Trench-Forearc Geology: Sedimentation and Tectonics on Modern and Ancient Active Plate Margins

Trench-Forearc Geology: Sedimentation and Tectonics on Modern and Ancient Active Plate Margins

edited by

Jeremy K. Leggett
Department of Geology, Royal School of Mines,
Imperial College of Science and Technology, London

1982

Published for
The Geological Society of London,
by Blackwell Scientific Publications
Oxford London Edinburgh
Boston Melbourne

Published by

Blackwell Scientific Publications
Osney Mead, Oxford OX2 OEL
8 John Street, London WCIN 2ES
9 Forrest Road, Edinburgh EH1 2QH
52 Beacon Street, Boston, Massachusetts 02108, USA
99 Barry Street, Carlton, Victoria 3053, Australia

First published 1982

DISTRIBUTORS

USA
Blackwell-Mosby Book Distributors
11830 Westline Industrial Drive
St. Louis, Missouri 63141

Canada
Blackwell-Mosby Book Distributors
120 Melford Drive, Scarborough
Ontario M1B 2X4

Australia
Blackwell Scientific Book Distributors
99 Barry Street, Carlton
Victoria 3053

British Library Cataloguing in Publication Data
Trench-forearc geology.
1. Geology, Structural
I. Leggett, Jeremy K.
551.1'36 QE601

ISBN 0-632-00708-7

Text set in 9/10 pt Linotron 202 Times, printed and bound in Great Britain at The Pitman Press, Bath

Contents

ASIA AND AUSTRALASIA

ATLANTIC

MEDITERRANEAN

MAKRAN OF IRAN AND PAKISTAN

CALIFORNIA

FOREARC TERRANES IN OROGENIC BELTS

FACIES, PETROLOGY AND MODELS

Preface

This book is a collection of papers on an aspect of plate tectonics of which our understanding is at present limited. In the mid-1970s, prior to the recent phase of IPOD active margin drilling, few geologists would have anticipated that at the start of the 1980s so many new questions concerning the nature of tectonic and sedimentary processes in forearc regions would have come to light.

1980 seemed to be a good time to synthesize current research and future problems. Several fascinating transects of active margins had been drilled in the preceding years, with largely enigmatic results. With the exception of drilling on the Barbados Ridge in early 1981, IPOD active margin activities were about to enter a period of abeyance. Several academic institutions had completed informative cruises on active margins, and numerous studies of emergent areas in modern forearc regions and of proposed ancient forearc terranes had come to fruition. For these reasons the British Sedimentological Research Group, a Specialist Group of the Geological Society of London, decided in 1979 to convene a three-day international conference on the theme of 'Trench and Forearc Sedimentation and Tectonics in Modern and Ancient Subduction Zones'. The meeting was held on 23–25 June 1980, at the Geological Society, Burlington House, Piccadilly, London. It was attended by more than 180 earth scientists from 17 countries.

Most of the contributions in this book were presented orally or as poster sessions. Many of the presentations were subjected to lengthy and often heated discussions. I consider it unfortunate that it has not proved possible to incorporate selections from those useful discussions in the final volume.

The ambitious title of the book is a concession to brevity rather than an exact description of the contents. The emphasis is on tectonics and sedimentation. However, in studies of forearc geology, perhaps more so than in any other topic in geology, it becomes difficult to consider tectonic and sedimentary phenomena independently. For this reason, these topics are discussed side by side in many of the papers. Metamorphic and igneous phenomena are also considered in several contributions.

The papers fall into three approximately equal categories: contributions with entirely new data, review papers on areas of special interest, and papers which combine reviews with new data and ideas. As at the original meeting, the Publications Committee of the Geological Society decided to allow a relatively high amount of review material so that the final volume would be a useful comprehensive and up-to-date reference for the geologist interested in this field.

The papers fall into 11 more or less natural groups, the first eight of which are defined on the basis of geography. These concern Japan, Central America, South America, the Aleutians, Asia and Australasia, the Atlantic, the Mediterranean, and the Makran of Iran and Pakistan. All these sections include data from both land-based and marine studies. Papers on the geology of the Franciscan and Great Valley terranes of California, in section 9, have been separated from papers on other possible ancient forearc terranes, in section 10, whose setting in orogenic belts makes their reconstruction more equivocal. In the final section are three papers on facies models, petrology of forearc sediments, and the origin of forearc-related ophiolites.

In editing this collection of papers I have tried to provide the interested geologist with a clear impression of the current state of research on forearc regions of active plate margins. If this goal has been achieved, he or she will hopefully be impressed with the enormous amount of data which has accrued over recent years, though should be equally aware that there are many puzzling variations in the geology of forearcs and that there are many disagreements in the interpretation of individual forearc areas. Variations in forearc tectonics will become clear in comparing, for example, the Late Cretaceous and Palaeogene forearc geology of land areas of SW Japan in the Shimanto Belt (Taira *et al.*) with the Neogene behaviour of the Honshu offshore forearc (von Huene & Arthur). Equally impressive are the differences in tectonic evolution seen along the same active margin in South America (Kulm *et al.*; Moberly *et al.*) and highlighted in two drilled transects of the Central American margin (Moore *et al.*; von Huene *et al.*). Differences in interpretation of the same area include the Hellenic Trench, where Kenyon *et al.* and Le Pichon *et al.* arrive at differing conclusions using different geophysical techniques, and the Franciscan Complex of California, where Blake *et al.* and Bachman/Aalto arrive at differing palaeogeographic interpretations using detailed mapping of different parts of the same terrane. The paper by

Wezel gives a salutory reminder that perhaps we are all barking up the wrong tree in adhering too closely to the tenets of plate tectonics.

I have preferred not to attempt the traditional editor's summary paper. My overall impression is that the recent flood of data on the forearc geology of active margins has created just as many, if not more, new problems than it has solved. This is good for our science, and augurs well for a stimulating future. If this collection of papers achieves one thing, I hope it will be to encourage the funding which will be necessary to tackle these problems, both at sea and on land, over the next few years.

Note on editorial procedure:
As editor I have followed wherever possible the editorial proceedures and format used in the *Journal of the Geological Society*. The papers are written by geologists whose specialities cover a variety of fields, and who hail from many different countries. I have tried to standardize scientific terminology, nomenclature, and spelling. I have made concessions to overseas authors regarding American spellings in some figure captions where I was so requested.

ACKNOWLEDGMENTS:

I am greatly indebted to many people. First, to the Council of the Geological Society: without their agreement to fund the conference, coverage of the subject in this book would not have been so complete, and the conference participants would not have enjoyed the meeting as much as they did. Second, to the staff at the Geological Society: they tolerated endless impossible deadlines in the administration of the conference. Third, to the contributors: all very busy men, who nonetheless found time to produce their manuscripts on time. Fourth, to referees and advisors too numerous to mention: they put in hours of their time to assist—Gwyn Thomas in particular. Fifth, to Mac (W. S. McKerrow) for all his support and encouragement. Finally, to my parents, Jessie and Sally: they know what for.

J. K. Leggett, Department of Geology,
Royal School of Mines, Imperial College,
London SW7 2BP.
June 1981.

JAPAN

The Shimanto Belt of Japan: Cretaceous-lower Miocene active-margin sedimentation

A. Taira, H. Okada, J. H. McD. Whitaker & A. J. Smith

SUMMARY: South-east of the two paired metamorphic belts of Kyushu, Shikoku and Honshu, and separated from them by the Chichibu and Sambosan belts and the Butsuzo Tectonic Line, lies a belt of mildly metamorphosed sedimentary rocks called the Shimanto Supergroup. Their area, extending from the Nansei Islands through Kyushu, Shikoku, Kii Peninsula, Akaishi and Kanto Mountains to the Boso Peninsula, is comparable in size with the combined Franciscan Formation and the Great Valley sequence in California. A Cretaceous lower group is flanked on the Pacific side by an upper group of Palaeogene to early Miocene age. The Shimanto sediments, mostly sandstones and mudstones, were deformed by cyclic subduction into open and isoclinal folds which close south-eastwards, and thrust slices which in places incorporate basaltic pillow lavas and radiolarian cherts.

Sandstones from Kyushu and Shikoku show marked changes of composition with stratigraphic position and the feldspar content can be used to distinguish Cretaceous from Tertiary sandstones. Data on sandstone petrography and palaeocurrents indicate that sediments of the Shimanto Supergroup were probably derived from the NW. Precambrian gneisses and older Mesozoic granitic rocks of the Korean Peninsula may have contributed sediments in Early Cretaceous times, when the Shimanto Terrane was located to the east of southern Korea. The unroofing of granites in the Inner Zone of SW Japan may also have contributed to the Cretaceous and later sedimentation. Quartz-rich, well-sorted sands in younger Shimanto sediments may have been recycled from older Shimanto formations.

The thick Shimanto sedimentary sequences were laid down in a variety of environments within a forearc basin on accretionary complex, trench-slope break, trench inner slope and trench settings; first in the Cretaceous and again, in a more southerly position, in the Palaeogene and early Miocene. In the shallower waters on the inner (arc) side of the Palaeogene forearc basin, coarsening- and thickening-upward deltaic sequences were deposited. In deeper water further offshore, submarine channel and fan complexes, base-of-slope slump deposits and red shales with cherts accumulated. Mélanges with basalts and slump olistostromes occupied the trench-slope break, while inner trench wall perched basins on accretionary basement were filled with coarsening-upward flysch, slump-olistostrome facies and mélanges with metabasalts: similar lithologies probably occupied a trench fill.

Sedimentation and deformation were controlled by intermittent cyclic subduction of the Kula Plate towards the north. Phases of tension and down-faulting, forming long, narrow intra-arc basins, alternated with those of compression. Comparable sedimentary environments are found today off SE Japan on the submarine terraces, trench-slope break, inner slope (including perched basins) and fill of the Nankai Trough.

To Western geologists, Japan is best known for its two pairs of metamorphic belts and for its igneous rocks, well summarized by Miyashiro (1973), and more recently by various authors in Tanaka & Nozawa (1977). Less well known, because the literature is mainly in Japanese, is the mildly metamorphosed Shimanto Belt lying to the SE of the Sambagawa metamorphic and Chichibu and Sambosan Belts and separated from them by the Butsuzo Tectonic Line (Fig. 1). Extending some 1800 km along strike from the South Nansei Islands to the Boso Peninsula and up to 70 km wide across the strike in the Kii Peninsula, the Shimanto Belt is comparable in area with the Franciscan Formation plus the Great Valley sequence of California.

The purpose of this paper is to summarize the structural and sedimentological features of the Shimanto Belt and to reconstruct the conditions of deposition of Shimanto rocks. We believe they accumulated through Cretaceous and lower Tertiary times in an active margin setting consisting of forearc basin on accretionary complex—trench-slope break—trench inner slope—trench environments (terminology of Dickinson & Seely 1979). Thrusting during accretion, with the production of mélanges bearing blocks of ocean-floor pillow basalts, hyaloclastites, cherts and sandstones was a feature of this active margin.

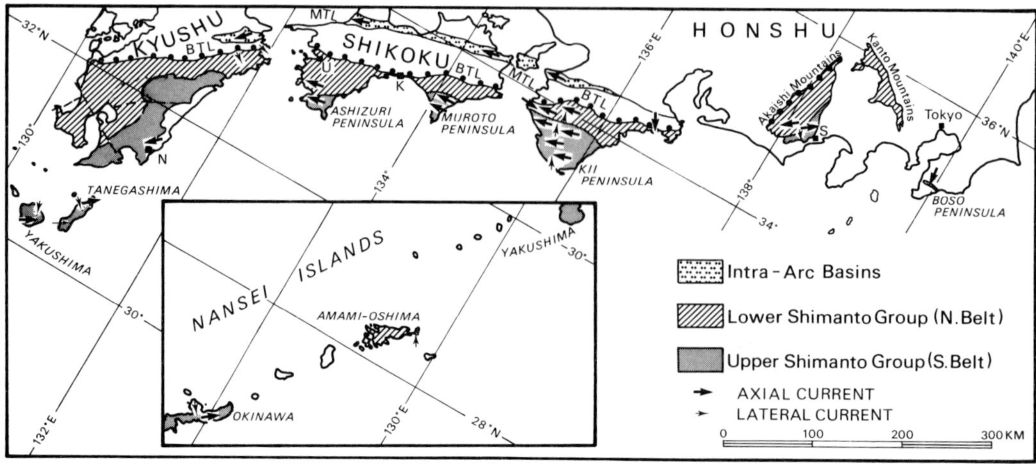

FIG. 1. Map of Japan to show the distribution of the Lower and Upper Shimanto Groups, with generalized palaeocurrents. K = Kochi; N = Nichinan; S = Shizuoka; U = Uwajima. BTL = Butsuzo Tectonic Line. MTL = Median Tectonic Line.

General geology of the Shimanto Belt

The Shimanto Supergroup is now exposed in the Nansei Islands, South Kyushu, South Shikoku, Kii Peninsula, Akaishi and Kanto Mountains and on the Boso Peninsula (Fig. 1). This long, narrow belt is divisible along its length into two major tectonostratigraphic units, a Northern Belt (Lower Shimanto Group) and Southern Belt (Upper Shimanto Group) (Teraoka 1979). The former is characterized by uppermost Jurassic to Cretaceous strata, the latter by Palaeogene and Lower Miocene rocks.

Both these groups are characterized by metabasalt and mudstone facies generally overlain by flysch facies, forming successions many kilometres in thickness (Kanmera 1976). These rocks are highly deformed, often into isoclinal folds, and there is important repetition by imbricate thrusting (Sakai 1978). There is general low-temperature regional metamorphism of the prehnite-pumpellyite and greenschist facies.

In the cross-sections of the Shimanto Belt, sediments tend to become younger oceanwards (Fig. 2) as does the unconformity between the highly deformed Shimanto strata and the overlying mildly deformed post-Shimanto sediments. A similar trend continues out of the Shimanto Belt into the present trench inner slope (Fig. 2, columns 5 & 6).

The lower metabasalts in both the Northern and Southern Belts are characterized by pillowed basalt and hyaloclastite (Kanmera 1976; Suzuki & Hada 1979; Tsuchiya *et al.* 1979). In the Southern Belt this facies contains gabbro, diabase and serpentinite. Recently, 'umber' deposits comparable to modern oceanic sediments have been found between pillows of basalt in the Mineoka Group in the Boso Peninsula (Tazaki *et al.* 1980). Large-scale olistostromes are common in both belts, but are better developed in the Southern Belt.

The overlying flysch facies in both belts is characterized by terrigenous turbidites alternating with mudstones, infrequently intercalated with acidic tuff layers. In the Northern Belt, ammonites and inoceramids have been found sporadically (Matsumoto & Okada 1978), while bivalves and gastropods occur rather commonly in some limited sequences in the Southern Belt. Trace fossils are prolific locally, e.g. in Shikoku (Katto 1960, 1964). Generally, however, fossils are rare and subdivision of the Shimanto Supergroup must use sandstone petrography.

Sandstones of the Northern Belt (Lower Shimanto Group) tend to be highly feldspathic in composition, less sorted and coarser grained than the quartz-rich, better sorted Upper Shimanto sandstones of the Southern Belt. In the mature arenites of the latter group, extremely well-rounded quartz grains abound. These sandstones are further characterized by the stable heavy mineral suite of zircon-tourmaline-rutile, and by a higher muscovite content. In both groups, microcline is a major constituent, but it decreases upwards. The total amounts of feldspars also decrease upwards: the Cretaceous sandstones of Kyushu, for example, have more than 35% feldspar, Tertiary ones

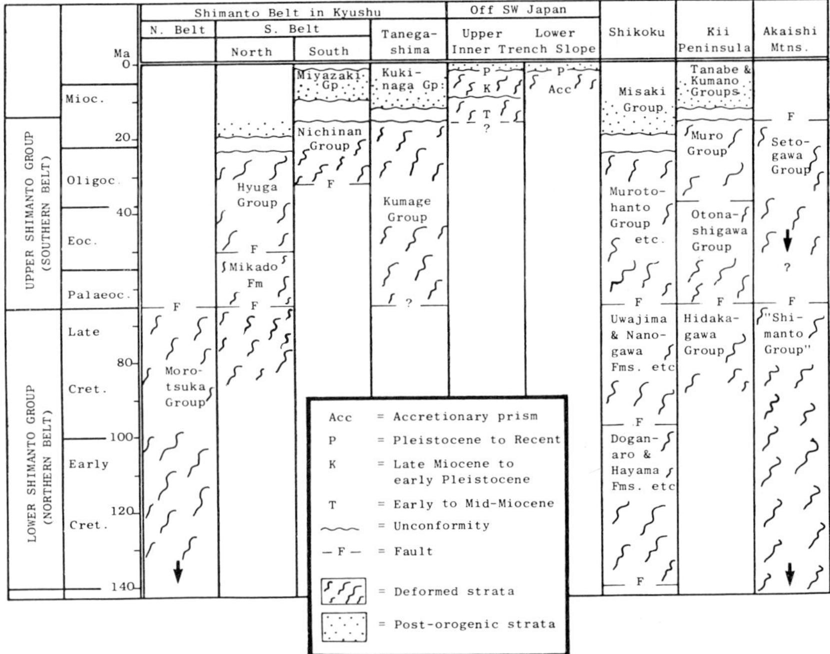

Fig. 2. Simplified stratigraphy of the Shimanto Supergroup and immediately overlying strata in four areas. Many formation names are omitted. The four Kyushu columns show the southward younging of the Shimanto rocks, which is continued offshore (next two columns, from Okuda *et al.* 1979). Acc, P, K and T are based on acoustic stratigraphy.

less than 35%. Moreover, the ratios between orthoclase and plagioclase and those between orthoclase and microcline alter with stratigraphy. Teraoka (1979), for example, found that the content of feldspar and the microcline/orthoclase ratios tend to decrease upwards from the Upper Cretaceous. Okada (1977) distinguished three petrographic zones in the Shimanto of Kyushu. Lower Shimanto Group (Cretaceous) sandstones (zone I) are characterized by ill-sorted, angular-grained feldspathic arenite containing more than 35% feldspar clasts. Sandstones of zones II and III are richer in quartz and the grains are better sorted and more rounded. These sandstones of zones II (lower sequence of the Upper Shimanto Group, lower Palaeogene), and III (upper sequence of the same group, upper Palaeogene to lower Miocene) contain 20–35% and 10–20% of feldspars respectively. Thus the Cretaceous, lower Palaeogene and upper Palaeogene in Kyushu are distinguishable from one another on the basis of sandstone petrography (Okada 1977; Teraoka 1977, 1979). Interestingly, similar strike-parallel petrographic zones have been delineated in a supposed ancient accretionary complex in the Ordovician and Silurian rocks of the Southern Uplands of Scotland by Floyd (1975): these zones can be traced along strike for up to 200 km (J. D. Floyd, pers. comm. 1980).

Regional geology of the Shimanto Belt

The best-studied areas (Shikoku and Kii) are described first, then the Akaishi, Kanto and Boso areas to the NE, and finally Kyushu and the Nansei Islands to the SW.

Southern Shikoku

In the southern half of the Island of Shikoku, the Shimanto Belt extends about 230 km along strike (Fig. 3): as in other regions, it is much folded and faulted. In eastern Shikoku, the triangle-shaped Muroto Peninsula provides excellent continuous exposures along the coastline while in the west, the type sections of the Shimanto Supergroup, here 70 km wide, are exposed along the Shimanto River.

(a) *The Lower Shimanto (Cretaceous) Belt*

The Cretaceous rocks of this belt are separated from a Jurassic subduction complex (the

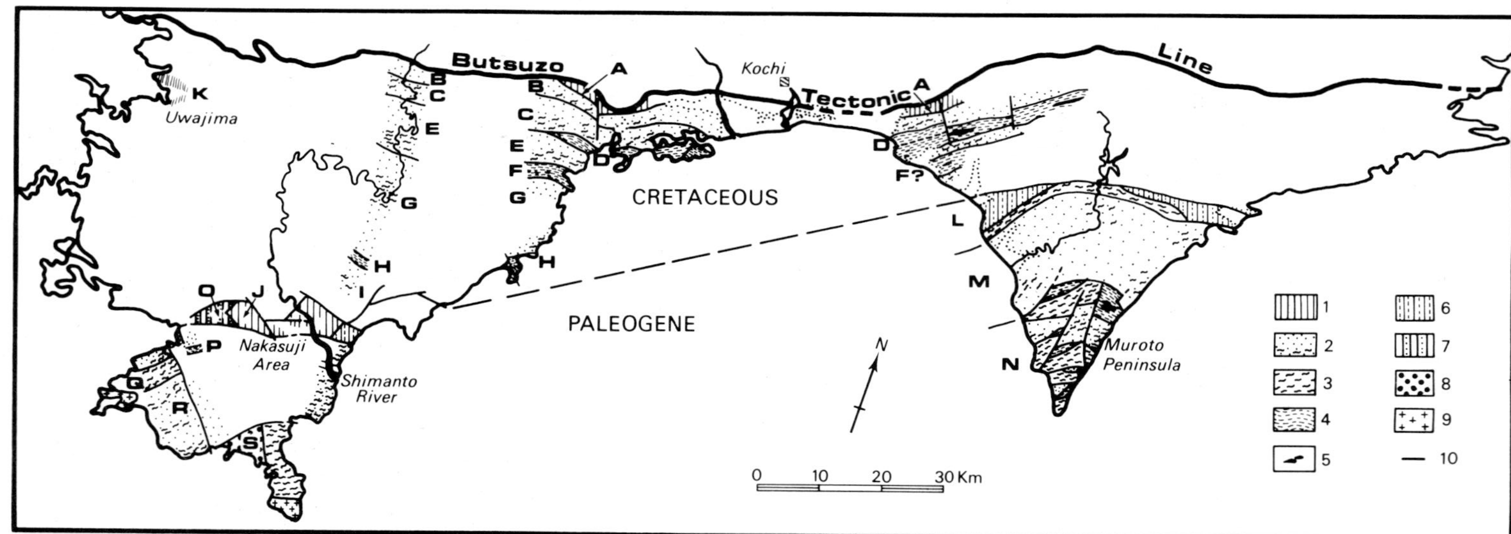

FIG. 3. Geological sketch map of the Shikoku Shimanto Belt.
(1) Cretaceous shallow marine facies; (2) turbidite facies (wave mark for slumping); (3) slump and olistostrome facies; (4) mélange facies; (5) basic volcanics; (6) Eocene conglomerate and sandstone; (7) Eocene and Oligocene shallow marine facies; (8) Miocene shallow marine facies; (9) Miocene granitic intrusives; (10) faults. (A) Neocomian to Turonian shallow marine facies (Doganaro Formation); (B) Albian turbidite facies (Hayama Formation); (C, E, G, I) Upper Cretaceous turbidite facies; (D, F, H) mélange facies; (J) Campanian to Maastrichtian shallow marine facies (Nakamura and Arioka Formations); (K) Turonian to lower Campanian shallow marine facies (Uwajima Group); (L) Ohyama-misaki Formation; (M) Naharigawa Formation; (N) Muroto Formation; (O) Palaeogene shallow marine facies (Hirata Formation); (P) Palaeocene? mélange facies; (Q) Eocene slump and olistostrome facies; (R) Eocene turbidite facies; (S) Miocene shallow marine facies (Misaki Formation). (Data from Katto *et al.* 1977; Katto & Taira 1978; Suzuki & Hada 1979; Sano *et al.* 1979; Katto *et al.* 1980; Taira *et al.* 1980.)

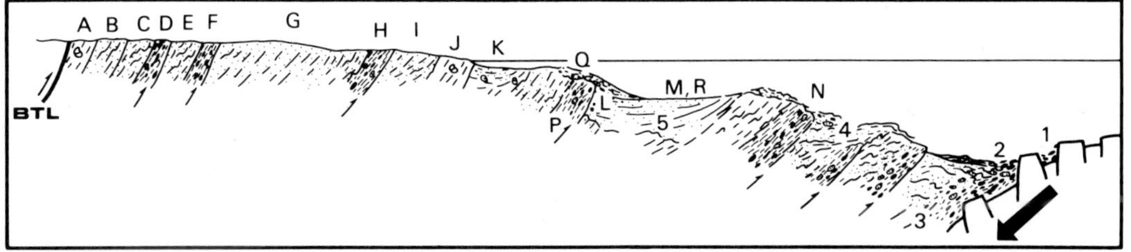

FIG. 4. Schematic reconstruction of the Shimanto cross-section during the time of Palaeogene subduction. Letters correspond with Fig. 3. (1) Extensional fracture of oceanic crust in the trench outer slope; (2) olistostrome of basalts and pelagic sediments mixed with trench sediments; (3) accreted mélange; (4) turbidite and slump deposits of perched basin on the trench inner slope; (5) turbidites in the accretionary forearc basin. BTL = Butsuzo Tectonic Line. (Based on Taira *et al.* 1980.)

Sambosan Group) by a major thrust fault, the Butsuzo Tectonic Line (Figs 3 &4) (Taira *et al.* 1979b). They can be divided into: (1) a shallow marine facies, (2) a turbidite facies and (3) a mélange facies (Figs 3 & 4) (Taira *et al.* 1979a; Suzuki & Hada 1979; Taira *et al.* 1980). Each facies is usually fault bounded.

(1) The *shallow marine facies* mainly occurs in three small areas: along the Butsuzo Tectonic Line (A of Figs 3 & 4); in the Uwajima area of western Shikoku (Figs 3K & 4K); and along the southern margin of the Cretaceous belt near the boundary with the Palaeogene (Nakasuji area) (Figs 3J & 4J).

Near the Butsuzo Tectonic Line in Kochi Prefecture the Doganaro Formation, composed of interbedded sandstone and mudstone, yields Neocomian brackish molluscan faunas, and Turonian fully marine types (Katto & Tashiro 1978). In the Uwajima area, the Supergroup consists mainly of the Lower Cretaceous (Albian) Kitanada Group and the Upper Cretaceous (Coniacian to Santonian) Uwajima Group (Teraoka & Obata 1975). The former is mainly pelitic, the latter is sandy with abundant molluscs. In the Nakasuji area, several localities yield upper Campanian to Maastrichtian molluscan faunas, serpulid limestones and *Zoophycus* trace fossils and there is intensive bioturbation (Tanaka 1977; Katto & Tashiro 1979b). This facies, of shallow marine origin, roughly defines the outer margin of the frontal arc area, which apparently migrated southwards through Cretaceous times.

(2) The *turbidite facies* forms a major part of the Cretaceous Shimanto Belt in Shikoku. Sandstone layers of various thicknesses interbed with black shales and occasional grey and red shales. Slumps and olistostromes are common. This facies generally shows an east-west structural trend with steeply dipping beds. While tight, closed folding occurs locally, the succession is repeated by imbricate faults which control the overall structural framework. Slaty cleavage sub-parallel to the bedding is ubiquitous.

Ammonites and inocerami have been found at only a few localities (Katto *et al.* 1980), but microfossils, especially radiolarians, have been found more recently. Although the structural framework cannot be dated in detail, an important result has been obtained (Katto *et al.* 1980; Taira *et al.* 1980): the oldest age found in this facies is Albian for the Hayama Formation which occupies the northernmost turbidite belt (B of Figs 3 & 4). Southwards, several belts of turbidite facies (C, E, G, I of Figs 3 & 4) show overall southward younging. However, in each individual turbidite belt (generally between 2000 and 5000 m in thickness) the age, based on radiolaria and the stratigraphic succession indicates a northward-younging trend. This compares with the model for modern accretionary wedges (Seely *et al.* 1974; Moore *et al.* 1979), supposed ancient analogues such as the Ordovician-Silurian Scottish Southern Uplands (Leggett *et al.* 1981) and the Franciscan of California (e.g. Bachman 1981).

In western Shikoku, the Coniacian to Campanian turbidite facies (Nonogawa Formation, G of Figs 3 & 4) is widely distributed.

(3) A zone of *mélange facies*, which separates the belts of turbidite facies in some places (D, F, H of Figs 3 & 4), consists of a chaotic mixture of blocks of sandstone, chert and basic volcanics in pervasively sheared shale matrix (Fig. 5d). The size and shape of the blocks is varied, but phacoidal and lenticular shapes with extensive foliation structures abound. The composition of the sandstone blocks is similar to the turbidite beds, though some differences are reported (Suzuki & Hada 1979). The mélange

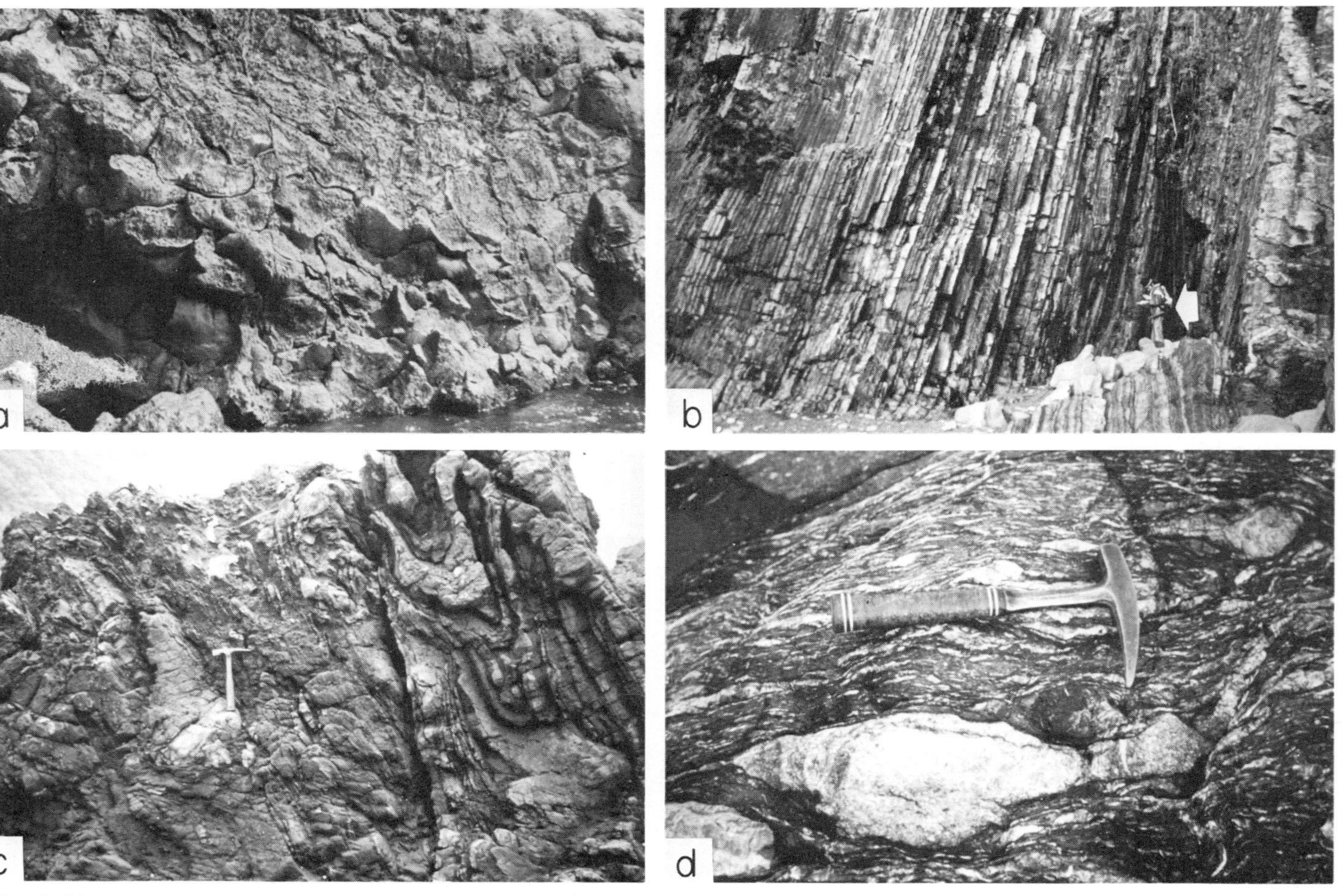

FIG. 5. (a) Huge block of basalt, Muroto Formation (Palaeogene) mélange complex. East side of Muroto Peninsula, Shikoku. The pillow structures show the block to be inverted. Height of cliff in view is about 12 m.
(b) Coherently bedded flysch sequence, west of Tatsukushi, western Shikoku. Arrow indicates figure for scale.
(c) Slump facies, Muroto Formation, west side of Muroto Peninsula.
(d) Close-up view of Cretaceous mélange facies, with phacoidal sandstone lenses. Okitsu coast SW of Kochi, Shikoku.

facies contains less K-feldspar and more siliceous matrix than the other facies. Blocks of basic volcanics, generally metamorphosed to the pumpellyite-prehnite facies, range in size from gravel to units of mappable size. Pillow structures are commonly observed. In some larger blocks, pillow lava is overlain by metalliferous black mudstone and red radiolarian chert. Radiolaria in such cherts are older than those of the argillaceous 'matrix' of the mélange and surrounding turbidite facies (Taira 1979; Nakaseko *et al.* 1979; Katto *et al.* 1980). For example, in belt D (see Figs 3 & 4), the facies contains Valanginian red radiolarian chert resting on basaltic pillow lava, but is embedded in a Coniacian to Santonian 'matrix'. In the southern mélange belt (H in Figs 3 & 4 and other localities), Albian red radiolarian chert blocks are found with basaltic lava blocks and this mélange facies is overlain by Coniacian to Campanian turbidites. These data indicate that the lavas and cherts within the mélange facies are emplaced older exotic blocks from an oceanic plate mixed with presumed arc- or continental margin-derived granitic clastic detritus.

(b) *The Upper Shimanto (Palaeogene) Belt*

The lithofacies of the Palaeogene rocks in Shikoku resemble the Cretaceous rocks, but the overall facies relationships are better established and their depositional setting can be deduced in more detail.

The representative Muroto-hanto Group of south-eastern Shikoku illustrates the relationships. Nannofossils, radiolarians and molluscs suggest an age from early Eocene to Oligocene (Katto & Tashiro 1979a). The group consists of three units (from north to south) : the Ohyama-misaki, Naharigawa and Muroto Formations (L, M, N of Figs 3 & 4) (Katto *et al.* 1961; Katto & Taira 1978).

The Eocene *Ohyama-misaki Formation* (Figs 3L & 4L) consists of conglomerates, sandstones and shales formed within a fining-upward depositional cycle. The conglomerate beds, which show channelling, scour-and-fill, large-scale grading and trough cross-bedding, contain large boulders of various lithologies interpreted as derived from the frontal arc areas including the Cretaceous Shimanto, Chichibu and Sambagawa Belts. Sandstones show graded bedding, parallel- and trough cross-bedding and abundant rip-up clasts typical of some turbidite sequences. This formation is interpreted as sumbarine channel- or canyon-fill conglomerates and sandstones, and over-bank, thin-bedded shales and sandstones, which possibly are part of an upper-fan facies (Katto & Taira 1978). It also contains slump and olistostrome deposits which may represent base-of-slope environments.

The *Naharigawa Formation* (M of Figs 3 & 4), containing Eocene nannofossils, radiolarians and molluscs, is a flysch sequence of interbedded turbidite sandstones (up to 10 m thick) and shales, with minor amounts of tuffs (Fig. 5b). Trace fossils include *Helminthoidea, Nereites, Paleodictyon* and *Spirorhaphe* suggesting deep-water deposition, a deduction supported by recent finds of traces of the last two genera on the surface of modern ocean sediments at depths ranging from 1436 to 3895 m for *Paleodictyon* and from 3358 to 5119 m for *Spirorhaphe* (Ekdale 1980). Palaeocurrent analysis indicates east to west dominated axial current directions (Fig. 1) (Katto & Arita 1966).

To the south the *Muroto Formation* (Figs 3N & 4N) is a mélange complex containing slump-olistostromes (Fig. 5a,c) and flysch. The mélange consists of blocks of ocean-floor hyaloclastite, basalt (Sugisaki *et al.* 1979), chert, tuff and sandstone in pervasively sheared shale and tuff matrix. The mélange material is not yet well dated, but radiolarians from the sheared grey shale indicate a Palaeocene and early Eocene age, the oldest age so far obtained in the Muroto-hanto Group.

Large slump-olistostrome deposits are common in the Muroto Formation, especially in a major zone ranging up to 700 m in thickness. The flysch facies consists of a coarsening upward sequence: shale, some of it red, intercalated with tuff passing up into numerous turbidite sandstone units. Radiolarians suggest an Eocene age for the post-mélange flysch facies.

Recent work on the Tsuro Formation at the south tip of the Muroto Peninsula has revealed Foraminifera and Radiolaria of lower Miocene age (Taira *et al.* 1980).

(c) *Sedimentation and tectonics*

The lithofacies and structural pattern of the Shikoku Shimanto Belt resembles other examples of forearc basin deposits resting on an accretionary complex (cf. Dickinson & Seely 1979, fig. 3, lower left). The overall southward younging trend of the Shimanto Supergroup shows that the accretionary process was dominant through Cretaceous and Palaeogene times.

The conditions of deposition are best documented in the Muroto-hanto Group (see Fig. 4), with the Ohyama-misaki and Naharigawa Formations interpreted as Eocene accretionary forearc basin sediments (5 of Fig. 4), the former representing submarine channel (or canyon) fill and slope sediments and the latter representing a possible submarine fan complex that filled a major part of the accretionary forearc basin (Katto & Taira 1978; Taira 1979). The Muroto Formation can be interpreted as trench-slope break to trench inner slope deposits and accretionary basement (subduction complex) (Taira 1979; cf. Bachman 1981). The broader mélange zone which contains the large metabasalt complex possibly makes up a basement high at the trench-slope break. The flysch and slump facies separated by the mélange zone with imbricated thrust faults are interpreted as small perched accretionary basins on the trench inner slope (4 of Fig. 4) similar to those described by Karig & Sharman (1975), Moore & Karig (1976) and Lewis (1980).

Similar arrangements of lithofacies can, with some variations, be deduced throughout the Palaeogene Shimanto Belt. In Western Shikoku, there is an extensive development of slump deposits containing blocks of shallow marine origin such as limestones containing large Foraminifera, and sandstones rich in molluscan fossils (Hiromi mélange) occurring in an upper slope setting (Figs 3Q & 4Q).

Similar features occur in Cretaceous sediments (see Fig. 4A–K); shallow marine facies suggest deposition south-eastwards of an exposed frontal arc area. Contemporaneous turbidites (e.g. the Nonogawa Formation) were probably filling an accretionary forearc basin (cf. the Naharigawa Formation during Palaeogene times). The other turbidite facies were either trench fill deposits or filled small perched basins on the trench inner slope. The abundant slump and olistostrome deposits within this facies indicate that there was much active slumping of the turbidite cover as well as gravity failure of accreted mélange basement.

The origin of the mélange facies (Fig. 5d) is not clear, but the following observations set some constraints: (1) Oceanic plate materials incorporated into the Shimanto terrane are a very small proportion of the whole, occurring only as blocks. There is no evidence of wholesale incorporation of oceanic crust itself into the Shimanto Supergroup. (2) The blocks in the mélange facies show variety of size and shape and are interpreted as olistostrome deposits. Bedded chert and sandstone blocks show slump folding. (3) The shale 'matrix' is often very sheared, with abundant quartz veins. In some places, it shows a schistose appearance. (4) The age of the mélange material indicates a southward younging trend, as for the turbidite facies, but the mélange facies contains blocks of older age.

These features of the mélange facies indicate original formation as olistostrome deposits, possibly in a trench (Taira *et al.* 1980). The oceanic plate materials such as basaltic pillow lava and radiolarian chert may have been derived from the faulted scarps of the oceanic crust or seamounts on the outer slope of the trench (see 1 and 2 of Fig. 4) as has been observed in the Peru–Chile Trench (e.g. Kulm *et al.* 1981). Such trench-filling olistostrome deposits may have been accreted to the arc front together with arc- or continental margin-derived clastic sediments (3 of Fig. 4). Most of the pelagic sediments and the oceanic lithosphere itself have probably been subducted.

Kii Peninsula

Exposures of the Shimanto Supergroup continue eastwards from Shikoku into the Kii Peninsula (Figs 1 & 6). There, between the Butsuzo Tectonic Line and the Pacific Ocean, thick accumulations of Shimanto Supergroup sediments are characterized by dark mudstones, turbidites and conglomerates, and the whole sequence is deformed (Fig. 6, sections). Here the Shimanto has been studied by the Kishu Shimanto Research Group led by T. Harata of Wakayama University and T. Tokuoka of Shimane University (Kishu Shimanto Research Group 1970, 1975; Harata *et al.* 1978). Central to much of their discussions of the Shimanto geology of the Kii Peninsula is their belief in the existence of a pre-Miocene ancient Kuroshio continent once lying to the south. This account owes much to the work of the Research Group but the conclusions here do not necessarily reflect the views of the group.

As in other areas, the geology of the Kii Shimanto Supergroup reflects a migration of a depo-centre towards the SE. The oldest Shimanto sediments are assumed to be Late Cretaceous on the basis of inoceramid fossils (*I.* (*Mytiloides*) *aff. labiatus* and *I.* (*Platyceramus*) *amakusensis*), though these may have been transported into the basin, whilst the youngest are lower Miocene in age on the basis of their molluscan fauna (see Kishu Shimanto Research Group 1970).

The Kii Shimanto Supergroup has been divided into three groups: the Hidakagawa, Otonashigawa and Muro Groups (see Fig. 2). Each group is separated by major reverse faults or tectonic lines with steep dips to the north (Fig. 6).

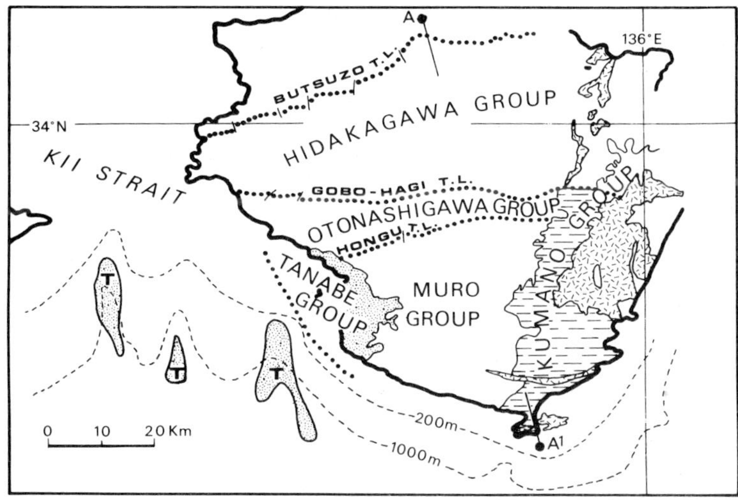

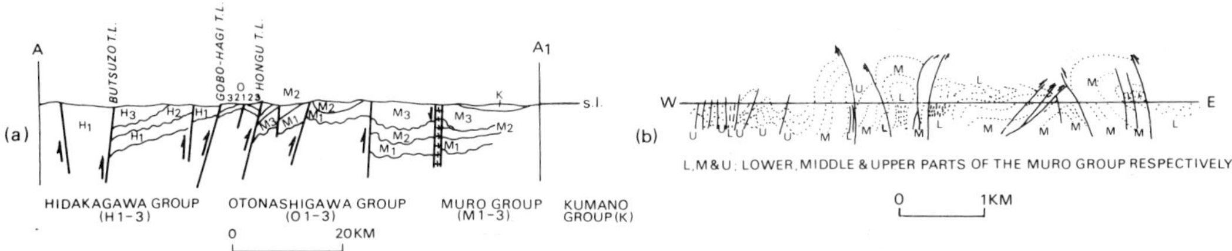

FIG. 6. Map and sections of the Kii Peninsula. The Hidakagawa, Otonashigawa and Muro Groups of the Shimanto are overlain unconformably by the Miocene Tanabe and Kumano Groups. Also shown are offshore exposures of the Tanabe Group (T) in submarine canyons in the outer Kii Strait. TL = Tectonic Line. Section (a) is along AA^1 (A.J.S.), (b) is a detailed E-W section through part of the Muro Group (after Suzuki 1975).

(a) *Hidakagawa Group*

Exposed between the Butsuzo Tectonic Line in the north (see Fig. 6) and the Gobo-Hagi Tectonic Line, it is composed of three formations, with complex structural relationships.

The oldest, the Nyunokawa Formation, is more than 2 km thick: it and contemporaneous formations are found near the Butsuzo and the Gobo-Hagi Tectonic Lines. The lower part is generally finer grained, composed of shales and muddy turbidites, while higher the turbidites become sandier and massive. Conglomerates, often coarse, are intercalated in the upper part and contain a wide variety of rock types: rhyolites are abundant, with granites, sandstones, cherts and some limestones. The succeeding Ryujin Formation is about 3 km thick and finer grained, being composed of shales and muddy turbidite sequences with only occasional more sandy sequences. Basalts with pillow lavas and rhyolitic tuffs occur.

The youngest formation is the Miyama Formation and it lies to the north of the preceding Ryujin Formation. More than 7 km thick, it is composed of massive, usually sandy, turbidites, which may be associated with basalts and radiolarian cherts, and subordinate shales.

Roughly contemporaneous with Hidakagawa sedimentation, turbidite deposition of the Izumi Group was taking place in a narrow basin north of the Median Tectonic Line (Fig. 1).

(b) *Otonashigawa Group*

Only differentiated from the succeeding Muro Group relatively recently, it is about 1.5 km thick (Hatenashi Research Group 1975) and is exposed between the Gobo-Hagi Tectonic Line and the Hongu fault. Though no reliable palaeontological evidence has been discovered, the Otonashigawa Group has been ascribed mainly to the Eocene. Again the lowest formation (the Uridani Formation) is predominantly shaly and again there are basalts and acid tuffs. The middle formation, the Lower Haroku Formation, is the most extensive formation and is somewhat sandier than the

Uridani Formation. The Upper Haroku Formation completes a coarsening upward sequence and in the southern outcrops includes the Kizekkyo conglomerates which contain clasts up to boulder size. Sandstone clasts predominate, with chert, granite, acid volcanics and limestones less abundant.

(c) *Muro Group*

The most southerly and youngest of the Kii Shimanto Supergroup is estimated to be 7.5–9 km thick, and fossils, though possibly transported, indicate an Oligocene to early Miocene age. Again the lowest formation, some 1200 m thick, is mainly fine grained whilst the middle formation is thicker and sandier with some intercalated conglomerates. The uppermost part is between 3 and 4 km thick and is of mixed composition: shaly horizons are associated with conglomerates, which have muddy and sandy matrices, and breccia-bearing mudstones occur. Large olistoliths, up to several metres across, occur near the southern limit of the peninsula (Fig. 7b). The clasts in the conglomerates are composed predominantly of acid volcanic rocks and also include cherts, orthoquartzites, sandstones, granites, and limestones. Spectacular slumps occur (Fig. 7a) and clastic dykes are common. The character of this formation suggests that it was deposited near its source, though many of the pebbles, including the orthoquartzites, are extremely well rounded: their possible origin is discussed later.

(d) *Sedimentation and tectonics*

Essentially, these Shimanto Supergroup successions show three sequences represented by the three groups, each coarsening upward and implying either a simple shallowing or an increased input from the source area. Each group is separated by some sort of hiatus, though it is not clear if this is solely due to tectonic events. The tectonic lines themselves must represent major planes of movement since they now effectively separate the three Shimanto groups (Fig. 6).

The palaeocurrent directions for the different groups (for details see Fig. 8) reveal a predominantly east to west transport, but both southerly and northerly directions are important. In the Muro Group, there are some west to east directions.

The Supergroup is much folded and faulted (Fig. 6, sections), with steeply dipping, vertical, and overturned strata common and much strike faulting. Harata *et al.* (1978) comment that there has been a 30–40% shortening in the Kii Peninsula south of the Butsuzo Tectonic Line. In the Muro Group, Suzuki (1975) draws attention to the synsedimentary tectonic structures, with major folds being regarded as slump structures, and the occurrence of clastic dykes. Much remains to be done in the Muro Group areas to determine the relationship between slump deposits and the early 'folding'. Here, much of the folding must have been penecontemporaneous with sedimentation for all the folding and much of the faulting was completed before the unconformable deposition of the sediments of the Tanabe and Kumano Groups of middle Miocene age. Palaeontological and radiometric dating clearly show that only a very short time separated these groups from the Muro Group.

The last events in the Shimanto succession of Kii reflect proximal sedimentary conditions including slumps, locally derived olistoliths, rounded quartzite pebbles, washouts, clean or relatively clean conglomerates and even ripple marks, all affected by penecontemporaneous folding. The Kishu Shimanto Research Group (1970) sees much of this as evidence of the close proximity of a now-lost southern 'Kuroshio continent'. However, recent work by Tokuoka & Okami (1979), although they still insist on a supply from the south, suggests to us that in the Palaeogene the orthoquartzite pebbles were recycled from the Lower Cretaceous Tetori Formation, lying to the north. A boulder of granitic gneiss from the Muro Group gave K-Ar ages of 63.3 Ma on biotite and 70.4 Ma on muscovite, suggesting that the provenance of the gneiss may be the Ryoke metamorphic belt (Shibata & Nozawa 1973).

Akaishi and Kanto Mountains

The Shimanto rocks of this deeply dissected mountain region have been described by Ogawa & Horiuchi (1978) and Ogawa (1981). The Akaishi Belt trends NNE-SSW and is separated from the Ryoke, Sambagawa, Chichibu and Sambosan Belts on the west by the Butsuzo Tectonic Line and from Miocene groups on the east by the Itoigawa-Shizuoka Line. Deformation is represented by large-scale recumbent folds in the west and a north-south trending anticlinorium in the east. Metamorphism is of the usual pumpellyite or greenschist facies. The Shimanto rocks, several kilometres thick, consist of turbidites, sandstones, mudstones and slump deposits, with local limestones and chert, together with basalts and rhyolitic tuffs in the lower and middle parts. The lower group, 'Shimanto s.s.' is Jurassic (in part) and Cretaceous in age (Fig. 2). The upper, Setogawa Group, is of Palaeogene age, probably passing up into the Miocene.

FIG. 7. (a) Large slump structure, Muro Group, S Kii. Arrow indicates figure for scale.
(b) Olistoliths in mudstone, Sarashikubi Formation, Muro Group, S Kii. The block to the left of Dr Shiki is approximately 4 m in diameter. The matrix forms the wave-cut platform in the foreground.
(c) Thinner bedded sandstones and shales within the thick cross-bedded sandstone facies, Kumage Group (Palaeogene), N of Nishino-omote, Tanegashima. Mainly shale (right) and mainly sandstone (left) forming a coarsening- and thickening-upward deltaic sequence.
(d) Slumping of plastically deformed packets of thin sandstones and mudstones, Kumage Group (Palaeogene), Minato, N Tanegashima. Compare the deformation of the Aberystwyth Grits, Wales (Davies & Cave 1976). Hammer scale arrowed.

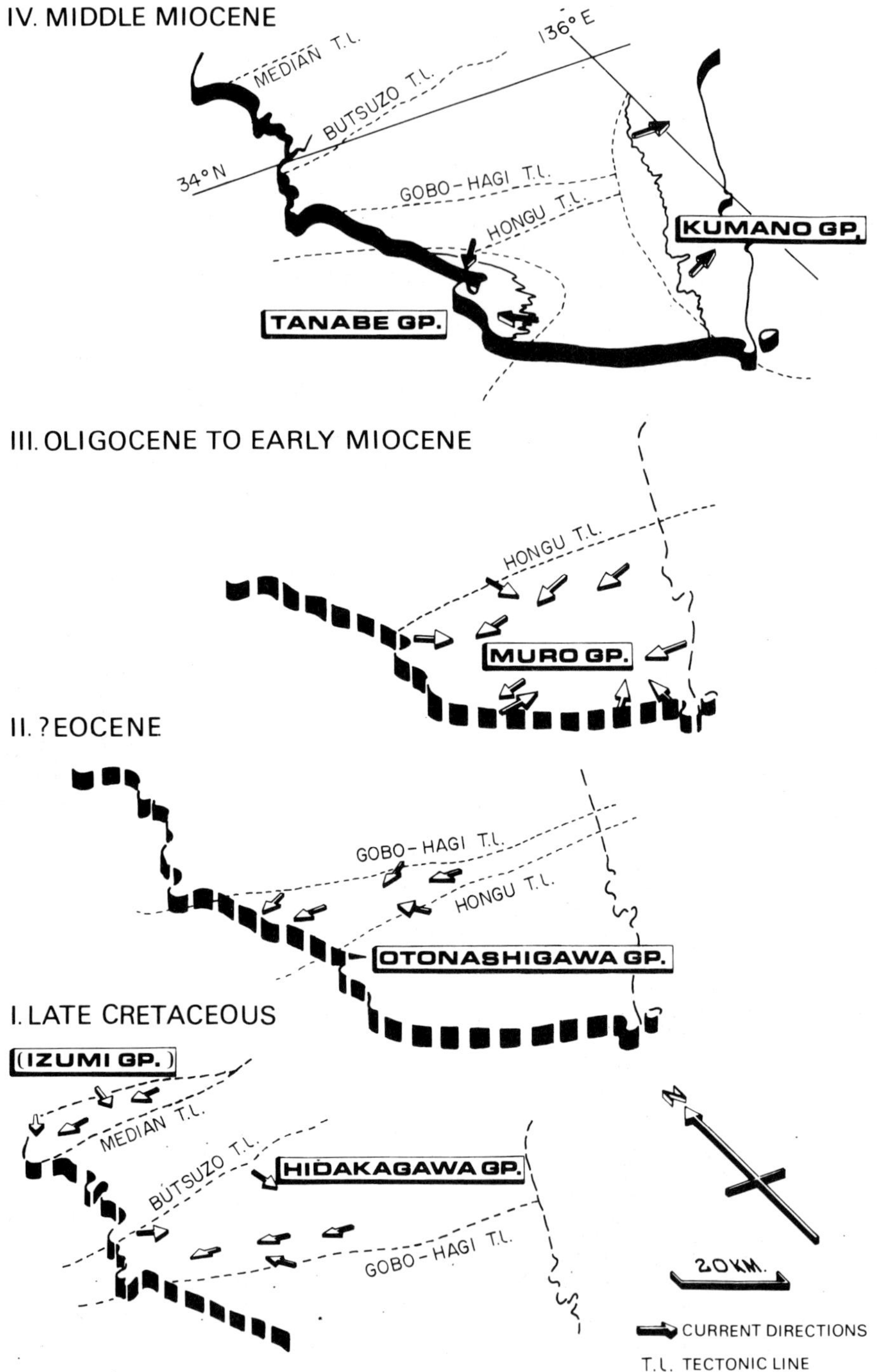

FIG. 8. Changing palaeocurrent patterns through the Kii Shimanto and up into the middle Miocene.

In the east-west-trending Kanto Belt, the northern Ogochi Group, Lower to Upper Cretaceous in age, consists of quartz/feldspar sandstones bearing *Inoceramus*, and mud-rich slumps. In the south, the Upper Cretaceous Kobotoke Group consists of sandstones, mudstones and turbidites forming sandy deltaic aprons marked by slumping: basalts, dacitic tuffs and chert are also present. There are no Palaeogene strata in the Kanto Mountains.

These Shimanto rocks are interpreted by Ogawa & Horiuchi (1978) as arc-trench gap and trench slope sediments that were later deformed by collision with the northward-moving Izu microcontinent, lying at the northern end of an aseismic ridge (Matsuda 1978): transcurrent movements were also important (Ogawa 1981).

Boso Peninsula

The Mineoka Group in the southern Boso Peninsula represents the easternmost development of Shimanto rocks. This relatively thin group (1000 m+) is made up of hemipelagic siliceous shale, flysch-type sandstones, shales and olistostromes, with basalts and serpentinite. Previously thought to be of Palaeogene age (Imai 1977), Miocene-type radiolaria have recently been found in the lower part of the Mineoka Group (K. Nakaseko, pers. comm. 1980).

We now return to the Shimanto developments SW of Shikoku.

Southern Kyushu

The Shimanto of Kyushu (Fig.9) is complicated in structure as shown in a map by Hashimoto (1962). In the Northern Belt, the Lower and Upper Cretaceous Morotsuka Group (Fig. 2) consists of phyllites, sandstones and mafic volcanic rocks with minor cherts in its lower part and mainly sandstones in its upper part (Sakai 1978). In the Southern Belt, the

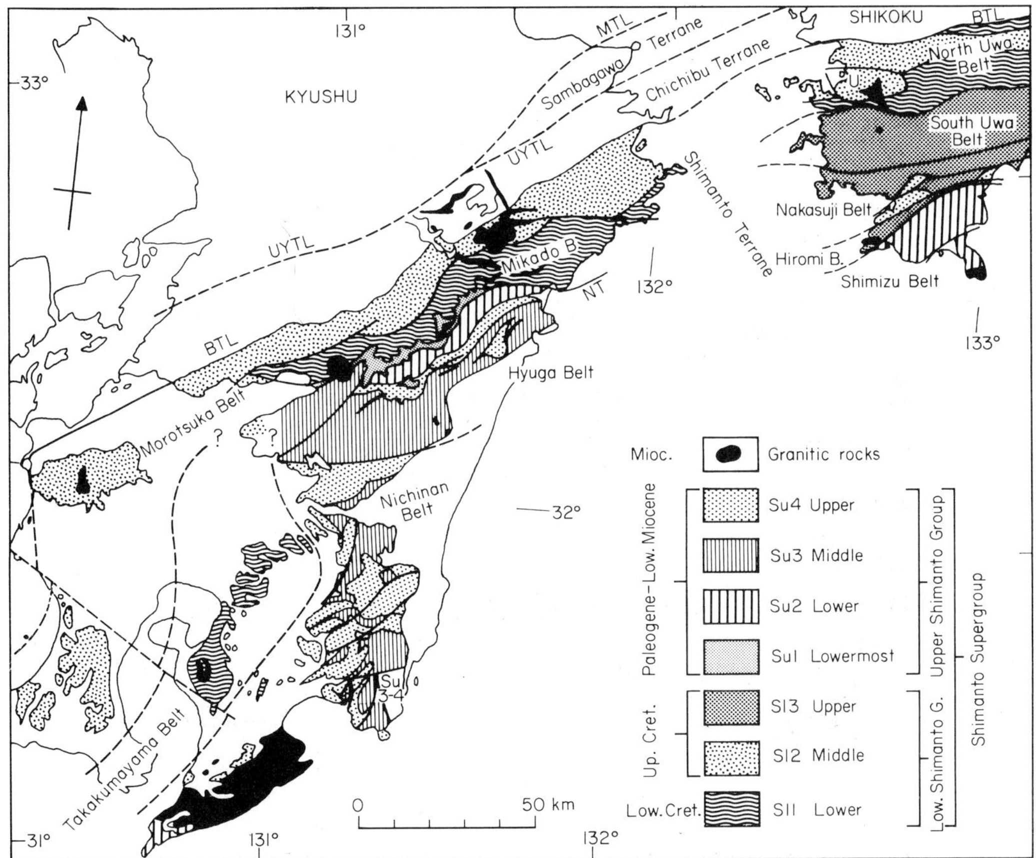

FIG. 9. Map of the Shimanto of Kyushu and West Shikoku. MTL = Median Tectonic Line; UYTL = Usuki-Yatsushiro Tectonic Line; BTL = Butsuzo Tectonic Line; NT = Nobeoka Thrust; Sl = Lower Shimanto; Su = Upper Shimanto. (Teraoka 1979, fig. 10, reproduced by permission of the author and the Geological Society of Japan.)

highly deformed Palaeogene Mikado Formation, of limited extent, is rich in pillowed basalts: it is separated from the Cretaceous Belt by a northward-dipping thrust. Other lower Tertiary rocks are the Hyuga Group (Eocene to early Miocene) in the north and the Nichinan Group, shaly in the lower part and sandy above, in the south: this group is late Eocene to early Miocene in age.

The petrographic work on the sandstones of Kyushu has already been described (p. 6). Further details on the stratigraphic division of the Shimanto of Kyushu and the distribution of sandstone compositions are given by Imai *et al.* (1975), Okada (1977), Teraoka (1977, 1979) and Teraoka *et al.* (1979). In the last three papers, the sandstone petrography of western Shikoku is also examined.

At an early stage of sedimentation of the Lower Shimanto Group, Precambrian gneisses and older Mesozoic granites in the Korean Peninsula and their equivalents on the Asian continent may have been important as the provenance of the Shimanto sediments. Yaskawa (1975) deduced from palaeomagnetic studies that in the Cretaceous, southern Japan was to the east of today's Korea. The Ryoke granites and acidic igneous rocks in the Inner Zone of SW Japan may have played increasingly important roles as source rocks for the Shimanto sediments at the later stages of development.

This is supported by palaeocurrent data, summarized in Fig. 1: i.e. current directions are mostly axial, but lateral currents from the north were also consistently present and, sporadically, currents came from the south. Some geologists (e.g. Harata *et al.* 1978), underlining the importance of the southerly currents have, as noted earlier, advocated a 'Kuroshio Palaeocontinent' composed of Pre-Cambrian rocks to the SE of the Shimanto basin, which contributed sediments that were deposited on Precambrian basement. We prefer that the Shimanto sediments of Kyushu, as in Shikoku and Kii, were deposited in basins analogous to modern accretionary forearc basins and partly to trench inner slope basins and trenches. Therefore, tectonic highs like the trench-slope break (Dickinson 1973), active during sedimentation, could have controlled the northward-flowing currents and may sometimes have provided a supply of new or recycled clasts (cf. the Oyashio 'landmass' in the DSDP Japan Trench transect (von Huene, Nasu *et al.* 1978)).

As in Shikoku and Kii, the greenstone facies associated with the flysch of Kyushu is of oceanic origin (Kanmera 1976; Suzuki & Hada 1979). Again the siliceous red and variegated claystones with abundant radiolarians (intensively studied by Nakaseko *et al.* 1979) are closely associated with metabasalts.

Northern Nansei Islands

The best studied Shimanto rocks, the Eocene to Oliogocene Kumage Group, are well exposed on Tanegashima (Okada & Whitaker 1974; Suzuki *et al.* 1979, Hayasaka *et al.* 1980). Rocks of similar age occur on Yakushima (Hashimoto 1956) and other smaller islands.

The Kumage Group occurs in tight folds, including isoclines, with parallel axial planes trending NNE-SSW, which are separated by many thrust planes. The group consists of three tectonostratigraphic units, separated from each other by thrusts. They comprise: (1) a flysch facies with variegated claystones, (2) a chaotic facies and (3) a thick cross-bedded sandstone facies.

(1) The *flysch facies*, dominant in southern Tanegashima and more than 2900 m thick, consists of 10–50 cm thick turbidities, normally interbedded with mudstones of similar thickness, with some thin tuffs and accompanied by radiolarian-bearing black, grey, green and red claystones near the base. The radiolarian assemblage is assigned to the *Thyrsocyrtis triacantha* Zone (Riedel & Sanfilippo 1974), i.e. latest early Eocene to earliest middle Eocene (Suzuki *et al.* 1979), but attempts to date the lutites palynologically have not been successful due to heat degradation (C. Downie, pers. comm. 1978). Within the flysch facies, some slumping occurs.

(2) The *chaotic facies*, more than 700 m thick in the SW of the island, is characterized by pebbles and blocks set in scaly claystones. The sedimentary clasts range from a few millimetres to tens of metres in diameter. A large block of pillowed basalt is also incorporated in claystone matrix surrounded by sandstone.

(3) The *thick cross-bedded sandstone facies*, dominant in northern Tanegashima and around Shimama in the SW, is more than 1300 m thick. This facies is characterized by well-sorted arenites in beds up to tens of metres thick. They are generally of medium sand grain-size and seem to be lacking in pebbles. Large-scale cross-bedding (Fig. 10a), ripple-marks, ripple-drift lamination, mega-ripples (Fig. 10b), parting step lineation and dewatering structures are well developed. The only sole markings recorded are very large flute casts and load casts. When slump structures occur, they are on a spectacular scale (Fig. 10c).

Between the sandstones are dark shales and

FIG. 10. Thick cross-bedded sandstone facies, Kumage Group (Palaeogene), Tanegashima. (a) Normal bedding at base of photograph, foreset bedding behind hammer; (b) Mega-ripple; (c) Large slump fold in massive sandstone.

mudstones bearing siderite nodules and trace fossils: thin coaly beds and lenses are occasionally found above seat-earths. Hayasaka *et al.* (1980) recently reported molluscan fossils suggesting an early Miocene age. The thin siltstones and intermediate sandstones (up to 1 m) which are also present in this facies show convolute lamination, slump structures, ripple-drift lamination, small-scale cross-bedding, thin clastic dykes and tubular trace fossils. Top surfaces sometimes show ripple marks and, at two localities, sand volcanoes (Okada & Whitaker 1979). Flute, groove, striation and gutter casts (Whitaker 1973) occur and, less commonly, longitudinal ridge-and-furrow, prod, bounce, brush and crescentic scour casts. From these sole markings and the foreset beds of the sandstones, a complex palaeocurrent pattern emerges (Okada & Whitaker 1974).

This facies shows thickening- and coarsening-upwards cycles (Okada 1971, 1973), (Fig. 7c) and well-developed slumps (Fig. 7d).

Environments of sedimentation

These three facies are interpreted as having been laid down in the following environments:
(1) The flysch deposits, with variegated radiolarian shale suggesting deep water, may have been deposited in upper to mid-slope perched basins on an inner trench wall.
(2) The chaotic deposits may be part of an accretionary complex beneath the flysch, forming a trench-slope break ridge, the pillowed basalt block having been incorporated into this sequence from the ocean floor during mélange formation: slumping off this ridge could account for disturbances in the other two facies.
(3) The thick cross-bedded sandstone sequence may be forearc basin deposits on the arc side of the trench-slope break ridge. The coals and sideritic nodules suggest a delta top environment, with the thick sandstones as shallow marine sandy delta deposits. The siltstones and mudstones, sometimes slumped, could have formed on an unstable delta-front slope. The variable palaeocurrents would support this interpretation.

If our conclusions are correct, it follows that facies (1) and (3) may be synchronous in their times of sedimentation and younger than facies (2). Unfortunately, fossils are too sparse or not sufficiently diagnostic in the Tanegashima Shimanto Rocks for confirming this suggestion.

Southern and Central Nansei Islands

On Okinawa, the Early Cretaceous Nago Formation of the Northern Belt, consisting of phyllites and mafic volcanic rocks, is in thrust contact with the Eocene Kayo Formation (850 m) of the Southern Belt, which consists of flysch-type sandstone-shale alternations. On the basis of *Nereites*-assemblage trace-fossils, Fukuda & Hayasaka (1978) consider their depth of formation to be 3500–5500 m. On Amami-oshima, the Northern Belt contains phyllites and pillowed basalts of the Naze Formation (1300 m) overlain by Upper Cretaceous flysch sandstones and shales of the Ogachi (2400 m) and Tatsugo (1000 m) Formations, that show coarsening- and thickening-upward cyclic sedimentation. These in turn are covered by Eocene sandstones and shales of the Wano Formation (800 m) (Ishida 1969; Sakai *et al.* 1977). This slightly metamorphosed sequence is severely deformed into isoclinal folds and SE-verging thrust faults (Kizaki 1978).

Cretaceous basins north of the Shimanto Belt

Although outside the Shimanto Belt, two types of basin lying to the north received Cretaceous sediments that may be related to the Cretaceous Shimanto (Teraoka 1977; Taira 1979; Taira *et al.* 1979a).

Forearc shelf basins

These occur within the Chichibu Belt and show three correlatable fining-upward depositional megacycles in the Lower Cretaceous and one in the Upper Cretaceous. Each cycle lasted 10–20 Ma, similar to cycles in the accretionary forearc basin, and is related to synchronous vertical movements of SW Japan.

Intra-arc basins

(i.e. between the volcanic and frontal arcs). These formed during the Late Cretaceous. The best-known are the Mifune, Goshonoura, Onogawa, Himenoura and the 300 km long Izumi Basins (Fig. 1) which, like the forearc shelf basins, had a life-span of about 10 Ma. Coarse turbidites accumulating by longitudinal accretion at 1–4 m/1000 yr gave enormous thicknesses of sediment up to 40 km thick in the Onogawa Basin (Teraoka 1970) and 50 km thick in the Izumi Basin (Suyari 1973) in fault-bounded troughs (compare the 25 km thick Devonian sediments of the Hornelen Basin, Norway (Steel & Gloppen 1980)). Synclinical in structure, these basins were filled from east to west but closed from west to east (Suyari 1973; Taira

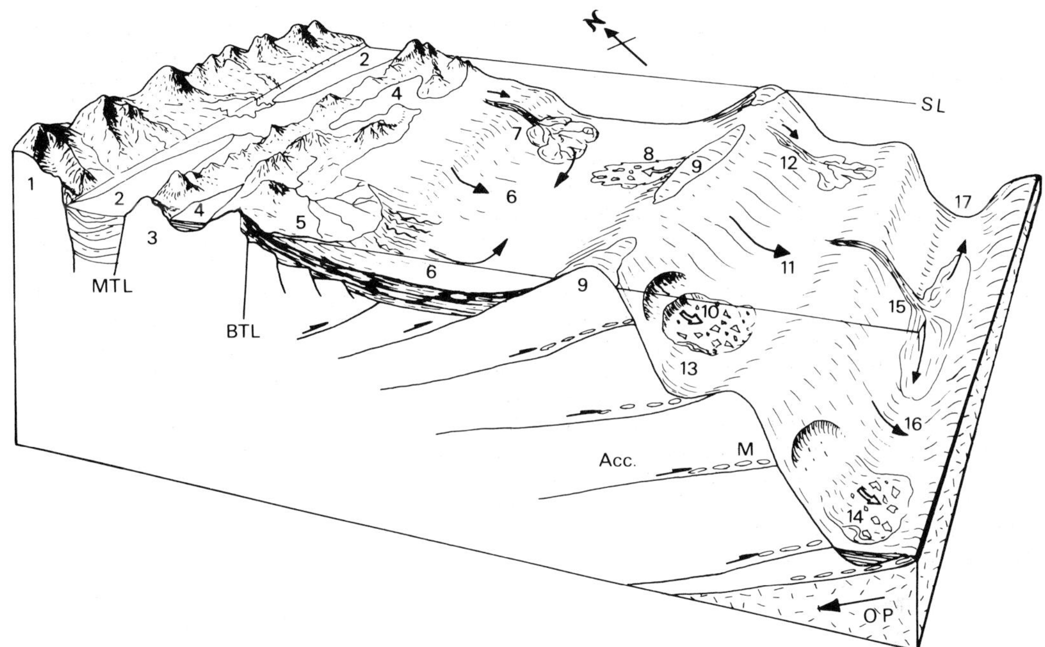

Fig. 11. Schematic block diagram showing inferred environments of deposition of the Lower Shimanto Group during the Cretaceous (not to scale) (after Niitsuma *et al.* 1979; Taira 1979; Taira *et al.* 1979a). (1) Seaward limit of Ryoke high T/P belt; (2) Intra-arc basins (Izumi, Onogawa, etc.), mainly Upper Cretaceous, with very thick longitudinally and rapidly deposited clastics; (3) Sambagawa low T/P and Chichibu Belts, with (4) small forearc shelf basins with thin fluvial and shallow marine sands and clays, mainly Lower Cretaceous; (5) to (8), Lower Shimanto accretionary forearc basin; (5) deltaic sandstones and clays; (6) flysch in deeper water with variable current directions; (7) submarine canyon, fan channel and fan system; (8) slumps and olistostromes from trench-slope break; (9) trench-slope break (partly submergent, partly emergent); (10) slumps and olistostromes on upper inner trench wall; (11) turbidites and (12) canyon-fan system contributing sediments to perched basin (13); (14) slumps and olistostromes on lower inner trench wall; (15) canyon-fan system and (16) deep-water turbidites feeding trench (17). MTL = Median Tectonic Line; BTL = Butsuzo Tectonic Line; Acc = Accretionary prism wedges separated by thrusts with mélange (M); OP = intermittently subducting oceanic plate. Conditions during the Palaeogene for the Upper Shimanto Group are visualized as being broadly similar, but with the Cretaceous Lower Shimanto deformed and exposed to erosion and new accretionary forearc basin, trench-slope break and trench developing further towards the right (SE).

1979). Filling was cyclical: the Himenoura Basin, for example, accumulating five cycles each of 1–2 Ma duration (Tashiro, Taira & Matsumoto, in preparation). The alternating tensional and compressional phases affecting these basins are attributed by one of us (Taira 1979) to subduction proceeding in cyclical manner. We shall return to this idea later, after considering the tectonic and sedimentary setting of the Shimanto Supergroup.

Evolution of the Shimanto Supergroup

Towards the end of Jurassic times, after northward subduction of the Kula Plate had accreted the Chichibu and Sambosan sediments on to the earlier paired metamorphic belts, a long, narrow, NE-SW trending forearc basin (or series of basins) developed above an accretionary complex to the SE of the Chichibu and Sambosan belts. Environments cited in the following account are numbered in Fig. 11. In these accretionary forearc basins (see Fig. 11–6) great thicknesses of Cretaceous sandstones and turbidites accumulated, fed by rivers draining the Eo-Japanese landmass to the north and the Korean region which then lay directly west of Kyushu (Yaskawa 1975). The shallow water sandstones that accumulated on the shelf and in deltas (5) were relatively angular and coarse grained, ill-sorted and rich in feldspars (especially microcline) but with K-feldspar content decreasing through Cretaceous times. Lateral and axial turbidity currents (6) and canyons and fans (7) were active. As subduction proceeded, accretionary wedges at the outer margins of the forearc basins continued to form a positive and active trench-slope break (9). This brought older mélanges and olistostromes (with large basaltic blocks and red chert derived from an oceanic trench or its faulted outer wall) into shallow waters or above sea-level, and slump masses (8) and turbidity currents were shed northwards into the forearc basins and southwards down the inner wall of a deep-sea trench (10), (11), (12). Terraced basins (13), perched on the inner trench wall, accumulated some of these slump deposits and turbidites while the rest of the clastic detritus remained on the slopes of the trench wall or reached the bottom of the trench (17) via turbidity currents (16), slumps (14) or through canyons feeding fans (15).

North of the Butsuzo Tectonic Line, Lower Cretaceous forearc shelf basins (Fig. 11–4) with thin fluvial and shallow marine sands and offshore muds rest on the Chichibu sequence and further north, beyond the Median Tectonic Line, Upper Cretaceous sediments of exceptional thickness occupy down-faulted intra-arc basins.

Towards the end of the Cretaceous, continued subduction had deformed, mildly metamorphosed and exposed some of the accretionary forearc basin deposits and created younger accretionary wedges towards the SE. Thus, new forearc—trench-slope break—trench inner slope—trench environments evolved, oceanward of the Cretaceous ones, and these lasted through the Palaeocene, Eocene and Oligocene into the early Miocene. The Palaeogene accretionary forearc basins were now fed mainly from the newly exposed Lower Shimanto (Cretaceous) rocks. In consequence, the Palaeogene forearc sandstones tend to be less angular, finer grained, better sorted and more quartz-rich than their Cretaceous equivalents, with less feldspars and, in particular, diminishing amounts of microcline upwards. Sandy deltas occupied the arc side of the forearc basin in the SW (Tanegashima area) while submarine fan complexes with channels built out to fill the major part of the central forearc basins (Shikoku region).

The Palaeogene trench-slope break consisted of a broad mélange zone containing a large metabasalt complex forming a basement high. This zone, like its predecessor, shed detritus north into the forearc basin and south on to the trench slope. On the trench inner slope, perched basins trapped flysch and slump deposits. Blocks in the slumps of western Shikoku and the later slumps of Kii are shallow-water sandstones and limestones that came to rest in upper trench slope basins.

During the early to middle Miocene, the Palaeogene Shimanto suffered the same fate as the Cretaceous Shimanto, becoming tightly folded, sliced into fault blocks, thrust into accretionary wedges and, in places, mildly

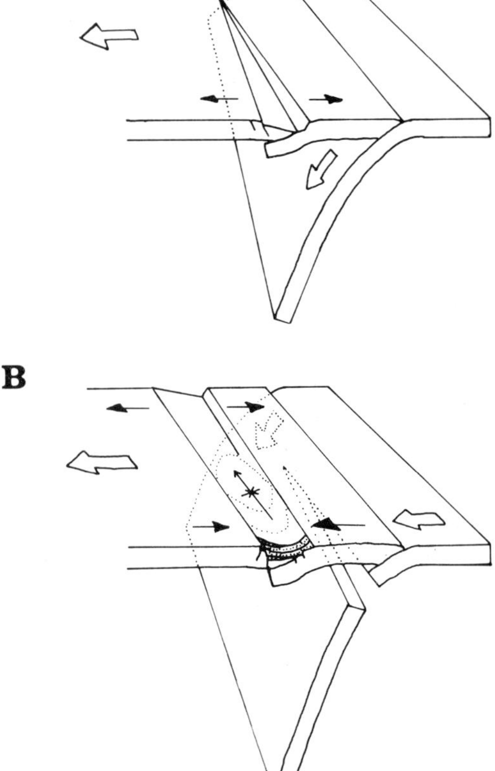

FIG. 12. (A) At the back of the diagram, compression results in accretionary prism developing on inner trench wall and keeping the intra-arc basin closed. At the front, the steepening dip of the downgoing slab gives relative tension, resulting in the opening of an intra-arc rift-type basin.
(B) At the back of the diagram, there is relative tension, as in (A), front. At the front, partial detachment of the slab begins a new cycle of subduction with compression closing the intra-arc basin. (From Taira 1979, fig. 6, with permission of *The Earth Monthly*, Kaiyo-shuppan Co., Tokyo.)

metamorphosed. Their uplift and erosion caused clastic sediments to be shed into new basins, such as those of the Miyazaki Group in southern Kyushu and the Tanabe and Kumano Groups of western and eastern Kii respectively (Fig. 2). Many of these Miocene deposits may be traced offshore (the Tanabe Formation being exposed in the walls of present-day submarine canyons, Fig. 9) and, with Pliocene and Pleistocene deposits, make up the fill of today's terraced forearc basins. The modern trench-slope break is flanked on the Pacific side by the inner wall of the Nankai Trough (Fig. 2, columns 5, 6) with basins perched on accretionary wedges (Inoue *et al.* 1977).

This shift SE of Shimanto forearc basin to trench environments from their Cretaceous positions, through Palaeogene positions, to their present sites, seems to have been effected at intervals by cyclic subduction (Niitsuma *et al.* 1979; Taira 1979; Taira *et al.* 1979a). This gave rise to alternating periods of compression (maximum subduction with accretionary prism development) and tension, which produced faulting and new intra-arc rift-type basins such as the Izumi, Onogawa and Mifune Basins on the arc side of the forearc basins ('retroduction' of Taira (1979), 'roll-back effect' of Dewey 1980). This idea, developed by one of us (A.T.) in several papers (Taira 1979; Katto & Taira 1978), could also account for simultaneous compression at one place and tension at another, as illustrated in Fig. 12 (from Taira 1979, fig. 6). Should subduction have been oblique, a lateral component of movement could have resulted perhaps in considerable strike-slip along the faults (such as the Median and Butsuzo Tectonic lines) bounding the intra-arc basins, as described by Lewis (1980) in New Zealand (Whitaker 1982).

The northward movement of the Kula Plate in Shimanto times changed to a westward movement of the Pacific Plate with transform motion parallel to the Shimanto trend in its eastern part. This, followed by the opening of the Shikoku backarc basin, the Pliocene-Recent renewed northward subduction under Shikoku and Honshu and the northward movement of the Izu Block to collide with the Akaishai & Kanto Shimanto Belts (Matsuda 1978) make a complicated but intriguing sequel to Shimanto events (Smith 1982).

ACKNOWLEDGMENTS: A.T. wishes to express his appreciation to Professor J. Katto, Drs M. Tashiro and M. Okamura and other faculty members of Kochi University for their encouragement. H.O. wishes to thank Professor S. Hayasaka and Dr Y. Teraoka for constructive comments and Messrs T. Arakawa and K. Suzuki for their help in the field. H.O. and A.T. are grateful for a Grant-in-Aid for Scientific Researches from the Ministry of Education, Japan (Grant No. 434041). J.H.McD.W. is grateful to the Royal Society for grants for two long visits to Japan, and to the many Japanese geologists, especially his co-authors Professor H. Okada and Dr A. Taira, who made these studies not only possible but enjoyable. A.J.S. is grateful to the Royal Society and the Japan Society for the Promotion of Science for grants for several visits to Japan, and to many Japanese geologists, particularly members of the Kishu Shimanto Research Group, for their guidance and companionship in the field. Professor Charles Downie kindly examined many specimens for palynomorphs. Constructive comments on the manuscript were made by Dr J. K. Leggett, Professor J. Tarney, Dr R. von Huene and a Japanese reviewer.

References

(J signifies that the paper is in Japanese: J,E indicates a summary in English).

BACHMAN, S. B. 1981. The Coastal Belt of the Franciscan: youngest phase of northern California subduction (this volume).

DAVIES, W. & CAVE, R. 1976. Folding and cleavage determined during sedimentation. *Sediment. Geol.* **15,** 89–133.

DEWEY, J. F. 1980. Abstract from GSL Meeting, London, June 23–25.

DICKINSON, W. R. 1973. Widths of modern arc-trench gaps proportional to past duration of igneous activity in associated magmatic arcs. *J. geophys. Res.* **78,** 3376–89.

—— & SEELY, D. R. 1979. Structure and stratigraphy of forearc regions. *Bull. Am. Assoc. Petrol. Geol.* **63,** 2–31.

EKDALE, A. A. 1980. Graphoglyptid burrows in modern deep-sea sediment. *Science*, **207,** 304–6.

FLOYD, J. D. 1975. *The Ordovician Rocks of West Nithsdale.* Thesis, PhD, Univ. St Andrews (unpubl.).

FUKUDA, Y. & HAYASAKA, S. 1978. Trace fossils from the Eocene Kayo Formation in Okinawa-shima, Ryukyu Islands, Japan. *Rep. Fac. Sci., Kagoshima Univ. (Earth Sci. Biol.)* **11,** 13–25.

HARATA, T., HISATOMI, K., KUMON, F., NAKAZAWA, K., TATEISHI, M., SUSUKI, H. & TOKUOKA, T. 1978. Shimanto geosyncline and Kuroshio Paleoland. *Suppl. J. Phys. Earth,* **26,** 357–66.

HASHIMOTO, I. 1956. Geological succession and structure of the undated strata in Yaku-shima, and some facts concerning the Kumage Group in the western part of Tanega-shima, Kagoshima Prefecture. *Rep. Earth Sci. Dep. Gen. Educ. Kyushu Univ.* **2,** 23–34.

—— 1962. The sedimentary complex of uncertain ages in South Kyushu. *Rep. Earth Sci. Dep. Gen Educ. Kyushu Univ.* **9,** 13–69 (J,E).

HATENASHI RESEARCH GROUP 1975. Stratigraphy and geologic structure of the Otonashigawa-Muro Group in the Kii Peninsula, southwest Japan. *Monogr. Assoc. Geol. Collab. Jpn* **19,** 157–66 (J,E).

HAYASAKA, S., FUKUDA, Y. & HAYAMA, A. 1980. Discovery of molluscan fossils and the paleoenvironmental aspects of the Kumage Group in Tanegashima, South Kyushu, Japan. *Prof. Saburo Kanno Mem. vol. Tsukuba Univ.* 59–70.

IMAI, I. 1977. Cretaceous and lower Tertiary Systems in the Shimanto Belt. *In*: TANAKA, K. & NOZAWA, T. (eds). *Geology and Mineral Resources of Japan 1*, 3rd edn, 207–13. Geol Surv. Japan, Hisamoto.

—— TERAOKA, Y. & OKUMURA, K. 1975. Structural division of the Shimanto Terrane of Kyushu, southwest Japan. *Monogr. Assoc. Geol. Collab. Jpn* **19,** 179–89 (J,E).

INOUE, E., ISHIBASHI, K., ISHIHARA, T., KINOSHITA, Y., JOSHIMA, M. & TAMAKI, K. 1977. Geological map off outer zone of southwest Japan, 1 : 1 million. *Mar. Geol. Map Ser.* **8.** Geol. Surv. Japan.

ISHIDA, S. 1969. Wano Formation (Eocene) in Amami-Oshima. Ryukyu Islands, Japan. *J. geol. Soc. Japan.* **75,** 141–56.

KANMERA, K. 1976. Comparison between ancient and modern geosynclinal sedimentary bodies. I, II, *Kagaku (Sci.)* **46,** 284–91, 371–8 (J).

KARIG, D. E. & SHARMAN III, G. F. 1975. Subduction and accretion in trenches. *Bull. geol. Soc. Am.* **89,** 377–89.

KATTO, J. 1960. Some problematica from the so-called unknown Mesozoic Strata of the southern part of Shikoku, Japan. *Sci. Rep. Tohoku Univ. Sendai, 2nd Ser. (Geol.)* **4.** 323–34.

—— 1964. Some sedimentary structure and problematica from the Shimanto Terrain of Kochi Prefecture, Japan. *Res. Rep. Kochi Univ. 13, Nat. Sci. I,* **6,** 14 pp.

—— & ARITA, M. 1966. Geology of the Muroto Peninsula, Shikoku, Japan. *Res. Rep., Kochi Univ. 15, Nat. Sci. I,* **8,** 59–63 (J,E).

—— KOJIMA, J., SAWAMURA, T. & SUYARI, K. 1961. *The Geology and Mineral Resources of Kochi Prefecure, with Geologic Map.* Kochi Prefecture (J).

—— SUYARI, T., KASHIMA, Y., HASHIMOTO, I., HADA, S., MITSUI, S. & AKOJIMA, I. 1977. *Surface Geologic Map of Shikoku.* Kochi Regional Forestry Office.

—— & TAIRA, A. 1978. Lithofacies and depositional environments of the Muroto-hanto Group. *Geol. News, Tokyo,* **287,** 21–31 (J).

—— & TASHIRO, M. 1978. A study on the molluscan fauna of the Shimanto terrain, southwest Japan, part 1: on the bivalve fauna of the Doganaro Formation in Susaki area, Kochi Prefecture. *Res. Rep. Kochi Univ. Nat. Sci.* **27,** 143–50 (J,E).

—— & TASHIRO, M. 1979a. A study on the molluscan fauna of the Shimanto terrain, southwest Japan, part 2: bivalve fauna from the Muroto-hanto Group in Kochi Prefecture, Shikoku. *Res. Rep. Kochi Univ. Nat. Sci.* **28,** 1–11.

—— & TASHIRO, M. 1979b. A study on the molluscan fauna of the Shimanto terrain, southwest Japan, part 3: on the bivalve fauna from the Arioka, Nakamura and Susaki Formations in Shimanto (Northern) terrain, Kochi Prefecture. *Res. Rep. Kochi Univ. Nat. Sci.* **28,** 49–58.

—— TASHIRO, M., TAIRA, A. & OKAMURA, M. 1980. Biostratigraphy of the Cretaceous Shimanto Belt in the Susaki area, Shikoku. *Geol. News, Tokyo,* **310,** in press.

KISHU SHIMANTO RESEARCH GROUP 1970. Sedimentological and paleontological studies of the Muro Group of the Kii Peninsula. *Mem. Fac. Educ. Wakayama Univ. Nat Sci.* **20,** 75–102 (J,E).

—— 1975. Development of the Shimanto geosyncline. *Monogr. Assoc. Geol. Collab. Jpn* **19,** 143–56. (J,E).

KIZAKI, K. 1978. Tectonics of the Ryukyu Island arc. *Suppl. J. Phys. Earth,* **26,** 301–7.

KULM, L. D., RESIG, J. M., THORNBURG, T. M. & SCHRADER, H. J. 1981. Cenozoic structure, stratigraphy and tectonics of the central Peru forearc (this volume).

LEE, S. M. 1974. The tectonic setting of Korea with relation to plate tectonics. *Tech. Bull. U.N. ESCAP, CCOP,* **8,** 39–53.

LEGGETT, J. K., MCKERROW, W. S. & CASEY, D. M. 1981. The anatomy of a Lower Palaeozoic accretionary forearc: the Southern Uplands of Scotland (this volume).

—— MCKERROW, W. S., MORRIS, J. H., OLIVER, G. J. H. & PHILLIPS W. E. A. 1979. The north-western margin of the Iapetus Ocean. *In*: HARRIS, A. L., HOLLAND, C. H. & LEAKE, B. E. (eds). *The Caledonides of the British Isles Reviewed.* Spec. Publ. geol. Soc. London, **8,** 499–512.

LEWIS, K. B. 1980. Quaternary sedimentation on the Hikurangi oblique-subduction and transform margin, New Zealand. *In*: BALLANCE, P. F. & READING, H. G. (eds). *Sedimentation in Oblique-slip Mobile Zones.* Spec. Publ. Int. Assoc. Sedimentol. **4,** 171–89. Blackwell Scientific Publications, Oxford, 232 pp.

MATSUDA, T. 1978. Collision of the Izu-Bonin arc with central Honshu: Cenozoic tectonics of the Fossa Magna, Japan. *Suppl. J. Phys. Earth*, **26,** 409–21.

MATSUMOTO, T. & OKADA, H. 1978. Evaluation of molluscan fossils from the Mesozoic of the Shimanto Belt. *Proc. Japan. Acad.* **54B,** 235–330.

MIYASHIRO, A. 1973. *Metamorphism and Metamorphic Belts.* Allen & Unwin, London.

MOORE, G. F. & KARIG, D. E. 1976. Development of sedimentary basins on the lower trench slope. *Geology*, **4,** 693–7.

MOORE, J. C., WATKINS, J. S., SHIPLEY, T. H., BACHMAN, S. B., BEGHTEL, F. W., BUTT, A.,

DIDYK, B. M., LEGGETT, J. K., LUNDBERG, N., McMILLEN, K. J., NIITSUMA, N., SHEPHARD, L. E., STEPHAN, J.-F. & STRADNER, H. 1979. Progressive accretion in the Middle America Trench, Southern Mexico. *Nature. London,* **281,** 638–42.

NAKASEKO, K., NISHIMURA, A. & KANNO, K. 1979. Study of radiolarian fossils from the Shimanto Belt. Spec. Issue *News Osaka Micropaleont.* **2,** 1–49.

NIITSUMA, N., TAIRA, A. & KITAZATO, H. 1979. History of the plate boundaries in the Japanese Islands. *Earth Monthly, Tokyo,* **1,** 193–205 (J).

OGAWA, Y. 1981. Tectonics of some forearc fold belts in and around the arc-arc crossing area in central Japan (this volume).

—— & HORIUCHI, K. 1978. Two types of accretionary fold belts in central Japan. *Suppl. J. Phys. Earth,* **26,** 321–36.

OKADA, H. 1971. A pattern of sedimentation in clastic sediments in geosynclines. *Mem. geol. Soc. Japan,* **6,** 75–82 (J,E).

—— 1973. Coarsening-upwards cycle. *Tohoku Univ. Sci. Rep., 2nd Ser. (Geol.)* **6,** 429–37 (J,E).

—— 1977. Preliminary study of sandstones of the Shimanto Supergroup in Kyushu, with special reference to "Petrographic Zone". *Sci. Rep., Dept Geol., Kyushu Univ.* **12,** 203–14 (J,E).

—— & WHITAKER, J. H. McD. 1974. Shimanto Group in Tanegashima off South Kyushu (Abs.). *81 Ann. meet. Geol. Soc Japan,* 201(J).

—— & WHITAKER, J. H. McD. 1979. Sand volcanoes of the Palaeogene Kumage Group, Tanegashima, southwest Japan. *J. geol. Soc. Japan.* **85,** 187–96.

OKUDA, Y., KUMAGAI, M. & TAMAKI, K. 1979. Tectonic development of the continental slope and its peripheral area off southwest Japan in relation to sedimentary sequences in sedimentary basins. *J. Japan Assoc. Petrol. Technol.* **44,** 279–90.

RIEDEL, W. R. & SANFILIPPO, A. 1974. Radiolaria from the southern Indian Ocean, DSDP Leg. 26. *Initial Rep. Deep Sea drill. Proj,* **26,** 771–813.

SAKAI, T. 1978. Geologic structure and stratigraphy of the Shimantogawa Group in the middle reaches of the Gokase River, Miyazaki Prefecture. *Sci. Rep., Dept Geol., Kyushu Univ.* **13,** 23–38 (J).

—— ONO, K., MOMOKI, Y., OTSUKA, H. & HAYASAKA, S. 1977. Geology of the northern part of Amami-Oshima. *Geol. Studies Ryukyu Islands,* **2,** 11–23 (J,E).

SANO, H., KANMERA, K., & SAKAI, T. 1979. Sediments associated with greenstones of the Shimanto terrain. *J. geol. Soc. Japan,* **85,** 435–44 (J,E).

SEELY, D. R., VAIL, P. R. & WALTON, G. G. 1974. Trench slope model. *In*: BURK, C. A. & DRAKE, C. L. (eds). *The Geology of Continental Margins,* 249–60. Springer-Verlag, New York.

SHIBATA, K. & NOZAWA, T. 1973. K-Ar ages of gravels of orthoquartzite and gneiss from the Muro Group. *Bull. geol. Surv. Japan,* **24,** 551–3.

SMITH, A. J. 1982. The Neogene to Recent geology of Japan and its surrounding seas. *Proc. Geof. Ass.* **93** (in press).

STEEL, R. & GLOPPEN, T. G. 1980. Late Caledonian (Devonian) basin formation, western Norway: signs of strike-slip tectonics during infilling. *In*: BALLANCE, P. F. & READING, H. G. (eds). *Sedimentation in Oblique-slip Mobile Zones.* Spec. Publ. Int. Assoc. Sedimentol. **4,** 79–103. Blackwell Scientific Publications, Oxford, 232 pp.

SUGISAKI, R., SUZUKI, T., KANMERA, K., SAKAI, T., & SANO, H. 1979. Chemical compositions of green rocks in the Shimanto Belt, southwest Japan. *J. geol. Soc. Japan,* **85,** 455–66.

SUYARI, K. 1973. On the lithofacies and the correlation of the Izumi Group of the Asan Mountain Range, Shikoku. *Sci. Rep., Tohoku Univ. Hatai Mem. Vol,* **6,** 489–95 (J,E).

SUZUKI, H. 1975. Deformed structures and deformational history of the Palaeogene Muro Flysch in southwest Japan. *Monogr. Assoc. Geol. Collab. Jpn,* **19,** 167–77 (J,E).

SUZUKI, K., NAKASEKO, K. & OKADA, H. 1979. Stratigraphy, structure of the Kumage Group in southern Tanegashima (Abs.) *86 Ann. meet. geol. Soc. Japan,* 134 (J).

SUZUKI, T. & HADA, S. 1979. Cretaceous tectonic mélange of the Shimanto belt in Shikoku, Japan. *J. geol. Soc. Japan,* **85,** 467–79.

TAIRA, A. 1979. Formation of sediment body in the arc-trench system and cyclic subduction model. *Earth Monthly, Tokyo,* **1,** 860–8 (J).

—— KATTO, J, & TASHIRO, M. 1979a. The Cretaceous and Cenozoic geologic development of southwest Japan and the tectonism of arc trench system. *Geol. News, Tokyo,* **296,** 27–40 (J).

——, ——, —— & OKAMURA, M. 1980. Geology and origin of the Shimanto Belt in Kochi Prefecture, Japan. *In*: TAIRA, A. & TASHIRO, M. (eds). *Geology and Paleontology of the Shimanto Belt.* Rinyakosaikai Press, Kochi, Japan (in press) (J,E).

—— NAKASEKO, K., KATTO, J., TASHIRO, M. & SAITO, Y. 1979b. New observations on the Sambosan Group in the western Kochi Prefecture. *Geol. News, Tokyo,* **302,** 22–35. (J).

TANAKA, K. 1977. Shimanto Supergroup in the Sukumo area, southwestern Shikoku. *Bull. geol. Surv. Japan,* **28,** 461–76 (J,E).

—— & NOZAWA, T. (eds) 1977. *Geology and Mineral Resources of Japan,* 3rd edn, **1,** 430 pp. Geol. Surv. Japan.

TAZAKI, K., INOMATA, M. & TAZAKI, K. 1980. Umbers in pillow lava from the Mineoka tectonic Belt, Boso Peninsula. *J. geol. Soc. Japan,* **86,** 413–6.

TERAOKA, Y. 1970. Cretaceous formations in the Onogawa Basin and its vicinity, Kyushu, Southwest Japan. *Rep. geol. Surv. Japan,* **237,** 87 pp. (J,E).

—— 1977. Comparison of the Cretaceous sandstones between the Shimanto Terrane and the Median Zone of Southwest Japan, with reference to the provenance of the Shimanto geosynclinal sediments. *J. geol. Soc. Japan,* **83,** 795–810 (J,E).

—— 1979. Provenance of the Shimanto geosynclinal sediments inferred from sandstone compositions. *J. geol. Soc. Japan,* **85,** 753–69 (J,E).

—— & Obata, I. 1975. Stratigraphy of the Upper Cretaceous Uwajima Group. *Mem. Nat. Sci. Museum, Tokyo,* **8,** 5–20 (J,E).

—— Okumura, K. & Imai, I. 1979. Sandstones of the Shimanto Supergroup in the Mimi-kawa area, Kyushu, southwest Japan—with reference to the structural division of the Shimanto terrane. *In*: Hara, I. *et al.* (eds). *The Basement of Japanese Islands,* 133–51. Prof. Hiroshi Kano Mem. Vol., Akita Univ. (J,E).

Tokuoka, T. & Okami, K. 1979. Orthoquartzite clasts and the problem on the basement rocks of the Japanese Islands. *In*: Hara, I. *et al.* (eds). *The Basement of Japanese Islands,* 601–23. Prof. Hiroshi Kano Mem. Vol., Akita Univ. (J,E).

Tsuchiya, N., Sakai, T. & Kanmera, K. 1979. Mode of occurrence and petrological characteristics of greenstones of the Shimanto Terrain in the Mimi River area, Kyushu. *J. geol. Soc. Japan,* **85,** 445–54 (J,E).

Von Huene, R., Nasu, N. *et al.* 1978. On Leg 57 Japan Trench transected. *Geotimes*, **23,** 16–20.

Whitaker, J. H. McD. 1973. 'Gutter casts', a new name for scour-and-fill structures: with examples from the Llandoverian of Ringerike and Malmöya, Southern Norway. *Nor. geol. Tidsskr,* **53,** 403–17.

—— 1982. Cretaceous-Palaeogene geology of south-west Japan. *Proc. Geof. Ass.* **93** (in press).

Yaskawa, K. 1975. Palaeolatitude and relative position of south-west Japan and Korea in the Cretaceous. *Geophys. J. R. astron. Soc.* **43,** 835–46.

A. Taira, Department of Geology, Kochi University, Akebono-cho 2–5–1, Kochi, 780, Japan.

H. Okada, Institute of Geosciences, Shizuoka University, Shizuoka, 422, Japan.

J. H. McD. Whitaker, Department of Geology, The University, Leicester LE1 7RH, England.

A. J. Smith, Department of Geology, Bedford College, Regent's Park, London NW1 4NS, England.

Sedimentation across the Japan Trench off northern Honshu Island

Roland von Huene & Michael A. Arthur

SUMMARY: The convergent margin along the International Program of Ocean Drilling (IPOD) Japan Trench transect was sampled by dredging, piston coring, and drilling by the Deep Sea Drilling Project (DSDP). This sampling was within an extensive network of single-and multichannel seismic reflection records. The Quaternary uplift of northern Honshu Island, climatic change, and lowered sea-levels, provided coarser grained sediment during the Quaternary than during the late Miocene and Pliocene. Within the Quaternary sediments, those in the forearc basin, in a slope basin, and on the trench floor are coarser grained and accumulated more rapidly than the clayey sediment from the trench slope. But overall, the sediment sampled from all environments is mainly clay and silt. Rates of accumulation varied greatly and prior to 3 Ma ago they were higher in trench slope basins than in the forearc basin. Tectonism along the Japan Trench transect must have changed the Miocene-Pliocene morphology by tilting some of the former trench slope basins; this may have actually triggered slumping and erosion rather than allowing sediment deposition.

The lower slope has a relatively large number of channels and a greater amount of slumped material then is found elsewhere; thus it may contain vertical and lateral changes of facies in sediments deposited over relatively short periods of time. Morphological features and sedimentation patterns on the lower slope are greatly affected by local tectonic structure which has changed significantly in the past 2–3 Ma. Slope basin and trench-floor deposits are distinguished from those of the main forearc basin by the relative abundance of material deposited by mass movement. Channels commonly cross the forearc basin and trench slope and provide the transport paths for coarse sediment to any part of the margin. No specific characteristics appear exclusively in a given environment, but the relative abundance of features can help identify broad environments of deposition.

Introduction

The Japan Trench transect, a 100 km wide corridor across the convergent margin off northern Honshu Island, was studied using a variety of geophysical methods, conventional sampling, and drilling by the *Glomar Challenger* during Legs 56 and 57 (Fig. 1). This effort was concentrated in the forearc area, where a network of reconnaissance single channel seismic reflection records (Tamaki *et al.* 1977) was obtained prior to the acquisition of five 24-channel CDP seismic reflection records (Nasu *et al.* 1980). Many dredge samples, several piston cores, and seven DSDP drill sites provided samples from the forearc basin, and across the Japan Trench. The distribution of sediment types recovered in samples combined with the structure shown in seismic records help outline a late Miocene to Quaternary history of sedimentation in the Japan Trench.

Geological structure across the Japan Trench transect consists of landward dipping reflections beneath strata paralleling the sea-floor. This pattern is similar to the structure seen in seismic records of other convergent margins. However, the Neogene sediment section off Honshu developed during regional subsidence of the outer forearc area which was once subaerially exposed and is now buried beneath 1–3 km of sediment (von Huene, Nasu *et al.* 1978). This contrasts greatly with certain other outer forearc areas which appear dominated by uplift (Moore, Watkins *et al.* 1979, 1981). A relatively flat terrace occupies the forearc which has subsided below the normal 200 m maximum depth of continental shelves; this is called the Sanriku deep-sea terrace (Nasu 1964). Seaward of the deep-sea terrace along the trench landward slope, is a 4000 m deep narrow terrace—the mid-slope terrace—which is structurally a slope basin and we use the morphological and structural terms interchangeably.

In this paper we summarize sample and seismic information to define the Neogene and Quaternary sediment facies in each morphotectonic environment. We describe the gross structure of the margin and examine the distribution of Quaternary sediment types and morphological features. Then we examine the succession of older sediments and relate this succession to tectonic history. Our observations of sediment facies and the sedimentary features associated with the late Neogene environments

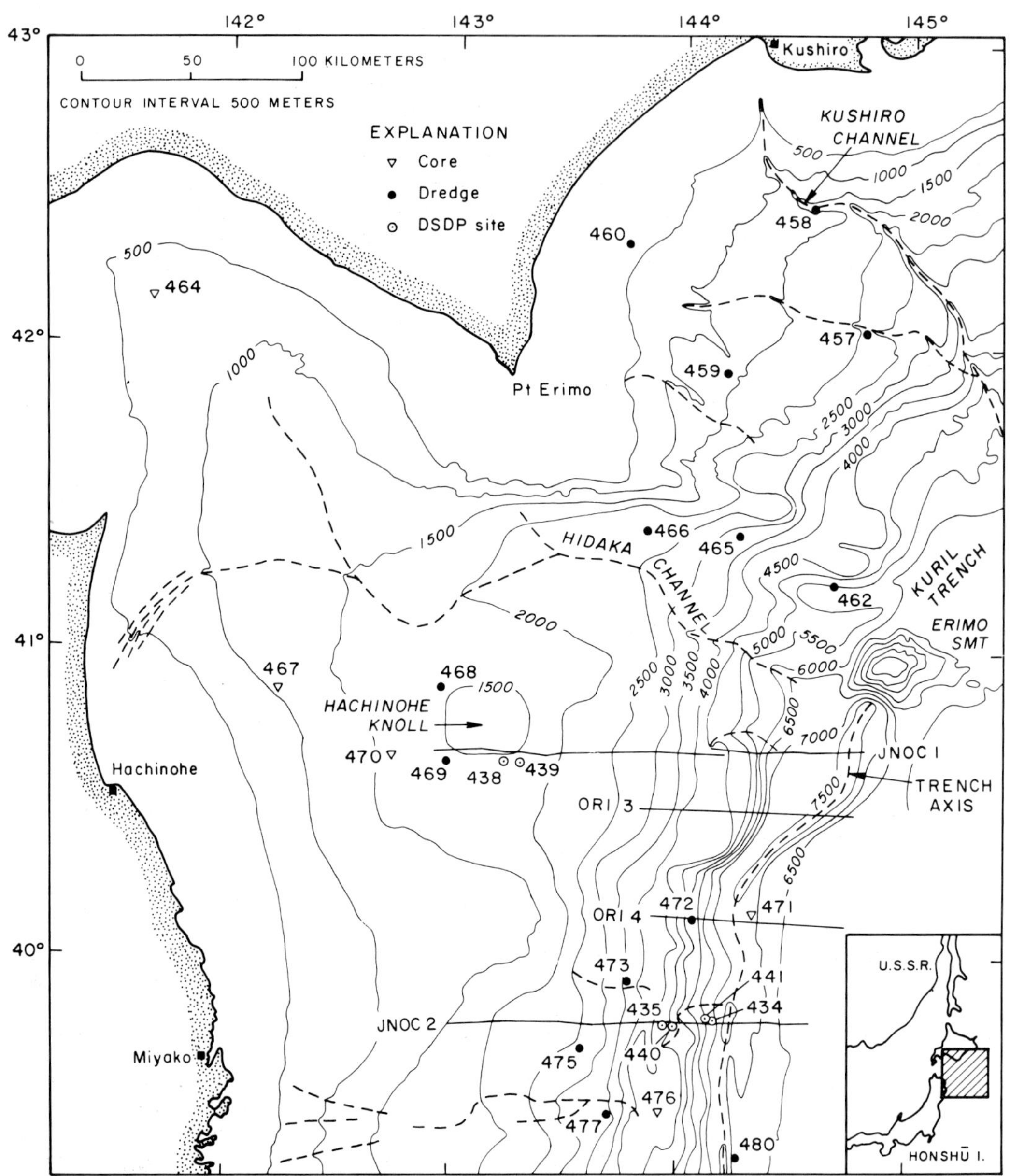

FIG. 1. Pacific side of northern Japan showing the coast of northern part of Honshu and southern Hokkaido Islands. Lines indicate 24-fold CDP seismic reflection records; circled dot indicates DSDP drill site; dot indicates location of dredge hauls; and triangle indicates location of piston core. Channels indicated by dashed lines.

lead to some predictions of features that may be found at convergent margins like the Japan Trench.

Structure of the Japan margin

The Japan margin off Honshu consists of the Neogene Ishikari-Hidaka forearc basin, a topographically simple trench landward slope, a trench floor with generally little ponded fill, and a block-faulted slope seaward of the trench (Fig. 2) (Mogi 1978; Ishiwada & Ogawa 1976). The configuration of the Ishikari-Hidaka basin is like that of most forearc basins, although it is in 1500–3000 m of water. In some places sediment of the basin makes a gradual transition to the trench slope (Fig. 2A); in other places it is marked by a sharp break in slope (Fig. 2B). Neogene sediment filling the Ishikari-Hidaka basin rests on an angular unconformity that can be followed throughout the seaward part of the basin.

The slope landward of the trench is divided into an upper and lower part by a narrow but continuous mid-slope terrace (Fig. 2). The upper slope is underlain by a continuation of the Ishikari-Hidaka basin sediment section and the unconformity at its base. The mid-slope terrace, formed by a narrow slope basin in which sediment has ponded, marks the seaward extent of the unconformity and a basic change in the structure of the margin (von Huene *et al.* 1980).

Seismic records that show the internal structure of the lower slope is badly obscured by diffractions (Fig. 2). On record JNOC-2, which shows the part of the slope where most of the drilling was done, the deep poorly defined reflectors dip landward. Overlying these reflectors are strata sub-parallel to the seafloor which are the slope deposits that were drilled at DSDP sites 435, 440, 441, and 434. The landward-dipping strata may have been penetrated only at site 434.

The trench floor is too deep to be drilled by *Glomar Challenger*. It has thin ponded sediment covering about 2 km of the trench floor that was sampled by piston coring. In record JNOC-1 (Fig. 2A) a large toe projects from the foot of the trench slope. This toe is interpreted as sediment transported by mass movement from the slope (Arthur *et al.* 1980; Nasu *et al.* 1980).

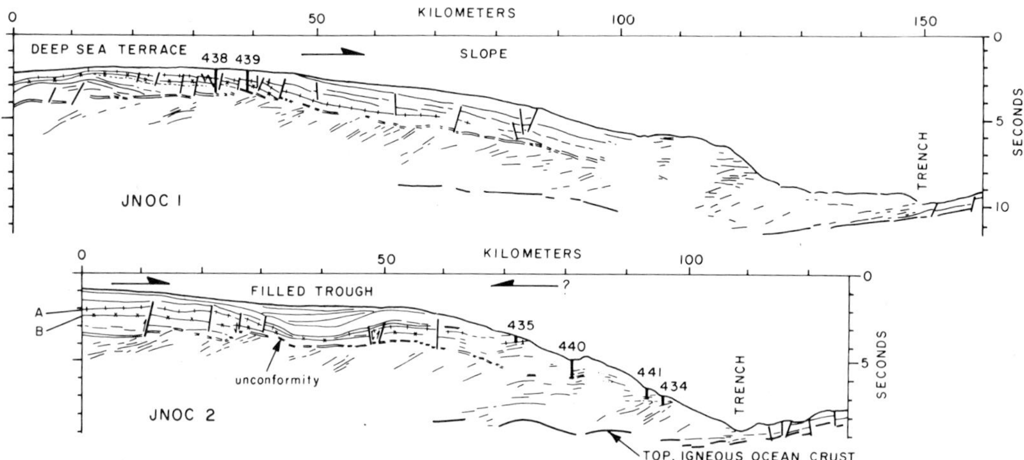

FIG. 2. Line drawings of 24-fold CDP seismic reflection records JNOC-1 (A) and JNOC-2 (B); locations of DSDP drill sites indicated by site numbers. Vertical axis is 2-way time (seconds) horizontal axis is distance (kilometres). Double line indicates angular unconformity.

The slope seaward of the trench consists of igneous oceanic basement overlain by a uniform layer of sediment that follows the topography (Fig. 2). Ocean crust and overlying sediment are broken by normal faults as they are flexed downward into the trench. The upper part of this sediment is composed of hemipelagic material from a terrigenous source.

Sampling of Quaternary sediment

Early studies

Descriptions of sediment samples from the northern Honshu and Hokkaido margins have

TABLE 1. *Dredged and cored samples after Yuasa et al. (1977); Inouchi et al. (1977); Hasegawa et al. (1977); Nasu et al. (1960); Nasu & Sato (1962). Samples in the vicinity of the Japan Trench transect are as shown in Fig. 1, and the lithologies penetrated at DSDP sites are shown in Fig. 3*

Station no.	Depth (m)	Lithology
		Deep sea terrace (Ishikari-Hidaka Basin)
443	1710–1700	Numerous cobbles and pebbles, and sandy silt. Clasts include sandstone, tuff, limestone, rhyolite, andesite, biotite granite, diorite, biotite hornfels, metavolcanic rock, and partly calcareous metamorphic rock. Age 1.65 to 2.1 Ma
444	1080	Olive-brown silt with pebbles of pumice, scoria, and siltstone
467	1150	377 cm core of silt, sandy silt, and clay broken by thin sand and volcanic sand beds
468	1740–1730	Medium volcanic sand, and cobbles of volcanic rock, weakly metamorphosed rock, and clinopyroxene-olivine basalt, pebbles of chert, pumice, shale, and volcanics. Dredge site in or near a canyon
469	1420–1415	Fine sand with pumice, and shale pebbles, dredged from side of Hachinohe Knoll
470	1650	376 cm core of clay with very thin sand layers from base of Hachinohe Knoll
479	1850	462 cm core of silt with six thin sand beds
E1T	1690	A dredge yielding sand, mud, and a few pebbles
E1P	1800	A core of mud with sand
		Trench upper slope
435	2800	Silty clay or silt, and pumice
437	3050–2850	Micritic limestone, sandy limestone, and tuffaceous sand clasts Limestone; pale bluish green in colour, with grains of plagioclase, hypersthene, augite, brownish glass, and rock fragments. Arenite; pale greyish brown with grains of quartz, plagioclase, clinopyroxine, and siliceous rock fragments. Dredged across a canyon. Age 0–0.65 Ma
439	2520–2500	Sandy silt, pebbles of chert, slate, and sandstone, sedimentary rock fragments, and scoria. Age 0–0.65 Ma
445	3120–3050	Sandy olive-brown silt with pebbles and gravels of chert, siltstone and pumice
448	2520–2500	Sandy silt with pebbles of granite, chert and siltstone. Age 0.4–2.1 Ma
473	3110–3050	Semiconsolidated silt, siltstone, and pebbles of chert, volcanic and metamorphic rocks. Age 0.9–1.4 Ma
475	2510	Clay and pumice dredged from top of slope
477	3330–3290	Silt, siltstone, and pumice dredged from outer lip of a small terrace
478	3400–3380	Silt, cobbles of siltstone and andesitic rock, pebbles of pumice, siltstone, igneous rock, and shale
EDS-2-S_1 and S_2	2300	Gravel of various rock types
EDS-2-B_1	430	Sandy mud from side of Kushiro Canyon
476	4770	556 cm core of clay with silt beds and thin layers of sand, volcanic sand, and plant fragments. Core from mid-slope terrace basin
440	4980	Silt, sedimentary rock fragments, gravels, pumice, and scoria dredged from mid-slope terrace. Age 1.65–2.1 Ma
		Trench lower slope
472	4870–4560	Clay including pebbles and semiconsolidated silt. Pumice and tuffaceous siltstone (?) fragment. Age 0.4–0.7 Ma
474	5280–5150	Silt, semi-consolidated siltstone, and siltstone
480	7000–6850	Silty clay, siltstone and a dolerite clast. Age 2.9–4.4 Ma
E2P	6935	324 cm core of blue green mud, and some red-brown mud

		Trench
434	7900	Pebbles of olivine basalt, chert, and mudstone. Corer was stopped by gravel
438	7300	475 cm core of clay with many thin sand and tuff beds and a 50 cm sand bed at base of core
447	7400	442 cm core of clay with thin beds of sand
471	7330	541 cm core of clay with four very thin tuffaceous sand and sand layers
JEDS-2-C_2	8005	85 cm core of mud with sand
		Trench seaward slope
446	5790–5740	Reddish-brown clay with manganese nodules and manganese oxide coated pebbles, and pale greenish grey clay
442	5400	464 cm core of silt, sand beds, clay and tuff layers
E2T	6700–7340	Sand and mud with numerous gravels were dredged along with contaminated red clay. Wood fragments and bamboo
E3P	5430	258 cm core of blue green mud mixed with red brown clay and some sand

been reported since the rebuilding of Japan's scientific capability after World War II (Table 1). Terrigenous sand, gravel, and mud were sampled from the Pacific side of northern Honshu and Hokkaido in the late 1950s by Nasu (Nasu *et al.* 1960; Nasu & Sato 1962). Of great interest to Nasu and his colleagues were the large amounts of gravel and boulders recovered by dredging (Nasu 1964), which, since the Leg 56–57 DSDP drilling, are thought to be Pleistocene ice-rafted materials. Interest in the Japan Trench increased with the seismic studies of Ludwig *et al.* (1966) and the area was one of the first recommended for drilling by the IPOD Active Margins Panel.

Sampling programme

In preparation for future IPOD drilling, the Geological Survey of Japan carried out a sampling programme within their network of reconnaissance geophysical tracklines (Honza 1977). Off northern Honshu, 28 sites were sampled and an additional 19 sites were sampled along the adjacent southern Kuril Trench. Sampling was by short dredge hauls, all of which recovered muddy sediment, and by piston coring (Table 1). Most dredge samples were taken from steep slopes, and most piston cores from flat areas. Almost all samples contain ice-rafted material, and the remaining material is predominantly silt and clay. The samples were examined for foraminifers, radiolarians, and nannofossils, and eight contained assemblages suitable for dating (Hasegawa *et al.* 1977).

Sediment samples from the southernmost Kuril Trench resemble material from the Japan Trench. Despite the narrow shelf without large sediment traps, and several well-defined submarine channels that would allow sediment to bypass the shelf and slope (Fig. 1), the two cores from the Kuril Trench are composed of clay interbedded with thin layers of fine sand interpreted as turbidites. One sample from a channel contains significant amounts of sand. Samples from the slope are mostly clay and silt (Yuasa *et al.* 1977; Inouchi *et al.* 1977).

Quaternary sediment samples were obtained by DSDP drilling at the edge of the deep-sea terrace (Sites 438, 439), on the upper slope (Site 435), on the mid-slope terrace (Site 440), on the lower slope (Sites 434, 441), and at the top of the seaward slope of the trench (Site 436) about 20 km east of the trench floor (Figs 1 & 2, Table 2). The sediment across the transect, regardless of location, is predominantly greenish-grey diatomaceous clay and silt which contains varying amounts of volcanic ash and scattered sand generally consisting of volcanic rock fragments, plagioclase, and quartz.

Forearc basin

Of the seven surface samples from the forearc basin, two dredge samples (468, 469) contain mainly sand (Yuasa *et al.* 1977; Inouchi *et al.* 1977). Both samples are near a topographic high, Hachinohe Knoll (Fig. 1), and one is in a canyon area. Three piston cores consist of clay and silt with some sand, and the remaining two dredge samples are silt and sandy silt.

Site 438, near the top of the trench landward slope, yielded about 30% muddy sand of Pleistocene age in the first 50 m penetrated. Muddy fine- to coarse-grained sand beds range in thickness from thin laminae to perhaps 10 m. The beds cannot be correlated between two holes that are about 800 m apart and so they appear to have a restricted lateral extent. However, sand is confined mainly to the Pleistocene and is a minor constituent in the underlying sediment. The section from 50 to 916 m consists of uniform diatomaceous clay and claystone of Pliocene to late early Miocene age and contains occasional pumice pebbles and ash layers (Fig. 3, Table 2).

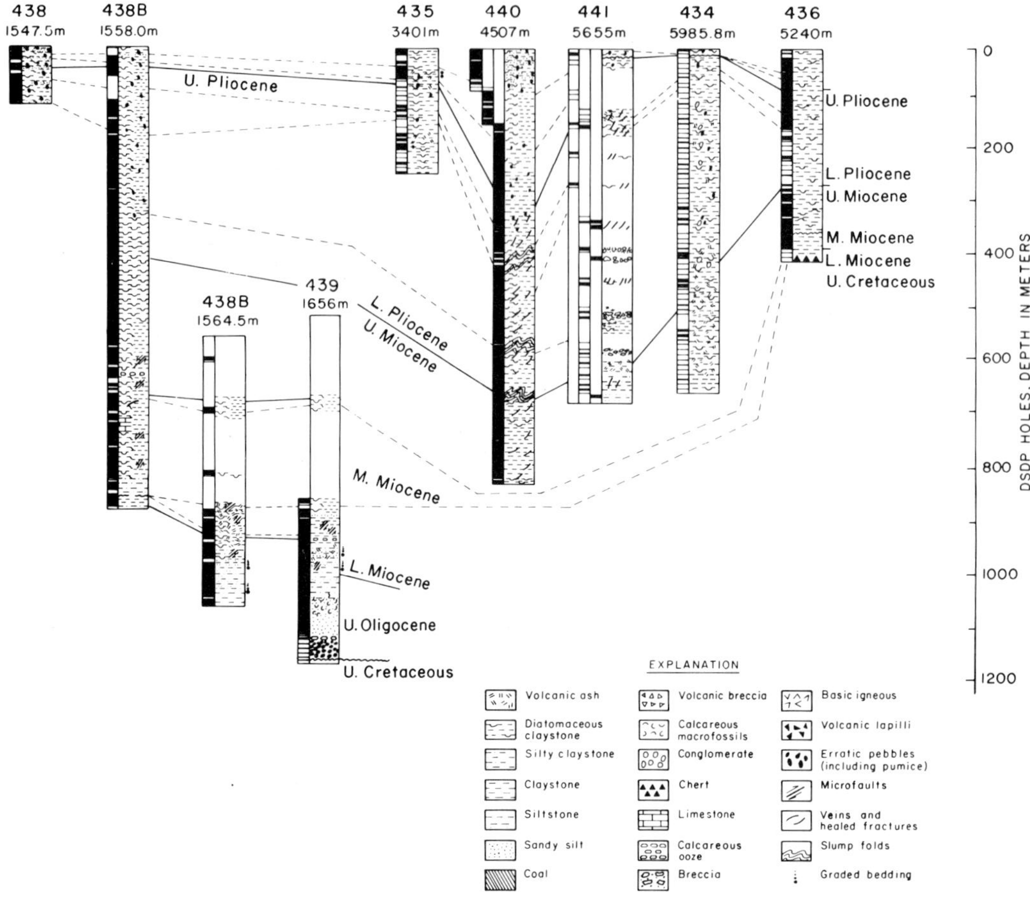

FIG. 3. Lithology of DSDP cores from Legs 56 and 57. Site 436 is not within the area of Figs 1 or 2.

The vertical distribution of sand is consistent with changes in sediment source inferred from Pleistocene sea-level changes and vertical tectonism. The shoreward flank of the forearc basin is a 20 km wide eroded platform (Mogi 1978) which rises 1000 m or more above the deep sea terrace. The adjacent coast is cut into a massif of Palaeozoic and Mesozoic rock that was uplifted during the Quaternary; short streams drain the massif area. During the Miocene and Pliocene the area of the massif was a lowland and much of the surrounding area was flooded (Chinzei 1966). Thus, during the pre-Quaternary period, local terranes were probably only of minor importance as sediment sources. During times of lowered sea-level the coast was further seaward and the landward flank of the Ishikari-Hidaka forearc basin was eroded (Fig. 4). This erosion may have been a major source of sediment.

Groups of small channels incised into the upper sediment layers are seen in seismic records from the area off Hachinohe and just south of Miyako (Figs 1 & 4) (Nakajima 1973; Sakurai *et al.* 1975; Mogi 1978). Most of these channels are prominent on the 2° slopes between the eroded coastal platform and the main part of the deep sea terrace. At station 468 (Fig. 1) one such channel may have been dredged, because the sample had thin layers of sand (Yuasa *et al.* 1977). In some places seismic records indicate broad overbank deposits associated with the channels (Fig. 4). Significant transport of sediment in channels is indicated by the coarse sediment that bypasses the shelf and is deposited in the mid-slope terrace basin and the trench. Certainly, the very high Quaternary rate of accumulation in the mid-slope basin—about 10 times that on the deep-sea terrace—suggests bypassing. Other evi-

TABLE 2. *Lithology*

Site	Lithological unit and sub-unit/ thickness (m)	Core	Depth (m)	Age	Lithology
434	1/101.5	434-1–434-11 434A-1 434B-1	0–101.5	Lower Pliocene to Pleistocene (lower Pleistocene missing)	Clayey diatomaceous ooze, largely uniform, pebbly (pumice), in part; greyish olive-green (5GY 4/2); sediment semilithified below Core 3
	2/199.5	434-12–434-33 434A-2	101.5–301.0	Lower Pliocene	Spicular diatomaceous mudstone and diatomaceous spicular mudstone, vitric in part, some thin (< 1 cm) ash layers. Scattered pebbles present (sedimentary and volcanic rocks). Sediment highly fractured in some intervals; greyish olive-green (5GY4/2), dusky yellow-green (5GY5/2), greyish-green (10G4/2), and light olive grey (5Y4/2) with abundant mottles
	3/160.5	434B-2–434B-18	295.5–456.0	Lower Pliocene (possible repetition of diatom zones)	Vitric diatomaceous mudstone and diatomaceous vitric mudstone; volcanic glass more abundant with diatoms and sponge spicules, minor terrigenous grains; some carbonate layers and pebbles; mostly olive-grey (5Y3/2) with slight colour variations to greenish-black (5GY2/1) and greyish olive-green (5GY3/2)
	4/153.0	434B-19–434B-34	456.0–609.0	Upper Miocene	Tuffite (30%–60% ash) and clayey tuffite generally without siliceous microfossils; no ash layers recovered; carbonate pebbles common; variable colour, mostly olive-grey (5Y3/2), dark greenish-grey (5GY4/1, 5G4/1) greenish-black (5GY2/1), and greyish olive-green (5GY3/2)
	5/>28.5	434B-35–434B-37	609–637.5	Upper Miocene	Vitric diatomaceous mudstone and diatomaceous vitric mudstone; vitric component less than 25%, diatoms as much as 25%; marlstone pebbles common; dark green-grey (5GY4/1) to greyish olive-green (5GY3/2)
435	1/244.5	435-1–435-11 435A-1–435A-16	0–149.5 149.5–244.5	Lower Pliocene to Pleistocene	Spicule-diatom ooze to diatom mud (diatoms: 10–40%); pebbles common (pumice, igneous rocks, argillite); (ash layers present; sediment commonly burrowed; lithified below about 225 m; greyish olive (10Y4/2) to dusky yellow-green (5GY4/2)
436	1/103	436-1–436-11	0–103	Upper Pliocene to Holocene	Vitric diatomaceous mud and diatomaceous mud (20% diatoms, 15% ash); olive-grey to greyish olive-green; ash layers and pockets

TABLE 2 (*continued*)

Site	Lithological unit and sub-unit/ thickness (m)	Core	Depth (m)	Age	Lithology
	2/62	436-12–436-18	103–165	Pliocene	Vitric muddy diatomaceous ooze; higher biogenic silica than Unit 1; thin ash beds or patches; dusky yellow-green to greyish-olive (5GY4/2), commonly mottled with brownish-grey (5YR5/1)
	3A/80	436-10–436-26	165–245	Pliocene	Diatomaceous vitric mud to stiff vitric diatomaceous mud and minor muddy diatomaceous ooze; irregular ash content
	3B/67	436-27–436-33	245–312	Upper Miocene	Diatomaceous mudstone and vitric mudstone; lithified at about 245 m; similar to 3A but colour changes downward from greenish-grey (5G6/1–5/1) to moderate yellow-brown (10YR5/2)
	4/48	436-34–436-38	312–360	Miocene	Radiolarian diatomaceous mudstone; bright moderate yellow-brown (10YR5/2, 6/4, 5/4). Transition to pelagic clay; diatoms and ash decrease downward
	5/19	436-39–436-40	360–379	Eocene and Oligocene (?)	Pelagic clay; very dark brown (10YR2/2-2/1), stiff, scattered burrows and lenses
	6/>18	436-41–436-42	379–?	Cretaceous	Pelagic clay with chert (poor recovery); brownish-black fragments of chert in dusky yellow-brown pelagic clay (10YR 2/2)
437	No sediment recovered				
438	1/107.0	438-1–438-12	0.0–107.0	Upper Pliocene to Pleistocene	Diatomaceous sandy, silty clay, and silt, pebbly (pumice and black greywacke [?]), dark grey, olive-grey, and dark olive-grey (5Y4/2-3/2, 3/1); numerous ash layers
	1/51.5	438A-1–438A-4	0.0–51.5	Pleistocene	Diatomaceous sandy, silty, pebbly clay, olive-grey to dark olive-grey (5Y3/2–4/2) in part; numerous ash layers
	2	438A-5–438A-80	51.5–817.0		
438 (cont.)	2A/312.5	438A-5–438A-32	51.5–363.0	Lower Pliocene through upper Pliocene	Homogeneous to burrow-mottled (lower part) clayey diatom ooze, diatom ooze, and diatomaceous clay (soft to firm), mostly olive-grey (5Y4/2) with dark and very dark grey intervals (5Y5/3); numerous ash horizons (green-grey to grey), occasional pebbles (pumice), and siliceous sponge remains
	2B/230.0	438A-33–438A-56	363.0–593.0	Upper Miocene to lower Pliocene (lowermost)	Diatomaceous claystone and clayey diatomite (silicified), burrow-mottled throughout, minor ash layers in upper portion. Most of sub-unit is olive-grey, dark olive-grey, grey to very dark grey. Sponge remains occur only in upper part. Lower part of sub-unit silicified. Rare pebbles (pumice)

TABLE 2. (*continued*)

Site	Lithological unit and sub-unit/ thickness (m)	Core	Depth (m)	Age	Lithology
	2C/224.0	438A-57–438A-80	593.0–817.0	Middle Miocene to lower upper Miocene	Mostly burrow-mottled claystone and diatomaceous claystone showing dewatering structures and some microfaults. Thin calcareous concretions and thin limestone beds common. Pebbles (pumice) and ash horizons rare. Sub-unit 2C is predominantly dark grey, grey, and grey-green (5Y4/2–5/1, and 5GY5/1–4/1)
	3/61.0	438A-80–438A-86	817.0–878.0	Lower Miocene to lower middle Miocene	Sandy claystone and diatomaceous claystone, vitric in part. Clayey sandstone and siltstone alternates with claystone. Burrow-mottling common. Colours mostly grey to dark grey and grey-green (5YG5/1–4/1 and 5GY4/1)
	2B	438A-1,438A-2	—	—	Overlap with Hole 438
	2C	438A-3, 438A-4	—	—	
	Continuous coring begins				
	3/766.5	438A-5–438A-11	849.5–916.0	Lower and [to] middle Miocene	Greyish olive-green (5GY3/2) and olive-grey to dark olive-grey (5Y4/2–3/1) claystone; intensely burrow-mottled (5Y 6/4 and 5Y4/3) concretionary limestone layers, uncommon thin tuff layers, pumice clasts and fragments, microfaults and dewatering veins, tension fractures, sharp transition to next unit below
				Hiatus	
	4/77.5	438A-12–438A-22 (Section 1)	916.0–993.5	Lower Miocene	Interbedded turbidites and silty claystone, greenish-black (5G2/1, 5Y4/1) to olive-grey, Bouma turbidite sequences are mostly 'ae' with a few 'ade' and 'acde'; medium-grain to fine- sand, silt grading upward to clay, sharp basal contacts, scoured; lithic wacke composition (micro-crystalline quartz lithic fragments), some mollusc fragments. A few tuff layers, mottled clayey siltstone at base-sharp transition to next unit below
	5/104.5	438A-22 (Section 2)–438A-32 (Section 1)	993.5–1098.0	Oligocene	Gray to olive-grey (N5 to 5Y4/1 and 2.5Y4/0) massive sandstone and siltstone moderately well sorted, friable, no sedimentary structures, abundant molluscan debris and articulated pelecypods; sandstone calcite-cemented in lowest part of unit. Wood chips, quartz, lithic fragments (sands are lithic arenites) common

TABLE 2. (*continued*)

Site	Lithological unit and sub-unit/ thickness (m)	Core	Depth (m)	Age	Lithology
	6/47.5	438A-32 (Section 2)–438A-37 (Section 1)	1098.0–1145.5	Oligocene (?)	Boulder to pebble conglomerate and breccia, mostly mud-supported, angular to sub-rounded clasts in clay-size matrix; boulder and smaller clasts of dacite (blue grey to greenish-grey) and pebbles and granules of very dark grey silicified silty claystone (similar to Unit 7). Boulders up to 65 cm in diameter. Several individual conglomerate units. Basal 60 cm is olive-grey (5Y4/1) plastic clay (firm)
				Hiatus	
	7	438A-37 (Section 2)–438A-39	1145.5–1156.5	Cretaceous	Dark to very dark grey (n4 to N2), well-indurated, silicified claystone and clayey siltstone with thin (<1 cm) calcareous graded silt beds; contorted into slump folds but some horizontal bedding
440 (cont.)	1	440-1–440-5	0.0–38.0	Upper Pleistocene and Holocene	Greenish-black to olive-grey (5G2/1 to 5Y4/2) clayey sand and gravel, poorly sorted, from dusky yellow-green to olive-grey (5GY5/2 to 5Y3/2) diatomaceous clay, sandy diatom ooze and greyish-olive (10Y4/2) diatomaceous silty and sandy clay; some graded bedding in coarser-grained intervals Rounded pebbles of various igneous and sedimentary rocks common. Some thin ash layers
	2	440-5–440-26	38.0–380.0	Lower Pliocene and upper Pliocene	Mostly dark olive-grey to olive-grey (5Y3/2–4/2) diatomaceous clay and (below 168 m) diatomaceous claystone. Randomly dispersed rounded pebbles common; burrow mottling increases downhole. Minor thin ash layers.
	3	440-27–440-71	380.0–808.0	Lower Pliocene and upper Pliocene	Generally olive-grey to dark olive-grey (5Y4/2–3/2) diatomaceous clay and (below 168 m) diatomaceous claystone. Zones of folded beds (due to mass movement), veins and microfaults. Thin (1 cm) ash layers common throughout
441	1	441-1–441-6 441A-1–441A-2	0.0–132.0	Upper Pleistocene and Pleistocene	Dark greenish-grey (5GY4/1) silty diatomaceous clay at top, predominantly olive-grey (5Y4/2) diatomaceous clay with minor ash layers; rounded pebbles up to 2 cm in diameter consisting of basalt, dacite, pumice, and some calcareous pebbles or concretions. Minor clayey silt or sand layers. Slightly mottled throughout. One graded silty tuffite noted at base of Unit 1

TABLE 2. (*continued*)

2	441-7–441-9 441A-3–441A-4 H1, H2[a]	132.0–380.0	Lower Pleocene and upper Pliocene	Dark green-grey (5GY4/1), dark olive-grey (5Y3/2), and olive grey (5Y4/1) diatomaceous claystone, minor clayey diatomite, vitric diatomaceous claystone, and a few thin tuff beds. Some calcareous mottles and concretions. Entire unit is characterized by fractures, parting, and some black veins (rehealed). Recovery is mostly in the form of <1cm chips (fissility) and very poor
3	441A-5	380.0–410.0	Lower Pliocene	Breccia of dark grey (N3) siltstone: rounded to angular siltstone fragments in more clay-rich matrix or gouge. Dewatering veins at top of unit. Several graded and ungraded very fine-grained to coarse-grained sandstone beds interbedded with claystone (5Y6/1). Bottom of unit chosen on basis of signature on geophysical log (FTC). Fair recovery
4	441A-6, 441A-7	410.0–504.0	Lower Pliocene	Mostly olive-grey (5Y3/2) diatomaceous claystone and claystone, some siltstone, possible redeposited claystone, minor tuff layers, and calcareous concretions; poor recovery and highly fractured material
5	441A-8–441A-15 H4	504.0–687.0	Upper Miocene and lower Pliocene	Heterogeneous lithology consisting of olive-grey (5Y3/2) claystone, greenish-black to greenish-grey (5G2/6–6/1) silty claystone, calcareous claystone and carbonate-cemented claystone breccia, to dark greenish grey (5GY4/1) resedimented claystone, 45° dip. Lower part of unit composed of greenish-black claystone (5G2/1–3/1) and olive-grey silty claystone (5Y4/1) with some light olive-grey (5Y5/1) silty tuffite and (5Y3/1) vitric silty claystone. Entire unit exhibits high fissility; intensely fractured except for breccia units.

[a]H signifies 'wash cores,' sediment cored through an interval longer than the core barrel. These are archived and sampled as a normal cores.

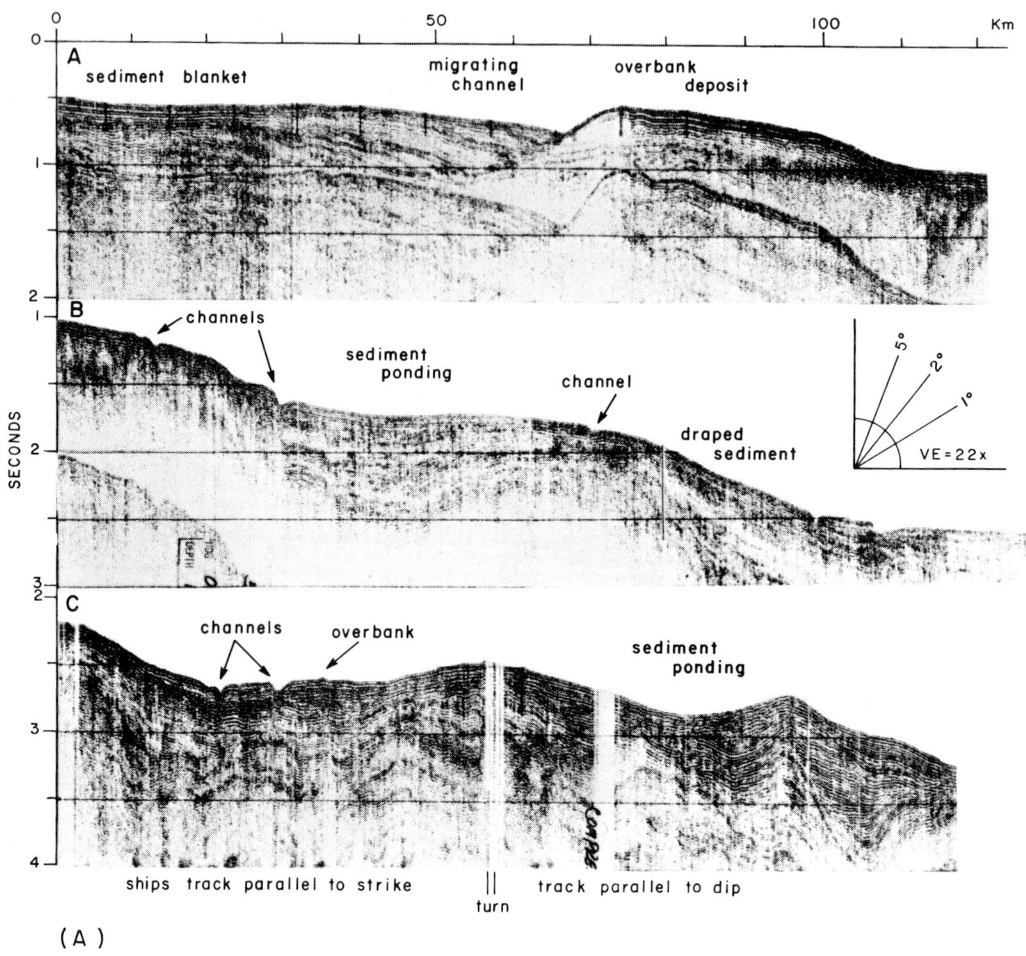

FIG. 4. (A) Examples of seismic reflection records along ships tracks paralleling the trend of the trench and coast off northern Honshu. Record A shows a migrating channel which is being filled and the associated overbank deposits which merge with draped deposits filling a trough. Part of this record parallels the dip of the slope and shows ponding in small tectonic features at the top of the trench slope.

dence for sediment bypassing is the displacement of Pleistocene microfauna from shelf and upper slope environments to depths of 4000 m and more at Sites 440 and 434 and the occurrence of calcareous microfauna below the carbonate compensation depth (CCD) in dredged samples from the trench lower slope (Hasagawa *et al.* 1977).

Three types of sediment deposits inferred from single-channel seismic records occur on the deep-sea terrace. First, sediment is ponded in lows behind a ridge (Fig. 4). Secondly, some ponded sediment has a convex-upward surface (Fig. 4) resembling the sediment lobes associated with channels just north of Hachinohe, which are interpreted as deep-sea fans (Sakurai *et al.* 1974, in Mogi 1978). The third type of sediment deposit seen in seismic records consists of layers conforming roughly to the topography, similar to the sediment blanket described by Inouchi & Tamaki (1977). The published seismic data are insufficient to map the distribution of each type of deposit but such a map might be made if all of the original published and unpublished records in this area were studied.

A change in sedimentation caused by tectonism is suggested by the prominent sediment pond in seismic record JNOC-2 (Fig. 2). Arthur *et al.* (1980) suggest that a former trough, along

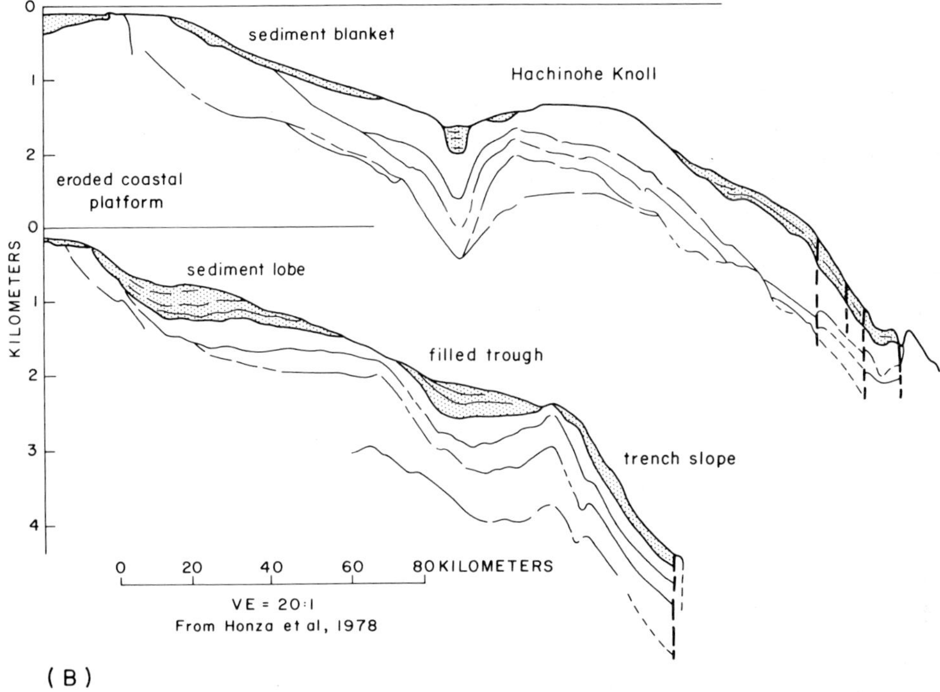

FIG. 4. (B) Line drawings of seismic records showing sediment lobe, filled trough, and sediment blanket from Honza *et al.* (1978).

which sediment was transported parallel to the slope, was disturbed by tectonism in the late Pliocene and Pleistocene. This trough extended for perhaps 200 km along the deep sea terrace from Hidaka Channel and emptied on to the trench slope just south of the location of seismic record JNOC-2. Arthur *et al.* (1980) suggest that continued uplift of the anticline that forms the seaward bank of the trough caused it to fill by lessening or reversing its gradient (Fig. 5). The trough probably passed over an outcrop of basement material that previously formed a high point in the unconformity surface. Immediately below the unconformity and beneath the trough, seismic refraction velocities of 6.3 km s^{-1} are reported (Nagumo *et al.* 1980), values suggesting that the high on the unconformity surface (Fig. 5A) is igneous rock. Sediment first transgressed over the high point and buried it (Fig. 5A). The region then began to tilt seaward and local uplift formed a 'shelf edge high' (deep-sea terrace edge), the relief of which was diminished by sedimentation (Fig. 5B). Further local uplift and regional tilting surpassed the rate of hemipelagic sedimentation required to maintain a level surface.

Despite abundant channelling and ponding which is commonly associated with sandy sediment, hemipelagic silt and clay appear to predominate on the deep-sea terrace from the sample information. The fine grain size may be exaggerated by a sampling bias because samples are located far from shore and mainly in the inter-channel areas. In the central Gulf of Alaska, for example, sand and gravel predominate within 25 km of the shore (Carlson *et al.* 1977); the sample closest to shore on the deep-sea terrace is about 90 km off Hachinohe. South of the transect area, near the Kuji River, Nasu (1964) reports abundant sand and gravel in the near shore area.

DSDP drill holes on the deep-sea terrace have been sited on structural highs where subsurface targets are at shallow depths. Thus, in the forearc basin, interchannel areas were sampled preferentially. Perhaps sampling closer to the eroded coastal platform or in channels would have produced more coarse material because the coarse sediment found in the slope basin and trench must have been transported from shoreward regions in channels. We suspect a bias toward fine-grained sediment in the sample data and have taken this into account in our conclusions.

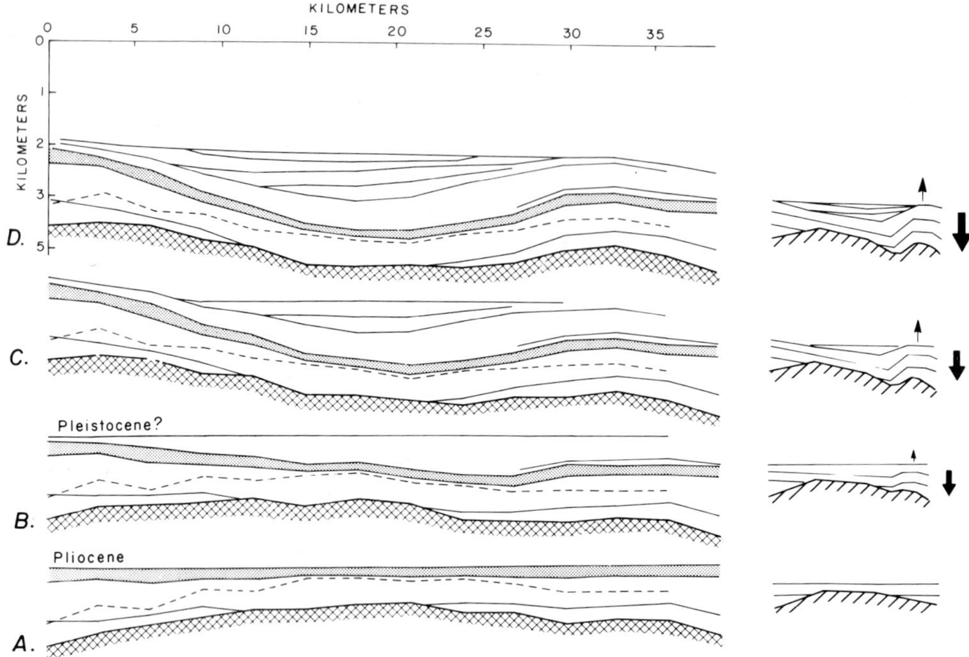

FIG. 5. Development and filling of trough in seismic record JNOC-2, Fig. 2. This sequence was made by flattening successively deeper horizons on a depth section of the seismic record. The dashed line over basement is a permissible alternate interpretation. We interpret a local uplift at the edge of the deep sea terrace, indicated by a simple arrow, superimposed on a regional tilt of the margin, indicated by the heavier arrow.

The information regarding the distribution of Quaternary sediment facies on the shelf and morphological features is summarized as follows—silt and clay dominate samples from interchannel areas well offshore but one sample, possibly from a channel, is mostly sand. Channels are observed in bathymetric and seismic records. Seismic data show sediment bulges, some of which have been interpreted as deep-sea fan lobes; they show some sediment ponding behind tectonic ridges, and a blanket of sediment in many other areas. Tectonism has changed local morphology, and the mode of sedimentation appears to have changed correspondingly.

Trench upper slope

The trench upper slope was sampled at nine dredge sites. As a whole the slope samples are finer grained than those from the forearc basin. Four samples consist of clay with varying amounts of silt, five samples are silt with sparse sand layers, and one consists of micritic limestone (Table 1).

Hesse (1977) examined the sedimentary structure in piston cores from the slope with x-radiographs. Cores from the upper slope appear enriched in sand; the sand grains are scattered throughout the muddy matrix. An undetermined amount of the sand and granule-size material is likely to be of ice-rafted origin.

Site 435 on the upper slope yielded diatomaceous silty clay with sparse sandy silt, one bed of sand, and volcanic ash from the top 40 m (Fig. 3 and Table 2). Site 435 has the same hemipelagic character as Site 438 except that the proportion of sand in the upper Pleistocene interval of 435 is 10% or less, far less than in samples from Site 438.

Quaternary sediment seen in seismic records down the upper slope is generally in a layer that follows the topography (Fig. 4). In the multichannel seismic records, older beds are truncated at the seafloor in some places and the Quaternary deposits may have been lost by slumping (Fig. 2) (Nasu *et al.* 1979). A well-defined slump is recorded on the upper slope by Nakajima (1973). It has a clear headwall scarp and a jumbled footwall. This slump block measures about 35 km in the downslope direction and is as much as 500 m (0.6 s) thick. Slumps in the Geological Survey of Japan seismic records are difficult to outline because of the high vertical exaggeration of the records (37:1). The progressive seaward loss of continuity in reflections indicates tectonic deformation, some of which can be interpreted as listric faulting (Nasu *et al.* 1980). Channels across the upper slope are seen in topographic contours and particularly in seismic and bathymetric records paralleling the strike of the slope (Fig. 4).

Seismic records made on transits to and from the area drilled by *Glomar Challenger* show large (10–20 km wide) canyons at 50–60 km spacings all along the Japan Trench upper slope (Fig. 4). Some are filled with sediment including slump debris, whereas others contain eroded areas. Filled depressions observed in the records may once have been major canyons. Inter-canyon areas commonly have a uniform blanket of sediment. From the scanty seismic records made parallel to the strike of the slope it is clear that seismic records parallel to the dip of the slope give an oversimplified impression of sediment processes on the trench slope. More data parallel as well as transverse to the slope are needed to quantify the relative importance of individual morphological features off Japan.

Slope basin (mid-slope terrace)

A single piston core from the slope basin consists of turbidite mud and massive mud broken by thinly laminated sand layers which are not megascopically graded (Inouchi *et al.* 1977). Rapid sedimentation in this environment means that deep penetration of cores is necessary to get a vertical succession of representative facies, such as was obtained at DSDP Site 440.

Site 440, on the trench mid-slope terrace, yielded muddy sand and gravel interbedded with silty clay in the first 40 m penetrated. Below this depth, there are very few sand beds in the diatomaceous clay and claystone section (Fig. 3 and Table 2). The composition of the sediment is much like that recovered at Site 438, but several zones of syndepositionally folded mud below 300 m probably originated from mass movement (Scientific Party, Leg 57, 1980). Despite its proximity to the trench, the total section is in a normal stratigraphic order and records slope sediment deposition in water more than 2000 m deep.

In the network of bathymetric and seismic records, the mid-slope terrace appears to be essentially continuous for about 250 km along the Japan Trench. Seismic records across the mid-slope terrace show ponded sediment in many places (Nakajima 1973, record VII; Nasu *et al.* 1979, record 78–4; Tamaki *et al.* 1977, records 18, 19, 22, 23, 24) and records with low vertical exaggeration show channels (Nakajima 1973, record VI). At DSDP Site 440 the ponding in the present slope basin began in the Pleistocene and was succeeded by sedimentation at a relatively rapid rate of accumulation (160 m Ma^{-1}) until the present.

Trench lower slope

The five samples dredged from the trench lower slope are dominantly silt and clay with some semiconsolidated siltstone. Most samples from the trench slope are no older than 2 Ma. Calcareous microfossils were present in all samples from the trench lower slope, and two samples contain abundant planktonic foraminifers and nannofossils. Their recovery from deep below the CCD probably indicates considerable down-slope transport. One lower slope sample is between 2.8 and 4.4 Ma old, which is much older than the other samples and suggests that it is a displaced fragment from an older upslope outcrop. At Sites 434 and 441 on the lower slope most of the Pleistocene section is missing owing to slumping and non-deposition. The beds sampled below the first 10 m are Pliocene or older. Some are re-sedimented claystone breccia which indicate even more down-slope mass movement than at Site 440 (Scientific Party, Legs 56 and 57, 1980).

Multichannel records across the trench lower slope show truncation of reflections at the seafloor (Nasu *et al.* 1979). The steep slope (4°–8°) and rugged terrain indicate little Quaternary sediment deposition; sediment erosion and slumping are typical, a conclusion confirmed in DSDP drill holes. Some channels continue from the upper slope across the lower slope, the most obvious one being Hidaka Channel. In the adjoining Kuril Trench, the well-developed Kushiro channel is also continuous across the entire slope (Mogi 1978).

Trench floor

Piston cores from the trench contain hemipelagic mud, mud turbidites, and some turbidite deposits composed of fine sand. Fine volcanogenic sand is apparently redeposited by bottom traction processes rather than turbidites (Hesse 1977).

The trench floor yielded somewhat coarser material than the trench slope. The three piston cores consist of clay with thin beds of fine sand, volcanic ash, and pumice. One core barrel apparently bounced off a gravel bed and another bottomed in a sand bed.

Except at its northernmost end, the trench floor generally has only a small amount of ponded sediment (Fig. 2). At the north end of the Japan Trench a large sediment mass associated with the Hidaka Channel fills the trench and displaces the trench axis seaward (Figs 1 & 2A) (Nasu *et al.* 1980; Arthur *et al.* 1980; Odonera & Honza 1977). The sediment mass may have been deflected southward by Erimo Seamount which is now on the trench floor. The sediment moving down the Hidaka Channel, perhaps augmented by slumps from the adjacent landward slope, where truncated beds crop out along steep slopes (Nasu *et al.* 1979), must have accumulated in the trench more rapidly than it could be subducted.

Trench seaward slope

Two piston cores from the trench seaward slope yielded red-brown clay, greenish clay, and silt with sand and tuff. The seamounts that were dredged appear to be covered with a layer of silt. Further examination of the sediment on seamounts is necessary in order to assess how much sediment on the margin falls through the water column (e.g. as from aeolian transport) as opposed to that transported by mass-flow or other means which presumably would not be able to climb the flank of a sediment.

Site 436, on the outer swell of the seaward slope, yielded about 300 m of diatomaceous silt and clay with much volcanic ash, underlain by a pelagic sequence. This Neogene section is largely hemipelagic material, similar to that sampled on the landward slope.

The seaward slope is covered by hemipelagic and pelagic sediment that conforms to the topography. It is possible that the dredge samples containing red clay were taken along fault scarps that exposed the pelagic lower part of the ocean section. Quaternary sediment at Site 436 is about 85 m thick, and its lower boundary cannot be resolved well in multichannel seismic records, probably because of the poor resolution of this technique at great water depths.

Summary of sediment sampled and rates of deposition

Many samples from the Japan Trench transect area are described only in general terms. Surface and near surface samples in conjunction with seismic and bathymetric data do, however, allow us to infer a very general distribution of Quaternary sediment facies in the Japan Trench Transect area. DSDP samples, which are from greater depths, show the vertical change in facies between the late Miocene and Quaternary.

The distribution of Quaternary sandy facies along the Japan Trench margin is similar to that in the areas discussed by Underwood & Karig (1980), Moore *et al.* (1981), Underwood *et al.* (1980), and Underwood & Bachman (1981). Sediment of the forearc basin, slope basin, and trench is characterized by coarser-grained material than is found in sediment of the slope. In the forearc basin only one of the eleven samples has no sand; in the mid-slope basin much sand and gravel was sampled; in the trench mostly mud was cored, but two of the three cores bottomed in sand or gravel. On the other hand, only four of nine samples from the trench upper slope had any trace of sand and no sand was reported from the trench lower slope. Overall however, *the great bulk of sediment sampled is clay and silt and discrete layers of sand are rare.* Facies characteristics are not distinct enough to identify the original environment without knowledge of sample location.

The morphological features and gross sedimentary structures are best seen in seismic lines parallel to the regional trend of the margin. Such records show sediment draping, ponding, lobate sediment masses (probably fans), channels with overbank deposits and slumps. Certain combinations of these features are more commonly observed in one environment than in others although most features can be found in all environments. Despite the large amount of seismic data off northern Honshu, it is difficult to quantify the relative occurrence of each type of morphological feature or sediment structure. From our study of the geophysical data, we conclude that the forearc basin consists mainly of ponded and draped sediment with common lobate masses, channels, and overbank deposits. Similar features are seen on the trench upper slope, but these channels and slumps are more common there. The trench slope basin and trench floor also show features seen in the forearc basin, but most of the sediment is ponded, slumps have been arrested, and channels are few. Details of the trench lower slope are less clearly resolved in seismic and bathymetric records. A single short line parallel to the margin shows a canyon with truncated beds and possibly some overbank deposits (Arthur *et al.* 1980). Large slump scars are inferred on the lower slope from morphology.

The *Glomar Challenger* made two seismic records parallel to the orientation of northern Honshu during transits to the Japan Trench transect. Features in the upper sediment layers of the forearc basin and the trench upper slope can be compared in the two records (Table 3),

TABLE 3. *Comparative amounts of various sedimentary structures along two* Glomar Challenger *seismic lines parallel to the regional topographic or tectonic trend*

Sedimentary structure	Forearc basin (%)	Trench upper slope (%)
Ponded sediment	45	10
Draped sediment	25	10
Channels and overbank	30	60
Slump	0	10
Unknown		10

but the comparison is tentative because three-dimensional structure can only be inferred from the two-dimensional seismic record. The records suggest, however, a trend from ponded and draped sediment bodies in the forearc basin toward more canyons and slumps on the upper slope.

Across the Japan Trench margin, Quaternary sediment has more coarse grained material than the underlying Pliocene and late Miocene sediment. The increased grain size which in Quaternary sediment includes gravel in many samples, may be attributed to ice rafting, to a lowering of the sea-level which caused the erosional platform along the coast (Honza *et al.* 1978; Nakajima 1973; Sakurai *et al.* 1975; Mogi 1978), and to Quaternary uplift of the coastal massif (Chinzei 1966). In contrast, Pliocene and Miocene sediment from the deep sea terrace (three holes at Sites 438 and 439) and the slope (four sites, of 15 holes) contains no discrete sand beds. Apparently relatively little coarse sediment was being supplied to the margin prior to the Quaternary when the adjacent part of Honshu was a lowland and partly submerged. The grain size differences in pre-Quaternary samples from across the margin are small; however, the number of sites are few for such a large area.

Despite the uniform lithostratigraphy from one DSDP drill site to another (Fig. 3), rates of pre-Quaternary sediment accumulation vary widely between the environments sampled. Simple plots of age versus depth indicate rates between 25 and 110 m Ma^{-1} at Sites 438 and 439 in the forearc basin, whereas higher rates of as much as 270 m Ma^{-1} are indicated at sites on the slope (Fig. 6). In contrast, rates of sediment accumulation seaward of the trench are between 14 and 50 m Ma^{-1}. There are major differences in the water contents of sediments at each of these sites, so various sediment components have been separated (Fig. 7). Rates of accumulation of biogenic silica are generally the same at all sites but are slightly higher in the Pliocene than at other times. The terrigenous component differs greatly between the forearc basin and the lower slope. Surprisingly, the lower slope sites (440, 441, 434) have Miocene and Pliocene accumulation rates roughly 1.5 to 3.5 times greater than sites in the forearc basin (438) and upper slope. Rates on the lower slope are about as high as the very high Pleistocene rate measured at Site 440 which probably corresponds to formation of the slope basin and sediment ponding. At Site 436 the accumulation rates of biogenic silica, volcanic ash, and total sediment increased toward the top of the core because the area of the site moved closer to the Japan margin.

The high rates of accumulation on the trench lower slope are probably due to a tectonic process, such as creation and disruption of basins on the lower slope, because it is difficult to otherwise explain higher rates of accumulation on the 5° trench slope than in the nearly flat forearc basin. The change from anomalously high rates, to the present lower rates of accumulation on the slope coincides with regional tectonic events on land, such as a change in the stress field (Nakamura & Uyeda 1981), and explosive volcanism (Cadet & Fujioka 1980). A compilation of the number of ash layers per million years indicate that and the middle Pliocene was a time of great airfall ash deposition. The pattern of accumulation of disseminated volcanic glass entrained in the sediment (Fig. 7) only partially reflects the trends inferred from discrete ash layers. The ash accumulation pattern of Site 436 also reflects the increased proximity to the volcanic source with time as shown by accumulation rates of terrigenous material. At the sites on the lower slope (440, 441, 434), the accumulation rates of ash follow the same trends as those of terrigenous sediment. The similarity suggests submarine transport of much of the volcanic glass, probably in dilute suspension turbidity currents. Further evidence of a change in rate or pattern of tectonism in the Pliocene is seen in the sequence of benthic foraminifers recovered from Site 438 and 439 (Keller 1980).

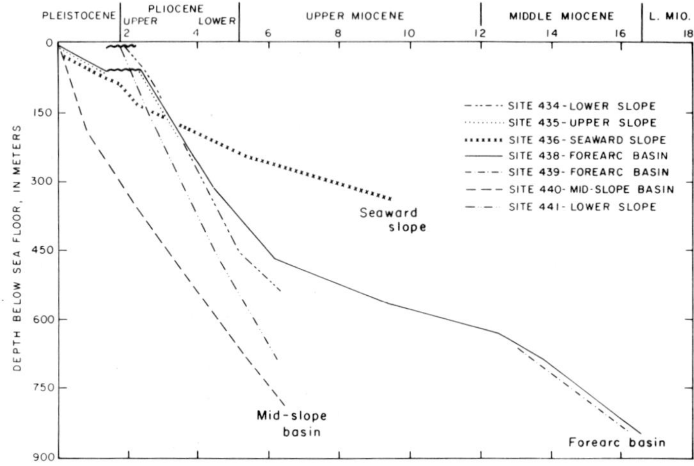

FIG. 6. Plots of sediment age versus depth in DSDP cores, uncorrected for compaction.

When analysed in a manner developed by van Hinte (1978), the benthic assemblages show rapid subsidence of the margin until about 3 Ma ago and then a rapid uplift (Fig. 8) (Arthur *et al.* 1980). The lithologies of rocks from which the foraminifers were recovered also indicate subsidence (von Huene *et al.* 1980).

Soon after the period of change from subsidence to uplift at Sites 438 and 439, and the Pliocene increase in explosive volcanism, basins on the lower slope were rearranged, sediment

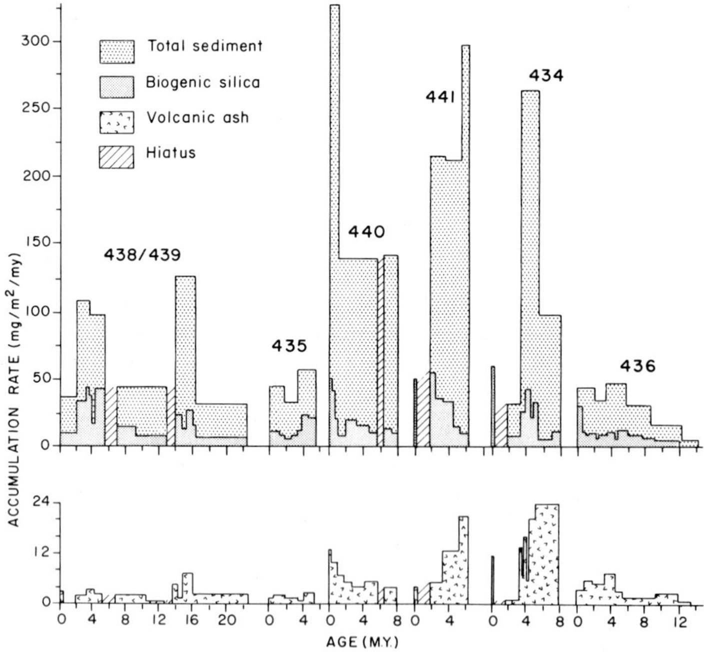

FIG. 7. Rates of sediment accumulation in DSDP cores, after Arthur *et al.* (1980). These are the same data as in Fig. 6 except that the water has been subtracted and components have been separated.

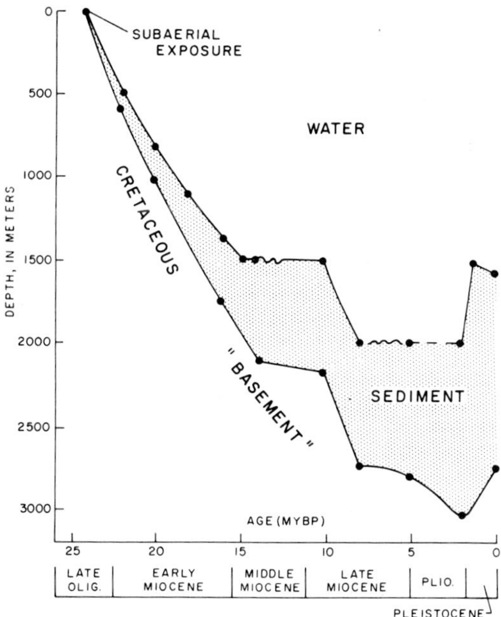

FIG. 8. Vertical movement at Sites 438 and 439 indicated by benthic foraminiferal assemblages and plotted in the manner described by Van Hinte (1978) (after Arthur *et al.* 1980 and Keller 1980).

was ponded in the mid-slope basin (Fig. 2), and the bank at the edge of the forearc basin in JNOC-2 was uplifted (Fig. 5). The stress field on Honshu also changed during this period (Nakamura & Uyeda 1981) and slow uplift began in the area of the coastal massifs (Chinzei 1966). We propose that during the Pliocene the lower slope sediment depocentres were rearranged in their present form. As the slope was tilted, failure resulting in mass movement caused the Pleistocene hiatus observed at Sites 434 and 441.

This Neogene tectonic history of the convergent margin off northern Honshu illustrates how the forearc basin area is tectonically more stable than the slope area. This relative stability is also shown by the depths of the erosional angular unconformity at the base of the Neogene section as seen in seismic records (Fig. 2). At its seaward end this unconformity has subsided as much as 6 km in Neogene time, a depth indicating very rapid rates of tectonism, whereas near the present coast of Honshu the unconformity is indistinct and there was very little subsidence and perhaps minor uplift. Sedimentation paths and depocentres must have shifted through time and caused changes in deposition and in facies associations. In the mid-slope area these changes are significant over 1 Ma periods or less whereas in the forearc basin area significant local changes took place over 2–3 Ma periods or longer. Thus it appears that the morphology which also influences sedimentation, changes at very rapid rates near the trench compared to areas near the arc.

One interesting feature in the Japan Trench is the large amount of material filling the trench floor opposite Hidaka Channel The widest part of this sediment mass, more than 20 km wide, almost bridges the trench floor. Perhaps this body represents a large deep-sea fan. The upper surface of the toe is marked by diffractions on seismic record JNOC 1 (Fig. 2), perhaps from slump blocks or fan channels. If a sediment mass bridges the trench, it could provide a transport path for coarse material seaward of the trench floor.

Conclusions

Although no sediment, sedimentary structure, or morphological feature is unique to any particular environment along the Japan Trench, certain features appear more common to one area relative to another. The samples are insufficient to establish sediment facies associations typical of each morphological element defined by the bathymetry or observed in seismic records. The high resolution seismic records however, are sufficient to make an estimate of the areas covered by each type of morphological feature or sediment deposit, as we have done in Fig. 9. Additionally, rearrangement of sediment paths and depocentres can be said with confidence to be more rapid near the trench than in the forearc basin. Our major conclusions are as follows:

(1) Neogene and Quaternary sediment on the Japan Trench convergent margin is mainly fine grained with rare discrete layers of sand.

(2) For specific morphologies and sediment structures seen in seismic records the facies associations cannot be determined without a great number of additional samples tied to the seismic and bathymetric data.

(3) Thick (200 m) hemipelagic sediment of terrigenous origin blankets the deep-ocean pelagic sediment of the seaward slope of the trench.

FIG. 9. Estimate of the frequency with which morphological features are found in various parts of the Japan Trench convergent margin, from seismic records and bathymetry.

(4) Tectonic disruption changed trench slope basins to unstable slopes, and unstable slopes to basins, in a 1–2 Ma period on the lower slope. The rearrangement of channels and shifts in depocentres were less abrupt (2–3 Ma) in the forearc basin. The *relative* rates of change in sedimentation corresponding to change of morphology may help to distinguish between broad environments in ancient convergent margins. This depends, however, on the ability to isolate tectonic and eustatic influences on sedimentation.

(5) Local sedimentation patterns at the leading edge of the convergent margin may at times be superimposed on sedimentary response to a regional tectonic event as is perhaps shown by increased volcanism. Such an event was associated with uplift of northern Honshu about 3.5 Ma ago and together with a change in climate and lowered sea-level, these factors increased the grain size of terrigeneous sediment supplied to the margin. Thus, the Pleistocene high-energy environments are easier to recognize by their sandier facies. During previous times the same sedimentary processes left a less diagnostic record in deposits of clay and silt. This change in sedimentation reinforces the well-known importance of tectonism in controlling sedimentary facies along convergent margins; at sea tectonism influences sediment depocentres and distribution systems, whereas on land it influences the character of the sediment source.

(6) The succession of litho- and bio-facies found off northern Honshu indicate massive subsidence of the margin; however, the pattern of sediment facies shows no distinct difference from margins that were uplifted.

(7) The coarse terrigenous material sampled may have been associated with transverse channels. Most of the coarse-grained sediment probably moves across the margin in such channels which deposit coarse-grained material in morphologically diverse areas of the margin. This opinion is shared by a number of other authors (Underwood *et al.* 1980; Underwood & Bachman 1981; Moore *et al.* 1981; Underwood & Karig 1980).

(8) Most morphological features and sedimentary structures are common to all parts of the margin, but ponded or draped sediment without many slumps is most characteristic of the forearc basin, slumps are most prevalent in slope basins and on trench floor, and canyons are most common on trench landward slopes. No environment appears to be distinguished by a single morphological feature, sedimentary structure, or facies association. The relative abundance of an assemblage of features, however, is suggestive of the bathymetric environment along the Japan convergent margin, especially when features in the different environments are compared. Given a single sample or even a small group of samples from one locality, it would be very difficult to establish depositional environment except in the case of deep-ocean pelagic sediment.

ACKNOWLEDGMENTS: We greatly appreciate the reviews of our early manuscripts by David Howell, Mike Fields, Arnold Bouma, Steve Bachman and Jerry Leggett.

References

ARTHUR, M. A., VON HUENE, R. & ADELSECK, C. 1980. Sedimentary evolution of the Japan forearc region off northern Honshu (DSDP Legs 56 and 57). *In: Initial Rep. Deep Sea drill. Proj.* **56** & **57,** pt 1. U.S. Govt Printing Office, Washington.

CADET, J. & FUJIOKA, K. 1980. Neogene volcanic ashes and explosive volcanism: Japan Trench Transect, leg 57. Deep Sea Drilling Project. *In: Scientific Party, Init. Rep. Deep Sea drill. Proj.* **56** & **57,** pt 2. 1027–42. U.S. Govt Printing Office, Washington.

CARLSON, P. R., MOLNIA, B. F., KITTELSON, S. C. & HAMPSON, JR, J. C. 1977. Map of distribution of bottom sediments on the continental shelf, northern Gulf of Alaska. Two sheets, scale, 1:500 000. *U.S. Geol. Surv. miscell. Field Studies Map MF 876.*

CHINZEI, K. 1966. Younger Tertiary geology of the Mabechi River Valley, northeast Honshu, Japan. *J. Fac. Sci. Univ. Tokyo,* Sec. II, **16,** 161–208.

HASEGAWA, S., SAKAI, T., OKAMURA, M. & TAKAYAMA, T. 1977. Age assignment of the siltstone fragments dredged. *In:* HONZA, E. (ed.). *Geological Investigation of Japan and Kurile Trench and Slope Areas GH 76–2 Cruise*, pp. 80–5. April–June 1976. Geol. Survey of Japan.

HESSE, R. 1977. Softex-radiographs of sliced piston cores from the Japan and southern Kurile Trench and slope areas. *In:* HONZA, E. (ed.). *Geological Investigation of Japan and Southern Kurile Trench and Slope Areas GH 76–2 Cruise,* pp. 86–108, April–June 1976. Geol. Survey of Japan.

HONZA, E. 1977. Outline of the cruise. *In:* HONZA, E. (ed.). *Geological Investigation of Japan and Southern Kurile Trench and Slope Areas GH 76–2 Cruise*, pp. 1–9, April–June 1976. Geol. Survey of Japan.

—— , TAMAKI, K. & MURAKAMI, F. 1978. *Geological map of the Japan and Kuril Trenches and the adjacent areas.* Geol. Survey of Japan. One sheet, 1:1 000 000.

INOUCHI, Y. & TAMAKI, K. 1977. 3.5 kHz echo sounder profiling survey. *In:* HONZA, E. (ed.). *Geological Investigation of Japan and Southern Kurile Trench and Slope Areas GH 76–2 Cruise*, pp. 17–20, April–June 1976. Geol. Survey of Japan.

—— , YUASA, M. & ODONERA, K. 1977. Cored materials. *In:* HONZA, E. (ed.). *Geological Investigation of Japan and Southern Kurile Trench and Slope Areas GH 76–2 Cruise*, pp. 78–9, April–June 1976. Geol. Survey of Japan.

ISHIWADA, Y. & ORGAWA, K. 1976. Petroleum geology of offshore areas around the Japanese islands. *UN-ESCAP. Bull. CCOP Tech.* **10,** 23–4.

KELLER, G. 1980. Benthic Foraminifera. *In: Init. Rep. Deep Sea drill. Proj.* **57.** U.S. Govt Printing Office, Washington.

LUDWIG, W. J., EWING, J. I., EWING, M., MURAUCHI, S., DEN, N., ASANO, S., HOTTA, H., HAYAKAWA, M., ASANUMA, T., ICHIKAWA, K. & NAGUCHI, I. 1966. Sediments and structure of the Japan Trench. *J. geophys. Res.* **71,** 2121–37.

MOGI, A. 1978. *An Atlas of the Seafloor Around Japan, Aspects of Submarine Geomorphology.* University of Tokyo Press. 96 pp.

MOORE, G. F. & KARIG, D. E. 1976. Development of sedimentary basins on the lower trench slope. *Geology,* **4,** 693–7.

MOORE, J. C., WATKINS, J. S., BACHMAN, S. B., BEGHTEL, F. W., BUTT, A., DIDYK, B. M., LEGGETT, J., LUNDBERG, N., MCMILLEN, K. J., NIITSUMA, N., SHEPHARD, L. E., SHIPLEY, T. H., STEPHAN, J. F. & STRADNER, H. 1979. The Middle America Trench off Mexico. *Geotimes*, **24,** 20–2.

—— , —— , MCMILLEN, K. J., BACHMAN, S. B., LEGGETT, J. K., LUNDBERG, N., SHIPLEY, T. H., STEPHAN, J.-F., BEGHTEL, F. W., BUTT, A., DIDYK, B. M., NIITSUMA, N., SHEPHARD, L. E. & STRADNER, H. 1981. Facies belts of the Middle America Trench and forearc region, northern Mexico—results from Leg 66 DSDP (this volume).

NAKAJIMA, T. 1973. Submarine topography off the southern part of Sanriku district. *Mag. Earth Sci.* **82,** 136–47.

NAKAMURA, K. & UYEDA, S. 1981. Stress gradiant in arc-back arc regions and plate subduction. *J. geophys. Res.* **85,** 6419–28.

NAGUMO, S., KASAHARA, J, & KORESAWA, S. 1980. OBS airgun seismic refraction survey near sites 441 and 434 (J-1A), 438 and 439 (J-12), and proposed site J-2B: legs 56 and 67, Deep Sea Drilling Project. *In: Scientific Party, Init. Rep. Deep Sea drill. Proj.* **56** & **57,** pt 1, 459–62. U.S. Govt Printing Office, Washington.

NASU, N. 1964. The provenance of the coarse sediments on the continental shelves and the trench slopes off the Japanese Pacific coast. *In:* MILLER, R. L. (ed.). *Papers in Marine Geology: Shepard Commemorative Volume.* New York, 531 pp.

—— , IIJIMA, A. & KAGAMI, H. 1960. Geological results in the Japanese Deep Sea Expedition in 1959. *Oceanogr. Mag.* **2,** 201–14.

—— , TOMODA, Y., KOBAYASHI, K., KAGAMI, H., UYEDA, S., NAGUMO, S., KUSHIRO, I., OZIMA, M., NAKAZAWA, K., TAKAYANAGI, Y., OKADA, H., MURAUCHI, S., ISHIWADA, Y. & ISHII, Y. 1979. *Multichannel Seismic Reflection Data Across the Japan Trench.* IPOD-Japan Basic Data Series, No 3, Ocean Research Institute, University of Tokyo, 22 pp.

—— , VON HUENE, R., ISHIWADA, Y., LANGSETH, M., BURNS, T. & HONZA, E. 1980. Interpretation of multichannel seismic reflection data, Legs 56 and 57, Japan Trench transect, Deep Sea Drilling Project. *In: Init. Rep. Deep Sea drill. Proj.* **56** & **57,** pt 1. U.S. Govt Printing Office, Washington.

—— , & SATO, T. 1962. Geological results in the Japanese Deep Sea Expedition in 1961 (JEDS-4). *Oceanogr. Mag.* **13,** 155–66.

ODONERA, K. & HONZA, E. 1977. Bathymetric survey. *In:* HONZA, E. (ed.). *Geological Investigation of Japan and Southern Kurile Trench and Slope Areas GH 76–2 Cruise,* pp. 10–60, April–June 1976, Geol. Survey of Japan.

SAKURAI, M., NAGANO, M., NAGAI, T., KATSURA, T., TOZAWA, M. & IKEDA, K. 1975. Submarine geology off the south coast of Hokkaido. *Hydrogr. Res.* **10,** 1–37. Hydrographic Office of Japan.

SCIENTIFIC PARTY, 1980. *Init. Rep. Deep Sea drill. Proj.* **56** & **57.** U.S. Govt Printing Office, Washington.

TAMAKI, K.,, INOUCHI, Y., MURAKAMI, F. & HONZA, E. 1977. Continuous seismic reflection profiling survey. *In:* HONZA, E. (ed.) *Geological Investigation of Japan and Southern Kurile Trench and Slope Areas GH 76–2 Cruise*, pp. 50–71, April–June 1976. Geol. Survey of Japan.

UNDERWOOD, M. B. & BACHMAN, S. B. 1981. Sedimentary facies associations within subduction complexes (this volume).

—— , —— & SCHWELLER, W. J. 1980. Sedimentary processes and facies associations within trench and trench-slope settings. *In:* FIELD, M. E., BOUMA, A. H. & COLBOURN, I. (eds). *Quaternary Depositional Environments on the Pacific Continental Margin*, 211–29. Pacif. Sec. Soc econ. Paleontol. Mineral.

UNDERWOOD, M. B. & KARIG, D. E. 1980. Role of submarine canyons in trench and trench-slope sedimentation. *Geology,* **8,** 432–6.

VAN HINTE, I. E. 1978. Geohistory analysis—application of micropaleontology and exploration geology. *Bull. Am. Assoc. Petrol. Geol.* **62,** 201–29.

VON HUENE, R., LANGSETH, M., NASU, N. & OKADA, S. 1980. Summary, Japan Trench Transect. *In: Init. Rep. Deep Sea drill. Proj.* **56** & **57,** pt 1. U.S. Govt Printing Office, Washington.

—— , NASU, N., ARTHUR, M., CADET, J. P., CARSON, B., MOORE, G. W., HONZA, E., FUJIOKA, K., BARRON, J. A., KELLER, G., REYNOLDS, R., SHAFFER, B. L., SATO, S. & BELL, G. 1978. Japan Trench transected on Leg 57. *Geotimes*, **23,** 16–21.

YUASA, M., IONUCHI, Y., ODONERA, K. & KIMURA, M. 1977. Rocks and sediments. *In:* HONZA, E. (ed.) *Geological Investigation of Japan and Kurile Trench and Slope Areas GH 76–2 Cruise*, pp. 72–7, April–June 1976, Geol. Survey of Japan.

ROLAND VON HUENE, U.S. Geological Survey, Menlo Park, California, U.S.A.
MICHAEL A. ARTHUR, U.S. Geological Survey, Denver, Colorado, U.S.A.

Tectonics of some forearc fold belts in and around the arc-arc crossing area in central Japan

Yujiro Ogawa

SUMMARY: The zonal structure of the old Honshu arc in central Japan is deflected and cut by the colliding younger Izu-Bonin arc causing peculiar types of accretionary fold belts. The Shimanto Fold Belt (s.s.), made up of Cretaceous flysch and pre-flysch sediments, shows a collisional fold pattern: arcward vergence in the inner side and oceanward vergence on the outer side. The minor structures are also different, with shear folds on the inner side of the belt and lens folds on the outer side. Structures in the inner side grew by flattening and those in the outer side by simple shearing. The divergent development of structural characteristics is explained by the differences in geothermal gradient and stress fields across the belt. The Izu-Bonin arc has a forearc fold belt situated between the Honshu and Izu-Bonin arcs. The stratigraphical, palaeogeographical and structural history of the area shows that it developed in a strike-slip regime associated with the transform fault between the Eurasian and Philippine Sea plates.

In the south-western part of the Mesozoic-Cenozoic Honshu arc of Japan, several accretionary fold belts have been tectonically juxtaposing since late Mesozoic times, making the area most appropriate for tectonic analysis of trench and forearc geology (Kimura 1974; Ogawa 1978; Kanmera *et al.* 1980). The accretionary fold belts are made of arc- or continent-derived flysch sediments which are underlain with both concordant and tectonic contacts by pre-flysch (pelagic) sediments and oceanic or ophiolitic rocks (Kanmera 1976; Sakai 1978; Sakai & Kanmera 1981). The old Honshu arc is cut and bent in its central part by collision of the Izu-Bonin arc which causes several peculiar types of accretionary fold belts (Ogawa & Horiuchi 1978) (Fig. 1).

In this paper I describe these belts in and around the arc-arc crossing area, especially the Cretaceous Shimanto Fold Belt (s.s.) in the Kanto Mountains and the Neogene Miura Fold Belt in the Miura and Boso Peninsulas (Fig. 1), I compare their tectonic features with those of the equivalent fold belts in other areas, and assess the tectonic significance of collisional and strike-slip tectonics in their evolution.

Tectonic setting

The old Honshu arc is composed chiefly of orogenic belts comprising middle or upper Palaeozoic to Mesozoic 'geosynclinal' sediments and their metamorphic equivalents. Paired metamorphic belts constitute the cores of the Akiyoshi and Sakawa Orogenic Belts (Fig. 1) (Kobayashi 1941; Miyashiro 1961; Kimura 1974). Each orogeny has a paroxism around 240 Ma (Permo-Triassic) and 140–110 Ma (Jurasso-Cretaceous) respectively (Ono 1980). The Sakawa Orogen forms the backbone of the Eo-Nippon (Eo-Japan) Cordillera (Kobayashi 1941). Several inliers expose older tectonic zones, made of remnant arcs or micro-continental fragments and dismembered ophiolite complexes which include high pressure metamorphic rocks dated at 380–240 Ma (Maruyama *et al.* 1978) beside 390 Ma granitic and gneissose rocks (Ishizaka 1972, the Rb-Sr age is recalculated by Miss R. Hamamoto 1980, pers. comm.).

The new Izu-Bonin arc, on the other hand, is a juvenile arc mostly comprising volcanogenic materials of Tertiary and Quaternary ages. Collision between the Izu-Bonin and Honshu arcs might have originated as long ago as the Cretaceous (Matsuda 1978; Ichikawa 1980). Voluminous basaltic to rhyolitic lavas and pyroclastics filled the intra-arc basins, and volcaniclastics filled the forearc basins. In the backarc basins volcanogenic materials are interbedded with terrigenous clastics supplied from the old arc. The Izu-Bonin arc runs through the old Honshu arc in the great tectonic graben called the 'Fossa Magna'.

The tectonic status of the central part of the old Honshu arc and the surrounding areas can be explained historically using the theory of plate tectonics as follows:

Before about 40 Ma, prior to the establishment of the present plate system around Japan, the eastern margin of the Eurasian Continent was underthrust by the NNW-subducting Kula or Pacific plate (Emperor Seamount direction)

(Uyeda & Miyashiro 1974). The parallel arrangement of the orogenic and accretionary fold belts of late Palaeozoic to early Cenozoic age were produced by welding and accretion of oceanic, continental and island arc materials during this long subduction phase (Uyeda & Miyashiro 1974; Kimura 1974; Ogawa 1978; Ono 1980; Kanmera *et al.* 1980).

When the spreading direction in the Pacific region changed to WNW (Hawaiian direction) in the late Palaeogene, about 40 Ma ago, the Pacific plate began to subduct along the Japan, Izu-Bonin and Mariana trenches, and consequently the Neogene to Quaternary Izu-Bonin volcanic arc was formed. The Philippine Sea plate was trapped to the west of the arc and backarc spreading occurred from 30 to 15 Ma ago, i.e. late Oligocene to middle Miocene (Kobayashi & Nakada 1978). In the Miocene, the Sea of Japan also began to open, associated with large-scale lateral faulting in the eastern and western parts of Japan (Otsuki & Ehiro 1978). The Eo-Nippon Cordillera was detached from the Eurasian Continent at this stage to form the detached type of island arc of Dickinson & Seely (1979). The Philippine Sea plate may have been separated from the Honshu arc by a strike-slip fault without any significant effects of subduction at this stage (Kobayashi & Nakada 1978). M. Takahashi (1980, pers. comm.) considers that a pulse of subduction of the hot Philippine Sea plate occurred in the middle Miocene, since considerable rhyolitic magmatism occurred on the forearc area of Southwest Japan around 14 Ma (Nakada & Takahashi 1979).

Throughout these subduction phases, from the Cretaceous to the present, a remarkable lateral bending of fold belts took place in central Japan. The west-east trending zonal structure in Southwest Japan became sharply bent and cut in the shape of the letter 'W' (Fig. 1). The en echelon arrangement of the internal fold axes in the Sambagawa Metamorphic Belt and the rotation of the Cretaceous Ryoke Metamorphic Belt and of the mylonite along the Median Tectonic Line in the western and eastern parts of the arc-arc crossing area indicate that northward inhomogenous compression effected ductile bending both on the western and eastern sides of the arc-arc crossing in the Cretaceous (Hara *et al.* 1980). The lateral drag of fold belts with northward convergence in the Akaishi Mountains (Fig. 1) indicates that plastic left-lateral dragging occurred due to large-scale lateral slip during the Cretaceous to Neogene (Kimura 1959, 1961). On the other hand, structural analyses in the Kanto Mountains (the eastern side of the arc-arc crossing area) by Kosaka (1979) indicate that the fold belts rotated rather rigidly in the Palaeogene to Neogene without significant lateral convergence or dragging of the fold belts. Vertical faults with minor right-lateral slippage occurred in the Neogene.

At present, the Philippine Sea plate borders on the Nankai Trough south of Southwest Japan. The plate boundary enters into the Suruga Trough or runs along the eastern edge of the Izu Peninsula, and then it turns sharply to the SE along the Sagami Trough, joining the Japan and Izu-Bonin trenches at the triple junction called the Japan Triple Junction in this paper (Fig. 1). Present tectonism around the north-eastern edge of the Philippine Sea plate is said to result from NW or NNW compression (Seno 1977; Matsuda 1978; Matsubara 1980), and lateral compression of the Honshu arc by the colliding Izu-Bonin arc has been effective since the Cretaceous.

Subduction and collision tectonics of the Shimanto Fold Belt in central Japan

General structure and stratigraphy

The structural character of the accretionary fold belts in the outer part of the Honshu arc is essentially imbricate (Kobayashi 1941; Kimura 1974; Ogawa 1978). The strata involved are broadly 'geosynclinal', comprising alternations of continent-derived flysch and pre-flysch argillaceous sediments, associated with tholeiitic and alkalic basalt, chert and limestone. The sediments in places are intercalated and intermingled by both sedimentary and tectonic deformation to form what may be termed sedimentary or tectonic mélanges. They are metamorphosed into greenschist or lower grade facies. Thrusts in the imbricate structure are mostly inclined inwards, but in the arc-arc crossing area folds on a variety of scales show both inward and outward inclination as will be described later.

A typical group of accretionary fold belts is well exposed in the Sambosan-Shimanto Terrane on the outer side of the Sakawa Fold Belt in Southwest Japan (Figs 1 & 4 bottom). In this terrane, several Mesozoic to Cenozoic fold belts, in which the strata become progressively younger outwards, are arranged parallel to the major zonal structure. The late Tertiary to Quaternary fold belts are restricted to the arc-arc crossing area.

The Cretaceous to early Miocene fold belts in

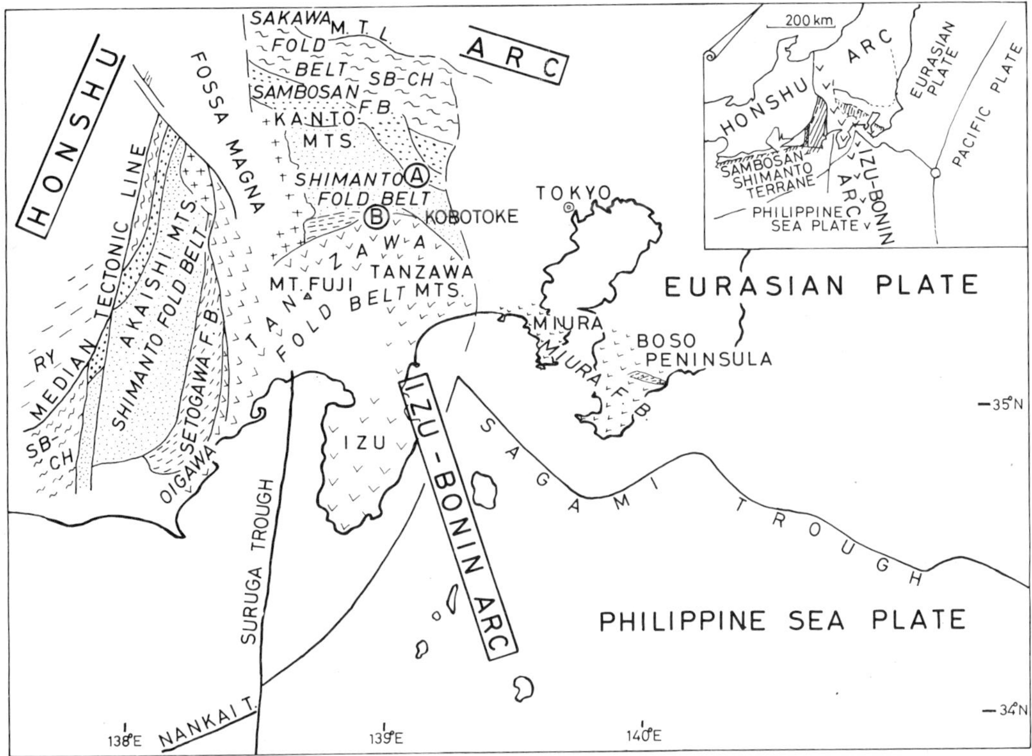

FIG. 1. Index maps of the forearc fold belts in the arc-arc crossing area in central Japan. From inner to outer: Sakawa Fold Belt (Ry, Ryoke Metamorphic Belt; Sb-Ch, Sambagawa-Chichibu Belt); Sambosan F. B.—coarsely dotted; Shimanto F. B.—finely dotted; Setogawa F. B.—dashed; Oigawa-Tanzawa F. B.—large v symbol; Miura F. B.—small v symbol. M.T.L.—Median Tectonic Line. A and B indicate the sites of cross-sections in Fig. 2.

the Shimanto Terrane are generally made up of flysch and pre-flysch sediments which have been explained as forearc and ocean floor deposits (Kanmera 1976; Sano *et al.* 1979; Sakai & Kanmera 1981; Taira *et al.* 1981). The flysch sediments are composed of thick sandy or muddy alternations, often turbiditic. Some large-scale slump or slide deposits are mappable as olistostrome beds (Kanmera 1977). The pre-flysch sediments are largely composed of argillaceous materials including fragments or strips of basalt, chert, limestone, as well as of sandstone. I formerly considered that the pre-flysch sediments were conformably overlain by, or intercalated to some extent with, the flysch sediments and that some basaltic volcanism occurred within the flysch basin (Ogawa & Horiuchi 1978). This I now consider doubtful since there is no strong evidence that the volcanism was *in situ*. Rather the basaltic rocks commonly accompany pelagic sediments (such as radiolarian chert, umber-like sediments and radiolarite) usually within a pebbly mudstone (Tsuchiya *et al.* 1979). The pre-flysch sediments may merit interpretation as accretionary prism basement, mixed by sedimentary and/or tectonic means, and the overlying terrigenous flysch as mantling lower trench slope sediments (Suzuki & Hada 1979; Sakai & Kanmera 1981; Taira *et al.* 1981).

Some limestone and chert blocks contained within the Cretaceous argillaceous pre-flysch sediments yield Triassic or Jurassic fossils, including conodonts, corals and radiolarians (Fujimoto & Suzuki 1969; Nishimiya & Yamagiwa 1973; Yamato-Ohmine Research Group 1979). Permian fusulinids are occasionally found (Fujimoto & Suzuki 1969). These blocks of older sediments are probably recycled materials derived as olistoliths from older (inboard) accreted strata of the Chichibu or Sambosan Terranes.

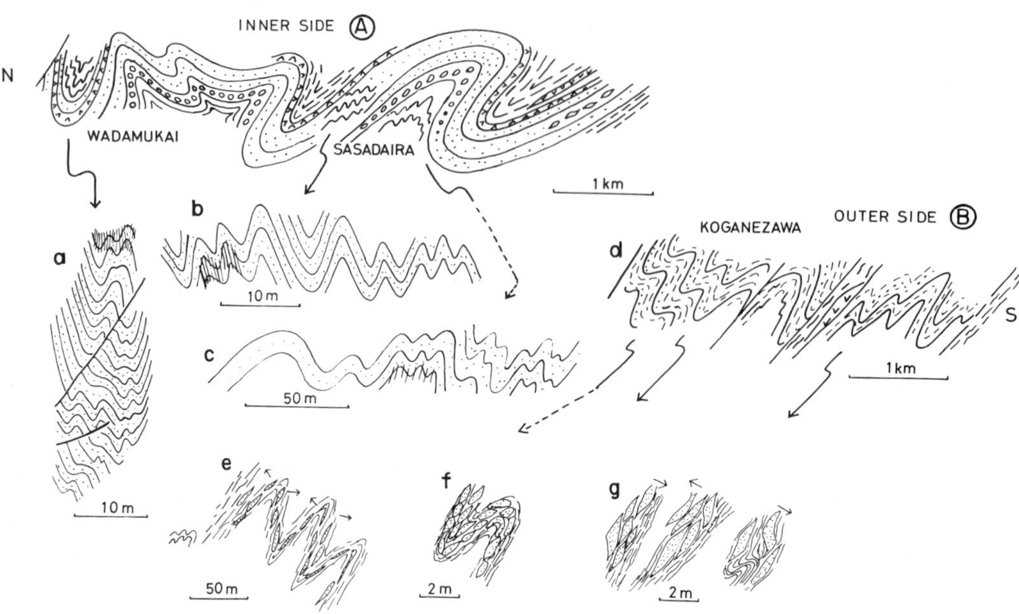

FIG. 2. Fold patterns in the Kobotoke Group of the southern part of the Shimanto Fold Belt in the Kanto Mountains around A and B in Fig. 1. Small arrows indicate the facing of beds.

Structural characteristics of the Cretaceous Kobotoke Group in the Kanto Mountains

The Cretaceous sediments of the north Ogochi and south Kobotoke Group in the Kanto Mountains are attributed to the products of a subduction regime. Large-scale olistostromes characterize the mélange zones, and are overlain by terrigenous turbidites. Greenstones within the pre-flysch facies are mostly large slabs or sheets of metabasalt and those within the flysch facies are volcaniclastic sediments such as volcanic conglomerates or sandstones.

Large-scale closed and overturned folds are developed in the flysch sediments of the Upper Cretaceous Kobotoke Group. Folds of various styles verge northward in the inner (north) side and verge southward in the outer (south) side (Figs 2 & 4). Structures with opposed vergence are well developed in the central part of the group. Similar fold patterns are also recognized in the Shimanto Fold Belt in the Akaishi Mountains (Matsushima 1978; Ogawa & Horiuchi 1978) and in the Chichibu Terrane in the Kanto Mountains (Takizawa 1979).

Differing development of fold styles (Ramsay 1963; Kimura 1968) is also distinguishable between the inner and outer sides of the Kobotoke Group outcrop. Terminology used in this account is summarized in Fig. 3. Flow folds characterized by schistosity are developed at the lower levels in the core of the fold belt. Metamorphic grade is commonly as high as greenschist facies, decreasing laterally or vertically into pumpellyite facies. In upper levels a series of shear folds and a series of lens folds are developed in the inner and outer zones respectively (Fig.3) Other structural characteristics are also different in the two sides (Fig. 4).

In the mid-depths of the inner side, W- or M-shaped symmetric folds are common and the fold style corresponds to the so-called shear folds, characterized by development of well-developed axial planar slaty cleavage (e.g. Fig. 2a). Typical folds in this region belong to Class 1C or Class 2 of Ramsay (1963). Associated bedding slip is occasionally recognizable in the interturbidite mudstones. In such cases the bedding slip plane is shown as a segregated vein of albite and quartz, and the vein is refolded on a microscale in the axial part of folds. This indicates that the folds were developed by bedding slip in the early stages of deformation and by flattening in the later stages. Sedimentary structures such as graded bedding or lamination are preserved, but sole marks are not.

In upper levels of the inner side, flexure folds are developed. Weakly developed slaty cleavage is observed in the muddy part, and bedding slip is only rarely observed (e.g. Fig. 2b).

In the mid-depths of the outer side, S- or Z-shaped asymmetric folds are common and the fold style corresponds to the so-called lens folds of Kimura (1968) (e.g. Fig. 2e,f,g). Brittle beds such as sandstone are fractured or faulted into lensoidal blocks and sometimes flow along the bedding slip plane. The blocks were originally deformed by extension or shear fracturing and then stretching and flowage or even rotation to form tectonic lenses in the later stages of folding. The axial parts of the folds are often recognized only by critical determinations of the facing of beds. Bedding slip and shearing are usually strong and some sedimentary structures are deformed and destroyed. The types of folds are difficult to determine but generally correspond to Class 1B or 1C. Vergence of the folds is mostly outward. Slaty cleavage is sometimes developed, but the texture is quite different from that in the shear folds of the inner side.

In upper levels of the outer side, flexural slip folds accompanied by remarkable bedding slip are commonly developed (e.g. Fig. 2f). No slaty cleavage occurs.

Tectonic significance of the differing fold styles on the inner and outer sides of the Shimanto Fold Belt

Similar development of differing fold styles on the inner and outer sides of the Shimanto Fold Belt is also well recognized in the Akaishi Mountains region (Ogawa & Horiuchi 1978).

In both the Akaishi and Kanto study areas fold styles can be compared in terms of brittle and ductile responses to deformation. The alternating sandstone and mudstone layers show different reactions according to two factors: mean ductility and ductility contrast (Uemura 1981). The mean ductility depends both on temperature and confining pressure (Handin & Hager 1957, 1958), i.e. it is a product of tectonic level. The ductility contrast depends chiefly on temperature when confining pressure is constant. Under relatively high temperature, the ductility contrast is low, while at lower temperature it is high, as shown by the experimental work of Handin & Hager (1958). Therefore, deformational mode in a horizon where the confining pressure is constant is determined by geothermal gradients (Ogawa & Horiuchi 1978; Kimura 1979) (Fig. 3).

When the ductility contrast is high, deformation during folding proceeds by pronounced

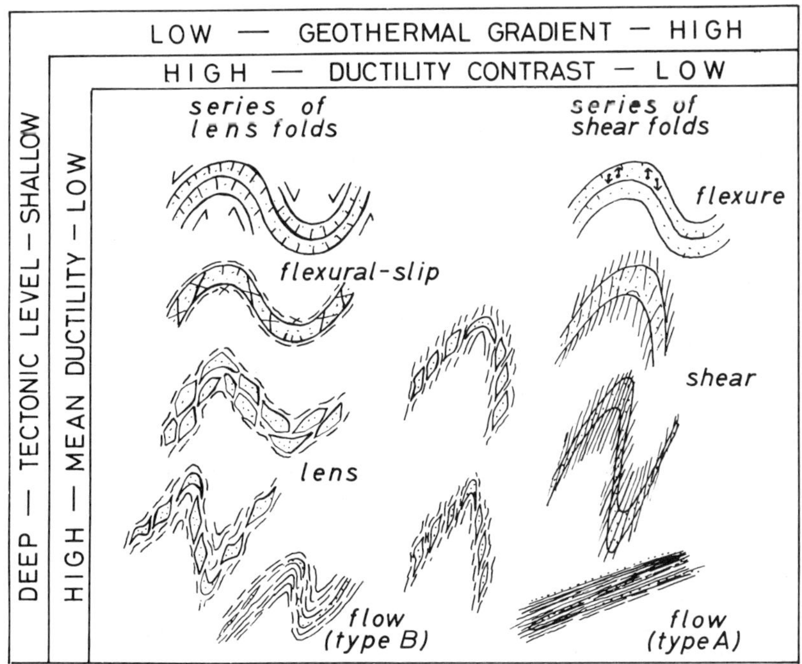

FIG. 3. Fold styles and their arrangement according to tectonic level in the Kobotoke Group, showing series of lens folds and shear folds. For further explanation, see text.

bedding slip, and grain boundary slip is confined within individual sandstone or mudstone beds. On the other hand, when the ductility contrast is weak, grain boundary slip occurs along planes which can penetrate through the bedding boundaries, and passive folding occurs. If bedding slip proceeds, with development of fractures or faults of the brittle layers, lensoidal blocks form giving rise to the characteristic lens fold style (Kimura 1968). Thus the series of lens folds is constituted by flow folds in the lowest tectonic level and lens folds and flexural slip folds at higher levels. The series of shear folds comprises flow folds, shear folds and flexure folds in upward succession (Fig. 3). These observations indicate that the lens fold series is the characteristic structural mode in areas of low geothermal gradient, and that the shear fold series is the characteristic mode in areas of high geothermal gradient.

In the Kobotoke Group, as described previously, the inner and outer sides of the outcrop belt are characterized respectively by series of shear folds and lens folds (Fig. 2). However, the assemblages of metamorphic minerals in the two sides do not show obvious critical differences indicative of different geothermal gradients. Vertical and lateral changes of metamorphic grade should be further studied.

In Japan, a similar divergent development of shear and lens folds is observed in the Sakawa Orogenic Belt. In the Ryoke high temperature belt, a shear fold zone is underlain by gneiss, while in the Sambagawa-Chichibu high pressure belt, lens folds are underlain by schist with sporadic development of shear folds (Ogawa 1978). This may indicate that the divergent development of fold-styles is possibly related to the paired metamorphism in the arc-trench system (Ogawa & Horiuchi 1978).

The tectonically divergent development of fold styles or series of folds and other deformational modes is as follows. Sediments of the Shimanto Fold Belt were deposited in trench and forearc areas—flysch sediments in the forearc trough or basin and pre-flysch sediments on the inner trench slope or in a trench. The pre-flysch sediments contain oceanic and pelagic materials, are covered by the flysch sediments and occupy the lower stratigraphic and structural position. They were accreted to the inner trench slope during subduction, giving rise to inward-dipping imbricate faults (Karig 1974; J. C. Moore & Karig 1976; G. F. Moore & Karig 1976). Sediments in the outer side of the Shimanto Fold Belt are much nearer to the subduction zone, and are pushed into and under the flysch sediments by simple shearing (Moore 1979), suffering deformation in a lower geothermal gradient. Faulting and fracturing occur and a series of lens folds is likely to result. On the other hand, sediments in the inner side of the belt are far from the subduction zone and much nearer to the volcanic front, and suffer from deformation by flattening in a higher geothermal gradient. Thus, steady-state flow occurs and a series of shear folds is likely to result.

In conclusion, sediments deposited in the trench and forearc area were progressively deformed in a subduction regime (Fig. 4). The vergence pattern across the belt around the arc-arc crossing area probably results from collision of the Izu-Bonin and Honshu arcs, giving inner vergence in the inner side of the belt and outer vergence in the outer.

Strike-slip tectonics of the Miura Fold Belt in the Miura and Boso Peninsulas

The stratigraphy and structure in the Miura and Boso Peninsulas is quite different from that in the other Shimanto terranes. This is because of the singular tectonic position of the area, which is situated on the sandwiched forearc basin between the Honshu and Izu-Bonin arcs (Fig. 1). Sediments were supplied both from the Izu volcanic arc to the west and from the Honshu arc to the north. The structural framework consists of gentle soft-sediment folds, forming curving outcrop patterns and arranged in NW–SE trending zones (Fig. 5). These indicate that, unlike other Shimanto forearc fold belts, the sediments did not suffer significant deformation or metamorphism, but rather were subjected to strike-slip tectonism throughout sedimentation. This tectonism was probably the result of movement on the transform fault between the Eurasian and Philippine Sea plates.

Stratigraphy and palaeogeography

The Palaeogene to Neogene rocks and sediments are divided from bottom to top into the Mineoka Group, Hota Group, Yabe Formation and Miura Group. The Miura Group forms the majority of the outcrop area, while the others are restricted to the vicinity of the Hayama-Mineoka Uplift Belt and the offshore (submerged) Okino-yama Uplift Belt (Geological Survey of Japan 1976; Kimura *et al.* 1976) (Fig. 5).

Mineoka Group

Outcropping within the structural horst in the

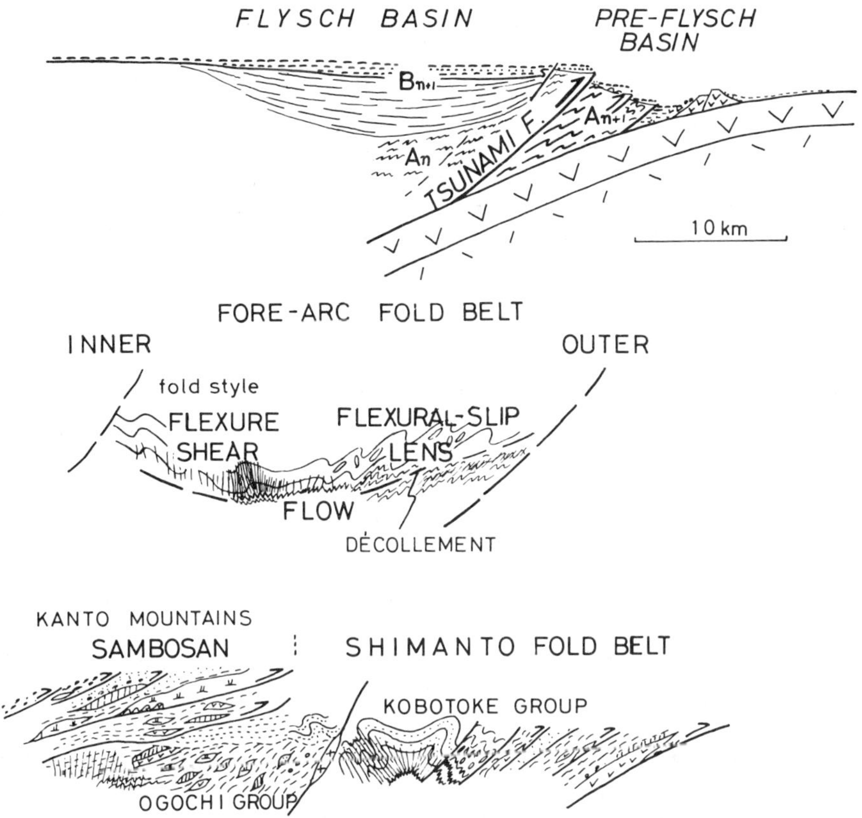

FIG. 4. Tectonic regime of the Shimanto Fold Belt in the Kanto Mountinas. Notice that pre-flysch sediments of A_n age are accreted under the flysch sediments of B_{n+1} age. Tsunami F. corresponds to the Tsunamigenic fault of Fukao (1979).

central part of the Boso Peninsula, the Mineoka Group is predominantly composed of ophiolitic rocks which comprise tholeiitic pillow lavas, dolerite intrusives, and hyaloclastites (Uchida & Arai 1978). Detrital or sheared serpentinites and pyroxene gabbro intrusives occur in the eastern part, with alternating siliceous and calcareous mudstones and turbidites in the western part. The sheared serpentinites are commonly intruded by hornblende gabbro and also contain blocks of hornblende gabbro, harzburgite, picrite and amphibolite schist (Kanehira *et al.* 1968; Kanehira 1976). Some ophiolitic rocks which crop out as knockers within serpentinite, mudstone or sandstone are emplaced as serpentinite mélange and olistostromes.

The amphibolite schist and hornblende gabbro or diorite give K-Ar ages of 38 Ma (late Eocene) and 14 Ma (middle Miocene) respectively (Yoshida 1974). Mudstone intercalated between the pillow lavas yields radiolarians probably of early Miocene age. Mudstone, alternating with serpentinite, yields early Miocene foraminifera (*Globigerinita unicava* and *Globigerinoides trilobus*) (Yoshida 1974).

Some rocks of the Mineoka Group, especially the amphibolite schist, pyroxene and hornblende gabbros, chert and claystone show weak but significant cataclastic or stilolitic textures. Only the amphibolite schist has undergone significant metamorphism and shearing. The

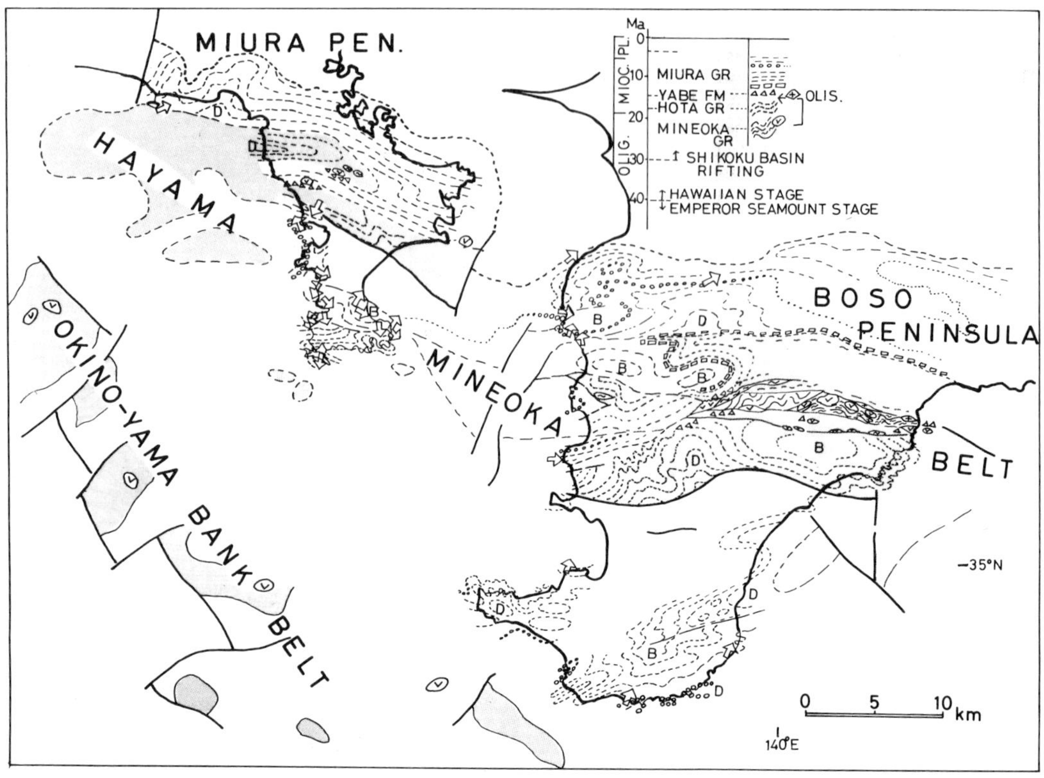

FIG. 5. General structural pattern in the Miura Fold Belt in the Miura-Boso Peninsulas. White arrows in the map indicate the palaeocurrents in the middle part of the Miura Group. D, dome or anticline; B, basin or syncline. For further explanation, see text.

others are of zeolite grade, the metamorphic minerals occurring in veins. These differing indicators of metamorphism indicate that the rocks were emplaced as tectonic blocks, incorporated in a serpentinite matrix. Subsequently gabbroic dykes were intruded into the breccias. The assemblage of rocks in the Mineoka Group seems to indicate trapped oceanic materials sandwiched in the forearc area in the early to middle Miocene (Fig. 6).

Hota Group

This group is composed of rhyolitic tuffaceous sediments and radiolarian-bearing siliceous or calcareous siltstone or claystone; it also contains some large blocks of altered tuff and chert probably from the Mineoka Group. The strata form gentle dome-basin structures. The upper part of the group yields lower-middle Miocene radiolarians such as *Cyrtocapisella cornuta, C. tetrapera* and *Stichocorys delmontensis.*

Yabe Formation

This formation is composed of basaltic to rhyolitic tuffaceous or pumiceous sediments which were derived from the west, i.e. from the Izu volcanic arc. These volcaniclastic rocks overlie the Mineoka and Hota Groups unconformably. In some areas blocks of pillow basalt are intercalated in the basal part of this formation (Kimura *et al.* 1976).

Miura Group

This group is widely distributed, and the stratigraphy, palaeogeography and structure are well known. The group is divided, in ascending order, into the Sakuma, Misaki, Hatsuse, Zushi and Ikego Formations. Foraminifera are of N8 to N21 (middle Miocene to early Pleistocene) age (Ikebe 1978). The group unconformably overlies the previously described units.

The Sakuma Formation only outcrops in the

vicinity of the Mineoka Group and has a characteristic basal conglomerate composed largely of Mineoka-derived volcanic or gabbroic rocks. In the Misaki and later formations, the lower part is generally composed of turbiditic alternations of scoriaceous sandstone or conglomerate and rhyolitic tuffaceous or pumiceous siltstone; the upper part is composed of rather coarse scoriaceous and pumiceous unsorted sediments which often show cross-bedding. Tuffaceous limestone or calcareous tuff occurs in the lower part. Scoriaceous or pumiceous materials are coarse and thick in the north-western and south-western coastal areas of the Miura-Boso Peninsulas and in these areas large-scale submarine slide and slump deposits are common (Ogawa & Horiuchi 1978; Kozima 1980).

In the middle horizon of the Miura Group, around the key tuff bed (called Ok by Mitsunashi & Yazaki 1958), remarkable conglomerates of various kinds are intercalated and traceable over a wide region (Fig. 5). In the southernmost part of the Boso Peninsula, greenish volcanic sandstone, containing pebbles of fresh terrestrially erupted basalt, is widely distributed. Cross-beds in the conglomerate indicate northward transport. Restricted distribution of the peculiar sediments and slump deposits as well as the northerly movement, indicate that the material was derived from a volcanic island sited to the south or SW. In the middle part of the peninsula, a greenish conglomerate contains basalt and gabbro pebbles, and it is interbedded within the scoriaceous and pumiceous beds. Cross-beds over a wide region indicate eastward transport, and grain size and thickness decrease eastward. In the western part of the Miura Peninsula, a conglomerate chiefly composed of angular blocks of basalt, scoria and pyroxene gabbro was deposited from various directions including NW and SW (Fig. 5).

These data indicate that much igneous material was supplied from the Izu volcanic arc in the west; sediment transport in general was from the south or SW in the southern outcrop area and from the NW or SW in the central and northern outcrop area (Fig. 5). At this time, in the late Miocene, basaltic and dacitic or rhyolitic volcanism occurred over a wide area of the Izu Peninsula, with associated large olistostromes of scoriaceous and pumiceous rocks.

In summary, the Neogene, Hota, Yabe and Miura Groups in the Miura Fold Belt are composed chiefly of volcanic materials deposited in the forearc basin of the Izu volcanic arc. They also include fragments of the Mineoka Group, an assemblage very similar to typical lithofacies of other parts of the Shimanto Terrane. The basin was probably dammed up temporarily by uplift of the Mineoka Group, caused by rotation of the Honshu arc during the Miocene in the transform zone between the Eurasian and Philippine Sea plates (Fig. 6).

Structures of the Miura Fold Belt and their tectonic significance

Structurally the Neogene strata in the Miura Fold Belt are characterized by gentle undulations giving rise to a dome-basin pattern. NW–SE trending faults bound the stratigraphic zones of the fold belt, while folds occur between each fault (Fig. 5). Sediments of the Miura Group are all poorly consolidated. The mechanical properties of fine siltstones of the Miura Group are characterized by 'visco-ductile' behaviour beyond 100 or 300 bars confining pressure in triaxial compression tests (Hoshino *et al.* 1972). This indicates that the sediments have never been subjected to a confining pressure over those values. The specific density ranges from 1.44 to 1.91 (Hoshino *et al.* 1972). No significant alteration nor metamorphism is recognized except in the rocks of the Mineoka Group. Volcanic glass and cristobalite or trydimite are common in the tuff beds, and montmorillonite or smectite is present in the fine sediments.

Although for the most part poorly consolidated and unmetamorphosed, the strata of the Miura and Boso Peninsulas show very complicated fault and fold systems. Generally the small-scale folds are slump structures. Strata from the Hota to Miura Groups are usually displaced laterally or vertically on the decimetre to metre scale. Small-scale faults usually cut these slump structures, but are commonly syn-sedimentary.

On the western coast of the Miura Peninsula, minor faults in the Miura Group show the following sequence of development (Ogawa & Horiuchi 1978; Ogawa 1980). The first-formed faults are normal and syn-sedimentary, being associated with slump beds and debris flows which accumulated along the fault scarps. Dewatering structures, called beard-like structure (Ogawa 1980) also formed at this stage. Next, thrust faults were formed by lateral compression, coincident with the present regional strike (Ogawa & Horiuchi 1978). These faults may represent the beginning of large-scale folding in a strike-slip regime. The first two types of faults usually show drag of strata along fault planes and show welding of fault planes. However, the last set of faults, vertical or oblique-slip faults, do not show such drag. This type of fault was developed at the same time as the large-scale folds.

These three fault systems in the Miura Group indicate that faults were generated during the formation of the sedimentary basin and continued during lateral compression at the same time as large-scale folds were developed. Sedimentation began in the south at first and gradually shifted northward (Mitsunashi 1973). The upper part of the group in each area does not show such strong deformation as the lower part. This indicates that deformation was occurring during sedimentation.

Discussion and conclusions

As I have shown, the area of the Miura Fold Belt has a different sedimentary and tectonic history to the Shimanto and Izu Terrane. In the Shimanto Terrane of the arc-arc crossing area, collisional tectonics occurred coincident with sedimentation of the Miura Terrane. Sediments in the latter area were deposited during strike-slip movements, and olistostromes, slumps and slide deposits are common. Moreover the fold axes are not parallel but *en echelon* or curving. Such evidence, as well as the changeable stress fields deduced from the fault patterns, indicates that the area was situated in the strike-slip

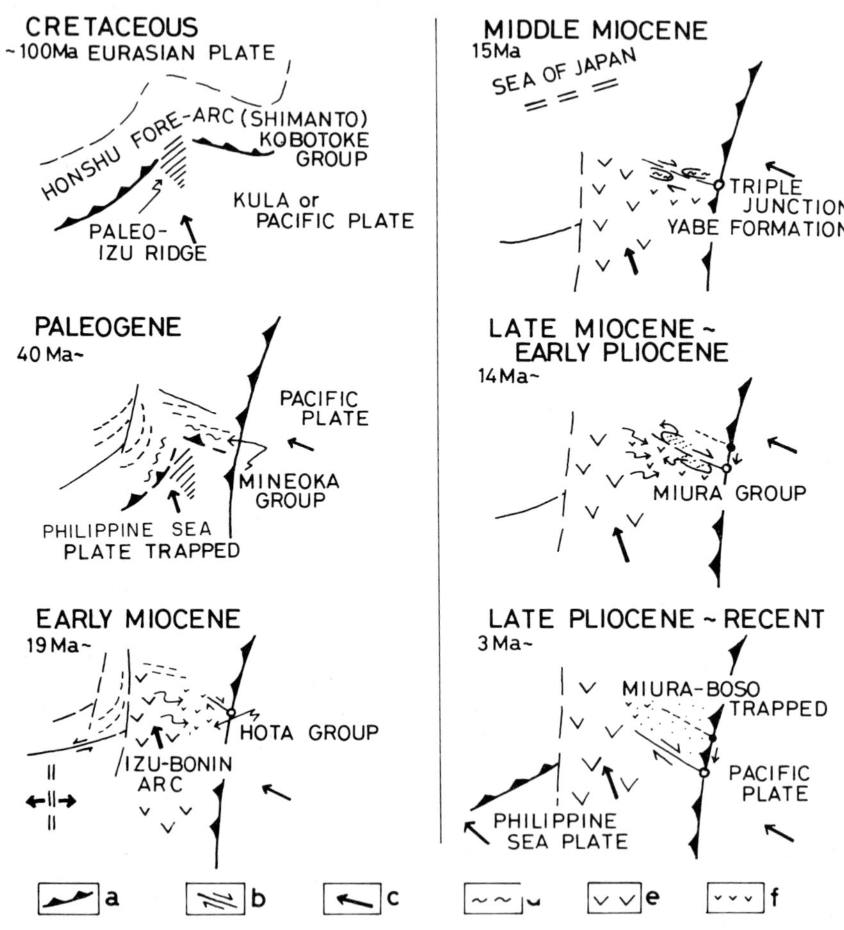

FIG. 6. Plate tectonic model of the arc-arc crossing area in central Japan from Cretaceous to Present. a, subduction zone; b, transform fault; c, direction of plate spreading; d, ophiolite; e, arc volcanism; f, volcaniclastics. Folded small arrows indicate sediment transport. Present radiolarian data indicate some Mineoka sediments are of lower Miocene and some Hota sediments are of middle Miocene age (see text).

regime between two different arcs.

Interestingly, the current stress field in the Sagami Trough, which is the boundary between the Eurasian and Philippine Sea plates, is generally of lateral-slip type, with a changeable component for thrust or normal faults between the co-seismic and inter-seismic times (Ando 1974; Matsuda *et al.* 1978; Seno 1980). Judging from the sedimentological analysis of the strata in the Miura-Boso Peninsulas, the Sagami Trough was not situated in its present position, but adjacent to the central part of the peninsulas from NW–SE during the Miocene and Pliocene. In the trough at that time numerous olistostrome, turbidity or debris flow units indicate near-continuous crustal movement along the transform fault.

The transform seems to have been formed when the system of plate spreading changed around 40 Ma ago. At the same time the crossing and colliding of the Izu-Bonin arc was taking place (Fig. 6). Some part of the oceanic crust (the ophiolitic Mineoka Group), and subsequently sediments of the middle Miocene Hota Group, Yabe Formation and a part of the Miura Group were also trapped in this transform zone. These sediments became welded to the Shimanto Terrane to form the Miura Fold Belt in the eastern side of the arc-arc crossing area (Fig. 6).

This trapping and welding may have been associated with the southward shifting of the Japan Triple Junction during Miocene and Pliocene times. The transform fault probably switched to its present position in the late Pliocene, after the last sediments of the Miura Group were deposited. Some tectonic erosion may have occurred during the triple junction migration (Matsubara 1980).

ACKNOWLEDGMENTS: I am grateful to Dr J. K. Leggett for inviting me to the Trench-Forearc Conference. Dr J. K. Leggett and Professor A. J. Smith are gratefully acknowledged for improvement of the manuscript. Professor K. Kanmera, Messrs S. Tonouchi, K. Fujioka and T. Sakai are thanked for their discussion. Professor K. Nakaseko determined the radiolarian names. Messrs K. Horiuchi, H. Taniguchi and K. Koga helped me in the field survey. This work was supported by grants from the Ministry of Education of Japan and the Takeda Scientific Foundation.

References

(J), in Japanese.

ANDO, M. 1974. Seismo-tectonics of the 1923 Kanto Earthquake. *J. Phys. Earth*, **22,** 263–77.

DICKINSON, W. R. & Seely, D. R. 1979. Structure and stratigraphy of fore-arc regions. *Bull. Am. Assoc. Petrol. Geol.* **63,** 2–31.

FUJIMOTO, H. & SUZUKI, M. 1969. Geology of the basin of the River Obora-gawa, a tributary of the River Arakawa. *Bull. Chichibu Mus. nat. Hist.* **15,** 1–18 (J).

FUKAO, Y. 1979. Tsunami earthquakes and subduction processes near deep-sea trenches. *J. geophys. Res.* **84,** 2303–13.

Geological Survey of Japan (ed). 1976. *Geological Map of Tokyo Bay and adjacent areas (1:100,000).* Geol. Survey of Japan.

HANDIN, J. & HAGER, R. V., JR 1957. Experimental deformation of sedimentary rocks under confining pressure: tests at room temperature. *Bull. Am. Assoc. Petrol. Geol.* **41,** 1–50.

—— & —— 1958. Experimental deformation of sedimentary rocks under confining pressure: tests at high temperature. *Bull. Am. Assoc. Petrol. Geol.* **42,** 2892–934.

HARA, I., SHOJI, K., SAKURAI, Y., YOKOYAMA, S. & HIDE, K. 1980. Origin of the Median Tectonic Line and its initial shape. *Mem. geol. Soc. Japan*, **18,** 27–49.

HOSHINO, K., KOIDE, H., INAMI, K., IWAMURA, S. & MITSUI, S. 1972. Mechanical properties of Japanese sedimentary rocks under high confining pressures. *Rep. geol. Soc. Japan*, **244,** 200 pp.

ICHIKAWA, K. 1980. Geohistory of the Median Tectonic Line of Southwest Japan. *Mem. geol. Soc. Japan*, **18,** 182–212.

IKEBE, N. 1978. Bio- and chronostratigraphy of Japanese Neogene, with remarks on paleogeography. *In*: HUZITA, K. (ed). *Cenozoic Geology of Japan*, pp. 13–34 (J).

ISHIZAKA, K. 1972. Rb-Sr dating on the igneous and metamorphic rocks of the Kurosegawa Tectonic Zone. *J. geol. Soc. Japan*, **78,** 569–75 (J).

KANEHIRA, K. 1976. Mode of occurrence of serpentinite and basalt in the Mineoka district, southern Boso Peninsula. *Mem. geol. Soc. Japan*, **13,** 43–50 (J).

—— , OKI, Y., SANADA, S., YABE, M. & ISHIKAWA, F. 1968. Tectonic blocks of metamorphic rocks at Kamogawa, southern Boso Peninsula. *J. geol. Soc. Japan*, **74,** 529–34 (J).

KANMERA, K. 1976. Correlation between the geosynclinal sediments of past and present. I & II. *Kagaku (Sci)*, **46,** 284–91, 371–6 (J).

—— 1977. General aspects and recognition of olistostromes in the geosynclinal sequence. *Monogr. Assoc. geol. Collab. Japan,* **20,** 145–59 (J).

—— , HASHIMOTO, M. & MATSUDA, T. (eds). 1980. *Geology of Japan.* Iwanami-shoten, Tokyo. 387 pp. (J).

KARIG, D. E. 1974. Evolution of arc systems in the western Pacific. *Ann. Rev. Earth planet. Sci.* **2,** 51–75.

KIMURA, M., YUASA, M., MASAI, Y. & KANIE, Y.

1976. Finding of pillow lava from the Oligo-Miocene formation of the Miura Peninsula. *Bull. geol. Soc. Japan*, **27**, 451–7 (J).

KIMURA, T. 1959. A sharp bend of the Median Tectonic Line and its relation to the Akaishi Tectonic Line—tectonic significances yielded by lateral faulting. *Jpn. J. Geol. Geogr.* **30,** 215–32.

—— 1961. The Akaishi Tectonic Line, in the eastern part of Southwest Japan. *Jpn. J. Geol. Geogr.* **32,** 119–36.

—— 1968. Some folded structures and their distribution in Japan. *Jpn. J. Geol. Geogr.* **39,** 1–26.

—— 1974. The ancient continental margin of Japan. *In*: BURK, C. A. & DRAKE, D. L. (eds). *The Geology of Continental Margin*, 817–29. Springer-Verlag, New York.

—— 1979. *The Japanese Islands: the Geologic History (II-A)*, 245–578. Kokinshoin, Tokyo (J).

KOBAYASHI, K. & NAKADA, M. 1978. Magnetic anomalies and tectonic evolution of the Shikoku inter-arc basin. *J. Phys. Earth Suppl.*, **26,** S391–402.

KOBAYASHI, T. 1941. The Sakawa orogenic cycles and its bearing on the origin of the Japanese Islands. *J. Fac. Sci. Univ. Tokyo. II*, **5,** 219–578.

KOSAKA, K. 1979. Cretaceous and post-Cretaceous faulting in the northeastern Kanto Mountains. *J. geol. Soc. Japan*, **85,** 157–76 (J).

KOZIMA, N. 1980. On the layers of disordered sedimentation in the Misaki Formation in the southwestern part of the Miura Peninsula, south Kanto, Japan (Part I). *J. geol. Soc. Japan*, **86,** 313–26 (J).

MARUYAMA, S., UEDA, Y. & BANNO, S. 1978. 208–240 M.Y. old jadeite-glaucophane schists in the Kurosegawa Tectonic Zone near Kochi City, Shikoku. *J. Jpn. Assoc. Mineral. Petrol. Econ. Geol.* **73,** 300–10(J).

MATSUBARA, Y. 1980. Izu Peninsula and Philippine Sea plate. *Chikyu* (*Earth Monthly*), **2,** 157–63 (J).

MATSUDA, T. 1978. Collision of the Izu-Bonin arc with the central Honshu—Cenozoic tectonics of the Fossa Magna, Japan. *J. Phys. Earth Suppl.*, **26,** S409–21.

—— , OTA, Y., ANDO, M. & YONEKURA, N. 1978. Fault mechanism and recurrence time of major earthquakes in southern Kanto district, Japan, as deduced from coastal terrace data. *Bull. geol. Soc. Am.* **89,** 1610–8.

MATSUSHIMA, N. 1978. Geology of the Southern Alps of Japan—geologic structure of the Shimanto belt in the Akaishi Mountains. *J. nat. Sci. Shimoina Educ. Soc.* (*Geol.*) **1,** 119–34 (J).

MITSUNASHI, K. 1973. Geologic development of south Kanto and Niigata sedimentary basins from Miocene to Pleistocene. *J. Assoc. geol. Collab. Japan*, **27,** 48–65 (J).

—— & YAZAKI, Y. 1958. The correlation between the Cenozoic strata in Boso and Miura Peninsulas by the pyroclastic key bed (No. 1). *J. Jpn. Assoc. Petrol. Technol.* **23,** 16–20 (J).

MIYASHIRO, A. 1961. Evolution of metamorphic belts. *J. Petrol.* **2,** 277–311.

MOORE, G. F. & KARIG, D. E. 1976. Development of sedimentary basins on the lower trench slope. *Geology*, **4,** 693–7.

MOORE, J. C. 1979. Variation in strain and strain rate during underthrusting of trench deposits. *Geology*, **7,** 185–8.

—— & KARIG, D. E. 1976. Sedimentology, structural geology, and tectonics of the Shikoku subduction zone, Southwest Japan. *Bull. geol. Soc. Am.* **87,** 1259–68.

NAKADA, S. & TAKAHASHI, M. 1979. Regional variation in chemistry of the Miocene intermediate to felsic magmas in the Outer Zone and the Setouchi province of Southwest Japan. *J. geol. Soc. Japan*, **85,** 571–82 (J).

NISHIMIYA, K. & YAMAGIWA, N. 1973. Coral fossils from the Kosuge Formation, Yamanashi Prefecture. *Trans. Proc. Palaeontol. Soc. Japan*, **89,** 15–23.

OGAWA, Y. 1978. Structural characteristics and tectonisms around the microcontinent in the outer margin of the Paleozoic-Mesozoic geosyncline of Japan. *Tectonophysics*, **47,** 295–310.

—— 1980. Beard-like veinlet structure as fracture cleavage in the Neogene siltstone in the Miura and Boso Peninsulas, central Japan. *Sci. Rep. Dep. Geol. Kyushu Univ.* **13,** 321–7 (J).

—— & HORIUCHI, K. 1978. Two types of accretionary fold belts in central Japan. *J. Phys. Earth Suppl.*, **26,** S321–36.

ONO, A. 1980. A model of the formation of the Ryoke-Sambagawa paired metamorphic belt. *J. Assoc. Jpn. Min. Petrol. econ. Geol.* **75,** 31–7 (J).

OTSUKI, K. & EHIRO, M. 1978. Major strike-slip faults and their bearing on spreading in the Japan Sea. *J. Phys. Earth Suppl.*, **26,** S537–55.

RAMSAY, J. G. 1963. *Folding and Fracturing of Rocks*. McGraw-Hill, New York. 568 pp.

SAKAI, T. 1978. Geologic structure and stratigraphy of the Shimantogawa Group in the middle reaches of the Gokase River, Miyazaki Prefecture. *Sci. Rep. Dep. Geol. Kyushu Univ.* **13,** 23–38 (J).

—— & KANMERA, K. 1981. Stratigraphy of the Shimanto Terrane and the tectono-stratigraphic setting of greenstones in the northern part of Miyazaki Prefecture, Kyushu. *Sci. Rep. Dep. Geol. Kyushu Univ.* **14,** 31–48 (J).

SANO, H., KANMERA, K. & SAKAI, T. 1979. Sediments associated with greenstone of the Shimanto Terrane. *J. geol. Soc. Japan*, **85,** 435–44 (J).

SENO, T. 1977. The instantaneous rotation vector of the Philippine Sea plate relation to the Eurasian plate. *Tectonophysics*, **42,** 209–26.

—— 1980. Changeable regional stress field. *Chikyu* (*Earth Monthly*), **2,** 146–54 (J).

SUZUKI, T. & HADA, S. 1979. Cretaceous tectonic melange of the Shimanto belt in Shikoku, Japan. *J. geol. Soc. Japan*. **85,** 467–79.

TAIRA, A., OKADA, H., WHITAKER, J. H. McD. & SMITH, A. J. 1981. The Shimanto Belt of Japan: Cretaceous-Lower Miocene sedimentation in forearc basin to deep-sea trench environments (this volume).

TAKIZAWA, S. 1979. Stratigraphy of the Chichibu Terrane in North Kanto Mountains, *In*: INAMORI, J. (ed.). *Biostratigraphy of Permian and Triassic Conodonts and Holotinrian Sclerites in Japan*, 89–102 (J).

TSUCHIYA, N., SAKAI, T. & KANMERA, K. 1979. Mode of occurrence and petrological characteristics of greenstones of the Shimanto Terrane in the Mimi River area, Kyushu. *J. geol. Soc. Japan*, **85,** 445–54 (J).

UCHIDA, T. & ARAI, S. 1978. Petrology of ultramafic rocks from the Boso Peninsula and the Miura Peninsula. *J. geol. Soc. Japan*, **84,** 561–70.

UEMURA, T. 1981. Deformation facies, series and grades. *J. geol. Soc. Japan*, **87,** 297–305.

UYEDA, S. & MIYASHIRO, A. 1974. Plate tectonic model and Japanese Islands: a synthesis. *Bull. geol. Soc. Am.* **85,** 1159–70.

YAMATO-OHMINE RESEARCH GROUP 1979. Triassic limestone blocks within the Shimanto Group in Shikoku. *J. geol. Assoc. Collab. Japan*, **33,** 59–61 (J).

YOSHIDA, A. 1974. Findings of foraminifera from Mineoka Mountains, Chiba Prefecture. *Chishitsu News, Geol. Surv. Japan,* **233,** 30–6 (J).

YUJIRO OGAWA, Department of Geology, Faculty of Science, Kyushu University, Hakozaki, Fukuoka 812, Japan.

Forearc geological structure of the Japanese Islands

Tsunemasa Shiki & Yoshibumi Misawa

SUMMARY: Under the forearc region off the Japanese Islands a conspicuous basal acoustic reflection plane can be traced at the surface of the acoustic basement from the oceanic plate into the bottom of the accretionary prism. Recent studies indicate this plane cannot be the upper boundary plane of the subducting oceanic plate, but must be the upper layer of the oceanic basaltic basement (or chert layer).

The tectonic activity of the Wadati-Benioff zone appears to be a thrust movement which cuts the boundary between the oceanic basement and the overlying sediments. Thrusts cut the basal reflector and extend into the accretionary prism. Large earthquakes which occur at intermediate and shallow depths under the continental shelf and slope are a direct manifestation of these thrust movements.

Thrust faulting also occurred in past geological time. Acidic and basic magma, possibly generated along the earthquake zone at intermediate depths, ascended along the thrust faults or associated normal faults. Intrusions occurred 15–16 Ma ago in the forearc region of Southwest Japan and 22 Ma ago in that of Northeast Japan; subsequently basins subsided landward of the intrusions. In some cases the forearc igneous rocks carried continental and oceanic crustal xenoliths. Good examples of such forearc igneous rocks, including xenoliths, occur in the Permian and Triassic structural belts of the Japanese Islands.

The model described here is specific to the forearc regions of the Japanese Islands. We envisage that different tectonic parameters such as convergence rate and sediment supply may engender different structural processes in other forearcs.

Introduction

The Japanese (or Honshu) Island Arc may be divided into two parts: the Northeast (or Tohoku) Arc and the Southwest Arc. The former is a typical island arc displaying the characteristic arrangement of a deep trench (in this case the Japan Trench), a gravity minimum arcward of the trench, earthquake hypocentres along a dipping Wadati-Benioff zone, with an inner volcanic arc and an outer sedimentary arc above the Wadati-Benioff zone. The Southwest Japan Arc, however, is not a typical arc: it lacks recent volcanism and a deep trench (though the Nankai Trough may be regarded as a trench) (Hoshino 1963) (Fig. 1).

The Japan Trench-Island Arc system is the best studied Pacific convergent margin. Careful re-examination of existing data and new and more detailed investigations of the forearc region of this system could lead to a better understanding of forearc sedimentation and tectonics in general. Shiki (1978) and Shiki & Misawa (1979, 1980) have attempted such an investigation, and presented a new model of the forearc structure of the Japanese Islands, especially that under the inner slopes of the Nankai Trough and the Japan Trench. Several major publications (e.g. Murauchi 1979; Nagumo 1980) have appeared since the Shiki and Misawa papers. In this contribution, we wish to develop and expand our previous ideas and to present a modified model of the forearc structure of the Japanese Islands.

The accretionary prism and basal acoustic reflection plane

Although a large number of studies have been made in the forearc region of the Honshu Arc, it is only in recent years that the precise structure of the region has been discussed. The discovery of the unexpectedly small size of the accretionary prism off Tohoku is one of the most important results of recent multi-channel seismic reflection surveys and DSDP-IPOD legs 56 and 57 (Nasu *et al.* 1978, 1979; von Huene & Arthur 1981). Another discovery is that no true pelagic ocean floor sediments were drilled in the continental slope of the Japan Trench forearc region (Okada & Sakai 1979; Langseth, Okada *et al.* 1978). In the case of the Southwest Japan forearc region, Leg 31 drilling and air-gun surveys indicate that most of the accretionary sediments under the lower continental slope (inner trench slope) comprise terrigenous and hemi-terrigenous materials (Karig, Ingle *et al.* 1975). In such examples, where trench turbidites are apparently the only materials accreted, the prism may be called a '*re-accretionary prism*' (Shiki & Misawa 1980).

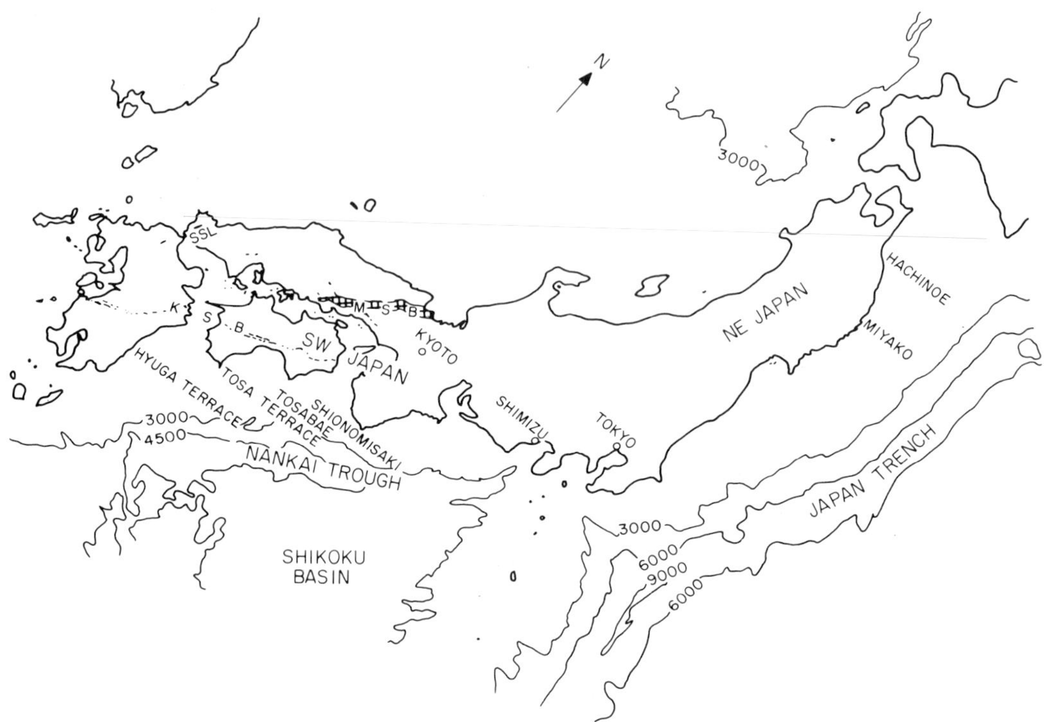

FIG. 1. Map showing the area discussed and Permian-Triassic structural belts in Southwest Japan.
MSB: Maizuru structural belt.
SSL: Su-etakegawa structural line.
KSB: Kurosegawa structural belt.

Another problem concerns the structure of the forearc region. A conspicuous acoustic reflection plane can be observed in multi-channel profiles, at the bottom of the accretionary prism under the lower continental slope of the Honshu Arc. In the profile across the Japan Trench shown by Beck *et al.* (1975) this major basal reflector is apparently flat. It is generally regarded as the slip plane between the oceanic plate and the continental plate. However, the reflection plane is traceable from the surface of the acoustic basement under the ocean floor arcwards under the inner trench slope. It therefore seems most likely that the relationship between the lower part of the accretionary prism above the reflection plane and the layer beneath the plane must be similar to that between the ocean bottom sediments above the reflection plane and the layer beneath the plane. Hence, the surface must be a non-conformity or unconformity within a stratigraphic section.

This non-conformable or unconformable stratigraphic relationship is more apparent in Southwest Japan. As is shown in Fig. 2(A), the acoustic basement, which can be regarded as basaltic basement there, is not flat under the Shikoku Basin and the Nankai Trough (Southwest Japan Trench) and uneven basement features can be traced under the inner slope of the Nankai Trough (Shiki & Misawa 1979). Other air-gun records by the Geological Survey of Japan (Inoue 1978; Okuda *et al.* 1979) show similar features more clearly (Fig. 2C). Such an uneven surface cannot be regarded as a plane of thrust slip occurring simply at the top of the ocean crust and below pelagic sediment of Layer 1.

More recent investigations (Langseth & Okada *et al.* 1978; Nasu *et al.* 1979) reveal that the basal reflection plane in the Japan Trench region is cut by thrust faults (Figs 3 & 4). This further supports interpretation of the basal reflection plane as a non-conformity or unconformity above the chert layer rather than a major décollement surface.

The thrusts which cut the basal reflection extend into the accretionary prism, some of them approaching the seafloor, forming an imbricate structure in the prism. Some of these

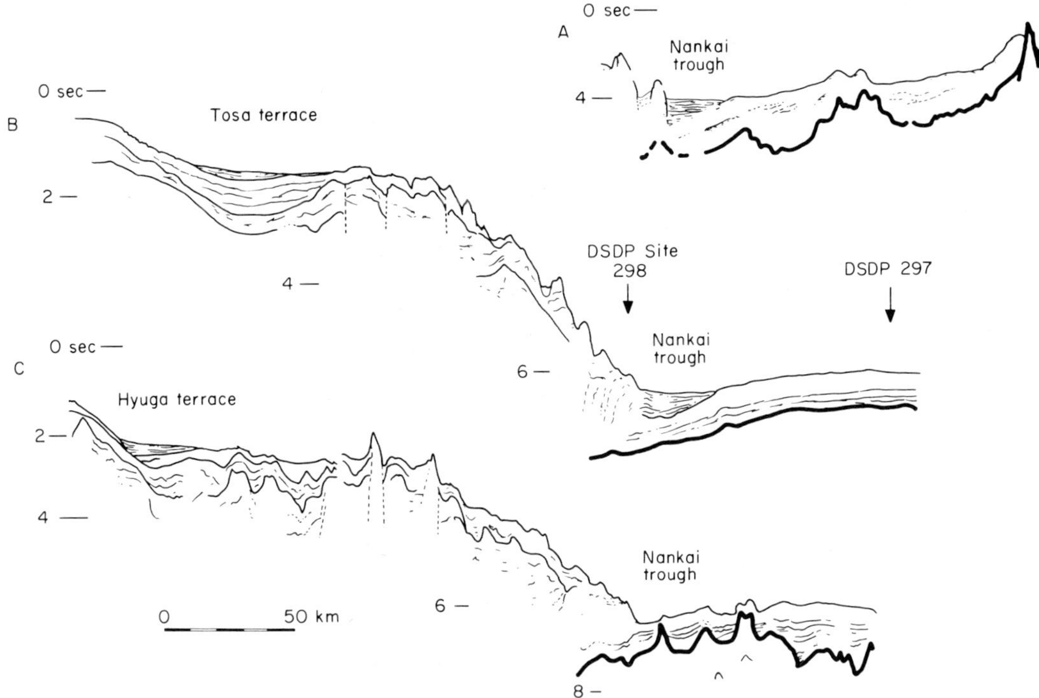

FIG. 2. Single-channel seismic reflection profile across Nankai Trough.
(A) Across central part of Nankai Trough (Misawa 1976).
(B) Off Kochi through Tosa Terrace and DSDP sites 297 and 298.
(C) Across western corner of Nankai Trough and Hyuga Terrace (Inoue 1978; Okuda *et al.* 1979).

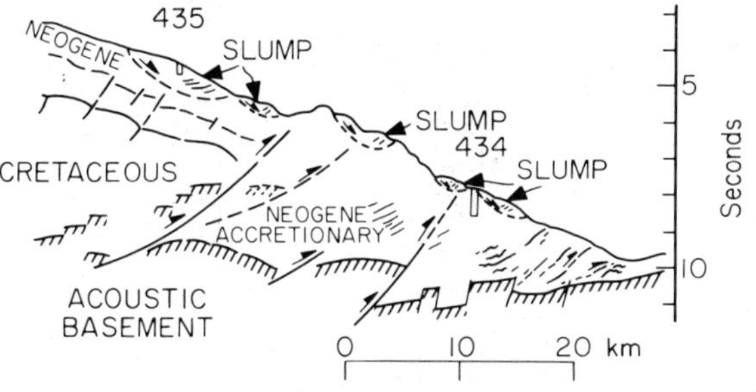

FIG. 3. Seismic reflection profile across the Japan trench and inner trench slope, off Miyako, Northeast Japan.
Note that the reflection plane under the accretionary prism is cut by thrust faults (after Leg 56 report; Langseth, Okada *et al.* 1978).

thrusts may unite at depth, as was suggested by Von Huene (1978) for the Aleutian forearc region.

Thrust faults and earthquakes

It has long been postulated that the Wadati-Benioff zone must extend along the upper surface of the oceanic lithosphere and appear at the trench bottom. However, recent detailed studies have yielded results somewhat different from this general conception.

For instance, Suzuki & Okada (1977) relate seismicity around Northeast Japan, to a fault plane which does not continue to the trench axis but extends into the continental crust, appearing at the sea-floor in mid-slope. A similar distribu-

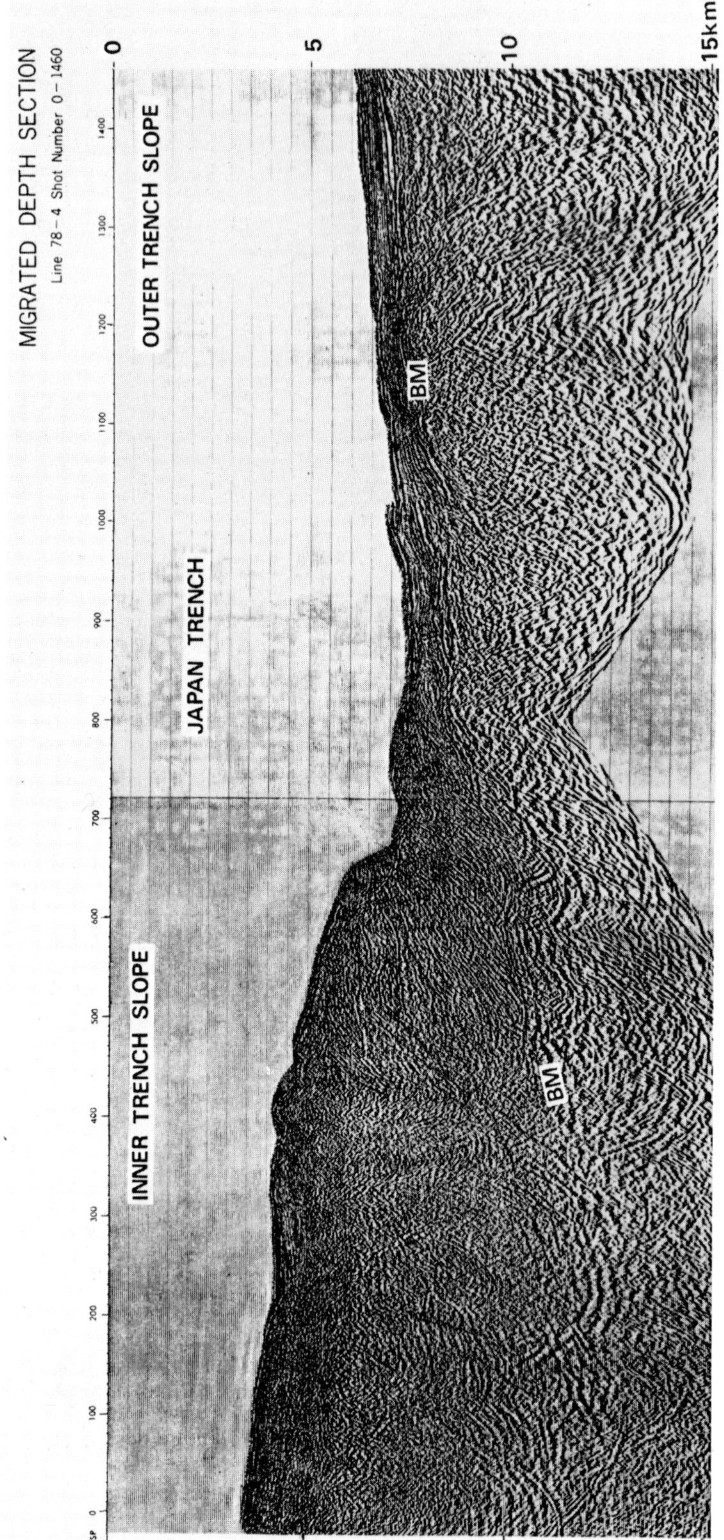

FIG. 4. Multi-channel seismic reflection record across Japan Trench (line 78–4) BM: acoustic basement (from Nasu *et al.* 1979)

tion of earthquake foci is seen in the figures of Hasegawa *et al.* (1978). Earthquake foci also occur, of course, along the upper surface of the lithosphere just under the trench. However, these earthquakes are different in character to the intermediate depth (30–60 km) earthquakes under the continental slope, as shown by Yoshii (1979).

Tectonic deformation associated with large earthquakes along trenches has been described and interpreted in terms of low angle thrust faulting between the continental block and ocean floor. Sawamura (1953) noticed such deformation along the Nankai Trough about 30 years ago. Plafker (1972) studied the thrusting of listric surfaces related to the 1964 Alaska earthquake. The main shock of the Tokachi-oki earthquake of 16 May 1968 ($M = 7.9$; depth '0 km') seems to have been caused by thrust motion along the upper plane of the dual deep seismic zones under the Japan Trench forearc region (Ichikawa & Kawamura 1971; Watanabe 1971). A thrust fault was also assumed from analyses of the data of the Miyagi-oki earthquake (12 June 1978; $M = 7.4$; depth 40 km), and was regarded as a branch of the Wadati-Benioff zone. Its extension seems to appear on the slope on the outer side of the structural high (Nagumo 1980).

It seems possible, therefore, that some of the large earthquakes are caused by motion on thrust faults which reach into the accretionary prism and appear at the mid-slope seafloor, linking at depth with the Wadati-Benioff zone.

Forearc igneous rocks

Present occurrence

One of the most important results of Leg 57 of the DSDP was the discovery of conglomerates and upper Oligocene course sandstones resting unconformably on Upper Cretaceous siltstones (Nasu *et al.* 1978; von Huene & Arthur 1981). It was assumed that the conglomerate pebbles (of acidic to intermediate volcanic rocks) originated from a palaeo-island (Oyashio Ancient Landmass) which was situated off Northeast Japan. Geochemical studies by Fujioka & Nasu (1978) indicate that the rocks might belong to a calcalkalic and tholeitiitic rock series of island arc-type. Since the location is so far east (i.e. oceanwards) of the present volcanic front of the Northeast Japan arc the finding surprised some geologists, and several speculative explanations were suggested for the occurrence of these rocks. However, in Southwest Japan, various igneous rocks such as basalt, diabase, gabbro, granophyre, and quartz-porphyry, occur in a forearc setting on the east coast of Honshu Island at Cape Shiono-Misaki and Cape Muroto-Misaki (Yoshizawa 1954a,b; Imoto *et al.* 1970; Yajima 1972a,b; Yajima & Fukuda 1976; Miyake 1981) and on Tosabae Bank, off Shikiku Island (Niino 1935). Hisatomi & Miyake (1981) clarified that the igneous activity at Shiono-Misaki was intimately associated with uplift and formation of a tectonic high which dammed sediments in the forearc basin.

The ages of the forearc igneous rocks from Northeast and Southwest Japan are different. The ^{40}Ar-^{39}Ar age of the Northeast Japan rocks is around 22 Ma (Ojima 1979), while the rocks of Southwest Japan are inferred to be 15–16 Ma old based on biostratigraphy (Hisatomi & Miyake 1981). The geological setting of both sets of rocks is similar; in both cases the intrusion or eruption of magma seems to be connected with the formation of a tectonic high and a forearc basin landward of the high.

Studies on the rocks of Shiono-Misaki and Muroto-Misaki (Yoshizawa 1954a,b; Yajima 1972a,b; Yijima & Fukuda 1976; Miyake 1981; Hisatomi & Miyake, in press) show a complicated occurrence of various rocks (acidic, and basic to ultrabasic), and poor K content. The linear outcrop distribution suggests the presence of a tectonic line; this has been called the 'Shiono-Misaki-Osumi line' (Nozawa 1975) and the 'Shiono-Misaki zone' (Hisatomi & Miyake 1981). It seems possible that the igneous material gained egress along thrusts or normal faults. Shiki & Misawa (1979) called these igneous rocks 'continental slope intrusive rocks' and Shiki & Misawa (1980) called them 'forearc igneous rocks'.

Shiki & Misawa (1979, 1980), have suggested that the occurrence of such igneous rocks on the continental slope may not just be a feature of the Japanese Islands. For example, gabbroic to granitic rocks from the structural high in the Kuril-Kamchatka forearc region, which intruded pre-Neogene deformed volcanogenic sediments (Gnibidenko *et al.* 1978) may be of similar origin. Boninite and other igneous rocks found in the Mariana Islands forearc on DSDP Leg 60 (Hussong & Uyeda *et al.* 1978) could be forearc igneous rocks, as might the boninites of the Bonin Islands (Ogasawara Islands) themselves.

Problems of generation

Although the rocks appear on the seafloor of the present forearc region, their intrusion or eruption occurred some time ago in all cases. The geological setting of the forearc igneous rocks of the Nankai Trough and Japan Trench suggest that the generation of the rocks possibly occurred in response to Tertiary tectonic movements such as the rise of the structural high and the depression of a forearc basin. The presence of the trench at the time of the generation of the rocks is not absolutely proven.

Clearly the genesis of such magma presents a serious problem. The heat flow flux is generally small at recent trenches and in adjacent zones; neither high temperatures nor magmatism can be expected. At 30–50 km beneath the sediments and continental crust, however, the temperature must be somewhat higher and magmatism may exceptionally occur. The study of Hayakawa & Iizuka (1979) on the relation between earthquakes and the process of partial melting is interesting; they note that a comparatively high heat flow zone is observed off South-western Japan, probably occurring in response to partial melting at the hypocentral depths of great earthquakes. Certainly the thermal gradient in these high heat flow zones shows that the temperature at hypocentral depths (30–40 km) can approach the melting point of wet peridotite, supporting the possibility of magma genesis in forearc regions under certain special geothermal and geological conditions. Such magma genesis may have occurred in the past, though no instances of contemporary forearc volcanism have been reported. Various kinds of lower crustal rocks including both oceanic and continental ones may melt, or be caught as xenoliths and rise to shallower crustal levels. It is natural that the presence of some zone of weakness makes the ascent of such magma and xenoliths easier. The imbricate thrusts mentioned above or other fractures associated with the thrusts could provide such planes of weakness. We believe that such a model explains the occurrence of Tertiary igneous rocks in the Japanese forearc region.

The Permian and Triassic structural belt of Japan and associated forearc igneous rocks

Shiki & Misawa (1979) pointed out that some examples of forearc igneous rocks may be found in much older Japanese geological provinces. Rocks in the Permian and Triassic Kurosegawa and Maizuru structural belts (Fig. 1) may be examples. The rocks of these belts are more complex than those of the younger forearc igneous rocks mentioned above. They include igneous rocks (such as trondhjemite, adamellite, quartz diorite, granophyre, gabbro, diabase), metamorphic rocks (such as amphibolite, amphibolite schist, and serpentinite) and Silurian sedimentary rocks (Hada *et al.* 1979; Igi 1973; Igi *et al.* 1979; Yamashita 1979). The acidic igneous rocks of the belts have semi-oceanic chemical characters, i.e. high Na/K ratios, high K/Rb ratios and low initial $^{87}Sr/^{86}Sr$ ratios (Ishizaka & Yanagi 1975). A protoclastic texture is another feature of many of these acidic rocks: such a texture can be caused by shear stress suffered at the time of intrusion. Although many of these rocks were dated as Permian to Triassic, others are more than 400 Ma old (Ishizaka 1972; Nishimura *et al.* 1976; Nishimura 1979; Nohda 1973; Shibata *et al.* 1977). Anatexis or remobilization of pre-Silurian or Precambrian rocks has been claimed in these cases (Tomita 1954; Noda 1961; Kano *et al* 1961; Matsumoto *et al.* 1962). The most significant feature of the geological setting of the Maizuru belt is its occurrence along the boundary between an outer uplifting province and subsiding 'geosynclinal' province of late Permian and Triassic age. Maruyama (1978) has interpreted the Kurosegawa belt as having between located in a palaeo-forearc, and the Maizuru belt appears to have a similar origin.

We do not maintain that the geological setting of the rocks of the Permian and Triassic structural belts is identical with that of the rocks of younger forearc igneous rocks. However, the resemblance in character (including complexity and occurrence along faults or tectonic lines) suggests some similarity in setting and geological significance.

Discussion and conclusions

The conspicuous basal acoustic reflection plane which can be traced under the accretionary prism from the oceanic plate must be the upper surface of oceanic Layer 2 (basaltic basement, or the chert layer). It is not a tectonic slip plane. Décollement resulting from subduction occurs along thrust faults which cut the basal reflector (Figs 5 & 6). Large earthquakes at intermediate and shallower depths under the continental shelves and slopes are the direct result of these fault movements (Figs 6 & 7). This model does not deny relative movement between the con-

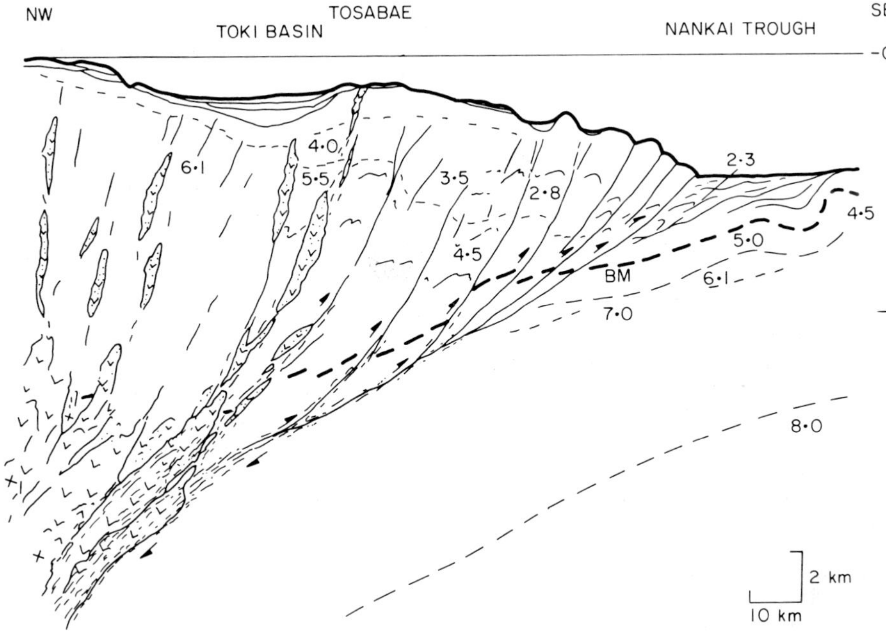

FIG. 5. Schematic structural profile across Nankai Trough off Central Shikoku, showing basal reflection plane (BM), thrust faults, forearc igneous rocks, and illustrative seismic velocities. Modified from Shiki & Misawa (1979), and based on references quoted therein.

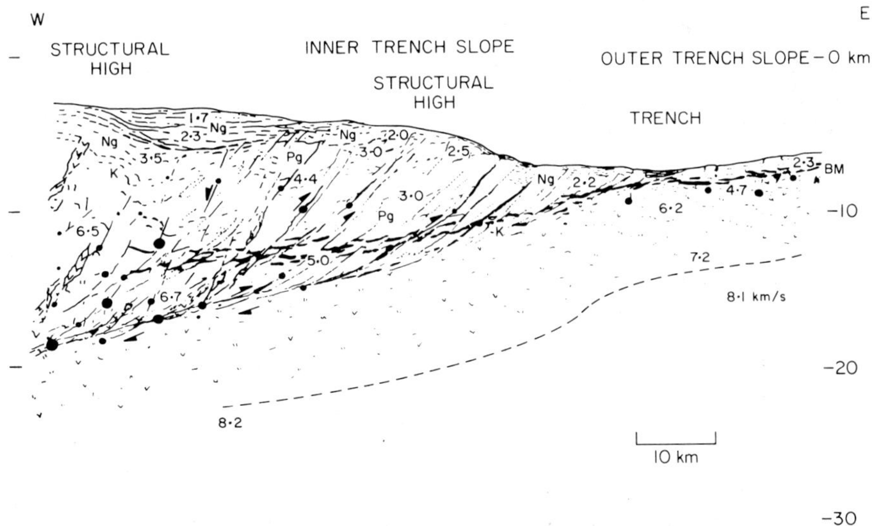

FIG. 6. Schematic structural profile across Japan Trench and inner trench slope off Hachinoe, Northeast Japan, showing basal reflection plane (= acoustic basement, BM), thrust faults, accretionary prism, and illustrative seismic velocities and earthquake foci. Revised from Shiki & Misawa (1979, 1980), and based on references quoted therein and Honza *et al.* (1978).

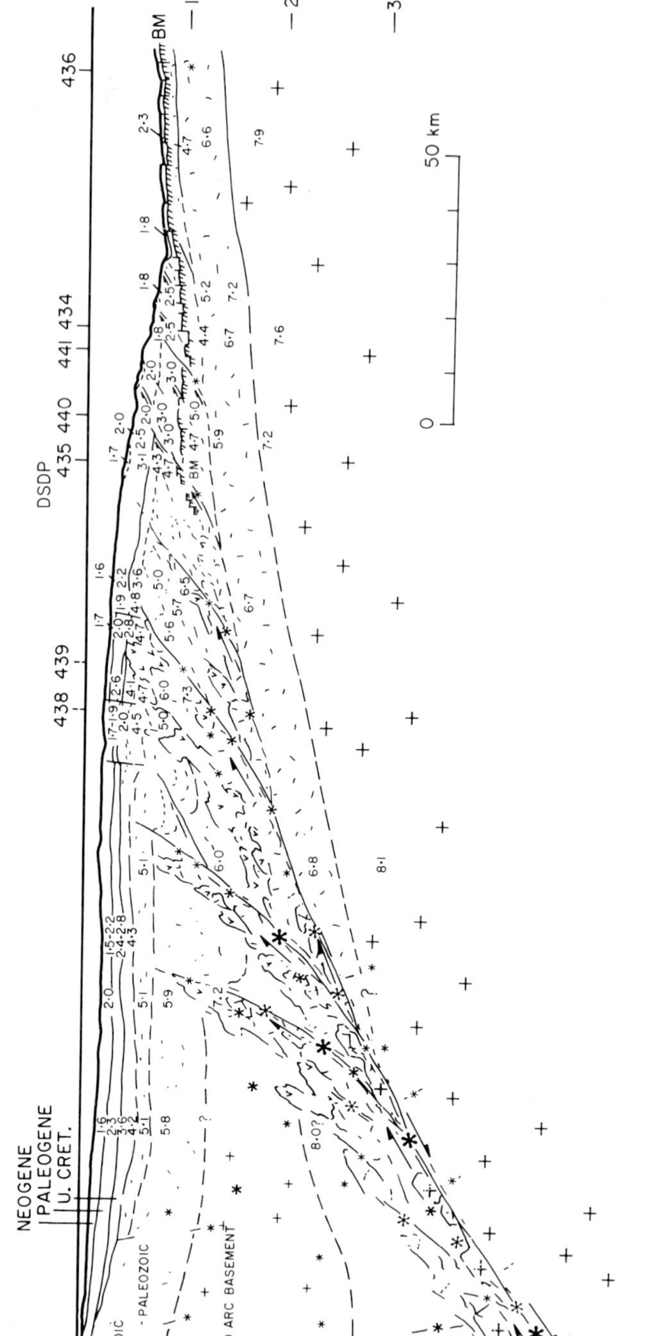

FIG. 7. Schematic structural profile through DSDP sites across Japan Trench and forearc region off Miyako, northeast Japan, showing acoustic basement (BM), thrust faults, accretionary prism, forearc igneous rocks, and idealized distribution of earthquake foci. Based on numerous sources quoted in Shiki & Misawa (1979, 1980). Seismic velocities are based mainly on collective study of seismic refraction information by Murauchi (1979) and on multi-channel reflection data in Nasu *et al.* (1979). Values of the velocity may not be accurate at deep horizons in each vertical sequence.

tinental lithosphere and the oceanic lithosphere. However, the mechanism of the movement in our model differs from that of current interpretions. Our mechanism, i.e. of thrust faulting which cuts the boundary between the continental lithosphere and the oceanic lithosphere, does not lead to such a rapid subsidence of the oceanic lithosphere as is held in current views on plate tectonics.

Although currently small, the size of the accretionary prism may have been variable in the past, depending not only upon speed of subduction of the oceanic plate, but also on factors such as the supply of sediment from land as well as the amount of sediment trapped in forearc basins.

Magmatism has occurred in the past at greater depths in the forearc. Acidic and basic magmas ascended from depth and intruded into some of the fault crush zones, sometimes carrying up continental and oceanic crustal xenoliths. Although currently-exposed forearc igneous rocks are likely to be old, magma genesis may still be taking place in certain circumstances under the forearc.

The model given here is based upon the results of many investigations and studies by many scientists, both at sea and on the Japanese mainland. Further studies, especially deeper drilling, the more precise determination of earthquake hypocentres and stress patterns, and closer observation of the heat flow flux are required to obtain a more accurate knowledge of the structure and tectonics of forearc regions of the world including those of the Japanese Islands.

ACKNOWLEDGMENTS: The writers wish to express their hearty thanks to Professors N. Nasu, A. Asano, S. Nagumo, Drs J. Kasahara, H. Kagami, and Mr K. Fujioka of the University of Tokyo, Dr A. Mizuno of the Geological Survey of Japan, Professors H. Aoki of the Tokai University and K. Nakazawa, Dr D. Shimizu, Messrs Y. Miyake and K. Hisatomi of the Kyoto University, and many others for their stimulating discussions and help. The writers are also indebted to Miss T. Imai for the preparation of the manuscript.

References

BECK, R. H., LEHNER, P., DIEBOLD, P., BAKKER, G. & DOUST, H. 1975. New geophysical data on key problems of global tectonics. *Proc. 9th World Petroleum Congress*, Tokyo, **2,** 3–32.

FUJIOKA, K. & NASU, N. 1978. Chemical composition of the volcanic rocks from the Oyashio Ancient Landmass. *Tectonophysics EOS Trans. Am. geophys. Union,* **59,** 1183.

GNIBIDENKO, H. S., KRASNY, M. L. & POPOV, A. A. 1978. Tectonics of the Kuril-Kamchatka deep-sea trench. *Tectonophysics EOS. Trans. Am. geophys. Union*, **59,** 1184.

HADA, S., SUZUKI, T., YOSHIKURA, S. & TSUCHIYA, N. 1979. The Kurosegawa Tectonic Zone in Shikoku and tectonic environment of the outer zone of Southwest Japan. *In: The Basement of the Japanese Islands.* Prof. Hiroshi Kano Memorial Volume, 341–68 (in Japanese with English abstract).

HASEGAWA, A., UMINO, N. & TAKAGI, A. 1978. Double-planed structure of the deep seismic zone in the northeastern Japan area. *Tectonophysics*, **47,** 43–58.

HAYAKAWA, M. & IIZUKA, S. 1979. A mechanism to explain the earthquakes around Japan by a process of partial melting. *In:* UYEDA, S., MURPHY, R. W. & KOBAYASHI K. (eds). *Geodynamics of the Western Pacific. Suppl. Issue, J. Phys. Earth*, 43–58.

HISATOMI, K. & MIYAKE, Y. 1981. Upheaval movement and igneous activity in the Shionomisaki area, Kii Peninsula, Southwest Japan. *J. geol. Soc. Japan* (in Japanese with English abstract).

HONZA, E., TAMAKI, K. & MURAKAMI, F. 1978. Geological map of the Japan and Kuril trenches and the adjacent areas. *Marine Geology Map Series, 1:1 000 000.* Geol. Surv. Japan.

HOSHINO, M. 1963. Southwest Japan Trench. *Mar. Geol. Tokyo*, **10,** 10–15 (in Japanese with English abstract).

HUSSONG, D. & UYEDA, S. *et al.* 1978. Leg 60 ends in Guam. *Geotimes,* **23,** 10, 19–22.

ICHIKAWA, M. & KAWAMURA, Y. 1971. Mechanism. *In: General Report on the Tokachi-oki Earthquake of 1968.* Eigaku Pub. Co., 33–8.

IGI, S. 1973. The metagabbros and related rocks of the 'Yakuno complex' in the inner zone, Southwest Japan. *Rep. geol. surv. Japan*, **248,** 1–39.

IGI, S., NAKAZAWA, K. & KURODA, Y. 1979. The basement complex in the Maizuru structural belt. *In: The Basement of the Japanese Islands.* Prof. Hiroshi Kano Memorial Volume, 143–52 (in Japanese with English abstract).

IMOTO, N., KANEKO, K., NAKAYAMA, I., YOSHITANI, A., TEMPAKU, T. & SUZUKI, H. 1970. The occurrence of so-called acidic igneous rocks in the north-eastern part of the Shionomisaki Peninsula, Wakayama Prefecture. *Bull. Kyoto Coll. Educ.* **37,** 91–7 (in Japanese with English abstract).

INOUE, E. (ed). 1978. *Investigation of the Continental Margin of Southwest Japan June–July 1975 (GH75–4 Cruise)*. Geol. Surv. Japan, 88 pp.

ISHIZAKA, K. 1972. Rb-Sr dating on the igneous and metamorphic rocks of the Kurosegawa tectonic zone. *J. geol. Soc. Japan*, **78,** 569–75 (in Japanese with English abstract).

ISHIZAKA, K. & YANAGI, T. 1975. Occurrence of oceanic plagiogranites in the outer tectonic zone, Southwest Japan. *Earth planet. Sci. Lett.* **27,** 371–7.

KANO, H., NAKAZAWA, K. & SHIKI, T. 1961. Consideration on the Permian back-grounds of the Maizuru district. *J. geol. Soc. Japan*, **67,** 463–75 (in Japanese with English abstract).

KARIG, D. E., INGLE, J. C., JR, *et al.* 1975. *Initial Rep. Deep Sea drill. Proj*, **31,** U.S. Govt Printing Office, Washington.

LANGSETH, H. & OKADA, H. *et al.* 1978. Near the Japan Trench, transects begun. *Geotimes,* **23,** 3, 22–7.

MARUYAMA, S. 1978. The dismembered ophiolite belt. *Earth Sci.* (Chikyu Kagaku), **32,** 317–20 (in Japanese with English abstract).

MATSUMOTO, T., NODA, M. & MIYAHISA, M. 1962. *Kyushu Region.* Asakura Pub. Co. (in Japanese).

MATSUZAWA, A., TAMANO, T., AOKI, Y. & IKAWA, T. 1980. Structure of the Japan Trench subduction zone, from multi-channel seismic-reflection records. *Mar. Geol.* **35,** 171–82.

MISAWA, Y. 1976. Submarine geology off the southwestern part of the Fossa Magna. *Mar. Sci./* monthly, **8,** 611–6 (in Japanese with English abstract).

MIYAKE, Y. 1981. Geology and petrology of the Shionomisaki igneous complex, Wakayama Prefecture, Japan. *J. geol. Soc. Japan*, **87,** 383–403 (in Japanese with English abstract).

MURAUCHI, S. 1979. Mechanism of subduction in Japan Trench. *Mar. Sci.*/monthly, **11,** 799–806 (in Japanese).

NAGUMO, S. 1980. Seismic activity and geological structure across the Japan Trench and forearc. *In:* SUGIYAMA, R., HAYAKAWA, M. & HOSHINO, M. (eds). *Earthquake.* Tokai Univ. Press, Tokyo, 25–40 (in Japanese).

NASU, N., HONZA, H., FUJIOKA, K. & SATO, T. 1978. Outline of drilling in Japan Trench by Glomer Challenger, Leg 57. *Bull. Japan Soc. Sedimentology,* **15,** 12–15 (in Japanese).

NASU, N., TOMODA, Y., KOBAYASHI, K., KAGAMI, H., UYEDA, S., NAGUMO, S., KUSHIRO, I., OZIMA, M., NAKAZAWA, K., TAKAYANAGI, Y., OKADA, H., MURAUCHI, S., ISHIWADA, Y. & ISHII, Y. 1979. Multi-channel seismic reflection data across the Japan Trench. *IPOD-Japan Basic Data Series, no. 3.* Ocean Res. Inst., Univ. Tokyo.

NIINO, H. 1935. On sounding at Tosabae. *Geogr. Rev. Japan*, **11,** 679–87 (in Japanese with English abstract).

NISHIMURA, S., KATSURA, K. & SASAJIMA, S. 1976. Fission track ages of zircon in conglomerates occuring in the Maizuru District. *J. geol. Soc. Japan,* **82,** 413 (in Japanese with English abstract).

NISHIMURA, Y. 1979. Metamorphic rocks of 300–400 My ages found in the inner zone of Southwest Japan. *In: The Basement of the Japanese Islands.* Prof. Hiroshi Kano Memorial Volume, 201–16 (in Japanese with English abstract).

NODA, M. 1961. Pre-Cambrian (?) rocks in the outer zone of Southwest Japan. *J. geol. Soc. Japan,* **67,** 346–9 (in Japanese with English abstract).

NOHDA, S. 1973. Rb-Sr Dating of the Yatsushiro granite and genesis, Kyushu, Japan. *Earth planet. Sci. Lett.* **20,** 140–4.

NOZAWA, T. 1975. A preliminary note on the distribution of granitic rocks on outer side of Southwest Japan, in Kii-Shikoku-Kyushu Area. *Assoc. Geol. Collaboration Japan Monogr. no. 19, Problems on geosynclines*, 209–12 (in Japanese with English abstract).

OJIMA, M. 1979. *History of the Earth.* Iwanami Pub. Co. (in Japanese).

OKADA, H. & SAKAI, T. 1979. Leg 56 deep sea drilling in the Japan Trench—1. *Mar. Sci.*/monthly, **11,** 756–62 (in Japanese).

OKUDA, Y., KUMAGAI, M. & TAMAKI, K. 1979. Tectonic development of the continental slope and its peripheral areas off Southwest Japan in relation to sedimentary sequences in sedimentary basins. *J. Japan Assoc. Petrol. Technol.* **44,** 279–90 (in Japanese with English abstract).

PLAFKER, G. 1972. Alaskan earthquake of 1964 and Chiliean earthquake of 1960, Implications for arc tectonics. *J. geophys. Res.* **77,** 901–25.

SAWAMURA, T. 1953. Relation between the activities of the outer earthquake zone in Southwestern Japan and the geologic structure and crustal movements of Shikoku and its vicinity. *Res. Rep. Kochi Univ.* **2,** 2, 1–46 (in Japanese with English abstract).

SHIBATA, K., IGI, S. & UCHIUMI, S. 1977. K-Ar ages of Hornblendes from gabbroic rocks in Southwest Japan. *Geoch. J.* **11,** 57–64.

SHIKI, T. 1978. Structure and tectonics under the continental slopes of the Honshu Arc. *Tectonophysics EOS Trans. Am. geophys. Union,* **59,** 1183.

SHIKI, T. & MISAWA, Y. 1979. A few problems concerning the structure under the continental slope and its origin. *Earth Sci.* (*Chikyu Kagaku*), **33,** 208–24 (in Japanese with English abstract).

SHIKI, T. & MISAWA, Y. 1980. Geologic structure and tectonics under the inner slopes of the Nankai Trough and the Japan Trench. *Geotectonics*, **6,** 98–109 (in Russian).

SUZUKI, S. & OKADA, H. 1977. Regional variation in Pn velocity and seismic activity in and around Northeastern Japan. *In: Subsurface Structure of Hokkaido and its Surroundings, and its Meanings in Geosciences, a Symposium,* 1–12 (in Japanese).

TOMITA, T. 1954. Geologic significance of the color of granite zircon, and the discovery of the Pre-Cambrian in Japan. *Mem. Fac. Sci. Kyushu Univ.* **40,** 135–61.

VON HUENE, R. 1978. A section across the Aleutian arc based on shallow and deep geophysical data. *Abstr. Int. Geodynamics Conference 'Western Pacific' and 'Magma Genesis',* Tokyo, 172.

von Huene, R. & Arthur, M. A. 1981. Sedimentation across the Japan Trench off northern Honshu Island (this volume).

von Huene, R., Nasu, N. *et al.* 1978. On Leg 57, Japan Trench transected. *Geotimes,* **23,** 4, 16–21.

Watanabe, H. 1971. Tsunami sources, aftershocks, and earthquake mechanism on the Tokachi-oki earthquake of 1968. *In: General Report on the Tokachi-oki Earthquake of 1968.* Eigaku Pub. Co., 137–51.

Yajima, T. 1972a. Petrology of the Murotomisaki gabbroic complex. *J. Japan Assoc. Min. Pet. econ. Geol.* **67,** 218–41.

Yajima, T. 1972b. Petrochemistry of the Murotomisaki gabbroic complex. *J. Japan Assoc. Min. Pet. econ. Geol.* **67,** 247–61.

Yajima, T. & Fukuda, H. 1976. On the Shionomisaki igneous complex, Wakayama Prefecture. *Mem. Fac. Educ. Saitama Univ.* **25,** 27–44.

Yamashita, N. 1979. A history of the researches on the Kurosegawa Tectonic Zone. *In: The Basement of the Japanese Islands.* Prof. Hiroshi Kano Memorial Volume, 299–318 (in Japanese with English abstract).

Yoshii, T. 1979. A detailed cross-section of the deep seismic zone beneath Northeastern Honshu, Japan. *Tectonophysics*, **55,** 349–60.

Yoshizawa, M. 1954a, b. On the gabbros of the Cape of Muroto, Shikoku Island, Japan. *Mem. Fac. Sci. Univ. Kyoto*, **20B,** 291–84, **21B,** 193–212.

Tsunemasa Shiki, Department of Geology and Mineralogy, Kyoto University, Kyoto 606, Japan.

Yoshibumi Misawa, Faculty of Marine Science and Technology, Tokai University, Shimizu 424, Japan.

CENTRAL AMERICA

Facies belts of the Middle America Trench and forearc region, southern Mexico: results from Leg 66 DSDP

J. Casey Moore, Joel S. Watkins, Kenneth J. McMillen, Stephen B. Bachman, Jeremy K. Leggett, Neil Lundberg, Thomas H. Shipley, Jean-Francois Stephan, Floyd W. Beghtel, Arif Butt, Borys M. Didyk, Nobuaki Niitsuma, Les E. Shephard & Herbert Stradner

SUMMARY: The Middle America Trench SE of Acapulco is flanked by a steep canyon-incised slope and narrow shelf, showing one of a variety of sedimentary facies patterns possible at convergent margins. Piston and drill cores from this region define eight facies belts including: (1) a pelagic facies of brown clay, (2) an outer slope mud facies, (3) a trench sand facies, (4) a foraminiferan-free facies on the lower slope, (5) a foraminiferan-bearing facies on the mid-slope, (6) a laminated mud facies on the upper slope, (7) a shelf facies of sand and mud, and (8) a canyon facies of sand and gravel. The superposition of trench and lower slope sediment during accretion results in a fining upward sequence reflecting a gradual uplift of the seafloor through the trench sediment-plume. The lower limit of the foraminiferan-bearing facies is defined by the absence of *in situ* calcareous foraminiferans and is controlled by the calcite compensation depth. The upper slope laminated mud facies probably reflects the depth range of the oxygen minimum zone.

In the Leg 66 area sedimentation rates are high in the trench and on the outer and lower slope, decrease on the mid-slope, and increase again on the shelf. On the inner shelf, waves and currents concentrate sand which funnels through a prominent submarine canyon, bypassing the mud-dominated slope and accumulating in the trench. A terrigenous sediment-plume generated by trench turbidity flow causes accelerated sediment accumulation to about 500 m above and 40 km seaward of the trench. The volume of material transported by the trench sediment-plume is five or six times greater than that moved by the shelf sediment-plume which supplies detritus to the shelf, upper slope and mid-slope environments.

Facies models for trench and trench-slope deposits are essential for the palaeogeographic interpretation of ancient convergent margin sequences, especially those involved in collision zones. Development of field criteria for the recognition of various sub-environments is especially desirable.

Available data from modern convergent plate boundaries and probable ancient equivalents indicate substantial variability in sedimentary deposits (Schweller & Kulm 1978; Underwood & Bachman 1981; Underwood *et al.* 1980 and references therein). The resultant facies mosaics reflect complex depositional systems developed in response to the diversity of morphotectonic patterns. We present a detailed facies model for a type of convergent margin that consists of a juvenile subduction complex emplaced against truncated older continental basement terrane.

Our facies model for the Middle America Trench and trench-slope off southern Mexico stems from the study of 21 piston cores (McMillen & Haines 1981; McMillen *et al.* 1981), eight Deep Sea Drilling Project Sites (Moore, Watkins *et al.* 1979), and associated geophysical and bathymetric data (Shipley 1981). The piston cores show the distribution of modern facies, whereas the drill cores record the distribution of facies belts back to early Miocene.

For convenience we have subdivided and informally named the marine sedimentary environments off southern Mexico as follows: (1) shelf environment (0–200 m); (2) upper slope environment (200–1000 m); (3) mid-slope environment (1000–3400 m); (4) lower slope environment (3400 m–trench floor or about 5000 m); (5) trench environment (any sediment ponded in trench axis); (6) outer slope environment (area from the outer margin of trench sediment pond, or the trench axis if unfilled, to seaward limit of hemipelagic sediment accumulation); (7) pelagic environment (seaward of limit of hemipelagic sediment accumulation).

Tectonic setting

The Pacific margin of southern Mexico is characterized by a rugged coastal mountain range, a narrow continental shelf, and a steep trench slope which lacks a forearc basin (Fig. 1). The

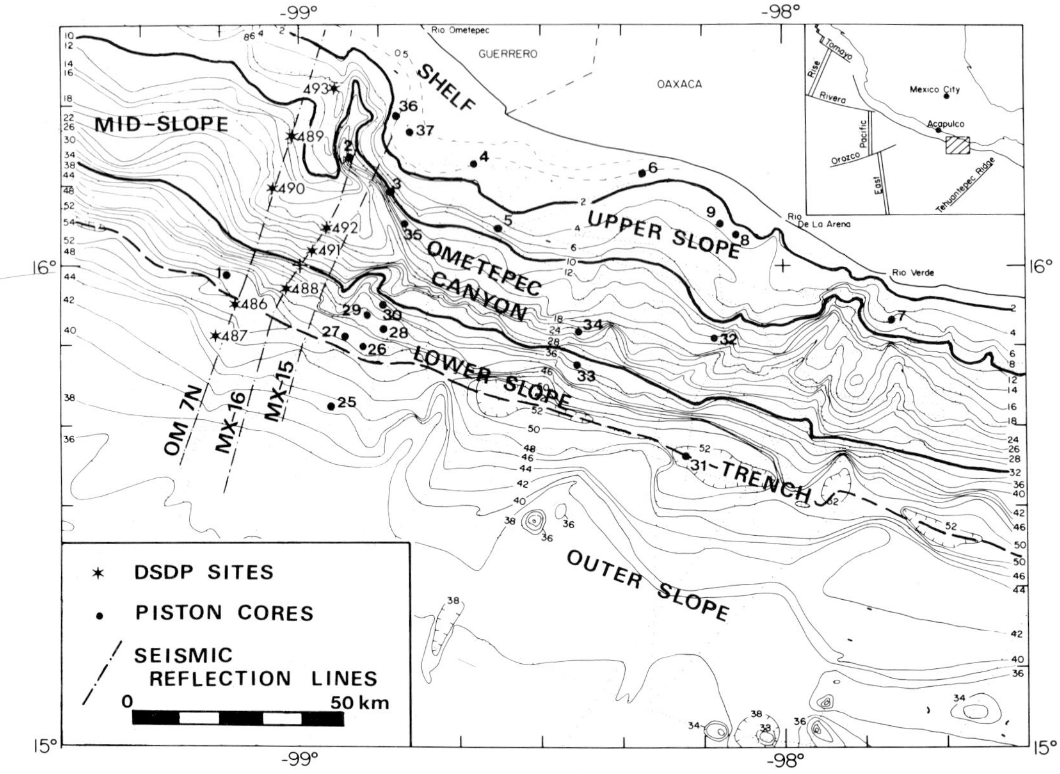

FIG. 1. Bathymetry and core locations. Bathymetry from Shipley (1981).

Middle America Trench in the Leg 66 area is about 5 km deep and marked by a series of discontinuous sediment ponds with a maximum of 625 m of turbidite fill overlying about 200 m of pelagic and hemipelagic sediment which rests on Miocene oceanic crust (Shipley 1981 and Fig. 2). In the Leg 66 drilling area the Cocos plate converges with the North American plate at about 7 cm yr^{-1} along an azimuth of 038°, or about 20° from perpendicular to the strike of the trench (Minster & Jordan 1978).

Plate tectonic reconstructions and the magmatic history of the continental margin both suggest that subduction has been occurring at least intermittently beneath southern Mexico about 100 Ma (Karig *et al.* 1978). An extensive zone of off-scraped deep-sea sediments and a broad forearc region might be expected along this margin in view of its long history of inferred subduction. However, Mesozoic and Palaeogene intrusions and associated basement rocks (Mejorada 1976) occur at the shoreline

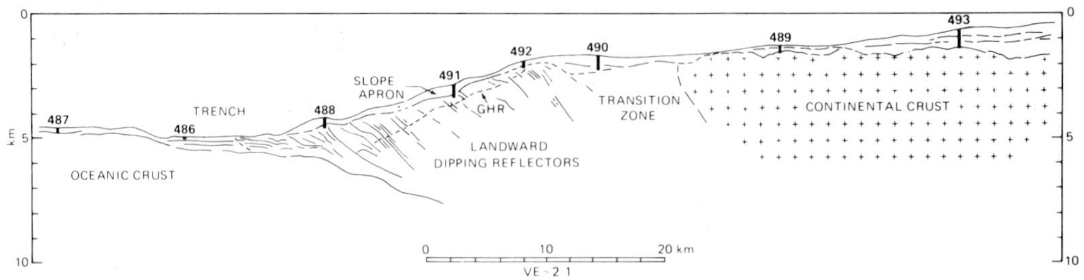

FIG. 2. Generalized cross-section along Deep Sea Drilling transect. Portion from Site 490 landward and 486 seaward taken along seismic line OM-7N, remainder along seismic line MX-16 (locations in Fig. 1).

and are shown, by magnetic, seismic and drilling data, to extend within 35 km of the trench (Karig *et al.* 1978; Shipley *et al.* 1980; Moore, Watkins *et al.* 1979). Neogene rocks both overlie the basement or are faulted against its seaward boundary, suggesting removal of any pre-existing Mesozoic and/or Palaeogene forearc and accretionary terrane. The present phase of accretion apparently began in the Neogene (Moore, Watkins *et al.* 1979) at the same time as magmatism in the Trans-Mexican volcanic belt (Cantagrel & Robin 1979).

Adjacent to the Leg 66 area coastal mountains of crystalline basement and intrusive rocks supply coarse quartzofeldspathic sediment to the marine environment while blocking volcaniclastic detritus from the Trans-Mexican volcanic belt situated 240 km NE of the trench (Enkeboll 1981; Bachman & Leggett 1981). The narrow shelf, steep trench slope cut by submarine canyons, and lack of a forearc basin all permit transport of large volumes of coarse clastic sediment into the trench.

Thus the tectonic elements controlling clastic sedimentary facies along southern Mexico include substantial sediment supply, a relatively rapid convergence rate, and direct input of sediment into the deep sea. Subduction zones with tectonic setting and sedimentary facies similar to southern Mexico may have occurred along western North America during the Mesozoic; the Cretaceous subduction zone bordering the Salinian block in California may be a prime example (Page 1981).

Interpretation of DSDP cores in the Leg 66 area

The interpreted drilling results are summarized below from the seaward-most to the landward-most site (Fig. 3). We interpret only those portions of the DSDP holes that directly pertain to the facies model; complete descriptions are provided elsewhere (Watkins, Moore *et al.* 1981).

Drilling on the outer trench slope at Site 487 penetrated Quarternary hemipelagic mud which covers Pliocene and upper Miocene brown pelagic clay that in turn overlies basaltic oceanic basement. Movement of the Cocos plate toward a terrigenous source apparently accounts for the superposition of hemipelagic over pelagic sediments. Lack of calcareous ooze directly above the oceanic crust suggests it originally formed below the local CCD.

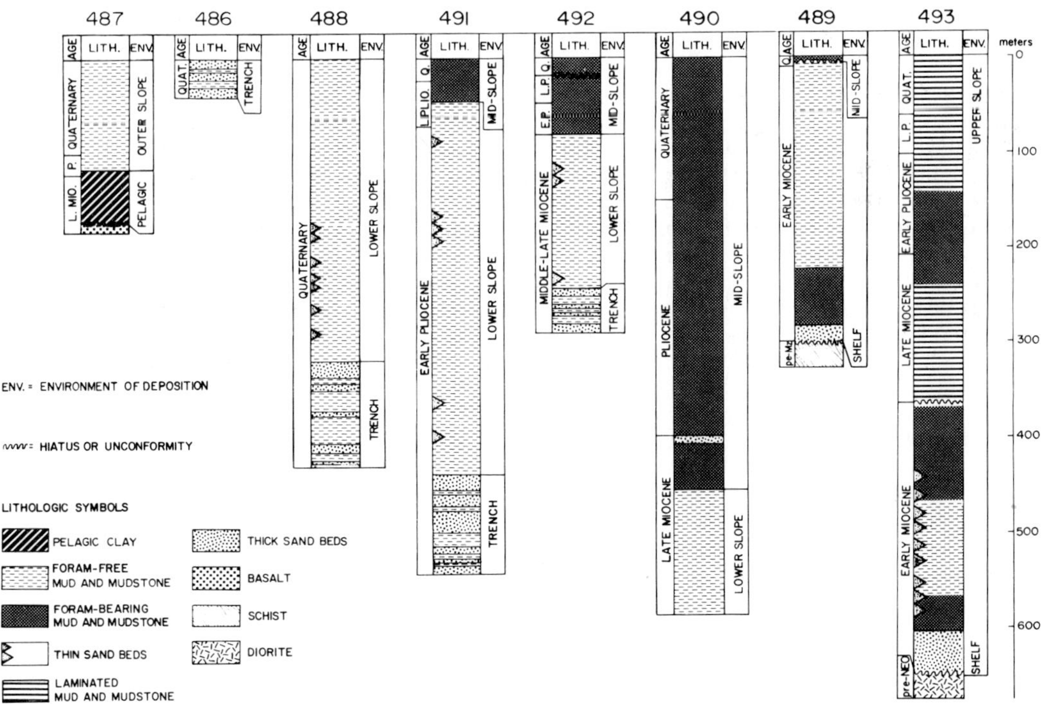

FIG. 3. Lithologic summary of DSDP sites. Interpretation of depositional environment locally omitted where controversial.

The upper Quaternary fine- to coarse-grained sands and muds at Site 486 confirm the sand-rich character of the modern trench fill as indicated by piston coring (McMillen & Haines 1981; McMillen *et al.* 1981). The sequence at the lower and mid-slope sites 488, 491 and 492 consists of hemipelagic mud underlain by mudstone and sand. Surface muds at Site 488 lack *in situ* calcareous foraminiferans indicating deposition below the CCD. On the mid-slope at Sites 491 and 492, hemipelagic mud, respectively below 44 and 80 m sub-bottom, was probably uplifted from below the CCD, as indicated by the absence of *in situ* calcareous foraminiferans (McMillen & Bachman 1981). The thick-bedded sand and mudstone units encountered at depth at Sites 488, 491 and 492 lithologically resemble the modern trench sediments, but could also have accumulated in a basin or submarine canyon on the lower slope. At each site the depth of the sand-rich units coincides with the depth of the zone of dipping reflectors (Fig. 2) which shows acoustic characteristics similar to the modern trench deposits (Shipley 1981). Neither buried submarine canyons nor slope basins are evident on a network of seismic lines across the trench lower slope. As such, we prefer to interpret the sand-rich unit as uplifted trench deposits.

In the mid-slope environment, drilling at Site 490 penetrated 589 m of mud and mudstone with local ash and thin silt and fine sand beds. At Site 490, rocks below 450 m lack *in situ* calcareous foraminiferans suggesting accumulation below the CCD (Fig. 3 and McMillen & Bachman 1981). We interpret the whole section as a slope deposit though its lower part (accumulated below the CCD) apparently was uplifted from the lower slope environment (Fig. 3).

Sites 489 and 493 are now respectively in the mid- and upper slope environments. Foraminiferan- and glauconite-bearing mud characterize surface deposition at Site 489 although at the base of the hole we cored a sand-rich unit that included articulated bivalves (Fig. 7C), scaphopods, echinoids, and shallow-water foraminiferans. Coring at Site 493 initially penetrated laminated muds but terminated in a sequence of muddy siltstone and sandstone overlying quartz dioritic basement. The muddy silt and sandstone contains foraminiferans indicative of a deep shelf environment.

Facies belts in the Leg 66 area

The facies belts described below are named for their depositional environments and defined by their lithology, sedimentary structures, fossil content, and sedimentation rates. The essential characteristics of each facies are summarized in Table 1 and augmented in the text by detailed references to specific DSDP sites and piston cores. To normalize for effects of compaction we quote sedimentation rates in units of mass per area per time (van Andel *et al.* 1975).

Pelagic facies

The Pliocene to upper Miocene brown clay in the lower portion of hole 487 constitutes an open-ocean facies having typical abyssal pelagic characteristics: a siliceous microfossil assemblage, authigenic phillipsite, a lack of terrigenous silt, and basal metal-enrichment (Leggett 1981). Accumulation rates range from 0.4 to 1.7 gm cm^{-2}/1000 yr and are the lowest in the Leg 66 area. In contrast to the pelagic facies at Site 487, modern pelagic deposits seaward of the Leg 66 area consist of siliceous and calcareous ooze (Horn *et al.* 1972).

Outer slope mud facies

The outer slope mud facies is distinguished by a green-grey colour, a terrigenous silt component, and a sedimentation rate higher than that of the pelagic facies. Pyrite, plant fragments and a modest organic carbon content (0.4–1.7%) reflect the reduced nature of the outer slope muds relative to the pelagic brown clays which lack pyrite and plant fragments and have organic carbon contents of 0.1–0.2%. Faint colour laminations and ash beds occur locally in cores from Site 487. Several 1 to 2 cm thick terrigenous sand beds were recovered from a piston core located 12 km seaward of and 700 m shallower than the trench axis (McMillen *et al.* 1981), but such sand beds were not drilled at Site 487.

While the concentration of terrigenous silt and sand-sized material in the outer slope mud is far greater than that in the pelagic clay, it is less than that of the inner slope mud (see below). The outer slope mud contains radiolarians but no foraminiferans except a probable redeposited fauna similar to that recovered in the trench. The sedimentation rate of the outer slope mud is 10 gm cm^{-2}/1000 yr, markedly higher than that of the pelagic mud but less than recorded on the lower slope. The outer slope mud is about 100 m thick at Site 487, 7 km seaward of the trench floor.

Trench sand facies

We sampled the trench sand facies in three piston cores and two DSDP holes (Site 486) in the modern trench and presumably at depth at DSDP Sites 488, 491, and 492 on the trench inner slope. In the Leg 66 area the trench sand facies occurs in a series of discontinuous sediment ponds whose localization is clearly related to major submarine canyons as well as irregularities on the oceanic plate (Fig. 4).

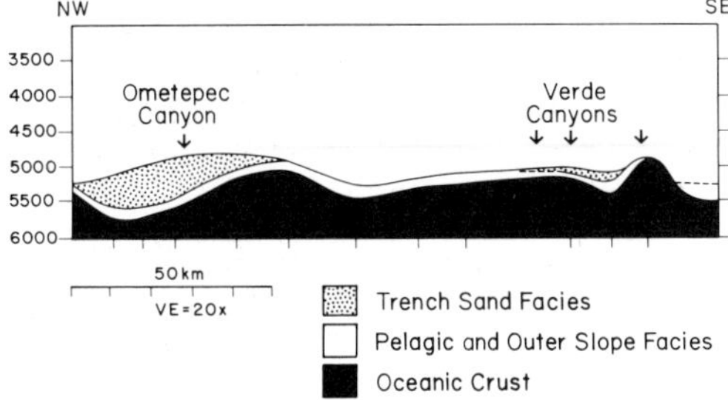

FIG. 4. Distribution of trench sand, pelagic and outer slope mud facies longitudinally along trench. Thickness of trench sediment taken at base of inner slope with corrections made for any thickening due to mild deformation (after Shipley 1981).

The major trench sediment pond occurs at the mouth of Ometepec submarine canyon in the Leg 66 area (Fig. 4). Two piston cores (26 and 27, Fig. 1) on the small fan at the mouth of this canyon consist predominantly of fine and coarse sand with minor mud. Similarly, the two holes drilled at Site 486 recovered predominantly fine- to very coarse sand (Fig. 5B) with minor mud.

The morphology of the lower slope where Ometepec Canyon enters the trench suggests levee development and a recent submersible dive in this area indicates that boulder-filled channels cut the fan (Lonsdale pers. comm. 1980). One seismic profile (MX-15, Shipley 1981; McMillen *et al.* 1981) shows a clearly defined axial channel about 400 m wide and 40 m deep bordering a fault-line scarp. Two small channels appear in the trench on seismic profile MX-16 9 km NW (down gradient) from MX-15; no channels cut the trench fill on seismic profile OM-7N 18 km NW of profile MX-15 (Shipley 1981; McMillen *et al.* 1981). Thus, a transition from channelized to non-channelized flow occurs within 27 km from the mouth of Ometepec Canyon to the vicinity of Site 486 lying on seismic line OM-7N. The gradient of the trench from the mouth of Ometepec Canyon towards Site 486 is 1/125 (0.5°) which is at the upper limit of slopes observed on submarine fans (Nelson *et al.* 1978) and an order of magnitude steeper than those from the continuous trench wedges of the Peru–Chile and eastern Aleutian trenches (Schweller & Kulm 1978; von Huene 1974).

A second sediment pond occurs in the trench off Rio Verde where a series of small canyons cut the trench slope (Fig. 4). One piston core from the north-western edge of this accumulation recovered mud with local graded beds showing Bouma Tc-e sequences (McMillen *et al.* 1981).

The inferred trench deposits comprising the basal units at Sites 488, 491, and 492 include sand beds locally coarser than those in the modern trench but overall the sequence is richer in mud. Sand ranges from fine- to very coarse-grained and granular; clasts locally reach 7 mm in diameter. At Site 488 sand layers inferred from accelerated drilling rates are as much as 11 m thick and appear to thicken upwards over an interval of 100 m. A semi-consolidated unit more than 6 m thick from Site 492 represents the thickest single sand bed we recovered; overall this bed appears massive (Fig. 5A), although it does grade up-section from very coarse to medium sand. The mud-rich sequences interlayered with the massive sand beds at Sites 488, 491, 492 locally show thin graded and laminated sand and silt beds (Fig. 5C).

Foraminiferal faunas in both modern trench deposits (Site 486) and their inferred ancient equivalents (Sites 488, 491, 492) include displaced inner shelf, outer shelf, upper slope, and mid-slope forms. Measurement of the sediment accumulation rate in the trench is precluded by poor biostratigraphic resolution. Using the dynamic steady-state model (e.g. Moore 1979) the calculated rate is about 322 gm cm^{-2}/1000 yr (about 2.6 km Ma^{-1}) for the area near seismic line OM-7N.

Foraminiferan-free mud facies: lower slope environment

The modern inner slope sediments include foraminiferan-free mud on the lower slope, foraminiferan-bearing mud on the mid-slope, and laminated mud on the upper slope (Figs 6 & 7). Each of these sediment types is defined as a separate facies although their distinction may be subtle.

The foraminiferan-free mud facies of the

2 cm
B
2 cm
C
A

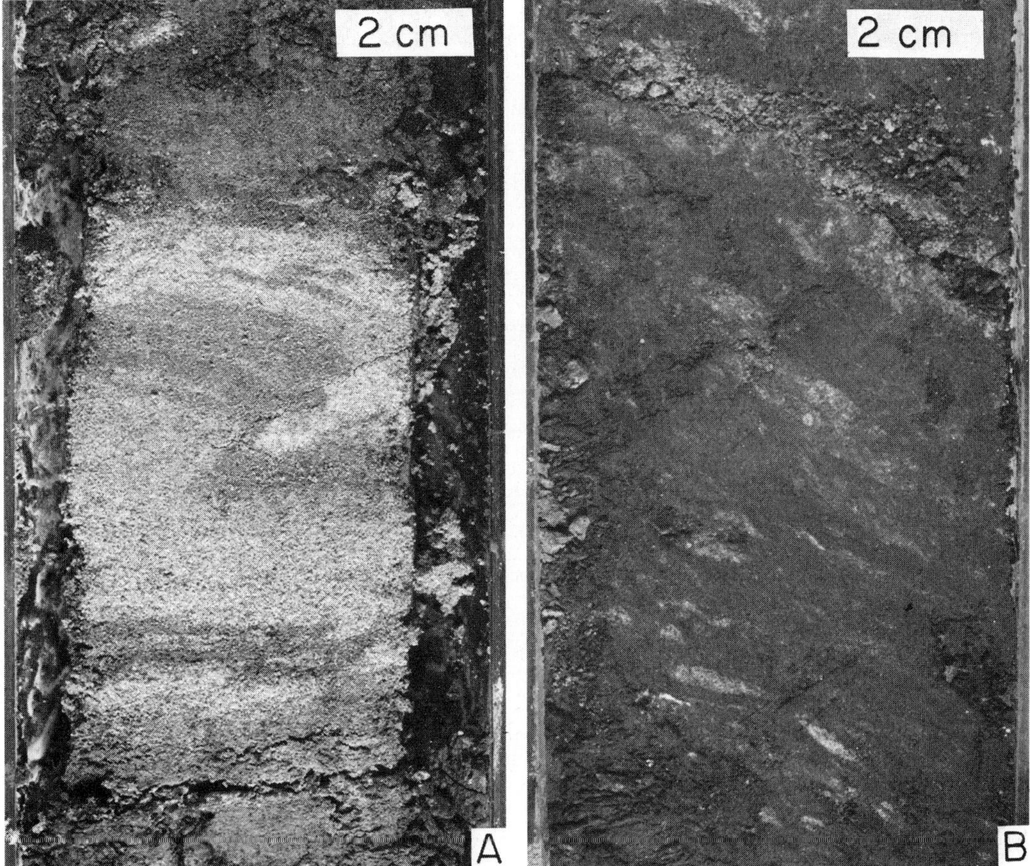

FIG. 6. Foraminiferan-free mud facies. (A) Thin sand bed from foraminiferan-free mud facies, Site 491, 167 m. This sand bed represents one of the thicker though photogenic sand beds of the foraminiferan-free mud facies. Most sand beds were less than 3 cm thick. (B) Disrupted silt beds, Site 488, 238 m.

modern trench lower slope has been sampled in five piston cores and at Site 488 and also encountered at depth at Sites 489, 490, 491, 492, and 493. Piston core 1 (Fig. 1) just upslope from the trench includes mud-rich turbidities with Tc-e Bouma sequences and a maximum sand bed thickness of 3 cm. Other piston cores show colour lamination and fine-grained sand beds up to 1 cm thick. Notably, not even the two cores from slope basins encountered any concentration of sand.

DSDP Site 488 penetrated 310 m of the foraminiferan-free mud facies before coring probable trench deposits (Fig. 3). The upper 170 m of the foraminiferan-free mud here consists of mud with local green and blackish-green laminations with isolated concentrations of plant matter. From 170 m to just above the trench sand facies at 310 m, locally graded fine sand and silt beds up to 3 cm thick punctuate the otherwise muddy section. Thin beds of silt and fine-grained sand, nowhere greater than 8 cm thick, make up a small percentage of the foraminiferan-free mud facies at Sites 491 and 492 (Fig. 3).

In summary, the foraminiferan-free mud facies is mud dominated with a few per cent of fine-grained, thin, sand beds (Fig. 6A). It is

FIG. 5. Trench sand facies. (A). Semiconsolidated 'massive' sand bed from 492, 280.5–290 m. Although the bed appears massive, it grades overall from medium to very coarse sand down-section (from left to right). (B) Very coarse sand, Site 486A, 22 m. (C) Laminated, locally graded fine sand and silt beds, Site 488, 383 m.

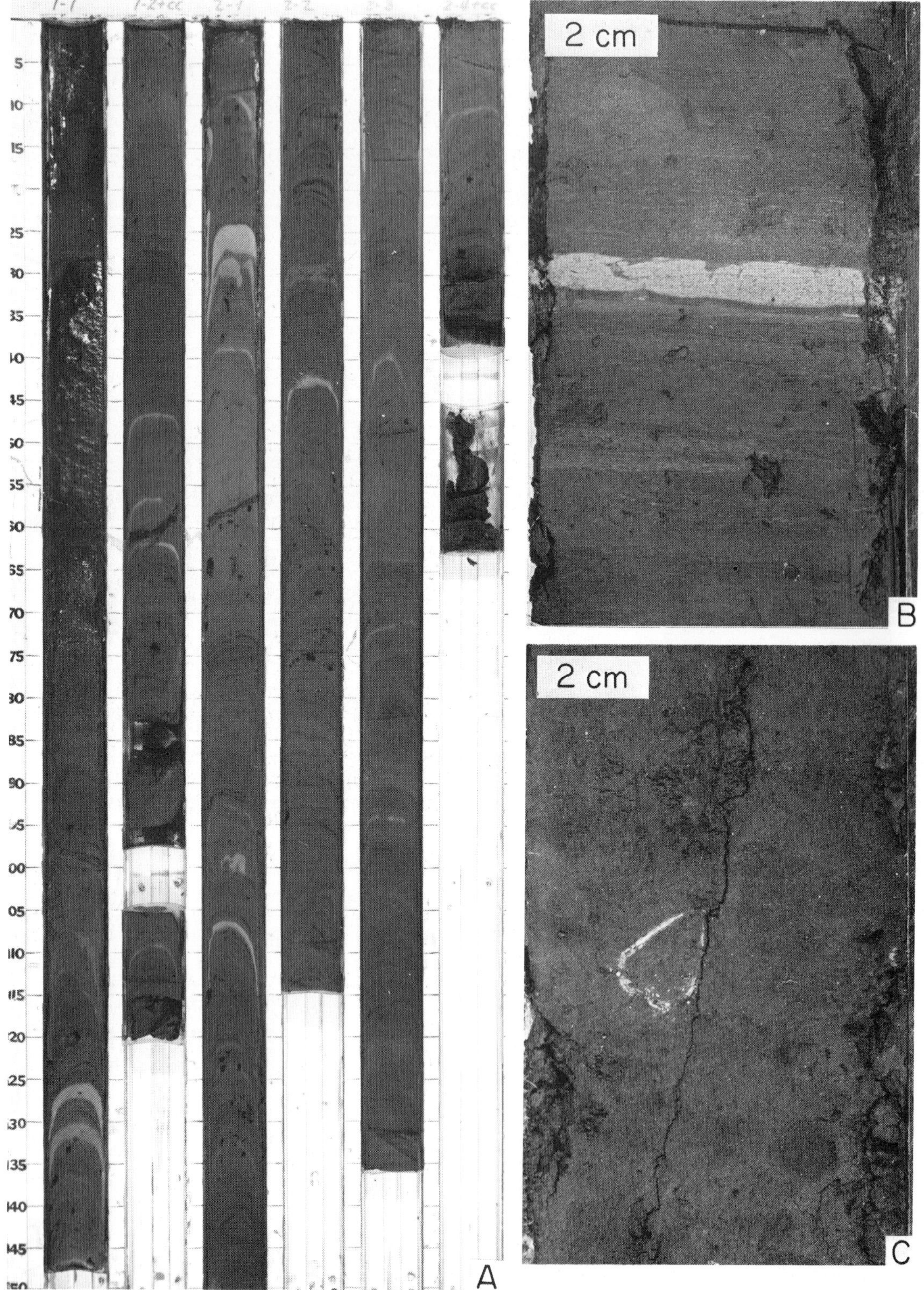
1-1
1-2+cc
2-1
2-2
2-3
2-4+cc
A
2 cm
B
2 cm
C

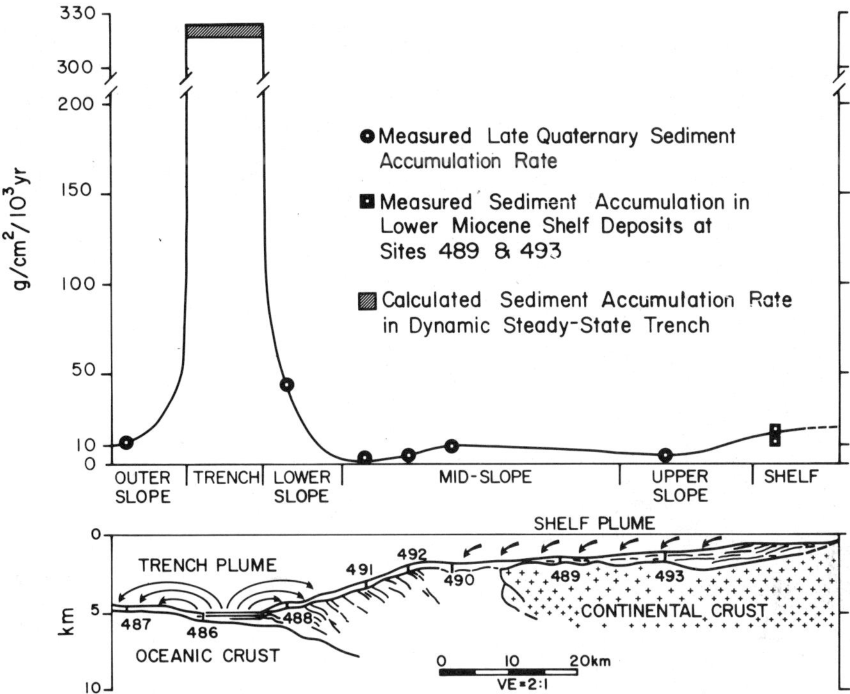

FIG. 8. Sediment accumulation rates plotted across continental margin.

distinguished from the foraminiferan-bearing and laminated mud facies by the absence of *in situ* calcareous foraminiferans. Both grading in the sand beds and rhythmic colour laminations with discernible grain-size variations suggest deposition in part by dilute turbidity currents. The thin sand beds occur in no recognizable megasequence although the foraminiferan-free mud facies does fine-upward overall. The sediment accumulation rate in the foraminiferan-free mud facies varies between 4 and 44 gm cm^{-2}/1000 yr and generally decreases away from the trench (Fig. 8).

The hemipelagic mud component of the foraminiferan-free mud facies is essentially indistinguishable from the hemipelagic mud of the outer slope mud facies. Except for a higher percentage of thin-bedded turbidities in the former, these two facies would be impossible to separate on the outcrop, though their associated rock types might allow their identification.

Foraminiferan-bearing mud facies: mid-slope environment

Three piston cores and four DSDP sites sample the foraminiferan-bearing mud facies. Piston cores 30, 32, and 34 (Fig. 1) consist predominantly of mud with local colour laminations and minor concentrations of foraminiferans, glauconite, radiolarians, and diatoms. Core 30 penetrated a slope terrace, whereas cores 32 and 34 were taken from the slope.

At Site 491 the foraminiferan-bearing mud facies is approximately 44 m thick and is characterized by a dominance of argillaceous material, the common occurrence of foraminiferans, and a complete lack of sand and silt beds (Fig. 3). Drilling at Site 492 penetrated about 80 m of foraminiferan-bearing mud which includes concentrations of foraminiferans ranging up to 10% of smear slides. Here, glauconite occurs as a dispersed component in the predominantly muddy section and in discrete beds up to 10 cm thick (Fig. 3).

The foraminiferan-bearing mud facies comprises the initial 450 m of sediment cored at Site 490. Concentrations of foraminiferans locally reach 5–15% in rare, thin sandy beds, whereas glauconite is a common dispersed component and also occurs in discrete beds up to 10 cm thick (Fig. 3). The foraminiferan-bearing mud

FIG. 7. Laminated mud and shelf facies. (A) Laminated mud, Site 493A, 0–10 m. (B) Laminated mudstone, Site 493, 278 m. (C) Articulated bivalve in silt of shelf facies, Site 489, 279 m.

facies at Site 490 includes recrystallized limestone beds and nodules. At Site 489 the foraminiferan-bearing mud facies is marked by the presence of both foraminiferans and glauconite, but is only 6 m thick, overlying a substantial hiatus.

Quaternary accumulation rates in the foraminiferan-bearing mud facies are low at Sites 491 and 492 (1.8 and 3.6 gm cm^{-2}/1000 yr, respectively) seaward of the mid-slope high, but modest at Site 490 (10 gm cm^{-2}/1000 yr) landward of the mid-slope high.

Laminated mud facies: upper slope environment

Five piston cores from about 400 to 1000 m water depth and DSDP Site 493 at 645 m water depth all sampled laminated muds. Laminations in cores 5, 7, 8 and 9 occur as contrasting colour bands, but in core 35 as thin-bedded, fine-grained, clean sand that locally shows channels and ripples (McMillen & Haines 1981). These thin sand layers may have been deposited by overbank flow from Ometepec Canyon. The near absence of bioturbation in all cores presumably accounts for the preservation of lamination.

Drilling at Site 493 penetrated Quaternary and upper Pliocene laminated sediments from 0 to 148 m (Fig. 3). Colour laminations are most common (Fig. 7A, B) though foram- and shell-rich sand laminations also occur. Between 234 and 372 m upper Miocene rocks display conspicuous colour laminations. The benthonic foraminiferal fauna in the upper Miocene rocks indicates accumulation under conditions of low dissolved oxygen (in the oxygen minimum zone). Except for a trace occurrence in one smear slide, glauconite is notably absent from the laminated muds. Sediment accumulated at about 4 gm cm^{-2}/1000 yr both in the Quaternary-upper Pliocene and upper Miocene laminated mud sections at Site 493.

Shelf facies

Modern shelf deposits were recovered in four piston cores and probable ancient shelf sequences were penetrated at DSDP Sites 489 and 493. Piston cores 6, 36 and 37 (Fig. 1) contain massive clean sand with molluscan shells, coral, and calcareous algae. A hemipelagic mud layer 1.8 m thick overlies an equal amount of shelly sand in piston core 6, and piston core 4 consists of mud with one thin sand bed.

A lower Miocene massive sand section at the base of Site 493 occurs unconformably on granitic intrusive rocks and fines upward through a predominantly muddy section. At Site 489 the basal sand-rich sequence also fines upward but is finer grained and thinner than the similar sequence at Site 493. Small articulated bivalves occur in the basal sequence at Site 489 (Fig. 7). The foraminiferans in the sand-rich sections of both Sites 489 and 493 indicate deposition in a deep shelf environment. Sediment accumulation rates in the sand-rich sequences are 4 and 6 gm cm^{-2}/1000 yr at Sites 489 and 493 respectively. The shelf facies deposits at both sites indicate an overall history of subsidence (McMillen & Bachman 1981).

Canyon facies

Ometepec Canyon and other smaller submarine canyons are the principal conduits supplying sediment to the trench in the Leg 66 area. Although we have not sampled these features in detail a summary of available data for Ometepec Canyon is warranted.

Piston cores 2, 3 and 35 penetrated deposits on the slope flanking Ometepec Canyon. Cores 2 and 3 contain layers of angular gravel clasts. Core 35, unique among upper slope surface samples, contains numerous thin sand laminae, possibly deposited in response to density current flow down Ometepec Canyon. Cores 26 and 27 from a small fan-shaped structure where the Ometepec Canyon intersects the trench contain predominantly fine- to very coarse-grained sand. Very coarse sediment up to boulder size occurs in channels at the canyon mouth (Lonsdale pers. comm. 1980).

Four seismic lines crossing Ometepec Canyon show a narrow rough axis with thin fill (100 m or less). A fifth seismic line displays a poorly stratified fill a maximum of 550 m thick and about 4 km wide that is cut by a channel approximately 300 m deep. The seismic data indicate that Ometepec Canyon cuts into both the crystalline basement complex and associated Neogene rocks.

Discussion of facies descriptions in the Leg 66 area

Non-diagnostic sedimentary components

Definitions of sedimentary facies described herein hinge on certain diagnostic sedimentary components. Non-diagnostic sedimentary components include siliceous microfossils, volcanic ash, and limestone beds and nodules. Radiolaria and diatoms occur in varying amounts in all facies although they are most conscpicuous in sediments deposited below the CCD. All sedimentary facies also include dispersed ash fragments and ash beds although sand derived from the crystalline basement directly onshore is far more abundant. Limestone beds occur locally in facies accumulated above the CCD and small calcareous nodules of probable diagenetic origin are present in sediments deposited both above and below the CCD.

Boundary between the foraminiferan-bearing and foraminiferan-free mud facies

The foraminiferan-free mud facies occurs down-hole at Sites 490, 491 and 492 where the sediments lack *in situ* calcareous foraminiferans. Diagenetic effects may have induced this down-hole transition from foraminiferan-bearing to foraminiferan-barren mud. However, at Sites 489 and 493 foraminiferans are abundant in surface cores, initially diminish down-hole, and then become common again at depth. This distribution of foraminiferans is probably not simply due to burial diagenetic effects but can be explained more simply by vertical motions of the drilling sites relative to the CCD (McMillen & Bachman 1981). A similar interpretation is preferred for the absence of foraminiferans at depth at Sites 490, 491 and 492.

The presence or absence of foraminiferans may be questioned as a criterion to separate middle and lower slope sedimentary environments. Sediments associated with the lower trench slope in the Leg 66 area might be defined alternatively by the presence of the included thin sand and silt beds. This definition is undoubtedly a more useful field criterion although whether or not the thin sand beds would be preferentially associated with the lower, middle, or upper slope sediments would depend on the dynamics of clastic sediment distribution along a particular trench slope. We use the absence of calcareous foraminiferans as an index to lower slope sediments since it is probably controlled by the CCD (see below) and less affected by trench slope physiography.

Controls on facies distribution

Oceanographic factors affecting facies distribution

Lithological differences between the facies belts in the Leg 66 area are controlled by variations in oceanic chemical parameters as well as physiography and current patterns. The downslope disappearance of *in situ* calcareous foraminiferans defines both the calcite compensation depth (CCD) (McMillen & Bachman 1981) and the boundary between the foraminiferan-free and foraminiferan-bearing mud facies. Clearly foraminiferans may be reworked below the CCD. As such, study of sediment deposited between turbidite flows would best define the boundary between the foraminiferan-free and foraminiferan-bearing mud facies. The brown clays comprising the pelagic facies in the Leg 66 area undoubtedly accumulated below the CCD but seaward of their present location, removed from the terrigenous influence of the continent. At other convergent margins one might encounter a pelagic facies composed of other types of open-ocean sediments including carbonate oozes (e.g. Site 495, Leg 67, von Huene, Aubouin *et al.* 1981).

In addition to the CCD, the oxygen minimum zone (OMZ) exerts a critical chemical control on the sediments accumulating in the Leg 66 area. The substantially diminished bioturbation in the OMZ results in the preservation of finely laminated sediments (Ingle 1975). In the Leg 66 area laminated sediment in piston cores and from shallow cores from Site 493 apparently records the approximate depth range of the modern OMZ.

The concentration of glauconite in the foram-mud facies occurs at greater water depths than those at which glauconite normally forms (McRae 1972). Since neither the laminated mud nor the shelf sand facies include significant amounts of glauconite, simple down-slope reworking of this mineral is apparently precluded. We therefore infer that glauconite is forming actively in the mid-slope region. The low sedimentation rates and moderately oxidizing bottom waters in the mid-slope environment would favour the *in situ* formation of glauconite (McRae 1972). Ross (1971) also noted a concentration of glauconite between 1 and 2 km water depth in the northern Middle America Trench slope off Manzanillo.

Control of facies by physical processes and physiography

The concentration of coarse material in the shelf sand facies undoubtedly results from the winnowing effects of waves and currents in this shallow environment. However, the occurrence of sand or mud in piston cores is not simply a function of water depth or distance from the shoreline and therefore must record the complex dynamics of this narrow shelf.

The presence of massive, coarse-grained sand

in the trench but not on the slope indicates sediment bypassing of the slope via Ometepec and other submarine canyons. Calcareous material is nearly absent in the modern trench sediments but common in the modern outer shelf sands. Dissolution below the CCD probably does not account for the lack of calcareous detritus in the trench sand since deposition occurs in pulses which shield the bulk of the sediment from the corrosive sub-CCD bottom waters. Rather, much of the trench sand may be derived from the inner shelf or littoral zone with any contribution from outer shelf calcareous sand being severely diluted. Ometepec Canyon does head in the inner shelf area and could trap sand being transported in this region. Hence, the submarine canyons in the Leg 66 area constitute a conduit for coarse sediment that bypasses the outer shelf as well as the slope.

The thin fine-grained sand beds that occur in piston core 25 seaward of the trench axis probably represent 'contourite' deposits formed through winnowing by bottom currents. Although the sediment bypassed into the trench probably creates a substantial terrigenous plume (see below) the absence of sand beds at Site 487, 7 km seaward of the trench, would seem to preclude trench-plume origin for sand beds in core 25, 12 km from the trench.

Effects of the trench sediment plume and the magnitude of bypassing

Submarine canyons facilitate sediment bypassing of the Middle America Trench slope as a whole (Underwood & Karig 1980). Data from the Leg 66 area provide a basis for quantifying the bypassing process in a specific locality. Sediment accumulation rates plotted along the generalized cross-section (Fig. 8) show a maximum in the trench, a minimum in the mid-slope region (Sites 491 and 492) and a secondary high on the shelf. This sediment distribution pattern may be simply modelled by a modest sediment-plume that emanates down-slope from the shelf area and a more significant plume that originates from the inner shelf, reaches the trench via Ometepec Canyon, and then flows laterally down the trench axis.

Herein, sediment-plume refers to an influx of sediment of a particular environment irrespective of transport process. The trench sediment-plume probably is initiated as a sediment gravity flow which may be characterized by low-density turbidity currents near its outer margins. Bottom currents may assist in re-distributing terrigenous sediment once it is introduced by the trench plume. Material in the shelf sediment-plume may be transported in sediment gravity-flows, by wave- and current-induced traction processes, or by grain-by-grain settling.

The presence of hemipelagic sediment considerably seaward of the trench axis (piston core 25) and the high sedimentation rate at Site 488 attest to the widespread influence of the trench sediment-plume beyond the trench floor. At Site 488 the trench sand facies is succeeded upsection by a sequence of thin-bedded turbidites which is in turn covered by mud at 172 m sub-bottom (*c.* 425,000 yr BP). We believe that the sand represents deposits of the trench floor and that the thin-bedded turbidites accumulated as Site 488 was progressively uplifted through the trench sediment-plume to its present water depth of 4259 m. By using an uplift rate of 400 m Ma^{-1} (McMillen & Bachman 1981) we deduce that the cessation of thin-bedded turbidite deposition (425,000 yr ago) occurred at a depth of 4429 m. If at this time the trench floor was at its present depth of 5000 m, then the thin-bedded turbidite-bearing plume rose 500–600 m above the trench floor. Within our limits of error, the apparent 500–600 m rise of the thin-bedded turbidite-bearing plume above its distributary channel is comparable to the 400 m upper limit of this phenomenon as determined elsewhere (Damuth & Embley 1979; Shipley 1978).

Alternatively the cessation of thin-bedded turbidite deposition at Site 488 could be due to the uplift isolation of this locality on a high shielded from a down-slope flux of fine sand. However, this explanation does not account for the lack of thin-bedded turbidites in the uppermost portions of Sites 491 and 492 (Fig. 3). The absence of sand beds at Site 487, 340 m above and 7 km seaward of the trench, apparently precludes the extension of the sand-bearing plume to that point.

By back-tracking at 6.5 cm yr^{-1} perpendicular to the trench (Minster & Jordan 1979) we deduce that the initial appearance of terrigenous detritus 2.5 Ma ago occurred when Site 487 was about 160 km seaward of its present location. For comparison, Horn *et al.* (1972) map the offshore limit of modern hemipelagic sedimentation about 500 km seaward of the trench. The sediment accumulation rate increases sharply in the hemipelagic sediment about 0.5 Ma ago when Site 487 was approximately 40 km seaward of the trench axis. Thus, we believe that the trench-plume strongly influences sedimentation within 40 km seaward of the trench axis, with secondary effects extend-

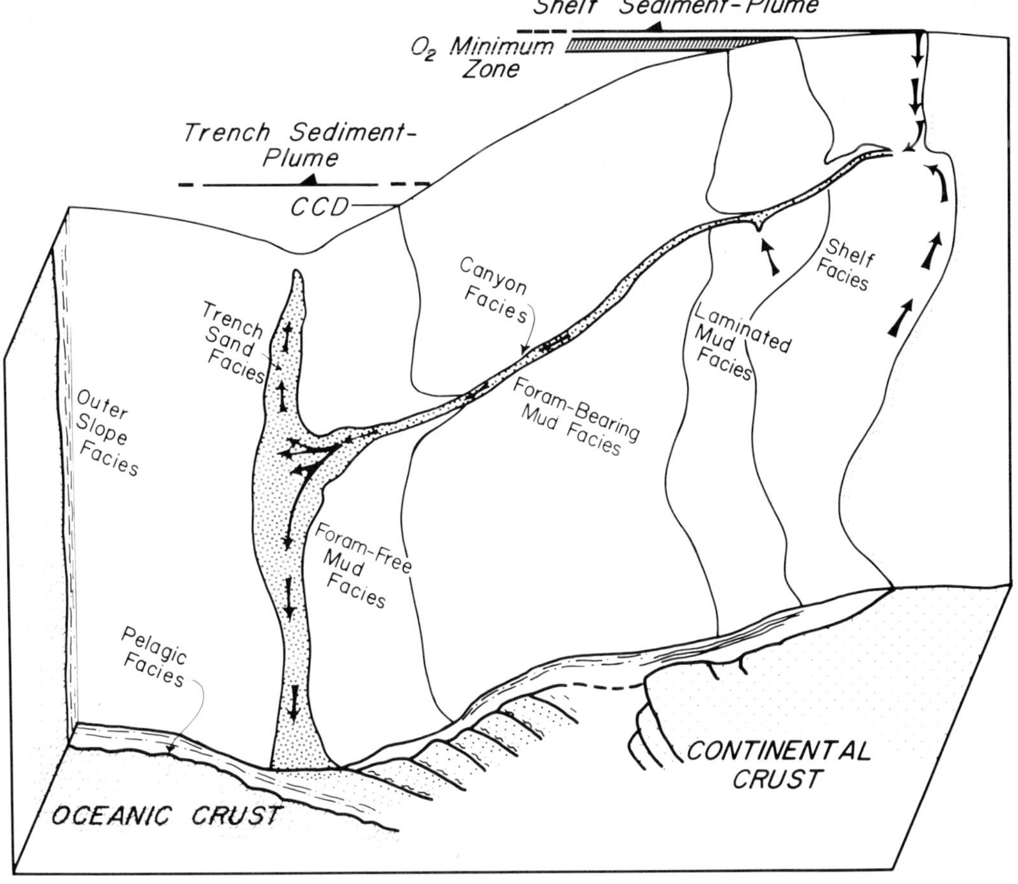

FIG. 9. Schematic representation of facies belts in Leg 66 area with possible controls. Note that facies boundaries are approximately parallel to bathymetric contours.

ing several hundred kilometres or more offshore.

The relative influence of the trench versus the shelf sediment-plume can be established by: (1) integrating the sediment accumulation-rate curve across the continental margin, and (2) measuring the volumes of material accumulated beneath the two plumes during the past 10 Ma. The first approach involves determining the area under the sediment accumulation-rate curve from the seaward edge of the cross-section to the lower-mid-slope boundary (trench-plume influence) and comparing this value to the area under this curve in the shelf, upper slope and mid-slope environments (shelf-plume influence). As such, the measured area of the sediment accumulation rate curve influenced by the trench plume is about six times larger than that affected by the shelf sediment-plume.

The volume of sediment accumulated under the shelf sediment-plume also can be estimated by measuring the cross-sectional area (on Fig. 2) of the sediment younger than 10 Ma that accumulated in the shelf, upper slope, and middle slope environments. Deposits of trench and outer slope have been structurally accreted beneath the lower slope. Similarly, the long-term contribution of the trench sediment-plume can be estimated from the cross-sectional areas (on Fig. 2) of sediment younger than 10 Ma that accumulated in the lower slope, trench and outer slope environments. The cross-sectional area of accreted deposits can be approximated by a triangle with corners at the base of the trench slope, Site 492, and at the intersection of

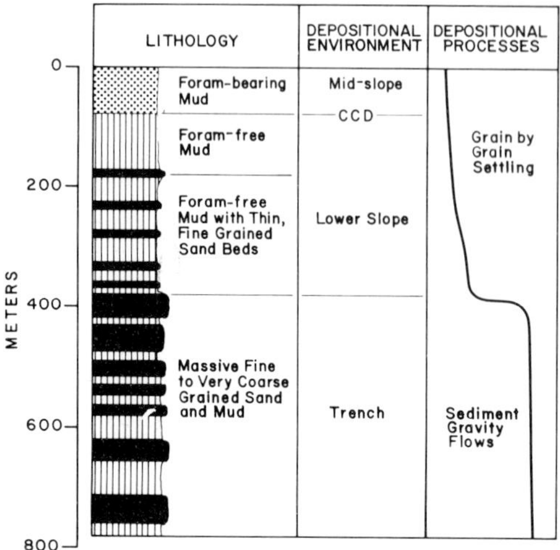

FIG. 10. Fining-upward sequence caused by uplift of seafloor from trench through lower slope to mid-slope environment.

the oceanic crust and a plane projected down along the dipping reflectors from Site 492 to the oceanic crust (Watkins *et al.* 1981). If sediments are accreted in the proportion that they enter the subduction zone, then 90% of this triangle is trench or outer slope sediment. As such, we find that the cross-sectional area of post-10 Ma accreted and overlying lower slope sediments is about five times larger than the cross-sectional area of existing shelf, upper slope and mid-slope sediments less than 10 Ma old. Note that the calculated volume of the accreted trench and outer slope sediments is a minimum since no adjustments were made for dewatering during accretion, nor for subduction of any of this material (Watkins *et al.* 1981).

Our two estimates of the magnitude of sediment bypassing are essentially equal and are probably valid for the cross-section along which they were taken. The magnitude of bypassing is large due to the influence of Ometepec Canyon. However, the volume of material in the shelf sediment-plume may also be large relative to other areas along the northern Middle America Trench due to the sediment input of Rio Ometepec near the landward end of the cross-section.

Secular and regional variation of the controls of facies distribution

Controls on the distribution of sedimentary facies may vary in both time and space along convergent margins. Hence, facies belts similar to those of the Leg 66 area may occur at a range of depths or even be absent elsewhere. For example, the CCD in the non-equatorial Pacific has varied more than 1 km in depth since Eocene times (van Andel *et al.* 1975) and thus probably shifted any boundary between the foraminiferan-free and foraminiferan-bearing mud facies along the Middle America Trench during the past 40 Ma. The oxygen minimum zone may vary in depth as well as intensity, thus allowing for changes in depth and width of, or even elimination of the laminated mud facies. Finally, the bypassing of the trench slope in the Leg 66 area constitutes but one of the many possible clastic sedimentation patterns along convergent margins (Underwood *et al.* 1980; Underwood & Bachman 1981), patterns that are primarily controlled by sediment supply and physiography.

Interpretation of ancient rock sequences

Predicted outcrop-scale characteristics of facies belts

One of our principal goals is the articulation of a facies model for the interpretation of ancient subduction complexes. Our descriptions may partially accomplish this end; in order to define further each facies we have outlined their expected outcrop characteristics (Table 1). The proposed outcrop-scale characteristics list lithology much as observed although we have inferred some types of sedimentary structures which may have been destroyed by drilling or not recorded in a single drill core. We use Walker's (1979) turbidite facies terminology (classical turbidites, massive sandstones, pebbly sandstones, etc.) which implies certain sedimentary structures and bedding characteristics not listed in Table 1.

While Table 1 lists the essential outcrop features of the various facies belts, some discussion of these characteristics is warranted. The blind application of turbidite fan models to trench and trench slope settings is clearly inappropriate; nevertheless, the sedimentary lithological associations and vertical sequences generated by the depositional elements of the fans may also occur within trenches and on trench slopes. For example, where Ometepec Canyon enters the trench the submarine channels would probably deposit coarse sedimentary sequences (possibly thinning- and fining-upward) that would be interbedded with finer grained levee and overbank deposits. However,

TABLE 1.

FACIES BELT	LITHOLOGY	DIAGNOSTIC FOSSIL/TRACE FOSSIL ASSEMBLAGES	SEDIMENT ACCUMULATION RATE $gm/cm^2/10^3$ yr.	POSSIBLE OUTCROP CHARACTERISTICS (Not covered under lithology)
Pelagic	Brown pelagic mud with authigenic phillipsite and lacking terrigenous silt.	Radiolarians, sponge spicules, fish teeth, some planktonic foraminiferans near base of section suggesting water depth near CCD.	1	
Outer Slope	Hemipelagic mud with rare thin ungraded sand or silt beds ("contourites").	Reworked benthonic foraminiferans from bathyal environment.	10	Erosion surfaces and lag deposits ("contourites") associated with reworking by bottom currents.
Trench Sand	Hemipelagic mud and massive sand beds, local gravel concentrations, thin-bedded turbidites.	Reworked benthonic foraminiferans predominantly from neritic environment, some reworked foraminiferans from bathyal environment.	322	Thinning and fining upward sequences where axial channel development occurs. Thin-bedded turbidites as levee and overbank deposits. Thickening coarsening upward sequences associated with axial lobes. Classical turbidites as sheet deposits down-current from lobes. Axial paleocurrent directions excepting where submarine canyons build small fans into trench.
Foraminiferan-free mud	Hemipelagic mud, with thin-bedded turbidites.	In-situ calcareous foraminiferans absent, radiolarians and diatoms present, diagnostic trace fossil assemblage: <u>Chondrites</u>-<u>Planolites</u>-<u>Zoophycus</u>-<u>Teichichnus</u>.	4-44 decreasing upslope away from trench	Slump folds, olistostromes, slope sequence of fines upward from contact with trench sand facies. Possible concentrations of turbidites in slope basins, showing predominantly axial paleocurrents.
Foraminiferan-bearing mud	Dominantly hemipelagic mud with minor glauconite, foraminifera conspicuous relative to radiolarian mud facies, rare sand beds, locally graded and/or glauconitic.	Benthonic foraminiferans of middle and lower bathyal zones, foraminiferans poorly preserved near lower limit of water depth of facies belt due to corrosion below lysocline.	2-10	
Laminated Mud	Hemipelagic mud commonly with color, shell rich, and foram rich laminations, thin sand and silt beds, locally graded.	Benthonic foraminiferans of upper and upper middle bathyal zone, foraminiferans indicative of low levels of dissolved oxygen.	4	
Shelf	Shelly sand, mud.	Articulated and fragmented bivalves, coraline fragments, calcareous algae, benthonic foraminiferans of outer neritic zone.	14-19	Sand both massive and cross-bedded, graded sand beds with hummocky cross-stratification.
Canyon	Sand, gravel, mud. Mud with thin sand beds on canyon wall.			Channelization conspicuous, very coarse-grained deposits in contrast to fine-grained sediments of adjacent slope and canyon wall.

instead of extending radially out across the seafloor the channels veer to the NW and die out in the trench axis. Thus, the vertical sequence produced by the channelized flow might be similar to that observed in a submarine fan but the palaeocurrent and lithofacies pattern would be different. Palaeocurrents would be predominantly axial rather than radial as in the case of a fan. Any transition from channels to depositional lobes would occur not on the fan at the canyon mouth but down the trench axis to the NW. Facies associations at variance to those observed on submarine fans also could develop on slopes and in slope basins (see Underwood *et al.* 1980 for discussion).

The trench slope transition: a diagnostic vertical sequence

Vertical stratigraphic sequences provide a critical tool for interpretation of ancient sedimentary deposits. Indeed, a classic fining-upward transgressive sequence occurs at Sites 489 and 493 in the upper and mid-slope region of the Leg 66 area. Another kind of fining-upward sequence generated by the superposition of lower slope and trench deposits constitutes a vertical succession of particular interest to geologists investigating convergent margins.

The vertical succession recording the shift from a trench to slope environment consists of coarse sand and mud topped by thin-bedded turbidites, covered by mud barren of foraminiferans, and finally succeeded by foram-bearing mud. Seismic profiles across the modern trench suggest that the basal sandy unit should not exceed 500 to 700 m in thickness. The thin-bedded turbidite unit ranges up to several hundred metres thick in the Leg 66 area, whereas the superjacent mud unit is a maximum of 172 m thick.

This fining-upward sequence apparently accumulates as the accretionary process moves a vertical section up off the trench floor, through the lower part of the sediment-plume (that deposits thin sand beds), then through the muddy portion of the trench sediment plume, and finally up through the CCD. This vertical sequence should develop at other convergent margins with substantial clastic influx and bypassing of sediment to a trench with axial turbidite flow.

Conclusions

Along the Middle America Trench in the Leg 66 area we discriminate: (1) a pelagic facies, (2) an outer slope mud facies, (3) a trench sand facies, (4) a foraminiferan-free mud facies on the lower slope, (5) a foraminiferan-bearing facies on the mid-slope, (6) a laminated mud facies on the upper slope, (7) a shelf sand facies, and (8) a canyon facies. During accretion superposition of the mid-slope, lower slope, and trench deposits results in a fining-upward sequence. The boundary between the foraminiferan-free and foraminiferan-bearing muds is defined by the CCD. Similarly, the oxygen mimimum zone controls the extent of the laminated muds on the upper slope. In the Leg 66 area, a prominent submarine canyon causes coarse sediment to bypass the outer shelf and slope. The volume of sediment bypassed is about five times that transported directly across the slope, which accounts for the high sediment accumulation rates in and adjacent to the trench. Bypassed sediment also appears to influence hemipelagic sedimentation several hundred kilometres seaward of the trench.

Acknowledgments: We acknowledge the U.S. National Science Foundation and similar agencies in France, Great Britain, Japan, West Germany, and the U.S.S.R. for support of the International Program of Ocean Drilling. The marine crew, drilling crew and technical staff of the *Glomar Challenger* were exceptionally cooperative in the collection and shipboard examination of the drill cores on Leg 66. The University of California at Santa Cruz supported the preparation of the manuscript. We thank Roland von Huene and Rick Stanley for reviews of various drafts of this manuscript.

References

Bachman, S. B. & Leggett, J. K. 1981. Petrology of Middle America Trench and trench-slope sands, Guerrero margin, Mexico: results from Leg 66 DSDP. *In:* Watkins, J. S., Moore, J. C. *et al.* (eds). *Init. Rep. Deep Sea drill. Proj.* **66.** U.S. Government Printing Office. Washington, in press.

Cantagrel, J. M. & Robin, C. 1979. K-Ar dating on eastern Mexican volcanic rocks—relations between the andesitic and alkaline provinces. *J. Volcan. Geotherm. Res.* **5,** 99–114.

Damuth, J. E. & Embley, R. W. 1979. Upslope flow of turbidity currents on the northwest flank of Ceara Rise: western Equatorial Atlantic. *Sedimentology,* **26,** 825–34.

Enkeboll, R. H. 1981. Sedimentary petrology and

provenance of sands and gravels from the Middle America Trench and trench slope. *J. sediment. Petrol.* in press.

HORN, D. R., HORN, B. M. & DELACH, M. N. 1972. Ferromanganese deposits of the North Pacific. *Techn. Rep. No. 1 NSF GX-33616.* Office of the International Decade of Ocean Exploration, National Science Foundation, Washington, 78 pp.

INGLE, J. C., JR 1975. Paleobathymetric analyses of sedimentary basins. *In:* DICKINSON, W. R. (ed.). *Current Concepts of Depositional Systems with Applications to Petroleum Geology.* 11–1–12. San Joaquin Geol. Soc., Bakersfield, California.

KARIG, D. E., CARDWELL, R. K., MOORE, G. F. & MOORE, D. G. 1978. Late Cenozoic subduction and continental margin truncation along the northern Middle American Trench. *Bull. geol. Soc. Am.* **89,** 265–76.

LEGGETT, J. K. 1981. Geochemistry of Cocos Plate pelagic and hemipelagic sediments, DSDP Site 487, Leg 66. *In:* WATKINS, J. S., MOORE, J. C., *et al.* (eds). *Initial Rep. Deep Sea drill. Proj.* **66.** U.S. Government Printing Office, Washington, in press.

MCMILLEN, K. J. & BACHMAN, S. B. 1981. Paleobathymetry and tectonics of the trench slope, Leg 66, DSDP. *In:* WATKINS, J. S., MOORE, J. C. *et al.* (eds). *Initial Rep. Deep Sea drill, Proj.* **66.** U.S. Government Printing Office, Washington, in press.

—— , ENKEBOLL, R. H., MOORE, J. C., SHIPLEY, T. H. & LADD, J. W. 1981. Sedimentation in different tectonic environments of the Middle America Trench, Southern Mexico and Guatemala (this volume).

—— & HAINES, T. R. 1981. Late Quaternary sediments of the southern Mexico margin. *In:* WATKINS, J. S., MOORE, J. C., *et al.* (eds). *Initial Rep. Deep Sea drill, Proj.* **66.** U.S. Government Printing Office, Washington, in press.

MCRAE, S. G. 1972. Glauconite. *Earth Sci. Rev.* **8,** 397–440.

MEJORADA, S. H. S. 1976. *Carta Geologica de la Republica Mexicana,* 4th ed. Instituto de Geologia U.N.A.M. Ciudad Universitaria.

MINSTER, J. B. & JORDAN, T. H. 1978. Present-day plate motions. *J. geophys. Res.* **83,** 5331–54.

MOORE, J. C. 1979. Variation in strain and strain rate during underthrusting of trench deposits. *Geology,* **7,** 185–8.

—— , WATKINS, J. C., SHIPLEY, T. H., BACHMAN, S. B., BEGHTEL, F. W., BUTT, A., DIDYK, B. M., LEGGETT, J. K., LUNDBERG, N., MCMILLEN, K. J., NIITSUMA, N., SHEPHARD, L. E., STEPHAN, J. F. & STRADNER, H. 1979. Progressive accretion in the Middle America Trench, Southern Mexico. *Nature,* **281,** 638–42.

NELSON, C. H., NORMARK, W. R., BOUMA, A. H. & CARLSON, P. R. 1978. Thin-bedded turbidites in modern submarine canyons and fans. *In:* STANLEY, D. J. & KELLING, G. (eds). *Sedimentation in Submarine Canyons, Fans, and Trenches,* 117–89. Dowden, Hutchinson & Ross, Stroudsburg, Pennsylvania.

PAGE, B. M. 1981. Southern coast ranges. *In:* ERNST, W. G. (ed.). *Geotectonic Development in California.* Prentice-Hall, New York.

ROSS, D. A. 1971. Sediments of the Northern Middle America Trench. *Bull. geol. Soc. Am.* **82,** 303–22.

SCHWELLER, W. J. & KULM, L. D. 1978. Depositional patterns and channelized sedimentation in active eastern Pacific trenches, *In:* STANLEY, D. J. & KELLING, G. (eds). *Sedimentation in Submarine Canyons, Fans, and Trenches.* 311–24. Dowden, Hutchinson & Ross, Stroudsburg, Pennsylvania.

SHIPLEY, T. H. 1978. Sedimentation and echo characteristics in the abyssal hills of the west-central North Atlantic. *Bull. geol. Soc. Am.* **89,** 397–408.

—— 1981. Regional geology of the southern Mexico continental margin. *In:* WATKINS, J. S., MOORE, J. C. *et al.* (eds). *Initial Rep. Deep Sea drill. Proj.* **66.** U.S. Government Printing Office, Washington, in press.

—— MCMILLEN, K. J., WATKINS, J. S., MOORE, J. C., SANDOVAL-OCHOA, J. H. & WORZEL, J. L. 1980. Continental margin of Guerrero and Oaxaca, Mexico. *Mar. Geol.* **35,** 65–82.

UNDERWOOD, M. B. & BACHMAN, S. B. 1981. Sedimentary facies associations within subduction complexes (this volume).

—— , & SCHWELLER, W. J. 1980. Sedimentary processes and facies associations within trench and trench slope settings. *In:* FIELD, M. E., BOUMA, A. H. & COULBURN, I. (eds). *Quaternary Depositional Environments on the Pacific Continental Margin.* Pacif. Sec. Soc. Econ. Palaeontol. Mineral. Tulsa.

—— & KARIG, D. E. 1980. Role of submarine canyons in trench-slope sedimentation. *Geology,* **8,** 432–6.

VAN ANDEL, T. H., HEATH, G. R. & MOORE, T. C., JR 1975. Cenozoic history and paleoceanography of the central Equatorial Pacific Ocean. *Mem. geol. Soc. Am.* **143,** 134 pp.

VON HUENE, R. 1974. Modern trench sediments. *In:* BURK, C. A. & DRAKE, C. L. (eds). *The Geology of Continental Margins.* 207–11. Springer-Verlag, New York.

VON HUENE, R., AUBOUIN, J., AZEMA, G., BLACKINGTON, G., CARTER, J., COULBOURN, W., COWAN, D., CURIALE, J., DENGO, C., FAAS, R., HARRISON, W., HESSE, R., HUSSONG, D., LADD, J., MUZYLOV, N., SHIKI, T., THOMPSON, P & WESTBERG, J. 1981. Leg 67 of the DSDP Mid-American Transect off Guatemala. *Bull. geol. Soc. Am.* in press.

WALKER, R. G. 1979. Turbidites and associated coarse clastic deposits. *In:* WALKER, R. G. (ed). *Facies Models,* **1,** 91–103. Geoscience Canada Reprint Ser.

WATKINS, J. S., MOORE, J. C. *et al.* 1981. *Initial Rep. Deep Sea drill. Proj.* **66.** U.S. Government Printing Office, Washington, in press.

WATKINS, J. S., MCMILLEN, K. J. & BACHMAN, S. B. 1981. Tectonic evolution of the Middle America Trench. *In:* WATKINS, J. S., MOORE, J. C. *et al.* (eds). *Initial Rep. Deep Sea drill. Proj.* **66.** U.S. Government Printing Office, Washington, in press.

J. CASEY MOORE, Earth Sciences Board, University of California, Santa Cruz, California 95064, U.S.A.

JOEL S. WATKINS, Gulf Research and Development Company, Pittsburg, Pennsylvania, U.S.A.

KENNETH J. MCMILLEN, Gulf Research and Development Co., P.O. Box 2038, Pittsburgh, Pennsylvania 15230, U.S.A.

STEPHEN B. BACHMAN, Department of Geological Sciences, Cornell University, Ithaca, New York, U.S.A.

JEREMY K. LEGGETT, Department of Geology, Imperial College of Science and Technology, London, England.

NEIL LUNDBERG, Earth Sciences Board, University of California, Santa Cruz, California, U.S.A.

THOMAS H. SHIPLEY, Scripps Institution of Oceanography, La Jolla, California, U.S.A.

JEAN-FRANCOIS STEPHAN, Département de Géotectonique, Université Pierre et Marie Curie, Paris, France.

FLOYD W. BEGHTEL, Phillips Research Center, Bartlesville, Oklahoma, U.S.A.

ARIF BUTT, Universität Tubingen, Tubingen, Federal Republic of Germany.

BORYS M. DIDYK, Empresa Nacional del Petroleo Concon, Chile.

NOBUAKI NIITSUMA, Department of Geosciences, Shizuoka University, Shizuoka, Japan.

LES E. SHEPHARD, Texas A & M University, College Station, Texas, U.S.A.

HERBERT STRADNER, Geologische Bundesanstalt, Vienna, Austria.

Tectonic processes along the Middle America Trench inner slope

T. H. Shipley, J. W. Ladd, R. T. Buffler & J. S. Watkins

SUMMARY: The relationship between sedimentation and structural variations observed in geophysical data provides insight into tectonic processes on the convergent Middle America Trench margin. We observe that some of the hemipelagic and pelagic oceanic sediments seem to continue relatively undeformed beneath the lowermost slope. Trench sediment bodies restricted to ponds in the vicinity of submarine canyons are incorporated into the lower slope by folding and thrust faulting. This offscraping (shallow accretion) at the base of the Middle America Trench slope appears at least partly dependent on the presence of trench fill. The internal structure of the so-called accretionary zone, the top of which is usually identified on seismic reflection profiles by a prominent zone of diffractions, is rarely resolved in the seismic reflection data. The shallow part of the accretionary zone along much of the margin may consist of deformed slope sediments since down-section increase in both folding and other deformation are observed in the seismic reflection and drill data. Beneath the undeformed and deformed slope sediments, offscraped trench fill may occur if trench fill was present in the ancient trench. The deeper part of the accretionary zone representing the bulk of the volume is not resolved in the reflection data but may consist of other sediments or crustal rocks added by an underplating process at deeper structural levels.

Evolution of ancient convergent continental margins is difficult to confirm in modern analogues. Exposures of ancient rocks provide most of our information on evolutionary processes. Recent seismic reflection and refraction surveys of modern margins, hampered by the local variability and abrupt transitions of major structures, are interpreted as evidence for sediment subduction, sediment and/or oceanic crust accretion or transitional (or steady-state) regimes. Every new investigation of a modern convergent margin reveals new variations. Perhaps one reason we have not been successful in completely understanding the variations between and within a single modern arc system is the difficulty in separating the accretionary processes from local complexities.

An investigation was designed for the modern continental margin adjacent to the Middle America Trench to address some of these problems. This margin is suited for a comparative study because of differences in the gross continental margin structure, physiography and sedimentation pattern along the arc (Fig. 1). A narrow trench-to-shoreline distance north of the North American-Caribbean plate boundary (near the intersection of the Tehuantepec Ridge trend with the shoreline in Fig. 1) contrasts with a wide trench-to-shoreline distance to the south (Molnar & Sykes 1969) while a wide and deep forearc basin extends along the shoreline from southern Mexico south to about the Nicoya Peninsula of Costa Rica.

However, the plate motion history of this margin, necessary for understanding the development of the accretionary zone, has been difficult to decipher. Spreading centre reorientation in the Miocene has complicated magnetic anomaly mapping and makes reconstruction of earlier history difficult with the present data base (Lynn & Lewis 1976; Schilt & Truchan 1976). Magnetic anomaly interpretations predict that Miocene ocean crust is now entering the trench with present plate motions producing slightly oblique convergence. The convergence rate increases southward from about 7 cm yr^{-1} off Mexico to more than 10 cm yr^{-1} off Costa Rica (Minster & Jordan 1978). The modern trench is bounded by transverse oceanic plate structural features; the Cocos Ridge on the south near Costa Rica and the Tamayo Fracture Zone off central Mexico.

The University of Texas Marine Science Institute undertook an investigation of the Middle American margin which focused on three specific areas for detailed study and included multichannel seismic reflection, refraction, coring, dredging, magnetics, bathymetry and sparker surveys. Each of the areas off Mexico, Guatemala and Costa Rica are topics of other investigations detailing the geology (Buffler & Watkins 1978, 1981; Ibrahim *et al.* 1979; Ladd *et al.* 1978; McMillen 1980; McMillen *et al.* 1981; Shipley *et al.* 1980; Shipley 1981). Deep sea drilling was conducted along transects in the Mexico and Guatemala areas based on these studies (Moore *et al.* 1979a, b; von Huene *et al.* 1980). This paper is a summary of these studies

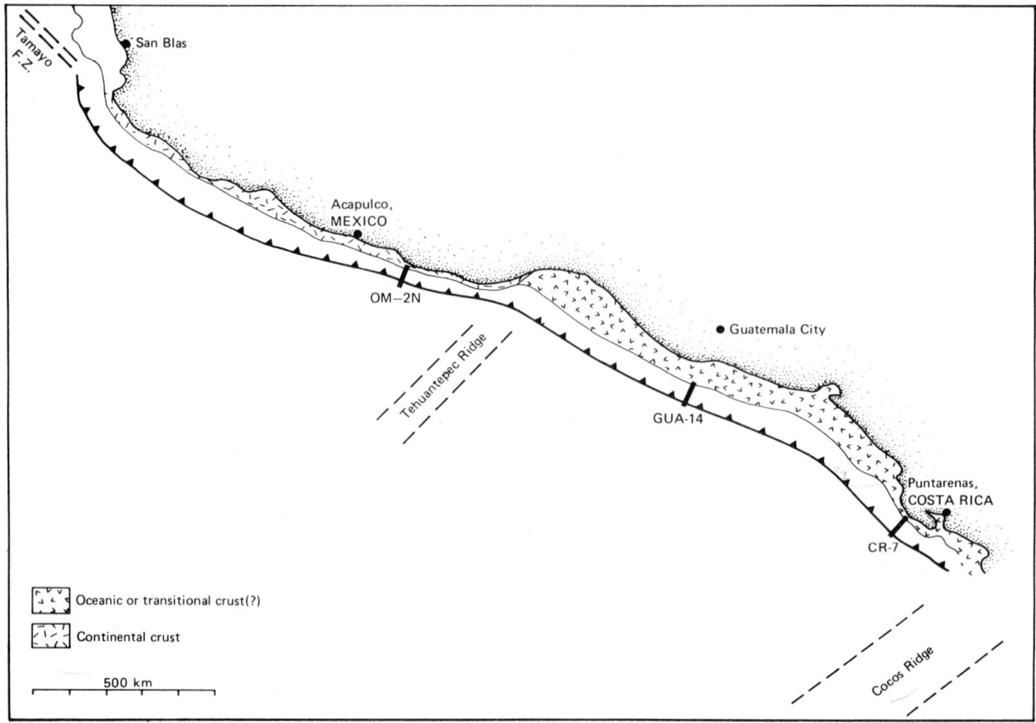

FIG. 1. Index map showing location of profiles in subsequent figures. Continental crust extends offshore north of the Gulf of Tehuantepec, oceanic or transitional crust inferred to the SE.

and attempts to document the significant structural variations and their causes. A single multichannel seismic reflection section from each area is presented in Figs 2–4.

Structural styles

Mexico

The narrow central Mexico shelf and truncation of Mesozoic and older terranes is evidence for an unusually abrupt continental/oceanic transition formed post-Cretaceous and pre-mid-Miocene. Exposed Cretaceous batholiths near the coast attest to an earlier arc which from several lines of evidence was probably truncated in the early Miocene although the mechanisms responsible for this feature are difficult to evaluate (e.g. erosion, oblique subduction) (Karig 1974a; Karig *et al.* 1978; Kesler 1973; Moore *et al.* 1979a; Schilt & Truchan 1976). Studies of the modern Mexico margin reveal recent sediment accretion at the base of slope (Karig *et al.* 1978; Moore *et al.* 1979a, b; Shipley *et al.* 1980).

The Mexican area reveals little variation in margin structure over the 150 km segment surveyed (Figs 2 & 5). Within this area the oceanic crust is overlain by pelagic and hemipelagic oceanic sediments, generally not more than 170 m thick. Recent normal faults, which mostly step down to the trench, offset the oceanic crust and most of the overlying sediments. The trench contains turbidite ponds only near two submarine canyons that are incised into and traverse the shelf. The slope section consists of numerous diffractions and short discontinuous reflections, while a zone of diffractions 0.5–1.0 sec sub-bottom in Fig. 2 defines the top of a slightly higher velocity section (the so-called accretionary zone). Farther to the NW near the Rio Ometepec Canyon, landward-dipping reflections generally 2–5 km in length down dip, occur in the section below the diffracting surface. Drilling data show they probably represent dipping interbedded sands and mudstones, presumably uplifted from the trench during the accretionary process (Fig. 6) (Moore *et al.* 1979b). Dips of the presumed bedding planes reach high values (>8°) within 4 km of the trench and remain high and variable farther up slope (Fig. 6). Landward of 1640Z (Fig. 2) is an upper slope sediment

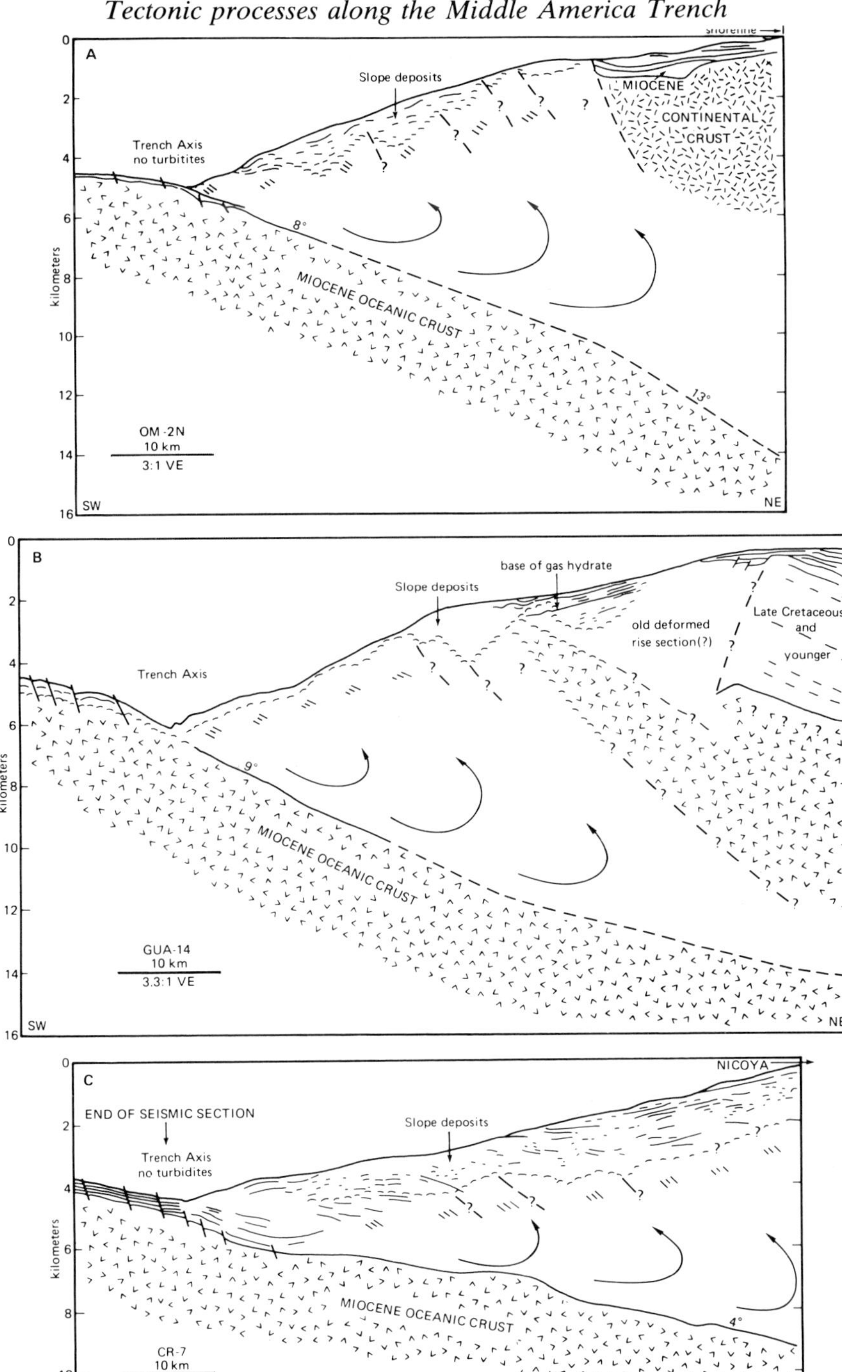

FIG. 5. Interpretive depth sections along seismic sections OM-2N, GUA-15 and CR-7. Structure deduced from adjacent profiles, drilling data of Moore *et al.* (1979a), Seely (1979), von Huene *et al.* (1980), and refraction data of Ibrahim *et al.* (1979), Mooney *et al.* (1975) and Shor & Fisher (1961). Short landward-dipping reflections represent schematic offscraped trench fill. Off parts of Mexico (Fig. 6) where large volumes of sediment were transported to the trench, these reflections would define a zone about 5 km in width down dip. Arrows depicting 'flow lines' of Cowan & Silling (1978), a possible underplating mechanism.

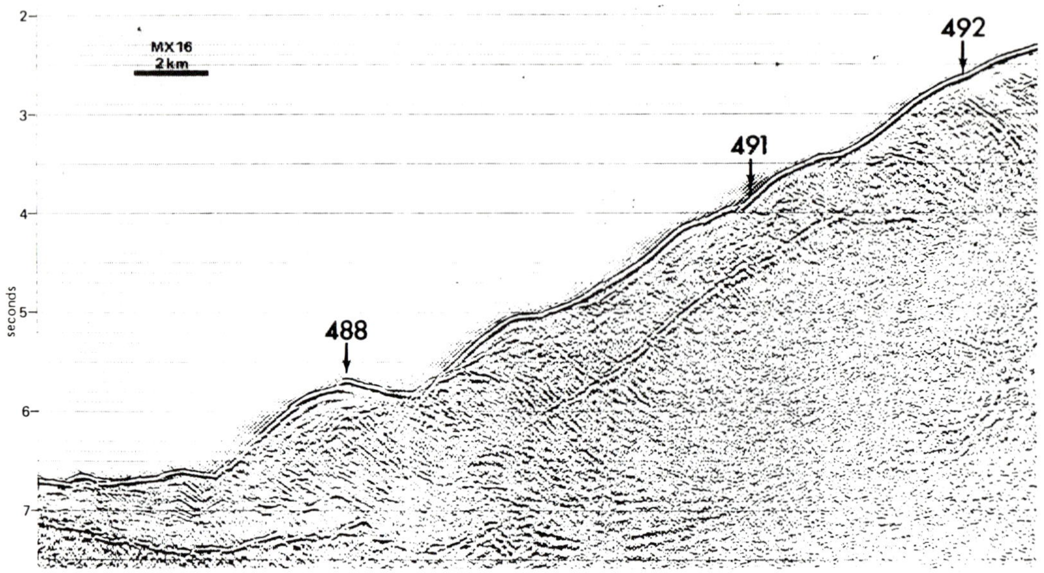

FIG. 6. Portion of migrated seismic reflection profile MX-16 near the Rio Ometepec Canyon. Three DSDP drill sites are shown. Note landward (NE) dipping reflections which are correlated with uplifted turbidites. Vertical exaggeration 5×.

package of variable thickness overlying (at 1 sec at 1440Z) the top of the continental crust which was confirmed from refraction data and by correlation to dredging and drilling in the northern part of the survey area. The shoreline is less than 3 km from the end of the section. Large shelf sedimentary basins are uncommon north of the Gulf of Tehuantepec and south of the Tamayo Fracture Zone. In the survey area shelf basins have widths up to 20 km and contain a maximum of 1 km of early Miocene or younger sediments (Moore *et al.* 1979a; Shipley 1981). Similar basins are reported and observed in other reflection data to the north (Karig *et al.* 1978; Ross & Shor 1965) and in unpublished University of Texas data to the south of the survey area.

Refraction profiles discovered no high velocities which might suggest accreted slices of oceanic crust within the slope (Shipley *et al.* 1980). The area between the seaward edge of continental crust 20–40 km offshore, and the base of the trench inner wall defines an accretionary zone less than 50 km wide (Fig. 5a). There is no evidence for large downfaulted continental blocks, and on some seismic reflection profiles the top of the oceanic crust extends to within 10 km of the edge of continental crust at a depth of 8 km (Shipley *et al.* 1980; Shipley 1981).

Small accretionary slope basins and ridges are generally present within 20 km of the trench and lack lateral continuity of more than 5 or 10 km. Many small channels originating in the upper or middle stope region are blocked by these ridge and basin structures. The zone of most bathymetric relief or roughness, which may be related to more intense deformation, is restricted largely to a width of less than 10 km with structures becoming more subdued beneath the slope sediments. In most profiles (Fig. 2) there are no discernible internal structures within the base of slope region and the slope sedimentary section is not seismically defined. This is repeated along most of the profiles in the area *except* near the Rio Ometepec Canyon. This is the same area where significant trench turbidite deposits are found.

The presence of a large submarine canyon on the slope within the survey area reveals both the influence of sedimentation parameters on structural development and the ability to identify such structures in seismic data. Profiles normal to and along the base of slope (Fig. 7) reveal changes in degree of deformation and structural style over short distances. In the vicinity of the Rio Ometepec (for a zone of about 50 of the 150 km of margin surveyed) the lower slope departs from a fairly consistent slope angle to abrupt scarps and trench subparallel ridges. This is also the only area where the internal structure is delineated by turbidite-related reflections which reveal anticlines, synclines and thrust faults (Fig. 7).

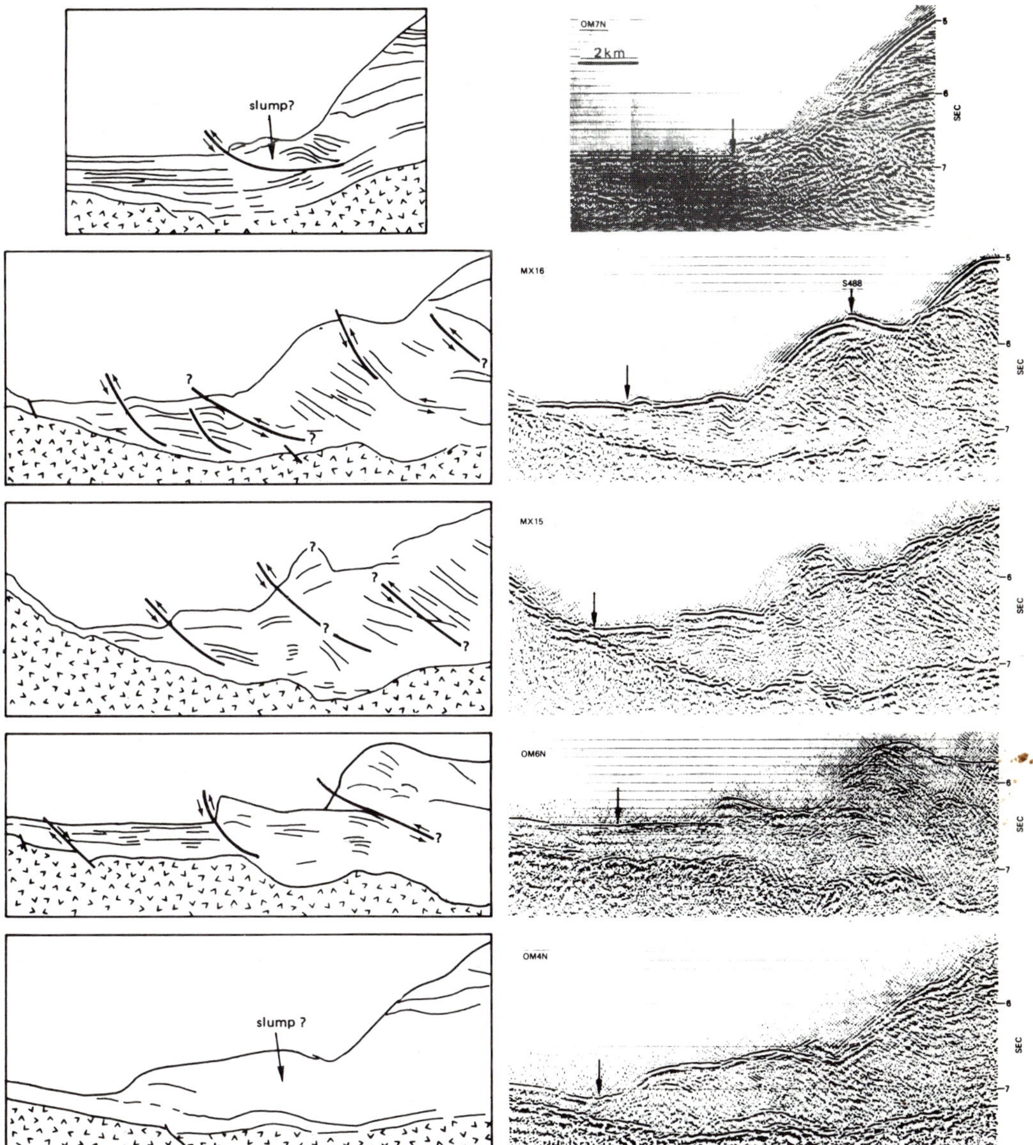

FIG. 7. Migrated 24-fold seismic reflection profiles normal to the trench axis in the vicinity of the Rio Ometepec Submarine Canyon of Mexico. The distance between adjacent profiles OM-7N, MX-16, MX-15, OM-6N and OM-4N is, respectively, 8, 8, 18 and 34 km. The Rio Ometepec Canyon reaches the trench floor between profiles MX-15 and OM-6N. Most turbidity currents flow to the NW (from MX-15 to MX-16 to OM-7N) and a higher portion of sand is expected in that direction. OM-6N contains less turbidite fill while OM-4N contains what is interpreted as a slump deposit. The oceanic section of pelagic and hemipelagic sediments are seismically distinguishable beneath the trench fill and seaward (left) on most of the profiles. Note the changes in slope and internal reflections, a result of deformation and incorporation of trench fill into the lower slope. Arrow notes trench axial channel or deepest point on crossing where channel is absent. Compare with Fig. 2, a profile approximately 60 km to the SE of the Canyon.

While the sedimentological aspects of the trench fill are not well known, seismically significant differences in the reflection response indicate significant changes in grain size or bedding thickness over short distances. For example, the difference between the acoustic response of the trench fill on OM-7N and OM-6N are striking (Fig. 7). OM-7N is 25 km down-axis (NW) from the Rio Ometepec entry point into the trench while OM-6N is up-axis (SE) about 8 km (Shipley 1981). OM-6N trench fill may consist of either coarse debris flows related to the canyon or finer-grained deposition from slumping. In either case the physical properties are significantly different from the coarse turbidites being deposited south of the canyon entry point. The production of gentle to tight folds and high angle to low angle thrusts might be explained by bedding competency related to the sediment variations in the trench. They do not seem related to an evolution of deformational styles with time since all are presumably of nearly the same age.

The variability of the lower slope structures and shallow accretion at the base of the slope seems related to the presence of trench fill. The sedimentological and reflection data are not dense enough to adequately determine whether the type of trench fill (e.g. sandy turbidites, muddy turbidites or slump material) or variations in the oceanic sediment physical properties influence the process or details of deformational style. However, the presence of these structures are striking evidence for incorporation of the trench sediments into the slope (accretion).

More than half of the existing profiles in the Mexico area show no base of slope bathymetric irregularities, no significant trench fill and thus little evidence for modern offscraping and shallow accretion of trench fill. Where significant trench sediments are not present in the modern trench much of the oceanic sediments seem to continue relatively undeformed at least initially beneath the slope (Fig. 2; Shipley 1981 figs 8 & 12). This implies that even in the absence of trench fill, offscraping (shallow accretion) of the pelagic and hemipelagic oceanic section is slow or inactive, or at least not volumetrically a very important process here.

Preserved landward-dipping reflections within the accretionary zone correlating with uplifted sands and muds suggest that ancient offscraping has been occurring near the Rio Ometepec since at least late Miocene (Fig. 6) (Moore *et al.* 1979b). The absence of landward-dipping reflections elsewhere on the margin implies that ancient offscraping has been confined to the Rio Ometepec region and that the supply to the trench has varied only slightly since the late Miocene.

While we do not deny that the lack of long (>2 km) dipping reflections in the slope may result from a lack of impedance contrasts in muddy turbidites and slumps, unless the stress/strain regime changes markedly over tens of kilometres along the trench, there is no reason to expect that we would not have detected ancient trench sand bodies elsewhere in the slope. This observation suggests long-term localization of the Ometepec submarine canyon, though the actual locus of deposition off the Rio Ometepec may have shifted back and forth up to 10 km depending on incoming plate relief and sediment and convergence rates. In the modern trench only a small part of the extensively surveyed area contains turbidite ponds thicker than 200 m or wider than 3 km, while most of the trench extending from Acapulco all the way to Costa Rica apparently contains very limited turbidite ponds based on a nearly continuous University of Texas seismic line down the trench axis. We observe evidence of folds near the base of slope only in regions with significant sediment ponds. We do not observe these structures to the SE in Mexico. We do not deny that offscraping may occur in the absence of significant trench fill but it may be volumetrically of little consequence here.

Within this survey area the shallow offscraped section apparently represents only a small portion of the accretionary zone. This is also supported by drilling on the Mexico slope which recovered slope deposits of early Pliocene age within a few hundred metres subbottom directly overlying the accretionary zone. A significant thickness of accretionary material (several kilometres) is inferred below the slope deposits and beneath the zone of landward-dipping reflections here (Fig. 5).

Guatemala

Investigations including the Guatemalan margin led Seely *et al.* (1974) to development of an imbricate thrust model for lower slope accretion of sediments and oceanic crust. Dickinson & Seely (1979) and Seely (1979) have gone on to develop a model for the entire slope and forearc structure off Guatemala. Studies by Ladd *et al.* (1978) and Ibrahim *et al.* (1979) with additional seismic reflection and refraction data confirmed the existence of oceanic crust at shallow depths within the slope. Recent drilling data at the base of slope do not support a presently active imbricate thrust regime but reveal significant subduction of sediments (von Huene *et al.* 1980).

Figure 3 is the seaward portion of a margin profile off Guatemala excluding the substantial forearc basin to the NE. Normal faults on the oceanic crust, usually separated by 3–5 km with a throw of about 75 m down to the trench, appear to offset recent sediments. Horst-graben structures are rare near the trench. The pelagic and hemipelagic oceanic section of Miocene, and younger sediments are about 450 m thick (von Huene *et al.* 1980). The muddy turbidite trench fill is notably restricted to the mouth of a canyon system and even then with less than 150 m of fill in localized axial basins (McMillen 1981; McMillen *et al.* 1981). The oceanic crust is not traceable landward of the trench in Fig. 3 but other profiles in the area reveal crust 18 km from base of slope.

The slope region, from the trench to about 0500Z (Fig. 3) is characterized by a thin slope blanket of low reflectivity sediments of variable thickness underlain by a strong zone of diffractions below which little structure is observed (accretionary zone). The upper slope region between 0500Z and 0100Z is more complex than off Mexico, consisting of a thickened slope blanket containing more short and discontinuous reflections.

A landward-dipping reflection about 8 km in length beneath the upper slope apron is similar to reflections found on other profiles in the Guatemala area (Fig. 3). The dipping reflections correlate with magnetic anomalies, high refraction velocities and in some places with near surface exposures of basalts, serpentinite and chert. These reflections are interpreted as oceanic crustal slices (Ibrahim *et al.* 1979; Ladd *et al.* 1978; McMillen 1981). Other sections show evidence for two or more such reflections suggesting small oceanic fragments. If these fragments are emplaced during initial or later subduction phases then the sediment overlying them may be older slope or rise sections.

Within the upper slope apron a strong reflection at 0.6 sec sub-bottom between 0230Z and 0330Z with inverted polarity correlates with the base of a gas hydrate layer (Shipley *et al.* 1979). Gas hydrated sediments were encountered in drilling on this transect but in an area where this reflection (the so-called BSR) is not present (von Huene *et al.* 1980).

The shelf section at the outer high (about 0010Z) consists of about 3 km of landward-dipping reflections which thickens toward the NE. Drilling data reveal that this section is as old as Late Cretaceous near the base (Seely 1979). The underlying basement was not reached but is presumed to be transitional ocean crust or Nicoya-type basement based on velocity characteristics (Ibrahim *et al.* 1979; Seely 1979).

The continental slope off Guatemala is smoother than that off Mexico but small-scale benches and ridges are common within 20 km of the trench (McMillen 1981; McMillen *et al.* 1981). Locally, near the San Jose Canyon, benches extend farther upslope (Ibrahim *et al.* 1979, fig. 7A). Reflection data reveal no evidence of internal structure at the base of slope, where the slope sediment cover is thinner and more discontinuous than off Mexico. While a single seismic line in this vicinity was interpreted by Seely *et al.* (1974) as containing landward-dipping reflections near the base of slope, none of our extensive survey data readily support this observation.

The lack of reflections (and thus definition of lower slope structure) could result from an absence of impedance mismatches. However, McMillen (1981) has argued, based on the bathymetry, present sedimentation regime, and bathymetric reconstructions related to the evolution of the outer high of Seely (1979) that little coarse detritus was able to broach the outer high and reach the trench for much of the history of this portion of the margin. Most present-day canyons do not traverse the wide forearc, and most fluvial sediments are trapped near shore. Sparsity of trench fill as off Mexico may limit the significance of sediments accreted by offscraping versus some other type of deeper mass addition process. Further, the Guatemala drilling programme discovered bedded deposits as old as Cretaceous near the base of slope, while Miocene oceanic crust is presently being subducted (von Huene *et al.* 1980).

Costa Rica

Seismic reflection studies of the Costa Rican margin along and just north and south of the Nicoya Peninsula reveal a complex slope morphology (Buffler & Watkins 1978, 1981). The peninsula, which is surrounded by a narrow shelf, contains Late Cretaceous oceanic crust and accretionary complex (Dengo 1962; Galli-Olivier 1979; Lundberg 1981). A seismic section beginning within 3 km of the shoreline of the Nicoya Peninsula and terminating in the trench axis is shown in Fig. 4. Seaward of the trench (not shown) normal faults with a throw down to the NE offset a highly reflective oceanic sediment section of pelagic sediments. The reflective oceanic sedimentary section and associated normal faults are identified up to

15 km beneath the lower slope (to 1630Z) and oceanic crust extends to the end of the profile (at 5.7 sec).

Most profiles show a thick slope blanket extending down the entire inner slope with internal reflections becoming more discontinuous down-slope. While the slope is fairly smooth and free of basins in Fig. 4 other profiles to the south reveal significant variations in the slope bathymetric profile. Between 0.8 and 2.5 sec sub-bottom beneath the slope sediments the top of the accretionary zone is defined by diffraction hyperbolas similar to the sections off Guatemala and Mexico. Again no structure is observed in the zone. The only complexities exist near the base of slope (1600Z–1700Z). Because the oceanic crustal reflection continues without significant interruption and with little relief in the time section, shallow occurrence of high velocity oceanic slices or blocks are probably not involved in this structure (Fig. 5c). The reflective sequence may be slope sediments and not uplifted oceanic sediments. Buffler & Watkins (1981) have not found evidence for shallow sediment accretion at the base of slope.

The origin of the thick slope sediment sequence is not obvious because the geography precludes a significant present near-shore sediment source. The present absence of turbidites in the trench precludes deposition on the slope of sediments during passage of recent downslope turbidity currents. The erosion and highly reflective nature of the shallow section (up to 0.3 sec sub-bottom) between 0000Z and 0200Z could be interpreted as evidence of erosion and redeposition related to shallow-water contour(?) currents. If the slope deposits are older then the modest deformation of the 2.0 sec thick sediment interval is anomalous considering the high convergence rate and expected high strain rate.

Well-defined oceanic sediments extending at least 15 km beneath the slope of Costa Rica reveal little evidence for compressional tectonics. Lower slope structures would be well-defined by the highly reflective oceanic sediment and mildly reflective slope apron if they existed. Reflection data in the slope section appear to confirm the absence of offscraping (shallow accretion) within the base of slope region. The gradual downslope decrease in reflection continuity in the slope sediments suggest that some sort of shallow deformation is more intense near the base of slope.

Discussion

The differences in major structural features within the upper slope and shelf region along the Middle America Trench margin represent a spectrum of convergent continental margin structures. While there have been changes in rates of convergence and obliquity of convergence along the margin with time, the entire margin has probably been subject to a convergence component since at least Miocene if not the Cretaceous.

The narrow shelf north of the Gulf of Tehuantepec is an unusual but not unknown feature of convergent continental margins. For example, parts of the Peru-Chile Trench have a similar narrow shelf (Karig 1974a; Kulm *et al.* 1977), where Hussong *et al.* (1976) favour a model of erosion of the continental crust by down-faulting in the middle slope region leading to a retreat of the continent. Our data do not resolve the mechanism of removal of the Cretaceous arc along the northern portion of the Middle America Trench but the process ended by early Miocene and no continental crust remains in the middle slope region (Moore *et al.* 1979a, b; Shipley *et al.* 1980; Shipley 1981).

Dickinson & Seely (1979) and Seely (1979) suggest that forearc basins are formed on transitional or oceanic crust. The differences in basin development along the Middle America margin may be related at least partly to the presence or absence of continental crust between the original trench suture and the continent. Variation in the regional sedimentation pattern does not seem large enough in itself to account for the basin evolution. A speculative possibility is that subduction along Central America and Mexico may have begun or restarted after significant left-lateral movement on the Motagua-Polochic Fault Zone. Such motion would produce an offset in the margin with oceanic crust trapped behind the suture south of the Gulf of Tehuantepec.

The oceanic crust within the accretionary zone of Guatemala at first appears unusual in terms of modern margins but will probably be more commonly recognized as more extensive investigations such as in the Middle America Trench are undertaken. The presence of oceanic crust within mélanges is well known; thus, the observations in Guatemala are not surprising except that there they seem to occur at shallow structural levels in an older(?) part of the accretionary zone. The incorporation of oceanic crust at this level might be related to initial ruptures of the oceanic crust. Sediments trapped above this old(?) crust may contain an earlier accretionary section or old rise sediments pre-dating subduction. The presence of old oceanic crustal slices would suggest that large-scale accretion of oceanic crust is episodic. A non-steady state regime might also explain a number of observations, particularly the old age of sediments at the base of slope in Guatemala and thick slope sediment in the trench off Costa Rica.

The Nicoya Peninsula Late Cretaceous mélange sequence is essentially an exposed outer arc high with a much smaller forearc sedimentary basin. The elevation of the outer high above sea-level may in part be related to increased uplift due to the shallow and presumably thickened oceanic crust of the Cocos Ridge currently being subducted (a hot spot trace, Hey 1977).

The presence or absence of base of slope structures, amounts of trench fill, the thickness of the slope sediments and internal deformation together reveal some of the active processes within the accretionary zone. Observations in the Middle America Trench off Mexico suggest that offscraping and shallow accretion of the type proposed by Seely *et al.* (1974) is controlled by the presence of trench fill. Four adjacent profiles near the Rio Ometepec Canyon (an average of 10 km apart) reveal a shallow subbottom zone of dipping reflections (less than 5 km in width) with inconsistent to constant dips over most of the upper slope (Fig. 6). These reflections are correlated with uplifted sands and muds and define a shallow zone of trench sediment fill offscraping that has occurred since the late Miocene (Moore *et al.* 1979b). The canyon has been a sediment source since at least the late Miocene. As the sections in Figs 2–4 show, significant trench fill has not been introduced in some other areas along the modern trench and are not observed as reflections within the slope.

If accretion at the base of slope is dependent on trench fill, then some accretionary complex mélanges, both modern and ancient, may contain large portions of sediments added in another fashion. Scholl & Marlow (1974) and Scholl *et al.* (1977) have emphasized the small amount of oceanic sediments in ancient subduction zones. Studies of modern and ancient subduction zones (Karig 1974b; Moore 1975) suggest selective shallow accretion of turbidites or indicate that oceanic sediments continue beneath the deformational front where the trench sediments are deformed and added to the base of slope (Karig & Sharman 1975; Kulm & Fowler 1974; Moore & Curray 1980; von Huene 1979). Thus transfer of pelagic and hemipelagic oceanic sections at the base of the slope to the accretionary zone may not always be an important process in outbuilding of the slope. Where trench fill is significant and has been for some time then offscraping becomes more important.

If our observations are valid in the Middle America Trench, then shallow trench fill even where present is volumetrically a minor accretionary component in the Middle America Trench and the accretionary zone consists mainly of material derived by a different process. One process might be deformation of the slope deposits. The accretionary zone has been defined seismically as the zone capped by diffractions and which contains few distinctive internal reflections (except near the Rio Ometepec Canyon). It is often associated with a distinct increase in velocity. The diffractions result partly from easily recognized large-scale irregularities in the slope. Diffractions may also be produced by a number of other geological affects dealing mainly with surface roughness or abrupt termination of reflecting horizons, neither of which appear applicable here.

Perhaps the diffraction 'horizon' is not a significant structural feature but an upward-moving deformational front within the slope sediments. Carson (1977) has shown that significant dewatering and strain hardening occur within the subduction zone. The gradual increase in fissility and other evidence of compaction and strain are observed progressively downhole in data within the slope deposits of the Nankai Trough (Karig *et al.* 1975); Japan (Arthur *et al.* 1980), and Mexico (Lundberg & Moore 1981; Moore *et al.* 1979b). Perhaps the zone of diffractions is related to some significant changes in physical properties resulting from the incipient strain. Drilling to date, however, has shown no obvious correlations, though it is hampered by poor recovery and shallow penetration.

While we observe a mechanism for incorporation of slope sediments into the seismically defined accretionary zone, much of the volume must still be related to sediment addition at a much deeper structural level. Another possible tectonic process which may be responsible for the large acoustically unresolvable portion of the accretionary zone is some type of mass addition by underplating. Watkins *et al.* (1981) have convincingly shown that underplating is a necessary process for Mexico mass balance calculations and fits a wide variety of other geological observations. Sediment budget calculations by Watkins *et al.* (1981) reveal that significant addition of material is required at depth beneath the slope section and landward-dipping reflections. Cowan & Silling (1978) have conducted model experiments which show that with a buttress, material will begin to flow upward in such a situation. Such a process does not preclude a zone of deformed slope sediments and a zone of offscraped landward-dipping trench fill. The underplating process may act beneath this section and in fact helps explain the observed uplift without further rotation of bedding planes.

Conclusions

As Dickinson & Seely (1979) and Seely (1979) have proposed, the presence of a large shelf basin along the margin of the Middle America Trench might perhaps be explained by the line of initial suture occurring seaward of the ocean-continent boundary trapping oceanic crust (as from the Gulf of Tehuantepec to Nicoya). In other areas the suture may have occurred right at the continental margin. Though highly speculative, perhaps the pre-Miocene history involved lateral movement in the Gulf of Tehuantepec which altered the margin configuration moving Mexico westward relative to Central America. The complexities in the southern part of the trench are probably partly related to subduction of thickened shallow oceanic crust.

Along the Middle America Trench margin at least three different tectonic processes may be responsible for accretion. At the base of slope offscraping seems dependent on the presence of trench fill. Oceanic sediments are not significantly involved in this process. A shallow imbricate thrust model generally fits such a sitution. The regional significance of such a model depends on the amounts of trench fill and appears fairly minor along most of the Middle America Trench south of Acapulco where trench fill is thin and discontinuous. Where offscraping is believed to have continued for any length of time (off the Rio Ometepec of Mexico) dips of presumed bedding planes reach high values within 4 km of the axis and remain high and variable at shallower water depths.

The composition of the seismically defined accretionary zone, particularly in the absence of landward-dipping reflections and shallow drilling penetration, make it difficult to resolve. Diffraction hyperbolas at the top of the zone correlate with an increase in velocities. Carefully documented changes in structure and incipient deformation observed in the drilling data show that deformation slowly increases downcore within slope deposits (Arthur *et al.* 1980; Lundberg & Moore 1981). Perhaps on some margins the shallow part of the accretionary zone represents increasingly deformed slope sediments which reach a stage where significant physical property changes produce a ragged deformational front responsible for the characteristic diffraction patterns.

While some sort of incorporation of slope sediments into the accretionary zone is likely, the processes of uplift and total volume of the accretionary zone suggest some sort of mass addition at deeper levels. The underplating proposed by Watkins *et al.* (1981) is probably a fundamental process. The absence of acoustic return from these zones could result from the mode of emplacement and associated bedding disruptions. Direct sampling will be required to verify and describe the details of the mass addition process. Underplating, offscraping and inplace slope deformation represent three different processes of tectonic accretion which will vary from place-to-place in volumetric significance even within one arc depending on the sedimentological and tectonic setting.

ACKNOWLEDGMENTS: This programme was conducted by the University of Texas Marine Science Institute in 1977 and 1978 and encompasses the effort of many individuals in the marine geophysical group including K. Griffiths, J. Kunselman and A. Roberts and their staffs. A number of colleagues, M. Houston, A. Ibrahim, G. Latham, K. McMillen, J. Moore, J. Shaub, R. von Huene, and J. Worzel, were involved in shipboard operations and later analysis which contributed greatly to the success of this programme. The paper was improved with thoughtful reviews by J. Leggett and anonymous Geological Society referees. This investigation was supported by the Oceanography section of the National Science Foundation Grant OCE-76-2330 and NSF-IPOD subcontracts CU-TEX 25907-2, -3. Contribution to Scripps Institution of Oceanography New Series.

References

ARTHUR, M. A., CARSON, B. & VON HUENE, R. 1980. Initial tectonic deformation of hemipelagic sediment at the leading edge of the Japan convergent margin. *In: Initial Rep. Deep Sea drill. Proj.*, Leg **56–57,** 569–613. U.S. Govt Printing Office, Washington.

BUFFLER, R. T. & WATKINS, J. S. 1978. Geologic structure of the continental margin off the

Nicoya Peninsula, Costa Rica. *Abstr. Prog. geol. Soc. Am.* **10,** 374.

—— & —— 1981. Geologic structure of the continental margin off the Nicoya Peninsula, Costa Rica, based on multifold seismic reflection data (in prep.).

CARSON, B. 1977. Tectonically induced deformation of deep sea sediments off Washington and northern Oregon: mechanical consolidation. *Mar. Geol.* **24,** 289–307.

COWAN, D. S. & SILLING, R. M. 1978. A dynamic, scaled model of accretion at trenches and its implications for the tectonic evolution of subduction complexes. *J. geophys. Res.* **83,** 5389–96.

DENGO, G. 1962. Tectonic-igneous sequences in Costa Rica. *In:* ENGEL, H. L. *et al.* (eds). *Petrologic Studies: a volume in honor of A. F. Buddington*, 131–61. Geol. Soc. Am.

DICKINSON, W. R. & SEELY, D. R. 1979. Structure and stratigraphy of forearc regions. *Bull. Am. Assoc. Petrol. Geol.* **63,** 2–31.

GALLI-OLIVIER, C. 1979. Ophiolite and island-arc volcanism in Costa Rica. *Bull. geol. Soc. Am.* **90,** 444–52.

HEY, R. 1977. Tectonic evolution of the Cocos-Nazca Spreading Center. *Bull. geol. Soc. Am.* **88,** 1404–20.

HUSSONG, D. M., EDWARDS, P. B., JOHNSON, S. H., CAMBELL, J. F. & SUTTON, G. H. 1976. Crustal structure of the Peru-Chile Trench: 8° S–12° S latitude. *In:* SUTTON, G. H. *et al.* (eds). *The Geophysics of the Pacific Ocean Basin and its Margin.* Geophys. Monogr. Am. Geophys. Union, **19,** 71–86.

IBRAHIM, A. K., LATHAM, G. V. & LADD, J. W. 1979. Seismic refraction and reflection measurements in the Middle America Trench offshore Guatemala. *J. geophys. Res.* **84,** 5643–9.

KARIG, D. E. 1974a. Tectonic erosion of trenches. *Earth planet. Sci. Lett.* **21,** 209–12.

—— 1974b. Evolution of arc systems in the western Pacific. *Earth planet. Sci. Ann. Rev.* **2,** 51–76.

—— , CARDWELL, R. K., MOORE, G. F. & MOORE, D. G. 1978. Late Cenozoic, subduction and continental margin truncation along the northern Middle America Trench. *Bull. geol. Soc. Am.* **89,** 265–76.

—— & SHARMAN, G. F. 1975. Subduction and accretion in trenches. *Bull. geol. Soc. Am.* **86,** 377–89.

—— *et al.* 1975. *Initial Rep. Deep Sea drill. Proj.* Leg **31.** U.S. Govt Printing Office, Washington, 927 pp.

KESLER, S. E. 1973. Basement rock structural trends in southern Mexico. *Bull. geol. Soc. Am.* **84,** 1059–64.

KULM, L. D. & FOWLER, G. A. 1974. Oregon continental margin structure and stratigraphy: a test of the imbricate thrust model. *In:* BURK, C. A & DRAKE, C. L. (eds). *The Geology of Continental Margins*, 261–83. Springer-Verlag, New York.

—— , SCHWELLER, W. J. & MASIAS, A. 1977. A preliminary analysis of the subduction processes along the Andean continental margin. *In:* TALWANI, M. & PITMAN, W. C. (eds). *Island Arcs, Deep Sea Trenches and Back-Arc Basins.* Am. Geophys. Union, M. Ewing Ser. **1,** 285–301.

LADD, J. W., IBRAHIM, A. K., MCMILLEN K. J., LATHAM, G. V., VON HUENE, R. E., WATKINS, J. S., MOORE, J. C. & WORZEL, J. L. 1978. Tectonics of the Middle America Trench. *Int. Symp. Guatemalan Earthquake and Reconstruction Process,* Feb. 4th, 1976, vol. 2.

LUNDBERG, N. 1981. Evolution of the slope landward of the Middle America Trench, Nicoya Peninsula, Costa Rica (this volume).

—— & MOORE J. C. 1981. Structural features of the Middle America Trench slope off southern Mexico, Deep Sea Drilling Project, Leg 66. *In:* MOORE, J. C. & WATKINS, J. S. *et al.* (eds). *Initial Rep. Deep Sea drill. Proj.*, Leg **66.** U.S. Govt Printing Office, Washington, in press.

LYNN, W. S. & LEWIS, B. T. R. 1976. Tectonic evolution of the northern Cocos Plate. *Geology*, **4,** 718–22.

MCMILLEN, K. J. 1981. San Jose Submarine Canyon, Middle America Trench, Guatemala: Its influence on slope and trench sedimentation, and tectonic accretion. *Mar. Geol.* (in press).

——, ENKEBOLL, R. H., MOORE, J. C., SHIPLEY, T. H & LADD, J. W. 1981. Sedimentation in different tectonic environments of the Middle America Trench, southern Mexico and Guatemala (this volume).

MINSTER, J. B. & JORDON, T. H., 1978. Present-day plate motions. *J. geophys. Res.* **83,** 5331–54.

MOLNAR, P. & SYKES, L. R. 1969. Tectonics of the Caribbean and Middle America regions from focal mechanisms and seismicity. *Bull. geol. Soc. Am.* **80,** 1639–84.

MOORE, G. F. & CURRAY, J. R. 1980. Structure of the Sunda Trench lower slope off Sumatra from multichannel seismic reflection data. *Mar. Geol. Res.* **4,** 319–40.

MOORE, J. C. 1975. Selective subduction. *Geology*, **3,** 530–2.

—— *et al.* 1979a. Middle America Trench. *Geotimes*, **24,** 20–2.

—— *et al.* 1979b. Progressive accretion in the Middle America Trench, southern Mexico. *Nature,* **281,** 638–42.

MOONEY, W. M., MEYER, R. P., HELSLEY, C. E., LOMNITZ, C. & LEWIS, B. T. R. 1975. Refracted waves across a leading edge: observations of Pacific shots in southern Mexico. *Trans. Am. geophys. Union,* **56,** 452.

ROSS, D. A. & SHOR, G. G. 1965. Reflection profiles across the Middle America Trench. *J. geophys. Res.* **70,** 5551–72.

SCHILT, F. S. & TRUCHAN, M. 1976. Plate motions in the northern part of the Cocos Plate. *Trans. Am. geophys. Union,* **57,** 333.

SCHOLL, D. W. & MARLOW, M. S. 1974. Sedimentary sequences in modern Pacific trenches and the deformed circum-Pacific eugeosyncline. *In:* DOTT, R. H. & SHAVER, R. H. (eds). *Modern and Ancient Geosynclinal Sedimentation.* Spec. Publ. Soc. econ. Paleont. Mineral, **19,** 193–211.

—— , —— & COOPER, A. K. 1977. Sediment subduction and offscraping at Pacific margins. *In:* TALWANI, M. & PITMAN, W. C. (eds). *Island Arcs, Deep Sea Trenches and Back-arc Basins.* Am. Geophys. Union, M. Ewing Ser. **1,** 199–220.

SEELY, D. R. 1979. The evolution of structural highs bordering major forearc basins. *Bull. Am. Assoc. Petrol. Geol.* **29,** 245–60.

—— , VAIL, P. R. & WALTON, G. F. 1974. Trench slope model. *In:* BURK, C. A. & DRAKE, C. L. (eds). *The Geology of Continental Margins*, 249–60. Springer-Verlag, New York.

SHIPLEY, T. H. 1981. Seismic facies and structural framework of the southern Mexico continental margin. *In:* MOORE, J. C. & WATKINS, J. S. *et al.* (eds). *Initial Rep. Deep Sea drill. Proj.*, Leg **66.** U.S. Govt Printing Office, Washington (in press).

—— , HOUSTON, M. H., BUFFLER, R. T., SHAUB, F. J., MCMILLEN K. J., LADD, J. W. & WORZEL, J. L. 1979. Seismic-reflection evidence for the wide-spread occurrence of possible gas hydrate horizons on continental slopes and rises. *Bull. Am. Assoc. Petrol. Geol.* **63,** 2204–13.

—— , MCMILLEN, K. J., WATKINS, J. S., MOORE, J. C., SANDOVAL-OCHOA, H. & WORZEL, J. L. 1980. Continental margin and lower slope structures of the Middle America Trench near Acapulco, Mexico. *Mar. Geol.* **35,** 65–82.

SHOR, G. G., JR. & FISHER, R. L. 1961. Middle America Trench: seismic-refraction studies. *Bull. geol. Soc. Am.* **63,** 2204–13.

VON HUENE, R. 1979. Structure of the outer convergent margin off Kodiak Island, Alaska, from multichannel seismic records. *Mem. Am. Assoc. Petrol. Geol.* **29,** 261–90.

—— *et al.* 1980. Leg 67: The Deep Sea Drilling Project Mid-America Trench transect off Guatemala. *Bull. geol. Soc. Am.* **91,** 421–32.

WATKINS, J. S. *et al.* 1981. Accretion, underplating, subduction and tectonic evolution of the Middle America Trench, southern Mexico: Results of Leg 66 DSDP, *Oceanologica Acta* (in press).

THOMAS H. SHIPLEY, Scripps Institution of Oceanography, University of California, La Jolla, California 92093, U.S.A.

JOHN W. LADD, Lamont-Doherty Geological Observatory of Columbia University, Palisades, New York 10964, U.S.A.

RICHARD T. BUFFLER, Marine Science Institute, University of Texas, Galveston, Texas 77550, U.S.A.

JOEL S. WATKINS, Gulf Science & Technology Company, P.O. Box 2038, Pittsburgh, Pennsylvania 15230, U.S.A.

Sedimentation in different tectonic environments of the Middle America Trench, southern Mexico and Guatemala

Kenneth J. McMillen, Robert H. Enkeboll, J. Casey Moore, Thomas H. Shipley & John W. Ladd

SUMMARY: Late Pleistocene to Holocene sediment facies and composition within and bordering the Middle America Trench offshore southern Mexico and Guatemala reflect two distinctly different tectonic provinces. The truncated Mexican margin with crystalline rocks onshore and a narrow shelf exhibits a locally thick sandy trench fill. Sand mineralogy of quartz, feldspar, and biotite matches the onshore source terrane. The Guatemalan margin with a volcanic terrane onshore and a wide forearc basin show smaller amounts of predominantly muddy trench fill. Sand composition of volcanic rock fragments, plagioclase, and heavy minerals accurately reflects the volcanic source. Trench fill in both areas corresponds to submarine canyon location, and extensive bypassing of the slope occurs. The outer shelf contributes little sediment to the Mexican slope and trench, the main source being in the littoral zone and inner shelf. Conversely, most Guatemalan slope and trench sand has come from the shelf. Trench fill correlates best with onshore geology, with similar slope sediments in both areas. The Mexican margin with locally thick trench fill displays clear evidence of accretion, the Guatemalan margin with meagre trench fill apparently has not accreted lower plate sediments recently. Possibly, larger volumes of trench fill encourage accretion.

Studies of modern sediment distribution and depositional processes on active margins provide an understanding of controls on the type and amount of sediment present in deep sea trenches, the sediment pathways to the trench, and trapping effects of the shelf and of slope basins. Description of sedimentary petrology can be related to source terrane and to variations in tectonic setting of the margin.

In this paper, we report on piston cores and seismic reflection profiles collected from two portions of the Middle America Trench on surveys by the University of Texas Marine Science Institute for IPOD drilling offshore southern Mexico on Leg 66, and offshore Guatemala on Leg 67 (Fig. 1). We describe trench, slope, and shelf sediments, infer depositional processes, and relate sedimentological and petrological variations to differences in tectonic setting.

Cores were collected with a Ewing-type piston corer of 20 or 40 ft (6.7–13.3 m) length and 2.5 in (6.4 cm) inside diameter fitted with polycarbonate liners. Cores were sectioned and stored aboard ship. Satellite navigation was used for all underway data collection and coring. Bathymetry, which was recorded on 3.5 kHz records, was corrected with Matthews Tables (Matthews 1939). Seismic reflection data were collected and processed using techniques described in Shipley *et al.* (1978). In the text, core numbers are preceded by an 'M' for Mexican cores and a 'G' for Guatemalan cores.

Laboratory analyses consisted of general descriptions of split cores noting colour, general texture, structures, and composition of the wet-sieved sand fraction. Micropalaeontological samples, taken roughly every 20 cm, were boiled in peroxide to remove organic matter and to clean fossil tests, wet-sieved through a 63 μm mesh screen and mounted as strewn slides. Petrological samples were taken from sand beds, soaked in peroxide to break down organic matter, wet-sieved through a 63 μm mesh screen, mounted in epoxy, and cut and ground for thin-section point-count analysis of 300–800 grains using the methods of Dickinson (1970). Slides were stained with concentrated sodium cobaltinitrate solution for potassium feldspars and amaranth solution for plagioclase feldspars. Radiography followed standard techniques of Bouma (1969).

Southern Mexico regional setting

The southern Mexico margin north of the Gulf of Tehuantepec differs from most active margins in several ways. The distance from trench to shoreline is small, generally less than 70 km (Fig. 2). The shelf is narrow and, as the land surface rises steeply from shore, there is no forearc basin. The trans-Mexican volcanic arc trends at an angle to the trench, and is located about 200 km landward of the Leg 66 drilling area. The area between the arc and shoreline

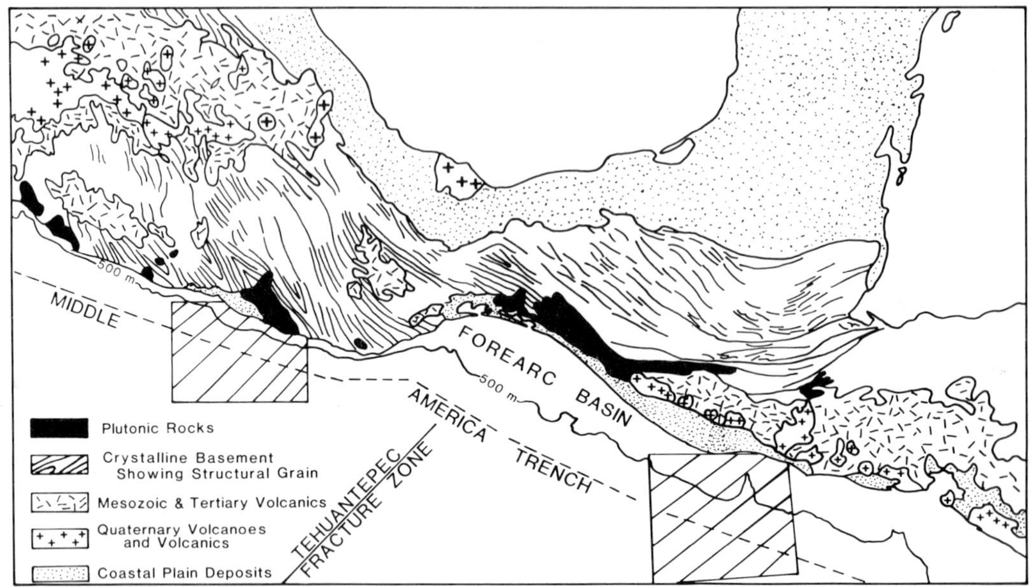

Fig. 1. Regional geology of Middle America with the location of study areas (cross-hatched). Note truncation of plutonic rocks and structural grain in Precambrian basement at the Mexican shoreline. From tectonic map of North America (King 1969).

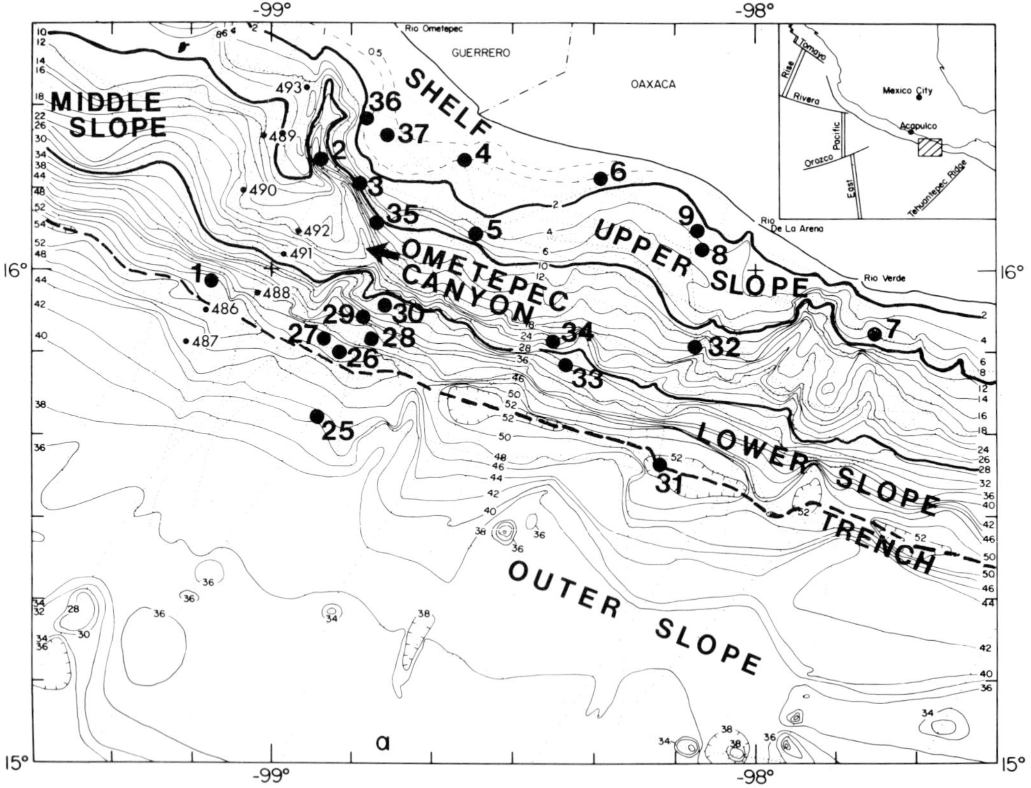

Fig. 2.(a) Bathymetry of the southern Mexico margin showing piston core locations and DSDP Leg 66 site locations (486–493).

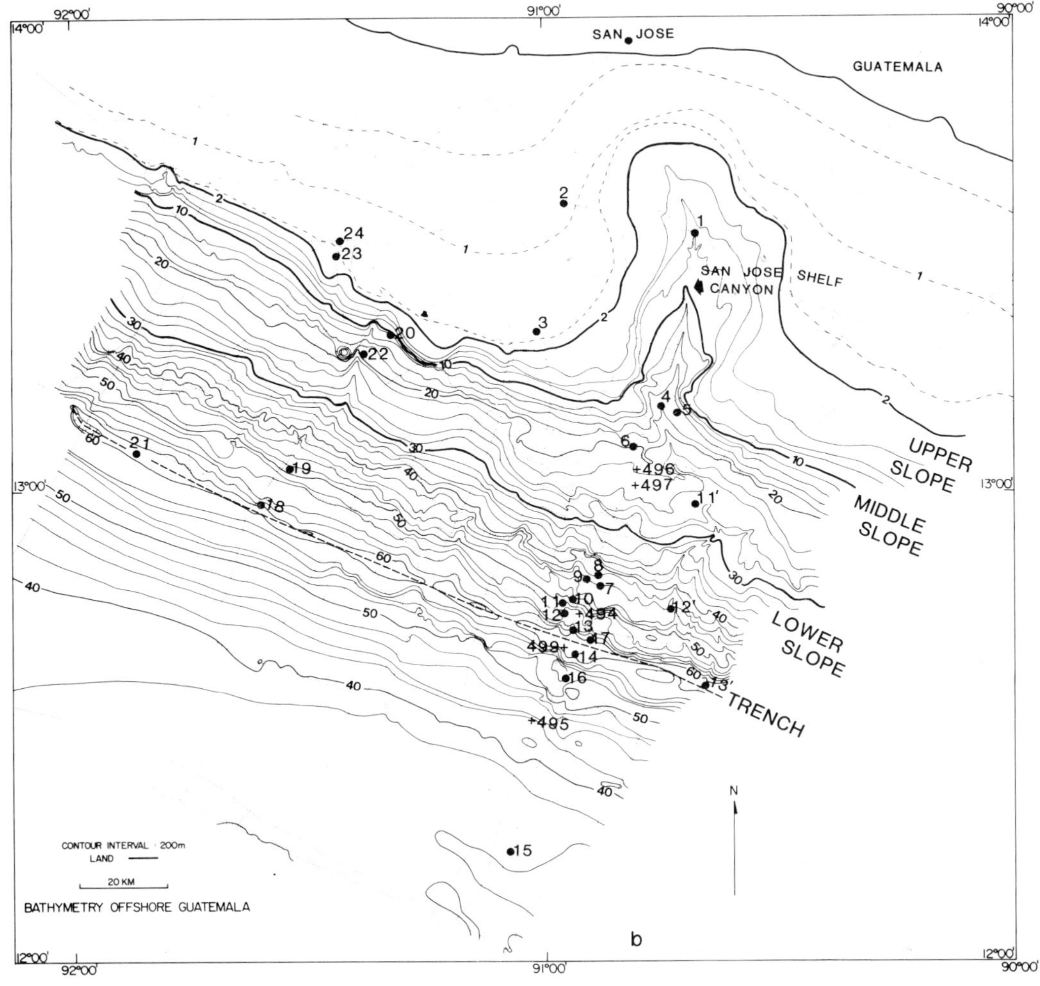

FIG. 2.(b) Bathymetry of the Guatemala margin showing piston core locations and DSDP Leg 67 site locations (494–499). Contour interval is 200 m.

contains a variety of continental rocks including Precambrian crystalline rocks, Palaeozoic and Mesozoic shallow-water and platform sedimentary rocks, and Mesozoic plutonic rocks (de Czerna 1965). Structural grain of many of these rocks trends offshore, crystalline outcrops occur at the shoreline (Fig. 1), and these continental rocks underlie the shelf and upper slope (Fig. 3a, Shipley *et al.* 1978, 1981) indicating margin truncation. Two theories to explain this truncation are 'tectonic erosion' (Karig 1974) and removal by strike-slip faulting (Karig *et al.* 1978). Results from Leg 66 drilling (Moore *et al.* 1979a) did not resolve the mechanism of truncation but show that it occurred prior to early Miocene.

Guatemala regional setting

The Guatemala Pacific margin typifies many active margins. Its wide (~ 120 km) forearc basin traps a large volume of sediments from the volcanic arc (Seely *et al.* 1974). The arc parallels the trench and is closer than in the Leg 66 area (Fig. 1). The forearc basin is probably floored by trapped oceanic crust (Seely *et al.* 1974; Seely 1978; Lundburg 1981) and filled by volcanic sediments to comprise a broad shelf and coastal plain. Oceanic crust probably occurs within the inner trench slope (Fig. 3b, Ibrahim *et al.* 1979). The actual zone of recent accretion is probably very small or non-existent (von Heune *et al.* 1980, 1981).

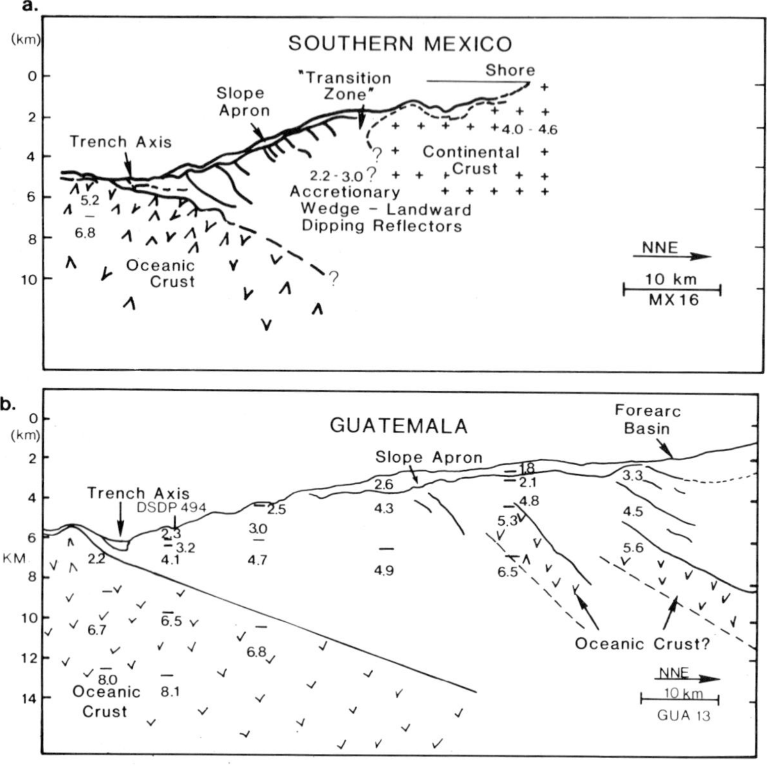

FIG. 3. Interpretive geological sections across (a) the Southern Mexico and (b) the Guatemala Pacific margins showing major reflectors, refraction and interval velocities, and underlying basement.

Southern Mexico bathymetry

Margin morphology and core locations are shown in Fig. 2(a) for the southern Mexico margin. Major indentations of the upper slope may be due to extensions of continental structures offshore (Shipley *et al.* 1978). The lower slope is steep with a few small terraces. Several submarine canyons cross the slope but only one, Ometepec Canyon, reaches the trench; others are blocked by uplifted or slumped masses. Trench sediment occurs as isolated basins with no apparent interconnection. Two large trench sediment basins are related to the locations of Ometepec Canyon and a series of canyons offshore from Rio de la Arena and Rio Verde (Fig. 2a). Multichannel profiles across the trench fill show evidence of fan channels, levees, and deformation of trench sediment on the inner trench wall (Fig. 4a–c). Landward-dipping seismic reflectors beneath the inner trench slope probably represent older, tilted and uplifted trench sediments (Moore *et al.* 1979b).

Guatemala bathymetry

The margin morphology offshore Guatemala is simpler than offshore Mexico (Fig. 2b). The shelf edge is fairly straight except for a major embayment associated with San Jose submarine canyon. The upper slope is steep facing the trench but is gentler within the embayment. The gentler gradient of the middle slope may be structurally controlled as it is underlain, at least in part, by emplaced oceanic crust (Ibrahim *et al.* 1979). San Jose Canyon deflects to the east as it crosses the middle slope and here shows levee-like depositional structure adjacent to the canyon. Several small terraces interrupt an otherwise steep lower slope. San Jose Canyon breaks up into smaller channels which are deflected around topographic ridges. The trench offshore Guatemala essentially lacks sediment except in one small pond fed by San Jose Canyon (Fig. 4d).

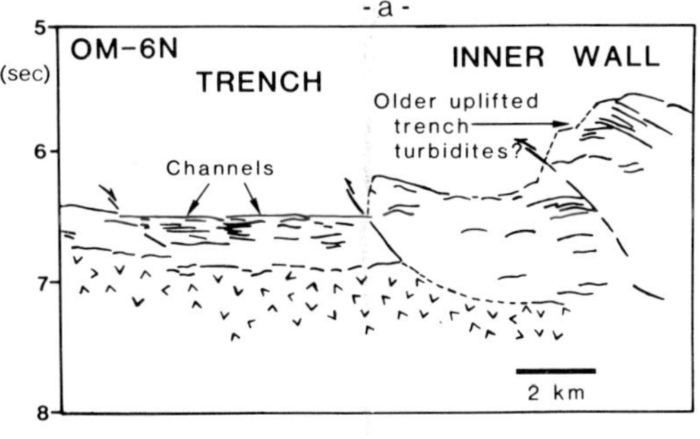

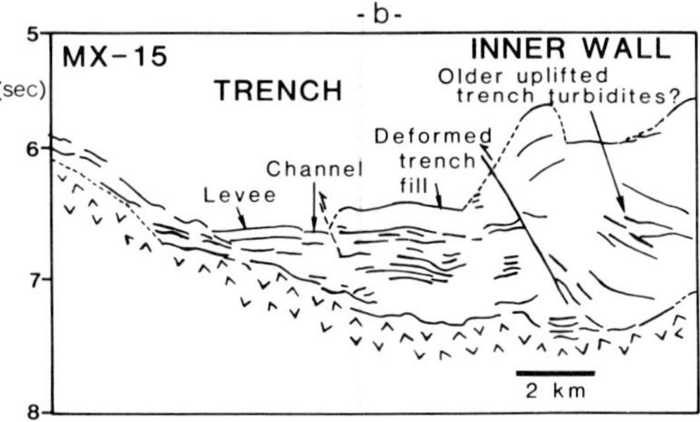

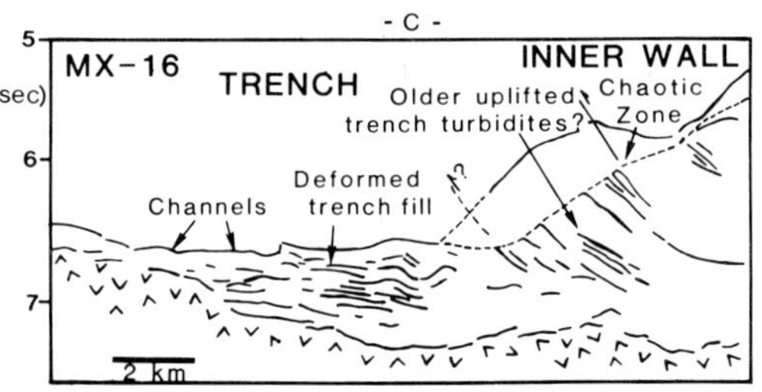

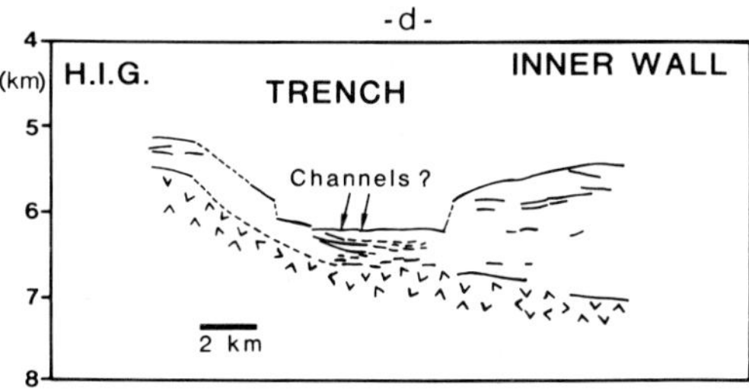

FIG. 4. Profiles across the Middle America Trench axis showing sediment fill: (a–c) offshore southern Mexico, and (d) Guatemala. Profiles (a–c) are from the University of Texas survey, profile (d) is from the Hawaii Institute of Geophysics.

Sediment distribution

Mexico

Trench sediments recovered in cores M26 and M27 consist of sand- to granule-size material with virtually no clay or silt matrix (Fig. 5a). Two other trench cores recovered sandy turbidites with c-e Bouma sequences (Fig. 6a).

A predominance of mud characterizes the slope sediments. Outer slope hemipelagic sediments, cored in M25, consist of hemipelagic mud with zones of 1–2 cm thick sand beds having sharp upper and lower contacts (Figs 5a & 6). These may be contourites deposited by bottom currents. The 700 m elevation of core M25 above the trench probably precludes sand deposition by turbidity currents, unless rapid vertical uplift has occurred.

Lower inner slope core M28 includes thin sand laminations (1 cm). Except for one thin sand bed in core M29, all other lower slope cores are devoid of sands. Middle slope cores are mud only except for gravels in Ometepec Canyon which may be derived from upslope or onshore. Upper slope sediments are also muddy except near Ometepec Canyon where rhythmically bedded sand and mud laminations occur with rippling and local channelling in core M35 (Fig. 6b). All other upper slope cores consist of laminated mud where laminations are recognized by colour contrasts (Fig. 5a). Outer shelf sediments recovered in cores M36 and M37 consist entirely of shelly sand, shelly sand capped by hemipelagic mud in core M6, and mud only in core M4. Laminations were not observed in any outer shelf cores.

Guatemala

Sand bed deposition at core G16 is similar to sand deposition at Mexican core M25: the sands have been deposited by bottom currents, or turbidity currents above the level of the trench. Subsequent rapid uplift analogous to uplifted turbidites in the Peru trench (Prince *et al.* 1974), or overtopping by large turbidity flows can explain sand deposition above trench level.

Trench sediments offshore Guatemala are mostly muddy to the 11 m cored depth (Fig. 5b). Coarse clastic deposits consist of thin sand and ash beds and thicker muddy ash horizons with convolute structures representing ashy turbidites in cores G14, G17 and G13. Most trench cores contain a great deal of interstitial gas which cracked open clayey sediment (Fig. 6d).

Inner and outer slope sediments are also nearly all muddy. Outer slope core G16 nearly 700 m above the trench axis, contains thin-bedded sands in the lower half of the core and core G15 on the outer rise contains ash beds.

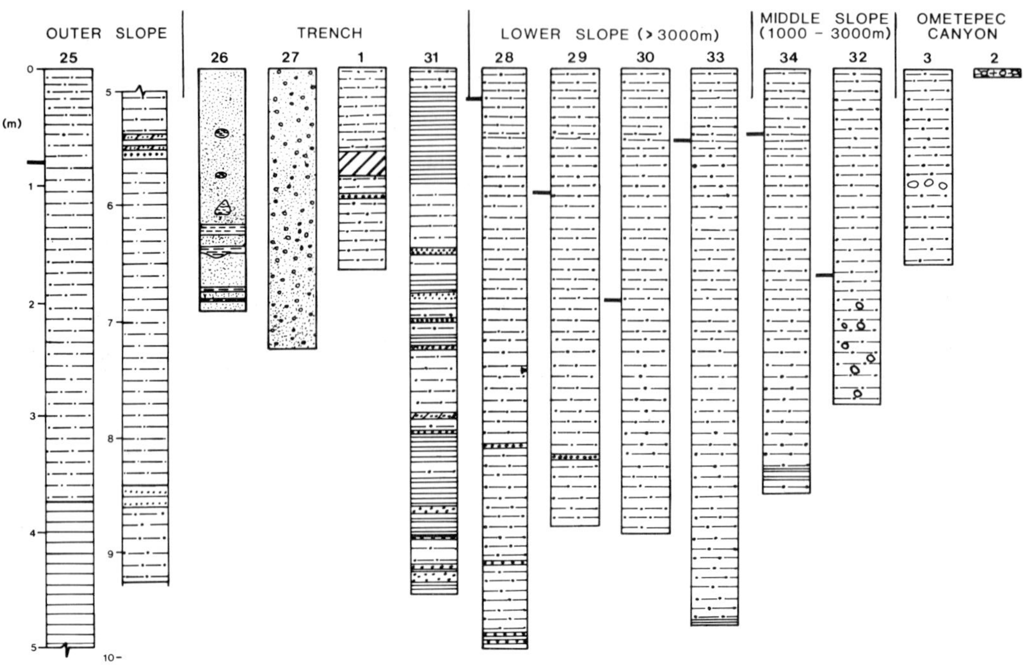

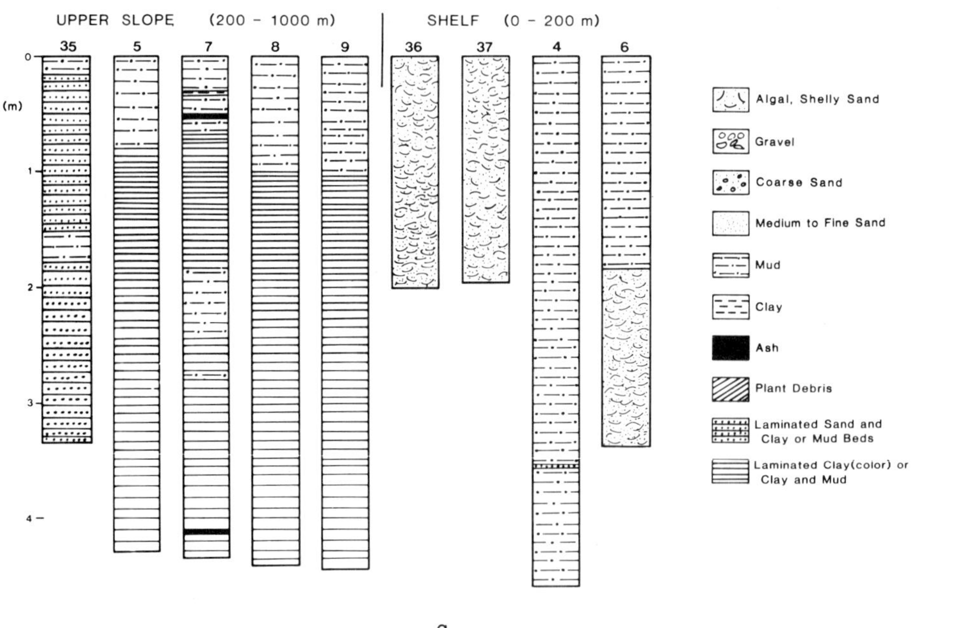

FIG. 5. Logs of all piston cores recovered from (a) the Mexico margin and (b) the Guatemala margin. Cores are grouped by environments, labelled in Fig. 2. In the text, cores from Mexico are prefixed by the letter 'M', and cores from Guatemala by the letter 'G'. The N/S (nasselarian–spumellarian) minimum is shown as a bar to the left of some core logs.

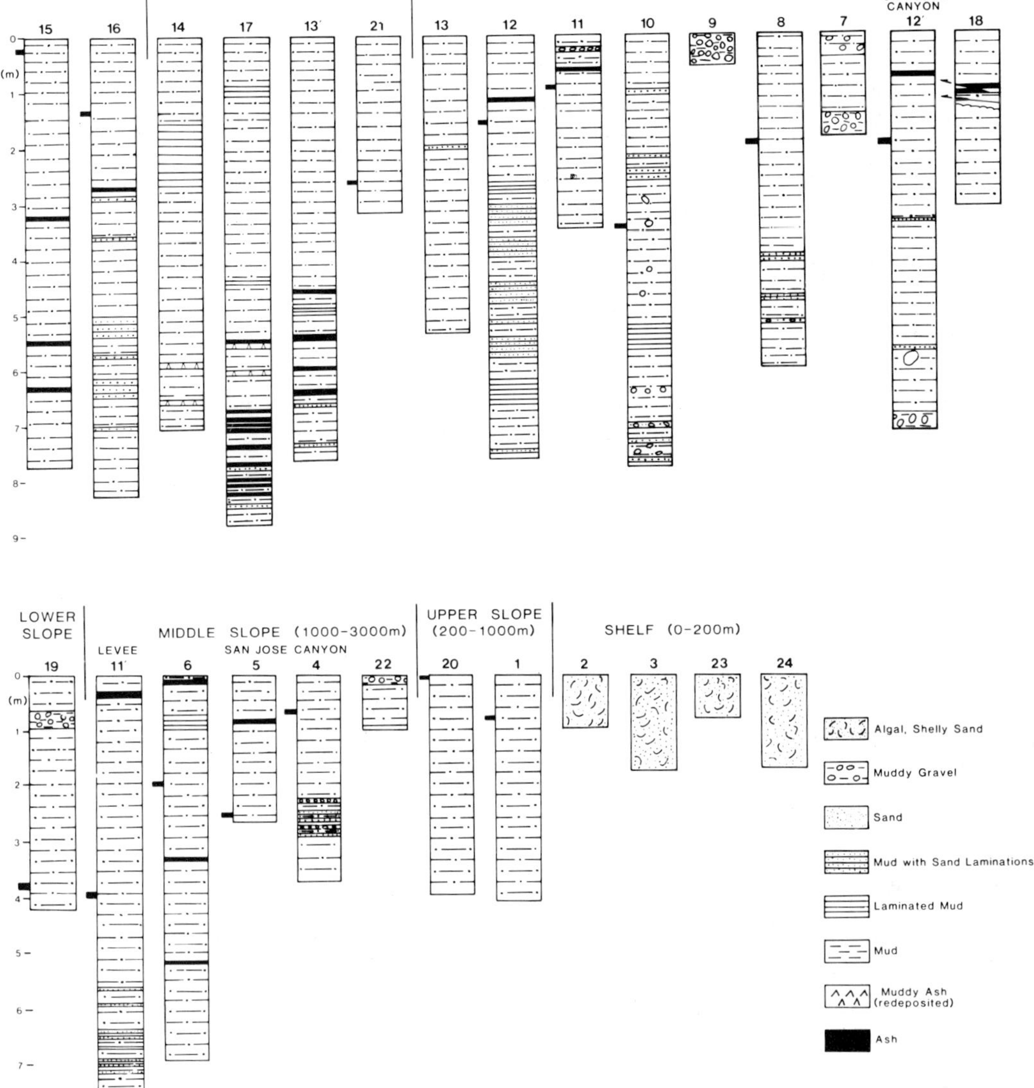

FIG. 5 (*continued*)

Inner slope cores contain four types of coarse clastic deposits:

1. thin-bedded, sandy turbidites (Fig. 6e) in lower slope terraces (in cores G8, G10, and G12);
2. thin-bedded sandy turbidites from the San Jose Canyon mid-slope levee (core G11) and from San Jose Canyon itself (cores G4 and G12′);
3. ash beds; and
4. coarse gravels (in cores G7, G9, G11, and G22 from lower slope scarps) and gravel clasts (from San Jose Canyon cores G4 and G12′) (Fig. 5b).

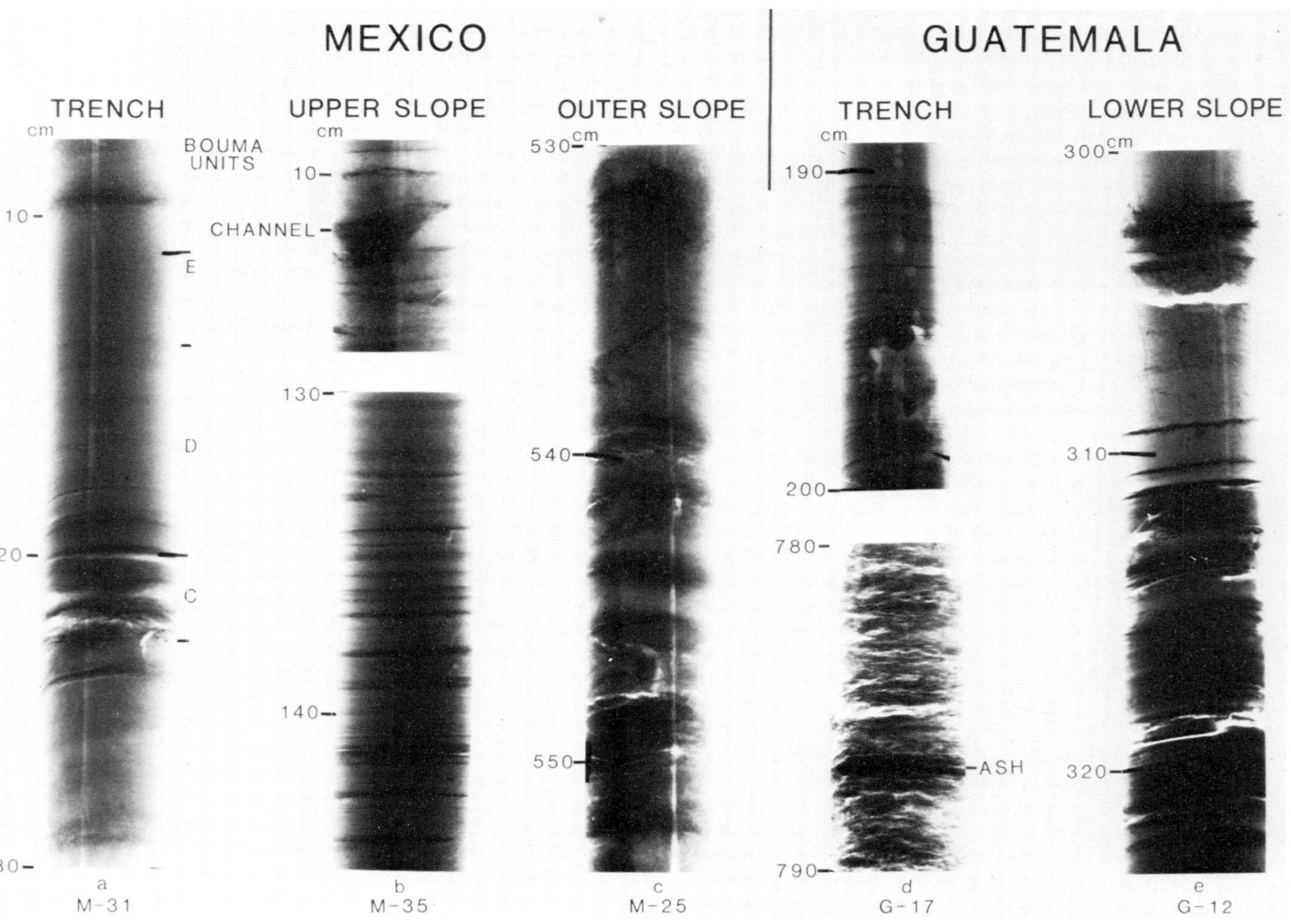

FIG. 6. Radiographs of cores showing distinctive structures: (a) turbidites from Mexican trench, (b) Mexican upper slope laminated sand, silt, and clay with ripples and channelling, (c) Mexican outer slope sand and clay deposits (contourites), (d) Guatemalan trench mud with gas cracks, and (e) Mexican inner-slope terrace thin-bedded turbidites. Photographs are contact prints of actual radiographs so density response is reversed.

Shelf sediments are sandy with much calcareous debris such as shells.

Micropalaeontology and correlation

Calcareous microfossils are unsatisfactory for Quaternary biostratigraphic zonation on the Middle America Pacific margins due to carbonate solution in response to the relatively shallow East Pacific calcite compensation depth (CCD) of 3.4 km (McMillen & Bachman 1981). Siliceous microfossils such as radiolarians should provide biostratigraphic control, but the commonly used Quarternary equatorial zonation of Nigrini (1971) is unsuitable because upwelling in this area has excluded the critical surface-dwelling index species (McMillen 1981). The extinction level of *Axoprunum angelinum* (Campbell and Clark) at 0.40 Ma (Hays 1970) provides a good datum in the NE Pacific (McMillen 1981; Dinkelman 1973). None of the piston cores we recovered contained *A. angelinum* so all cores are of late Quaternary age less than 0.40 Ma old. The base of core M-32 contains mud with late Miocene-age radiolarians. This sediment could be older slope deposits or reworked.

Recognition of the Pleistocene–Holocene boundary, based on the presence of radiolarian-rich sediments above the boundary and foraminiferal-rich sediments below, has been attempted by Duncan *et al.* (1970). As Bernard & McManus (1973) point out, however, this change from foraminiferal- to radiolarian-rich sediments up-core is controlled by changes in the CCD, and is time-transgressive up-slope. Nevertheless, the foraminiferal-radiolarian boundary, which can be identified offshore Guatemala (McMillen 1979), does provide a rough indication of the Pleistocene–Holocene boundary.

The ratio of nassellarian (cone-shaped) to spumellarian (spherical) radiolarians, or the N/S ratio, has been shown to correlate roughly with the down-core foraminiferal increase offshore Guatemala (McMillen 1979). Since variations in the N/S ratio represent faunal changes due to changing oceanographic conditions rather than solution changes, maxima and minima in the N/S ratio should be isochronous over a small area. A N/S minimum, which occurs just below the level of foraminiferal increase, is correlative from core to core and is usually located near a time-parallel ash horizon. McMillen (1979) reasoned that this N/S minimum was due to the late Pleistocene lowering of sea-level, occurring at 15,000 yr bp (Milliman & Emery 1968), as the N/S ratio is low in modern nearshore environments (Casey & Bauer 1976). Since this N/S minimum is probably isochronous, it can be used to compare relative rates of sedimentation in different environments and is plotted in Fig. 5.

Sedimentation rates estimated using the 15 000 yr N/S minimum are seen to be highest in ponded areas of trench sediment and lowest on the slope. Near submarine canyons such as Ometepec and San Jose Canyons, the rate increases. Upper slope sediments off Mexico were deposited at a much higher rate than comparable sediments off Guatemala. Sedimentation rates in portions of the trench with no sediment ponds are low off Guatemala (cores G18 and G21) and similar to slope rates. Holocene to late Pleistocene slope sediments are generally muddy and thin sand beds show up only in the Pleistocene section. A notable exception is upper slope core M35 with laminated sand and mud in the Holocene or late Pleistocene.

Petrology

We have investigated coarse-fraction constituents of sands and muds, and mineralogy of the terrigenous portion of sand beds. Coarse-fraction constituent study gives clues as to the origin of lower slope and trench sands. Sediments delivered by river to the Middle America margins consists of mud, terrigenous sand, and plant material. Shelf sediments receive a significant contribution of calcareous debris in the form of shell fragments, foraminiferal tests, and algae. Trench and lower slope sands offshore Mexico contain mostly terrigenous grains, especially in the thick-bedded, coarse-grained sands at the base of Ometepec Canyon. These coarse-grained sands were derived from a littoral drift or fluvial source, with little contribution of shelf sands. Sand beds on the Guatemala slope and trench often contain calcareous shelly debris from molluscs, ostracodes, and foraminifers; this indicates a source on the shelf.

Terrigenous petrology of sands differ considerably between the two areas (Fig. 7, Enkeboll 1981). Mexican sands contain abundant quartz, feldspar, and metamorphic rock fragments. The modal composition for fine-grained sands is $Q_{31}F_{50}L_{19}$, whereas it is $Q_{41}F_{22}L_{37}$ for coarse-grained sands. The chert to total quartz ratio is 0.05 in fine-grained sands and 0.04 in coarse-grained sands. The plagioclase to total feldspar ratio is 0.53 in fine-grained sands and 0.43 in coarse-grained sands. Biotite is the commonest mafic constituent.

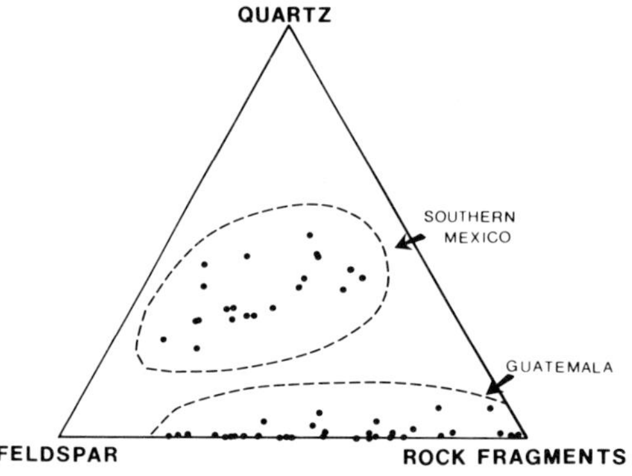

FIG. 7. Compositional comparison of sands from the trench, trench slope, and shelf from the Mexico and Guatemala margins.

Guatemalan sands, all fine-grained, are rich in pyroclastic material and poor in quartz; the average composition is $Q_1F_{43}L_{56}$. The plagioclase to total feldspar ratio is 0.99. Volcanic rock fragments constitute 99% of all rock fragments. Pyroxene is the dominant mafic mineral. In addition to these sands, thick gravel beds also occur on the lower Guatemalan slopes. These gravels have unusual lithic constituents consisting of chert, mudstone, zoelite-facies argillite, and prehnite-pumpellyite-facies metabasite.

Discussion

Sediments of the southern Mexico and Guatemalan portions of the Middle America Trench margin can be compared lithologically, petrologically, and on the basis of their accumulation rates. Trench sediments differ most between the two areas. The Mexican Trench contains Holocene coarse- to medium-grained, clean sand or thick-bedded sandy turbidites, whereas the Guatemalan trench contains muddy sediments with late Pleistocene or early Holocene thin-bedded sandy or ashy turbidites. Sediment fill within the trench offshore Mexico is thicker and more extensive, showing evidence of large channels and levees, and local tectonic deformation. Guatemalan trench fill is much less extensive with only a few small channels and no compelling evidence for deformation within the trench.

Trench sediment lithology and texture correlate well with nearshore and onland geology. The Mexican margin lacks a good forearc basin, has a narrow shelf, highlands of crystalline rock immediately onshore, and numerous submarine canyons. Sandy trench fill reflects the crystalline source terrane. Guatemala has a wide forearc basin underlying a wide shelf and fluvial coastal plain, volcanic highlands onshore, and few submarine canyons. Trench fill is muddy due to sand trapping on the shelf and breakdown of unstable volcanic products. Middle and lower slope sediments are similar in both areas, consisting mainly of slowly deposited hemipelagic mud with a few thin sand beds. Upper slope sediments differ betweeen Mexico and Guatemala. Laminated sediments deposited at a fairly high rate by slope currents occur offshore Mexico in or near the oxygen-minimum zone. Hemipelagic upper slope sediments offshore Guatemala have been deposited at a slow rate and no effect of the oxygen-minimum zone is seen. Erosion of the Mexican slope at slightly greater depths at DSDP Site 489 (Moore *et al.* 1979a) suggests that considerable variability can exist in upper slope deposition. Shelf sediments are similar in both areas, consisting of terrigenous and bioclastic calcareous sand.

The volume of trench fill differs between both areas as well. While neither area has especially thick fill, the portion of the Guatemalan trench studied has considerably less fill than the Mexican area studied. The trench fill offshore Guatemala is both thinner and less extensive laterally. The wide forearc basin of the Guatemalan margin may capture more sediment and result in a smaller quantity actually being delivered to the trench than offshore Mexico.

On the other hand, the higher convergence rate across the Guatemalan margin (10 cm yr^{-1}, Molnar & Sykes 1969) than across the Mexican margin (about 7 cm yr^{-1}, Minster & Jordan 1978) would also tend to reduce the volume of trench sediment at any given moment (see Schweller & Kulm 1978).

Sediment bypassing of the slope is a significant feature of both margins; however, the role of the shelf in supplying sediment varies. Offshore Mexico, trench sediment is derived mainly from fluvial or possibly littoral-drift sources with a little contribution from the outer shelf, as is shown by differing coarse fraction composition between terrigenous trench sands and terrigenous and bioclastic sands of the outer shelf. Trench sediments here bypass both slope and shelf with little residence time on the shelf. Offshore Guatemala high sedimentation rates in the trench indicate that the muddy trench-fill largely bypasses the slope, although local slope deposition does occur in a levee near San Jose Canyon and in lower slope basins. Similarities in the coarse-fraction composition of trench, slope, and shelf sands (all sands contain bioclastic as well as terrigenous debris) suggests that these sands have been derived from the shelf. Most sands accumulate in the forearc basin. Sediments are suspended during flood or by wave action and carried over the shelf edge. The San Jose Canyon embayment probably acts as a catchment basin for this muddy sediment and funnels it down slope and into the trench.

Before considering the possible tectonic effects of varying trench wedge size, we need to evaluate possible variations in the trench wedge through time. Factors which affect the volume of a trench wedge include: the convergence rate (which determines the rate that the trench wedge is deformed or consumed), and the rate at which sediments are supplied to the trench to build up the trench wedge (Karig & Sharman 1975; Schweller & Kulm 1978). Convergence varies along the Middle America Trench with a rate of roughly 7 cm yr^{-1} off Southern Mexico (Minster & Jordan 1978) and roughly 10 cm yr^{-1} off Guatemala (Molnar & Sykes 1969). Sediment input to the trench is controlled by sediment load of rivers (reflecting climate and topographic relief), the presence of sediment-trapping areas such as forearc and slope basins, and relative sea-level, which controls whether rivers discharge into an estuary, on to the shelf, or directly on to the shelf edge and the heads of submarine canyons.

With these variables in mind, estimates of past amounts of trench fill based on present-day amounts can be hazardous. The Pleistocene Middle America Trench could have had more fill because: (1) the sea-level was lower and rivers discharged nearer the shelf edge, and (2) the Pleistocene climate could have been wetter and resulted in greater runoff.

Even with these possible causes for increased sedimentation, the Pleistocene trenches offshore Guatemala and Mexico are inferred to have not contained a great deal more sediment than at present. Current convergence rates result in roughly 1 km of underthrusting at the trench over the last 10 000 yr, or since the beginning of the Holocene, a time when trench fill would have been greater than at present. Yet the width of trench fill offshore both Mexico and Guatemala is 5–6 km, so most trench fill should consist of relict Pleistocene sediments.

Therefore, we infer that trench fill has been small offshore Guatemala during the late Quaternary, and that there have been appreciably greater amounts of fill offshore Mexico than offshore Guatemala.

Long-term extrapolation of the amount of trench fill offshore Guatemala is possible using Seely's (1978) model for development of the forearc basin. The basin initially formed from oceanic crust trapped landward of a tectonic ridge of oceanic crust. Sediments sampled in the Esso Petrel #1 well within the basin document continued shallowing since the Palaeocene. It is probable that nearly all coarse-grained sediments delivered to the margin were trapped within this forearc basin so that the trench has remained virtually sediment-free throughout the Tertiary. In fact, now that the forearc basin is essentially filled at the present time, sediments might even be spilling over into the trench at a greater rate than in the past.

The relative amounts of trench fill do seem to be related to the amount of accretion taking place on the margins. The Southern Mexico margin, with the sizable accretionary wedge abuting on to continental basement, has been accreting trench material since the Miocene (Moore *et al.* 1979a). The Guatemala margin, on the other hand, seems to have very little or no active accretion taking place (von Heune *et al.* 1981), perhaps a direct consequence of the meagre trench fill. In both areas, the thickness of pelagic cover on the subducted oceanic plate is small (Fig. 4) so differences in the amount of accretion are probably due to the different amounts of trench fill present in these two areas.

Conclusions

We can make the following generalizations about sedimentation along the margins associated with the Middle America Trench:

1. Trench sediment type and amount correlates well with the tectonic environment of the margin. Relatively thick, sandy trench fill occurs where truncated crystalline highlands occur close to shore and where the shelf is very narrow offshore Mexico. Thinner, muddy trench fill occurs where volcanic highlands occur far from shore and where a large forearc basin intervenes between the volcanic arc and shelf edge offshore Guatemala.
2. Trench fill correlates well with the position of major submarine canyons that supply sediment to the trench, a point already emphasized by Underwood & Karig (1980).
3. Slope sediments are similar in both areas and consist mainly of muddy, hemipelagic

sediments. Bypassing of the slope occurs in both areas, although there is local deposition near large submarine canyons.

4. Sedimentation on a narrow shelf, such as occurs offshore Mexico, does not provide significant detritus for removal to the trench, as much greater amounts of sediment are delivered there directly from rivers. Conversely, the wide shelf offshore Guatemala contributes most of the sediment to submarine canyons and the trench.
5. Relative amounts of margin accretion may be controlled by the amount of trench fill in the Middle America Trench, as ocean plate sediment thickness coming into the subduction zone is so low (Fig. 4). The thicker trench fill offshore Mexico may result in the much greater amount of observed accretion there than off Guatemala, where trench fill is small and probably has been small in the past.

ACKNOWLEDGMENTS: Cores and seismic data were collected on NSF contract CU-TEX-2S907-3 to the University of Texas Marine Science Institute. We wish to thank the captain and crew of the RV *Idea Green* and the assistance of Helen Kim, John Kumselman and staff, and Fran Sebak. Acknowledgment is made to the donors of the Petroleum Research Fund, administered by the American Chemical Society, for partial support of Enkeboll and Moore's participation in this study. Don Hussong allowed us to use the line drawing from the Hawaii Institute of Geophysics profile across the Guatemala trench.

References

BERNARD, W. D. & MCMANUS, D. A. 1973. Planktonic foraminiferan-radiolarian stratigraphy and the Pleistocene-Holocene boundary in the northeast Pacific. *Bull. geol. Soc. Am.* **84,** 2097–100.

BOUMA, A. H. 1969. *Method for the Study of Sedimentary Structures.* Wiley, London, 450 pp.

CASEY, R. E. & BAUER, M. A. 1976. A seasonal study of Radiolaria and Foraminifera in the water overlying the south Texas outer continental shelf. *Abstr. Prog. geol. Soc. Am.* **8–1,** 11.

DE CZERNA, A. 1965. Reconocimiento geologico en la Sierra Madre del Sur de Mexico, entre Chilpaningo Y Acapulco, Estado de Guerrero. *Bol. Inst. Geo, Univ. Nac. Aut. Mexico,* **62,** 76 pp.

DICKINSON, W. R. 1970. Interpreting detrital modes of graywacke and arkose. *J. sediment. Petrol.* **40,** 695–707.

DINKELMAN, M. G. 1973. Radiolarian stratigraphy: Leg 16 Deep Sea Drilling Project. *Initial Rep. Deep Sea drill. Proj.* **16,** 747–813.

DUNCAN, J. R., FLOWER, G. A. & KULM, L. D. 1970. Planktonic foraminiferan-radiolarian ratios and Holocene-late Pleistocene deep-sea stratigraphy off Oregon. *Bull. geol. Soc Am.* **81,** 561–6.

ENKEBOLL, R. H. 1981. Sedimentary petrology and provenance of sands and gravels from the Middle America trench and trench slope: Guatemala and southern Mexico. *Bull. Am. Assoc. Petrol. Geol.*, in press.

HAYS, S. D. 1970. Stratigraphy and evolutionary trends of Radiolaria in North Pacific deep-sea sediments. *Mem geol. Soc. Am.* **126,** 185–218.

IBRAHIM, A. K., LATHAM, G. V. & LADD, J. W. 1979. Seismic refraction and, reflection measurements in the Middle America Trench offshore Guatemala. *J. geophys Res.* **84–10,** 5643–9.

KARIG, D. E. 1974. Tectonic erosion of trenches. *Earth planet. Sci. Lett.* **21,** 209–12.

—— , CARDWELL, R. K., MOORE, G. F., & MOORE, D. G. 1978. Late Cenozoic, subduction and continental margin truncation along the northern Middle America Trench. *Bull. geol. Soc. Am.* **89,** 265–76.

—— & SHARMAN, G. F. 1975. Subduction and accretion in trenches. *Bull. geol. Soc. Am.* **86,** 377–89.

KING, P. B. 1969. *Tectonic Map of North America.* U.S. Geol. Survey.

LARSON, R. L. & CHASE, C. G. 1970. Relative velocities of the Pacific, North America and Cocos Plates in the Middle America Trench region. *Earth planet. Sci. Lett.* **7,** 425–8.

LUNDBERG, N. 1981. Evolution of the slope landward of the Middle America Trench, Nicoya Peninsula, Costa Rica (this volume).

MCMILLEN, K. J. 1979. Radiolarian ratios and the Pleistocene-Holocene boundary. *Trans. Gulf Coast Assoc. geol. Soc.* **29,** 298–301.

—— 1981. Radiolarians from the Southern Mexico active margin, DSDP Leg 66. *Initial Rep. Deep Sea drill. Proj.* **66,** in press.

—— & BACHMAN, S. B. 1981. Bathymetry and vertical tectonics of Leg 66 DSDP sediments. *Initial Rep. Deep Sea drill. Proj.* **66,** in press.

MATTHEWS, D. J. 1939. *Tables of the Velocity of Sound in Pure Water and Sea Water for Use in Echo-sounding and Sound-ranging.* The Hydrographic Department, Admiralty, London, 52 pp.

MILLIMAN, J. D. & EMERY, K. O. 1968. Sea levels during the past 35 000 years. *Science,* **162,** 1121–3.

MINISTER, J. B. & JORDON, T. H. 1978. Present-day plate motions. *J. geophys. res.* **83-B11,** 5331–54.

MOLNAR, P. & SYKES, L. R. 1969. Tectonics of the Caribbean and Middle America regions from focal mechanisms and seismicity. *Bull. geol. Soc. Am.* **80,** 1639–84.

MOORE, J. C., WATKINS, J. S., BACHMAN, S. B., BEGHTEL, F. W., BUTT, A., DIDYK, B., LEGGETT, J. K., LUNDBERG, N., MCMILLEN, K. J., NIITSUMA, N., SHEPHARD, L. E., SHIPLEY, T. H.,

Stephan, J. F. & Stradner, H. 1979a. Middle America Trench. *Geotimes,* **24,** 20–2.

——, ——, Shipley, T. H., Bachman, S. B., Beghtel, F. W., Butt, A., Didyk, B., Leggett, J. K., Lundberg, N., McMillen, K. J., Niitsuma, N., Shephard, L. E., Stephan, J.-F. & Stradner, H. 1979b. Progressive accretion in the Middle America Trench southern Mexico. *Nature. London,* **281,** 638–42.

Nigrini, C. A. 1971. Radiolarian zones in the Quaternary of the equational Pacific Ocean. *In:* Funnell, B. M. & Riedel, W. R. (eds). *The Micropalaeontology of Oceans*, 443–61. Cambridge University Press.

Prince, R. A., Resig, J. M., Kulm, L. D. & Moore, T. C. 1974. Uplifted turbidite basins on the seaward wall of the Peru Trench. *Geology,* **2,** 607–11.

Schweller, W. J. & Kulm, L. D. 1978. Depositional patterns and channelized sedimentation in active eastern Pacific trenches. *In:* Stanley, D. J. & Kelling G. (eds). *Sedimentation in Submarine Canyons, Fans, and Trenches,* 311–24. Dowden, Hutchinson & Ross, Stroudsburg, Pennsylvania.

Seely, D. R. 1978. The evolution of structural highs bordering major forearc basins. *In:* Watkins, J. S., Montadert, L. & Dickerson, P. W. (eds). *Geological and Geophysical Investigations of Continental Margins. Mem. Am. Assoc. Petrol. Geol.* **29,** 245–60.

—— , Vail, P. R. & Walton, G. G. 1974. Trench slope model. *In:* Burk, C. A. & Drake, C. L. (eds). *The Geology of Continental Margins*, 249–60. Springer-Verlag, New York.

Shipley, T. H., McMillen, K. J., Watkins, J. S., Moore, J. C., Sandoval-Ochoa, H. & Worzel, J. L. 1978. Continental margin and lower slope structures of the Middle America Trench near Acapulco, Mexico. *Mar. Geol.* 65–82.

—— , Ladd, J. W., Buffler, R. T. & Watkins, J. S. 1981. Tectonic processes along the Middle America Trench inner slope (this volume).

Stewart, R. 1978. Neogene volcaniclastic sediments from Atka Basin, Aleutian Ridge. *Bull. Am. Assoc. Petrol. Geol.* **62,** 87–97.

Underwood, M. B. & Karig, D. E. 1980. Role of submarine canyons in trench and trench-slope sedimentation. *Geology*, **8,** 432–6.

von Heune, R., Auboin, J., Azema, J., Blackinton, G., Carter, J., Coulbourn W., Cowan, D., Curiale, J., Dengo C., Faas, R., Harrison, W., Hesse, R., Hussong, D., Ladd, J., Muzylor, N., Shiki, T., Thompson, P. & Westberg, J. 1980. Leg 67: The Deep Sea Drilling Project Mid-America Trench Transect off Guatemala. *Bull. geol. Soc. Am.* **91,** 421–32.

—— , —— , —— , —— , —— , —— , —— , —— , —— , —— , —— , —— , —— , —— , —— , —— , —— & —— , 1981. A summary of Deep Drilling Project Leg 67 shipboard results from the Mid-America Trench transect off Guatemala (this volume).

K. J. McMillen, Gulf Science and Technology Co. P.O. Box 2038, Pittsburgh, Pennsylvania 15230, U.S.A.

R. H. Enkeboll, H. Esmaili and Associates, Inc., 2718 Telegraph Avenue, Suite 200, Berkeley, California 94705, U.S.A.

J. C. Moore, Earth Sciences Board, University of California, Santa Cruz, California 95064, U.S.A.

T. H. Shipley, Scripps Institution of Oceanography, La Jolla, California 92093, U.S.A.

J. W. Ladd, Lamont-Doherty Geological Observatory, Palisades, New York 10964, U.S.A.

A summary of Deep Sea Drilling Project Leg 67 shipboard results from the Mid-America Trench transect off Guatemala

Roland von Huene, Jean Aubouin, Jacques Azema, Grant Blackinton, Jerry A. Carter, William T. Coulbourn, Darrel S. Cowan, Joseph A. Curiale, Carlos A. Dengo, Richard W. Faas, William Harrison, Reinhard Hesse, Donald M. Hussong, John W. Ladd, Nikita Muzylov, Tsunemasa Shiki, Peter R. Thompson & Jean Westberg

SUMMARY: The Middle America Trench off Guatemala was transected by 24-channel seismic reflection surveys, seismic-refraction surveys, and drilling with the *Glomar Challenger*. The drilling was done at three sites on the oceanic Cocos plate and four sites on the Caribbean plate. These plates converge at about 10 cm yr^{-1}. At all drill sites sediment of upper Miocene to Quaternary age is almost entirely hemipelagic mud with interbedded thin volcanic ash, except in the trench where mud and fine sand turbidites less than 400 000 yr old are ponded. However, the underlying rocks are very different. On the oceanic Cocos plate a basal chalk sequence of lower and middle Miocene age is overlain by a thin section of abyssal clay. At a site only 3 km landward of the trench axis where drilling penetrated the slope deposits we recovered a Cretaceous to lower Miocene claystone sequence resting on a section containing igneous rock of continental affinity. A large net subduction of sediment along with ocean crust has occurred during the present (Miocene-Quaternary) episode of subduction and perhaps parts of the continental framework have been subducted as well. However, no current model satisfactorily explains the surprising occurrence of Cretaceous-Miocene claystone at the foot of the trench slope.

The Mid-America Trench transect off Guatemala is the second of two geophysical and drilling transects recommended for the International Program of Ocean Drilling (IPOD) across the convergent margin of southern Mexico and Central America. The first, a transect off Oaxaca, Mexico, was drilled on Leg 66 of the *Glomar Challenger* to investigate a convergent margin where the leading edge is a subduction complex tectonically accreted during the Neogene, and where a truncated Precambrian and Palaeozoic continental framework extends oceanward to the mid-slope (Moore, Watkins *et al.* 1979, 1981; McMillen *et al.* 1981). Off Guatemala our principal objective was to study a margin where geophysical data indicated continuous accretion and imbrication (Seely *et al.* 1974). Seely and his colleagues made a case for the imbricate thrust model that is widely accepted, and this model is supported by the multichannel seismic reflection site surveys (Ladd *et al.* 1978; Ibrahim *et al.* 1979).

This paper is patterned after a more complete report prepared on the preliminary results of Leg 67, where the data from the cores are discussed in greater detail (von Huene, Aubouin *et al.* 1981).

Geological setting

Investigations of global plate motions and geophysical studies provide a convincing case for normal convergence of the Cocos and Carribbean plates along the Mid-America Trench off Guatemala (Fig. 1) (Minster & Jordan 1978; Jordan 1975; Molnar & Sykes 1969). The rate of convergence is estimated at 10 cm yr^{-1} since the late Miocene and the compressional tectonism implied by plate convergence is compatible with the geology of the trench slope deduced largely from geophysics (Seely *et al.* 1974; Ladd *et al.* 1978; Ibrahim *et al.* 1979). The oceanic crust continues beneath the trench landward slope and is overlain by a thick sequence of landward-dipping reflections (Fig. 2). Near the top of the trench slope strata deduced from landward-dipping reflections become the seaward flank of a forearc basin that is filled with as much as 8 km of sediment. This thick sediment probably underlies the coast above which tower the volcanoes of the magmatic arc. From a drill hole at the edge of the shelf, Seely (1979) documented Palaeocene and Eocene uplift of the shelf edge and showed that the Cenozoic section is broken by a series of hiatuses, one of the most visible ones in seismic

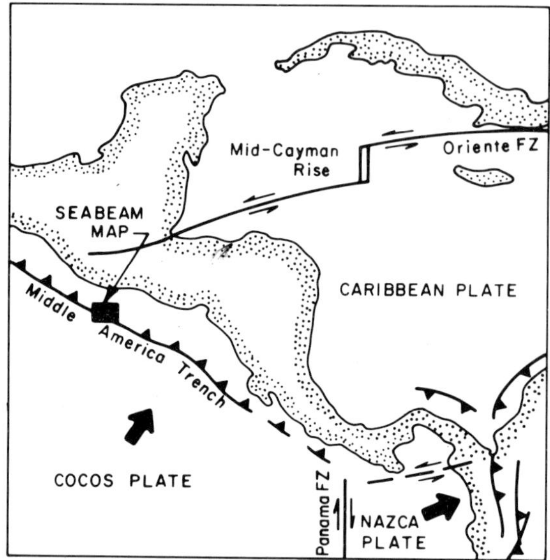

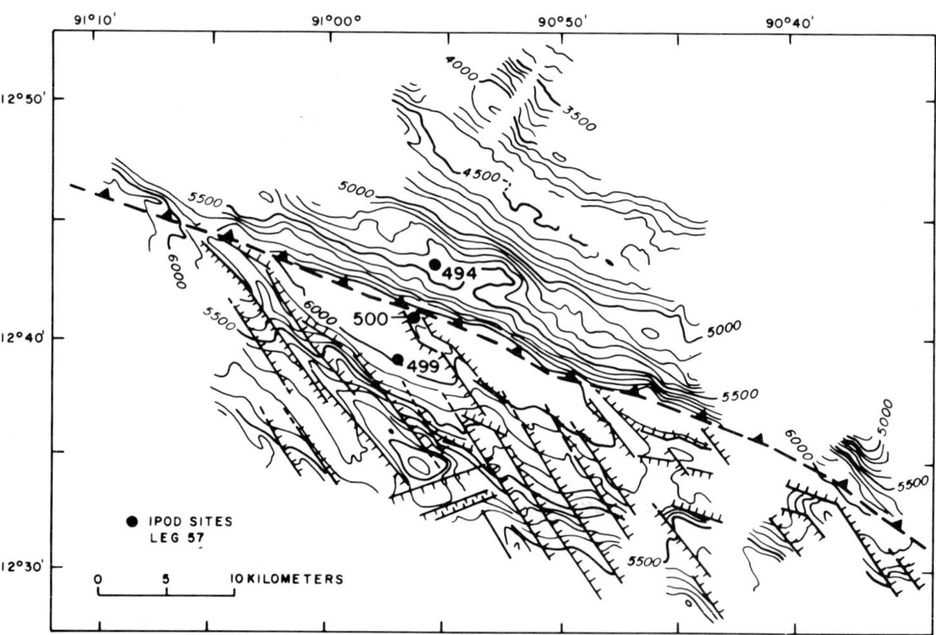

Fig. 1. Plate tectonic diagram of the Central American area showing the relation of the Caribbean and Cocos plates along the Mid-America Trench. The Caribbean and the Cocos plate are converging at 10 cm yr^{-1} across the trench. The lower part of the figure is a bathymetric map made with a Seabeam instrument (after Renard *et al.* 1980).

records corresponding to a Miocene angular unconformity. Seely *et al.* (1974) interpret the structure of the Guatemalan margin as one of stacked and tilted imbricate slices, based on pervasive landward dips in seismic records, as well as a series of benches on the lower slope

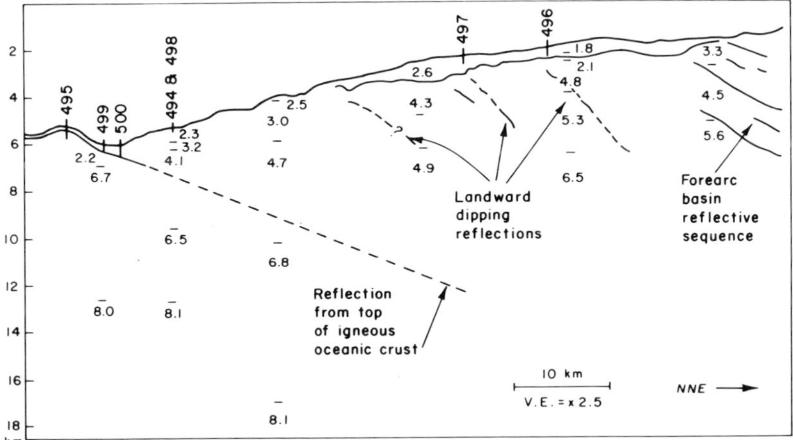

FIG. 2. Summary of seismic-reflection and refraction studies showing velocity structure and major reflectors. The location of sites drilled by *Glomar Challenger* on Leg 67 are indicated.

which may be likely places for the surfacing of large thrust faults.

Seabeam bathymetry

The first results from a Seabeam survey by the RV *J. Charcot* (Renard *et al.* 1980) made after the Leg 67 drilling, revealed an unexpected topographic trend of horst and graben on the trench seaward slope. The Seabeam instrument installed on the Charcot maps bathymetry in a swath with width that is about 75% of the water depth (Renard & Allenou 1979). Individual swaths are not distorted by inter-ships-track navigational discrepancies and the effects of diffractions from steep slopes are minimized. The preliminary map shown in Fig. 1 will later be adjusted to the limit of resolution which is less than 10 m (Renard & Allenou 1979).

The Seabeam bathymetry indicates that structure on oceanic crust strikes about 30° from the trend of the Mid-America Trench axis. This is also the trend of regional oceanic magnetic anomalies (Fig. 1). A typical horst and graben topography, commonly seen as ocean crust is flexed down into the trench, is very apparent in the Seabeam bathymetry. Ridges formed by horsts are not completely covered by sediment ponded in the trench axis and these ridges divide the trench fill into small isolated basins (Fig. 1). The trench fill basins terminate landward against a linear escarpment that forms the beginning of the trench landward slope. Site 500 on the trench floor was drilled along one of the ridges.

The trench landward slope is stepped into a sequence of benches and the lowest was well surveyed. It continues for at least 30 km with minor relief. The next higher bench is shown for 20 km of its extent. Site 494 on the first bench appears to be on a secondary low nose trending diagonally across the bench.

Drilling results

Cocos Plate

The oceanic Cocos plate forms the seaward slope of the Mid-America Trench off Guatemala. Ocean basement is overlain by a thin sediment blanket that is uniformly 200–300 m thick. As the ocean crust is flexed down into the trench the crust breaks into a sequence of horst and graben. The downward flexure forms a trench slope that drops about 2000 m into the trench in the area of the drilled transect. Site 495 is on a horst about 22 km seaward of the trench axis and 1925 m above it (Fig. 2). The sediment follows topography (except where faulted) with uniform thickness, a feature often interpreted in seismic reflection records as characteristic of pelagic sediment.

The sediment section penetrated at Site 495 (Fig. 3) records the early and middle Miocene path of this site as it passed northward through the equatorial carbonate belt, subsiding down the east flank of the East Pacific Rise, into a zone of normal slow pelagic deposition. Then during the late Miocene it drifted into a hemipelagic environment influenced by a terrigenous source (rather than the pelagic

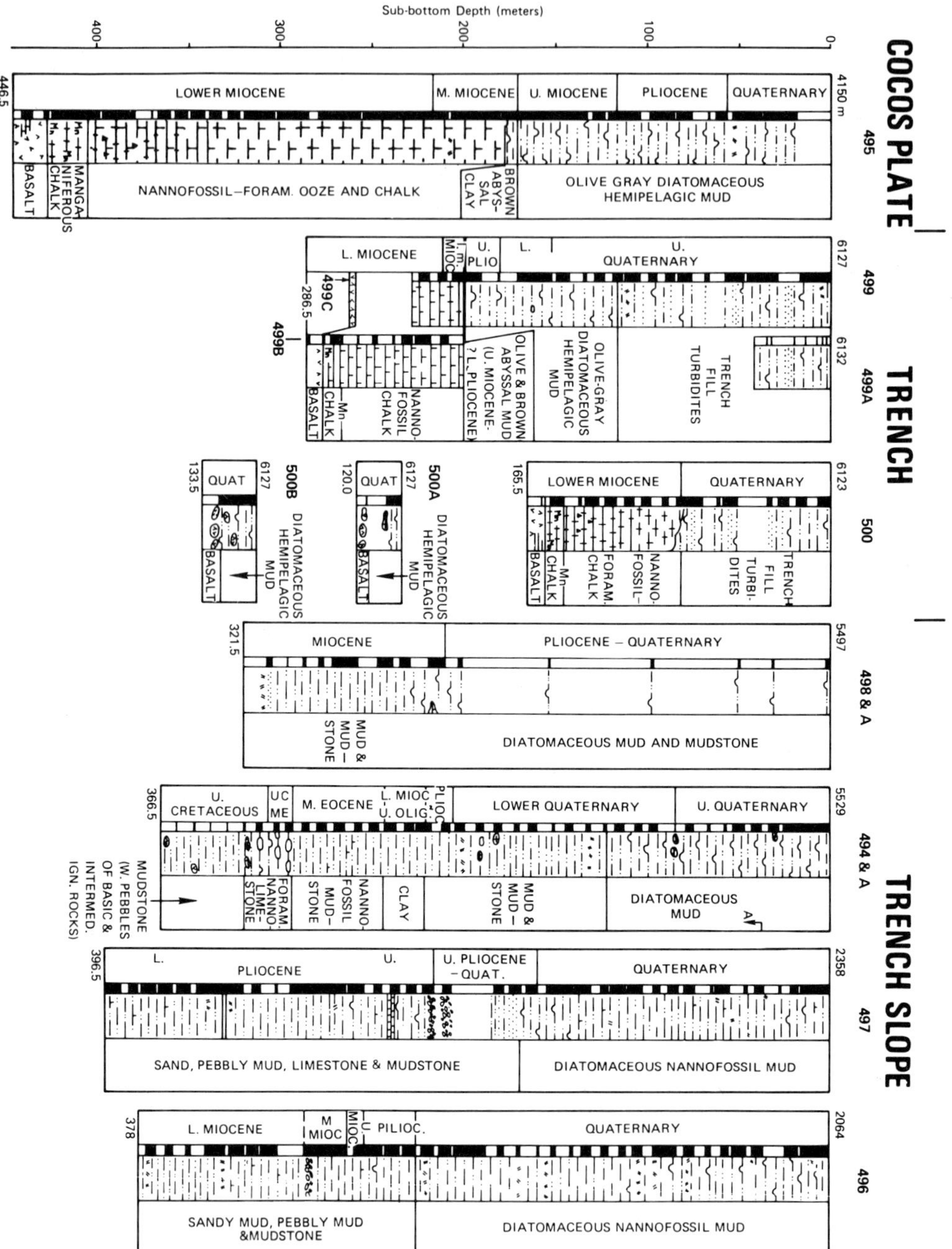

FIG. 3. Simplified bio- and litho-stratigraphy of sites drilled on Leg 67 (from von Huene, Aubouin *et al.* 1981).

environment previously inferred from seismic records). The upper hemipelagic section indicates that a surprisingly large amount of sediment was transported a great distance seaward across the trench. If present rates of convergence have been constant during deposition of the hemipelagic section the lowest beds were deposited when the site was about 900 km seaward of the trench. Such great distances of sediment transport are not seen in the present pattern of hemipelagic sedimentation as recorded at other DSDP sites in the area (van

Andel, Heath *et al.* 1973; Hays, Cook *et al.* 1972).

Trench floor

The trench floor has ridges separated by intermittent flat areas which are underlain by horizontal seismic reflections (Fig. 2). These reflections are presumed to represent ponded sediment which is in turn underlain by weak reflections that dip gently landward and are presumed to be the deep ocean sediment section riding into the trench axis on the Cocos Plate.

Sites 499 and 500 were drilled on the trench floor and they include seven holes most of which bottomed in basalt or basalt rubble at various depths (Fig. 3). The trench fill, as seen in seismic records, is represented by an upper Quaternary (slightly less than 400 000 yr old) sequence of alternating muddy and sandy turbidites with a microfossil assemblage transported from the shelf and slope. This turbidite sequence is underlain by a lower Miocene to Quaternary oceanic sequence like the one recovered at Site 495 on the Cocos plate, but about half as thick. At Site 500 this section is cut by a normal fault that probably formed during the development of the horst and graben structure of the trench seaward slope. The sediment beneath the trench floor is compacted normally and shows few signs of compressional deformation even against the trench landward slope.

The drilling results indicate that the trench is more complex than would be expected from the seismic records. Basalt and basalt rubble occur at a variety of levels, probably from a basalt basement structure not revealed by the seismic records. The trench fill has no pronounced lateral facies differences transverse to the trench axis as might be expected from axial turbidity current channels (Piper *et al.* 1973; von Huene 1974).

Trench lower slope

The trench lower slope has a series of benches. The lowest of these is 3 km from the trench floor, about 500 m above it, and extends more than 30 km parallel to the trench. In our previous article (von Huene *et al.* 1980) we refer to bathymetry from a detailed survey by the *Kana Keoki* that suggests re-entrants in the terrace. A Seabeam survey (Renard *et al.* 1981) has shown that the bench is a linear feature without re-entrants. This type of topographic feature is often interpreted as having been formed by the emergence of a large thrust fault at the seafloor. Below the seafloor is a weak sequence of reflections that dip gently landward. These in turn overlie a strong reflection from the top of the igneous oceanic crust. When the vertical axis of the seismic reflection record is converted from time to depth the beds of the reflective sequence are nearly horizontal and have a thickness of about 1200 m. At Site 494 on the bench just above the trench floor, we recovered a 366 m sequence of rock in normal stratigraphic succession that ranges in age from Late Cretaceous to Quaternary (Fig. 3). The Upper Cretaceous rock is about 900 m above the top of igneous oceanic crust (Fig. 4).

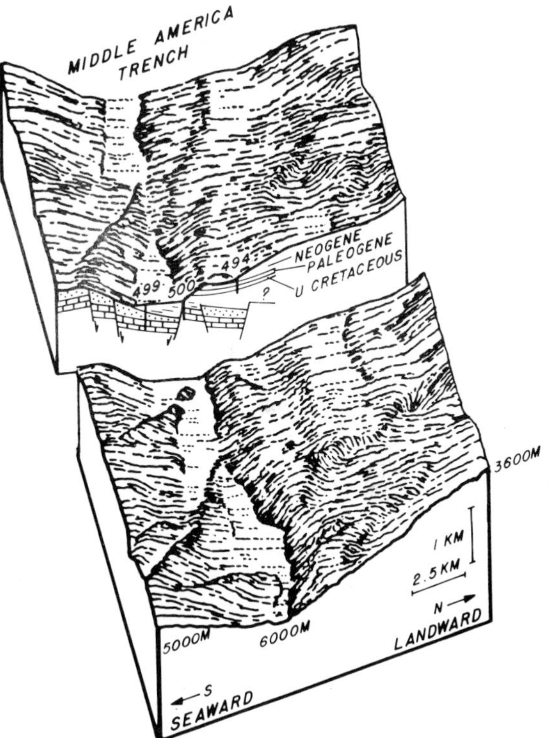

Fig. 4. Diagram of interpreted subsurface geology across the Mid-America Trench off Guatemala based on geophysics and Leg 67 drilling. The topography is after the Seabeam survey in Fig. 1 and was drawn by Tau Alpha.

At Site 494 upper Miocene to Quaternary slope deposits cover a sequence similar to that reported by Seely (1979) from the edge of the continental shelf. The environments represented by this sequence, the nature of hiatuses, and their relation to the sequence of the shelf are interpreted in down-hole order as follows:

1. A Pliocene and Quaternary deposit consisting principally of sediment transported from the shelf and upper slope to this site.
2. An upper Miocene unconformity coincident with the widespread upper Miocene unconformity on the shelf.
3. A sequence of distal, terrigenous, hemipelagic clay that accumulated during the early Miocene on seafloor near the CCD and at rates an order of magnitude less than age-equivalent deposits on the adjacent shelf (3 versus 100 m my^{-1}).
4. An upper Eocene-Oligocene hiatus which is the age-equivalent of a widespread unconformity in the adjacent shelf section but of unknown origin in the section drilled.
5. An Eocene hemipelagic clay deposited below the foraminiferal CCD but above the nannofossil CCD at rates at least five times less than age-equivalent shelf deposits.
6. A hiatus of unknown origin but of an age that is represented by thick, widespread Palaeocene sediments below the adjacent shelf.
7. An Upper Cretaceous claystone that accumulated in an open ocean environment above the CCD at depths about equivalent to, but at rates apparently less than age-equivalent rocks below the present shelf.
8. A contact of unknown origin between claystone and igneous rocks.
9. Igneous rock originally of basaltic and andesitic composition and atypical of igneous oceanic crust; the nature of the rock body sampled is unknown.

This sequence shows microfracturing from tectonic stress beginning with rock at the base of the Pleistocene, but if there are thrust faults cutting the drilled section they are not large enough to give major discontinuity in rock ages.

At Site 498, only the slope section (olive-grey mudstone) appears to have been penetrated. We interpret the difference in thickness of the Miocene sections of olive-grey mudstone as reflecting relief on the upper Miocene unconformity.

Trench middle slope

Sites on the trench middle slope were positioned to penetrate the slope deposits and to sample the underlying landward-dipping reflections. We planned to test whether the reflectors are part of an imbricated stack resulting from subduction as has been inferred by Seely *et al.* (1974), Ladd *et al.* (1978) and Ibrahim *et al.* (1979). Drilling was terminated short of the primary objectives for safety considerations because we encountered gas hydrate. Slope deposits, as interpreted from seismic reflection records, unconformably drape the rock that makes up the bulk of the continental margin (Fig.2). The slope deposits thin toward the trench lower slope.

Site 497 was drilled in olive-grey mud with varying biogenic and vitric tuffaceous components (Fig. 3). It is essentially a uniform section of mixed hemipelagic and terrigenous muds interspersed with a thin pebbly mudstone that represents mass movement. A similar lithology was penetrated at Site 496 where the upper sequence is biogenic mud of Quaternary age underlain by biogenic sandy mudstone of Pliocene and Miocene age.

Sediments recovered from the mid-slope area are rich in terrigenous detritus and contain volcanic ash beds, and yield a microfossil assemblage displaced from upslope. The foraminiferal assemblage at Site 496 indicates subsidence of the site in Miocene time from shelf to lower bathyal depths. This subsidence is coincident with a subsidence reported by Seely (1979) from the adjacent shelf edge.

Conclusions

Geophysical data from the Mid-America Trench transect off Guatemala contain nothing unusual when compared with geophysical data from other convergent margins although the records do not resolve a great deal of structure at depth along the trench landward slope, as is seen for instance along the companion transect off Mexico (Moore, Watkins *et al.* 1979; Shipley 1981). The results from drilling off Guatemala also reveal nothing unexpected in the post-middle Miocene sediment. The unexpected results occur in the one drill hole that sampled below the cover of slope deposits, returning rock that is much older than that anticipated from previous geophysical work. Perhaps another unexpected aspect is the scant evidence for compressional deformation. However, it is difficult to detect structure geophysically that has less than a few hundred metres amplitude in very deep water and structure defined from the study of isolated drill cores is no longer than about 10 cm. Thus the lack of more compressional structure can be attributed to a lack of information at the appropriate scale.

The ocean crust entering the Mid-America Trench off Guatemala has a Miocene section that clearly records northward passage of the Cocos Plate first beneath the equatorial carbonate belt and then into proximity with an upper Miocene to Present terrigenous source. The distance of Site 495 from land when the terrigenous sediment first reached it is problematical, but is certainly hundreds of kilometres if plate reconstructions for this area are correct. As the ocean plate was flexed down into the trench, the crust was tensionally faulted into horst and graben that persist beneath the trench until hidden from view under the trench landward slope.

The trench floor is distinguished from the trench seaward slope by the ponded turbidites. The ponded sediment contains a microfauna less than 400 000 yr old, attesting to the youth of the present trench floor. The youthful sediment together with active seismicity and continental volcanism, are strong arguments for rapid convergence of oceanic and continental crust. Nonetheless, the failure to find any increase in compressional structure or even initial tectonic consolidation of trench fill and underlying oceanic sediment at the foot of the trench landward slope is puzzling.

On the trench landward slope, a cover of lower Miocene to Present sediment drapes the underlying rocks which are generally obscured in seismic reflection records. Few DSDP holes in any convergent margin have penetrated deeply into the zone of obscured reflections so commonly observed in convergent margins. In seismic reflection records, the upper layer consists of reflections paralleling the seafloor and the obscured zone occasionally reveals landward-dipping reflections (see for instance, Moore, Watkins *et al.* 1979). The upper layer is presumed to be slope deposits. These reveal more small-scale deformation at depths of about 200 m than was seen in the sediment of the trench at Site 500 almost at the trench landward slope.

At Site 494, only 3 km landward from the trench floor, we penetrated the slope deposits and found rock that is surprisingly old and overconsolidated. The sedimentary section rock here was first deposited in water shallower than the CCD and then at depths at or just below the CCD. The section has distal continental affinities, that is to say, it received terrigeneous sediment but probably not trench slope deposits. From the shipboard studies we are unable to constrain the original environments of deposition much more than to infer that they are of a base of slope or open ocean provenance near a terrigeneous source. The Miocene section at Site 494 originated seaward of those deposits recovered from the mid-slope at sites 496 and 497. The 494 section was deposited close enough to land or entirely at latitudes north of the carbonate zone of high productivity not to be inundated by carbonate ooze.

Tectonic interpretation

An interpretation of our shipboard data in the context of previous geological studies is difficult in particular because the data from Site 494 do not fit the commonly applied steady-state model of tectonic accretion (Fig. 4); nor do any of the less often invoked models seem to apply without much modification. Three end-member models or explanations, all of which may apply in part, include the slumping of a large block, subduction of sediment without accretion, or tectonic erosion.

A slump that displaced the large bench on which Site 494 was drilled would have originated in an environment seaward of the shelf edge (Seely 1979) and of Sites 496 and 497 based on the slope sediment facies. Since there are no slump scars as large as the proposed slump block, and since the shallowed unconformity is of early Pliocene age, early Pliocene or previous slumping seems to be required. Although a slump block would alleviate the problematical absence of compressional structure at the foot of the trench slope and provide a means of transporting an older accreted section from upslope to the present trench floor, slumping does not explain the absence of accretion in the Pliocene and Quaternary. During the minimum period of accretion allowable by slumping 4 Myr ago, 280 km of ocean crust has been subducted and the totally dewatered volume of oceanic sediment on 280 km of ocean crust greatly exceeds the maximum space available for undrilled accreted material. Thus slump explanation also requires considerable subduction of sediment.

The concept of sediment subduction implies disposal of a great amount of sediment somewhere down the Benioff zone since the present convergence began (early Miocene). Perhaps the rather surprising section at Site 494 is part of an elevated block brought in on the Cocos plate from the west. The block would have arrived in the pre-middle Miocene based on the oldest slope deposits at Site 498 and the youngest anomalous deposits at Site 494. There is no evidence in the post-middle

Miocene slope deposits at Sites 494 and 498 for collision between a crustal fragment and the Caribbean plate. Thus the total sediment subducted must involve all of the post-middle Miocene sediment on the oceanic crust that was subducted. Although possible, such a history raises the question of mechanism and the ultimate disposition of so much sediment.

Tectonic erosion is a third concept that might be applied to explain the drill data. Since the vector of plate convergence is normal to the trench axis little strike-slip rifting seems possible and tectonic erosion would require subduction of the continental framework (previously accreted rock) a difficult process to conceptualize. However, tectonic erosion is an efficient way to truncate the base of the slope despite the conceptual problems of stuffing rock of less density beneath those of greater density or the abrasion of the continental framework on a massive scale.

These explanations are briefly discussed here to provide some focus for subsequent study, despite their problematical aspects.

References

HAYS, J. D., COOK, H. *et al.* 1972. *Initial Rep. Deep Sea drill. Proj.* **9,** U.S. Govt Printing Office, Washington. 1205 pp.

IBRAHIM, A. K., LATHAM, G. V. & LADD, J. 1979. Seismic refraction and reflection measurements in the Middle America Trench offshore Guatemala. *J. geophys. Res.* **84,** 5643–9.

JORDAN, T. H. 1975. The present day motions of the Caribbean Plate. *J. geophys. Res.* **80,** 4433–9.

LADD, J. W., IBRAHIM, A. K., MCMILLEN, K. J., LATHAM, G. V., VON HUENE, R. E., WATKINS, J. S., MOORE, J. C. & WORZEL, J. L. 1978. Tectonics of the Middle America Trench offshore Guatemala. *In: Int. Symp. Guatemala February 4 Earthquake and Reconstruction Process,* Guatemala City, May 1978.

MCMILLEN, K. J., EUKEBOLL, R. H., MOORE, J. C., SHIPLEY, T. H. & LADD, J. W. 1981. Sedimentation in different tectonic environments of the Middle America Trench, southern Mexico and Guatemala (this volume).

MINSTER, J. B. & JORDAN, T. H. 1978. Present-day plate motions. *J. geophys. Res.* **83,** 5331–4.

MOLNAR, P. & SYKES, L. R. 1969. Tectonics of the Caribbean and Middle America regions from focal mechanisms and seismicity. *Bull. geol. Soc. Am.* **80,** 1639–84

MOORE, J. C., WATKINS, J. S., BACHMAN, S. B., BEGHTEL, F. W., BUTT, A., DIDYK, B. M., LEGGETT, J. K., LUNDBERG, N., MCMILLEN, K. J., NIITSUMA, N., SHEPARD, L. E., SHIPLEY, T. H., STEPHAN, J. F. & STRADNER, H. 1979. The Middle America Trench off Mexico. *Geotimes,* **24,** 20–2.

—— , —— , MCMILLEN, K. J., BACHMAN, S. B., LEGGETT, J. K., LUNDBERG, N., SHIPLEY, T. H., STEPHAN, J.-F., BEGHTEL, F. W., BUTT, A., DIDYK, B. M., NIITSUMA, N., SHEPARD, L. E. & STRADNER, H. 1981. Facies belts of the Middle America Trench and forearc region, southern Mexico: results from Leg 66 DSDP (this volume).

PIPER, D. J. W., VON HUENE, R. E. & DUNCAN, J. R. 1973. Late Quaternary sedimentation in the active Eastern Aleutian Trench. *Geology,* **1,** 19–22.

RENARD, V., AUBOUIN, J., LONSDALE, P. & STEPHAN, J. F. 1980. Premiers resultats d'une etude de la fosse d'Amerique centrale au sondeur multifaisceaqux (Seabeam). *C. r. Seances Acad. Sci. Paris,* **291,** 137–42.

SEELY, D. R. 1979. Geophysical investigations of continental slopes and rises. *In:* WATKINS, J. S., MONTADERT, L. & DICKERSON, P. W. (eds). *Geological and Geophysical Investigations of Continental Margins.* Mem. Am. Assoc. Petrol. Geol. **29,** 245–60.

SEELY, D. R., VAIL, P. R. & WALTON, G. G. 1974. Trench slope model. *In:* BURK, C. A. & DRAKE, C. L. (eds). *Geology of Continental Margins,* 261–83. Springer-Verlag, New York.

VAN ANDEL, T. H., HEATH, G. R. *et al.* 1973. *Init. Rep. Deep Sea drill. Proj.* **16,** 949 pp. U.S. Govt Printing Office, Washington.

VON HUENE, R. E. 1974. Modern trench sediments. *In:* BURK, C. A. & DRAKE, C. L. (eds). *The Geology of Continental Margins,* 207–11. Springer-Verlag, New York.

VON HUENE, R. E. & AUBOUIN, J. 1981. Leg 67: The Deep Sea Drilling Project Mid-America Trench transect off Guatemala. *Bull. geol. Soc. Am.* in press.

ROLAND VON HUENE, U.S. Geological Survey, Menlo Park, California 94025, U.S.A.

JEAN AUBOUIN, Département de Geologie Structurale, Université Pierre et Marie Curie, 4, Place Jussieu, Paris 75230, France.

JACQUES AZEMA, Département de Geologie Structurale, Université Pierre et Marie Curie, 4 Place Jussieu, Paris 75230, France.

GRANT BLACKINTON, Department of Geology and Geophysics, University of Hawaii, Honolulu, Hawaii 96822, U.S.A.

Jerry A. Carter, Department of Geology and Geophysics, University of Hawaii, Honolulu, Hawaii 96822, U.S.A.

William T. Coulbourn, Scripps Institution of Oceanography, La Jolla, California 92093, U.S.A.

Darrel S. Cowan, Department of Geological Sciences, University of Washington, Seattle, Washington 98195, U.S.A.

Joseph A. Curiale, School of Geology and Geophysics, University of Oklahoma, Norman, Oklahoma 73019, U.S.A.

Carlos A. Dengo, Center for Tectonophysics, Texas A & M University, College Station, Texas 77843, U.S.A. and ICAITI, Guatemala.

Richard W. Faas, Department of Geology, Lafayette College, Easton, Pennsylvania 18042, U.S.A.

William Harrison, School of Geology and Geophysics, University of Oklahoma Norman, Oklahoma 73019, U.S.A.

Reinhard Hesse, McGill University, Department of Geological Sciences, Montreal M3A 2A7, Canada, and Technische Universität, München, Federal Republic of Germany.

Donald M. Hussong, Department of Geology and Geophysics, University of Hawaii, Honolulu, Hawaii 96822, U.S.A.

John W. Ladd, Lamont-Doherty Geological Observatory, Palisades, New York 10964, U.S.A., and Marine Science Institute, University of Texas, Galveston, Texas 77550, U.S.A.

Nikita Muzylov, Geological Institute of USSR Academy of Sciences, Moscow, U.S.S.R.

Tsunemasa Shiki, Department of Geology and Mineralogy, Kyoto University, Kyoto 606, Japan.

Peter R. Thompson, Lamont-Doherty Geology Observatory, Palisades, New York 10964, U.S.A.

Jean Westberg, Scripps Institution of Oceanography, La Jolla, California 92093, U.S.A.

Evolution of the slope landward of the Middle America Trench, Nicoya Peninsula, Costa Rica

Neil Lundberg

SUMMARY: The Nicoya Peninsula of Costa Rica represents an uplifted portion of the trench slope break landward of the Middle America Trench, and is composed of the Nicoya Complex and its sedimentary cover. The Nicoya Complex is upper Mesozoic oceanic crust, showing the effects of continued igneous activity and deformation probably due to the Late Cretaceous initiation of subduction in an oceanic region. The sedimentary cover reflects uplift and progressive deformation throughout the evolution of this intra-oceanic arc-trench system.

Sediments of the Nicoya Complex are open-ocean deposits, comprising locally derived sedimentary breccias made up of recycled Nicoya Complex material as well as more typically pelagic radiolarian cherts, black shales and deep-water limestones. The basal unit of the sedimentary cover is the mainly hemipelagic Campanian Sabana Grande Formation, which was probably deposited on the juvenile trench slope. In the Sabana Grande Formation, basal siliceous mudstones give way up-section to the foraminifer-rich calcareous mudstones, representing the passage of the sediment surface up relative to the CCD. Interbedded conglomerates and sedimentary breccias reflect recycling of Nicoya Complex material, apparently by uplift and erosion of the forearc basement. The Sabana Grande Formation is overlain by the ?Campanian to Palaeocene Rivas and Las Palmas Formations which consist of thick, mainly volcanogenic, thin- to thick-bedded turbidites and massive and/or pebbly sandstones characteristic of mid-fan facies associations. These rocks were probably deposited in a forearc basin situated landward of a structural high. In contrast, Palaeocene turbidites which crop out to the SW and seaward of the proposed high are thin-bedded, associated with redeposited hemipelagic mudstones, and interpreted as trench slope deposits. This thin-bedded turbidite/hemipelagic mudstone unit is in turn overlain by Palaeocene or Eocene calcareous mudstones and Eocene siliceous mudstones, both of which are interpreted to represent continued deposition on the trench slope, respectively above and below the CCD. These ?Palaeocene and Eocene mudstones are overlain unconformably by Eocene to ?Quaternary shallow-water clastic and carbonate deposits, reflecting rapid Eocene tectonic uplift. The general trend of shallowing in depositional environments through time is paralleled by an overall decrease in intensity of structural deformation up-section, indicating progressive deformation in this forearc terrane.

The Middle America Trench is becoming one of the world's best known trenches, following two recent Deep Sea Drilling transects, yet the processes operating during the development of the inner trench slope remain controversial. The results of IPOD Leg 66 off southern Mexico (Moore *et al.* 1979) indicate progressive incorporation of trench deposits into the lower trench slope, supporting the idea of sediment accretion through some version of the popular imbricate-thrust model (Seely *et al.* 1974; Karig 1974a; Karig & Sharman 1975). However, Leg 67 results off Guatemala (von Huene, Aubouin *et al.* 1980) are not readily explained by a simple imbricate-thrust model. Drilling off Guatemala, as well as the transects off Japan (Scientific Party 1980) and across the Mariana Trench (Hussong, Uyeda *et al.* 1978), suggest little or no sediment accretion accompanying subduction. These drilling results support instead a process of sediment subduction, in which most or all trench and pelagic deposits are subducted to at least shallow levels along with the oceanic lithosphere, which itself may even tectonically erode and subduct upper plate material (Karig 1974b; Scholl *et al.* 1980; Murauchi & Ludwig 1980). These contrasting drilling results suggest considerable variation in processes associated with subduction, even along the strike of a single trench. The critical questions concern the evolution of the trench slope, and the fate both of sediments deposited in this environment and of the subducting oceanic plate with its mantle of pelagic, hemipelagic, and trench deposits.

While deep-sea drilling provides otherwise unavailable data on active subduction zones, the information is limited to about the upper kilometre of sediment, and drilling sites comprise one-dimensional data sets which must be extended laterally by geophysical data. Onland geological studies of presumed uplifted portions of active trench slopes are therefore useful for

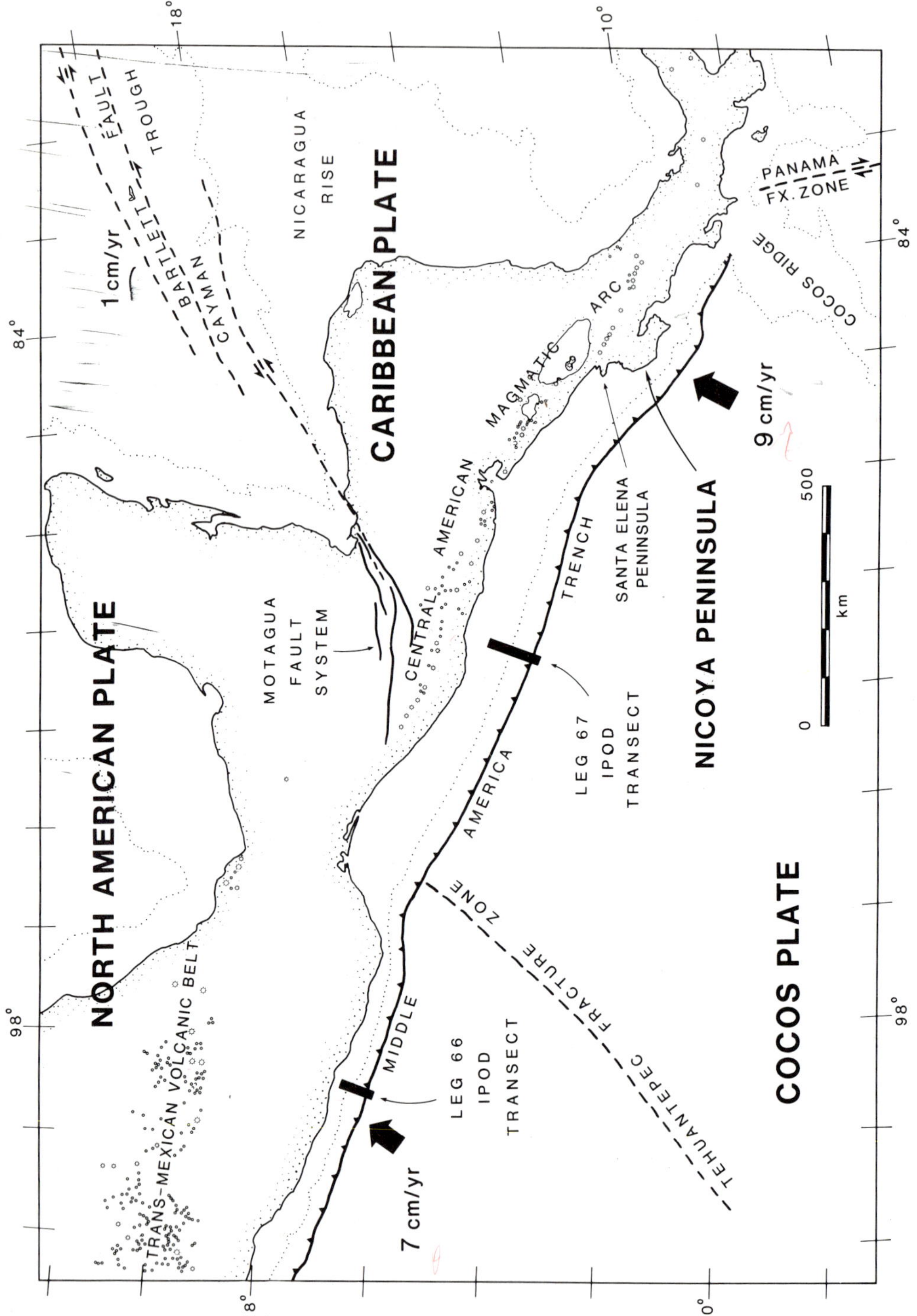

FIG. 1. Tectonic setting of the Nicoya Peninsula, Costa Rica. Starred circles represent volcanoes, dotted line is the 2000-m isobath. After King (1969); present-day plate motions from Minster & Jordan (1978).

comparison, although by their emergent nature these terranes are somewhat removed from the locus of tectonic activity within the lowermost trench slope.

The Nicoya Peninsula of Costa Rica, 600 km SE of the IPOD Leg 67 transect off Guatemala and 300 km NW of the Cocos Ridge (Fig. 1), constitutes an emergent portion of the structural high landward of the Middle America Trench. This paper reports preliminary results of an onland sedimentological and structural study of the Nicoya Peninsula, and the consequent implications for the evolution of the slope landward of the Middle America Trench.

Tectonic setting and geological background

Judging from the record of intermediate-composition volcanic activity, rocks of the Nicoya Peninsula apparently have been near the site of more or less continuous plate convergence since Late Cretaceous times. Thick sequences of Upper Cretaceous and Palaeocene volcaniclastic strata were deposited on the Nicoya Complex beginning in the early Campanian (Fisher & Pessagno 1965; Galli-Olivier & Schmidt-Effing 1977; Galli-Olivier 1979). Eocene to Quaternary volcanic rocks are currently exposed in the Costa Rican magmatic arc (Dengo 1962a; Pichler & Weyl 1973, 1975).

The Cocos Plate is currently being consumed beneath Costa Rica at 9 cm yr^{-1}, along an azimuth of N29E (Minster & Jordan 1978; Fig. 1). The broad shelf landward of the Middle America Trench south of the Tehuantepec Ridge (Fig. 1) comprises a large forearc basin, about 100 km wide and at least 9 km deep, bounded by a prominent structural high which underlies the shelf edge (Seely *et al.* 1974; Ladd *et al.* 1978; Seely 1979). This structural high is similar to the antiformal structure of the Nicoya Peninsula (Dengo 1962a), in which a cover of Upper Cretaceous to ?Quaternary marine sediments are exposed on the landward and seaward flanks of a structural high (Fig. 2). This antiformal high is cored by the basement unit of the Nicoya Peninsula, the ?Upper Jurassic to Upper Cretaceous Nicoya Complex. The Nicoya Complex is thought to represent the basement of all southern Central America (south of Nicaragua) as well, upon which an Upper Cretaceous through Quaternary magmatic arc has been built (Case 1974). Broadly similar terranes extend down the west coast of South America into Ecuador (Pichler *et al.* 1974; Goossens *et al.* 1977), into the Caribbean region (Donnelly 1975), and north along the Pacific offshore of Central America at least as far as Guatemala (Ladd *et al.* 1978).

Nicoya Complex

The Nicoya Complex was originally described as 'an intensely folded rock series consisting of basalt, sedimentary units and intrusives' by Dengo (1962a), who interpreted it as uplifted oceanic crust. Because of the complex history of continued volcanism, sedimentation and deformation in this region, however, more recent studies have adopted a variety of definitions of the Nicoya Complex (see Kuijpers 1980 for a review). In addition, the Nicoya Complex has been shown to be more complicated than a simple piece of oceanic crust.

The Nicoya Complex has been divided into two major units by several workers (De Boer 1979; Kuijpers 1980; Schmidt-Effing *et al.* 1981). The lower unit has generally been described as massive or pillow basalt, with little or no intercalated sediment, overlain by a sequence of siliceous sediments. The age of these sediments is predominantly Early Cretaceous (latest Jurassic (Galli-Olivier 1977) or earliest Cretaceous to Hauterivian or possibly Barremian, according to Schmidt-Effing *et al.* (1981); and Berriasian to Aptian according to Baumgartner, in Kuijpers (1980)). Most workers have agreed that the lower unit of the Nicoya Complex represents the upper level of uplifted oceanic crust, and some (Galli-Olivier 1979; De Boer 1979) have interpreted the serpentinized peridotite of the Santa Elena Peninsula, 30 km north of the Nicoya Peninsula (Fig. 1) as the associated 'layer 3' ultramafic rocks.

The upper unit of the Nicoya Complex is more heterogeneous, however, and its interpretation is disputed. De Boer (1979) has described an upper unit of pillow lava and volcanic agglomerate, intercalated with chert, siliceous limestone, and tuffaceous sediment. He has interpreted this unit, along with gabbros and diorites which intrude the lower unit of the Nicoya Complex, as an early volcanic arc, built on oceanic crust represented by the lower Nicoya Complex. Schmidt-Effing (1979) has subdivided an upper unit, composed of mainly submarine basalt flow material with irregular inclusions of sedimentary rocks, into six 'subcomplexes'. These are defined by differing ages of sedimentary rocks associated with the more prevalent volcanic rocks, both those incorporated as xenoliths and those deposited on volcanic units, providing respectively maximum and minimum ages of volcanism. The minimum age of latest volcanism in these six subcomplexes ranges from Maastrichtian (latest Cretaceous) to Eocene, and the thickness of the upper Nicoya Complex in general is estimated at 2–3 km. Schmidt-Effing (1979) and Schmidt-Effing *et al.* (1981) have interpreted the upper unit as an oceanic plateau, similar to the Ontong–Java Plateau, built on oceanic crust represented by the lower Nicoya Complex. Kuijpers (1980), on the other hand, has described a thin (300+ m) upper unit which consists mainly of mostly basalt and ophitic diabase, commonly including gabbroic rocks and less commonly 'plagiogranite'. He has reported a Cenomanian or early Santonian (late Cretaceous) age for a radiolarite intercalation in the upper part of this unit, which he has interpreted as a nappe of younger oceanic crust which has been thrust over the older oceanic crust of the lower Nicoya Complex. Galli-Olivier (1977, 1979) has also interpreted the Nicoya Complex as structurally complicated

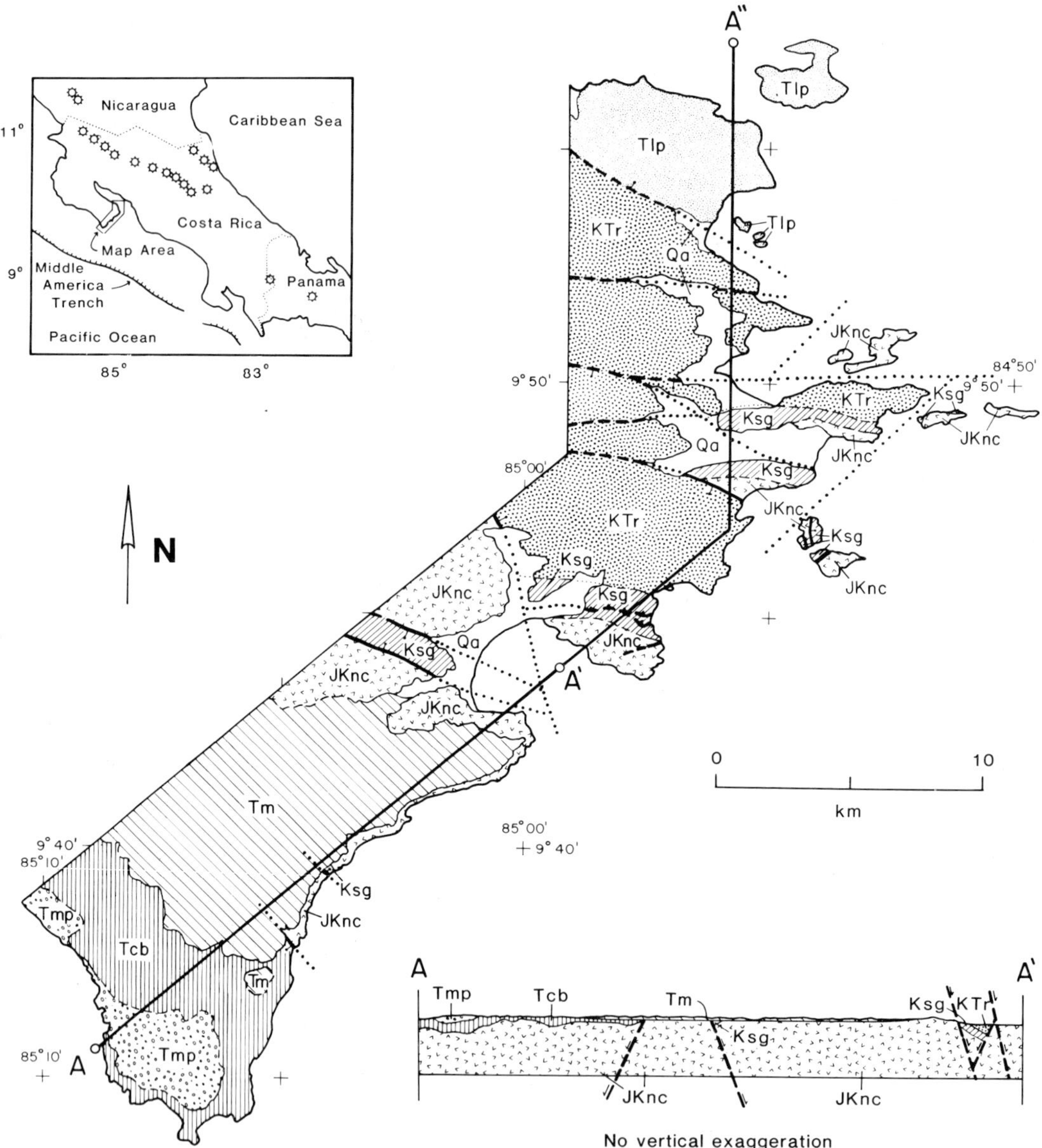

FIG. 2. Generalized geological map and cross-section of the south-eastern portion of the Nicoya

oceanic crust, although he did not recognize major subdivisions within it. He has interpreted the Nicoya Peninsula as an accretionary prism formed of slabs of oceanic crust progressively sliced off a subducting lithospheric plate, implying the presence of imbricated thrust-bounded packages of rock (Galli-Olivier & Schmidt-Effing 1977; Galli-Olivier 1979).

These various interpretations of the upper part of the Nicoya Complex are difficult to evaluate on the basis of field relations alone. Chemical analyses of the volcanic rocks are potentially useful, but unfortunately few analyses are available. Analyses of three samples of Nicoya Complex basalt by Pichler & Weyl (1975) suggest mid-ocean ridge origins, using Pearce & Cann's (1973) empirically-derived Ti versus Zr discrimination technique. More recent compilations of chemical analyses have questioned the validity of this method, however, for ancient (and often altered) volcanic rocks (Garcia 1978; Hill 1979); and moreover, Pichler & Weyl (1975) did not differentiate major units within the Nicoya Complex. Additional chemical analyses are currently in progress, in a study integrating more recent ideas of a multistage evolution of the Nicoya Complex (H. Wildberg, pers. comm. 1980) and may constrain its origin. Interestingly, there is no clear distinction between a lithologically homogeneous lower unit and a more heterogeneous upper unit of the Nicoya Complex in the south-eastern Nicoya Peninsula. In this area at least minor pieces of deeper, older levels of the Nicoya Complex, including radiolarian cherts of late Jurassic or early Cretaceous age (D. Jones, pers. comm. 1980) have been juxtaposed with the more widespread exposures of shallower, younger portions of the Nicoya Complex and of the sedimentary cover. Both younger and older terranes have relatively abundant associated sediments, and both include radiolarian cherts, sedimentary breccias and black shales. Gabbroic rocks have intruded volcanic breccia as well as basalt; both layered and non-layered gabbro are present.

Notwithstanding the controversy over the origin of the upper Nicoya Complex outlined above, the Nicoya Complex represents at least in part oceanic crust of late Jurassic and/or early Cretaceous age. Sediments of the Nicoya Complex document a history of highly variable

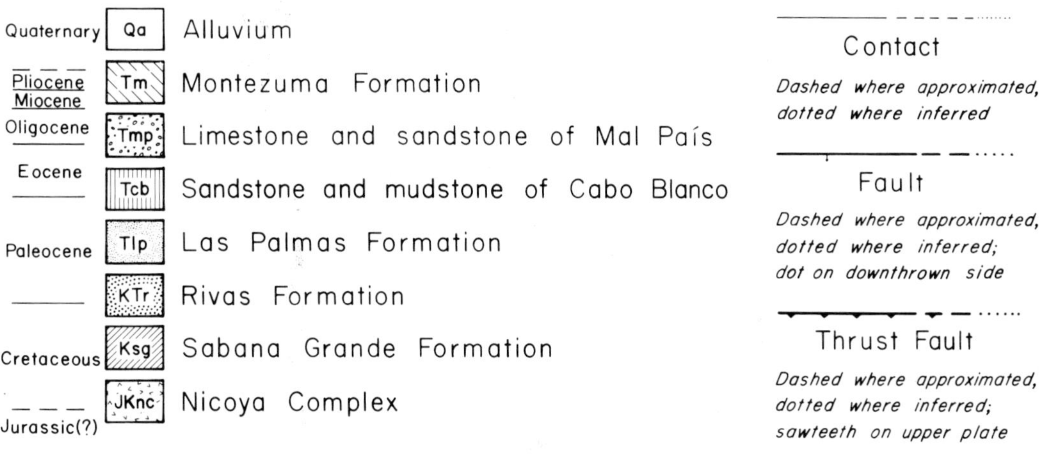

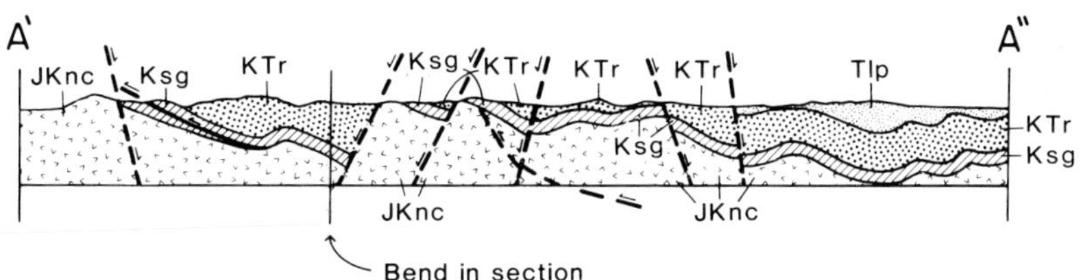

Peninsula. Alluvium and water not shown on cross-section.

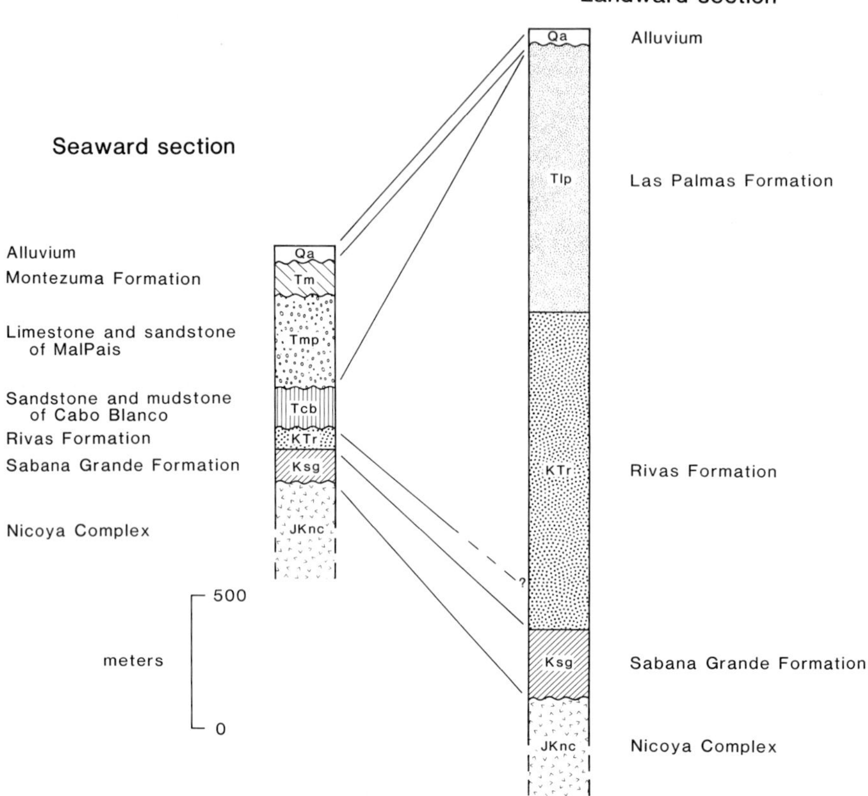

FIG. 3. Stratigraphic columns for seaward (SW) and landward (NE) flanks of antiformal high of the SE Nicoya Peninsula. Symbols refer to legend to Fig. 2.

deep-sea sedimentation, and an analysis of the sedimentary cover offers insights on the subsequent development of the terrane which now forms the structural high landward of the Middle America Trench.

Sedimentology

In order to expand the available data base I have mapped a strip along the SE coast of the Nicoya Peninsula, across the structural strike of the margin (Fig. 2). The study area includes considerable exposures of the sedimentary cover and mainly shallow levels of the Nicoya Complex, apparently due to down-to-the-SE offset along a series of NW-trending faults which cut the Nicoya Peninsula (De Boer 1979). Specifically I use lateral and vertical facies changes and regional unconformities, integrated with major contrasts in structural style, to constrain the evolution of this trench slope. Because of abundant faulting, the reconstructed thicknesses I report I consider to be minimum values. Petrographic characterization of units is based on reconnaissance study; more detailed summaries of sedimentary petrography and of measured sections are currently being compiled.

Nicoya Complex

The sediments of the Nicoya Complex are mainly pelagic deposits and coarse sedimentary breccias, and contain no significant continent-derived quartzo-feldspathic detritus. Pelagic sediments are predominantly layers and pods of radiolarian chert, with minor black shale and deep-water limestone, while the sedimentary breccias represent reworked Nicoya Complex material.

Although the Nicoya Complex includes abundant volcanic breccias and locally tectonic breccias as well, the sedimentary breccias constitute a key lithology. They are typically

FIG. 4. Sedimentary breccia of the Nicoya Complex, including principally light-weathering basalt clasts and dark-weathering chert clasts.

formed of angular to sub-rounded clasts of basalt and lesser amounts of chert, with minor limestone and gabbro (Fig. 4). The small amount of matrix generally comprises sand and gravel apparently made up of broken clasts, although occasionally the clasts are set in chert or (very rarely) limestone. Beds of sedimentary breccia are locally interbedded with argillaceous chert, and typically occur between basalt and the overlying sedimentary units as well as interbedded with basalt at deeper levels in the Nicoya Complex. Clasts are locally derived, as demonstrated by a unique sedimentary breccia which overlies gabbro and is itself composed largely of gabbroic clasts. These breccias reflect considerable relief in a deep-sea environment, which could be related to a spreading ridge, an aseismic rise or seamount, a fracture zone, or a trench. Pelagic sediments and sedimentary breccias are intercalated with basalt on Isla Negritos Fuera, the easternmost island in the study area (Fig. 2), in a section both underlain and overlain by basalt and volcanic breccia, indicating a complex history of volcanism and sedimentation.

Sabana Grande Formation

In several localities sedimentary breccias of the uppermost Nicoya Complex are concordantly overlain by the Sabana Grande Formation, a lithologically heterogeneous unit of early to late Campanian age (Galli-Olivier & Schmidt-Effing 1977; S. Hart & S. Percival, written comm. 1980). The most complete section of the Sabana Grande Formation (Fig. 5) has a minimum thickness of 220 m. The true thickness is probably somewhat more, as abundant faults have apparently removed minor sections. Green radiolarian-rich siliceous mudstone at the base of the Sabana Grande Formation is interbedded upsection with beds of volcanic ash (Fig. 5). Ash beds are commonly 5–10 cm thick, and rarely exceeded 30 cm in thickness. Siliceous mudstone gives way upsection to calcareous mudstone rich in calcareous foraminiferal tests, representing the passage of the sediment surface up through the calcite compensation surface. Volcanic ash beds remain prevalent, and distinctive interbedded conglomerates contain clasts of Nicoya Complex basalt and chert as well as smaller clasts of the underlying siliceous mudstone, in a matrix of foraminiferan-rich calcareous mudstone. The calcareous mudstone is overlain by a 10 m section of red shale and very thin-bedded green volcanogenic turbidites. This thin clastic section grades up-section to a pink to white limestone, indicating a gradual cessation of clastic input. In one locality, however, sedimentary breccias similar to those of the underlying Nicoya Complex are interbedded at the base of the limestone, suggesting continued uplift and erosion of the forearc basement. The limestone is up to 80 m thick and is very widespread in the southeastern Nicoya Peninsula, marking the top of the Sabana Grande Formation. The limestone is late Campanian in age (Schmidt-Effing 1979), medium to thin-bedded, and composed largely of planktonic calcareous foraminiferan tests, coccoliths, and fine-grained volcanogenic detritus. This white, pelagic limestone grades upsection through pink argillaceous limestone to shale and interbedded thin turbidites of the overlying Rivas Formation, reflecting a gradual re-introduction of clastic detritus to this area.

Rivas Formation

The Rivas Formation is composed of clastic turbidites which range in age from late Campanian or early Maastrichtian (Late Cretaceous) to Palaeocene (Galli-Olivier & Schmidt-Effing 1977; S. Hart & S. Percival, written comm.

1980). Reconstruction of numerous fault blocks along the best exposed and most complete section of the Rivas Formation suggests a thickness of 1200+ m (Fig. 6). In general this unit begins with a shale-rich section with thin-bedded turbidites, overlain by more sand-rich sequences of thick-bedded turbidites. These are overlain in turn by thick massive sandstones which are commonly amalgamated and infrequently pebbly, interbedded with shale and very thin-bedded turbidites. The thin turbidite beds at the base of the Rivas Formation are generally less than 30 cm thick with sharp bases and gradational tops. They do not form thinning or thickening megasequences, and correspond to Walker's (1978) basin plain and lower fan facies associations, probably representing distal turbidites (Nilsen 1980). The overlying thick-bedded turbidites are generally 10–120 cm thick, representing discrete depositional events, and in many cases form cycles of thickening- and coarsening-upward sandstone beds. They probably represent progradational processes similar to those which form depositional lobes in mid-fan environments of submarine fans (Walker 1978, 1980). Palaeocurrent data (Fig. 6) and a preliminary reconstruction of basin geometry, however, suggest deposition by axial flow down an elongate basin sub-parallel to the present-day trench, suggesting that a fan geometry is probably not applicable to this unit. However, the identification of fan facies associations remains a useful exercise in order to characterize turbidite deposits (Nilsen 1980), and comparison with fan models allows an initial interpretation of large-scale changes in depositional processes.

The massive sandstones which overlie the thick-bedded turbidites are up to 12 m thick and generally show evidence of amalgamation of sands deposited by more than one turbidite event. These sandstones and the interbedded shale and thin-bedded turbidites probably represent channel-fill and interchannel deposits

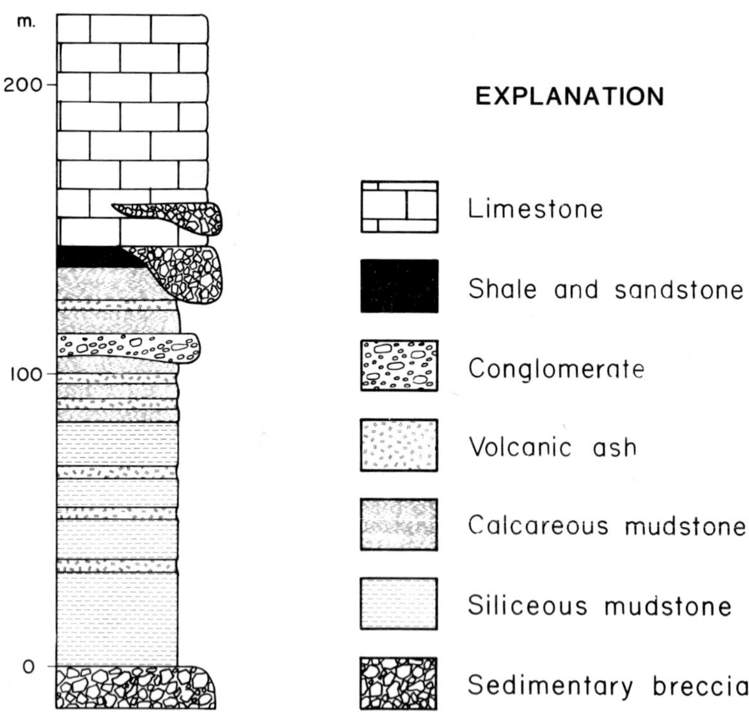

FIG. 5. Generalized section of the Sabana Grande Formation. Uppermost beds of the underlying Nicoya Complex are shown to emphasize the concordant depositional contact and the similarity between sedimentary breccias interbedded in Sabana Grande limestone and the Nicoya Complex. Bed thicknesses not to scale.

respectively (Walker 1978). Further upsection in the landward portion of the study area this unit commonly exhibits thinning- and fining-upward cycles, in which massive sandstones give way gradually to thin-bedded turbidites and shale, probably representing episodes of channel filling and abandonment (Walker & Mutti 1973; Walker 1978). The massive sandstones with interbedded shale and thin beds are more similar to mid-fan than upper fan channel deposits, in that they generally lack truly coarse material but clearly include thinning- and fining-upward megasequences (Nilsen 1980).

Although upper fan facies associations were not identified in the study area, they are present in Rivas Formation exposures along the west coast of the Nicoya Peninsula, NW of the study area. These exposures are located seaward of the antiformal high of the Nicoya Peninsula, and the rocks dip predominantly seaward. The Rivas Formation beds in this area are composed of thick (up to 30+ m), massive, amalgamated sandstones with clear evidence of channelling, overlain by coarse conglomerates interbedded with abundant shale and thin-bedded turbidites. These constitute an upper fan facies association, with the coarse material representing fill of feeder channels and the fine-grained beds representing channel-levee deposits (Walker 1978). Large, well-rounded extraformational clasts in the conglomerates (including basalt clasts up to 2 m across) suggest that the coarse detritus is fed by a major canyon which cut through any bathymetric highs present landward of this area.

Turbidites of the Rivas Formation thus form an overall progradational sequence, with channelled sandy deposits overlying prograding lobe deposits (Walker 1980). These overlie thin-bedded, probably distal turbidites similar to those of lower fan and basin plain environments, which in turn overlie hemipelagic to pelagic deposits of the Sabana Grande Formation. The Rivas Formation is volcanogenic,

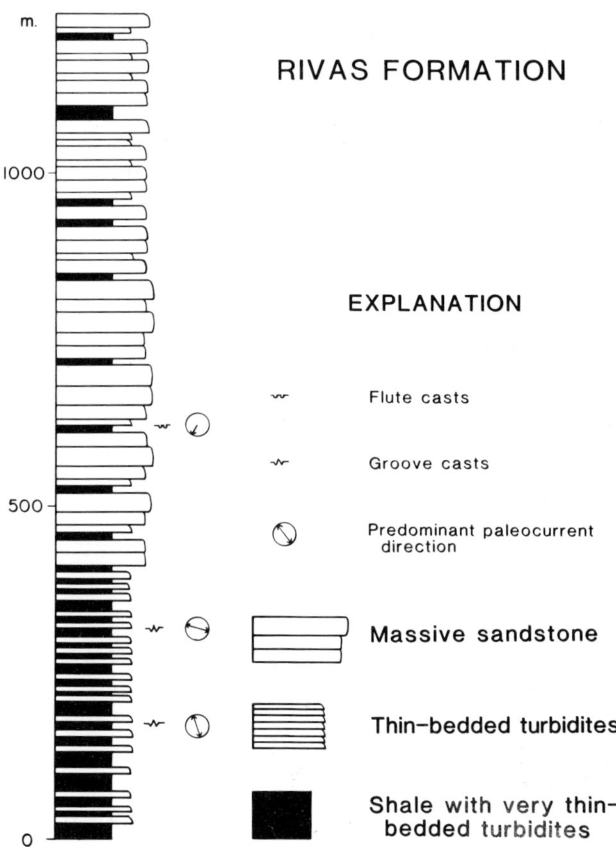

FIG. 6. Generalized section of the Rivas Formation. Individual bed thicknesses not to scale, though sand/shale ratios are representative.

with both intermediate and mafic volcanic detritus, and reflects substantial erosion of the volcanic arc which was presumably located to the NE (present coordinates).

Las Palmas Formation

The Rivas Formation is overlain by the Las Palmas Formation, a thick accumulation of Palaeocene to possibly lower Eocene turbidites (Fig. 7). These turbidites are rich in carbonate lithic fragments as well as volcanogenic detritus, probably reflecting reworking of shallow-water limestones as old as Palaeocene which are exposed NW of the study area (Dengo 1962b; Schmidt-Effing 1979). The basal contact of the Las Palmas Formation with the Rivas Formation is gradational, as minor carbonate detritus is present in the upper Rivas Formation. The Las Palmas Formation is over 1000 m thick at the landward end of the study area, and composes a second turbidite wedge which prograded over the generally similar Rivas Formation. The lower portion of the Las Palmas Formation is dominated by thin- to thick-bedded turbidites which locally form thickening- and coarsening-upward cycles, commonly ending in thick, massive sandstones. The upper Las Palmas Formation is composed of mainly thick, massive, and amalgamated sandstones, interbedded with shale and thin-bedded turbidites. Thus prograding depositional-lobe features are overlain by channel-system facies, as in the Rivas Formation, although fan facies associations are less well-defined and the Las Palmas Formation lacks deposits corresponding to basin plain and lower fan environments. Again, however, a fan geometry may not be applicable to this unit.

Sandstone and mudstone of Cabo Blanco

In the seaward portion of the study area, the thick turbidite sequences of the Rivas and Las Palmas Formations are not present. Instead, thin (200–300 m) sequences of thin-bedded turbidites overlie the Nicoya Complex and Sabana Grande Formation with probable angular unconformity, although vestiges of the lowermost Rivas Formation are preserved

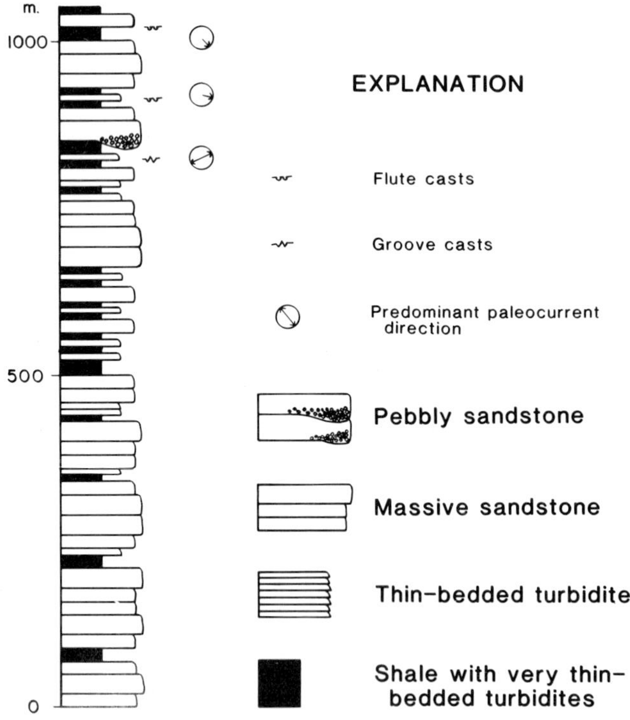

Fig. 7. Generalized section of the Las Palmas Formation. Bed thicknesses not to scale, though sand/shale ratios are representative.

locally (Fig. 3; these remnants are too small to be portrayed on the geological map in Fig. 2). The thin-bedded turbidites have yielded a Palaeocene assemblage of nannoplankton (S. Hart & S. Percival, written comm. 1980) and so are correlative with the Rivas and/or Las Palmas Formations. These turbidites are generally 5–15 cm thick, rarely exceeding 30 cm in thickness, and the intercalated fine-grained sediment is commonly rich in biogenic material. The absence of thick Palaeocene turbidites corresponding to down-current facies of the channel systems found landward strongly suggests the presence of a bathymetric high which shielded this area from the bulk of the clastic detritus. At the seaward edge of the study area these turbidites are overlain by at least 150 m of radiolarian-rich siliceous mudstone, and along the SE coast extending closer to the centre of the study area they are underlain and overlain by at least 50 and 125 m, respectively, of foraminiferan-rich calcareous mudstone. The turbidites and mudstones together make up the unit informally called the sandstone and mudstone of Cabo Blanco (Fig. 8). The mudstones are uniformly thin- to medium-bedded and individual beds are subtly graded. They probably represent locally-derived hemipelagic material redeposited by minor turbidity currents. The siliceous and calcareous mudstones are very similar except for the virtual absence of calcareous foraminiferan tests in the former. The siliceous mudstone is of lower or middle Eocene age (R. Schmidt-Effing, pers. comm. 1980) and was probably deposited below the CCD. The calcareous mudstone is of Palaeocene or Eocene age (R. Schmidt-Effing, pers. comm. 1980) and was deposited above the CCD. Although broadly similar in age as well as appearance, the exact age relationship between these two units must await more detailed dating.

I interpret the hemipelagic mudstones of this unit as slope deposits, and the thin-bedded turbidites as the basin fill of small slope basins. These slope basins were probably fed by minor canyons which incised the landward high.

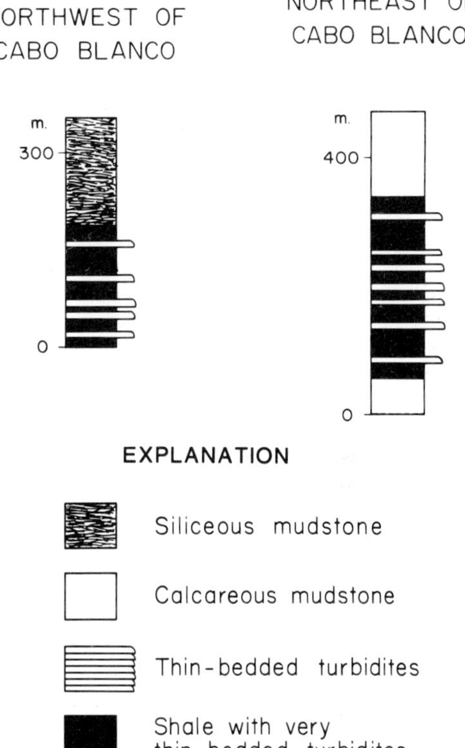

FIG. 8. Generalized sections of the sandstone and mudstone of Cabo Blanco (southernmost point of study area; see Fig. 2). Bed thicknesses not to scale, though sand/shale ratios are representative.

Limestone and sandstone of Mal País

In the seaward portion of the study area the generally deep-water deposits of the Nicoya Complex, the Sabana Grande and Rivas Formations, and the sandstone and mudstone of Cabo Blanco are overlain by shallow-water sediments. These range in age from Eocene to probably Quaternary, and generally overlie the older section with angular unconformity. The shallow-water deposits form two distinct units, the limestone and sandstone of Mal País and the Montezuma Formation (Fig. 3). The limestone and sandstone of Mal País is only exposed at the seaward (SW) end of the study area, and includes late Oligocene sandstones reported by Dengo (1962b). A distinctive white limestone of highly variable thickness (up to several tens of metres) is present at the base of this unit in several localities. The limestone is made up of larger foraminiferans and is Eocene in age (R. Schmidt-Effing, pers. comm. 1980). The remainder of this unit is mainly sandstone, the lower portion of which is fine- to medium-grained and unstratified. The lack of stratification is attributed to complete reworking of the sediment by bioturbation. Upsection the sandstone becomes gradually coarser grained and stratified, with planar-stratified sandstone giving way in turn to moderate-angle cross-stratified sandstone interbedded with conglomerate. The upper portion apparently represents a higher-energy environment than does the lower portion, but certainly the abundant larger foraminiferans and the prevalence of moderate-angle cross-stratification both indicate a shallow-water origin (Bandy 1964; Walker 1979).

Montezuma Formation

The Montezuma Formation ranges in age from Miocene (Dengo 1962a) to probably Quaternary (S. Hart & S. Percival, written comm. 1980) and unconformably overlies the Nicoya Complex, the Sabana Grande and Rivas Formations, and the sandstone and mudstone of Cabo Blanco. The Montezuma Formation reaches a maximum thickness of 120 m along the sea-cliffs of the south-eastern coastline of the study area, and thins northward to taper out against a series of hills composed of the underlying Nicoya Complex (Fig. 2). Commonly a 1–3 m thick basal conglomerate of the Montezuma Formation directly overlies basalt of the Nicoya Complex. The conglomerate, composed of sub-rounded to sub-angular basalt and chert clasts in a coarse sandstone matrix, is unstratified, and generally matrix supported. The remainder of the Montezuma Formation is predominantly sandstone. Sandstone in the lower portion is typically medium- to coarse-grained, pebbly in places, planar- or cross-stratified, and is commonly interbedded with pebbly and/or shell-rich horizons. In contrast, sandstone in the upper Montezuma Formation is generally fine-grained, unstratified, devoid of pebbles and shells, and well burrowed.

The lower portion of the Montezuma Formation represents a high energy depositional environment in which material as coarse as medium- to coarse-pebbly sand was deposited as migrating bedforms. The interbedded pebbly and shelly horizons probably represent lag deposits of intermittent storms (Johnson 1978). The lack of stratification in the upper portion is probably due to biological reworking of the sediment, and this massive sandstone represents a quieter, lower energy environment, in which the sedimentation rate was lower than the rate of bioturbation. The Montezuma Formation is clearly a shallow-water deposit, probably a transgressive sequence which formed over a possibly subaerial erosional surface. In this interpretation, the stratified sands with interbedded pebbly and shelly horizons represent deposition in a nearshore environment, and overlying finer-grained bioturbated sands a less energetic, probably offshore environment (Harms *et al.* 1975, chapter 5; Seilacher 1978). Significant tectonic uplift must be invoked to explain the superposition of these shallow-water deposits directly over siliceous mudstones presumably deposited below the CCD, and deep-sea rocks of the Nicoya Complex.

Structural geology

The Nicoya Complex is structurally complex on a large scale, although it is not a mélange terrane. Folding and faulting, including some thrust faulting, has resulted in a juxtaposition of fault blocks which is generally difficult to reconstruct, and stratigraphic sections are difficult to piece together. However, I found no evidence of large-scale nappe emplacement or of the imbricate stacking of thrust-bound slices of ocean floor. The Nicoya Complex may instead represent in originally coherent piece of oceanic crust, of a somewhat atypical nature as outlined earlier. The Nicoya Complex has suffered significant shortening, however, especially prior to deposition of the sedimentary cover.

The overlying sedimentary units have largely not been affected by this complex juxtaposition of fault blocks, although significant thrusting has involved the Sabana Grande Formation and the lower part of the Rivas Formation. The structural style of the lower part of the sedimentary cover, including all units stratigraphically below the shallow-water deposits, is generally characterized by large-scale folding with seaward vergence, with fold patterns becoming tighter in more seaward exposures. Dips range up to steeply overturned in local zones of intense deformation, and faulting is generally dip-slip along moderate to high-angle surfaces. Major Tertiary normal faults separate large landward-dipping slabs floored by the Nicoya Complex landward of the present antiformal high (Fig. 2). The shallow-water deposits are considerably less deformed than the older rocks. The limestone and sandstone of Mal País exhibits broad open folds with bedding dips up to 45–50° and normal faults, while the Montezuma Formation is virtually flat-lying and is affected by only very minor faults.

Evolution of the trench slope

Sediments of the Nicoya Complex reflect predominantly pelagic deposition. The sedimentary breccias intercalated with basalt probably formed at a spreading ridge, an aseismic rise, or a fracture zone. Certainly the Nicoya Complex does not represent a piece of simple oceanic crust, with a straightforward stratigraphy of sediments over ridge-generated basalt, but the field relations seen in the south-eastern portion of the Nicoya Peninsula could be explained in various ways: by a significant ridge jump (e.g. Handschumacher 1976) or by proximity to a ridge-ridge transform, elevating portions of the sea-floor above the CCD and extruding basalt over pelagic deposits, as well as by the formation of an aseismic rise (Schmidt-Effing 1979) or by the initiation of the island arc (De Boer 1979). Although drilling results from the Mariana Trench suggest that arc rocks may underlie most or all of an intra-oceanic forearc region (Hussong, Uyeda *et al.* 1978; Meijer 1980), none of the Nicoya Complex rocks have been shown to be arc related (Pichler & Weyl 1975). De Boer (1979) has also reported younger volcanic rocks interbedded in Palaeogene sediments of the Nicoya Peninsula, and suggested that they represent arc activity in this region. Alternative origins for near-trench magmatism, however, have been suggested for other forearc terranes. Echeverria (1980) has proposed the leaking of basaltic magma through extensional cracks in down-bending oceanic lithosphere, and the intersection of a fracture zone and a trench as two possible origins of gabbroic sills in the Mesozoic Franciscan Complex of California. These gabbros intruded turbidite deposits after their initial incorporation into an accretionary wedge but before their subduction to depths producing blueschist-grade metamorphic assemblages (Mattinson & Echeverria 1980). Reid & Gill (1980) have proposed a ridge-trench encounter to explain the interbedding of basalt and andesite with trench or near-trench turbidites of the Palaeocene Ghost Rocks Formation of Kodiak Island. Thus near-trench volcanism may be related to oceanic heat sources rather than to arc processes, although the hypothesis of a ridge-trench encounter is not supported in the Nicoya Peninsula case by the remaining magnetic anomalies of the Cocos Plate (Handschumacher 1976).

I suggest that much of the deformation of the Nicoya Complex and the formation of the uppermost sedimentary breccias in this unit were due to the Mesozoic initiation of the Middle America Trench in this region (Fig. 9, stage 1). The deformation, and these breccias post-date the bulk of pelagic deposition, and pre-date arc-derived detritus. Additionally these breccias mark a contrast in structural style, between the structurally complex basement and its generally intact sedimentary cover, which overlies these upper breccias concordantly. Because there is no record of a pre-existing arc or continental massif in Central America south of Nicaragua, nor of any other rocks as old as the Nicoya Complex, this subduction zone apparently developed in the open ocean, well removed from continental sediment sources. A major suture has been suggested by De Boer (1979) and Gose *et al.* (1980) between southern Central America (including Costa Rica), with its oceanic basement, and the continental 'Chortis' block of northern Central America. De Boer (1979) has proposed a Late Cretaceous suture marked by the ultramafic rocks of the Santa Elena Peninsula of northern Costa Rica (Fig. 1) on the basis of regional geology and aeromagnetic data, while Gose (1980) has suggested a Tertiary suture north of southern Nicaragua using palaeomagnetic data. The question of a pre-existing arc in southern Central America remains open, however, because the basement of the magmatic arc in Costa Rica is largely covered by Cenozoic volcanic deposits (Dóndoli *et al.* 1968; see also Schmidt-Effing 1979).

Schmidt-Effing (1979) has suggested that the Sabana Grande Formation represents continued deposition in an open ocean environment, and the relative shallowing of the sediment surface is caused by submarine volcanic build-ups which form an aseismic rise represented by the upper part of the Nicoya Complex. This interpretation explains the complex lateral variations in both the Sabana Grande Formation and the Nicoya Complex, but it is difficult to reconcile with the concordant contact between the Sabana Grande Formation and the overlying Rivas Formation, and the deposition of a thin sequence of turbidites just before the upper limestone of the Sabana Grande Formation. This interpretation would also predict the presence of vesicular basalt in the upper part of the Nicoya Complex, especially under the sediments reflecting shallow-water deposition. However, the shallowest age-equivalent deposits on the Nicoya Peninsular, upper Campanian Rudist-bearing reefal carbonates, overlie non-vesicular basalt and coarse-grained gabbro in at least one locality, suggesting tectonic uplift into the photic zone. An alternative interpretation has been suggested by De Boer (1979), in which the upper part of the Nicoya Complex represents the initial island arc built on oceanic crust, and the Sabana Grande Formation may thus represent back-arc basin sediments. Additional geochem-

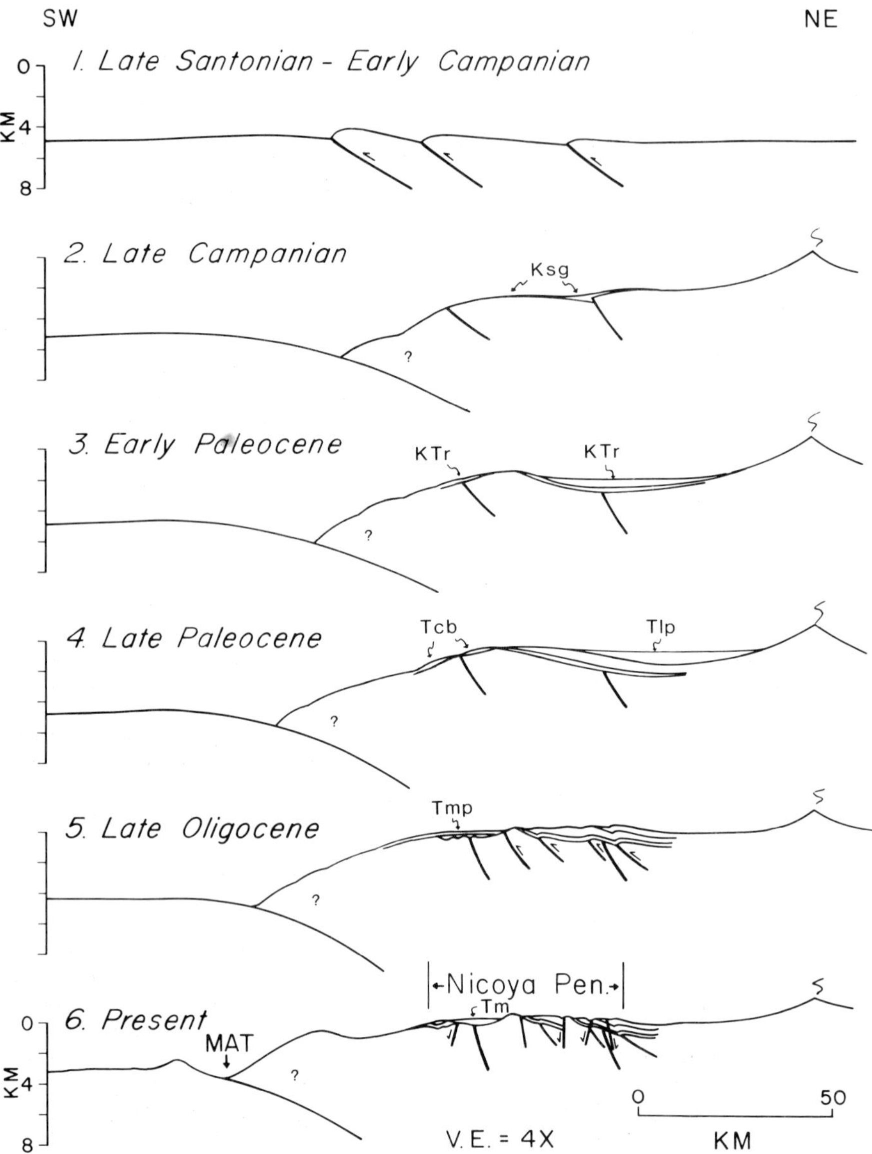

FIG. 9. Proposed model for the evolution of the Nicoya Peninsula. Each diagram represents a time slice recording the development of a major unit (see text for explanation). Symbols for units refer to legend of Fig. 2. Present-day bathymetric profile off Nicoya Peninsula from seismic data courtesy of R. T. Buffler, University of Texas Marine Science Institute.

ical data integrated with detailed mapping of field relations are needed to evaluate these models for the origin of the Nicoya Complex. Nevertheless, I suggest that lateral variations can be expected on a juvenile trench slope as well, and that the Sabana Grande mudstones resemble hemipelagic muds that have been recovered on the present-day slope landwards of the Middle America Trench (Moore *et al.* 1981). Thus the Sabana Grande Formation may simply represent deposition on the juvenile trench slope, with the newly formed island arc, presumably now covered by the Cenozoic volcanic arc, providing the clastic component of the mudstones and turbidites, as well as the ash layers (Fig. 9, stage 2). The passage of the sediment surface up through the CCD recorded in this unit probably represents uplift of the early trench slope, because of the relative scarcity of calcareous sediment below this level and the abundance of them above it. Alternatively this may reflect sediment accumulation through the CCD on a stationary basement, or a secular lowering of the CCD. The recycling of Nicoya Complex and other material to form interbedded conglomerates and sedimentary breccias is probably due to uplift and erosion of the oceanic basement in the forearc region. The widespread, thick, and relatively pure pelagic limestone at the top of this unit may represent uplift of the Nicoya Peninsula terrane in Campanian time, preventing clastic detritus from reaching this area.

The Rivas and Las Palmas Formations probably were deposited in a large forearc basin (Fig. 9, stage 3). The age-equivalent thin-bedded turbidites in the seaward portion of the study area I believe represent deposition in small basins on the trench slope, which must have been shielded from the voluminous clastic detritus of the forearc basin by a structural high in approximately the same position as the present-day anticlinal high of the Nicoya Peninsula (Fig. 9, stage 4). Sparse palaeocurrent data from the thick landward sections of the Rivas and Las Palmas Formations suggest significant longitudinal transport, supporting the existence of a structurally-restricted elongate forearc basin. In addition, the presumed angular unconformity below the seaward Palaeocene beds suggests tectonism and perhaps uplift in the seaward area at this time. I interpret the siliceous and calcareous mudstones which overlie thin-bedded turbidites in the seaward region to represent continued deposition on the trench slope, respectively below and above the CCD, in the absence of significant clastic input from the arc. This shut-off of clastic detritus probably reflects more efficient local shielding by bathymetric highs, as Eocene volcanic and clastic sequences have been reported elsewhere in this portion of the margin (Dengo 1962a; Pichler & Weyl 1973, 1975).

The Cenozoic shallow-water deposits indicate uplift of much of the area to shelf depths during the Eocene, and by this time the older strata had been folded and faulted (Fig. 9, stage 5). Palaeogene shallow-water deposits were broadly folded and normal faulted by Miocene time, and shallow-water to emergent conditions have prevailed through to the present (Fig. 9, stage 6).

Conclusions

The foregoing observations and interpretations suggest a straightforward scenario for the evolution of this portion of the Middle America Trench margin (Fig. 9). The Nicoya Peninsula is an uplifted trench slope break composed of the Nicoya Complex and its sedimentary cover. The Nicoya Complex is upper Mesozoic oceanic crust, upon which younger volcanic rocks of as yet uncertain tectonic affinity were extruded. I suggest that the deformation of the Nicoya Complex was due to the initiation of subduction during the Late Cretaceous in an oceanic region. The sedimentary cover reflects uplift and progressive deformation of the trench slope through the evolution of this intra-oceanic arc-trench system. The schematic diagrams of selected time periods in Fig. 9 depict the formation of the major units in the study area, and the area labelled 'Nicoya Peninsula' in the final diagram shows the lateral extent of onland data on which this interpretation is based. The early stages illustrated in Fig. 9 depict a subduction zone initiating in simple ocean crust, and are based in part on Weissel & Anderson's (1978) model of an initiating subduction zone northeast of the Caroline Plate. These early stages may be especially subject to revision, if for example this present-day forearc region is underlain in part by arc rocks as suggested by De Boer (1979).

Major differences in lithology and timing of events suggest the presence of a variety of terranes with differing geological histories even within the Costa Rican forearc region (Galli-Olivier & Schmidt-Effing 1977; Stibane *et al.* 1977; Schmidt-Effing 1979). Thus caution should be exercised in extrapolating this model and the data on which it is based to the allegedly correlative terranes in Columbia and Ecuador, the Caribbean region, and the Guatemalan trench slope, although it is noteworthy that the sequence of events in the vicinity of the shelf-edge high off Guatemala outlined by Seely (1979) is broadly similar.

ACKNOWLEDGMENTS: Acknowledgment is made to the Donors of the Petroleum Research Fund (PRF 9219-AC2), administered by the American Chemical Society, for the partial support of this research. I thank the Mobil Foundation for supporting much of the field research, and S. Hart and S. Percival of Mobil Exploration and Producing Services Inc. for nannofossil and palynological age determinations. The Instituto Geográfico Nacional of Costa Rica provided significant logistical support, and the Patent Fund of the University of California provided partial travel support. R. Schmidt-Effing kindly provided microfossil age determinations and a preprint and D. Jones provided radiolarian age determinations. I thank C. Rodríguez for enthusiastic and capable assistance in the field, and H. Gursky, H. Wildberg, and M. Streben for stimulating discussions. I am indebted to the family of Sr. Ing. Mario Rodríguez, Chap Porter and Isabel Vásquez de Porter, Fritz Schutt, and many other people of Costa Rica for their generous hospitality. Finally I thank J. C. Moore for his unflagging encouragement and assistance throughout this project.

References

BANDY, O. L. 1964. Foraminiferal biofacies in sediments of Gulf of Batabano, Cuba, and their geologic significance. *Bull. Am. Assoc. Petrol. Geol.* **48,** 1666–79.

CASE, J. E. 1974. Oceanic crust forms basement of eastern Panama. *Bull. geol. Soc. Am.* **85,** 645–52.

DE BOER, J. 1979. The outer arc of the Costa Rican Orogen (oceanic basement complexes of the Nicoya and Santa Elena Peninsulars). *Tectonophysics,* **56,** 221–59.

DENGO, G. 1962a. Tectonic-igneous sequence in Costa Rica. *In:* ENGEL, A. E. J., JAMES, H. L. & LEONARD, B. F. (eds). *Petrologic Studies: a volume to Honor A. P. Buddington,* 133–61. Geol. Soc. Am., Boulder.

—— 1962b. *Estudio Geológico de la Región de Guanacaste, Costa Rica.* Instituto Geográfico Nacional, San José, Costa Rica, 112 pp.

DÓNDOLI, C., DENGO, G. & MALAVASSI, E. 1968. *Mapa geológico de Costa Rica. Scale 1:700,000.* Instituto Geográfico Nacional, San José, Costa Rica.

DONNELLY, T. W. 1975. The geological evolution of the Caribbean and Gulf of Mexico—some critical problems and areas. *In:* NAIRN, A. E. M. & STEHLI, F. G. (eds). *The Ocean Basins and Margins*, vol. III. *The Gulf of Mexico and the Caribbean*, 663–89. Plenum Press, New York.

ECHEVERRIA, L. M. 1980. Oceanic basaltic magmas in accretionary prisms: the Franciscan intrusive gabbros. *Am. J. Sci.* **280,** 697–724.

FISHER, S. P. & PESSAGNO, E. A. 1965. Upper Cretaceous strata of northwestern Panama. *Bull. Am. Assoc. Petrol. Geol.* **49,** 433–44.

GALLI-OLIVIER, C. 1977. Edad de emplazamiento y periodo de acumulación de la ofiolita de Costa Rica. *Cienc. Tech. Univ. Costa Rica*, **1,** 81–6.

—— 1979. Ophiolite and island arc volcanism in Costa Rica. *Bull. geol. Soc. Am.* **90,** 444–52.

—— & Schmidt-Effing, R. 1977. Estratigrafia de la cubierta sedimentaria supra-ofiolítica cretácica de Costa Rica. *Cienc. Tech. Univ. Costa Rica*, **1,** 87–96.

GARCIA, M. O. 1978. Criteria for the identification of ancient volcanic arcs. *Earth Sci. Rev.* **14,** 147–65.

GOOSSENS, P. J., ROSE, W. I. JR. & FLORES, D. 1977. Geochemistry of tholeiites of the Basic Igneous Complex of northwestern South America. *Bull. geol. Soc. Am.* **88,** 1711–20.

GOSE, W. A., SCOTT, G. R. & SWARTZ, D. K. 1980. The aggregation of Mesoamerica: paleomagnetic evidence. *In:* PILGER, R. H. JR. (ed). *The Origin of the Gulf of Mexico and the Early Opening of the Central North Atlantic Ocean.* Proceedings of a symposium at Louisiana State University, Baton Rouge, Louisiana, March 3–5 1980, pp. 51–4.

HANDSCHUMACHER, D. W. 1976. Post-Eocene plate tectonics of the Eastern Pacific. *In:* SUTTON, G. H., MANGHNANI, M. H. & MOBERLY, R. (eds). *The Geophysics of the Pacific Ocean Basin and its Margin: a volume in Honor of George P. Woollard.* Monogr. Am. Geophys. Union, **19,** 177–202.

HARMS, J. C., SOUTHARD, J. B., SPEARING, D. R. & WALKER, R. G. 1975. Depositional environments interpreted from primary sedimentary structures and stratification sequences. Short Course, Soc. Econ. Paleontol. Mineral., Tulsa, **2,** 161 pp.

HILL, M. D., 1979. *Volcanic and plutonic rocks of the Kodiak-Shumagin Shelf, Alaska: subduction deposits and near-trench magmatism.* Thesis, PhD, Univ. California, Santa Cruz, 259 pp. (unpubl.).

HUSSONG, D., UYEDA, S., BLANCHET, R., BLEIL, U., ELLIS, C. H., FRANCIS, T. J. G., FRYER, P., HORAI, K., KLING, S., MEIJER, A., NAKAMURA, K., NATLAND, J., PACKHAM, G. & SHARASKIN, A. 1978. Leg 60 ends in Guam. *Geotimes,* **23,** 19–22.

JOHNSON, H. D. 1978. Shallow siliciclastic seas. *In:* READING, H. G. (ed.). *Sedimentary Environments and Facies*, 207–58. Elsevier, New York.

KARIG, D. E. 1974a. Evolution of arc systems in the western Pacific. *Ann. Rev. Earth planet. Sci.* **2,** 51–75.

—— 1974b. Tectonic erosion at trenches. *Earth planet. Sci. Lett.* **21,** 209–12.

—— & SHARMAN, G. F. III. 1975. Subduction and accretion in trenches. *Bull. geol. Soc. Am.* **86,** 377–89.

KING, P. B. 1969. *Tectonic Map of North America. Scale 1:5,000,000.* U.S. geol. Surv. Washington.

KUIJPERS, E. P. 1980. The geologic history of the Nicoya Ophiolite Complex, Costa Rica, and its geotectonic significance. *Tectonophysics,* **68,** 233–55.

LADD, J. W., IBRAHIM, A. K., MCMILLEN, K. J., LATHAM, G. V., VON HUENE, R. E., WATKINS, J. S., MOORE, J. C. & WORZEL, J. L. 1978. Tectonics of the Middle America Trench offshore Guatemala. *Int. Symp. Guatemala 4 February Earthquake and Reconstruction Process*, **2,** Guatemala City, May 1978.

MATTINSON, J. M. & ECHEVERRIA, L. M. 1980. Ortigalita Peak gabbro, Franciscan Complex: U-Pb dates of intrusion and high-pressure—low-temperature metamorphism. *Geology*, **8,** 589–93.

MEIJER, A. 1980. Primitive arc volcanism and a boninite series: examples from western Pacific island arcs. *In:* HAYES, D. E. (ed). *The Tectonic and Geologic Evolution of Southeast Asian Seas and Islands*. Am. Geophys. Un. Monogr. **23,** 271–82. Washington.

MINSTER, J. B. & JORDAN, T. H. 1978. Present-day plate motions. *J. geophys. Res.* **83B,** 5331–54.

MOORE, J. C., WATKINS, J. S., BACHMAN, S. B., BEGHTEL, F. W., BUTT, A., DIDYK, B. M., LEGGETT, J. K. LUNDBERG, N., MCMILLEN, K. J., NIITSUMA, N., SHEPHARD, L. E., SHIPLEY, T. H., STEPHAN, J. F. & STRADNER, H. 1979. Progressive accretion in the Middle America Trench, southern Mexico. *Nature. London,* **281,** 638–42.

MOORE, J. C. *et al.* 1981. Facies belts of the Middle America Trench and forearc region, southern Mexico: results from Leg 66 DSDP (this volume).

MURAUCHI, S. & LUDWIG, W. J. 1980. Crustal structure of the Japan Trench: the effect of subduction of ocean crust. *In: Scientific party.* Initial Rep. Deep Sea drill. Proj. **56, 57,** 463–9. U.S. Govt Printing Office, Washington.

NILSEN, T. H. 1980. Modern and ancient submarine fans: discussion of papers by R. G. Walker and W. R. Normark. *Bull. Am. Assoc. Petrol. Geol.* **64,** 1094–101.

PEARCE, J. A. & CANN, J. R. 1973. Tectonic setting of basic volcanic rocks determined using trace element analyses. *Earth planet. Sci. Lett.* **19,** 290–300.

PICHLER, H., STIBANE, F. R. & WEYL, R. 1974. Basischer Magmatisms und Krustenbau im sudlichen Mittelamerika, Kolumbein und Ecuador. *Neues Jahrb. Geol. Palaontol. Monatshefte*, **2,** 102–26.

—— & WEYL, R. 1973. Petrochemical aspects of Central American magmatism. *Geol. Rdsch.* **62,** 357–96.

—— & WEYL, R. 1975. Magmatism and crustal evolution in Costa Rica (Central America). *Geol. Rdsch.* **64,** 457–75.

REID, M. & GILL, J. B. 1980. Near trench volcanism, Kodiak Island, Alaska: implications for ridge-trench encounter. *Abstr. Prog. geol. Soc. Am.* **12,** 148–9.

SCHMIDT-EFFING, R. 1979. Alter und Genese des Nicoya-Komplexes, einer ozeanischen Palaokruste (Oberjura bis Eozan) im sudlichen Zentralamerika. *Geol. Rdsch.* **68,** 457–94.

—— , GURSKY, H.-J., STREBIN, M. & WILDBERG, H. 1981. The ophiolites of southern Central America with special reference to the Nicoya Peninsula (Costa Rica). *Trans. 9th Caribbean Geol. Conf.*, Santo Domingo, Dominican Republic.

SCHOLL, D. W., VON HUENE, R., VALLIER, T. L. & HOWELL, D. G. 1980. Sedimentary masses and concepts about tectonic processes at underthrust ocean margins. *Geology*, **8,** 564–8.

SCIENTIFIC PARTY 1980. *Initial Rep. Deep Sea drill. Proj.* **56, 57.** U.S. Govt Printing Office, Washington.

SEELY, D. R. 1979. The evolution of structural highs bordering major forearc basins. *In:* WATKINS, J. S., MONTADERT, L. & DICKERSON, P. W. (eds). *Geological and Geophysical Investigations of Continental Margins*. Mem. Am. Assoc. Petrol. Geol. **29,** 245–60.

—— , VAIL, P. R. & WALTON, G. G. 1974. Trench slope model. *In:* BURK, C. A. & DRAKE, C. L. (eds). *The Geology of Continental Margins*, 249–60. Springer-Verlag, New York.

SEILACHER, A. 1978. Use of trace fossils for recognizing depositional environments. *In:* BASAN, P. B. (ed.). *Trace Fossil Concepts.* Short Course Soc. econ. Palaeontol. Mineral. Tulsa, **5,** 185–201.

STIBANE, F. R., SCHMIDT-EFFING, R. & MADRIGAL, R. 1977. Zur stratigraphisch-tektonischen Entwicklung der Halbinsel Nicoya (Costa Rica) in der Zeit von Ober-Kreide bis Unter-Tertiar. *Giessener geol. Schriften* **12,** 315–58.

VON HUENE, R., AUBOUIN, J., AZEMA, J., BLACKINTON, G., CARTER, J. A., COULBOURN, W. T., COWAN, D. S., CURIALE, J. A., DENGO, C. A., FAAS, R. W., HARRISON, W., HESSE, R., HUSSONG, D. M., LADD, J. W., MUZYLOV, N., SHIKI, T., THOMPSON, P. R. & WESTBERG, J. 1980. Leg 67: The Deep Sea Drilling Project Mid-America Trench transect off Guatemala. *Bull. geol. Soc. Am.* **91,** 421–32.

WALKER, R. G. 1978. Deep-water sandstone facies and ancient submarine fans: models for exploration for stratigraphic traps. *Bull. Am. Assoc. Petrol. Geol.* **62,** 932–66.

—— 1979. Shallow marine sands. *In:* WALKER, R. G. (ed.). *Facies Models*, 75–89. Geol. Assoc. Can., Toronto.

—— 1980. Modern and ancient submarine fans: reply. *Bull. Am. Assoc. Petrol. Geol.* **64,** 1101–8.

—— & MUTTI, E. 1973. Turbidite facies and facies associations. *In:* MIDDLETON, G. V. & BOUMA, A. H. (eds). *Turbidites and Deep-water Sedimentation*, 119–57. Short Course, Pacif. Sect. Soc. econ. Paleontol. Mineral, Tulsa.

WEISSEL, J. K. & ANDERSON, R. N. 1978. Is there a Caroline Plate? *Earth planet. Sci. Lett.* **41,** 143–58.

NEIL LUNDBERG, Earth Sciences Board, University of California, Santa Cruz, California 95064, U.S.A.

SOUTH AMERICA

Cenozoic structure, stratigraphy and tectonics of the central Peru forearc

L. D. Kulm, J. M. Resig, T. M. Thornburg & H.-J. Schrader

SUMMARY: The central Peru forearc (6°–14°S) is characterized by a variety of structural and tectonic styles. Continental shelf basins, Sechura, Salaverry, and East Pisco, are floored by a block faulted Precambrian to Palaeozoic massif and accumulated as much as 3 km of Cenozoic and probably Mesozoic sediment. The massif rises to form an Outer Shelf High (OSH) covered with disrupted Neogene strata. Structural and stratigraphic relationships suggest several pulses of uplift and subsidence for the massif with the most recent uplift detected on the positive massif features.

Upper slope basins are cradled between the OSH and a seaward Upper Slope Ridge (USR) of deformed sediment. Late Eocene and younger clastic sediments, with minor limestone beds, occupy the eastern Trujillo Basin, whereas late Miocene to Pleistocene dolomicrite, micrite and glauconitic micrite are prominent in the western part of the basin. Both the Trujillo and Yaquina Basins (7°–9°30′S) exhibit severe internal disruption caused by compressional stresses. To the south, the Lima Basin (9°30′–13°S) contains similar carbonate lithologies, but this basin and underlying massif have subsided 275 to 500 m Ma^{-1} since Pliocene–Pleistocene time, indicating tensional stresses.

Geological and geophysical data show the subducting plate is rupturing by thrust faulting of layer 2 basalts which may form a sediment-basalt mélange in the overlying subduction complex. Because the seaward boundary of the arc massif is not well-defined, several forearc models are presented which allow the subduction complex to range in width from 15 to 60 km.

The Peru forearc off Lima (12°S) exhibits many of the tectonic characteristics of the Japan forearc off Honshu, especially the subsiding slope basin and contemporaneous accreting subduction complex.

The Andean convergence zone along the west coast of South America is often considered the classical continental arc. It displays all the structural-tectonic elements (subduction zone, arc massif, forearc basins and volcanic arc) that characterize an arc (Seely 1979) as well as the modern-day seismicity and volcanism. The rate of convergence is 10 cm yr^{-1} (Minster *et al.* 1974).

Recent studies show that the Peru continental margin and trench are characterized by a variety of crustal structures, stratigraphies, and tectonic regimes (Johnson & Ness 1981; Thornburg & Kulm 1981; Kulm *et al.* 1977, 1981a,b; Shepherd & Moberly 1981; Schweller *et al.* 1981; Coulbourn & Moberly 1977; Hussong *et al.* 1976; Mordojovich 1974; Scholl *et al.* 1970). Precambrian and Palaeozoic strata crop out in the coastal regions and may, in some areas, occur in close proximity to the Peru–Chile Trench, giving the impression that the margin is truncated. A number of forearc basins of various sizes are distributed over the margin, some resting upon this ancient arc massif and some resting upon a deformed subduction complex farther seaward. The geology of the convergence zone may be further complicated by aseismic ridges and fracture zones which collide with the margin (Nur & Ben Avraham 1981).

This study describes the geology of the Peru forearc and shows the variability and complexity that exist within one of the most extensively studied areas of the forearc extending from 6° to 14°S latitude (Fig. 1). We also contrast and compare the crustal structure and stratigraphy of the major forearc features and relate them to the tectonic framework of the region.

Crustal structure

Basement structures and distribution of basins

Continental margin

Recent geological and geophysical studies indicate that the sedimentary basins on the continental shelf are underlain by Precambrian and Palaeozoic crystalline rock and metasediment (Kulm *et al.* 1977, 1981a,b; Thornburg & Kulm 1981). This crystalline basement beneath the shelf exhibits compressional velocities ranging from 5.7 to 6.1 km s^{-1} and densities from 2.72 to 3.0 g cm^{-3}. Mantle velocities (8.2 km s^{-1}) occur at a depth of 30 km in the Salaverry Basin (Fig. 2; Hussong *et al.* 1976).

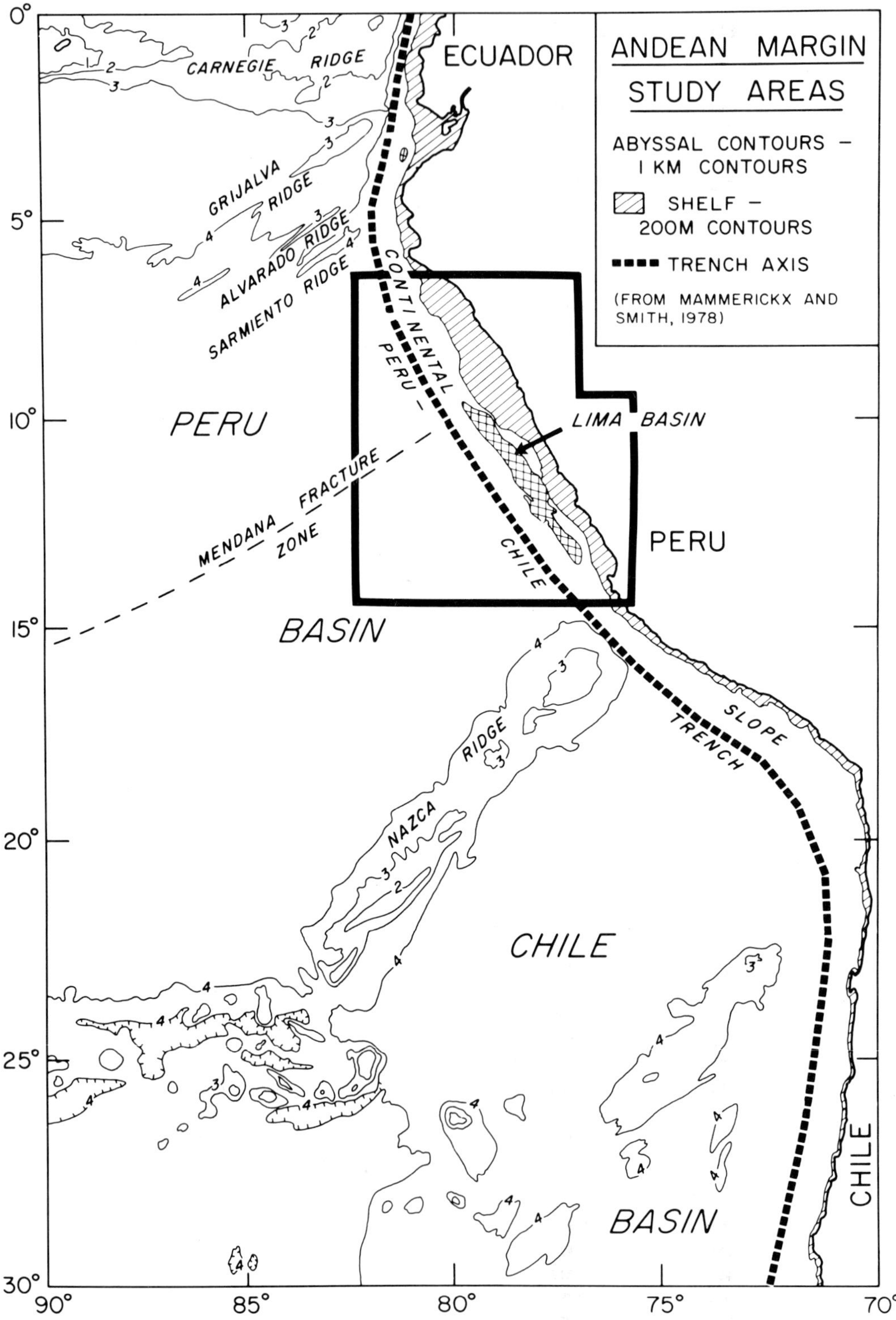
ANDEAN MARGIN STUDY AREAS
ABYSSAL CONTOURS – 1 KM CONTOURS
SHELF – 200M CONTOURS
TRENCH AXIS
(FROM MAMMERICKX AND SMITH, 1978)
ECUADOR
CARNEGIE RIDGE
GRIJALVA RIDGE
ALVARADO RIDGE
SARMIENTO RIDGE
CONTINENTAL
PERU –
PERU
LIMA BASIN
MENDANA FRACTURE ZONE
CHILE
PERU
BASIN
SLOPE
TRENCH
NAZCA RIDGE
CHILE
BASIN
CHILE
0°
5°
10°
15°
20°
25°
30°
90°
85°
80°
75°
70°

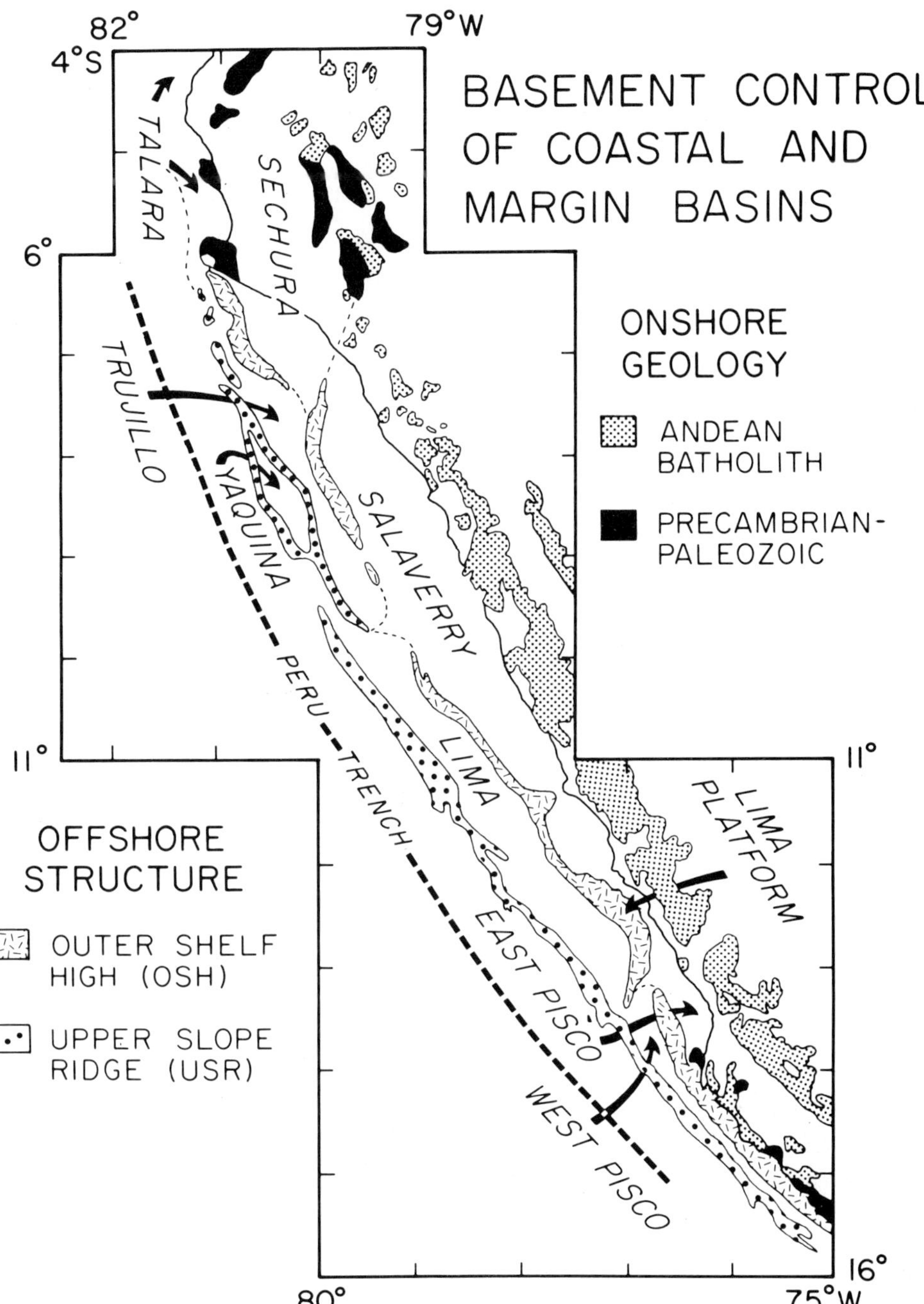

FIG. 2. Geological features onshore and offshore Peru (after Thornburg & Kulm 1981). See text for explanation.

FIG. 1. Location map of Peru continental margin and adjacent Nazca plate. Heavy-lined box is study area.

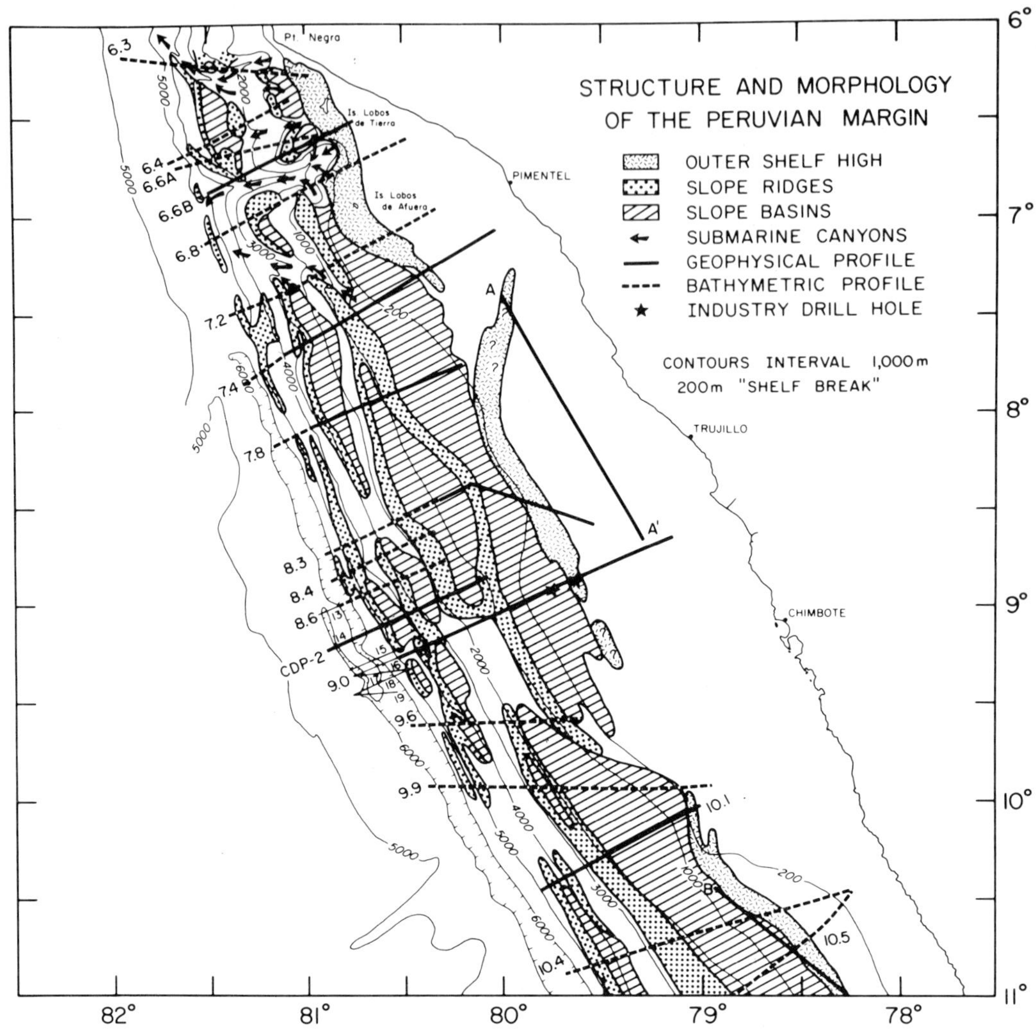

FIG. 3. Crustal structure and morphology of Peru margin. Note petroleum industry drill holes (stars) on line 9.0 (after Thornburg & Kulm 1981). Refer to Fig. 2 for names of features. Numbered profiles are locations of geophysical profiles displayed in following figures.

Structural ridges form an Outer Shelf High (OSH) which is traced along the continental margin and rises above sea-level as mountain ranges in north-western (6°S) and southern (14°S) Peru (Figs 2 & 3). Foliated metamorphic rocks were recovered in drill holes on the OSH at 9°S and from exposures on small islands that occupy the outer shelf. These basement ridges form the boundaries of the continental shelf basins (Thornburg & Kulm 1981). From north to south, these basins are the Sechura, Salaverry and eastern Pisco Basins (Fig. 2).

Structural ridges also form the foundation for the basins on the continental slope (Figs 2 & 3; Thornburg & Kulm 1981). Three upper-slope basins, Trujillo, Lima and West Pisco, lie between OSH and a prominent Upper Slope Ridge (USR) of deformed sediments (Figs 2 & 3). A fourth and smaller basin, the Yaquina Basin, cradles between the USR and another structural ridge to the west. Numerous smaller basins occur within an anastomosing network of structural ridges on the middle and lower continental slope (Fig. 3).

Trench

The Peru–Chile Trench is characterized by

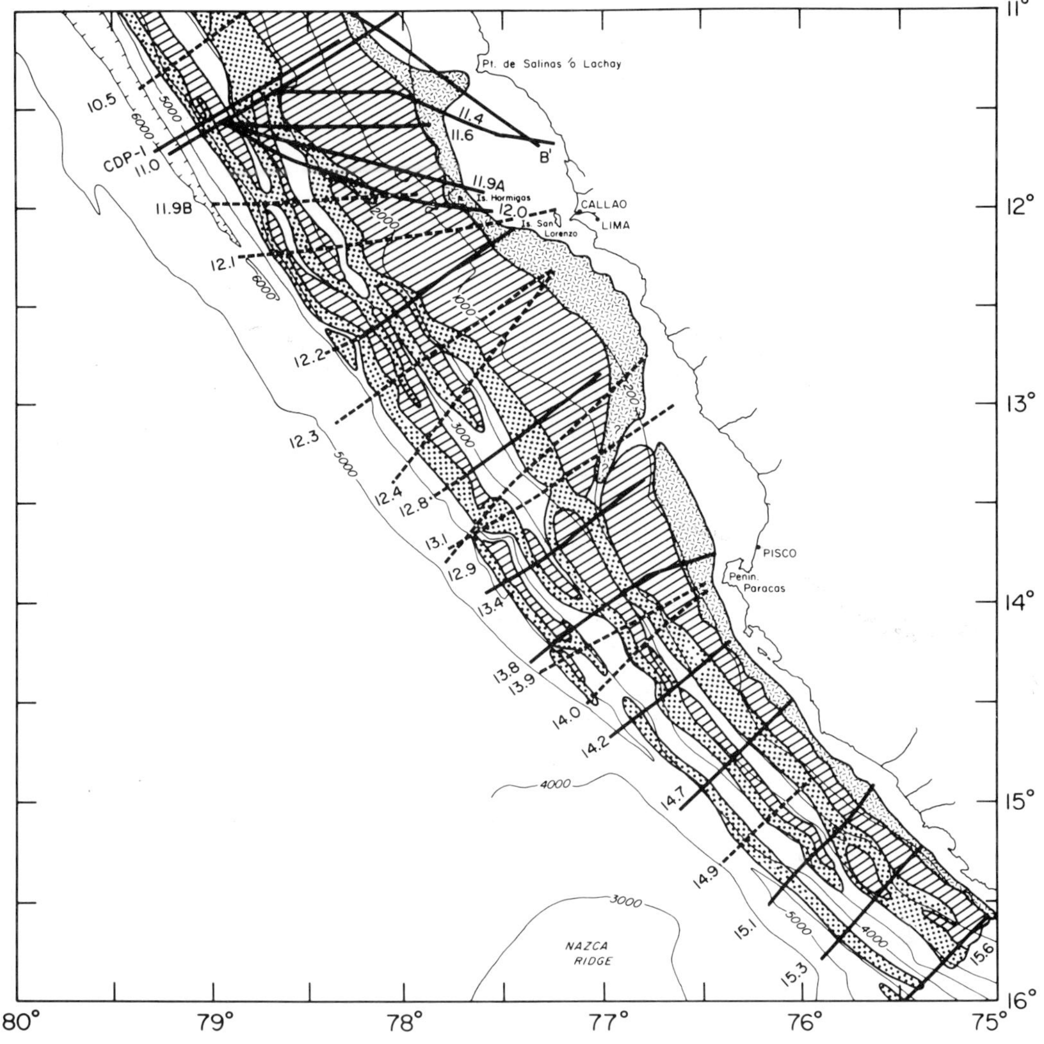

Fig. 3. (continued).

two different systems of faulting on the descending oceanic slab (Schweller *et al.* 1981). Bending of the oceanic plate produces extensional stress and brittle fracture of the basaltic crust, creating step faults and grabens on the seaward trench slope. Compressional stress resulting from plate convergence can be transmitted seaward from beneath the overriding continental plate through the oceanic plate, emerging as thrust faults within the oceanic crust near the trench axis (Fig. 4). Large ridges consisting of upper layer 2 basalts occur between 7°40′S and 9°30′S and signify the presence of the thrust regime in this area of the trench off central Peru (Fig. 5; Schweller *et al.* 1981; Kulm *et al.* 1981a; Prince & Kulm 1975). Axial turbidites are disrupted and occur on the 900 m high ridge, indicating uplift of 4–22 cm yr^{-1} of the oceanic crust along reverse faults inclined from 30° to 60° (Prince & Schweller 1978).

To the south, from 9°30′S to 14°00′S, the compressional regime changes to an extensional regime with step faults predominating in the oceanic crust of the trench (Fig. 6). The step faults are simple offsets with the downward displacement on the landward side of the fault with apparent steep dips landward (Schweller *et al.* 1981).

Comparison of two multichannel depth sections across the two tectonic regimes in the trench at a low vertical exaggeration (Fig. 7), shows the large scale of crustal rupture that characterizes the 60 km long thrust ridge from 8°55′ to 9°28′ (Fig. 5) and the relatively small

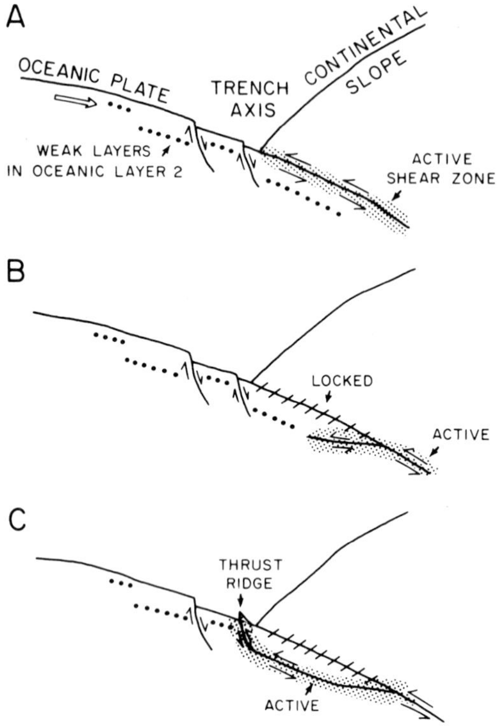

FIG. 4. Schematic diagram showing formation of a basaltic thrust ridge within trench due to locking of portion of main underthrust surface. Uplift and landward tilting of axial turbidites caused by upward curving of leading edge of thrust fault in upper oceanic layer (after Schweller *et al.* 1981).

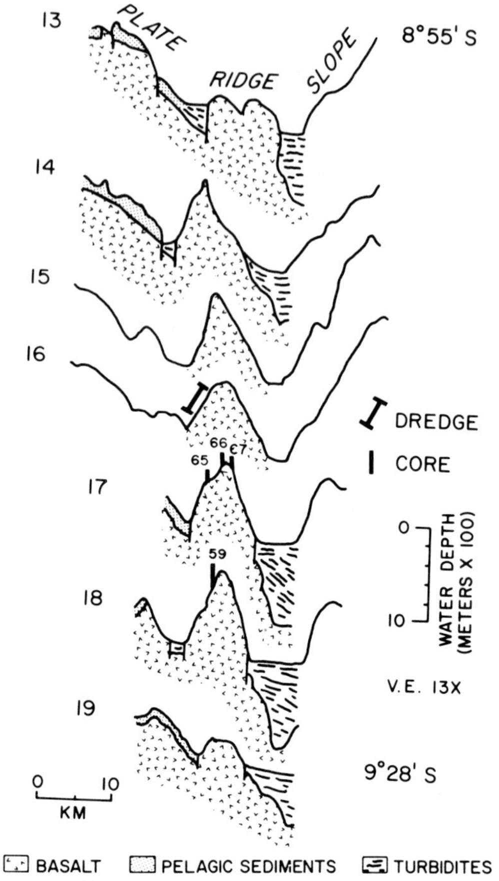

FIG. 5. Schematic diagram showing structure and morphology of basaltic ridge in trench axis between 8°55′ and 9°28′S. Cores are numbered on ridge. Diagram from Kulm *et al.* (1981a).

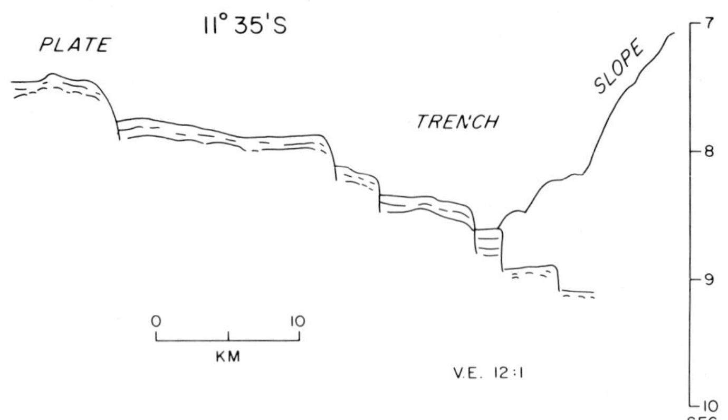

FIG. 6. Line drawing of single channel reflection record on ocean plate showing step faults as described in Schweller *et al.* (1981) for the Peru–Chile Trench.

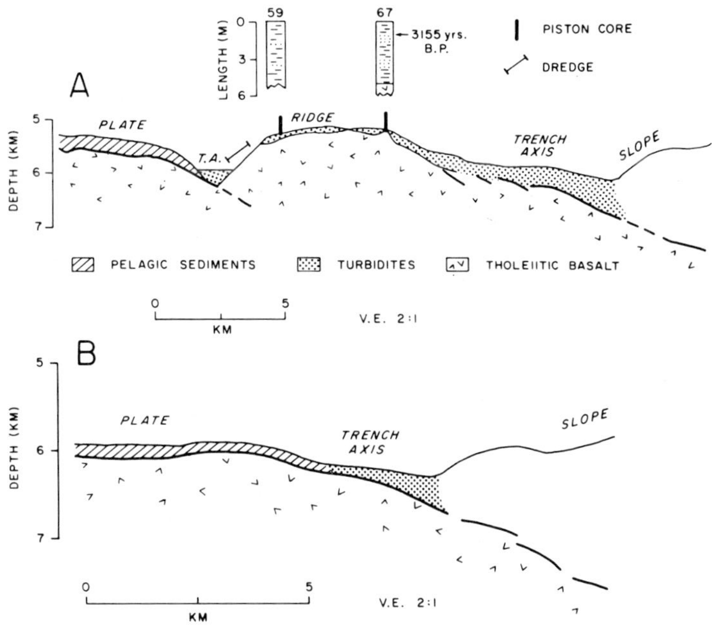

FIG. 7. Geological cross-sections constructed from multichannel seismic records CDP-2 (A) and CDP-1 (B) across the Peru Trench at 9°S (see Fig. 8A) and 12°S (see Fig. 8B), respectively. See Fig. 3 for location of lines.

amount of sediment within the trench axis in both trench sections. About 200–250 m of pelagic and hemipelagic sediment is entering the trench from the oceanic plate with 170–850 m of continent derived turbidites being deposited over these sediments.

Massif and subduction complex

While the arc massif on the continental shelf is identified in drill holes and in island exposures, the nature of the subduction complex on the continental slope is virtually unknown in the Andean forearc. Only geophysical data and one set of geological observations indicate that a complex is present. Multichannel seismic reflection records trace the subducting oceanic crust landward beneath the continental slope at least 50–60 km (Figs 8, 9 & 10) before the signal becomes obscured by the bottom-multiple reflection. The slab dips initially from 8° to 10° as it slides beneath the outer arc. A wedge-shaped diffracting mass overlies the oceanic crust; this is a typical geophysical signature of a disturbed subduction complex (Seely 1979).

The basaltic ocean crust is rupturing within the trench (Fig. 7A) and beneath the subduction complex as shown in the multichannel depth sections (Figs 8, 9 & 10). Manipulation of the overlying velocity structure cannot remove these offsets of rupture structures underlying the complex in either depth section (Kulm *et al.* 1981a; Hussong *et al.* 1976). The broken basaltic slabs may be inherited as a result of thrust faulting within the trench (Kulm *et al.* 1981a) or apparently rupture later within the subduction complex from the oceanic crust underthrusting itself (Hussong *et al.* 1976).

The generalized velocity structure of the central Peru margin is shown in Fig. 11 and can be divided into three different structures. The lowest-velocity material (2.0–3.6 km s^{-1}) forms a 15 km wide wedge-shaped mass across the lowermost continental slope. Velocities in this range are commonly found in the most recently formed dewatered part of the subduction complex (Kulm, von Huene *et al.* 1973). The only geological evidence of continental accretion is documented opposite the Nazca Ridge (Fig. 1) where pelagic sediment from the ridge and turbidites from the trench are accreted to the lowermost continental slope (Kulm *et al.* 1974). High-velocity material (5.0–5.6 km s^{-1}) occurs landward of this wedge and extends to the mid- to upper-slope region. Such material may comprise dewatered and highly cemented trench-slope turbidites and hemipelagic deposits and a mélange of these

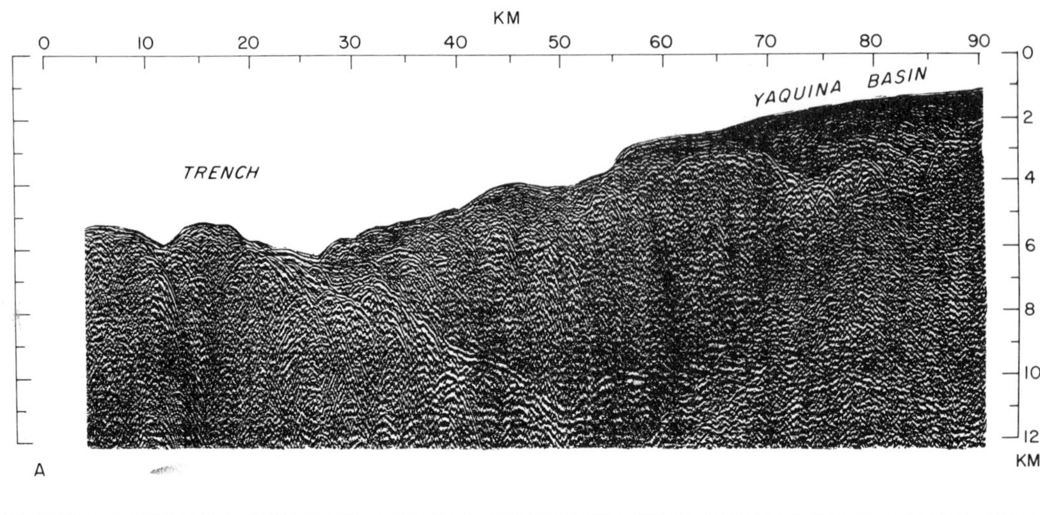
KM
0 10 20 30 40 50 60 70 80 90
TRENCH
YAQUINA BASIN
0 2 4 6 8 10 12
KM
A

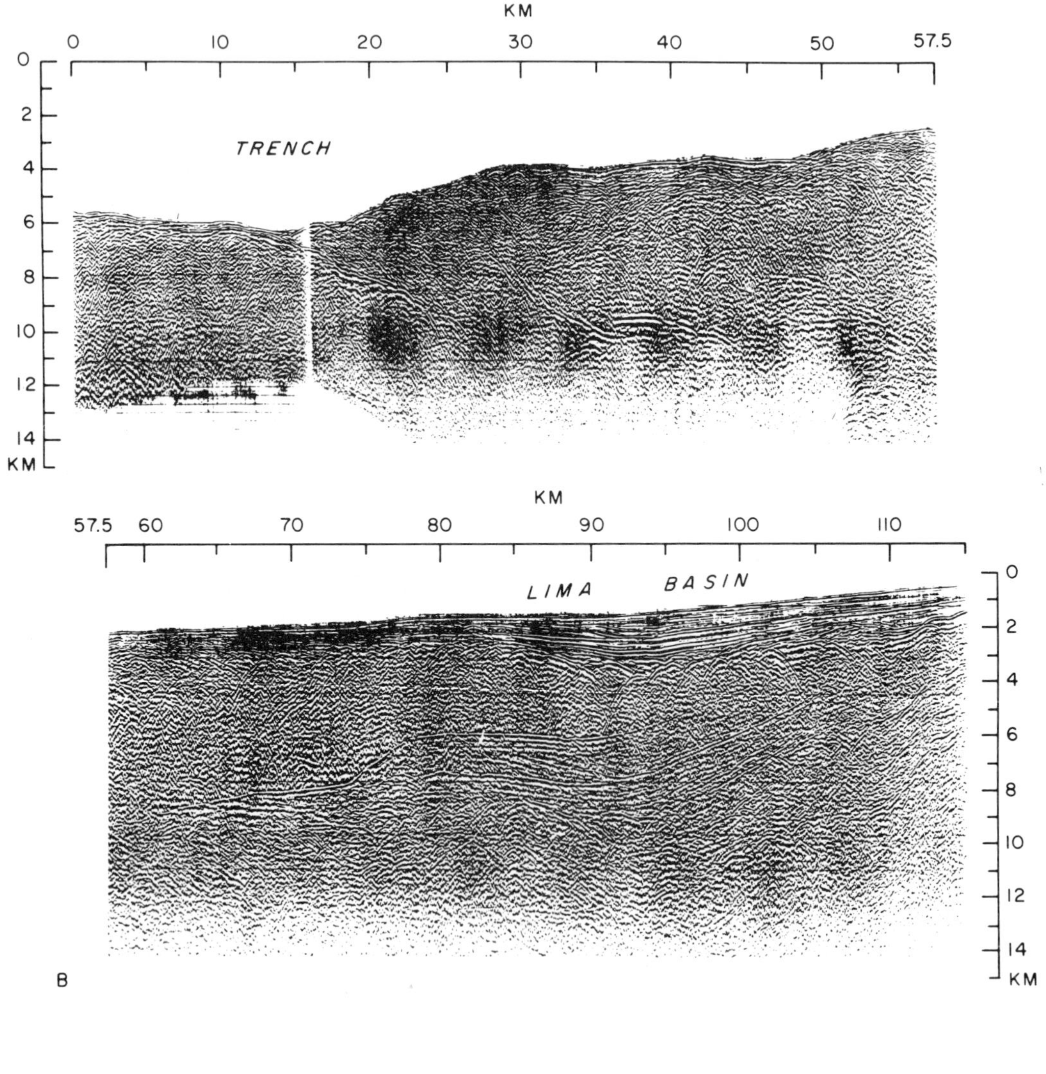

KM
0 10 20 30 40 50 57.5
TRENCH
0 2 4 6 8 10 12 14
KM
KM
57.5 60 70 80 90 100 110
LIMA BASIN
0 2 4 6 8 10 12 14
KM
B

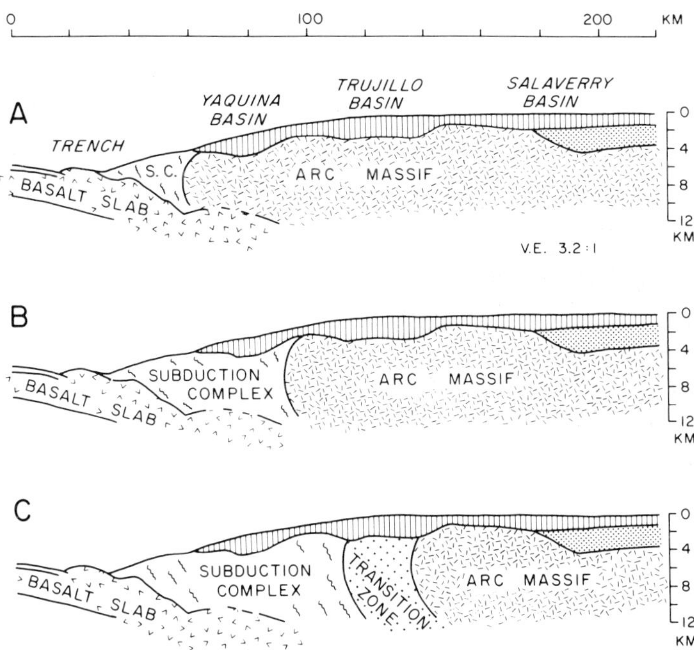

FIG. 9. Alternative explanations of geological cross-section at 9°S. Data from multichannel depth section (Fig. 8A), seismic refraction line at 9°S (Fig. 3; Jones 1981; Hussong *et al.* 1976), drill holes and dredge data (Kulm *et al.* 1981b). Dotted pattern is inferred Mesozoic strata based on seismic refraction data. Lined pattern is Cenozoic strata in forearc basins. See text for explanation.

sediments mixed with slivers of the underthrusting basaltic slab (Fig. 12; Kulm *et al.* 1981a). Basalt slivers should substantially increase the velocities of this portion of the complex. Landward, the deeper structure is characterized by velocities of 5.7–5.9 km s^{-1} or higher material which is known to consist of foliated metamorphic strata typical of the arc massif (Kulm *et al.* 1981a). These structures are mantled by slope sediment in the range of 1.5–3.0 km s^{-1}.

If largely geophysical criteria are used to identify the position of the arc massif-subduction complex boundary, several models are feasible. There is little doubt that the low-velocity mass occupying the toe of the continental slope consists of sediment (Figs 9, 10 & 11) and accretion by some sort of thrust faulting is strongly implied at 9°S during the most recent subduction history (Figs 4, 7A & 12) (Kulm *et al.* 1981a). Accretion is also suggested for the same material at 12°S (Hussong *et al.* 1976), especially since the basalt slab appears to be underthrusting itself (Figs 8B & 10).

There is also little doubt that the highest velocity material is the Palaeozoic arc massif which extends from the shoreline to at least the edge of the continental shelf (Figs 2, 3 & 11). But the massif's position between the small accretionary-sediment wedge and shelf edge is uncertain. If we assume that the 5.0–5.6 km s^{-1} material is the subduction complex (Fig. 11), the massif terminates in the vicinity of the upper continental slope. On the other hand, if this material is part of the foliated metamorphic strata associated with the massif elsewhere, then the massif extends to the lower slope and the subduction complex is indeed quite small

FIG. 8. (A) Non-migrated multichannel depth section CDP-2 of trench and continental slope at 9°S (after Kulm *et al.* 1981a). Vertical exaggeration 3:1. Note position of Yaquina Basin on upper continental slope. See Fig. 3 for location.
(B) Non-migrated multichannel depth section CDP-1 of trench and continental slope at 12°S. Note location of Lima Basin on the upper continental slope. Vertical exaggeration 3:1. See Fig. 3 for location. This is a depth section of the time section shown by Hussong *et al.* (1976, fig. 8). Record processing described in Kulm *et al.* (1981a).

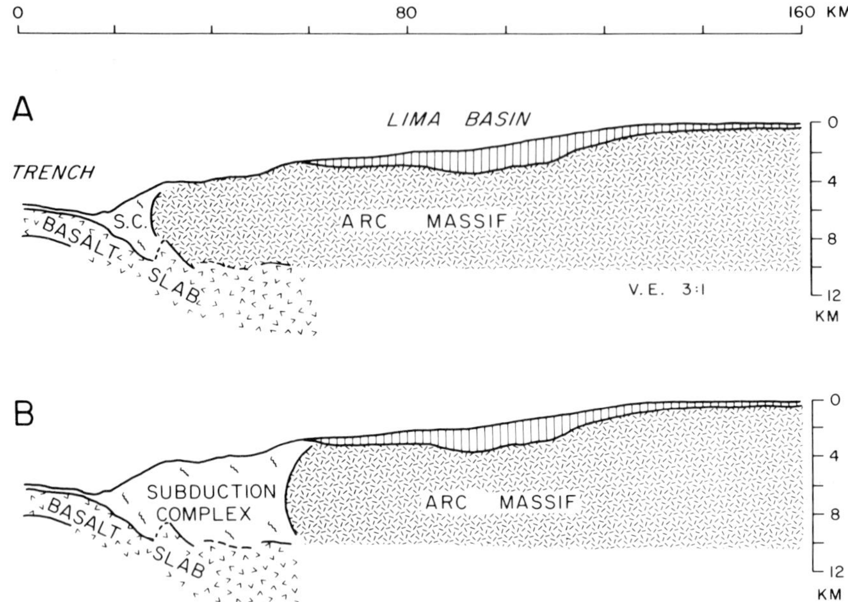

FIG. 10. Alternative explanations of geological cross-section at 12°S. Data from multichannel depth section (Fig. 8B), seismic refraction data at 12°S (Hussong *et al.* 1976) and dredge data (Kulm *et al.* 1981b). Lined pattern is Cenozoic strata in Lima forearc basin. See text for explanation.

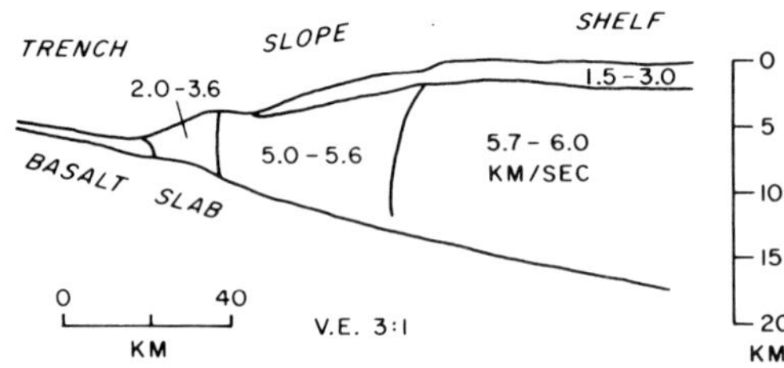

FIG. 11. Generalized velocity structure of central Peru margin. Data compiled from multichannel velocity spectra (Kulm *et al.* 1981a) and seismic refraction information (Hussong *et al.* 1976; Jones 1981).

(Figs 9 & 10). At 9°S, Kulm *et al.* (1981a) favour the intermediate position (i.e. upper slope, Fig. 9B) and at 12°S Hussong *et al.* (1976) favour the lower position (i.e. lower slope, Fig. 10A) for the arc massif. Additional discussion of these models is presented in the following section.

Geological history of forearc basins

Continental shelf basins

From the limited information available on the geology of the forearc basins off central Peru (Travis *et al.* 1976; Kulm *et al.* 1977; Thornburg & Kulm 1981), it is clear that these basins have undergone different depositional and tectonic histories which are related, in part, to the basement features described previously. Extensive normal faulting characterizes the subaerial outcrops of the coastal basins, and geophysical data (Travis *et al.* 1976) indicate that irregular horst and graben basement features occur on the adjacent continental shelf. Cenozoic subsidence of the Sechura, Salaverry and East Pisco Basins (Fig. 2) on this block-faulted arc massif is indicated by the deposition of approximately 1000–3000 m of marine clastic sediments (shale, siltstone, and sandstone) with minor interbedded carbonate, phosphate, and glauconite deposits. Diatomite is especially prominent in the East Pisco Basin. Large-scale gravity slumping and sliding and shale flowage in the younger strata are detected on seismic reflection sections (e.g. Fig. 13, 9°S [east]) in the Salaverry Basin.

The southern end of the Salaverry Basin shallows on a feature called the Lima Platform (Fig. 2), which has either undergone only minor subsidence or has experienced uplift and subaerial erosion during late Cenozoic time because only a few hundred metres of sediment currently rest upon this massif block (Fig. 14).

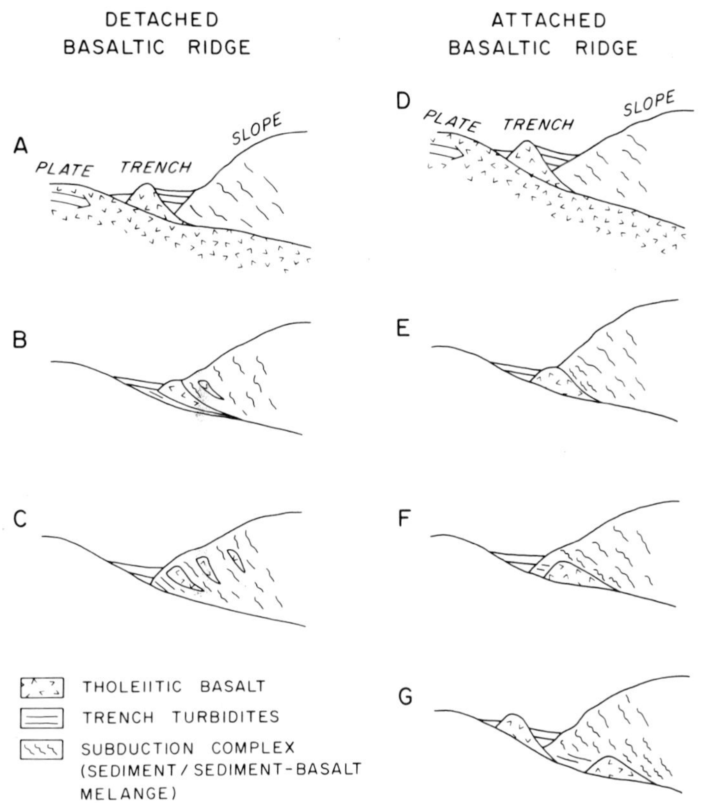

FIG. 12. Transfer of ocean plate basaltic slivers to subduction complex by thrust faulting (detached basaltic ridge) forming sediment-basalt mélange. Evolution of two basaltic ridges as shown in cross-section (Fig. 9). They remain essentially coupled to the subducting slab (attached basaltic ridge). See Fig. 4 for origin of thrust features.

Upper slope basins

The upper slope Lima Basin is located immediately seaward of the Lima Platform and contains a maximum of 2 km of sediment resting largely upon the strongly reflecting and high velocity arc massif (Figs 8B, 10 & 14). The Lima Basin lithologies consist of late Miocene to Pleistocene siltstone, calcareous siltstone, micrite, dolomicrite and sheared dolomicrite which were dredged at two locations near 11°30′S (Fig. 15, D-1 and D-46, Kulm *et al.* 1981b). These same strata are shown in the upper 1300 m of the multichannel section (CDP-1) located in the vicinity of the dredge sites (Figs 3 & 15). All of these rocks were originally deposited in water depths of 100–500 m and now lie at water depths of 837–1201 m (D-1) and 1639–2263 m (D-46) which indicates subsidence of about 500 m during the Pleistocene for the shallow deposits and about 1100 m during the Pliocene–Pleistocene for the deeper rocks. Minimum rates of subsidence are calculated at 500 and 275 m Ma^{-1}, respectively, for the metamorphic arc massif and overlying sedimentary deposits (Kulm *et al.* 1981b).

In the Lima Basin strata are truncated on the seafloor in water depths ranging between about 100 and 1800 m (Fig. 14). A poleward-flowing undercurrent presently impinges on the outer continental shelf and upper slope in water depths ranging between 100 and 400 m (Thornburg & Kulm 1981). If the undercurrent was present in the same position during the Pliocene and Pleistocene, truncated strata found deeper than 400 m today also suggest subsidence of the continental margin. Whereas subsidence characterizes the most recent tectonic movements in the southern Lima basin, compressional stress is also inferred (Thornburg & Kulm 1981).

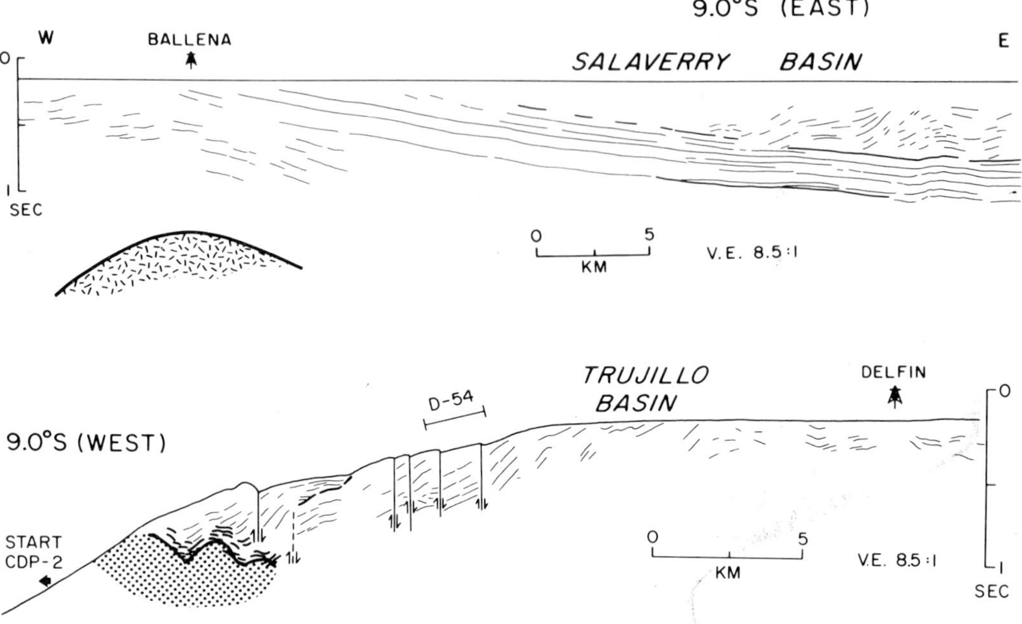

FIG. 13. Line drawing of single channel seismic record across Salaverry and Trujillo Basins at 9°S (see Fig. 3 for location). Line pattern is Outer Shelf High and dotted pattern deformed sediment of Upper Slope Ridge (Fig. 3). Dredge (D-54) taken across fault blocks. Note slumping and movement of shales in 9°S (east) younger sediments (Travis *et al.* 1976). Drill holes shown by towers.

Normal faults penetrate the sediment surface along the seaward edge of the metamorphic Outer Shelf High (OSH), but high-angle reverse faults in the basin deposits suggest compressional stress. This stress pre-dates or is in part synchronous with basin subsidence (Thornburg & Kulm 1981; Johnson & Ness 1981).

To the north, the upper slope Trujillo Basin contains more than 3 km of post-Eocene shale, siltstone, marlstone, minor thin limestone beds, and in the basal part, late Eocene sandstone (Travis *et al.* 1976). This section was initially called the Salaverry Basin section by Travis *et al.* (1976) but Thornburg & Kulm (1981), using additional structural data, define it as the Trujillo Basin section. The latter author's interpretation of the Trujillo Basin section is shown in Fig. 16; the Delfin drill hole penetrated the section described above but cannot be shown for proprietary reasons. Seismic data suggest older lower Tertiary strata at greater depth in this basin (Thornburg & Kulm 1981).

Single channel reflection data indicate that the Trujillo Basin deposits are more structurally disturbed than the Lima Basin deposits (cf. Figs 13, 14, 15 & 17) which accounts for the large volume of Neogene sheared dolomicrite recovered from the reverse (?) fault block at dredge sites 54 (Fig. 13) and 59 (Fig. 17) west of the drill holes. An orientated brecciation suggests a compressive origin for these rocks. Undeformed Neogene dolomicrite, glauconitic micrite, and phosphorite were also recovered in these dredges. The intense deformation of the sheared rocks occurred during late Pliocene–early Pleistocene time if radiometric dating of the breccia and enclosing cement are correct (Kulm *et al.* 1981b). The limited amount of palaeodepth information obtained at dredge sites 54 and 59 (Figs 13 & 17) indicates deposition in today's water depths.

Discussion

The central Peru margin is characterized by its diverse structural, stratigraphic, and tectonic elements. Andean subduction occurred throughout most of Mesozoic and Cenozoic time, and no doubt has created and destroyed numerous forearc elements in its 150 Ma history. Portions of the Cenozoic record are examined in this study and compared along approximately 900 km (6°–14°S) of the Peru margin.

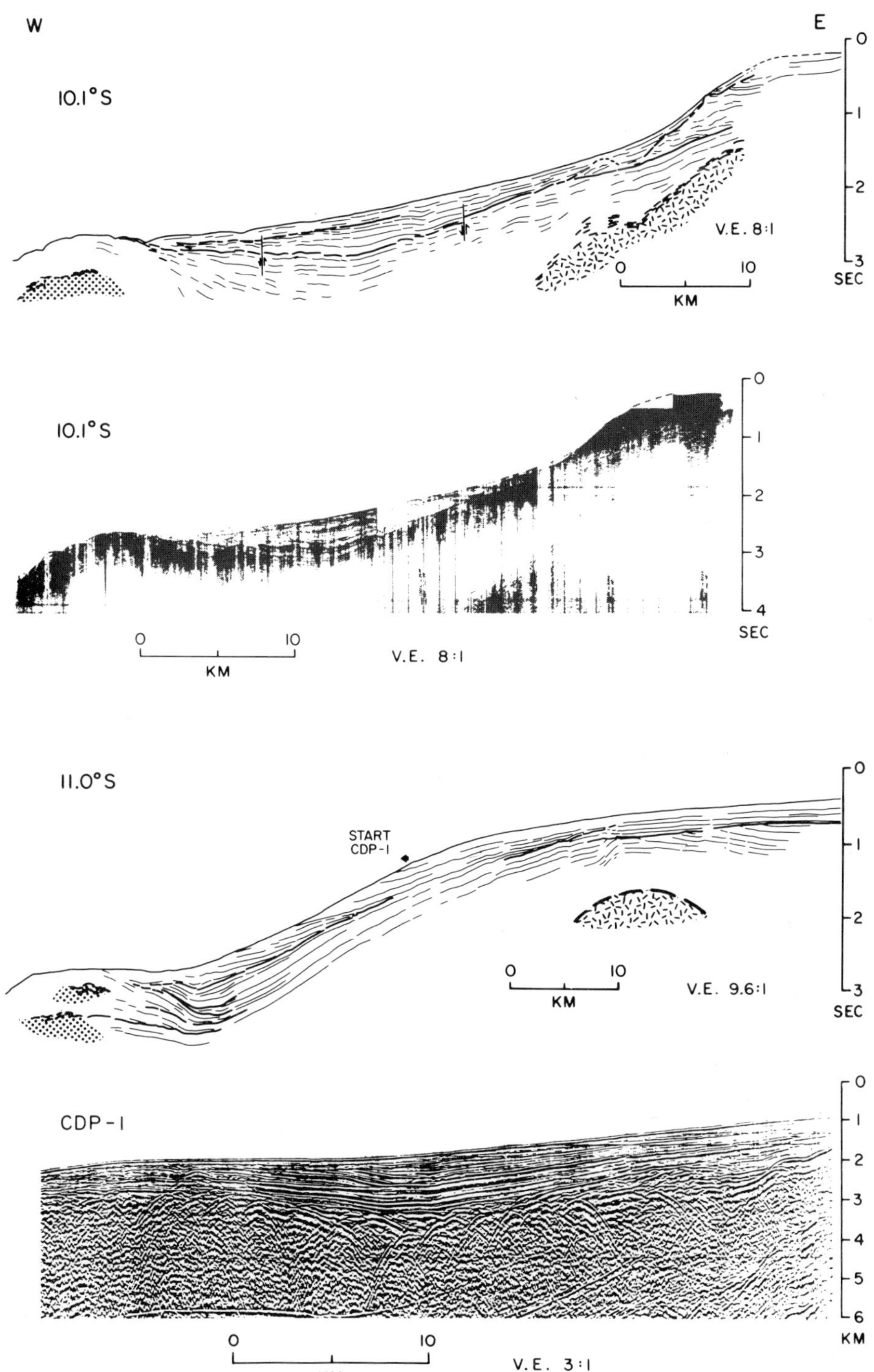

FIG. 14. Line drawings of single channel seismic records across Lima Basin (10.1°S, 11.0°S) and single channel record (10.1°S) and multichannel depth section CDP-1 showing Lima Basin deposits near 11.0°S single channel line. Line pattern is OSH and dotted pattern deformed sediment of USR (Fig. 3). See Fig. 3 for location.

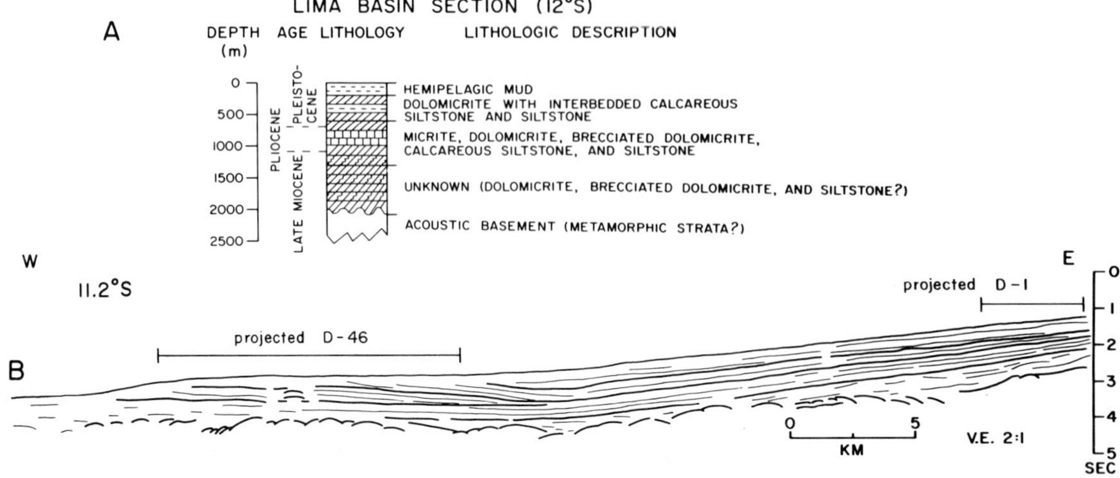

FIG. 15. (A) Stratigraphic section of Lima Basin near 12°S (modified from Kulm *et al.* 1981b). Section based upon dredge data described in text. (B) Line drawing of time section of multichannel seismic line CDP-1 across Lima Basin. See Figs 2 & 3 for location.

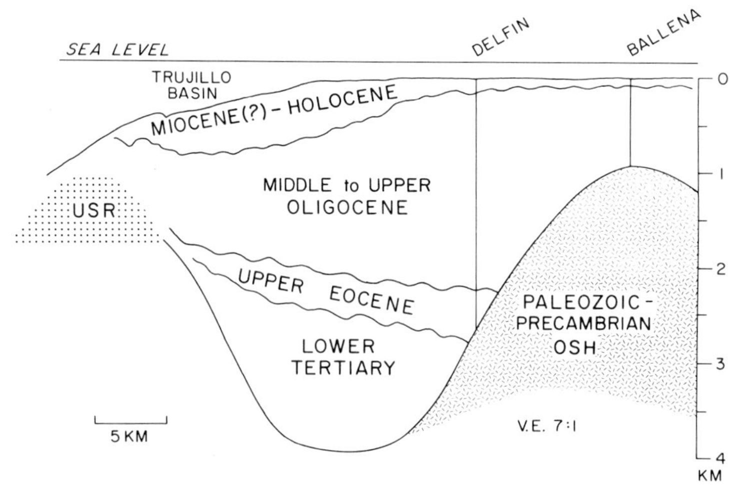

FIG. 16. Reconstruction of stratigraphic section at 9°S in Trujillo Basin (Figs 2 & 3). See Figs 3 (stars) & 13 (towers) for location of drill holes. Dredge data (Figs 13 & 17) also used in section. Diagram after Thornburg & Kulm (1981).

Massif tectonics and basin deposition

The block-faulted arc massif underlies the shelf and probably upper slope forearc basins. The Outer Shelf High (OSH) and Lima Platform are the elevated expressions of the massif, and form the seaward boundaries and configurations of the shelf forearc basins (Figs 2 & 3). Limited sampling and extrapolation with onshore geology suggest these foliated metamorphic strata are Palaeozoic–Precambrian and contain a complex history of deformation that is difficult to decipher.

Mesozoic strata probably occur in the deeper portions of the shelf basins, particularly in the Salaverry Basin where 4.6–5.2 km s^{-1} material (Fig. 9, dotted pattern) is sandwiched between the high velocity massif and much lower velocity (1.5–2.6 km s^{-1}) Cenozoic deposits (line pattern) (Jones 1981).

Correlation with onshore stratigraphy suggests that the region of the shelf basins was emergent and positive until late Eocene or post-Eocene time, when the Cenozoic marine sediment record begins in the shelf basin, unconformably overlying the deformed Mesozoic

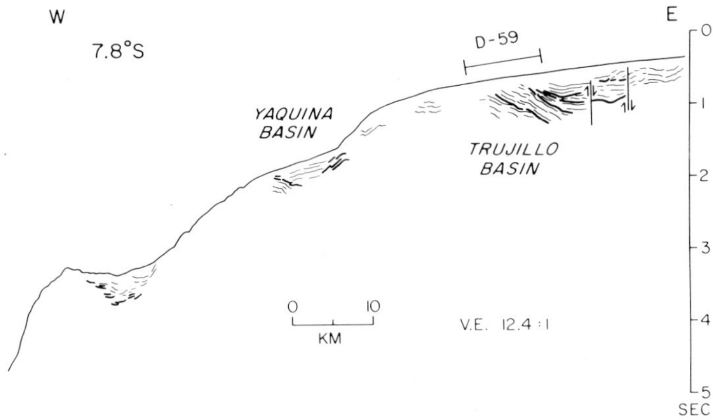

FIG. 17. Line drawing of single channel seismic record at 7.8°S across the Trujillo and Yaquina Basins (see Fig. 3 for location). Dredge D-59 shown over rock outcrops in Trujillo Basin.

sediments or pre-Mesozoic OSH (Travis *et al.* 1976). Sedimentation during the early Tertiary may have been more continuous in the upper slope Trujillo Basin, however, and the exposed massif (OSH) may have provided a sediment source (Figs 2 & 16). One can speculate that the OSH was a Mesozoic volcanic arc (Masias 1976): Mesozoic stratigraphy records the influence of such a volcanic source (references cited in Thornburg & Kulm 1981). Mesozoic dykes, which cut the Precambrian/Palaeozoic complex of correlative coastal Cordillera outcrops (Travis *et al.* 1976), may represent the core of such a feature.

In middle Cenozoic times, the OSH subsided along with the shelf basins to allow the accumulation of several hundred metres of sediment. Subsequent uplift in middle to late Miocene time deformed and truncated the Palaeogene sediments (Thornburg & Kulm 1981). The bulk of the stratigraphic section in the Trujillo Basin appears to pre-date this event, implying that this deformational episode caused a pronounced reduction in the Trujillo sedimentation which followed. In the neighbouring upper slope Lima Basin, however, 1.3 km of the 2 km thick sedimentary section may be of late Miocene or younger age (see *Geological History of Forearc Basins*), implying that inception of the predominantly Neogene deposition may have resulted from these same middle to late Miocene tectonic movements. Thus, a major change in structural style near 9.5°S apparently separated the basins at this time. This zone of structural transition has persisted throughout younger episodes of deformation also, causing the neighbouring basins to respond in distinctly different ways of subsequent tectonic movements.

Nature of subduction complex

The structure and stratigraphy of the continental slope is complex. As shown previously, the arc massif extends seaward some distance to abut against a subduction complex of largely unknown composition. In both the 9°S and 12°S areas, the minimum size of the subduction complex is a 15 km wide wedge (Figs 9A & 10A). Considering this relatively small accretionary prism (25 km^2), Hussong *et al.* (1976) calculated that during 150 Ma of subduction most of the sediment from the oceanic plate (750 km^2) and the trench (unknown volume) was subducted beneath rather than accreted to the continental margin. Using a wider prism of 40 km (Fig. 10B) or 60 km (Fig. 9B), it is still unlikely that a large volume of the potentially available plate and trench sediment has been accreted to the margin during this period.

This subduction of plate and trench sediment and vague evidence of downfaulting in the multichannel seismic line at 12°S (Fig. 8B) led Hussong *et al.* (1976) to conclude that the continental slope and outer shelf is actually subsiding, which suggests tectonic removal of the continental plate (i.e. the subduction complex and the arc massif). Kulm *et al.* (1981b) confirmed this large-scale subsidence on the upper to middle slope at 12°S during Neogene time as will be discussed later.

Rupture of subducting plate versus evolution of subduction complex

While the subduction complex may not have historically swept up all the sediments entering the subduction zone, it has nonetheless played an important role in recent subduction-related tectonics. Crustal rupture of the subducting oceanic slab is well documented off central Peru (Figs 5, 7 & 8A). Using the 10 cm yr^{-1} convergence rate (Minster *et al.* 1974), all the rupture zones have evolved within the trench and beneath the first 15 km of the subduction complex during the past 0.5 Ma, with the trench rupture occurring during the past few thousands or tens of thousands of years. Inclusion of basaltic material into the subduction complex is strongly implied from geological and geophysical considerations (Fig. 12; Kulm *et al.* 1981a). These data suggest that the complex is actively accreting material, even at 12°S, where tectonic removal of the continental plate is postulated by Hussong *et al.* (1976). However, this relationship is similar to the Japan forearc where accretion along the lower continental slope with large-scale subsidence immediately landward on the middle to upper slope is well documented (von Huene, Nasu *et al.* 1978). The data presented in this paper suggest the Peru margin, between 9°30′ and 14°S, is experiencing a tectonic regime of accretion (uplift) and subsidence similar to that of the Japan forearc.

Continental accretion seems even more likely at 9°S, because of the large basaltic thrust ridge in the trench axis and slab rupture beneath the subduction complex.

Tectonics of upper slope basins

The sedimentary deposits in basins on the continental slope offer important clues about the tectonic framework and processes affecting the forearc. The Lima Basin (Figs 2 & 3) has undergone from 500 to 1100 m of subsidence during the past 4 Ma at 12°S (Kulm *et al.* 1981b), probably over much of its 520 km length (Thornburg & Kulm 1981). Subsidence occurred without intensive deformation of the basin deposits (Fig. 14) in contrast to the Trujillo and Yaquina Basins to the north (Figs 8A, 13 & 17). Nevertheless, the Lima Basin exhibits subsurface high-angle reverse faults (Johnson & Ness 1981; Thornburg & Kulm 1981) possibly caused by compression which uplifted the Upper Slope Ridge (Figs 2 & 3) forming the seaward boundary of the basin. The compression either predated or was, in part, synchronous with basin subsidence. Our geographically limited palaeodepth information indicates that the greatest rates of subsidence (500 m Ma^{-1}) occurred during the Pleistocene, suggesting that the Lima Basin has shifted from a compressional to extensional regime in its relatively recent geological history.

Neogene strata in the Trujillo and Yaquina Basins (Figs 2 & 3) show abundant large-scale structural evidence of compression including faults and folds (Figs 8A & 13), as well as the highly sheared dolomicrite. The entire Yaquina Basin sequence is disturbed to a depth of 2 km, including the acoustic basement. Age dating of the Trujillo Basin dredged lithologies (Kulm *et al.* 1981b) indicates that the Pleistocene section is thinner here than in the Lima Basin; sediment trapping in the Salaverry Basin to the east of the OSH (Fig. 2), and subsequent deposition in the upper slope basins may be responsible.

We cannot rule out a moderate amount of subsidence in the Trujillo and Yaquina Basins because our palaeodepth information was obtained only at water depths between 202 and 430 m, the approximate depth of the original environment of deposition. (Several attempts were made to recover lithologies at deeper water depths, but each one failed. The dolomicrites are probably present in the Yaquina Basin because the dredges—one was lost—contacted very resistant outcrops there, typical of the ones dredged successfully.) The disturbed structures of these basins, however, are more typical of compressional than extensional tectonics, at least in the upper slope region.

A major structural change occurs near latitude 9°30′S between the highly disturbed Trujillo and relatively undisturbed Lima Basins (Thornburg & Kulm 1981). The two basins are offset here as is the Upper Slope Ridge of deformed sediments (Figs 2 & 3). This boundary coincides with the strike of the Mendana Fracture Zone on the Nazca Plate which intersects the margin at this geographical location (Fig. 1). The basaltic ridge (Fig. 5) also terminates near the fracture zone. Unfortunately, we have no geophysical data parallel to the continental slope to determine whether the fracture zone does influence the structure of the margin.

Comparison of Peru and other forearcs

The central Peru forearc has many similarities to other circum-Pacific forearcs, but it is quite dissimilar in other aspects. The Peru continental arc massif consists of metamorphic strata similar to that drilled off southern Mexico (Moore, Watkins *et al.* 1979), which places very old crystalline rocks beneath the continental shelf and upper continental slope. Truncation of the continental block has been proposed on the basis of Palaeozoic and older rocks found near the shoreline and at the edge of the margin. The seaward limit of the massif can be defined by drilling where the sedimentary section is relatively thin, but it is often difficult to define solely from geophysical measurements. Subsidence of the inner shelf massif is well documented off Mexico and strongly inferred off central Peru in the Salaverry Basin on the inner shelf (Fig. 2). Interestingly, the outer edge of the massif off Mexico subsided perhaps as much as 3000 m and then began a pulse of uplift about 19 Ma ago (Moore, Watkins *et al.* 1979). The Outer Shelf High on the Peru margin also appears to be experiencing a current cycle of uplift which followed the subsidence of this former topographic high. Data from both of these margins indicate that the massif is an unstable block that periodically undergoes vertical isostatic adjustments. The block-faulted basement of the Peru massif (Travis *et al.* 1976) and numerous unconformities reflect these movements.

The Lima Basin on the upper continental slope is subsiding at the rate of 275–500 m Ma^{-1} since late Miocene time. Geophysical data suggest the underlying basement is the metamorphic massif (Fig. 10). A similar rate of subsidence was first reported for the Japan forearc basin deposits laid down on an Upper Cretaceous, steeply dipping, black silicified claystone (Arthur *et al.* 1980). The Japan forearc subsided from 18 to 4 Ma ago with subsequent uplift. The subsidence was in part contemporaneous with the formation of a small accretionary wedge at the toe of the continental slope off Japan (Langseth, Okada *et al.* 1978; von Huene, Nasu *et al.* 1978). Low-velocity material at the toe of the slope at 12°S off Peru (Fig. 11) and the underthrusting of the basaltic slab upon itself (Figs 9 & 10) show that a similar subduction complex probably exists here. Both forearcs are characterized by fast convergence rates (8–10 cm yr^{-1}). The similarities of the Neogene tectonic histories of these two forearcs is striking and warrants further study.

The forearc basins on the continental slope off central Peru exhibit various degrees of internal deformation with the Yaquina Basin being the best example of extreme deformation (Fig. 8A). In fact, all slope basins are faulted to some extent which indicates the entire continental slope is undergoing structural adjustments in response to compressional or tensional stresses. We do not have enough structural and stratigraphic control to determine whether these stresses co-exist or follow one another in some rational fashion.

Conclusions

The central Peru margin and trench are characterized by a variety of structural, stratigraphic, and tectonic styles. The area studied extends from 6° to 14°S latitude and covers the indentation in the coastline lying between the Palaeozoic and Precambrian coastal mountains of southern and northwestern Peru (Fig. 2).

Three forearc basins, Sechura, Salaverry, and eastern Pisco Basins, rest upon a Palaeozoic–Precambrian metamorphic massif whose surface is undulatory and block faulted. Portions of the massif, especially the Outer Shelf High (OSH) and Lima Platform are elevated above the deeper parts of the massif and apparently have been uplifted in Neogene time creating folds and faults in the overlying strata. While the massif must have subsided to allow the accumulation of as much as 3.0 km of Mesozoic and Cenozoic sediment, it has probably been influenced by several pulses of uplift and subsidence during its long history.

Three additional forearc basins, Yaquina, Trujillo and Lima, occupy the middle to upper continental slope and are bounded on the east by the OSH and on the west by an upper slope ridge of deformed sediment. Drill holes in the eastern Trujillo Basin contain late Eocene sandstone with overlying post-Eocene shale, siltstone, marlstone, and minor thin limestone beds, whereas the western Trujillo Basin and possibly the Yaquina Basin to the west contain late Miocene to Pleistocene micrite, dolomicrite, and glauconitic micrite dredged from rock outcrops. The Lima Basin has similar carbonate lithologies.

The Trujillo Basin and particularly the Yaquina Basin are experiencing severe internal disruption which is expressed by fault and fold structures in seismic records and by sheared dolomicrite obtained in the dredges. Compressional stresses best explain the structural style of these basins. In contrast, the Pliocene to Pleistocene sediments in the Lima Basin indicate subsidence at a rate of 275–500 m Ma^{-1} and thus tensional stresses. An earlier phase of uplift and compression is proposed for the Lima Basin because of the presence of subsurface reverse faults.

Multichannel seismic records and geological data at 9°S and 12°S show that the subducting oceanic plate within the trench (9°S) and beneath the overriding plate (9°, 12°S) is rupturing by thrust faulting of the upper layer 2 basalts. Slivers of basalt may be incorporated into the overlying sediments of the subduction complex forming a sediment-basalt mélange.

Several models are proposed which place the subduction complex-massif interface from 15 to 60 km from the trench axis depending upon which set of geophysical criteria are used to define these features. The subduction complex is at least a 15 km wide wedge and the massif extends seaward to the uppermost continental slope in all models.

The central Peru forearc at 12°S has many of the tectonic characteristics of the Japan forearc off Honshu, Japan. The severe internal deformation of the Peru upper slope basins at 9°S seems to be rather unusual with respect to other circum-Pacific forearc basins in a similar position.

ACKNOWLEDGMENTS: This research was sponsored by the Office of International Decade of Ocean Exploration, National Science Foundation, under grants GX 28675, IDOE71-04208, and OCE76-05903. Hawaii Institute of Geophysics Contribution No. 10601.

References

ARTHUR, M., VON HUENE, R. & ADLESECK, C., JR. 1980. Sedimentary evolution of the Japan fore-arc region off Honshu, legs 56 and 57, Deep Sea Drilling Project. *In:* SCIENTIFIC PARTY (eds). *Initial Rep. Deep Sea drill. Proj.* **56** & **57,** Part I, 521–68. U.S. Govt Printing Office, Washington.

COULBOURN, W. T. & MOBERLY, R. 1977. Structural evolution of forearc basins off southern Peru and northern Chile. *Can. J. Earth Sci.* **14,** 102–16.

HUSSONG, D. M., EDWARDS, P. B., JOHNSON, S. H., CAMPBELL, J. F. & SUTTON, G. H. 1976. Crustal structure of the Peru-Chile Trench: 8°S–12°S latitude. *In*: SUTTON, G. H., MANGHNANI, M. H. & MOBERLY, R. (eds). *The Geophysics of the Pacific Ocean Basin and its Margin.* Geophys. Monogr. Am. geophys. Union, **19,** 71–86.

JOHNSON, S. H. & NESS, G. E. 1981. Shallow structures of the Peruvian margin 12°S–18°S. *In*: KULM, L. D., DYMOND, J., DASCH, E. J. & HUSSONG, D. M. (eds). *Nazca Plate: crustal formation and Andean convergence.* Mem. geol. Soc. Am. 154, in press.

JONES, P. R. 1981. Crustal structures of the Peru continental margin and adjacent Nazca plate, 9°S latitude. *In*: KULM, L. D., DYMOND, J., DASCH, E. J. & HUSSONG, D. M. (eds). *Nazca Plate: crustal formation and Andean convergence.* Mem. geol. Soc. Am. 154, in press.

KULM, L. D., PRINCE, R. A., FRENCH, W., JOHNSON, S. & MASIAS, A. 1981a. Crustal structure and tectonics of the central Peru continental margin and trench. *In*: KULM, L. D., DYMOND, J., DASCH, E. J. & HUSSONG, D. M. (eds). *Nazca Plate*: *crustal formation and Andean convergence*. Mem. geol. Soc. Am. 154, in press.

—— , RESIG, J. M., MOORE, T. C., JR & ROSATO, V. J. 1974. Transfer of Nazca Ridge pelagic sediments to the Peru continental margin. *Bull. geol. Soc. Am.* **85,** 769–80.

—— , SCHWELLER, W. J. & MASIAS, M. 1977. A preliminary analysis of the geotectonic processes of the Andean continental margin, 6° to 45°S. *In*: TALWANI, M. & PITMAN, W. C. (eds). *Problems in the Evolution of Island Arcs, Deep Sea Trenches, and Back-arc Basins*. Am. geophys. Union, M. Ewing. Ser. **1,** 285–301.

—— , SCHRADER, H.-J., RESIG, J. M., THORNBURG, T. M., MASIAS, A. & JOHNSON, L. 1981b. Late Cenozoic carbonates on the Peru continental margin: Lithostratigraphy, biostratigraphy, and tectonic history. *In*: KULM, L. D., DYMOND, J., DASCH, E. J. & HUSSONG, D. M. (eds). *Nazca Plate*: *crustal formation and Andean convergence*. Mem. geol. Soc. Am. 154, in press.

—— , VON HUENE, R. *et al.* 1973. *Initial Rep. Deep Sea drill. Proj.* **18.** U.S. Govt Printing Office, Washington, 1077 pp.

LANGSETH, M., OKADA, H. *et al.* 1978. Near the Japan Trench: transects begun. *Geotimes*, **23,** 22–6.

MASIAS, A. 1976. *Morphology, shallow structure, and evolution of the Peruvian continental margin, 6° to 18°S.* Thesis, MS, Oregon State Univ., Corvallis, 92 pp.

MINSTER, J. B., JORDAN, T. H., MOLNAR, P. & HAINES, E. 1974. Numerical modelling of instantaneous plate tectonics. *Geophys. J. R. astron. Soc.* **36,** 541–76.

MOORE, J. C., WATKINS, J. S. *et al.* 1979. Off Mexico: Middle American Trench. *Geotimes*, **24,** 20–2.

MORDOJOVICH, D. 1974. Geology of a part of the Pacific Margin of Chile. *In*: BURK, C. A. & DRAKE, C. L. (eds). *The Geology of Continental Margins*, 591–8. Springer-Verlag, New York.

NUR, A. & BEN-AVRAHAM, Z. 1981. Consumption of aseismic ridges and volcanic gaps in South America. *In*: KULM, L. D., DYMOND, J., DASCH, E. J. & HUSSONG, D. M. (eds). *Nazca plate: crustal formation and Andean convergence*. Mem. geol. Soc. Am. 154, in press.

PRINCE, R. A. & KULM, L. D. 1975. Crustal rupture and the initiation of imbricate thrusting in the Peru–Chile Trench. *Bull. geol. Soc. Am.* **86,** 1639–53.

—— & SCHWELLER, W. J. 1978. Dates, rates and angles of faulting in the Peru–Chile Trench. *Nature, London*, **271,** 743–5.

SCHOLL, D. W., CHRISTENSEN, M. N., VON HUENE, R. & MARLOW, M. S. 1970. Peru–Chile Trench sediments and sea-floor spreading. *Bull. geol. Soc. Am.* **81,** 1339–60.

SCHWELLER, W. J., KULM, L. D. & PRINCE, R. A. 1981. Tectonics, structure and sedimentary framework of the Peru–Chile Trench. *In*: KULM, L. D., DYMOND, J., DASCH, E. J. & HUSSONG, D.

M. (eds). *Nazca Plate: crustal formation and Andean convergence*. Mem. geol. Soc. Am. 154, in press.

SEELY, D. R. 1979. The evolution of structural highs bordering major forearc basins. *In*: WATKINS, J. S., MONTADERT, L. & DICKERSON, P. W. (eds). *Geological and Geophysical Investigations of Continental Margins*. Mem. Am. Assoc. Petrol. Geol. **29,** 245–60.

SHEPHERD, G. L. & MOBERLY, R. 1981. Coastal structure of the continental margin, northwest Peru and southwest Ecuador. *In*: KULM, L. D., DYMOND, J., DASCH, E. J. & HUSSONG, D. M. (eds). *Nazca Plate: crustal formation and Andean convergence*. Mem. geol. Soc. Am. 154, in press.

THORNBURG, T. M. & KULM, L. D. 1981. Sedimentary basins of the Peru continental margin: structure, stratigraphy, and Cenozoic tectonics from 6°S to 16°S latitude. *In:* KULM, L. D., DYMOND, J., DASCH, E. J. & HUSSONG, D. M. (eds). *Nazca Plate: crustal formation and Andean convergence*. Mem. geol. Soc. Am. 154, in press.

TRAVIS, R. B., GONZALES, G. & PARDO, A. 1976. Hydrocarbon potential of coastal basins of Peru. *In*: HALBOUTY, M. T., MAHER, J. C. & LIAN, H. M. (eds). *Circum-Pacific Energy and Mineral Resources*. Mem. Am. Assoc. Petrol. Geol. **25,** 331–8.

VON HUENE, R., NASU, N. *et al*. 1978. Japan Trench transected. *Geotimes*, **23,** 16–20.

L. D. KULM, T. M. THORNBURG & H.-J. SCHRADER, School of Oceanography, Oregon State University, Corvallis, Oregon 97331, U.S.A.

J. M. RESIG, Hawaii Institute of Geophysics, 2525 Correa Road, Honolulu, Hawaii 96822, U.S.A.

Forearc and other basins, continental margin of northern and southern Peru and adjacent Ecuador and Chile

R. Moberly, G. L. Shepherd & W. T. Coulbourn

SUMMARY: The continental margin between the Peru–Chile Trench and the west coast of South America is the type example of the Andean margin, where one of the two converging blocks of lithosphere is oceanic and the other is continental. Along Peru and Chile, basement at the coast and under the landward side of forearc basins is dominantly metamorphic and plutonic. Apparently, the edge of continental crust has been stoped away during subduction. The seaward side of basins may be structurally high dams of sediments accreted and deformed by subduction, or they may be horsts of continental basement. Subsurface mapping in the on-shore and off-shore oilfields of the forearc basin of north-western Peru has shown a mosaic of faulted blocks of tensional origin, overlain by low-angle gravity glides. Details of this faulting, however, cannot be seen on single- or multi-channel seismic records of forearc basins, even where they cross the fields. Off northern and southern Peru, forearc basins are separated from one another along strike by culminations of the basement; across strike there may be more than one forearc basin. The pattern of culminations and depressions may be inherited from structures of the oceanic crust and trench. Deeper seismic reflectors of the basins generally dip landward and converge seaward, and show progressive migration shoreward of the centres of deposition. Sediment movement into the basins and down the slopes is by turbidity currents and slumps. Subduction is an efficient process, removing about four-fifths of the trench sediments.

The Progreso Basin under the Gulf of Guayaquil owes its origin to a transform fault, the Dolores-Guayaquil megashear, oblique to the trench. It is a pull-apart basin or rhombochasm opened in the wake of the northward movement of a small plate. A forearc basin at Talara and the adjacent Progreso Basin have substantial hydrocarbon accumulations.

The evolution of the Nazca oceanic lithospheric plate was chosen as a topic for study in the U.S. programme of the Geodynamics Project (Drake 1973), and a large part of that research was conducted in the Nazca Plate Project of the International Decade of Ocean Exploration, a part of the U.S. National Science Foundation. The broad multidisciplinary programme included participants from the United States as well as South American institutions, and in particular the universities of Hawaii and Oregon State.

Hawaii Institute of Geophysics concentrated its marine surveys of the convergent margin in two areas (Fig. 1), between 2° and 7°S, where the coastline of South America is closest to the Peru–Chile Trench (Shepherd 1979), and between 18° and 23°S, where the coastline and trench change their strike at the Arica Bight (Coulbourn 1977). The purpose of this article is to compare and contrast the tectonic and sedimentary processes between the two areas, and to show the degree to which our interpretations support or need to modify some currently held concepts of processes at convergent plate boundaries.

Comprehensive reviews of Andean geology are by Weeks (1947), Jenks (1976), Bellido (1969) and Gansser (1973). The tectonic evolution of the Nazca Plate in the Cenozoic has been described by Herron (1972), Handschumacher (1976), and Mammerickx *et al.* (1980). Geophysical investigators of the convergent zone between the Andes and the oceanic plate, *i.e.* of the Peru–Chile Trench and the continental margin, include Fisher & Raitt (1962),

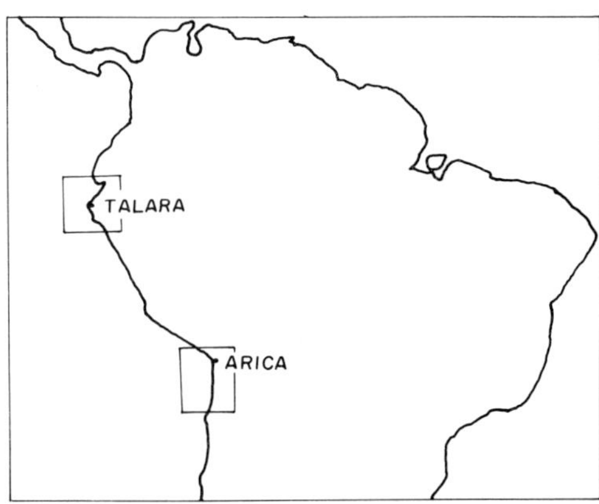

FIG. 1. Areas of the South American margin described in this report.

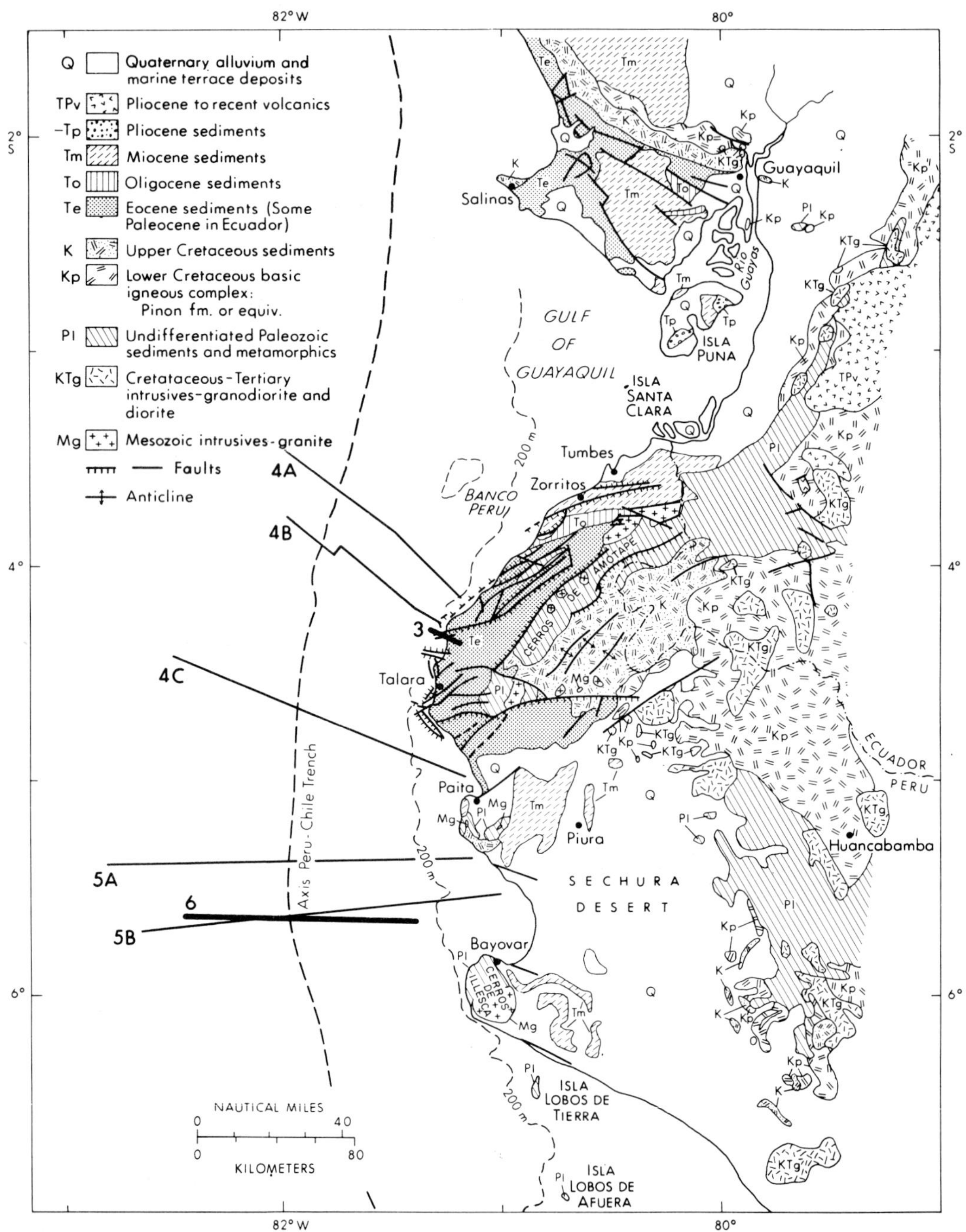

FIG. 2. Tectonic setting of NW Peru and adjacent Ecuador. Horsts of pre-Cenozoic metamorphic and instrusive rocks in Cerros de Amotape, the headlands at Paita and Bayovar, and Lobos Islands. Location of Figs 3 (oilfield subsurface), 4 and 5 (single-channel seismic lines), and 6 (multi-channel seismic line) are shown. Map is simplifed from Shepherd (1979).

Hayes (1966), Hussong *et al.* (1976), Kulm *et al.* (1977), and Lonsdale (1978).

Basins of the convergent margin

Sedimentary processes in the basins of coastal western South America and the adjacent Pacific differ in degree and detail from similar processes in a number of more familiar areas. The climate is exceptionally arid, so there are no major streams entering the Pacific along the Peru and northern Chile coasts. Recently uplifted coastal terranes of unconsolidated sediment, and ash from Andean volcanoes provide abundant detritus for the intermittent streams that do exist. Upwelling and the Humboldt Current give waters too cold to allow reef growth but with nutrients to support a high level of organic productivity. A result is the layer of oxygen-depleted waters which intercepts the continental slope, allowing phosphorites to form (as they did in the Miocene of Sechura) and organic matter is preserved in a band along the margin (Burnett 1977). The steep slopes of the margin allow slumping at various scales; abundant large and small earthquakes may trigger slumps and turbidity currents.

Typical surface sediments between the coast and the trench are diatomaceous muds. They are green and grey as a result of the near-reducing conditions, and have epiclastic and pyroclastic volcanic components. Glauconite pellets are abundant, and phosphorite nodules and early diagenetic pyrite are common locally. Submarine canyons, basins, and the trench also contain turbidite sands, identified by modest grading and benthonic calcareous foraminifers displaced from shallower depth. On the oceanic plate, because of the depth of water, dissolution of microfossil tests has occurred, and the red-brown to yellow muds contain planktonic nannofossils and benthonic agglutinated foraminifers. The environment and the lithogenous, biogenous, and hydrogenous sources of sediment are given in more detail by Rosato *et al.* (1975), Coulbourn (1977), Shepherd (1979), and Burnett *et al.* (1980).

The regions of this report, as well as the Peruvian margin between them (Masias 1975; Kulm *et al.* 1981), have forearc, trench-slope, and trench basins to trap sediment. We describe mainly the forearc basins, but also include a description of a 'pull-apart' basin and comment on some other aspects of sedimentation.

Forearc basins of NW Peru

In coastal NW Peru, old metamorphic and igneous rocks are exposed within 80 km of the axis of the Peru–Chile Trench, and important petroleum resources exist in a forearc basin. Generally speaking, oilfields and crystalline basement so close to a trench are unusual features of arc-trench gaps.

The economic importance of the coastal basins of NW Peru has resulted in their intense study. Much information from the oil fields remains unpublished, but an introduction to the geology of the Coastal Province and adjacent border of the High Cordilleran Province is in the local studies and regional reviews of such investigators as Travis (1953), Fischer (1956), M. Paredes (1958), J. Paredes (1966), Cossio & Jaen (1967), Ruegg (1967), Martinez (1970), Paz (1974), Morris & Aleman (1975) and Shepherd (1979).

The sediments of the forearc basins overlie or surround horsts, of a basement of Palaeozoic and Mesozoic sedimentary, metasedimentary, and igneous rocks. The oldest of these are exposed discontinuously in low massifs curving south from the Cerros de Amotape to Cerros de Illesca and to islands offshore (Fig. 2). Although not exposed here, similar Palaeozoic metasedimentary rocks lie on Precambrian metamorphic and or intrusive rocks nearby in southern Ecuador (Campbell 1975), the Cordillera Central (Hosmer 1959), and in southern coastal Peru (Bellido 1969). Thick volcanic rocks and volcaniclastic sediments of Triassic, Jurassic, and Late Cretaceous ages lie east of the coastal massifs. Within the massifs and in the thick Mesozoic section are dioritic, granodioritic, and granitic intrusions of the Andean coastal batholith. The intrusions are mainly of Palaeocene and Eocene age (Giletti & Day 1968) and east of the massifs, but some Mesozoic granites are exposed at the coast near Bayovar and Paita.

Massive normal faulting began in Late Cretaceous time and continued through the Cenozoic into the Quaternary, providing the framework of the landward sides of the forearc basins. They have filled with Cretaceous and Cenozoic detrital sediments of paralic and marine facies. The section near Talara is about 9000 m thick (Travis 1953). Unconformably above 150 m of Aptian-Albian limestones are about 1000 m of Senonian-Maestrichtian shales with subordinate sandstones and conglomerates, as much as 2000 m of Danian shales and lenticular sandstones, and a thick Eocene section. The Eocene is predominantly silty, micaceous shale, but also contains oil-bearing, feldspathic sandstones and a few conglomerates. Surface mapping and subsurface data from

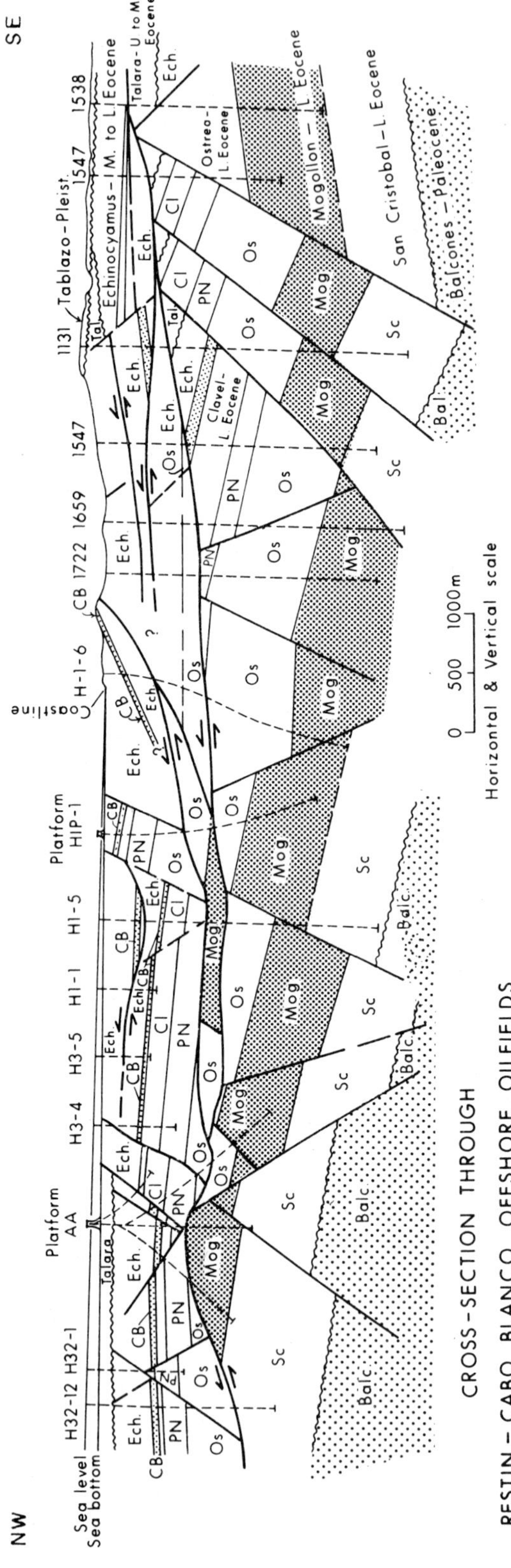

FIG. 3. Off-shore and on-shore section through an oilfield 15 km NE of Talara, Peru, near Cabo Blanco. For location see Fig. 2. Eocene and Palaeocene section is broken by high-angle normal faults, and later cut by low-angle gravity slides. This is the inner edge of a forearc basin. Coastal oilfields extend 50 km either side of Talara. Fields between Zorritos and Isla Santa Clara (reference Figs 2 & 13) are in Progreso Basin. Section courtesy of Belco Petroleum Co.

nearly 10 000 wells show exclusively normal and low-angle gravity faulting, rare minor warpings, and no folds. Gravity slides with rubble, resembling olistostromes, extend seaward at angles of a few degrees (Baldry 1938). Many of the master faults cut basement. The subsurface information has been extended to the offshore oilfields (Fig. 3). The entire region is being broadly uplifted, amounting to more than 300 m since the early Quaternary.

The seaward edge of the forearc basins is beyond the narrow continental shelf, at one of the topographic breaks of the slope. Typically, the edge is not at the most prominent of the breaks. Also, the edge is the locus of the inflection in free-air gravity-anomaly profiles, suggesting that denser rocks may underlie the trench-slope break. Examination of single-channel (Figs 4 & 5) and multi-channel (Fig. 6) seismic profile records shows major faults, but does not show the mosaic of block faults and olistostrome-like slides known to exist from offshore drilling (Fig. 3). The minor structures on the geophysical profiles look like folds, and slides may be indistinguishable from packets of sediment. We believe that the apparent difference between coastal-zone structures and those farther offshore is due to the different investigative techniques (seismic versus subsurface).

The bathymetry, sub-bottom profiling, and bottom samples show mass movement by slumps and slides down the slopes, and turbidity currents and mass movement in submarine canyons to be the important processes of sediment transport. Canyons generally head at about 100 m depth, and extend to the trench at about 5000 m, and thereby cross different tectonic domains. They have cut through consolidated sedimentary rocks of the upper slope as is evident on the sub-bottom reflection records (Fig. 5). They can be traced bathymetrically across the lower slope, but sub-bottom reflections cannot be resolved there. If the lower slope is underlain by an accumulating wedge of tectonized sediments, the structural grain of that rising wedge is crossed by canyons. Thus it is apparent that the canyon-cutting processes were sufficiently efficient to maintain a canyon in its course rather than allow it to be deflected by wedge morphology. Canyons are essentially free of sediment. Thus canyon-cutting and transport in canyons are active geological processes (Shepherd & Moberly 1981).

The complexly faulted oil pools in the coastal fields of Peru are of 'giant' status, by virtue of having produced more than 1.3×10^8 tonnes of oil. Tar seeps and tar sands constitute an enormous potential resource of energy that is sub-economic at the present time. Most of these on-shore and off-shore fields and seeps are near Talara in the landward part of the forearc basin.

Elsewhere in the world oilfields are uncommon in arc-trench gaps. Those of Barbados and Timor are non-commercial. The St Elena fields of coastal SW Ecuador (1.5×10^7 t cumulative), and off-shore fields east of Trinidad and west of Palawan are of commercial value. Poor quality of reservoirs, low geothermal gradients for maturation of organic-rich sediments to hydrocarbons, and complex geology have tended to bias explorationists away from convergent margins.

Forearc basins of southern Peru–northern Chile

The region of southernmost Peru and northernmost Chile is characterized by a major change of strike of the Peru–Chile Trench, by great ruptures in the oceanic lithosphere sinking under South America (Rodriguez *et al.* 1976) and by active volcanism, in contrast to gaps in the line of active volcanoes to the north and south. The change in strike at Arica also marks a change in rupture length of great earthquakes (Kelleher & McCann 1976). East of Arica and Inquique the Andes are of their greatest width.

As is the case for NW Peru, rocks characteristic of continental crust are exposed along the coast of southern Peru and northern Chile (Farrar *et al.* 1970; Paredes & Megard 1972; Frutos & Feraris 1973; Pitcher 1974; Audebaud *et al.* 1973). The Peruvian Coastal Cordillera from near Pisco (14°S) to within the study area (Fig. 7) has at the coast mainly thick metasedimentary and metavolcanic sections of Precambrian age. Farther south and beyond the area of study the Peruvian and Chilean coasts are bordered chiefly by granitoid intrusives of Permian through Palaeogene age, and locally by middle Palaeozoic to late Mesozoic sedimentary and metasedimentary rocks, or Mesozoic (submarine) and Cenozoic volcanic rocks.

Large forearc basins, with about 1.5 sec (perhaps 2.9 km) of moderately deformed sediment in them, lie under the bight between 17° and 20.5°S. Farther south and farther off-shore the forearc and trench-slope basins are smaller, thin, and discontinuous; some have moderately deformed sediment but most show little evidence of deformation (Fig. 7). Of 29 crossings of the trench, 13 show no sediment. Sediments are thin on the oceanic plate. Sediments of the region have been studied by Schmalz (1958),

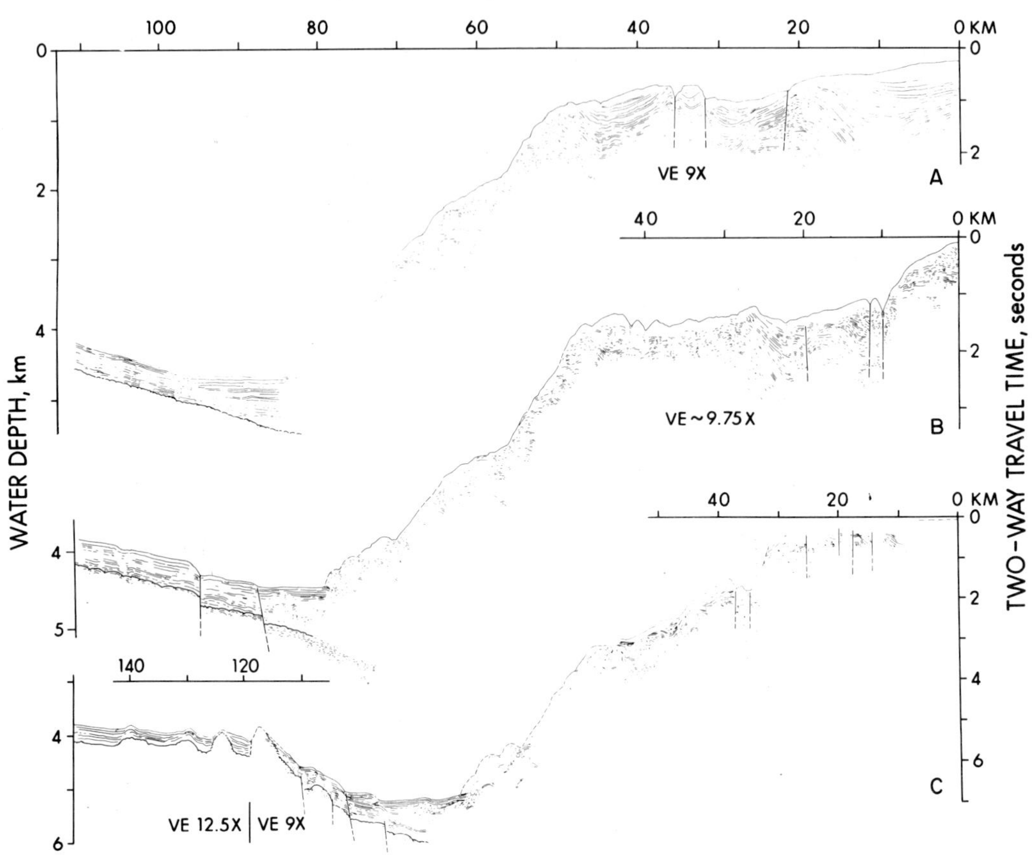

FIG. 4. Tracings of single-channel reflection records. For location see Fig. 2. Line A (top) shows no faulting of sediments and basement of oceanic plate (left). Trench floor is flat with turbidites. Record of inner trench wall is poor. Minimum free-air gravity anomaly is at about 80 km and maximum is at 50 km. Faults between 30 and 40 km have the same trend as faults on other profiles on SE edge of Banco Peru. Basin between 20 and 30 km is NE of Talara Forearc Basin. Basin between 0 and 15 km is SW end of Progreso Basin. Line B (middle) is typical of many profiles in the area: Oceanic basement and its pelagic cover are faulted; trench has an apparent mixture of pelagic, turbidite, and slump deposits; inner trench wall offers unresolvable acoustic reflections suggesting deformed sediments; sub-bottom structure and gravity inflections indicate that the small trench-slope break at 63 km (rather than the break at 45 km) is the 'structural high' marking the outer limits of the Talara Basin; canyons cut the upper slope; some reflectors of the Talara forearc basin are broken indicating high-angle faults, but most reflections suggest folds or unconformities. The inner end of line B overlaps Fig. 3 within the Cabo Blanco oilfields. Line C shows slumps at (and in?) the trench. The trench-slope break at 60 km marks the seaward edge of the Paita forearc basin.

FIG. 5. Tracings of single-channel reflection records. For locations, see Fig. 2. Lines A (top) and B (bottom) show Sarmiento Ridge, one of several fracture zones on the Nazca Plate that strike NE into the trench. On a structural basis, the trench-slope break is picked at 60 km in each profile. Apparently, the seaward edges of the sediments in the Sechura forearc basin are slumping away from the upper slope. Internal reflections suggest folds or unconformity-bounded packets of sediment. Eastern end of each line is within 15 km of test well VIRU 4X-1.

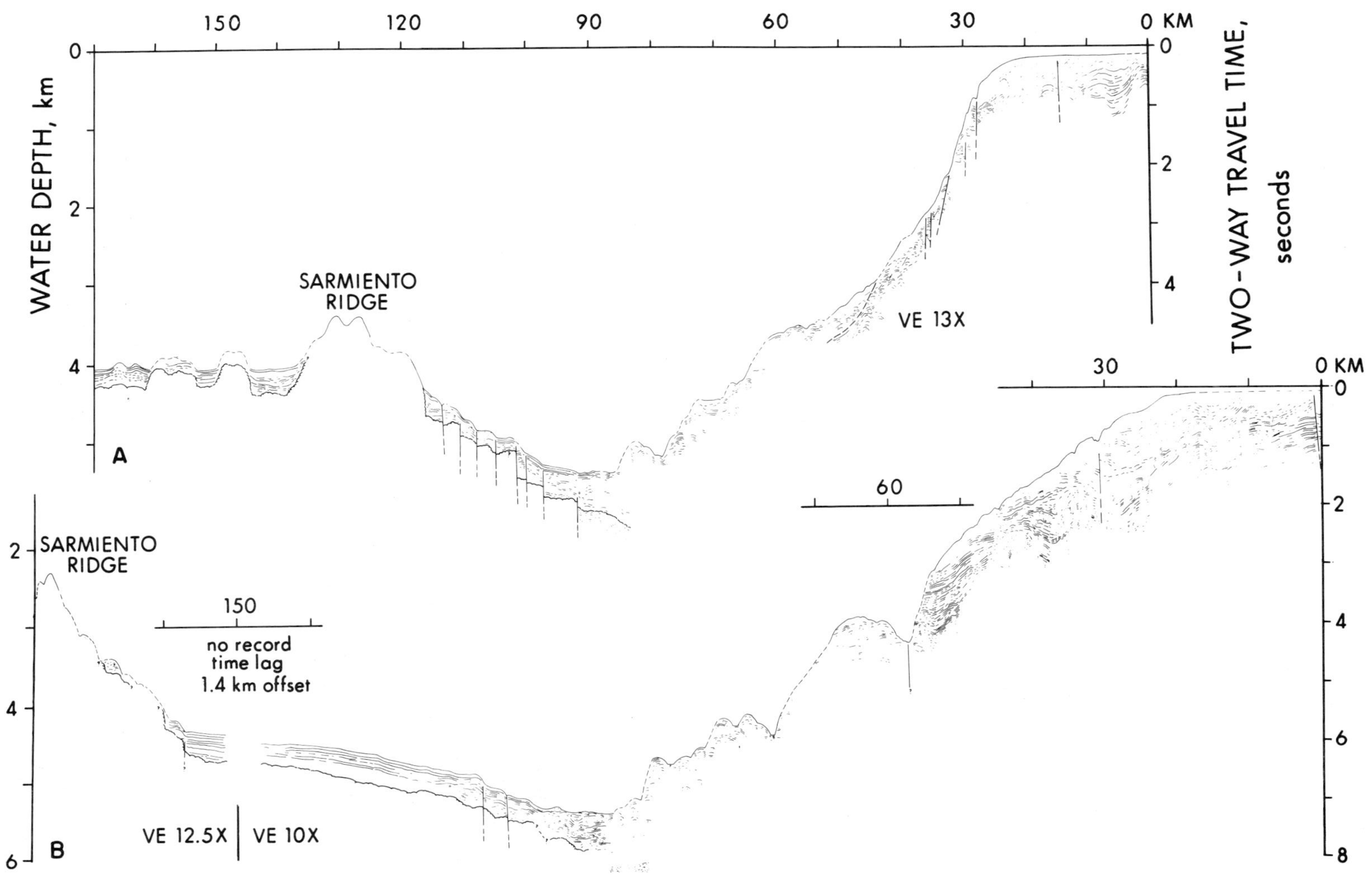

WATER DEPTH, km
TWO-WAY TRAVEL TIME, seconds
150
120
90
60
30
0 KM
0
2
4
SARMIENTO RIDGE
VE 13X
A
30
0 KM
60
2
4
6
8
SARMIENTO RIDGE
150
no record
time lag
1.4 km offset
VE 12.5X
VE 10X
B

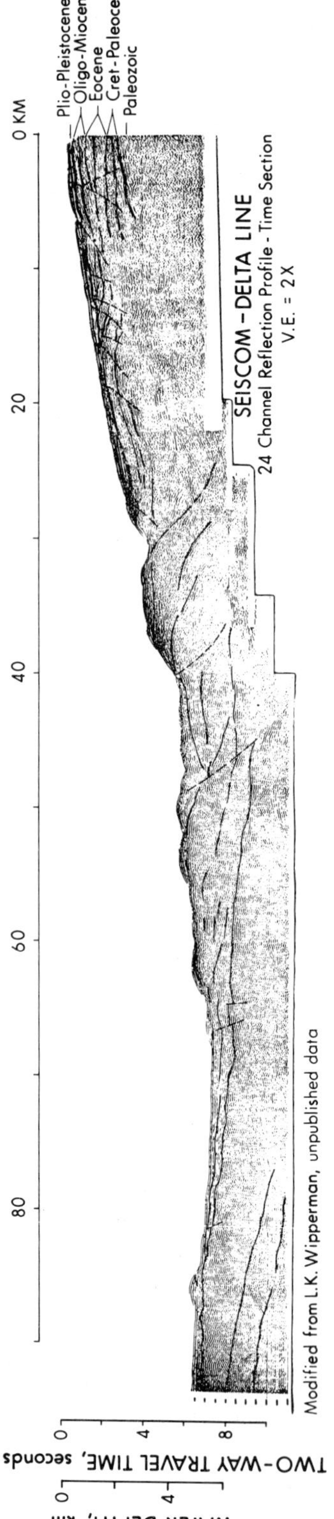

Fig. 6. Multi-channel, common-depth-point seismic reflection profile, obtained by Seiscom Delta Co. for Nazca Plate Project. Line nearly lies along middle of Fig. 5(B), for comparison. Lowest reflector at left (about 9.5 sec) probably is oceanic Moho. Highest sub-bottom reflector at left (at about 6.7 sec) is probably top of oceanic basement, and can be traced under trench and lower slope to 46 km, possibly 37 km (offset). Trench-slope break is about 35 km. Incoherent reflectors between trench and the break are accreted sediment. Coherent but faulted reflectors between the break and the end of the profile are in the Sechura Basin. Interpretation of the age and thickness of the section is projected from VIRU 4X-1, which had a velocity survey. Further discussion is in Shepherd & Moberly (1981).

Zen (1959), Bandy & Rodolfo (1964), and Rosato *et al.* (1975).

The three largest forearc basins overlap or are separated from one another by constrictions, as controlled by oblique trends of the structurally high ridges of the more highly deformed acoustic basement (Fig. 7). The basins lie over gravity minima (Fig. 8). Sediments within the Arequipa, Arica, and Iquique forearc basins can be separated into packets between successive reflectors. The thickest parts of the packets shift landward with time (Fig. 9), showing that the structural high has had a long intermittent history of growth (Coulbourn & Moberly 1977).

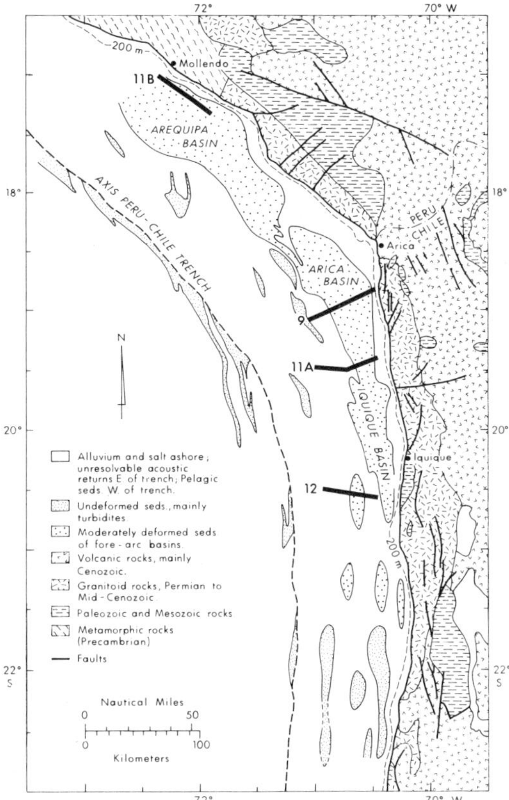

FIG. 7. Tectonic setting of the Arica Bight, coastal southern Peru and northern Chile. Coast is lined with Mesozoic batholiths and other crystalline rocks characteristic of continental basement. Shown offshore are forearc, slope, and trench basins. Locations of Figs 9, 11 and 12 are shown. Maps is simplified from Coulbourn (1977).

The origin of most cored sediment samples from the Arica Bight could be deduced easily, but some samples seemed anomalous in location, texture, or microfauna content. A set of 54 samples was subjected to a variety of statistical methods to solve for relationships between sedimentological properties and physical environment. Variables, methods, computations, and results are in Coulbourn (1980). In general there are four clusterings of sediment. One group of samples is characterized by agglutinated foraminifers, nannofossils, red to yellow colours, and dominantly clay size. Most samples come from the Nazca-plate side of the trench, and the microfossils and seismic profiles show them clearly to be pelagic. Two samples clustered with them were recovered several kilometres landward of the trench, however, at depths of 4700 and 3700 m, and presumably indicate incorporation of pelagic sediment in an uplifted tectonic wedge.

The other three clusters are characterized by grey-green colours and relatively high contents of fecal pellets, glauconite, biotite, glass, and detrital grains. They are hemipelagic. One cluster is found at a range of sites over forearc basins and the inner trench slope, and represents unconsolidated hemipelagic mud that has not been resedimented. Another cluster is also found within the forearc basins, and in addition was recovered from the axis of a submarine canyon, sites in the trench axis, and two sites slightly above the trench on faulted blocks of the Nazca Plate descending into the trench. Displaced benthonic faunas and abundant sand-sized glauconite and detrital grains indicate these to be turbidites. Presumably the two anomalous samples are of hemipelagic turbidites that crossed the trench axis where fault-block ridges and depressions of the oceanic plate intercept the trench floor off Peru (Fig. 10), and thence were channelled to the south in trench-slope depressions that have southward gradients less than that of the axis of the trench off Chile. The final cluster is a small one, of olive-grey semi-indurated mudstone sampled on a traverse over a structural high in about 1000 m of water (Fig. 11A). It is an outcrop of the tectonic wedge, which in most places is veiled by hemipelagic mud.

Reflection records, bathymetry, and sampling indicate that whereas off northern Peru slumps and turbidity currents seem to be of subequal importance in sediment transport, off southern Peru and northern Chile both are present but turbidity currents appear to be dominant. Some slumps are seen on sub-bottom reflection profiles, especially on the upper slope adjacent to structural highs (Coulbourn & Moberly 1977). Low-angle glides can

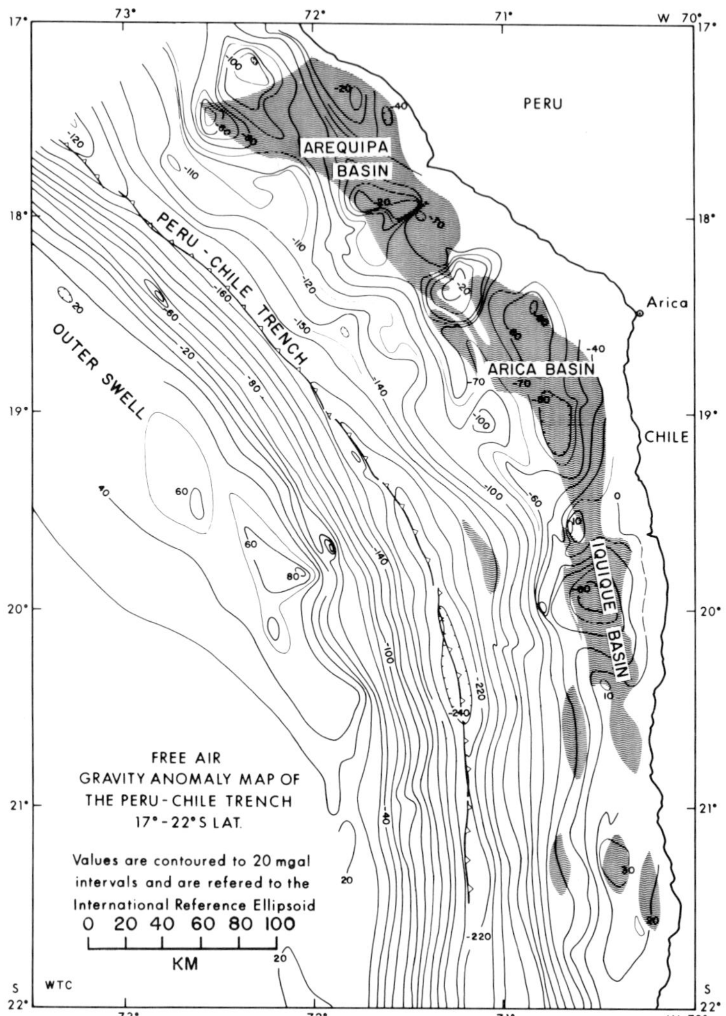

FIG. 8. Free-air gravity-anomaly map of the Arica Bight area. Gravity minimum lies virtually along trench axis. The forearc basins lie along a band of alternating gravity maxima and minima, suggesting that the basins are localized along fault-bounded depressions of a segmented basement, and that some of the faults trend at high angles to the coast. From Coulbourn (1977).

be interpreted from single-channel records of the basins (Fig. 12), using the analogy from northern Peru (Fig. 3). Perhaps some of the boundaries of the packets of sediments (Fig. 9 and Coulbourn & Moberly 1977) are low-angle glide surfaces rather than structurally deformed turbidite bedding or unconformities.

Extensive submarine canyons, recordings of off-shore cable breaks (Benest 1899), and displaced benthonic foraminifers indicate that turbidity currents fill the forearc basins. After the basins overflow, transport continues downslope. Off southern Peru the canyons strike south, at an acute angle to coast and trench, and cut into sediments of the Arequipa and Arica forearc basins (Fig. 11B). Canyons are less conspicuous off Chile, and the trench is deeper, perhaps due to less sediment supply.

Progreso Basin, northern Peru–southern Ecuador

The eastern part of the Gulf of Guayaquil is underlain by the Progreso Basin. From southern Ecuador the basin exis plunges SE to the eastern end of the Isla Puna and thence con-

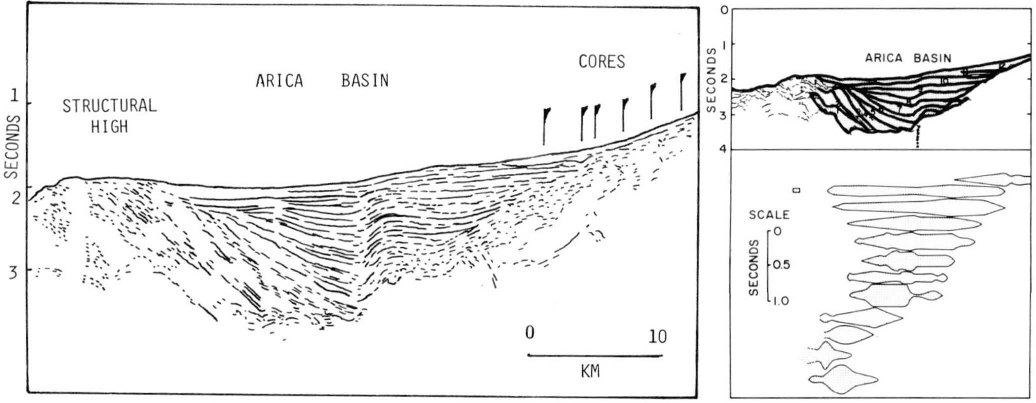

FIG. 9. Tracing of a seismic reflection profile through the Arica Basin. For location see Fig. 7. Packets of sediment (shown at right) may indicate successive migration of the centre of deposition landward as the structural high (trench-slope break) is forced upwards by accretion of a tectonized wedge of sediments between it and the trench (after Coulbourn & Moberly 1977, fig. 10). Alternatively, reflectors may be a mixture of bedding and low-angle glide planes. Symbols indicate location of cores that obtained sediments and faunas indicative of hemipelagic sedimentation.

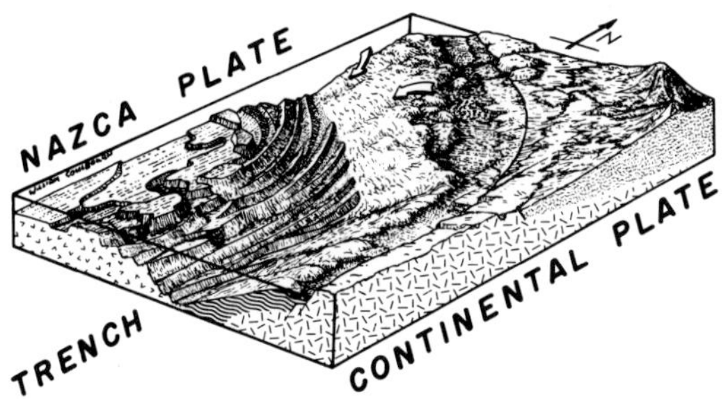

FIG. 10. Perspective of the Arica Bight, looking NW. Ridges from the Nazca Plate descend into the Peru–Chile Trench. Although nearly parallel to the Chilean part, they intercept the Peru Trench at an acute angle and may channel turbidites into the oceanic side of the trench slope. The principal sediment supply across the fore-arc basins, slope, and trench apparently comes from the Peruvian margin.

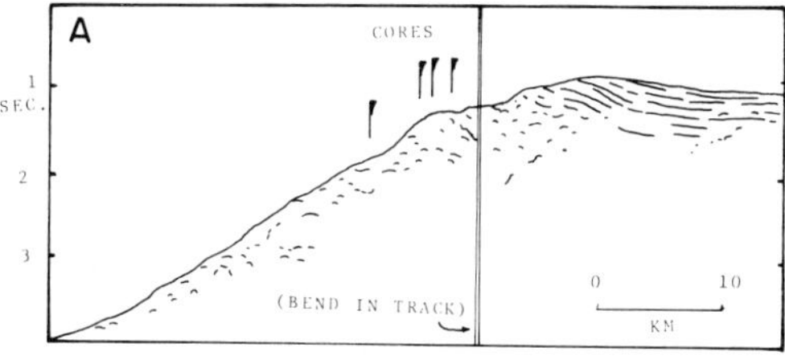

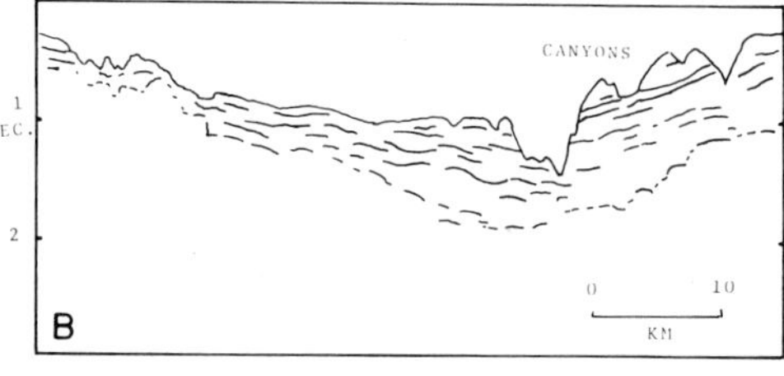

FIG. 11. Profiles with evidence of geological processes in the Arica Bight. Simplified from Coulbourn (1977). For location see Fig. 7. Line A (top) crosses a thin forearc basin (here, the arbitrary division between the Arica and Iquique basins). The main structural high was cored (symbols), and yielded indurated sediment that probably is the inner edge of the accreted tectonized wedge. Line B (bottom) shows canyons near the coast of Peru that are cutting sediments of the Arequipa forearc basin. Canyons with terraces and levees in the Arequipa Basin have been illustrated elsewhere (Coulbourn & Moberly 1977).

tinues plunging SW past Isla Santa Clara (Fig. 2). Gravity and seismic surveys suggest that the centre of the basin lies 20 km west of Isla Santa Clara (Fig. 13). The Zorritos oilfields on its south flank produce from Miocene sandstones, and gas discoveries have been made in the Gulf of Guayaquil. The basin holds more than 6000 m of Miocene and younger sediments. Its north and south margins are marked by steep gradients in the free-air gravity anomalies (Fig. 13), and the south margin has growth faults along it. The eastward embayment of the basin

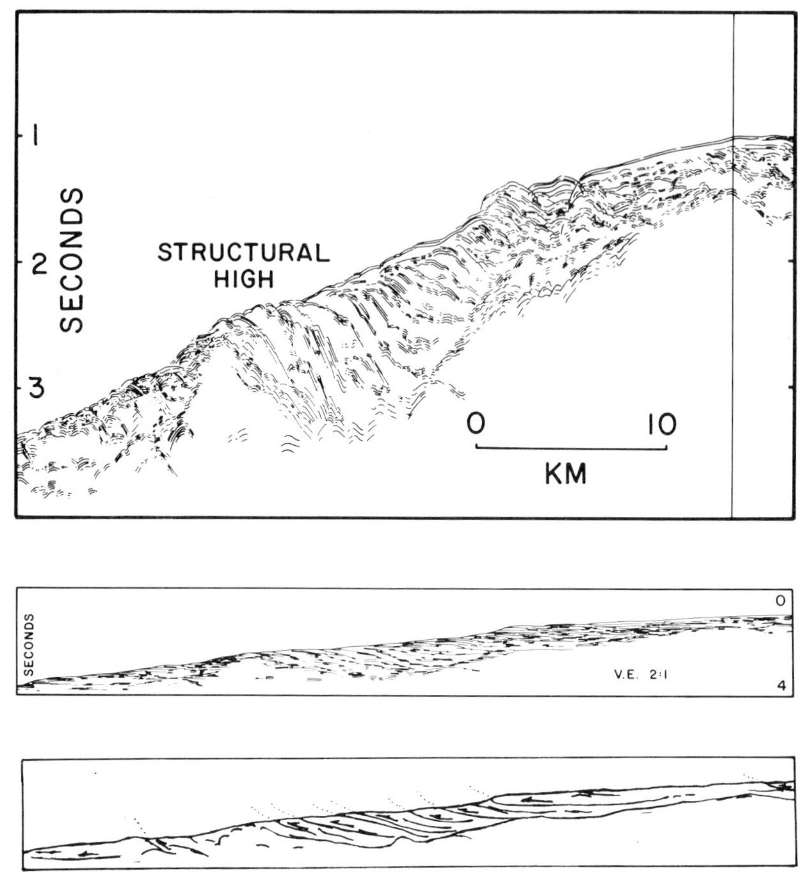

FIG. 12. Tracing of a seismic reflection profile of a small forearc basin seaward of the Iquique Basin, showing reflectors generally converging toward the structural high. For location, see Fig. 7. Below, the tracing has been photographed through a strip camera so that the vertical exaggeration of the topography is reduced to 2:1. Whether or not any reflectors are low-angle glide faults is an unanswered question; a possible interpretation is shown below.

contrasts with the strike of other South American coastal basins, which follow the trend of the Andes and trench. According to Faucher & Savoyat (1973) the basin continues landward into the Jambeli Graben, where about 9000 m of Cenozoic sediments lie. Several investigators have proposed that the Gulf of Guayaquil and Progreso Basin have formed in response to right-lateral motion on the Dolores-Guayaquil megashear (Marchant 1961; Malfait & Dinkelman 1972; Goosens 1973; Faucher & Savoyat 1973; Shepherd & Moberly 1975), or to interaction between the Tumbes-Guayana (left lateral) and Dolores-Guayaquil megashears (Campbell 1975).

In a different interpretation, Lonsdale (1978) terms the Progreso Basin a mature example of a trench-slope basin, believing it to be structurally and genetically similar to basins forming today between thrust slices on the inner slope of trenches. In this case, however, the basin has received a thick fill, and has been sufficiently uplifted that its landward part is exposed. Although Henderson & Evans (1980) also term the Progreso a trench-slope basin, they differ from Lonsdale in interpreting the present tectonic division between it and the Guayas Valley Basin, which they term a forearc basin, as dating only from the late Pliocene.

The Dolores-Guayaquil megashear (Fig. 14A) is traced on land from north-western Colombia to the latitude of Guayaquil, Ecuador. It separates a terrane of Mesozoic and Cenozoic mafic volcanic and marine sedimentary rocks in coastal Colombia and Ecuador, which probably formed as oceanic

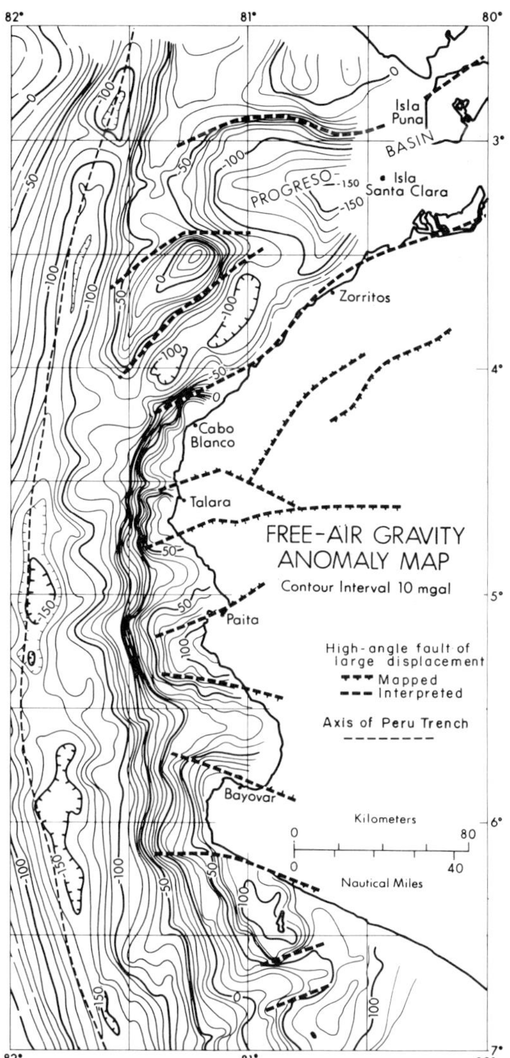

FIG. 13. Free-air gravity-anomaly map of the South American continental margin off NW Peru and SW Ecuador. Gravity minimum lies along and landward of the trench axis. High-angle faults bounding the Gulf of Guayaquil and Banco Peru are seen on reflection records or inferred from the gravity. They are splays of the Dolores-Guayaquil megashear, and outline the Progreso Basin. Other high-angle faults bound massifs or horsts of Palaeozoic metasedimentary and Mesozoic plutonic basement rocks. They are inferred to continue seaward and mark the lateral limits of the forearc basins along the coast.

crust, from a terrane of Mesozoic and other metamorphic and plutonic rocks, the continental crust of northern South America. Geological evidence for the fault zone separating two types of crust, and its right-lateral movement, is supported by seismicity, earthquake first-motion, and gravity studies, as summarized by Case *et al.* (1971) and Campbell (1974), who projected the fault SW under the south edge of the Gulf of Guayaquil. Geophysical evidence for the extension under the Gulf into the Peru Trench was presented by Shepherd & Moberly (1975, 1981).

According to our interpretation, the Dolores-Guayaquil megashear is a transform fault ending at a trench-trench-transform triple junction near 4°S. It is the eastern boundary of a small plate uplifted from oceanic crust in coastal Ecuador and Colombia, trapped against the continent when the convergent margin jumped from a former position at the megashear to its present location at the Colombia-Ecuador trench. Progreso Basin is a rhombochasm-like basin (pull-apart basin of Crowell 1974) caused by right-lateral movement on the megashear (Fig. 14B).

The river Guayas and its many tributaries are the principal source of sediment for the Gulf of Guayaquil and the Progreso Basin. Presumably the ancestral Guayas was the past source as well. The climate-controlled gradient of sediment supply probably existed during the Cenozoic, favouring a northern source. Moreover, the former subduction zone or suture which became the Dolores-Guayaquil megashear would have provided an intermontane lowland trapping runoff (Potter 1978) and directing it laterally.

A number of present-day island arcs and other young orogenic belts are cut at acute angles by major fault zones that are active today or have been in the recent past. Several of the faults control the position of rivers. Pull-apart basins with thick sediments may be a common feature of ancient convergent margins.

Trench and trench-slope basins

Canyons on the slopes as well as samples of displaced faunas and turbidite sands, and profiler records of slumps and of pelagic-capped fault blocks that descend into the trench all show that the Peru–Chile Trench is a sedimentary basin. The basin, however, does not accommodate great thicknesses of sediment. Cross-sections of profiles rarely show as much as 1 sec (two-way travel time; perhaps 800 m) of sediment in the trench, and sediment is nil in many crossings.

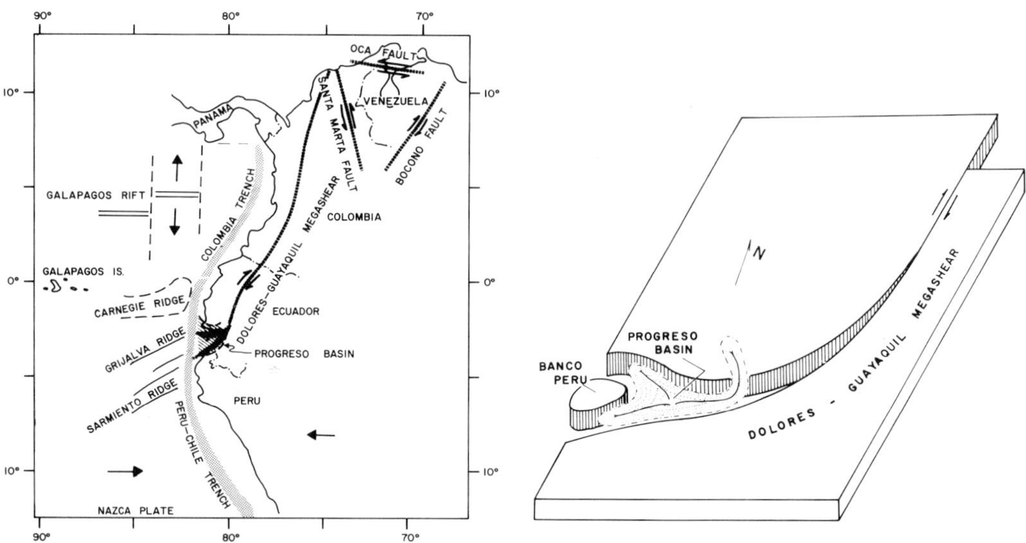

FIG. 14. Regional tectonic trends of NW South America, showing Dolores-Guayaquil Megashear as a transform fault separating a small plate between the Colombia Trench and the South American Plate. The transform fault may be nearly inactive today, but its earlier right-lateral movement may have left the Progreso Basin as a rhombochasm or pull-apart basin.

Probably more than half of the sediment is pelagic, entering from the oceanic plate, and less than half is hemipelagic, redeposited as turbidites and slumps or deposited directly as a rain of suspended particles. Pyroclastic and authigenic contributions are easily identified but modest. About four-fifths or more of the trench sediment is removed by subduction (Shepherd & Moberly 1981), leaving a small fraction to enter an accretionary prism landward of the trench. The accretionary prism may be more substantial where a thicker section of pelagic sediment can be provided to the trench, as for example the carbonate oozes preserved on the Nazca (Kulm *et al.* 1974) and Carnegie (Lonsdale 1978) ridges.

Geological processes along the Andean margin

For the most part, our observations off South America support current concepts of sedimentary and tectonic processes during subduction. In some instances, however, we believe modifications of those concepts are needed. The question is, how typical of recent and ancient arc-trench gaps is the Andean margin?

Sedimentation

The mixture of sedimentary sources and tectonic processes off Peru today shows that caution is necessary in palaeoenvironmental reconstructions of ancient subduction zones where exposures are incomplete and samples are few.

Submarine canyons are efficient avenues of sediment supply for forearc basins, trench-slope basins, and trenches. As the forearc basins are lifted, canyons may cut them, reworking sediment and carrying it to greater depths. Mass movement in the form of small slumps and immense low-angle slides lead to olistrostrome-like structures. They may gravitate to the same levels as the tectonic mélange of the accretionary wedge, leading to a possible source of confusion in interpretation.

The hydrocarbon accumulations in the forearc basins and a pull-apart basin in northern Peru and southern Ecuador show that not all arc-trench gaps are unsatisfactory tectonic and sedimentary settings for large oil and gas fields. Perhaps they are exploited off Peru because they are partly exposed on land, unlike most young forearc basins, so that their tar seeps are plainly visible. Perhaps they have received enough sediment to generate and mature hydrocarbons, but are not so old or deformed that the oil is lost.

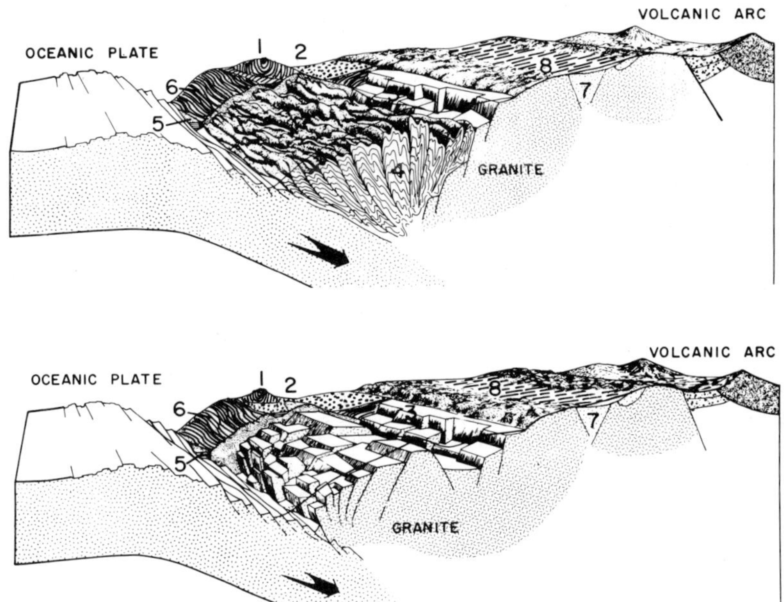

FIG. 15. Alternative hypotheses for the relative roles of accretion and erosion or attrition by subduction. Legend: (1) structural high, (2) forearc basin, (3) block-faulted continental crust under all or part of forearc basins, (4) tectonic mélange of the accretionary prism, (5) trench-axis turbidites, (6) deformed and dewatered hemipelagic sediments of the continental margin, (7) Mesozoic sediment, (8) coastal batholith. An earlier version of this illustration appeared in the 1977 JOIDES report on the Future of Scientific Ocean Drilling.

Shallow deformation

Across the Andean margin, evidence of young compressive tectonics is restricted to: (1) lithospheric convergence shown by earthquakes, (2) the Subandean Province between the Cordillera and the shields, and (3) a small prism immediately landward of the trench.

In the case of (1), the planes and first-motion studies of the shallow to deep earthquakes dipping under South America show the convergence. Burchfiel & Davis (1976) believe that thermal weakening of the overriding edge of the continental plate led to its middle Miocene to recent deformation, thickening, and uplift in response to the horizontal compressive stress at the shallow-dipping (10–15°) Benioff zone. (2) Foreland-type folding and thrust faulting of the sedimentary prism characterizes the Subandean Province.

For (3), the inner wall of the trench, morphotectonics and interpretation of seismic reflection profiles suggest a belt of tight folds and imbricate thrust faults, but there is no proof of it. An alternative to a compressive-structure interpretation of the profiles in the Arica Bight where pelagic and hemipelagic sediment supply is so scanty, is that some ridges of the belt may in part be ridges of oceanic basalt crust clipped off the top of the descending lithosphere (Coulbourn & Moberly 1977).

Tensional tectonics prevail in all other segments of a wide traverse across this convergence zone, as may be the general case for convergent plates (Elsasser 1971; Moberly 1972). Kanamori (1971) has suggested that the landward and subducting plates nearly decouple from time to time, along the thrust-fault interface between. The topography of an outer rise may result from flexure of the oceanic plate without buckling from horizontal forces (Caldwell *et al.* 1977, for a contrary view see Watts *et al.* 1976), even though an interpretation off central Peru suggests buckling (Hussong *et al.* 1976). Normal faulting in the shallow arch of the outer ridge clearly is tensional (Coulbourn 1977), and so is the step-like fault-block pattern from the outer ridge down the outer wall of the trench (Shepherd 1979; Coulbourn & Moberly 1977).

Although the single- and multi-channel seismic reflection records might suggest folds (Figs

5 & 6), the subsurface information (Fig. 3) and one interpretation of seismic records (Fig. 12) show gravity faulting rather than compressive folding. Some faults are high-angle ones that cut into basement (not listric), and others are immense low-angle glides. Farther on-shore and through the central Andes broad uplift and normal faulting has prevailed into recent times (Cobbing & Pitcher 1972; Gansser 1973).

Tectonic erosion and accretion during subduction

The efficiency of the subduction process has an obvious control on the sediments and shallow structures in trench and forearc basins. We have pointed out that most of the sediment entering the trench is subducted. We believe that continental crust is removed as well. Coastal Peru and northern Chile have plutonic and metamorphic rocks at the coast, exposed 80–140 km from the trench axis, and interpretations of the style of deformation, the position of gravity-anomaly inflections, and velocities along multi-channel and ASPER seismic lines suggest that the edge of continental crust may be as close as 20–40 km to the trench axis off NW Peru (Shepherd & Moberly 1981), 50 km at 12°S (Hussong *et al.* 1976), and about 50 km off the Arica Bight. Moreover, structures transverse to the coast strike seaward (and help to delimit the lateral limits of individual forearc basins; Shepherd 1979; W. S. Pitcher pers. comm. to W.T.C. 1975). Mesozoic plutons now at the coast could not have formed by any reasonable pattern of geotherms and dipping plate so close to the trench. All of these observations strongly suggest that subduction has removed by stoping the underside and edge of continental crust.

Karig (1974) thought that tectonic erosion could be by lateral translation. The terrane of coastal Middle America, Colombia, and Ecuador resembles old oceanic crust, not continental (Case *et al.* 1971; Schmidt-Effing 1979). Mesozoic batholiths and Precambrian and Palaeozoic metasedimentary rocks line the Chilean shoreline for 3000 km south of the region studied. We do not know where lateral slices could have gone.

The past decade has been one of revolutionary interpretations of the origin of much of western Canada, the U.S. Pacific North-west, and Alaska, as a collage of large and small blocks of 'exotic terranes' that have been swept against the North American continent by plate convergence (e.g. Davis *et al.* 1978). The Andes and indeed all mountain belts will have to be re-examined to determine whether or not they have had a similar history. To date, no such exotic blocks have been discovered in the Andes, but if they are, the granitic rocks at the coast might be explained as a result of a rifted margin: origin of a spreading centre within some Mesozoic granitic crust, movement of a block or blocks with a Pacific passive or trailing margin to a subduction zone along nuclear South America, and then welding of those blocks to South America and the initiation of a new subduction zone along the present trench. Until there is clear field evidence of exotic terranes from mapped suture zones and palaeomagnetic-pole calculations in the Andes, we prefer the alternative that subduction has removed the edge of the crust.

Fig. 15 shows two extreme interpretations for the tectonic control of basins in the arc-trench gap. One is that the accretion of a wedge of tectonized mélange makes the structural high separating the forearc basin from the remainder of the slope. Imbricate slices of the mélange trap hemipelagic sediments in slope basins and may deform and dewater them. Moore *et al.* (1980) present a current interpretation of this tectonic type for Sumatra.

The figure also shows an interpretation where accretion is slight and subcrustal erosion is dominant. The structural high is a horst of continental crust over which sediments are draped. The trench slope under which structure is so difficult to resolve by seismic reflection methods is shown as a mosaic of fault-sliced blocks of continental crust entering the trench. Their movement, and a modest amount of accreted pelagic sediment or oceanic crust slivers moving up over them, deforms and dewaters the hemipelagic blanket. This example may exist off Peru, Chile, and Mexico. Or, more likely, truth may lie somewhere between the two extremes.

ACKNOWLEDGMENTS: This research was supported by National Science Foundation grant ID071–04207–A04. We appreciate technical assistance by HIG staff on board R/V *Kana Keoki* and in the laboratories. Conversations with Donald Hussong, the late George Woollard, LaVerne Kulm, Johanna Resig, William Woodard, Larry Wipperman, Roland von Huene, Leo Ocola, Hector Ayon, Fernando Feraris, O. Zevallos, and others have helped to shape the evolution of our thoughts about the South American margin. So also have the results of the 1977–1979 *Glomar Challenger* drilling of Pacific active margins. Hawaii Institute of Geophysics Contribution No. 1091.

References

AUDEBAUD, E., CAPDEVILA, R., DALMYRAC, B., DEBELMAS, J., LAUBACHER, G., LEFEURE, C., MAROCCO, R., MARTINEZ, C. MATTAUER, M., MEGARD, F., PAREDES, J. & TOMASI, P. 1973. Les traits geologiques essentiels des Andes Centrales (Perou-Bolivie). *Rev. Geog. phys. Geol. dynamique,* **15,** 73–114.

BALDRY, R. A. 1938. Slip-planes and breccia zones in the Tertiary rocks of Peru. *Q.J. geol. Soc. London,* **94,** 347–58.

BANDY, O. L. & RODOLFO, K. 1964. Distribution of foraminifera and sediments: Peru–Chile Trench area. *Deep Sea Res.* **11,** 817–37.

BENEST, H. 1899. Submarine gullies, river outlets and fresh water escapes beneath the sea level. *Geog. J.* **14,** 394–413.

BELLIDO, E. 1969. Sinopsis de la geologia del Peru. *Bol. Serv. Geol. Min. Lima,* **22.**

BURCHFIEL, B. C. & DAVIS, G. A. 1976. Compression and crustal shortening in Andean-type orogenesis. *Nature. London,* **260,** 693–4.

BURNETT, W. C. 1977. Geochemistry and origin of phosphorite deposits from off Peru and Chile. *Bull. geol. Soc. Am.* **88,** 813–23.

—— , VEEH, H. H. & SOUTAR, A. 1980. U-series, oceanographic, and sedimentary evidence in support of recent formation of phosphate nodules off Peru. *In:* BENTOR, Y. K. (ed.). *Marine Phosphorites.* Spec. Publ. Soc. econ. Paleontol. Mineral. Tulsa, **29,** 61–71.

CALDWELL, J. G., TURCOTTE, D. L., HAXBY, W. F. & KARIG, D. E. 1977. Thin elastic plate analysis of outer rises (abs.). *In:* TALWANI, M. & PITMAN, W. C. III (eds). *Island Arcs, Deep Sea Trenches, and Back-arc Basins.* Am. geophys. Union, M. Ewing Ser. **1,** 467.

CAMPBELL, C. J. 1974. Ecuadorian Andes. *In:* SPENCER, A. M. (ed.). *Mesozoic-Cenozoic Orogenic Belts.* Spec. Publ. geol. Soc. London, **4,** 725–32.

—— 1975. Ecuador. *In:* FAIRBRIDGE, R. W. (ed.). *The Encyclopedia of World Regional Geology, Part 1, Western Hemisphere,* 261–270. Dowden, Hutchinson & Ross, Stroudsburg, U.S.A.

CASE, J. E., DURAN, L. G., LOPEZ, A. R. & MOORE, W. R. 1971. Tectonic investigations in western Colombia and eastern Panama. *Bull. geol. Soc. Am.* **83,** 2685–712.

COBBING, E. & PITCHER, W. 1972. Plate tectonics and the Peruvian Andes. *Nature (phys. Sci)* **240,** 51–3.

COSSIO, A. & JAEN, H. 1967. *Geologia de los Cuadrangulos de Puemape, Chocope, Otuzco, Trujillo, Salaverry y Santa.* Bol. Serv. Geol. Miner. Peru, no. 17.

COULBOURN, W. T. 1977. *Tectonics and sediments of the Peru–Chile Trench and continental margin at the Arica Bight.* Thesis, PhD, Univ. Hawaii, Honolulu (unpubl.).

—— 1980. Relationship between the distribution of foraminifera and geologic structures of the Arica Bight, South America. *J. Paleontol.* **54,** 696–718.

—— & MOBERLY, R. 1977. Structural evidence for the evolution of fore-arc basins off South America. *Can. J. Earth Sci.* **14,** 102–16.

CROWELL, J. C. 1974. Origin of late Cenozoic basins in southern California. *In:* DICKINSON, W. R. (ed.). *Tectonics and Sedimentation.* Spec. Publ. Soc. econ. Paleontol. Mineral. Tulsa, **22,** 190–204.

DAVIS, G. A., MONGER, J. W. H. & BURCHFIEL, B. C. 1978. Mesozoic construction of the Cordilleran 'collage' central British Columbia to central California. *In:* HOWELL, D. G. & MCDOUGALL, K. A. (eds). *Mesozoic Paleogeography of the Western United States.* Pacif. Sec., Soc. econ. Paleontol. Mineral. Tulsa, 1–32.

DRAKE, C. L. (ed.) 1973. *U.S. Program for the Geodynamics Project: Scope and Objectives.* Natn. Acad. Sci., Washington.

ELSASSER, W. M. 1971. Sea-floor spreading as thermal convection. *J. geophys. Res.* **76,** 1101–12.

FARRAR, E., CLARK, A. H., HAYNES, S. J., QUIRT, G. S., CONN, H. & ZENTILLI, M. 1970. K:Ar evidence for the post-Paleozoic migration of granitic intrusion foci in the Andes of northern Chile. *Earth planet. Sci. Lett.* **10,** 60–6.

FAUCHER, B. & SAVOYAT, E. 1973. Equisse geologique des Andes de l'equateur. *Rev. Geog. phys. Geol. dynamique* **15,** 115–42.

FISCHER, A. G. 1956. Desarrollo geologico del noroeste Peruano durante el Mesozico. *Bol. Soc. geol. Peru,* **30,** 117–90.

FISHER, R. L. & RAITT, R. W. 1962. Topography and structure of the Peru–Chile Trench. *Deep Sea Res.* **9,** 423–43.

FRUTOS, V. & FERARIS, F. 1973. *Mapa Tectonico de Chile.* Instit. de Invest. geol., Santiago.

GANSSER, A. 1973. Facts and theories on the Andes. *J. geol. Soc. London,* **129,** 93–131.

GILETTI, B. J. & DAY, H. W. 1968. Potassium-argon ages of igneous rocks in Peru. *Nature. London,* **220,** 570–2.

GOOSENS, P. J. 1973. Characteristicas estructurales de la parte noroeste del margin continental de America del Sur relaconadas con la tectonica de placas (abs.). *Int. Union Geod. Geophys. Conf. on Geodynamics, Lima.*

HANDSCHUMACHER, D. W. 1976. Post-Eocene plate tectonics of the Eastern Pacific. *In:* SUTTON, G. H., MANGHNANI, M. H. & MOBERLY, R. (eds). *The Geophysics of the Pacific Ocean Basin and its Margin [Woollard Volume].* Monogr. Am. geophys. Union, **19,** 177–202.

HAYES, D. E. 1966. A geophysical investigation of the Peru–Chile Trench. *Mar. Geol.* **4,** 309–51.

HENDERSON, W. G. & EVANS, C. D. R. 1980. Ecuadorian subduction system: discussion. *Bull. Am. Assoc. Petrol. Geol.* **64,** 280–2.

HERRON, E. M. 1972. Sea-floor spreading and the Cenozoic history of the east-central Pacific. *Bull. geol. Soc. Am.* **83,** 1671–92.

HOSMER, H. L. 1959. *Geology and structural development of the Andean system of Peru.* Thesis, PhD, Univ. Michigan, Ann Arbor (unpubl.).

HUSSONG, D. M., EDWARDS, P. B., JOHNSON, S. H., CAMPBELL, J. F. & SUTTON, G. H. 1976. Crustal structure of the Peru–Chile Trench: 8°–12°S latitude. *In:* SUTTON, G. H., MANGHNANI, M. H. & MOBERLY, R. (eds). *The Geophysics of the Pacific Ocean Basin and its Margin* [*Woollard Volume*]. Monogr. Am. geophys. Union, **19,** 71–86.

JENKS, W. F. (ed.) 1956. Handbook of South American geology. *Mem. geol. Soc. Am.* **65,** 378 pp.

KANAMORI, H. 1971. Great earthquakes at island arcs and the lithosphere. *Tectonophysics*, **12,** 187–98.

KARIG, D. E. 1974. Tectonic erosion at trenches. *Earth planet. Sci. Lett.* **21,** 209–12.

KELLEHER, J. & MCCANN, W. 1976. Buoyant zones, great earthquakes, and unstable boundaries of subduction. *J. geophys. Res.* **81,** 4885–96.

KULM, L. D., RESIG, J. M., MOORE, T. C., JR. & ROSATO, V. J. 1974. Transfer of Nazca Ridge pelagic sediments to the Peru continental margin. *Bull. geol. Soc. Am.* **85,** 769–80.

——, ——, THORNBURG, T. M. & SCHRADER, H. J. 1981. Cenozoic tectonics and stratigraphy of a fast-convergence zone—Peru forearc region (this volume).

——, SCHWELLER, W. & MASIAS, A. 1977. A preliminary analysis of the subduction process along the Andean continental margin, 6° to 45°S. *In:* TALWANI, M. & PITMAN, W. C. III (eds). *Island Arcs, Deep Sea Trenches, and Back-arc Basins.* Am. geophys. Union, M. Ewing Ser. **1,** 285–301.

LONSDALE, P. 1978. Ecuadorian subduction system. *Bull. Am. Assoc. Petrol. Geol.* **62,** 2454–77.

MALFAIT, B. I. & DINKELMAN, M. G. 1972. Caribbean tectonic and igneous activity and the evolution of the Caribbean plate. *Bull. geol. Soc. Am.* **83,** 251–71.

MAMMERICKX, J., HERRON, E. & DORMAN, L. 1980. Evidence for two fossil spreading ridges in the southeastern Pacific. *Bull. geol. Soc. Am.* **91,** 263–72.

MARCHANT, S. 1961. A photogeological analysis of the western Guayas Province, Ecuador. *Q. J. geol. Soc. London,* **117,** 215–32.

MARTINEZ, M. 1970. Geologica de basamento Paleozoic en las Montanas de Amotope y posible origen del petroleo en rocas paleozoicas del noroeste del Peru. *Primir Congreso Latinoamer. Geol.* **2,** 105–38.

MASIAS, J. A. 1975. *Morphology, shallow structure, and evolution of the Peruvian continental margin, 6° to 18°S.* Thesis, MS, Oregon State Univ., Corvallis (unpubl.).

MOBERLY, R. 1972. Origin of lithosphere behind island arcs, with reference to the western Pacific. *In:* SHAGAM, R. (ed.). *Studies in Earth and Space Sciences* [*Hess Volume*]. Mem. geol. Soc. Am. **132,** 35–55.

MOORE, G. F., BILLMAN, H. G., HEHANUSSA, P. E. & KARIG, D. E. 1980. Sedimentology and paleobathymetry of Neogene trench-slope deposits, Nias Island, Indonesia. *J. Geol.* **88,** 161–80.

MORRIS, R. C. & ALEMAN, A. 1975. Sedimentation and tectonics of middle Cretaceous Copa Sombero Formation in northwest Peru. *Bol. Soc. geol. Peru,* **48,** 49–64.

PAREDES, J. 1966. *Estratigrafia del Paleozoico del noroeste Peruano.* Comision Carta Geologica Nacional del Peru (unpubl. report).

—— & MEGARD, F. 1972. *Carte Structurale schematique des Andes de Perou* (privately distributed).

PAREDES, M. P. 1958. *Terciario de la Brea y Parinas y area de Lobitos.* Thesis, PhD, Univ. Sciences et Techniques de Languedoc, Montpellier (unpubl.).

PAZ, M. H. 1974. Estudio estratigrafico del subsuelo de los yacimentos Zapatal-La Tuna-Coyonitas-Ronchudo Petroles del Peru, operaciones noroeste, Talara (abs.). *III Cong. Peru de Geol., Soc. geol. del Peru.*

PITCHER, W. S. 1974. The Mesozoic and Cenozoic batholiths of Peru. *Pacif. Geol.* **8,** 51–62.

POTTER, P. E. 1978. Significance and origin of big rivers. *J. Geol.* **86,** 13–33.

RODRIGUEZ, R. E., CABRÉ, R. & MERCADO, A. 1976. Geometry of the Nazca Plate and its dynamic implications. *In:* SUTTON, G. H., MANGHNANI, M. H. & MOBERLY, R. (eds). *The Geophysics of the Pacific Ocean Basin and its Margin* [*Woollard Volume*]. Monogr. Am. geophys. Union, **19,** 87–103.

ROSATO, V. J., KULM, L. D. & DERKS, P. S. 1975. Surface sediments of the Nazca Plate. *Pacif. Sci.* **29,** 117–30.

RUEGG, W. 1967. El margen suroriental la cuenca Para-Andina de Sechura en el noroeste del Peru. *Bol. Soc. geol. Peru,* **40,** 99–121.

SCHMALZ, R. F. 1958. *A technique for quantitative model analysis by X-ray diffraction and its application to modern sediments of the Peru–Chile Trench.* Thesis, PhD, Univ. Harvard, Cambridge (unpubl.).

SCHMIDT-EFFING, R. 1979. Alter und Genese des Nicoya-Komplexes, einer ozeanischen Palaokruste (Oberjura bis Eozan) in sudlichen Zentralamerika. *Geol. Rdsch.* **68,** 457–94.

SHEPHERD, G. L. 1979. *Shallow crustal structure and marine geology of a convergence zone, northwest Peru and southwest Ecuador.* Thesis, PhD, Univ. Hawaii, Honolulu (unpubl.).

—— & MOBERLY, R. 1975. Southern extension of the Dolores-Guayaquil megashear across the continental margin of northwest Peru and the Gulf of Guayaquil (abs.). *Trans. Am. geophys. Union,* **56,** 442.

—— & —— 1981. Coastal structure of the continental margin, northwest Peru and southwest Ecuador. Unpublished manuscript accepted for: KULM, L. D. *et al.* (eds). *Nazca Plate,* Monogr. geol. Soc. Am.

TRAVIS, R. B. 1953. La Brea-Parinas oil field, northwest Peru. *Bull. Am. Assoc. Petrol. Geol.* **37,** 2093–118.

WATTS, A. B., TALWANI, M. & COCHRAN, J. R. 1976. Gravity field of the north-west Pacific Ocean at its margin. *In:* SUTTON, G. H., MANGHNANI, M.

H. & MOBERLY, R. (eds). *The Geophysics of the Pacific Ocean Basin and its Margin [Woollard Volume]*. Monogr. Am. geophys. Union, **19**, 17–34.

WEEKS, L. G. 1947. Paleogeography of South America. *Bull. Am. Assoc. Petrol. Geol.* **31,** 1194–241.

ZEN, E. 1959. Mineralogy and petrology of marine bottom sediment samples off the coast of Peru and Chile. *J. sediment. Petrol.* **29,** 513–39.

R. MOBERLY, Hawaii Institute of Geophysics, University of Hawaii, Honolulu, Hawaii 96822, U.S.A.

G. L. SHEPHERD, Cities Service East Asia, Inc., 606 Cathay Bldg., Mount Sophia, Singapore 9.

W. T. COULBOURN, Geologic Research Division, Scripps Institution of Oceanography, La Jolla, California 92093, U.S.A.

The geology of the western part of the Borbón Basin, North-west Ecuador

C. D. R. Evans & J. E. Whittaker

SUMMARY: The western part of the Borbón Basin, North-west Ecuador, contains some 5 km of predominantly thin-bedded Tertiary mudstones with rare shallow-water sandstones, resting on a basement of Cretaceous basalts. The basin is floored by thin, discontinuous, turbiditic carbonates of middle Eocene age, which pass up into a 3 km thick sequence of mudstones, cherty at the base, extending into the mid-Miocene. Shallow-water sandstones accumulated during the late Miocene and early Pliocene and were succeeded by deeper water mudstones, which in the Pliocene, are over 1 km thick and contain resedimented sandstone units. The strata are folded into NE-SW-trending structures, parallel to the mid-slope basement high centred on the Río Verde area. Stratigraphic evidence suggests that much of this folding, and the emergence of the high, post-dates the late Pliocene (*c*.2.9 Ma).

Since the middle Eocene the area has occupied the arc-trench gap between the Ecuador–Colombia oceanic trench, and the volcanic arc of the Andes; it now lies on the oceanward side of the mid-slope basement high.

Along the west coast of Ecuador and Colombia, between the Andes and the Pacific Ocean, lies a belt of lowlands varying in width from 60–200 km (Fig. 1c). This is the site of the Bolivar 'geosyncline' (Nygren 1950), a thick accumulation of Tertiary sediments resting on a Mesozoic, largely basaltic basement. Nygren (1950) recognized within the 'geosyncline' a number of basins or 'deeps', three of which, the South Guayaquil (or Progreso), the Manta (or Manabí) and the Borbón, occur in the coastal lowlands of Ecuador (the Costa). This paper deals with the geology of the western part of the Borbón Basin (Fig. 2).

The International Ecuadorian Petroleum Company (I.E.P.C.) drilled three wells in the Borbón Basin, including Camarones-1 in the mapped area, during the 1940s. These wells, and unpublished oil company reports provided the basis for the regional stratigraphy (Stainforth 1948, 1968; Canfield 1966). Between 1966 and 1972 the Institut Français du Pétrole (I.F.P.) examined in detail some sections in the basin and data from their Río Verde traverse (Faucher *et al.* 1970) are included in Fig. 1(a).

The area consists of undulating ground rising up to 400 m above sea-level: much of the northern coastal strip and valleys are covered by grassland but primary forest persists inland. Apart from the coast, new road cuttings and along the deeper valleys, exposures are few and deeply weathered. The area was mapped in 1976–77 as part of a collaborative project between the Institute of Geological Sciences, London, and the Dirección General de Geología y Minas, Quito.

Stratigraphy

A 5 km-thick section, extending from Cretaceous to upper Pliocene, is exposed in the western part of the Borbón Basin. The older rocks are exposed only in the core of the Río Verde anticline; the younger Miocene and Pliocene strata crop out over most of the area with the youngest strata forming a belt along the coast. The stratigraphic nomenclature follows Faucher *et al.* (1970); however, changes are introduced in the upper Miocene and Pliocene units, mainly after Bristow (1976b). The units were dated (J.E.W.) using planktonic foraminifera recovered from over 150 samples collected across the area.

Piñón Formation

Thirty metres of green, basaltic breccia of the Piñón Formation are exposed in the core of the Río Verde anticline, near Businga (Cuadro 82. Faucher *et al.* 1970). Camarones-1 well (I.E.P.C. 1944) terminated in 248 m of grey, fine-grained, basaltic agglomerate, pumiceous tuffs and lavas which were assigned to the formation on lithological grounds.

Outside the mapped area, the nearest major outcrops of the formation are 150 km SW of Esmeraldas in the Montañas de Jama (Fig. 2) where they form a complex, at least 2000 m thick, of massive, fine-grained gabbros with brecciated, altered, pillow basalts near the top of the section (Evans & Arguello 1977).

Feininger (1977) states that '. . . the chemical composition, lithology and structure of the

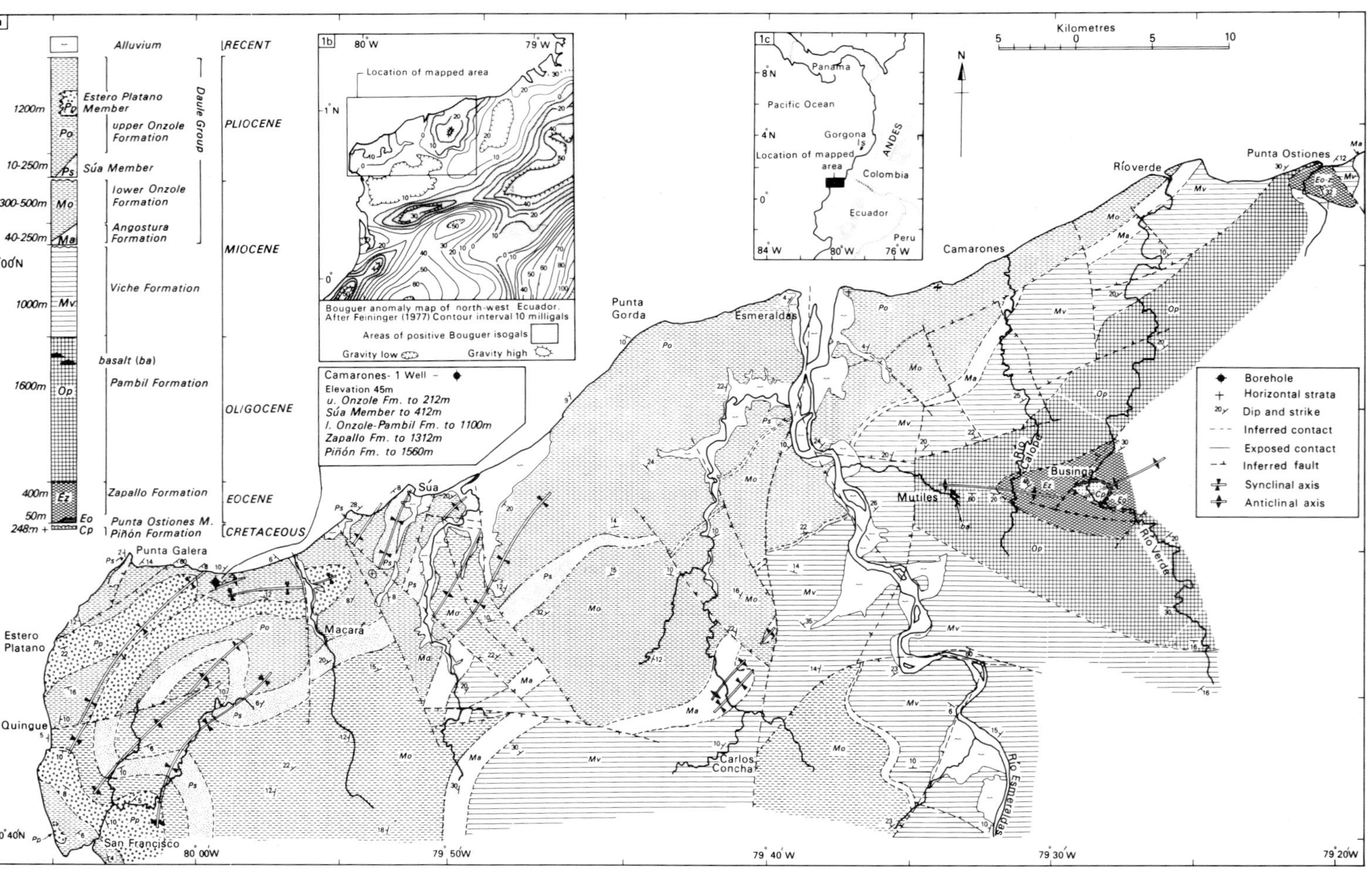

FIG. 1. (a) Geological map of the western part of the Borbón Basin.
(b) Gravity map of the Borbón Basin, after Feininger (1977).
(c) Map showing the location of the mapped area in South America.

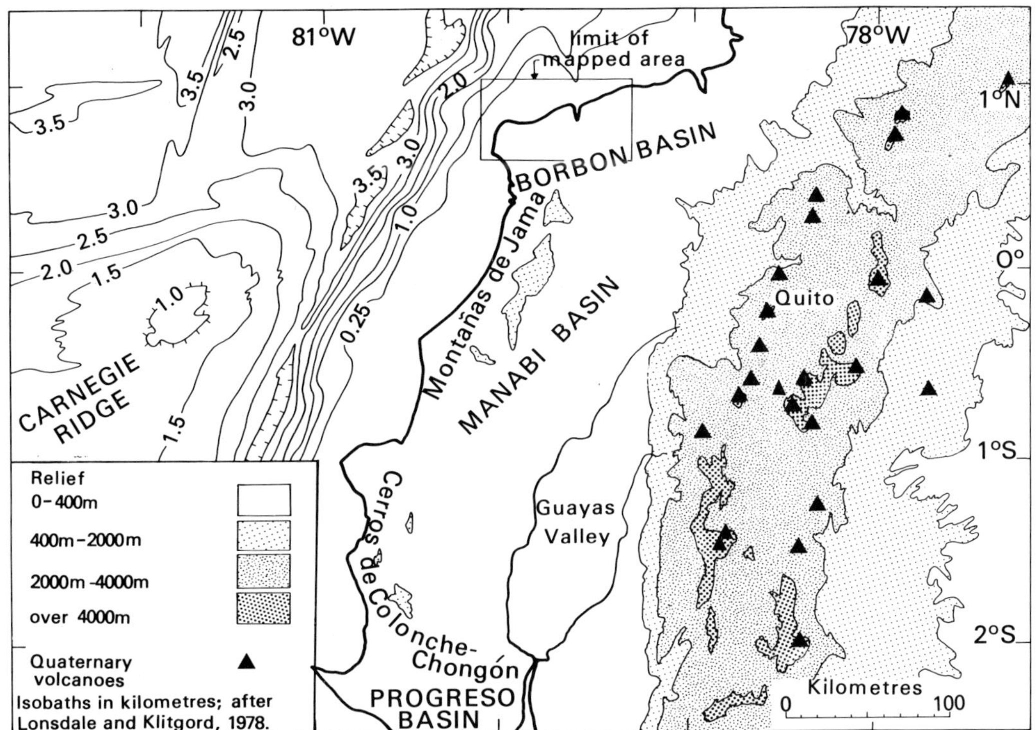

FIG. 2. Physiography of western Ecuador and the adjacent Pacific Ocean.

formation all point to its oceanic orgin' and suggests that it is '. . . a remnant of ocean floor of Cretaceous age'. However, Henderson (1979) believes that the formation represents the tholeiitic, primitive, oceanward part of a volcanic arc more fully represented to the east in the Western Cordillera of the Andes.

In southern Ecuador, the Piñón Formation is overlain by the Senonian-Maastrichtian, Cayo Formation (Bristow 1976a) which, over part of the northern Costa, has been removed by pre-middle Eocene erosion (Evans & Arguello 1977). The oldest radiometric date for the formation is 110±10 Ma, Aptian (Goosens & Rose 1973).

Zapallo Formation

Across much of the Costa the Zapallo Formation, or its equivalent, has at its base a thin discontinuous limestone unit, named in this area, the Punta Ostiones Member. At Punta Ostiones it consists of about 50 m of brown-grey, silicecus biomicrites and porcellanites, in beds 5–50 cm thick, interbedded with thinner, green, calcareous sandstone and marl. Silica occurs as irregular banks, nodules and veins of chert, and as a matrix in pseudo-breccias (A. F. Wilkinson, pers. comm.). Rich faunas of plantonic foraminifera of mid-Eocene age are found with radiolaria in the porcellanites, and calcareous algae and bryozoans occur in the biomicrites. Near Businga, the member consists of 40 m of bioclastic limestone and it is absent from the Camarones-1 well. The unsorted nature of the biomicrites and the commonly worn condition of the fauna and flora, suggest that the limestones were originally deposited in shallow water at the edge of the basin, and redeposited by slumping into deeper water where the porcellanites accumulated. The member passes transitionally up into the siliceous mudstones which comprise the bulk of the Zapallo Formation.

The formation, which is well exposed in the river north of Businga and in the road cuttings east of Mutiles, is at least 400 m thick. It consists of green-grey, hard, bioturbated, in part silicified mudstones with thin sandstone and tuffitic beds. Dykes, lenses and rare beds of dark brown chert are found, often fractured and traversed by clear quartz veins. In the

Camerones-1 well the formation, which is 212 m thick, contains in the lower part thin cherty limestones (I.E.P.C. 1944).

A re-interpretation of planktonic foraminifera from the formation, listed in Stainforth (1948) and Bristow & Hoffstetter (1977), suggests a mid-late Eocene age, and deposition in bathyal depths.

Pambil Formation

In the Río Verde the formation is 1600 m thick (Faucher *et al.* 1970) and consists of green-grey, bioturbated, silty mudstones with local flaggy bedded, fine-grained sandstones.

Near Mutiles, two basalt domes break the sequence. They are at least 50 m thick and over 400 m long, with an outer crust, up to 8 m thick, of brecciated basalt traversed by a network of zeolite veins. The domes are concordant and the absence of vesiculation suggests emplacement beneath a considerable depth of water.

The Pambil Formation is entirely Oligocene in age. Its lower part is developed in a primarily radiolarian facies; the overlying beds contain a rich foraminiferal fauna indicative of a late Oligocene age, and deposition in outer shelf to bathyal depths.

Viche Formation

The Viche Formation, which has a thickness in excess of 1000 m (Faucher *et al.* 1970), consists of a uniform sequence of green-grey to brown, well-bedded, blocky, bioturbated mudstone with thin ribs of sandstone, locally rich in molluscs and fish scales. Palaeontological evidence indicates that the formation extends from the early to lower mid-Miocene. The absence of lowermost Miocene sediments may be attributed to incomplete sampling or a non-sequence at this stratigraphic level. The benthic foraminifera suggest deposition in bathyal depths and give no suggestion of shallowing towards the top of the unit.

Daule Group

Angostura Formation
North of Carlos Concha and east of the mouth of the Río Verde, the Angostura Formation rests with a sharp, conformable contact on the underlying Viche mudstones. The formation is between 40–250 m thick and consists of fine- to coarse-grained, rusty-orange sandstones with lenses of bioclastic debris and conglomerates near the base, and a gradational contact up into the overlying mudstones of the lower Onzole Formation. The poorly sorted sandstones are lithic-feldspathic greywackes and contain abundant, angular quartz, fresh, zoned plagioclase, red-brown biotite and green hornblende grains.

The basal samples of the overlying formation contain a late Miocene fauna and the age of the Angostura Formation is probably mid- to late Miocene with a possible depositional hiatus indicated for much of the mid-Miocene.

Onzole Formation
West of the Río Esmeraldas, the Onzole Formation is divided into a lower and an upper (= Punta Gorda Formation *auct*) unit separated by sandstones of the Súa Member. These sandstones thin rapidly eastwards and are absent east of the river where the formation can only be divided on faunal grounds. A second sandstone sequence (Estero Platano Member) occurs towards the top of the formation in the west of the area.

The lower Onzole Formation is between 300–500 m thick and composed of uniform, silty mudstones with thin, poorly graded, fine ribs and rare beds of sandstone. The late Miocene fauna found in the mudstones indicates, in the lower part, deposition in shallow water but thereafter a gradually deepening environment.

At Súa the member consists of about 250 m of well bedded greywacke sandstones which thin eastwards to about 10 m in the Río Esmeraldas. The orange to yellow-grey, bioturbated sandstones are coarse- to fine-grained, with pebbly lenses, powdery siltstone intercalations and locally large, rounded, calcareous concretions. Lenses of worn and broken molluscs and lignite-rich bedding planes are common. The lower contact varies from an undulating unconformity to a planar disconformity and the contact with the overlying mudstones is gradational. The member contains shallow water, benthic foraminifera and the rich, planktonic foraminifera in the immediately overlying mudstones place the member within the early Pliocene.

The upper Onzole Formation consists of about 1200 m of mudstones with thin-bedded turbidites and rare, thicker, massive sandstones and conglomerates. In the west, the formation includes about 300 m of re-sedimented sandstones and mudstones (the Estero Platano Member) which thin rapidly eastward and are not defined east of Súa. The formation is well-exposed along the coast where it forms cliffs of dark, green-grey, well bedded, faintly graded mudstones, often intensely bioturbated. Discontinuous partings and thin beds of rusty, fine-grained sandstone and sharply defined beds of white tuff occur on a decimetre scale.

In contrast to the lower part of the formation, the microfaunas in the upper Onzole Formation are very rich, well-preserved and wholly of Pliocene age; the benthic foraminifera suggest deposition in bathyal depths. The minimum age for the formation is fixed at early Zone 21, lower late Pliocene (*c*.2.9 Ma).

Interbedded with the thin-bedded turbidites, especially in the east, are dark brown, medium-grained sandstones up to 1 m thick with unconformable bases and convoluted, internal laminations. They contain a rich fauna of shallow water molluscs and have been transported by turbidity flows into deeper water. Lenses of pebbly sandstone and conglomerates up to 50 m thick with step-like, erosive, lower contacts occur near Esmeraldas. Locally the basal 1–3 m contains rounded mudstone boulders up to 1 m in diameter set in a pebbly mud matrix which passes up into an irregular sequence of sands and pebbles with trains of mudstone blocks. These coarse redeposited sediments have flowed into the deep water basin, via a submarine canyon, and are preserved in a series of channels cut into the thin-bedded turbidites of the basin floor.

Matrix-supported mudstone breccias up to 5 m thick are found in the formation east of Punta Galera. They contain rounded to angular, commonly sheared blocks of local and exotic mudstone up to 1 m across set in a muddy matrix, locally rich in sandy lenses. These breccias are the products of slumps which have slid into the basin floor from the submarine canyon walls.

At the type locality, on the road between Estero Platano and Quinque (Fig. 1a), the Estero Platano Member consists of about 300 m of rusty-brown, very regularly bedded, fine- to medium-grained, well-sorted sandstones separated by thin mudstones. In the lowest exposures the individual sandstone beds are about a metre thick and form over 75% of the succession. The thickness of the sandstone beds and the sandstone/mudstone ratio decreases gradually up the unit which, at the top, passes transitionally into the normal, thin-bedded turbidite sequence.

The absence of coarse detritus and the well-sorted nature of the sandstones suggest a source distinct from that of the conglomeratic lenses which are found in the formation. The sands of the Estero Platano Member are similar to the beach deposits along the north coast: the conglomerates are similar to the material on the floor of the Río Esmeraldas which debouches into the submarine canyon immediately offshore at present (Fig. 2).

Quaternary deposits

Parts of the coastal area, at an elevation of up to 60 m above sea-level, are covered by up to 4 m of grey, fine sand with gravel and bioclastic lenses. These are typical of the Tablazo Formation (Bosworth 1922) a raised beach deposit of Pleistocene or younger age common along the coast of Ecuador and northern Peru.

Isolated patches of river gravel, stranded up to 25 m above the present river level, are found along the banks of the Río Esmeraldas along with a number of low river terraces (Fig. 1a).

Stratigraphical comparison of the western and eastern parts of the Borbón Basin

Stainforth (1968), using data largely from the eastern part of the basin, identified an axial and a peripheral stratigraphic sequence and the study area is typical of the former. The peripheral sequence, developed along the eastern margin of the basin, near the Andean foothills, is distinguished by having a series of shallow-water formations interdigitated with the deep-water mudstones. The oldest three shallow-water units, the Santiago, Playa Rica, and the 'lower Angostura' are not recorded in the western part of the basin. Along the Andean foothills the Santiago is developed above the basal Punta Ostiones Member; the Playa Rica separates the Pambil and Zapallo Formations and the 'lower Angostura' rests unconformably on the Pambil Formation. The 'upper Angostura' sandstones of Stainforth (1968), equivalent to the Angostura Formation in this paper, are a sequence '. . . moving slowly basinwards over the Viche Shales' (Stainforth 1968); the base of the formation in the east is dated mid-Miocene but in the west it is mid- to late Miocene. In the eastern part of the basin the overlying Onzole Formation is of shallow marine character while in the west the analogous lower Onzole Formation was deposited in a deepening basin.

The early Pliocene event, which led to the uplift and the deposition of the Súa Member, is recognized across most of the Costa. Much of the eastern Borbón, the Manabí and Progreso Basins remained positive areas after the uplift. However, subsidence recommenced in the western part of the Borbón Basin where the upper Onzole Formation accumulated in outer shelf to bathyal depths. The post-late Pliocene uplift of the western part of the basin was associated with the emergence of the Río Verde anticlinal area and the development, to the east, of a new basin where 500 m of sands and gravel (Cachabi Formation) accumulated during the Pleistocene (Bristow & Hoffstetter 1977). This situation persists with thick recent alluvium accumulating in the eastern basin and the western part of the basin continuing to rise.

Structure

The study area lies on the north-western margin of the gravity low which delimits the Borbon Basin (Fig. 1b), but the youngest strata are exposed near the coast, well to the north of the area of thickest accumulation of Tertiary sediments, as defined by the gravity data. An ENE-WSW-trending, fault-controlled gravity high, which separates the Borbón Basin from the Manabí Basin to the south, forms the southern margin to the basin.

The strata are folded into a series of open, NE-SW-trending structures with wavelengths upwards of 2 km and sinuous axial traces which are parallel to the main structural feature of the area, namely: the anticline which brings the older formations to crop in the Río Verde area. This feature, which extends the length of the Costa from the Cerros de Colonche–Chongon northwards to the Río Verde area (Fig. 2), obliquely crosses the axis of the Borbón Basin as defined by the gravity low. Stratigraphic evidence indicates that the feature, a mid-slope basement high, formed a positive topographic feature in the Manabí and Borbón Basins only after the late Pliocene.

The coastal sections (Fig. 1a) show that the region is divisible into structurally simple areas (Esmeraldas to Punta Gorda) and discrete more complex areas (around Súa). The faults mostly trend between 330° and 360° with some, east of Esmeraldas, parallel to the coastline. All are normal faults with no evidence of thrusting.

In the Camarones-1 well slickensided fractures are reported at a depth of 500 m and a poorly developed cleavage, parallel to bedding, below 1000 m. The Pliocene mudstones on the west coast, near Quinque, display a fine, slightly irregular pattern of dark, linear veins (dewatering structures?) developed at a steep angle to the bedding.

Tectonic history and palaeogeography

The Pacific coastal lowlands of Ecuador and Colombia occupy the 'arc-trench gap' (Dickinson 1971) between the offshore bathymetric trench—the site of the oceanic Nazca Plate subducting under the South American Plate (Lonsdale & Klitgord 1978)—and the andesite volcanoes of the Andes (Fig. 2). The gap is about 240 km wide in Ecuador, compared with a characteristic value of 250 ± 50 km for continental margin settings (Dickinson 1971).

Lonsdale (1978) divided the Costa into a mid-slope basement high (the coastal range of hills) with a forearc basin to the east (the Manabí Basin) and the trench-slope basin to the west (the Progreso Basin). In the northern Costa (Fig. 2) this sub-division is applicable only to the post-late Pliocene; the mid-slope basement high had no major topographic expression prior to this time.

The tectonic history of the northern Costa is divisible into pre- and post-mid-Eocene phases with a further three-fold sub-division of the latter into mid-Eocene–mid-Miocene, late Miocene–late Pliocene and post-late Pliocene to Quaternary phases.

Pre-mid-Eocene

Feininger (1977) suggested that the basement of the Costa, the Piñón Formation, was a remnant of ocean floor which '. . . now crops out owing to its having been isolated by a westward jump of the continental border subduction zone in Early Tertiary time'. On gravity evidence (Fininger 1977), the boundary between the pre-early Tertiary continental and oceanic crust lies approximately at the base of the western Andes, about 180 km east of the present axis of the Ecuador trench. However, Henderson (1979) considered the Piñón a tholeiitic oceanward part of the arc and suggested that the co-linearity of the Cretaceous to Eocene and post-Eocene volcanic arc indicated that no change had occurred in the position of the oceanic trench in the early Tertiary.

The event which resulted in uplift and erosion of the Costa and the establishment of the present forearc regime is dated between late Palaeocene—the youngest date for basaltic activity in the Cayo Formation (Goosens & Rose 1973)—and the early mid-Eocene—the age of the Punta Ostiones Member. Limestones of the latter unit rest unconformably on the weathered top of the Cretaceous basement (I.E.P.C. 1944) with no evidence for a totally tectonic contact between the two.

Mid-Eocene–mid-Miocene

During this interval the sequence was dominated by a thin-bedded turbidite facies, deposited in a forearc basin, whose main axis of deposition trended ENE-WSW. The occurrence of cherts only in the older strata indicates an initially deep basin floor which became shallower with time. In the late Eocene to Oligocene, sediment accumulation in the mapped area, which is on the north-western margin of the main deposition axis, was about 100 m/Ma, compared to about 60 m/Ma in the present Panama Basin (Lonsdale 1978). As the basin infilled, the shallow-water sequences associated with the uplift of the basin floor, extended further oceanward (westwards). Thus the earliest shallow-water formations were restricted to the eastern, peripheral parts of the basin and the upper Miocene Angostura Formation was the first to extend into the western part of the Borbón Basin. The Angostura Formation was also the first shallow-water sequence developed west of the Montañas de Jama in the Manabi Basin (Evans & Arguello 1977).

The basaltic activity in the upper part of the Pambil Formation was contemporaneous with the breakup of the Farallon Plate into the Nazca and Cocos Plates 27 Ma ago (late Oligocene) (Lonsdale & Klitgord 1978). The effect of the breakup of the plate on the local stratigraphy was minimal; the facies remained unchanged although the rate of sedimentation during the early Miocene was about double that for the late Eocene–Oligocene.

Late Miocene–late Pliocene

During this interval, between 8 and 2.9 Ma, two shallow-water sequences transgressed across the basin to its oceanward margin, interrupting the normal thin-bedded turbidite sequence. The start of the interval coincided with a reduction in the rate of subduction (Lonsdale & Klitgord 1978) under the forearc and therefore changes in the stresses on the basement.

Sedimentation was more rapid than during the earlier part of the basin's history with a rate of 500 m/Ma in some areas. No facies variations in the Pliocene units can be attributed directly to the present mid-slope basement high, though the very rapid accumulation of sediment, probably in localised fault-controlled basins, may have been complementary to the initial rise which eventually resulted in the mid-slope basement high centred on the Río Verde area.

Post-late Pliocene–Quaternary

The youngest Pliocene strata were deposited in outer shelf to bathyal depths and, in the Esmeraldas–Súa area, a gently dipping succession over 1200 m thick is now exposed. The exposure of the Súa Member in this area represents uplift and erosion of at least 1500 m (water depth plus thickness of strata removed by erosion) since the deposition of the unit. Geomorphological evidence indicates that the main uplift of the Andes was contemporaneous with the uplift of the coastal range of hills (Campbell 1974). With the rise of the mid-slope basement high the axes of deposition in the Costa moved eastwards into the Guayas Valley (Fig. 2) and eastern Borbón Basin where thick alluvium and shallow-water sequences accumulated during the Quaternary (Bristow & Hoffstetter 1977). This sequence of events is recognized in the coastal lowlands northwards into Panama, although parts of the forearc in Colombia remain submerged with the offshore islands (Gorgona Island), equivalent to the mid-slope basement high in Ecuador (Fig. 1c).

Conclusions

The Tertiary stratigraphy of the western part of the Borbón Basin is typical of the forearc in the northern Costa and southern Colombia (Nygren 1950). The sequence can be summarized as follows: thin discontinuous limestones which floor the basin pass up into thin-bedded turbidites which alternate, in the upper part, with shallow-water sandstones. The sequence is typical of that in an idealised residual forearc basin '. . . where strata lie depositionally on oceanic or transitional crust trapped between the arc massif and the subduction zone (Dickinson & Seely 1979). The Borbón Basin is interesting because of the alternation of shallow- and deep-water sedimentation. The fluctuations in the depth of the basin floor are caused by regional tectonic events, such as rates and direction of subduction which influence the tectonic stresses on the basement of the arc-trench gap. The regional character of the events is indicated by the contemporaneous changes in the Tertiary stratigraphy of the forearc from northern Peru to Venezuela (Stainforth 1968).

Karig & Sharman (1975) propose a model of forearc accretion which includes the oceanward movement and vertical growth of the trench-slope break as the basin matures. However, in the Borbón Basin the similarity of tectonic styles and facies of the Eocene–Oligocene and late Pliocene strata implies no significant migration of the basin margin since the mid-Eocene. A similar situation occurs off Guatamala, where the position of the shelf-edge has remained static since the end of the Cretaceous (Seely 1979). Growth of the trench slope break as the basin infills appears to have been vertical in the Borbón Basin. During the early part of the basin's history the trench slope break was obviously deep and the basin may have sloped, almost without a break, from the shore into the oceanic trench. As the basin infilled and shallow water sequences edged oceanwards the trench slope break rose to become a pronounced topographic feature separating the forearc basin from the lower trench slope into the oceanic trench.

Gravity data indicate that the bulk of the Tertiary sediments accumulated in a basin whose shape was controlled primarily by faults in the basement and not the presumed arc-trench geometry. The balance of influence changed during the Pliocene and the shape of the later basins was related to the present arc-trench system which cut obliquely the earlier dominant trend. However, these basement faults continue to influence the geology of the area as is shown by the fault which forms the southern boundary of the Borbón Basin. There is an abrupt change in the strike of the arc where it intercepts the Andean Massif (Fig. 2) and an offset in the alignment of the Quaternary volcanoes. Thus mega-structures in the basement, which cross the present arc-trench geometry at an oblique angle, may divide the arc and forearc into sectors with differing patterns of sedimentation, uplift and igneous activity.

ACKNOWLEDGMENTS: The authors are grateful to the personnel of the Dirección General de Geología y Minas in Quito for the support during the fieldwork especially Ing. L. Cevallos and Sr. C. Cardenas. Thanks are also due to the rest of the staff, past and present, of the British Mission in the D.G.G.M. especially Mr W. G. Henderson and Dr A. F. Wilkinson.

References

BOSWORTH, T. O. 1922. *Geology of the Tertiary and Quaternary Periods in the Northwest Part of Peru (with an account of the Palaeontology by H. Woods, J. W. Vaughan, J. A. Cushman, and others).* MacMillan, London, 434 pp.

BRISTOW, C. R. 1976A. The age of the Cayo Formation, Ecuador. *Newslr. Stratig.* **4,** 169–73.

—— 1976b. The Daule Group, Ecuador. *Newslr. Stratig.* **5,** 190–200.

—— & HOFFSTETTER, R. 1977. *Lexique Stratigraphique International, 4, Fasc.* **5a** 2nd ed. Centre National de la Recherche Scientifique, Paris.

CAMPBELL, C. J. 1974. Ecuadorian Andes. *In:* SPENCER, A. M. (ed). *Mesozoic–Cenozoic Orogenic Belts, Data for Orogenic Studies.* Spec. Publ. geol. Soc. London, **4,** 725–32.

CANFIELD, R. W. 1966. *Reporte Geológico de la Costa Ecuatoriana.* Ministerio de Industrias y Comercio, Quito.

DICKINSON, W. R. 1971. Clastic sedimentary sequences deposited in shelf, slope and trough settings between magmatic areas and associated trenches. *Pacif. Geol.* **3,** 15–30.

—— & SEELY, D. R. 1979. Structure and stratigraphy of forearc regions. *Bull. Am. Ass. Petrol. Geol.* **63,** 2–31.

EVANS, C. D. R. & ARGUELLO, C. 1977. *Geological Map of Sheet 10 (Jama) 1:100,000.* Dir. Gen. de Geol. y Minas, Quito.

FAUCHER, B., JOYES, R., MAGNE, F., SIGAL, J., VERNET, R., GRANJA, V. (J.C.), GRANJA, B. (J.C.), CASTRO, R. & GUEVARA, G. 1970. *Estudio general de la Cuenca de Esmeraldas.* Instituto Frances de Petrole. Servic. Nat. de Geol. Miner, Quito (unpubl.).

FEININGER, T. 1977. Bouguer anomaly map of Ecuador. *Inst. Georg. Militar,* Quito.

GOOSENS, P. J. & ROSE, W. I. 1973. Chemical composition and age determinations of tholeiitic rocks in the Basic Igneous Complex, Ecuador. *Bull. geol. Soc. Am.* **84,** 1043–52.

HENDERSON, W. G. 1979. Cretaceous to Eocene volcanic arc activity in the Andes of northern Ecuador. *J. geol. Soc. London.*, **136,** 367–78.

I.E.P.C. 1944. Composite log, Camarones-1 Well, International Ecuadorian Petroleum Company. *Dir. Hidrocarburos,* Quito (unpubl.).

KARIG, D. E. & SHARMAN, G. F. 1975. Subduction and accretion in trenches. *Bull. geol. Soc. Am.* **86,** 377–89.

LONSDALE, P. 1978. Ecuadorian subduction system. *Bull. Am. Assoc. Petrol. Geol.* **62,** 2454–77.

—— & KLITGORD, K. D. 1978. Structure and tectonic history of the eastern Panama Basin. *Bull. geol. Soc. Am.* **89,** 981–99.

NYGREN, W. E. 1950. Bolivar geosyncline of northwestern South America. *Bull. Am. Assoc. Petrol. Geol.* **34,** 1998–2006.

SEELY, D. R. 1979. Evolution of structural highs bordering major forearc basins. *In:* WATKINS, J. S., MONTADERT, L. & DICKERSON, P. W. (eds). *Geological and Geophysical Investigations of Continental Margins.* Mem. Am. Assoc. Petrol. Geol. **29,** 245–60.

STAINFORTH, R. M. 1948. Applied micropalaeontology in coastal Ecuador. *J. Paleontol.* **22,** 133–51.

—— 1968. Mid Tertiary diastrophism in Northern South America. *In:* SAUNDERS, J. B. (ed.). *Trans. Fourth Caribbean Geol. Congr.* 159–77, 1965. Port of Spain, Caribbean Printers, Arima.

C. D. R. EVANS, Institute of Geological Sciences, Ring Road, Halton, Leeds LS15 8TQ, England.

J. E. WHITTAKER, British Museum (Natural History), Cromwell Road, London SW7 5BD, England.

ALEUTIANS

Ancient plate boundaries in the Bering Sea region

M. S. Marlow, A. K. Cooper, D. W. Scholl & H. McLean

SUMMARY: Plate tectonic models of the Bering Sea suggest that the abyssal Bering Sea Basin is underlain by oceanic crust, a supposition supported by refraction and magnetic data. The oceanic crust is thought to be a remnant of the Kula(?) plate that was isolated within what is now the Bering Sea when the proto-Aleutian arc began to form between the Alaska Peninsula and Kamchatka in late Mesozoic or earliest Tertiary times. Prior to the formation of the Aleutian arc, the Kula(?) plate moved NW, directly underthrusting eastern Siberia; the plate's eastern edge either obliquely underthrust or slid past the Bering Sea margin along a transform boundary.

The Koryak Range in eastern Siberia is composed in part of mélange units that include Palaeozoic and Mesozoic allochthonous blocks juxtaposed within a matrix of Cretaceous sedimentary rocks. Structural trends suggest that these blocks were accreted into the Koryak area from the south along an ancient subduction zone formed by underthrusting of the Kula(?) plate.

The base of the Bering Sea continental margin that extends from eastern Siberia to the Alaska Peninsula—the so-called Beringian margin—is underlain by a thick (7–10 km) sedimentary section along the base of the slope. Rocks dredged from the basement exposed farther up the slope (1500–2000 m deep) include shallow-water Upper Jurassic sandstone that is unconformably overlain by shallow-water Eocene to Miocene diatomaceous mudstone. Fauna in the dredge samples indicate that the shelf edge has subsided several kilometres since late Palaeogene time, perhaps in response to the cessation of motion relative to the adjacent oceanic plate and subsequent sediment loading of the oceanic plate.

Uplift of the former plate boundary exposed in the Koryak Range occurred principally in late Cenozoic time, and collapse of the adjacent plate boundary, the Beringian margin, began in earliest Tertiary time and has continued to the present. Both tectonic events occurred after the site of active plate collision shifted south to near the present Aleutian Trench. We are uncertain as to why these two ancient, yet adjacent former plate boundaries should behave so differently, i.e. why one area was folded and uplifted while the other was extensionally deformed and subsided, both apparently in response to the cessation of convergent or strike-slip plate motion.

We examine two postulated Mesozoic plate boundaries in the Bering Sea; a SE facing convergent boundary through the Koryak Range in eastern Siberia, and a SW facing obliquely underthrust or strike-slip boundary, the Beringian margin, connecting Alaska and Siberia (Fig. 1).

A terrane of complexly deformed rocks that include mélange sequences and ultramafic masses exposed in the Koryak Range of eastern Siberia is thought to record Mesozoic to earliest Tertiary convergence between the Kula(?) and North American plates (Scholl *et al.* 1975). The range was deformed and in part uplifted in Cenozoic time. The adjacent Beringian margin, striking at right angles to the Koryak margin, has undergone substantial subsidence in Cenozoic time. The Beringian margin is underlain by a broadly folded sequence of shelf and upper slope sedimentary and volcanic (evidence is volcanic-rich detritus) units. These units are Jurassic and Cretaceous in age and presumably are not typical of subduction complexes.

Geological data from eastern Siberia

A generalized cross-section that traverses the north-eastern Koryak Mountains (Fig. 2) is adapted from McLean (1979a), who in turn abstracted the section from the original work of Aleksandrov *et al.* (1976). The eastern Koryak Range is underlain by structurally juxtaposed blocks of different age and lithology. McLean (1979a) compares the complex rock fabric of the Koryak Range to the Franciscan assemblage in California as described by various workers (e.g. Bailey *et al.* 1964, p. 5; Hamilton 1969; Bailey & Blake 1969; Ernst 1970; Page 1970, 1978) and to the McHugh and Uyak Complexes of south-western Alaska (Clark 1973; Connelly 1978; Tysdal & Case 1976; Tysdal *et al.* 1977). Rock types common to all three areas include chert (red, grey, and green), siliceous shale, siltstone, fine-grained greywacke, tuff, spilite, pillow basalt, blueschist, and ultramafic rocks. The age of the greywacke in the Koryak Range is estimated as either Triassic to Early Cre-

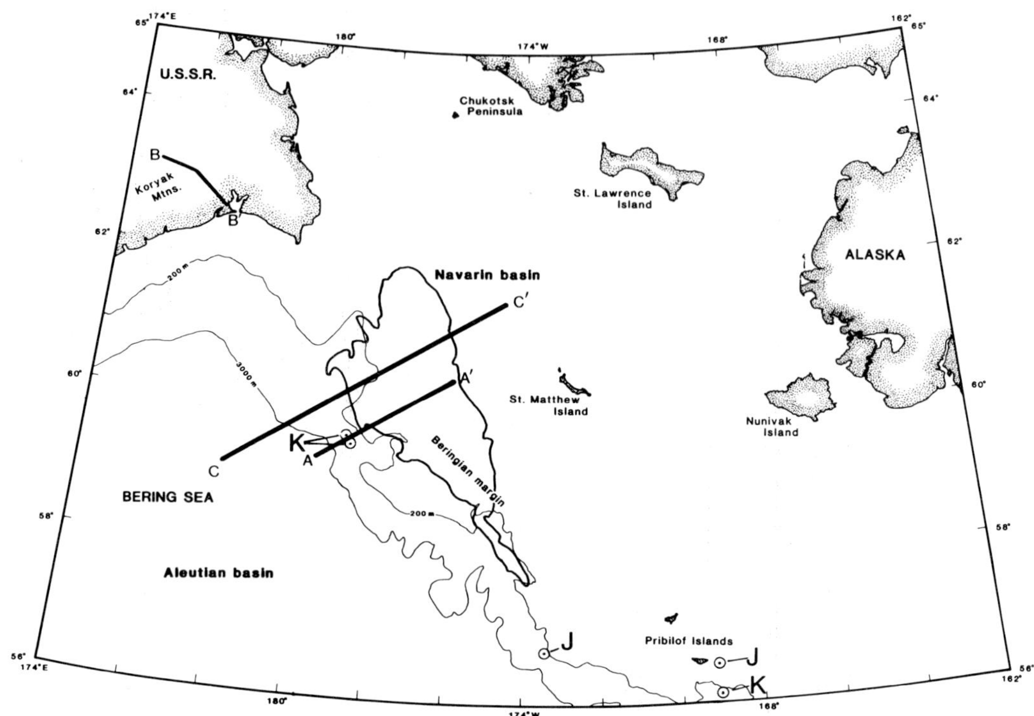

FIG. 1. Location of seismic-reflection profile A-A′ across the Beringian margin shown in Figs 3 and 4. Line B-B′ shows location of the generalized structure cross-section shown in Fig. 2. Line C-C′ shows location of hypothetical section shown in Fig. 6. Encircled dots indicate sites where Jurassic (J) and Cretaceous (K) sedimentary rocks were dredged from the margin (Marlow *et al.* 1979b; Vallier *et al.* 1980; Marlow & Cooper 1981). Albers equal area projection.

taceous (Valanginian) or Late Jurassic (Tithonian) to Early Cretaceous (Hauterivian) (McLean 1979a).

Large slab-like rock bodies occur within the Koryak Mountains. These slabs are separated by northward-dipping thrust faults and include olistostrome and mélange sequences (Fig. 2). At the base of many of the tectonic slabs and sheets are ultramafic masses consisting of pyroxenite, peridotite and serpentinite associated with gabbro and diorite. These mafic and ultramafic assemblages may be part of an ophiolite suite (Aleksandrov *et al.* 1976). The assemblages are unconformably overlain by either siliceous sedimentary rocks of Palaeozoic age or volcanic and terrigenous deposits of Mesozoic age. Exotic blocks of serpentinite mélanges and olistostromes (*sensu* Silver & Beutner 1980) include shallow-water Devonian Carboniferous, and Lower and Upper Permian limestones. Aleksandrov *et al.* (1976) suggest that all Palaeozoic deposits in the Koryak Mountains are allochthonous units that by the end of the Mesozoic were tectonically derived from the south, in the vicinity of the present-day Bering Sea. Thrusting of the allochthonous units apparently continued into early Palaeogene time (Aleksandrov *et al.* 1976; McLean 1979a).

Early Cretaceous magmatism in the Koryak Mountains coincided with the accumulation of olistostromes to the south toward the Bering Sea, in front of rising tectonic slabs (Aleksandrov *et al.* 1976). Compression and foreshortening of the Koryak Range apparently continued into Palaeocene time as evidenced by isoclinally folded rocks of Palaeocene age exposed on the SE side of the range (Aleksandrov *et al.* 1976; McLean 1979a). These younger deformed rocks were derived as reworked debris from older, previously deformed units of the subduction complex present in the Koryak Mountains.

Seismic-reflection data from the Beringian margin

A 1620 km long seismic-reflection profile, A-A′ (Fig. 1), transects the north-western Beringian margin west of St Matthew Island. An

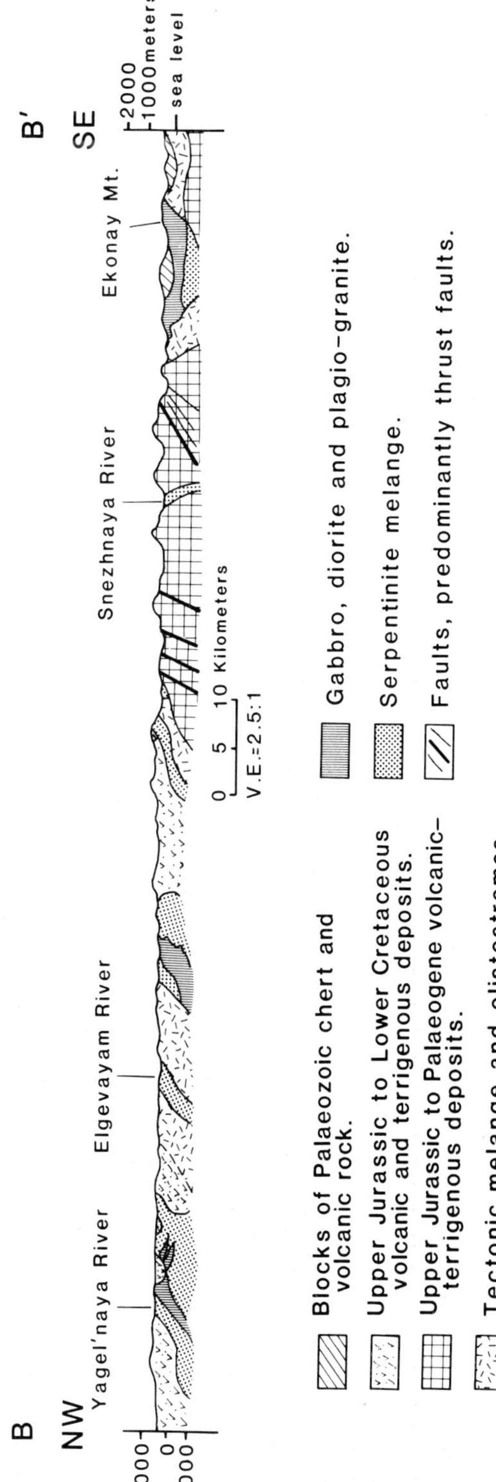

FIG. 2. Generalized structure cross-section through the north-eastern Koryak Range (after Aleksandrov *et al.* 1976, p. 294 and McLean 1979a, p. 1473). Line of section is located on Fig. 1.

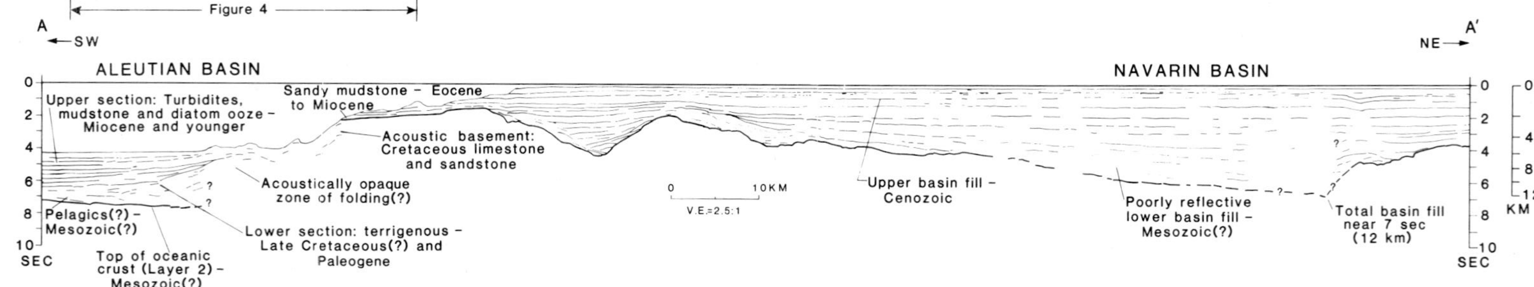

FIG. 3. Interpretative drawing of 24-channel seismic reflection profile A-A′ across the northern Beringian margin. For location of profile see Fig. 1. Travel time, in seconds, is two-way time. Detailed segment noted is shown in Fig. 4.

interpretive drawing of profile A-A′ (Fig. 3) shows that the south-western end of the profile crosses the eastern edge of the Aleutian Basin in water more than 3000 m (4.0 s) deep, where 4–5 km of undeformed sedimentary strata overlie a distinct acoustic basement. Magnetic and refraction-velocity data indicate that the basement is oceanic crust of Mesozoic age (Kula(?) plate: Cooper *et al.* 1976a, b). Immediately overlying the basement is a highly reflective section a few hundred metres thick. Other profiles (not shown) show that this layer blankets and in-fills the irregular surface of igneous oceanic crust, suggesting that the infilling layer is a pelagic deposit, possibly similar to the cherty units that blanket and smooth the surface of Mesozoic igneous crust in the Pacific (Ewing *et al.* 1968). Overlying the pelagic section are moderately to poorly reflective strata that thin and dip gently to the SW away from the margin. On the basis of this geometry and of close proximity to the margin, these deposits are interpreted to be mainly Upper Cretaceous(?) and Palaeogene terrigenous debris. The upper section beneath the Aleutian Basin along profile A-A′ may be correlative with the section at DSDP Hole #190, where Creager, Scholl *et al.* (1973) drilled through more than 600 m of turbidites, mudstone, and diatomaceous ooze of late Miocene and younger age. Strata equivalent in age and lithology presumably underlie the basin along the south-western end of profile A-A′.

Below the lower part of the continental slope the rock sequence is acoustically characterized by scattered and discontinuous reflectors (Figs 3 & 4). Oceanic basement was not resolved here. In contrast, a strong basal reflector, a sub-shelf basement, can be traced from the NE seaward to the middle of the continental slope where the basement crops out. Dredged rocks and the flatness of the subshelf basement indicate that beneath the slope the shelf basement is a wave-base unconformity cut across broadly folded rock sequences of Jurassic and Cretaceous age. These dredge and seismic-reflection data, discussed more fully below, attest that the subshelf basement beneath the slope has subsided at least 1500–1700 m since about Eocene time.

Basement beneath the shelf can be followed to the NE below the thick sedimentary section filling Navarin basin (Fig. 3). Near the north-eastern end of the profile strata in the basin are nearly 12 km (7 s) thick. Reflections from the upper basin fill are strong, continuous, and flat. Apparent breaks in the continuity of these reflectors are associated with columnar 'wipe-out' zones beneath short discontinuous reflectors in the upper few hundred metres of the section. The acoustic 'wipe-out' zones are thought to be produced by shallow accumulations of gas that mask the lateral continuity of deeper reflectors (Marlow & Cooper 1981). Dredge data indicate that the upper 3–4 km of the beds in the basin are younger than early Eocene in age. Deeper strata are poorly reflective, diverge in dip from the overlying strata, and near the base of the section may be Cretaceous or older in age (Marlow *et al.* 1976; Marlow 1979).

Dredge data

Dredging in 1978 sampled rocks of the subshelf basement that crops out on the continental slope (Figs 1 & 3; Marlow *et al.* 1979a, b). Rocks recovered near profile A-A′ include muddy limestone deposited in shallow water (based on fossil flora), siltstone, and mudstone of Late Cretaceous (Campanian and Maestrichtian) age. Farther south and west of the Pribilof Islands, shallow-water Upper Jurassic sandstone was dredged from the continental slope (based on the presence of *Buchia rugosa,* Marlow *et al.* 1979a, b; Marlow & Cooper 1981). All of the Mesozoic basement rocks recovered to date are generally highly indurated but are not significantly sheared or fractured.

Additional rock samples from the continental slope in the vicinity of profile A-A′ include mudstone and sandy mudstone that yield diatoms and foraminifers of Eocene, Oligocene, and Miocene age (Marlow *et al.* 1979a, b). These Tertiary samples were recovered from the acoustically layered section overlying rocks of the subshelf basement (Fig. 3). The fossils recovered from both the subshelf basement and the overlying rocks and the presence of wave-cut terraces now 1–2 km deep attest that the outer-shelf and upper-slope region have subsided 1–2 km since Eocene time. The exceptionally thick sedimentary fill in the adjacent Navarin basin implies that locally areas of the shelf may have subsided 10–12 km since late Mesozoic time (Marlow 1979).

Discussion

A postulated plate model for the evolution of the Bering Sea and the Beringian margin (Fig. 5) proposed by Scholl *et al.* (1975) suggests that the Beringian margin was initially the site of oblique convergence or strike-slip (transform)

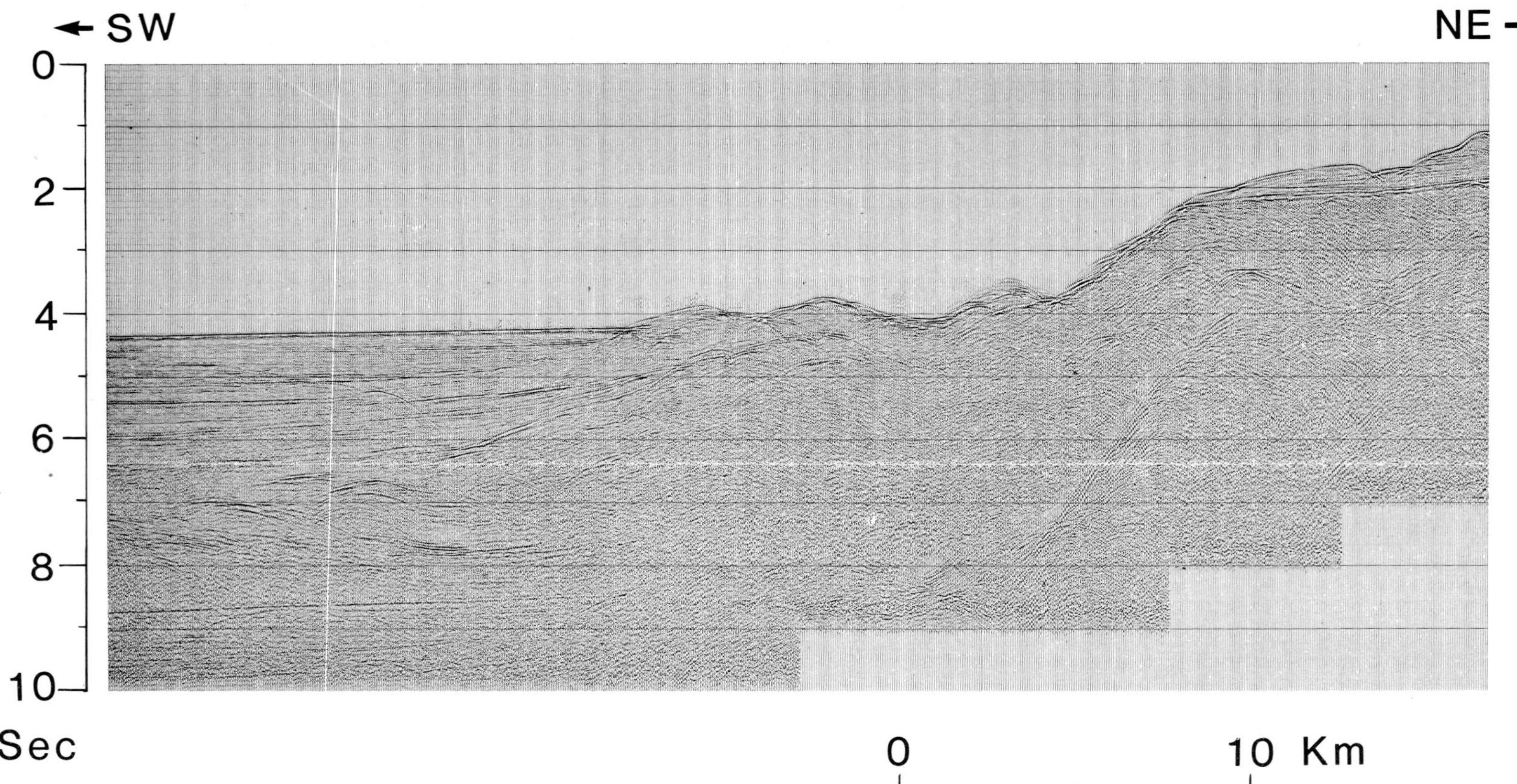

FIG. 4. Twenty-four channel seismic-reflection record used in constructing part of the interpretative drawing shown in Fig. 3.

motion between the Kula (?) and North American plates. Scholl *et al.* (1975) further argue that the formation of the Aleutian arc by earliest Tertiary time would have shifted the site of plate interactions south to an ancestral Aleutian Trench. The shift in the site of plate convergence would have tectonically isolated both eastern Siberia (Koryak Mountains) and the Beringian margin from plate-bounding interactions with the Kula(?) plate, and, at the same time, trapped a piece of Kula(?) plate in a proto-Bering Sea.

Koryak Range

The structural and lithological similarities between the mélange units in the Koryak foldbelt and the McHugh and Uyak Complexes of south-western Alaska noted by McLean (1979a) support Burk's (1965) original comparison of the two areas. The similarities between the Alaskan complexes and the Koryak Foldbelt suggests that the deformed rocks of the Koryak Mountains are a subduction complex formed by the underthrusting of the Kula(?) plate beneath the North American or Eurasian plate in Mesozoic to earliest Tertiary time.

Allochthonous blocks of Palaeozoic and Mesozoic age, associated serpentinite mélange, and northward-dipping thrust faults all suggest Mesozoic to earliest Tertiary foreshortening of the Koryak Range by horizontal compression. The ultramafic complexes in association with cherts, spilites, and pillow basalts in the range may be remnants of either oceanic crust, perhaps fragments of former Kula(?) plate, or island-arc crust. Convergence along the continental margin had apparently ceased in or by earliest Tertiary time when the ancestral Aleutian arc began to form and the convergence zone jumped to the south along an ancient Aleutian Trench (Scholl *et al.* 1975). We note, however, that the Siberian margin has undergone repeated uplift and broad compressional deformation in the Cenozoic, the strongest episode of Cenozoic tectonism occurred in middle Miocene through early Pliocene time (Gladenkov 1964; Tilman *et al.* 1969; Drabkin 1970). The Koryak area may, therefore, still be part of a broad plate boundary, separating but continuing to reflect very limited (<100 km) interaction between North America and Eurasia(?) (Fig. 5; see fig. 1 of Zoneshayn *et al.* 1978).

Beringian margin

The plate model proposed by Scholl *et al.* (1975) and Cooper *et al.* (1976b) predicts that the Beringian margin could be underlain by a subduction complex of oceanic and slope deposits accreted to the North American plate by oblique underthrusting of the Kula(?) plate (Fig. 5). Even if the margin was formerly the site of substantial transform or strike-slip motion between the two plates, the ancient margin should nevertheless have been the site of deposition of deep-water slope deposits (for example, as is true of the present Queen Charlotte transform boundary of the NE Pacific off Chigagof and Baranof Islands; von Huene *et al.* 1978). During transform motion between the two plates, slivers of allochthonous crustal fragments transported from the south could have been tectonically incorporated into what is now the bedrock framework of the Beringian continental slope.

Dredging in 1978 and earlier sampling work by Hopkins *et al.* (1969) have not yielded evidence of a former active or underthrust plate margin (McLean 1979b; Marlow & Cooper 1981). Rocks recovered thus far from the subshelf basement beneath the margin are indurated and broadly deformed but they are not part of mélange sequences nor are they pervasively sheared or fractured as might be expected along an ancient plate boundary of convergent character. Fossils in the basement samples indicate deposition in shallow water or upper slope environments.

Oceanic crust can be traced on seismic reflection profiles to within 15–20 km of the shallow-water basement rocks of continental affinity that underlie the continental slope. If deformed lower continental slope or oceanic deposits exist beneath the margin, they must be small in volume. Alternatively, large volumes of oceanic beds were subducted or thrust to the NE beneath the continental slope in Mesozoic time (Scholl *et al.* 1975).

We should note here that our sample and seismic-reflection data attest only that the Cenozoic Beringian margin has been a subsiding one since about Eocene time. In Late Cretaceous time the outer shelf was underlain by a structural high that included broadly deformed Jurassic and Cretaceous sedimentary and igneous (based on detritus in Mesozoic sandstone) rocks that accumulated in shallow-water regions. Tectonic models that suggest strike-slip or oblique convergence along the Beringian margin in Mesozoic time are based on regional geological reconstructions and plate reconstructions that are based on magnetic anomalies (Scholl *et al.* 1975; Cooper *et al.* 1976a).

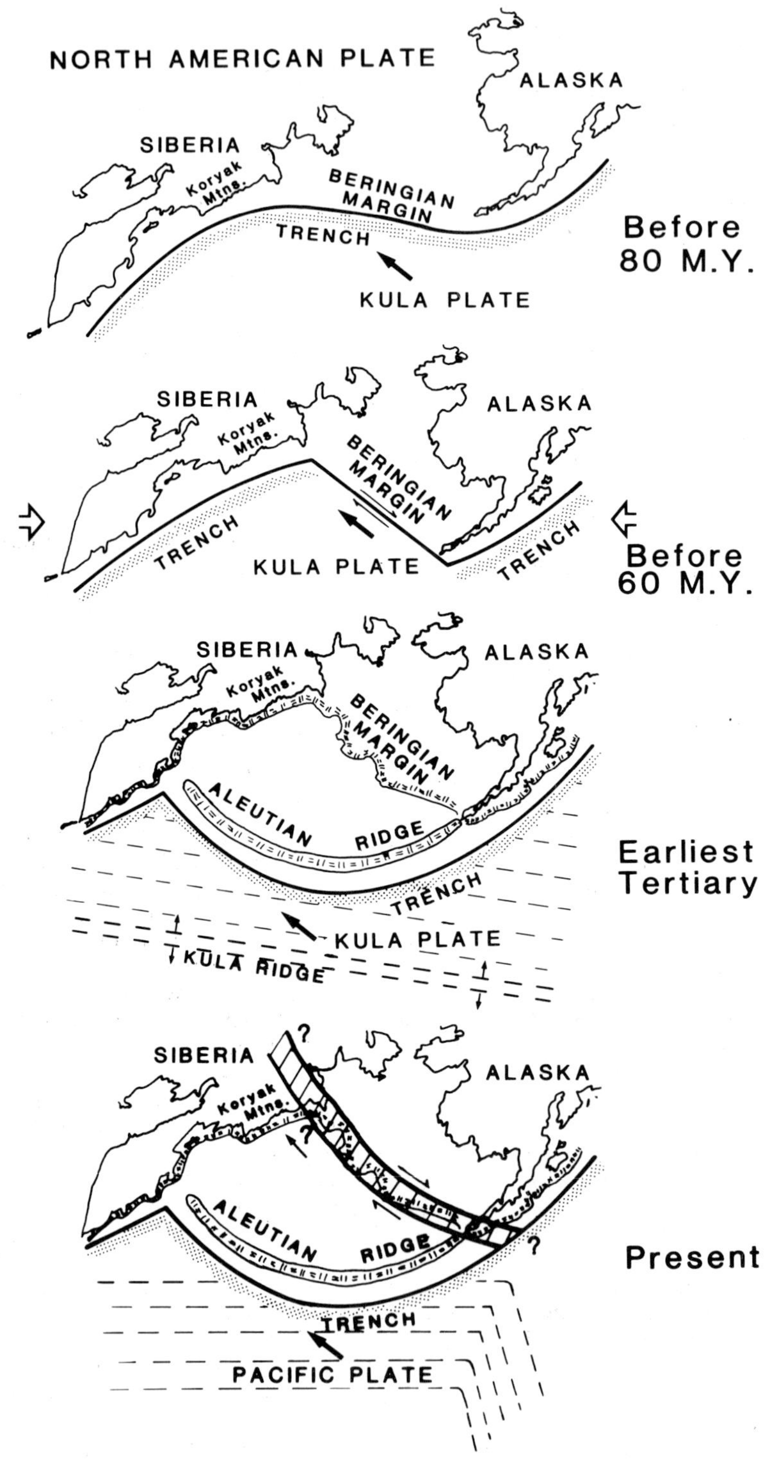
NORTH AMERICAN PLATE
ALASKA
SIBERIA
Koryak Mtns.
BERINGIAN MARGIN
TRENCH
KULA PLATE
Before 80 M.Y.
SIBERIA
ALASKA
Koryak Mtns.
BERINGIAN MARGIN
TRENCH
KULA PLATE
TRENCH
Before 60 M.Y.
SIBERIA
ALASKA
Koryak Mtns.
BERINGIAN MARGIN
ALEUTIAN RIDGE
TRENCH
KULA PLATE
KULA RIDGE
Earliest Tertiary
SIBERIA
ALASKA
Koryak Mtns.
ALEUTIAN RIDGE
TRENCH
PACIFIC PLATE
Present

Conclusions and speculations

Serpentinite mélange, olistrostromes, ultramafic complexes, and compressionally deformed terrigenous sequences exposed in the Koryak Mountains of eastern Siberia are presumed to be a subduction complex of Mesozoic age (McLean 1979a). The converging lithospheric plates are thought to have been the Kula(?) and North American or Eurasian plates.

Following cessation of rapid convergence between the two plates in late Mesozoic or earliest Tertiary time, the Koryak Range underwent repeated uplift, principally during the middle Miocene and early Pliocene. Parts of the coastal area subsided deeply, especially in the early Tertiary. Cenozoic deformation, differential uplift, and subsidence of the Siberian margin may in part be attributable to the buoyant uplift of sediment subducted beneath the margin during earlier periods of plate convergence.

In contrast to the uplifted Koryak Range in eastern Siberia, the adjacent Beringian margin to the SE is a foundered continental margin that appears to be underlain by shallow-water sedimentary and igneous rocks of Jurassic and Cretaceous age. The initial collapse of the margin in Late Cretaceous or early Tertiary time could be ascribed to subsidence along wrench faults that are associated with obliquely underthrust plate boundaries (Fitch 1972). But the continued collapse of the margin in middle and late Cenozoic time must be attributed to other factors. One mechanism for collapse might be the cooling of continental lithosphere that had been heated by the underthrusting action of oceanic crust.

Underthrusting in the Mesozoic would have resulted in the juxtaposition of thick low-density continental crust against thin ocean crust; cessation of underthrusting would allow tensional stresses to develop in the continental crust (Bott 1971; Bott & Dean 1972). The stresses in the continental crust cause an oceanward creep of lower crustal rocks beneath the outer shelf toward the outer margin and Aleutian Basin. Outflow of the lower crust would contribute to subsidence of the margin and to extensional deformation of the outer shelf.

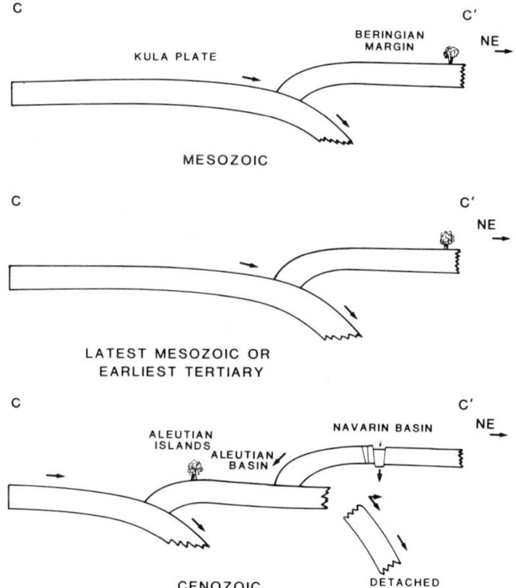

FIG. 6. Postulated plate motion between the Kula(?) plate and Beringian margin (North American plate) during the Mesozoic. For location of cross-section see Fig. 1. By latest Mesozoic or earliest Tertiary time, formation of the Aleutian arc isolates a remnant of Kula(?) plate in the proto-Bering Sea and also isolates the Beringian margin. The detachment of a piece of Kula(?) plate beneath the Beringian margin and its subsequent sinking causes the subsidence and extensional deformation of the margin throughout the Cenozoic.

Another mechanism that could result in the collapse of convergent plate boundaries is the detaching of a segment of subducted oceanic plate (Blanton 1977; Uyeda & Kanamori 1979). Fig. 6 is a sketch adapted from Blanton (1977) illustrating the oblique subduction of Kula(?) plate beneath the Beringian margin. Following formation of the Aleutian arc and isolation of the margin, a slab of the subducted plate is mechanically detached from the rest of the plate underlying the adjacent Aleutian basin. According to Blanton, after detachment of the

FIG. 5. Generalized sketch map of late Mesozoic and earliest Tertiary plate motions along the Beringian margin. Prior to 80 Ma, the margin was obliquely underthrust by the Kula(?) plate. Between 80 and 60 Ma, oroclinal bending of western Alaska and subsequent clockwise rotation of the margin resulted in change in plate motion to transform or strike-slip movement between the North American and Kula(?) plates (W. Patton written comm. 1979). Formation of the Aleutian arc by earliest Tertiary time isolated a piece of Kula(?) plate, thus forming the Bering Sea as the site of subduction shifted south to near the present Aleutian Trench. After Scholl *et al.* (1975) and Marlow & Cooper (1981). Speculative boundary between the North American-Eurasian plates shown by diagonal hachures.

subducted slab the continental margin bowed upward, then ruptured extensionally and began to subside.

A final speculation as to the cause of Cenozoic collapse of the Beringian margin is based on plate reconstructions for the recent movements between the North American and Eurasian plates by Zonenshayn *et al.* (1978, fig. 1). Their work indicates that the North American-Eurasian plate boundary passes through the Koryak Mountains (eastern Siberia)–Beringian margin area. In their model, the Koryak Mountains (eastern Siberia) could have been the site of limited Cenozoic convergence along a broad and diffuse boundary separating the North American-Eurasian plates (Fig. 5). The Beringian margin could also have been the site of limited (few tens of kilometres) Cenozoic transform or strike-slip motion extending from the Koryak Mountains south-eastward along the margin and intersecting the Aleutian Trench near Unimak Pass (Fig. 5). Subsidence of the margin during the Cenozoic could be ascribed to wrench faulting associated with strike-slip motion.

In summary, we note that our model for the late Mesozoic to early Tertiary evolution of the Bering Sea does not satisfactorily explain the continued Cenozoic tectonism along the deactivated and supposedly quiescent former plate boundaries within the Bering Sea. The continued uplift of the Koryak Range and the steady subsidence of the Beringian margin remain difficult to explain in a plate-tectonic model for the North Pacific and Bering Sea during middle and late Cenozoic time. Our explanations, problematical as they may be, are offered here to stimulate discussion and further study.

ACKNOWLEDGMENTS: We thank Michael Fisher and Roland von Huene for their thoughtful and thorough reviews.

References

ALEKSANDROV, A. A., BOGDANOV, N. S., BYALOBZHESKIY, S. G., MARKOV, M. S., TILMAN, S. M., KHAIN, V. YE. & CHEKHOV, A. D. 1976. New data on the tectonics of the Koryak highlands. *Geotectonics*, **9,** 292–9.

BAILEY, E. H. & BLAKE, M. C. JR. 1969. Tectonic development of western California during the late Mesozoic. *Geotectonics*, **3,** 148–54.

BAILEY, E. H., IRWIN, W. P. & JONES, D. L. 1964. Franciscan and related rocks, and their significance in the geology of western California. *Bull. Calif. Div. Mines Geol.* **183,** 177 pp.

BLANTON, S. L. 1977. Geology of the Bering shelf. *In:* SISSON, A. (ed). *The Relationship of Plate Tectonics to Alaskan Geology and Resources.* E1–E6, 1977 Symp. Proc. Alaska Geol. Soc. Anchorage, Alaska.

BOTT, M. H. P. 1971. Evolution of young continental margins and formation of shelf basins. *Tectonophysics*, **11,** 319–27.

—— & DEAN, D. S. 1972. Stress systems at young continental margins. *Nature (phys. sci.)* **235,** 23–5.

BURK, C. A. 1965. Geology of the Alaska Peninsula island arc and continental margin. *Mem. geol. Soc. Am.* **99,** 250 pp.

CLARKE, S. H. B. 1973. The McHugh Complex of south-central Alaska. *Bull. U.S. geol. Surv.* **1372-d,** D1–D11.

CONNELLY, W. 1978. The Uyak Complex, Kodiak Islands. Alaska: a Cretaceous subduction complex. *Bull. geol. Soc. Am.* **89,** 755–69.

COOPER, A. K., MARLOW, M. S. & SCHOLL, D. W. 1976a. Mesozoic magnetic lineations in the Bering Sea marginal basin. *J. geophys. Res.* **81,** 1916–34.

——, SCHOLL, D. W. & MARLOW, M. S. 1976b. Plate tectonic model for the evolution of the Bering Sea basin. *Bull. geol. Soc. Am.* **87,** 1119–26.

CREAGER, J. S., SCHOLL, D. W. *et al.* 1973. *Init. Rep. Deep Sea drill. Proj.* **19,** 913 pp. U.S. Govt Printing Office, Washington.

DRABKIN, I. YE (ed). 1970. Severo-Vostok SSSR, Geologischeskoye Opisaniye. *In: Geologiya SSSR, Tom 30, Kniga, 2.* Izd. Nedra, Moscow (Geology of USSR, northeast USSR).

ERNST, W. G. 1970. Tectonic contact between the Franciscan melange and the Great Valley sequence—crustal expression of late Mesozoic Benioff zone. *J. geophys. Res.* **75,** 886–902.

EWING, J., EWING, M., AITKEN, T. & LUDWIG, W. 1968. North Pacific sediment layers measured by seismic profilings. *In: The Crust and Upper Mantle of the Pacific Area.* Geophys. Monogr. **12,** 147–73. Am. geophys. Union, Washington.

FITCH, T. J. 1972. Plate convergence, transcurrent faults, and internal deformation adjacent to southeast Asia and the western Pacific. *J. geophys. Res.* **77,** 4432–60.

GLADENKOV, IU. B. 1964. O tektonike vostochnoi chasti Koryakskogo Nagor'ya. *Akad. Nauk SSSR Geol. Inst. Trudy.* **113,** 7–23 (Tectonics of the eastern part of the Koryak Uplands).

HAMILTON, W. 1969. Mesozoic California and the underflow of Pacific mantle. *Bull. geol. Soc. Am.* **80,** 2409–30.

HOPKINS, D. M. *et al.* 1969. Cretaceous, Tertiary, and early Pleistocene rocks from the continental margin in the Bering Sea. *Bull. geol. Soc. Am.* **80,** 1471–80.

MARLOW, M. S. 1979. Hydrocarbon prospects in Navarin basin province, northwest Bering Sea shelf. *Oil Gas J.* October 29, 190–6.

—— & COOPER, A. K. 1981. Mesozoic and Cenozoic structural trends beneath the southern Bering Sea shelf. *Bull. Am. Assoc. Petrol. Geol.*, in press.

—— , —— , SCHOLL, D. W., VALLIER, T. L. & MCLEAN, H. 1979b. Description of dredge samples from the Bering Sea Continental Margin. *Open-File Rep. U.S. geol. Surv.* **79–1139,** 5 pp.

—— , SCHOLL, D. W., COOPER, A. K. & BUFFINGTON, E. C. 1976. Structure and evolution of Bering Sea shelf south of St. Lawrence Island. *Bull. Am. Assoc. Petrol. Geol.* **60,** 161–83.

—— , —— , —— & JONES, D. L. 1979a. Mesozoic rocks from the Bering Sea; the Alaska-Siberia connection. *Abstr. Prog. geol. Soc. Am.* **11,** 90.

MCLEAN, H. 1979a. Review of petroleum geology of Anadyr and Khgatyrka basins, USSR. *Bull. Am. Assoc. Petrol Geol.* **63,** 1467–77.

—— 1979b. Pribilof segment of the Bering Sea continental margin: a reinterpretation of Upper Cretaceous dredge samples. *Geology,* **7,** 307–10.

PAGE, B. M. 1970. Sur-Nacimiento fault zone of California. *Bull. geol. Soc. Am.* **81,** 667–90.

—— 1978. Franciscan melanges compared with olistostromes of Taiwan and Italy. *Tectonophysics,* **47,** 223–46.

ROEDER, D. H. 1975. Tectonic effects of dip changes in subduction zones. *Am. J. Sci.* **275,** 252–64.

SCHOLL, D. W., BUFFINGTON, E. C. & MARLOW, M. S. 1975. Plate tectonics and the structural evolution of the Aleutian-Bering Sea region. *In:* FORBES, R. B. (ed). *Contributions to the Geology of the Bering Sea basin and Adjacent Regions. Spec. Pap. geol. Soc. Am.* **151,** 1–32.

SILVER, E. A. & BEUTNER, E. C. 1980. Melanges. *Geology,* **8,** 32–4.

TILMAN, S. M., BELYI, V. F., NIKOLAEVSKII, A. A. & SHILO, N. A. 1969. Tektonika Severo-Vostoka SSSR. *Akad. Nauk SSSR Sibirsk.* Otdeleniye, Severo-Vostoch. Kompleks, Nauchno Issled. Inst. Trudy, **33,** 78 pp. Magadan (Tectonics of the northeast of USSR).

TYSDAL, R. G. & CASE, J. E. 1976. The McHugh Complex in the Seward quadrangle, south-central Alaska. *In:* BLEAN, K. M. (ed). *The United States Geological Survey in Alaska: accomplishments during 1976.* Circ. U.S. geol. **751-B,** B48–9.

—— , —— , WINKLE, G. R. & CLARK, S. H. B. 1977. Sheeted dikes, gabbro, and pillow basalt in flysch of coastal southern Alaska. *Geology,* **5,** 377–83.

UYEDA, M. S. & KANAMORI, H. 1979. Back-arc opening and the mode of subduction. *J. geophys. Res.* **84,** 1049–61.

VALLIER, T. L., UNDERWOOD, M. B., JONES, D. L. & GARDNER, J. V. 1980. Petrography and geological significance of Upper Jurassic rocks dredged near Pribilof Islands, southern Bering Sea continental shelf. *Bull. Am. Assoc. Petrol. Geol.* **64,** 945–50.

VON HUENE, R., SHOR, G. S. JR & WAGEMAN, J. 1978. Continental margins of the eastern Gulf of Alaska and boundaries of tectonic plates. *In:* WATKINS, J. S., MONTADERT, L. & DICKERSON, P. W. (eds). *Geological and Geophysical Investigations of Continental Margins. Mem. Am. Assoc. Petrol. Geol.* **29,** 273–90.

ZONENSHAYN, L. P., NATAPOV, L. M., SAUOSTIN, L. A. & STAVSKII, A. P. 1978. Recent plate tectonics of northeastern Asia in connection with the opening of the North Atlantic and the Arctic Ocean Basins. *Oceanology,* **18,** 550–5.

M. S. MARLOW, A. K. COOPER, D. W. SCHOLL & H. MCLEAN. U.S. Geological Survey, 345 Middlefield Road, Menlo Park, California 94025, U.S.A.

The Chugach Terrane, a Cretaceous trench-fill deposit, southern Alaska

Tor H. Nilsen & Gian G. Zuffa

SUMMARY: The Chugach terrane of southern Alaska extends for approximately 2000 km along the margin of the Gulf of Alaska. A seaward flysch facies of the terrane, the Chugach flysch terrane, represents the fill of a Late Cretaceous trench and consists of structurally deformed turbidites with some mafic volcanic rocks. It is intruded by anatectic granitic plutons of early Tertiary age. The Chugach flysch terrane in most places is bounded to the north by a landward-dipping thrust fault separating it from mélange of the Chugach terrane at least as young as Late Cretaceous. To the south the flysch terrane contains oceanic volcanic rocks and is bounded by faults that separate it from Palaeogene turbidites or upper Mesozoic metavolcanic rocks.

Interpretations of folds and faults in SW Alaska suggest that the Chugach terrane was deformed during NW-directed subduction. Palaeocurrents indicate primarily westward flow along the axis of the outcrop belt, and secondary southward transport. Turbidite facies associations indicate an east-to-west progression from inner-fan to middle-fan, outer-fan, fan-fringe and basin-plain deposits down the axis of the outcrop belt, and a bounding slope facies association to the north. Rock-fragment petrography of sandstone samples from the Chugach terrane indicates derivation from a magmatic arc that was increasingly dissected eastward and from an older subduction complex. The magmatic arc and adjacent shallow-marine forearc basin deposits are located to the north in southern Alaska. Palaeomagnetic data from the Chugach terrane and adjacent deposits indicate original deposition far to the south of the present latitude of Alaska. The entire magmatic arc-forearc basin-trench complex migrated northward in early Tertiary time, when it was probably oroclinally bent and accreted as a mesoplate to Alaska prior to the middle Miocene.

Although modern deep-sea trenches adjacent to both continental margins and island arcs are abundant and widely distributed, very few good examples of well-preserved and extensive ancient trench-fill deposits have been described in the geological literature. Most examples described are of obducted fragments or slices of trench-fill deposits rather than the entire fill of a major trench system. The reason for the lack of good examples is clearly the structural deformation imposed on trench-fill deposits, which generally are partially or wholly subducted, disrupted by lateral faulting, or intensively deformed by plate-margin tectonism.

The purpose of this paper is to describe the general setting, geological history, and palaeogeographic framework of part of the Chugach terrane, inferred to be a Cretaceous trench-fill complex accreted to southern Alaska. The Chugach terrane may represent the longest, most continuous, and best preserved ancient trench-fill sequence in the world. After more than 10 years of detailed study of various parts of the terrane, the overall size, character and significance of this remarkable sequence are beginning to emerge.

Modern trenches may extend for thousands of kilometres and can be filled by sediment derived from various sources and transported to the trench by various routes. Continental-margin trenches are most likely to contain thick sediment fills because the magmatic arc and uplifted older terranes form larger source areas. Both the thickness and geometry of trench fills depend upon the rate of convergence across the trench and whether sediment transport is longitudinal or lateral (Schweller & Kulm 1978). Some trenches may be starved of sediment derived from the continental margin because most if not all of the sediment may be deposited in forearc basins (the arc-trench gap) or in major basins on the trench slope (Scholl & Marlow 1974). Because of extensive tectonic deformation and submarine landsliding, continuous submarine canyons may not extend across the landward trench slope to funnel sediments from the magmatic arc to the trench floor (Underwood & Karig 1980). Some trench-fills may consist mostly of hemipelagic sediment deposited along the continental margin and pelagic sediments carried into the trench by the subducting oceanic plate rather than turbidites derived from the continental margin.

The Chugach terrane was named by Berg *et al.* (1972) for outcrops of greywacke, argillite, slate, conglomerate, volcanic rocks, chaotic mélanges, and granitic plutons in the Chugach Mountains of southern Alaska. They suggested that the terrane extended south-eastward to Baranof and Chichagof Islands and divided it into two units thought to be of different age. Their older Chugach terrane cropped out landward of the younger Chugach terrane and consists of a polydeformed and regionally metamorphosed assemblage of phyllite, metagreywacke, quartzite, metachert, greenstone, amphibolite, and ultramafic rocks originally thought to be as old as Permian and Triassic. It originally included some rocks currently assigned to Wrangellia, a separate accreted terrane to the NE (Jones & Silberling 1979). The outboard younger Chugach terrane consists of an exterior belt of greywacke and slate, generally very thick and unfossiliferous, also extending south-eastward to Baranof and Chichagof Islands. Fossils from the younger Chugach terrane are chiefly of Late Cretaceous age. The Chugach terrane was extended by mapping to SW Alaska (Plafker *et al.* 1977; Moore & Connelly 1979; Nilsen & Moore 1979), where the younger Chugach terrane had previously been mapped over a wide area as the 'slate and graywacke belt' (Burk 1965).

As mapped at present, the Chugach terrane crops out continuously, except where covered by large glaciers and marine waters of the Gulf of Alaska, for about 2000 km from Baranof Island in the SE to Sanak Island in the SW (Fig. 1). The outcrop belt is as wide as 100 km, but is highly variable where affected by Cenozoic strike-slip faulting, particularly in SE Alaska.

Moore (1972, 1973a) suggested that the outcrop belt of the Chugach terrane may extend NW for another 1200 km from the Sanak Islands along the edge of the shelf of the Bering Sea to the Koryak Mountains of SE Siberia. However, because Upper Cretaceous rocks have not been recovered by recent dredging or detected by geophysical studies of the Bering Sea shelf edge, the extension of the Chugach terrane to the NW is now considered unlikely (Cooper & Marlow 1979; McLean 1979).

Because faunal evidence now indicates that both the older and younger Chugach terranes of Berg *et al.* (1972) contain fossils as young as Late Cretaceous, the distinction of the two parts of the terrane by age has become less desirable. For the purposes of this paper, I divide the Chugach terrane into the Chugach mélange terrane, formerly part of the 'older Chugach terrane', and the Chugach flysch terrane, formerly the 'younger Chugach terrane'. The Chugach flysch terrane will be of principal interest here, because it appears to represent a late Mesozoic trench-fill.

The age of the Chugach flysch terrane has been established as Late Cretaceous, primarily Maestrichtian (Jones & Clark 1973), on the basis of scattered megafossils. Brew & Morrell (1979) reported Early and Late Cretaceous fossils from parts of SE Alaska.

The Chugach flysch terrane is bounded by major faults both landward and seaward throughout most of its extent. The Eagle River fault, called the Uganik fault on Kodiak Island by Moore (1978), bounds the Chugach terrane on the north. It generally dips landward, except in SE Alaska, and separates the Chugach flysch terrane from the Chugach mélange terrane, known as the Uyak Complex on Kodiak Island (Connelly 1978), the McHugh Complex in the Chugach Mountains and adjacent areas (Clark 1973), the mélange facies of the Yakutat Group east of Yakutat, the Tarr Inlet suture zone rocks in the Glacier Bay National Monument (J. Decker, pers. comm. 1980), and the upper part of the Kelp Bay Group near Baranof Island (Loney *et al.* 1975). The Chugach mélange terrane contains blocks of many different lithologies and ages. Blocks of chert from the McHugh Complex contain radiolarians of Triassic, Jurassic, and Cretaceous as young as Cenomanian (Karl *et al.* 1979; G. R. Winkler, pers. comm. 1980); megafossils in the matrix are as young as Maestrichtian (Plafker *et al.* 1977).

The Border Ranges fault forms the northern margin of the Chugach flysch terrane where the Eagle River fault is not present and extends for most if not all the outcrop length of the Chugach terrane. The Border Ranges fault dips landward and locally juxtaposes Palaeozoic and Mesozoic metamorphic rocks with the Cretaceous units (MacKevett & Plafker 1974; Plafker *et al.* 1976, 1977; Plafker & Campbell 1979).

To the south, the Chugach flysch terrane is separated from Palaeogene turbidites and mafic volcanic rocks by the Contact fault, in most places a high-angle landward-dipping reverse fault. In SE Alaska, the Queen Charlotte and Fairweather faults form the SW margins of the Chugach flysch terrane and related late Mesozoic(?) volcanic rocks (Plafker *et al.* 1977; Plafker & Campbell 1979). These faults are right-lateral strike-slip faults that record Cenozoic offsets.

North of the Chugach terrane are extensive outcrops of non-marine and shallow- to deep-marine Upper Cretaceous sedimentary rocks (Fig. 1). These units, of Campanian and Maestrichtian age, are common in parts of the Alaska Peninsula, on the NW side and beneath the

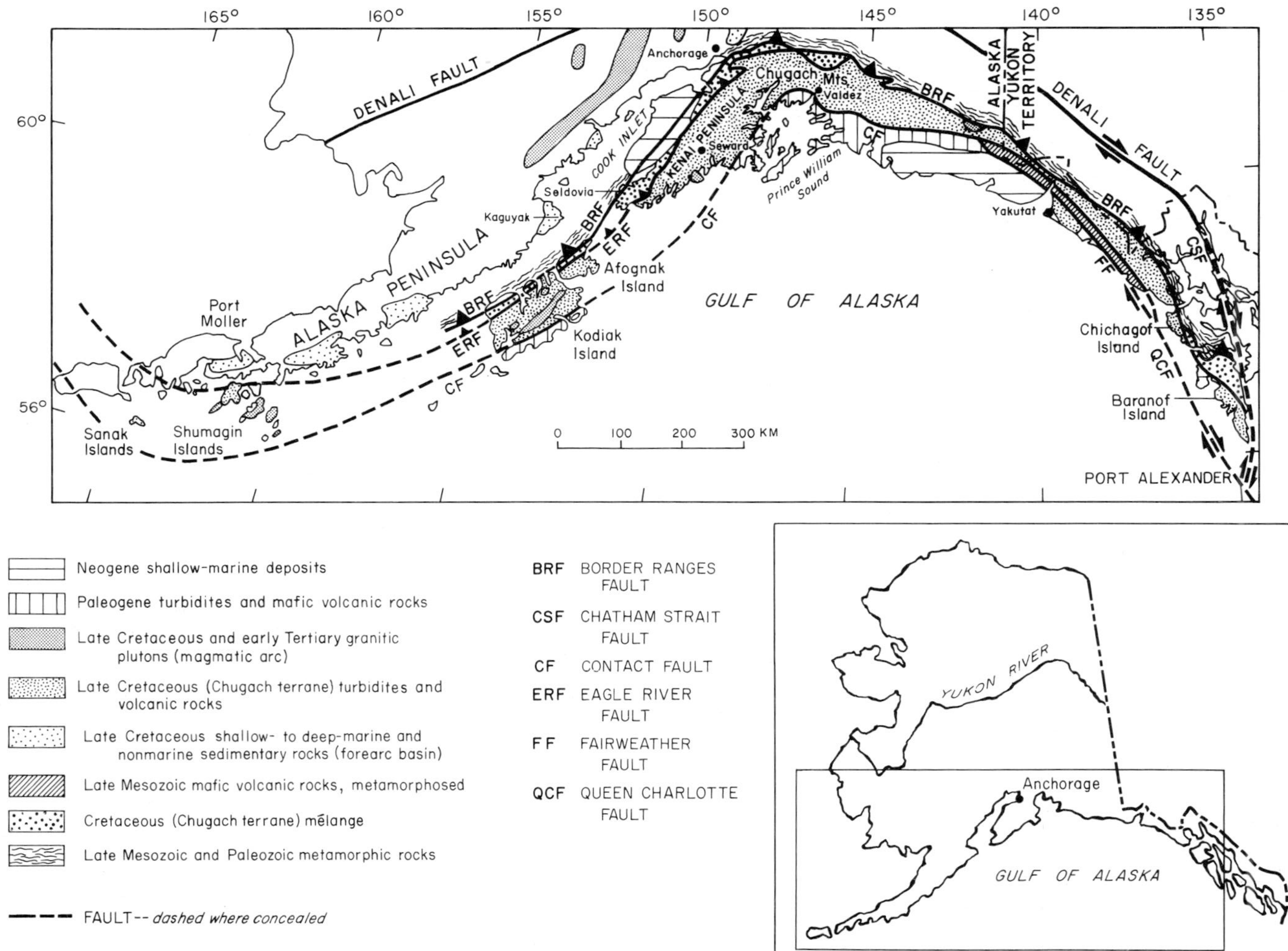

FIG. 1. Distribution of Chugach terrane and major faults along margins of the Gulf of Alaska. Geology modified from Beikman (1974a,b, 1975a,b), Plafker *et al.* (1977), and Plafker & Campbell (1979).

Cook Inlet, and in the Matanuska Valley NE of Anchorage (Grantz 1964; Jones & Detterman 1966; Mancini *et al.* 1978; Detterman 1978; Magoon *et al.* 1978, 1980). These clastic sedimentary rocks, generally transported southward and derived from volcanic and plutonic sources, are forearc basin or arc-trench gap deposits coeval with and situated landward of the Chugach terrane.

North of the forearc basin deposits is a belt of Upper Cretaceous to lower Tertiary quartz diorite to granite plutons that form part of a large magmatic arc (Fig. 1). The plutons, called the Iliamna-McKinley phase of the Alaska Range-Talkeetna Mountains batholithic complex by Hudson (1979a,b,c), have yielded potassium-argon ages of 83 to 55 Ma (Reed & Lanphere 1973). Plots of K_2O trends by Moore & Connelly (1979) from data presented by Reed & Lanphere (1974) for the Late Cretaceous and early Tertiary Alaska Range-Talkeetna Mountains batholiths in SW Alaska suggest that these plutons were generated in response to NW-directed subduction. In SE Alaska, Cretaceous and early Tertiary granitic plutons are abundant landward of the Border Ranges fault (Brew *et al.* 1978). The magmatic arc may be bounded on the north in Central and SW Alaska by the Denali fault system, which may be a major Cenozoic right-lateral strike-slip system.

Within the Chugach flysch terrane and on its south flank are extensive but discontinuous belts of volcanic rocks that are locally highly deformed and metamorphosed adjacent to plutons. On Kodiak Island, basalts and andesites have been included in the Palaeocene and Eocene Ghost Rocks Formation (Moore 1969; Nilsen & Moore 1979; Reid & Gill 1980). In the Prince William Sound area diabasic sheeted dykes, gabbros, and basalts crop out within both the Chugach flysch terrane and the Palaeogene Orca Group to the south (Tysdal *et al.* 1977). In SE Alaska, late Mesozoic oceanic basalts form a mappable terrane along the SW margin of the Chugach flysch terrane (Plafker & Campbell 1979). Volcanic rocks are interbedded with the Sitka Graywacke on Chichagof Island (J. Decker, pers. comm. 1980). These various volcanic and plutonic rocks are clearly of two distinct ages, the first coeval with deposition of the Maestrichtian Chugach flysch terrane and the second coeval with deposition of Palaeogene turbidites. These rocks, partly oceanic in character, indicate the involvement of ocean floor with deposition and deformation of the Chugach flysch terrane.

Southern Alaska has been reinterpreted in the last few years as a collage of various terranes that have been transported northward and accreted to Alaska. Many different Mesozoic terranes have been recognized in British Columbia (Monger *et al.* 1972), SE Alaska (Berg *et al.* 1978), and central and southern Alaska (Jones & Silberling 1979; Csejtey 1979). Palaeomagnetic data, largely from rocks associated with the Chugach terrane, indicate that it was originally deposited far to the south of its present position (Hillhouse & Gromme 1977; Stone & Packer 1979); during the Cenozoic the terrane moved northward, eventually reaching its present latitude and accreting to Alaska. However, the process, timing, and tectonic effects of the accretion process on the Chugach flysch terrane and related rocks are not well understood.

The Chugach flysch terrane has previously been considered to be a trench deposit on the basis of tectonostratigraphic considerations (Burk 1965; Moore 1972, 1973a,b; Budnik 1974a,b; Nilsen & Bouma 1977; Connelly 1978; Nilsen & Moore 1979; Moore & Connelly 1979). However, regional petrographic, palaeocurrent, and facies studies have never been completed, so that most palaeogeographic and palaeotectonic generalizations have not been substantiated by geological data.

Stratigraphy

The Chugach flysch terrane includes a number of stratigraphic units of somewhat differing characteristics. Most published descriptions note that it consists of deformed, thick, repetitively interbedded greywacke and shale or slate with local conglomerate, limestone, and volcanic rocks.

On the Sanak and Shumagin Islands, the Chugach flysch terrane consists of the Shumagin Formation. Moore (1973a) estimated 3–4 km as a minimum thickness after having measured sections as thick as 1470 m on the Sanak Islands and 4100 m on the Shumagin Islands. Scattered fossils indicate a Maestrichtian age (Jones & Clark 1973). On the Sanak Islands, pillow lava and bedded chert are locally in depositional contact with the turbidites.

On Afognak and Kodiak Islands, the Chugach flysch terrane consists of the Kodiak Formation. Nilsen & Moore (1979) estimated a stratigraphic thickness of 5000 m and noted no volcanic rocks or chert within it. Scattered fossils indicate a Maestrichtian age (Jones & Clark 1973).

In the Chugach Mountains and Kenai Peninsula, the Chugach flysch terrane has been mapped as the Valdez Group by Clark (1972) and Tysdal & Plafker (1978), but no stratigraphic thicknesses have been estimated for it. Maestrichtian megafossils are present at scattered localities (Jones & Clark 1973). The Valdez Group includes local interbedded tuff and pillow basalt, and on the Resurrection Peninsula near Seward is intruded by basalt sills, sheeted dykes, gabbro and serpentinized dunite (Tysdal *et al.* 1977).

The Chugach flysch terrane SE of the Yakutat area has been mapped as part of the Yakutat Group (see Plafker & Campbell 1979) and as an unnamed unit east of the Fairweather fault (Brew & Morrell 1979). Much of the Yakutat Group crops out west of the Fairweather and Contact faults, outside the main outcrop belt of the Chugach flysch terrane (Fig. 1), and could be considered to be a separate terrane from the Chugach terrane. No thicknesses or fossils have been reported for this area, which is regionally metamorphosed.

On Chichagof, Baranof, and adjacent islands, the Chugach flysch terrane has been mapped as the Sitka Graywacke (Loney *et al.* 1975). No thickness for it has been estimated. Fossils include only Early Cretaceous pelecypods transported and redeposited in debris flows at one or two localities. The Sitka Graywacke contains local bodies of greenstone, greenschist and chert.

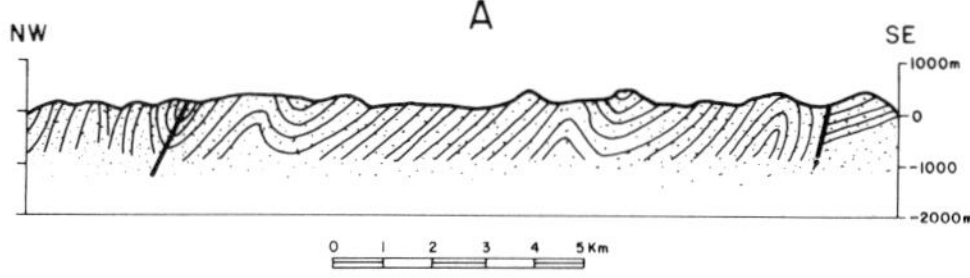

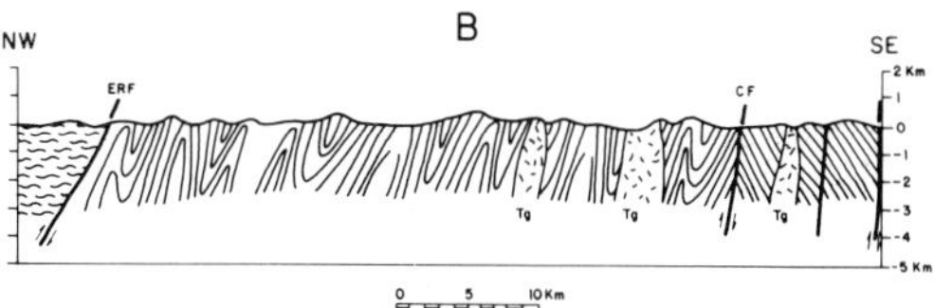

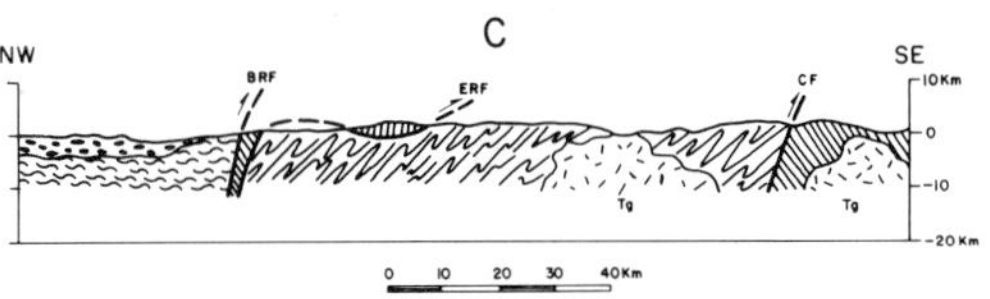

FIG. 2. Cross-sections showing general structure of Chugach terrane and relations with bounding terranes. Map symbols and abbreviations as in Fig. 1 except for Tg, Tertiary granitic plutons.
(A) NW–SE cross-section of part of Shumagin Islands (from Moore 1973b, Fig. 8).
(B) NW–SE cross-section of Kodiak Island (from von Huene *et al.* 1978).
(C) NW–SE cross-section from Anchorage area to Prince William Sound (from Plafker *et al.* 1977).

Structure

Throughout its extent, the Chugach flysch terrane is highly folded and faulted (Fig. 2). Strata generally dip landward, except in SE Alaska, and strike parallel to the regional trend of the belt. Faults and axial planes of folds also dip landward and strike parallel to the regional trend. Moore (1973b) demonstrated that on the Sanak and Shumagin Islands the Chugach flysch terrane was initially deformed in a partially lithified state and developed an axial plane slaty cleavage. A last phase of deformation involved uplift and landward tilting along high-angle faults.

The Chugach flysch terrane on Kodiak Island consists chiefly of landward-dipping homoclinal sequences of turbidites with stratigraphic tops toward land, separated by high-angle landward-dipping reverse faults. Because of this geometry, Moore (1969) originally postulated a thickness of 30,000 m for the Chugach flysch terrane. Directions of underthrusting on Kodiak Island were determined at 332° ± 11° for the Uyak Complex (Moore & Wheeler 1978), 334° ± 7° along the Uganik or Eagle River fault (Moore 1978), and 340° for the Chugach flysch terrane (Moore & Bolm 1977). These underthrust directions are perpendicular to the present general strike of the Chugach flysch terrane, forearc basin sequence, and Alaska Range-Talkeetna Mountains batholithic complex.

Granitic batholiths, informally referred to as the Sanak-Baranof belt, intrude the Chugach flysch terrane in many areas, and have been dated by the potassium-argon method as 59.9 Ma on the Sanak Islands, 54–64 Ma on the Shumagin Islands, 58 Ma on Kodiak Island, 47–52 Ma in the Chugach Mountains, and 43–48 Ma on Baranof Island (see summaries in Hudson 1979a,b). These batholiths have been interpreted to be products of anatectic melting of the subduction complex rather than arc magmatism (Hill & Morris 1977; Hudson *et al.* 1979). Regional metamorphism of Late Cretaceous age with penetrative deformation typifies the Chugach flysch terrane in the Yakutat area (Hudson *et al.* 1977).

The structural framework of the Chugach terrane suggests early deformation in a trench setting followed by regional uplift and tilting, at least in SW Alaska, along landward-dipping reverse faults. In parts of SE Alaska, the terrane was regionally metamorphosed. The entire belt of turbidites was intruded by anatectically derived granitic rocks produced by syndeformational melting of the sedimentary pile. These plutons decrease in age eastward.

Facies associations

Turbidite facies can be most simply grouped on the basis of vertical sequence into five facies associations: slope, inner-fan, middle-fan, outer-fan, and basin-plain (Mutti & Ricci Lucchi 1972, 1975; Nilsen 1977). These facies associations have been delineated for parts of the Chugach flysch terrane, particularly on Chichagof Island, the Kenai Peninsula, and Kodiak Island (Fig. 3).

On western Chichagof Island in SE Alaska, Decker *et al.* (1979) recognized an inner-fan facies association toward the SE, a middle-fan facies association toward the NW, and a slope facies association between the two and along the NE margin of the belt of outcrops of the Sitka Graywacke. The inner-fan facies association extends SE to Baranof Island, where it forms the type area of the Sitka Graywacke. It consists chiefly of massive sandstone and conglomerate with a maximum clast size of about 25 cm. The middle-fan facies association consists of channelized bodies of massive sandstone surrounded by thicker sequences of thin-bedded turbidites and shaly channel-margin and interchannel deposits. The slope facies association consists of thick sequences of hemipelagic mudstone with thin-bedded siltstone and mudstone turbidites and local olistostromes and slide blocks of shallow-marine sandstone.

Between Chichagof Island and Yakutat, the Chugach flysch terrane consists largely of argillite and greywacke with smaller amounts of conglomerate and mafic volcanic rocks (Brew & Morrell 1979, table 15; Plafker & Campbell 1979). Although Brew & Morrell (1979) suggested that these deposits consisted mostly of outer-fan facies, J. Decker (pers. comm. 1980) has interpreted these deposits as an inner-fan channel and interchannel facies association. The incorporation of conglomeratic units within thick sequences of argillite and thin-bedded turbidites suggests the presence of channelized fan deposits and possibly slope deposits.

On the Kenai Peninsula between Anchorage and Seward, Nilsen & Bouma (1977) reported the presence of a slope facies association to the NW and a middle-fan facies association to the SE in the Valdez Group. They suggested that part of the slope facies association included large olistoliths and olistostromes previously mapped as part of the McHugh Complex. Remapping of the Chugach flysch terrane-McHugh Complex contact by Tysdal & Case (1977) had resulted in placing some of these olistostrome deposits within the Chugach terrane. Budnik (1974a,b) and Mitchell (1979, 1980) also described facies associations in the Valdez Group, referring to slope, middle-fan, outer-fan, and basin-plain deposits.

On NE Kodiak Island, Nilsen & Moore (1979) divided the Kodiak Formation into two facies associations, a landward slope association and a seaward basin-plain association. The slope association contains abrupt changes in thickness and grain size of deposits, conglomerate deposited in channels and canyons, small trench-slope basins, abundant slumps and synsedimentary folds, and thick pelitic sections. The small slope basins contain well-developed thin accumulations of little-deformed deep-sea fan facies associations fed by small canyons and fan channels that are enveloped by more highly deformed fine-grained slope deposits. These little-deformed slope-basin deposits are typically more coarse-grained than the adjacent trench-fill deposits to the south and have palaeocurrent directions more indicative of lateral transport. The basin-plain association contains, in contrast, classical fine- to medium-grained sandstone turbidites that extend for the entire outcrop length without perceptible changes in thickness or grain size and little organization of beds into thickening- or thinning-upward megasequences, although locally outer-fan lobe and fan-fringe deposits can be distinguished.

On the Shumagin and Sanak Islands, Moore (1973a) reported the presence of 35% massive beds of sandstone, 35% thin-bedded turbidites, 30% intermediate bed types, and less than 0.5% conglomerate and tuff. Conglomerates contain clasts as large as 20 cm and are described as pebbly mudstone olistostromes. Slumps are not uncommon and one as thick as 135 m is reported; directions of slumping are sub-parallel to the palaeocurrent flow directions. Although Moore (1973a) did not interpret facies associations, his descriptions suggest the presence of slope and fan or basin-plain

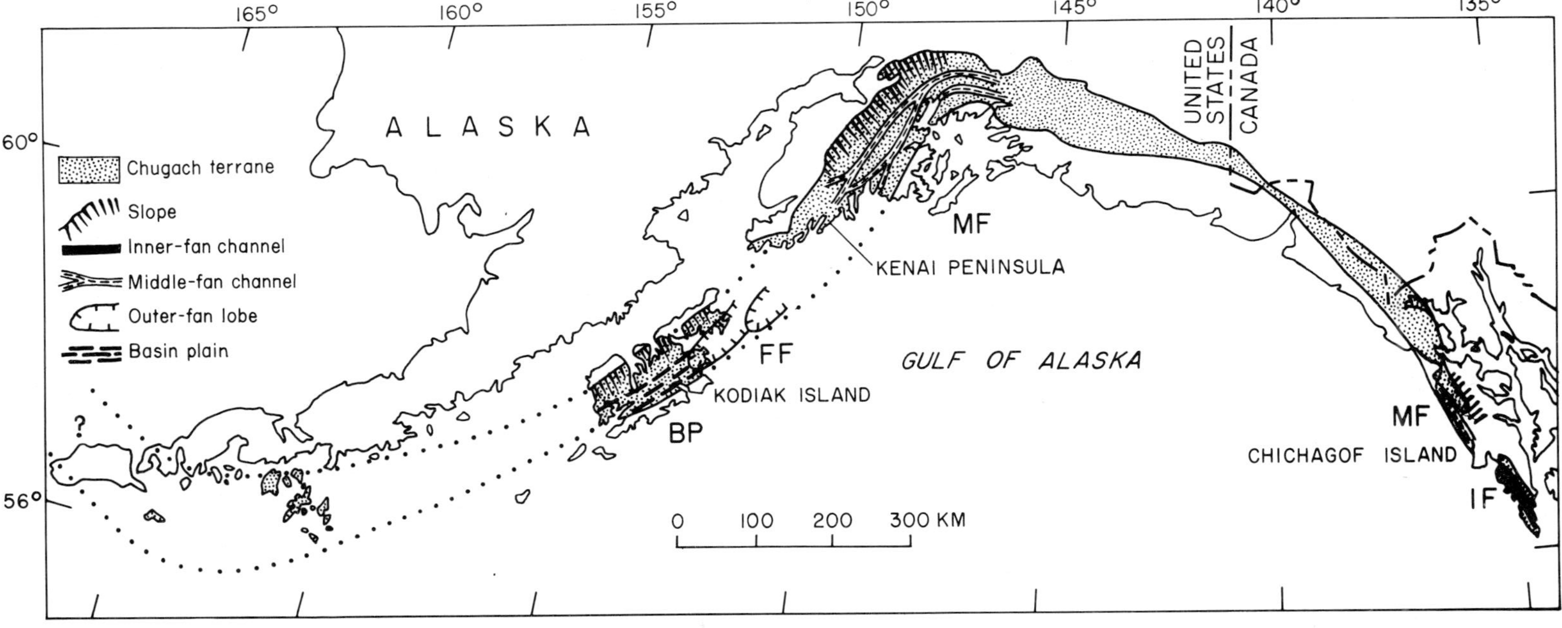

Fig. 3. Generalized distribution of facies associations from the Chugach terrane. Abbreviations (from east to west); IF, inner fan; MF, middle fan; FF, fan fringe; BP, basin plain.

facies associations in these islands. No major channelling was noted, so that the bulk of the deposits may represent outer-fan or basin-plain deposits.

In summary, preliminary analyses of facies associations in the Chugach flysch terrane suggest the presence coevally of an inboard slope facies association, locally difficult to distinguish from the adjacent older mélange terrane because of its extensive synsedimentary deformation, and an outboard inner-fan to basin-plain sequence that appears, on the basis of published interpretations, to be most proximal to the SE and distal to the SW (Fig. 3). The clearly developed down-axis changes in facies associations of the trench-fill deposits suggest fill by a radiating fan system characterized by a distributary system rather than by a single large channel depositing a wedge, such as that described by Schweller & Kulm (1978). The relations suggest penecontemporaneous slope deposition adjacent to a large, elongate, outbuilding fan system in a narrow trough, most probably a trench. Sediments appear to have been derived chiefly from a point source at the SE end of the trough. Although not fully demonstrated to date by facies studies, the longitudinal system of trench filling was undoubtedly supplemented by lateral infilling from canyons that cut across the inboard trench slope to produce mixed sediments.

Petrography

Burk (1965) inferred that the western part of the 'slate and greywacke' terrane, later included in the Chugach terrane, was derived chiefly from a volcanic source area. From study of sandstone of Shumagin and Sanak Islands, Moore (1973a) concluded that the Chugach flysch terrane in those areas consisted primarily of litharenite derived chiefly from andesitic volcanic rocks, with minor contributions from plutonic, sedimentary, and metamorphic rocks. Connelly (1978) reported that samples of the Kodiak Formation from the NW edge of the outcrop belt on Kodiak Island were more feldspathic than samples of the Chugach flysch terrane to the west. Winkler (in Nilsen & Moore 1979) determined that the composition of Kodiak Formation sandstone samples from other parts of Kodiak Island is similar to that obtained by Moore (1973a) farther west.

Mitchell (1979, 1980) reported for the Valdez Group south of Anchorage a litharenitic sandstone composition, 60% volcanic, 30% metamorphic, and 10% plutonic types. Sandstone samples of the Chugach flysch terrane from the Yakutat area are primarily arkose (G. R. Winkler, pers. comm. 1980), further suggesting that these rocks may form a separate terrane from the Chugach flysch terrane. In the Chichagof Island area, Chugach flysch terrane sandstone samples are primarily volcanogenic lithic arenite, probably derived chiefly from a magmatic arc provenance (J. Decker, pers. comm. 1980).

In order to examine the lateral variability in composition of the Chugach flysch terrane, Zuffa *et al.* (1980) studied the rock-fragment petrography of about 150 samples of medium- to coarse-grained sandstone from Baranof Island in SE Alaska to Sanak Island in SW Alaska. The rock fragments of about 60 samples were divided into 26 categories, and 100 or 200 grains were counted on each thin section. The resulting data indicate that two main source types supplied sediment to the Chugach flysch terrane.

The first source area shed argillite, siltstone, volcanic wacke, chert, limestone, basalt, basaltic andesite, and andesitic felsite, in varying amounts in different areas. This source area probably consisted of an active volcanic arc, older accreted terranes, and an emergent or submerged subduction complex. The second source area shed quartz, plagioclase feldspar, andesitic felsite, microlitic rock fragments, and potassium feldspar. This source area was probably a dissected arc terrane.

A ternary plot of polycrystalline quartz lithic fragments, volcanic and metavolcanic lithic fragments, and sedimentary and metasedimentary lithic fragments clearly indicates an arc-orogen source field for the Chugach flysch terrane (Fig. 4A). The plot also indicates an increasing degree of dissection of the arc-orogen eastward. The petrographic changes suggest lateral as well as longitudinal input of sediments to the trench. Sandstone from the Yakutat area contains the highest amount of K-feldspar, which together with abundant plutonite fragments (Fig. 4B), is indicative of a fairly distinct magmatic arc provenance and supports the inference that these rocks either form a separate terrane or were derived from a completely different source area.

Conglomerate clasts in the Chugach flysch terrane include granitic rocks, chert, limestone, quartzite, greywacke, argillite and silicic volcanic rocks in SE Alaska (Brew & Morrell 1979); granitic rocks, chert, greenstone, limestone, argillite and greywacke in southern Alaska (Tysdal & Case 1977); and chert, volcanic rocks, greywacke, mudstone and granitic rocks in SW Alaska (Moore 1973a). Thus, the con-

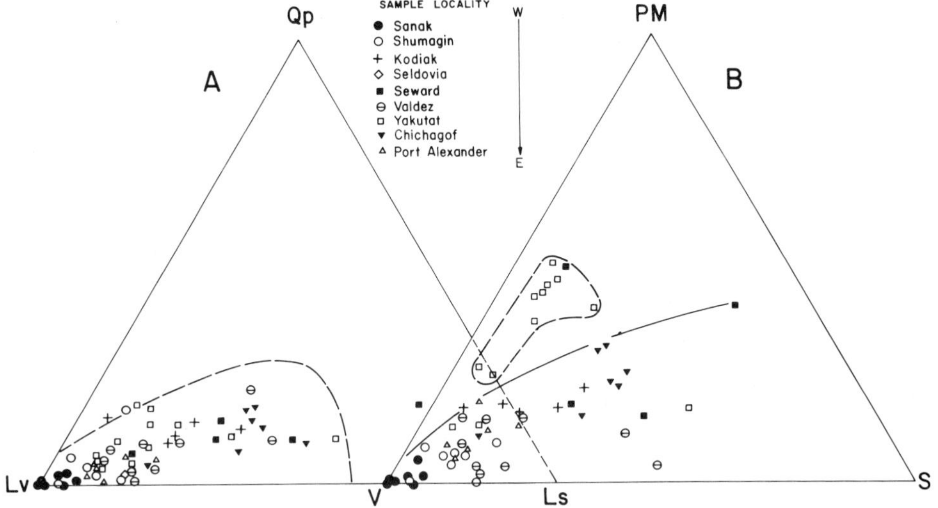

FIG. 4. Ternary diagrams showing rock-fragment petrography of Chugach terrane.
(A) Plot of polycrystalline quartz (Qp), volcanic and metavolcanic lithic fragments (Lv), and sedimentary and metasedimentary lithic fragments (Ls). Dashed line defines arc-orogen source field as proposed by Dickinson & Suczek (1979).
(B) Plot of all rock-fragment varieties: polycrystalline quartz-plutonic-gneiss-schist-phyllite (PM), felsite-microlitic grains—ophitic grains—vitrophyric grains—undetermined cherty grains—hypabyssal volcanic (V), and sandstone—siltstone—limestone—argillite—chert—shale—slate (S). The three poles in a continental margin-arc-trench geometry may indicate magmatic arc roots (PM), volcanic-arc (V), and subduction complex (S) sources. Dashed line around Yakutat area samples indicates greater influence of magmatic arc source in this part of Chugach terrane. Dashed line with arrow indicates from left to right eastward increasing dissection of magmatic arc and increasing amounts of sedimentary detritus.

glomerates reflect the same two major sources as the sandstone petrography.

Discussion and conclusions

The deformation history, petrography, palaeocurrents, facies associations, and regional stratigraphic and structural relations suggest that the turbidites of the Upper Cretaceous Chugach flysch terrane represent a trench-fill deposit. Inclusions of pillow basalt, sheeted dykes, serpentinized dunite, and chert suggest that the turbidites were deposited on ocean floor and were subsequently offscraped on the upper plate of the subduction zone. Regional evidence for the trench depositional setting is indicated by the presence progressively to the north of coeval: (1) slope deposits with south-trending submarine canyons and trench-slope basins, (2) remnants of a mélange zone separated from the Chugach flysch terrane by a landward-dipping major fault zone, (3) south-transported forearc basin deposits that form a belt parallel to the outcrop trend of the Chugach flysch terrane and contain abundant shallow-marine and non-marine sequences, and (4) granitic intrusions that form a major subparallel magmatic arc. The complete arc-trench system is at present bounded by major faults, mostly right-lateral strike-slip faults of Cenozoic age.

The style of folding, faulting, and metamorphism is typical of deposits emplaced into a subduction zone. The filling of the trench was by two separate processes, longitudinal and lateral. The major palaeocurrent trends and facies associations indicate an axial east-to-west change from proximal to distal turbidite sedimentation, suggesting the development of very elongate and highly channelized deep-sea fans within the narrow trench (Fig. 5). The principal source areas for these axial deposits were at the eastern end of the arc-trench system. Lateral infilling from the north is suggested by the presence of south-directed palaeocurrents and a slope facies associations along the northern edge of the outcrop belt (Fig. 6). The at least

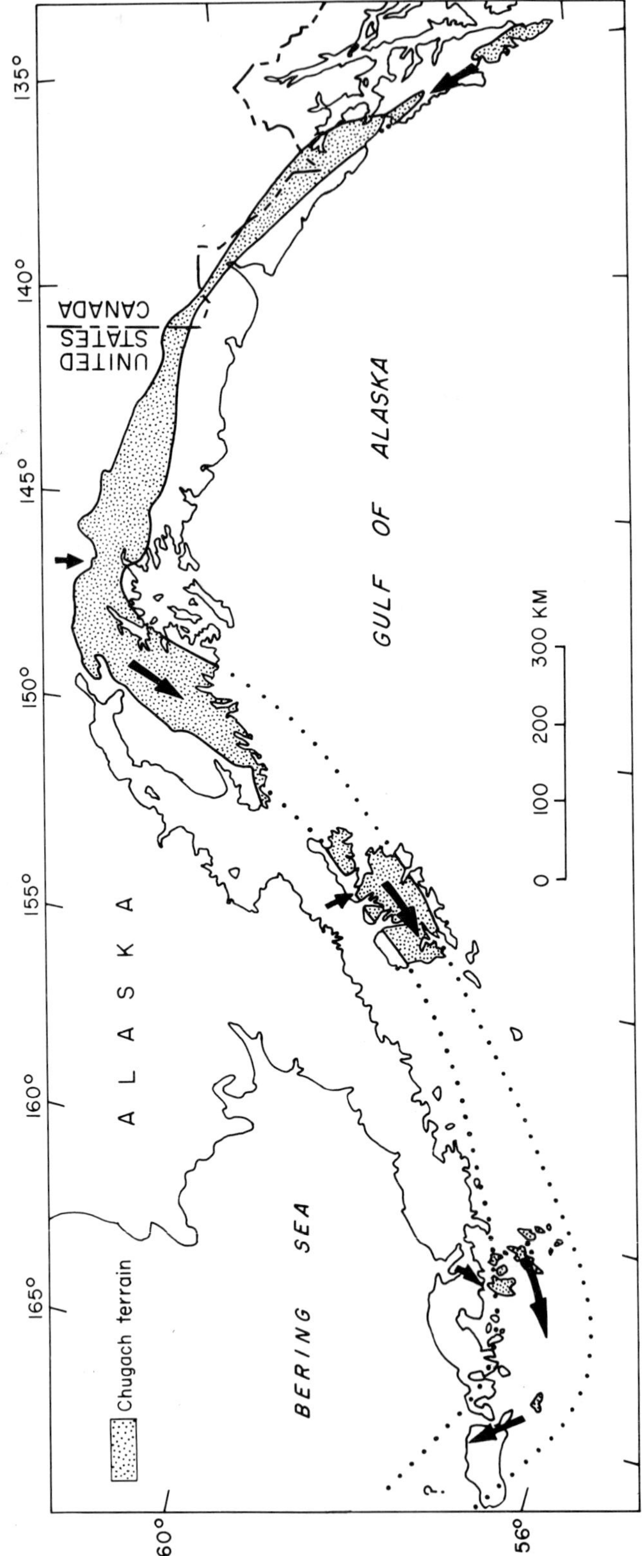

FIG. 5. Generalized palaeocurrent directions from the Chugach terrane. Longer arrows indicate longitudinal flow, shorter arrows lateral flow.

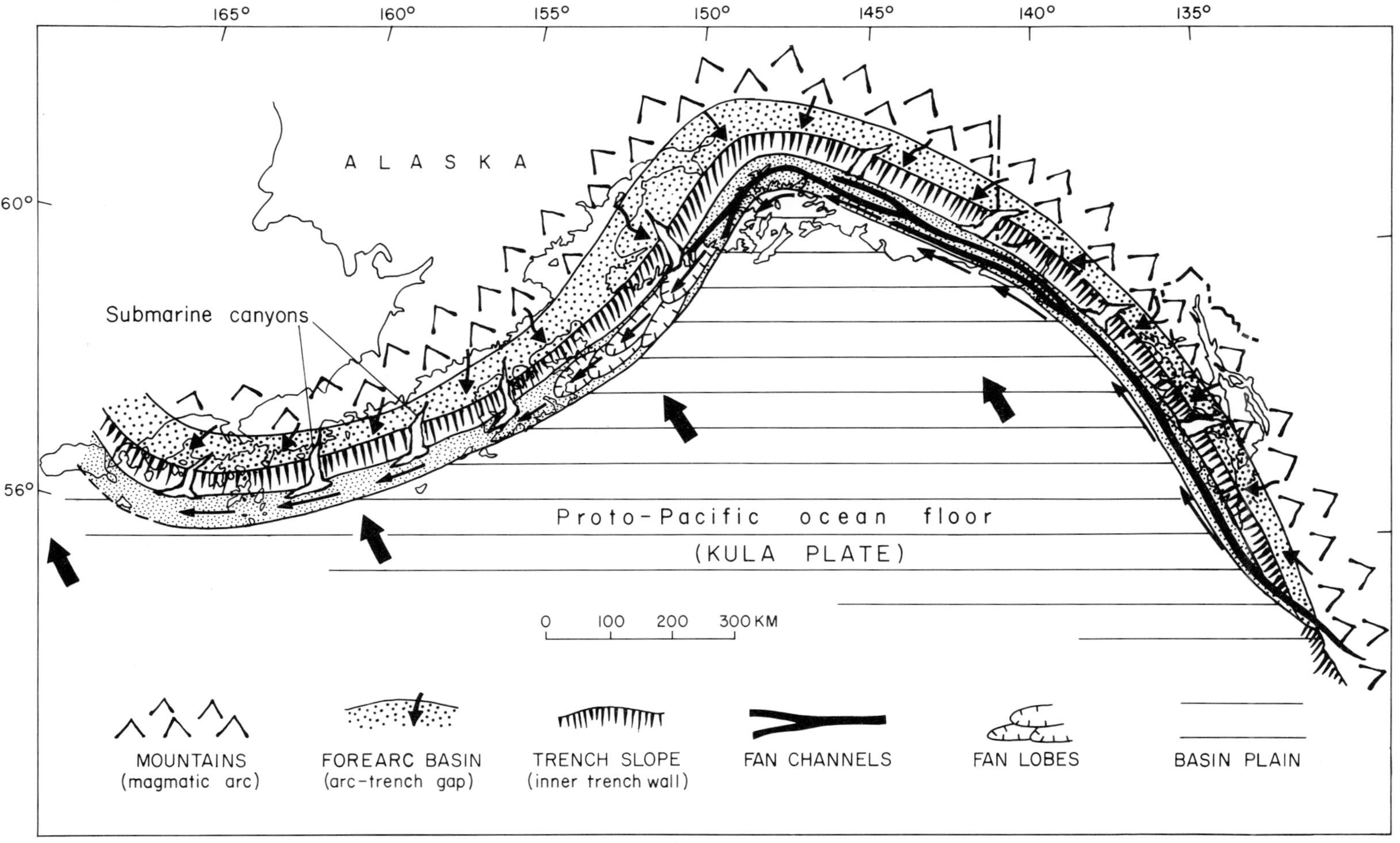

FIG. 6. Palaeogeographic map of Chugach terrane showing magmatic-arc source area, forearc basin, trench slope, trench and proto-Pacific ocean floor. Large arrows show approximate direction of seafloor movement. Small arrows show principal directions of sediment transport. Incised areas on trench slope and forearc basin indicate submarine canyons, some of which transport sediment to the trench floor. The Late Cretaceous trench is shown in its present position without palinspastic restoration to a more southerly position, as indicated by palaeomagnetic data, or straightening, as indicated by structural studies of the southern Alaska orocline.

partly coeval Chugach mélange terrane, although severely deformed and metamorphosed, could in part represent deformed olistostromes emplaced by large-scale submarine landsliding on the trench slope.

The trench formed in response to north-directed subduction within the framework of present distribution of the trench-fill, forearc basin, and magmatic arc. It is clear that the Chugach terrane and its related Late Cretaceous mélange terrane, forearc basin and magmatic arc constitute a coherent larger terrane that moved northward as a single unit. Oceanic plate reconstructions for the Pacific suggest that the Kula plate was moving northward and being subducted beneath Alaska during the Late Cretaceous. However, palaeomagnetic data from both Jurassic and Eocene rocks in southern Alaska indicate that these units formed at latitudes considerably south of the present 56–60° N (Stone & Packer 1976, 1979). If valid, these data require that the Chugach terrane and its associated belts form a large, mostly continental plate that accreted to southern Alaska by middle Miocene time, prior to deposition of the Yakataga Formation. The accretion must have occurred long after the Late Cretaceous subduction zone had ceased to be active.

The Chugach terrane is oroclinally bent about an axis extending through the Prince William Sound area. The time of bending is not known with certainty, and it remains possible that the Chugach terrane originally filled an arcuate trench. The terrane in SE Alaska may have been partly a zone of right-lateral slip rather than orthogonal subduction, as is the case at present. Palaeomagnetic data are required from both limbs of the Chugach flysch terrane, SE and SW Alaska, to test for post-depositional rotation of the limbs.

Within the framework of available data, it appears most reasonable that the Chugach flysch terrane and its associated belts formed a large NE-SW-trending continental fragment along the southern margin of which the Kula plate was subducting during the Late Cretaceous. Subsequent northward migration of the large block resulted in its probable post-Eocene, pre-middle Miocene accretion to Alaska. On its northward or possibly north-eastward trek, the eastern end of the block impinged on the west coast of North America, resulting in oroclinal bending of the plate during its subsequent attachment to Alaska. Continued drag along the eastern part of the block resulted in slivering of the terrane along numerous Cenozoic strike-slip faults, especially in SE Alaska. The present Aleutian trench and arc system was superimposed on the Chugach block in Cenozoic time; this arc system contains volcanic rocks as old as Eocene, yielding an overlap in the timing for arrival of the Chugach block and initiation of the new arc system.

The Chugach terrane and related belts, which might be called the Chugach mesoplate, form part of a series of upper Mesozoic subduction complexes along the west coast of North America (Jones *et al.* 1978). The term 'mesoplate' is used to indicate a larger plate constructed of previously accreted smaller plates generally referred to as 'microplates'. The terrane has clearly been post-depositionally transported northward, and formed a major continental plate within the proto-Pacific Ocean, much as the Salinian-Peninsula-Baja California plate west of the San Andreas fault will in the future. It probably formed the much-debated source for the Aleutian abyssal plain turbidites (Stewart 1976). A speculative source for the Chugach mesoplate may have been the Pacific Northwest of the contemporaneous United States, where Palaeocene rocks are missing from the coastal area and the extensive north-trending late Mesozoic Franciscan assemblage and Great Valley sequence arc complex of California is abruptly truncated (Nilsen & McKee 1979). Our thinking about the tectonic evolution of western North America must account for large-scale translation of continental mesoplates, which in the present case includes an outstanding example of a trench-fill sequence.

ACKNOWLEDGMENTS: We thank J. Decker, J. C. Moore, and G. R. Winkler for helpful reviews of this paper. Zuffa's work on rock-fragment petrography was made possible by the contributions of thin sections by a large number of geologists, who are individually acknowledged in Zuffa *et al.* (1980).

References

BEIKMAN, H. M. 1974a. Preliminary geologic map of the southwest quadrant of Alaska. *U.S. geol. Surv. Misc. Field Studies Map MF-611*, scale 1:1,000,000.

—— 1974b. Preliminary geologic map of the southeast quadrant of Alaska. *U.S. geol. Surv. Misc. Field Studies Map MF-612*, scale 1:1,000,000.

—— 1975a. Preliminary geologic map of southeastern Alaska. *U.S. geol. Surv. Misc. Field Studies Map MF-673*, scale 1:1,000,000.

—— 1975b. Preliminary geologic map of the Alaska Peninsula and Aleutians: *U.S. geol. Surv. Misc. Field Studies Map MF-674*, scale 1:1,000,000.

Berg, H. C., Jones, D. L. & Coney, P. J. 1978. Map showing pre-Cenozoic tectonostratigraphic terranes of southeastern Alaska and adjacent areas. *Open-File Rep. U.S. geol. Surv. 78–1085.*

—— & Richter, D. H. 1972. Gravina-Nutzotin belt—tectonic significance of an upper Mesozoic sedimentary and volcanic sequence in southern and southeastern Alaska. *Prof. Pap. U.S. geol. Surv.* **800-D,** D1–24.

Brew, D. A., Johnson, B. R., Ford, A. B. & Morrell, R. P. 1978. Intrusive rocks of the Fairweather Range, Glacier Bay National Monument, Alaska. *Circ. U.S. geol. Surv.* **772–B,** B88–90.

—— & Morrell, R. P. 1979. Correlation of the Sitka Graywacke, unnamed rocks in the Fairweather Range, and Valdez Group, southeastern Alaska. *Circ. U.S. geol. Surv.* **804–B,** B123–5.

Budnik, R. T. 1974a. *The geologic history of the Valdez Group, Kenai Peninsula, Alaska—deposition and deformation at the Late Cretaceous consumptive plate margin.* Thesis, PhD, Univ. California at Los Angeles, 139 pp.

—— 1974b. Deposition and deformation along an Upper Cretaceous consumptive plate margin, Kenai Peninsula, Alaska. *Abstr. Prog. geol. Soc. Am.* **6,** 150.

Burk, C. A. 1965. Geology of the Alaska Peninsula—island arc and continental margin. *Mem. geol. Soc. Am.* **99,** 250 pp.

Clark, S. H. B. 1972. Reconnaissance bedrock geologic map of the Chugach Mountains near Anchorage, Alaska. *U.S. geol. Surv. Misc. Field Studies Map MF–350*, scale 1:250,000.

—— 1973. The McHugh Complex of south-central Alaska. *Bull. U.S. geol. Surv.* **1372-D,** D1–11.

Connelly, W. 1978. Uyak Complex, Kodiak Islands, Alaska: a Cretaceous subduction complex. *Bull. geol. Soc. Am.* **89,** 755–69.

Cooper, A. K. & Marlow, M. S. 1979. Geologic and geophysical cruise across the outer Bering continental margin. *Circ. U.S. geol. Surv.* **804–B,** 132–4.

Csejtey, B., Jr 1979. Regional significance of tectonics of the Talkeetna Mountains, south-central Alaska. *Circ. U.S. geol. Surv.* **804–B,** B90–2.

Decker, J., Nilsen, T. H. & Karl, S. 1979. Turbidite facies of the Sitka Graywacke, southeastern Alaska. *Circ. U.S. geol. Surv.* **804-B,** B125–9.

Detterman, R. L. 1978. Interpretation of depositional environments in the Chignik Formation, Alaska Peninsula. *Circ. U.S. geol. Surv.* **772-B,** B62–3.

Dickinson, W. R. & Suczek, C. R. 1979. Plate tectonics and sandstone composition. *Bull. Am. Assoc. Petrol. Geol.* **63,** 2164–82.

Grantz, A. 1964. Stratigraphic reconnaissance of the Matanuska Formation in the Matanuska Valley, Alaska. *Bull. U.S. geol. Surv.* **1181-I,** I1–33.

Hill, M. D. & Morris, J. D. 1977. Near-trench plutonism in southwestern Alaska. *Abstr. Progr. geol. Soc. Am.* **9,** 436–7.

Hillhouse, J. W. & Gromme, C. S. 1977. Paleomagnetic poles from sheeted dikes and pillow basalt of the Valdez(?) and Orca Groups, southern Alaska. *EOS*, **58,** 1127.

Hudson, T. 1979a. Plutonism and regional geology in southern Alaska. *U.S. Circ. geol. Surv.* **804-B,** B86–8.

—— 1979b. Mesozoic plutonic belts of southern Alaska. *Geology*, **7,** 230–4.

—— 1979c. Calc-alkaline plutonism along the Pacific rim of southern Alaska. *Open-File Rep. U.S. geol. Surv.* **79–953,** 31 pp.

—— & Plafker, G. 1979. Paleogene anatexis along the Gulf of Alaska margin. *Geology,* **7,** 573–7.

—— , —— & Peterman, Z. E. 1977. Paleogene anatexis in flyschoid rocks along the Gulf of Alaska margin. *Abstr. Prog. geol. Soc. Am.* **9,** 439.

Jones, D. L., Blake, M. C., Bailey, E. H. & McLaughlin, R. J. 1978. Distribution and character of upper Mesozoic subduction complexes along the west coast of North America. *Tectonophysics*, **47,** 207–22.

—— & Clark, S. H. B. 1973. Upper Cretaceous (Maestrichtian) fossils from the Kenai-Chugach Mountains, Kodiak and Shumagin Islands, southern Alaska. *J. Res. U.S. geol. Surv.* **1,** 125–36.

—— & Detterman, R. L. 1966. Cretaceous stratigraphy of the Kamishak Hills, Alaska Peninsula. *Prof. Pap. U.S. geol. Surv.* **500-D,** D53–8.

—— & Silberling, N. J. 1979. Mesozoic stratigraphy—the key to tectonic analysis of southern and central Alaska. *Open-File Rep. U.S. geol. Surv.* **79–1200,** 37 pp.

Karl, S., Decker, J. & Jones, D. L. 1979. Early Cretaceous radiolarians from the McHugh Complex, south-central Alaska. *Circ. U.S. geol. Surv.* **804-B,** B88–90.

Loney, R. A., Brew, D. A., Muffler, L. J. P. & Pomeroy, J. S. 1975. Reconnaissance geology of Chichagof, Baranof, and Kruzof Islands, southeastern Alaska. *Prof. Pap. U.S. geol. Surv.* **792,** 105 pp.

MacKevett, E. M., Jr & Plafker, G. 1974. The Border Ranges fault in south-central Alaska. *J. Res. U.S. geol. Surv.* **2,** 323–9.

McLean, H. 1979. Pribilof segment of the Bering Sea continental margin: a reinterpretation of Upper Cretaceous dredge samples. *Geology*, **7,** 307–10.

Magoon, L. B., Egbert, R. M. & Petering, G. 1978. Upper Jurassic and Cretaceous rocks of the Kamishak Hills-Douglas River area, lower Cook Inlet. *Circ. U.S. geol. Surv.* **772-B,** B57–9.

—— , Griesback, F. B. & Egbert, R. M. 1980. Nonmarine Upper Cretaceous rocks, Cook Inlet, Alaska. *Bull. Am. Assoc. Petrol. Geol.* **64,** 1259–66.

Mancini, E. A., Deeter, T. M. & Wingate, F. H. 1978. Upper Cretaceous arc trench gap sedimentation on the Alaska Peninsula. *Geology*, **6,** 437–9.

Mitchell, P. A. 1979. *Geology of the Hope-Sunrise (gold) mining district, north-central Kenai Peninsula, Alaska.* Thesis, MS, Univ. Stanford, 123 pp.

—— 1980. Genesis of the Valdez Group south of Turnagain Arm, Alaska. *Abstr. Prog. geol. Soc. Am.* **12,** 142.

Monger, J. W. H., Southey, J. G. & Gabrielse, H.

1972. Evolution of the Canadian Cordillera. *Can. J. Earth Sci.* **8,** 259–78.

Moore, G. W. 1969. New formations on Kodiak and adjacent islands, Alaska. *Bull. U.S. geol. Surv.* **1274-A,** A27–35.

—— & Bolm, J. B. 1977. Orientation of Late Cretaceous and early Tertiary subduction, Kodiak Islands, Alaska. *Abstr. Prog. geol. Soc. Am.* **9,** 1099–100.

Moore, J. C. 1972. Uplifted trench sediments—southwestern Alaska-Bering Shelf edge. *Science*, **175,** 1103–5.

—— 1973a. Cretaceous continental margin sedimentation, southwestern Alaska. *Bull. geol. Soc. Am.* **84,** 595–614.

—— 1973b. Complex deformation of Cretaceous trench deposits, southwestern Alaska. *Bull. geol. Soc. Am.* **84,** 2005–20.

—— 1978. Orientation of underthrusting during latest Cretaceous and earliest Tertiary time, Kodiak Islands, Alaska. *Geology*, **6,** 209–13.

—— & Connelly, W. 1979. Tectonic history of the continental margin of southwestern Alaska, Late Triassic to earliest Tertiary. *In:* Sisson, A. (ed). *The Relationship of Plate Tectonics to Alaskan Geology and Resources*, H1–29. Alaska geol. Soc. Symp.

—— & Wheeler, R. L. 1978. Structural fabric of a melange, Kodiak Islands, Alaska. *Am. J. Sci.* **278,** 739–65.

Mutti, E. & Ricci Lucchi, F. 1972. Le torbiditi dell' Appennine settentrionale—introduzione all' analisi di facies. *Mem. Soc. geol. Ital.* **11,** 161–99.

—— & —— 1975. Turbidite facies and facies associations. *In: Examples of turbidite facies and facies associations from selected formations of the northern Apennines.* Ninth Int. Congr. Sedimentology, Field Trip Guidebook, **A11,** 21–36, Nice, France.

Nilsen, T. H. 1977. Turbidite facies and sedimentation patterns. *In*: Nilsen, T. H. (ed). *Late Mesozoic and Cenozoic Sedimentation and Tectonics in California*, Short Course, San Joaquin geol. Soc. 39–52.

—— & Bouma, A. H. 1977. Turbidite sedimentology and depositional framework of the Upper Cretaceous Kodiak Formation and related stratigraphic units, southern Alaska. *Abstr. Prog. geol. Soc. Am.* **9,** 1115.

—— & McKee, E. H. 1979. Paleogene paleogeography of the Western United States. *Pacif. Coast Paleogeogr. Symp. Soc. econ. Paleontol. Mineral.* **3,** 257–76.

—— & Moore, G. W. 1979. Reconnaissance study of Upper Cretaceous to Miocene stratigraphic units and sedimentary facies, Kodiak and adjacent islands, Alaska. *Prof. Pap. U.S. geol. Surv.* **1093,** 34 pp.

Plafker, G. & Campbell, R. B. 1979. The Border Ranges fault in the Saint Elias Mountains. *Circ. U.S. geol. Surv.* **804-B,** B102–4.

—— , Jones, D. L., Hudson, T. & Berg, H. C. 1976. The Border Ranges fault system in the Saint Elias Mountains and Alexander Archipelago. *Circ. U.S. geol. Surv.* **722,** 14–6.

—— , Jones, D. L. & Pessagno, E. A., Jr. 1977. A Cretaceous accretionary flysch and melange terrane along the Gulf of Alaska margin. *Circ. U.S. geol. Surv.* **751-B,** B41–3.

Reed, B. L. & Lanphere, M. A. 1973. Alaska-Aleutian Range batholith: chronology, chemistry, and relation to circum-Pacific plutonism. *Bull. geol. Soc. Am.* **84,** 2583–610.

—— & —— 1974. Chemical variations across the Alaska-Aleutian Range batholith. *J. Res. U.S. geol. Surv.* **2,** 343–52.

Reid, M. & Gill, J. B. 1980. Near trench volcanism, Kodiak Island, Alaska: Implications for ridge-trench encounter. *Abstr. Prog. geol. Soc. Am.* **12,** 148–9.

Scholl, D. W. & Marlow, M. S. 1974. Sedimentary sequences in modern Pacific trenches and the deformed circum-Pacific eugeosyncline. *In*: Dott, R. H. & Shaver, R. H. (eds). *Modern and Ancient Geosynclinal Sedimentation. Spec. Publ. Soc. econ. Paleontol. Mineral. Tulsa*, **19,** 193–211.

Schweller, W. J. & Kulm, L. D. 1978. Depositional patterns and channelized sedimentation in active eastern Pacific trenches. *In:* Stanley, D. J. & Kelling, G. (eds). *Sedimentation in Submarine Canyons, Fans and Trenches,* 311–24. Dowden, Hutchinson & Ross, Stroudsburg, Pennsylvania.

Stewart, R. J. 1976. Turbidites of the Aleutian Abyssal Plain—mineralogy, provenance, and constraints for Cenozoic motion of the Pacific plate. *Bull. geol. Soc. Am.* **87,** 793–808.

Stone, D. B. & Packer, D. R. 1976. Tectonic implications of Alaska Peninsula paleomagnetic data. *Tectonophysics*, **37,** 183–201.

—— & —— 1979. Paleomagnetic data from the Alaska Peninsula. *Bull. geol. Soc. Am.* **90,** 545–60.

Tysdal, R. G. & Case, J. E. 1977. The McHugh Complex in the Seward quadrangle, south-central Alaska. *Circ. U.S. geol. Surv.* **751-B,** B48–9.

—— , —— , Winkler, G. R. & Clark, S. H. B. 1977. Sheeted dikes, gabbro, and pillow basalt in flysch of coastal southern Alaska. *Geology*, **5,** 377–83.

—— & Plafker, G. 1978. Age and continuity of the Valdez Group, southern Alaska. *Bull. U.S. geol. Surv.* **1457-A,** A120–4.

Underwood, M. B. & Karig, D. E. 1980. Role of submarine canyons in trench and trench-slope sedimentation. *Geology*, **8,** 432–6.

von Huene, R., Moore, J. C. & Moore, G. W. 1978. Cross section of Alaska Peninsula-Kodiak Island-Aleutian Trench. *Map Chart Series, geol. Soc. Am.* **MC-28A**.

Zuffa, G. G., Nilsen, T. H. & Winkler, G. R. 1980. Rock-fragment petrography of the Upper Cretaceous Chugach terrane, southern Alaska. *Open-File Rep. U.S. geol. Surv.* **80–713,** 28 pp.

The Chugach Terrane, a Cretaceous trench-fill deposit

Tor H. Nilsen, U.S. Geological Survey, Branch of Western Environmental Geology, 345 Middlefield Road, Menlo Park, California 94025, U.S.A.
Gian G. Zuffa, Dipartimento di Scienze della Terra, Universita della Calabria, 87030 Castiglione Scalo (Cosenza), Italy.

Structural evolution of coherent terranes in the Ghost Rocks Formation, Kodiak Island, Alaska

Tim Byrne

SUMMARY: The Late Cretaceous to Palaeocene Ghost Rocks Formation, Kodiak Island, Alaska, comprises two structural styles: multiply deformed mélange and relatively coherent terranes of turbidite deposits which together constitute a probable ancient accretionary complex. The mélange consists almost exclusively of orientated, elongate sandstone boudins which lie in a foliated pelitic matrix. In the coherent terranes, structural histories, rock types and hornfels metamorphism delineate two NE trending units: Unit A, and a probably younger Unit B. Unit A can be subdivided into two gradational structural belts, a seaward conjugate fold belt and a landward spaced cleavage belt. The conjugate fold belt contains 'S' and 'Z' type kink folds that often intersect to form conjugate fold sets. The folds of this fold belt are hornfelsed by a sill which intruded partially lithified sediments; the fold belt therefore also formed when the sediments were partially lithified. Analyses of the fold axial surfaces indicate a shortening direction of 319°, when later folding is removed. Sill emplacement and hornfels metamorphism interrupted the conjugate fold deformation and resulted in a change in style of deformation in Unit B to broad, open to tight folds associated with a spaced cleavage. The cleavage strikes 046° and dips approximately vertically, fanning divergently about fold axial surfaces. The approximately coaxial conjugate folding and folding with cleavage in unit A constitute D_1. Unit B generally contains sub-horizontal beds except for a large-scale, overturned, thrusted fold, which emplaced mélange to the NW and above Unit B. Slickenlines, fold axes and poles to bedding indicate this unit was shortened along a trend of 334°. This deformation is termed D_2 and is correlated elsewhere in the Ghost Rocks Formation to thrust faults which place mélange to the NW and under the coherent terranes.

The structural histories of the coherent terranes in the Ghost Rocks Formations show that accretionary prism deformation can involve coaxial deformations and the tectonic deformation of partially lithified sediments. The structural histories also show that the Palaeocene deformation axes in the Ghost Rocks Formation have not rotated out of their original sub-horizontal orientations.

The structural evolution of accretionary terranes presents an outstanding problem in convergent margin geology. Structural models of deformation and accretion at convergent margins based on marine geophysics (Seely *et al.* 1974; Karig & Sharman 1975), scale models (Cowan & Silling 1978; Seely 1977) and on-land geological studies (Cowan 1974; Suppe 1973; Moore 1973; Moore & Karig 1979) have been enlightening but have been most successful in showing the wide variety of unconstrained parameters. Moreover, neither geophysical profiles or scaled models are directly applicable to outcrop-scale problems. In order to understand the geological consequences of plate convergence and plate reorganizations, and to interpret ancient subduction complexes accurately, detailed structural studies and appropriate kinematic interpretations of ancient accretionary terranes are essential. The Late Cretaceous to early Palaeocene Ghost Rocks Formation, Kodiak Islands, Alaska, provides an excellent opportunity to resolve convergent margin processes as it is well-exposed, young in age, only mildly metamorphosed and may record events associated with the later stages of Kula Plate subduction.

Tectonic setting

The continental margin of SW Alaska has experienced repeated episodes of subduction-related magmatism and accretion since at least the early Mesozoic. While magmatism has been concentrated on the Alaskan Peninsula, and is continuing today, accretion of deep-sea rocks has progressively broadened the continental shelf to the SE. The Kodiak Islands constitute an emergent portion of this continental shelf and expose NE-trending belts of deep-sea rocks that dip NW and young progressively toward the present day Aleutian trench (Fig. 1; Moore & Connelly 1979; Moore 1969). The Ghost Rocks Formation (Armentrout 1979a; Moore 1969) comprises the next to youngest sequence of deep-sea deposits in the Kodiak Islands and is in fault contact with older rocks to the NW

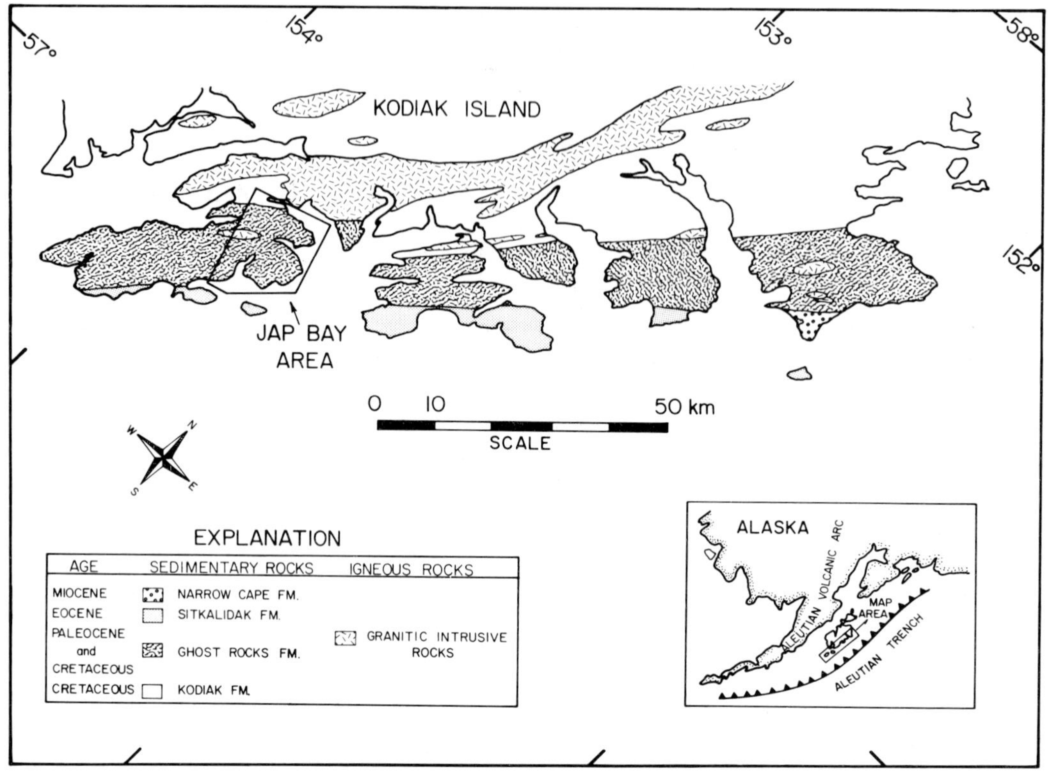

FIG. 1. Generalized geological map of SE Kodiak Island, Alaska, after Moore (1967).

and younger rocks to the SE. While the Kodiak margin may have lain south of its present position (Stone & Packer 1979) at the time when these rocks were deformed, geochemical trends (Reed & Lanphere 1974) in the Alaska-Aleutian Range batholith clearly indicate plate convergence was toward the NW during the time of deposition and accretion of the Ghost Rocks Formation.

The Ghost Rocks Formation

The Ghost Rocks Formation consists of complexly deformed sandstone and shale with interbedded volcanic rocks and scattered hypabyssal intrusions. The sedimentary rocks are predominantly thin (1–5 cm) to thick (tens of metres) bedded, mostly massive sandstones with interbedded shales of similar thicknesses. Pebbly mudstone and conglomerate also occur locally. Although tectonic deformation has severely dismembered almost all original stratigraphy, sedimentary structures in the sandstones and shales indicate turbidite deposition and the occurrence of massive sandstones, pebbly mudstones and conglomerates suggest deposition in channelized environments.

Fossiliferous limestones are locally associated with the igneous rocks in the Ghost Rocks Formation. The igneous rocks are clearly interbedded with the sedimentary rocks and include pillow lava, pillow breccia, tuff, dykes and sills of andesitic to basaltic composition (Reid & Gill 1980; Hill 1978). The limestones typically overlie or are interbedded with the volcanic rocks and contain planktonic foraminifera of Palaeocene age at some localities, and of Late Cretaceous age at others.

Silicic igneous rocks also intrude the Ghost Rocks Formation. These consist of quartz dioritic to tonalitic plutons that probably represent satellitic intrusions of the Kodiak batholith. The plutons are associated with zones of iron-staining and zeolitization, are chilled at their margins, and have thermally altered the sedimentary rocks of the Ghost Rocks Formation. The pluton exposed NW of Jap Bay (Fig. 2) is similar to other early Tertiary intrusive rocks of the Ghost Rocks Formation and has been isotopically dated by the Mobil Oil Company. The ages of this pluton (62.6 ± 0.6, 62.1 ± 0.6: K-Ar, biotite, and 63 ± 3: Rb-Sr isochron defined by biotite and plagioclase, Armentrout, pers. comm. 1978) are slightly older but not greatly different from the 57 to 60 Ma ages determined from the main body of the Kodiak batholith (Armentrout 1979b, and written comm. 1979) and probably record the time of pluton emplacement. Locally, the Ghost Rocks pluton is cross-cut by discrete zones of low grade alteration and cataclasis suggesting that some deformation post-dates the emplacement of the pluton.

The Ghost Rocks Formation crops out in a belt 160 km long and 15 km wide along the SE side of the Kodiak Islands (Fig. 1). During three seasons of fieldwork, I examined most exposures of this formation. The conclusions presented here, however, are derived from a detailed study in the Jap Bay area and regional studies of the bays and headlands to the NW (Fig. 2).

Jap Bay area

Detailed mapping in the vicinities of Jap, Kiavak and Kaiugnak Bays (Fig. 2) reveals the presence of two dominant structural styles within the Ghost Rocks Formation. The structural styles are broadly differentiated as mélange and coherent terranes, which occur as mappable belts that strike NE, parallel to the structural grain of Kodiak Island. The coherent terranes are sedimentary deposits that may be analogous to the slope basin deposits recognized in many modern convergent margins. As such, the deformation history of the coherent terranes record not only the kinematics of specific events in the history of the accretionary prism but perhaps also events present in the more complexly deformed mélange terranes. The fundamental contribution of this study is the conclusion that the deformation axes in the coherent terranes have not rotated from their original sub-horizontal orientation since Palaeocene time. Systematic imbrication and underthrusting of younger more seaward deep-sea rocks is not indicated.

Mélange terranes

Areas designated as mélange (Fig. 2) are characterized by strongly foliated and discontinuous argillite-sandstone units in which structures and included blocks are unresolvable at a mapping scale of 1:25 000. Sandstone and argillite are the dominant rock types but conglomerate, limestone, greenstone and tuffaceous units also occur, both interlayered in the matrix and as inclusions. On a mesoscopic scale the foliation occurs only in the pelitic lithologies and distinctively anastomoses around the more competent sandstone and greenstone lithologies, resulting in a 'phacoids in a matrix' appearance at most outcrops.

The limestones in the mélange terrane contain planktonic foraminifera of Late Cretaceous age (Poore, written comm. 1980; Rau, written comm. 1980; see section B-B′ Fig. 3 for location) and are intimately associated with the volcanic rocks, suggesting that the volcanic rocks are of Late Cretaceous age, also.

Detailed mapping of the mélange (1:300) in the vicinity of Jap Bay shows that the steep, NW-dipping foliation crosscuts and locally transposes an earlier foliation. The earlier foliation is defined by the parallel alignment of elongate sandstone boudins that plunge gently SW. The intersection of the two foliations locally produces a pencil cleavage that plunges gently to the SW, parallel to the sandstone boudins.

The remaining area, designated as undifferentiated (Fig. 2), was not mapped in detail but reconnaissance mapping shows it contains both mélange (as described above) and relatively more coherent, but still unresolvable, units of interbedded sandstone and pelite with minor greenstone and conglomerate (section C-C′ Fig. 5). These relative coherent areas are lithologically indistinguishable from unit A in the coherent terranes.

Coherent terranes

The structure of the coherent terranes contrasts strikingly with that of the mélange terranes. The coherent terranes are characterized by gently to tightly folded turbidite deposits. Pinch-and-swell structures and boudinage are rare and individual sandstone beds can be traced for tens of metres. Tectonic cleavage, where present, penetrates both the pelitic and sandstone lithologies.

Critical to understanding the tectonic significance of the coherent terranes is knowing their relationship to the underlying mélange. Studies of modern convergent margins suggest that relatively less-deformed slope sediments become incorporated into more deformed, accreted trench deposits through time, resulting

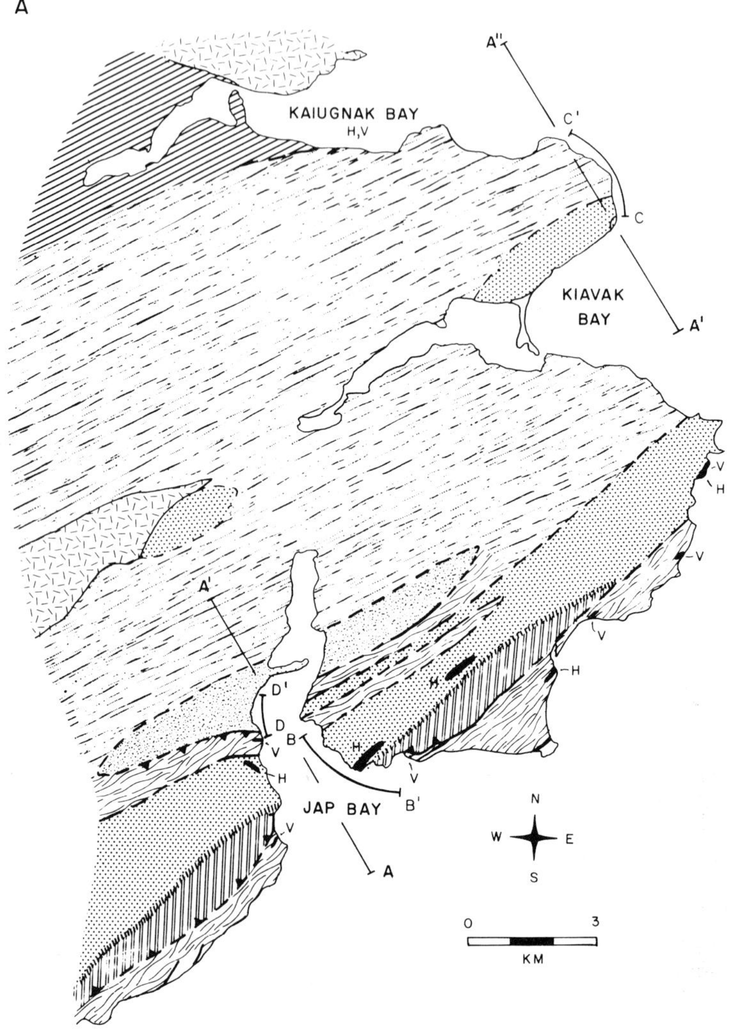

FIG. 2.(A) Geological map and (B) cross-section of the Jap Bay area, generalized from coastline (excellent exposure) and inland (poor exposure) traverses.

in a transition from more-deformed to less-deformed sedimentary sequences (Lundberg & Moore 1981). In the Jap Bay area the contacts between the mélange and coherent terranes, where they are exposed, are either moderately dipping thrust faults or the relationships are ambiguous and the contacts could be deformed depositional or tectonic boundaries. Structural studies (see below) indicate, however, that the thrust faults are some of the latest structural features in the Ghost Rocks Formation. Thus, no direct evidence as to the primary relationship between the coherent and mélange terranes is available. To help resolve this problem a detailed study of the structural history of the mélange terranes is in preparation so that it

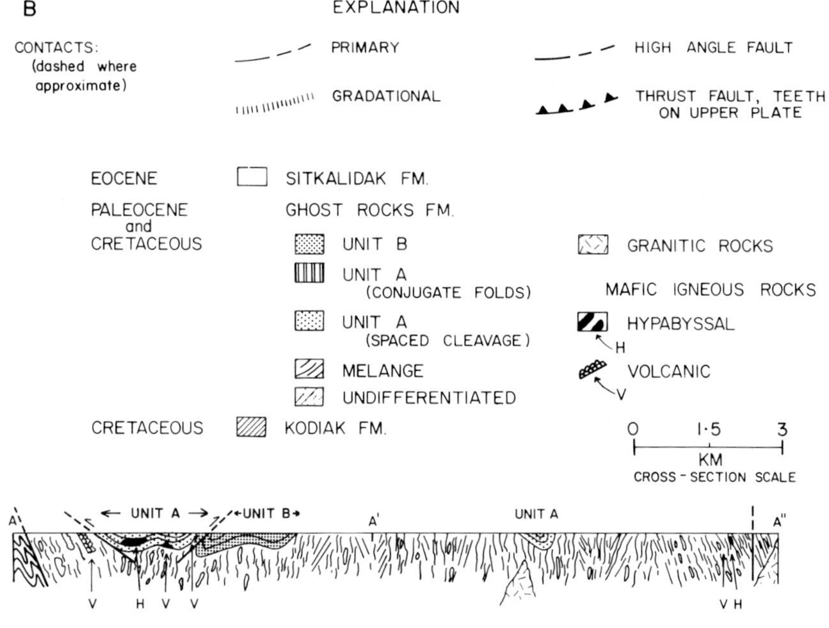

Fig. 2.(B)

can be compared with the structural histories of the coherent terranes.

The coherent terranes can be subdivided into two stratigraphic units based on contrasting lithologies and structural and metamorphic histories; Unit A and a relatively younger Unit B.

Unit A

Unit A contains a wide variety of rock types including conglomerate, pebbly mudstone, thick bedded sandstone (tens of metres thick), thin bedded sandstone (less than 10 cm thick),

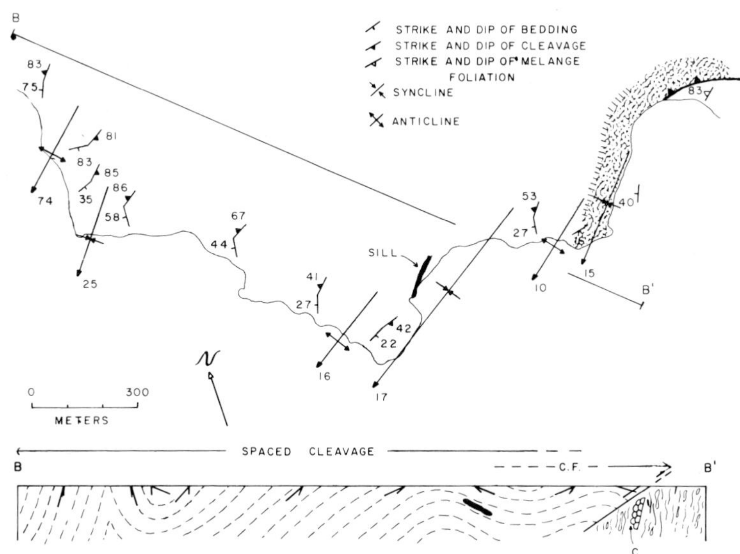

Fig. 3. Geological map and cross-section of Unit A. See Fig. 2 for location and symbols. Fold axes were determined from more data than shown in maps. In cross-section long and short lines represent apparent dip directions of bedding and cleavage, 'C' shows late Cretaceous fossil locality and 'CF' shows extent of conjugate fold belt.

massive argillite and interbedded extrusive and hypabyssal igneous rocks. Overall, the lithological character of Unit A is strikingly similar to that of the mélange and undifferentiated terranes, and the gross lithologies alone cannot be used to distinguish these terranes. Areas of Unit A therefore may represent parts of a coherent terrane that were folded or thrust into the mélange, relatively less deformed remnants of an originally coherent stratigraphic sequence, or both.

Unit A appears to be early Palaeocene in age. This unit is the most extensive coherent terrane exposed in the Jap Bay area and is lithologically and structurally similar to rocks that crop out locally along the entire length of the Ghost Rocks Formation. These similar and probably correlative rocks contain Palaeocene fossils at two localities (Nilsen & Moore 1979; Poore, written comm. 1978), suggesting that Unit A may be Palaeocene in age also. Thus, if Unit A is gradational with the mélange, the original sedimentary sequence extended from Late Cretaceous to early Palaeocene.

All the sedimentary rocks in Unit A have been mildly recrystallized by a hornfels metamorphism related to emplacement of the hypabyssal rocks. The recrystallization is most intense 1 to 2 m below and above the andesitic sill that crops out at the mouth of Jap Bay (Figs 2 & 3). Here, the pelites and sandstones are very well indurated, and break conchoidally. At greater distances from this sill, a tectonic cleavage in the pelites cross-cuts the hornfels texture but the indurated texture indicative of recrystallization is still present in the pelites and sandstones. In thin section, the recrystallization is recorded by fine-grained epidote and chlorite porphyroblasts, patches of stilpnomelane, and veins of prehnite and quartz. Near the contact, the epidote porphyroblasts are especially conspicuous in thin section because they occur in pelitic layers as large subhedral crystals.

The andesitic sill appears to have intruded partially lithified sediments. In most areas the upper and lower contacts between the sill and the sedimentary rocks are sharp and regular. Locally, however, the contact consists of a several metre thick zone of angular to rounded greenstone fragments in a pelitic matrix. The fragments typically have chilled margins and vary from a few centimetres to several metres in size. Pelitic dykes usually less than 5 mm thick also commonly penetrate both the greenstone fragments and the sill for several tens of centimetres. The fragmentation and dyking are interpreted as being results of explosive brecciation and steam-mud injections related to interaction of the sill with partially lithified sediments. These field relations indicate that hornfelsing apparently accentuated the diagenetic processes that cause lithification. A change in the degree of lithification of the sediments after sill emplacement is also reflected by a change in structural styles.

Deformation styles within Unit A define two mappable structural belts that grade into each other, a seaward belt characterized by conjugate folds, and a landward belt characterized by a spaced cleavage (Figs 2 & 3). The structures of the conjugate fold belt are hornfelsed by the andesite sill exposed at the mouth of Jap Bay. Thus, this fold belt also is interpreted to have developed when the sediments were partially lithified. Both the sill and the conjugate folds are locally cross-cut by the cleavage of the spaced cleavage belt. Unit A therefore was deformed by two modes of deformation, conjugate folding in partially lithified sediments and spaced cleavage development in lithified sediments. Together, these structural belts document a progressive deformation in the Jap Bay area that I have termed D_1.

Conjugate fold deformation

The folds of the conjugate fold belt occur in a wide variety of styles and forms. Typically, they are disharmonically and/or polyclinally folded and locally sandstone layers are faulted, boudinaged or show pinch-and-swell structures (Fig. 4). The mesoscopic scale disruption of the folded layers is consistent with the deformation of relatively weak, partially lithified sediments as indicated by the intrusive character of the andesite sill. The folds are not chaotic, however, and generally form 'S' or 'Z' type asymmetric fold sets which locally intersect to form conjugate folds (Fig. 4A & B). Symmetrical folds are also common (Fig. 4C).

The direction of maximum shortening during deformation of the conjugate fold belt can be determined from either individual conjugate folds or from the average orientation of the 'S' and 'Z' asymmetric fold sets (Gay & Weiss 1974; Tobisch & Fiske 1976). The individual conjugate folds yield a shortening direction that is subparallel to bedding and trends 326° when later folding is removed. The extension direction is orientated normal to bedding. The shortening direction in these folds bisects either the acute (e.g. Fig. 4B) or obtuse (e.g. Fig. 4A) angle formed between the fold axial surfaces, although the obtuse bisectrix is more common. The average orientation of the 'S' and 'Z' fold sets yields a shortening direction that is subparallel to bedding and trends 319°, when later

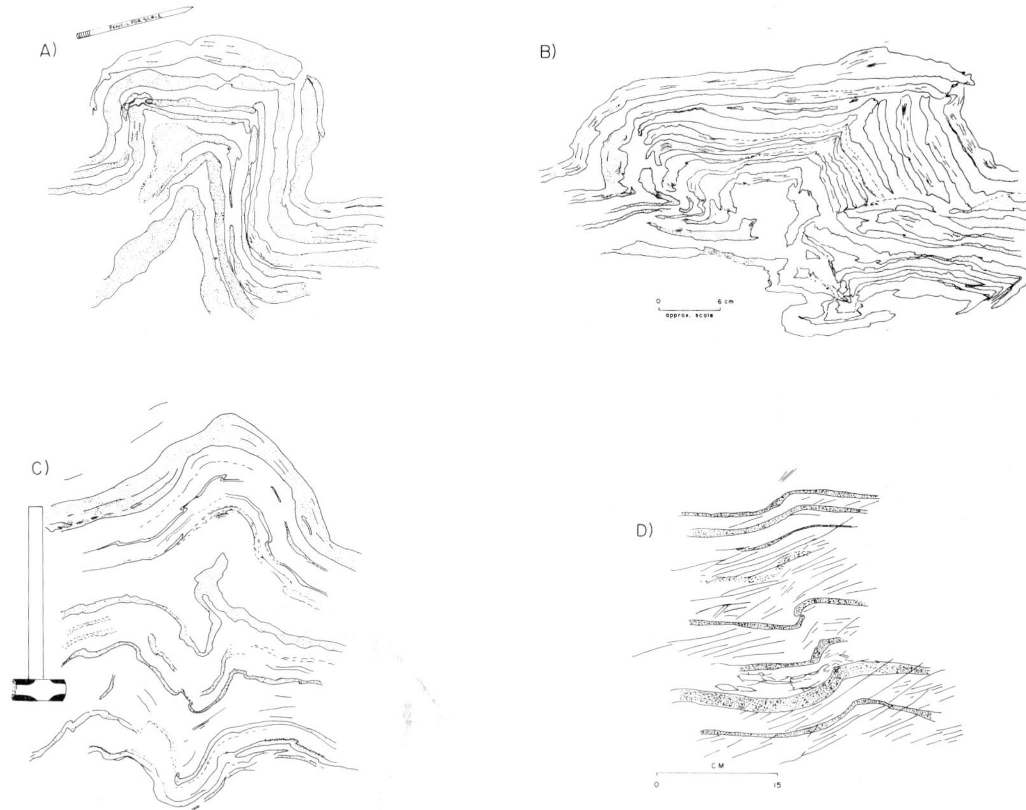

FIG. 4. Sketches from photographs of conjugate (A and B) and symmetrical (C) folds that formed in partially lithified sediments in Unit A. Lines within sand layers indicate traces of bedding. Abundant quartz filled fractures in sand layers are not shown. (D) Sketch of spaced cleavage cross-cutting axial surface of asymmetric fold of conjugate fold belt. Bedding is shown right side up in all folds and fold axes and cleavage surfaces are approximately normal to plane of sketches.

folding is removed. The extension direction is normal to bedding. These results are in good agreement with the results from the individual conjugate folds and indicates that both types of data can be used to determine the direction of maximum shortening. These combined data (Fig. 6A) show that on a regional scale the shortening direction trended sub-horizontal about 319° when the conjugate fold belt was deformed. In rotating the fold axial surfaces around the later fold axes, I assume that bedding was horizontal prior to conjugate folding. This appears to be a valid assumption because the sediments were only partially lithified when they were deformed and there is no evidence of an earlier deformation. This assumption is also supported by the consistency in the orientation of the axial surfaces after they have been rotated so that bedding is horizontal (Fig. 6A).

An important problem in interpreting the tectonic significance of this fold belt is defining its possible origin; did the folds form from gravity-induced slumping or from tectonic processes? Although there is no direct way of answering this question, three field observations suggest the folds are of tectonic origin. First, the northern contact of the conjugate fold belt is gradational. The asymmetric folds grade laterally into undeformed rocks of approximately the same stratigraphic position. There is no discontinuity suggestive of a zone of slump folds. Moreover, there is no evidence within the belt to suggest that the folds were surficial features. Secondly, the development of the folds post-dates both the compaction of the sediments (i.e. post-dates 50% volume loss by dewatering) and the formation of calcareous concentrations. Thus, if the folds formed during gravitational sliding, the slumped unit was of sufficient size to cut deeply into a compacted and partially lithified sequence. Finally, the conjugate fold deformation is nearly coaxial with a later deformation that is clearly of tectonic origin. The simplest interpretation for this belt is that it records the initial stages of a tectonic deformation that produced, after sill emplacement and lithification, broad folds associated with a spaced cleavage (see below).

Spaced cleavage deformation

The spaced cleavage belt of Unit A is characterized by broad (hundreds of metres) open to tight folds and a 5 mm spaced cleavage best developed in pelitic layers (Fig. 3). The cleavage typically fans divergently (Hobbs *et al.* 1976) around the axial surfaces of related mesoscopic and macroscopic folds although areas of axial planar cleavage are also locally present. On a regional scale the cleavage has an average sub-vertical orientation, striking 046° (Fig. 6B). Near the conjugate fold belt, the spaced cleavage cross-cuts the axial surfaces of the conjugate folds (Fig. 4D) indicating it post-dates these structures.

On a microscopic scale, the spaced cleavage occurs in two gradational modes: (1) as localized zones of concentrated, preferentially orientated phyllosilicates, and (2) as a pervasive but poorly developed fabric defined by the parallel alignment of fine-grained phyllosilicates. The phyllosilicates that define both modes of cleavage are very fine-grained (less than a few microns) relative to the phyllosilicates that parallel bedding and occur as individual grains and composite stringers. The cleavage trace in the phyllosilicates parallels the trace of the tectonic cleavage. Clusters and trains of opaques also define both fabrics. The cleavage folia often cross-cut or abut against the conspicuously large phyllosilicates that parallel bedding. This relationship and the strong disparity in size between bedding and cleavage phyllosilicates indicate that the cleavage phyllosilicates are not reorientated detrital grains. Instead, they have apparently grown at the expense of the detrital micas through dissolution-neocrystallization processes. Evidence for dissolution of detrital grains is also shown by the sutured contacts between well-rounded chert pebbles in conglomeratic units. Not all of the strain was accommodated through dissolution-neocrystallization, however, as many of the detrital quartz grains are also undulatory indicating internal deformation of the quartz grains.

Although the spaced, divergent cleavage is not geometrically analogous to many slaty cleavages that parallel axial surfaces of folds and define the plane of maximum flattening (Wood 1974; Ramsay 1967), the average orientation of the spaced divergent cleavage parallels related fold axial surfaces and probably also defines the plane of maximum shortening in the cleavage belt. The direction of maximum shortening indicated by the cleavage trends sub-horizontally 336° (Fig. 6B). The direction of extension during cleavage development, as indicated by quartz fibres around pyrite framboids and crystals, has a sub-vertical orientation. These axes of the deformation are coaxial with the directions of shortening and extension determined from the conjugate fold belt.

Thus, D_1 is interpreted to have been a coaxial, progressive deformation that started with the localized development of mesoscopic-scale conjugate folds in partially lithified sediments (Fig. 7A). Sill emplacement apparently lithified the sediments and interrupted this deformation, after which the mode of deformation consisted of open to tight large-scale folds with an associated divergent spaced cleavage (Fig. 7B). During both phases of D_1 the shortening axes were sub-horizontal and trended about 318°. Significantly, the shortening axis of the latest phase of this deformation is still horizontal.

Unit B

Unit B rock types and structural styles are unique in the Jap Bay area and in the Ghost Rocks Formation. The belt consists of medium bedded (10–40 cm thick beds) sandstone interbedded with similar thicknesses of shale. The sandstones and shales are rhythmically bedded and sole markings and graded beds indicate turbidite deposition. Stratigraphic sections show no systematic cycles, however. Distinctively, this unit is not hornfelsed and contains only rare thick-bedded (greater than 1 m thick) sandstones; conglomerate, pebbly mudstone and igneous rocks are absent. Moreover, Unit B was apparently not deformed by the multiple deformations of D_1. Thus, although no basal contacts were observed, the contrasting lithologies and structural and metamorphic histories suggest that Unit B was deposited after deformation and metamorphism of Unit A.

Although Unit B has not been metamorphosed, it has been thermally altered. Seven shale samples from this unit, examined by Mobil Oil Company, indicate a degree of organic maturation (average R_0 = 2.15) 3 to 4 times higher than that recognized in the Eocene Sitkalidak

Formation (Moore & Allwardt 1981). Considering the age and tectonic setting of the Ghost Rocks Formation sediments, the high reflectance values are explained most simply by the thermal alteration associated with the emplacement of the early Palaeocene pluton exposed to the NE of Jap Bay, suggesting that Unit B is of earliest Palaeocene age.

Structural geology

Unit B is generally deformed into outcrop-scale asymmetric folds displaying landward or seaward vergence. Bedding-parallel slickensides indicate that folding occurred through flexural slip and that the direction of slip was normal to the fold axes. Stereographic plots of fold axes and poles to bedding indicate that the direction of maximum shortening was orientated about 334° (Fig. 6C). This direction of shortening is different from that of D_1 and, along with the contrasting lithologies and metamorphic histories, suggests that Unit B records a separate deformation, D_2, younger than D_1.

On the west side of Jap Bay, D_2 is represented in part by a large-scale thrusted fold which placed mélange to the NW (landward) and above Unit B (section D-D′, Fig. 5). Smaller-scale landward-verging asymmetric folds, unrelated to the larger fold, are also present. The north-eastern end of Unit B on this side of Jap Bay (not shown in Fig. 5) is poorly exposed but the limited exposures suggest that this area is more complexly deformed than the area of Fig. 5.

On the east side of Jap Bay, beds of Unit B dip homoclinally 40–45° to the NW and are locally folded by small-scale asymmetric folds that consistently verge to the SE (seaward). No evidence of NW vergence and overthrusting of mélange was observed on the east side of Jap Bay. Thus, Jap Bay appears to define, and may be the topographic expression of, a north-trending tear fault separating NW verging structures on the west from SE verging structures on the east.

Elsewhere in the Jap Bay area, D_2 is represented by thrust faults (Fig. 7C). Near the mouth of Jap Bay, the contact between mélange and Unit A is a moderately dipping thrust fault which places mélange to the NW and below Unit A (Figs 2 & 3). In the mélange nearest the thrust contact (less than 0–15 m from the contact) subsidiary thrust faults pervasively cross-cut the mélange fabric. The faults are sharp planar discontinuities generally spaced about 10 cm apart. Drag of the mélange fabric along the faults indicates underthrusting of mélange on all of the NW dipping faults. Rare SW dipping faults show evidence of NW overthrusting of mélange. A stereographic plot of slickenlines (Fig. 6D), commonly present on both sets of fault surfaces, indicates a shortening direction of 336°. Mélange is also thrusted below Unit A across Jap Bay (Fig. 2) but no kinematic data were obtained from this fault. However, minor thrust faults occur within Unit A in this area and show north-westerly dips and slickenline trends parallel to the shortening direction recognized in Unit B.

The timing of D_2 deformation is ambiguous but landward-verging structures like those in Unit B have been documented elsewhere in the Kodiak Islands only in the Eocene Sitkalidak Formation. Moore & Allwardt (1981) show that the Sitkalidak Formation on Sitkinak Island represents a thick, obducted trench fill sequence that was accreted to the continental margin in late Eocene time. Thus, the landward-verging structures of D_2 in the Ghost Rocks Formation may be a local response within this formation to obduction and accretion of the Sitkalidak Formation.

Structural synthesis and discussion

The coherent terranes in the Ghost Rocks Formation in the area of Jap Bay record two major tectonic deformations, D_1 and D_2 (Fig. 7). D_1 is recorded in Unit A and was initiated in partially lithified sediments. After sill emplacement, D_1 progressed coaxially into a deformation that involved lithified sediments. D_2 caused large-scale folding in Unit B and thrust mélange to the NW above and below Unit B and below Unit A.

Critical to understanding the tectonic significance of the coherent terranes is knowing the tectonic setting in which they were deposited and deformed: do the terranes represent accreted trench or slope basin deposits? Unit B is lithologically, metamorphically and structurally unique relative to other units in the Ghost Rocks Formation and appears to be the youngest unit in this formation, suggesting that it is a slope basin deposit (Fig. 7B). A more seaward, trench environment is possible but unlikely because there is no evidence in the deformation histories of the coherent terranes that indicates Unit B was thrust into its present position.

The structural history of Unit A suggests it was also deposited in a slope basin environment (Fig. 7A). Recent studies of modern convergent margins suggest that trench sediments become imbricated and tilted landward as they

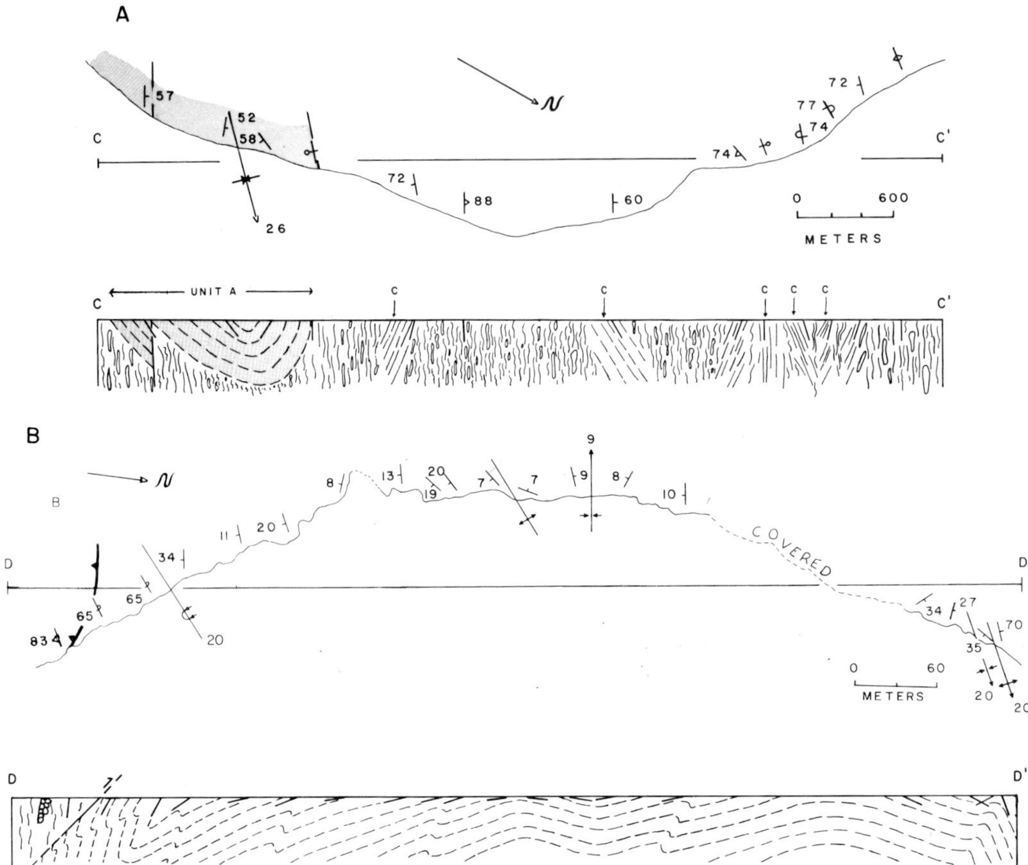

FIG. 5. Geological maps and cross-sections of undifferentiated terrain and Unit A (section C-C′) and Unit B (section D-D′). See Figs 2 and 3 for locations and symbols. Small-scale folds in D-D′ are exaggerated for clarity. 'C' in section C-C′ shows coherent areas in undifferentiated terrane (see text).

are deformed (Moore *et al.* 1979). In Unit A of the Ghost Rocks Formation the sub-horizontal deformation axes that deformed this unit are still in their sub-horizontal orientation. This unit apparently was not imbricated and tilted landward during deformation as would be expected in a trench environment. Unit A is therefore most simply interpreted as a slope basin deposit. Alternatively, trench environments may be more complicated than inferred from studies of modern convergent margins and Unit A may represent the remnants of complexly deformed trench deposits. A detailed study of the deformation history of the mélange terranes (in preparation) could resolve this problem. In either case, the deformation history of Unit A is significant because it indicates that uplift of the Ghost Rocks Formation occurred without imbrication and landward tilting of younger more seaward deposits (Fig. 8).

FIG. 6. Stereographic plots and contour diagrams for D_1 (A and B) and D_2 (C and D). A and B respectively show poles to axial surfaces in conjugate fold belt and poles to cleavage in spaced cleavage belt. Closed circle and square indicate directions of maximum shortening and extension, respectively. (C) Poles to bedding and fold axes (shown as open circles) in Unit B. (D) Slickenlines (arrows) and poles to fault planes (+) of thrust faults subsidiary to large thrust fault that separates Unit A from mélange. SE-plunging slickenlines are on SE dipping fault planes.

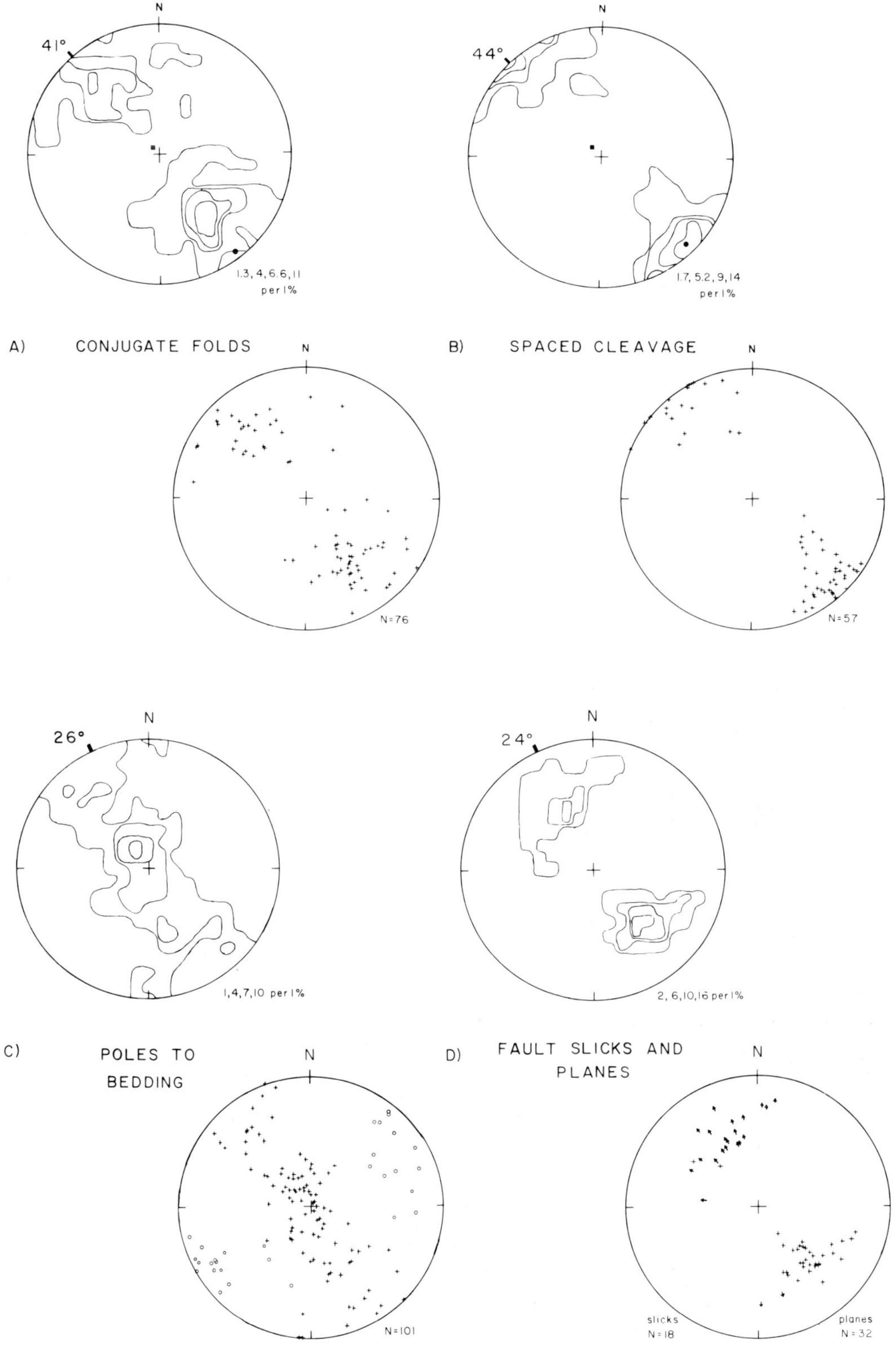
N
41°
1.3, 4, 6.6, 11
per 1%
N
44°
1.7, 5.2, 9, 14
per 1%
A) CONJUGATE FOLDS
N
N=76
B) SPACED CLEAVAGE
N
N=57
N
26°
1, 4, 7, 10 per 1%
N
24°
2, 6, 10, 16 per 1%
C) POLES TO BEDDING
N
8
N=101
D) FAULT SLICKS AND PLANES
N
slicks
N=18
planes
N=32

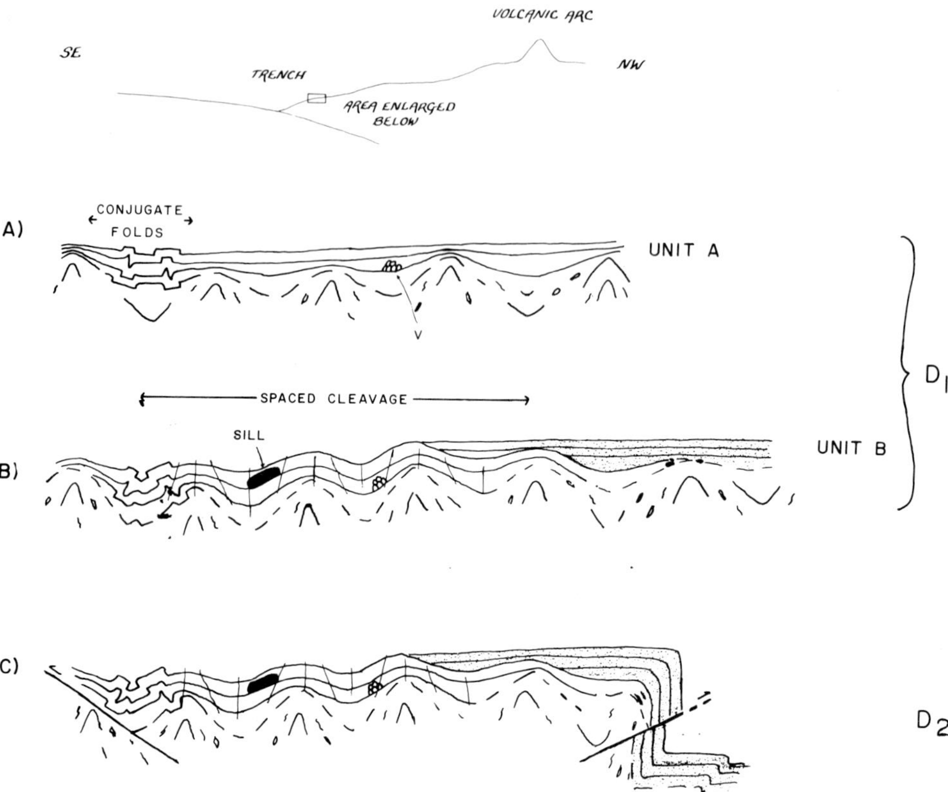

FIG. 7. Interpretative deformation history of the coherent terranes in the Ghost Rocks Formation. Unit A is inferred to grade into more complexly deformed units but not necessarily into mélange with depth (see text). (A) Sub-horizontal shortening in Unit A causes development of conjugate fold belt in partially lithified sediments. Size and extent of conjugate folds are exaggerated for clarity. (B) Unit A is intruded and hornfelsed by a shallow level sill which accentuated lithification. This resulted in a change in style of deformation during continued sub-horizontal shortening. Broad folds with an associated cleavage cross-cut the sill and the conjugate folds. (C) D_2 resulted in NW-directed overthrusting and underthrusting in Units B and A, respectively. (NW-directed underthrusting also occurred in Unit B but is not shown in this diagram.)

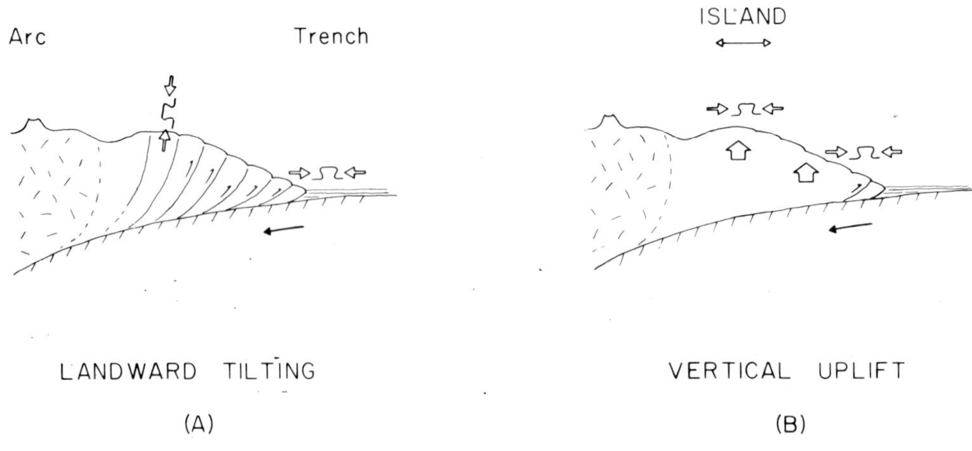

ACKNOWLEDGMENTS: This paper is part of a PhD thesis completed at the University of California, Santa Cruz, and I greatly appreciate the stimulating environment the people there provided. Mary Reid's dedicated assistance and encouragement both in and out of the field, and Casey Moore's inspiring enthusiasm played significant roles in completing this paper. Gene Gonzales still makes better thin sections than I, and it is greatly appreciated. Casey Moore, Othmar Tobisch, Ed Beutner and George Moore provided comments on early drafts of this paper. John Armentrout and Mobil Oil Company graciously provided the isotopic ages, the vitrinite reflectance measurements and first introduced me to the geology of Kodiak Island. National science Foundation grant EAR 77–07836, Mobil Oil Corporation, Exxon Company, Union Oil, Amoco Production, U.S. Geological Survey and the U.S. Coast Guard provided financial support.

References

ARMENTROUT, J. 1979a. Cenozoic stratigraphy, Kodiak Island archipelago, Alaska. *Abstr. geol. Soc. Am. Progr.* **11,** 66.

—— 1979b. Paleocene plutonism in southern Alaska (abstr.). *Proc. 14th Pacif. Sci. Congr.* Khabarovsk, U.S.S.R.

BACHMAN, S. B. 1978. A Cretaceous and early Tertiary subduction complex, Menodocino coast, northern California. *In:* HOWELL, D. G. & McDOUGALL, K. A. (eds). *Mesozoic Paleogeography of the Western United States*, 419–29. Pacif. Sec., Soc. Econ. Paleontol. Mineral. Tulsa.

COWAN, D. S. 1974. Deformation and metamorphism of the Franciscan Subduction Zone Complex northwest of Pacheco Pass, California. *Bull. geol. Soc. Am.* **85,** 1623–34.

—— & SILLING, R. M. 1978. A dynamic, scaled model of accretion at trenches and its implications for the tectonic evolution of subduction complexes. *J. geophys. Res.* **83,** 5389–96.

GAY, N. & WEISS, L. 1974. The relationship between principle stress directions and the geometry of kinks in foliated rocks. *Tectonophysics,* **21,** 287–300.

HILL, M. D. 1978. *Volcanic and plutonic rocks of the Kodiak-Shumagin Shelf, Alaska: subduction deposits and near-trench magmatism.* Thesis, PhD, Univ. California, Santa Cruz.

HOBBS, B., MEANS, W. & WILLIAMS, P. 1976. *An Outline of Structural Geology.* Wiley, New York, 571 pp.

KARIG, D. E. & SHARMAN, G. F. III. 1975. Subduction and accretion in trenches. *Bull. geol. Soc. Am.* **86,** 377–89.

LUNDBERG, N. & MOORE, J. 1981. Structural features of the Middle America trench slope off southern Mexico. *In:* MOORE, J. *et al.* (eds). *Initial Rep. Deep Sea drill. Proj.* **66,** (in press).

MOORE, G. 1967. Preliminary geologic map of Kodiak Island and vicinity, Alaska. *Open File Rep. 271,* U.S. geol. Surv. Reston, Virginia.

MOORE, G. F. & KARIG, D. E. 1979. Structural geology of Nias Island, Indonesia: implications for subduction zone tectonics. *Am. J. Sci.* **280,** 193–223.

MOORE, G. W. 1969. New Formations of Kodiak and adjacent islands, Alaska. *Bull. U.S. geol. Surv.* **1274-A,** 428–35.

MOORE, J. C. 1973. Complex deformation of Cretaceous trench deposits, southwestern Alaska. *Bull. geol. Soc. Am.* **84,** 2005–20.

—— & ALLWARDT, A. 1981. Progressive deformation of a Tertiary trench slope, Kodiak Islands, Alaska. *J. geophys. Res.* (in press).

—— & CONNELLY, W. 1979. Tectonic history of the continental margin of southwestern Alaska: Late Triassic to earliest Tertiary. *In: The Relationship of Plate Tectonics to Alaska Geology and Resources,* H1–29, Alaska Geol. Soc., Anchorage, Alaska.

—— *et al.* 1979. Progressive accretion in the Middle America Trench, southern Mexico. *Nature, London,* **281,** 638–42.

NILSEN, T. & MOORE, G. 1979. Reconnaissance study of Upper Cretaceous to Miocene stratigraphic units and sedimentary facies, Kodiak and adjacent islands, Alaska. *Prof. Pap. U.S. geol. Surv.* **1093.**

RAMSAY, J. G. 1967. *Folding and Fracturing of Rocks.* McGraw-Hill, New York, 568 pp.

REED, B. L. & LANPHERE, M. A. 1974. Chemical variations across the Alaska-Aleutian Range batholith. *J. Res. U.S. geol. Surv.* **86,** 819–29.

REID, M. R. & GILL, J. B. 1980. Near trench volcanism, Kodiak Island, Alaska: implications for ridge-trench encounter. *Abstr. Progr. geol. Soc. Am.* **12,** 148.

SEELY, D. R. 1977. The significance of landward vergence and oblique structural trends on trench inner slopes. *In:* TALWANI, M. & PITMAN, W. C., III (eds). *Island Arcs, Deep Sea Trenches, and Back-arc Basins,* 187–98. Am. geophys. Union M. Ewing Ser. **1.**

FIG. 8. Two models for uplift of the Ghost Rocks Formation. In both, opposing arrows show shortening directions inferred from *conjugate fold* and spaced *cleavage* belts. In imbricate thrust model (A) (after von Huene 1979; Seely *et al.* 1974) sub-horizontal deformation axes formed in the area of the lower slope or trench become tilted landward as younger more seaward rocks are accreted. (B) shows the model required by the structural history of Unit A. In this model deformation axes remain in their sub-horizontal orientation as the prism rises vertically.

—— , VAIL, P. R. & WALTON, G. G. 1974. Trench slope model. *In:* BURK, C. A. & DRAKE, C. L. (eds). *The Geology of Continental Margins*, 249–60. Springer-Verlag, New York.

STONE, D. B. & PACKER, D. R. 1979. Paleomagnetic data from the Alaska Peninsula. *Bull. geol. Soc. Am.* **90,** 545–60.

SUPPE, J. 1973. Geology of the Leech Lake Mountain-Ball Mountain Region, California. *Univ. Calif. Publ. Geol. Sci.* **107,** 82.

TOBISCH, O. & FISKE, R. 1976. Significance of conjugate folds and crenulations in the central Sierra Nevada, California. *Bull. geol. Soc. Am.* **87,** 1411–20.

VON HUENE, R. 1979. Structure of the outer convergent margin off Kodiak Island, Alaska, from multichannel seismic records. *In:* WATKINS, J., MONTADERT, L. & DICKERSON, W. (eds). *Geological and Geophysical Investigations of Continental Margins*. Mem. Am. Assoc. Petrol. Geol. **29,** 261–72.

WOOD, D. S. 1974. Current views of the development of slaty cleavage. *Ann. Rev. Earth Sci.* **2,** 1–35.

TIM BYRNE, Earth Sciences Board, University of California, Santa Cruz, California 95064, U.S.A.

ASIA AND AUSTRALASIA

Sedimentation in the Sunda Trench and forearc region

Gregory F. Moore, Joseph R. Curray & Frans J. Emmel

SUMMARY: Sedimentation in the Sunda Trench and forearc region is dominated by the transport of terrigenous detritus from orogenic and volcanic terranes around the margins of the north-eastern Indian Ocean. Quartzose Himalayan detritus is transported into the Bay of Bengal and down the trench axis as far south as the Sunda Strait. Prior to development of the outer-arc ridge, the majority of arc-derived sediment bypassed the forearc basin and was deposited in the trench. Nearly all sediment derived from the arc terranes of Java and Sumatra during the Neogene has been trapped in the forearc basin and has not reached the trench.

Hemipelagic sedimentation dominates on the lower part of the inner trench slope. Calcareous microfossils and volcanic ash are the dominant constituents of these hemipelagic deposits. Higher on the slope, terrigenous and hemipelagic sediments accumulate in large trench-slope basins.

The Sunda forearc basin is a series of smaller individual basins separated from each other by transverse highs, isolating the respective sedimentary sequences. Off north Sumatra, quartzose detritus accumulates, whereas off Java, volcaniclastic sediments predominate.

Long-distance transport of sediment down the trench axis, damming of sediments behind the outer-arc ridge, and the segmented nature of the forearc basin leads to the juxtaposition of sediment bodies with different provenances.

Sedimentation along convergent margins reflects a complex interaction between tectonic and depositional processes. Sediments are deposited seaward of the trench, in the trench axis, on the inner trench slope, and in the forearc basin. These depositional environments are continuously modified as the accretionary prism grows. Sediment provenance and dispersal are also complicated. For example, terrigenous detritus may be derived from a nearby arc terrane, it may be transported long distances down the trench axis, or it may be carried into the trench on the subducting oceanic plate. Biogenic and volcanogenic detritus are also important components of convergent margin sediments.

Convergent margin sedimentation is poorly understood, partly because of the paucity of studies of modern margins, but also because of the great variability among convergent margins around the world (Dickinson & Seely 1979; Underwood *et al.* 1980). The purpose of this paper is to outline recent sedimentation patterns and discuss the development of stratigraphic relationships along one convergent margin: the Sunda Arc of the north-eastern Indian Ocean. In this paper, we define the depositional framework of the Sunda Trench and forearc region, outline the patterns of sediment dispersal and provenance, and infer the characteristics of the forearc deposits. Our results serve to illustrate the complexity of sedimentary facies and spatial relationships that develop in forearc regions.

This study is primarily based on the analysis of seismic reflection, 3.5 kHz and bathymetric data collected by ships of the Scripps Institution of Oceanography since 1968. Our data were supplemented by seismic data collected by the Lamont-Doherty Geological Observatory (published by Hamilton 1979), the U.S. Naval Oceanographic Office (Bowles *et al.* 1978), and Union and Mobil Oil Companies (see Karig *et al.* 1980a). These geophysical data comprise a regional reconnaissance study of the entire Sunda Arc, as well as two detailed transect studies, and include approximately 30 000 km of track (see Moore *et al.* 1980b for locations of most tracks). Bottom samples from the trench and lower trench slope are insufficient to define sediment types and facies distributions precisely. The geophysical data, however, allow inferences to be made concerning depositional processes. Data from several exploration wells on the landward margin of the forearc basin off central and north Sumatra provide insights into forearc basin development. Our sedimentological studies of uplifted Tertiary trench and trench slope strata exposed on Nias Island, west of Central Sumatra (Moore 1979; Moore *et al.* 1980a), provide additional data on forearc depositional processes.

Regional setting

The Sunda Arc is continuous for more than 5000 km from Burma in the NW to Sumba Island in the SE (Figs 1 & 2). It is a continental-margin arc-trench system with a ridged forearc (Dickinson & Seely 1979) and marks the site of active subduction of the Indian (or Australian) Plate beneath the Eurasian (or China) Plate. The arc is bounded on the NW by the continent-continent collision zone of the Himalayas and on the SE by the incipient continent-arc collision zone of the Banda Arc. Subduction has apparently been occurring along the Sunda Arc, at least intermittently, since the Permian (Katili 1973). The most recent episode of subduction may have begun in late Oligocene time (Karig *et al.* 1979; Hamilton 1979).

Neogene sedimentation in the northern part of the Sunda Trench has been dominated by detritus eroded from the Himalayan collision belt (Moore *et al.* 1974). Accretion of these thick sediments to the Sunda margin during the Neogene has led to outbuilding of a wide accretionary prism and upbuilding of a high outer-arc ridge (Curray & Moore 1974; Hamilton 1979; Karig *et al.* 1979, 1980b).

Depositional environments

Deposition in the Sunda forearc region occurs in four distinct environments: the Indian Plate, the trench, the trench inner slope (defined as the area between the trench and the outer-arc ridge), and the forearc basin. The types of sediments deposited, and the depositional processes that operate in each environment differ significantly as we demonstrate below.

Indian Plate

Terrigenous, hemipelagic, and pelagic sediments that are deposited on the Indian Plate seaward of the Sunda Trench are carried into the trench by plate convergence. In the Bay of Bengal, deposition is dominated by transport of terrigenous detritus onto the Bengal Fan (Curray & Moore 1971). Deposits in the Bay of Bengal are predominantly turbidites that are greater than 16 km thick west of Burma and thin markedly to the south (Fig. 3; Curray & Moore 1971, 1974). During low stands of sea-level, great amounts of sediment are delivered to the Bengal Fan. In contrast, during the present high stand of sea-level, most of the sediment derived from the Himalayas is trapped in the rapidly subsiding delta and on the inner shelf, and relatively little sediment is transported to the fan (Moore *et al.* 1974).

Off the Nicobar Islands, the Ninetyeast Ridge has recently impinged on the trench and cut off the flow of turbidites to the Nicobar Fan (Fig. 2; Curray & Moore 1974; McDonald 1977). Pelagic deposits are now accumulating on the Nicobar Fan (Bowles *et al.* 1978). Prior to the mid-Pleistocene, however, sediments flowed down the Nicobar Fan as far south as the latitude of the Sunda Strait (Fig. 2). At DSDP site 211, west of the Sunda Strait (Fig. 2), Upper Pliocene-Quaternary pelagic silicic ooze overlies Pliocene silt turbidites. At the time of deposition of the Pliocene turbidites, approximately 200–300 km less convergence of the Indian Plate and the Sunda Trench had occurred, and turbidites were transported directly to this part of the Nicobar Fan from Ganges-Brahmaputra sources (Curray & Moore 1974; McDonald 1977). These sediments are underlain by brown clay of unknown age and Lower Cretaceous nannofossil ooze (von der Borch, Sclater *et al.* 1974).

Off Java, mainly pelagic and hemipelagic sediments have been deposited on the Indian Plate (Cook *et al.* 1978; Heirtzler *et al.* 1978). The thickness of pelagic and hemipelagic sediments seaward of the trench varies from 200 to 400 m (unpublished data from S.I.O. cruise *Rama*, September 1980).

Sunda Trench

West of the Andaman Islands, where the Bengal Fan is being subducted, the trench has little morphological expression and is less than 3000 m deep (Fig. 4A). A more distinct morphological depression appears off north Sumatra, where the thickness of Nicobar Fan sediment decreases, the trench deepens to 5000 m, the trench floor narrows, the outer trench slope steepens, and a trench sediment wedge is developed (Fig. 4C). The trench has relatively steep inner and outer slopes south and southeast of Java, and is greater than 6000 m deep (Fig. 1). It has several isolated troughs that are greater than 6500 m deep and are separated by bathymetric sills (Fig. 3). The width of the trench floor varies markedly south of Java, and locally, a trench sediment wedge is well-developed (Fig. 4D).

A major submarine channel (Fig. 4A,B) transports sediment from the northern end of the trench to at least 10°N. The channel is not in the trench axis as in the eastern Aleutian Trench (von Huene 1974), but is 10–50 km seaward of

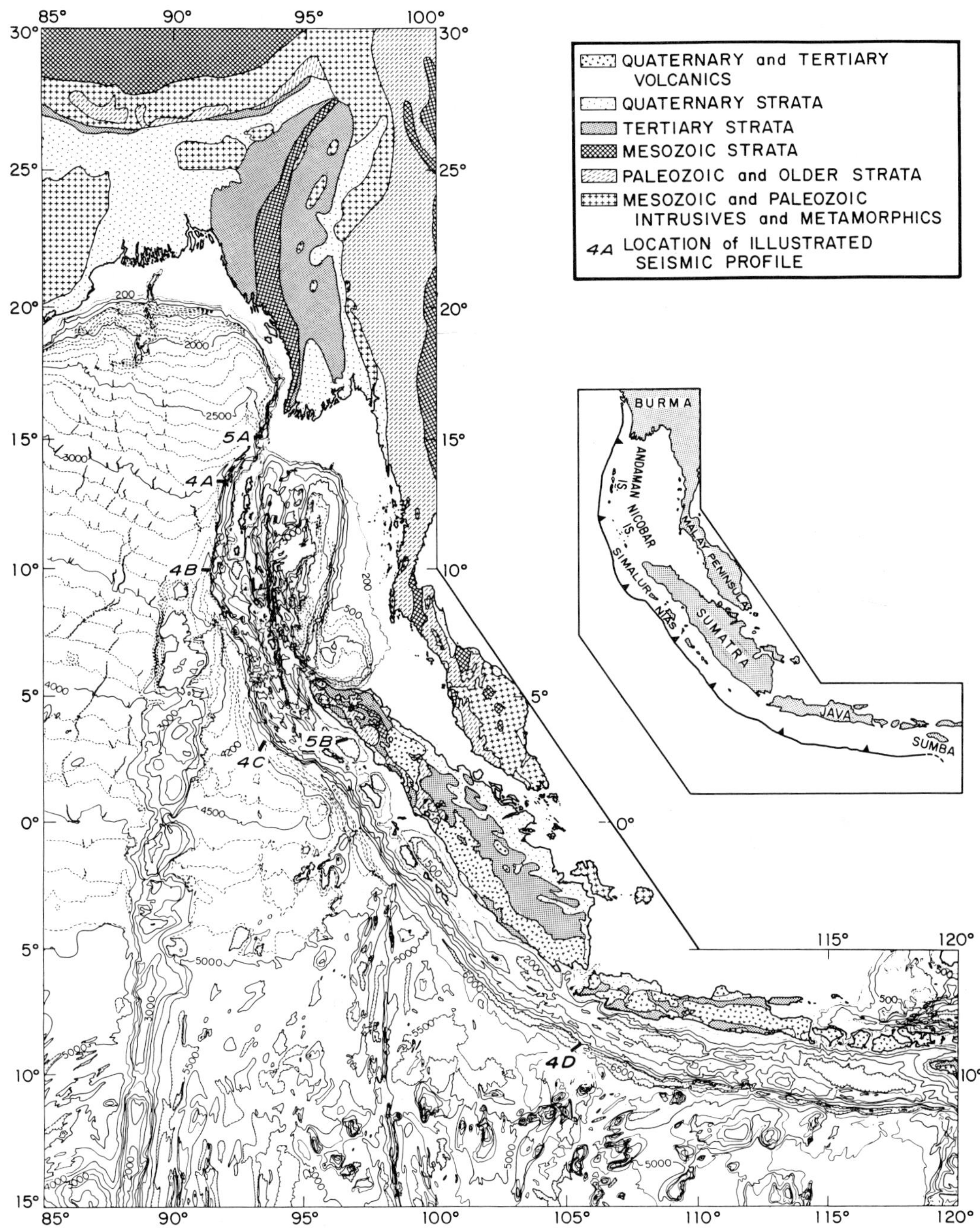

FIG. 1. Regional bathymetry of the north-east Indian Ocean (from Curray *et al.* 1981) and simplified geological map of landmasses (from United Nations 1971; Hamilton 1979; and Page *et al.* 1979). Locations of seismic profiles of Figs 4 and 5 are also indicated.

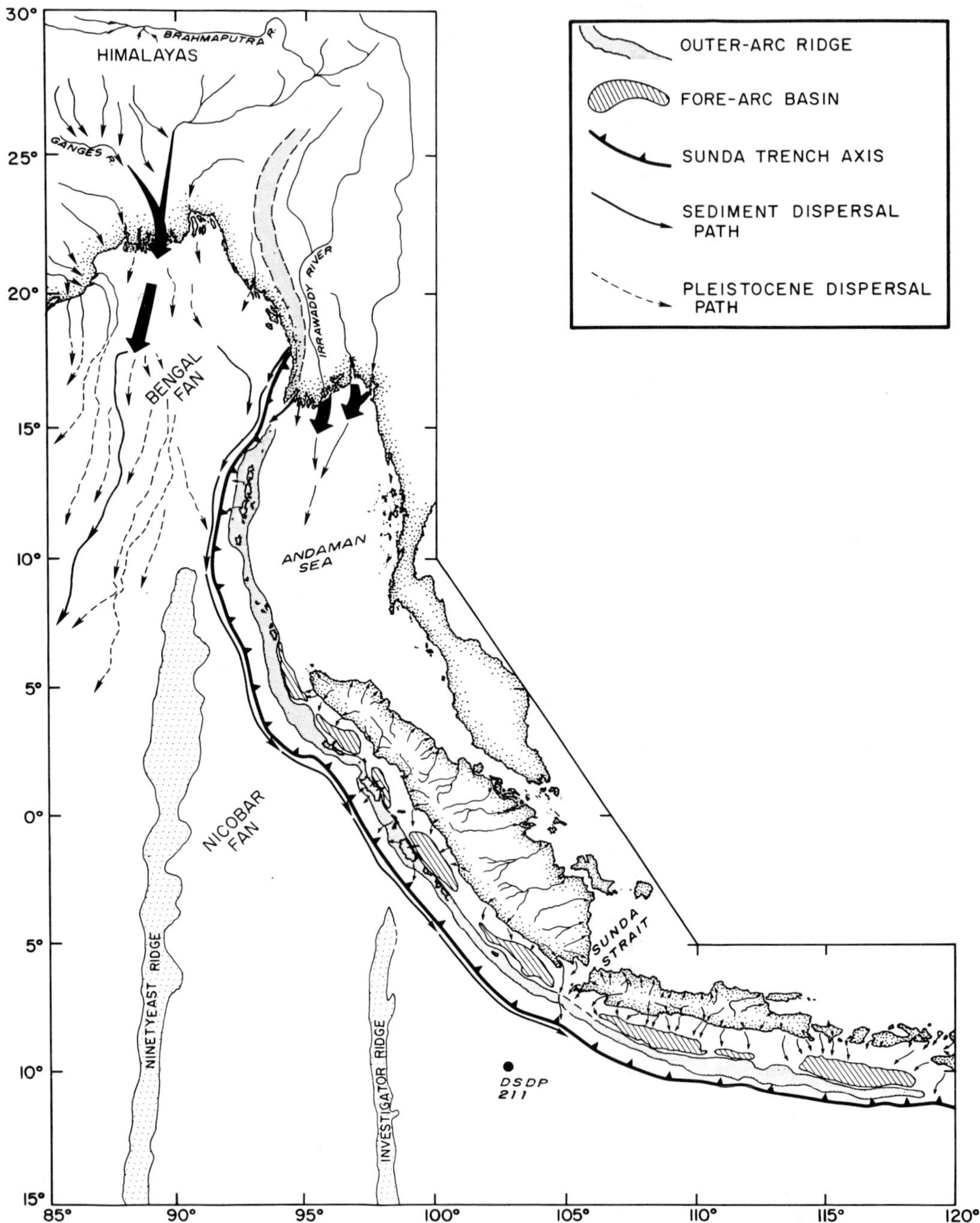

FIG. 2. Tectonic map showing major sediment dispersal paths to the Sunda forearc region. Solid arrows indicate present dispersal paths; dashed arrows are dispersal paths that were active during the Pleistocene low stands of sea-level.

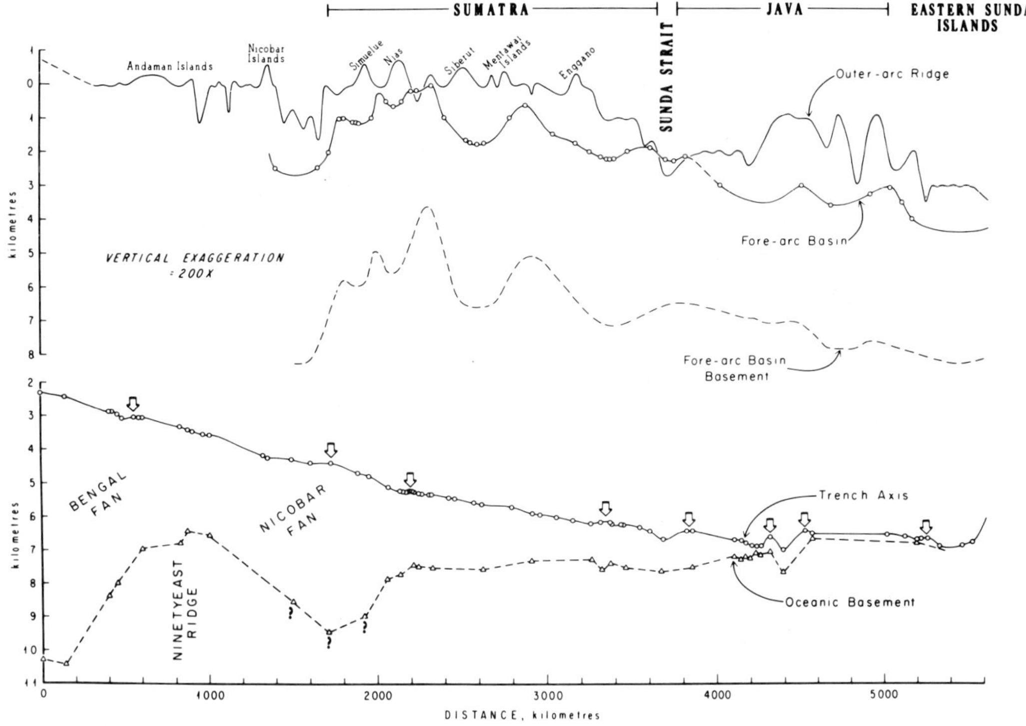

FIG. 3. Depth profiles along the Sunda forearc region from south-west Burma to Sumba Island. Lower diagram is an axial profile along the trench. Depth to oceanic basement based on seismic reflection and refraction measurements (triangles are data points). Trench axis depths are from echo-sounder crossings (circles). Arrows are locations of sills that may block the flow of turbidites. Upper diagram is an axial profile along the forearc basin. Depth to forearc basin basement from Curray *et al.* (1977), Hamilton (1979), Kieckhefer *et al.* (1980) and unpublished S.I.O. seismic refraction data. Depth of forearc basin from echo-sounder crossings (circles). Depth of outer-arc ridge measured from regional bathymetric map (Curray *et al.* 1981).

the base of the inner trench slope as in the Peru–Chile Trench (von Huene 1974; Schweller & Kulm 1978). This channel, probably fed by a canyon from the Bangladesh shelf (Figs 1 & 2), has been filled SW of Burma by a large submarine slide and associated debris flows (see below). West of the Andaman Islands, another major channel from the Bengal Fan intersects the trench at 12°N and trends southward (Fig. 2).

Transport of large amounts of sediment down the trench axis is indicated by the broad trench floor west of Sumatra (Fig. 1) and the relatively smooth axial gradient from Burma to the south tip of Sumatra (Fig. 3). Himalayan detritus may be transported as far south as the Sunda Strait approximately 3000 km from the head of the trench. Hamilton (1979) interprets quartzose, micaceous, silty turbidites recovered from the trench axis at 6°S (Anikouchine & Ling 1967) as having been transported from the Himalayas. The rough topography and variable width of the trench axis south of Java suggest that sediments are not transported along the trench axis in this area at present.

Sediments from Sumatra and Java are at present prevented from reaching the trench by the outer-arc ridge along most of the arc. Arc-derived sediments generally are trapped in the forearc basin (Hamilton 1979). An exception is in the area of the Sunda Strait, where the Sumatran Fault cuts across the forearc region (e.g. Hamilton 1979; Curray *et al.* 1981). Turbidity currents from south Sumatra and west Java apparently flow down the canyon formed along the fault and into the trench axis (Figs 1 & 2).

The morphotectonic setting of the Sunda Arc was probably very different during the early and mid-Tertiary. Sumatran detritus is found in

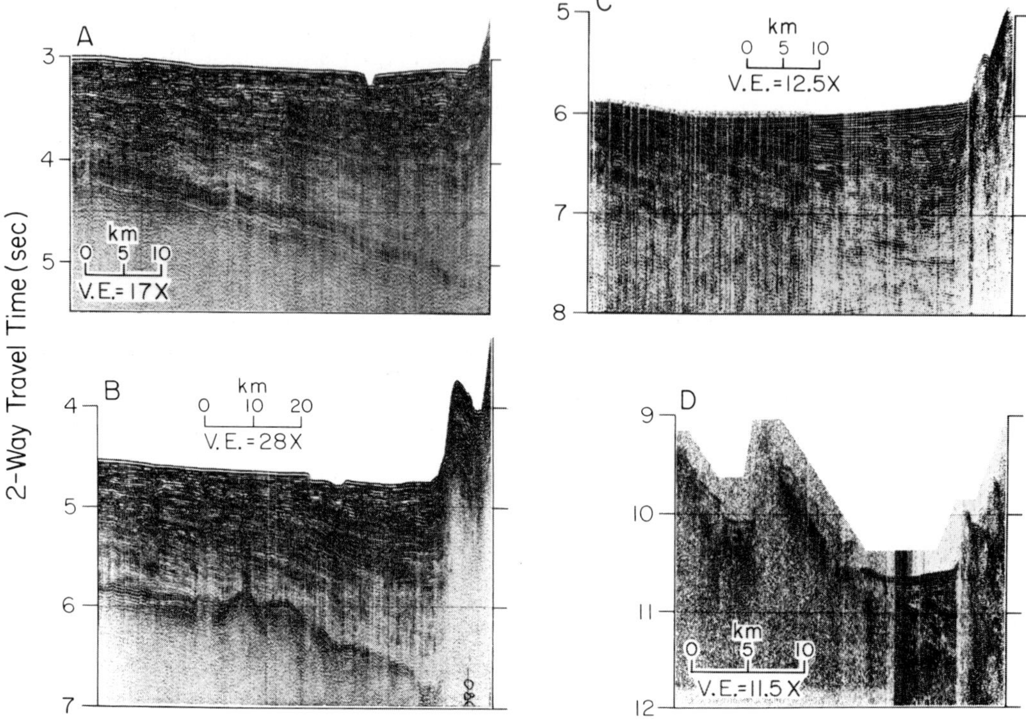

FIG. 4. Seismic reflection profiles across the Sunda Trench (locations shown in Fig. 1). (A) Profile west of Andaman Islands shows thick sediments of Bengal Fan with channel west of trench axis. (B). Profile west of Nicobar Islands shows wider channel with levees. Acoustic basement is the Ninetyeast Ridge. (C) Profile off North Sumatra shows seaward-dipping trench wedge and onlapping relationship of trench sediments on to Indian plate sediments. (D) Profile off Java shows oceanic pelagic section onlapped by trench wedge with seaward-dipping surface.

uplifted upper Oligocene/lower Miocene trench deposits on Nias Island, suggesting that at least some arc-derived sediments were bypassing the forearc basin and flowing into the trench (Moore 1979). Prior to the Miocene, the outer-arc ridge was probably much smaller in size because thick fan sediments had not yet been accreted (Curray & Moore 1974; McDonald 1977). The time of formation of a morphological outer-arc ridge probably varied along the arc, occurring first in the north and later in the SE.

Many seismic reflection and 3.5 kHz profiles across the trench show a seaward-dipping seafloor at the base of the landward trench slope (Fig. 4B,C). The echo character of this sediment apron (type IIB of Damuth & Hayes 1977) indicates that it is formed of large amounts of bedded sands and silts. Several crossings of the apron show hummocky topography (echo type IIIC) characteristic of debris flows or small slumps. These sediment aprons are widespread along the trench south of 2°N (Bowles *et al.* 1978), suggesting that small-scale slumping and redeposition of accreted material and surficial slope sediments through mass flows and turbidites is common. In areas of the trench off Java that are isolated from the axial transport of turbidites, mass flow deposits may form a substantial portion of the trench wedge. Small fans emanating from minor canyons on the lower trench slope (Karig *et al.* 1980b) probably direct turbidity currents and debris flows of slumped hemipelagic slope sediments and uplifted trench sediments to the trench. The fans SW of Nias have a mounded, chaotic seismic character (Karig *et al.* 1980b) and extend across the trench axis, possibly blocking or disrupting the flow of turbidites down the trench axis (Fig. 3).

A large sediment slide (over 2000 km^2) has been identified in the trench off Burma (Moore *et al.* 1976). This slide has transported a huge volume (over 900 km^3) of sediment which may have come from a westerly Pleistocene lobe of the Irrawaddy River delta. Other slump masses tens of kilometres wide have not been identified elsewhere in the Sunda Trench (Hamilton 1979; Moore *et al.* 1980b), suggesting that slumping of large-scale blocks (several kilometres in size) is not an important mechanism of sediment transport to the Sunda Trench.

Trench inner slope

Sediments accumulate on the landward trench slope between the trench and the outer-arc ridge. The inner slope west and NW of Sumatra is characterized by a ridge and trough morphology, and the width of the troughs generally increases upslope (Moore & Karig 1976; Karig *et al.* 1980b). The basins are continuous along strike for only a short distance (a few tens of kilometres), and the basin floors generally exhibit an irregular topography. Many small canyons traverse short sections of the slope, providing a mode of transport of sediment into the basins (Karig *et al.* 1980b). The lower trench slope off Java is steeper than off Sumatra, and slope basins are not as well-developed.

Slope basin sediments are thin near the base of the slope, but are as thick as 2–3 km near the outer-arc ridge. Over many of the ridges, 3.5 kHz profiles show evenly stratified reflectors (echo type IB), indicating a uniform pelagic or hemipelagic drape. The echo character of the basin sediments ranges from prolonged (type IIA) to bedded (type IIB), suggesting that the strata are a mix of hemipelagic sediments and bedded silts and sands (turbidites).

Basins near the top of the slope have been filled, causing sediment to spill over into the deeper basins (Fig. 5A). As small basins fill with sediment they combine to form larger basins (Fig. 5A). Some basins near the top of the slope contain up to 1.5 km of layered sediments. The shallow reflectors are generally continuous across the basins, but the deeper reflectors are discontinuous (Karig *et al.* 1980b) and many unconformities are visible (Fig. 5A). We believe that the discontinuous nature of the older sediments is due primarily to deposition in smaller, discontinuous basins lower on the slope and secondarily to slumping. The unconformities are due to tilting and folding contemporaneous with sedimentation.

Piston cores taken on the lower slope off Nias (Karig *et al.* 1980b) indicate that hemipelagic sediments are presently accumulating at a high rate. Volcanic ash layers are common in the cores. The emergent islands of the outer-arc ridge west and north of Sumatra are supplying detritus at present from the uplifted accretionary prism to the inner trench slope. For example, there are several large rivers that drain the interior of Nias and discharge large amounts of sediment on to the trench slope. We identified several canyons on the Nias shelf that lead into trench slope basins (unpublished data from Expedition *Rama*, October 1980). The transport of accretionary prism material to the trench slope has varied considerably along the arc. It has probably been occurring since the Oligocene in Burma, whereas it is a fairly recent phenomenon off Sumatra, because the islands of the outer-arc ridge have only been emergent for a few million years. Accretionary prism material is probably not being transported to the trench slope off Java, because the outer-arc ridge is below sea-level and blanketed by hemipelagic deposits. Data from Nias (Moore *et al.* 1980a) suggest that redeposition of slumped hemipelagic sediments into the slope basins is important near the base of the inner trench slope. In the past, submarine canyons probably cut across the forearc basin and delivered arc-derived sediment from Java and Sumatra to the slope basins.

Forearc basin

Between the outer-arc ridge and the active arc terrane from the eastern Sunda Arc to north Sumatra is a series of forearc basins (Figs 1–3). The basins are 50–100 km wide and 150–500 km long. The water depth of the basins off Java is 3500–4000 m, but is only 500–2000 m off Sumatra. They are distinct depositional basins that are separated from each other by transverse structural highs. The sedimentary fill in each basin is thus expected to be different in composition from adjacent basins because of the differences in source terranes along the arc (see below). The sediment fill in the basins is up to 6 km thick (Fig. 3). The maximum sediment thickness in each basin underlies the present deep basins (Hamilton 1979), and the basin sediments onlap the structural highs, indicating that the structural separation of the basins has persisted for a long period of time.

Sediments deposited in the forearc basins are derived primarily from the arc terranes at their landward boundaries. Submarine canyons are common on the slope from the arc terrane into the forearc basins, and are apparently the main dispersal paths for arc-derived sediments. Canyons are more numerous along the Java margin than along the Sumatra margin (Fig. 2). In addition to the arc-derived terrigenous and

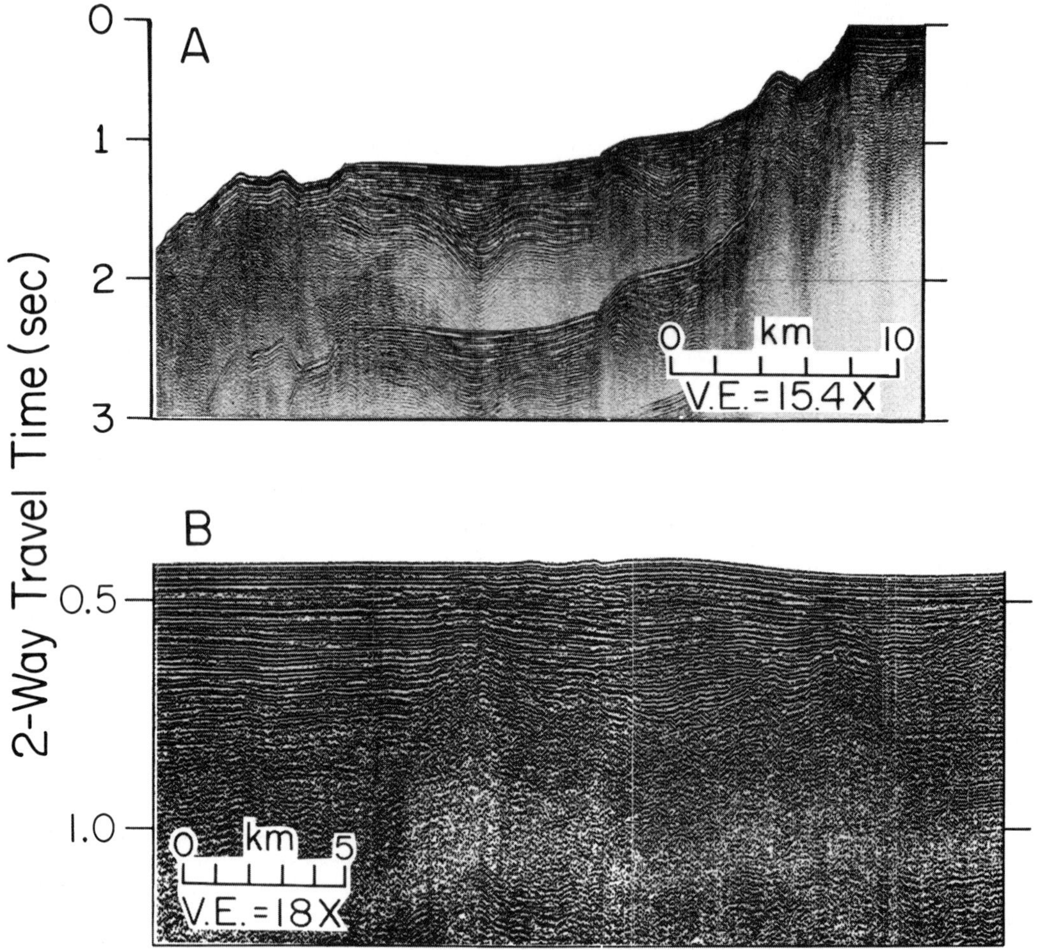

FIG. 5. (A) Seismic profile across trench-slope basin west of the Andaman Islands showing several periods of deformation as indicated by many unconformities. Upper basin has been filled and sediments are now spilling into lower basin. (B) Seismic profile across small fan in forearc basin landward of Simalur Island showing characteristic concave-upward surface and chaotic internal seismic facies. Location of profiles shown in Fig. 1.

volcaniclastic sediment, shallow-water carbonate debris is an important component of the forearc basin sediments.

Surprisingly, we have identified only a few small submarine fan complexes in the forearc basin. This may be because the morphological fans are smaller than the resolution of our seismic data. One fan, in the basin off Simalur Island (Figs 1 & 2), has a characteristic convex-upward morphology and chaotic seismic character (Fig. 5B). Seismically identifiable buried fan and channel complexes are also uncommon. The flat floors of the basins and the reflection configuration of the sedimentary fills suggest that deposition is mainly by turbidity currents that spread laterally across the basin floors.

The forearc terrane north of Sumatra is anomalous. There is at present no forearc basin between the Nicobar Islands and Burma (Fig. 1). A forearc basin probably existed prior to the Miocene, but it has been structurally disrupted by the opening of the Andaman Sea (Curray *et al.* 1981). The morphotectonic outer-arc ridge becomes subaerial as the Indoburman Ranges of Burma, and the forearc basin is the western trough of the Central Valley of Burma. The sediments in the basin are many kilometres thick and probably have been derived and transported longitudinally within the trough from the Himalayan terrane.

Sediment provenance

Terrigenous detritus is at present the major constituent of the sediments being deposited in the Sunda Trench and forearc region. Volumetrically, the most important provenance is the uplifted basement terrane of the Himalayas. The uplifted crystalline basement rocks are eroded rapidly and the detritus is carried by the confluent Ganges–Brahmaputra River system into the Bay of Bengal (Curray & Moore 1971), where it is dispersed southward on the Bengal deep-sea fan system (Curray & Moore 1974; Curray *et al.* 1981).

Distal turbidites of late Miocene-Pleistocene age were recovered at DSDP Site 211 west of Sumatra (von der Borch, Sclater *et al.* 1974). Detailed point counts of these turbidites (Ingersoll & Suczek 1979) document the petrological characteristics of sediments derived from a Himalayan provenance: they are micaceous and highly quartzose (quartz-feldspar-lithic modes average Q_{57} F_{28} L_{14}), and lithic types are dominated by metamorphic fragments (metamorphic-volcanic-sedimentary lithic populations average Lm_{87} Lv_4 Ls_9). Other petrological studies of Himalayan sediments confirm their quartzose, metamorphic nature (e.g. Raju 1967; Thompson 1974; Mallik 1976, 1978). The range of modal compositions for Himalayan detritus is shown in Fig. 6.

Terrigenous detritus eroded from the arc terrane of Sumatra is transported to the west coast by rivers and delivered to the forearc region by submarine canyons. Although no quantitative point count information exists on the character of Sumatran detritus, inferences can be made based on knowledge of the rocks exposed on Sumatra. In north and central Sumatra, Palaeozoic schists, gneisses and granites, Mesozoic volcanic and volcaniclastic rocks, and quartzose Tertiary strata are exposed (Page *et al.* 1979). The Quaternary volcanoes of north and central Sumatra lie NE of the drainage divide, so recent volcanic detritus is transported mainly to the backarc region (Fig. 2). Thus, detritus delivered to the forearc region off north and central Sumatra reflects primarily a metamorphic and sedimentary provenance. Such a provenance should contribute quartzose detritus with low ratios of plagioclase to total feldspar (P/F) and low abundance of volcanic lithic clasts (Fig. 5; Dickinson 1970; Crook 1974; Schwab 1975; Potter 1978; Dickinson & Suczek 1979; Ingersoll & Suczek 1979). Because there are several ophiolite bodies exposed in north Sumatra (Page *et al.* 1979), a small amount of ophiolitic detritus would be expected to be delivered to the forearc basin.

In south Sumatra, Tertiary and Quaternary volcanic rocks make up a large portion of the exposures (Hamilton 1979; Fig. 1). Mesozoic and Tertiary sedimentary rocks also contribute detritus to the forearc region. Sediments derived from southern Sumatra are probably less quartzose and more volcanic-rich than those of north and central Sumatra.

Java is composed largely of Quaternary volcanics and Tertiary sedimentary rocks (Hamilton 1979; Fig. 1). Palaeogene strata are quartzose, but Neogene strata are volcanic-rich (Hamilton 1979). Sediment derived from Java should therefore be dominated by volcanic lithics and plagioclase feldspar with subsidiary quartz. The islands east of Java are composed mainly of volcanics and volcaniclastic sediments. Detritus derived from the eastern is-

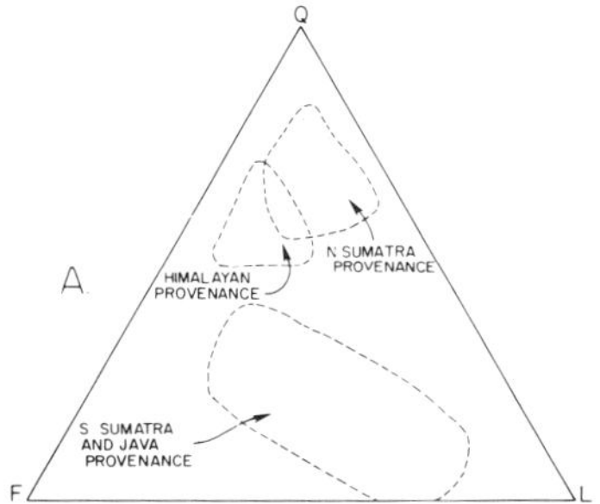

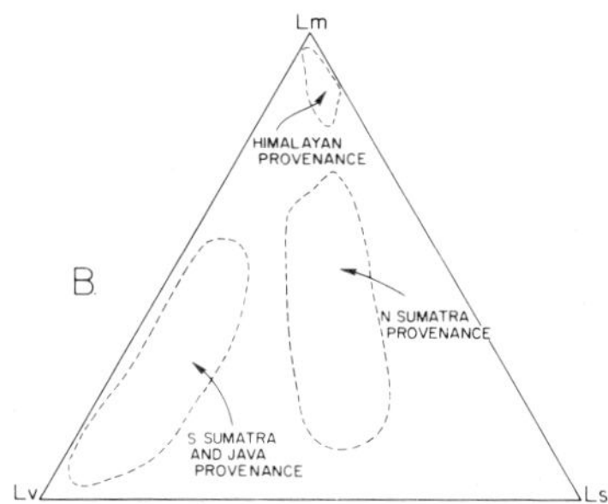

FIG. 6. Detrital modes of sands transported from various provenances to Sunda forearc region. Himalayan modes are from detailed point counts (see text for references); Java and Sumatra modes are inferred from the lithologies exposed in the source terranes (Page *et al.* 1979; Hamilton 1979) and from studies in other arcs (Dickinson & Suczek 1979). (A) Quartz-feldspar-lithic diagram. (B) Metamorphic-volcanic-sedimentary lithic fragment diagram.

lands of the Sunda Arc is probably highly lithic-rich and low in quartz, is dominated by volcanic lithic clasts and has high P/F ratios (Fig. 6).

Facies relationships

Due to differences in sediment transport to the different depositional environments of the forearc region and to changes in character of these environments along strike, sediments of different facies and composition are deposited adjacent to each other. These facies relationships are best illustrated with cross-sections from the Indian Plate, across the trench, trench inner slope, and forearc basin. However, because of the variations along the arc, any cross-section is valid only for describing one narrow transect. We will present here three schematic cross-sections from transects inferred to be representative of large areas of the arc (Fig. 7).

Andaman section

The oceanic crust entering the trench off the Andaman Islands is inferred to be mid-Cretaceous in age (e.g. Curray *et al.* 1981) and is probably overlain by a thin Cretaceous pelagic section. Overlying the pelagics are Upper Cretaceous-Palaeocene strata that probably represent a continental rise sediment wedge that developed off India prior to its collision with Asia (Curray & Moore 1974). These sediments are overlain by a sequence of Eocene to mid-Miocene turbidites of the ancestral Bengal Fan, which is in turn overlain by approximately 1.5 km of Upper Miocene to Pliocene turbidites of the Bengal Fan (Moore *et al.* 1974), including large channel and levee complexes (Curray & Moore 1974). Quaternary trench fill is essentially a part of the Bengal Fan, and consists dominantly of turbidites of Himalayan provenance.

Young sediments deposited on the inner

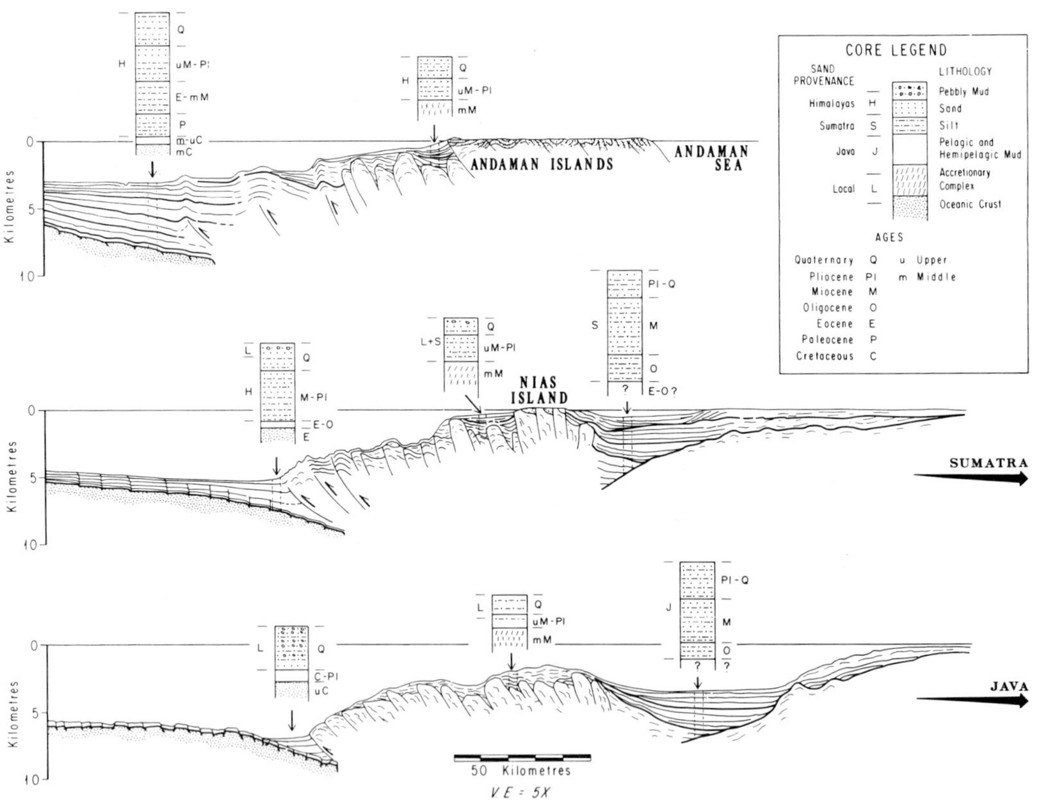

FIG. 7. Schematic depth sections across the Sunda forearc region off the Andaman Islands (top), central Sumatra (middle) and central Java (bottom). Columns are inferred stratigraphic sequences developed in various depositional settings. Structure is based on data presented by Curray *et al.* (1977), Karig *et al.* (1979), and Moore *et al.* (1980b).

trench slope off the Andaman Islands are probably a mixture of terrigenous debris derived from the exposed ophiolitic and sedimentary terranes on the Andaman Islands (e.g. Karunakaran *et al.* 1968) and infrequently some terrigenous detritus delivered to the trench slope by western distributaries of the Irrawaddy River of Burma. Because the strata exposed on the Andaman Islands are uplifted Bengal Fan deposits, the composition of the slope sediment derived from the Andamans should be similar to the sediment that has accumulated in the trench with the addition of minor ophiolitic debris. Sedimentary rocks which may be older slope deposits exposed on the Andaman Islands are hemipelagic and pelagic deposits interbedded with coarse strata that contain abundant ophiolitic debris.

Sumatra section

The age of the crust being subducted off Sumatra is uncertain, but is inferred to be Eocene from marine magnetic data (Sclater & Fisher 1974). Oceanic crust is probably succeeded by Eocene to Oligocene pelagic sediments (Fig. 7), then by a 1 km thick sequence of Nicobar Fan turbidites which is overlain in turn by the trench deposits. Both the Nicobar Fan and trench turbidites reflect a Himalayan provenance. Sediment cores from the Indian Plate contain a large component of terrestrial clay that has probably been transported from India (Venkatarathnam & Biscaye 1973) and considerable amounts of volcanic ash (Ninkovich 1979). Carbonate comprises less than 10% of the sediment because most of the Indian Ocean floor is below the carbonate compensation depth (Kolla *et al.* 1976).

Very thin slope sediments are deposited on top of accreted trench and fan sediments at the base of the slope off Sumatra. Slope sediments are dominantly hemipelagic clays and mud turbidites, because the flow of arc-derived sediment is cut off by the outer-arc ridge. Higher on the slope, turbidites carrying terrigenous detritus from the emergent islands on the outer-arc ridge are important. As inferred from Nias Island data (Moore *et al.* 1980a), a hypothetical section through a slope basin on the upper part of the trench slope has thin hemipelagics at the base overlain by a thicker sequence of terrigenous turbidites (Fig. 7).

The basement of the Sunda forearc basins is believed to be either thinned continental crust (Kieckhefer *et al.* 1980), thickened oceanic crust (Curray *et al.* 1977; Hamilton 1979), or an old subsided and highly metamorphosed accretionary complex (Karig *et al.* 1980a). The sedimentary fill in the basins records several periods of sedimentation and deformation. Palaeogene strata fill a structural low formed landward of a palaeo-outer-arc ridge. Uplift and erosion of the entire margin occurred at the end of the Oligocene (Karig *et al.* 1980b), producing a regional unconformity. During the Neogene, turbidite sedimentation has dominated. The forearc basin strata off Sumatra are 4–6 km thick and are dominantly turbidites derived from Sumatra. They range in age from Palaeogene to Recent. In the central Sumatra area, the terrigenous turbidites are quartzose with andesitic detritus present in varying amounts (Fig. 6).

Java section

Upper Cretaceous oceanic crust is at present being subducted off Java (Heirtzler *et al.* 1978). It is overlain by a thin (300–400 m) section of pelagic deposits that are Cretaceous to Pleistocene in age. As off Sumatra, volcanic ash is an important component in the pelagic sediments, and carbonate contents are less than 10%. The pelagic section is overlain by Quaternary trench deposits. Trench sediments south and SE of Java are less than 1 km thick, and several isolated areas are nearly devoid of trench fill (Fig. 3). The trench off Java is probably starved of terrigenous sand, so the trench deposits are mostly hemipelagic silts and muds and redeposited material that was accreted to the inner trench slope and subsequently slumped back into the trench.

The inner trench slope is blanketed by a hemipelagic slope sediment section that is over 1 km thick on the outer-arc ridge. Unlike the slope deposits off Sumatra, those off Java probably do not contain large amounts of terrigenous material from the outer-arc ridge, because the ridge is not subaerial along the Java section of the arc.

The forearc basin deposits off Java are 4–5 km thick and generally display similar sedimentation patterns to those in the Sumatra forearc basins. Compositionally, however, the Java deposits should differ markedly from those off central and north Sumatra. The sediments probably are dominantly volcanic-rich turbidites from Java, reflecting their arc provenance.

Discussion and conclusions

Sedimentation patterns in the Sunda forearc region vary considerably both spatially and temporally. Transport of Himalayan detritus along the trench axis dominates in the north. As the amount of sediment transported from the north decreases, hemipelagic sediments become more important components of the trench wedge. The isolated nature of the forearc basins and the changes in provenance along the arc lead to marked spatial changes in composition of sediment that is deposited in the basins. At the present time, most of the coarse sediment derived from Java and Sumatra is trapped in the forearc basins. Prior to the Neogene, the outer-arc ridge was a much smaller feature, and arc-derived detritus was able to bypass the forearc basin and flow into the trench. Because the time of formation of the outer-arc ridge varied along the arc, this change in sedimentation patterns also varied temporally along the arc.

A subsidiary effect of this change in tectonic regime was a corresponding change in sedimentation rates in the trench and forearc basins. Prior to uplift of the present outer-arc ridge, sedimentation rates were relatively high in the trench and relatively low in the forearc basin. After uplift of the ridge, sedimentation rates must have decreased in the trench and increased in the forearc basins. Until filling of the basins is complete, this pattern should continue.

Coarse turbidites that have been longitudinally transported are deposited in the trench off north Sumatra, while finer-grained sediments may be deposited in the adjacent forearc basin. A similar juxtaposition of facies in Mesozoic deposits of California (Franciscan Complex and Great Valley Series; Blake & Jones 1974) might be the result of deposition in a similar tectonic setting (Hamilton 1978). Off south Sumatra, quartzose turbidites derived from the Himalayas are deposited in the trench, whereas volcanic-rich turbidites derived from Sumatra are deposited in the forearc basin, producing a juxtaposition of facies of markedly different compositions. Off Java, hemipelagic sediments dominate both in the trench and on the lower trench slope, so offscraped sediments might be indistinguishable from the overlying slope sediments.

ACKNOWLEDGMENTS: Most of the S.I.O. data used in this paper were collected with funds from the Office of Naval Research as part of a large-scale study of tectonic and sedimentary processes in the northeastern Indian Ocean. Data acquisition in the central Sumatra and Java areas and preparation of this paper were supported by the IDOE/SEATAR Office of the National Science Foundation under Grants OCE76-24101, OCE79-18185 and OCE79-20482. We thank Dr Fred Hehuwat of the Indonesia National Institute of Geology and Mining for his efforts in coordinating field studies of the Sunda Arc. Reviews by M. A. Arthur, S. B. Bachman, K. J. McMillen, J. C. Moore, and T. H. Shipley improved the manuscript.

References

ANIKOUCHINE, W. A. & LING, H. 1967. Evidence for turbidite accumulation in trenches in the Indo-Pacific region. *Mar. Geol.* **5,** 141–54.

BLAKE, M. C., JR & JONES, D. L. 1974. Origin of the Franciscan melanges in northern California. *In*: DOTT, R. H., JR & SHAVER, R. H. (eds). *Modern and Ancient Geosynclinal Sedimentation.* Spec. Publ. Soc. econ. Paleontol. Mineral. Tulsa, **19,** 345–57.

BOWLES, F. A., RUDDIMAN, W. F. & JAHN, W. H. 1978. Acoustic stratigraphy, structure, and depositional history of the Nicobar Fan, eastern Indian Ocean. *Mar. Geol.* **26,** 269–88.

COOK, P. J., VEEVERS, J. J., HEIRTZLER, J. R. & CAMERON, P. J. 1978. The sediments of the Argo Abyssal Plain and adjacent areas, Atlantis II Cruise, Northeast Indian Ocean. *BMR J. Aust. Geol. Geophys.* **3,** 113–24.

CROOK, K. A. W. 1974. Lithogenesis and geotectonics: the significance of compositional variation in flysch arenites (graywackes). *In*: DOTT, R. H., JR & SHAVER, R. H. (eds). *Modern and Ancient Geosynclinal Sedimentation.* Spec. Publ. Soc. econ. Paleontol. Mineral. Tulsa, **19,** 304–10.

CURRAY, J. R., EMMEL, F. J., MOORE, D. G. & RAITT, R. W. 1981. Structure, tectonics, and geological history of the northeastern Indian Ocean. *In*: NAIRN, A. E. M. & STEHLI, F. G. (eds). *Ocean Basins and Margins: the Indian Ocean.* (in press).

—— & MOORE, D. G. 1971. Growth of the Bengal Deep-Sea Fan and denudation in the Himalayas. *Bull. geol. Soc. Am.* **82,** 563–72.

—— & —— 1974. Sedimentary and tectonic process in the Bengal Deep-Sea Fan and geosyncline. *In*: BURK, C. A. & DRAKE, C. L. (eds). *The Geology of Continental Margins*, 617–27. Springer-Verlag, New York.

—— , SHOR, G. G., JR, RAITT, R. W. & HENRY, M. 1977. Seismic refraction and reflection studies of crustal structure of the Eastern Sunda and Western Banda Arcs. *J. geophys. Res.* **82,** 2479–89.

DAMUTH, J. E. & HAYES, D. E. 1977. Echo character of the east Brazilian continental margin and its relationship to sedimentary processes. *Mar. Geol.* **24,** 73–95.

DICKINSON, W. R. 1970. Interpreting detrital modes

of graywacke and arkose. *J. sediment. Petrol.* **40,** 695–707.

DICKINSON, W. R. & SEELY, D. R. 1979. Structure and stratigraphy of forearc regions. *Bull. Am. Assoc. Petrol. Geol.* **63,** 2–31.

—— & SUCZEK, C. A. 1979. Plate tectonics and sandstone composition. *Bull. Am. Assoc. Petrol. Geol.* **63,** 2164–82.

GRAHAM, S. A., DICKINSON, W. R. & INGERSOLL, R. V. 1975. Himalayan-Bengal model for flysch dispersal in Appalachian-Ouachita system. *Bull. geol. Soc. Am.* **86,** 273–86.

HAMILTON, W. 1978. Mesozoic tectonics of the western United States. *In*: HOWELL, D. G. & MCDOUGALL, K. A. (eds). *Soc. econ. Paleontol. Mineral. Pacif. Coast Paleogeogr. Symp.* **2,** 33–70.

—— 1979. Tectonics of the Indonesian Region. *Prof. Pap. U.S. geol. Surv.* **1078,** 345 pp.

HEIRTZLER, J. R., CAMERON, P. J., COOK, P. J., POWELL, T., ROESER, H. A., SUKARDI, S. & VEEVERS, J. J. 1978. The Argo Abyssal Plain. *Earth planet. Sci. Lett.* **48,** 21–31.

INGERSOLL, R. V. & SUCZEK, C. A. 1979. Petrology and provenance of Neogene sand from Nicobar and Bengal Fans, DSDP Sites 211 and 218. *J. sediment. Petrol.* **49,** 1217–28.

KARIG, D. E., LAWRENCE, M. B., MOORE, G. F. & CURRAY, J. R. 1980a. Structural framework of the forearc basin, NW Sumatra. *J. geol. Soc. London*, **137,** 77–91.

—— , SUPARKA, S., MOORE, G. F. & HEHANUSSA, P. 1979. Structure and Cenozoic evolution of the Sunda Arc in the central Sumatra region. *Mem. Am. Assoc. Petrol. Geol.* **29,** 223–37.

—— , MOORE, G. F., CURRAY, J. R. & LAWRENCE, M. B. 1980b. Morphology and shallow structure of the trench slope off Nias Island, Sunda Arc. *In*: HAYES, D. E. (ed.). *The Tectonic and Geologic Evolution of Southeast Asian Seas and Islands*. Geophys. Monogr. Washington, **23,** 179–208.

KARUNAKARAN, C., RAY, K. K. & SAHA, S. S. 1968. Tertiary sedimentation in the Andaman-Nicobar Geosyncline. *J. geol. Soc. India*, **9,** 32–9.

KATILI, J. A. 1973. Geochronology of west Indonesia and its implication on plate tectonics. *Tectonophysics*, **19,** 195–212.

KIECKHEFER, R. M., SHOR, G. G., JR, CURRAY, J. R., SUGIARTA, W. & HEHUWAT, F. 1980. Seismic refraction studies of the Sunda trench and forearc basin. *J. geophys. Res.* **85,** 863–90.

KOLLA, V., BE, A. W. H. & BISCAYE, P. E. 1976. Calcium carbonate distribution in the surface sediments of the Indian Ocean. *J. geophys. Res.* **81,** 2605–16.

MCDONALD, J. M. 1977. *Sediments and structure of the Nicobar Fan, northeast Indian Ocean*. Thesis, Ph.D. Univ. Calif., San Diego, 148 pp. (unpubl.).

MALLIK, T. K. 1976. Shelf sediments of the Ganges delta with special emphasis on the mineralogy of the western part, Bay of Bengal, Indian Ocean. *Mar. Geol.* **22,** 1–32.

—— 1978. Mineralogy of deep-sea sands of the Indian Ocean. *Mar. Geol.* **27,** 161–76.

MOORE, D. G., CURRAY, J. R. & EMMEL, F. J. 1976. Large submarine slide (olistostrome) associated with Sunda Arc subduction zone, northeast Indian Ocean. *Mar. Geol.* **21,** 211–26.

—— , —— , RAITT, R. W. & EMMEL, F. J. 1974. Stratigraphic-seismic section correlations and implications to Bengal Fan history. *In*: VON DER BORCH, C. C., SCLATER, J. G. *et al.* (eds). *Initial Rep. Deep Sea drill. Proj.* **22,** 403–12.

MOORE, G. F. 1979. Petrography of subduction zone sandstones from Nias Island, Indonesia. *J. sediment. Petrol.* **49,** 71–84.

—— , BILLMAN, H. G., HEHANUSSA, P. E. & KARIG, D. E. 1980a. Sedimentology and paleobathymetry of Neogene trench-slope deposits, Nias Island, Indonesia. *J. Geol.* **88,** 161–80.

—— , CURRAY, J. R., MOORE, D. G. & KARIG, D. E. 1980b. Variations in geologic structure along the Sunda Forearc, northeastern Indian Ocean. *In*: HAYES, D. E. (ed.). *The Tectonic and Geologic Evolution of Southeast Asian Seas and Islands.* Geophys. Monogr. Washington, **23,** 145–60.

NINKOVICH, D. 1979. Distribution, age, and chemical composition of tephra layers in deep-sea sediments off Western Indonesia. *J. Volcanol. geotherm. Res.* **5,** 67–86.

PAGE, B. G. N., BENNETT, J. D., CAMERON, N. R., BRIDGE, D. MCC., JEFFERY, D. H., KEATS, W. & THAIB, J. 1979. A review of the main structural and magmatic features of northern Sumatra. *J. geol. Soc. London*, **136,** 569–79.

POTTER, P. E. 1978. Petrology and chemistry of modern big river sands. *J. Geol.* **86,** 423–49.

RAJU, A. T. R. 1967. Observations on the petrography of Tertiary clastic sediments of the Himalayan foothills of North India. *Bull. Oil nat. Gas Comm.* Delhi, **4,** 5–16.

SCHWAB, F. L. 1975. Framework mineralogy and chemical composition of continental margin-type sandstone. *Geology*, **3,** 487–90.

SCLATER, J. G. & FISHER, R. L. 1974. Evolution of the east-central Indian Ocean, with emphasis on the tectonic setting of the Ninetyeast Ridge. *Bull. geol. Soc. Am.* **85,** 683–702.

THOMPSON, R. W. 1974. Mineralogy of sands from the Bengal and Nicobar Fans, Sites 218 and 211, eastern Indian Ocean. *In*: VON DER BORCH, C. C., SCLATER, J. G. *et al.* (eds). *Initial Rep. Deep Sea drill. Proj.* **22,** 711–3.

UNDERWOOD, M. B., BACHMAN, S. B. & SCHWELLER, W. J. 1980. Sedimentary processes and facies associations within trench and trench-slope settings. *In*: FIELD, M. E., BOUMA, A. H. & COLBURN, I. (eds). *Quaternary Depositional Environments on the Pacific Continental Margin*, 211–29. Pacif. Sec. Soc. econ. Paleontol. Mineral. Tulsa.

UNITED NATIONS 1971. *Geologic map of Asia and the Far East. Scale 1:5,000,000*. 2nd ed.

VENKATARATHNAM, K. 1974. Mineralogical data from

sites 211, 212, 213, 214, and 215 of DSDP Leg 22 and origin of noncarbonate sediments in the equatorial Indian Ocean. *In*: von der Borch, C. C., Sclater, J. G. *et al* (eds). *Initial Rep. Deep Sea drill. Proj.* **22,** 489–501. U.S. Govt Printing Office, Washington.

—— & Biscaye, P. 1973. Clay mineralogy and sedimentation in the eastern Indian Ocean. *Deep Sea Res. Oceanogr. Abstr.* **20,** 727–38.

von der Borch, C. C., Sclater, J. G. *et al.* 1974. *Initial Rep. Deep Sea drill. Proj.* **22,** U.S. Govt Printing Office, Washington, 890 pp.

von Huene, R. 1974. Modern trench sediments. *In*: Burk, C. A. & Drake, C. L. (eds). *Geology of Continental Margins,* 207–11. Springer-Verlag, New York.

Gregory F. Moore, Joseph R. Curray, Frans J. Emmel, Geological Research Division, Scripps Institution of Oceanography, La Jolla, California 92093, U.S.A.

Development of the North Island Subduction System, New Zealand

Gerrit J. van der Lingen

SUMMARY: Subduction of the Pacific plate underneath the North Island of New Zealand began near the beginning of the Neogene, when the Tonga-Kermadec Subduction System propagated southwards into the New Zealand continental crustal block, together with the southwards migration of the relative pole of rotation between the Pacific and Australian plates. Many geological and geophysical aspects of the North Island, both on land and offshore, are in agreement with models of subduction systems from elsewhere. Several NE-trending zones can be distinguished: Taupo Volcanic Zone (volcanic arc), Axial Ranges (exposed part of arc massif), East Coast Depression (forearc basin), Coastal Ranges (structural high of subduction complex), continental shelf and slope (trough-inner-slope of subduction complex). A west-dipping Benioff zone underlies the volcanic arc and arc massif. A negative gravity anomaly coincides with the surface extension of the Benioff zone. However, no subduction trench exists below the negative gravity anomaly. *Sensu stricto,* Γhe Hikurangi Trough, to the east of the subduction complex, is not a subduction ɾrench.

The forearc basin and subduction complex are riding passively on top of the Pacific plate to the east of the Benioff zone. The forearc basin initially was shallow marine, but subsequently was filled with coarse debris shed from the rising arc massif, and is now above water. The subduction complex consists of highly deformed autochthonous Cretaceous-Palaeogene strata, possibly deposited in a pre-subduction continental borderland. Accreted oceanic sediments have not been recorded, but may be present in the lower part of the subduction complex, or, to a large degree, may have been subducted. During the formation of the subduction complex, small basins were created by elongate narrow structural ridges parallel to the trough-inner-slope. Several are now exposed in the Coastal Ranges structural high. Their sedimentary fills comprise sediment gravity flow deposits, hemipelagic muds, and arc-derived ash. Similar sedimentary facies have been cored in present-day offshore slope basins. Some sediment gravity flows reach the Hikurangi Trough via submarine canyons, bypassing the slope.

A directional, morphological and geophysical break exists between the North Island and Kermadec subduction systems, probably caused by the Vening Meinesz Fracture Zone.

Early-upper Miocene rhyolitic tuff beds in trough-inner-slope flysch basins are the first sign of volcanic arc activity. They cannot have come from the Quaternary Taupo Volcanic Zone, but were probably derived from the upper Miocene Coromandel Arc further to the NW. As subduction is oblique, strike-slip movement has moved the forearc region east of the Axial Ranges from a position closer to the Coromandel Arc to its present situation.

The present-day boundary between the Pacific and Australian plates passes through the New Zealand continental crustal block. To the north of New Zealand the plate boundary follows the structurally relatively simple Tonga-Kermadec Volcanic Arc system, and to the south the more complicated Macquarie Ridge system, which joins the present-day spreading ridge between the Australian-Pacific and Antarctic plates in a triple junction (Christoffel 1971; Christoffel & Falconer 1972; Hayes & Talwani 1972; Weissel & Hayes 1972). Nowadays it is generally accepted that the plate boundary in the New Zealand area consists of two opposing subduction systems, the North Island and Fiordland ones, connected by the Alpine Fault system (Fig. 1; Walcott 1978, 1979; Hatherton 1978; Oliver & Coggon 1979).

The New Zealand crustal block once formed part of Gondwana. It started to separate from the Australian-Antarctic block in the Campanian (Christoffel & Falconer 1972; McDougall & van der Lingen 1974; Weissel & Hayes 1977). Spreading between New Zealand and Australia ceased *c.* 60 Ma ago, but spreading between New Zealand and Antarctica is still continuing.

Plate tectonic reconstructions based on marine geophysical data go back to the Cretaceous (Fig. 1; Pittman *et al.* 1968; Molnar *et al.* 1975; Weissel *et al.* 1977; Barron & Harrison 1979). The further back in time, the greater the degree of uncertainty in the reconstructions.

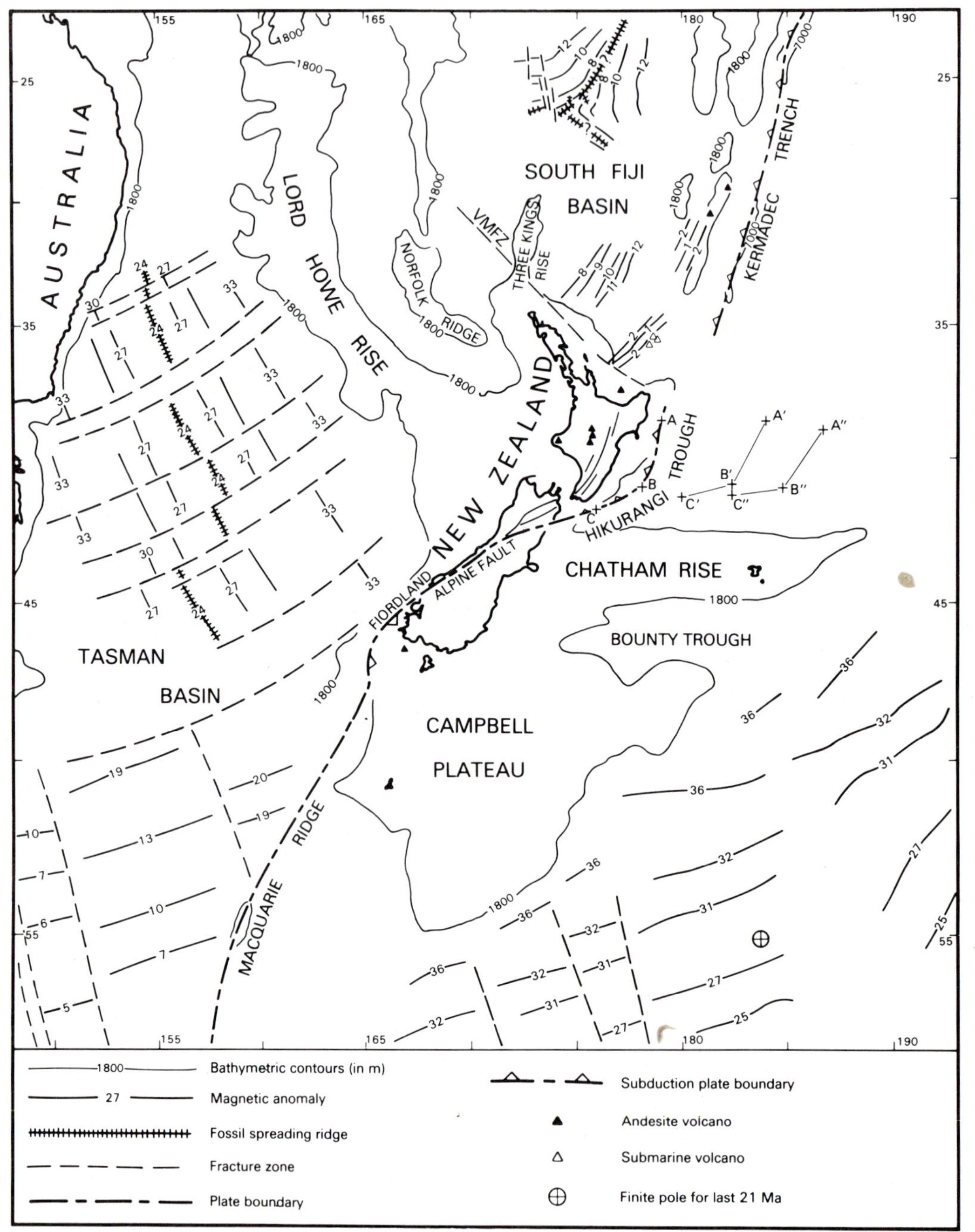

FIG. 1. Present-day plate tectonics setting of New Zealand in the Southwest Pacific. The line A-B-C connects three arbitrary points on the Pacific plate. A′-B′-C′ is the position of this line 10 Ma ago, and A″-B″-C″ 21 Ma ago. VMFZ—Vening Meinesz Fracture Zone. Seafloor magnetic anomaly data after Christoffel & Falconer (1972), Weissel & Hayes (1972, 1977), and Malahoff *et al.* (1981).

Trying to match the geology of New Zealand with the various 'geometric' reconstructions has provided problems, many of which still await solution. However, for the last 38 Ma (since magnetic anomaly 13), it has been possible to determine the positions of the relative poles of rotation between the Pacific and Australian plates with a reasonable degree of reliability (Walcott 1978, 1979). During this period the pole migrated southwards, from a position near the Bounty Trough to its present position at latitude −62.0° ± 1.8°, longitude 174.3°±1.8° (Chase 1978; see also Walcott 1979). The position of this migrating pole, close to the plate boundary through New Zealand, has resulted in a very complex structural history of the New Zealand crustal block. The southwards pole-shift was accompanied by a gradual southward propagation of subduction into the North Island of New Zealand. This subduction is oblique to the main geophysical, structural, and volcanic elements (Figs 1 & 2; Ballance & Reading 1980). The further southward the pole migrated, the larger the compressional (subduction) component and the smaller the strike-slip component of plate movement became.

The volcanic arc characteristics of the North Island were first described in detail by Dickinson & Hatherton (1967), Hatherton & Dickinson (1968), and Hatherton (1969, 1970a, b). Since then the overall picture has been changed in minor detail only.

Estimates of the time of initiation of subduction underneath the North Island vary. Ballance (1976) suggested 20 Ma, and Hayward (1979) 22 Ma, both based on the first occurrence of island-arc type volcanics. Sameshima (1975) gives 19 Ma, calculated from spreading rates and the maximum depth of the Benioff zone. Walcott (1978), using similar, but more up-to-date geophysical data, calculated at least 15 Ma.

A speculative reconstruction of the subduction history of the North Island was made by Ballance (1976), who distinguishes six successive, geographically separate volcanic arcs; the Waitakere Arc (20–15 Ma), the Northland Arc (18–15 Ma), the Coromandel Arc (15–6 Ma), the Whitianga Arc (6–3 Ma), the Tauranga Arc (3–0.75 Ma) and the present-day Taupo Arc (from 0.75 Ma, but others have suggested that the Taupo Arc had started somewhat earlier, e.g. Challis 1978; Cole 1979). The first five, now extinct, arcs are aligned in a NNW direction, whereas the Taupo Arc (or Taupo Volcanic Zone) trends NNE, close to the direction of the Tonga-Kermadec Arc. Ballance (1976) suggested that the east coast area of the North Island has been part of the forearc region of all six volcanic arcs, and that these arcs rotated anticlockwise away from their position above the North Island Benioff zone by a combination of NNE-trending dextral strike-slip movement and sinistral displacement along the Vening Meinesz Fracture Zone (Fig. 1; van der Linden 1967).

This paper presents a review and discussion of current ideas on the North Island Subduction System. A brief description of its various geological zones and geophysical elements will first be given, together with a discussion of their likely position and role in the subduction system, followed by a summary and discussion. The volcanic arc terminology of Seely (1979) is followed as closely as possible.

The North Island Subduction System

The forearc region of the North Island Subduction System is bounded to the west by a line of andesite volcanoes of the Taupo Volcanic Zone, and to the east by the Hikurangi Trough (Fig. 2). Its maximum width is 270 km. From west to east the forearc region can be subdivided into the following zones; Axial Ranges, East Coast Depression, Coastal Ranges, and the continental shelf and slope.

Taupo Volcanic Zone

Volcanic rocks of the Taupo Volcanic Zone are calc-alkaline, and consist of rhyolites, andesites, dacites and basalts (Cole 1979—Dr Cole's paper appeared after completion of the manuscript of my paper). Volumetrically, rhyolite extrusives form the dominant component (approximately 16 000 km^3; Challis 1978). Widespread ignimbrite sheets and thick ash deposits have blanketed the surrounding area and partly filled a tectonic graben in Permian to Jurassic sedimentary rocks of the Torlesse Supergroup (Figs 2 & 3). Rhyolitic volcanism commenced *c.*1 Ma ago. Andesite volcanism began in the Late Quaternary, about 50 000 BP (Cole 1979); although volumetrically far less important (about 800 km^3), it forms spectacular active volcanoes. A few dacite domes and some isolated basalt flows are volumetrically insignificant.

The Taupo Volcanic Zone is situated approximately 120–150 km above the Benioff zone (Fig. 4). Based on geochemical evidence, Ewart *et al.* (1977; see also Cole 1979) considered the andesitic lavas of the Tonga-Kermadec-North Island volcanic arc to be derived from Benioff

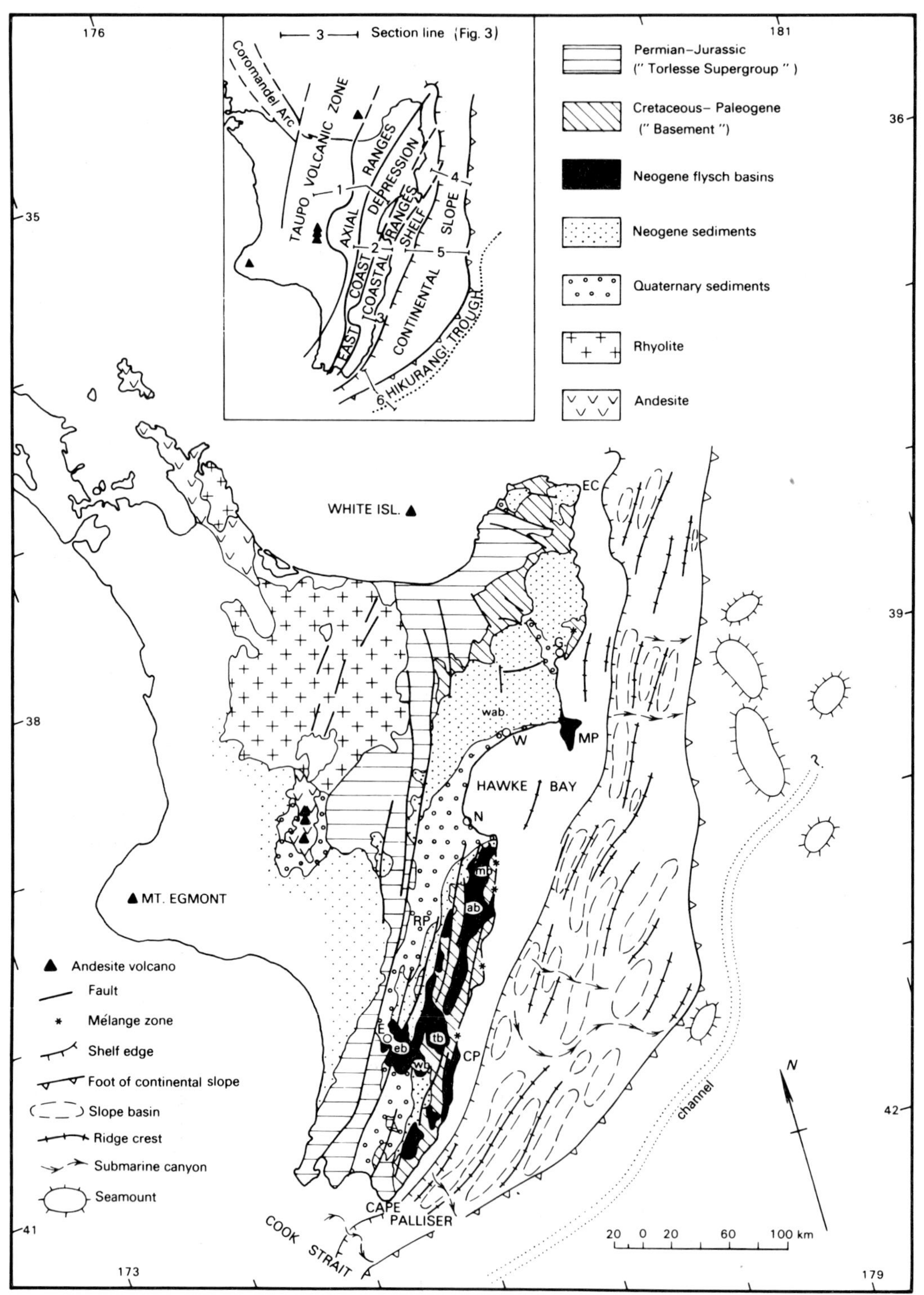
Section line (Fig. 3)
Coromandel Arc
TAUPO VOLCANIC ZONE
AXIAL RANGES
EAST COAST
COASTAL RANGES
DEPRESSION
SHELF
CONTINENTAL SLOPE
HIKURANGI TROUGH
Permian–Jurassic ("Torlesse Supergroup")
Cretaceous–Paleogene ("Basement")
Neogene flysch basins
Neogene sediments
Quaternary sediments
Rhyolite
Andesite
WHITE ISL.
HAWKE BAY
MT. EGMONT
CAPE PALLISER
COOK STRAIT
channel
Andesite volcano
Fault
Mélange zone
Shelf edge
Foot of continental slope
Slope basin
Ridge crest
Submarine canyon
Seamount
20 0 20 60 100 km

zone melting, with extensive modification by low-pressure crystal fractionation processes. They believe the rhyolites of the Taupo Volcanic Zone to be the product of fusion of the crustal continental rocks caused by the magmas rising from the Benioff zone.

Another Quaternary andesitic volcanic area, the Egmont Volcanic Chain, exists in the southwest of the North Island, well outside the Taupo Volcanic Zone (Fig. 2; Challis 1978). It consists of three volcanoes, of which Mt Egmont has been active in historic time. These andesite volcanoes have a higher K-content than those of the Taupo Volcanic Zone. This is in agreement with the theory that the K-content is directly related to the depth of the Benioff zone below the volcanoes (Hatherton & Dickinson 1968). The Benioff zone is situated between 150 and 250 km below Mt Egmont (Adams & Ware 1977).

Axial Ranges

The Axial Ranges consist of complexly deformed, well-indurated geosynclinal sediments of the Permian to Jurassic Torlesse Supergroup.

During much of the Tertiary, the Axial Ranges were submerged. Parts started to emerge at the beginning of the Pliocene. In late Pliocene time some parts of the ranges had acquired substantial relief (Grant-Taylor 1978).

Apart from forming the Axial Ranges and the floor of the Taupo Volcanic Zone graben, the Torlesse rocks also underlie at least part of the East Coast Depression to the east of the Ranges (Beu *et al.* 1980). The eastern boundary of the Torlesse rocks is not known.

In a volcanic arc context, the Torlesse rocks can be regarded as constituting the Arc Massif.

East Coast Depression

In between the Axial Ranges and the Coastal Ranges exists a structural depression (Figs 2 & 3). From Cook Strait to Wairoa, just north of Hawke Bay, this depression is filled with Pliocene-Quaternary sediments. This part of the depression is called the 'East Coast Inland Depression' by Beu *et al.* (1980). It started to subside in mid-Pliocene time, when the Axial and Coastal Ranges at either side began to emerge. The depression was never very deep, and shallow-marine sediments were deposited until the middle of the Pleistocene (Grant-Taylor 1978), after which it filled with abundant gravels from the rapidly rising ranges, and ceased to be marine. The basement of the depression is known only from a few drill holes sunk in the Ruataniwha Plains to the SW of Hawke Bay (Fig. 2). They proved the existence of Torlesse rocks below mid-Pliocene sediments (Grant-Taylor 1978).

In the Eketahuna area (Fig. 2), mid-Pleistocene to Quaternary sediments unconformably overlie Cretaceous to lower Pliocene sediments (including Pliocene flysch sediments), which, in turn overlie Torlesse rocks (Neef 1974). It may be that, structurally, the Eketahuna area originally belonged to the Coastal Ranges, later to be incorporated in the East Coast Inland Depression.

No Quaternary basin sediments occur north and NW of Wairoa but may have been removed by erosion, middle Miocene to Pliocene sediments form a broad syncline, called the 'Wairoa Basin' (Figs 2 & 3; Grindley 1960). To the south this syncline disappears underneath the Quarternary strata of the East Coast Inland Depression. Modern sedimentological data on the Wairoa Basin are sparse, but Miocene sediments seem to be mainly shelf sands and mudstones with minor algal limestones, grading into shallow-marine Pliocene sediments (Grindley 1960). North of a line due east through Gisborne, structures become very complicated. Tertiary strata have been involved in several phases of décollement tectonics, as a result of uplift of areas to the NE (Ridd 1964; Stoneley 1968).

In a volcanic arc context, the East Coast Inland Depression probably represents a forearc basin. The Wairoa Basin may represent an earlier phase of forearc basin development, which gradually propagated southwards, together with the southwards migrating subduction system. I have adopted the term 'East Coast Depression' for both the East Coast Inland Depression and the Wairoa Basin.

Coastal ranges

A series of small Neogene flysch basins, bounded by narrow zones of highly deformed

FIG. 2. Geological map of the North Island Subduction System. ab—Akitio Basin; CP—Castlepoint; E—Eketahuna; eb—Eketahuna Basin; EC—East Cape; G—Gisborne; mb—Makara Basin; MP—Mahia Peninsula; N—Napier; RP—Ruataniwha Plains; tb—Tawhero Basin; W—Wairoa; wab—Wairoa Basin; wb—Whareama Basin. Insert: zonations of the North Island Subduction System, and locations of section lines (Fig. 3). Offshore data after Lewis (1980), and Katz & Wood (1981).

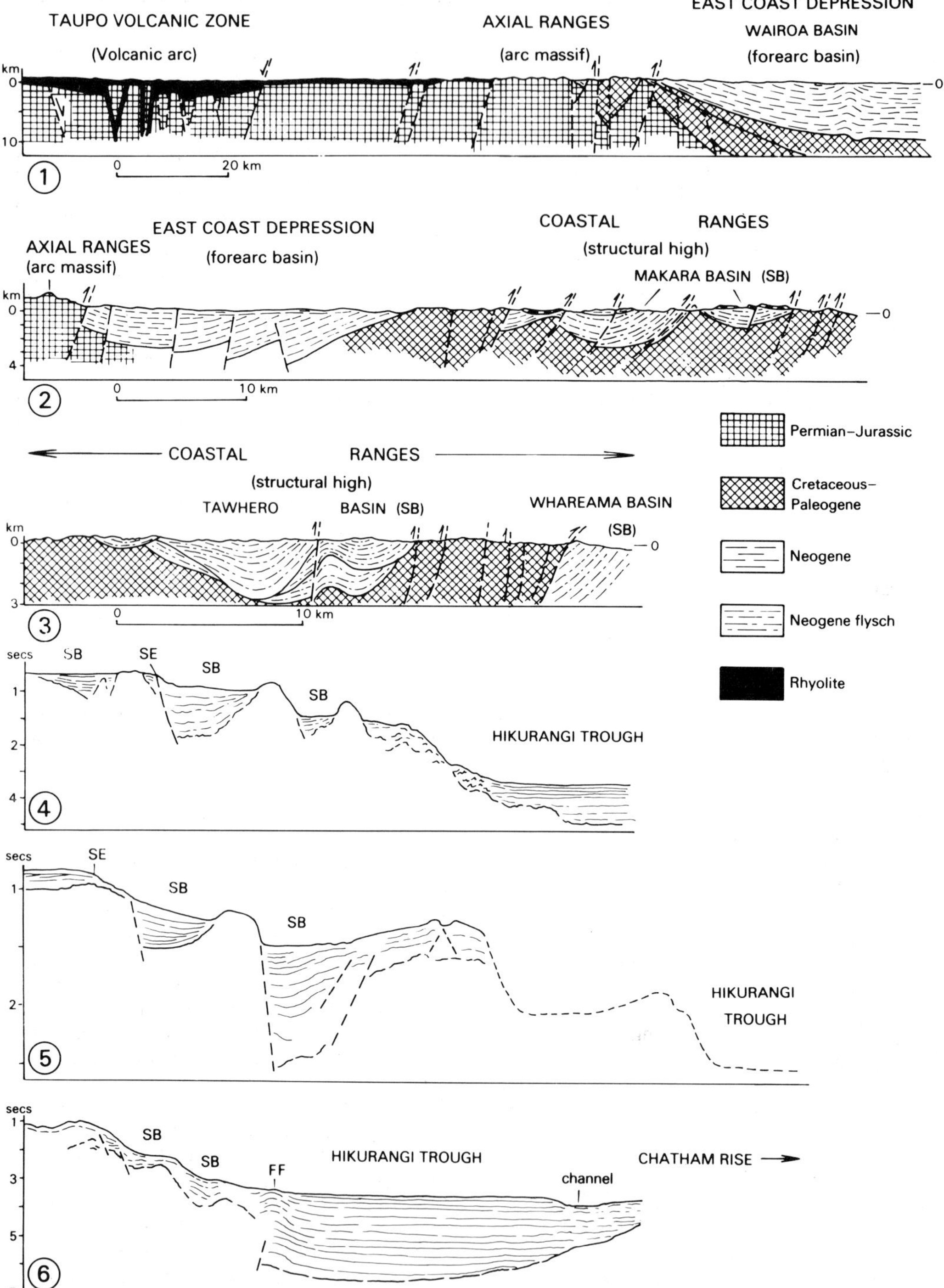

FIG. 3. Sections across the North Island Subduction System. For locations, see insert of Fig. 2. Section 1 after Grindley (1960): section 2 after van der Lingen & Pettinga (1980) and Grant-Taylor (1978); section 3 after Johnston (1975); sections 3–6 are tracings of offshore seismic profiles recorded by Mobil Oil Corp. Vertical scale is two-way travel time. SB—slope basin; SE—shelf edge; FF—frontal fold.

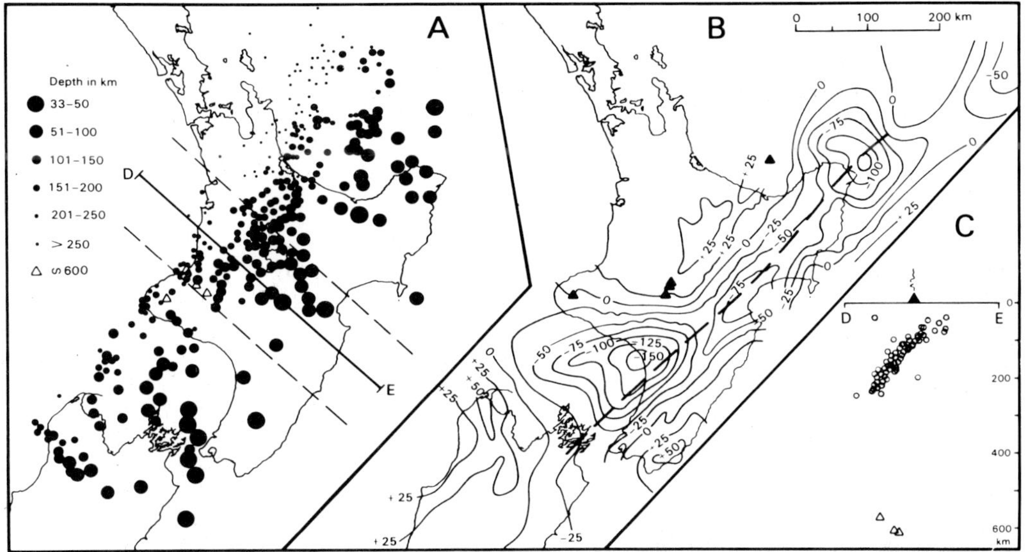

Fig. 4. Geophysical maps of the North Island Subduction System. (A) Deep earthquake foci. After Walcott (1979). (B) Isostatic gravity anomalies and projected surface extension of the Benioff zone (dashed line). After Hatherton & Syms (1975), and Reilly *et al.* (1977). C—cross-section D-E (Fig. 4A), showing the Benioff zone and the cluster of very deep earthquake foci. Projected earthquake foci are from area between dashed lines in (A). After Adams & Ware (1977).

Cretaceous-Palaeogene rocks, is exposed in a coastal belt between Cape Palliser and Hawke Bay (Fig. 2). In recent years, one of the flysch basins, the Makara Basin, has been studied in detail in the light of modern ideas about deep-sea sedimentation processes and plate tectonics (van der Lingen & Pettinga 1980). Some of the other basins have been described in the context of general mapping only: the Akitio Basin (Lillie 1953), the Whareama and Tawhero Basins (Johnston 1975), and the Eketahuna Basin (Neef 1974).

Based on tectonic setting and basin-fill lithologies, van der Lingen & Pettinga (1980) interpreted the upper Miocene Makara Basin as a basin which originated on the trough-inner-slope (for reasons to be explained later, the term 'trough' will be used instead of 'trench') of a subduction complex, in accordance with the model proposed by Moore & Karig (1976). The basin measures about 20 by 30 km, and is divided in half and bounded by narrow zones of highly deformed (in part mélange) Cretaceous-Palaeogene rocks (so-called 'thrust zones'; Fig. 3). Thrust faults associated with these thrust zones dip to the west. The basin sediments consist of flysch strata (mass-flow deposits: turbidites to grain flows, derived from upslope), hemipelagic mudstones (normal slope sedimentation) pebbly mudstones (debris flow deposits, derived from the thrust zones), and rhyolitic tuff beds (as derived from the volcanic arc) (Fig. 5). Microfauna indicate an outer-shelf to upper bathyal palaeodepth for the basin.

Descriptions from the literature and my reconnaissance fieldwork suggest that the other flysch basins have similar settings. Several more mélange zones have been found (Figs 2 & 5) and no doubt others exist elsewhere. A striking feature everywhere in the Coastal Ranges is the pronounced unconformity between the highly deformed Cretaceous-Palaeogene 'basement' and the Neogene sediments (Fig. 3). This unconformity is interpreted as marking the onset of subduction in this area.

The flysch of the Makara Basin does not contain typical submarine-fan facies (Mutti & Ricci-Lucchi 1972), such as thickening- and thinning-upward, or coarsening- and fining-upward sequences (Fig. 5; van der Lingen & Pettinga 1980). The same is true for flysch strata observed in other flysch basins (Fig. 5). This feature may be typical for trench-inner-slope basins in general, and could be a function of the small size of the basins, in which submarine fans cannot fully develop.

North of Hawke Bay, relationships between flysch basins and 'basement' are less clear,

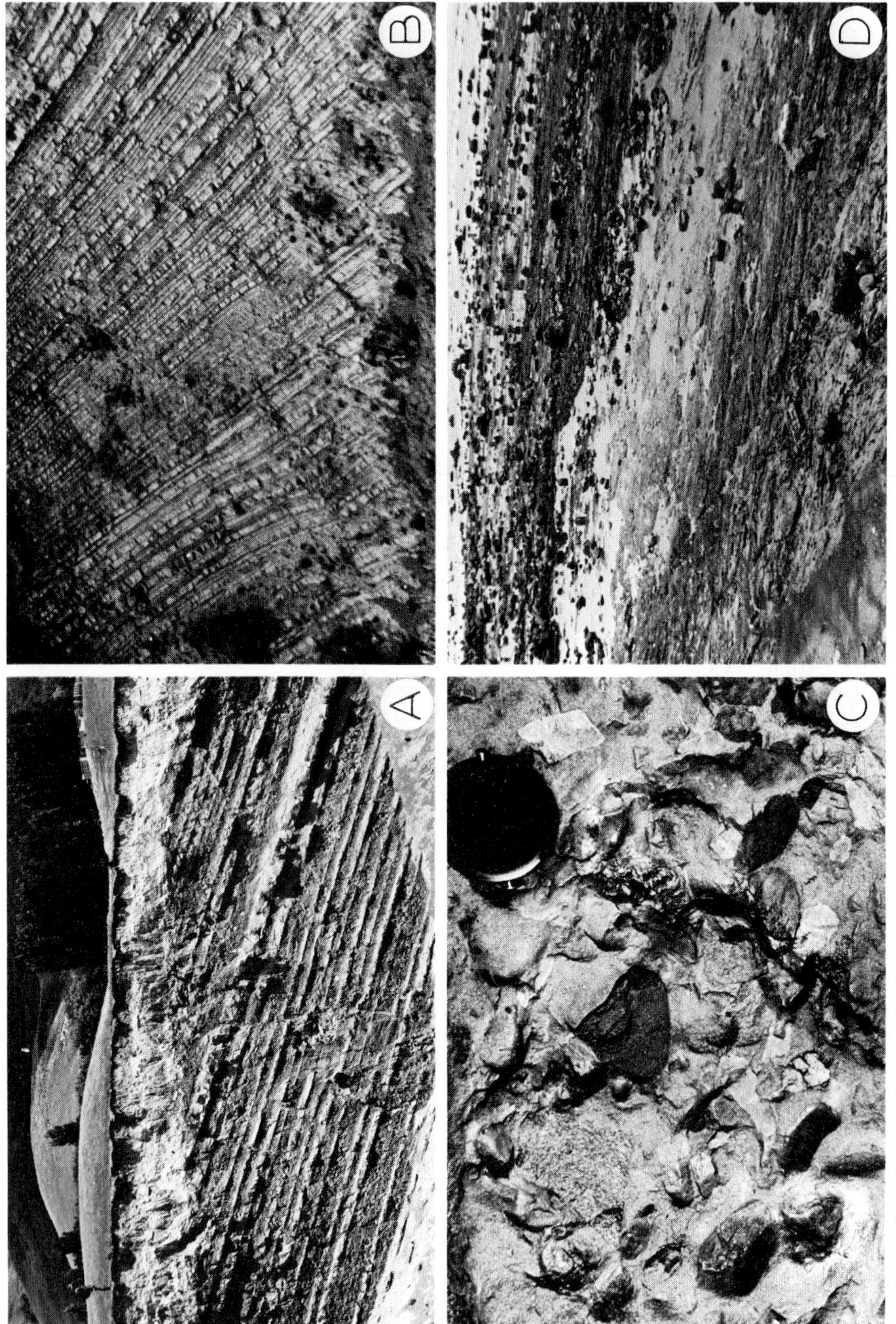

FIG. 5. Field photographs from the Coastal Ranges. (A) Flysch strata in the Makara Basin. (B) Flysch strata in the Whareama Basin. (C) Pebbly mudstone (debris flow) intercalated in flysch strata of the Makara Basin. Pebbles are of Cretaceous-Tertiary lithologies. Diameter of lens cap is 55 mm. (D) Mélange zone exposed on shore platform, north of Castlepoint. Loose blocks represent a variety of Upper Cretaceous and Palaeogene lithologies, left behind on the shore platform as erosional remnants.

partly due to the earlier mentioned décollement tectonics. But flysch sequences similar to those in the Makara Basin are exposed on Mahia Peninsula, and Cretaceous-Palaeogene mélange zones, surrounded by décollement units, have been described north of Gisborne (Ridd 1964). On Fig. 2 I have therefore tentatively extended the boundary line between the East Coast Depression and the Coastal Ranges north of Hawke Bay.

By analogy with the Makara Basin, the other Neogene flysch basins in the Coastal Ranges are interpreted as (fossil) trough-inner-slope basins as well.

Because of the intense deformation of the Cretaceous-Palaeogene basement rocks, it is difficult to reconstruct the palaeogeography of the east coast area at the onset of subduction. A reconstruction has only been attempted for the Lower Cretaceous, by Moore & Speden (1979). But lithological characteristics of Upper Cretaceous and Palaeogene sedimentary rocks can provide some clues as to palaeoenvironments. All Cretaceous-Palaeogene sediments were deposited in a marine environment (Kingma 1971). In Upper Cretaceous time the dominant lithologies were glauconitic sandstones, carbonaceous and siliceous mudstones, conglomerates, and flysch sediments. In Palaeocene-Eocene time sedimentation was generally slow, and the sediments fine-grained. Locally, restricted circulation conditions prevailed (e.g. the 'Waipawa Black Shales'). Typical lithologies are massive mudstones and bentonites. Depth of deposition was probably shelf to upper bathyal. In the Oligocene there is a gradual change to more rapid, coarser-grained sedimentation, accompanied by a greater variety of environments, probably shelf with localized deeper basins. A possible modern analogue for this environment could be the continental borderland that exists offshore from California.

The area of the Coastal Ranges was uplifted to shallow-marine depth in the early Pliocene, forming a 'structural high' (Seely 1979). Barnacle and coquina limestones were deposited unconformably on this high and along the borders of the East Coast Depression, which started to form in this same period (Beu *et al.* 1980). The Coastal Ranges finally emerged in the Pleistocene.

Continental shelf and slope

The structure of the offshore area is known from seismic profiles (Figs 2 & 3; Lewis 1971, 1980; Katz 1975; Katz & Wood 1981), and data on lithologies come from a limited number of piston and gravity cores and dredge samples (Lewis & Kohn 1973; Katz, 1975).

Morphologically, the slope consists of NE aligned narrow, elongate ridges and basins, cut perpendicularly by several submarine valleys and canyons (Fig. 2; van der Linden 1968; Lewis & Kohn 1973).

Seismic profiles show that the ridges are actively growing structures and that sediments are being ponded in small basins formed by the ridges (Fig. 3). Similar structures exist underneath the planed shelf. Attitudes of faults are difficult to determine on the profiles, which has led to different structural interpretations. Lewis (1980) interprets the highs as imbricate thrust zones controlled by west-dipping faults, in accordance with generally accepted ideas about the development of subduction complexes (Seely 1979). Katz & Wood (1981), in contrast, interpret the highs as being controlled by tectonics caused by diapiric movements of overpressured clays; the influence of overpressured clays has been suggested also for the Makran subduction complex (White 1979). Whatever the interpretation, the seismic profiles are very similar to those from trench-inner-slopes of other subduction complexes (e.g. the Sunda Arc: Karig *et al.* 1979, the Makran Arc: White 1979, and the Aleutian Arc: von Huene & Shor 1969 and Grow 1973).

In situ deformation of slope basement lithologies by diapiric movements seems well established (Katz & Wood 1981) but compressional deformation no doubt also plays an important role. For instance, as in other subduction complexes (e.g. the Makran Arc, White 1979, 1981), several profiles show a so-called 'frontal fold', representing the first compressional fold of trough (trench) sediments at the foot of the continental slope (Fig. 3).

Slope basin and trough sediments consist of hemipelagic muds, turbidites, and tephra layers. One dated ash layer (3400 yr) was encountered in several slope basins as well as in the Hikurangi Trough (Lewis & Kohn 1973). In basins high up on the slope, up to nine turbidites exist above the ash layer, while in the trough only two turbidites occur above the same ash layer. This suggests that most downslope-moving mass flows are being trapped in slope basins, but that some reach the trough (most probably via submarine canyons).

The sediment types filling the slope basins thus seem to be similar to those of the Neogene flysch basins in the Coastal Ranges, apart from pebbly mudstones, which were not encountered in the short piston cores.

Dredge samples have shown that the structural highs consist of older strata (Katz 1975; Lewis 1980).

The continental shelf and slope thus display many of the features typical of trench-inner-slopes of subduction systems elsewhere. Together with the Coastal Ranges they form a subduction complex (Seely 1979).

Hikurangi Trough

Morphologically, the Hikurangi Trough is not a trench (Katz 1974). It only represents the deepest area where the gently dipping abyssal plain of the Pacific plate meets the foot of the continental slope. In this configuration it has acted as a valley along which sediment gravity flows are being channelled. Most profiles across the trough show a shallow channel (Fig. 3). The average water depth of the trough is 3000 m.

Although the North Island Subduction System is generally regarded as a continuation of the Tonga-Kermadec system, there is no bathymetric continuity between the Kermadec Trench and the Hikurangi Trough (Katz 1974; Katz & Wood 1981). The area directly south of the termination of the Kermadec Trench is characterized by several seamounts, probably consisting of oceanic igneous material (Fig. 2; Katz & Wood 1981).

'Classical' subduction trenches (like the Tonga-Kermadec Trench) are characterized by two associated geophysical features: the projected continuation of the Benioff zone cuts the surface near the trench, and a negative gravity anomaly is situated above the trench (Hatherton 1969). However, in the North Island this geophysical association occurs about 200 km to the west of the Hikurangi Trough (Fig. 4). *Sensu stricto*, the Hikurangi Trough therefore cannot be considered a subduction trench.

Benioff zone

The Benioff zone underneath the North Island has been described in detail by Hatherton (1970a, b) and Adams & Ware (1977). A well-defined seismic zone, dipping at about 50° to the west, has a maximum depth of 350 km at its northern end and gradually shallows along strike to about 200 km at its southern end, where it terminates underneath the northern part of the South Island (Fig. 4). This shallowing and termination has been explained as being a function of the position of the pole of rotation in relation to the plate boundary (see Walcott 1978). Earthquake foci shallower than 50 km show a diffuse pattern extending from the top of the Benioff zone (at about 50 km depth) to the Hikurangi Trough (Adams & Ware 1977).

A small cluster of very deep earthquakes was recorded, at about 600 km depth (Fig. 4; Adams & Ware 1977). It has been suggested that these earthquakes originated in a detached lithospheric slab (Barazangi *et al.* 1973; Christoffel & Calhaem 1973).

The Benioff zones of the North Island and Kermadec systems are not continuous (Eiby 1977). In a transitional zone between 35° and 37° south latitude, the strike changes from N35.5E (North Island) to N18E (Kermadec). In that transitional zone is a gap in deep earthquake foci, where the Kermadec Benioff zone shallows from 500 to only 200 km. In this gap the earthquakes are also of smaller magnitude.

Gravity anomalies

Gravity anomaly maps of the North Island have been produced by Reilly *et al.* (1977). The isostatic anomalies show a negative belt extending from East Cape, via Hawke Bay to Cook Strait (Fig. 4). Its maximum approximately coincides with the surface intersection of the Benioff zone. Its southern boundary roughly coincides with the termination of the Benioff zone.

A positive gravity anomaly covers the Taupo Volcanic Zone. A large part of the Coastal Ranges is also covered by a positive anomaly, which extends eastwards over the Hikurangi Trough (Fig. 4).

There is no continuity with the negative gravity anomaly belt of the Kermadec system. Like the two Benioff zones, the two belts differ in strike (Hatherton 1970a). The North Island negative belt ends in a circular low with a minimum of −130 mgal, just north of East Cape, and is separated from the Kermadec system by a saddle of only −25 mgal (Fig. 4; Hatherton & Syms 1975). This saddle lies in the transitional zone between the two Benioff zones. There is also a dextral offset between the two anomaly belts.

The gravity anomalies thus have a spatial relationship with the Taupo Volcanic Zone and the Benioff zone which is typical for subduction systems.

Discussion

That subduction of the Pacific plate underneath the North Island takes place is now fairly well established. Calculations derived from strict geometric plate tectonic reconstructions suggest that subduction commenced near the beginning of the Neogene. Subduction along the Tonga-Kermadec Arc began somewhat earlier, about 25 Ma ago, and it is thought that subduction propagated southwards from there into the New Zealand continental crustal block, following a southward migration of the relative pole of rotation.

Many geological and geophysical aspects of the North Island, both on land and offshore, are in agreement with characteristics of subduction systems in other parts of the world. Several NE-trending geological zones can be interpreted as representing specific elements of a subduction system. *Volcanic arc*: the Taupo Volcanic Zone; *arc massif*: rocks of the Torlesse Supergroup, forming the Axial Ranges, and underlying the Taupo Volcanic Zone and part of the East Coast Depression; *forearc basin*: the East Coast Depression; *subduction complex*: the Coastal Ranges (*structural high*), and continental shelf and slope (*trough-inner slope*). A Benioff zone underlies the volcanic arc and arc massif, and a negative gravity anomaly exists near the projected surface extension of the Benioff zone.

In some aspects, however, the North Island Subduction System differs from subduction systems elsewhere. There is no subduction trench coinciding with the negative gravity anomaly and the projected surface extension of the Benioff zone. The forearc basin and subduction complex are situated to the east of a line at which the subducting plate starts its descent into the mantle. For a distance of over 200 km, the forearc basin and subduction complex are riding passively on top of the WSW-wards moving Pacific plate. Where the Pacific plate disappears underneath the subduction complex, the continental slope and abyssal plain form the Hikurangi Trough. In relation to the Benioff zone and the negative gravity anomaly, the Hikurangi Trough is not a subduction trench, but in relation to the subduction complex it could be considered as such. This poses the problem where exactly to draw the plate boundary.

Although the Coastal Ranges, shelf and continental slope are interpreted as a subduction complex, no evidence has been found on land that oceanic sediments, scraped off the Pacific plate, have been incorporated into the subduction complex. Lithological characteristics suggest that we are dealing with sediments originally deposited in a continental borderland setting. Continuity in structural style from the Coastal Ranges to the continental shelf and slope suggests that the same may be true for the rocks underlying the offshore slope basins and forming the ridges. It therefore seems likely that the subduction complex for the larger part is being formed by sediments from a pre-existing borderland that, passively riding on the Pacific plate, is being squeezed together. The borderland area was probably much wider originally than the present-day subduction complex. This process, of course, can also be regarded as a form of accretion. It is possible that sediments deposited in the Hikurangi Trough and on the Pacific Plate are being accreted to the lower part of the subduction complex, adjacent to the trough. Alternatively, substantial amounts of sediments could have been subducted, as has been suggested for other subduction systems (e.g. Moore *et al.* 1979). The necessary data are lacking to solve this problem.

The style of deformation of the subduction complex is, however, similar to that of 'classical' subduction complexes. The 'accreting' rocks are being strongly deformed, often into mélange zones, forming narrow, elongate thrust ridges on the trough-inner-slope, creating small flysch basins in which sediments moving downslope are being ponded. Some sediment gravity flows reach the Hikurangi Trough via submarine canyons, bypassing the slope.

Most authors have considered the North Island and Kermadec subduction systems as being one continuous system. Although they are closely related, there is a definite break between the two. The Hikurangi Trough does not form a continuation of the Kermadec Trench, and their relative positions within the two systems is quite different. In a transitional zone north of East Cape the strike direction changes, and breaks exist between the Benioff zones and the negative gravity anomaly belts. The gravity belts are also offset dextrally. As this transitional zone is situated along the direction of the Vening Meinesz Fracture Zone, it seems likely that the break between the two systems is an expression of this fracture zone, dividing the Pacific plate into two parts.

Authors have used two different sets of data to determine the onset of subduction below the North Island, the depth of the Benioff zone (Sameshima 1975; Walcott 1978), and the first occurrence of island arc-type volcanics (Ballance 1976; Hayward 1979). The latter is the less reliable, as the relationship of the volcanics with the present-day west-dipping Benioff zone is purely speculative. The Waitakere Arc, for instance, could well be related to the detached slab at 600 km depth. It is unlikely that this detached slab originally belonged to the west-dipping Benioff zone, as it is situated to the east of the lower end of the Benioff zone (Fig. 4C). It could well be a remnant of an originally east-dipping Benioff zone. This would require a 'flip' in the subduction direction, but similar flips have been suggested for segments further to the NW along the same plate boundary, e.g. the New Hebrides and Solomon Islands systems (Coleman 1975). This hypothesis requires further testing.

A timing problem is presented by the rhyolitic tuff beds in the Neogene flysch basins of the Coastal Ranges. It is assumed that these tuffs are derived from a volcanic arc. In which case the Taupo Volcanic Zone cannot be the source, as the first tuff beds recorded are of early-late Miocene age, whereas the rhyolitic eruptions of the Taupo Volcanic Zone did not start until the Pleistocene. Of the several Neogene volcanic arcs proposed by Ballance (1976), the Coromandel Arc was active when the tuff beds were deposited. In a speculative reconstruction, Ballance (1976) juxtaposes the Coastal Ranges opposite the Coromandel Arc in late Miocene time. Recent fission-track dating of some Coromandel rhyolites gave a late Miocene age (Rutherford 1978). But more work, especially geochemical, is required to test this hypothesis.

As subduction is oblique, a strike-slip component has to be taken into account in any spatial reconstruction. Over the last 10 Ma, about 190 km of dextral strike-slip parallel to the strike of the Benioff zone, has taken place. This brings the subduction complex closer to the Coromandel arc back in time. Any rotational movement, however, of the early Neogene Arcs or the forearc region, remains highly speculative. Opening of the South Fiji Basin cannot be invoked, as spreading in this basin had already terminated before the end of the Oligocene (Watts *et al.* 1977; Malahoff *et al.* 1981).

According to Walcott (1978), the strike-slip movement was mainly taken up by faults along the eastern side of the Axial Ranges. The forearc region to the east of the Axial Ranges thus moved southwards during the Neogene, while being compressed due to the gradually increasing compressional component of Pacific plate movement.

ACKNOWLEDGMENTS: The author benefited from discussions with Drs R. I. Walcott, F. J. Davey, I. Speden, and R. A. Wood. Messrs R. A. Cook and T. Montague assisted with the interpretation of the seismic reflection profiles. Mr B. D. Field helped with the reconnaissance fieldwork. The figures were drafted by Mr E. T. H. Annear. The photographs were developed and printed by Mr A. Downing and Miss Carol Hulse. Miss Christina Johnstone typed the various versions of the manuscript. Drs J. D. Bradshaw, S. D. Weaver, R. I. Walcott, M. G. Laird and H. R. Katz, and Mr T. Montague critically read the manuscript and made suggestions for its improvement.

References

ADAMS, R. D. & WARE, D. E. 1977. Subcrustal earthquakes beneath New Zealand; locations determined with a laterally inhomogeneous velocity model. *N.Z. J. Geol. Geophys.* **20,** 59–83.

BALLANCE, P. F. 1976. Evolution of the Upper Cenozoic Magmatic Arc and plate boundary in northern New Zealand. *Earth planet. Sci. Lett.* **28,** 356–70.

BALLANCE, P. F. & READING, H. G. (eds). 1980 *Sedimentation in Oblique-slip Mobile Zones.* Spec. Publ. Int. Assoc. Sediment. **4.**

BARAZANGI, M., ISACKS, B. L., OLIVER, J., DUBOIS, J. & PASCAL, G. 1973. Descent of lithosphere beneath New Hebrides, Tonga-Fiji, and New Zealand: evidence for detached slabs. *Nature. London,* **242,** 98–101.

BARRON, E. J. & HARRISON, C. G. A. 1979. Reconstructions of the Campbell Plateau and the Lord Howe Rise. *Earth planet. Sci. Lett.* **45,** 87–92.

BEU, A. G., GRANT-TAYLOR, T. L. & HORNIBROOK, N. DE B. 1980. The Te Aute limestone facies, Poverty Bay to northern Wairarapa. *Misc. Series Map. No. 13.* N.Z. geol. Surv., Dept Sci. Industr. Res., Wellington.

CHALLIS, G. A. 1978. Upper Quaternary volcanism. *In:* SUGGATE, R. P., STEVENS, G. R. & TE PUNGA, M. T. (eds). *The Geology of New Zealand,* **2,** 651–70.

CHASE, C. G. 1978. Plate kinematics: the Americas, East Africa and the rest of the world. *Earth planet. Sci. Lett.* **37,** 253–68.

CHRISTOFFEL, D. A. 1971. Motion of the New Zealand Alpine Fault deduced from the pattern of sea-floor spreading. *Bull. R. Soc. N.Z.* **9,** 25–30.

—— & CALHAEM, I. M. 1973. Upper mantle viscosity determined from Stokes' law. *Nature (phys. Sci.)* **243,** 51–3.

—— & FALCONER, R. K. H. 1972. Marine magnetic measurements in the south-west Pacific Ocean and the identification of new tectonic features. *In:* HAYES, D. E. (ed). *Antarctic Oceanology II, the Australian-New Zealand Sector.* Am. geophys. Union. Ant. Res. Ser. **19,** 197–209.

COLE, J. W. 1979. Structure, petrology, and genesis

of Cenozoic volcanism, Taupo Volcanic Zone, New Zealand—a review. *N.Z. J. Geol. Geophys.* **22,** 631–57.

Coleman, P. J. 1975. On island arcs. *Earth Sci. Rev.* **11,** 47–80.

Dickinson, W. R. & Hatherton, T. 1967. Andesite volcanism and seismicity around the Pacific. *Science,* **157,** 801–3.

Eiby, G. A. 1977. The junction of the main New Zealand and Kermadec seismic regions. *Int. Symp. Geodyn. SW Pacific, Noumea, New Caledonia, 1976,* 167–78. Technip, Paris.

Ewart, A., Brothers, R. N. & Mateen, A. 1977. An outline of the geology and geochemistry, and the possible petrogenetic evolution of the volcanic rocks of the Tonga-Kermadec-New Zealand island arc. *J. Volcan. Geotherm. Res.* **2,** 205–50.

Grant-Taylor, T. L. 1978. The geology and structure of the Ruataniwha Plains. *In:* Smale *et al.* (eds). *Geology and Erosion in the Ruahine Range.* Rep. N.Z. geol. Surv. **G20,** 31–61.

Grindley, G. W. 1960. Sheet 8 Taupo, 1st ed. *Geological Map of New Zealand 1:250 000.* Dept Sci. Industr. Res., Wellington.

Grow, J. A. 1973. Crustal and upper mantle structure of the central Aleutian arc. *Bull. geol. Soc. Am.* **84,** 2169–92.

Hatherton, T. 1969. The geophysical significance of calcalkaline andesite in New Zealand. *N.Z. J. Geol. Geophys.* **12,** 436–59.

—— 1970a. Upper mantle inhomogeneity beneath New Zealand: surface manifestations. *J. geophys. Res.* **75,** 269–84.

—— 1970b. Gravity, seismicity, and tectonics of the North Island, New Zealand. *N.Z. J. Geol. Geophys.* **13,** 126–44.

—— 1978. Astride two plates—an account of current tectonics in New Zealand. *Endeavour,* **2,** 180–5.

—— & Dickinson, W. R. 1968. Andesite volcanism and seismicity in New Zealand. *J. geophys. Res.* **73,** 4615–19.

—— & Syms, M. 1975. Junction of Kermadec and Hikurangi negative gravity anomalies (note). *N.Z. J. Geol. Geophys.* **18,** 753–6.

Hayes, D. E. & Talwani, M. 1972. Geophysical investigation of the Macquarie Ridge Complex. *In:* Hayes, D. E. (ed.). *Antarctic Oceanology II, the Australian-New Zealand Sector.* Am. Geophys. Union, Ant. Res. Ser. **19,** 211–34.

Hayward, B. W. 1979. Eruptive history of early to mid Miocene Waitakere Volcanic Arc, and palaeogeography of the Waitemata Basin, northern New Zealand. *J. R. Soc. N.Z.* **9,** 297–320.

Johnston, M. R. 1975. Sheet N159 and part sheet N158 Tinui-Awaitoitoi, 1st ed. *Geological Map of New Zealand 1:63 360.* Dept Sci. Industr. Res., Wellington.

Karig, D. E., Suparka, S., Moore, G. F. & Hehanussa, P. E. 1979. Structure and Cenozoic evolution of the Sunda Arc in the central Sumatra region. *In:* Watkins, J. S., Montadert, L. & Dickerson, P. W. (eds). *Geological and Geophysical Investigations of Continental Margins*, Mem. Am. Assoc. Petrol. Geol. **29,** 223–37.

Katz, H. R. 1974. Margins of the south-west Pacific. *In:* Burk, C. A. & Drake, C. L. (eds). *Geology of Continental Margins,* 549–65. Springer-Verlag, Berlin.

—— 1975. Ariel Bank off Gisborne. An offshore late Cenozoic structure, and the problem of acoustic basement on the East Coast, North Island, New Zealand. *N.Z. J. Geol. Geophys.* **18,** 93–107.

—— & Wood, R. A. 1981. Submerged margin east of the North Island, New Zealand, and its petroleum potential. *CCOP/SOPAC, Techn. Bull. 3.*

Kingma, J. T. 1971. Geology of Te Aute subdivision. *Bull. N.Z. geol. Surv.* **70,** 173 pp.

Lewis, K. B. 1971. Growth rate of folds using tilted wave-planed surfaces: coast and continental shelf, Hawkes' Bay, New Zealand. *In: Recent Crustal Movements.* Bull. R. Soc. N.Z. **9,** 225–31.

—— 1980. Quaternary sedimentation on the Hikurangi oblique-subduction and transform margin, New Zealand. *In:* Ballance, P. F. & Reading, H. G. (eds). *Sedimentation in Oblique-slip Mobile Zones.* Spec. Publ. Int. Assoc. Sediment. **4,** 171–89.

—— & Kohn, B. P. 1973. Ashes, turbidites, and rates of sedimentation on the continental slope off Hawke's Bay. *N.Z. J. Geol. Geophys.* **16,** 439–54.

Lillie, A. R. 1953. Geology of the Dannevirke subdivision. *Bull. N.Z. geol. Surv.* **46,** 156 pp.

McDougall, I. & van der Lingen, G. J. 1974. Age of the rhyolites of the Lord Howe Rise and the evolution of the south-west Pacific Ocean. *Earth planet. Sci. Lett.* **21,** 117–26.

Malahoff, A., Feden, R. H. & Fleming, H. S. 1981. Magnetic anomalies and tectonic fabric of marginal basins north of New Zealand. *J. geophys. Res.*, in press.

Molnar, P., Atwater, T., Mammerickx, J. & Smith, S. M. 1975. Magnetic anomalies, bathymetry, and the tectonic evolution of the South Pacific since the late Cretaceous. *Geophys. J. R. astron. Soc.* **40,** 383–420.

Moore, G. F. & Karig, D. E. 1976. Development of sedimentary basins on the lower trench slope. *Geology,* **4,** 693–7.

Moore, J. C. *et al.* 1979. Off Mexico. Middle America Trench. Rept. Leg 66 IPOD. *Geotimes,* Sept. 1979. 20–2.

Moore, P. R. & Speden, I. 1979. Stratigraphy, structure and inferred environments of deposition of the Early Cretaceous sequence, eastern Wairarapa, New Zealand. *N.Z. J. Geol. Geophys.* **22,** 417–33.

Mutti, E. & Ricci-Lucchi, F. 1972. Le torbiditi dell'Apennino settentrionale: introduzione all'analisi de facies. *Mem. Soc. geol. Ital.* **11,** 161–99.

Neef, G. 1974. Sheet N153 Eketahuna, 1st edn. *Geological Map of New Zealand 1:63 360.* Dept Sci. Industr. Res., Wellington.

Oliver, G. J. H. & Coggon, J. H. 1979. Crustal

structure of Fiordland, New Zealand. *Tectonophysics,* **54,** 253–92.

PITTMAN, III, W. C., HERRON, E. M. & HEIRTZLER, J. R. 1968. Magnetic anomalies in the Pacific and sea-floor spreading. *J. geophys. Res.* **73,** 2069–85.

REILLY, W. I., WHITEFORD, C. M. & DOONE, A. 1977. North Island, 1st ed. *Gravity Map of New Zealand, 1:1 000 000, Isostatic Anomalies.* Dept Sci. Industr. Res., Wellington.

RIDD, M. F. 1964. Succession and structural interpretation of the Whangara-Waimata area, Gisborne, N.Z. *N.Z. J. Geol. Geophys.* **7,** 279–98.

RUTHERFORD, N. F. 1978. Fission-track age and trace element geochemistry of some Minden Rhyolite obsidians. *N.Z. J. Geol. Geophys.* **21,** 443–8.

SAMESHIMA, T. 1975. Silica indices of volcanoes in and around New Zealand with reference to volcanic zones in the North Island. *N.Z. J. Geol. Geophys.* **18,** 523–39.

SEELY, D. R. 1979. The evolution of structural highs bordering major forearc basins. *In:* WATKINS, J. S., MONTADERT, L. & DICKERSON, P. W. (eds). *Geological and Geophysical Investigations of Continental Margins.* Mem. Am. Assoc. Petrol. Geol. **29,** 245–60.

STONELEY, R. 1968. A lower Tertiary décollement on the East Coast, North Island, N.Z. *N.Z. J. Geol. Geophys.* **11,** 128–56.

VAN DER LINDEN, W. J. M. 1967. Structural relationships in the Tasman Sea and south-west Pacific Ocean. *N.Z. J. Geol. Geophys.* **10,** 1280–301.

—— 1968. Cook bathymetry, N.Z. Oceanogr. Map Chart. *Oceanic Series 1:1 000 000.* Dept Sci. Industr. Res., Wellington.

VAN DER LINGEN, G. J. & PETTINGA, J. R. 1980. The Makara Basin: a Miocene slope-basin along the New Zealand sector of the Australian-Pacific obliquely convergent plate boundary. *In:* BALLANCE, P. F. & READING, H. G. (eds). *Sedimentation in Oblique-slip Mobile Zones.* Spec. Publ. Int. Assoc. Sediment. **4,** 191–215.

VON HUENE, R. & SHOR, JR. G. G. 1969. The structure and tectonic history of the eastern Aleutian trench. *Bull. geol. Soc. Am.* **80,** 1889–902.

WALCOTT, R. I. 1978. Present tectonics and Late Cenozoic evolution of New Zealand. *Geophys. J. R. astron. Soc.* **52,** 137–64.

—— 1979. Plate motion and shear strain rates in the vicinity of the Southern Alps. *In:* WALCOTT, R. I. & CRESSWELL, M. M. (eds). *The Origin of the Southern Alps.* Bull. R. Soc. N.Z. **18,** 5–12.

WATTS, A. B., WEISSEL, J. K. & DAVEY, F. J. 1977. Tectonic evolution of the South Fiji marginal basin. *In:* TALWANI, M. & PITMAN III, W. C. (eds). *Island Arcs, Deep Sea Trenches and Back-arc Basins.* Am. Geophys. Union, M. Ewing Ser. **1,** 419–27.

WEISSEL, J. K. & HAYES, D. E. 1972. Magnetic anomalies in the southeast Indian Ocean. *In:* HAYES, D. E. (ed). *Antarctic Oceanology II: the Australian-New Zealand Sector.* Am. geophys. Union, Ant. Res. Ser. **19,** 165–96.

—— 1977. Evolution of the Tasman Sea reappraised. *Earth planet. Sci. Lett.* **36,** 77–84.

——, HAYES, D. E. & HERRON, E. M. 1977. Plate tectonics synthesis: the displacements between Australia, New Zealand and Antarctica since the late Cretaceous. *Mar. Geol.* **25,** 231–77.

WHITE, R. S. 1979. Deformation of the Makran continental margin. *In:* FARAH, A. & DE JONG, K. A. (eds). *Geodynamics of Pakistan,* 295–304. Geol. Surv. Pakistan, Quetta.

WHITE, R. S. 1981. Deformation of the Makran accretionary sediment prism in the Gulf of Oman (north-west Indian Ocean) (this volume).

GERRIT J. VAN DER LINGEN, Sedimentology Laboratory, New Zealand Geological Survey, University of Canterbury, Christchurch, New Zealand.

ATLANTIC

The Barbados Ridge Complex: tectonics of a mature forearc system

G. K. Westbrook

SUMMARY: The Barbados Ridge Complex is the wide accretionary sediment pile associated with the Lesser Antilles island arc. Its width is so great (>200 km) that a trench no longer exists on the oceanward side of it. This development is a product of the age of the system (>50 Ma) and the thickness of sediment on the ocean floor (0.8 km in the north, >4 km in the south). The northward decrease in elevation of the sediment pile and the variation in the style of initial deformation at the leading edge of the pile are related to the northward change in sediment thickness and type. The region in which deformation is prevalent has a westward limit just west of the axis of the minimum negative Bouguer gravity anomaly. In the south the axis of this minimum is coincident with the Barbados Ridge (an outer 'sedimentary arc'). The deformed rocks of the accretionary pile are overlain by later sediments, which show varying degrees of deformation, and often occupy small basins.

In the northern part of the complex, the relief and structure of the accretionary pile are complicated by ridges and troughs running east-west across the general trend of structures. These appear to be related to variations in the relief of the subducted oceanic basement.

The Barbados Ridge Complex is a region with complicated bathymetry lying along the eastern margin of the Lesser Antilles Island arc in the eastern Caribbean (Fig. 1). It takes its name from the northerly trending Barbados Ridge which lies 150 km east of the arc. The island of Barbados is its most elevated part. The complex generally deepens towards the east, and most bathymetric features have a northerly trend, except between latitudes 13°30′N and 15°30′N where easterly trending ridges and troughs occur. The complex becomes broader and more elevated towards the south (Fig. 2).

The considerable quantity of geophysical information obtained in the region shows that the complex is composed of sedimentary rocks above a linear depression in the igneous basement (Fig. 3). The axis of this depression, which reaches a depth of 20 km beneath Barbados, lies up to 50 km west of the eastern limit of seismicity associated with the subduction zone beneath the Lesser Antilles, and is where the crystalline crust of the Atlantic Plate passes beneath the crystalline crust of the Caribbean Plate. The axis is strongly out of isostatic equilibrium, being depressed beneath its equilibrium position. The eastern part of the complex is above its equilibrium position and corresponds to the outer trench rise of most island arcs which do not have such a broad sediment complex. The existence of metamorphosed sediments beneath the Barbados Ridge (7–10 km beneath Barbados) is indicated by seismic velocities of around 5.0 km sec^{-1} and magnetic anomalies (Westbrook 1975). Another possible contributory cause of the magnetic anomalies in the vicinity of Barbados may be diagenetic magnetite formed in sediments from reduction of hydrated ferric oxides by petroleum microseepage (Donovan *et al.* 1980); hydrocarbons are found in Barbados.

The overall thickness of the complex increases towards the south, and broadly reflects a similar southward thickening of undeformed sediment on the Atlantic ocean floor. Along the eastern margin of the complex, the sediments of the Atlantic ocean floor are progressively deformed as they pass into the complex, eventually losing any structure that can be recognized from those seismic reflection sections that are currently available. The deformed sediment is overlain by more recent sediment which thickens towards the west, and is absent in the easternmost part of the complex. North of 14°N this overlying layer shows gentle deformation. South of 14°N, the Tobago Trough, a well-developed basin containing a thickness of at least 3 km of undeformed sediment, lies between the Barbados Ridge and the volcanic island arc. Similar, but smaller, basins lie to the east of the Barbados Ridge. On the island of Barbados, strongly folded and faulted Eocene flysch (the Scotland Group) is overlain by pelagic sediments of upper Eocene to Miocene age (Saunders 1979). Pleistocene to recent coral rock covers most of the island. The trend of deformational structures on Barbados is generally NE.

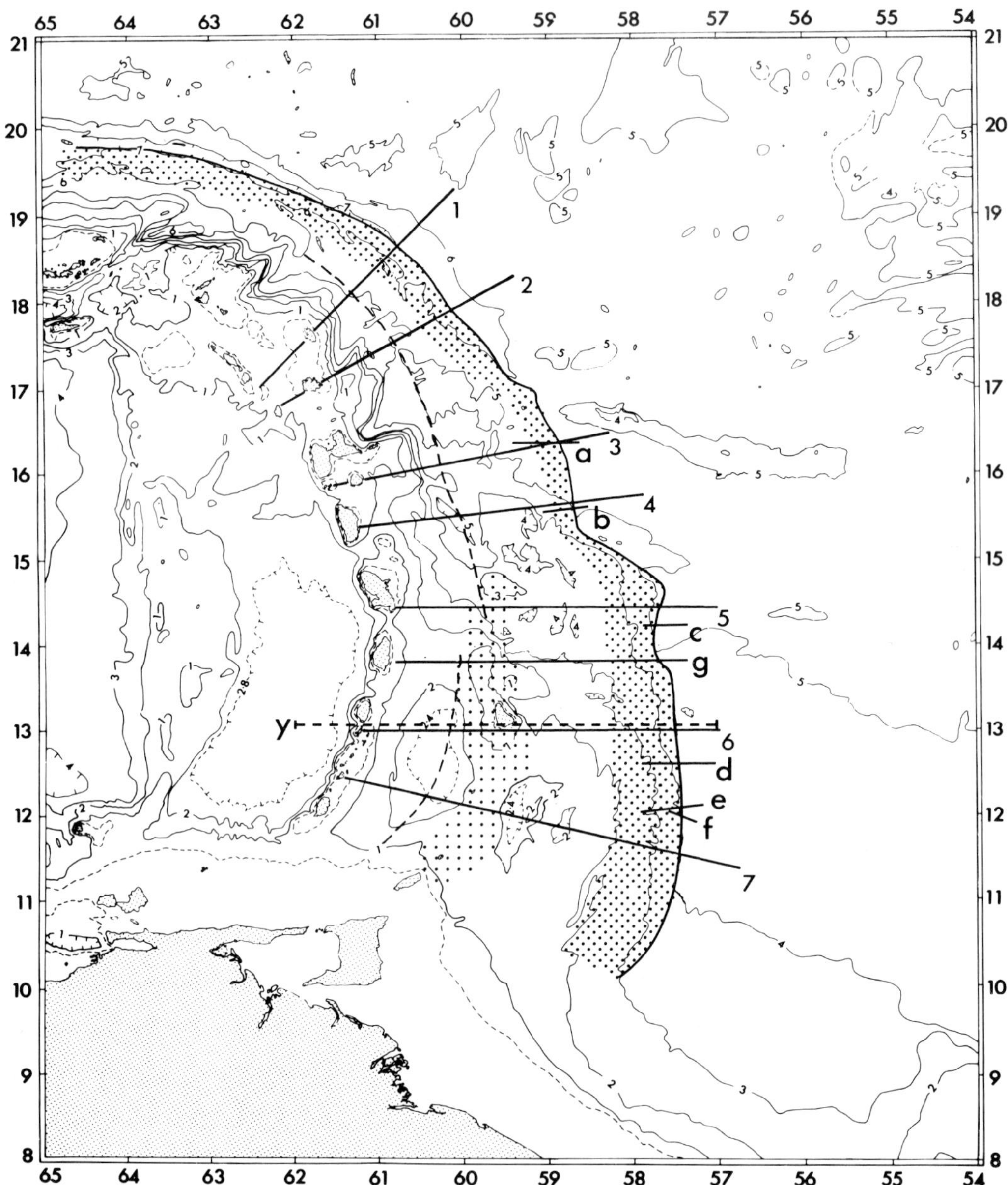

FIG. 1. Eastern Caribbean showing: bathymetry from U.S.G.S. Geologic-Tectonic Map of the Caribbean Open File 75–146, & Kearey *et al.* (1975) (isobaths at 1 km intervals); lines of profiles 1–7 of Fig. 2; the eastern limit of the Barbados Ridge Complex (solid black line); the region of the initial slope of the complex (close stipple); the western limit of deformation (dashed line); the Barbados Ridge (open stipple); the line of section for Fig. 3 (y); the lines of the seismic reflection sections of Fig. 5 (g) and Fig. 6 (a,b,c,d,e,f).

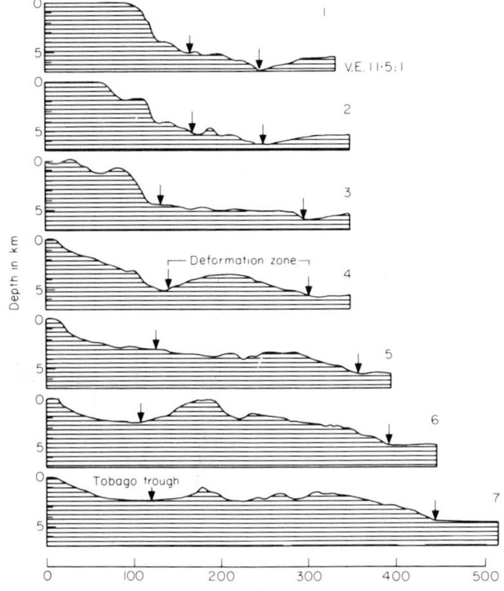

FIG. 2. Bathymetric profiles across the Barbados Ridge Complex showing the variation in its width and shape. The left edge of each profile ends at the active volcanic island arc. The arrows mark the limits of the region in which deformation can be seen on seismic reflection sections. The positions of the profiles are shown on Fig. 1.

Eastern margin

Between 17°30′N and 12°N the distance of the eastern edge of the complex from the Lesser Antilles increases from 200 to 450 km. Similarly the distance between the axis of the Bouguer gravity anomaly minimum (lying over the line of subduction of the Atlantic crystalline crust) beneath the Caribbean crystalline crust) and the eastern edge increases from 95 to 265 km. The increase in width of the complex is gradual, except at 15°N where the edge runs WNW for about 90 km because of the influence of a ridge in the oceanic basement. South of this change in strike of the edge of the complex, the margin is typically in the form of a slope rising from the Atlantic ocean floor by about 1.5 km in 50 km then flattening off or becoming less steep. North of the change in strike, the slope is more gentle, 0.5 km in 50 km, until the Puerto Rico Trench is reached, where it becomes steeper again (Fig. 2).

The deformation of the sedimentary horizons of the Atlantic ocean floor at the eastern margin of the Barbados Ridge Complex was first reported by Chase & Bunce (1969), who drew an analogy with the sandbox experiment of Hubbert (1951) to explain the mechanism of thrusting and folding. Westbrook *et al.* (1973) subsequently suggested that the deformation was brought about by successive thrusts emanating from a master décollement close to the sediment/basement interface accompanied by shearing parallel to bedding planes, and producing a reverse stratigraphy of thrust slices in the complex which is characteristic of models published for other forearc regions (Seely *et al.* 1974; Karig & Sharman 1975). This was incorporated later into an evolutionary model for the whole of the complex by Westbrook (1975).

The style of deformation varies considerably along the eastern margin (Peter & Westbrook 1976; Mascle *et al.* 1977; Biju-Duval *et al.* 1978). In the south it takes the form of gently asymmetric east-facing folds (amplitude 0.5 km; wavelength 8 or 9 km) riding on thrusts dipping westward at 20° which probably have a listric form at depth, presumably becoming parallel to the basement as they run into a major décollement (Fig. 6d,e,f). North of 15°30′N the deformation of sediment is so intense and/or chaotic that no structure can be recognized in it from seismic reflection sections. This deformed material, however, overlies undeformed bedded sediment on a plane of discontinuity (probably a décollement) which lies at shallow depth within the undeformed sedimentary section (Fig. 6a,b). A much smaller part of the ocean sediments is involved in the initial deformation (a layer 0.2 km thick as opposed to 1.5–2 km in the south). The plane of discontinuity extends at least 30 km beneath the deformed sediment, and how it does so without being deformed does pose some problems (Peter & Westbrook 1976; Mascle *et al.* 1977). If the deformed material was essentially a submarine slide formed by the gravitational collapse of the frontal slope of the complex it would explain the lack of disruption beneath the discontinuity and the gentler gradient of the frontal slope in the region where this discontinuity exists. However, reflection profiles run by Lamont-Doherty (Ewing *et al.* 1974) and NOAA (Peter & Westbrook 1976 and unpublished data) show that this discontinuity exists over such a wide area (between 15°30′N and 16°40′N) that a submarine slide, although possible, is an unlikely explanation. Given that the plane of discontinuity is likely to be a décollement, then the presence of a particularly weak horizon above the décollement, probably with a more competent (more lithified) sequence of sediments lying beneath it, is re-

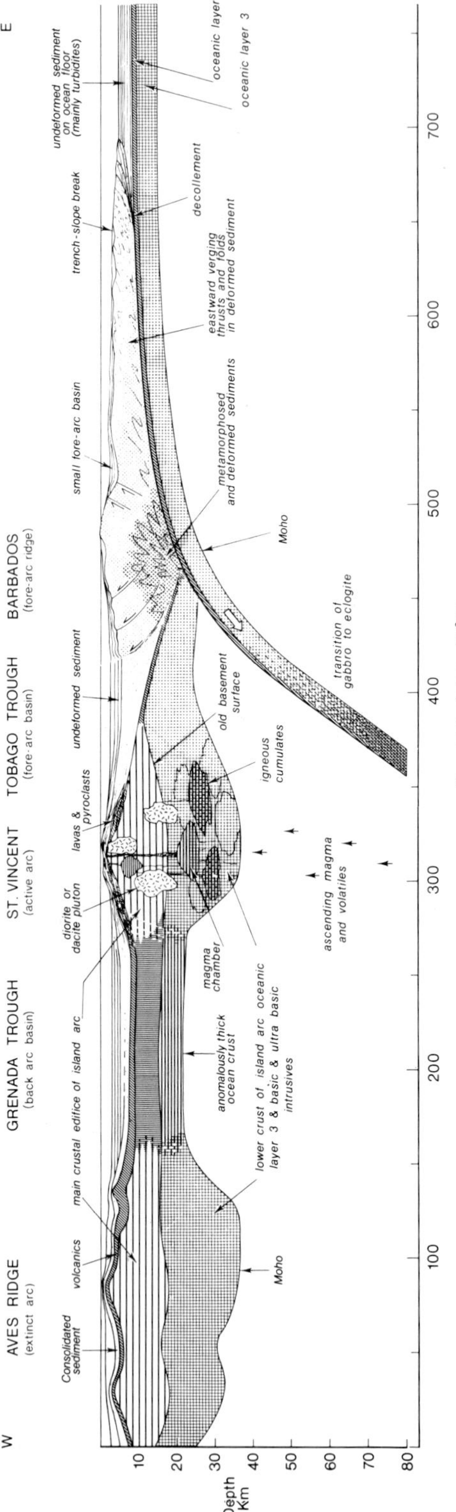

FIG. 3. Interpretative cross-section of the Lesser Antilles and the Barbados Ridge Complex; from Westbrook (1975) and Boynton *et al.* (1979). Position of section is shown in Fig. 1 (line y).

quired to enable it to persist so far without disruption.

There is a general correlation between the thickness of sediment on the Atlantic Ocean floor and style of deformation. The thinner the sediment the more intense is the deformation, because the stresses are distributed in a smaller volume of material. This feature of the Barbados Ridge Complex has been used by Moore (1979) in developing a model for deformation in forearc sediments, but thickness is not the only factor. Lithological variation also plays a part, as discussed above for the region north of 15°30′N where the development of a décollement high in the sedimentary column results in only one-quarter of the sediment being involved in the deformation at the front of the sediment pile. The sediments in the north are further from continental sources than those in the south (Bunce *et al.* 1971; Peter & Westbrook 1976) and likely to have a higher clay content. Consequently they may deform in a far less competent fashion, especially if pore water is trapped in them, causing them to become overpresssured.

Western margin

The sediments of the forearc region wedge out against the Lesser Antilles island arc. Up to 120 km from the active arc the sediments are completely undeformed except for some occasional slumping. South of 14°N this zone of undeformed sediments exists in the Tobago Trough which can be followed round to the south-west passing beneath the Venezuelan continental margin as far as the island of Margarita (Feo-Codecido 1977). The thickness of undeformed sediment exceeds 6 km to the east of Grenada. North of 14° the undeformed sediment does not occupy a bathymetric basin except opposite Dominica where the so-called Lesser Antilles Trench lies at the foot of the slope down from the arc. Much of the sediment comes from the island arc, but a proportion of clay and silt size material comes from the Amazon and Orinoco (Keller *et al.* 1972). The sediments become deformed as they pass eastward into the Barbados Ridge Complex. The limit of deformation lies just to the west of the axis of the negative Bouguer gravity anomaly (Fig. 4). In the east of the Tobago Trough the deformation has been accompanied by considerable uplift to form the Barbados Ridge. The deformation is in the form of gentle westward facing folds, although on the western flank of the Barbados Ridge there is the implication that they are the near surface expression of westward directed thrusts (Westbrook 1975; Mascle *et al.* 1981). Almost without exception, the youngest sedimentary horizons are affected by this deformation.

Southern margin

In the south the Barbados Ridge Complex abuts the continental margin of South America. Although much of it has been explored for petroleum, the structure of the region where the two provinces merge is not well known.

East of Trinidad, the deformed margin of the Barbados Ridge Complex passes into the undeformed passive margin of eastern South America. The area of transition is characterized by diapiric structures, which can be seen on reflection profiles (Lowrie & Escowitz 1969; Ewing *et al.* 1974; Mascle *et al.* 1981). Whether these result from salt as on parts of the African Margin or from overpressured shales, such as those producing mud volcanoes on Trinidad (Arnold & Macready 1956) is not known, but the latter seems more likely.

In the west, as mentioned above, the forearc basin beneath the Tobago Trough continues south-westward until it pinches out between the metamorphic rocks of the Venezuelan Coast Range and the south-westward extension of the volcanic arc.

In the region of Tobago there is a sharp drop in bathymetry from the continental shelf down to the Barbados Ridge and the continuity of structure from the Barbados Ridge into the shelf is very unclear, although some authors claim it exists (Weeks *et al.* 1971). Seismic refraction and reflection show that rocks with a high seismic velocity occur close to the surface at the southern end of the Barbados Ridge (Ewing *et al.* 1957; Edgar *et al.* 1971); these could be the Cretaceous metamorphic rocks that crop out on Tobago and Trinidad. These rocks deepen towards the north. The trend of structures in Trinidad suggest that even if there is no direct continuity of structures between the Barbados Ridge Complex and the South American margin, then they have been produced in a similar tectonic situation. There is no very obvious evidence for any significant transcurrent faulting along the northern margin of the present continental shelf.

Variation in structure along the Barbados Ridge Complex

The overall elevation of the complex decreases towards the north and the Barbados Ridge does

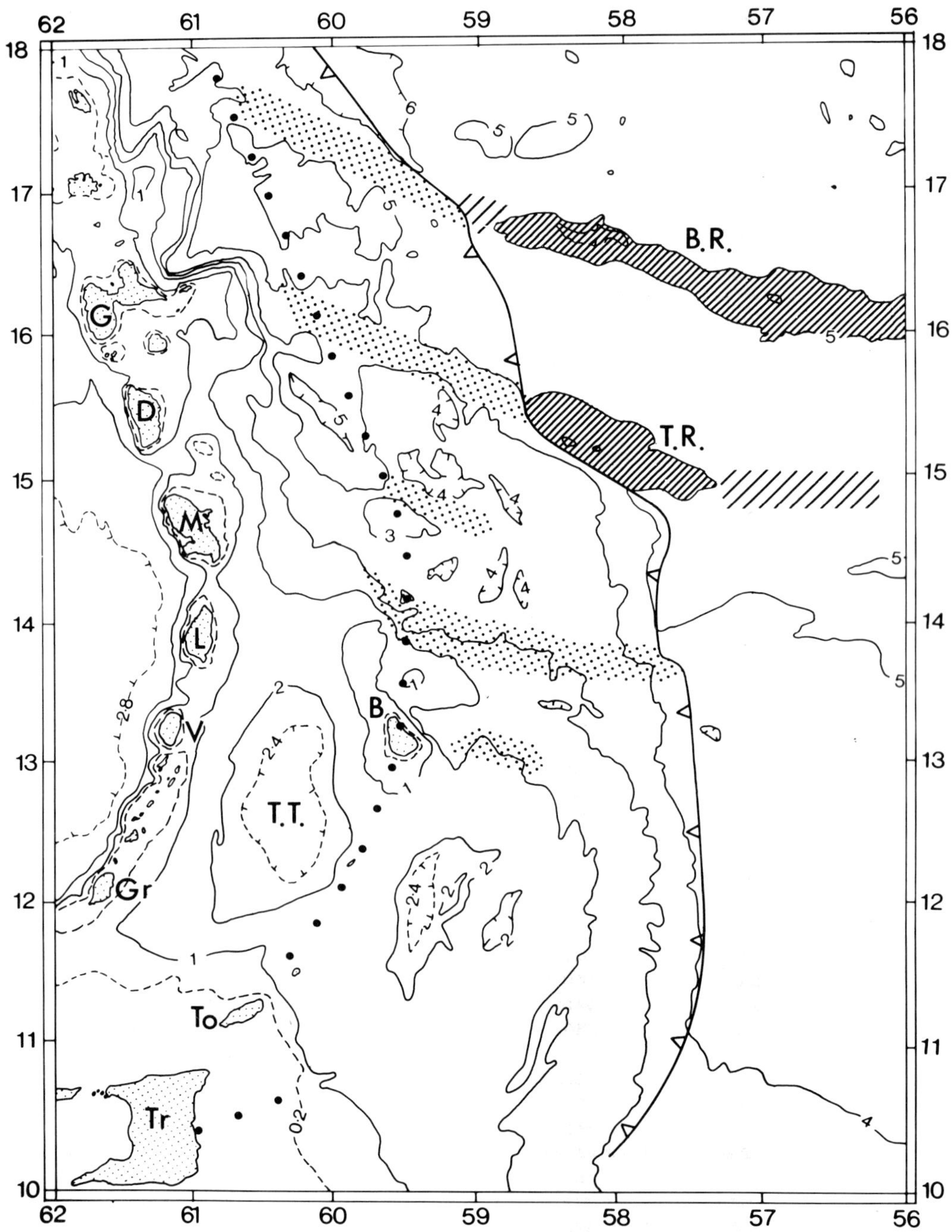

FIG. 4. Lateral variations in the Barbados Ridge Complex. Northward facing slopes ('steps') are stippled. Ridges in the oceanic basement are shown in close diagonal hatching where they form bathymetric features and in open hatching where they are known from seismic profiles to exist beneath the ocean floor. Hatching is not continued where the ridges extend beneath the Barbados Ridge Complex, as these areas are already stippled. The eastern margin of the complex is shown by the solid line with triangles on the overthrusting side. The large dots mark the axis of the negative Bouguer gravity anomaly. B.R.—Barracuda Ridge; T.R.—Tiburon Rise; G—Guadeloupe; D—Dominica; M—Martinique; L—St Lucia; B—Barbados; V—St Vincent; T.T.—Tobago Trough; Gr—Grenada; To—Tobago; Tr—Trinidad.

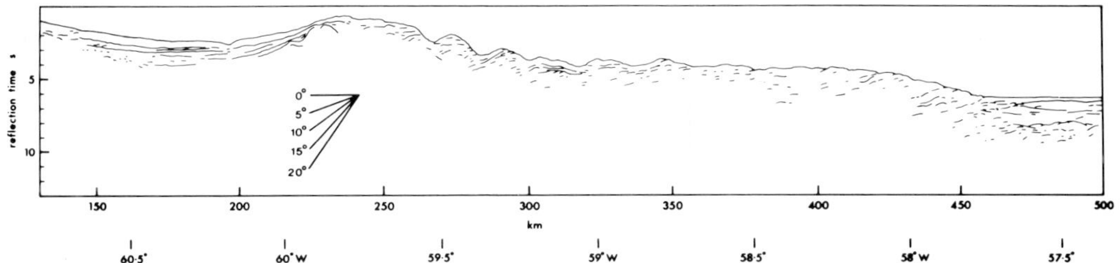

FIG. 5. Line drawing of seismic reflection section across the Barbados Ridge Complex (g in Fig. 1) from Westbrook (1975).

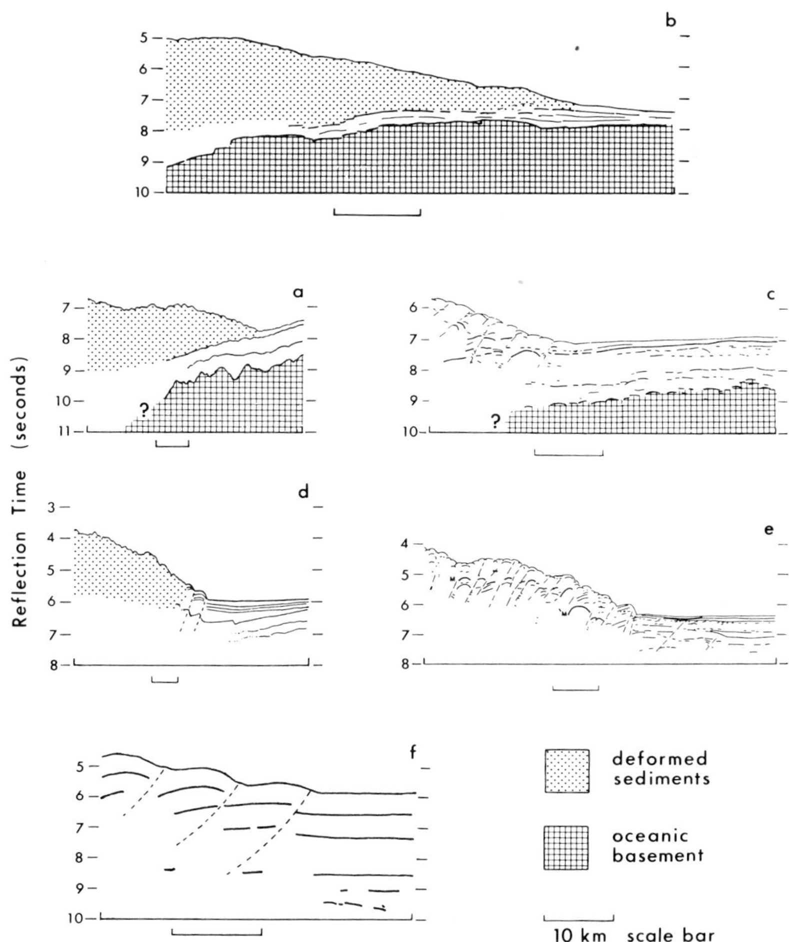

FIG. 6. Line drawings of seismic reflection sections across the eastern margin of the Barbados Ridge Complex. (a) and (d) from Peter & Westbrook (1976), (b) and (f) from Biju-Duval *et al.* (1978), (c) and (e) from Chase & Bunce (1969). The zones which are stippled on sections (a), (b) and (d) strongly reflect seismic energy, but the pattern of reflections lacks coherency. Pecked lines show the interpreted positions of thrusts.

not continue further north than 14°50′N (Fig. 2). This appears to be principally related to the northward decrease in thickness of the ocean floor sediments (>4 km at 12°N; 1 km at 16°N). This northward decrease in elevation is not uniform however (Fig. 4). South of 13°N the elevation of the complex is fairly uniform at around 2000 m, east of the Barbados Ridge, it then descends at 13°N and 13°50′N to 3500 m, the biggest drop coming at 13°50′N. It descends again at 15°50′N to around 5000 m and at 17°10′N to 5600 m. These 'steps' which all have a WNW trend are associated with ridges in the oceanic basement (Peter & Westbrook 1976). The northernmost of these is the Barracuda Ridge which extends at least 100 km beneath the complex (Schubert 1974), the next south is the Tiburon Rise at 15°20′N. Both these ridges have a parallel trend (WNW) and run beneath the 'steps' in the complex, which have the same trend. At 13°55′N the ridge is not expressed as a bathymetric feature on the seafloor east of the arc, but its presence is shown on seismic reflection profiles and by a linear positive gravity anomaly (Westbrook 1975).

The ridges can influence the complex in various way. (Fig. 7). They control the thickness of sediment on the ocean floor and consequently provide lateral control on the amount of accreted sediment in the complex. When the ridges first encounter the complex, their effect must be to uplift material above them.

After some time this initial effect will be reversed by the accretion of thicker sediment from the troughs flanking the ridges. With turbidite sediments coming from the south, sediment thickness south of each ridge will tend to be greater than to the north, and this would lead to a northward decrease in accreted sediment across each ridge.

Another important effect is produced by the difference between the trend of the ridge and the direction of convergence of the ridge with the complex. As the ridge moves obliquely into the complex the sediment will tend to be swept to one side by its 'plough' action. Earthquake first motions and the orientation of Caribbean plate boundaries indicate that the direction of convergence between the Caribbean and the Atlantic is easterly (Jordan 1975; Tomblin, pers. comm.). The trend of the ridges is 290°, which give a difference in angle of 20°. This produces a southward sweeping action which builds up material to the south of the ridge producing compressional structures semi-parallel to the ridge (Fig. 8). The cross-stress produced by the ridge comes from the reaction normal to its surface, which depends on the angle of slope of the ridge and the angle between the strike of the ridge and the direction of convergence.

A significant aspect of this process is that structures with a trend oblique to the main structural trend are produced without any change in the overall geotectonic setting. No north-south regional stresses are required other than the reaction to the stresses produced at the ridges. Between 13°50′N and 15°20′N the complex is dominated by WNW to west trending ridges and troughs superimposed upon the northerly trending features characteristic of the southern part of the complex. The northernmost ridge is that associated with the Tiburon Rise. The sediment in this ridge shows no clear reflecting horizons by comparison with the ridges and troughs south of it, implying recent deformation by the oblique convergence of the Tiburon rise (Peter & Westbrook 1976, line M). The ridge east of Martinique may also be associated with a basement ridge, but there is no clear geophysical evidence to support its existence, except that the NW flank of this ridge is the site of a positive gravity anomaly which locally modifies the form of the main negative gravity anomaly. The blanket of undeformed sediment which overlies the tectonized rocks of the complex in this region of cross-trends is displaced vertically by topography (Peter & Westbrook 1976). The troughs are not filled by more recent sediment and the displaced sediment layer shows no variations in thickness related to topography, implying fairly recent movement. By contrast, in the area south of 13°30′N, the blanket of overlying sediment is undisturbed and the troughs are sites of recent sedimentation.

The buried basement ridge at 13°55′N is probably responsible for the relatively high elevation of Barbados and the portion of the Barbados Ridge just north of it, compared with the southern part of the ridge. It is possible that it has been a significant feature since the late Eocene producing the southward dipping palaeoslope on which the Oceanic Formation of Barbados was deposited (Lohmann 1973).

Development of the Barbados Ridge Complex

The Barbados Ridge Complex is the widest example of an accretionary sediment pile associated with an island arc. This implies that some features of accretionary sediment piles, which may only be nascent in many arcs, should

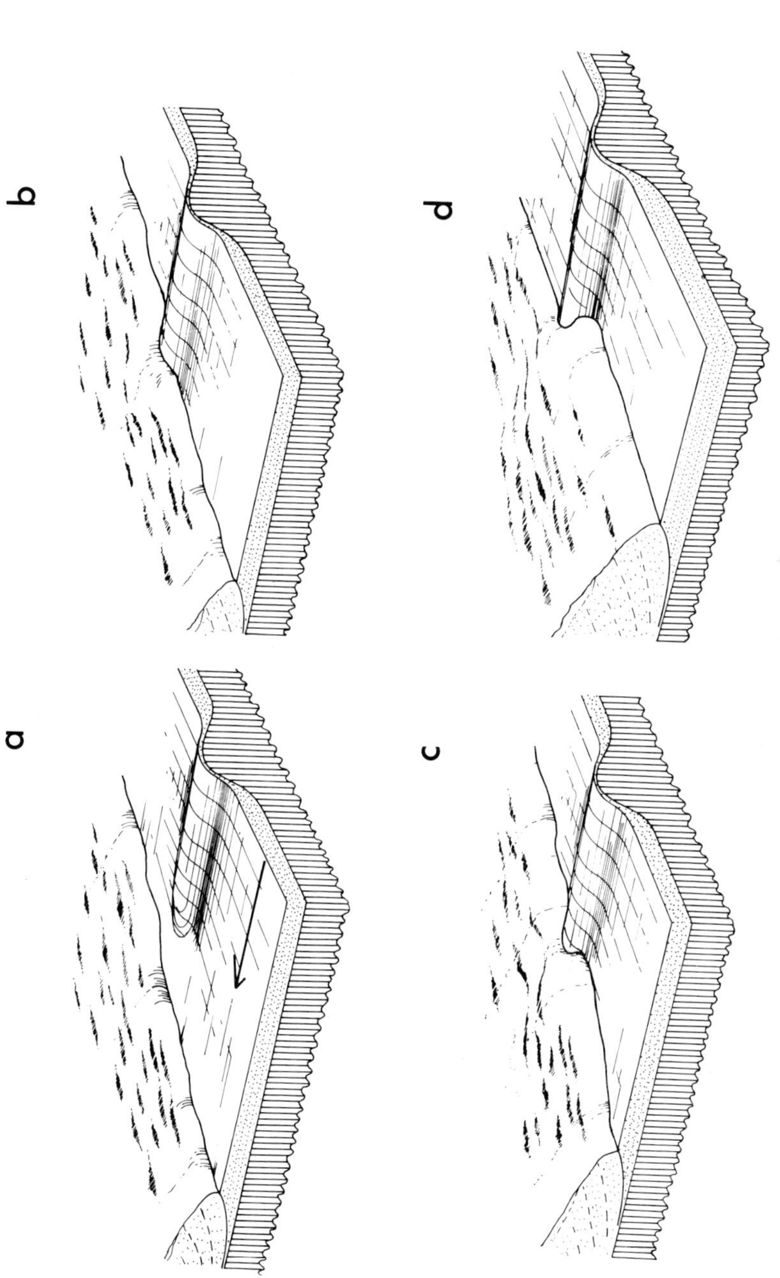

Fig. 7. The influence of a ridge in oceanic basement upon a forearc complex:
(a) Ridge approaching the edge of the forearc complex. The direction of convergence is the same as the strike of the ridge.
(b) After coming beneath the edge of the complex the ridge locally uplifts the part of the complex above its crest.
(c) After further subduction the effect of the relief of the ridge is outweighed by the greater contribution of sediment to the complex from either side of the ridge which builds the complex further out than over the ridge, where little sediment is added.
(d) If there is a greater thickness of sediment on one side of the ridge than the other, because of its acting as a barrier to turbidites, then one side of the complex will grow out further than the other, leading to a change in strike of the front across the ridge.

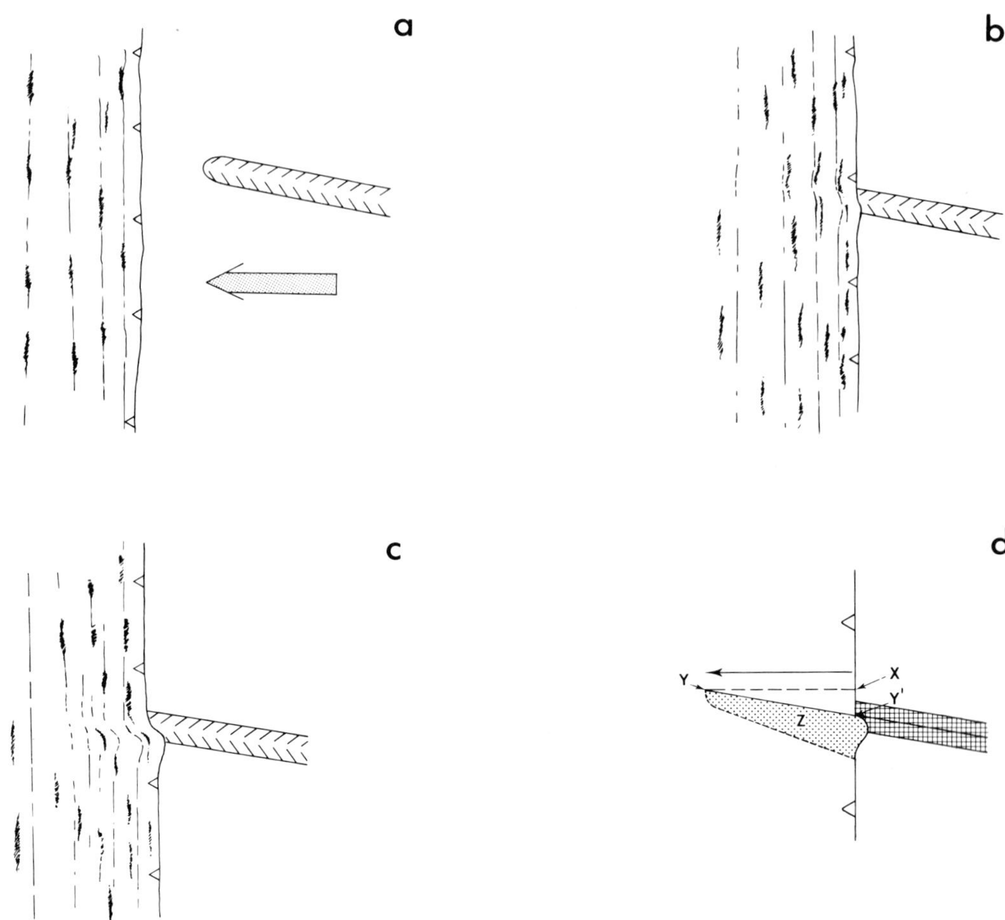

FIG. 8. The effect of a ridge in oceanic basement converging obliquely with a forearc complex:
(a) Ridge approaching the edge of the complex in a direction oblique to that of the strike of the ridge.
(b) Shortly after passing into the complex the ridge begins to distort the form of the leading edge as the point of entry of the ridge migrates along it.
(c) After further subduction, the southward sweeping action of the ridge has built up material on its southern side. There should also be some depletion on the northern side, but this would not produce such an obvious change in strike in the leading edge of the complex.
(d) Diagram showing the main elements of the situation. Y marks the end of the ridge. Y′ marks the point at which the crest of the ridge passes under the leading edge of the complex. This has migrated along the leading edge, from X, the point at which the ridge entered the complex. Z is the zone in which most of the compressional structures produced by the ridge exist.

be fully developed in the Barbados Ridge Complex. One of these is the trench-slope break, the break of slope at the top of the inner trench-wall which is very often the highest point of an accretionary sediment pile. The outer sedimentary ridges of arcs such as the Java-Sumatra, have also been described as the trench-slope break (Karig *et al.* 1980). The Barbados Ridge, however, is clearly not the equivalent of a trench-slope break, since it only exists in the southern part of the complex, and even in the region where it exists the topography and structure of the area between it and the front of the complex is so varied that it is conceptually difficult to identify the whole of it as the inner trench wall. The 'trench-slope break' of the Barbados Ridge complex lies 50–60 km west of its eastern edge (Figs 1 & 2). It is interpreted as representing the maximum elevation attained by material involved in the initial accretion process before gravitational forces nullify the upward movement of material caused by the horizontal convergence. The opposition of gravitational body forces to tectonically induced forces allows the complex to build forward rather than be pushed upward and backward. On a large scale the deformation of the pile can be considered as a viscous or plastic process and therefore the position of the trench slope break and the angle of the trench slope will be a function of gravitational body forces, tectonically induced stresses and the rheology of the rocks. This last is the most difficult to assess, because it varies with lithology, and the amount of deformation and metamorphism. Following from this, one can speculate that variations in the angle of the trench slope on the eastern margin of the complex are related to lithological variations in the material being accreted. Between the Tiburon Rise and the Barracuda Ridge where only the topmost sediment has been stripped away from the undeformed sedimentary column and intensely deformed, the angle of the trench slope is less steep than it is further south.

The Barbados Ridge has been formed by the uplift produced by the horizontal compressional forces in the complex. These arise principally from shear stresses applied to the base of the sediment pile by the subducting ocean crust which are opposed by the reaction of the island arc and consequently achieve a maximum value immediately above the line along which the ocean crust of the Atlantic is subducted beneath the crystalline crust of the Caribbean. A wide accretionary pile is subject to more horizontal compression than a narrow one, because shear stress has been applied over a greater surface area. Also, a purely frictional model for imparting the shear stress at the base of the accretionary complex is probably inadequate, as some shear stress will be transmitted by viscous shear through deforming sediment. This time-dependent element in the deformational process implies that an increase in subduction rate would increase the amount of compressional stress acting on the accretionary complex.

Comparison with other arcs

If the Barbados Ridge Complex is taken as a model, then which other forearc systems possess a sedimentary ridge that is a separate feature from the trench-slope break? Off Sumatra a major break of slope lies 30 km from Nias Island and 50 km from the trench (Karig *et al.* 1980), and further north along the same arc system, the break of slope is 80 km from the Nicobar Islands and again 50 km from the trench (Weeks *et al.* 1967). The Nicobar Islands and Nias are the equivalents of Barbados. In the eastern Aleutians where the major break of slope occurs 50–70 km from the trench, the position of Barbados is analagous with that of Kodiak Island 100 km from the break of slope. The presence of a large forearc basin between Kodiak Island and the break of slope does not detract from the comparison. Forearc basins occur between the Barbados Ridge and the eastern edge of the complex. Although in Dickinson & Seely's (1979) classification of forearcs, the Barbados Ridge Complex is cited as an example of a narrow ridged forearc, it should be classed as a broad ridged forearc. It is implicit in the evolutionary model of Westbrook (1975) that narrow ridged forearcs evolve into broad ridged forearcs.

It is common for geological models of forearc ridges (e.g. Dickinson & Seely 1979; Seely 1979) to include rises of oceanic crust beneath the ridges. The geophysical evidence is strongly against the existence of a rise of oceanic crust beneath the Barbados Ridge (Westbrook 1975; Boynton *et al.* 1979), and is ambivalent for its existence beneath the forearc ridge of the Java Sumatra arc system (Curray *et al.* 1977; Karig *et al.* 1980; Kieckhefer *et al.* 1980).

Barbados and history of the complex

Barbados, the geology of which is succinctly summarized by Saunders (1979) is an important source of information on the nature of the rocks comprising the Barbados Ridge Complex, but is also a source of problems. The Scotland Formation (Pudsey & Reading 1981) comprises a group of sediments of early Eocene (perhaps Palaeocene) to middle Eocene age which can be generally described by flysch. Current opinion is that these rocks were deposited in deep water. Pudsey & Reading propose that they were deposited as a deep sea fan in the former trench of the Lesser Antilles arc. While there is a strong likelihood of this, it is also possible that the fan was formed on the ocean floor and engulfed later by the subduction complex. The very small amount of material in the formation derived from a volcanic source lends some support to the latter view.

The Scotland Formation is quite strongly deformed. Most of this deformation is tectonic, but some synsedimentary gravity-driven deformation has been recognized. Exploration wells show that there is tectonic repetition of the Scotland Formation, at least four times to a depth of at least 4.5 km (Baadsgaard 1960).

The strike of this deformation is NE, rather than the northerly trend expected from general consideration of the shape of the arc and the direction of subduction. Herrera & Spence (1964) suggested that this trend was the result of emplacement of the Scotland Formation as gravity slides from the south. The difficulty with this explanation is that after sliding into the trench the material underwent relatively little deformation. The possibility that the deformation reflects a southerly component in the direction of subduction since the Cretaceous is supported by some plate tectonic reconstructions of the area (Malfait & Dinkleman 1972; Ladd 1976) for the period from Late Cretaceous until late Eocene. Subsequent tectonic rotation of the small area of Barbados is a possible explanation which could be tested palaeomagnetically. The Oceanic Formation (middle Eocene to Oligocene) and the Bissex Hill Formation (early Miocene) which overlie the Scotland Formation are also deformed, although by no means to the same extent as the Scotlands. They show the continuing influence of tectonism, comparatively far from the leading edge of the complex.

Reconstructions of plate motions and the stratigraphy of Trinidad suggest that Jurassic and Cretaceous rocks should have been present on the ocean floor during the early history of the arc, but they are not present in Barbados as might be expected from many models of subduction complexes. Maybe these older sediments were not scraped off, but taken further down the subduction zone in the initial stages of its formation, or exist further west in the complex.

Barbados has undergone considerable vertical movement since the late Eocene when the Oceanic Formation began to be deposited on the Scotlands at a depth of 3500 m (Saunders 1979). In the early Miocene during the deposition of the Bissex Hill Formation the water depth can have been as little as 300 m (Steineck & Murtha 1981). In the middle Miocene during deposition of the Conset Marl the water depth was between 1000 and 1500 m (Steineck & Murtha 1981). The seismic stratigraphy of the Tobago Trough suggests that Barbados has risen about 3 km relative to the trough's centre since the Miocene. The vertical displacements are controlled by the subduction process, which over a long term would tend to increase the elevation of Barbados. On a shorter term a slowing of subduction would result in a rise of the complex as the negative isostatic imbalance was reduced, whereas an increase in rate would initially depress the surface. A possible scenario for the history of these movements is shown in Fig. 9.

The great volume of material in the Barbados Ridge Complex is undoubtedly a consequence of the proximity of the Lesser Antilles to sources of terrigenous sediment from the South American continent. How much was directly deposited in a trench in front of the arc earlier in its history and how much was deposited on the ocean floor can only be conjectured. A considerable proportion, however, must be continental slope and rise deposits from the northern margin of South America, incorporated as the Lesser Antilles moved eastward along the margin (Fig. 10) (Chase & Bunce 1969). How far the Lesser Antilles have migrated past South America is undetermined, but if one follows Jordan (1975) this could be virtually the whole length of the Caribbean. (This is illustrated by Pudsey & Reading 1981.) The tectonic history of northern Venezuela suggests that it might not have moved as far as this, however (Bell 1972; Maresch 1974). In the region currently occupied by the Barbados Ridge Complex, there must once have existed a thick sedimentary cone built in front of a proto-Orinoco River system. The collision of the complex with this cone must have had a significant influence on the development of the complex. This collision could have begun at any time between the end of the Eocene and the middle of the Miocene, depending on the model chosen for the development of the eastern Caribbean (Fig. 9f).

Maturity of the Barbados Ridge Complex

Can the Barbados Ridge Complex be considered to be a mature member of a sequence of forearc evolution? Listed below are variables which can influence the development of forearc complexes.

(1) Age of the subduction zone.
(2) Rate of subduction.
(3) Thickness of sediment:
  (a) on ocean plate to be subducted,
  (b) in trench at the foot of the forearc.
(4) Lithologies of sediments on (a) and in (b).
(5) Shape of the basement surface.
(6) Sedimentation directly on to the forearc complex.

It is implicit in most models of accretionary complexes that there will be growth but will this growth always proceed in the same way? Looked at in simple terms the volume of sediment in the complex should be the product of

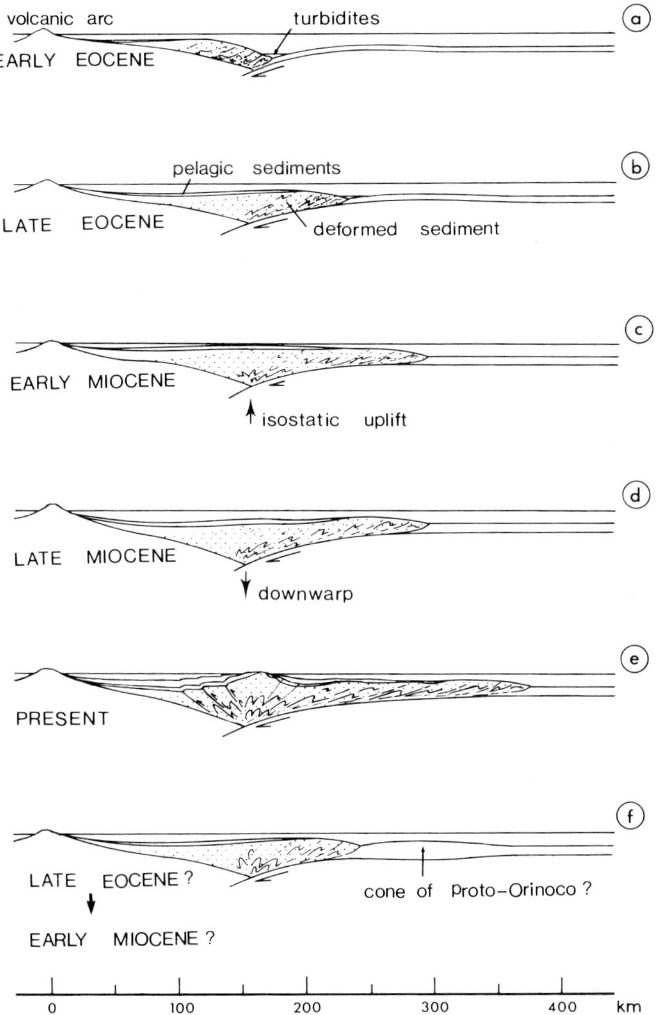

Fig. 9. Evolution of Barbados Ridge Complex.
(a) Early Eocene: youthful stage in the evolution of the forearc which had a trench possibly with direct deposition of Scotland Formation sediments into it.
(b) Late Eocene: region in which Barbados now is, no longer in active accretionary zone and receiving pelagic sedimentation of Oceanic Formation. Elevated position keeps it free of terrigenous sediments. The forearc may also have been far from a source of continentally derived material at this time. This stage of development began in the mid-Eocene.
(c) Early Miocene: a hiatus in subduction has allowed the central part of the forearc complex to rise by isostatic rebound. During this stage the Bissex Hill Formation of Barbados was laid down.
(d) Late Miocene: following the recommencement of subduction in the middle Miocene the central part of the forearc has become depressed by the reimposition of the negative isostatic imbalance. Some of the Conset Marl of Barbados was probably laid down at the beginning of this stage, as the water began deepening.
(e) Present: continued accretion and consequent widening of the forearc have imposed more stress on the centre, and caused it to rise forming the Barbados Ridge and Tobago Trough. Present rate of uplift is about 400 m Ma^{-1}.
(f) End Eocene—early Miocene?: at some time during the period, the Barbados Ridge Complex encountered what was probably quite a thick cone of sediment produced by some proto-Orinoco delta. This must have produced a rapid widening of the complex.
Note: The hiatus in subduction referred to in (c) and (d) may have been a considerable slowing rather than a cessation.

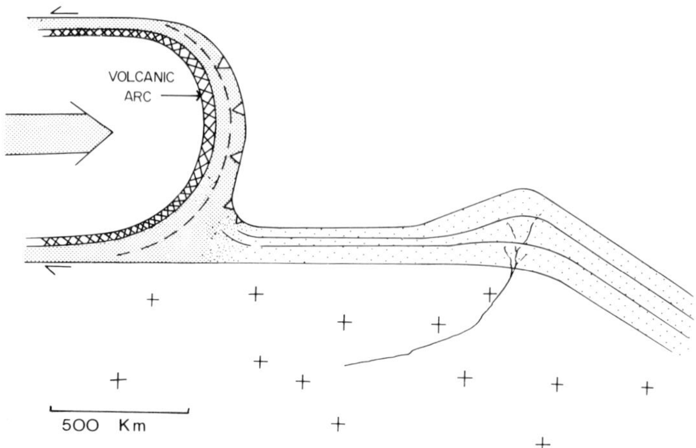

FIG. 10. Migration of the Lesser Antilles island are eastward along the northern margin of South America during the Tertiary. The continental slope deposits of the South American margin become assimilated in the forearc complex. Eventually the sediment cone of the proto-Orinoco becomes incorporated in the complex. The front of the complex is shown by a solid line with triangle. The line of subduction of crystalline ocean crust beneath crystalline crust of Caribbean Plate is shown by a dashed line.

variables (1), (2) and (3), plus (6), but other effects complicate this. Dewatering and compaction of accreted sediment will reduce its volume. If the mean density of the Barbados Ridge Complex is taken to be 2.35 Mg m^{-3} and the mean density of sediment before addition is 1.9 Mg m^{-3}, then the volume of sediment in the complex is 0.66 of its original volume. Only part of the sedimentary section on the subducted oceanic lithosphere may be accreted in the forearc complex, the remainder being taken down further into the crust and mantle. This is probably strongly dependent on lithology; mature lithified rocks of biogenic origin being far more likely to remain attached to the oceanic lithosphere than overlying terrigenous sediments which were probably only added to the section close to the subduction zone, under most circumstances (Moore 1975). In some situations such as Peru-Chile it is possible that the forearc complex has been 'eroded' at its base by the subducting lithosphere (Karig 1974a). As well as governing the rate at which sediment enters the complex, subduction rate is also important in controlling stress imparted to a forearc complex, and consequently affecting its shape and strain within it.

The shape of the basement must have a strong influence particularly on the early development of the complex. The upthrusting of a slice of oceanic crust beneath which accreted sediment is 'understuffed' (Seely 1979) produces quite a different evolution, with a fixed position of the outer structural high, from that proposed by Karig (1974b) in which sediment accretes in a depression above the oceanic basement and the trench slope break migrates outward. Evidence for Seely's model exists from forearcs associated with continental margins, e.g. Middle America, but not for intra-oceanic arcs. The Barbados Ridge Complex does not follow Seely's model.

Given the variety of situations in which arcs occur and the episodic nature of geological history it is obvious that the development of a universal model for the evolution of forearcs which will predict what will occur under any circumstances during the evolution of forearcs may be an intractable problem. However, in a situation where there is a plentiful supply of terrigenous sediment and a moderate rate of subduction, growth of the forearc complex is a well-established consequence of continued subduction, it is reasonable to suppose that the Barbados Ridge Complex is a mature forearc complex, representing a state towards which forearcs will evolve, given sufficient time.

Conclusions

(1) The Barbados Ridge Complex is an accretionary forearc complex of broad ridged type. Its width and thickness are a consequence of its long history (50 Ma)

and the thickness of terrigenous sediment on the ocean floor.

(2) The changes in elevation along the complex are broadly related to variation in sediment thickness on the ocean floor, and are locally controlled by ridges in the oceanic basement.

(3) Oceanic basement ridges influence the position of the leading edge of the complex and its thickness by laterally controlling the amount of sediment accreted, and by deforming accreted material when they pass into the complex at an angle oblique to their strike.

(4) Style of deformation and accretion at the leading edge of the complex are dependent upon sediment thickness and lithology. Thicker sediment gives a more open style of deformation. More competent lithologies in the lower part of the sedimentary section can restrict the amount of sediment accreted, and confine significant deformation to the upper part of the section.

(5) Increase in the width of the complex, and also perhaps, an increase in subduction rate, has increased the horizontal stress in the complex. This has produced the uplift of the Barbados Ridge and the westward verging structures on its western flank.

(6) Isostatic response to variations in subduction rate has superimposed changes in the elevation of the Barbados Ridge upon the long-term uplift produced tectonically.

ACKNOWLEDGMENTS: Some of the ideas concerning basement ridges in this paper were originally discussed by George Peter and myself some years ago, although not presented in our paper of 1976. I am grateful to Carol Pudsey and Harold Reading for a preprint of their paper on Barbados. Harold Reading and Casey Moore gave helpful reviews of this paper.

References

ARNOLD, R. & MACREADY, G. A. 1956. Island-forming mud volcanoes in Trinidad, British West Indies. *Bull. Am. Assoc. Petrol. Geol.* **40,** 2749–58.

BAADSGAARD, P. H. 1960. Barbados, W.I. Exploration results 1950–1958. *Int. Geol. Congr. Rep. 21st Session,* **18,** 21–7.

BELL, J. S. 1972. Geotectonic evolution of the southern Caribbean area. *In: Studies in Earth and Space Sciences* (*Hess Vol*). Mem. geol. Soc. Am. **132,** 367–86.

BIJU-DUVAL, B., MASCLE, A., MONTADERT, L. & WANNESON, J. 1978. Seismic investigations in the Colombia, Venezuela and Grenada basins, and on the Barbados Ridge for future IPOD drilling. *Geol. Mijnbouw* , **57,** 105–16.

BOYNTON, C. H., WESTBROOK, G. K., BOTT, M. H. P. & LONG, R. E. 1979. A seismic refraction investigation of crustal structure beneath the Lesser Antilles island arc. *Geophys. J.R. astron. Soc.* **58,** 371–93.

BUNCE, E. T., PHILLIPS, J. D., CHASE, R. L. & BOWIN, C. O. 1971. The Lesser Antilles Arc and the eastern margin of the Caribbean Sea. *In:* MAXWELL, A. E. (ed). *The Sea,* **4,** part II, Wiley, New York.

CHASE, R. L. & BUNCE, E. T. 1969. Underthrusting of the eastern margin of the Antilles by the floor of the western North Atlantic Ocean, and the origin of the Barbados Ridge. *J. geophys. Res.* **74,** 1413–20.

CURRAY, J. R., SHOR, G. G., RAITT, R. W. & HENRY, M. 1977. Seismic refraction and reflection studies of crustal structure of the eastern Sunda and western Banda arcs. *J. geophys. Res.* **82,** 2479–89.

DICKINSON, W. R. & SEELY, D. R. 1979. Structure and stratigraphy of forearc regions. *Bull. Am. Assoc. Petrol. Geol.* **63,** 2–31.

DONOVAN, T. J., RANDAL, L. F. & ROBERTS, A. A. 1980. Aeromagnetic detection of diagenetic magnetite over oil fields. *Bull. Am. Assoc. Petrol. Geol.* **63,** 245–8.

EDGAR, N. T., EWING, J. I. & HENNION, J. 1971. Seismic refraction and reflection in the Caribbean Sea. *Bull. Am. Assoc. Petrol. Geol.* **55,** 838–70.

EWING, J., EWING, M., WINDISCH, C. & ATHEN, T. 1974. *Underway Marine Geophysical Data in the North Atlantic June 1961–January 1971. Parts E and F: Seismic Reflection Profiles.* Lamont-Doherty Survey of the World Ocean (Talwani, M. (ed)).

EWING, J. I., OFFICER, C. B., JOHNSON, H. R. & EDWARDS, R. D. 1957. Geophysical investigations in the eastern Caribbean: Trinidad shelf, Tobago Trough, Barbados Ridge, Atlantic Ocean. *Bull. geol. Soc Am.* **68,** 897–912.

FEO-CODECIDO, G. 1977. Un esbozo geologico de la plataforma continental Margarita-Tobago (A geological outline of the Margarita-Tobago continental shelf). *Bol. Geol* (*Venezuela*), **3,** 1923–46.

HERRERA, R. C. & SPENCE, J. 1964. *The geology and oil prospects of the Scotland Group of sediments at Barbados*. Sinclair and B.P. Explorations Inc. (unpubl. report).

HUBBERT, M. K. 1951. Mechanical basis for certain familiar geologic structures. *Bull. geol. Soc. Am.* **62,** 355–72.

JORDAN, T. H. 1975. The present-day motions of the Caribbean plate. *J. geophys. Res.* **80,** 4433–9.

KARIG, D. E. 1974a. Tectonic erosion at trenches. *Earth planet. Sci. Lett.* **21,** 209–12.

—— 1974b. Evolution of arc systems in the Western Pacific. *Ann. Rev. Earth planet. Sci.* **2,** 51–75.

—— , LAWRENCE, M. B., MOORE, G. F. & CURRAY, J. R. 1980. Structural framework of the forearc basin, NW Sumatra. *J. geol. Soc. London,* **137,** 77–91.

—— & SHARMAN, G. F. III. 1975. Subduction and accretion in trenches. *Bull. geol. Soc. Am.* **86,** 377–89.

KEAREY, P., PETER, G. & WESTBROOK, G. K. 1975. Geophysical maps of the eastern Caribbean, *J. geol. Soc. London,* **131,** 311–21.

KELLER, G. H., LAMBERT, D. N., BENNETT, R. H. & RUCKER, J. B. 1972. Mass physical properties of Tobago Trough sediments. *Trans. Sixth Caribbean Geol. Conf.* 405–8.

KIECKHEFER, R. M., SHOR, G. G. JR. CURRAY, J. R., SUGIARTA, W. & HEHUWART, F. 1980. Seismic refraction studies of the Sunda Trench and forearc basin. *J. geophys. Res.* **85,** 863–89.

LADD, J. W. 1975. Relative motion of South America with respect to North America and Caribbean Tectonics. *Bull. geol. Soc. Am.* **87,** 969–76.

LE PICHON, X. & ANGELIER, J. 1979. The Hellenic arc and trench system: a key to the neotectonic evolution of the Eastern Mediterranean area. *Tectonophysics,* **60,** 1–42.

LEWIS, J. & ROBINSON, E. 1976. A revised stratigraphy of the Lesser Antilles. *Trans. Seventh Caribbean Geol. Conf.* 339–44.

LOHMANN, G. P. 1973. *Palaeo-oceanographic interpretation of the Oceanic Formation (Eocene-Oligocene), Barbados, West Indies,* Thesis, PhD, Brown Univ. (unpubl.).

LOWRIE, A. & ESCOWITZ, E. (eds) 1969. *Kane 9.* Global ocean floor analysis and research data series, **1,** U.S. Naval Oceanographic Office.

MALFAIT, B. T. & DINKLEMAN, M. G. 1972. Circum-Caribbean tectonic and igneous activity and the evolution of the Caribbean plate. *Bull. geol. Soc. Am.* **83,** 251–72.

MARESCH, W. V. 1974. Plate tectonics origin of the Caribbean Mountain system of northern South America: discussion and proposal. *Bull. geol. Soc. Am.* **85,** 669–82.

MARLOW, M. S., GARRISON, L. E., MARTIN, R. G., TRUMBULL, J. V. A. & COOPER, A. K. 1974. Tectonic transition zone in the northeastern Caribbean. *J. Res. U.S. geol. Surv.* **2,** 289–302.

MASCLE, A., BIJU-DUVAL, B., LETOUZEY, J., MONTADERT, L. & RAVENNE, C. 1977. Sediments and their deformations in active margins of different geological settings. *Int. Symp. Geodynamics in South-west Pacific Noumea (New Caledonia) 27 August–2 September 1976,* 327–44. Technip, Paris.

MASCLE, A., LAJAT, D. & NELY, G. 1981. Sediment deformation linked to subduction and to argillokinesis in the southern Barbados Ridge from multichannel seismic surveys. *Trans. 4th Latin American geol. Congr.* (in press).

MOORE, J. C. 1975. Selective subduction. *Geology,* **3,** 530–2.

—— 1979. Variation in strain and strain rate during underthrusting of trench deposits. *Geology,* **7,** 185–8.

PETER, G. & WESTBROOK, G. K. 1976. Tectonics of southwestern North Atlantic and Barbados Ridge Complex. *Bull. Am. Assoc. Petrol. Geol.* **60,** 1078–106.

PUDSEY, C. J. & READING, H. G. 1981. Sedimentology and structure of the Scotland Group, Barbados (this volume).

SAUNDERS, J. B. 1979. Field Trip I(J), Barbados. *Field Guide: 4th Latin American Geol. Congr. 1979,* 71–88.

SCHUBERT, C. 1974. Seafloor tectonics east of the northern Lesser Antilles. *Abstr. Seventh Caribbean Geol. Conf.*, 62–3.

SEELY, D. R. 1979. The evolution of structural highs bordering major forearc basins. *In:* WATKINS, J. S., MONTADERT, L. & DICKERSON, P. W. (eds). *Geological and Geophysical Investigations of Continental Margins.* Mem. Am. Assoc. Petrol. Geol. **29,** 245–60.

—— , VAIL, P. R. & WALTON, G. G., 1974. Trench slope model. *In:* BURK, C. A. & DRAKE, C. L. (eds). *The Geology of Continental Margins,* 249–60. Springer-Verlag, New York.

STEINECK, P. L. & MURTHA, G. 1981. Foraminiferal palaeobathymetry of Miocene rocks, Barbados, Lesser Antilles. *Trans. 4th Latin American geol. Congr.* (in press).

VON HUENE, R. E. 1979. Structure of the outer convergent margin of Kodiak Island, Alaska, from multichannel seismic records. *Mem. Am. Assoc. Petrol. Geol.* **29,** 261–72.

WEEKS, L. A., HARBISON, R. N. & PETER, G. 1967. Island arc system in the Andaman Sea. *Bull. Am. Assoc. Petrol Geol.* **51,** 1803–15.

—— , LATTIMORE, R. K. & HARBISON, R. N. 1971. Structural relations among Lesser Antilles, Venezuela, and Trinidad-Tobago. *Bull. Am. Assoc. Petrol. Geol.* **55,** 1741–52.

WESTBROOK, G. K. 1975. The structure of the crust and upper mantle in the region of Barbados and the Lesser Antilles. *Geophys. J.R. astron. Soc.* **43,** 201–42.

—— 1976. The structure and evolution of the Barbados Ridge. *Trans. Seventh Caribbean geol. Conf. 1974,* 321–8.

—— , BOTT, M. H. P. & PEACOCK, J. H. 1973. The Lesser Antilles subduction zone in the region of Barbados. *Nature (phys. Sci.)* **244,** 18–20.

G. K. WESTBROOK, Department of Geological Sciences, University of Durham, Science Laboratories, South Road, Durham DH1 3LE, England.

Sedimentology and structure of the Scotland Group, Barbados

C. J. Pudsey & H. G. Reading

SUMMARY: The lower to middle Eocene Scotland Group is about 1700 m thick at outcrop and can be divided into: (A) The Lower Scotland Formation consisting of (1) the Walkers Member (500 m): graded turbidite sands and clays, arranged in sand-dominated and clay-dominated packets; (2) the Morgan Lewis Member (200 m): almost all clay, with two sand intervals; (B) the Upper Scotland Formation consisting of (3) the Murphys Member (about 350 m): a thickening and coarsening upward sequence including slumped beds; (4) the Chalky Mount Member (0–250 m): coarse and conglomeratic sands with large slide blocks and minor interbedded muds; (5) the Mount All Member (about 500 m): coarse and fine turbidite sands and clays, much disturbed and displaying no overall sequence. The group was deposited by mass-flow processes in a deep marine environment, probably in a deep-sea trench or a submarine fan.

Current flow was from the SW and mineral composition supports derivation from South American metamorphic rocks. Clay mineralogy shows an upward trend from mainly smectite to mainly kaolinite.

An ENE structural trend is present throughout the island and is anomalous in the present plate-tectonic setting adjacent to a north-south trending island arc. Deformation increases in complexity up the succession; this may be due to sedimentary slumping, partly a result of increasing disturbance by rising mud diapirs. This locally culminated in extrusion of hydrocarbon-rich mud—the overlying Joes River Formation. The Barbados Ridge probably comprises clastic sediments deposited both in a trench and on the Atlantic Ocean floor.

The island of Barbados in the eastern Caribbean is the only exposed part of a major north-south structure known as the Barbados Ridge (Fig. 1). The Ridge is a bathymetric feature extending 400 km northwards from Tobago, about 100 km wide at Barbados and rising 4–5 km from the Atlantic Ocean floor to the east (Westbrook 1981). Its highest part coincides with linear negative free-air and Bouguer gravity anomalies (Kearey *et al.* 1975; Bowin 1976) and a small magnetic anomaly (Westbrook 1975). The Ridge also marks the eastern limit of seismicity associated with a west-dipping Benioff zone (Sykes & Ewing 1965), i.e. the eastern boundary of the 'Caribbean plate' (Molnar & Sykes 1969). Fault-plane solutions for earthquakes along the east and north margins of this plate imply that it is today moving east with respect to North America and the central Atlantic: the rate of convergence of the Caribbean and Atlantic plates is $\frac{1}{2}$ to 2 cm yr^{-1} depending on the method of calculation (Molnar & Sykes 1969). Caribbean plate motion with respect to South America is less well known but appears to include a compressional component: a number of small blocks may be behaving as microplates in a broad zone of deformation along the north coast of Colombia and Venezuela (Jordan 1975; Bowin 1976).

Seismic reflection and refraction profiles, combined with gravity data, show that the Barbados Ridge is underlain by a thick (20 km beneath Barbados) slice of low-velocity, acoustically opaque material before oceanic crust is reached (Westbrook 1975). The presence of faint reflectors dipping shallowly west along the eastern margin of the ridge led Chase & Bunce (1969) to propose an origin by accumulation of crumpled and thrust-faulted sedimentary material in the southern part of the Antilles trench (sic). This idea of an accretionary prism has been elaborated by Westbrook (1975, 1981), Bowin (1976), Dickinson & Seely (1979) and Speed (1979) among others; such an origin is certainly to be expected from the position of the ridge just east of the Lesser Antilles island arc. Alternatively the ridge has been interpreted as an *in situ* northward extension of the continental rise of S America (Senn 1947; Meyerhoff & Meyerhoff 1972), and as a huge gravity slide into a trench from the west (Daviess 1971).

Most of Barbados is covered by the Coral Rock of Pleistocene to Recent age. An inlier on the NE coast, the Scotland District (Fig. 2) exposes a highly deformed succession of terrigenous clastic sediments (Scotland Group) overlain by much less deformed pelagic sediments (Oceanic Group). In the south of the inlier the Joes River Formation mudflow outcrops between these two groups. Foraminifera and molluscs from the Scotland Group yield an age of lower to middle Eocene and possibly

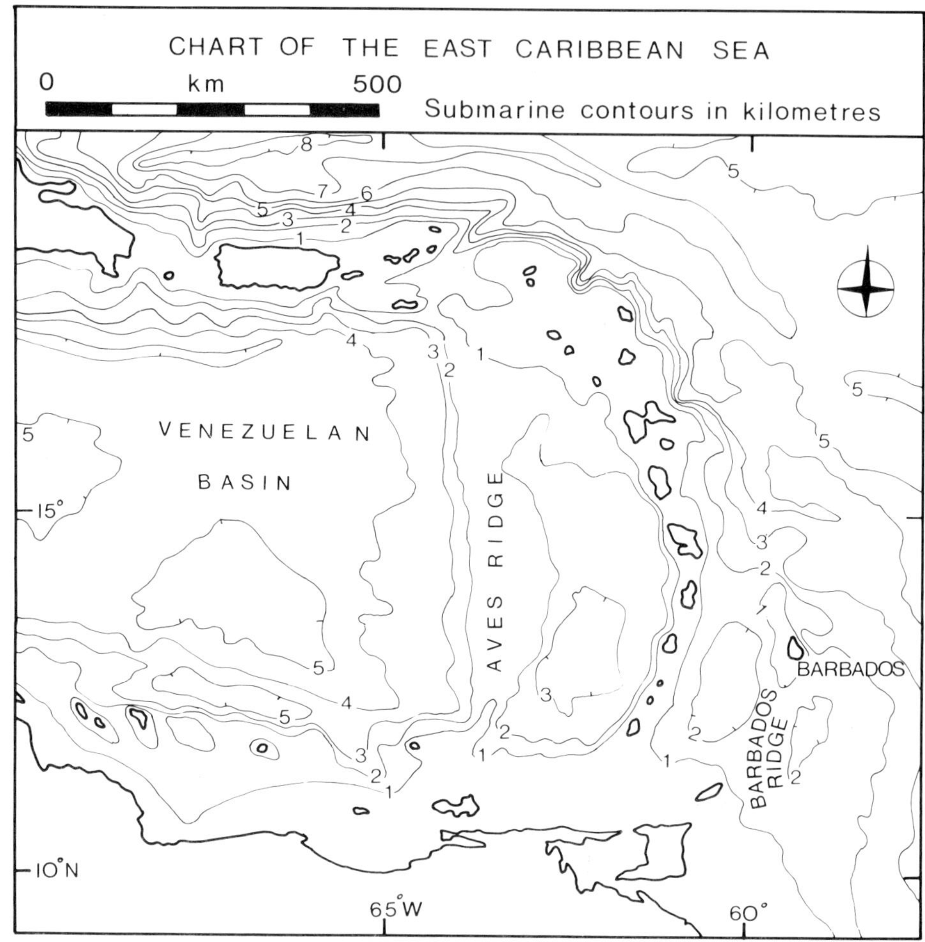

FIG. 1. Location map of Barbados. Bathymetry from Hurley (1966); Edgar & Saunders *et al.* (1973); Westbrook (1975); Ladd & Watkins (1978).

include upper Eocene forms (Trechmann 1925; Senn 1940, 1947; Cizancourt 1948; Caudri 1972). The Joes River Formation contains blocks of Palaeocene limestone (Caudri 1972) but its matrix has not been accurately dated. The Oceanic Group is largely Oligocene but its base could be as old as middle Eocene (see Lohmann 1974 for summary). It has been argued (e.g. Baadsgaard 1959; Speed 1979) that the Oceanic and Scotland Groups are, in part, time-equivalents, separated everywhere by a tectonic contact. We maintain that the contact is unconformable in at least a few places (Jukes-Brown & Harrison 1891, p. 271, appendix) though the unconformity may have been a plane of movement during subsequent deformation.

In the Scotland District the Tertiary rocks occur in a series of ENE-trending fault-bounded blocks (Fig. 2). The major faults are not exposed and their nature can only be inferred. The occurrence of a few fault slices of steeply dipping Oceanics and their presence in boreholes which penetrate $4\frac{1}{2}$ km of stacked Oceanic/Scotland sediments led Speed (1979) to propose that the Scotland District was a pile of totally unrelated fault slices. Sedimentary sequences and structural styles vary between blocks, but the five lithostratigraphic members of Senn (1940) have been mapped and correlated from block to block (Fig. 3). Pudsey (1979) has presented a table of systematic stratigraphy.

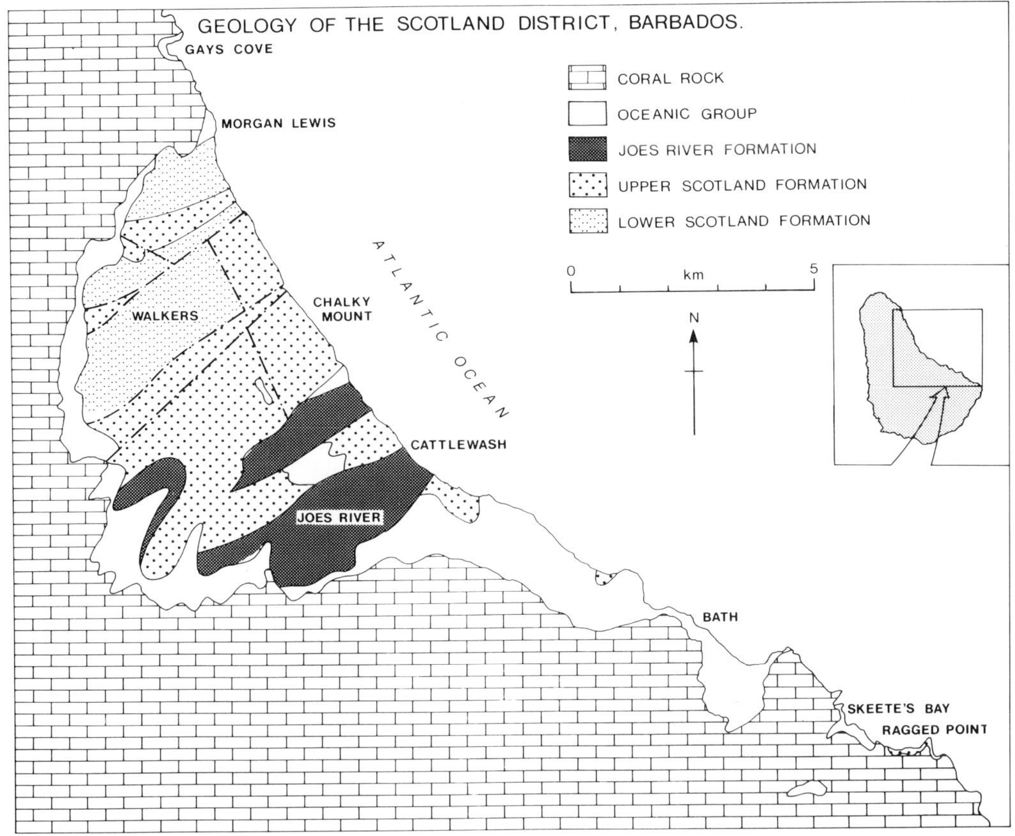

FIG. 2. Simplified solid geology of the Scotland district. Coral Rock, Oceanic and Joes River outcrops from Baadsgaard (1959) and Saunders (1965). Marls of Bissex Hill etc., included within Oceanic Group. Inset map shows location on Barbados.

Sedimentology

Lower Scotland Formation

Walkers Member

The member consists of alternating graded sands and laminated clays, in turn arranged in packets dominated by sand and packets dominated by clay.

Sand bed thickness ranges continuously from a few centimetres to 2 m. Most sands over 10 cm thick are medium or coarse-grained, smoothly graded to fine (Fig. 6A); there are occasional granular basal lags. Fine, poorly graded sand constitutes most of the thinner beds. The bases of sands are sharp (rarely erosive) and, where exposed, about half the beds bear sole marks: flutes, grooves, load casts, trails and prod marks. Most sands are pale yellowish-grey to white, well-sorted, quartzose and non-micaceous. Internally, thicker beds are parallel-laminated, often with ripple cross-laminated or convolute tops. Thinner sands are rippled or convolute throughout. The few unlaminated sands exhibit dish structures in their middle portions. The sole markings and ripples define a very consistent palaeocurrent direction from the SW and west.

The tops of sands grade into the overlying clay over a thickness of ½ to 1 cm. The clays are dark grey to brown, thickly laminated, non-silty and ferruginous. Clay units are a few centimetres to 2 m thick, rarely up to 5 m; they generally contain very thin graded silts and very fine sands, indistinctly parallel or ripple cross-laminated. Beds of all thicknesses are laterally continuous across most outcrops.

On a scale of tens of metres vertically, there are quite distinct sandy and clayey packets of beds, sandy packets containing about 50% sand and clayey packets 10% or less. Sand-

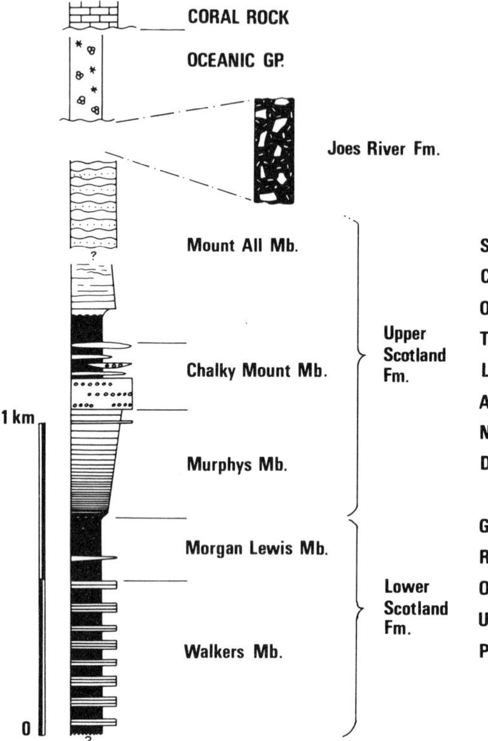

FIG. 3. Stratigraphic terminology used in this paper, with approximate thicknesses and a diagrammatic guide to lithology. The thicknesses of the Mount All Member and the Joes River Formation have not been accurately determined, on account of the complex structure of the former and the lack of bedding in the latter. The Oceanics seem to be much thicker in some subsurface wells (Baadsgaard 1959), but this may be due to tectonic repetition. Conformable contacts have been documented between each pair of members except for part of the Mount All Member. In particular the Morgan Lewis-Murphys transition has been mapped for several kilometres.

dominated packets in the type area of the Walkers Member (for location see Fig. 9) may be correlated over several hundred metres (Fig. 4) with no apparent interconnection between them.

Throughout the member there is no overall change in grain size of the sands, in bed thickness or packet thickness. The only trend discernible is that the clayey packets contain slightly fewer sands towards the top.

Morgan Lewis Member

This member consists almost entirely of clay with very thin graded silts and fine sands, similar to the clay packets in the Walkers Member (Fig. 6B). The clays are dark purplish-grey to brown, thickly laminated and ferruginous to the point of containing stringers of ovoid iron concretions. A few of the silt beds are completely cemented by iron oxy-hydroxides ('clay-ironstones') and trails are well-preserved on their bases. Scattered fine sands up to 12 cm thick bear sole marks indicating current flow from the SW. Isolated coarse sand beds from 10 cm to 1 m thick occur sporadically: they have stepped erosive bases and some contain clay flakes and fine sand clasts.

Two laterally impersistent sand intervals are present in the type area, each about 25 m thick. The lower (and better exposed) contains *c.* 50% sand in massive and parallel-laminated beds up to 2 m thick. The sands are fine-grained, moderately sorted, poorly graded and slightly carbonaceous. A few sole marks indicate current flow from the NW.

Upper Scotland Formation

Murphys Member

The base of the member is a thickening and coarsening upward sequence, the sand:shale ratio increasing from 1:20 to 1:3 in a thickness of 10–15 m. The bulk of the Murphys Member is quite variable both laterally and vertically, and does not show the good packeting developed in the Lower Scotland Formation. Overall it becomes sandier upwards, sand:shale ratio being 2:1 at the top; the modal grain size remains constant at fine sand. In most measured sections half the sands belong to a distinctive laminated carbonaceous facies, variously coloured red, grey, yellowish or black due to combinations of carbon and iron oxides. Both carbonaceous and non-carbonaceous fine sands are moderately sorted, micaceous and rather silty, usually poorly graded though the bases of beds are sharper than their tops. Erosion is virtually absent. There are a few medium and coarse graded sands but hardly any sole marks, except for sporadic load casts and trails. Sand bed thickness, commonly 10–40 cm, ranges up

FIG. 4. Measured sections in type area of the Walkers Member (for location see Fig. 9), showing lateral correlations between sandy packets of beds. Part of two sections are enlarged to demonstrate bed-by-bed correlation. Arrows indicate current flow directions.

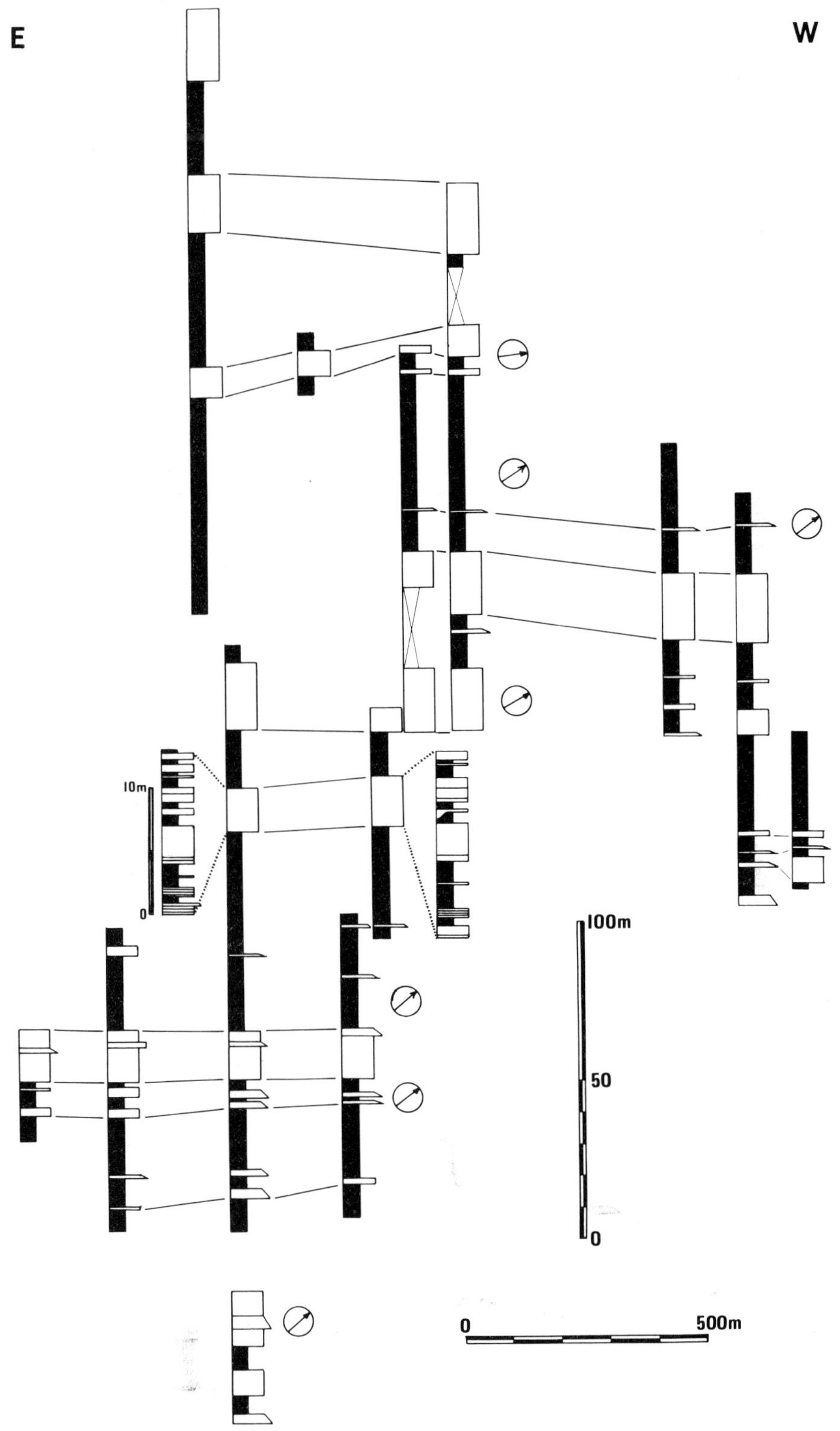
E
W
10m
0
100m
50
0
0
500m

to 2 m. The thick beds are commonly composite, consisting of several parallel or ripple cross-laminated units stacked together and varying slightly in composition or type of lamination. Palaeocurrents, almost all from ripple cross-lamination, are again consistently from the SW.

The clays are much siltier than those in the Lower Scotland Formation and there are no clay-ironstones. Very thin silt and fine sand interbeds are carbonaceous and ripple or trough cross-laminated. The clay is almost black and pyritic when fresh, weathering brownish grey and growing gypsum crystals on the surface. In the northern Scotland district a few very thin beds of silty cream-coloured chalk were found—the same lithology as parts of the Oceanic Group.

Two conglomerates occur in the Murphys Member; the better exposed is 3–4 m thick and mappable for 1½ km along strike. Small and medium pebbles are dominantly of well-rounded hard micritic limestone and clay-ironstone, with minor quartzite, shale and vein quartz, and locally numerous shell fragments (bivalves, gastropods and scaphopods) in a fine sand matrix.

A principal feature of this member is synsedimentary deformation, including normal faulting (Fig. 6C), slumped packets of beds a few metres thick, disruption of sand beds by clay injection, and olisthostrome units up to 10 m thick consisting of rafts of coarse sand up to 4 m × 30 cm in a red shaly matrix. Measurement of slump fold and fault orientations yielded no consistent palaeoslope direction.

Chalky Mount Member

This member is distinguished by the very coarse grain size, massive nature and lateral discontinuity of the sediments: it shows great variation between outcrops and only the main section will be considered in detail here. In the type area (Fig. 9) the member has been divided into the Chalky Mount Summit Beds (lowest 100 m) and the Chalky Mount North Beds (metre 100 to top) (Pudsey 1979). The former consist of thick, amalgamated, coarse to granular clayey white sands. Some beds are well-graded (Fig. 6D) and a very few have single sets of tabular cross-bedding at the top. Silty mud interbeds are thin and impersistent because of widespread scouring. There is no upward change in bed thickness or grain size and the unit has a sharp top.

In the eastern part of the type area the Chalky Mount North Beds (Fig. 5) consist of very distinct graded packets of coarse sands and conglomerates up to 20 m thick, separated by carbonaceous silty clays. Westwards the sands become thinner and finer and there is much ripple cross-lamination. The most westerly sections comprise clays containing large isolated blocks of very coarse white sands as well as a few coarse turbidite sand beds, fine rippled sand being unimportant except near the base. The clays are contorted beneath some of the sand blocks.

In contrast to the view of Senn (1940) the Chalky Mount Member does not thicken consistently to the SE. The few palaeocurrent measurements (Fig. 5) show that both coarse and fine sands were deposited by currents flowing from the SW and west. Chalky Mount outcrops exhibit many slump features, particularly synsedimentary normal faulting with a downthrow to the north. Trace fossils are rare, but there are stellate impressions at the top of the Chalky Mount Summit Beds, and two beds containing vertical burrows in other localities.

Conglomerates are prominent in the eastern part of the type area: large rounded clay and fine sand intraformational clasts are most common, followed by micritic limestone, iron concretions, quartzite and vein quartz (see Speed 1979). Baadsgaard (1959) reports an Upper Cretaceous age for a limestone pebble.

In the SE of Barbados there is no Oceanic Group between the Coral Rock and the local Chalky Mount Member, and the coarse and granular sandstones are cemented by calcite.

Mount All Member

The member has such a complex structure that it is not easy to derive a coherent sedimentary picture: the thickness estimate of 500 m is very tentative. Above the Chalky Mount Member immediately north of the Chalky Mount type area, the Mount All Member begins with 75 m of thickly laminated brown and grey clays with very thin graded silts and no iron concretions. Above these clays and at Mount All, it consists of thin to thick, fine to coarse sands interbedded with red-brown silty clays (Fig. 6E). The interbedding is usually random though packets of thin and thick sands are well-developed in one or two places. The sands have sharp or erosive bases, gradational tops, and are quite commonly cemented by calcite or silica. Sorting varies from moderately good to poor. The fine sands are micaceous and carbonaceous (though less so than the Murphys sands) and internally parallel or convolute-laminated: the coarse sands are quartzose, ferruginous, clayey and well to poorly graded. Some contain dish structures and others water

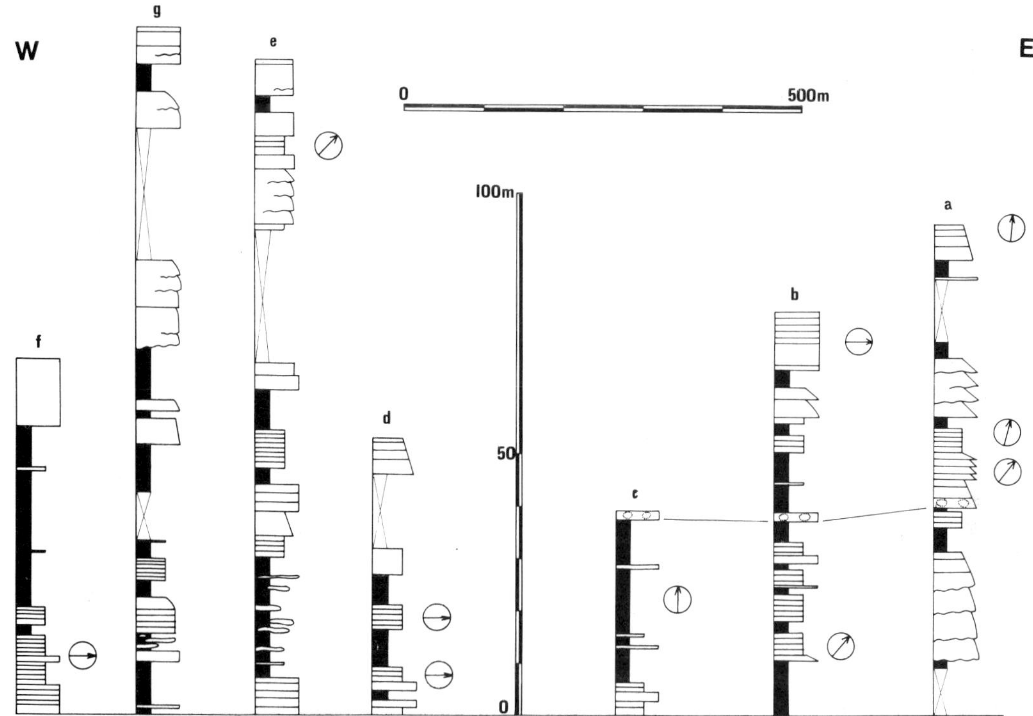

FIG. 5. Measured sections 3a to 3f at Chalky Mount (for location see Fig. 9), showing lateral variation in the Chalky Mount North Beds. Datum at base is top of 100 m thick, sheet sand. There is a marker bed containing huge concretions at metre 40 in sections a, b and c. Note extreme discontinuity of sand bodies in sections e, f and g. Units of sand rafts near base of sections e and g are similar to olisthostromes described in Murphys Member. Sections are overlain by thick clay unit at base of Mount All Member.

escape features. Overall sand : shale ratio is about 2:1. Sole structures are uncommon and include flutes modified by load casting, and frondescent marks. Few current measurements were made, on account of the difficulty of unfolding the steeply plunging structures, but current flow seems to have been from the southern quadrant.

At least some, and probably most, of the complex folding in this member can be ascribed to slumping very shortly after sedimentation (see below under 'Structure'). No consistent palaeoslope direction has been determined.

The top of the Member is not seen and there are no interbeds of Oceanic lithology.

Interpretation

The Lower Scotland Formation is entirely turbiditic and hemipelagic and we assume it to have been deposited in deep water. The Upper Scotland Formation is more variable, but still mainly turbiditic; it is also probably a deep-water deposit. Foraminifera show the whole Scotland Group to be marine. Because of its very coarse grain size and the presence of a shallow water benthic fauna, the Chalky Mount Member has often been interpreted as a shallow marine deposit (e.g. Trechmann 1925; Baadsgaard 1960). We believe that the member is entirely a deep-water mass-flow deposit for the following reasons: (1) position within a turbidite succession; (2) presence of grading and poor sorting of the coarse sands; (3) absence of wave ripples and of evidence for emergence, and almost total lack of cross-bedding in the sands; (4) shallow-water fossils occur only in conglomerates and most of the shells are fragmented: shells were resedimented along with pebbles; (5) extreme rarity of trace fossils of any kind and only two instances of vertical burrowing in all the sections examined (see below).

The Scotland Group on Barbados is part of

FIG. 6(A) Walkers Member, type area. Parallel-laminated medium-grained sand bed with fine-grained top showing water-escape structures, overlain by thinner parallel and ripple cross-laminated fine-grained sand bed. Slightly erosive base of sand is unusual. Scale: hammer 32 cm.

(B) Morgan Lewis Member, type area. Monotonous clays with very thin graded slits (lighter) and one clay-ironstone (darker, at top). Scale: 1 m of tape extended, lens cap 5.5 cm.

the Barbados Ridge and any interpretation must take into account the size, shape and position of the Ridge as well as the sedimentary features of the miniscule outcrop area. Thick piles of deep-sea clastic sediments may accumulate in the following environments (Rupke 1978):

Basin plains
- 1 Open oceans
- 2 Deep-sea trenches
- 3 Enclosed seas
- 4 Borderland basins
- 5 Miscellaneous small basins

Turbidite fans
- 6 Abyssal cones
- 7 Deep-sea fans
- 8 Short-headed delta front fans
- 9 Continental rise fans

Some of these may be eliminated in the case of the Scotland Group.

(1) is unlikely because of the thickness (nearly 2 km cf. a few hundred metres in modern examples) and the lenticularity within the Upper Scotland Formation.

(3) is not possible because there are no land areas enclosing the Ridge.

(4), (5) and (8) involve relatively small sediment bodies.

(9) can probably be eliminated because an essential characteristic of such accumulations is extensive reworking by indigenous bottom currents. The Scotland Group sediments show very little evidence for reworking.

The three remaining possibilities of a deep-sea trench, an abyssal cone and a deep-sea fan must all be considered likely (see 'Regional significance'). Certain features of the submarine fan model (e.g. Walker 1978; Normark 1978) are consistent with the Scotland Group sequences (Fig. 7). For instance the non-erosive, laterally continuous, rather regularly spaced, sand packets of the Walkers Member may represent fan lobes (cf. Mutti *et al.* 1976). During Morgan Lewis deposition sand supply became finer and more intermittent: the Murphys Member is the result of renewed progradation, the source now supplying more carbonaceous and finer sand. The uncommon and thin chalk beds are more likely to be carbonate turbidites shed from an adjacent swell than to represent pelagic background sedimentation. If the latter were so, one would expect many more such thin beds, or the clay should have an appreciable carbonate content.

The Chalky Mount Member is interpreted as channel deposits though the small size of the outcrop area makes it difficult to distinguish between a series of small distributary channels, and one major channel containing a complex network of smaller braided channels. Both associations have been described from the mid-fan regions of modern submarine fans. The sheet-like non-erosive Chalky Mount Summit beds perhaps represent a proximal fan lobe or channel mouth bar. The rare cross-bedding could have been formed by the tails of turbidity currents or by down-channel storm-driven currents (Shepard *et al.* 1979). The slide blocks

---

(C) Synsedimentary normal faulting in fine laminated sands of Murphys Member. Dark laminae are concentrations of carbonaceous material. Fault movement of the base of the bed occurred before deposition of the top was completed, as uppermost lamination is undisturbed. Scale: lens cap 5.5 cm.

(D) Chalky Mount Member, type area. Very thick, well-graded pebbly sands. Younging direction to right. Scale: hammer 32 cm.

(E) Mount All member, type area. Thin to thick fine and coarse-grained sands interbedded with silty clays. Younging direction to left. Scale: hammer 37 cm.

(F) Joes River Formation, Joes River. Blocks of sand in structureless mud matrix. The largest block is of very coarse sand graded to medium and its younging direction is to the right. Scale: hammer 37 cm.

(G) Syncline in clays of Morgan Lewis Member. Note simplicity of fold style and general lack of parasitic folds even in this incompetent lithology. Scale: figure 1.6 m.

(H) Folding in lower part of Mount All member. Overall younging direction is upwards and to the right. An almost upright anticline at left is followed by a slightly faulted syncline with parasitic minor folds, and at right a recumbent anticline. Folding took place at a stage when the sands were sufficiently lithified neither to thin nor thicken around the hinges of folds. Scale: hill is about 40 m high.

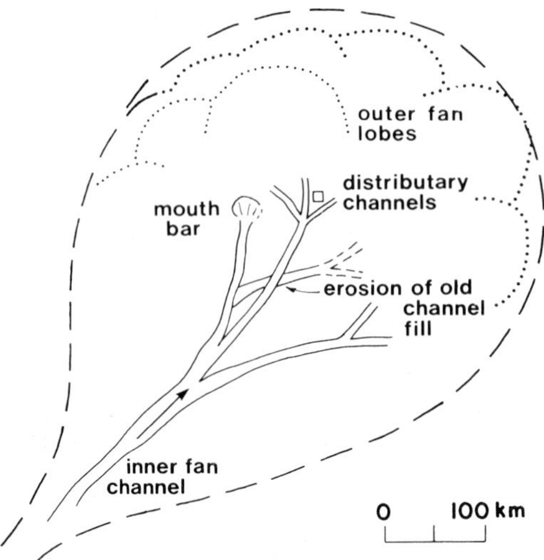

FIG. 7. Some features of the submarine fan model applicable to the Scotland Group. Small square represents position of Barbados (relative to the model) during Chalky Mount deposition.

probably moved only a short way downslope following erosion of a previously deposited channel fill (Fig. 7).

After the hiatus in sand supply represented by the thick clays at the base of the Mount All Member, progradation was renewed. The very thick laminated fine sands, seen both in the Mount All and Murphys Members, are clearly not the products of single classical turbidity currents. They may have been laid down by longer-lasting and gently pulsating river-generated turbidity currents (Heezen *et al.* 1964). The lack of organisation of the Mount All Member as a whole may be a consequence of the very frequent disturbances which triggered slumping: stable conditions were never maintained for long enough for a consistent sediment dispersal pattern to develop.

Petrography and provenance

The composition of the conglomerates has been described above. In thin section, Scotland Group sandstones are poorly to moderately well sorted with a high matrix content. Most grains are angular, though large quartz grains are well-rounded. A number of quartz grains show corroded edges particularly where the rock is locally calcite-cemented. Compositionally, the sands belong to the category of quartz wackes of Dott (1964). Of the large (>1 mm) quartz grains, a majority show marked strain shadows and a few are polycrystalline, but about 30% have uniform extinction and some approximate to the hexagonal form of β quartz. The latter are probably of volcanic origin (Todd & Folk 1957), the former plutonic or metamorphic (Blatt *et al.* 1972). Feldspar is uncommon and heavily altered: we did not find widespread microcline as reported by Velbel (1980). The heavy minerals in the sands were derived from a metamorphic terrane (Matley 1932; Hedberg, in Senn 1940).

The two external sediment source areas were the metamorphics on the north coast of South America and the volcanics of the Lesser Antilles arc, with the former predominating. This is in agreement with the conclusions of Velbel (1980). A local source is postulated for the conglomerates since they contain no metamorphic pebbles.

Clay mineralogy

The clay mineralogy of 37 samples was determined using X-ray diffractometry and the results are summarized in Fig. 8. The mineralogy

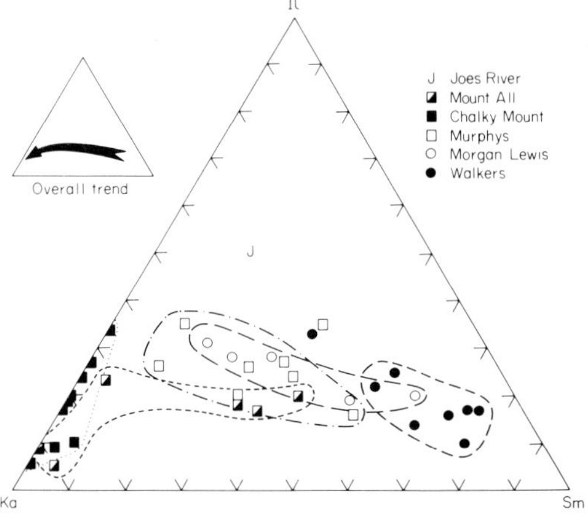

FIG. 8. Clay mineralogy of the Scotland Group. Ka = kaolinite, Il = illite, Sm = smectite. Minor components (chlorite and vermiculite) not plotted.

of each member of the Scotland Group is more or less distinct. The Walkers and Chalky Mount clays are the most readily identifiable, the former by their high smectite content (up to 75%) and the latter by the virtual absence of smectite and abundance of well-crystalline kaolinite. The Morgan Lewis clays plot in a very narrow field, all having Ka:Il close to 3:2. The Murphys and Mount All both show a rather wide range in clay composition. The general trend is a decrease in smectite and increase in kaolinite during Scotland sedimentation, accompanied by an increased illite contribution in the middle (see inset in Fig. 8). Only one sample of Joes River clay was analysed: it bears little relation to the Scotland clays, being much richer in illite.

We believe that most of these clays are turbiditic rather than hemipelagic because they frequently have a gradational lower boundary with the c or d division of a turbidite sand, they are not bioturbated, planktonic microfossils are extremely rare (Senn 1940; Baadsgaard 1959), and interbedded clay bands of two or more distinct colours are not seen (cf. Rupke & Stanley 1974; Hesse 1975).

The obvious source for the smectite is the volcanic Lesser Antilles arc, and for the kaolinite the deeply weathered tropical soils of northern South America. Kaolinite flocculates very easily in sea-water and is in that sense a proximal clay mineral; it is therefore not surprising to find it in clays interbedded with proximal turbidites. The smectitic Lower Scotland clays may be hemipelagic, or more probably redeposited ashfalls. Most of the samples contain a trace of chlorite, three containing 10% or more: this may be derived from metamorphics. The effects of diagenesis can probably be neglected since the clays are quite young and have never been deeply buried; even smectite, which is particularly unstable, remains in large quantities.

The progressive decline in smectite and increase in kaolinite may be attributed to either or both of (a) lessening of volcanic activity during Scotland Group deposition, (b) swamping of the volcanic contribution as the continent-derived sediments prograded into the basin.

Palaeoecology

Shelly fauna of the Chalky Mount Member

The diverse molluscan fauna of a few conglomerate beds was described and illustrated by Trechmann (1925) who dated the assemblage as middle/upper Eocene. During the present study small collections were made from Murphys and Chalky Mount conglomerates, and Trechmann's collection of nearly 100 species in the British Museum was re-examined briefly. The fauna consists largely of worn and broken specimens, and the evidence for redeposition in deep water is now substantial (see above). All the specimens are fully marine. The taxa represented are all thought to have lived in shallow water (<100 m) deep) and are characteristic of sandy substrates. Many were shallow burrowers. There is a marked lack of bored and encrusted shells. This may indicate burial by sand soon after death, or frequent slumping and turbidity flow of the sediment into water too deep for any epifauna to live.

The source area for a small part of Scotland sedimentation was a shallow sandy marine shelf, well-oxygenated and with an abundant food supply, supporting a very diverse benthic molluscan fauna. Mixing of the pebbly debris with shell debris probably took place during resedimentation, as there are no byssally attached molluscs.

No palaeoecological studies have been undertaken on the larger foraminifera of the Scotland Group.

Trace fossils

Most traces are post-depositional trails on the bases of thin turbidite sand beds. Those in the Lower Scotland Formation include *Palaeodictyon*, *Spiroraphe*, *Scolicia* and other grazing trails typical of quiet deep-water environments (Seilacher 1978). A small form of *Chondrites* occurs in clay-ironstones. In the Upper Scotland Formation there are two isolated beds containing *Tisoa*-like vertical pipes, and stellate traces cf. *Haentzschelinia* occur on the tops and bases of a few coarse sands, in addition to nondescript sinuous grazing trails beneath Murphys and Mount All Member sands. Crimes (1977) has described a similar association from an Eocene deep-sea fan in Spain: he postulates that 'shallow-water' trace-makers may live in the mid-fan region as turbidity currents periodically bring in oxygen and food from the shelf (see also Corbo 1979). Thus the trace fossil evidence is consistent with the deep marine interpretation.

The Joes River Formation

This curious unit presents stratigraphical problems: it has not been accurately dated and at outcrop it is not clear to what extent it overlies, interfingers with, or is faulted against the Scotland Group. It is a mélange *sensu lato* containing blocks of fine to coarse sand of all sizes up to tens of metres long in a massive, often sheared, gritty mud matrix. The whole formation is impregnated with heavy black oil. Senn (1940) was able to detect a faint stratification and estimate a maximum thickness of 400 m.

The sand inclusions are matrix-supported, vary in shape from equant blocks to elongate rafts (Fig. 6F) and occur in random orientations. Some are well-graded and all are similar to Scotland Group sands, from which they could have been derived. There are also small (few centimetres) angular blocks of Lower Scotland-type clay-ironstones, some of which appear slickensided. Senn (1940) mentions blocks of sandy foramineral limestone of Palaeocene age, which do not outcrop elswhere on the island. Herrera & Spence (1964) in a detailed account, describe a locality with very large sand blocks up to 300 m long and parallel to the regional strike, in a contorted but definitely bedded mud matrix.

The formation has been variously interpreted as a fault breccia (Baadsgaard 1959), as a submarine slide deposit (Hess 1938; Daviess 1971; Herrera & Spence 1964) and as the product of sedimentary volcanism (Senn 1940). Baadsgaard's theory is rejected because of the thickness and outcrop width of the Joes River Formation (Fig. 2) and because there are major faults in the northern Scotland district where it is not present. Submarine sliding is more likely but does not explain the oil content, which is much higher than in the exposed Scotland Group. Lithologically the Joes River Formation strongly resembles mud volcano deposits described from Trinidad by Higgins & Saunders (1967, 1974) and Arnold & Macready (1956) where the mud matrix is often sheared during extrusion and large sand blocks can be detached from the walls of the vent and brought to the surface. Generation of methane during bacterial decay of organic matter assists mud injection by rendering it more fluid (Hedberg 1974) and the extruded mud is often very hydrocarbon-rich. We therefore concur with Senn's interpretation, which is supported by the contrast in mineralogy between Joes River and Scotland clays (see above); one would expect more illite in clays brought up from greater depth.

The mobile mud would of course, be subject to downslope movement after extrusion, just as the Joes River Formation is much disturbed by land-slipping at the present day. The formation is highly significant in interpreting the structure of the Scotland district (see below).

Structure

The ENE structural trend is constant not only at outcrop (Figs 2 & 9), but in the subsurface throughout the island. This trend is anomalous in the present-day tectonic setting adjacent to a north-south trending island arc. Major strike faults separate discrete slices in the Scotland district. These faults are not normally exposed and their nature and exact positions can only be inferred. Their attitudes vary from low to high angle (Speed 1979) and dips are to both north and south. There is outcrop evidence for low-angle north-dipping reverse faults in the northern part of the area.

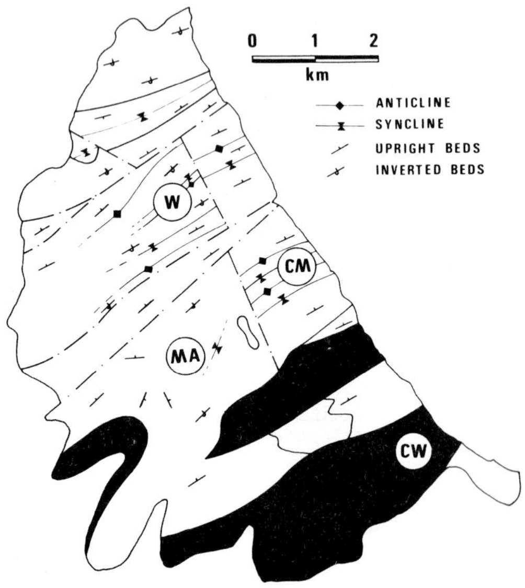

Fig. 9. Structural map of the Scotland District showing fold axes and bedding attitudes. Encircled letters are localities. W = Walkers Member type area; CM = Chalky Mount type area; MA = Mount All type area; CW = Cattlewash, where a large faulted slice, now enclosed within the Joes River Formation, exposes flat-lying Oceanic Group unconformable on gently dipping Upper Scotland Formation.

In general, the structure simplifies with increasing stratigraphic depth, this trend being modified by variations in competence of the rocks. There is a progression upwards from simple, laterally persistent, gently east-plunging folds in the Walkers Member to very complex, steeply plunging structures in the Mount All type area. In the Lower Scotland Formation there are open to tight, upright to overturned, angular folds with wavelengths of tens to hundreds of metres (Fig. 6G & 10). The axial traces of the folds undulate slightly between ENE and NE and the poles to bedding show some scatter about a NNW–SSE girdle (Fig. 10): the initially almost planar axial surfaces have been refolded by later folds with near-vertical axes and steep NW-striking axial planes. This is similar to the results of Speed's work on the Chalky Mount

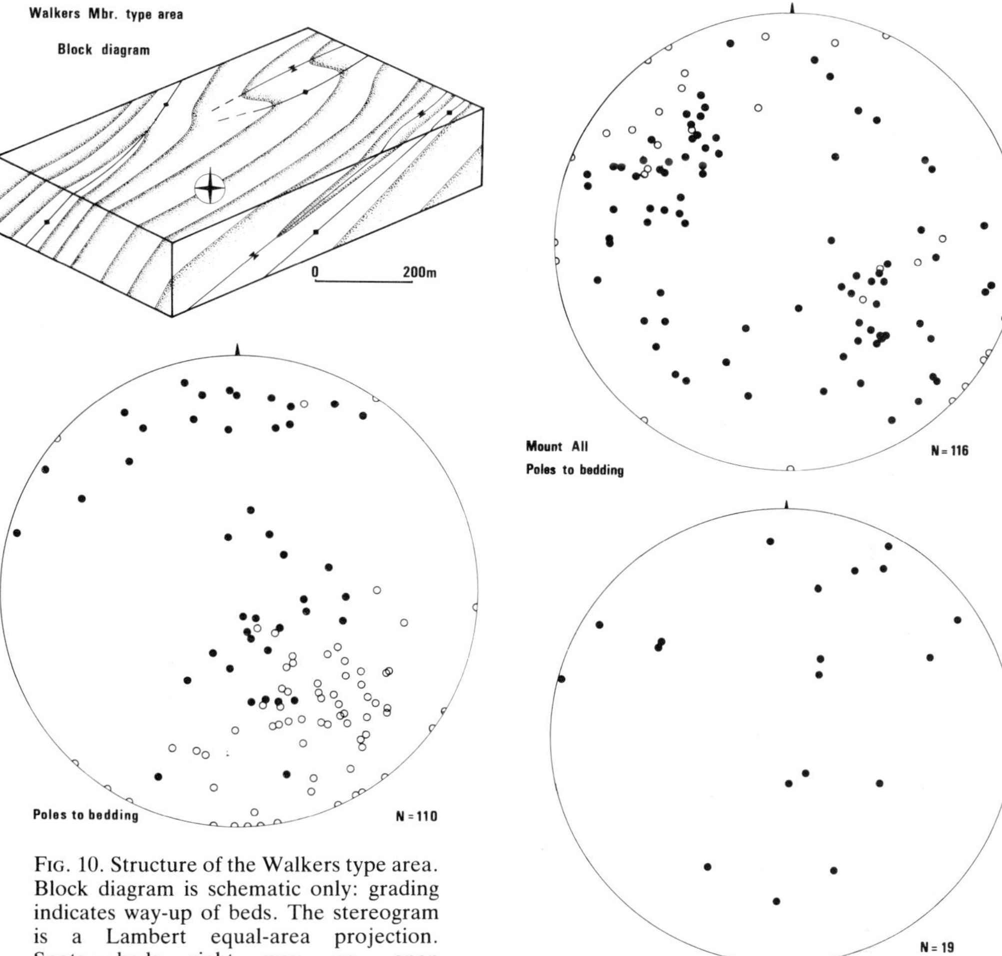

FIG. 10. Structure of the Walkers type area. Block diagram is schematic only: grading indicates way-up of beds. The stereogram is a Lambert equal-area projection. Spots = beds right way up, open circles = inverted beds.

FIG. 11. Structure of the Mount All type area. Stereograms are Lambert equal-area projections. In the upper diagram open circles = inverted beds. The only mesoscopic folds plotted are those where the plunge could be measured directly at outcrop.

area (Speed 1979). Towards the top of the stratigraphic succession, folds become more complex, with parasitic and disharmonic minor structures (Fig. 6H), though the intensity of deformation (amount of shortening) probably does not vary much. The Mount All area shows widely scattered poles to bedding (Fig. 11) and mesoscopic fold axes plunging to the north, NE, SE, south and SW. There is very little inversion of beds, but few fold axial traces are mappable for more than 100 m, and it is possible that there is considerable faulting in this heavily forested area.

There is no cleavage in the Scotland Group and all deformation took place at a very high structural level. Much was synsedimentary, slumping frequently occurring while the sediments were poorly lithified. The overlying Oceanic Group is virtually undeformed except for faulting and tilting of blocks, so the main folding took place before the middle of the upper Eocene. At the top of the Scotland Group it is not possible or useful to separate 'tectonic' from 'sedimentary' deformation. Evidently local slopes were generated on the seafloor during and after Scotland deposition. Possible causes include: (a) bank collapse into channels or sliding of levee material towards the interchannel area (Mutti 1977); (b) slumping down the front of the aggrading sediment pile, or down the frontal slope of an advancing accretionary wedge; (c) mud diaprism. Local slopes generated by the upward movement of overpressured mud are a common cause of slumping on deltas (Roberts *et al.* 1976).

Although progradational slumping and bank collapse may have been responsible for some small-scale faulting and the sand blocks in the Chalky Mount Member, there is no evidence that the superficial folds are related to channel margins and the random fold patterns conflict with the regularity of the palaeoflow indicators. We therefore think that mud diapirism is the most important factor. As distal, organic-rich muds become buried by an advancing sediment pile, their impermeability quickly results in overpressuring by water and hydrocarbon gases (Hedberg 1974; Chapman 1973, 1974). Diapirs form at a spacing related to the thickness of the stratigraphic units involved, to the viscosity contrast and to any basement structures. They may be circular, or linear parallel to the depositional strike (Chapman 1974). Slopes are formed in varying directions around their flanks and slumping and faulting occur as the diapirs continue to rise. In some circumstances they may break through to the surface. We believe that increasing upward disturbance by rising mud diapirs locally culminated in extrusion on to the seafloor of hydrocarbon-rich mud—the Joes River Formation.

Diapiric structures have been identified on seismic reflection profiles in the SE part of the Barbados Ridge complex, and are indicated on the USGS Geologic-Tectonic map of the Caribbean. Although the Scotland district is too small, and now too severely faulted, to be able to delineate individual diapirs and their associated radial and concentric structures, the present distribution of Joes River Formation outcrops (Fig. 2) suggests they were elongated parallel to strike.

Regional significance

In terms of plate-tectonic theory the Barbados Ridge should be an accretionary prism of deformed and thrust-faulted trench and ocean floor sediment, analogous to the Mentawai Islands (Karig *et al.* 1980). However, there are major inconsistencies between Barbados and the accretionary prism model: the structural trend is at a high angle to that of the arc, there being no outcrop evidence for major north-south structures; and pelagic sediments *overlie* turbidites unconformably.

As pointed out above (p. 299) the Scotland Group sediments may have been deposited in a deep-sea trench, an abyssal cone fed by a broad delta front, or a deep-sea fan fed by a submarine canyon. Although tectonic accretion may have led to the construction of the large Barbados Ridge from sediment deposited entirely within a narrow trench, it is more likely that sediments originally deposited on the Atlantic Ocean floor were also incorporated (Fig. 12). The occurrence of packets of sand and clay, plus channel sequences, indicates that there was at least sometimes an organized sediment dispersal system, i.e. some kind of cone or fan: though this may of course have been a very elongate 'fan' within a trench, rather than an unconfined one on open ocean floor. The consistent palaeocurrent directions support an elongate fan. Without knowing more about the thicker, more proximal end of the Barbados Ridge, it is impossible to say whether the source was a submarine canyon or a broad delta-front.

In Fig. 12 the east margin of the Caribbean plate has been drawn west of its present position with respect to South America. How far west is not known: it is clearly inappropriate to extrapolate the present convergence rate back 50 or 60 Ma, and the Caribbean and South American plates may have been coupled together at some time during the Tertiary (Jordan 1975). J. F. Dewey (pers. comm.) has calculated, from magnetic lineations in the Atlantic, that 50 Ma ago North and South America were converging very slowly, with the Caribbean plate squeezed sideways between them. Very oblique convergence along a complex fault system between the Caribbean and South America might have resulted in uplift and sediment supply to the newly initiated (Malfait & Dinkelman 1972) Lesser Antilles Trench to the north. Either the now inactive Aves Ridge, or the first volcanic islands of the Lesser Antilles chain, could have supplied the volcanic quartz detritus and the smectitic clays.

The prodigious change in sedimentary facies from Scotland Group to Oceanic Group could be due to one or more of: (1) cessation of uplift of source as convergence between North and South America changed to sideways motion (J. F. Dewey, pers. comm.); (2) strike-slip movement of depositional area eastwards away from the delta source (cf. Howell *et al.* 1980); (3) development during strike-slip movement of deep trenches such as the Venezuela Trench (Ladd & Watkins 1979) which trapped terrigenous clastics coming from the south; (4) uplift of the sediment pile during accretion of further slices of ocean-floor sediment under-

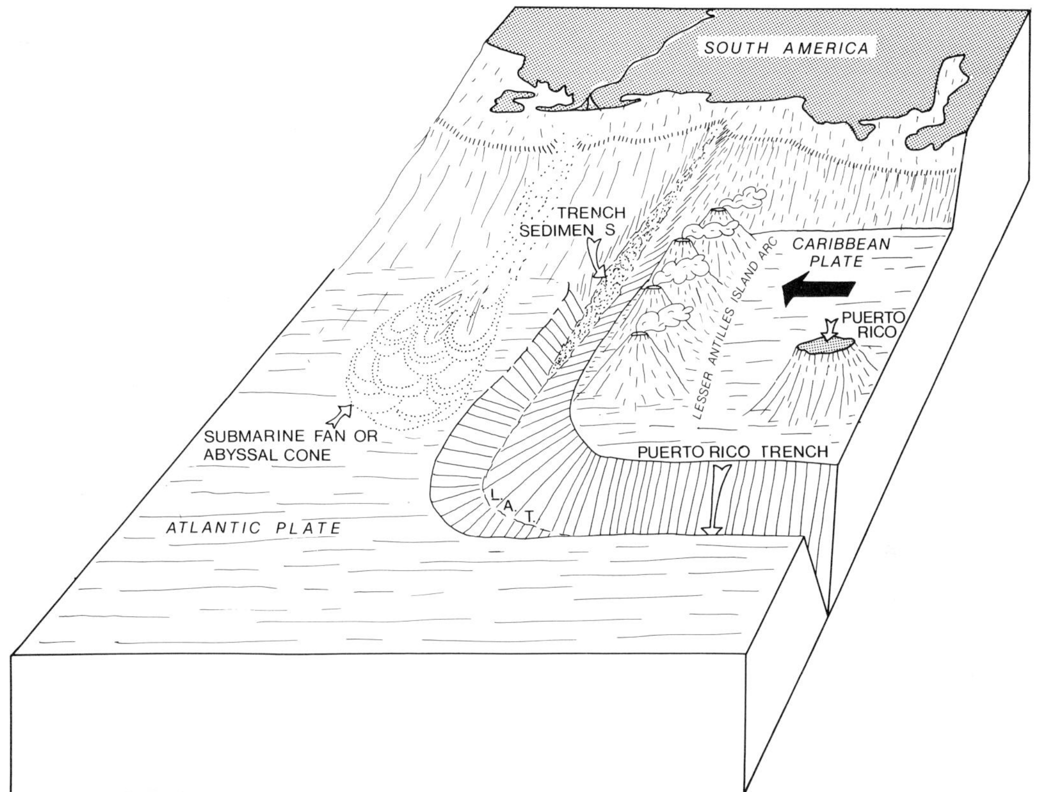

FIG. 12. Block diagram of the eastern Caribbean, reconstruction for the middle Eocene (about 50 Ma ago). View is to south towards the north coast of South America. Caribbean plate moving east with respect to South America (black arrow). Both trench sediments and ocean-floor fan sediments pile up to construct Barbados Ridge. L.A.T. = Lesser Antilles Trench.

neath it (Moore & Karig 1976; Dickinson & Seely 1979). Turbidity currents then went round the sides and only pelagic sediments could accumulate on top (Westbrook 1974). This is consistent with the gradual shallowing during Oceanic Group deposition reported by Lohmann (1974) and with continuing uplift through the Pleistocene and Recent, witnessed by Coral Rock deposition in a series of terraces.

Transverse trends occur in many active island arcs (summary in Ranneft 1979). At the end of the Barbados Ridge, an east-west cross-warp separates the calc-alkaline volcanics of the Guadelupe-St Lucia sector of the arc from the predominantly basaltic Grenada-St Vincent sector (Rea 1981). This feature and the ENE structures on Barbados may be related to oceanic transform faults in the downgoing Atlantic slab (Peter & Westbrook 1976). One Caribbean reconstruction (J. F. Dewey, pers. comm.) shows an east-west trending spreading ridge from the Mid-Atlantic Ridge to Central America during part of Cretaceous time. This plate boundary had become inactive by the mid-Eocene but probably remained as a topographic feature on the seafloor. Westbrook (1981) gives a full discussion. Subduction involving any such east-west irregularity introduces many complications not predicted by a simple two-dimensional subduction-accretion model.

ACKNOWLEDGMENTS: Field work was financed by the U.K. Ministry of Overseas Development and the Ministry of Trade, Barbados. C. J. Pudsey acknowledges a N.E.R.C. research studentship.
The manuscript benefited by discussions with J. F. Dewey, D. Karig, J. B. Saunders and G. K. Westbrook.
John Cooper, British Museum (NH) helped with the shells and W. J. Kennedy with trace fossils.
We thank R. Holland, P. Jackson, G. Collins and D. Relton for assistance and O.D.M. for permission to publish.

References

Arnold, R. & Macready, G. A. 1956. Island-forming mud volcano in Trinidad, British West Indies. *Bull. Am. Assoc. Petrol. Geol.* **40,** 2748–58.

Baadsgaard, P. H. 1959. *Barbados geological review and evaluation.* Unpubl. report, Gulf Oil Co., 132 pp.

Blatt, H., Middleton, G. V. & Murray, R. C. 1972. *Origin of Sedimentary Rocks.* Prentice-Hall, Englewood Cliffs, New Jersey, 634 pp.

Bowin, C. O. 1976. Caribbean gravity field and plate tectonics. *Spec. Pap. geol. Soc. Am.* **169,** 79 pp.

Caudri, C. M. B. 1972. The larger foraminifera of the Scotland District of Barbados. *Eclog. geol. Helv.* **65,** 221–34.

Chapman, R. E. 1973. *Petroleum Geology: a Concise Study.* Elsevier, Amsterdam, 304 pp.

—— 1974. Clay diapirism and overthrust faulting. *Bull. geol. Soc. Am.* **85,** 1597–1602.

Chase, R. L. & Bunce, E. T. 1969. Underthrusting of the eastern margin of the Antilles by the floor of the western North Atlantic Ocean, and the origin of the Barbados Ridge. *J. geophys. Res.* **74,** 1413–30.

de Cizancourt, M. 1948. Nummulites de i'ile de Barbade. *Mém. Soc. géol. Fr.* **57,** Tome 27, Fasc. 1.

Corbo, S. 1979. Vertical distribution of trace fossils in a turbidite sequence, Upper Devonian, New York State. *Palaeogeogr. Palaeoclimatol. Palaeoecol.* **28,** 81–101.

Crimes, T. P. 1977. Trace fossils of an Eocene deep-sea sand fan, north Spain. *In*: Crimes, T. P. & Harper, J. C. (eds). *Trace Fossils 2.* Spec. issue Geol. J. **9,** 71–90.

Daviess, S. N. 1971. Barbados: a major submarine gravity slide. *Bull. geol. Soc. Am.* **82,** 2593–601.

Dickinson, W. R. & Seely, D. R. 1979. Structure and stratigraphy of forearc regions. *Bull. Am. Assoc. Petrol. Geol.* **63,** 2–31.

Dott, R. H., Jr 1964. Wacke, greywacke and matrix—what approach to immature sandstone classification? *J. sediment. Petrol.* **34,** 625–32.

Edgar, N. T., Saunders, J. B. *et al.* 1973. *Initial Rep. Deep Sea drill. Proj.* **15,** U.S. Govt Printing Office, Washington, 1137 pp.

Hedberg, H. D. 1974. Relation of methane generation to undercompacted shales, shale diapirs and mud volcanoes. *Bull. Am. Assoc. Petrol. Geol.* **58,** 661–73.

Heezen, B. C., Menzies, R. J., Schneider, E. D., Ewing, W. M. & Granelli, N. C. L. 1964. Congo Submarine Canyon. *Bull. Am. Assoc. Petrol. Geol.* **48,** 1125–49.

Herrera, R. C. & Spence, J. 1964. *The geology and oil prospects of the Scotland Group of sediments at Barbados.* Unpubl. report, Sinclair and B. P. Explorations Ltd.

Hess, H. H. 1938. Gravity anomalies and island arc structure with particular reference to the West Indies. *Proc. Am. philos. Soc.* **79,** 71–96.

Hesse, R. 1975. Turbiditic and non-turbiditic mudstone of Cretaceous flysch sections of the East Alps and other basins. *Sedimentology*, **22,** 387–416.

Higgins, G. E. & Saunders, J. B. 1967. Report on 1964 Chatham mud island, Erin Bay, Trinidad, West Indies. *Bull. Am. Assoc. Petrol. Geol.* **51,** 55–64.

—— & —— 1974. Mud volcanoes—their nature and origin. *Verl. naturforsch. Ges. Basel*, **84,** 101–52.

Howell, D. G., Crouch, J. K., Greene, H. G., McCulloch, D. S. & Vedder, J. G. 1980. Basin development along the late Mesozoic and Cainozoic California margin: a plate tectonic margin of subduction, oblique subduction and transform tectonics. *In*: Ballance, P. F. & Reading, H. G. (eds). *Sedimentation in Oblique-slip Mobile Zones.* Spec. Publ. int. Assoc. Sediment. **4,** 43–62.

Hurley, R. J. 1966. Geological studies of the West Indies. *Can. geol. Surv. Pap.* **66–15,** 139–50.

Jordan, T. H. 1975. The present-day motions of the Caribbean plate. *J. geophys. Res.* **80,** 4433–9.

Jukes-Browne, A. J. & Harrison, J. B. 1891. The geology of Barbados. Part I—the Coral-rocks of Barbados and other West-Indian islands. *Q. J. geol. Soc. London*, **47,** 197–250.

Karig, D. E., Lawrence, M. B., Moore, G. F. & Curray, J. R. 1980. Structural framework of the fore-arc basin, NW Sumatra. *J. geol. Soc. London*, **137,** 77–91.

Kearey, P., Peter, G. & Westbrook, G. K. 1975. Geophysical maps of the eastern Caribbean. *J. geol. Soc. London*, **131,** 311–21.

Ladd, J. W. & Watkins, J. S. 1979. Tectonic development of trench-arc complexes on the northern and southern margins of the Venezuela Basin. *In*: Watkins, J. S., Montadert, L. & Dickerson, P. W. (eds). *Mem. Am. Assoc. Petrol. Geol.* **29,** 363–71.

Lohmann, G. P. 1974. *Paleo-oceanography of the Oceanic Formation, Barbados.* Thesis, PhD, Univ. Brown.

Malfait, B. T. & Dinkelman, M. G. 1972. Circum-Caribbean tectonic and igneous activity and the evolution of the Caribbean plate. *Bull. geol. Soc. Am.* **83,** 251–71.

Matley, C. A. 1932. The old basement of Barbados: with some remarks on Barbadian geology. *Geol. Mag.* **69,** 366–73.

Meyerhoff, A. A. & Meyerhoff, H. A. 1972. Continental drift, iv: the Caribbean "plate". *J. Geol.* **80,** 34–60.

Molnar, P. & Sykes, L. R. 1969. Tectonics of the Caribbean and Middle America regions from focal mechanisms and seismicity. *Bull. geol. Soc. Am.* **80,** 1639–84.

Moore, G. F. & Karig, D. E. 1976. Development of sedimentary basins on the lower trench slope. *Geology*, **4,** 693–7.

Mutti, E. 1977. Distinctive thin-bedded turbidite facies and related depositional environments in the Eocene Hecho Group (South-central

Pyrenees, Spain). *Sedimentology*, **24,** 107–31.

—— , NILSEN, T. H. & RICCI LUCCHI, F. 1976. Outer fan depositional lobes of the Laga Formation (Upper Miocene and Lower Pliocene), east-central Italy. *In*: STANLEY, D. J. & KELLING, G. (eds). *Sedimentation in Submarine Canyons, Fans and Trenches*, 210–23. Dowden, Hutchinson & Ross, Stroudsburg, Pennsylvania.

NORMARK, W. R. 1978. Fan valleys, channels and depositional lobes on modern submarine fans: characters for recognition of sandy turbidite environments. *Bull. Am. Assoc. Petrol. Geol.* **62,** 912–31.

PETER, G. & WESTBROOK, G. K. 1976. Tectonics of the south-western North Atlantic and Barbados Ridge complex. *Bull. Am. Assoc. Petrol. Geol.* **60,** 1078–106.

PUDSEY, C. J. 1979. Stratigraphy and sedimentology of the Scotland Group, Barbados. *Trans. 4th. Latin American Geol. Congr.* (in press).

RANNEFT, T. S. M. 1979. Segmentation of island arcs and application to petroleum geology. *J. Petrol. Geol.* **1,** 35–53.

REA, W. J. 1981. The Lesser Antilles. *In:* THORPE, R. S. (ed). *Orogenic Andesites*, 167–85. Wiley, New York.

ROBERTS, H. H., CRATSLEY, D. W. & WHELAN, T. 1976. Stability of Mississippi delta sediments as evaluated by analysis of structural features in sediment borings. *Offshore Tech. Conf. Paper No. OTC 2425*, 14 pp.

RUPKE, N. A. 1978. Deep clastic seas. *In*: READING, H. G. (ed.). *Sedimentary Environments and Facies*, pp. 372–415. Blackwell Scientific Publications, Oxford.

—— & STANLEY, D. J. 1974. Distinctive properties of turbiditic and hemipelagic mud layers in the Algero-Balaeric Basin, western Mediterranean Sea. *Smithson. Contrib. Earth Sci.* **13,** 40 pp.

SAUNDERS, J. B. 1965. *Field trip guide.* Barbados.4th Caribbean Geol. Congress, Trinidad, 443–9.

SEILACHER, A. 1978. Use of trace fossil assemblages for recognising depositional environments. *In*: BASAN, P. B. (ed.). *Trace Fossil Concepts*. Short Course Soc. econ. Paleontol. Mineral. Tulsa, **5,** 167–81.

SENN, A. 1940. Paleogene of Barbados and its bearing on history and structure of Antillean-Caribbean region. *Bull. Am. Assoc. Petrol. Geol.* **24,** 1546–610.

—— 1947. Die Geologie der Insel Barbados B.W.I. (Kleine Antillen) und die Morphogenese der umliegende marinen Grossformen. *Eclog. geol. Helv.* **40,** 199–222.

SHEPARD, F. P., MARSHALL, N. F., MCLOUGHLIN, P. A. & SULLIVAN, G. G. 1979. *Currents in Submarine Canyons and other Sea Valleys.* Studies in Geology, **8,** Am. Assoc. Petrol. Geol. Tulsa, Oklahoma, 173 pp.

SPEED, R. C. 1979. New views on the geology of Barbados. *Trans. 4th Latin American Geol. Congr.* (in press).

SYKES, L. R. & EWING, M. 1965. The seismicity of the Caribbean region. *J. geophys. Res.* **70,** 5065–74.

TODD, T. W. & FOLK, R. L. 1957. Basal Claiborne of Texas: record of Appalachian tectonism during Eocene. *Bull. Am. Assoc. Petrol. Geol.* **41,** 2545–66.

TRECHMANN, C. T. 1925. The Scotland Beds of Barbados. *Geol. Mag.* **62,** 481–504.

VELBEL, M. A. 1980. Petrography of subduction zone sandstones—discussion. *J. sediment. Petrol.* **50,** 303–4.

WALKER, R. G. 1978. Deep-water sandstone facies and ancient submarine fans: models for exploration for stratigraphic traps. *Bull. Am. Assoc. Petrol. Geol.* **62,** 932–66.

WESTBROOK, G. K. 1974. The structure and evolution of the Barbados Ridge. *7th Caribbean Geol. Congr.* 321–8.

—— 1975. The structure of the crust and upper mantle in the region of Barbados and the Lesser Antilles. *Geophys. J. R. astron. Soc.* **43,** 201–42.

—— 1981. Barbados ridge complex: tectonics of a mature fore-arc system (this volume).

C. J. PUDSEY & H. G. READING, Department of Geology and Mineralogy, Oxford OX1 3PR, U.K.

Appendix

The base of the Oceanic Group

Biostratigraphic evidence is not yet clear enough to confirm or disprove any age overlap between the Scotland and Oceanic Groups. We believe that the Oceanic/Scotland contact is at least partly unconformable because:

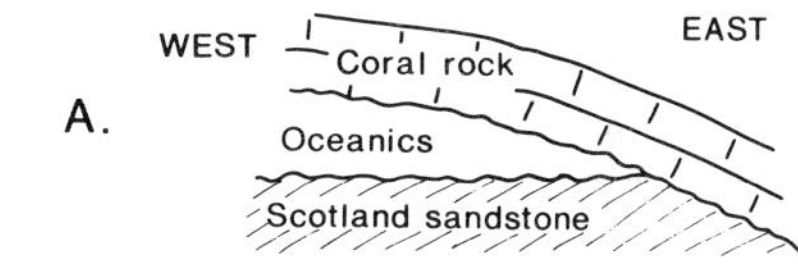

FIG. 13. (A) Double unconformity Coral rock/Oceanics/Scotland sandstones at Skeete's Bay; from Jukes-Browne & Harrison (1891).
(B) Measured section at base of Oceanic Group, Cattlewash. Section is horizontal and right way up. Murphys Member consists of interbedded sand and clay: Cattlewash Member consists of muddy cherts.

(a) Jukes-Brown & Harrison (1891, p. 217) figure a section at Skeete's Bay showing the double unconformity (Fig. 13A). Unfortunately this locality is now entirely overgrown. These authors were certainly able to recognize a tectonic contact between the Oceanic and Scotland Groups in other places.
(b) A new locality at Cattlewash (Barbados grid. ref. 67346185) shows the short section in Fig. 13B). This is measured in a large block within the Joes River Formation (Fig. 9) but subsequent fragmentation and incorporation of the block into a mudflow does not invalidate the point that the base of the Oceanic Group was at least partly unconformable on the Scotland Group. In many places the unconformity has been a plane of movement during deformation.

Subduction and tectonics on the continental margin off northern Spain: observations with the submersible *Cyana*

Jacques-André Malod, Gilbert Boillot, Raymond Capdevila, Pierre-Alain Dupeuble, Claude Lepvrier, Georges Mascle, Carla Müller & Josette Taugourdeau-Lantz

SUMMARY: Short-term latest Mesozoic–early Cenozoic convergence between the Iberian and European plates transformed the passive north Spanish margin into an active margin with southward subduction of the oceanic crust of the Bay of Biscay. This margin became passive again when subduction ceased in late Eocene or Oligocene time. Nine dives with the submersible *Cyana* allowed us to sample the rocks outcropping on the Le Danois Bank slope. A Mesozoic sequence includes shelf limestones of the Dogger to the Lower Cretaceous, and Lower Cretaceous detrital sediments (sandstones, conglomerates, pelites); it overlies a typical Palaeozoic Asturian basement.

The main objective of the diving campaign was to study the structure of these strata in order to elucidate deformation of the continental wall of an active margin during its early stages of evolution, before the formation of an accretionary prism. The occurrence in Lower Cretaceous pelites of a weak schistosity in a reversed fold limb gives evidence of compressive structures, probably northward-verging overfolds. Additional evidence of compressive tectonics is the dislocation of quartz grains and dolomite crystals in Lower Cretaceous sandstones and Jurassic dolomites. These observations suggest that the continental slope in the Le Danois Bank area has a predominantly imbricate structure. Deformation and uplift of the northern Spanish margin appear to be related to the early stages of subduction of the European plate under Iberia.

The northern and southern continental margins of the Bay of Biscay are very different. The continental slope off Brittany is in general gently inclined and 50–100 km wide, whereas the northern Spanish one is narrow (25 km) and steep (Brenot & Berthois 1962; Laughton *et al.* 1975). A deep sediment-filled trough characterized by a large negative gravity anomaly lies at the base of the northern Spanish margin (Fig. 1). According to Sibuet & Le Pichon (1971) and Le Pichon *et al.* (1971) this is a fossil trench associated with an Eocene subduction zone. Convergence between the Iberian and European plates, during the latest Mesozoic and the early Cenozoic, and the resulting southward subduction of the oceanic crust of the Bay of Biscay, have transformed the passive northern Spanish margin into an active one. Short term subduction at this time is put forward to explain the deformation and uplift of the rocks on the northern Spanish continental shelf (Boillot *et al.* 1971, 1973; Lamboy & Dupeuble 1975; Boillot & Capdevila 1977; Boillot *et al.* 1979). The margin became passive again when the convergence and the subduction ceased, owing to collision between the Iberian and European plates along the Pyrenees. The margin assumed its current physiography after further phases of normal faulting, transcurrent faulting and subsidence.

Much new data including seismic reflection profiles and rock samples have been collected in the area of the Le Danois Bank, a marginal plateau located on the northern Spanish margin (Fig. 1). The Le Danois Bank was a part of the deep margin before the Pyrennean orogeny and culminates now at a depth of 450 m. Boillot *et al.* (1979) proposed that the impressively steep slope of this bank is the result of an imbricate structure, a hypothesis which we examine further in this paper.

We used the submersible *Cyana* during July 1979 to address three problems on the Le Danois Bank:

the early Mesozoic stratigraphy,
the nature of the basement,
the style of deformation.

We surveyed two geological cross-sections of the northern slope of the bank between 3000 and 500 m depth during nine dives. Fifty-four samples of rocks (most of them taken *in situ*), numerous photographs and video-records were collected (Malod *et al.* 1980; Capdevila *et al.* 1981). We concentrate here on the structural observations.

Mesozoic stratigraphy and palaeogeography

A continuous survey of the bedrock is not possible, because of local soft mantling sedi-

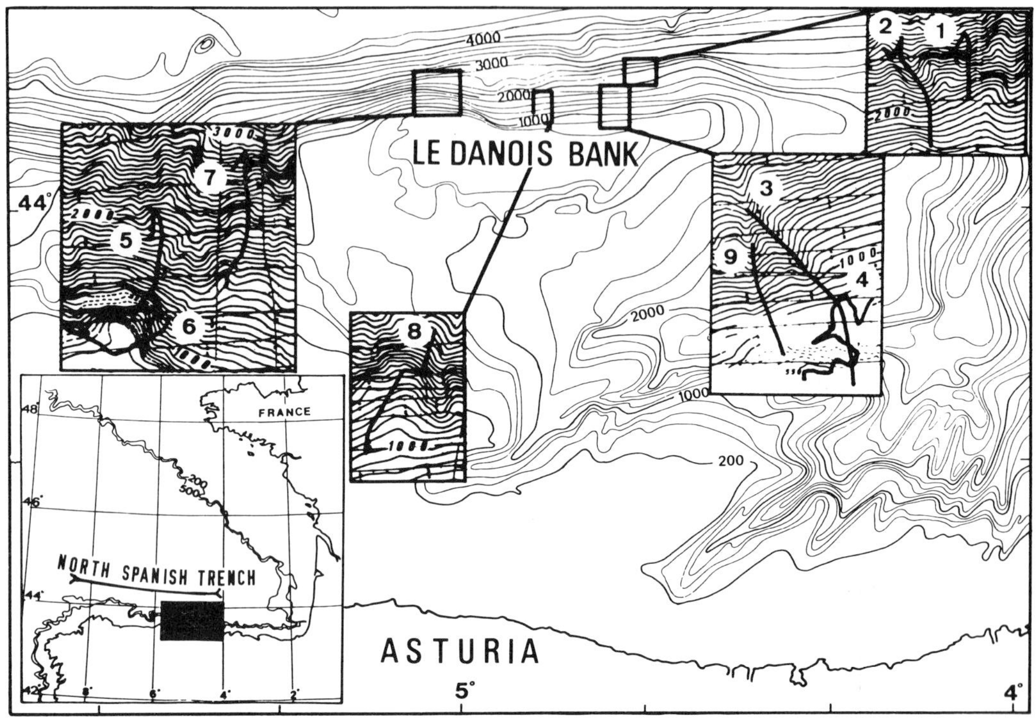

FIG. 1. Location of the *Cyana* dives on the Le Danois Bank.
The dives are numbered from 1 to 9. Insert: location of the chart with respect to the Bay of Biscay and the fossilized north Spanish trench.

ments. However, rock outcrops are numerous and we can draw an accurate geological cross-section for each dive, plotting lithologies by means of our numerous samples. We found three stratigraphic units in different depth zones. These were located respectively during dives 1, 2, 3, 4, 9, dives 5, 6, 7 and dive 8, and are summarized in Fig. 2.

Submarine cliffs associated with Upper Jurassic outcrops show clearly on a bathymetric chart of the Le Danois Bank slope (Malod & Vanney 1981). These cliffs are remarkably parallel, extending several kilometres along the slope at nearly uniform depths. The observed sedimentary units are therefore probably of good lateral continuity. Despite the existence of faults, oblique to the general east-west orientation of the slope, we consider that the succession of sedimentary units in Fig. 2 may be representative for all the studied area.

The lower unit (I in Fig. 2) occurs between 3000 and 2000 m depth. Kimmeridgian to Portlandian limestones in decimetre-scale beds, sometimes with marly interbeds, occur at the base. They are succeeded by massive limestones, commonly slightly dolomitized, ranging in age from Portlandian to earliest Cretaceous (perhaps Valanginian). Turbidite-like Hauterivian sandstones rich in plants debris follow, and are capped by Barremian pollen-rich pelites. These pelites and a previous sample of Aptian age (H.76 DR.09 in Boillot *et al.* 1979) dredged at the same level on the slope are comparable in lithology to black shales sampled in Deep Sea Drilling Project holes 398 (Sibuet & Ryan *et al.* 1979) and 400, 402 (Montadert & Roberts *et al.* 1979).

The intermediate unit (II in Fig. 2) occurs between 2000 and 1500 m depth in the west, and between 2000 and 600 m in the east. This unit consists of azoic dolomites succeeded by thick pollen-rich Aptian micaceous sandstones. In the upper part of the unit one or more levels of conglomerate indicate coarse detrital sedimentation of unknown but probably Aptian to Albian age. Because of their stratigraphical position and because Portlandian to earliest Cretaceous limestones of unit I are slightly dolomitized, we suppose that dolomites in unit II are of Late Jurassic to earliest Cretaceous age.

The upper unit (III in Fig. 2) occurs from 1500 m depth to the top of the bank in the west

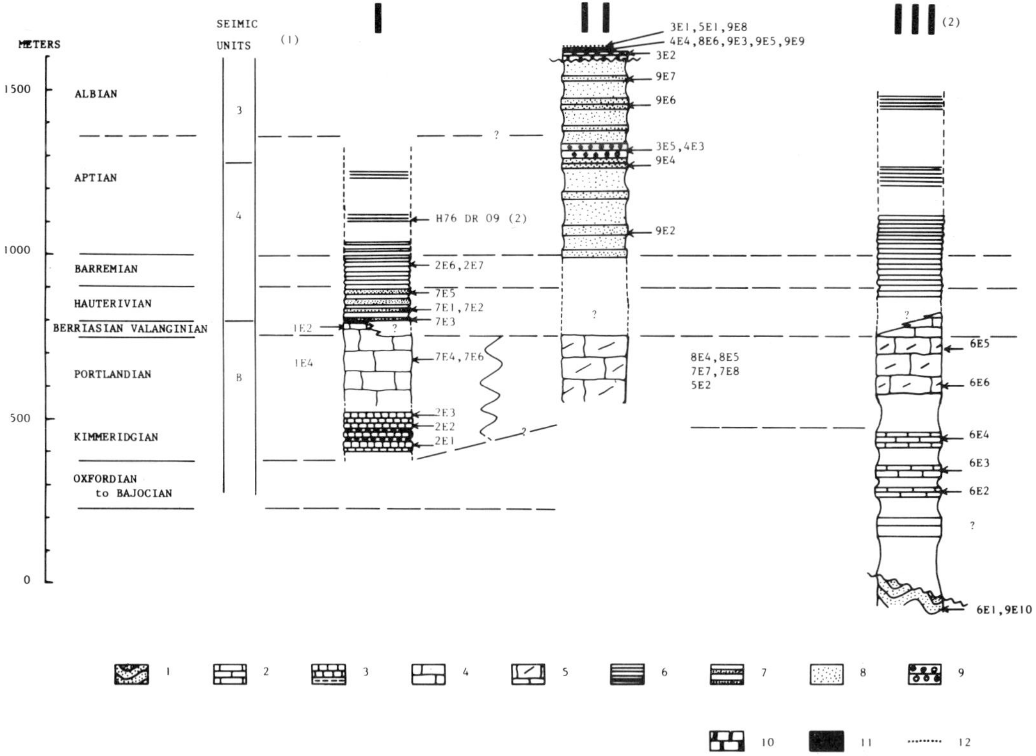

FIG. 2. The three stratigraphic units of the Le Danois Bank.
1: Cambrian quartzite; 2: limestone in decimetre-scale beds; 3: limestone with interbedded marl; 4: massive limestone; 5: dolomite; 6: pelite; 7: sandstone (turbiditic); 8: micaceous sandstone; 9: conglomerate; 10: early Tertiary reefal limestone; 11: Aquitanian limestone and Pliocene marl; 12: Quaternary mud.
Sample locations, with numbers, are shown. (1) After Montadert *et al.* (1979); (2) partly after Boillot *et al.* (1979).

and only on the top of the bank in the east. This unit begins with *in situ* pre-Mesozoic basement, here consisting of a two-mica quartzite of probable Cambrian age. Above it massive strata of undetermined nature crop out. Limestones (Bajocian to Oxfordian) and dolomites complete the succession observed with the submersible. Samples cored and dredged on the top of the bank show that the upper part of this unit consists partly of grey marls of Early Cretaceous age (Boillot *et al.* 1979).

Late Jurassic and earliest Cretaceous deposits indicate platform-type sedimentation and are similar in all three stratigraphic units. Later Early Cretaceous facies are more detrital and contemporary deposits differ within the three stratigraphic units. The transformation occurred in Hauterivian time, probably as a result of the initial rifting of the margin. We assume that during the Cretaceous opening of the Bay of Biscay the evolution of the north and south margins were very similar. On this basis, the profile OC.412 recorded by Montadert *et al.* (1979) on the Armorican margin has been taken as a model for the structure of the north Spanish margin (Fig. 3A, in which we have removed the post-Aptian layers to draw the Aptian reconstruction). The Early Cretaceous sediments are located in narrow basins between tilted blocks of the stretched continental crust. In this way the post-Hauterivian facies contrasts between the three stratigraphic units are interpreted as being the result of sedimentation in different basins.

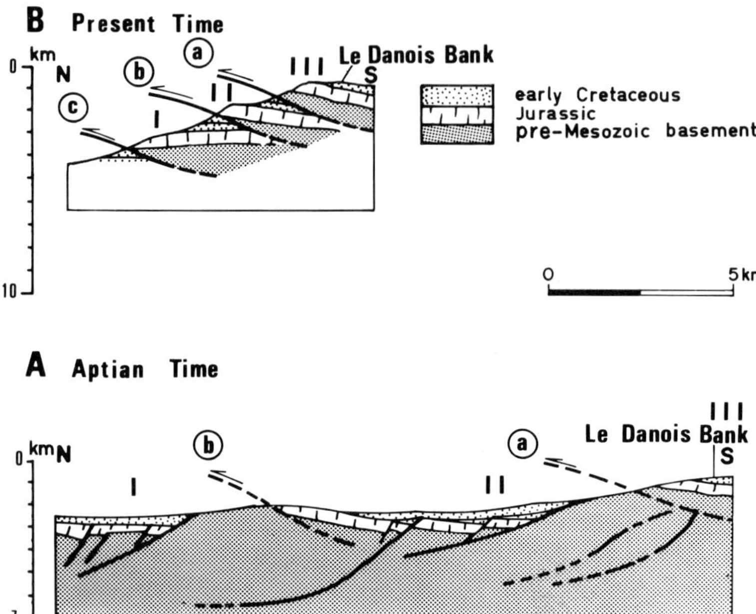

FIG. 3. A sketch of the geological evolution of the studied area.
Stratigraphic sequences I–III as in Fig. 2.
(A) Aptian time: the cross-section is based on a profile of the Armorican margin (OC.412 of Montadert *et al.* 1979). On this profile, the post-Aptian layers have been removed and the thickness of the Jurassic is based on diving data. a and b are the positions of future overthrusts.
(B) Present time: the slope of the Le Danois Bank is characterized by an imbricate structure. a, b and c represent overthrusts.

Structural observations

No obvious folds have been observed in any dives. Notwithstanding the narrow field of view from the submersible, it appears to us that the outcropping layers mainly form a homoclinal sequence dipping gently to the south. However, we have observed local N, E and SE dips. Our structural observations may be summarized as follows:

(1) An outcrop of Barremian pelites shows two cleavages (Fig. 4). The more inclined one is parallel to bedding; the other, an apparently horizontal fabric, is interpreted as a weak schistosity. It suggests that the outcrop belongs to the reversed limb of a fold, perhaps associated with a thrust fault.

(2) At a microscopic scale, some rocks show other evidence of a compressional tectonic event. Dolomite crystals in the Jurassic and quartz grains in the Early Cretaceous sandstones are often broken and crushed.

(3) The most important observation is probably the tectonic repetition of the Mesozoic sedimentary sequences up-slope (Fig. 2). It may be caused by normal faulting or by thrust faulting (Fig. 3). The schistosity and the microscopic breakage of crystals in some rocks are strong arguments for the second hypothesis. Moreover, the close proximity of the three different sedimentary units and the facies contrasts between them (Fig. 2) argue against normal faulting. They may be more easily explained by superposition in an imbricate structure (Fig. 3B). The overthrusts may have taken place either at the base of the Mesozoic or in a plastic level of the pre-Mesozoic basement.

(4) Normal faults (Fig. 5) and probable strike-slip faults, also occur. They explain the limited repetition of strata with the same facies, but not, in our opinion, the repetition of units where the Lower Cretaceous has a different lithology.

How can we explain the coexistence of extensional and compressive structures? The structural evolution of the margin has been

FIG. 4. Outcrop of Barremian pelites showing two cleavages.
The more inclined fabric is bedding-parallel; the other apparently horizontal fabric is probably a weak schistosity (dive 1, 17h09, 2299 m depth, scale bar: about 30 cm).

FIG. 5. Normal fault slickenside with vertical striae in Tithonian limestones (dive 7, 13h15, 2355 m depth, scale bar: about 20 cm).

described by Boillot *et al.* (1973) and Lamboy & Dupeuble (1975), using a detailed survey of the Asturian and Galician continental shelves. The deformation of this part of the margin, situated 60 km south of Le Danois Bank, is polyphase. During the latest Cretaceous and the early Eocene, compression induced folding, uplift and erosion by subaerial processes. At the end of the Oligocene, after the Palaeogene transgression, extensional (horst and graben) structures appeared. Finally NE and NW strike-slip faults cut the margin.

On the Le Danois Bank, the Eocene hiatus does not permit us to work out such a precise history. Aquitanian neritic limestones lie on a flat erosion surface on top of the bank, resting unconformably on deformed Mesozoic strata. We ascribe the main compressive deformation on the slope of the Le Danois Bank to the latest Cretaceous-early Eocene tectonic phase and the normal or transcurrent faults to later Oligocene and Miocene tectonic phases.

Conclusions

The north Spanish margin has been shortened, deformed and uplifted during the latest Cretaceous and early Eocene, coincident with convergence between the Iberian and European

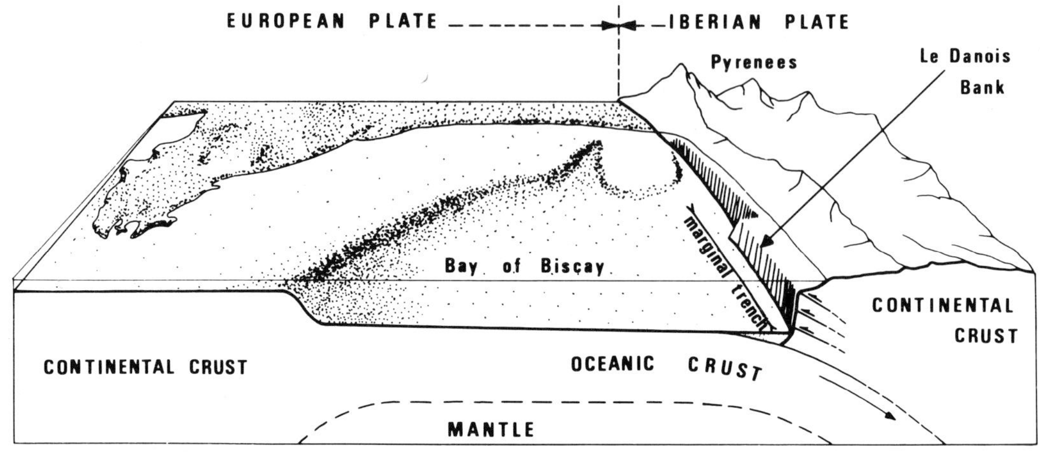

FIG. 6. Reconstruction showing how Le Danois Bank, near the junction of the European and Iberian plates, acquired an imbricate structure during the latest Cretaceous–early Eocene by contemporaneous subduction of the oceanic crust of the Bay of Biscay under Spain.

plates (Fig. 6). The Cretaceous stable margin was telescoped by thrusting resulting from southward subduction of the oceanic lithosphere of the Bay of Biscay under Spain. The resulting imbricate structure is completely different from that of an accretionary prism, involving deformed pieces of the old stable margin, not trench and ocean floor material.

This subduction seems anomalous in several respects:

(a) The lack of calc-alkaline volcanism. This may be explained by the small amount of oceanic lithosphere involved in the subduction (less than 130–150 km according to Boillot & Capdevila 1977).

(b) The importance of the deformation of the margin, despite the brevity of the subduction. At the beginning of subduction (−65 Ma) the oldest oceanic lithosphere of the Bay of Biscay was 45 Ma old, according to Montadert *et al.* (1979) (seafloor spreading under the Bay of Biscay began at −110 Ma). Consequently resistance to the initiation of subduction may have been great, because of the relative buoyancy of the young lithosphere (Molnar & Atwater 1978). The rather intense deformation of the stable north Spanish margin is probably a consequence of this 'forced subduction'.

ACKNOWLEDGMENTS: We particularly acknowledge the Spanish Government who authorized the diving campaign. We are grateful to all the people and organizations who made the cruise and the dives possible: the Commander and crew of the R.V. *Nadir*; Commander Caillart, responsible for the *Cyana* team, and Y. Morel, in charge of logistic operations. We especially acknowledge Messrs Kientzy, Leroux, Arnoux and Nivaggioli, pilots and co-pilots of the *Cyana*. The following organizations supported the diving campaign and subsequent analysis of the data: Centre National pour l'Exploitation des Océans, Université P. & M. Curie de Paris, Université de Rennes, Université de Rouen, Centre National de la Recherche Scientifique and Comité d'Etudes Pétrolières Marines.

Contribution no. 128 of the 'Groupe d'Etude de la Marge Continentale' (University of Paris and C.N.R.S., ERA 605).

References

BOILLOT, G. & CAPDEVILA, R. 1977. The Pyrenees: subduction and collision? *Earth planet. Sci. Lett.* **35,** 151–60.

—— , DUPEUBLE, P. A., HENNEQUIN-MARCHAND, I., LAMBOY, M. & LEPRETRE, J. P. 1973. Carte géologique du plateau continental nord-espagnol entre le Canyon de Capbreton et le Canyon d'Aviles. *Bull. Soc. geol. Fr.* **XV,** 367–91.

—— , LAMBOY, M., D'OZOUVILLE, L. & SIBUET, J. C. 1971. Structure et histoire géologique de la marge continentale au Nord de l'Espagne (entre 4° et 9°W). *In: Histoire structurale du Golfe de Gascogne.* V. 6.1–52. Technip, Paris.

—— & MALOD, J. A. 1979. Subduction and tectonics on the continental margin off northern Spain. *Mar. Geol.* **32,** 53–70.

BRENOT, R. & BERTHOIS, L. 1962. Bathymétrie du secteur atlantique du banc Porcupine (Ouest de l'Irlande) au cap Finistère (Espagne). *Rev. Trav. Inst. Pêches Mar.* **26,** 219–46.

Capdevila, R., Boillot, G., Lepvrier, C., Malod, J. A. & Mascle, G. 1981. Les formations cristallines du Banc Le Danois. *C.r. Acad. Sci. Paris.*

Lamboy, M. & Dupeuble, P. A. 1975. Carte géologique du plateau continental nord-ouest espagnol entre le Canyon d'Aviles et la frontière portugaise. *Bull. Soc. geol. Fr.* **XVII,** 442–61.

Laughton, A. S., Roberts, D. G. & Graves, R. 1975. Bathymetry of the northeast Atlantic: Mid-Atlantic Ridge to northwest Europe. *Deep Sea Res.* **22,** 791–810.

Le Pichon, X., Bonnin, J., Francheteau, J. & Sibuet, J. C. 1971. Une hypothèse d'évolution tectonique du golfe de Gascogne. *In: Histoire Structurale du Golfe de Gascogne,* VI. 11.1–44. Technip, Paris.

Malod, J. A., Boillot, G., Capdevila, R., Dupeuble, P. A., Lepvrier, C., Mascle, G., Müller, C. & Taugourdeau-Lantz, J. 1980. Plongées en submersible au Sud du Golfe de Gascogne: stratigraphie et structure de la pente du Banc Le Danois. *C.r. Somm. Soc. geol. Fr.* **3,** 73–6.

—— & Vanney, J. R. 1980. Etude morphologique de la pente du banc Le Danois (marge continentale nord-espagnole) d'après un relevé bathymétrique au sondeur multifaisceaux et des observations en submersible Cyana. *Ann. Inst. Océanogr.* **56,** 73–83.

Molnar, P. & Atwater, T. 1978. Interarc spreading and cordilleran tectonics as alternates related to the age of subducted oceanic lithosphere. *Earth planet. Sci. Lett.* **41,** 330–40.

Montadert, L., Roberts, D. G. *et al.* 1979a. *Initial Rep. Deep Sea drill. Proj.* **48,** U.S. Govt Printing Office, Washington.

——, De Charpal, O., Roberts, D. G., Guennoc, P. & Sibuet, J. C. 1979b. Rifting and subsidence processes. *In:* Talwani, M., Hay, W. & Ryan, W. B. F. (eds). *Deep Drilling Results in the Atlantic Ocean: continental margins and paleoenvironment.* M. Ewing Ser. **3,** 154–86.

Sibuet, J. C. & Le Pichon, X. 1971. Structure gravimétrique du Golfe de Gascogne et le fossé marginal nord-espagnol. *In: Histoire Structurale du Golfe de Gascogne*, VI. 9.1–18. Technip, Paris.

——, Ryan, W. B. F. *et al.* 1979. *Initial Rep. Deep Sea drilling Proj.* **47,** U.S. Govt Printing Office, Washington.

J. A. Malod, G. Boillot, C. Lepvrier, G. Mascle & J. Taugourdeau-Lantz, Université P. & M. Curie, 4 place Jussieu, 75230 Paris Cedex 05, France.

R. Capdevila, Université de Rennes, 35042 Rennes Cedex, France.

P. A. Dupeuble, Université de Rouen, 76130 Mont-Saint-Aignan, France.

C' Müller, BEICIP, 232, avenue Napoléon Bonaparte, 92500 Rueil-Malmaison, France.

MEDITERRANEAN

Subduction in the Hellenic Trench: probable role of a thick evaporitic layer based on Seabeam and submersible studies

X. Le Pichon, P. Huchon, J. Angelier, N. Lybéris, J. Boulin, D. Bureau, J. P. Cadet, J. Dercourt, G. Glaçon, H. Got, D. Karig, J. Mascle, L. E. Ricou & F. Thiebault

SUMMARY: A Seabeam survey of the Hellenic Trench in 1978 was followed by a submersible survey in 1980. Fifteen dives were completed between 1500 and 3000 m covering 58 km on the bottom and obtaining 48 samples. Thus, the Hellenic Trench is probably now the most intensively studied trench as previous work included seismic reflection, dredging, coring, bottom photography, side-scan sonar and drilling. Because of the high sedimentation rate and the relatively shallow depth, we did not expect outcrops except along faults and in canyons. Active tectonics with numerous faults and folds have been observed. But we have also discovered vertical cliffs with hard rock exposure over a depth range of 1000–1500 m. In this paper, we discuss the nature of the hard rock outcrops observed both on the inner wall and on two hills that we interpret as diapirs. We conclude that at least part of these hard rocks are probably Messinian evaporites and concentrate on the tectonic role of this evaporitic layer in the Hellenic subduction. It is likely that the evaporites do not subduct but tend to pile up in the trench, to form a floating evaporitic basin.

We call Aegea the landmass which includes southern Greece, the Aegean sea and Western Anatolia. Aegea is overriding the Mediterranean seafloor toward the south-west along the Hellenic Trench (McKenzie 1978; Le Pichon & Angelier 1979; Angelier 1979; Le Pichon & Angelier 1981). As a contribution to the HEAT program (Hellenic Arc and Trench: Le Pichon & Hsü 1979), we have surveyed parts of the Hellenic Trench using multi-beam echo-sounding (Seabeam) in 1978 (Le Pichon *et al.* 1979a) and then the *Cyana* SP-3000 submersible in 1979. Fifteen dives were conducted between water depths of 1500 and 3000 m (Le Pichon *et al.* 1979b, 1980; Huchon *et al.* 1981); the three diving zones are located on Fig. 1. A single dive was made in zone 1. Eleven dives were made in zone 2 in the NW–SE Ionian branch of the Hellenic Trench, which is approximately perpendicular to the direction of subduction. Three dives were made in zone 3, in the Strabo Trench, one of two parallel trenches forming the SW–NE Levantine branch. This branch makes an angle of only 20° with the direction of subduction and is thus a nearly pure transform portion of trench. These areas had previously been mapped using Seabeam on board *Jean Charcot* (Le Pichon *et al.* 1979a; Lyberis *et al.* 1981) and had been the object of numerous sea-surface cruises (e.g. Got *et al.* 1977; Le Quellec *et al.* 1980).

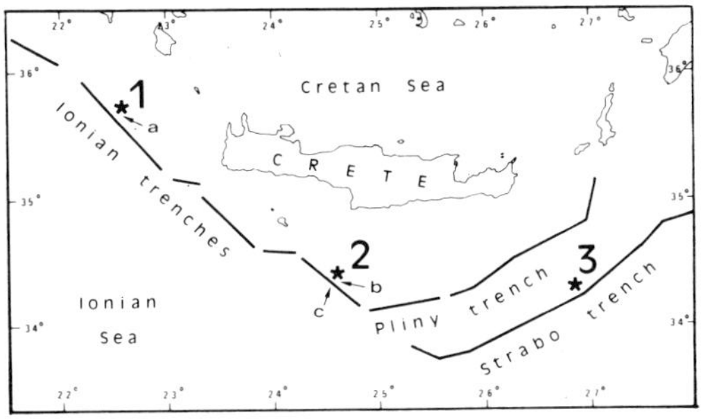

FIG. 1. Location map. 1,2,3: diving zones. a,b,c: locations of Figs 3, 5 & 6 respective-

The principal aim of these systematic detailed studies was to characterize the strain pattern over the whole length of the Hellenic subduction zone, using an interpretation of the fine-scale bathymetry, as given by Seabeam, and the submersible observations of tectonics on the seafloor. Le Pichon *et al.* (1981) have shown that the resulting superficial strain pattern is different over the inner wall, where it is purely extensional (σ_1 vertical) and over the main part of the trench floor area where it is purely compressional (σ_3 vertical). Thus the trench floor is the locus of a major discontinuity in the strain pattern. The Aegean extensional zone (Aubouin & Dercourt 1965; Angelier 1979) extends over the whole of the Hellenic arc at the foot of the inner wall. The trench floor, on the other hand, belongs to a compressional zone which may extend seaward to a large part of the Mediterranean ridge (Le Pichon & Angelier 1981). The deformation in the trench floor

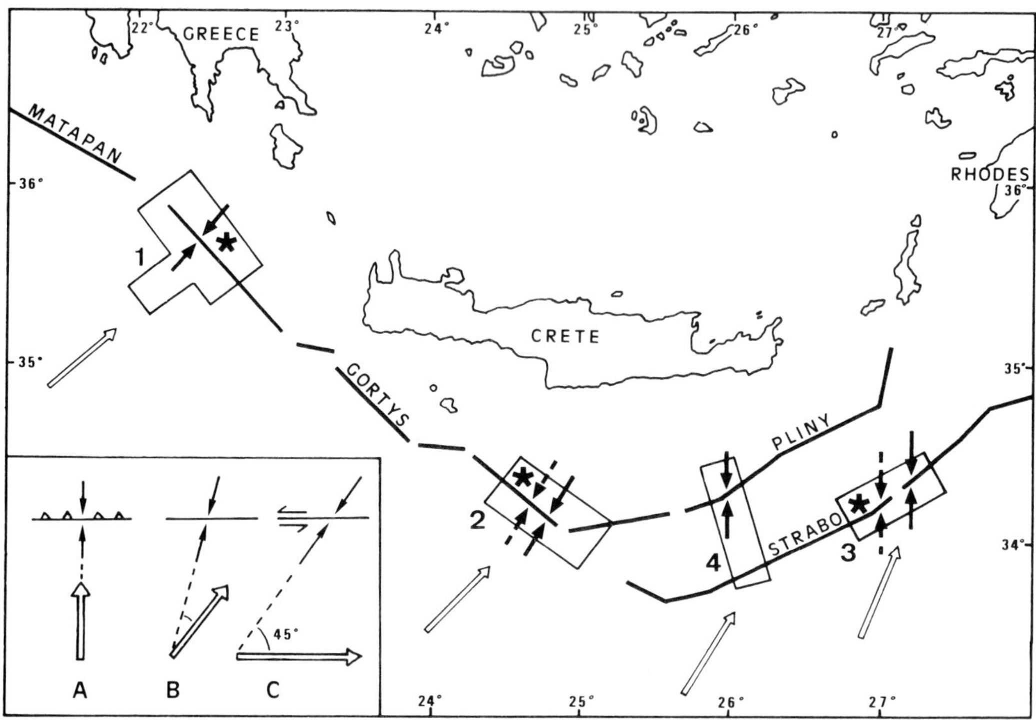

FIG. 2. Stress-strain relationship along the Hellenic subduction zone after Le Pichon *et al.* (1981). 1,2,3 and 4: areas mapped with Seabeam. Open arrows: horizontal projection of the Africa–Hellenic arc slip vector according to Le Pichon & Angelier (1979). Black arrows: direction of the axis of maximum shortening σ_1 based on Seabeam analysis. Dashed arrows: direction of σ_1 based on the analysis of tectonic structures observed during submersible dives. A,B,C: theoretical relation between stress and relative motion (schematic).

area is compatible with subduction of the Mediterranean seafloor toward the north-east, σ_1 being orientated 030° (perpendicular to the trench) along the Ionian branch and N 0° (oblique to the trench) along the Levantine branch (Fig. 2).

In this paper, we wish to emphasize a unique characteristic of the Hellenic subduction zone, which is the presence of a thick evaporitic layer, and to discuss observations which seem to suggest that the evaporites are tectonically thickened below the trench. They consequently tend to act as a decoupling layer between the Plio–Quaternary sedimentary cover and the deeper more compact pre-Messinian sediments.

Observations of consolidated sedimentary outcrops

Prior to our submersible survey, it was feared that all tectonic activity might be hidden by the high sedimentation rate (about 10 cm/1000 yr: Hsü & Ryan 1974; Got *et al.* 1977). However, our observations indicate that tectonic features are frequent within the unconsolidated to semi-consolidated sedimentary cover (Le Pichon *et al.* 1981) but also that hard rock outcrops exist over large portions of the trench system. Specifically, extensive hard rock outcrops have been observed in two areas: in the Matapan site (area 1 in Fig. 1) and in the Poseidon site (area 2 in Fig. 1).

The Matapan site

The only dive in the Matapan area was made on the upper portion of the inner wall, about 2000 m above the 4600 m deep trench floor, which is too deep for *Cyana*. Fig. 3 shows the geological section along the dive. The inner wall consists of a series of steps with sub-vertical cliffs facing the trench floor, each a few metres high. Three of these cliffs are several tens of metres up to about 100 m high and have been interpreted as major fault scarps. Prob-

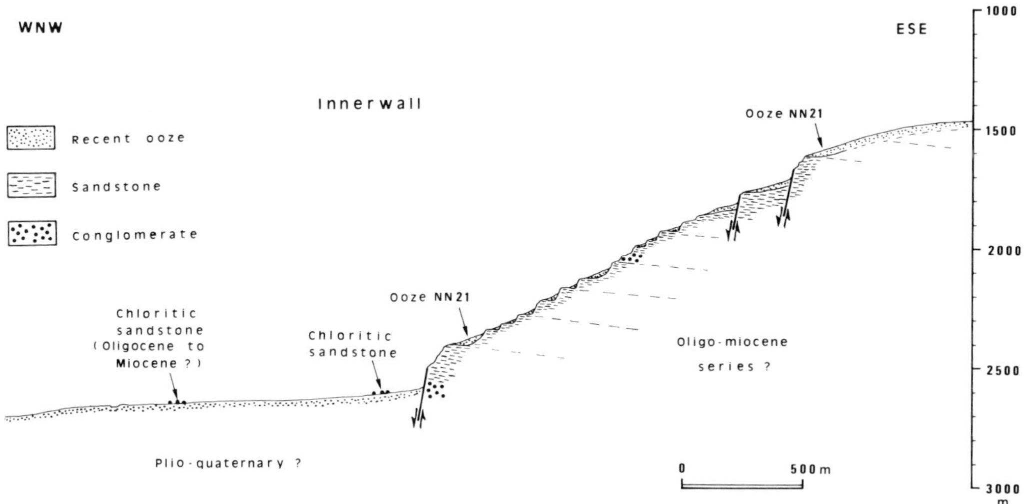

FIG. 3. Geological section across a part of the inner wall of the Matapan Trench (site 1; see Fig. 1).

able slickensides were observed within the massive rocks which outcrop on the cliffs; they are compatible with pure normal faulting. Two samples collected at the foot of this cliff consist of sandstones containing plant debris. One of them has been dated by the nannoflora (Muller, pers. comm. 1980) as late Oligocene to early Miocene. Their composition suggests that they result from the erosion of rocks similar to those which outcrop in the Peloponnesus, immediately to the north of site 1 (Chamley, pers. comm. 1980).

Although we have no indications as to the nature and composition of the lower inner wall in the Matapan area, the upper part of the inner wall is most probably the result of extensional block faulting within an Oligo–Miocene detrital sedimentary series.

The Poseidon site

Poseidon Trench (2 in Fig. 1) is much shallower than Matapan Trench, in part because of the presence of a complex of circular hills rising from the trench floor. The largest one we called Pollux (P in Fig. 4); it is 5 km in diameter and rises 1000 m above the 3000 m deep trench floor. A smaller hill, named Castor, which occurs immediately to the SE, is only 500 m high (C in Fig. 4). The hills are asymmetrical, the steeper side facing seaward. They divide the trench floor into two portions of unequal width, the narrowest portion to the south. Actually, Castor very nearly encroaches upon the outer rise, leaving only a narrow corridor, a few metres wide. The Seabeam survey has demonstrated that such hills are a common feature of the

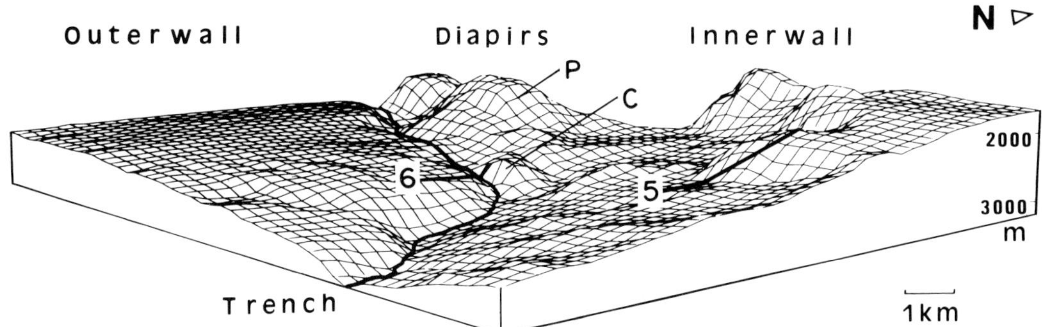

FIG. 4. Three-dimensional diagram of the diving zone of site 2. The observer looks to the NW from the surface of the sea, 30 km away from the centre of the diagram. Vertical exaggeration: ×2. The thick line indicates the axis of the trench. C: Castor hill. P: Pollux hill. 5,6: position of sections 5 and 6.

Hellenic Trench (Le Pichon *et al.* 1979a; Lyberis *et al.* 1981), and seismic reflection suggests that they are diapiric (Le Quellec *et al.* 1980). Because the depth at the axis of the trench is less than 3000 m, we were able to investigate two complete sections from the outer rise to the inner wall, one crossing Pollux hill and the other crossing Castor hill.

Detailed portions of the Castor section are shown in Figs 5 & 6 and general interpretation is shown in Fig. 7. The section is about 10 km long. From south to north, the outer rise on the seaward side has a maximum slope of 7° and progressively deepens from 2300 m to about 3000 m at the axis of the narrow corridor. The straight steep southern flank of Castor has a slope of about 40°. The northern flank has a semi-circular outline and an average slope of 10°–20°. The northern portion of the trench floor is about 3 km wide and 2600–2800 m deep. Finally, the inner wall has an average slope between 25° and 45° and rises to less than 1500 m.

Extensive hard rock outcrops have been observed both on the flanks of Castor and Pollux hills and on the inner wall. These outcrops occur on the faces of sub-vertical cliffs a few metres to a few tens of metres high, separated by portions of 25°–45° slopes covered by a Pteropod–Foraminifer ooze of Holocene or uppermost Pleistocene age (see Fig. 5).

A common characteristic of these outcrops is that they are affected by erosion resulting from dissolution. However, the intensity of erosion is extremely variable from thin micro-grooves to deep grooves (Fig. 8). In quite a few cases, the dissolution has been sufficiently intense to isolate columns or pillars a few centimetres to a few tens of centimetres in diameter (Fig. 9). In

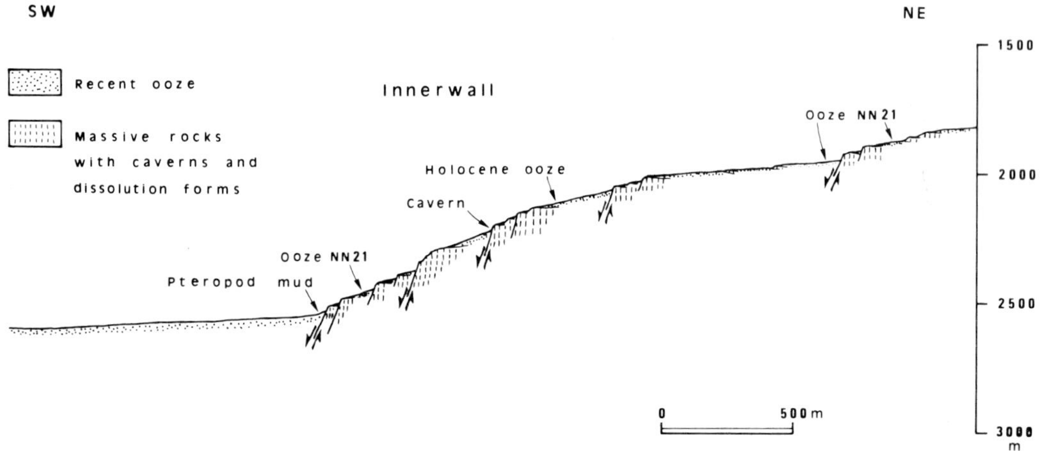

FIG. 5. Geological section across the lower part of the inner wall of the Poseidon trench (site 2; see Fig. 1).

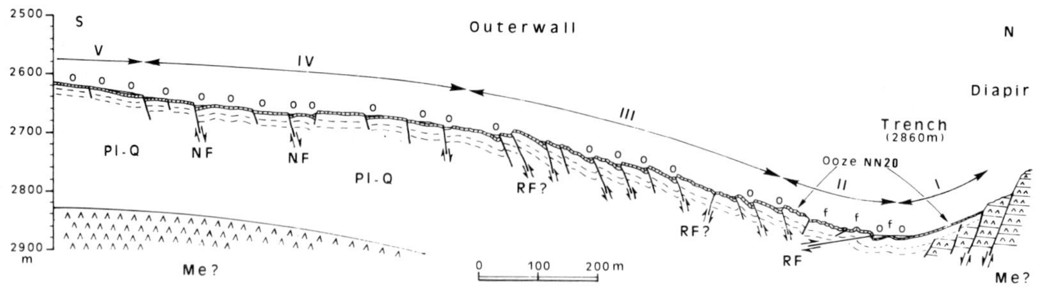

FIG. 6. Geological section across the outer wall to the base of Castor diapir, after Le Pichon *et al.* (1981) (site 2, see Fig. 1). Pl-Q: Plio–Quaternary, with overlying crust. Me: Messinian. O: ooze. F: fold. RF: reverse fault. NF: normal fault. I: base of diapir. II: zone of overthrusting. III: zone of intense shortening. IV: zone of limited extension. V: zone with little tectonic activity. NN20 is the calcareous nannoplankton biozone recognized in the crusts.

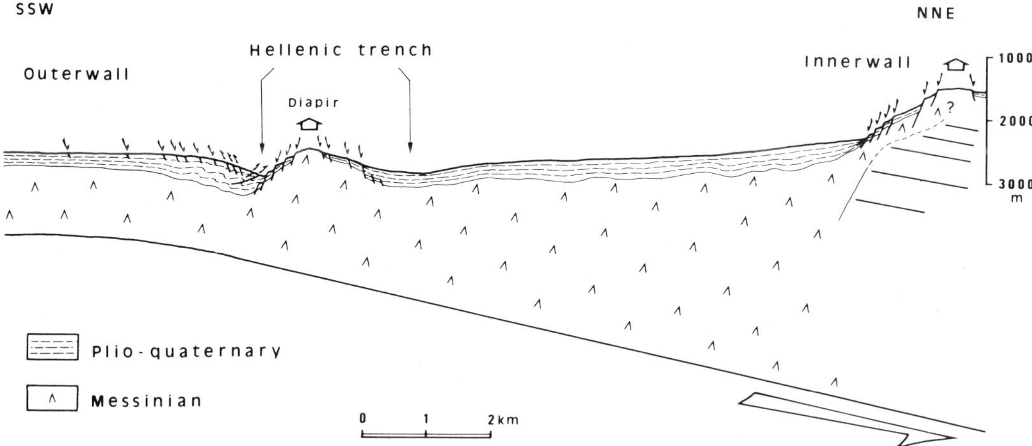

FIG. 7. Interpretative geological section across the Hellenic Trench, from the outer wall to the inner wall (site 2; see Fig. 1).

FIG. 8. Sub-vertical cliff, deeply affected by dissolution, at the foot of Castor diapir, site 2,—2956 m. Note that gouges are less developed at the foot of the cliff, indicating uplift of the diapir with respect to the talus.

The four numbers at the bottom left of each photograph (Figs 8–19) are, from top to bottom:
hour (e.g. 4400 = 14 h 40.0. Fig. 8);
heading of the submersible in degrees. This makes an angle of 20° to the right with the axis of the camera (e.g. 0038 = 038° (submersible) = 018° (camera);
uncorrected depth in metres (e.g. 2781 = 2956 m after conversion of pressure to water depth; altitude in decimetres with regard to the bottom (e.g. 0043 =4.3 m).
The scale of the photographs is variable, from 2 to 3 m in width for the foreground. For example, in Fig. 10, the pillars are 1 m high.

FIG. 9. Intersection of two fault scarps trending 110° (on the right) and 040° (on the left) on the inner wall, site 2,—2310 m. The dissolution is here sufficiently intense to isolate columns or pillars (see text).

FIG. 10. Cavern with pillars supporting the sloping roof formed by indurated slope formation. Inner wall, site 2,—2370 m.

one extreme case, the erosion has carved a cavern about 3 m deep and 1 m high, with numerous pillars supporting the sloping roof (Fig. 10).

The roof of the cavern is actually an indurated slope formation. Such slope formations have been observed frequently (Figs 11 & 12). They abut against the sub-vertical scarps and

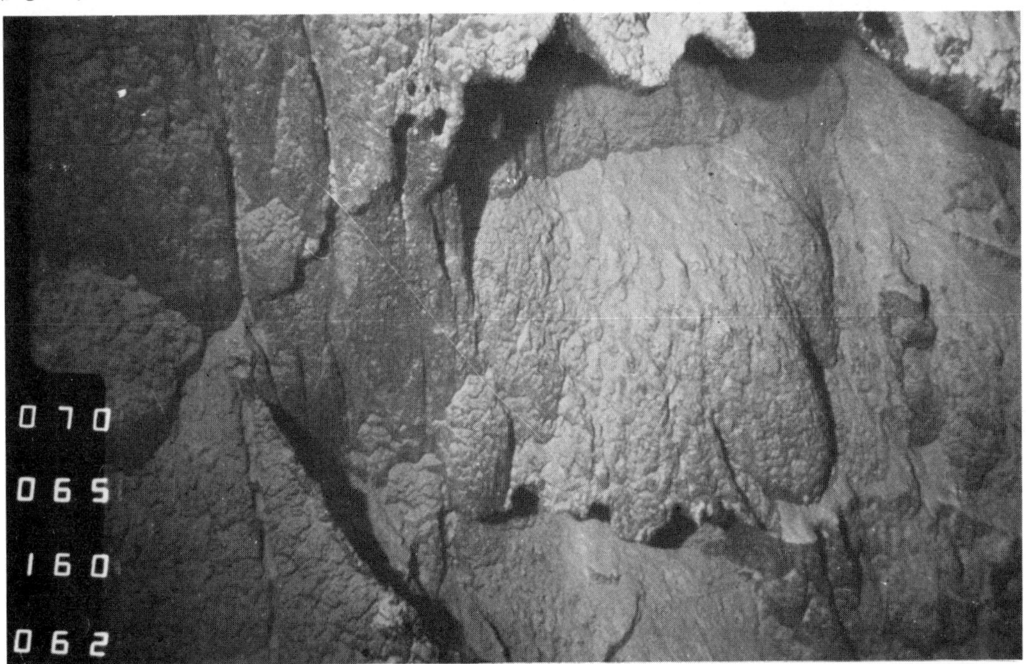

FIG. 11. Drapes formed by indurated slope formation, hanging from the vertical cliff. Inner wall, site 2,—2311 m.

FIG. 12. Dissolved drapes with festoons and holes. In the foreground, the mechanical arm of *Cyana*. Base of Castor diapir, site 2,—2937 m.

have a slope of 25°–45°, similar to the slope of the sedimentary talus, forming drapes which are obviously affected by dissolution (as shown by the common occurrence of circular holes a few centimetres in diameter). Some of these holes are still complete while others have been eaten away by erosion (Figs 10, 11 & 12). We suggest that these features resulted from a process of induration in the uppermost layer of the sedimentary talus. When the talus slumped they were left hanging in mid-water, which explains the occasional occurrence of a series of several such drapes at 1 m intervals.

An interesting point is that a fossil fauna of sponges and scleractinian corals, still in life position, covers the hard rock surfaces above 2200 m depth. Dating gave an age of 18,200 ± 500 yr for one of the scleractinian corals (Delibrias, pers. comm. 1980). It is thus probable that this fauna is part of the cold water Mediterranean fauna which disappeared at the end of the last glaciation (Zibrowius, 1980). Some indurated slope formations include sponges and consequently may be less than 20,000 yr old.

Unfortunately, the evidence we have on the nature of the rocks observed is meagre. Despite repeated attempts at sampling, we were unable to break and retrieve any sample of either the sub-vertical scarps or the indurated slope formation, except for a small piece of oolitic dolomitic limestone, which was deposited in very shallow water and has been subjected to subaerial erosion several times (Purser, pers. comm. 1980). We have not been able to date this sample. More surprising, perhaps, has been our inability to sample fallen debris on the sedimentary talus at the base of the cliffs; but such talus is lithified early by interstitial slope sediment and is equally resistant to sampling.

The other type of evidence we have is the observation of a scarp a few metres high and a few tens of metres long at the base of Pollux hill which has a thin and wavy stratification (Fig. 13). The thin beds stand out in relief because they are less affected by dissolution (Fig. 14). In one location, the horizontal stratification is interrupted by a pocket of breccias which has apparently filled a synsedimentary hole produced by dissolution (Fig. 15).

Finally a polygenetic conglomerate containing pebbles of Upper Miocene to Messinian pelagic limestone has been sampled at the foot of Pollux hill, on the top of a talus accumulation. The matrix is of Lower Pliocene age. However, although it is unlikely that the pebbles have been transported a long distance, their exact origin is unknown.

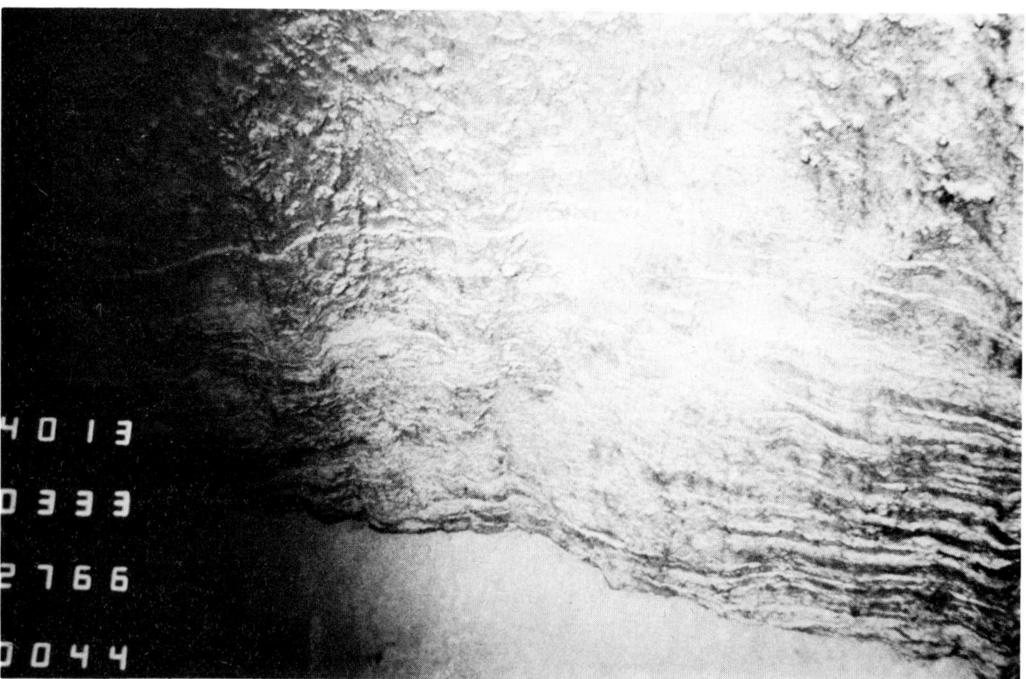

FIG. 13. Formation with wavy thin bedding. Foot of Castor diapir, site 2,—2940 m.

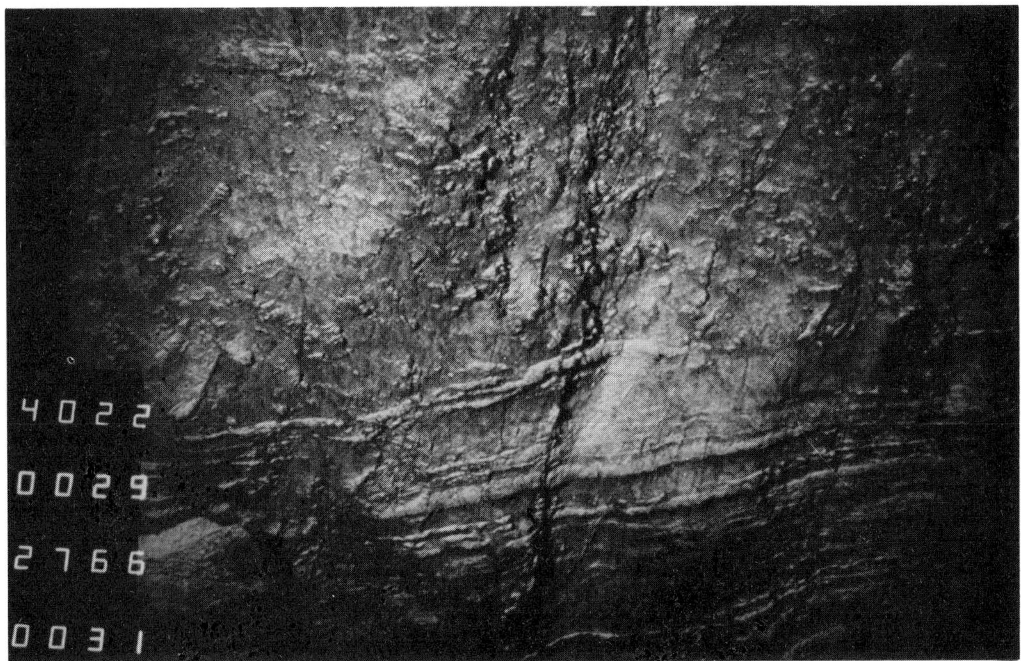

FIG. 14. Thin beds standing out in relief by differential dissolution. Castor diapir, site 2,—2942 m.

FIG. 15. Pocket of breccias interrupting the horizontal bedding. Castor diapir, site 2,—2812 m.

Discussion of the lithology of consolidated sedimentary outcrops

Although we cannot yet reach firm conclusions concerning the lithology of these hard rock outcrops, the available evidence leads to the following tentative deductions.

There is a striking difference between the Oligo–Miocene sandstones observed on the upper part of the inner wall in the Matapan area and the hard rock formations in the Poseidon area. In the Matapan area, we did not find evidence for intensive erosion by dissolution. In the Poseidon area there is strong evidence for major erosion by dissolution, although the intensity of erosion varies widely from one place to another. We conclude that the lithology observed in the Poseidon area includes a large variety of interstratified rocks, ranging from highly soluble ones to those which are barely affected by dissolution.

The erosional forms observed are always vertical which indicates that the dissolution process has been controlled by gravity and that there has been no major tectonic tilting since. Furthermore, the scarps of the inner wall are cut by 030°-trending corridors which most probably correspond to normal faults, perhaps with a small strike-slip component (Le Pichon *et al.* 1981). The walls of these corridors are affected by the same dissolution as the main scarps (see Fig. 9). Finally, we noted that the slope formations, which may be quite recent, are strongly affected by dissolution. This admittedly circumstantial evidence suggests that the process of erosion by dissolution is geologically recent and perhaps still active in deep water. A subaerial erosion could date from the Messinian desiccation stage (Hsü & Ryan 1974), but this old age is quite difficult to reconcile with the above evidence. A much younger age would imply an implausibly rapid subsidence rate.

The stratification observed in Figs 13, 14 & 15 strongly resembles the type of thin bedding observed in evaporitic rocks, either gypsum or halite. The small piece of shallow water dolomitic limestone was sampled a few tens of metres above these beds. It thus seems quite reasonable to hypothesize that the hard rock formations observed in the Poseidon area belong, at least in part, to the Messinian evaporitic series and include beds of gypsum or even halite as well as dolomitic limestone, which would explain the variable intensity of erosion by dissolution.

Tectonics related to Castor and Pollux hills

As noted previously, the sub-vertical cliffs observed on the flanks of Castor and Pollux probably correspond to fault scarps. However, in contrast to the faults of the inner wall, their trend changes continuously and follows the general shape of the circular hills. The scarps at the base of the hills commonly show a zone about 1 m high which has been only recently exposed to water and consequently is not covered by the ubiquitous thin dark concretionary film which covers older rock exposures. At other places, the base of the scarp over a similar height is much less affected by dissolution than the zone immediately above (Fig. 8). Both facts indicate upward motion of the hill with respect to the adjacent sedimentary fill and suggest that this motion is geologically recent.

On the flank of Pollux, between 2700 and 2200 m, a possible tectono-sedimentary mélange has been observed. It consists of imbricated marly plates mixed with large conglomeratic blocks. The sedimentary formation observed on the summit of Pollux appears to be conglomeratic. This may further indicate upward movement of the hills. In this interpretation, Castor and Pollux are active diapirs which lift, occasionally on their flanks and perhaps on their summits, portions of the sedimentary fill of the trench (which apparently includes a tectono-sedimentary mélange).

In this respect, we need to discuss briefly the tectonics of the outer rise, to the south of Castor and Pollux, investigated in detail during three different dives and described by Huchon *et al.* (1981) and Le Pichon *et al.* (1981). A cross-section extending south of Castor across the narrow southern corridor and the lower outer rise is shown in Fig. 6. It is representative of an east-west zone at least 1 km wide and indicates active shortening of the upper and semi-consolidated sedimentary layer, σ_1 being close to north-south (Figs 16–19).

Ritsema (1974) has given a reverse fault solution for a 6.4 magnitude earthquake (December 17 1952) with an epicentre in this general area. This evidence confirms that the zone is active and that the activity is compressional.

The amount of shortening measured is about 150 m; the actual amount may be somewhat larger depending on the amount of overthrusting of the outer rise over the trench floor (see Fig. 6), but probably does not exceed 500 m. The tectonic observations are fairly precise because the deformation is clearly inscribed in

Fig. 16. NNE-facing overthrust of marls and crusts of the outer wall on the recent ooze of the axis of the trench. South of Castor diapir, site 2,—2985 m.

Fig. 17. Small folds with broken axis in a crust of biozone NN20, near the overthrust of Fig. 16, site 2,—2866 m.

FIG. 18. Small fold with broken axis (detail), site 2,—2865 m.

FIG. 19. Tectono-sedimentary mélange exposed on a fault scarp. South of Pollux diapir, site 2,—2683 m.

the indurated crust which covers the semi-consolidated sediments in this area. This crust is the result of lithification of the calcareous hemipelagic sediment from the top downward (Allouc, pers. comm. 1980). The age of the sediments forming the crust is 200–400 000 yr (biozone NN20; nannoplankton determined by C. Müller). The age of the deformation is unfortunately unknown but is much more recent than formation of the crust because broken portions of crust are not covered by the dark oxidized film.

A similar but less intensive shortening was observed on the outer rise 10 km to the north along the Pollux section. Here, in addition, semi-consolidated marls exposed over several metres on a fault scarp have the structure of a sedimentary mélange (Fig. 19). We conclude that the upper sedimentary cover of the outer rise is actively shortened and that a tectono-sedimentary mélange is probably the result of this shortening. The intensity of shortening is greater on the Castor section, where the outer rise impinges on the diapir, than on the Pollux section, which is separated from the diapir by a portion of trench 1 km wide. Consequently, it is probable that the upward movement of the diapirs blocks the progression of at least the upper part of the sedimentary layer toward the north and is the cause of the localization of the shortening in this area.

Tectonic role of the evaporites in the subduction zone

Glomar Challenger drilling in the Mediterranean has demonstrated that the Messinian Stage is associated with a sedimentary evaporitic complex both in the Western and Eastern basins (Hsü & Ryan 1974; Hsü *et al.* 1978). Seismic reflection investigations confirm that evaporites, including halite beds, exist in the Poseidon area below a few hundred metres of Plio–Quaternary pelagic sediments (Biju-Duval *et al.* 1978). It is thus difficult to avoid the conclusion that the two diapirs Castor and Pollux are the product of tectonic thickening of the evaporites below the trench, with local extrusions. This would explain the asymmetry of the diapirs, with the steep side facing seaward. We noted previously that the Seabeam survey suggests that such diapirs are common within the Hellenic trench (Lyberis *et al.* 1981).

Fig. 7 shows a hypothetical model of the Poseidon trench. At the level of Castor, we show a thickness of 1 km for the Messinian series and 300 m for the Plio–Quaternary pelagic cover immediately south of the trench. Le Pichon & Angelier (1979) have estimated the rate of subduction as about 3–4 cm yr^{-1} in this area which would represent 150–200 km since the Messinian. The amount of evaporites shown in Fig. 7 thus corresponds to less than 1 Ma of subduction if none of them are subducted. However, this reasoning assumes that the thickness of evaporites was uniform, which is doubtful. In any case, it is clear that most of the sediments, possibly including part of the evaporites, are subducted and that the present situation is not a steady-state one.

It is likely that a sizeable portion of the deformed sedimentary cover as well as the diapiric hills will be later incorporated into the inner wall as suggested in Fig. 7. We are unable to estimate this amount. The tectonic configuration of the layers underlying the trench may thus be extremely complex and may have little resemblance to the schematic representation of Fig. 7.

Finally, we shall briefly discuss the results obtained in the Strabo trench. We noted previously that the motion there is nearly parallel to the trench which, in detail, can be interpreted as a series of Riedel shears within a general left-lateral strike slip zone with a limited component of subduction (Lyberis *et al.* 1981; Huchon *et al.* 1981). No hard rock outcrops have been observed in the immediate vicinity of the floor of the trench although large exotic boulders suggest that hard rock outcrops exist further upslope. This has been confirmed by dredging (Ariane 1979).

On the other hand, Biju-Duval *et al.* (1978) showed that the Strabo trench is the southern limit to a large and thick evaporitic basin which is limited to the north by the Pliny trench (see Fig. 2). This trapping of an evaporitic basin between two trenches suggests that the evaporitic basin was built by tectonic thickening in front of Pliny trench and that Strabo trench is the result of a recent migration of the subduction zone. As a result the thickness of evaporites entering the Strabo trench is now much less (see Biju-Duval *et al.* 1978). In addition, the actual rate of subduction is low because of the mostly transform nature of the motion. Consequently, the tectonic style of the Strabo trench is quite different from that of the Poseidon trench.

Conclusion

Detailed Seabeam and submersible studies have enabled us not only to decipher the regional strain pattern but also to discover and map extensive hard rock outcrops which had not previously been detected from sea-surface studies.

A remarkable character of these outcrops in the Poseidon area, but not in the investigated part of the Matapan area, is the presence of variable but often very intense erosion by dissolution. This fact, coupled with the similarity in stratification of some outcrops with gypsum or halite beds, leads us to suggest that the Messinian evaporitic complex is outcropping in the Poseidon area.

Seismic reflection has shown that only a few hundred metres of pelagic Plio–Quaternary sediments cover the Messinian formation seaward of the trench. Consequently we suggest that the Messinian evaporites are tectonically thickened below the trench and may occasionally extrude through the thin sedimentary cover. The Hellenic trench may thus be a subduction zone where the tectonic style is dominated by the presence of a thick evaporitic layer which is difficult to subduct.

References

ANGELIER J. 1979. *Néotectonique de l'arc égéen*, Thesis, Société geologique du Nord, Publication no 3, 417 pp.

ARIANE 1979. Résultats de dragages sur la bordure externe de l'arc hellénique (Méditerranée orientale). *Mar Geol.* **32,** 291–310.

AUBOUIN, J. & DERCOURT, J. 1965. Sur la géologie de l'Egée: regard sur la Crète (Grèce). *Bull. Soc. geol. Fr.* **VII,** 789–821.

BIJU-DUVAL, B., LETOUZEY, J. & MONTADERT, L. 1978. Structure and evolution of the Mediterranean basins. *In:* HSÜ, K. J. *et al.* (eds). *Initial Rep. Deep Sea drill. Proj.* **42,** 951–84. U.S. Govt Printing Office, Washington.

GOT, H., STANLEY, D. J. & SOREL, D. 1977. Northwestern Hellenic arc: concurrent sedimentation and deformation in a compressive setting. *Mar. Geol.* **24,** 21–36.

HSÜ, K. J., MONTADERT, L., BERNOUILLI, D., CITA, M. B., ERICKSON, A., GARRISON, R. C., KIDD, R. B., MELIERES, F., MÜLLER, C. & WRIGHT, R. 1978. History of the Mediterranean salinity crisis. *In:* HSÜ, K. J., MONTADERT, L. *et al.* (eds). *Initial Rep. Deep Sea drill. Proj.* **42,** 1053–78. U.S. Govt Printing Office, Washington.

—— & Ryan, W. B. F. 1974. Deep sea drilling in the Hellenic trench. *Geol. Soc. Greece*, **10,** 81–9.

HUCHON, P., ANGELIER, J., LE PICHON, X., LYBERIS, N. & RICOU, L. E. 1981. Les structures tectoniques observées en plongée dans les fosses helléniques *C. r. somm. Soc. geol. France*, in press.

LE PICHON, X. & ANGELIER, J. 1979. The Hellenic arc and trench system: a key to the neotectonic evolution of the eastern Mediterranean area. *Tectonophysics*, **60,** 1–42.

—— & —— 1981. The Aegean Sea, *Proc. R. Soc. London*, in press.

——, ——, AUBOUIN, J., LYBERIS, N., MONTI, S., RENARD, V., GOT, H., HSÜ, K., MART, Y., MASCLE, J., MATTHEWS, D., MITROPOULOS, D., TSOFLIAS, P. & CHRONIS, G. 1979a. From subduction to transform motion: a Seabeam survey of the Hellenic trench system. *Earth planet. sci. Lett.* **44,** 441–50.

——, ——, BOULIN, J., BUREAU, D., CADET, J. P., CHAPEL, A., DERCOURT, J., GLAÇON, G., GOT, H., KARIG, D., LYBERIS, N., MASCLE, J., RICOU, L. E. & THIÉBAULT, F. 1980. Importance des formations attribuées au Messinien dans les fossés de subduction helléniques: observations par submersible. *C. r. Acad. Sci. Paris*, **290D,** 5.

——, ——, ——, ——, ——, DERCOURT, J., GLAÇON, G., GOT, H., KARIG, D., LYBERIS, N., MASCLE, J., RICOU, L. E. & THIÉBAULT, F. 1979b. Tectonique active dans le fossé de subduction hellénique: observations par submersible. *C. r. Acad. Sci. Paris*, **289D,** 1225.

——, HUCHON, P., ANGELIER, J., LYBERIS, N., BOULIN, J., BUREAU, D., CADET, J. P., DERCOURT, J., GLAÇON, G., GOT, H., KARIG, D., MASCLE, J., RICOU, L. E. & THIÉBAULT, F. 1981. Active tectonics in the Hellenic trench. *Oceanol. Acta*, **SP4,** 273–81.

LE PICHON, X. & HSÜ, K. J. 1979. HEAT, *Episodes*, **1,** 3–6.

LE QUELLEC, P., MASCLE, J., GOT, H. & VITTORI, J. 1980. Seismic structure of southwestern Peloponnesus continental margin. *Bull. Am. Assoc. Petrol. Geol.*, **64,** 242–63.

LYBÉRIS, N., ANGELIER, J., LE PICHON, X. & RENARD, V. 1981. Interprétation d'un fossé de subduction à partir des levés bathymétriques au sondeur multifaisceaux 'Seabeam': l'exemple du fossé hellénique. *C. r. somm. Soc. geol. France*,

MCKENZIE, D. P. 1978. Active tectonics of the Alpide-Himalayan belt: the Aegean Sea and surrounding regions (tectonics of Aegean region). *Geophys. J. R. astron. Soc.* **55,** 217–54.

RITSEMA, A. R. 1974. The earthquake mechanisms of the Balkan region. *R. Netherl. Meteorol. Inst., De Bilt, Sci. Rep.* **74–4,** 36 pp.

ZIBROWIUS, H. 1980. Thanatocoenose pleistocene profonde à spongiaires et Scléractiniaires dans la fosse hellénique. *XXVIIe Congrès CIESM, Cagliari*, October 9–18.

X. le Pichon, P. Huchon, J. Angelier, N. Lybéris, J. Boulin & D. Bureau, Département de Géotectonique, Université Pierre et Marie Curie, 75230 Paris Cedex 05, France.

J. P. Cadet, Département des Sciences de la Terre, Université d'Orleans, 45045 Orléans, France.

J. Dercourt, Laboratoire de Stratigraphie, Université Pierre et Marie Curie, 75230 Paris Cedex 05, France.

G. Glaçon, Département de Géotectonique, Université Pierre et Marie Curie, 75230 Paris Cedex 05, France.

H. Got, Centre de Recherches en Sédimentologie, Université de Perpignan, 66000 Perpignan, France.

D. Karig, Department of Geological Sciences, Cornell University, New York 14853, U.S.A.

J. Mascle, Laboratoire de Géologie dynamique, Université et Marie Curie, 75230 Paris Cedex 05, France.

L. E. Ricou, Laboratoire de Géologie historique, Université Paris Sud, 91000 Orsay, France.

F. Thiébault, Laboratoire de Géologie structurale, Université de Lille, 59000 Lille, France.

Detailed tectonic trends on the central part of the Hellenic Outer Ridge and in the Hellenic Trench System

N. H. Kenyon, R. H. Belderson & A. H. Stride

SUMMARY: Extensive new sonograph coverage in the Eastern Mediterranean allows a re-assessment of the complex relief outside the Hellenic Arc. The trenches to the SE and SW of Crete are similar in their plan view appearance, each having predominantly sinuous or curved structures that are interpreted as the surface expression of major thrusts. Thrusting, possibly imbricate, is also suggested on the central, shallowest part of the Hellenic Outer Ridge where interlocking convex-outward, curved structures occur, along with other faults and folds. These are thought to relate to décollement-type displacement within or beneath evaporite sediments. The structural style deduced from the morphology is consistent with a continuing outward radial push of the Hellenic Arc to the SW, S and SE, rather than with a migration to the SW and strike-slip on the SE side of the arc as proposed by other workers.

The region considered here includes the central portion of the Hellenic Trench System and Hellenic Outer Ridge in the vicinity of Crete (Figs 1–3). The nature and significance of these two major relief features are still incompletely understood. The Outer Ridge may have been initiated during the late Miocene as it seems that the Messinian evaporates thin onto the 'Upper Plateau' region (Biju-Duval *et al.* 1978). Because the Outer Ridge evolved mainly during the Plio–Quaternary it is probably related largely to the latest phase in the outward growth of the tectonically more disturbed Hellenic Trench System.

The present consensus is that the relief of the Outer Ridge is largely tectonic in origin, although there is disagreement about the mechanisms involved in its development. Major thrusting has been proposed within the Hellenic Outer Ridge (Finetti 1976; Biju-Duval *et al.* 1978) while Kenyon & Belderson (1977) proposed a regional décollement mechanism to explain the folding and faulting along the whole length of the Outer Ridge. In addition, there are areas of postulated submarine salt karst within the tectonised ground (Belderson *et al.* 1978).

The new data consist of greatly increased sonograph coverage over both the Outer Ridge and the Hellenic Trench System, together with additional narrow-beam echo-sounder profiles and sub-bottom profiler data. These are used to re-assess our earlier conclusions (Stride *et al.* 1977) as well as to make a broad comparison with the data from the multi-beam echo-sounder 'Seabeam' (Le Pichon *et al.* 1979). We contest the proposal by Le Pichon *et al.* that compression has been dominant to the SW of Crete while strike-slip motion has been dominant to the SE of Crete.

RRS *Discovery* surveys of 1977 and 1979

The two cruises of RRS *Discovery* in 1977 and 1979 doubled the area of ground previously examined in the Eastern Mediterranean by means of Gloria, the long-range side-scan sonar (Fig. 1). The new coverage included large parts of the Hellenic Trench System as well as an extensive survey of the central portion of the Hellenic Outer Ridge. Both cruises made use of the improved two-way looking Mark 2 Gloria 'fish', which in this instance viewed a 30 km wide swathe of seafloor. Single-channel sub-bottom profiling was also used, and narrow-beam echo-sounding provided more realistic profiles of the surface of the floor than can be obtained from conventional echo-sounding.

In analysing the sonographs it must be borne in mind that linear features oriented parallel to the ship's track show up most clearly and small linear features at right angles to the track may not show up at all. Thus to be sure of detecting all structural trends it is necessary to examine the whole region on two sets of courses at an angle to one another. Inspection of Fig. 1 will show that, although the majority of courses are roughly parallel to the Hellenic Arc, crossing tracks are spread throughout the region and so most morphological trends should have been recognised.

Salt karst

The existence of salt karst of submarine origin on the Hellenic Outer Ridge was postulated by Belderson *et al.* (1978) to account for the presence of sub-circular to irregular patches of rough ground that have a conspicuous appear-

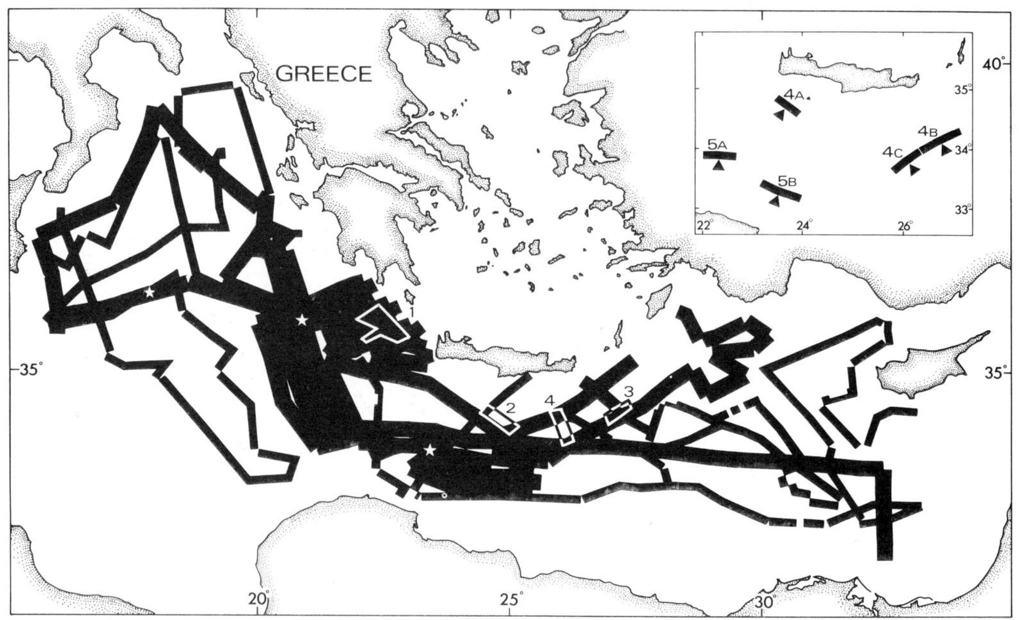

Fig. 1. Gloria long-range side scan sonar coverage in the Eastern Mediterranean up to 1979, the approximate limits of the Seabeam surveys (boxes) and the location of Scripps Institution of Oceanography deep-tow sites (stars). The inset figure shows the location and direction of view of the sonar beam for sonographs in Figs 4 & 5.

ance on 'fixed gain' sonographs. The more recent side-scan sonar coverage has revealed a few more of the distinctive sub-circular patches of rough ground, interpreted as dissolution surfaces above diapirs. Such rough floors are confined to the inner portion of the Hellenic Outer Ridge (Fig. 2). West of Crete two groups of these sub-circular salt karst bodies are connected by prominent linear faults, or salt 'dykes' (Belderson *et al.* 1978). A detailed

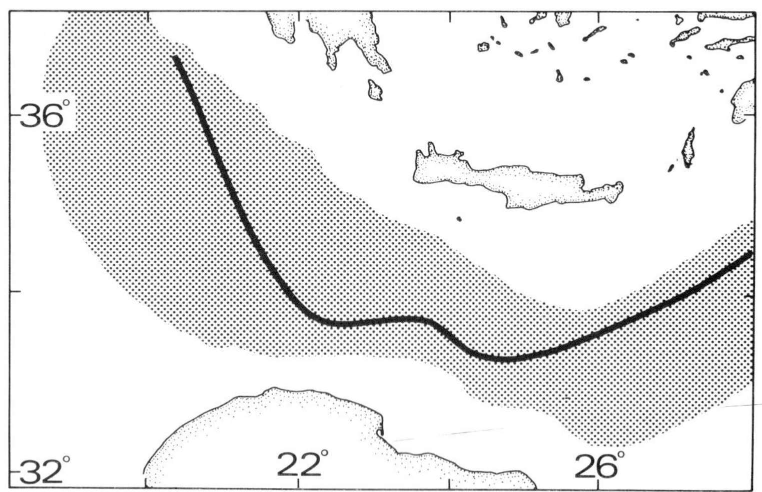

Fig. 2. Sub-circular to irregular shaped areas of rough floor (mostly pits) attributed to salt karst are confined to a belt on the internal portion of the Hellenic Outer Ridge, north of the limit shown. (Hellenic Outer Ridge shaded).

between the two areas postulated by Le Pichon & Angelier (1979), who proposed a thrust motion for those located SW and west of Crete and a predominant strike-slip motion along the Pliny and Strabo Trenches located SE of Crete.

The high ground behind the trenches is blocky, with the main structural trends made evident on bathymetric charts. The sonographs show, in addition, many relatively short, straight scarp-like features, both to the east and west of Crete, but trends are neither so consistent nor so numerous as the structural trends on the Hellenic Outer Ridge. They are considered to be normal faults, which is consistent with earlier interpretations of normal faulting both here and on Crete (e.g. Angelier 1978, fig. 5). In one area to the west of Crete there are several more sharply defined straight faults with only minor throw; these are presumably the youngest ones recognizable at the scale of the sonographs.

Hellenic Outer Ridge

The new data re-emphasize the presence of a dominant lengthwise tectonic trend that is traceable along virtually the whole of the Hellenic Outer Ridge (Figs 3 & 6), including some areas where there was previously little or no sonograph cover. The features comprising this trend are known to consist of folds along the southern edge of the Outer Ridge, and elsewhere are interpreted as folds, largely faulted and slumped along their strike (Stride *et al.* 1977). The localized kink in the lengthwise tectonic trend on the south edge of the Hellenic Outer Ridge (Fig. 5B) is attributed to compression of the Outer Ridge against a projecting spur at the base of the African continental margin (as was previously suggested for similar relief located further west by Stride *et al.* 1977, fig. 14). This is analogous, but on a much smaller scale to the way the Outer Ridge wraps around both sides of the foot of the Cyrenaican continental slope (Fig. 3). One notably anomalous area lies just south of the Hellenic Trench System to the west of Crete. Here a bundle of probable folds bends round to become transverse to the Outer Ridge. By chance, the only Seabeam coverage of the Outer Ridge reported on by Le Pichon *et al.* (1979, fig. 3) included a part of this area and gave hints of its anomalous trends.

Tectonic cross-trends are particularly well-developed between Crete and Cyrenaica, with a tendency to decrease in importance outwards. The arrangement of these may be interpreted as implying a conjugate shear system developed in relation to an outwards radial push (Fig. 6).

Between Crete and Cyrenaica there are adjacent areas of flat and intensely deformed ground. The large flat northern part of the Hellenic Outer Ridge, to the SW of Crete ('Lower Plateau') is intersected in its western half by a substantial ridge of deformed ground (Stride *et al.* 1977, fig. 4) perhaps underlain by a thrust. South-east and SW of the Lower Plateau there are some much narrower bands of relatively flat floor that alternate with probable fold bundles, possibly underlain by thrusts. South of the Lower Plateau on the shoalest part of the Outer Ridge ('Upper Plateau') there are numerous tectonic trends with relief of up to 130 m. These are frequently curved and more widely spaced than elsewhere on the Outer Ridge (Fig. 5a); they are interpreted as probable outcrops of thrusts.

Morphological indications of compression between the Hellenic Arc and the Hellenic Outer Ridge

It is generally agreed from geological studies on land that there has been outward migration of the Hellenic Arc since the Mesozoic. However, a more recent assessment, while supporting thrusting on the SW side of the arc suggests by first-motion studies of earthquakes (during the short period of available observations) that there is NE to SW strike-slip motion on the eastern side of that arc (Le Pichon & Angelier 1979). This proposed contrast between the eastern and western sides of the arc needs to be examined in relation to the the structural trends in the Hellenic Trench System and the adjacent Outer Ridge.

FIG. 4. Sonographs (located on Fig. 1) showing the morphological similarity of the trenches (A) to the SW of Crete (where there is supposed thrusting), and (B) to the SE of Crete (where there is supposed predominant strike-slip motion according to Le Pichon & Angelier 1979). The narrow cleft which defines each trench axis lies in shadow cast by the outer wall. Echo-sounder profiles show that the outer wall is about 550 m high near the west end of (A), 300 m high near the centre of (B), and 330 m high near the east end of (B). Sonograph (C) shows overlapping curved and mainly convex outward-facing structures interpreted as the early stages in the development of a new trench system just south of the main Strabo Trench. Note that sonograph (C) is continuous with sonograph (B). White arrows show the supposed motion. The black arrowheads point in the direction of view of the sonar beam as indicated on Fig. 1.

The trench system outside Crete is notable for approximately NW-trending major structural lines in the west and approximately NE-trending ones in the east. The partial sediment fill in the western examples obscures their deeper relief, so that they contrast with the narrow cleft-like appearance of the eastern examples, where the deepest portions are still largely sediment-free. However, the almost sediment-free trenches more immediately to the SE and SW of Crete are remarkably similar in appearance, as shown on Fig. 4. It is likely that there is progressive outward younging of trenches, from the Pliny Trench to the narrower and less deep Strabo Trench and then to the interconnected group of proposed thrusts immediately to the south of the Strabo Trench, where the newest trench of all seems to be developing. At the same time the block of ground between the Pliny and Strabo trenches is probably being uplifted.

On the Hellenic Outer Ridge, the two sets of cross-trends which cut the Outer Ridge-parallel folds (Fig. 6) are interpreted as strike-slip faults related to a conjugate shear system which results from regional compression between the Hellenic Arc and the foot of the African continental slope. The fact that such a pattern is found both SW and SE of Crete implies that, even if some strike-slip motion is now occurring in the trenches SE of Crete, a substantial amount of compression is also being transmitted across the trenches onto the Outer Ridge. In any case the Outer Ridge and folds parallel to it are as well-developed SE of Crete as they are to the SW of Crete and belie any long-term difference in origin between the two sides.

Thrusts or incipient thrusts are considered to lie beneath the bundles of folds separating smooth ground in the south-eastern part of Fig. 3, while other well-developed thrusts (Fig. 6) may lie beneath the ridge that intersects the Lower Plateau, as well as beneath the southern edge of the Hellenic Outer Ridge and beneath a step located above it to the south of the Upper Plateau (Stride *et al.* 1977, figs 6 & 12). In addition, outcrops of thrusts, possibly in places imbricate and overlapping, are suggested by the numerous curved, interlocking structural trends on the 'Upper Plateau'. For example, the two thrusts in the western half of Fig. 5(B) correspond to the two that outcrop in profile in fig. 15 of Finetti (1976). Further east they merge into a single thrust. The net effect of such thrusts should be to cause some thickening of the sediments in this, the shoalest, part of the Hellenic Outer Ridge. Alternatively, its elevation might be due to arching up of the basement. A seismic refraction line shot from Crete to Egypt across the Outer Ridge to the east of the 'Upper Plateau' shows the crust beneath the ridge to be 27 km thick, of which the sedimentary cover accounts for about 8 km, with some evident uparching of the basement (J. Makris, pers. comm.).

Thrusting involving even supposed Mesozoic sediments within parts of the Outer Ridge has already been interpreted on sub-bottom profiles (Finetti 1976; Biju-Duval *et al.* 1978), although such data are not entirely convincing on their own. We propose shallow thrusts which we think may be related to a regional décollement or a series of décollement horizons within or at the base of the Messinian evaporites (Kenyon & Belderson 1977). However, we do not exclude some deeper, higher angle thrusts below the Upper Plateau. Messinian salt will have contributed to the complexity of the deformation.

In conclusion, the broad regional coverage with sonographs does not lend support to the view that there is significant contrast between the shape of the bottoms of the trenches to the SE and SW of Crete, as has been suggested by Le Pichon & Angelier (1979) and Le Pichon *et al.* (1979), but rather points to reasonable similarity in their morphology. The conflict between the structural data presented and interpreted here and the first-motion studies (implying transform motion to the SE of Crete) might perhaps be resolved if the proposed transform motion is younger than the trench and Outer Ridge relief. Certainly the first-motion studies of this region are only available for some earthquakes of the past few years, and there is no reason to suppose that they should be relevant for the very much longer period of time needed to build present relief. Moreover, Le Pichon & Angelier (1979, fig. 2 and table II) did not make use of three horizontal projections of slip vectors which indicated thrusting normal to the trench system east of Crete (south of Karpathos), which diverged by about 90° from others that were used preferentially by them in the vicinity. Clearly it will be necessary to obtain a much longer time series of first-motions for the region before there can be any certainty about whether the proposed transform motion east of Crete is typical of present deformation in the region.

Acknowledgments: Mr Kenyon thanks Professor F. N. Spiess and Dr W. B. F. Ryan for the opportunity to sail on R.V. *Melville* so as to explore in detail, with the deep-tow system, some of the postulated salt karst revealed by Gloria sonographs.

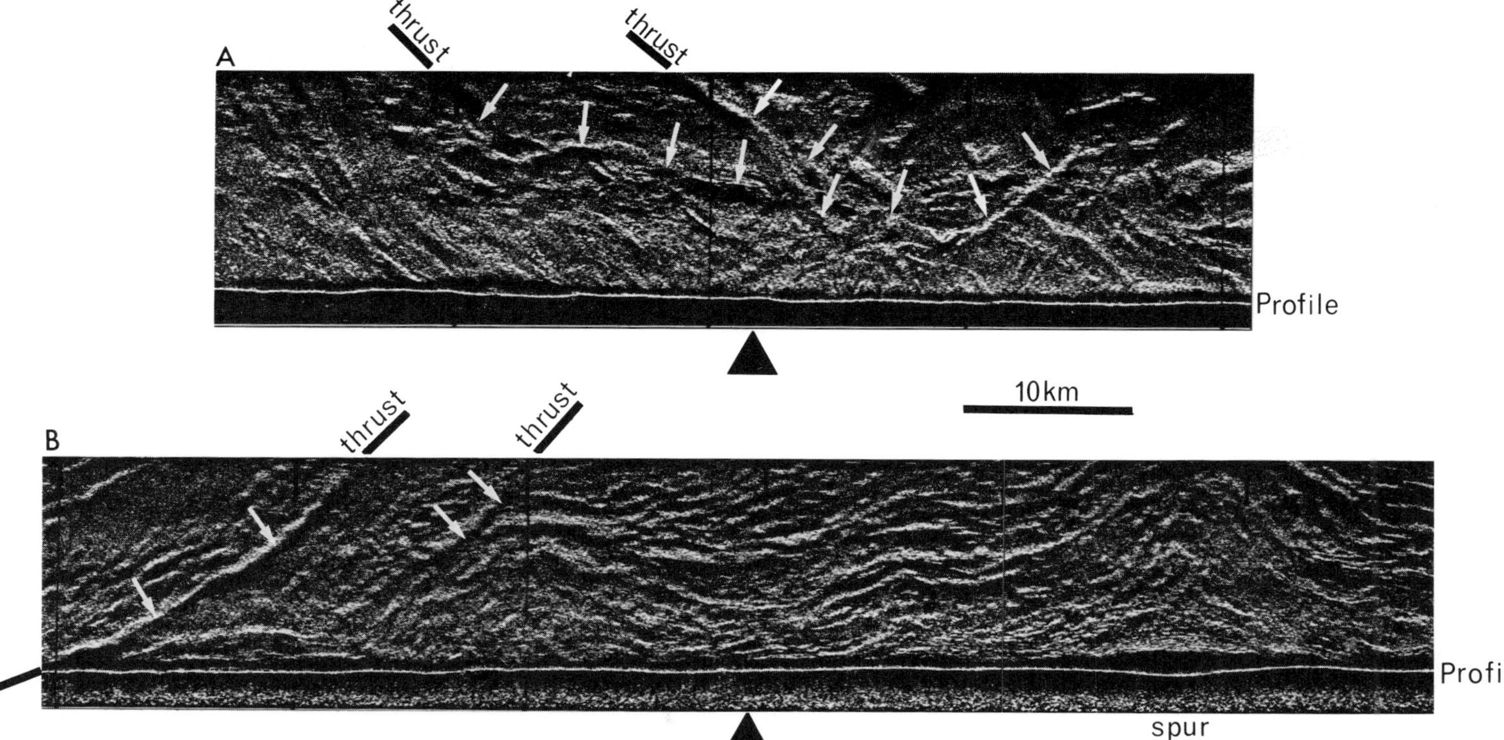

Fig. 5. (A) Sonograph (located in Fig. 1) showing curved tectonic trends, with relief of up to about 130 m, on the shoalest part of the Hellenic Outer Ridge. They are interpreted as the surface of two overlapping thrust sheets. White arrows show the supposed motion. These thrusts may be seen in depth on a sub-bottom profile of Finetti (1976, fig. 15). The black arrowhead points in the direction of view of the sonar beam as indicated on Fig. 1. (B) Sonograph on the south side of the Hellenic Outer Ridge (located in Fig. 1) showing ridge-parallel tectonic trends, with a general relief of about 40 m, wrapped around a spur at the base of the African continental slope. White arrows show the supposed movement direction of two thrust sheets with curved fronts. The black arrowhead points in the direction of view of the sonar beam as indicated on Fig. 1.

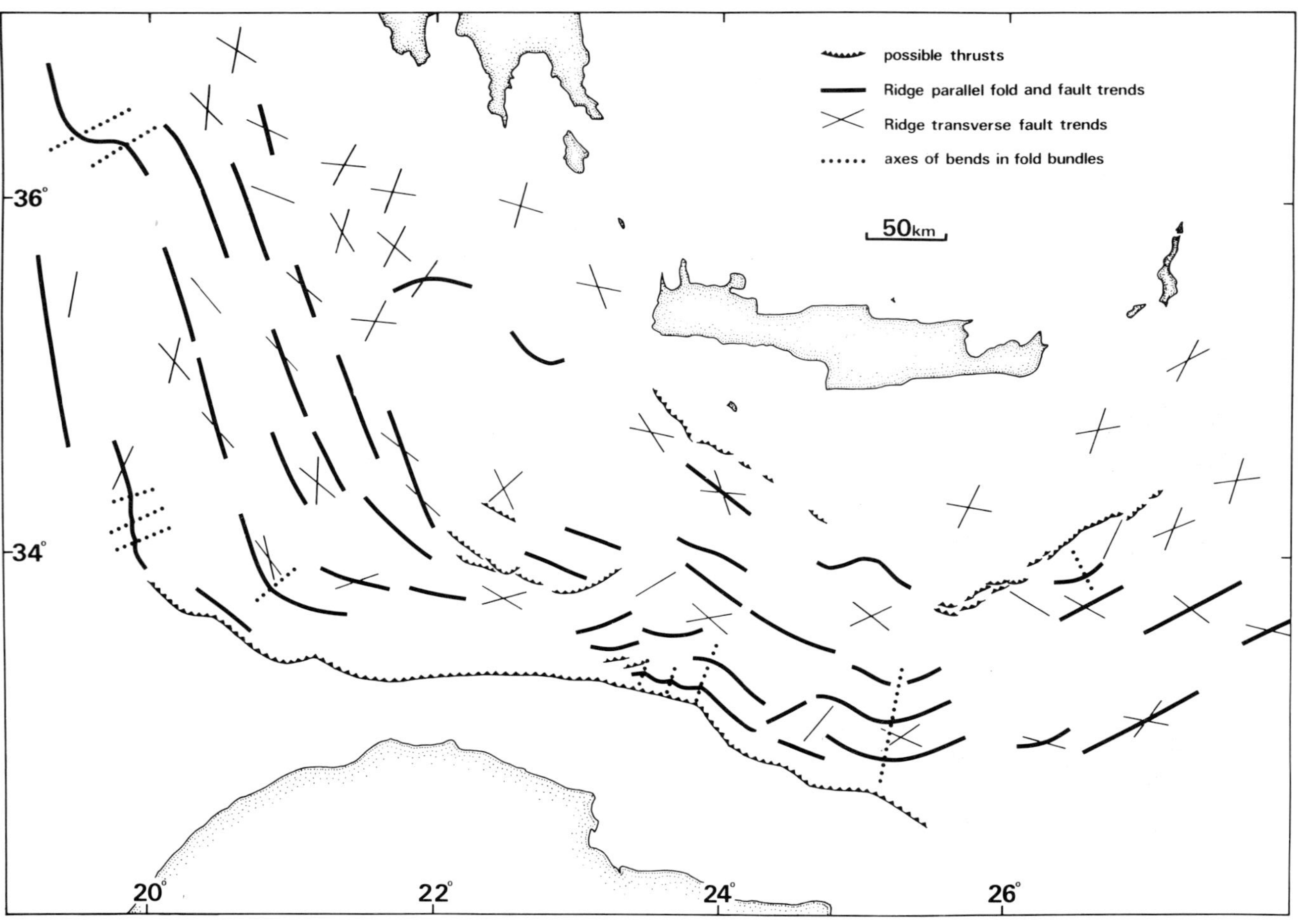

FIG. 6. Simplified structural map showing the curved outcrops of some of the most probable thrusts observed on sonographs in the Hellenic Trench System and on the Hellenic Outer Ridge, together with a summary of the Outer Ridge-parallel fold and fault trends and the conjugate system of strike-slip fault trends. The principal stress trajectories will be normal to the ridge-parallel fold trends and bisect the angle between the conjugate system of supposed strike-slip faults and lie parallel with the axes of bends in fold bundle trends. This is consistent with a scheme of radial stress. The possible thrust bounding the southern margin of the Hellenic Outer Ridge is thought to represent the front of a décollement within the Outer Ridge. The continuity of this feature is based on additional bathymetric data.

References

ANGELIER, J. 1978. Tectonic evolution of the Hellenic Arc since the late Miocene. *Tectonophysics*, **49,** 23–36.

BELDERSON, R. H., KENYON, N. H. & STRIDE, A. H. 1978. Local submarine salt-karst formation on the Hellenic Outer Ridge, eastern Mediterranean. *Geology*, **6,** 716–20.

BIJU-DUVAL, B., LETOUZEY, J. & MONTADERT, L. 1978. Variety of margins and deep basins in the Mediterranean. *In:* WATKINS, J. S. *et al.* (eds). *Geological and Geophysical Investigations of Continental Margins*. Mem. Am. Assoc. Petrol. Geol. **29,** 293–317.

FINETTI, I. 1976. Mediterranean Ridge; a young submerged chain associated with the Hellenic Arc. *Boll. Geofis. teor. appl.* **19,** 31–65.

FOOSE, R. M. 1977. Structural lineaments and tectonics of the Mediterranean Basin. *In:* BIJU-DUVAL, B. & MONTADERT, L. (eds). *Structural History of the Mediterranean Basins*, 221–7. Technip, Paris.

KENYON, N. H. & BELDERSON, R. H. 1977. Young compressional structures of the Calabrian, Hellenic and Cyprus Outer Ridges. *In:* BIJU-DUVAL, B. & MONTADERT, L. (eds). *Structural History of the Mediterranean Basins*, 233–40. Technip, Paris.

LE PICHON, X. & ANGELIER, J. 1979. The Hellenic arc and trench system: a key to the neotectonic evolution of the Eastern Mediterranean. *Tectonophysics*, **60,** 1–42.

—— , AUBOUIN, J., LYBERIS, N., MONTI, S., RENARD, V., GOT, H., HSÜ, K., MART, Y., MASCLE, J., MATTHEWS, D., MITROPOULOS, D., TSOFLIAS, P. & CHRONIS, G. 1979. From subduction to transform motion: a Seabeam survey of the Hellenic Trench System. *Earth planet. Sci. Lett.* **44,** 441–50.

—— *et al.* 1981. Subduction in the Hellenic trench: probable role of a thick evaporite layer based on Seabeam and submersible studies (this volume).

STRIDE, A. H., BELDERSON, R. H. & KENYON, N. H. 1977. Evolving miogeanticlines of the East Mediterranean (Hellenic, Calabrian & Cyprus Outer Ridges). *Philos. Trans. R. Soc. London*, **284A,** 255–85.

N. H. KENYON, R. H. BELDERSON & A. H. STRIDE, Institute of Oceanographic Sciences, Wormley, Godalming, Surrey, England.

The structure of the Calabro-Sicilian Arc: result of a post-orogenic intra-plate deformation

F. C. Wezel

SUMMARY: Morphotectonic and geological data show that this arc constitutes a post-orogenic ring-like system concentric round the Marsili 'backarc' basin. Peripheral concentric features comprise the 'Aeolian andesitic ring' and the arcuate troughs and ridges characteristic of the 'forearc' region. The Hyblean horst-like plateau and the Strait of Sicily represent the block-faulted foreland of the Sicilian segment of the Calabro-Sicilian Arc. The present shape of the arc has largely resulted from Plio-Quaternary vertical tectonic movements, superimposed on compressive structures caused by previous phases of tectonic deformation. The effect can be visualized as an outward-moving topographic wave producing inversion of tilting movements. This kinematic process (termed 'krikogenesis') is postulated as being related to midplate domal upwarp, dilatation and cauldron-like collapse of the Marsili block in the core of the arcuate system. Volcanicity appears to be associated with various stages of this geological history. Rather than originating by collision of two rigid plates, present-day morphological, volcanic and seismic configurations seem to be due to interaction of intra-plate 'arches', activated by local mantle movements.

Along with the Hellenic and Cyprus arc systems, the Calabro-Sicilian Arc is considered by many workers (e.g. Biju-Duval *et al.* 1978) to be a Mediterranean counterpart of Pacific island arcs. Subduction of Ionian Sea lithosphere should be responsible for the calc-alkaline volcanicity of the Aeolian Islands and for behind-arc spreading which created the Tyrrhenian Basin (backarc basin). However, a possible difference with western Pacific systems arises from the fact that Mediterranean arcs must be interpreted as the result of motion of small plates, rather than of a general descent of African lithosphere beneath Europe (McKenzie 1972; Alvarez *et al.* 1974).

The purpose of this paper is to show that the recent tectonics and volcanism of the Calabro-Sicilian region more likely resulted from intraplate deformation (here termed '*krikogenesis*') instead of from the interaction and horizontal convergence of two lithospheric plates. This different view arises from detailed land and marine studies, mainly of the Sicilian sector, and from morphostructural analysis of the Tyrrhenian Sea floor.

Morphotectonic configuration of the Tyrrhenian Sea floor

Structural and morphological observations led to the identification, in different physiographic zones and depths, of several arcuate bathymetric structures on the Tyrrhenian Sea floor (see morphological map of Savelli & Wezel 1979 and revised colour version 1980). The structures constitute a nearly complete circle or ellipse, but more commonly encompass a semicircle or less (Fig. 1). These peculiar and previously undescribed arcuate structures surround composite crustal blocks that, according to their diameter, I will name: *local-blocks* (up to 10′ of latitude or 18.5 km), *minor-blocks* (10′–1° or 18.5–111.1 km), *meso-blocks* (1°–4° or 111.1–444.5 km) or '*small ocean basin*'-*blocks* (4°–7° or 444.5–777.8 km). Elsewhere, outside the Mediterranean region, blocks with a diameter greater than 7° (*geo-blocks*) are of the same size as certain 'lithospheric plates' (e.g. the Tibet Plateau 20 degree-block can be compared in size with the Cocos plate).

Larger Tyrrhenian circular structures are of fairly uniform size, having a roughly constant diameter of between 1° to 2°, averaging 150 km (Fig. 1). These nearly equidimensional meso-blocks are subdivided morphologically into minor component units.

The 'backarc' region of the Calabro-Sicilian Arc (Marsili Basin)

The arc constitutes a *multi-ringed system* consisting of a 'backarc' region, concentrically surrounded by an inner ring of volcanic islands and seamounts, and by an outer series of post-orogenic 'forearc' basins and ridges.

The core of the Calabro-Sicilian circular structure, the most impressive of the Tyrrhenian rings, is represented by the Marsili Basin,

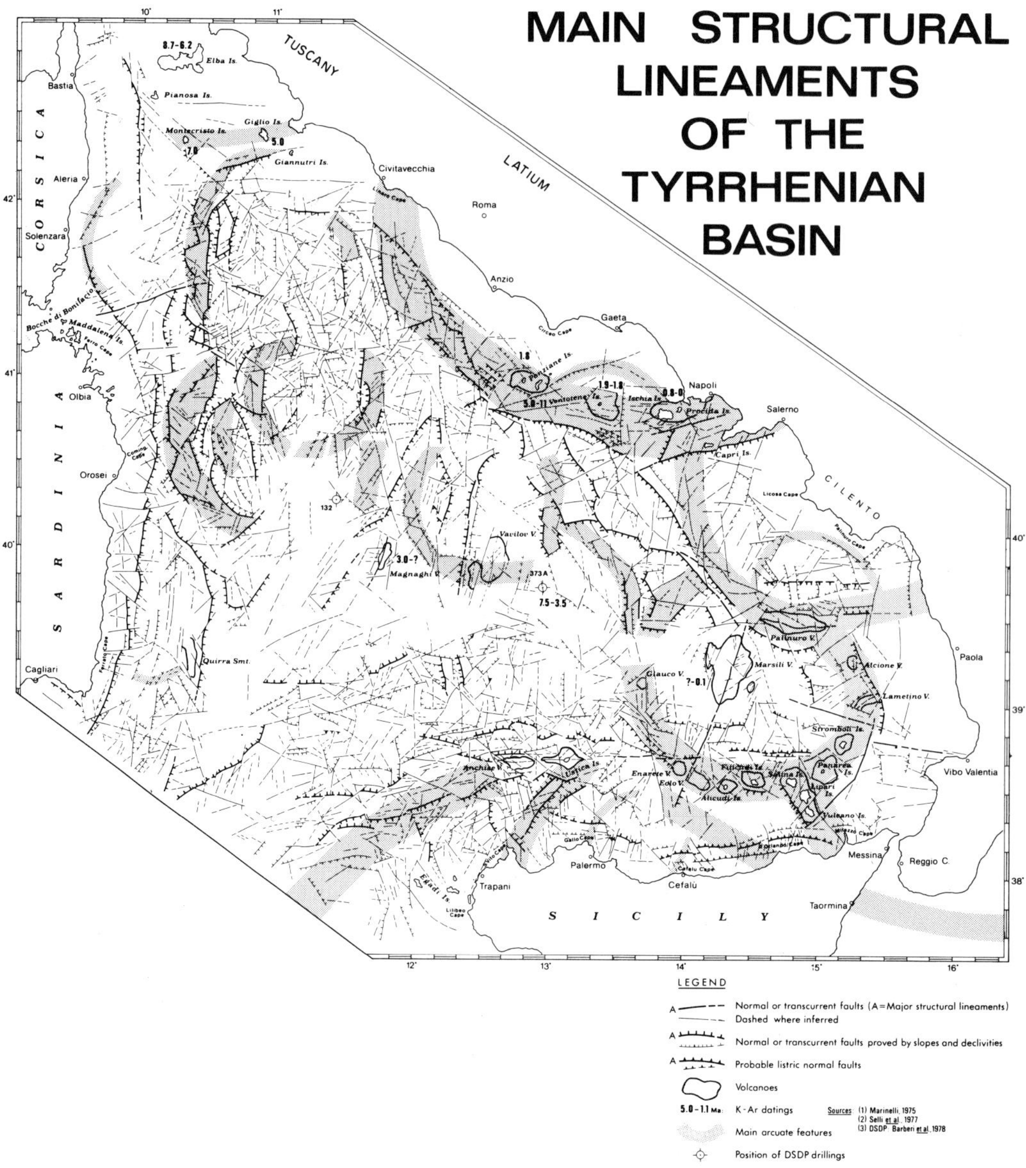

FIG. 1. Major tectonic lineaments in the Tyrrhenian Sea floor deduced from a morphological analysis. Note that the constituent circular blocks ('krikogens') show a peripheral suture rim with a width commensurate with size of the block. Circular fragmentation affected both submarine and continental regions. C.G. = Campidano Graben.

1° in diameter. This nearly circular depression, with a maximum depth of 3400 m, has in the centre the huge Marsili volcanic seamount, consisting of a ridge of extrusive rock, situated on a major fault trending N17°E. The seamount has 2900 m of relief, with its base lying at a depth of about 4000 m according to seismic data (Selli *et al.* 1977).

Dredge samples include tholeiitic, alkaline (Del Monte 1972) and calc-alkaline basalts (Maccarrone 1970; Selli *et al.* 1977), perhaps indicating different stages of igneous activity (P. L. Rossi, pers. comm. 1980). According to Selli *et al.* (1977) the volcano displays in its uppermost portion a calc-alkaline affinity above a base with tholeiitic affinity. Radiometric data suggest a 0.1–0.2 Ma (late Pleistocene) age for the more recent eruptives of Marsili seamount, while the onset of volcanism possibly took place in Pliocene times between 4 and 2 Ma ago (Selli *et al.* 1977).

The Marsili Basin is characterized by significant crustal attenuation (crust 16–18 km thick: Morelli *et al.* 1975), by extensional tectonics, and by the presence of an extremely strong local thermal field (maximum north–south diameter of about 40′) in the southern sector of Marsili volcano; here in the region of 39°N heat flow values are up to 11 HFU (Della Vedova & Pellis 1979). These geophysical data suggest the existence at a shallow crustal depth (about 18 km) of hot mantle material (magma chamber?), especially beneath the seismically active eastern half of the basin. Converging on this area of abnormally elevated heat flow are several major transverse (radial) lineaments such as the Sangineto, Catanzaro and the Giardini-Lipari lines (Fig. 2). In contrast, the western aseismic half of the Basin has a Bouguer gravity-high anomaly of +250 mGal, centred over the deepest basinal area ('*Marsili mascon*').

The Aeolian calc-alkaline ring

Alvarez *et al.* (1974, p. 312) were apparently the first to draw attention to '. . . the remarkably circular pattern of the south-easternmost Tyrrhenian, with the inner ring of the Aeolian Islands—M. Palinuro and the outer ring of the southern Apennines-Calabria-northern Sicily mountains concentric about the Marsili volcano . . .'. This peculiar circular feature was called 'anello delle Eolie' by Selli (see Selli *et al.* 1977, p. 228), who demonstrated the calcalkaline nature of rocks from six volcanic seamounts belonging to the ridge (Figs 1 & 2).

The start of the Aeolian volcanism shifted eastward during the Quaternary. Lava composition evolves from calc-alkaline to shoshonitic suites (Barberi *et al.* 1974). Stromboli is still active with persistent moderate explosive activity. Though the available radiometric data indicate a Quaternary age for the older Aeolian activity (see Marinelli 1975), it is however possible that earlier volcanic activity took place from the middle Pliocene, after sedimentation of the Trubi Formation (Selli *et al.* 1977; Bacini Sedimentari 1980).

A quantitative interpretation of magnetic and gravity anomalies along the Aeolian arc axis shows the continental nature of the underlying basement. Its morphology is characterized by horst-and-graben structures which probably control the eruptive activity (Iacobucci *et al.* 1977). Present-day tensional faulting is also shown by numerous faults cutting the metamorphic basement and volcanic successions, and by geodetic measurements (Nappi *et al.* 1977). The 20 km thick intermediate continental crust in the western (older and inactive) sector of the arc includes several low-velocity layers, while in the eastern (active) sector the crust contains only a single low-velocity layer at a depth of about 10 km (Morelli *et al.* 1975).

The annular volcanic ridge is limited on the concave side by inward-sloping normal faults, sometimes of listric type (see Fig. 2 and Savelli & Wezel 1979), indicating tilting of stepped blocks towards the 'forearc' region. The circular shape of the wall ridge and the evidence from some seismic reflection profiles (e.g. profile MS 3 in Bacini Sedimentari 1980) for a vertical boundary fault suggest a spectacular huge *cauldron-like collapse* of the Marsili region along ring fractures (Fig. 3). This may perhaps be connected with subsidence of magma in a magma chamber underlying the region.

In summary, Aeolian andesitic volcanicity mainly occurred in a region that seems to have been subjected to extensional stress. Eruptive centres are chiefly located on the intersections of crustal fractures. Volcanic features have resulted from long-active phases of magma ascent with formation of volcanic cones, followed by stages of volcano-tectonic collapse possibly due to the emptying of a magma chamber.

In general it has been postulated that the change of composition of erupted lavas toward the concave side of the arc may be related to the thickness and the structure of the crustal basement (e.g. stable, 'orogenized' or labile: see Wezel 1977). The Aeolian calc-alkaline eruptive centres are presumably related to crustal fractures cutting an 'orogenized' basement consisting of folded flysch-type units, similar to those outcropping in the Sicilian mountain chain (see Bacini Sedimentari 1980). Additionally, geochemical data indicate contamination of some andesitic magmas (Lipari Island) with flysch-type sediments metamorphosed under conditions of low-pressure/high-temperature (Marinelli 1975, p. 210).

The 'forearc' region

The convex side of the arc system faces towards the Calabro-Sicilian continental region. It consists essentially of troughs and ridges (Fig. 2)

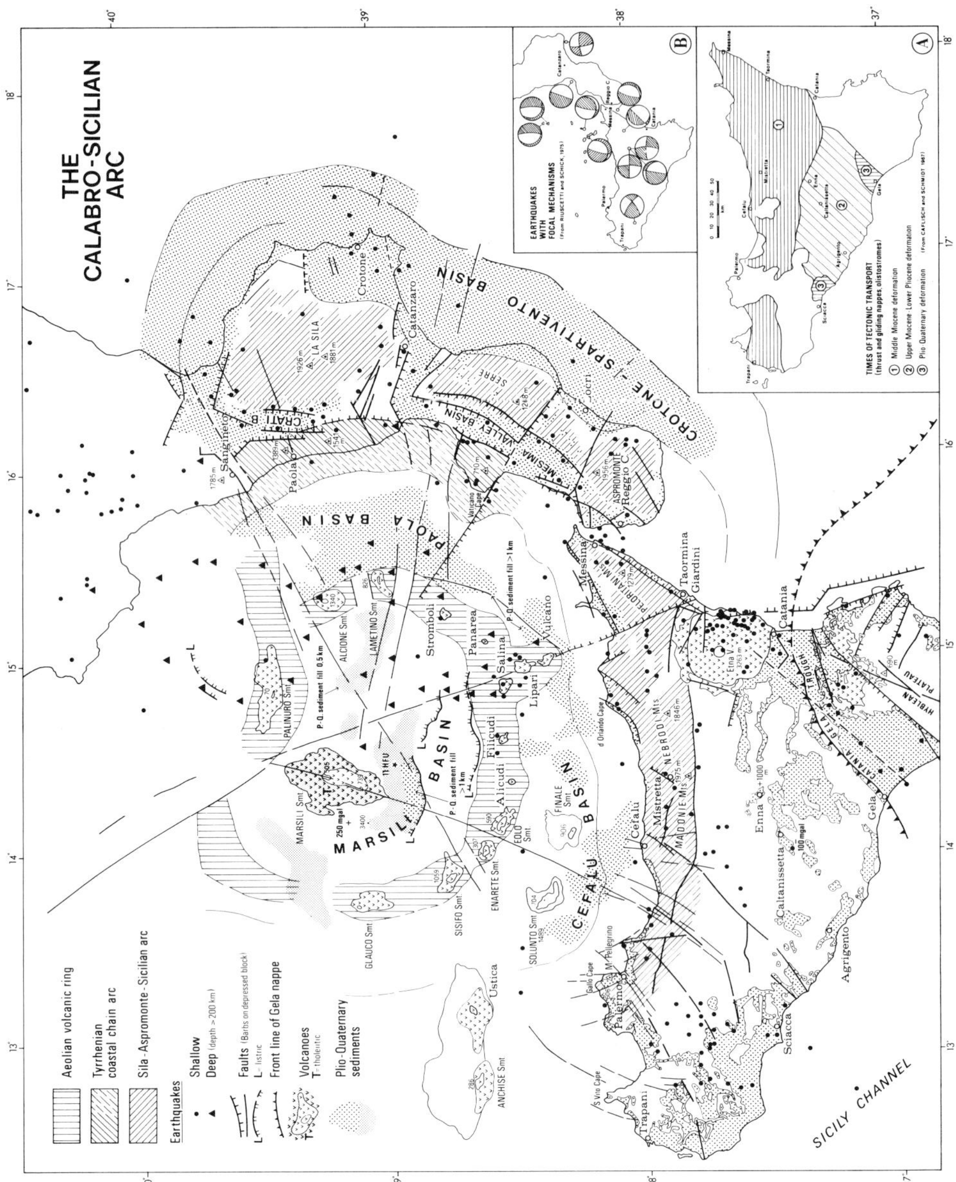

FIG. 2. Simplified geological map of Calabro-Sicilian Arc showing the concentric regional distribution of selected key tectonic, volcanic and seismic features (derived from published data, mainly the 'Structural Model of Italy' and the 'Carta Tettonica d'Italia'). (A) Shows the outward lateral migration, from middle Miocene to late Pleistocene, of compressive deformation, concomitantly with a shifting of the foredeep axis. (B) Fault-plane solutions for earthquakes along the arc denoting dip-slip faulting and left-lateral strike-slip motion. Note the lack of thrust events.

FIG. 3. Diagrammatic and grossly simplified cross-sections across Sicilian segment of the Calabro-Sicilian Arc, showing its 'krikogenetic' evolution. (A) (early-middle Miocene); development, uplift and erosion of volcanic arc; infilling of peripheral geosyncline with 'internal' preflysch and flysch sediments, partially volcaniclastic. (B) (late Miocene): inversion of regional relief; outward growth and subsidence of Marsili block; tectonization of preflysch and flysch sediments; Hyblean submarine

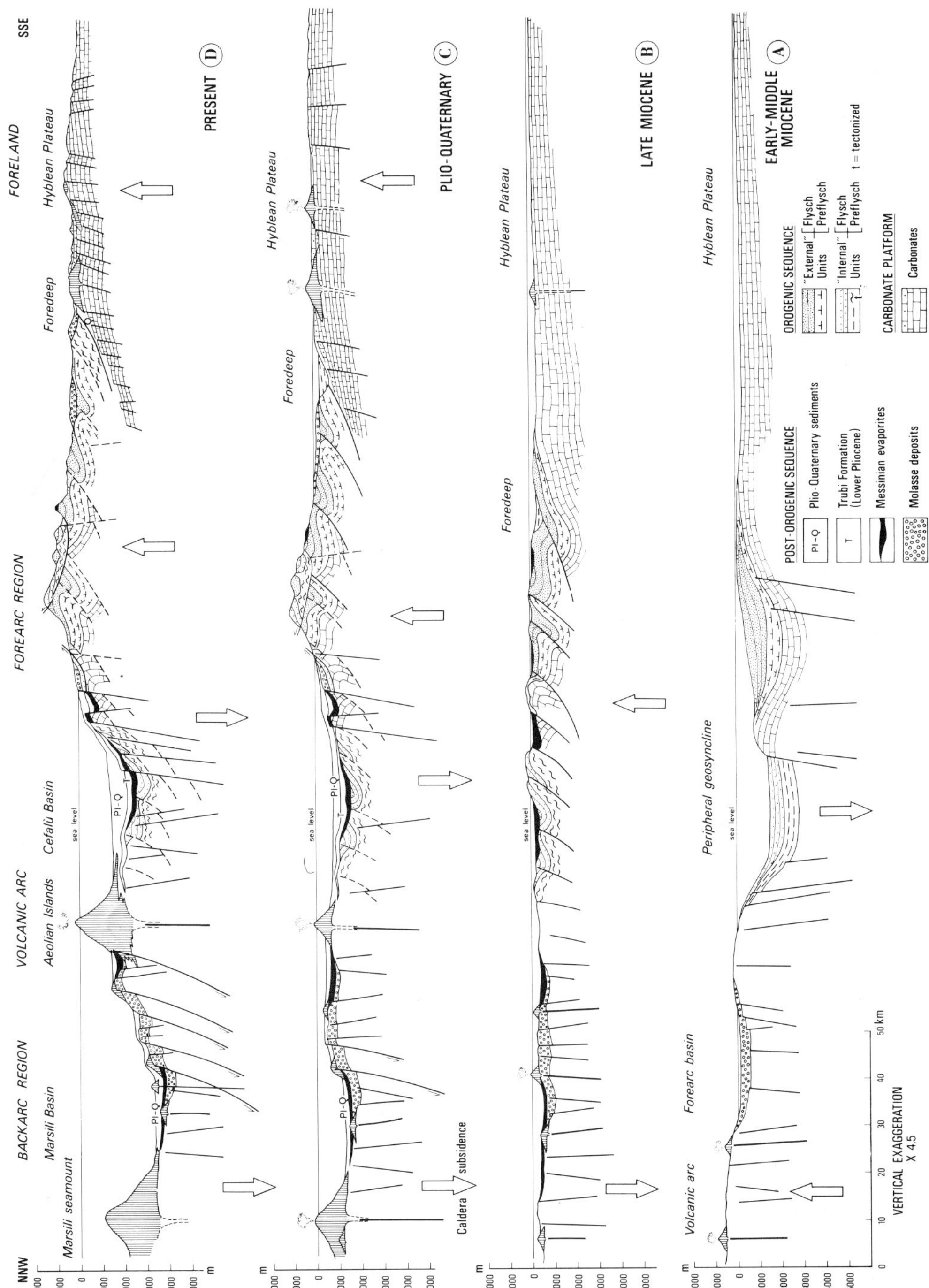

volcanic activity. (C) (Plio-Quaternary): caldera-like collapse of Marsili block; complementary activity of Marsili volcano and Aeolian andesitic ring; development and filling of Cefalù Basin; faulting and uplift of deformed orogenic sediments; outward migration of uplift axis and foredeep; activation of foreland 'arch' and development of 'outer ridge'. (D) (Present): continued collapse of, and sedimentation in, Marsili and Cefalù basins; further uplift and erosion of mountain arc; 'subduction' of the Hyblean 'arch' ('krikogen').

that developed in outwards progression during the Plio-Quaternary.

The Cefalù and Paola 'outer-arc' troughs

Filled by over 1000 m of Plio-Quaternary sediments (up to 1900 m in Sicily and 4000 m in Calabria) the Cefalù and Paola basins represent major depositories for sediment accumulation along the present Tyrrhenian margin (Fig. 2). The morphology of the basement is complex, giving rise to several depocentres separated by structural highs. In the Cefalù Basin the basement consists (see Bacini Sedimentari 1980) of the same tectonic units (lower Miocene flysch sediments) which constitute the Sicilian mountain chain (cf. Ogniben 1960; Wezel 1974). The undeformed Cefalù Basin succession, lying unconformably on the tectonized flysch strata starts with middle Miocene sediments of likely Tortonian age. In Tortonian and younger strata all the structures shown by seismic reflection profiles are *extensional* and indicate foundering of the Cefalù Basin, with distinct episodes of subsidence (Bacini Sedimentari 1980). In the initial stage a fault-controlled subsidence was dominant, but later it was replaced by regional subsidence lacking conspicuous fault control.

Thus it appears that the upper Miocene and Plio-Quaternary basin-fill deposits overlying tectonically emplaced units are entirely *post-orogenic* and clearly indicate a tensional regime following tectonic compression.

The mountain arc

The arcuate mountain chains represent an overall arch-shaped ('geanticline') ridge outcropping in Calabria and Sicily (Fig. 2). In Calabria the mountain arc comprises two uplift axes separated by the intramontane Crati and Mesima valleys which are graben-like depressions (Ogniben 1973). The western axis (i.e. the Tyrrhenian Coastal Chain and its southern extension, Fig. 2) was arched and normal-faulted during Plio-Quaternary times. The total post-orogenic vertical displacement along the steeply dipping fault surfaces amounts to more than 500 m (Carrara & Zuffa 1976). The eastern axis, centred on the crystalline highlands of the Sila, Serre and Aspromonte (Fig. 2), was uplifted and tilted eastward during the Pleistocene.

Remnants of a lower Pleistocene erosional peneplain can be traced from the Ionian coast up to almost 2000 m (Ergenzinger *et al.* 1978). The fragmentation of this surface occurred by tilting toward the Ionian side and by stepwise downfaulting towards the Tyrrhenian side.

The Sicilian segment of the mountain arc (i.e. the Peloritani-Nebrodi-Madonie chain: Fig. 2) was a horst structure bordered by mainly east-west orientated normal faults in Pleistocene times (Di Geronimo *et al.* 1979). The magnitude of uplift is manifested spectacularly by the presence of the Peloritani of lower Pleistocene marine sediments at an altitude of more than 1000 m (see Bonfiglio 1970).

The present-day structural configuration of the mountain arc results from intense Plio-Quaternary vertical tectonic movements (tilting and faulting). These are superimposed on previous compressive structures (thrust and gliding nappes: see Ogniben 1960, 1973) produced by preceding phases of orogenic deformation, which migrated progressively towards the external (foreland) zones (see Fig. 2A). Ancient thrust arcs internal to the trench, such as this one, should not be confused with modern 'accretionary prisms' created by converging plates. Calabro-Sicilian arcuate chains owe their present structure not to lateral compressive stress but to a post-orogenic tensional regime.

The external (foreland) zones of the Calabrian orogenic arc appear to extend to the east beneath the Ionian seafloor and constitute a thrust pile ('accretionary prism'; see Biju-Duval & Montadert 1977), much wider in its surface expression than the Sicilian one. The thrust front, passing through Gela and Catania in Sicily, might be expected to continue in the Ionian Sea (Fig. 2) and to be traced southward as much as about 36°35′N on the 17th meridian.

The foredeep

The Catania-Gela trough represents the emerged portion of the trench (Fig. 2). The submarine trench below the Ionian Sea is not a pronounced bathymetric feature but it is marked by local and discontinuous deeps at about 3000 m. The 'narrow incipient' Calabrian trench . . . 'has a flat floor underlain by up to 800 m of layered (and faulted) fill' (Belderson *et al.* 1974).

The Catania-Gela foredeep contains only some hundred metres of Pleistocene marls and sands. Boreholes have shown the presence of the 'Gela Nappe' in the foredeep inner wall. This gliding nappe was emplaced from the north in lower Pleistocene times (see Ogniben 1960; Di Geronimo *et al.* 1979).

The foreland region

In Sicily the foreland is represented by the Hyblean Plateau (carbonate platform), up to 986 m above sea-level (Fig. 2). On the whole it exhibits a horst-like structure (directed NE-SW), being dissected by normal faulting. The plateau shows a close correlation with a gravity high which was considered by Vecchia (1954, p. 14) to correspond with a 'simatic bulge'. Towards the Catania-Gela foredeep the platform sequence includes several Plio-Quaternary volcanic units with an average thickness of 500 m. Volcanic rocks consist largely of alkali-basalts with some interbedded tholeiitic lavas (cf. Marinelli 1975, p. 211).

This foreland plateau is broadly analogous (cf. Fig. 3) to the 'outer swell' or 'outer ridge', external to deep submarine trenches, interpreted in terms of plate tectonics as resulting from flexing in the lithosphere because of downward bending of the rigid, subducting oceanic plate. A further analogy can be made between recent Hyblean lavas and some oceanic basalts (from areas of 'smooth basement'), suggesting in the Hyblean case derivation from cracks and fissures along foundering continental blocks rather than from oceanic crust.

The submerged portion of the Sicilian carbonate platform below the Sicily Channel is characterized by Siculo-Tunisian graben-like troughs (up to 1700 m deep) resulting from Plio-Quaternary block-faulting, with contemporaneous volcanic activity (Pantelleria and Linosa alkali-basalts).

Tectonic evolution of the arc region

I argue from the data presented that the Calabro-Sicilian region constitutes a *post-orogenic multi-ringed system* concentric round the Marsili meso-block. Spatial concentricity is produced by an arcuate series of swell-shaped upheavals (horsts) and basin-shaped depressions which become younger and more widely spaced with increasing distance from the Marsili cauldron-like structure.

Field studies in Sicily indicate a southward migration of orogenic deformation (thrust and gliding nappes, olistostromes) from the lower Miocene to the lower Pleistocene (Fig. 2A), concomitant with the formation and outward shift of a foredeep axis with rate of progression in the early Miocene of about 2 to 10 km Ma^{-1} (Wezel 1975). My as yet unpublished stratigraphical research suggests that in early Miocene times a step-like shift of deformation occurred every 2 or 3 Ma. As deformation progressed toward the external zones (in the south), foundering phases occurred in the internal (Tyrrhenian) zones (Fig. 3). Tectonic movements can be visualized as a lateral migration of the 'positive uplift axis' or still better as an outward-moving *topographic wave* producing reversal in the direction of tilting movements.

Clastic material for the lower Miocene internal (Sicilide) flysch sediments exposed in the northern coastal zone or drowned beneath the present-day 'outer-arc' troughs, was undoubtedly derived from a sourceland immediately to the north (cf. Ogniben 1960; Wezel 1974). This probably corresponds to the Marsili block, now sunk to depths of more than 3000 m beneath the sea. The presence of calc-alkaline volcanic detritus in some internal flysch sediments (e.g. Tusa Flysch) represents direct evidence for derivation from an uplifted andesitic volcanic island arc considered to lie to the north in early Miocene times (Wezel 1974) and which has now disappeared beneath the Tyrrhenian Sea (Fig. 3A).

The beginning of the inversion of regional morphological relief probably occurred during middle Miocene times, and was probably connected with an intense tectonic phase affecting the Sicilide flysch strata. Outward growth of the Marsili block has probably caused the closing of the contiguous foredeep and the tectonization of its flysch sediments (Fig. 3B). The pulsatory expansion of the block was subsequently followed by its breaking up and collapse.

Intra-plate krikogenetic fragmentation

From the available geological evidence it seems likely that the Marsili block, the central core of the Calabro-Sicilian arc system, was affected early by positive and later by negative vertical tectonic movements which initiated the outward-migrating topographic wave.

This case history is used to propound a model for the evolution of similar arc systems elsewhere. Empirical data suggest some distinct stages in the morphotectonic evolution of a circular block or '*krikogen*'. The new term '*krikogenesis*', derived from the Greek words, *κρίκος*, for ring and from *γένεσις*, for formation, is proposed to define the tectonic movements which create the arcuate (island and mountain belt) structures so widespread on terrestrial (e.g. the Australasian island arcs) and planetary surfaces. The relief originating from krikogenetic movements results primarily from extensional stress. Krikogens generally exhibit radial and concentric deformation patterns, with magmatic activity often accompanying the deformation.

The envisaged time-sequence of geological events comprises the following stages (Fig. 3):

Stage 1 ('dome and trough' structure)—after the post-paroxysmal phase, krikogenesis develops a crustal mosaic of blocks (circular in shape) of different size, bounded by arcuate and asymmetric troughs. The width of the troughs is commensurate with the diameter of blocks. Shallow seismic activity (less than 40 km deep) is largely concentrated in trenches but also occurs in central elevated areas of blocks. The pattern of faulting is radial and, subordinately, concentric. At present this stage may be recognized in the Calabro-Sicilian Arc either in the foreland (e.g. the Hyblean Plateau) or in the 'forearc' (e.g. the Silan block in Calabria) continental regions. The presence of positive gravity anomalies corresponding to such blocks indicates crustal thinning, possibly above locally ascending mantle material.

Such an episode of development may have occurred during Plio-Quaternary times in the case of the 'Sardinian block', at present a giant 'semi-dome'. The central-eastern part of the island was intermittently uplifted and tilted westward. This led to phases of erosion and subsequent infilling of western peripheral troughs (e.g. the Campidano graben) with continental clastics (Cocozza *et al.* 1974). Associated alkaline volcanic activity developed along fissures roughly contemporaneously.

Stage 2 ('axial collapse zone')—continued arch-shaped uplift may lead to an axial collapse zone with graben structures. Tension across the 'geanticline' arch provides lines of weakness exploited by active arc volcanism. At the periphery of larger blocks a submarine trench ('geosynclinal trough') can originate. Turbidites derived from the volcanic arc and from older uplifted rocks fill the trench with flysch-type sediments.

This stage of development may have occurred during early Miocene times in the case of the Marsili block (Fig. 3A) and in late Miocene and Plio-Quaternary times for the 'Etruscan block' (Tuscan Archipelago, Umbro-Tuscan and northern Latium regions: see northern part of Fig. 1). In the latter area block-faulting and uplift of the arch were accompanied by the emplacement of plutons, volcanism and metallogenesis resulting from hydrothermal processes (cf. Marinelli 1975). Latium alkaline potassic volcanism originated in an elongate collapse zone (rift valley-type) within a late Pliocene updomed region (Locardi *et al.* 1977), which continues southward a 'metallogenic-volcanic' arch (in the southern Tuscan region) flanked by rifts. Intense volcanicity occurred during the culmination of the uplift phase.

Within the larger 'Pelagian Sea krikogen', the Sicilian Channel was an arch-like feature which collapsed during the Pliocene (Burollet & Winnock 1979). The isolation of the Pantelleria, Linosa and Malta troughs occurred by rifting in Quaternary times accompanied by alkali basaltic volcanism. These rifted arch basins are located within a region where a thinned crust is suggested by positive Bouguer anomalies (Vecchia 1954).

Stage 3 ('cauldron-type collapse')—in response to the vast effusion of lava the central area of krikogens collapses. Surface subsidence was probably caused by sinking of the magma column in the magma chamber. The sinking crust produces volcano-tectonic depressions like that of the Marsili Basin. Mainly because of the coalescence of minor basins (Savelli & Wezel 1979), their size tends to increase with time, giving rise to 'small ocean basins' like the Tyrrhenian Sea (see Fig. 1).

In such basins the floor comprises a mosaic of collapsed blocks, at different stages of development as manifested by their seismic activity. Deep earthquakes (hypocentral depths more than 200 km) are probably related to the vertical sinking of the older, more evolved, krikogens. In fact, fault-plane solutions for three Tyrrhenian deep-focus shocks (depth of 280–300 km) show dip-slip faulting along vertical fault planes (see Fig. 2B and Riuscetti & Schick 1975). The complete lack of foci in the 40–200 km depth interval, the absence of correlation between shallow and deep shocks and the overall lack of thrust events are arguments against the presence of a Benioff zone (Riuscetti & Schick 1975). Furthermore, the low level of seismicity of the Calabro-Sicilian Arc is consistent with post-orogenic tectonic activity.

The vast caldera-like subsidence was probably either accompanied or immediately succeeded by the outward-moving topographic wave, producing the lateral shift of the foredeep axis. The uplift of the mountain arcs compensates the sinking of the Marsili block ('*touche-de-piano*' tectonics).

In Plio-Quaternary times the foreland 'arch' began to evolve (Fig. 3C). It is mainly represented in the present Sicily Strait by grabens, banks and volcanic activity, indicative of a Pliocene phase of fracturing (stage 1) and a Quaternary phase of intra-arch collapse (stage 2). Thus, it appears that the present structural configuration of the Sicilian 'forearc' region may be considered equivalent to that of an 'inter-arch zone' lying between intra-plate arches at different stages of development.

Conclusions

For the reasons illustrated in the previous sections it seems most unlikely that the Calabro-Sicilian Arc has resulted from descent of lithosphere along a Benioff zone. The concentric and radial shape of the system in plan view suggests that the active tectonic force is vertical. Present-day morphological structure is entirely post-orogenic and indicative of intracontinental extensional tectonics.

The geological history of the area shows an outward step-like growth of the orogenic system beginning from a central circular block ('krikogen'), presently forming the Marsili Basin and almost completely encircled by the 'Aeolian andesitic ring'. It is postulated that this block suffered 'krikogenetic' movements (and igneous processes) consisting of Miocene arching and rifting stages, followed in the Pliocene by large-scale cauldron collapse, possibly triggering outward-moving 'touche-de-piano' movements which structured the 'forearc' region. Foundering phases occurred in the internal (hinterland) zones as deformation and migration of the foredeep axis progressed outwards in time toward the external (foreland) zones.

Beginning in Messinian times the 'foreland' block (Sicily Channel) was also activated, exhibiting the early two stages of evolution (i.e. arching and rifting), accompanied by Plio-Quaternary volcanic activity. The thrusting of the Sicilian orogenic system over the 'foreland' block is related to the rate of subsidence of the latter (i.e. to its relative stage of 'krikogenetic' development). It seems difficult to account for this intra-continental contact, manifested by the Catania-Gela foredeep, by appealing to an active Benioff zone.

Also the underlying 'foreland' circular block ('krikogen') is an intra-plate feature whose tectonic history might be attributable to local mantle movements.

'Krikogenetic' movements are thought to be induced by thermal regimes of mantle dilatational nuclei ('DIN' of Wezel 1977). Stage 1 is related to the regional upwelling, stage 2 to subsequent thermal expansion of the 'DIN', while stage 3 manifests thermal contraction.

The fact that the whole Tyrrhenian floor (Fig. 1) and the Italian region appear to have fractured in post-orogenic times in different circular blocks assembled like a mosaic, emphasizes the importance and the spreading of intraplate 'krikogenetic' deep-seated deformation in shaping present-day geomorphological and structural configurations. I believe that this tectonic process is particularly efficient in developing the topographic features of the 'backarc' and the 'forearc' regions located within previous Alpino-Mediterranean mountain belts.

ACKNOWLEDGMENTS: I should like to thank Drs J. Leggett, A. Robertson and A. Smith, for their many helpful suggestions and for revising the English. The study of the Tyrrhenian Sea was part of an Italian programme funded by the Consiglio Nazionale delle Ricerche ('P.F. Oceanografia e Fondi Marini').

References

ALVAREZ, W., COCOZZA, T. & WEZEL, F. C. 1974. Fragmentation of the Alpine orogenic belt by microplate dispersal. *Nature, London,* **248,** 309–14.

BACINI SEDIMENTARI 1980. Dati geologici preliminari sul Bacino di Cefalù (Mar Tirreno). *Ateneo Permense, Acta Nat.* **16,** 3–18.

BARBERI, F., INNOCENTI, F., FERRARA, G., KELLER, J. & VILLARI, L. 1974. Evolution of Eolian arc volcanism (southern Tyrrhenian Sea). *Earth planet. Sci. Lett.* **21,** 269–76.

BELDERSON, R. H., KENYON, N. H. & STRIDE, A. H. 1974. Calabrian Ridge, a newly discovered branch of the Mediterranean Ridge. *Nature, London,* **247,** 453–4.

BIJU-DUVAL, B., LETOUZEY, J. & MONTADERT, L. 1978. Structure and evolution of the Mediterranean basins. *Initial Rep. Deep Sea drill. Proj.* **42,** 951–84.

—— & MONTADERT, L. 1977. Introduction to the structural history of the Mediterranean basins. *Int. Symp. Struct. Hist. Med. Basins,* 1–12. Technip, Paris.

BONFIGLIO, L. 1970. Facies biodetritica tardoplioceni-ca nei Peloritani a 1250 metri di altitudine. *Boll. Soc. geol. Ital.* **89,** 499–506.

BUROLLET, P. F. & WINNOCK, E. 1979. Essai de synthèse. *In:* BUROLLET, P. F., CLAIREFOND, P. & WINNOCK, E. (eds). *La Mer Pélagienne. Géol. Médit.* **6,** 321–7.

CAFLISCH, L. & SCHMIDT DI FRIEDBERG, P. 1967. L'evoluzione paleogeografica della Sicilia e sue relazioni con la tettonica e la naftogenesi. *Mem. Soc. geol. Ital.* **6,** 449–74.

CARRARA, A. & ZUFFA, G. G. 1976. Alpine structures in northwestern Calabria, Italy. *Bull. geol. Soc. Am.* **87,** 1229–46.

COCOZZA, T.,JACOBACCI, A., NARDI, R. & SALVATORI, I. 1974. Schema stratigrafico-strutturale del Massiccio Sardo-Corso e minerogenesi della Sardegna. *Mem. Soc. geol. Ital.* **13,** 85–186.

DELLA VEDOVA, B. & PELLIS, G. 1979. Risultati delle misure di flusso di calore eseguite nel Tirreno sud-orientale. *Atti Conv. Sci. Naz. P.F. Oceanografia e Fondi Marini CNR (Roma 5–7 marzo 1979)*, **2,** 693–712.

DEL MONTE, M. 1972. Il vulcanesimo del Mar Tirreno: nota preliminare sui vulcani Marsili e Palinuro. *G. Geol.* **38,** 231–52.

DI GERONIMO, I., GHISETTI, F., LENTINI, F. & VEZZANI, L. 1979. Lineamenti neotettonici della Sicilia orientale. *Mem. Soc. geol. Ital.* **19,** 543–9.

ERGENZINGER, P., GÖRLER, K., IBBEKEN, H., OBENAUF, P. & RUMOHR, J. 1978. Calabrian Arc and Ionian Sea: vertical movements, erosional and sedimentary balance. *In:* CLOSS, H., ROEDER, D. & SCHMIDT, K. (eds). *Alps, Apennines, Hellenides,* 359–73. Schweizerbart'sche, Stuttgart.

IACOBUCCI, F., INCORONATO, A., RAPOLLA, A. & SCARASCIA, S. 1977. Basement structural trends in the volcanic islands of Vulcano, Lipari and Salina (Eolian Islands, Southern Tyrrhenian Sea) computed by aeromagnetic and gravimetric data. *Boll. Geof. Teor. Appl.* **20,** 49–61.

LOCARDI, E., LOMBARDI, G., FUNICIELLO, R. & PAROTTO, M. 1977. The main volcanic groups of Latium (Italy): relations between structural evolution and petrogenesis. *Geol. Romana,* **15,** 279–300.

MCKENZIE, D. 1972. Active tectonics of the Mediterranean region. *Geophys. J. R. astron. Soc.* **30,** 109–85.

MACCARRONE, E. 1970. Notizie petrografiche e petrochimiche sulle lave sottomarine del Seamount 4 (Tirreno Sud). *Boll. Soc. geol. Ital.* **89,** 159–80.

MARINELLI, G. 1975. Magma evolution in Italy. *In:* SQUYRES, C. H. (ed). *Geology of Italy.* Earth Sci. Soc. Libyan Arab Republic, Tripoli, **1,** 165–219.

MORELLI, C., GIESE, P., CASSINIS, R., COLOMBI, B., GUERRA, I., LUONGO, G., SCARASCIA, S. & SCHÜTTE, K. G. 1975. Crustal structure of Southern Italy. A seismic refraction profile between Puglia-Calabria-Sicily. *Boll. Geof. Teor. Appl.* **17,** 183–210.

NAPPI, G., BONASIA, V., FERRI, M., MONTAGNA, S. & PINGUE, F. 1977. Primi risultati sulla sorveglianza del complesso vulcanico Lipari-Vulcano (Tirreno meridionale). *Boll. Soc. geol. Ital.* **95,** 967–80.

OGNIBEN, L. 1960. Nota illustrativa dello Schema geologico della Sicilia nord-orientale. *Riv. Min. Siciliana,* **11,** 183–212.

—— 1973. Schema geologico della Calabria in base ai dati odierni. *Geologica Romana,* **12,** 243–585.

RIUSCETTI, M. & SCHICK, R. 1975. Earthquakes and tectonics in Southern Italy. *Boll. Geof. Teor. Appl.* **17,** 59–78.

SAVELLI, D. & WEZEL, F. C. 1979. Morfologia e stile tettonico del Bacino Tirrenico. *Atti Conv. Sci. Naz. P.F. Oceanografia e Fondi Marini CNR (Roma 5–7 marzo 1979),* **2,** 729–38.

SELLI, R., LUCCHINI, F., ROSSI, P. L., SAVELLI, C. & DEL MONTE. 1977. Dati geologici, petrochimici e radiometrici sui vulcani centro-tirrenici. *J. Geol.* **42,** 221–46.

VECCHIA, O. 1954. Lineamenti geofisici e geologia profonda nella Sicilia ed aree circostanti. *Riv. Geof. Appl.* **15,** 3–34.

WEZEL, F. C. 1974. Flysch successions and the tectonic evolution of Sicily during the Oligocene and early Miocene. *In:* SQUYRES, C. H. (ed). *Geology of Italy.* Earth Sci. Soc. Libyan Arab Republic, Tripoli, **2,** 105–27.

—— 1975. Diachronism of depositional and diastrophic events. *Nature, London,* **253,** 255–7.

—— 1977. Widespread manifestations of Oligocene-Lower Miocene volcanism around western Mediterranean. *Int. Symp. Struct. Hist. Med. Basins,* 287–301. Technip, Paris.

FORESE CARLO WEZEL. Istituto di Geologia, Università degli Studi, 61029 Urbino, Italy.

MAKRAN OF IRAN AND PAKISTAN

Deformation of the Makran accretionary sediment prism in the Gulf of Oman (north-west Indian Ocean)

Robert S. White

SUMMARY: The offshore Makran of Pakistan and Iran in the Gulf of Oman forms the seaward margin of a folded and faulted accretionary sediment wedge which extends several hundred kilometres inland across the onshore Makran. Sediment is being scraped off the oceanic Arabian plate as it subducts shallowly northward beneath the Eurasian plate. The submerged front of the accretionary prism has been traced along the entire width of the Makran continental margin, from the Strait of Hormuz in the west to the boundary with the Indian plate in the east. The structure of a 70 by 90 km portion of the deformed sediment belt was mapped in detail by a grid of continuous seismic reflection profiles. This survey demonstrates that the folds are laterally continuous and are well lineated, that the deformed belt appears to be migrating southward at a rate of about 10 km every million years and that folding of the initially undisturbed abyssal plain sediments occurs in the southernmost, or 'frontal' fold. Mapping of the three-dimensional structure of the frontal fold shows that it systematically increases then decreases in height along strike over a distance of about 80 km. Folding is initially restricted to the uppermost $2\frac{1}{2}$ km of the sediment pile, and there is local evidence of small-scale shale diapirism. The frontal fold is then incorporated into the accretionary prism by uplift along a thrust fault which flattens out within the sedimentary section, possibly in overpressured shales. There is little subsequent deformation of the uplifted material or of the sediment which infills the basins between the fold ridges until some 70 km north of the frontal fold when further rapid uplift occurs, eventually elevating the sediment above sea-level. I attribute this second phase of uplift to major imbricate thrust faulting extending down to the oceanic basement; most of the shortening and thickening of the sediment wedge occurs in this region. The relatively simple pattern of uplifted folds and intervening basins in the frontal 70 km of the accretionary wedge results from a balance of compressive and normal stresses on the thick sediment section. This simple deformational style of the seaward part of the accretionary prism can be traced eastwards until it becomes chaotically disrupted by the consumption of a basement ridge which protrudes to just beneath the seafloor.

The Makran continental margin of Pakistan and Iran in the Gulf of Oman, north-west Indian Ocean stretches from the Strait of Hormuz in the west to near Karachi in the east and forms the seaward edge of a large accretionary sediment prism extending hundreds of kilometres inland. This paper examines the style of deformation in the offshore Makran where the thick flat-lying sediments of the Gulf of Oman become folded and faulted in the initial stages of their incorporation into the accretionary prism. Previous work on RRS *Shackleton* leg 3/75 in the central part of the Gulf of Oman demonstrated that the Makran continental margin comprises a series of fold ridges and intervening basins (White & Klitgord 1976; White 1979b); similar structures were found extending westwards to the Strait of Hormuz on RV *Atlantis II* leg 96(13) (White & Ross 1979). I report preliminary results from RRS *Shackleton* leg 1/80 in January 1980 which completes the eastern end of the Makran margin survey off the coast of Pakistan (Fig. 1).

The Makran accretionary wedge lies above the oceanic part of the Arabian plate which is subducting shallowly northwards beneath the continental Eurasian plate at a convergence rate of about 50 mm yr^{-1} (see inset to Fig. 1). The easternmost boundary of the wedge is formed by an active right-lateral strike-slip fault system extending along the Ornach Nal and Chaman faults (Abdel-Gawad 1971). The western end of the Makran accretionary prism is at the Strait of Hormuz where collision between the Arabian and Eurasian plates changes to a continent-continent type (White & Ross 1979). On the south-eastern side of the Gulf of Oman the Arabian plate is separated from the Indian plate by the Murray Ridge (Barker 1966) which, as our recently acquired seismic profiles indicate, is a tensional feature.

Preliminary results from a reversed seismic refraction line above the Gulf of Oman abyssal plain show that normal oceanic crust lies beneath the 6 to 7 km thick pile of flat or gently dipping sediments. Although there are no rec-

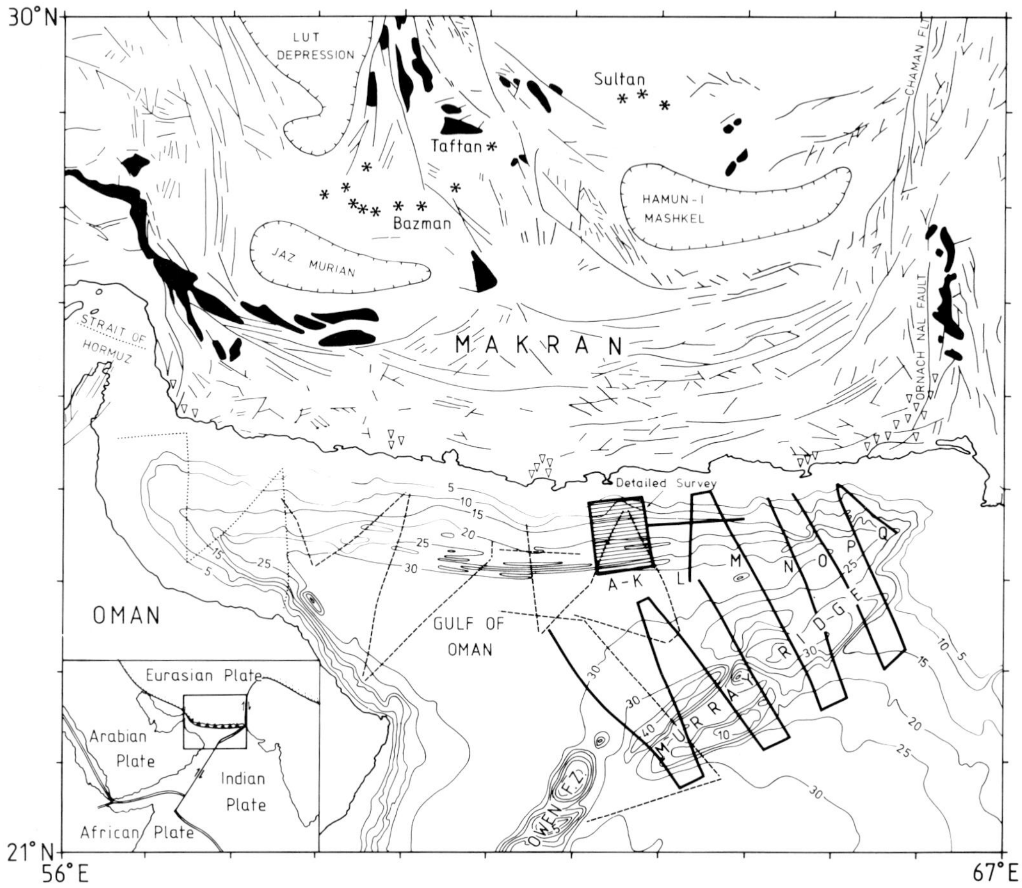

FIG. 1. Map showing location of continuous seismic profiles (all with coincident gravimetric, bathymetric and magnetic measurements), in the Gulf of Oman, north-west Indian Ocean. Dotted tracks from RV *Atlantis II* leg 96(13), dashed tracks from RRS *Shackleton* leg 3/75 and solid tracks from RRS *Shackleton* leg 1/80. Detailed survey tracks are shown expanded in Fig. 2. Stars show location of volcanic centres, triangles the positions of active mud vents, solid areas mark ophiolitic assemblages and 'coloured mélanges' and thin lines show the major faults on land. Information compiled from Ahmed (1969), Barker (1966), Berberian (1976), Hudson *et al.* (1954), Jacob & Quittmeyer (1979), Kazmi (1979), Nowroozi (1976) and Snead (1964). Inset shows general setting of the area with the main plate boundaries.

ognizable seafloor spreading magnetic anomaly lineations beneath the Gulf of Oman by which the crust can be dated, its age has been estimated by comparing the heat flow measured at a number of stations within the basin to the standard curves published by Parsons & Sclater (1977).

Only rather broad limits can be placed on the age of the oceanic crust because the sedimentation rates and history are poorly known and so the correction factor to convert from the observed heat flow through the seafloor to the heat flow through the buried basement is subject to considerable uncertainty. Even though the measured heat flow values within the basin show little scatter the best limits that can be placed on the age of the crust if all possible sources of error are considered are that it was formed between 70 to 120 Ma ago (Hutchison *et al.*, in press). This means that the crust is at least as old as and may be older than the large ophiolite sequences emplaced in Oman. Crustal creation during the Cretaceous quiet zone also explains the absence of seafloor-spreading magnetic lineations.

Accreted sediments have been uplifted and exposed in the onshore Makran as thick, faulted flysch sequences (see e.g. Hunting 1960,

Arthurton *et al.* and McCall & Kidd 1981 for discussions of the geology of the Makran). Between four to five hundred kilometres north of the coast is a chain of Cenozoic volcanic and plutonic rocks of andesitic to rhyolitic composition (Fig. 1) which may represent a volcanic arc (Farhoudi & Karig 1977). There are large offsets in the volcanic chain such as to the east of the Jaz Murian depression at about 60°E, and Dykstra & Birnie (1979) have suggested that the subduction zone may be segmented. However, we have found no significant offsets anywhere in the offshore Makran structures. There is no well-developed Benioff zone, although the seismicity appears to deepen somewhat to the north in a way consistent with a shallowly subducting plate (Jacob & Quittmeyer 1979).

The Makran subduction zone exhibits no topographic trench, and the accretionary wedge is unusually wide; the very thick sediment pile on the subducting plate is probably responsible for these abnormal features.

Interpretation of seismic reflection profiles

The continuous seismic reflection profiles illustrated in this paper are unprocessed except for the application of linear time variable gain triggered by the seafloor return and of band pass filtering from 5 to 70 Hz. They exhibit a number of features which we briefly review to prevent possible misinterpretation.

a. Bubble pulse

A single 2.6 litre (160 cubic inch) airgun fired once every 12 sec was used as a sound source. It generates a quarter second long wavetrain which causes multiple returns from every interface and obscures other arrivals within this period. Higher resolution of the structure immediately beneath the seafloor was achieved by making a separate record of the signal after high pass filtering at 100 Hz to remove the dominant bubble pulse frequency.

b. Water multiple

The prominent return at twice the travel time to the seafloor (marked 'm' on Fig. 3) is an artefact generated by an extra bounce in the water layer.

c. Migration

The illustrated profiles are unmigrated time sections; to turn them into true depth sections they must be migrated and the normal incidence travel times multiplied by the seismic wave velocity in the sediment. This has been done for the frontal fold (Fig. 10).

d. Gas hydrate/free gas reflection

Accumulations of free gas, which have very low impedance to seismic waves and consequently give rise to large-amplitude reflections are common within the uppermost 700 m of sediment beneath both the abyssal plain and the continental margin of the Gulf of Oman (there is a good example at the south end of profile C, Fig. 3, marked 'g'). The compelling evidence that these anomalously strong reflectors are generated by free gas is reported by White (1977b). Furthermore, the free gas is not found in stratigraphic or structural traps but forms a bottom simulating reflector (BSR) by accumulating beneath an impermeable sediment plus gas hydrate layer which is stable in the topmost few hundred metres of sediment. White's (1979a) evidence that the BSR's are caused by free gas trapped beneath gas hydrate has been confirmed in several ways: the measured temperature gradient above a prominent gas reservoir (Hutchison *et al.,* in press) agrees with the temperature gradient predicted from the depth and the pressure and temperate (P–T) conditions of the free gas to hydrate phase change, the marked increase in the seismic velocity of hydrated sediments found in laboratory experiments (Stoll & Bryan 1979) has been observed *in situ*, and many more examples of BSR's mirroring the P-T conditions of the hydrate phase change have been recorded on our recently acquired profiles. This confirmation that the BSR's are caused by trapped gas is important for our purposes beause it means that some of the strongest reflectors on the seismic profiles bear no relation to the structure but simply record particular P-T conditions, and indeed frequently cross-cut major structural boundaries (several such reflectors are marked 'g' on Figs 3 & 9).

Detailed survey across fold belt

A detailed grid survey of a 70 × 90 km area extending from the Gulf of Oman abyssal plain onto the Makran continental margin (Fig. 1) shows that the sequence of ridges and intervening basins is strongly lineated parallel to the coast. The survey consists of 11 profiles perpendicular (A through K, Fig. 2), and 10 profiles parallel (lines 1 to 10, Fig. 2), to the strike of the fold belt. Gravimetric, bathymetric and magnetic measurements were made in addition to a continuous seismic reflection profile along each line. The profiles were navigated using a combination of moored radar transponder, satellite and land radar fixes, which were interpolated using measurements of the speed

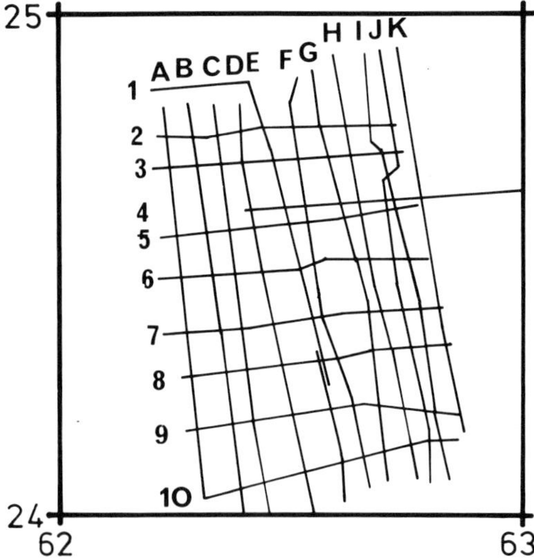

FIG. 2. Location map showing positions of grid of continuous seismic reflection profiles in the detailed survey area over the offshore Makran accretionary belt. Gravimetric, bathymetric and magnetic measurements were also made along these tracks.

through the water from an electro-magnetic log.

Profile C (Fig. 3) from the western part of the detailed survey exhibits the typical sequence of ridges and intervening basins seen along the entire Makran continental margin (see White & Klitgord 1976 and White & Ross 1979 for profiles from further west off the Iranian coast). Gently dipping sediments beneath the abyssal plain in the south are folded to a maximum amplitude of about 400 m in the 'frontal fold' (at 24° 10′ N on Fig. 3), then uplifted by faulting to a height of about $1\frac{1}{4}$ km above the abyssal plain. Basins between the uplifted folds subsequently become filled by sediment, the amount of infill increasing towards the coast until eventually the crests of the folds are completely buried. The sediment is derived partly from reworking of the material of the fold ridges although the absence of slump accumulations at the base of the ridges suggests that they are well consolidated and that most of the basinal infill comes from the surrounding land masses of Arabia and the Makran. Most of the adjacent land is desert and much of the material is wind-borne (Stewart *et al.* 1965). In the lower part of the fold belt the ridges form effective barriers to downslope sediment movement and

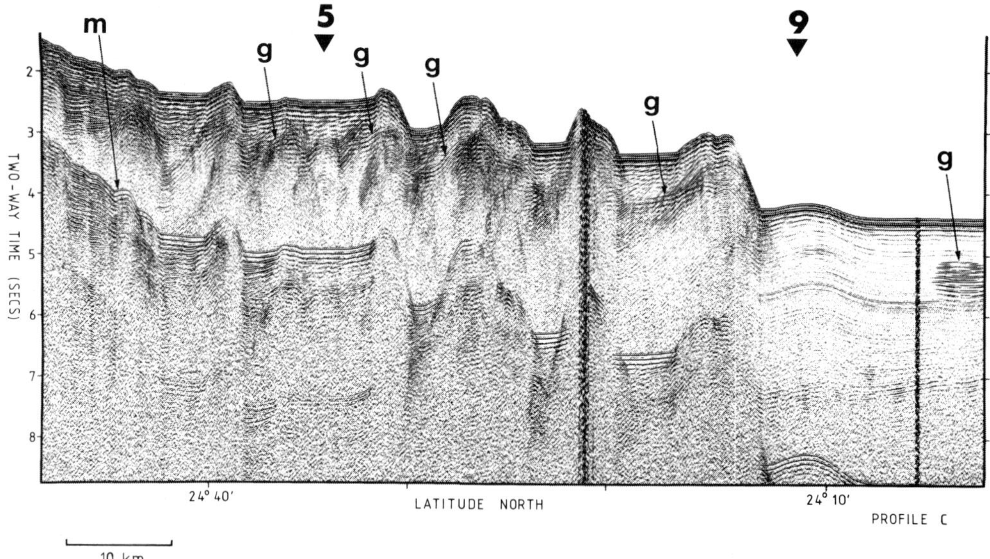

FIG. 3. Continuous seismic reflection profile along line C perpendicular to the strike of the fold belt (see Fig. 2 for location). Arrowheads mark intersection with profile 5 (Fig. 5), and profile 9 (Fig. 9). 'm' marks water multiple and 'g' denotes prominent returns caused by free gas trapped beneath gas hydrate. The profiling system comprised a single 2.6 litre (160 cubic inch) airgun fired once every 12 sec and a Geoméchanique hydrophone streamer. Linear time variable gain triggered by the seafloor return and 5–70 Hz band pass filtering has been applied to the signal. Vertical exaggeration at seafloor is approximately 7:1. Heavy vertical lines are noise caused by interference.

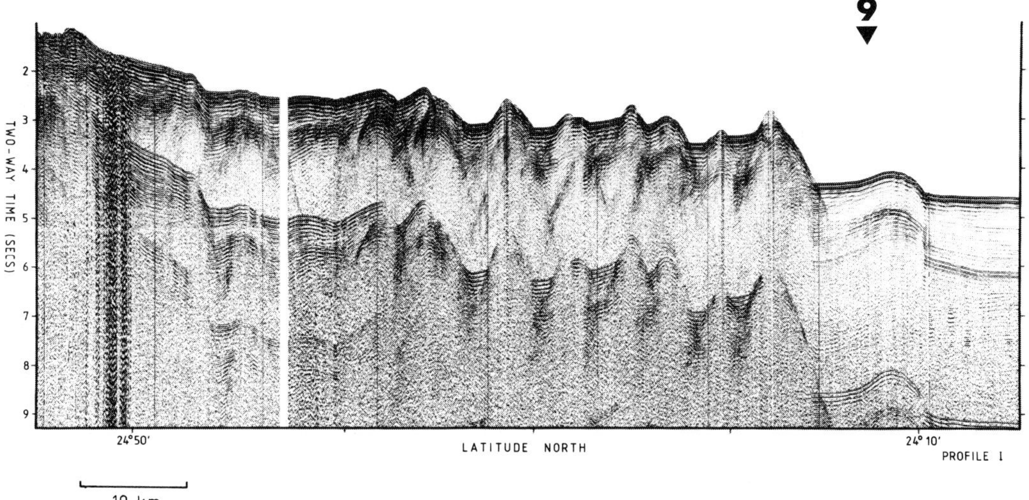

FIG. 4. Continuous seismic reflection profile along line I perpendicular to the strike of the fold belt (see Fig. 2 for location). Arrowhead marks intersection with profile 9 (Fig. 9). Vertical exaggeration at seafloor is approximately 7:1. Vertical lines are one-hourly marks superimposed on the record.

the seafloor is often at different levels in adjacent basins. On profile I (Fig. 4), the southernmost basin at 24° 19′ N is at a higher level than the next landward basin at 24° 22′ N, presumably as a result of recent uplift of the outermost ridge.

Profiles across the inter-ridge basins (e.g. Figs 3 & 4) show the infilling sediment to be little disturbed and in particular there is no evidence of buckling. I infer that in the southernmost 60 km of the fold belt there is no major shortening other than in the frontal fold, although as suggested below further major deformation and faulting commences approximately 60 to 70 km landward of the frontal fold. Horizons within the basinal sediments dip gently landward and the dips increase with depth; this structure may be attributed partly to the unconformable deposition of sediment on the landward side of the crest of the frontal fold as it develops (discussed further below), and partly to tilting of the sediment ponds as the frontal fold is uplifted by a thrust fault which flattens with depth. Landward tilts of up to $\frac{1}{2}$° in the seafloor of the basins are common (see, e.g. Fig. 4 and White & Klitgord 1976), so if we assume that the infilling sediment was deposited horizontally this indicates that recent post-depositional tilting has occurred.

Profiles parallel to the strike along the inter-ridge basins in the frontal 60 km of the margin exhibit remarkably little disturbance of the infilling sediment (e.g. profile 5, Fig. 5). There is no evidence of folding, of diapirism or of faulting of the basinal sediments until a major region of uplift near the coast is reached.

The bathymetric map of the detailed survey area illustrated in Fig. 6 demonstrates the linearity of both the scarp generated by the uplifted ridge behind the frontal fold (at approximately 24° 13′ N), and of the ridges and basins lying behind this. In the northernmost part of the survey (north of 24° 47′ N), is a steep slope rising rapidly towards sea-level; unlike the lower part of the margin this is cut by deep transverse valleys which feed terrigenous sediment on to the upper part of the fold belt.

Individual ridges can be traced along strike across the entire detailed survey area as can be seen in the bathymetric profiles of the north-south survey lines illustrated in Fig. 7. Some of the ridges die out in the vicinity of profile G, giving in detail a different sequence of ridges and basins in the west of the area (e.g. profile C, Fig. 3), than in the east (e.g. profile I, Fig. 4). In Fig. 8 the north-south bathymetric profiles are shown superimposed to emphasize the degree of similarity of shapes and depths of the ridges and intervening basins. The lower part of the figure shows the heights of all the ridge crests where they are not buried by secondary sediment. After the initial uplift of the outermost ridge to an elevation of $1\frac{1}{4}$ km above the abyssal plain the ridge crests remain at roughly the same height, increasing in elevation by an average of less than 1:100 (about 0.6°), until rapid uplift occurs some 60 to 70 km landward of the frontal fold.

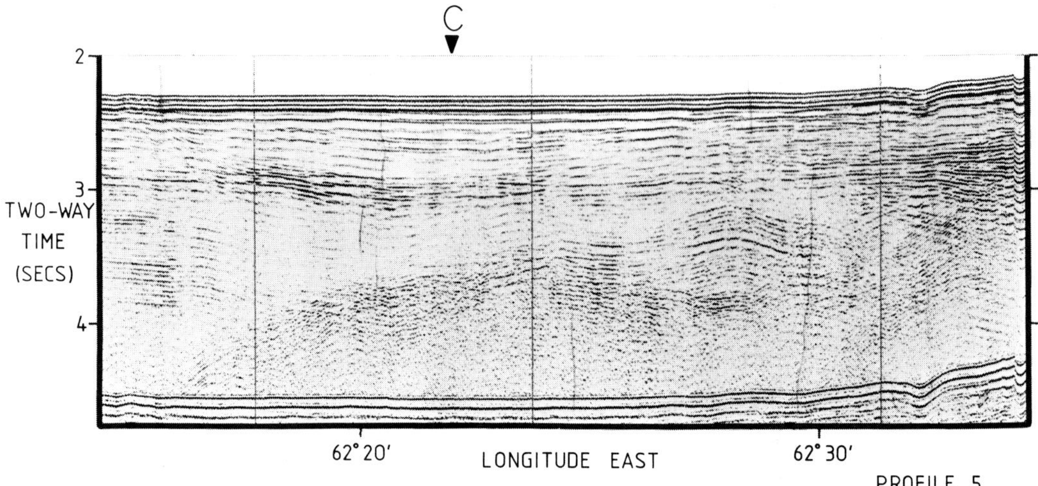

FIG. 5. Part of continuous seismic reflection profile along line 5 parallel to the strike of the fold belt (see Fig. 2 for location) showing flat-lying sediments. Arrowhead marks intersection with line C (Fig. 3). Vertical exaggeration at seafloor is approximately 7:1. Vertical lines are one-hourly marks superimposed on the record.

The frontal fold

The frontal fold is sufficiently gently folded to allow us to investigate its structure by tracing reflectors into the fold from beneath the abyssal plain. This is not possible with the ridges futher on to the Makran continental margin because they are too highly deformed. I have chosen three prominent horizons (marked by arrowheads on Fig. 9) which can be seen on every crossing, and correlated them between profiles using two tie lines orientated along strike; they are profile 10 which runs approximately east-west across the abyssal plain sediments just south of the fold, and profile 9 (illustrated in Fig. 9), which is positioned just north of the crest of the frontal fold. A very strong reflection from the gas hydrate/free gas boundary also occurs at 4.7 to 4.8 sec near the crest of the fold and can be traced continuously for more than 40 km ('g' on Fig. 9). Although some reflectors are visible from deeper into the sedimentary section (e.g. on Fig. 3), we have not mapped them because they are insufficiently clear to correlate between profiles using the tie lines. The time sections were migrated into true depth sections (Fig. 10) using sediment velocities calculated from variable angle sonobuoy profiles in the Gulf of Oman (reported by White 1977a and Closs *et al.* 1969). The main features of the frontal fold are:

a. The amplitude of the fold varies along strike from 120 m on profile A in the west to 400 m by profile F. Near profile G the fold becomes much broader and subsequently decreases in amplitude to 240 m by profile K in the east (e.g. Fig. 7).

b. The fold is asymmetric in shape with the steepest limb facing seaward. The seaward side often exhibits a crumpled surface (e.g. Figs 3 & 11), which may be tectonic in origin but may also be caused in part by sediment slumping off the seaward slope.

c. Sediment accumulates in the depression between the crest of the frontal fold and the adjacent uplifted ridge. Much of the infilling sediment is probably derived from the steep seaward-facing scarp of the elevated ridge. As the frontal fold develops the secondary sediment is deposited with increasing angle of unconformity against the landward limb of the fold. The ponded sediment accentuates the skewness of the seafloor expression of the frontal fold (e.g. Fig. 10).

d. With the exception of the uppermost sedimentary unit which is about $\frac{1}{2}$ km thick the top $2\frac{1}{2}$ km of sediment is deformed in parallel folds (Fig. 10). Only the topmost $\frac{1}{2}$ km thick sediment unit thins appreciably as it is traced from beneath the abyssal plain into the frontal fold. The fold at depth is generally broader than the seafloor ridge.

e. Deep reflectors within the sedimentary column, where they can with confidence be traced from the abyssal plain into the frontal

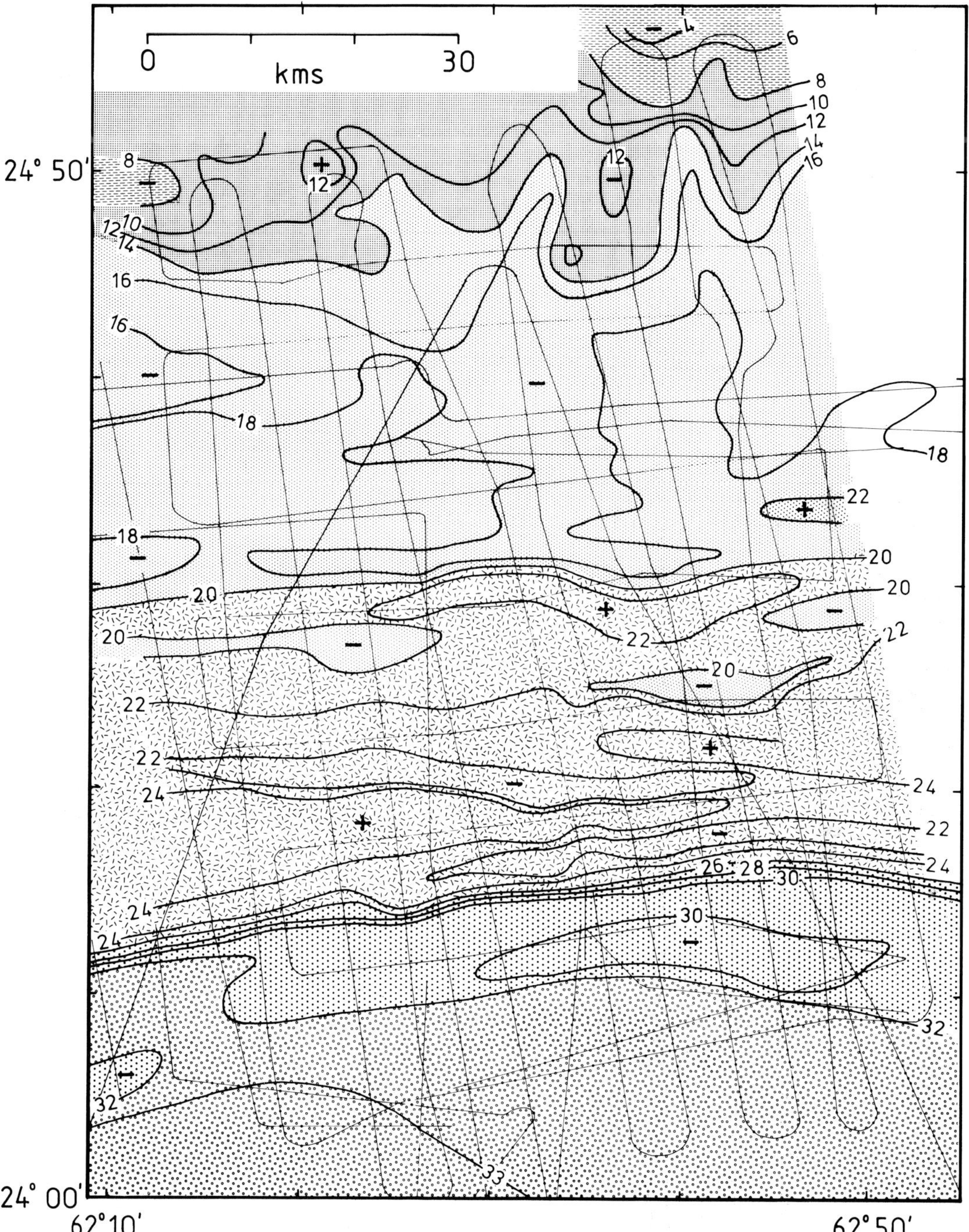

FIG. 6. Bathymetric map of the detailed survey area over the offshore Makran accretionary belt Contours in hundreds of metres, corrected for variations in the speed of sound in the water. Shading is in 600 m intervals. Track control shown by light lines; '−' and '+' mark bathymetric highs and lows, respectively. Note the well-lineated sequence of fold peaks and intervening basins which is particularly clear on the lower part of the slope where it is not masked by overlying sediment. Cross-cutting channels are visible on the upper portion of the slope.

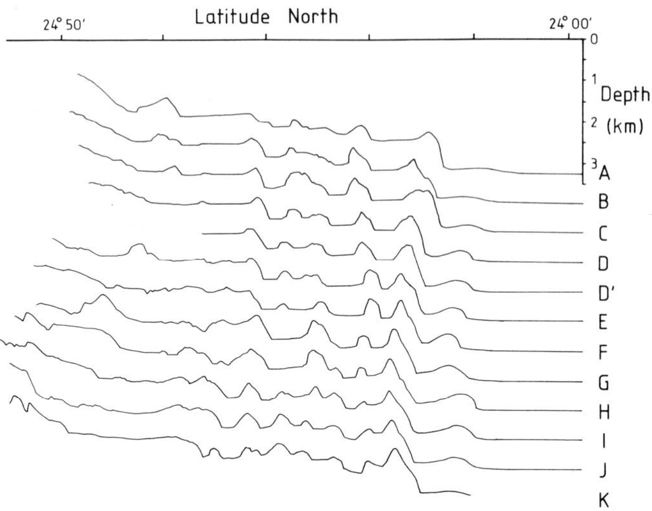

Fig. 7. Bathymetric profiles from detailed survey lines projected on to north-south axis shown along top of figure. See Figs 1 and 2 for location of profiles. Vertical exaggeration is 7.7:1. Note the increase in height of the frontal fold over profiles A to G, then the change in width and decrease in height over profiles H to K.

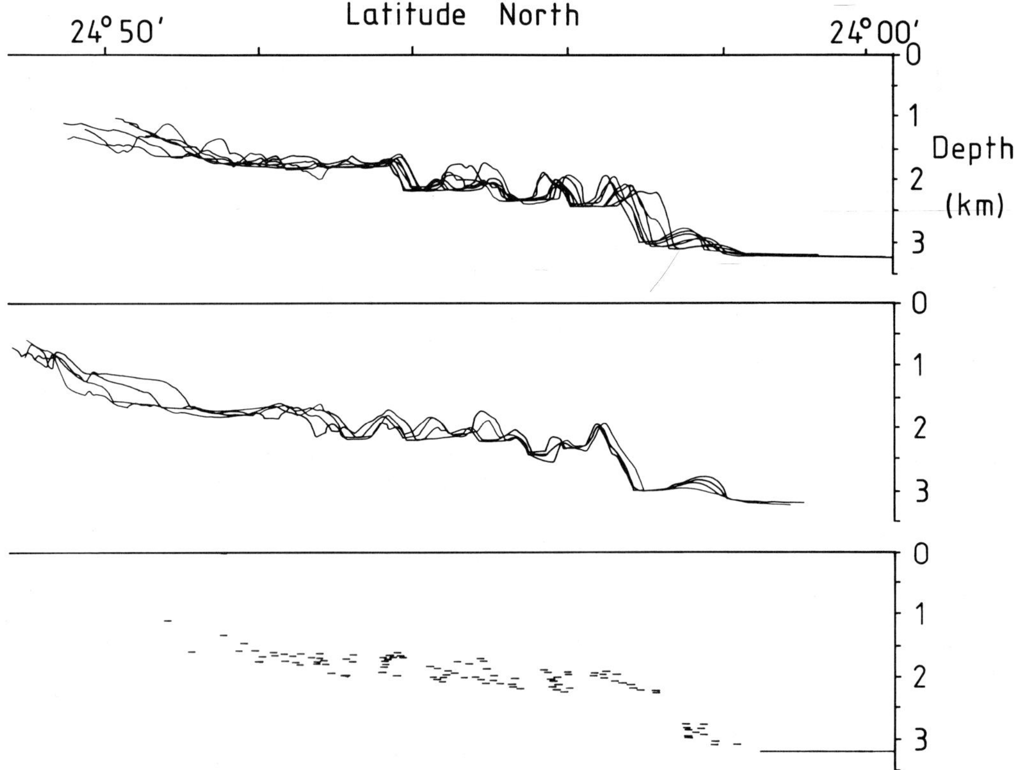

Fig. 8. Top two diagrams show stacked bathymetric profiles from detailed survey lines projected on to north-south axis (as in Fig. 7). Bottom diagram shows depths of crests of ridges in the detailed survey where they are not buried beneath sediment. Vertical exaggeration is 7.7:1.

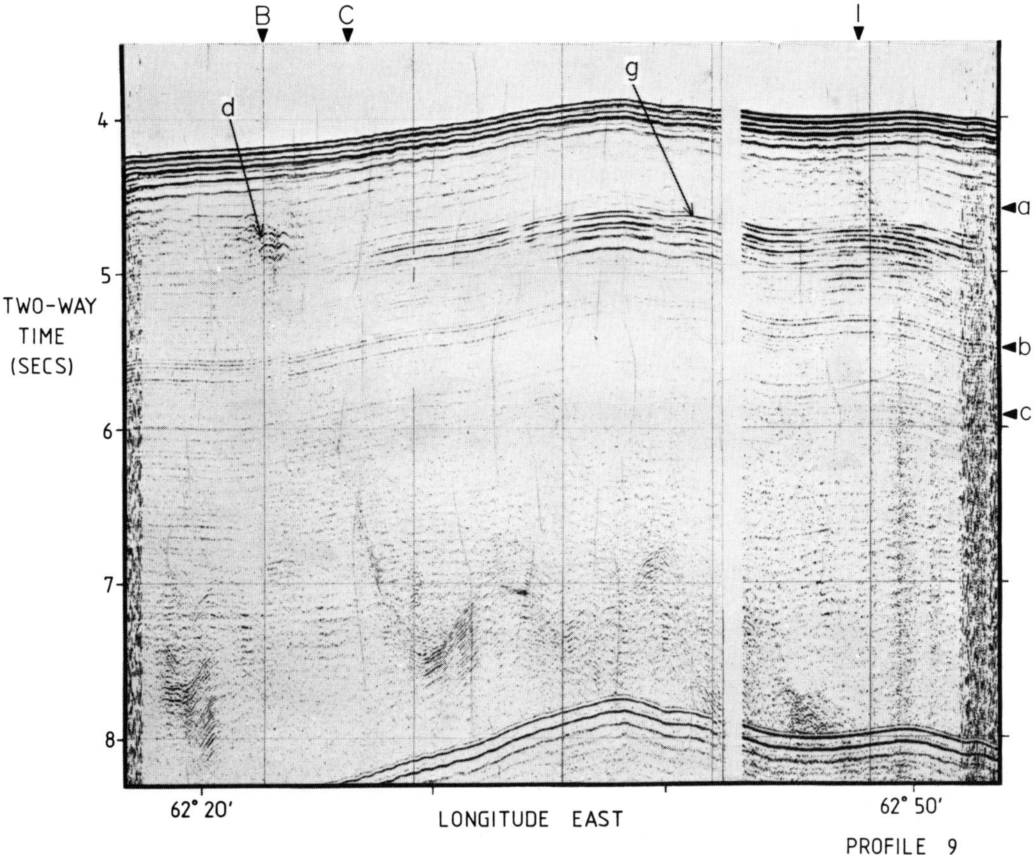

FIG. 9. Continuous seismic reflection profile along line 9 orientated parallel to the strike of the fold belt approximately 1 km north of the crest of the frontal fold (see Figs 2 and 6 for location). Arrowheads mark intersection with profiles B (Fig. 11), C (Fig. 3) and I (Fig. 4). Note the inferred shale diapir marked 'd' also seen on profile B (Fig. 11), the strong bottom simulating reflector marked 'g' caused by free gas trapped beneath a gas hydrate layer and the lateral continuity of reflectors. The reflectors marked a, b, c are those used to construct the depth sections illustrated in Fig. 10. Some of the deep returns are side echoes from the adjacent ridge. Vertical exaggeration at seafloor is approximately 15:1. Vertical lines are one-hourly marks.

fold (which is generally only where the amplitude of the fold is small), appear to dip beneath the overlying $2\frac{1}{2}$ km of folded material without themselves becoming buckled (see, e.g. Figs 3, 11 and profiles from further west in White 1977a).

f. On some profiles the beds are upturned where they abut against the adjacent uplifted ridge (e.g. profile B, Fig. 11).

Mechanisms of fold formation and emplacement

In the previous two sections I have described the observed features of the offshore Makran fold belt and have alluded to the likely emplacement mechanisms. I now look in more detail at the processes involved in transferring the thick undeformed sediment pile from the subducting Arabian plate to the folded and faulted accretionary wedge of the Makran using the observational constraints. In so doing, two points are worth bearing in mind: firstly that, apart from the frontal fold where we can see reflectors from deep within the sedimentary section, we only have details of the surficial deformation from which to infer the deep structure; secondly that the entire offshore belt is controlled ultimately by gravity forces acting on the subducting plate and the overlying sediments.

One of the main questions concerns the extent to which diapirism rather than crustal shortening controls the formation of the charac-

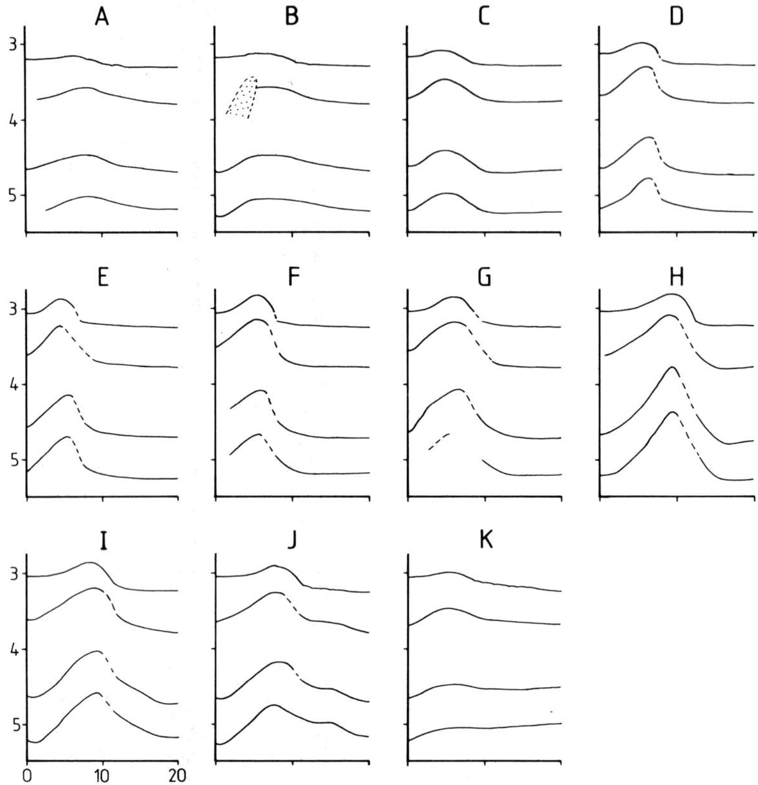

FIG. 10. Migrated depth sections of the upper portion of the frontal fold in the detailed survey area (see Fig. 2 for location). The horizons shown were correlated between profiles using profiles 9 (Fig. 9) and 10 as tie lines. Dotted area on profile B marks a shale diapir (see Fig. 11). Depths (vertical) and distances projected perpendicular to the strike (horizontal), are in kilometres, with north on the left of each section and distance 0 km at the foot of the scarp marking the seaward edge of the adjacent uplifted ridge. Vertical exaggeration is 10:1.

teristic ridge and basin morphology of the offshore Makran. Salt diapirism cannot be responsible for the structures because no large outcrops of salt are found anywhere on land in the Makran, nor do the reflection and refraction profiles at sea show any evidence of the distinctive reflectors and seismic velocities produced by salt. Clay diapirism is, however, much more likely in the thick, rapidly accumulated flysch deposits of the Gulf of Oman. Chapman (1974) has suggested that thick regressive sequences present ideal conditions for the generation of a folded and overthrust belt across a continental margin and Buffler *et al.* (1979) have attributed part of the Mexican Ridge province in the western Gulf of Mexico entirely to downslope overthrusting; the sequence of ridges and basins in the Mexican Ridges has some similarities in general size and appearance to those on the Makran continental margin. Other, smaller-scale evidence that diapiric clay is present in the Makran is given by the widespread occurrence of active mud volcanoes along the coast (see Fig. 1 and Ahmed 1969, Snead 1964), which are frequently triggered by earthquakes and may be accompanied by outgassing (Sondhi 1947).

We have found only one example of a probable clay diapir in all our reflection profiles across the Makran continental margin; this has formed within the frontal fold on profile B (illustrated in Fig. 11), at a depth of about 250 m beneath the seafloor. The diapir extends to about 700 m depth, the southern edge being marked by a curved fault surface which is concave landward. Deeper reflectors show an inflexion beneath the diapir which is probably an artefact caused by low seismic velocities in the over-pressured diapiric clays. Profile 9 (Fig. 9) crosses the diapir perpendicularly and shows that it is of limited extent, with a diameter of about $2\frac{1}{2}$ km.

Despite the evidence for localised diapiric clay intrusions I believe that diapirism is not primarily responsible for the formation of the fold ridges on the Makran margin. The diapir on profile B discussed above is much smaller than, and lies above, the folding in the frontal fold. The larger-scale buckling of the frontal fold nowhere exhibits an increase in amplitude with depth as would be expected if it were driven by a deep diapiric intrusion. Furthermore, there is no indication that any of the elevated ridges to the north of the frontal fold show any signs of the continued growth in amplitude that might be expected if they were cored by diapiric clay. Considering the overall morphology of the Makran continental margin it is unlikely that it is generated by the type of down-slope overthrusting controlled by clay diapirism postulated by Chapman; the rather flat 60 to 70 km wide bench terminated in the south by a $1\frac{1}{4}$ km scarp down to the frontal fold and the abyssal plain is atypical of deformation driven by down-slope slumping, which generally dies out gradually towards the seaward edge.

I suggest that as the undeformed abyssal plain sediments are carried northwards they undergo the following sequence of events:

(*a*) First the upper part of the sediment pile is folded in the frontal fold. The layer immediately beneath the seafloor is thinned over the crest of the fold possibly because it is unconsolidated and thus is buckled with a smaller amplitude than deeper lithified layers (Biot 1961) but perhaps mainly because lack of turbiditic sedimentation on the elevated crest of the fold preferentially thickens the topmost layer over the adjacent abyssal plain. This also causes the bathymetric expression of the folding to be narrower than at depth by burying the seaward edge of the fold. The observation that only the upper $2\frac{1}{2}$ km of sediment become folded suggests that there is a décollement zone about $2\frac{1}{2}$ km beneath the seafloor which may lie in a mud unit weakened by abnormally high pore pressures. Overpressured clays, which we suspect are present in the uppermost 700 m of sediment from the evidence of the diapir on profile B, may exist as deep as 3 km without becoming gravitationally unstable (Chapman 1974); they may then assist folding by providing interfacial lubrication (Biot 1961). The consistent asymmetry of the frontal fold may be explained by its growth in a region where one of the principal axes of the stress ellipsoid is not vertical (Price 1967), because the stress field has been locally modified by the step topography of the adjacent fold belt (see, e.g. King 1978).

(*b*) The frontal fold is next uplifited by a thrust fault which is concave towards the land and flattens out within the lower sediments. The frontal fold must be uplifted by a fault rather than simply by continued folding because the maximum height the frontal fold attains anywhere along the entire Makran margin is only 400 m (see, e.g. White & Klitgord 1976) whilst the next landward ridge is $1\frac{1}{4}$ km above the abyssal plain. It may be that the fold reaches a peak amplitude of only 400 m because the sediments involved have by then become sufficiently consolidated by the folding to support the thrust fault. The upturned beds at the landward edge of some of the frontal folds may be caused by drag folding on the footwall beneath the thrust fault.

The thrust elevates the fold by a very consistent $1\frac{1}{4}$ km, and then movement along the fault ceases and a new frontal fold starts to form in the flat-lying abyssal plain sediments to the south. There is thus a delicate balance between folding and faulting of the sediments in the frontal fold. The deformation reverts to folding when the additional normal stress across the thrust fault imposed by the weight of the uplifted ridge exceeds about 2×10^7 N m^{-2} (i.e. 200 bars; the additional stress caused by the displacement of $1\frac{1}{4}$ km of water by uplifted sediment). As we shall see in the next section this balance, which depends on the relative magnitudes of the vertical and horizontal stresses and on the strength of the thick pile of sediments, is easily upset by basement peaks projecting into the sedimentary section.

(*c*) Over the next 60 to 70 km towards the coast there is little further deformation other than a small amount of additional tilting resulting from minor movement on the thrust faults. From the very small average bathymetric slope of only 0.6° I infer that the subducting plate is slipping beneath the fold belt along weak horizons within the sediments (Chapple 1978), in a similar manner to that reported from the frontal 50 km of the Barbados accretionary wedge (Westbrook 1975).

(*d*) To the north of this gentle slope the fold belt becomes abruptly and rapidly uplifted by further thrust faults. These faults probably extend down to or even into the basement rocks. Uplift rates on the coast are typically 1.5 to 2.0 mm yr^{-1} increasing towards the east (Vita-Finzi 1975, 1979; Page *et al.* 1979). Extensive seismicity accompanies the faulting; it was reported that the magnitude 8.3 earthquake of 28

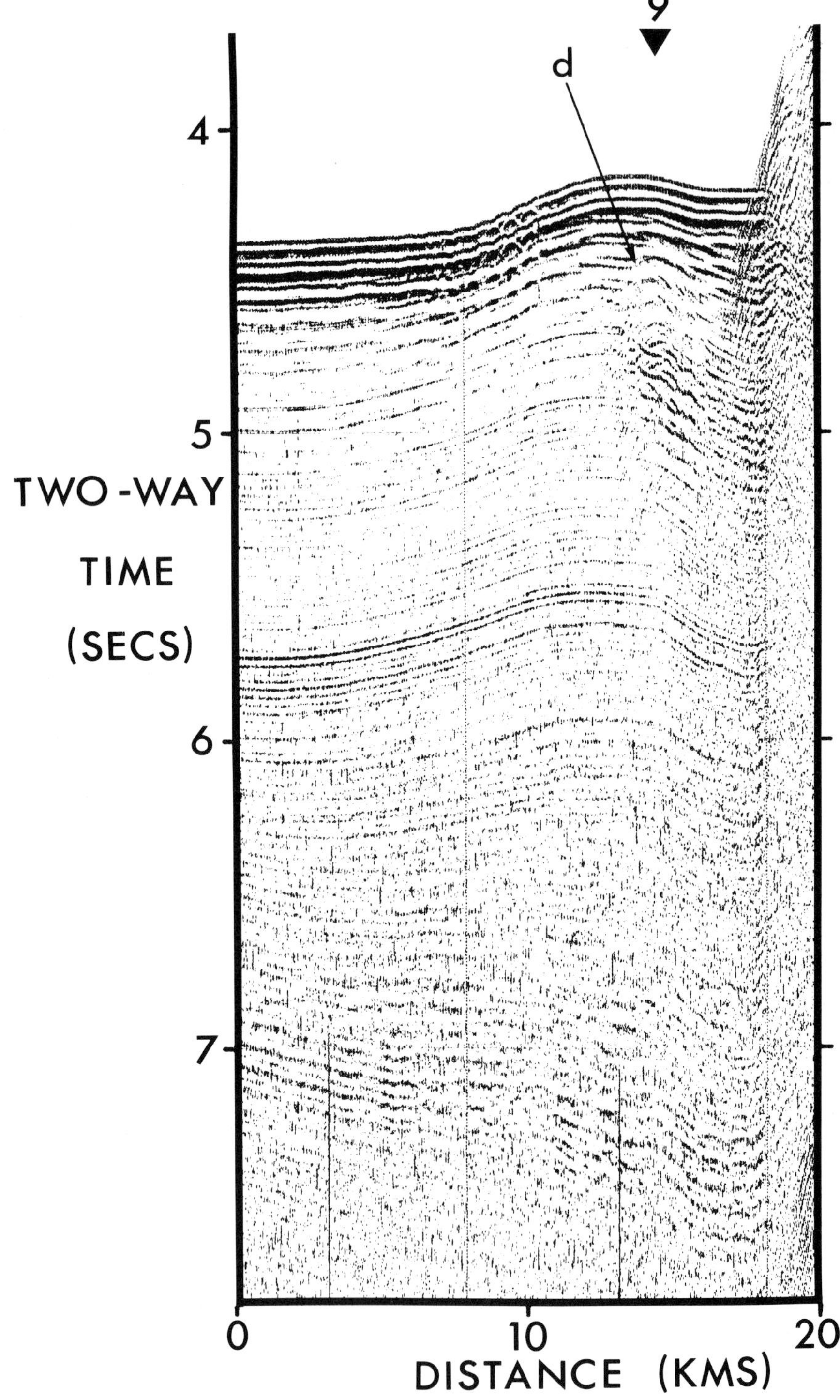

PROFILE B

November 1945 created a 2 m high fault scarp (Page *et al.* 1979). The pattern of incised streams and wave-cut terraces suggests that uplift occurs episodically along the Makran coast.

There is no evidence of any major basement discontinuity beneath the abyssal plain south of the fold belt at the longitude of profile G near which the pattern of uplifted ridges changes within the detailed survey area. This of course does not mean that there is not a basement feature now buried beneath the accretionary fold belt which was responsible for those perturbations in the ridge lineations. However, it is more likely that even given a thick sediment pile of great lateral extent then individual ridges will be of limited length because the sediment section is insufficiently strong to transmit stresses an indefinite distance laterally. The detailed survey area did not extend far enough along strike to follow an individual uplifted ridge over its entire length. However, an estimate of the lateral extent of the frontal fold can be made from 19 roughly equally spaced crossings over a distance along strike of 80 km over which interval the amplitude increases from about 10 m in the west to a maximum of 400 m and then decreases again to 240 m in the east. This lateral extent of over 80 km gives a length to width ratio for the frontal fold in excess of 8:1.

Although we cannot directly measure the rate of deformation, we can estimate it from two different approaches, as below:

(i) Geological mapping in the onshore Makran (Ahmed 1969) indicates southward regression of the marine-fresh water sedimentary boundary by 250 km since the Oligocene (i.e. about 6 to 10 mm yr^{-1}). If we assume that the deformational front is moving southwards at the same speed as the coastline, then a new fold must be added to the edge of the accretionary wedge approximately once every million years.

(ii) Alternatively we can calculate the fold growth rate provided we know the sedimentation rate and we assume that the uppermost sediment layer is thin over the frontal fold solely as a result of non-deposition on the developing fold. The ratio of the amount by which the uppermost layer is thinned to the amplitude of the fold is remarkably constant at 0.5 for all profiles across the frontal fold (Fig. 10 and White 1977a). Our best estimates of the sedimentation rate are 0.54 mm yr^{-1} over the past 15 Ma derived by summing the thicknesses of formations now exposed on land (from Ahmed 1969), and 0.50 mm yr^{-1} over the past 0.02 Ma measured by radio-carbon dating a piston core in the western Gulf of Oman abyssal plain (Stoffers & Ross 1979). These rates suggest that it takes about 0.4 Ma for the frontal fold to grow to its maximum height of 400 m. If we allow for some of the sedimentation being aeolian rather than turbiditic, and thus deposited uniformly over both the fold and the abyssal plain, our estimate of the time taken for the fold to develop is proportionately greater. If some of the reduction in thickness is due to tectonic thinning, the time taken for the fold to develop could be less.

The convergence rate of the Arabian plate beneath the Eurasian plate is about 50 mm yr^{-1}. Only part of this is taken up by deformation and southward migration of the seaward edge of the continental margin; much of the shortening of the sedimentary section occurs by internal deformation and thickening by imbricate thrusting commencing more than 60 km further north.

Eastward extent of the Makran fold belt

The offshore Makran fold belt was traced eastwards towards the triple junction near Karachi between the Indian, Arabian and Eurasian plates. Deformed ridge and basin topography is found along the entire continental margin, but it becomes increasingly irregular as the Murray Ridge impinges on the Makran margin. Similarly, the simple open fold and basin structure is perturbed in the west where the Makran margin is in collision with the opposing continental margin of Oman (White & Ross 1979).

On profile M (illustrated in Fig. 12), to the east of the detailed survey, is a large basement ridge only 10 km south of the fold belt. This ridge causes large gravity and magnetic anomalies (Taylor 1968). The basement, where it can last be seen adjacent to the fold belt, is only about 2½ km beneath the seafloor, and it has already caused the pattern of ridges and basins to be much less simple than within the detailed survey area (Figs 3, 4).

As the basement ridge passes obliquely beneath the Makran fold belt in the east of the Gulf of Oman it generates chaotic seafloor

FIG. 11. Continuous seismic reflection profile across frontal fold on line B (see Fig. 2 for location). Note the small inferred shale diapir marked by 'd' cutting through the upper sediments. The shale diapir is also seen on profile 9 (Fig. 9), which intersects at the position marked by an arrowhead. Vertical exaggeration at seafloor is approximately 7:1.

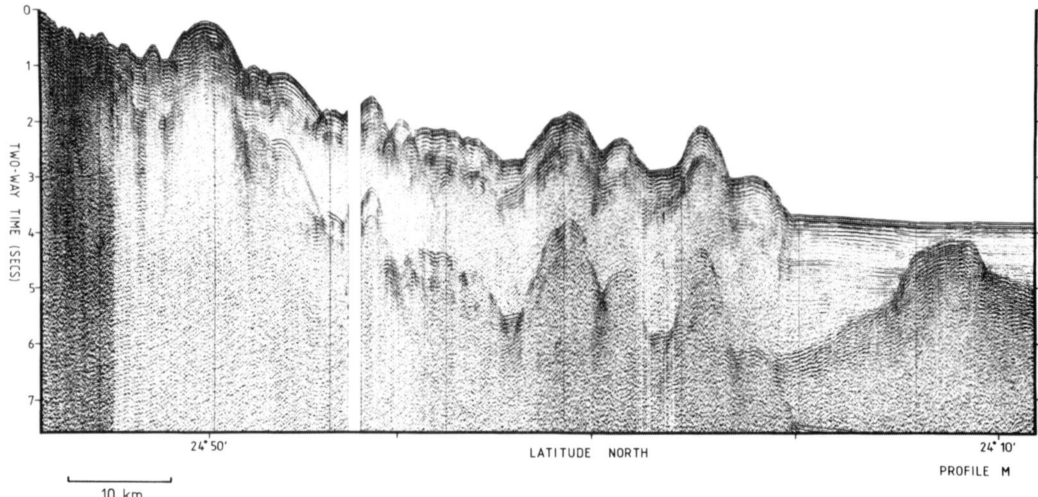

FIG. 12. Continuous seismic reflection profile along line M approximately perpendicular to the strike of the fold belt (see Fig. 1 for location). Vertical exaggeration at seafloor is approximately 7:1. Note the prominent offshore basement peak near 24° 12′ N.

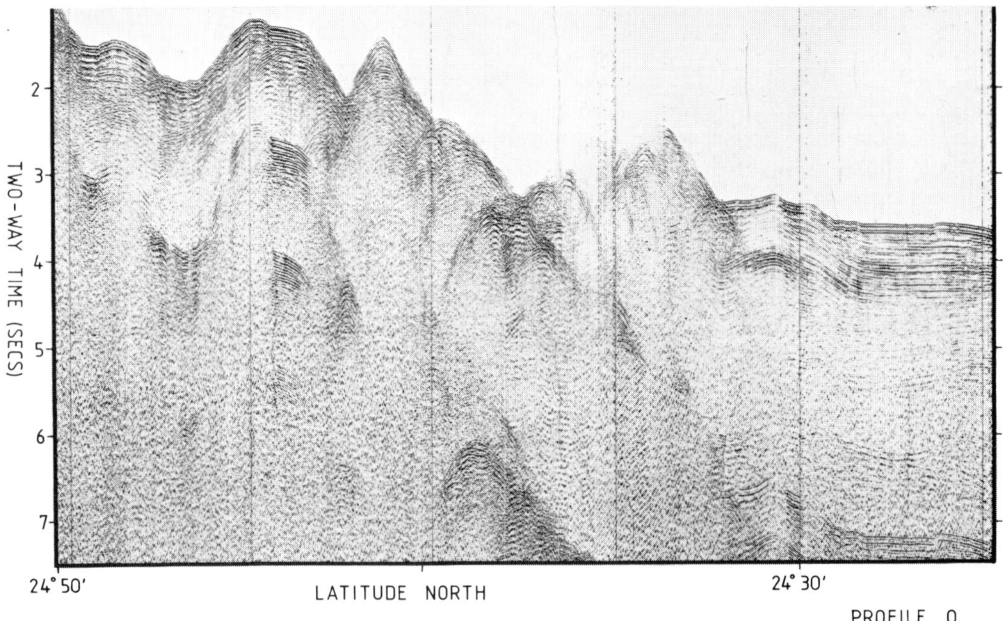

FIG. 13. Continuous seismic reflection profile along line O near the eastern end of the fold belt (see Fig. 1 for location). Vertical exaggeration at seafloor is approximately 7:1. The accretionary fold belt has now consumed the basement ridge seen in Fig. 12, generating chaotic seafloor relief.

relief (e.g. profile O, Fig. 13). The sediments are still being deformed and incorporated in the accretionary wedge by thrust faults, but the slope of the continental margin is here much steeper than to the west and the faulting extends into the basement. The initial stages of open folding and thrusting of the thick sediment pile (stages 'a' through 'c' of the previous section), have been bypassed and the shallow basement causes immediate rapid uplift and imbricate thrusting such as is seen near the coast in the detailed survey area (stage 'd').

Conclusions

The systematic, well lineated pattern of uplifted folds and intervening basins at the seaward margin of the offshore Makran is the result of scraping a very thick pile of sediment off a shallowly dipping oceanic plate at a convergent margin. The oceanic crust of the subducting plate is Cretaceous in age and the great width of the accretionary prism extending across the continental margin and into the Makran of Pakistan and Iran is the result of continued subduction from the Late Cretaceous or early Tertiary until the present. However, the pattern of plate interactions in this region has not always been as simple as it is now, as is attested to by the extensive Late Cretaceous ophiolites and mélanges emplaced on the adjacent continental margins of Oman, Iran and Pakistan and by the large strike-slip faults east of the Jaz Murian and Lut depressions. Nevertheless there are no major offsets in the present seaward margin of the accretionary wedge.

Material from the flat-lying abyssal plain is added to the seaward edge of the accretionary belt by the upper $2\frac{1}{2}$ km first becoming folded above a décollement zone within the sedimentary section and then elevated by a thrust fault which flattens out within the sediments. There is no further major deformation of the elevated ridges until some 60 to 70 km to the landward when they are rapidly uplifted until they are raised above sea-level; this is achieved by large-scale imbricate thrust faulting extending to the basement and accompanied by major seismicity.

The simple and relatively gentle deformation observed over the seaward part of the accretionary wedge will not be preserved in the geological record because in the process of emplacement above sea-level the sediment becomes further faulted and stacked up in thrust slices in a similar manner to that reported from many other subduction zones. The occurrence of large seamounts or basement ridges on the subducting plate also disrupts the open fold ridge and basin structure, as is seen in the eastern part of the Makran continental margin.

ACKNOWLEDGMENTS: Many people have contributed to this work by generously giving their time and their expertise both at sea and in Cambridge. Although there are too many to list by name, I am indebted to them all and wish to acknowledge their assistance. In particular, I am grateful to Dr K. E. Louden who was Chief Scientist on RRS *Shackleton* leg 1/80, to Dr D. H. Matthews who initiated my study of the Gulf of Oman and to Drs G. C. P. King and D. P. McKenzie, amongst many others, for their comments on the work. I thank the captain, officers and crew of the RRS *Shackleton* on leg 1/80 for navigating us through blissfully calm seas. The Natural Environment Research Council funded this project and provided personal support through a Postdoctoral Fellowship.

Department of Earth Sciences, Cambridge University, Contribution No. ES 59.

References

ABDEL-GAWAD, M. 1971. Wrench movements in Baluchistan Arc and Relation to Himalayan-Indian ocean tectonics. *Bull. geol. Soc. Am.* **82,** 1235.

AHMED, S. S. 1969. Tertiary geology of part of south Makran, Baluchistan, West Pakistan. *Bull. Am. Assoc. Petrol. Geol.* **53,** 1480–99.

BARKER, P. F. 1966. A reconnaisance survey of the Murray Ridge. *Philos. Trans. R. Soc. London,* **259A,** 187–97.

BERBERIAN, M. 1976. Contribution to the seismotectonics of Iran (Part II). *Rep. Geol. Surv. Iran,* **39,** 516.

BIOT, M. A. 1961. Theory of folding of stratified viscoelastic media and its implications in tectonics and orogenesis. *Bull. geol. Soc. Am.* **72,** 1595–620.

BUFFLER, R. T., SHAUB, F. J., WATKINS, J. S. & WORZEL, J. L. 1979. Anatomy of the Mexican Ridges, southwestern Gulf of Mexico. *In:* WATKINS, J. S., MONTADERT, L. & DICKERSON, P. W. (eds). *Geological and Geophysical Investigations of Continental Margins.* Mem. Am. Assoc. Petrol Geol. **29,** 319–27.

CHAPMAN, R. E. 1974. Clay diapirism and overthrust faulting. *Bull. geol. Soc. Am.* **95,** 1597–602.

CHAPPLE, W. M. 1978. Thin-skinned fold-and-thrust belts. *Bull. geol. Soc. Am.* **89,** 1189–98.

CLOSS, H., BUNGENSTOCK, H. & HINZ, K. 1969. Ergebnisse seismischer Untersuchungen im nordlichen Arabischen Meer ein Beitrag zur Internationalen Indischen Ozean Expedition. *Meteor forsch.* **2C,** 28.

DYKSTRA, J. D. & BIRNIE, R. W. 1979. Segmentation of the Quaternary subduction zone under the Baluchistan region of Pakistan and Iran. *In:* FARAH, A. & DEJONG, K. A. (eds). *Geodynamics of Pakistan*, 319–23. Geol. Surv. Pakistan, Quetta.

FARHOUDI, G. & KARIG, D. E. 1977. Makran of Iran and Pakistan as an active arc system. *Geology,* **5,** 664–8.

HUDSON, R. G. S., MCGUGAN, A. & MORTON, D. M. 1954. The structure of the Jebel Hagab area, Trucial Oman. *Q. J. geol. Soc. London,* **60,** 121–57.

HUNTING SURVEY CORPORATION 1960. *Reconnaissance Geology of Part of West Pakistan.* Maracle Press, Toronto.

HUTCHISON, I., LOUDEN, K. E., WHITE R. S. & VON HERZEN, R. 1981. Heat flow and age of the Gulf of Oman. *Earth planet. Sci. Lett.*, in press.

JACOB, K. H & QUITTMEYER, R. C. 1979. The Makran region of Pakistan and Iran: trench-arc gap with active plate subduction. *Geodynamics of Pakistan,* 305–18. Geol. Surv. Pakistan, Quetta.

KAZMI, A. H. 1979. *Preliminary Seismotectonic Map of Pakistan 1:2,000,000.* Map Ser. Geol. Surv. Pakistan, Quetta.

KING, G. C. P. 1978. Geological faults: fracture, creep and strain. *Philos. Trans. R. Soc. London,* **288A,** 197–212.

NOWROOZI, A. A. 1976. Seismotectonic provinces of Iran. *Bull. seism. Soc. Amer.* **66,** 1249–76.

PAGE, W. D., ALT, J. N., CLUFF, L. S. & PLAFKER, G. 1979. Evidence for the recurrence of large-magnitude earthquakes along the Makran coast of Iran and Pakistan. *Tectonophysics,* **52,** 533–47.

PARSONS, B. & SCLATER, J. G. 1977. An analysis of the variation of ocean floor bathymetry and heat flow with age. *J. geophys. Res.* **82,** 803–27.

PRICE, N. J. 1967. The initiation and development of asymmetrical buckle folds in non-metamorphosed competent sandstones. *Tectonophysics,* **4,** 173–201.

SNEAD, R. 1964. Active mud volcanoes of Baluchistan, West Pakistan. *Geog. Rev.* **54,** 545–60.

SONDHI, V. P. 1947. The Makran earthquake, 28th November, 1945. The birth of new islands. *Indian Miner.* **1,** 146–54.

STEWART, R. A. PILKEY, O. H. & NELSON, B. W. 1965. Sediments of the northern Arabian Sea. *Mar. Geol.* **3,** 411–27.

STOFFERS, P. & ROSS, D. A. 1979. Late Pleistocene and Holocene sedimentation in the Persian Gulf—Gulf of Oman. *Sediment. Geol.* **23,** 181–208.

STOLL, R. D. & BRYAN, G. M. 1979. Physical properties of sediments containing gas hydrates. *J. geophys. Res.* **84,** 1629–34.

TAYLOR, P. T. 1968. Interpretation of the north Arabian Sea aeromagnetic survey. *Earth. planet. Sci. Lett.* **4,** 232–6.

VITA-FINZI, C. 1975. Quaternary deposits in the Iranian Makran. *Geog. J.* **141,** 415–20.

—— 1979. Contributions to the Quaternary geology of southern Iran. *Rep. Geol. Miner. Surv. Iran,* **47.**

WESTBROOK, G. K. 1975. The structure of the crust and upper mantle in the region of Barbados and the Lesser Antilles. *Geophys. J. R. astron. Soc.* **43,** 201–42.

WHITE, R. S. 1977a. Recent fold development in the *Gulf of Oman. Earth planet. Sci. Lett.* **36,** 85–91.

—— 1977b. Seismic bright spots in the Gulf of Oman. *Earth planet. Sci. Lett.* **37,** 29–37.

—— 1979a. Gas hydrate layers trapping free gas in the Gulf of Oman. *Earth planet. Sci. Lett.* **42,** 114–20.

—— 1979b. Deformation of the Makran continental margin. *In:* FARAH, A. & DEJONG, K. A. (eds). *Geodynamics of Pakistan,* 295–304. Geol. Surv. Pakistan, Quetta.

—— & KLITGORD, K. D. 1976. Sediment deformation and plate tectonics in the Gulf of Oman. *Earth planet. Sci. Lett.* **32,** 199–209.

—— & ROSS, D. A. 1979. Tectonics of the western Gulf of Oman. *J. geophys. Res.* **84,** 3479–89.

ROBERT S. WHITE, Bullard Laboratories, Department of Earth Sciences, Cambridge University, Madingley Rise, Madingley Road, Cambridge CB3 0EZ, England.

The Late Cretaceous–Cenozoic history of western Baluchistan Pakistan—the northern margin of the Makran subduction complex

R. S. Arthurton, A. Farah & W. Ahmed

SUMMARY: The study area lies some 400 km N of the Makran coast, and extends westwards for some 450 km from the Chaman Fault, a major sinistral transform fracture zone which marks the suture between the Eurasian and Indian plates. It lies to the north of the Makran ranges of Tertiary flysch, and flanks the southern margin of a stable, aseismic area known as the Dasht-i-Margo block.

The Makran region as a whole has been interpreted as an accretionary prism of Late Cretaceous to Holocene age, resulting from the northward subduction of oceanic lithosphere. The geological record of the study area is consistent with such an interpretation, for the period Late Cretaceous–Palaeocene at least.

Rapid northward subduction during the Late Cretaceous gave rise to profuse andesitic volcanism on the sub-parallel, ENE-trending arc massifs of the Chagai Hills and Ras Koh geanticlines. The waning of this volcanism during the Maestrichtian coincided with the accumulation and deformation of flysch sediments in a trench which extended from Mirjawa to the south of Ras Koh; also with thick forearc sedimentation in the Dalbandin Trough, between the geanticlines.

In late Palaeocene times, deformed flysch—interpreted as a subduction complex—was elevated to form a structural high in the Mirjawa range. In the western Ras Koh, the flysch, along with ultramafic slices, became emplaced against the arc massif to form a new landmass. This emplacement may have been a result of a late Palaeocene collision between the northward-migrating Indian continental plate and the continental lithosphere of the Dasht-i-Margo block, the Ras Koh flysch belt taking the brunt of the initial impact.

The principal forearc faults have regular, arcuate traces which, in the south-west, parallel the structural grain of the Mirjawa flysch belt. Movement on the faults influenced sedimentation in the forearc, particularly during the Eocene and Miocene. There is evidence of tensional transform movement in Eocene times. The forearc faults were especially active during the Pliocene. Movement across them during this period was compressional and associated with major folding, particularly in the Dalbandin Trough. The trend of this folding may imply dextral transform displacement within the trough.

This paper is concerned with the Late Cretaceous and Cenozoic evolution of an area that is widely considered to include the contemporary volcanic arc and forearc components of a consuming plate margin (Farhoudi & Karig 1977; Sillitoe 1978; Jacob & Quittmeyer 1979). The study area (Fig. 1) extends from the Afghan frontier southwards to the Makran, and from the Iranian frontier in the west of the Chaman Fault in the east. The Chaman Fault is one of the principal fractures in a zone of left-lateral transcurrent displacement according to Auden (1974) active throughout the Tertiary, and according to Sillitoe (1978) initiated in or before early Miocene times.

Present knowledge of the area is due largely to Jones (1960), following the pioneer work of Vredenburg (1901). Jones established the present lithostratigraphy, though his mapping was modified by Bakr & Jackson (1964). He also identified four principal structural components including the Ras Koh geanticline, and the North Chagai Arch (here referred to as the Chagai Hills geanticline) (Fig. 1), the latter being interpreted as an area of stable shelf underlain by continental basement. Between these two structural highs he recognized a major depression, filled with Cenozoic marine and continental sediments, the Mirjawa–Dalbandin Trough (Fig. 1). The Makran region to the south was regarded as lying within the Baluchistan geosyncline.

Auden (1974) included the area within his Dasht-i-Margo zone, the northern part of which is a poorly exposed, aseismic region termed the 'Dasht-i-Margo Block'. This 'Block' has been referred to as the 'Afghan Block' (Jacob & Quittmeyer 1979); and forms part of the Cenozoic Iran–Afghanistan microcontinent of Sillitoe (1978). Auden termed the area of the present study the 'Chagai Eruptive Zone' in recognition of the recurrent volcanism that has

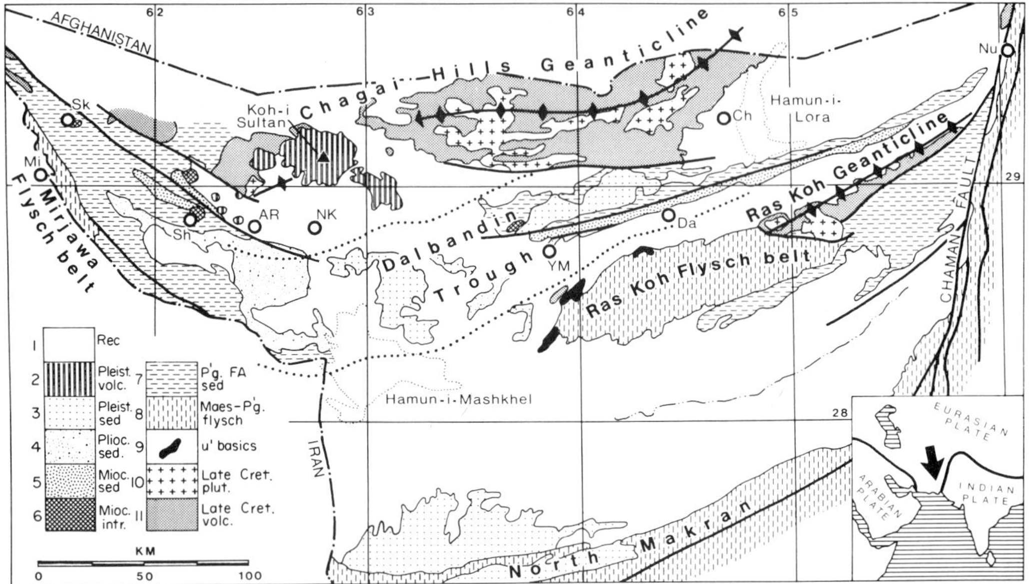

FIG. 1. Geological map of study area showing the principal structural components. (1) Recent and undivided; (2) Pleistocene volcanics; (3) Pleistocene sediments; (4) Pliocene sediments; (5) Miocene sediments; (6) Miocene intrusives; (7) Palaeogene forearc sediments; (8) Maestrichtian–Palaeogene flysch; (9) ultrabasic igneous rocks; (10) Late Cretaceous–Palaeogene plutonic igneous rocks; (11) Late Cretaceous volcanic rocks. AR Alam Reg; Ch Chagai; Da Dalbandin; Mi Mirjawa; NK Nok Kundi; Nu Nushki; Sh Shor Nalla; Sk Saindak; YM Yak Mach.
Sources: Jones (1960), Bakr & Jackson (1964), and recent geophysical and geological fieldwork by or for the Geological Survey of Pakistan. Limits of Dalbandin Trough (heavy dots) based on aeromagnetic data. Light dots show limits of present topographical depressions.

affected particularly the Chagai Hills and northeastern Ras Koh from Late Cretaceous to Quaternary times (Jones 1960). He placed the Makran ranges within his 'Tertiary Flysch Trough', but included the Mirjawa and Ras Koh ranges, as well as the present topographic depression of Hamun-i-Mashkhel (Fig. 1), within the Chagai Eruptive Zone to the north.

More recently, the entire Makran deformed belt and the contemporary depressions to the north, including the Hamun-i-Mashkhel, were interpreted by Farhoudi & Karig (1977) as an accretionary prism of Late Cretaceous or early Tertiary to Holocene age. This accretion results from the northward subduction of oceanic lithosphere (Stoneley 1974). White (1979, 1981) demonstrates that flysch sediments continue to be added to the accretionary prism today, with undeformed abyssal plain sediments becoming abruptly involved in frontal fold deformation some 150 km south of the Makran coast. The chain of Quaternary volcanoes stretching from south-central Iran through the study area into southern Afghanistan has been interpreted as the volcanic arc associated with this subduction (Farhoudi & Karig 1977; Jacob & Quittmeyer 1979).

Sillitoe's (1978) suggestion that left-lateral transcurrent faulting affected the southern margin of the Chagai belt (Ras Koh–Dalbandin area) during the late Cenozoic, perhaps causing eastward translation of the Ras Koh, has been developed by Lawrence *et al.* (1981), who have linked such movement to left-lateral displacement on the Chaman Fault.

In the following account, the stratigraphy and structural history of the area as determined by Jones (1960) is revised incorporating data collected during detailed surveys by field parties of the Geological Survey of Pakistan. The principal structural features of the area are reassessed and recent accounts of the geodynamic evolution of this part of Pakistan are discussed in the light of the new data. Particular attention is focused on the position and behaviour of the presumed margin of continental lithosphere during the Late Cretaceous and Palaeogene, and on the structural behaviour of the forearc zone throughout the Cenozoic. Lithostratigraphic terms and formation ages are those used by Jones (1960) unless otherwise stated.

The structural components

The North Chagai Arch (Jones 1960) we regard as having geanticlinal status, and this structural feature is here termed the 'Chagai Hills geanticline' (Figs 1 & 2). It has a history of which is well-defined both by aeromagnetic data (A. Farah *et al.*, in preparation) and geological evidence. It extends from a point about 100 km ENE of Dalbandin to the northern part of Hamun-i-Mashkhel. The Dalbandin Trough may be regarded as a forearc basin (*sensu* Seely *et al.* 1974) in relation to the arc massif of the Chagai Hills geanticline. A further forearc basin (Saindak–Alam Reg) (Fig. 2) abuts against the south-western end of the Chagai Hills geanticline.

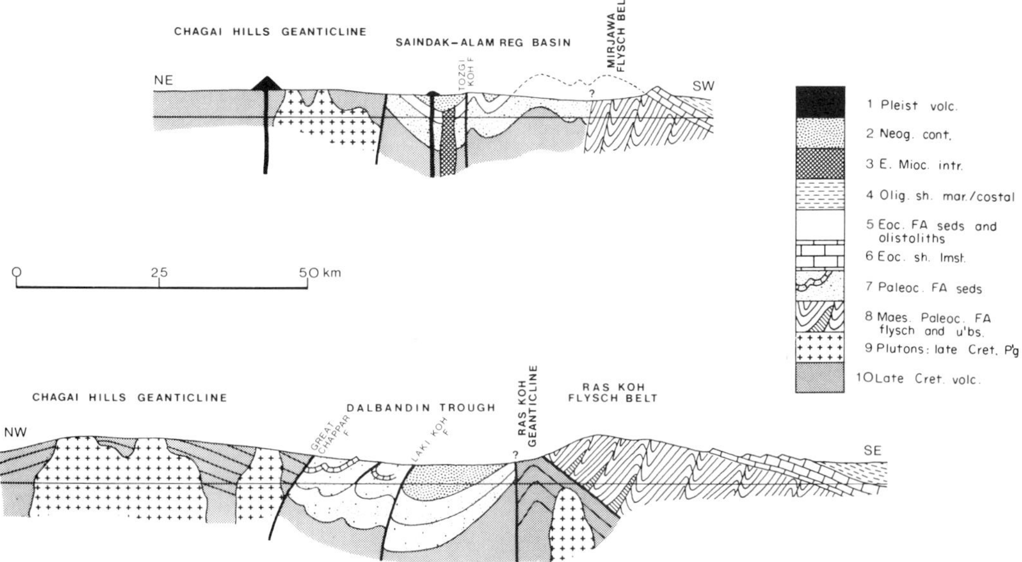

FIG. 2. Generalized horizontal sections from the Chagai Hills geanticline (a) across the Saindak–Alam Reg basin and Mirjawa flysch belt, and (b) across the Dalbandin Trough and Ras Koh flysch belt. (1) Pleistocene volcanics; (2) Neogene continental sediments; (3) early Miocene intrusions; (4) Oligocene shallow marine/coastal plain sediments; (5) Eocene forearc sediments with olistoliths; (6) Eocene shallow-water limestones; (7) Palaeocene forearc sediments, including limestones; (8) Maestrichtian/Palaeocene deformed flysch, with tectonically emplaced ultrabasics; (9) Plutons of Late Cretaceous/Palaeogene age; (10) Late Cretaceous volcanics.

recurrent volcanism and plutonism. In the context of the arc-trench system proposed by Farhoudi & Karig (1977), this component may be classified as the continental margin 'arc massif' (Dickinson & Seely 1979). The geanticline is clearly defined along the length of the Chagai Hills to a point some 50 km NE of Nok Kundi, where the structure plunges to the west. Its northern limb re-emerges from Quaternary cover to the west of Koh-i-Sultan, but the structure terminates abruptly south-westwards some 30 km WNW of Nok Kundi.

The Mirjawa–Dalbandin Trough (Jones 1960) is here renamed the 'Dalbandin Trough' (Figs 1 & 2). Cenozoic sediments several kilometres thick fill this structural depression,

The Ras Koh geanticline (Jones 1960) (Figs 1 & 2) is well-defined from near the Chaman Fault in the east to a point some 40 km E of Dalbandin where its surface expression is abruptly terminated. Aeromagnetic evidence (A. Farah *et al.*, in preparation) suggests that a comparable structure may extend from the western end of the Ras Koh range to southern Hamun-i-Mashkhel. The Ras Koh geanticline is intruded by plutons, though these are of minor extent compared to those of the Chagai Hills geanticline (Auden 1974). There is evidence of considerable Cretaceous volcanicity.

Two new structural units are defined. The south-western part of the Mirjawa range is termed the 'Mirjawa flysch belt', and the western part of the Ras Koh range, the 'Ras Koh flysch belt'. These belts may be linked under upper Cenozoic sediments to the south of Hamun-i-Mashkhel. Both are interpreted as belonging to the 'subduction complex' (Dickinson & Seely 1979) of the supposed arc-trench system.

Geological history

The interrelationship of formations and the ages of the principal magmatic and tectonic events are summarized in Table 1. A palaeogeographical history is suggested in Fig. 3.

glomerate suggest that much of the eruption was subaerial. No eruptive centres are known, though evidence of these may have been destroyed by the intrusion of batholiths. The basement to these volcanics is nowhere exposed.

The thickest known sequence is that of the Sinjrani Volcanic Group near Alam Reg in the western part of the Chagai Hills geanticline (Figs 1 & 3a), where at least 10 km of andesitic lavas, tuffs and volcanic conglomerates underlie the Maestrichtian Humai Formation (Arthurton *et al.* 1979). Similar components make up the Sinjrani Volcanic Group at the eastern end of the Chagai Hills near Chagai village, where a

TABLE 1. *Ages and interrelations of lithostratigraphic units; dates of igneous and tectonic events within the study area, and regional geodynamic events. Asterisk indicates proposed new formation name*

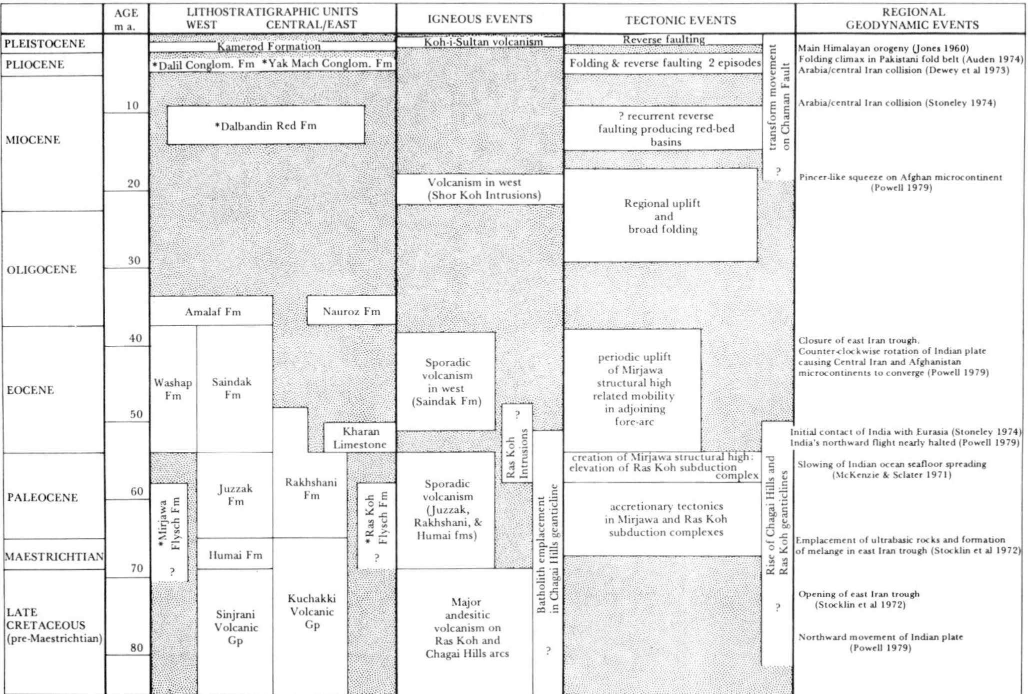

	AGE m a.	LITHOSTRATIGRAPHIC UNITS WEST	LITHOSTRATIGRAPHIC UNITS CENTRAL/EAST	IGNEOUS EVENTS	TECTONIC EVENTS	REGIONAL GEODYNAMIC EVENTS
PLEISTOCENE		Kamerod Formation	Kamerod Formation	Koh-i-Sultan volcanism	Reverse faulting	Main Himalayan orogeny (Jones 1960)
PLIOCENE		*Dalil Conglom. Fm	*Yak Mach Conglom. Fm		Folding & reverse faulting 2 episodes; transform movement on Chaman Fault	Folding climax in Pakistani fold belt (Auden 1974); Arabia/central Iran collision (Dewey et al 1973)
MIOCENE	10, 20	*Dalbandin Red Fm	*Dalbandin Red Fm	Volcanism in west (Shor Koh Intrusions)	? recurrent reverse faulting producing red-bed basins; ?	Arabia/central Iran collision (Stoneley 1974); Pincer-like squeeze on Afghan microcontinent (Powell 1979)
OLIGOCENE	30	Amalaf Fm	Nauroz Fm		Regional uplift and broad folding	
EOCENE	40, 50	Washap Fm; Saindak Fm	Kharan Limestone	Sporadic volcanism in west (Saindak Fm); Ras Koh Intrusions ?	periodic uplift of Mirjawa structural high related mobility in adjoining fore-arc	Closure of east Iran trough. Counter-clockwise rotation of Indian plate causing Central Iran and Afghanistan microcontinents to converge (Powell 1979); Initial contact of India with Eurasia (Stoneley 1974); India's northward flight nearly halted (Powell 1979)
PALEOCENE	60	*Mirjawa Flysch Fm; Juzzak Fm	Rakhshani Fm; *Ras Koh Flysch Fm	Sporadic volcanism (Juzzak, Rakhshani, & Humai fms)	creation of Mirjawa structural high: elevation of Ras Koh subduction complex; accretionary tectonics in Mirjawa and Ras Koh subduction complexes; Rise of Chagai Hills and Ras Koh geanticlines	Slowing of Indian ocean seafloor spreading (McKenzie & Sclater 1971)
MAESTRICHTIAN	70	?; Humai Fm	?			Emplacement of ultrabasic rocks and formation of melange in east Iran trough (Stocklin et al 1972)
LATE CRETACEOUS (pre-Maestrichtian)	80	Sinjrani Volcanic Gp	Kuchakki Volcanic Gp	Major andesitic volcanism on Ras Koh and Chagai Hills arcs; Batholith emplacement in Chagai Hills geanticline ?	?	Opening of east Iran trough (Stocklin et al 1972); Northward movement of Indian plate (Powell 1979)

Late Cretaceous

Andesitic volcanism dominated the area during the Late Cretaceous. Andesite lavas, tuffs and volcanic conglomerates several kilometres thick accumulated mostly in marine environments as indicated by intercalations of marine limestones and conglomerate, the latter including boulders of 'Hippuritic Limestone' (Vredenburg 1901). However, local red sediments and extensive beds of volcanic con-

thickness of about 4 km has been estimated. In the north-eastern Ras Koh range, the broadly equivalent Kuchakki Volcanic Group is estimated to be at least 6 km thick though, according to Jones (1960), the proportion of lavas is less than that in the Sinjrani volcanics. Again, no eruptive centres are recorded.

Jones (1960) suggested that the Sinjrani and Kuchakki volcanics might be continuous under the Dalbandin Trough but this suggestion is not supported by the aeromagnetic evidence (A. Farah *et al.*, in preparation). Their southern extension is also speculative. Small inliers of the volcanics were mapped in the northern Mirjawa range (Jones 1960); a further possible outcrop occurs at the extreme western end of the Ras Koh range. In the Makran, there are no records of Cretaceous volcanics *in situ*.

In the Chagai Hills geanticline, the Late Cretaceous volcanism was associated with the widespread intrusion of granite and granodiorite batholiths. This plutonism must have been accompanied at least locally by substantial upheaval and erosion. Pebbles of these intrusives have been recorded (Jones 1960) in the basal conglomerate of the overlying Humai Formation.

The Humai Formation, which overlies the Sinjrani Volcanic Group around much of the periphery of the Chagai Hills geanticline, reflects an abrupt reduction in the intensity of volcanism. It includes shallow-water limestones (Humai Limestone). Underlying these limestones in peripheral sections north of Alam Reg are several hundred metres of turbidites, including limestone conglomerates. By contrast, an outlier of Humai Limestone at Sor Koh, nearer the crest of the structure, rests directly on the Sinjrani volcanics. This lateral variation in the sequence is interpreted as indicative of uplift of the geanticline, resulting in widespread shoaling and the erosion of earlier-formed limestones from the crest of the structure.

In the Ras Koh range, the limestone-bearing upper part of the Kuchakki Volcanic Group was regarded by Jones (1960) as equivalent to the Humai Formation (Table 1). Although Jones recorded no indication of a possible shallow-water origin for these limestones, it is probable that both the Ras Koh geanticline and the Dalbandin Trough to the north were established structural features by this time (Fig. 3b).

Palaeocene

During the Palaeocene, sediments several kilometres thick and largely free from volcanics accumulated in the Dalbandin Trough and Saindak–Alam Reg basin (Figs 2 & 3b). The Juzzak Formation in the west comprises interbedded mudstones and turbiditic, volcaniclastic sandstones with some thin laminated limestones. It includes lenses of conglomerate with limestone debris including olistoliths (Fig. 5), and volcanic boulders. Andesite lavas occur locally within the sequence in the Mirjawa range, and tuff breccias mark the top of the formation near Alam Reg. The broadly equivalent Rakhshani Formation of the Dalbandin area is similar to the Juzzak in its lower part, but the presence in the upper part of the partly reefal Chapper Limestone, the oncolitic Gat-i-Hamun Limestone, and cross-bedded sandstones indicates that shallow-water conditions were temporarily established. Limestone conglomerate lenses and olistoliths occur north of Yak Mach, whereas coarse conglomerates including debris of intrusives comparable to those of the Chagai Hills, limestones of 'Humai' type, and quartzites of more distant derivation occur within the Palaeocene sediments north of Dalbandin. The conglomerates within these trough sediments point to the continuing uplift and erosion of the Chagai Hills geanticline throughout the Palaeocene. The quartzites may have been transported from far to the north, around the eastern end of the geanticline.

The plutonism in the Chagai Hills geanticline, which was associated with the active volcanism of the Late Cretaceous, recurred or perhaps continued well into the Palaeogene. Sediments of the Juzzak Formation on the north-west flank of the geanticline are themselves intruded by diorite, followed in places by grandodiorite and granite. Palaeogene plutonism occurred also in the Ras Koh geanticline. There, according to Jones (1960), syenite and diorite intrusions cut the Kuchakki volcanics and the Rakhshani Formation, and were thus dated as post-Palaeocene. Evidence for a post-Middle Eocene age for the intrusions as claimed by Bakr (1958) was not confirmed by Jones (1960), who found no indication that the intrusives cut Eocene sediments. The Ras Koh Intrusions are clearly defined as positive anomalies on the regional aeromagnetic map (A. Farah *et al.*, in preparation). A zone of similar anomalies extending WSW from western Ras Koh towards Hamun-i-Mashkhel may mark an extension of the intrusive belt and perhaps also the Ras Koh geanticline under thin upper Cenozoic cover in this tract (Fig. 3b).

Palaeocene sediments were shown by Bakr & Jackson (1964) to comprise a discrete structural unit which forms the western part of the Ras Koh range and is here termed the Ras Koh flysch belt (Fig. 1). In the same area, Jones (1960) had previously mapped outcrops of the Rakhshani Formation and Pishi Group, considered to be similar in part to the Oligocene flysch of northern Makran. The Ras Koh flysch belt has an arcuate strike and comprises a cleaved and highly folded and thrust succession (Ahmed 1961; Bakr 1963, 1965).

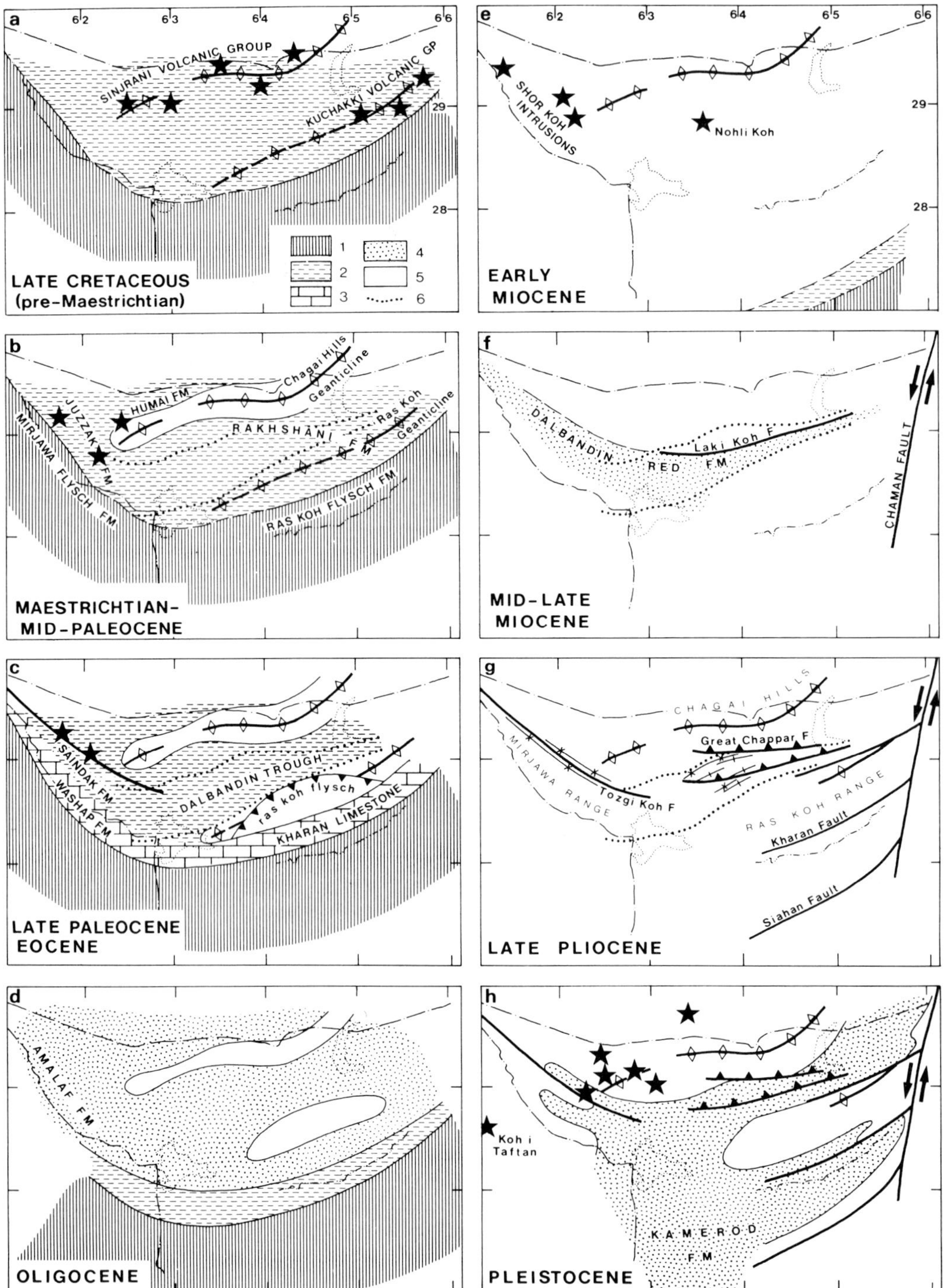

FIG. 3. Palaeogeographical history of study area from Late Cretaceous to Pleistocene times. (1) Flysch sediments; (2) marine sediments/volcanics of forearc; (3) shallow-water limestones; (4) coastal plain/continental sediments; (5) undifferentiated and erosional areas; (6) limits of Dalbandin Trough. Stars indicate volcanoes.

The Ras Koh flysch belt includes the Bunap Intrusions, bedding-parallel lenses up to some 15 m wide and about 1.5 km long of sheared ultrabasic rocks, including pyroxenite and peridotite, and commonly containing chromite (Ahmed 1956; Jones 1960). Jones took the view that the host strata were newly deposited when 'invaded' by these ultrabasics. In the context of the proposed arc-trench model (Farhoudi & Karig 1977), the ultrabasic slices may be considered to have been emplaced tectonically into the flysch deposits of a subduction complex. The host sediments were classified as Rakhshani Formation by Jones (1960), but these differ both in lithology and in tectonic style from the Rakhshani of eastern Ras Koh and the Dalbandin Trough. In view of the confusion over the classification of the western Ras Koh sediments, they are, as a whole, referred to here as the 'Ras Koh Flysch Formation'.

A similar deformed flysch facies of Maestrichtian–Palaeocene age forms much of the south-western part of the Mirjawa range (Fig. 4). It comprises mudstones with an axial-plane cleavage and turbiditic, volcaniclastic sandstones. There are rare lenses of conglomerate including limestone and volcanic debris, which must have been transported across contemporaneous Juzzak sediments from a source in the Chagai Hills geanticline. Unlike the Ras Koh Flysch, there is no record of the occurrence of ultrabasic intrusions. Jones (1960) included these sediments in the Juzzak Formation. We propose that the new term 'Mirjawa Flysch Formation' be applied to this deformed flysch facies. Like the Ras Koh Flysch Formation, the Mirjawa Flysch is believed to owe its deformation largely to the penecontemporaneous incorporation of its sediments into an accretionary fold belt. The precise sedimentary and structural relationships of the Mirjawa Flysch and the relatively undeformed Palaeocene sediments of the Saindak–Alam Reg basin (Juzzak Formation) are unclear (Fig. 2).

The Mirjawa and Ras Koh Flysch Formations are probably of the same general age and the oldest exposed accretionary sediments of the Makran arc-trench system. A major tectonic phase, perhaps in part associated with further accretion to the south, affected both of these flysch formations late in the Palaeocene. In the Mirjawa range, the Mirjawa Flysch was uplifted

FIG. 4. Cleaved and highly folded flysch sediments of the Mirjawa Flysch Formation (Maestrichtian–Palaeocene) between Mirjawa and Saindak (see Fig. 1 for locations). View to north.

FIG. 5. Open-folded mudstones and turbidite sandstones of the Juzzak Formation (Palaeocene) near Shor Nalla. View to south-east from olistolith of probable Maestrichtian limestone.

to form a structural high which defined the south-western limit of the adjoining (Saindak–Alam Reg) forearc basin (Table 1).

By contrast, in the Ras Koh range, deformation of the Ras Koh Flysch against the Ras Koh geanticline produced a complex thrust relationship (Figs 1, 2 & 3c). The Ras Koh flysch belt with its arcuate fold traces and included ultrabasic slices, was emplaced against the rocks of the geanticline, and the combined structures formed a new landmass along the length of the Ras Koh range (see below).

Eocene

The upper age limit for the emplacement of the Ras Koh flysch belt is fixed by the existence of relatively undeformed early Eocene shallow-water limestones (Kharan Limestone) overlying the deformed flysch of southern Ras Koh (Fig. 3). Comparable limestones, the Washap and Robat formations, were mapped by Jones (1960) in the Mirjawa range, while in eastern Iran shallow-water limestones of late Palaeocene to early-Middle Eocene age overlap mélange with pronounced unconformity (Stöcklin *et al.* 1972).

The northwards thinning of the Kharan Limestone was interpreted by Jones (1960) as indicating an Eocene landmass along the Ras Koh. By contrast, the Washap limestones appear to have formed as a shoal accumulation on the Mirjawa structural high described above (Fig. 3c). This structural high separated the Saindak–Alam Reg forearc basin from a deep flysch basin which extended into eastern Iran, there termed the 'Eastern Flysch Trough' by Stöcklin (1974). The existence of a further Eocene structural high, within the forearc basin, is indicated by a narrow zone of shallow-water limestones within the largely turbiditic Saindak Formation, flanking the Tozgi Koh Fault (Table 1, Figs 2 & 3). Carbonate sediment, as well as lithified Eocene limestone debris including boulders and olistoliths, was introduced by gravity slide and turbidity flow to the forearc basin both from the Tozgi Koh high and probably also from the Mirjawa high. It seems likely that lithified carbonate debris from the Mirjawa structural high also spilled off into the flysch trench to the south-west, there becoming deformed in mélange. Blocks of similar limestone of Lutetian (Middle Eocene) age are known in the mélange of eastern and south eastern Iran (Gansser 1960).

Two other features of the Saindak Formation are of interest in connection with the possible Eocene activity of the Tozgi Koh Fault, implied by the existence of its flanking structural high. The first is the abrupt thickness variation within the formation adjacent to the fault; at Shor Nalla (Fig. 1) the formation thickens along strike from 900 to 3000 m over a distance of 5 km. The second is the volcanic component of the formation: partly tuffs and thin andesite flows, but, more particularly, volcanic conglomerates up to several hundred metres thick, and well seen both at Saindak and at Shor Nalla. The volcanic conglomerates are locally mixed with limestone blocks and boulders, and, like the components of the limestone debris layers, appear to have been transported by gravity sliding from structural highs.

Taken together, the features of the Saindak Formation may be interpreted as indicating Eocene instability on the line of the Tozgi Koh Fault (Fig. 3), perhaps arising from contemporaneous transform movement (see Reading 1980). The incidence of associated volcanics could point to a tensional element in such a transform system.

Oligocene

The Eocene sediments described above are the youngest within the study area to have been correlated palaeontologically. The approximate dating of the succeeding Cenozoic sediments has been achieved partly by the correlation of tectonic events (Table 1) which have been dated elsewhere, and partly by a regional comparison of lithofacies. Only one radiometric date is available, for an igneous event at Saindak in the early Miocene (Sillitoe & Khan 1977). The chronostratigraphy of the younger part of the Cenozoic is therefore speculative.

The Amalaf Formation of the Mirjawa range was tentatively assigned an Oligocene age (Jones 1960) on the basis of faunal identification by Vredenburg (1901). It marks a widespread shallowing of the forearc basin in the Saindak area (Fig. 3d). The red and green, cross-bedded sandstones and mudstones were interpreted by Jones (1960) as estuarine-fluviatile, marking the end of the marine conditions. This interpretation is supported by the view of Stocklin (1974) that, in late Eocene to Oligocene time, a north-south trending flysch trough in eastern Iran (the Eastern Flysch Trough) closed up, thus integrating the Iranian and Afghan microplates. Powell (1979) has suggested that this convergence may have been caused by the counter-clockwise rotation of the Indian plate while in contact with the Afghan microcontinent, an event which he dates as late Eocene (40 Ma).

Miocene

In the Makran, the incorporation of contemporaneous flysch into the accretionary fold belt continued during the Oligocene and Miocene, resulting in a continuing southerly migration of the trench. According to Ahmed (1969) the Miocene shoreline was probably in the north Makran. In the present study area, however, a substantial break in the sedimentary record occurs between the Oligocene Amalaf Formation and the Dalbandin Red Formation of supposed mid-late Miocene age (see below).

During this hiatus two important events are recorded in the west. One was the regional uplift of the forearc and broad folding of its sediments. The other was a volcanic episode marked now by the intrusive roots (Shor Koh Intrusions) of andesitic volcanic centres penetrating Palaeogene forearc sediments, mostly in the Saindak–Shor Nalla area close to the Tozgi Koh Fault, but including one in the Dalbandin Trough at Nohli Koh (Figs 1 & 3e) immediately N of the Laki Koh Fault. Intrusive rocks at one of the centres, near Saindak, have been dated at 18–22 Ma (early Miocene) by Sillitoe & Khan (1977). The intrusion of the andesitic Tanki Sill of the Mirjawa range belongs to this phase of volcanism. This sill is notable both for its wide distribution, and for its lateral penetration from the open-folded forearc sediments of the Juzzak Formation to the more intensely deformed and the cleaved Mirjawa Flysch. This relationship confirms the juxtaposition of the Mirjawa Flysch and the Juzzak Formation in early Miocene times.

Following the main phase of Oligo-Miocene erosion of the forearc, a sedimentary basin was re-established to the north of the Mirjawa and Ras Koh ranges. Continental sediments, for which the new name 'Dalbandin Red Formation' is proposed, accumulated in the Dalbandin trough and Saindak–Alam Reg basin (Fig. 3f). In the type section, 40 km ENE of Dalbandin, red conglomerates pass up through cross-bedded, probably fluviatile, red and grey-green sandstones to red and green mudstones. The mudstones carry thin dolomitic layers. They are partly gypsiferous and of probable lacustrine origin. The basal conglomerates in the vicinity of the Laki Koh Fault (Fig. 3f) include much coarse, locally derived material, such as blocks of Palaeocene limestone north of Yak Mach (Fig. 1); this is taken as an indication of contemporaneous activity on the fault. Conglomerates higher in the sequence contain well rounded, far-travelled clasts, and include a high proportion of quartzites. The sequence is at least 3000 m thick against the Laki Koh Fault in the Dalbandin area (Figs 2 & 3f), and it is probable that contemporaneous activation of this and similar faults controlled the accumulation of the formation.

The extent of this continental 'red bed' sedimentation is poorly known. The most westerly occurrence of the Dalbandin Red Formation is in the Amalaf syncline at Shor Nalla (Fig. 1), where it rests unconformably on the Saindak Formation and on Shor Koh Intrusions. The age range of these red beds is speculative. Jones (1960) included them in the Kamerod Formation of Plio-Pleistocene age (see below). They were, however, clearly involved in the major folding and faulting episode ascribed to the main Himalayan Orogeny, dated as late Pliocene–early Pleistocene (Jones 1960; Auden 1974). On lithological grounds, it seems probable that the Dalbandin Red Formation is the counterpart of the Mio-Pliocene Upper Red Formation of Iran (Stöcklin 1971). Similar red beds on the eastern marginal parts of the Lut block in east Iran have been assigned a probable Mio-Pliocene age (Stöcklin *et al.* 1972). A mid-late Miocene age for the Dalbandin Red Formation seems likely, in view of the early Miocene date for the Saindak intrusion, which pre-dates the red beds.

Pliocene

In the study area, Pliocene deformation occurred in two main phases. These were separated by a period during which there was substantial erosion, followed, and perhaps locally accompanied, by the further accumulation of conglomerate. In the Shor Nalla area of the Mirjawa range, the earlier phase caused reverse faulting and related folding about NNW-to-WNW trending axes, parallel to the strike of the Mirjawa flysch belt (Fig. 3g). This earlier phase involved strata up to and including the Dalbandin Red Formation. In the same area, the succeeding conglomerates, which were also included by Jones (1960) in the Kamerod Formation (see below) and for which the name Dalil Conglomerate Formation is proposed, rest with marked angular unconformity on older Cenozoic formations. These conglomerates have an extensive outcrop immediately north of Koh-i-Dalil railway station (35 km west of Alam Reg); they include mostly subangular debris, but also scattered well-rounded pebbles more typical of the conglomerates of the Dalbandin Red Formation. The Dalil conglomerates are themselves folded, though considerably less intensely than the Dalbandin Red Formation nearby. Folded conglomerates believed to be of similar age, and for which the name Yak Mach Conglomerate Formation is proposed, are seen in the Dalbandin Trough, in an anticline immediately NE of Yak Mach. The stratigraphic relationship of the Yak Mach conglomerates to the adjoining Dalbandin Red Formation is not clear.

The sediments of the eastern part of the Dalbandin Trough suffered particularly as a result of the two Pliocene tectonic phases. There are two manifestations of the deformation (Figs 2 & 3g). One is folding with axial traces which swing from SW–NE west of Yak Mach to E–W north of Dalbandin. The other is major reverse faulting and associated folding along arcuate fractures, convex to the south. The Great Chappar and Laki Koh faults are the most important fractures for which there is clear field evidence (Fig. 3). As suggested by Jones (1960), and later by Lawrence *et al.* (1981), a similar reverse fault (Ahmad Wal Fault) may define the present northern boundary of the Ras Koh range (Fig. 2). The aeromagnetic and gravity data (A. Farah *et al.*, in preparation) provide definite support for a major fracture on this line. Along parts of their traces, these faults are low-angle thrusts; for example at Laki Koh, 8 km N of Yak Mach, tightly folded Palaeocene limestone of the Rakhshani Formation rests upon a low-dipping surface of vertical conglomerates of the Dalbandin Red Formation. The alignment of the axial traces of the folds in the Dalbandin trough reflects the general alignment of the trough itself (Fig. 3g), confined between the Chagai Hills geanticline and the Ras Koh massif. To the west of Dalbandin, the axial traces become markedly discordant to the trace of the Laki Koh Fault. Both the folding and the faulting are indicative of horizontal shortening between the geanticlines. The extent to which this compression may have been accompanied by transcurrent movement on the faults is unclear. It is arguable that the discordance of the fold axial traces with the Laki Koh fault in the Dalbandin area is best explained in terms of a dextral shear system. However, NNW-trending minor folds which affect Dalil conglomerates immediately S of the Tozgi Koh fault, 6 km S of Alam Reg (Fig. 1), may be taken as an indication of sinistral displacement across the fault. The case for major sinistral movement in this part of the Makran convergence zone (Lawrence *et al.* 1981) remains speculative.

The sediments of the succeeding Kamerod Formation (see below), regarded by Jones (1960) as Plio-Pleistocene, overlie the Yak Mach and Dalil conglomerates with angular unconformity. Their deposition ended a period of widespread erosion on the forearc which must have both accompanied and followed the later of the Pliocene tectonic episodes.

Pleistocene

The Kamerod sediments are buff silts of lacustrine and perhaps loessic origins, capped by fan gravels composed mostly of subangular volcanic debris (Fig. 3h). They are well-preserved north of the Laki Koh Fault where they form terraced surfaces at a number of levels up to about 50 m above the present alluvial pavement. In the west near Nok Kundi (Fig. 1), the formation is seen in the terraces south of the Tozgi Koh Fault (Fig. 3g), and is preserved as isolated patches north of that fault under conformable flows of andesite belonging to the Koh-i-Sultan Volcanic Group (Table 1, Fig. 1).

There has been confusion over the application of the term Kamerod Formation to the study area. Jones (1960) included not only sediments fitting the above Kamerod description, but also the strongly folded (Dalbandin) red beds of Shor Nalla, and part of the strongly deformed 'Dalbandin Assemblage' of the Dalbandin trough. Arthurton *et al.* (1979) followed Jones by classifying what is here termed the Dalil Conglomerate Formation as Kamerod, but applied the term 'Quaternary Older Alluvium' to the undeformed silts and gravels. These classifications we now consider to be erroneous, not only because the Dalbandin red beds show little resemblance to those of the Kamerod type area, but also because they and the Dalil Conglomerate clearly pre-date the later Pliocene tectonic phase.

The Koh-i-Sultan volcanics were erupted from a number of vents in the north-western part of the study area (Fig. 3h). They form part of a Quaternary volcanic arc extending west-south-westwards to south-central Iran (Gansser 1971). The principal eruptive centres are Kuh-i-Taftan and Bazman in Iran, and Koh-i-Sultan (Fig. 1). Besides these main centres, there are, in the study area, many smaller vents. Their distribution broadly coincides with the western part of the Chagai Hills geanticline. No vents are known in the Mirjawa range, though there is a cluster just north of the Tozgi Koh Fault, near Alam Reg (Fig. 1).

Further movement on the major faults within the forearc zone occurred during the Pleistocene. Kamerod sediments have been displaced across the Tozgi Koh Fault at Alam Reg, and across the Laki Koh Fault north of Yak Mach. Near Yak Mach railway station, the Kamerod sediments have been gently warped about a pre-existing WSW-trending axis. During the later part of the Pleistocene, the Kamerod sediments were deeply dissected and, over wide areas, completely eroded. Travertine deposits, post-dating but related to the Koh-i-Sultan volcanism (Sillitoe 1978), occur in a number of places on the Chagai Hills geanticline, mostly around the western end of the structure, north of Alam Reg.

Discussion

The oldest exposed parts of the Makran subduction complex in Pakistan are of Maestrichtian–Palaeocene age, in the Mirjawa and Ras Koh Flysch belts. The flysch deposits, which are highly deformed, and associated locally with ophiolitic slices, are juxtaposed against thick, broadly folded piles of probable Senonian volcanics, intruded by plutons, in the Chagai Hills and Ras Koh geanticlines (Figs 1 & 2). The junction between the flysch belts and volcanics is taken as defining the southern limit of the Dasht-i-Margo block in Maestrichtian–Palaeocene times (Fig. 3b). It is possible that older parts of the subduction complex underlie the volcanics; if so, the major faults of this part of the forearc, such as the Great Chappar and Laki Koh faults (Fig. 3), may be inherited from fractures dating from earlier accretionary deformation of the subduction complex.

The extent to which the Late Cretaceous–Recent magmatic and tectonic events in the study area were dictated by external geodynamic events is open to debate (Table 1). The collision of the northward-moving Indian subcontinent against the southern margin of Eurasia was dated as Late Cretaceous–early Palaeogene by Powell & Conaghan (1973), though initial contact in Pakistan was placed as late as early Eocene by Stoneley (1974). The Ras Koh flysch belt, which flanked the continental lithosphere of Dasht-i-Margo to the south-east, may have suffered severely as a result of the collision. The late Palaeocene thrusting of the Ras Koh flysch against the volcanics of the Ras Koh geanticline may have been caused by this collision (Figs 1, 2 & 3c). The subsequent, more stable conditions in which the early Eocene Kharan Limestone accumulated may reflect the late Palaeocene slowing or stopping of seafloor spreading in the Indian Ocean (McKenzie & Sclater 1971).

Sedimentation in the forearc basins of Saindak–Alam Reg and the Dalbandin Trough appears to have been influenced from Eocene times, at least, by movement on major faults within the forearc. Both the Eocene Saindak Formation and the Miocene Dalbandin Red Formation accumulated under such influence (Fig. 3), and evidence is cited herein favouring tensional transcurrent movement controlling the accumulation of the Saindak Formation.

The possibility of major transcurrent movement having occurred on the forearc faults has been raised by Sillitoe (1978), who suggested that the Ras Koh has been transported eastwards to its present position by sinistral transcurrent movement. Lawrence *et al.* (1981) support this idea, and favour a Pliocene age for the event, which they relate to the collision of the Arabian and central Iranian continental slabs (Table 1). Lawrence *et al.* (1981) have demonstrated a history of sinistral transcurrent movement on faults further south in the subduction complex. Sillitoe (1978) has claimed sinistral displacement of some 450 km on the Chaman Fault since early Miocene times at least; the Chaman Fault forms the eastern boundary of both the Dasht-i-Margo block and the Makran subduction zone (Fig. 1).

Counter to the proposal for eastward translation of the Ras Koh during the Pliocene, the supposed Pliocene structures in the Dalbandin Trough indicate dextral, rather than sinistral, displacement across the Laki Koh Fault (Fig. 3). However, sinistral displacement across the Tozgi Koh Fault is indicated by minor folds of NNW trend near Alam Reg (Fig. 1).

Whatever the pros and cons of sinistral, as opposed to dextral, transcurrent movement on the forearc faults, it is clear that these fractures have had a long history of activity up to recent times. Their regular, arcuate trends (Figs 1 & 3) suggest that the faults at outcrop are the superficial manifestation of major faults within the underlying northern part of the subduction complex, or its transition to the continental lithosphere of the Dasht-i-Margo block. The repeated activation of these faults highlights a fundamental structural weakness of even the oldest part of the Makran accretionary prism.

ACKNOWLEDGMENTS: Much of the new information contained in this paper is the result of fieldwork by officers of the Geological Survey of Pakistan, whose invaluable contribution to this paper is gratefully acknowledged. One of us (R.S.A.) is indebted to R. G. Davies and K. A. de Jong for an introduction to the study area. The work was partly supported by the U.S. National Science Foundation under Grant INT-76-22304, and has arisen in part from the engagement of one of us (R.S.A.) by the Ministry of Overseas Development, U.K. The authors acknowledge the major contribution in fieldwork by W. J. Barclay, and are grateful to A. J. Reedman for discussion and critical reading of the manuscript. This paper is published by permission of the Director, Institute of Geological Sciences, U.K.

References

AHMED, S. S. 1969. Tertiary geology of part of south Makran, Baluchistan, West Pakistan. *Bull. Am. Assoc. Petrol. Geol.* **53,** 1480–99.

AHMED, W. 1956. Geological report of Kambran-Rayo and Nag-Parshad-Charkohan Kullan, western and eastern Raskoh. *Geol. Surv. Pakistan (unpubl. report).*

—— 1961. Geology of the eastern Ras Koh range, Chagai and Kharan districts. *Geol. Surv. Pakistan (unpubl. report).*

ARTHURTON, R. S., SARWAR ALAM, G., ANISUDDIN-AHMAD, S. & SAEED IQBAL, 1979. Geological history of the Alamreg-Mashki Chah area, Chagai District, Baluchistan. *In*: FARAH, A. & DE JONG, K. A. (eds). *Geodynamics of Pakistan*, 325–31. GEOL. SURV. PAKISTAN, QUETTA.

AUDEN, J. B. 1974. Afghanistan–West Pakistan. *In*: SPENCER, A. M. (ed.). Mesozoic–Cenozoic Orogenic belts. Data for Orogenic studies. *Spec. Publ. geol. Soc. London*, **4,** 235–53.

BAKR, M. A. 1958. Geology of part of western Ras Koh Range, Chagai District, Kalat Division. *Pak. J. Sci.* **10,** 227–36.

—— 1963. Geology of the western Ras Koh Range, Chagai and Kharan Districts, Quetta and Kalat Divisions, West Pakistan. *Rec. geol. Surv. Pak.* **10,** 28 pp.

—— 1965. Structure of the Western Ras Koh Range, Chagai and Kharan Districts, Quetta and Kalat Divisions, West Pakistan. *Pak. J. Sci.* **17,** 1–17.

—— & JACKSON, R. O. 1964. *Geological map of Pakistan 1:2,000,000.* GEOL. SURV. PAKISTAN.

DEWEY, J. F., PITMAN, W. C. III, RYAN, W. B. F. & BONNIN, J. 1973. Plate tectonics and the evolution of the alpine system. *Bull. geol. Soc. Am.* **84,** 3137–80.

DICKINSON, W. R. & SEELY, D. R. 1979. Structure and stratigraphy of forearc regions. *Bull. Am. Assoc. Petrol. Geol.* **63,** 2–31.

FARHOUDI, G. & KARIG, D. E. 1977. Makran of Iran and Pakistan as an active arc system. *Geology*, **5,** 664–8.

GANSSER, A. 1960. Ausseralpine Ophiolithprobleme. *Eclog. geol. Helv.* **52,** 659–80.

—— 1971. The Taftan volcano (SE Iran). *Eclog. geol. Helv.* **64,** 319–34.

JACOB, K. H. & QUITTMEYER, R. L. 1979. The Makran region of Pakistan and Iran: trench-arc system with active plate subduction. *In:* FARAH, A. & DEJONG, K. A. (eds). *Geodynamics of Pakistan*, 303–17. Geol. Surv. Pakistan, Quetta.

JONES, A. G. (ed.) 1960. *Reconnaissance geology of part of West Pakistan: a Colombo Plan Co-operative Project.* Report by Govt of Canada for Govt of Pakistan, Toronto. 550 pp.

LAWRENCE, R. D., KHAN, S. H., DEJONG, K. A., FARAH, A & YEATS, R. S. 1981. Thrust and strike slip fault interaction along the Chaman transform zone, Pakistan. *In:* MCCLAY, K. R. & PRICE, N. J. (eds). *Thrust and Nappe Tectonics*, 363–70. Geol. Soc. London. Spec. Publ. 9.

MCKENZIE, D. & SCLATER, J. G. 1971. The evolution of the Indian Ocean since the late Cretaceous. *Geophys. J. R. astron. Soc.* **24,** 437–528.

POWELL, C. MCA. 1979. A speculative tectonic history of Pakistan and surroundings: some constraints from the Indian Ocean. *In*: FARAH, A. & DEJONG, K. A. (eds). *Geodynamics of Pakistan*, 5–24. Geol. Surv. Pakistan, Quetta.

—— & CONAGHAN, P. J. 1973. Plate tectonics and the Himalayas. *Earth planet. Sci. Lett.* **20,** 1–12.

READING, H. G. 1980. Characteristics and recognition of strike-slip fault systems. *Spec. Publ. Int. Assoc. Sedimentol.* **4,** 7–26.

SEELY, D. R., VAIL, P. R. & WALTON, G. G. 1974. Trench-slope model. *In*: BURK, C. A. & DRAKE, C. L. (eds). *The Geology of Continental Margins*, 249–60. Springer-Verlag, New York.

SILLITOE, R. H. 1978. Metallogenic evolution of a collisional mountain belt in Pakistan: a preliminary analysis. *J. geol. Soc. London*, **135,** 377–87.

—— & KHAN, S. N. 1977. Geology of the Saindak porphyry copper deposit, Pakistan. *Trans. Instn Ming Metall.* **86,** B27–42.

STÖCKLIN, J. 1971. Stratigraphic lexicon of Iran, Part 1: Central, North and East Iran. *Rep. geol. Surv. Iran*, **18,** 338 pp.

—— 1974. Possible ancient continental margins in Iran. *In*: BURK, C. A. & DRAKE, C. L. (eds). *The Geology of Continental Margins*, 873–87. Springer-Verlag, New York.

—— , EFTEKHAR-NEZHAD, J. & HUSHMAND-ZADEH, A. 1965. Geology of the Shotori Range (Tabas area, East Iran). *Rep. geol. Surv. Iran*, **3,** 69 pp.

—— , —— & —— 1972. Central Lut reconnaissance, East Iran. *Rep. geol. Surv. Iran*, **22,** 62 pp.

STONELEY, R. 1974. Evolution of the continental margins bounding a former southern Tethys. *In*: BURK, C. A. & DRAKE, C. L. (eds). *The Geology of Continental Margins*, 889–903. Springer-Verlag, New York.

VREDENBURG, E. W. 1901. A geological sketch of the Baluchistan desert and part of eastern Persia. *Mem. geol. Surv. India*, **31,** 179–302.

WHITE, R. S. 1979. Deformation of the Makran continental margin. *In*: FARAH, A. & DEJONG, K. A. (eds). *Geodynamics of Pakistan*, 295–304. Geol. Surv. Pakistan, Quetta.

—— 1981. Deformation of the Makran accretionary sediment prism in the Gulf of Oman (north-west Indian Ocean) (this volume).

RUSSELL S. ARTHURTON, Institute of Geological Sciences, Ring Road, Halton, Leeds LS15 8TQ, England.

ABUL FARAH, Geological Survey of Pakistan, P.O. Box 15, Quetta, Pakistan.

WAHIDUDDIN AHMED, Geological Survey of Pakistan, P.E.C.H. Society, Karachi, Pakistan.

The Makran, Southeastern Iran: the anatomy of a convergent plate margin active from Cretaceous to Present

G. J. H. McCall & R. G. W. Kidd

SUMMARY: The inland geology of the Iranian Makran, long neglected by geologists because of its lack of hydrocarbon potential, is now reasonably well known following a regional mapping programme carried out on behalf of the Geological and Mineral Survey of Iran. The mountain range can be divided into seven geotectonic provinces forming an arc round the late Pliocene epeirogenic Jaz Murian Depression at the southern end of the Lut block. From the edge of the Jaz Murian outwards to the S these provinces are: (1) a marginal basin in which well-preserved ophiolites formed and deep-water pelagic sediments were deposited from Jurassic to Palaeocene; (2) a narrow zone of continental crust of Palaeozoic metamorphics capped by shelf limestones of mainly Cretaceous age but including Carboniferous, Permian, Jurassic and Palaeocene developments; (3) a zone of ophiolitic mélange, the renowned Coloured Mélange; (4) a zone of immensely thick Eocene-Oligocene flysch; (5) a similar zone of Oligocene-Miocene flysch; (6) a southern zone of Miocene neritic to molassic sediments and (7) a Miocene-early Pliocene neritic zone W of the Zendan fault. The structure is dominated by steep inward-dipping reverse faults forming schuppen. These developed in a series of events which climaxed in the late middle Miocene with some faults continuing to be active to the present day with continuing uplift. Intense periclinal folds of dominantly chevron style are developed in the flysch, synchronous with the faulting. Over substantial areas in the flysch deformation has been so intense that a mélange has developed, resembling some Franciscan mélanges more closely than the Coloured Mélange. It contains exotic blocks which are tectonically intruded from the Coloured Mélange, which forms the basement to the flysch zones.

The most significant discovery of this programme is the continuation into the Makran of the Sanandaj-Sirjan zone in the form of the Bajgan–Dur-Kan complexes (province 2 above), thus forming a sliver of continental crust that stretches from the Bitlis Massif in Turkey to the Makran. This separates the two ophiolitic developments of the Makran: to the south the Coloured Mélange, the trench margin sequence of a north-dipping subduction zone formed mainly in Late Cretaceous-Palaeocene time, and to the north a belt of well-preserved Cretaceous-Palaeocene ophiolites formed in a marginal basin. The latter ophiolites may be the equivalent of the island arc of this subduction system. The Makran was not involved with the third group of ophiolites in this region, those of the Oman, Neyriz and Kermanshah, which were emplaced as a result of collision of the Arabian continental margin with an intra-oceanic NE-dipping subduction zone in the Campanian. Any association of these ophiolites with the Coloured Mélange along the Zagros is due to the Pliocene collision of Arabia and Iran.

Following uplift of the Inner Makran in the late Palaeocene northward subduction has continued to the present day with related andesitic volcanism to the north and migration of the trench to the south following tectonic events in the Oligocene and middle Miocene, so that the trench is now 300 km from the present andesitic volcanism. The products of Eocene to Present subduction, including the immense deformed flysch deposits, are thus superimposed on the condensed Cretaceous-Palaeocene subduction system.

Introduction

The Iranian Makran comprises the 150 km wide area of SE Iran between the Jaz Murian Depression and the coast (Fig. 1). The east-west trending mountain range rises from the coast to around 6000 feet and in a few places to over 8000 feet before dropping down again into the Jaz Murian. Within Iran it is 600 km long from the coast at Minab in the west to the Pakistan border in the east.

Apart from the excellent pioneer work of Harrison & Falcon (1936) and Harrison *et al.* (1935–36) and some unpublished oil company studies, the geology of the Makran was long neglected because of its lack of hydrocarbon potential. However, most of the Iranian Makran mountain range has now been mapped by Paragon-Contech (an Iranian-Australian joint venture) on contract for the Geological and Mineral Survey of Iran. The mapping covered four full and two half 1:250,000 quadrangles (1°

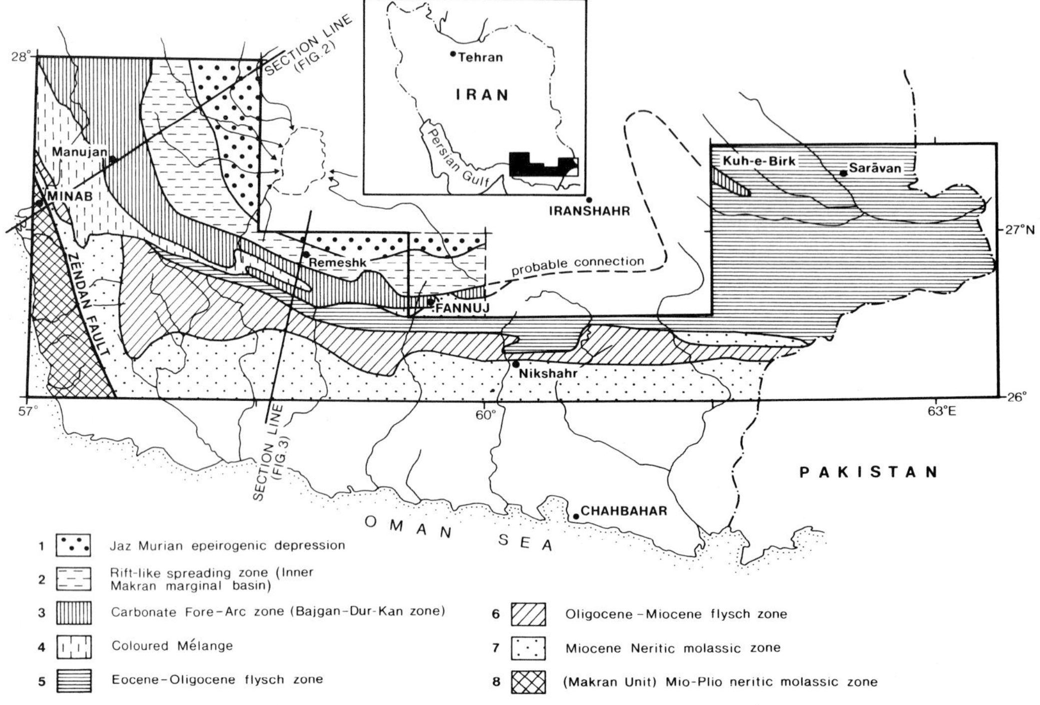

FIG. 1. Sketch map of the Iranian Makran showing the eight geotectonic provinces and the lines of the sketch sections (Figs 2 & 3).

by $1\frac{1}{2}°$). In addition, nine maps ($\frac{1}{2}°$ by $\frac{1}{2}°$) of special interest were mapped on a 1:100,000 scale and a 1:500,000 tectonic map accompanied the overall report. In all an area of over 80,000 km^2 (about the size of Scotland) was mapped in 2 years using helicopters and more than 30 geologists. The maps and reports with full details of the geology will be published shortly by the Geological and Mineral Survey of Iran (McCall 1981).

This paper is intended to relate the broad results of this programme to the global tectonics of this area. The mapping was a group project but the conclusions in this paper are those of the two authors alone. G. J. H. McCall was senior consultant to this project and compiled the reports (McCall 1981) and R. G. W. Kidd contributed to it in the second field season, and wrote the section on the plate tectonics and evolution of the region in the final report.

Geology of the Iranian Makran

There are eight geotectonic provinces (including the Jaz Murian) in the Iranian Makran (McCall 1981)—these are shown in Fig. 1. Sketch sections across the Makran mountain range are shown in Fig. 2 (Minab sector—NW) and Fig. 3 (Fannuj sector—centre). The lines of these sections are on Fig. 1. The structure is dominated by steep inward dipping (toward the Jaz Murian) faults forming schuppen. Thus the relationships between the geotectonic provinces (apart from the Jaz Murian) are largely tectonic. From the Inner Makran out towards the coast these provinces are:

(1) Jaz Murian Depression

This is a late Pliocene epeirogenic depression 300 km in length in an east-west direction and over 100 km across north-south. The only rocks are superficial: alluvial fans of gravel, silt plains, playa lakes, dunes and salt pans.

(2) Inner Makran spreading zone

This is a zone of rifting occupied by largely undeformed ophiolites (as defined by the 1972 Penrose Conference). There are three distinct ophiolites. One is tholeiitic and resembles the Troodos Complex, Cyprus. It is Early Cretaceous to early Palaeocene in age and consists

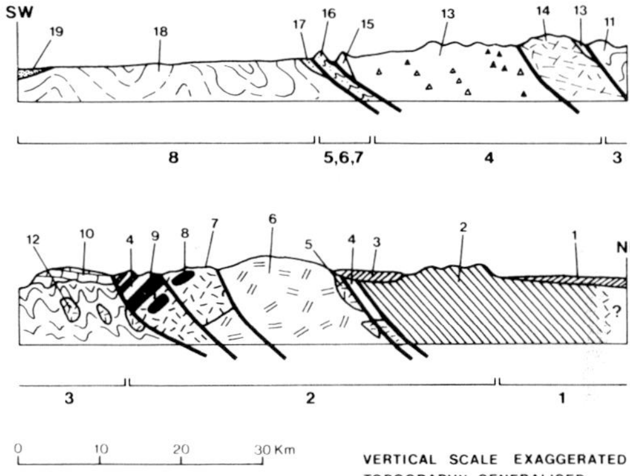

FIG. 2. Sketch section across the Makran in the Minab sector along the line shown in Fig. 1. Geotectonic province numbers are shown in heavy type beneath the section. Other numbers indicate geology as follows: (1) gravel and silt fans of the Jaz Murian depression; (2) Lower-Upper Cretaceous calc-alkaline ophiolite: mainly sheeted dykes of intermediate to acid composition, screens of pillow lavas, pelagic sediments, gabbro and trondhjemite; (3) lower-upper Eocene proximal turbidites, acid welded tuffs, diabase and diorite sills and shelly limestone, all overlying ophiolite sequence; (4) Lower Cretaceous-lower Palaeocene pillow lavas of outer tholeiitic ophiolite with pelagic limestones; (5) serpentinite; (6) low-level cumulate gabbro of outer ophiolite; (7) high-level non-cumulate gabbro; (8) trondhjemite at top of 7; (9) sheeted diabase dykes; (10) Dur-Kan Complex: mainly Cretaceous but including Permian and Jurassic shelf carbonates structurally overlying 11; (11) Bajgan Complex: greenschist and amphibolite facies Palaeozoic metamorphics; (12) tectonic enclaves of ultrabasics with chromite; (13) Coloured Mélange: Lower Cretaceous-lower Palaeocene block-to-block mélange; (14) ultrabasic tectonites, partly layered with chromitite cumulates; (15) Eocene-Oligocene flysch; (16) upper Oligocene-lower Miocene flysch plus some neritic, gypsiferous Miocene sediments; (17) Burdigalian reefal limestone; (18) Makran unit: upper Miocene-lower Pliocene neritic to molassic sediments; (19) coastal mudflats.

of cumulate gabbros overlain by high-level gabbro, trondhjemite, diabase sheeted dykes, pillow lavas with copper shows and pelagic sediments. The other two ophiolites are distinctly calc-alkaline and may represent separate parts of the same ophiolite sequence. One of these, of Early to Late Cretaceous age, has only the upper parts exposed and consists mainly of sheeted dykes which are largely dacitic-rhyolitic and include micro-trondhjemite. It also contains screens of pillow lavas and of gabbro and trondhjemite. The third ophiolite, of Early Cretaceous to early Palaeocene age, consists of fragmented plutonics (ultrabasics, troctolite, cumulate gabbro, high-level gabbro and trondhjemite) folded into immense flexures, and diabase dykes above (some sheeted), grading up into pillow lavas overlain by pelagic sediments.

The dyke trends in these ophiolites are roughly parallel to the strike of the ophiolites as a whole—following the curvature of this Inner Makran spreading zone round from a north-south trend in the NW of the area to east-west near Fannuj. This suggests that these ophiolites were formed in approximately their present positions and have not been tectonically emplaced from elsewhere. Indeed their present width of about 50 km in the NW to as little as 15 km near Remeshk (see Fig. 1) may not necessarily be a great deal less than their original width, as apart from steep bounding reverse faults they are relatively undeformed. Near Fannuj, however, there is a local development of metamorphics including blueschists in a mélange within the Spreading Zone. This includes Cretaceous *Globotruncana* limestone and radiolarites, and metamorphics which may be as old as Palaeozoic.

(3) Bajgan—Dur-Kan zone

This is a narrow but continuous zone of continental crust. In the NW of the area where it is up to 40 km wide (see Figs 1 & 2) it consists of Palaeozoic metamorphics (the Bajgan Complex). Eastward the Bajgan Complex is overlain by a sequence of shelf limestones of the Dur-Kan Complex (see Figs 1 & 3), which are predominantly of Early Cretaceous to early Palaeocene age, but have tectonic inclusions of Carboniferous, Permian and Jurassic shelf limestones. This continental sliver thins eastward and appears to pass into the Kuh-e-Birk range through a sigmoid flexure and thence into Pakistan. To the NW of the mapped area it can be traced up through the Sanandaj-Sirjan zone (see Fig. 4) as far as the Bitlis Massif in Turkey (Stocklin 1977). It has been suggested that it may be an Alpine-type nappe (Gansser pers. comm.) but the structure of this region is one of steep schuppen not of low-angle thrusts and a nappe thousands of kilometres long seems improbable. It must be considered a global geotectonic entity and the continuation of the Sanandaj-Sirjan zone into the Makran is impor-

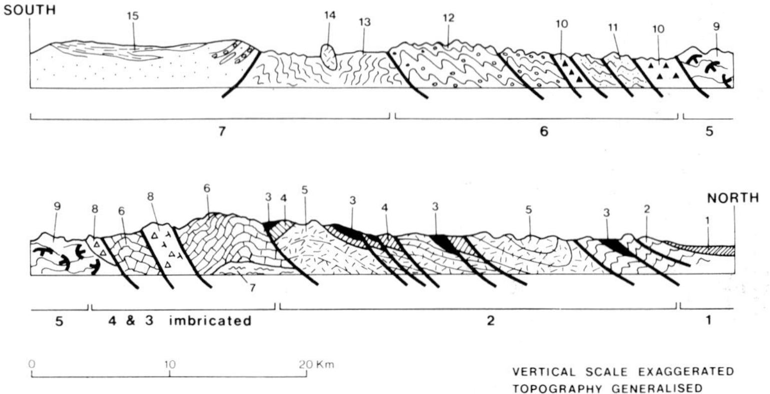

FIG. 3. Sketch section across the Makran in the Fannuj sector along the line shown in Fig. 1. Geotectonic province numbers are shown in heavy type beneath the section. Other numbers indicate geology as follows: (1) superficial gravel fans and dunes of Jaz Murian depression; (2) mainly greenschists and blueschists with Lower-Upper Cretaceous pelagic limestones and radiolarites; (3) pillow lavas and pelagic sediments; (4) diabase dykes, some sheeted; (5) fragmented and folded ophiolitic plutonics: layered cumulates at base (ultrabasics, troctolite, gabbro), non-cumulates above (gabbro, trondhjemite); (6) Dur-Kan complex: mainly Cretaceous, but including Permian and Jurassic, shelf carbonates; (7) Bajgan Complex of Palaeozoic metamorphics structurally overlain by 6; (8) Coloured Mélange; (9) lower-upper Eocene distal flysch and minor pelagic and shelf limestones dislocated throughout with exotic blocks; (10) tectonic dislocation and protrusion mélange, 80% flysch, 20% exotics; (11) lower-middle Miocene proximal flysch, deep water facies; (12) upper Oligocene-lower Miocene mainly distal flysch; (13) lower-middle Miocene neritic sediments, mainly fine gypsiferous mudstones; (14) exotic raft to harzburgite; (15) middle-upper Miocene neritic sediments; conglomerates and sandstone in the north, sandstone, siltstone and mudstone in the south, lateral facies changes to deltaic and estuarine facies with some evaporites.

tant in the consideration of any plate tectonic model for the region.

(4) Coloured Mélange Zone

This is a tectonic, ophiolitic, block-to-block mélange (Gansser 1974) consisting of serpentinite, other ultrabasic and basic ophiolitic rocks, pillow lavas, pelagic limestones, radiolarites and distal turbidites. It also contains minor andesite, rhyolite, rhyolitic welded tuff, trachyte and exotic components (metamorphics and Lower Cretaceous reefal limestone). There is no true matrix: block boundaries are sheared and often serpentinite acts as a 'lubricant'. The mélange is not all chaotic: much of it consists of stacked slabs that are the right way up and consistently dip steeply inward toward the NE. The fossil ages are enigmatic in that the radiolarites are Jurassic-Coniacian while the pelagic limestone (*Globotruncana*) is Cenomanian-Maastrichtian, yet the radiolarites and pelagic limestones sometimes appear to form an interbedded sequence. The youngest rocks are early Palaeocene biomicrites. The pillow lavas and sediments are commonly interbedded.

We suggest that the Coloured Mélange was formed in the trench of a NE- to north-dipping subduction zone by scraping off fragments of the downgoing plate, possibly mixing these fragments with small bits of the overriding plate. Whether there is a contribution to the mélange from the overriding plate depends upon establishing the derivation of the exotic components of the mélange (shallow water limestones and metamorphics). This question has not yet been resolved.

The outcrop of Coloured Mélange is about 30 km wide in the NW of the area but thins rapidly to a few kilometres and pinches out to the south of Remeshk (see Fig. 1), again reappearing in the area not mapped in this project to the south of Iranshahr. However, exotics within the flysch of the sediments of provinces 5, 6 and 7 indicate that the Coloured Mélange extends 70 km of the south of its area of outcrop beneath these sediments (see below).

The age of onset of mélange formation cannot be determined but the abundance of Cenomanian-Maastrichtian rocks suggests that much of it did not start forming a mélange until Maastrichtian time. The involvement of early

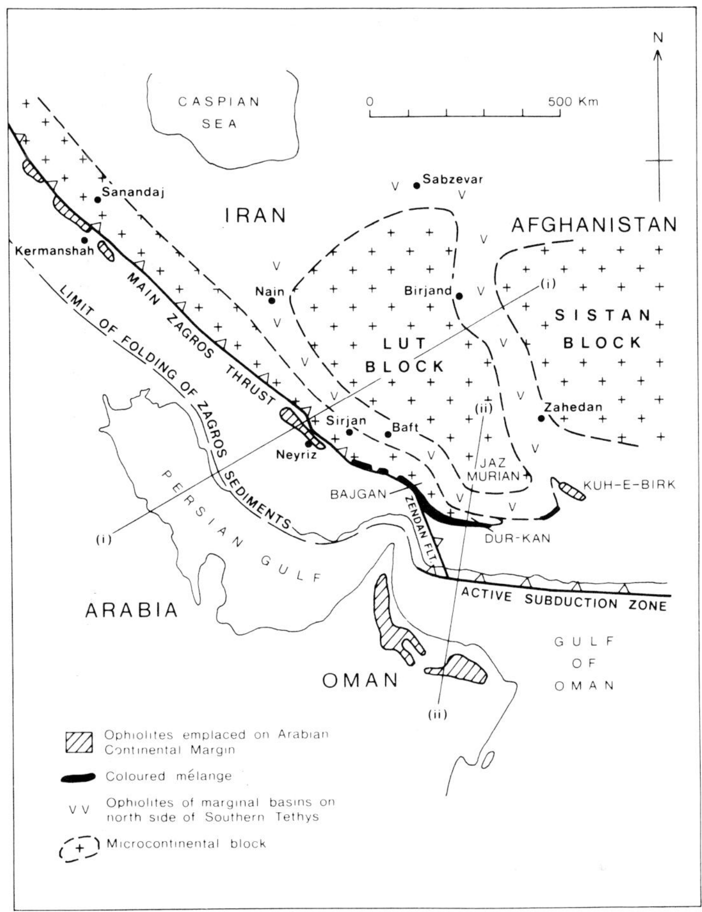

FIG. 4. The three distinct ophiolite developments in the Makran and surrounding regions with the associated microcontinent of Southern Tethys; (i) and (ii) are section lines shown in Fig. 5.

Palaeocene rocks indicates that some mélange formation continued into the Palaeocene. This event ended by the late Palaeocene when there was an abrupt change in palaeography mainly involving uplift of geotectonic provinces 2, 3 and 4. These provinces are patchily obscured by minor younger shelf limestone and flysch but for the most part marine deposition ceased in the Inner Makran in the Palaeocene.

(5) Eocene-Oligocene flysch

To the south of the Coloured Mélange Zone is an immensely thick flysch sequence. It is mainly distal, calcareous, turbiditic flysch, with thousands of repetitions of classic, complete or partial, Bouma sequences. It has beautifully preserved trace fossils and classic flute and tool marks. Continuous sequences may be more than 10,000 m thick and certainly occasional unfaulted sections several thousand metres thick may be traversed in the field. However, for the most part the flysch is highly folded and faulted. There are box, kink, isoclinal and open folds but by far the predominant type are periclines of chevron style. There is a slight south-vergence so that synclines have long south limbs and short north limbs, thus there is an overall northward dip of the strata as a result of the folding. Folds are developed on all scalds up to a few kilometres across but frequently both limbs and axial regions are cut out by north-dipping, steep, reverse faults. There is dominant faulting out of the anticlines and preservation of the synclines. This contrasts with the Zagros style of dominant preservation of anticlines that occurs west of the Zendan fault south of Minab in geotectonic province 8 (Fig. 1).

Commonly the flysch has been so tectonized that it has become dislocated and often has developed into a mélange. This mélange is distinct from the Coloured Mélange being very similar to some of the Franciscan mélanges (Cowan 1978, 1981). It contains exotic blocks derived from the underlying Coloured Mélange basement and also blocks of Eocene pelagic and shelf limestones. It has been referred to as wildflysch but it is clearly not. It is very similar to the exotic flysch of the Arakan, Burma (Brunnscheiler 1966). The exotic blocks were all emplaced tectonically, being squeezed up from the basement like pips. Some large blocks are composed of Coloured Mélange itself, not just of blocks that were once constituents of Coloured Mélange.

In the east of the project area the Eocene-Oligocene Flysch Zone widens out to over 100 km in width. In this region the flysch is less commonly tectonized to a mélange and contains less exotic components. In the north of the eastern part of the project area the flysch appears to grade up into a series of shallow-water basins.

The style of folding described above is similar to that described by White (1977, 1981) off the coast of the Makran. In the inland Makran the folding has been many times more intense and has developed on much smaller scales as well as on a large scale. As a result of the intense deformation it is not possible to demonstrate that there was syn-sedimentary deformation of basins as White has described off the coast.

(6) Oligocene-Miocene flysch

Further uplift occurred in this region in the mid-Oligocene. The trough of flysch deposition shifted to the south (and to the SW in the extreme west of the area). The Oligocene to earliest Miocene flysch is mainly distal, while the early to middle Miocene flysch is proximal. There is no evidence of any co-existing shelf limestone during this time, unlike in the Eocene.

The deformation of the Oligocene-Miocene flysch is similar to that of the Eocene-Oligocene flysch. There is similar development of mélange with exotic blocks some of which are of Eocene flysch. The pronounced spatial realtionship of dislocation and mélange formation to Miocene faults suggests that this extreme tectonism may be predominantly a Miocene event.

(7) Miocene Neritic Zone

In the south of the area rapid shallowing to platform sea conditions commenced in the Aquitanian producing a trough to the north in which flysch deposition persisted in the late middle Miocene between the emerging southern area and emergent provinces 2 to 5. Shallow shelf evaporites and reefal limestones were formed in the south. The latter have yielded important and diverse coral collections and accompanying foraminiferal faunas. In the late middle Miocene when the flysch trough finally filled up, coarse detritus spread over the entire southern area. These middle to upper Miocene estuarine, deltaic and shallow shelf sediments range from conglomerate to fine gypsiferous mudstones. There are local developments of fluviatile fanglomerate (true molasse) at the top of this sequence. This sequence forms the chain of immense open synclines (see Fig. 3) which are so conspicuous on the *Landsat* imagery. These were incorrectly interpreted as uplifted deepwater sequences of the trench slope by Farhoudi & Karig (1977). In general the neritic sediments are contorted where they were thinly bedded and incompetent and are folded into large open folds where they are thickly bedded and competent.

(8) Mio-Pliocene Neritic Zone

In the project area rocks of this zone (the Makran unit) (Huber 1952; Stocklin 1952) are only exposed west of the Zendan Fault (see Fig. 1). This sequence is the youngest marine sequence of the Makran and includes gypsiferous mudstones, deltaic sandstones and estuarine conglomerates with minor fluviatile conglomerates at the top. The term Makran (or Mekran) has been loosely applied by palaeontologists to virtually any neritic sediments encountered near the Makran coast in Iran and Pakistan which range in age through the entire Miocene and Pliocene. Here the name has no stratigraphic significance outside the limited zone described. Equivalent sediments probably also occur along the south coast of the Makran (Stocklin 1952; Anon 1962) outside the area mapped in this project.

West of the Zendan fault these sediments are highly folded and even Pliocene beds may be vertically disposed. The synclines are frequently faulted out so that the anticlines are preferentially preserved. This is the opposite of the dominant process within tectonic provinces 5, 6 and 7. The different styles of folding may be due to the different nature of the basement: to the west of the Zendan fault the basement is probably continental whereas to the east of the Zendan fault it is oceanic or comprises oceanic lithologies broken up into a mélange.

Plate tectonics and the evolution of the Makran

Figure 4 shows the continents, microcontinents, ophiolites and ophiolitic mélanges that make up the essential tectonic units and sutures of the Makran and surrounding region. The plate tectonic development of this region will be demonstrated by means of diagrams (Fig. 5) that show the relationships of the different tectonic units across two sections (see Fig. 4): (1) from Arabia NW across the Persian Gulf, the Zagros mountains including the Neyriz ophiolite, the Zagros Crush Zone, the Sanandaj-Sirjan belt, the Baft-Nain belt, the Lut block, the Birjand-Zahedan belt into the Afghanistan-Sistan block, and (2) from Arabia including the Oman ophiolite north across the Oman Sea, the Makran including the flysch belt, Coloured Mélange zone, Bajgan-Dur-Kan zone and Inner Makran ophiolite zone into the Lut block.

The exact positions in the Palaeozoic of the microcontinental fragments that now form Iran are not likely ever to be established but the similarities of the Palaeozoic sequences of India, Iran and Arabia (Stocklin 1977) suggest that along with India, most of Iran was once part of Gondwanaland and separated from Gondwanaland during the Triassic (Fig. 5a).

By middle Jurassic time a substantial Southern Tethys had developed in this region. Subduction appears to have commenced along the Sanandaj-Sirjan zone, indicated by andesites of

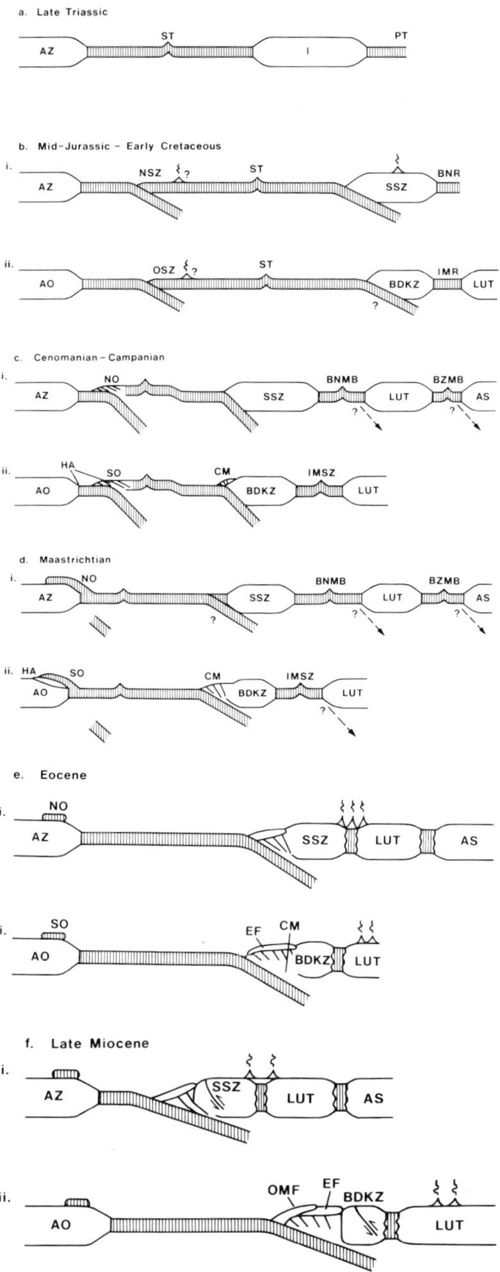

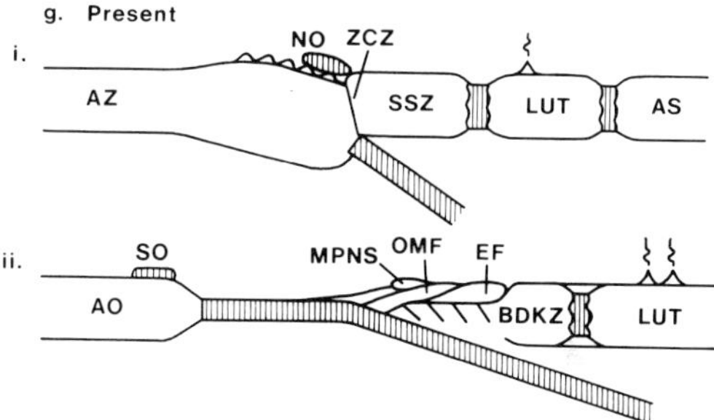

FIG. 5. Sketch sections along lines (i) and (ii) in Fig. 4 showing reconstructions of the probable relationships of continents, micro-continents, ridges, subduction zones and back-arc basins from Triassic to the present:
AO Arabia (Oman), AS Afghanistan—Sistan Block, AZ Arabia (Zagros), BDKZ Bajgan-Dur-Kan zone, BNMB Baft-Nain marginal basin, BNR Baft-Nain rift, BZMB Birjand-Zahedan marginal basin, CM Coloured Mélange, EF Eocene flysch, HA Hawasina, I Iran, IMR Inner Makran rift, IMSZ Inner Makran Spreading Zone, LUT Lut Block, MPNS Miocene-Pliocene neritic and molassic sediments, NO Neyriz ophiolite, NSZ Neyriz subduction zone, OMF Oligo-Miocene flysch, OSZ Oman subduction zone, PT Palaeozoic Tethys, SO Semail ophiolite, SSZ Sanandaj-Sirjan Zone, ST Southern Tethys, ZCZ Zagros Crush Zone.

this age (Fig. 5b(i)). There is no direct evidence of any subduction in the Makran at this time although Jurassic pelagics in the Inner Makran (geotectonic province 2) indicate the development of a rift (Fig 5b(ii)). During Early Cretaceous time or possibly a little later, an intra-oceanic NE-dipping subduction zone developed. It was as a result of collision of the Arabian continental margin with this subduction zone that the Kermanshah, Neyriz and Oman ophiolites were emplaced in the Campanian.

Fig. 5(c) shows the situation in Cenomanian to Campanian time. The continental margin of Arabia was about to collide with the NE-dipping subduction zone. The Hawasina thrust slices of the Oman were being stacked up in this subduction zone—the distal uppermost sheet first followed by successively lower and more proximal sheets. Finally, in Campanian time the collision was completed with the Semail ophiolite being emplaced on top of the stacked Hawasina continental margin sequence. The oceanic crust that was subducted would have been mainly Jurassic and Triassic, yet the Semail ophiolite is Cenomanian to Coniacian (Glennie *et al.* 1973) and hence came from the ocean to the NE of the subduction zone—where it was part of a forearc, backarc or even represents the island arc itself. The origin of the Neyriz and Kermanshah ophiolites (Fig. 5c(i)) is identical to that of the Oman ophiolite. The only difference is that the Zagros (regarded as part of the Arabian continental margin to the SW of the Zagros Crush Zone) collided with Iran along the Zagros Crush Zone during the Pliocene.

Fig. 5(d) shows the situation in the Maastrichtian when the ophiolites have been emplaced. There was ocean to the NE of both the Oman and Zagros; this ocean may have still been spreading at this time. There was certainly no collision of Arabia with the Makran region during the Campanian.

In the Makran the Inner Makran spreading zone was actively spreading from Early Cretaceous until the early Palaeocene. If this was a subduction-related process then it follows that there was some subduction of oceanic crust in a NE or northward direction throughout this period as indicated in Fig. 5(b(ii), c(ii) and d(ii)). The Baft-Nain, Birjand-Zahedan and Sabzevar belts were also spreading during the Cretaceous after initiation in the Jurassic. In the Makran the age of formation of the Coloured Mélange as opposed to the age of its constituent rocks cannot be precisely dated. Mélange formation may have commenced as early as Early Cretaceous or even Jurassic with late addition of younger rocks. However, the great abundance of Campanian-Maastrichtian rocks suggests that most of it was not formed until Maastrichtian or even Palaeocene. Thus the Coloured Mélange was mainly formed after completion of the emplacement of the Oman ophiolites. Two possible causes for the production of the Coloured Mélange are: (1) collision of the intra-oceanic subduction zone with the Arabian continental margin in the Campanian would cause a major rearrangement of plate boundaries in this region probably leading to an increased rate of subduction in the Makran, and (2) complete subduction of the older oceanic crust would result in attempted subduction of young oceanic crust that would be standing much higher than the old crust and so would be more easily scraped off the downgoing plate.

The emplacement of the Oman, Neyriz and Kermanshah ophiolites appears to coincide with changes in the relative motion between the African and Eurasian plates (Dewey *et al.* 1973) from predominantly compressional with a slight sinistral component in the region of Arabia to dextral strike slip. However, the existence of several plates and more than one subduction system without preservation of any oceanic crust between Africa/Arabia and Eurasia makes it impossible to reconstruct the motion along a single plate boundary from the motions of the major continental plates. The probable age of Coloured Mélange formation is coincident with the breaking away of India from Gondwanaland and its rapid acceleration northward.

There are no typical andesitic rocks in the Iranian Makran and it is suggested that the Inner Makran spreading zone may be the equivalent of the arc. As described above the orientation of the sheeted dykes along the length of the zone and the distinctly calcalkaline nature of the inner (more northward or north-eastward) ophiolites support this suggestion. It is possible that 'arc' rocks may lie beneath the Jaz-Murian depression (see Fig. 1) or have been tectonically removed, but in the former case the ophiolites would be in front of the arc (rather than being back-arc) and in the latter it is surprising that no trace should remain when the ophiolites are not severely deformed. There are small amounts of possible 'arc' rocks in the Coloured Mélange but these could equally well originate from oceanic island volcanoes or simply be unusual ocean-floor rocks. It is possible to construct a more complicated plate tectonic model involving many microplates. However, in this paper we wish to put forward the simplest model compatible with the known geology although accepting that whole sections might have completely vanished.

In this Makran subduction system the Bajgan-Dur-Kan Zone (geotectonic province 3 in Fig. 1) formed a forearc of shelf limestones on continental basement. This forearc zone extends to the NW and appears to have been emergent during the Cretaceous in the NW of the project area and in the Sanandaj-Sirjan belt. Since there is a continuation of the Makran to the NW in the Sanandaj-Sirjan and Baft-Nain belts it is inferred that subduction also took place along the line of the Zagros Crush Zone where there are remnants of Coloured Mélange. This is supported by the presence of andesites and flysch of Cretaceous age along the Sanandaj-Sirjan belt.

The Baft-Nain, Sabzevar and Inner Makran ophiolite troughs closed up by the late Palaeocene (Fig. 5e) so that the Sanandaj-Sirjan and Bajgan–Dur-Kan microcontinents had coalesced with the Lut block to form the Central and East Iran block of Takin (1972). In the Inner Makran there was uplift and compression of the Coloured Mélange, Bajgan–Dur-Kan and Inner Makran spreading zones. It is possible that there was some minor subduction of the latter zone indicated by the blueschists near Fannuj.

By the end of the Eocene the Birjand-Zahedan trough had also closed, joining the Afghan-Sistan Block to the Central and East Iran microcontinent to form a larger continental block. The resulting tectonic situation was a simple north-dipping subduction zone in the Makran with a single continental mass to the north (Fig. 5e(ii)). There was substantial Eocene andesitic volcanism to the north of the project area, which provided the source for the huge flysch deposits of geotectonic province 5. This flysch was deposited on an already accreted prism of Coloured Mélange up to 70 km wide, or possibly partly on oceanic crust, parts of which were later detached from the subducting plate and tectonically intruded into the flysch. It might be possible to distinguish between these alternative hypotheses for the nature of the basement to the flysch by detailed study of the exotics within the flysch. However, since the oceanic crust subducted during and subsequent to the Eocene may be similar to that which formed the Coloured Mélange in the Maastrichtian or Palaeocene this might prove difficult.

Along the Zagros Crush Zone there also appears to have been substantial NE subduction of oceanic crust in the Eocene (Fig. 5e(i)). The main evidence for this subduction is the great volume of Eocene volcanics 150 km NE of the Crush Zone in a parallel belt more than 1600 km long. There are also minor accumulations of Eocene flysch indicating the presence of a trough along the Crush Zone at this time. These may be remnants of a large volume of Eocene flysch that has now vanished by subduction or by overthrusting of Central Iran on to the Zagros. However, it is possible that the Sanandaj-Sirjan zone was mainly emergent and blocked off the supply of flysch from the volcanics to the north.

In the Oligocene there was some adjustment to the north-dipping subduction zone in the Makran. This resulted in a southward shift of the trench and uplift of the Eocene flysch, which then provided part of the source of the Oligocene-Miocene flysch. This event may be related to a readjustment of plate motions at this time which include the cessation of sea-floor spreading in the Indian Ocean (McKenzie & Sclater 1971). The further readjustment of plate motions when Indian Ocean sea-floor spreading recommenced in the middle Miocene approximately coincides with renewed uplift and yet further southward migrations of the trench in the Makran. The andesitic volcanics to the north, the tectonics, and the flysch deposition, however, are all consistent with the existence of a north-dipping subduction zone in the Makran throughout the Tertiary. The discrete tectonic events probably relate to changes in the rate and/or direction of subduction.

Fig. 5(f) shows the relationship between Arabia and Central Iran in the late Miocene. Arabia/Zagros is about to collide with Central Iran while there is still a substantial width of ocean separating Oman and the Makran. Not until the Pliocene collision (Stocklin 1977) did Arabia/Zagros have any direct relationship with Central Iran. This collision resulted in the juxtaposition of two separate subduction systems: (1) the subduction system represented by the Neyriz and Kermanshah ophiolites and related rocks which had been sitting passively on the Arabian/Zagros continental margin since their emplacement in the Campanian, and (2) the subduction system that had been active along the SW edge of the Sanandaj-Sirjan belt from the Jurassic through to the Tertiary. Due to the deep embayment in the Arabian (plus the Zagros) continental margin represented by the Gulf of Oman, Arabia has not yet collided with the Markan (Fig. 5g(ii)). In the Zagros the continued covergence resulted in intense folding of the Arabian continental platform sediments and thickening and possible subduction even of the continental basement. Along the Zagros Crush Zone all but a few remnants from the trench zone of subduction system (2) above have been obliterated by overthrusting of Central Iran from the NE. This effect can be seen in the NW of the project area near Minab (Fig. 1) where the immense flysch belt of the Makran has been thinned against the Zendan fault (which is a continuation of the south side of the Zagros Crush Zone). Going further NW into the Zagros (see Fig. 4) the Coloured Mélange Zone also thins except for a few remnants so that the Arabian continental margin is juxtaposed against the Sanandaj-Sirjan belt (the NW continuation of the Bajgan–Dur-Kan Zone).

Subduction in the Makran is continuing at the present day with deformation of the sediment of the Gulf of Oman off the Makran coast where the subducting plate initially bends down only about 1° (White & Klitgord 1976; White 1977). Further inland the subducting plate must bend down more steeply, but the initial low dip is related to the unusually great (250 km) separation between the trench and the 'arc'. From the Eocene onwards the 'trench' has migrated southwards while the 'arc' has remained approximately stationary. In the process a substantial accretionary prism of sediment has built up. Accretion of material from the downgoing plate clearly occurred during the Late Cretaceous and Palaeocene and may have occurred since (see discussion above).

If subduction continues as at present, Oman will collide with the Makran and this accretionary prism may be overthrust from the north and obscured as has occurred along the Zagros Crush Zone (assuming such an accretionary prism was once developed there). The result will be a single suture from Turkey to Pakistan, but one marking the disappearance of at least two subduction systems.

ACKNOWLEDGMENTS: This paper was prepared at the suggestion of the organizers of this symposium to present an up-to-date model of the tectonics of this region based on the recent work carried out on behalf of the Geological and Mineral Survey of Iran. The full descriptions of the geology, only summarized here sufficiently to support the plate tectonic model, will be published by the Geological and Mineral Survey of Iran. The contributions of the survey and of the geologists who worked on this regional mapping programme are acknowledged.

References

ANON, 1962. *Unpublished report on the Geology of the Southern Makran,* AGIP, Milan, filed at the Geological and Mineral Survey of Iran, Tehran.

BRUNNSCHEILLER, R. O. 1966. On the geology of the Indo-Burman ranges. *J. geol. Soc. Aust.* **13,** 137–94.

COWAN, D. S. 1978. Origin of blue-schist bearing chaotic rocks in the Franciscan complex, San Simeon, California. *Bull. geol. Soc. Am.* **89,** 1415–23.

—— 1981. Deformation of partly dewatered and consolidated Franciscan sediments near Piedras Blancas Point, California (this volume).

DEWEY, J. F., PITMAN III, W. C., RYAN, W. B. F. & BONNIN, J. 1973. Plate tectonics and the evolution of the Alpine system. *Bull. geol. Soc. Am.* **84,** 3137–80.

FARHOUDI, G. & KARIG, D. E. 1977. Makran of Iran as an active arc system. *Geology,* **5,** 664–8.

GANSSER, A. 1974. The ophiolitic melange, a world wide problem, and Tethyan examples. *Ecolog. geol. Helv.* **67,** 479–507.

GLENNIE, K. K., BOEUF, M. G. A., HUGHES-CLARKE, M. W., MOODY-STUART, M., PILAAR, W. F. H. & REINHARDT, B. M. 1973. Late Cretaceous nappes in the Oman Mountains and their geologic evolution. *Bull. Am. Assoc. Petrol. Geol.* **57,** 5–27.

HARRISON, J. V., & FALCON, N. L. 1936. Geology of the Coastal Makran. *Unpubl. maps*, filed at the Nat. Iranian Oil Co., Tehran and the Geol. Soc. Lond.

—— , —— ALLISON, A., HUNT, J. A., MALING, P. B. & MCCALL, R. J. C. 1935–36. Geology of the Landward Makran. *Unpubl. maps,* filed at the Nat. Iranian Oil Co., Tehran and the Geol. Soc. Lond.

HUBER, H. 1952. Geology of the Western Coastal Makran area. *Unpubl. report*, Iranian Oil Co., filed at the Iranian National Oil Co., Tehran, No. GR 91B.

MCCALL, G. J. H. 1981. Compiler of reports of Geological and Mineral Survey of Iran:
(1) Report on East Iran Project Area No. 1.
(2) Explanatory text of the Minab quadrangle map, 1:250,000.
(3) Explanatory text of the Tahenii quadrangle map 1:250,000.
(4) Explanatory text of the Fannuj quadrangle map 1:250,000.
(5) Explanatory text of the Pishin quadrangle map 1:250,000.
(6) Explanatory text of the Nikshahr quadrangle (southern half), 1:250,000.
(7) Explanatory text of the Saravan quadrangle (southern half), 1:250,000 (in press).

MCKENZIE, D. P. & SCLATER, J. G. 1971. The evolution of the Indian Ocean since the Late Cretaceous. *Geophys. J. R. astron. Soc.* **24,** 437–528.

STOCKLIN, J. 1952. Geology of the Central Coastal Makran Area. *Unpubl. report,* Iranian Oil Co., filed at the National Iranian Oil Co., Tehran, No. GR 91C.

—— 1977. Structural correlation of the Alpine ranges between Iran and Central Asia. *Mem. h. Ser. geol. Fr.* **8,** 333–53.

TAKIN, M. 1972. Iranian geology and continental drift in the Middle East. *Nature. London,* **235,** 147–50.

WHITE, R. S. 1977. Recent fold development in the Gulf of Oman. *Earth planet. Sci. Lett.* **36,** 85–91.

—— & KLITGORD, K. 1976. Sediment deformation and plate tectonics in the Gulf of Oman. *Earth planet. Sci. Lett.* **32,** 199–209.

WHITE, R. S. 1981. Deformation of the Makran accretionary sediment prism in the Gulf of Oman (this volume).

G. J. G. MCCALL, 57 Venns Lane, Hereford, U.K.

R. G. W. KIDD, RSMAS-MGG (Marine Geology and Geophysics), University of Miami, 4600 Rickenbacker Causeway, Miami, Florida 33149, U.S.A.

CALIFORNIA

The Coastal Belt of the Franciscan: youngest phase of northern California subduction

Steven B. Bachman

SUMMARY: The Coastal Belt of the Franciscan complex represents the latest Cretaceous to middle Tertiary part of the accretionary prism of an arc-trench system in northern California. On the Mendocino coast, both deformed slope and trench deposits and much less deformed slope basin deposits are recognized. Deformation increases with depth in the slope deposits, towards the contact with the accretionary prism. In the accretionary prism, bands of deformed rocks from tens of metres to kilometres in width suggest major zones of shearing. Stratigraphic tops of beds face eastward, yet palaeontological ages young westward; imbricate stacking is suggested by these relationships. Stratigraphical tops and palaeontological ages indicate that the slope basin deposits are in normal stratigraphic sequence.

Facies in the accretionary prism may represent trench fill, abyssal plain, feeder channel, and small fan deposits that were accreted to the leading edge of the continent and may also include slope deposits that were deformed during progressive deformation of the accretionary prism. The less deformed slope deposits may have been deposited as slope basin turbidites, small fans, and channels. South- and south-westward-directed palaeocurrent indicators and sandstone petrology suggest sediment transport from the Klamath Mountains and Oregon across the shelf and down the slope and/or along the trench axis.

Geological setting

The Franciscan complex of California and Oregon is an ancient subduction complex ranging in age from Jurassic to Miocene or younger. The Franciscan complex is divided into three belts on the basis of characteristic lithologies and structures; the Coastal Belt is the westernmost of these units (Bailey *et al.* 1964). The Coastal Belt is exposed in the northern Coast Ranges of California and extends from near Cape Mendocino in the north to as far south as the San Francisco Bay area (Fig. 1). East of it is the Central Belt mélange.

The Central Belt of the Franciscan is a mélange unit that includes blocks of sandstone, greenstone, chert, blueschist, eclogite, amphibolite, and serpentinite incorporated in a sheared shale and sandstone matrix (Bailey *et al.* 1964; Berkland *et al.* 1972; O'Day 1974). Central Belt rocks range in age from Jurassic to Cretaceous (Blake & Jones 1974), but near the contact with the Coastal Belt in this area, only Lower and Upper Cretaceous strata are present (O'Day 1974; Kramer 1976).

The Coastal Belt consists of interbedded sandstones, siltstones, shales, and minor conglomerates. Volcanic rocks, limestone, and chert occur locally. Laumontite is common in veins and on grain overgrowths in the sedimentary rocks. The presence of only zeolite facies metamorphism indicates that the Coastal Belt experienced shallower burial than some of the rest of the Franciscan complex.

In the central area of Coastal Belt exposures, the Coastal Belt is in contact on the east with either the Franciscan Central Belt mélange or

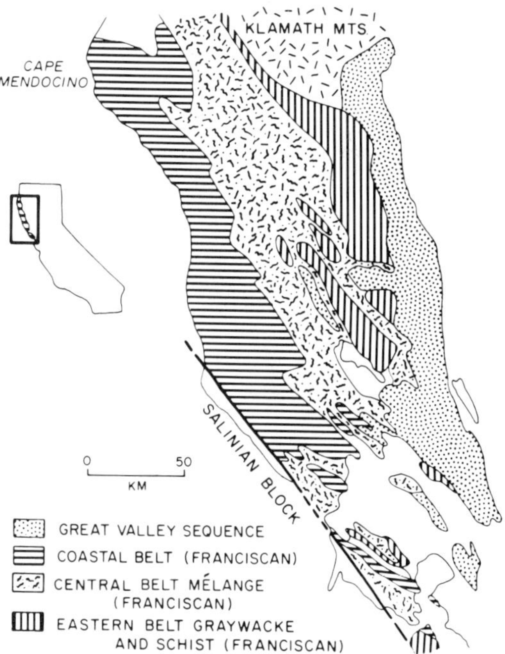

FIG. 1. General geology of northern California Coast Ranges after Blake & Jones (1978).

the Leggett peridotite. This eastern contact has been decribed variously as a high angle shear zone (Ogle 1953; O'Day 1974), an east-dipping, high angle reverse fault (Kramer 1976, 1977), or a more recent landslide depositional contact (Kleist 1974a). A broad band of current earthquake activity parallel to the contact suggests that an earlier reverse and/or strike-slip fault zone separating the Coastal and Central Mélange belts may be reactivated in sympathy with movement along the San Andreas fault. The San Andreas fault is located 7–17 km offshore and is the present western boundary of the Coastal Belt.

Coastal Belt structure

Analytical techniques

Because normal structural mapping techniques were of limited value in the complexly deformed and stratigraphically disrupted parts of the Coastal Belt, a different type of structural mapping scheme was used to resolve some of the complexities. This technique is described in more detail elsewhere (Bachman 1979). The technique is based on analysis of the styles of deformation within the Coastal Belt which range from sheared, boudinaged, and stratigraphically disrupted beds to continuous beds with little structural disruption. The less deformed rocks are commonly involved in local, open folds and minor faulting. Accordingly, the styles of deformation are classified on a numerical scale, with the end members as described above. Each outcrop was designated and mapped according to its tectonic style, both on small- and large-scale maps.

Style 1

This end member includes the sheared, boudinaged, and stratigraphically discontinuous beds such as those in Fig. 2. The majority of these beds consist of dismembered sandstone and mudstone, but locally contain conglomerates, limestones, and volcanic rocks. In all cases, the mudstone is a matrix that encases the other material. Individual beds cannot be traced laterally for more than 100 m or so, but on an outcrop scale, sandstone/mudstone ratios are relatively constant laterally, giving a gross sense of stratigraphy. Typically, the sandstones are deformed into elongate saucer-shaped or bladed phacoids or boudins. The edges of the boudins commonly are sheared, with broken sandstone grains up to several millimetres in thickness forming the circumference of the boudins. The ends of boudins are commonly sheared as well, with shale intruding along small fractures.

Anastomosing shear surfaces disrupt the mudstones so that no internal bedding is evident. The mudstone matrix of these sheared rocks in some instances contains extensive fractures formed parallel to the shear and boudin foliation. Where fracturing is pervasive, the

FIG. 2. Structural style 1 deformation with sandstone boudins in sheared mudstone matrix. Stratal continuity is extensively disrupted.

fine-grained matrix has a scaly appearance. The long axes of the boudins in style 1 areas are generally aligned with the shear fabric in the shales to form a foliation. This foliation is sub-parallel to the shear surfaces in the matrix and the elongation of the boudins is in the plane of shearing. The foliation in the boudins and the surrounding matrix generally dips steeply eastward.

The elongation of boudins that is typical in style 1 deformation is best measured quantitatively in conglomerate beds. The aspect ratios (long dimension ÷ short dimension) of chert and limestone clasts in undeformed conglomerate beds and in the adjacent deformed parts of these beds change from 1.5:1 to between 2 and 3:1. This elongation of well-rounded chert clasts strongly suggests a tectonic cause for many of the boudin shapes (Bachman 1978a).

Style 2

The rocks of this tectonic style show many of the deformation features of style 1, but strata have greater lateral continuity (Fig. 3). Pinch and swell and some boudinage are present, but bedding units can be traced laterally. Generally, the mudstone matrix is little or moderately deformed, with some traces of original bedding features often evident.

Style 3

This tectonic style includes the least-deformed rocks, which are continuous beds with little structural disruption (Fig. 4). Intrastratal slump folds, local open folding, and minor faulting are the only evidence of deformation in these rocks.

Structural units

Mapping and contouring the tectonic styles in the Coastal Belt reveal several scales of deformation geometry. On the largest scale (Fig. 5), the Coastal Belt can be divided into two units (Kramer 1976; Bachman 1978a), a west-

Fig. 3. Structural style 2 deformation. Individual beds are often discontinuous, but overall stratal continuity of units is preserved.

Fig. 4. Structural style 3 rocks. Stratal continuity has not been disrupted.

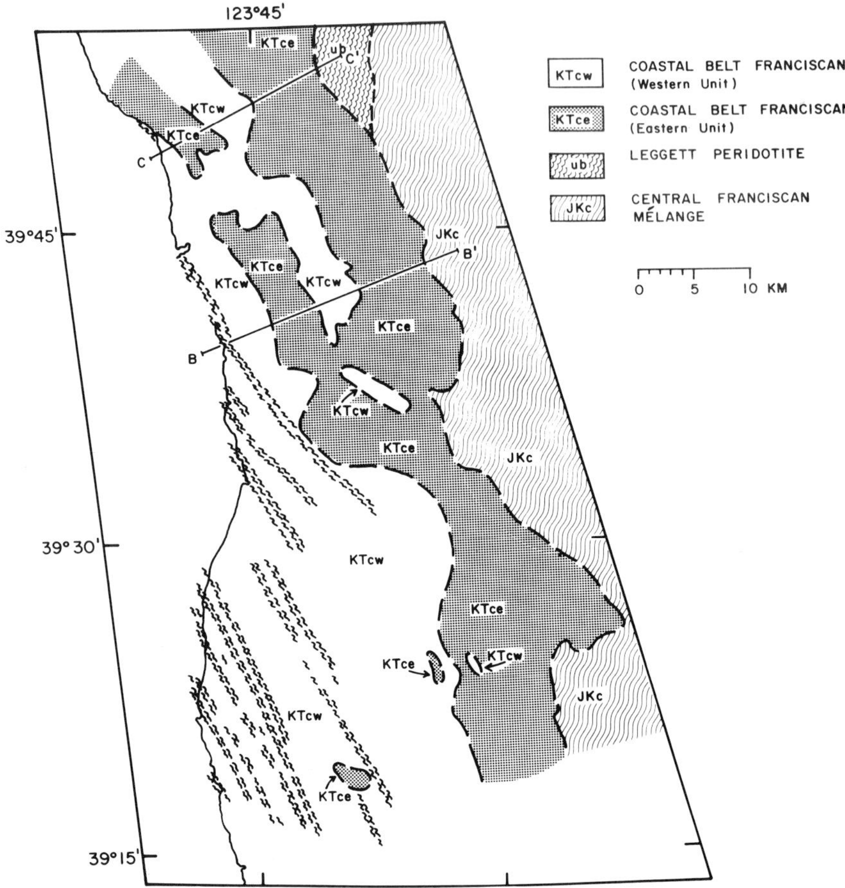

FIG. 5. Structural units in the Coastal Belt. Western Unit is primarily style 1 and 2 rocks with minor style 3, while Eastern Unit is primarily style 3 rocks. Linear bands of short wavy lines are inferred shear zones represented by style 1 rocks. See text for definition of structural styles.

ern unit of style 1, 2 and minor style 3 rocks, and a more eastern unit of primarily style 3 rocks. Locally, these two units alternate across strike (Fig. 5). Smaller scale maps (Fig. 6) show more detailed structure.

Western Mélange Unit

A western unit exposed along the coast and adjacent inland areas (Fig. 5) is recognized by the abundance of tectonic style 1 and 2 rocks. In current usage, the tectonic style of this unit would categorize it as a mélange. The remarkable aspect of the Western Unit is the inhomogeneity across strike in structural style over a distance of tens to hundreds of metres (Figs 5 & 6). Bands of sheared rocks several kilometres wide envelop less-sheared rocks (Fig. 5). Within the less-sheared rocks, narrower bands of deformed rocks tens of metres wide occur (Figs 6 & 7). In a few localities, wide bands of sheared rocks also appear to consist of sets of these narrow deformed bands as well. These narrow bands are characterized by a transition zone with a progressive increase in the degree of deformation from more-coherent beds into the boudinaged zone over a width of 10 to 100 m (Figs 6 & 7). The narrow bands of highly sheared rocks probably represent individual fault zones of the Coastal Belt.

The structural trend in the Mélange Unit is NW–SE. Beds dip predominantly to the east at a high angle or are slightly overturned. Stratigraphic tops of beds face eastward. The overall age progression in the Mélange Unit youngs westward from Cretaceous to Palaeogene, however, suggesting imbricate stacking of sedimentary units.

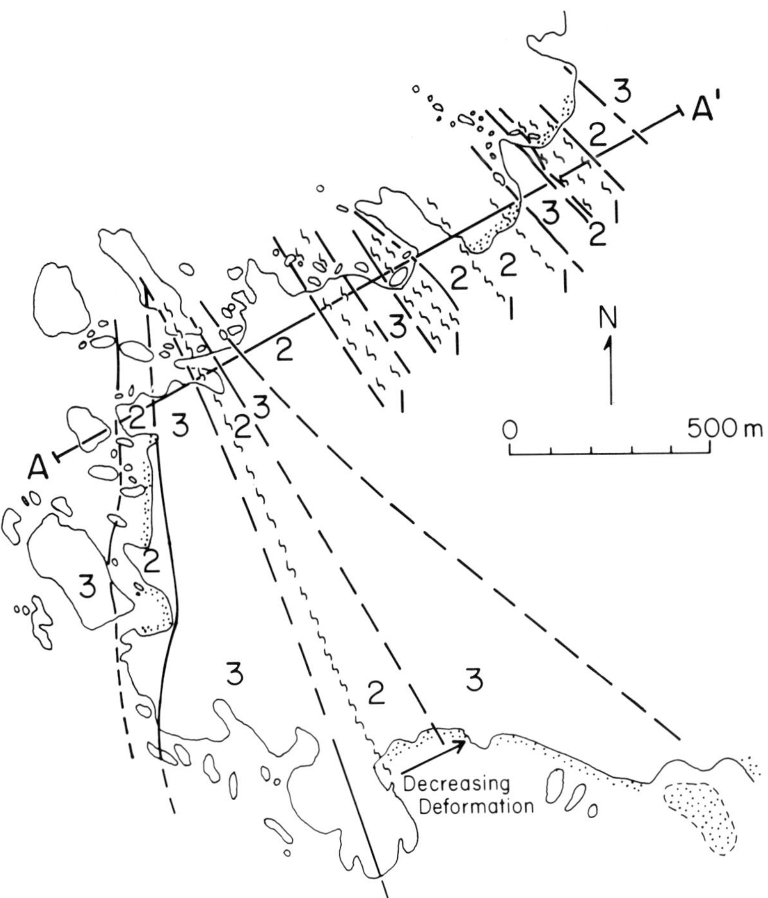

Fig. 6. Structural styles at Mendocino Headlands, showing deformation changes across strike at more detailed scale than Fig. 5. See text for definition of structural styles. Section A-A′ is Fig. 7.

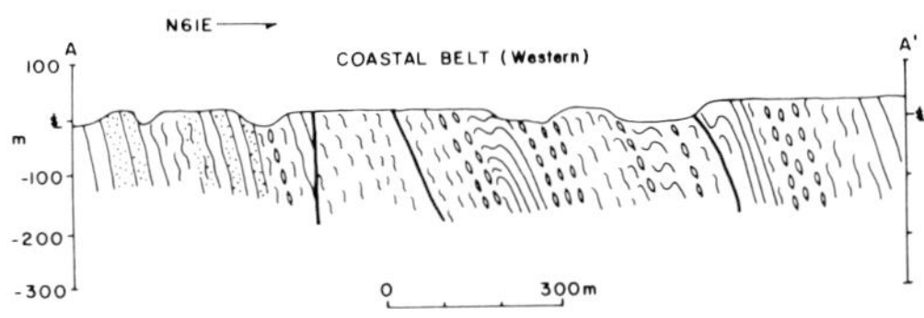

Fig. 7. Structure section across Fig. 6. Elongate pods represent structural style 1 rocks, short wavy lines represent structural style 2 rocks, and straight or folded continuous lines represent structural style 3 rocks. Stippled pattern is massive sandstone.

Eastern Unit

An eastern unit in the Coastal Belt (Fig. 5), defined by large areas of style 3 deformation, has been deformed locally into open folds, both synclines and eastward-dipping homoclines (Fig. 8). Unlike the Mélange Unit, the eastern beds are in a normal stratigraphic sequence; they young in the direction of stratigraphic facing. The degree of deformation increases with stratigraphic depth in the Eastern Unit, as well as towards the contact with the Mélange Unit. The contact between the units ranges from abrupt to gradational; it varies from gradational over a distance of hundreds of metres,

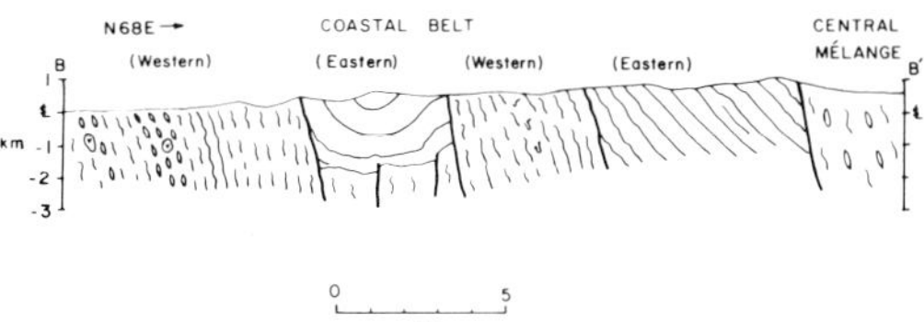

Fig. 8. Structure section B-B′ referenced in Fig. 5. Patterns are the same as Fig. 7.

with gradually increasing deformation of the Eastern Unit toward the contact with the Mélange Unit, to faulted (Kramer 1976) with a complex combination of Quaternary faulting (brecciated zone) and smaller-scale alternation of tectono-stratigraphic units (Bachman 1979).

Faults

In addition to the shear zones recognized in the Mélange Unit, discrete fault planes are also evident in both the Mélange and Eastern Units. The orientation of several hundred of these faults was measured (Fig. 9) in the Mélange Unit.

The low-angle north- and NE dipping faults (trend 1, Fig. 9) are offset by faults of other trends and thus appear to be the oldest faults. Where it is apparent, the sense of movement on faults with these trends is underthrusting of western blocks under eastern blocks. The relative ages of the north- and NE dipping faults are not known; the two trends were never seen with cross-cutting relationships.

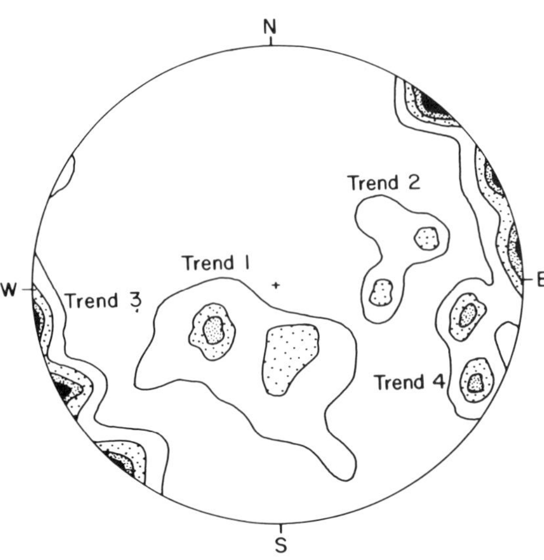

FIG. 9. Lower hemisphere equal area projection of poles to fault planes, Mélange Unit. Densities of orientations are contoured as 1, 3, 5, and 7% of total population in 1% of hemisphere area.

The moderately NW dipping faults (trend 2, Fig. 9) offset the north- and NE dipping faults (trend 1), but are in turn offset by the other fault trends. The sense of displacement of these faults is not evident.

High-angle NW and NE trending faults (trends 3 and 4) are the most abundant faults in the Coastal Belt. They are the youngest faults recognized. Slickensides indicate that strike-slip movement predominates, with offset beds and conglomerate clasts suggesting right-lateral movement. Fault gouge, up to tens of centimetres thick, is common along these faults.

The origin of the shear foliation in the Mélange Unit apparently is unrelated to the San Andreas fault system. The foliation, although subparallel to the NW trending San Andreas fault, is clearly offset by the youngest faults (trends 3 & 4) that may be sympathetic to the San Andreas fault. The shear foliation also appears to pre-date the north- and NE dipping faults in most areas.

Folds

Folding on several scales can be recognized in the Coastal Belt. On the largest scale, kilometre- and larger-wavelength NW trending folds expose mélange and volcanic rocks in the cores of anticlines with less-deformed Eastern Unit rocks on the flanks of the folds (Fig. 10). Smaller-scale folding is more difficult to recognize except in well-exposed areas; outcrop-scale

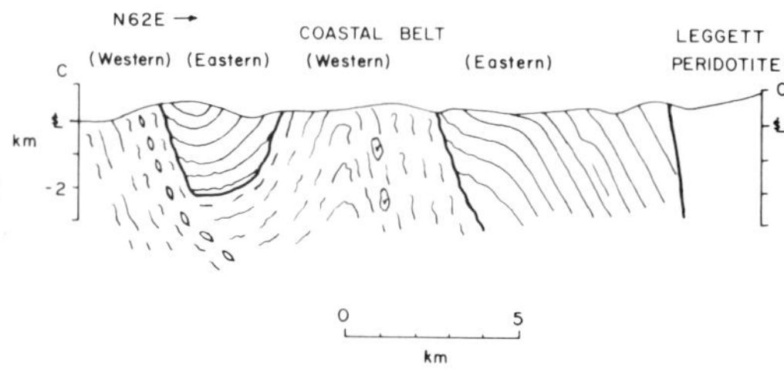

FIG. 10. Structure section C-C′ referenced in Fig. 5. Patterns are the same as Fig. 7.

folds in well-exposed mélange rocks fall into two groups. Folds with SE plunging axes often verge to the SW. The plunge of these folds varies from less than 5° to 30°. The second group of folds plunge 10°–50° to the NE. Two of these folds verge to the NW; the others are upright. Many of these smaller-scale folds are isoclinal.

Folds in the mélange are commonly separated by faults and shear zones. The limbs of isoclinal folds are commonly sheared, leaving overturned limbs boudinaged (Bachman 1978a) or fold hinges floating in the sheared matrix.

The folding and shearing appear to be contemporaneous. Early-formed fractures are folded during recurring folding and movement along the shear zones. Locally, sand has flowed into the mudstone matrix in the isoclinal fold hinges.

The latest period of folding, presumably Quaternary in age, slightly tilted the uplifted marine terrace deposits of probable Plio-Pleistocene age exposed along the coast (Kramer 1976). The youngest, NW-trending faults offset these terraces and truncate all the earlier folding.

Soft sediment deformation

The wispy, pull-apart, and somewhat chaotic appearance of many Coastal Belt mélange rocks suggests deformation of somewhat unconsolidated sediments (Kleist 1974b). The elongation of chert clasts, thrust stacking origin of some boudins (Bachman 1979; Bachman & Allmendinger 1980), and lateral extent of deformed rocks indicate that the boudinage and pull-apart features are probably tectonic rather than sedimentary in origin. However, intrastratal folding, chaotic beds, and the swirled mixing of sediments were probably caused by slumping and gravitational flowage on the margin slope (cf. Cowan 1981).

In the Eastern Unit, the only soft sediment deformation features apparent are slump folds, load structures, and flame structures.

Coastal Belt stratigraphy

Sandstone and shale comprise the vast majority of Coastal Belt stratigraphy. Minor amounts of conglomerates, volcanic rocks, limestone and chert are present also. Sandstone to shale ratios exceed 1:1 in most areas; in local areas, they approach 8:1.

Limestones occur as clasts in conglomerates, as blocks in volcanic rocks, as thin, discontinuous beds or boudins in shale units, or as lensoid- to ball-shaped concretions. Grey and brick-red limestone blocks up to 1 m across are found within the volcanic rocks described below. The limestones have been recrystallized and are now primarily sparite. The contacts between the limestones and igneous rocks are somewhat irregular and have the appearance of intrusive contacts, with limestone blocks floating in the igneous rock, although no chilled margins are evident in the igneous rocks. The brick-red limestones may owe their colour to mineralization during igneous activity. Secondary formation of carbonate from pore-water precipitation after the formation of the volcanic rocks cannot be ruled out.

Volcanic rocks are present in the Mélange Unit in blocks as large as 100–200 m across. They are arranged in somewhat linear bands parallel to the tectonic fabric. The volcanic rocks include pillow lavas, pillow breccias, and hyoloclastics with isolated pillows. One such sequence is 10 m thick. As already described, limestone blocks are prevalent in the rocks. O'Day (1974) reported that chemical analyses of some of the rocks are equivalent to the compositions of spilite, albite diabase, and labradorite diabase. Some of the rocks are porphyritic, with albite, olivine, and augite usually replaced by carbonate and magnetite. The groundmass is chloritized with minor amounts of pumpellyite and pyrite. Quench crystals are common. Zeolite and silica veins have been partially replaced and are cut by calcite veins.

The contact between the igneous rocks and the Coastal Belt sediments is usually obscured, but in several places shearing along a fault zone is evident. Within 6–8 m of the contact, both sediments and volcanic rocks are brecciated, with a reduction in sand grain size along discrete shear zones. The sheared volcanic material along the contact contains slickensided pillows and sandstone and shale boudins. The igneous rocks may be displaced exotic blocks tectonically emplaced into the Coastal Belt sediments.

The thickness of the Coastal Belt has been variously estimated as up to 7500 m (O'Day 1974), up to 10 000 m (Kleist 1974a), and 6000 m (Kramer 1976). In convergent margin depositional settings such as subduction complexes, deformation and sedimentation are often synchronous; consequently, the thickness of tectono-stratigraphic units cannot be measured directly. The structural thickness of stacked imbricate sheets can far exceed the stratigraphic thickness. The structural thickness of the Mélange Unit may exceed 20 km (Fig. 8). Kramer's 6000 m calculation is the apparent thickness of the homoclinal Eastern Unit west of Willits. With structural repetition primarily by open folding, this 6000 m represents a maximum present stratigraphic thickness of the Eastern Unit. The thickness of sediment lost to erosion is unknown. Thermal alteration of organic matter in the Eastern Unit (Bachman 1978a, 1979) suggests that several kilometres of overburden of unknown age has been removed by erosion.

The age of the Coastal Belt has been determined using a combination of palynomorph assemblages (Evitt & Pierce 1975; Kleist 1975;

Bachman 1978a, 1979) and foraminifera (Bachman 1978a, 1979; McLaughlin *et al.* 1979). These fossil data suggest that the Mélange Unit rocks range in age from Late Cretaceous on the east side to Oligocene or Miocene on the west side. Stratigraphic tops of beds face eastward, suggesting imbricate stacking of sedimentary units. Even younger Coastal Belt rocks may be found west of the Mendocino coast area, because subduction continued into the Plio-Pleistocene (Atwater & Molnar 1973). The Eastern Unit ranges in age from Cretaceous to Eocene. The youngest fossil assemblages occur in the structurally highest positions, suggesting that normal stratigraphic relationships exist in the Eastern Unit. In all cases where the resolution of palaeontological ages permits interpretation, the Eastern Unit is younger than the underlying or immediately adjacent mélange.

Facies analysis

I have used the facies classification scheme for turbidites and related resedimented deposits (e.g. Walker & Mutti 1973; Bouma & Nilsen 1978) to characterize Coastal Belt sediments. However, in most cases it is not appropriate to interpret the associations, or groupings, of these facies using a deep-sea fan facies association model: submarine fan facies terminology is only appropriate if the geometry of a fan complex can be reasonably documented. Instead a convergent margin sedimentation and facies model (Bachman 1979; Underwood *et al.* 1980; Underwood & Bachman 1981) is used for interpretation of facies associations.

Mélange Unit facies

Bedding disruption restricts measurement of detailed stratigraphic sections to less-deformed areas, although sequences of facies are very often on a scale smaller than the size of the coherent blocks (tens to hundreds of metres thick) within the long linear shear zones. Thus, gross stratigraphic patterns, traceable laterally, remain despite the shearing. The sub-parallel orientation of shear surfaces with bedding may aid preservation of these patterns.

The full range of turbidite and associated facies is present in the Western Mélange Unit. The thick, massive sands of facies A and B are particularly abundant in this unit, as well as the Eastern Unit. Sandstone/mudstone ratios in the Mélange Unit range from somewhat less than one to as high as 16. The overall ratio is approximately 2 or 3.

The beds of facies A range from 2 m to 30 m or more in thickness. They are mostly medium or coarse sandstone, but minor conglomerates and pebbly sandstones also are present. The beds commonly are amalgamated; locally no internal features or textural changes are evident in the beds. Facies A conglomerates range from clast supported to matrix supported; both disorganized and organized beds occur.

Channelling of facies A rocks into the underlying sediments is not uncommon. Rip-up clasts of the underlying sediments are often found near the base of the A beds.

Facies B sandstone beds are difficult to distinguish from those of facies A in many cases; dish structures, diffuse laminations, fluid-escape pipes and more frequent shale partings are indicative of facies B rocks when present. Facies B beds are up to 6 m thick, but the thicker beds are amalgamated. Commonly, facies A and B rocks are found in the same locality. Channellized facies A rocks are often associated with facies B, C, and D rocks (Fig. 11). These facies may occur in megasequences (Ricci-Lucchi 1975), with beds thinning and fining upwards.

Facies E overbank and levee deposits also occur with some of the A, B, C, D associations (Fig. 11), or alone with facies A rocks. Facies F slumps and pebbly mudstones also are associated with thick facies A rocks (Fig. 11). The pebbly mudstones range in thickness from 1.5 m to 8 m. A section that contains the largest volume of fine-grained rocks in the Western Unit has the following facies sequence:

$>$20 m A, B $\rightarrow$ 100 m G, D $\rightarrow$ 10 m A, B $\rightarrow$ 80 m G, minor D $\rightarrow$ $>$100 m D, C, B (coarsening- and thickening-upward sequence).

Typical facies B beds in the Mélange Unit are associated with facies C beds. These facies may be separated by facies G mudstones, or interbedded with facies D rocks, usually in a thinning-upward sequence. Facies B, C, F and G rocks are also associated with channels, grooves, and mudstone rip-up clasts up to 4 cm by 30 cm in size at the base of B beds. The facies F slumps have sets of discrete glide planes at the base of each slump that cut across lower chaotic beds. Flame structures and ball and pillow loading are common in the sandstone units.

Facies B rocks may also have the association B, C, D, E and minor F. The facies E levee deposits have abundant starved ripples and very low-angle reverse fault blocks that are interpreted as levee slumps (T. Nilsen, pers. comm.). Small intrastratal folds also indicate

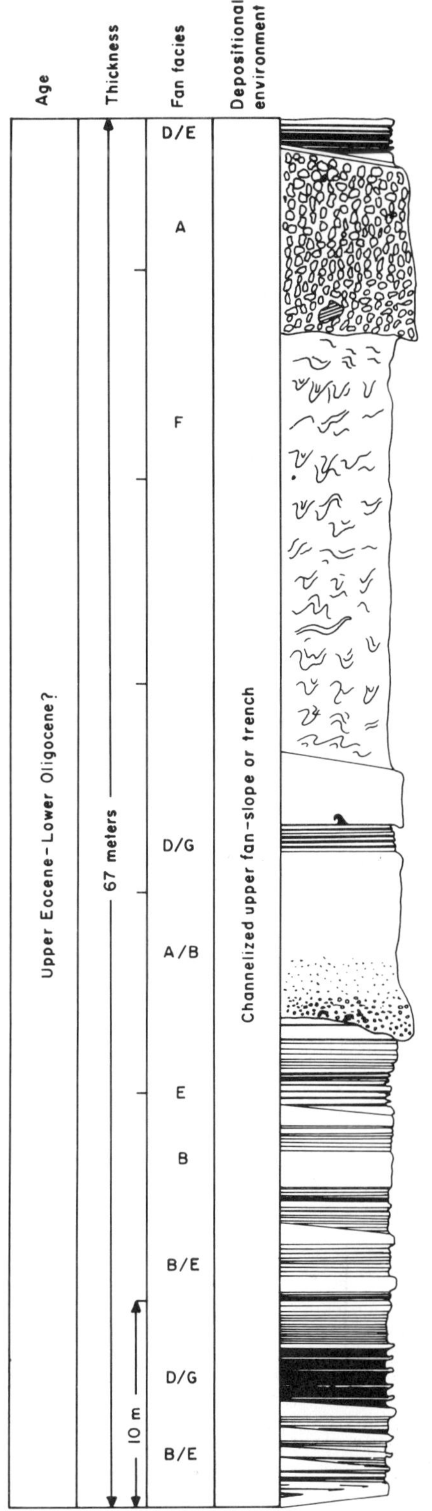

FIG. 11. Representative stratigraphic section in a portion of the Mélange Unit.

slumping of these levee deposits. Levee deposits and facies B rocks are also found in association with C and D beds in the following sequence:

$$C, D \rightarrow B, E \rightarrow C.$$

Eastern Unit facies

Facies relationships in the Eastern Unit are clearer because the rocks are little deformed. along the South Fork of the Eel River, lateral exposures can be traced over several kilometres as the river meanders. The full range of turbidite and associated facies also is present in the Eastern Unit. Thick massive sands of facies A and B are abundant. Sandstone/mudstone ratios of sequences in the Eastern Unit range from 2.5 to 13. Amalgamated sandstone beds, with thin shale partings, approach 85 m in thickness (Fig. 12).

Facies A conglomerates are similar to those in the Mélange Unit. In one area, a 25 m thick facies A sandstone onlaps against an erosional surface cut several metres into overbank deposits. Elsewhere, a channel cuts into folded, disrupted sandstone beds that were apparently part of a slump.

Dune cross-bedding up to 55 cm thick is common within levee and overbank deposits (facies E). Locally, a sequence of graded sand, parallel laminated sand, convolute laminations, and silt having the appearance of a Bouma T_{a-d} sequence is deposited on an inclined dune bed surface. This antecedent has not been reported elsewhere in turbidite facies and the process of formation is not clear.

A 5 m thick facies F pebbly mudstone that fills a channel cut into lower beds contains a basal layer of mudstone rip-up clasts overlain by blocks of deformed sandstone in the matrix. The sandstone clasts apparently were eroded from only semi-lithified beds. A common association is channellized facies A and/or B with facies E that thin- and fine-upward to C and/or D facies rocks (Fig. 13) and may correspond to progressive channel fill and abandonment. The sequence may be capped by facies D and G (Fig. 13) and the fining-upward sequence may go directly from A, E, D, G. A slight modification of this trend involves the sequence channellized A, G, E, B, C with a thinning-upward sequence. The facies C and D beds are laterally continuous over tens of metres; exposures preclude tracing them further. Facies A beds may also be encased in hemipelagic deposits, giving a G-A-G sequence.

Several thickening- and coarsening-upward sequences are evident in the Eastern Unit. Laterally continuous facies C and D sediments are overlain by G and E beds, then covered by thick channelled facies A and B beds.

Similar to this sequence is the coarsening-upward sequence facies B, C, D, G overlain by channellized facies A and B rocks. The Eastern Unit consists of many mixed facies as well: A, B, C with minor D, E; A, B, F, C, D; A, F, G, E with some B; and A, B, C, E, G.

FIG. 12. Thick massive sandstones of Eastern Unit. Note automobile in lower left for scale.

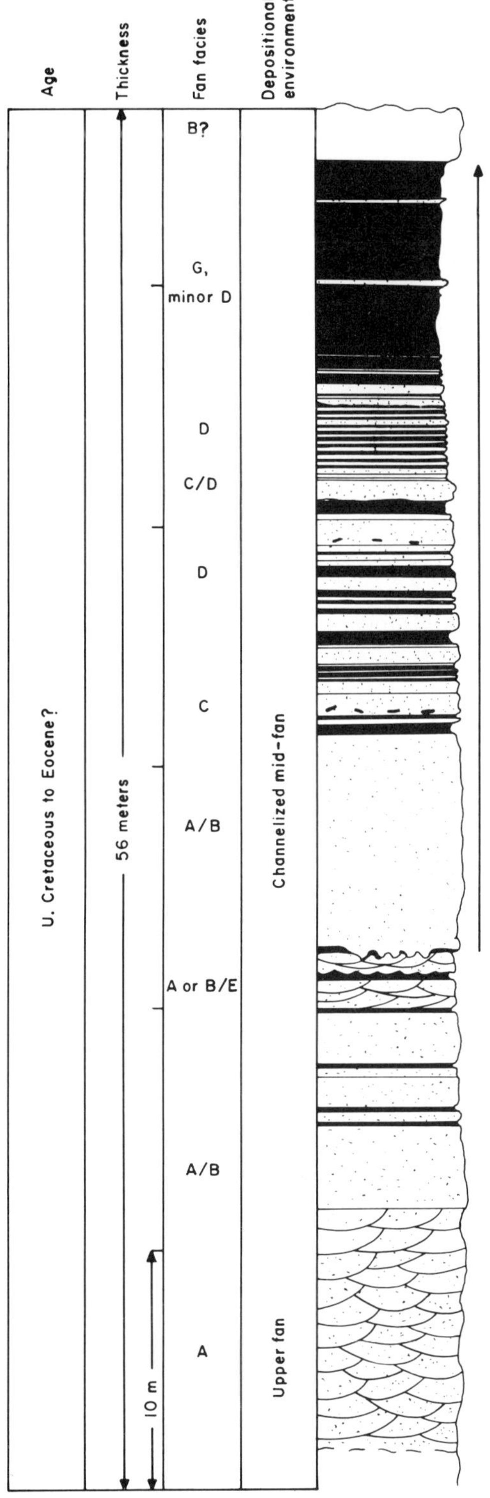

FIG. 13. Stratigraphic section in a portion of the Eastern Unit.

Sedimentary petrology

Using the petrographic techniques of Dickinson (1970) and Graham *et al.* (1976), 99 sandstones were analysed (Bachman 1979). In addition, a detailed study of quartz types aided in provenance interpretation. Triangular diagrams of modal percentages of the various mineralogical components were compared to Dickinson & Suczek's (1979) tectonic provenances for sandstones; this comparison indicates a mixed tectonic provenance for Coastal Belt sandstones.

Data from Tertiary sandstones are plotted in two separate groupings (Fig. 14). Most samples, from both Mélange and Eastern Units, are quartz- and feldspar-rich; some mélange samples are lithic-rich and plot nearer the L pole. Cretaceous sandstones, not shown, display similar groupings.

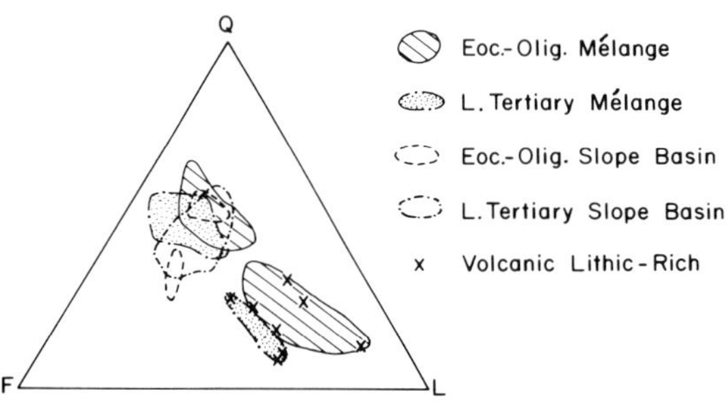

FIG. 14. Triangular diagram of modal percentage of quartz, feldspar, and lithic fragments in Coastal Belt sandstones of Tertiary age.

The anomalous Mélange Unit rocks are abundant in volcanic lithic fragments and plot in the volcanic-arc to transitional-arc fields of each of Dickinson & Suczek's (1979) diagrams. These volcanic-rich sandstones are all located in the Mélange Unit in somewhat linear outcrop patterns. These sandstones also contain small amounts of polycrystalline quartz types such as polygonized, partly recrystallized (bimodal), and schistose texture quartz that are most likely from metamorphic terranes; volcanic-rich conglomerate matrices contain clasts of varying lithologies. Thus, the volcanic lithic-rich sandstones appear to be derived from a distinct volcanic source, and other detrital material from non-volcanic sources was mixed locally. This volcanic source contributed detritus sporadically from Late Cretaceous into the Oligocene or later. The Eocene-Oligocene part of the Eastern Unit appears to have a compositional sub-field with more feldspar and less total quartz than the bulk of the rocks; these rocks plot within the plutonic arc provenance field of Dickinson & Suczek.

The remaining rocks appear to have a mixed provenance. Volcanic detritus is present in virtually all the rocks in varying amounts. Most polycrystalline quartz grains are finely crystalline with metamorphic textures. Chert is a ubiquitous component. Quartz-mica schist fragments and detrital epidote are not uncommon; blueschist fragments are present, but rare.

The variety of conglomerate clasts includes widely varying proportions of red, green, and black chert, sandstones of varying compositions, mudstone and siltstone, volcanic rocks of varying compositions, plutonic rocks, limestone, and vein quartz. Sandstone clasts vary in composition within any one conglomerate. Many of the sandstone clasts are similar in composition to the sandstone matrix. In other cases, compositions are dissimilar; a Mélange Unit conglomerate matrix enriched in volcanic lithic fragments commonly contains sandstone clasts with low volcanic content. The sandstone clasts have elongate or irregular shapes, suggesting erosion of only partly lithified sediments.

The volcanic clasts are for the most part porphyritic and range in composition from acidic to basaltic. The majority of the clasts are probably andesitic. More mafic clasts commonly have remnants of rinds suggesting a pillow origin. The volcanic clasts are included in sandstone matrices of varying compositions; there is no correlation between volcanic lithic-rich matrices and the relative abundance of volcanic clasts of any composition.

Palaeocurrent analysis

The amount of deformation in the Mélange Unit precludes a straightforward analysis of palaeocurrent data. Only data from the less-deformed rocks (style 2 and 3) were used for analysis. Measured palaeocurrent indicators include flute casts, groove and tool casts, and ripple and dune cross-laminations. The directions of overturning of intrastratal folds were measured as palaeoslope indicators. The direction of overturning of flame structures may indicate current directions or down-slope directions. Fig. 15 shows the calculated palaeocurrent directions. Grooves and symmetric ripples are plotted with bimodal directions.

Sole marks are poorly preserved in the Mélange Unit because of shearing along the

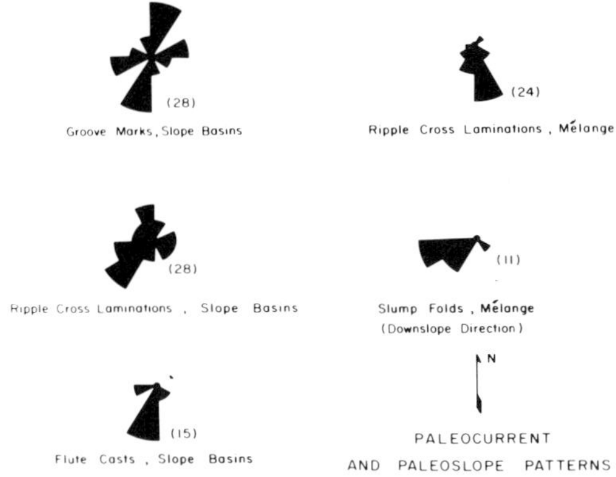

FIG. 15. Palaeocurrent and palaeoslope measurements from the Coastal Belt. The number of measurements is shown in parentheses.

bases of the beds. Only two or three groove marks could be measured; the small number are statistically insignificant. The much more abundant ripples suggest transport to the south. Less than one-fourth of the ripple directions were measured on inferred overbank deposits; current directions of flows that overtop channels would be expected to flow in divergent directions from the channel (Walker 1977), but generally downslope. The general sense of current flow in the Mélange Unit apparently was to the south and SW.

Flute casts, groove casts, and ripple laminations all suggest current direction was toward the SW in the Eastern Unit (Fig. 15). The minor NW directed ripple indicators may represent local turbidity flow towards the NW or may be divergent overbank currents.

Palaeogeography and tectonics of the Coastal Belt

Reconstructed Late Mesozoic and early Tertiary plate motions indicate oblique subduction of the Farallon plate beneath the North American plate (Coney 1981), with rates of convergence varying from 14 cm yr^{-1} in the period 80 to 45 Ma to 8 cm yr^{-1} after 45 Ma. The increased convergence rate may have resulted in flattening of the dip of the Benioff zone, eastward shift (Coney & Reynolds 1977) or temporary disappearance of the volcanic arc, and eastward shift or disappearance of the locus of blueschist metamorphism (Bachman 1978b).

Interpretation of structural styles and trends

The Mélange Unit is probably a Late Cretaceous to late Eocene or younger accretionary prism, with the youngest sediments towards the seaward side of the prism (Bachman 1978a). The Eastern Unit has many of the characteristics of trench slope basins (e.g. Moore & Karig 1976). It is much less deformed than the Mélange Unit, slightly younger than the adjacent mélange, and its contacts with the mélange are partly faulted, partly gradational in degree of deformation. It is unlikely that many segments of the Eastern Unit were juxtaposed with the melange after cessation of subduction because of the complex structural patterns shown in Figs 5 & 10, the complementary age relationships of the Eastern Unit overlying the mélange, and the partly gradational contacts between units. However, such a juxtaposition of portions of the outcrops by either later thrust or strike-slip faulting cannot be entirely ruled out.

The linear patterns of boudinaged and sheared rocks in the Mélange Unit probably represent shear zones in the accretionary prism (Figs 5 & 6). The pattern of age determinations in the unit suggests imbricate stacking; these sheared rocks may be wide zones of imbricate thrusts along which sediments were detached from the oceanic plate and accreted to the leading edge of the overriding continental plate. The shear foliation now dips steeply to the east or is slightly overturned; if landward rotation during progressive uplift and deformation of the accretionary prism is taken into account, the imbricate thrust zones dipped gently towards the east before tilting. Palaeontological age resolution does not permit identification of individual thrust sheets in the mélange.

Isoclinal folds in the Mélange Unit probably developed contemporaneously with the shearing. The seaward-verging folds are similar in orientation to those recognized developing near the toe of modern accretionary prisms, immediately after sediments are detached from the oceanic plate (Seely *et al.* 1974; Kulm & Fowler 1974; Moore, J. C. & Karig 1976). Slump folds in the Mélange Unit, presumably formed on the oversteepened, tectonically active trench slope, are also overturned seaward (SW).

The volcanic and intrusive rocks in the Coastal Belt are all located in the most sheared parts of the mélange, forming linear bands parallel to the shear foliation. Volcanic and intrusive rocks can be introduced into the accretionary prism as offscraped fragments of seamounts, aseismic ridges, oceanic fault blocks, or as intrusions into the trench or prism along leaky transforms or triple junctions (see discussion in Bachman 1978a). The association of carbonates with volcanic blocks suggests that some of the igneous rocks are offscraped seamounts or oceanic crust formed above the CCD.

The pattern of Eastern Unit exposures along section B-B′ (Fig. 5) indicates basin erosional remnants exposed between outcrops of mélange rocks. These Eastern basin deposits (Figs 8 & 10) may be either erosional remnants of a large slope basin that once covered much of the Western Mélange Unit accretionary prism, or they may have been deposited as several discrete basins, each ponded behind a ridge of the uplifted accretionary prism. On modern trench slopes, several ponded basins may eventually fill and overtop an uplifted accretionary ridge to form a larger basin (G. F. Moore & Karig 1976).

Other tectonic settings for part of the Eastern Unit could be possible if the postulated oblique subduction along the coast was decoupled into a subduction zone/transcurrent fault pair evident in some western Pacific subduction complexes (Fitch 1972; Walcott 1978; Lewis 1981). Part of the Eastern Unit would then have been deposited in a basin similar to the present continental borderland basins of southern California (e.g. Crowell 1976). However, no evidence of older large-scale strike-slip faults within the Eastern Unit has been reported. Borderland basins and marginal ocean basins must be enclosed on the seaward side by an uplifted basement terrane older than the accumulated sediments. No rocks older than Upper Cretaceous Coastal Belt rocks have been found on the seaward side of the Eastern Unit.

Interpretation of Coastal Belt lithofacies

With the aid of a covergent margin sedimentation and facies model (Bachman 1979; Underwood *et al.* 1980; Underwood & Bachman 1981), the depositional setting of many of the Coastal Belt sediments can be postulated. Many of the Mélange Unit sedimentary facies associations represent channellized flow, with local channel cutting and thinning- and fining-upward sequences.

The areal distribution and current direction indicators of these sandstones are parallel to the margin. Both trench and slope environments can have channellized sediment dispersal patterns parallel to the margin (Underwood & Bachman 1981), representing either filling of the trench or filling of elongate slope basins. The deformation and imbricate stacking sequence of these Mélange Unit rocks suggests an origin in the trench or on the lower slope where deformation is greatest; lower slope basins are generally not filled with thick, coarse clastics (see Underwood & Bachman 1981), so the channellized mélange sandstones are most likely trench deposits. No definitive criteria are present for interpreting many of the other Mélange Unit sediments. The absence of any thick pelagic or hemipelagic deposits suggests that any fine-grained slope or trench deposits were overwhelmed by coarse clastics and that abyssal plain deposits may have been selectively subducted (Moore 1975) and/or underplated beneath the margin at a level deeper than the present erosional surface.

Eastern Unit sedimentary facies associations resemble well-developed slope basin and channel facies (Underwood & Bachman 1981). The thick channel sands may have been deposited in feeder channels that led into the trench and distribution networks within the slope basins; the outcrop pattern of several conglomerate bodies suggest transport at an angle to the margin. The SW directed palaeocurrent indicators in the Eastern Unit probably represent flow in basins parallel to the margin, with local feeder channels at higher angles to the margin.

The Eastern Unit also has well-developed coarsening- and thickening-upward sequences that can be interpreted as prograding depositional lobes of small fans developed in slope basins. The small amount of typical slope and canyon facies (i.e. G → A → G), particularly hemipelagic muds, suggests that large amounts of coarse clastic material were available from the source area and that sedimentation rates were fairly high. Continuing fieldwork in the northern section of the Coastal Belt exposures confirms the presence of more typical thick hemipelagic slope muds to the north of the sand-rich slope basins. Poor palaeontological age resolution and structural complexities preclude sedimentation rate calculations. High energy flow in the basins could have prevented low energy deposition, and the fine material was transported to deeper water (Bachman 1978a). High sand/mud ratios are also present in another inferred Franciscan trench slope basin in central California (Howell *et al.* 1977).

Interpretation of provenance and sediment distribution

It has been shown that transport of sediments

in the Coastal Belt was generally toward the south and SW. Several rock types, including metamorphic, plutonic and volcanic rocks, and recycled oceanic sediments, contributed detritus to the Coastal Belt. Inferred trench deposits are locally very volcanic lithic-rich. Source areas previously suggested for Coastal Belt sediment include the Sierra Nevada, Klamath Mountains, and uplifted older Franciscan terranes (O'Day 1974; Kleist 1974; Kramer 1976).

The Klamath mountains to the NE of the Coastal Belt contain source rocks that would account for virtually all Coastal Belt detritus. The abundance of metamorphic rock types over plutonic types in the Klamaths would also account for the predominance of metamorphic quartz types in the Coastal Belt. The direction of palaeocurrent indicators also suggests transport from the NE, the present position of the Klamath Mountains relative to the Coastal Belt assuming no large-scale lateral transport of Coastal Belt rocks. Lower Tertiary current directions in streams draining the Klamath Mountains were towards the south (Dickinson *et al.* 1979). Thus, the Klamath Mountains must be considered as a possible major source for Coastal Belt sediments. However, large-scale lateral translations parallel to the margin may have separated the Coastal Belt from source rocks similar in composition to the Klamath Mountains, but at large distances from the present site of the Coastal Belt. Palaeomagnetic studies by K. Verosub and myself indicate resetting of remnant magnetism after folding, possibly at slightly elevated temperatures (100°C or more) for a long period of time (Bachman 1979). Thus, there are at present no data suggesting large lateral translations of Coastal Belt rocks.

The source for the volcanic lithic-rich sandstones has been proposed to be a western andesitic arc that has subsequently been subducted in part (Beutner 1977), based on one DSDP site drilled off northern California that penetrated volcanic rocks similar in composition to andesite. This hypothesis is unlikely based on several arguments: (1) no other evidence, sedimentary or tectonic, exists to suggest a western arc; (2) the DSDP site is west of the San Andreas fault, so the rocks formed as much as 1000 km south of their present position relative to the Coastal Belt; (3) if the arc was 1000 km long so that the northern end could contribute detritus to the Coastal Belt, fragments of the arc would now be exposed all along the coast of California (it is unlikely that the entire arc would later be subducted); (4) a western arc is not compatible with the evidence for a west-facing Coastal Belt subduction complex; and (5) volcanic detritus is found in rocks of Late Cretaceous, early Tertiary, and Eocene to Oligocene age, suggesting that the source was stable over a long period of time. However, arc volcanism was widespread in central and southern Oregon during Coastal Belt deposition (Snyder *et al.* 1976) and the area was topographically high (Axelrod 1968). In the lower Tertiary forearc basin of western Oregon, volcanic detritus was transported long distances northward from this southern Oregon arc (Chan & Dott 1980). It is probable that these volcanic arc rocks also shed detritus towards the trench; during volcanic pulses, currents may have brought this detritus similar distances south along the trench to the site of Coastal Belt deposition. In the central part of Coastal Belt exposures, the volcanic-rich sandstones are restricted to possible trench environments in the Western Mélange Unit.

Discussion

The Coastal Belt subduction complex shares many structural and stratigraphic relationships with other convergent margins. Imbricate stacks younging oceanward, mélange terrane, and slightly younger less-deformed rocks representing trench slope basins are all present. The structural mapping technique used here has further defined the geometry of deformation within this subduction complex.

The recognition of trench and trench slope deposits in ancient subduction complexes is a difficult problem. Using structural and sedimentary facies models it is possible to determine the depositional setting of some of the less-deformed subduction complex rocks. When the rocks are highly deformed and depositional sequences become fragmented, these criteria are less useful. The sandstone petrology, however, may reflect differing source areas and distribution patterns of sediments deposited on the abyssal plain-trench and trench slope. In the Coastal Belt, the volcanic lithic-rich sandstones of the Western Mélange Unit may have received volcanic detritus by long distance transport along the trench. Thus, with the proper sedimentological setting and lack of extensive stratigraphic dismemberment, the complex structural and stratigraphic relationships in subduction complexes may be partly reconstructed.

References

ATWATER, T. & MOLNAR, P. 1973. Relative motion of the Pacific and North American plates deduced from sea-floor spreading in the Atlantic, Indian and South Pacific oceans. *In:* KOVACH, R. L. & NUR, A. (eds). *Proc. Conf. Tectonic Problems San Andreas Fault System.* Univ. Stanford Publ. geol. Sci. **13,** 136–48.

AXELROD, D. I. 1968. Tertiary floras and topographic history of the Snake River Basin, Idaho. *Bull. geol. Soc. Am.* **79,** 713–34.

BACHMAN, S. B. 1978a. A Cretaceous and early Tertiary subduction complex, Mendocino coast, northern California. *In:* HOWELL, D. G. & MCDOUGALL, K. A. (eds). *Mesozoic Paleogeography of the Western United States. Pacif. Coast Paleogeogr. Symp.* Pacif. Sec. Soc. econ. Paleontol. Mineral. **2,** 419–30.

—— 1978b. Latest Cretaceous and early Tertiary subduction in northern California—fundamental changes from Late Mesozoic. *Abstr. Prog. geol. Soc. Am.* **10,** 361–2.

—— 1979. *Sedimentation and margin tectonics of the Coastal Belt Franciscan, Mendocino coast, northern California.* Thesis, PhD, Univ. California, Davis, 166 pp. (unpubl.)

—— & ALLMENDINGER, R. 1980. Subduction complex deformation related to thrust ramps—a possible model. *Abstr. Prog. geol. Soc. Am.* **12,** 380–1.

BAILEY, E. H., IRWIN, W. P. & JONES, D. L. 1964. Franciscan and related rocks, and their significance in geology of Western California. *Bull. Calif. Div. Mines Geol.* **183,** 177 pp.

BERKLAND, J. O., RAYMOND, L. A., KRAMER, J. C., MOORES, E. M. & O'DAY, M. 1972. What is Franciscan? *Bull. Am. Assoc. Petrol. Geol.* **56,** 2295–302.

BEUTNER, E. C. 1977. Evidence and implications of a Late Cretaceous-Paleogene island arc and marginal basin along the California coast. *Abstr. Prog. geol. Soc. Am.* **9,** 389.

BLAKE, M. C. JR & JONES, D. L. 1974. Origin of Franciscan melanges in northern California. *In:* DOTT, R. H. JR & SHAVER, R. H. (eds). *Modern and Ancient Geosynclinal Sedimentation.* Spec. Publ. Soc. econ. Paleontol. Mineral. Tulsa, **19,** 345–57.

—— & —— 1978. Allochthonous terranes in northern California?—a reinterpretation. *In:* HOWELL, D. G. & MCDOUGALL, K. A. (eds). *Mesozoic Paleogeography of the Western United States.* Pacif. Sec., Soc. econ. Paleontol. Mineral. Pacif. Coast Paleogeogr. Symp. **2,** 397–400.

BOUMA, A. H. & NILSEN, T. H. 1978. Turbidite facies and deep-sea fans—with examples from Kodiak Island, Alaska. *Proc. Offshore Technol. Conf.* **1,** 559–70.

CHAN, M. A. & DOTT, R. H., JR 1980. Deep-sea fan deposition in an Eocene forearc basin, western Oregon. *Abstr. Prog. geol. Soc. Am.* **12,** 401.

CONEY, P. J. 1981. Mesozoic-Cenozoic Cordilleran plate tectonics. *Spec. Pap. geol. Soc. Am.* (in press).

—— & REYNOLDS, S. J. 1977. Cordilleran Benioff zones. *Nature. London,* **270,** 403–6.

COWAN, D. S. 1981. Deformation of partly dewatered and consolidated Franciscan sediments near Piedras Blancas Point, California (this volume).

CROWELL, J. C. 1976. Implications of crustal stretching and shortening of coastal Ventura basin, California. In: *Aspects of the geologic history of the California continental borderland.* Pacif. Sec. Misc. Publ. Am. Assc. Petrol. Geol. **24,** 365–82.

DICKINSON, W. R. 1970. Interpreting detrital modes of graywacke and arkose. *J. sediment. Petrol.* **40,** 695–707.

—— & SUCZEK, C. A. 1979. Plate tectonics and sandstone compositions. *Bull. Am. Assoc. Petrol. Geol.* **63,** 2164–82.

—— , INGERSOLL, R. V. & GRAHAM, S. A. 1979. Paleogene sediment dispersal and paleotectonics in northern California. *Bull. geol. Soc. Am.* **90,** 1458–528.

EVITT, W. R. & PIERCE, S. T. 1975. Early Tertiary ages from the coastal belt of the Franciscan Complex, northern California. *Geology,* **3,** 433–6.

FITCH, T. J. 1972. Plate convergence, transcurrent faults, and internal deformation adjacent to Southeast Asia and the western Pacific. *J. geophys. Res.* **77,** 4432–60.

GRAHAM, S. A., INGERSOLL, R. V. & DICKINSON, W. R. 1976. Common provenance for lithic grains in Carboniferous sandstones from Ouachita Mountains and Black Warrior Basin. *J. sediment. Petrol.* **46,** 620–32.

HOWELL, D. G., VEDDER, J. C., MCLEAN, H., JOYCE, J. M., CLARK, S. H. JR & SMITH, G. 1977. Review of Cretaceous geology, Salinian and Nacimiento blocks, Coast Ranges of central California. *In:* HOWELL, D. G. (ed). *Pacif. Coast Paleogeogr. Field Trip Guide.* Pacif. Sec. Soc. econ. Paleontol. Mineral. **2,** 1–46.

KLEIST, J. R. 1974a. *Geology of the Coastal Belt, Franciscan Complex, near Ft. Bragg, California.* Thesis, PhD, Univ. Texas, Austin, 133 pp.

—— 1974b. Deformation by soft-sediment extension in the Coastal Belt, Franciscan Complex. *Geology,* **2,** 501–4.

—— 1975. Variation and distribution of sandstone composition in the Coastal Belt, Franciscan Complex, near Ft. Bragg, California. *Abstr. Prog. geol. Soc. Am.* **7,** 335–6.

KRAMER, J. C. 1976. *Geology and tectonic implications of the Coastal Belt Franciscan, Ft. Bragg-Willits area, northern Coast Ranges, California.* Thesis, PhD, Univ. California, Davis, 128 pp.

—— 1977. Cenozoic tectonics (plate motions) of the Coastal Belt Franciscan, northern California Coast Ranges. *Abstr. Prog. geol. Soc. Am.* **9,** 448–9.

KULM, L. D. & FOWLER, G. A. 1974. Oregon continental margin structure and stratigraphy: a test of the imbricate thrust model. *In:* BURK, C. A. & DRAKE, C. L. (eds). *Geology of Continental Margins*, 261–83. Springer-Verlag, New York.

LEWIS, K. B. 1981. Quaternary sedimentation of the Hikurangi oblique-subduction and transform Margin. *In:* BALLANCE, P. F. & READING, H. G. (eds). *Sedimentation in Oblique-slip Mobile Zones.* Spec. Publ. int. Assoc. Sediment. **4,** 171–89. Blackwell Scientific Publications, Oxford.

MCLAUGHLIN, R. J., KLING, S. A., POORE, R. Z., MCDOUGALL, K., BEUTNER, E. C. & OHLIN, H. N. 1979. Post-Middle Miocene microplate accretion of Franciscan Coastal Belt rocks to northern California. *Abstr.Prog.geol. Soc. Am.* **11,** 476–7.

MOORE, G. F. & KARIG, D. E. 1976. Development of sedimentary basins on the lower trench slope. *Geology,* **4,** 693–7.

MOORE, J. C. 1975. Selective subduction. *Geology,* **3,** 530–2.

—— & KARIG, D. E. 1976. Sedimentology, structural geology, and tectonics of the Skikoku subduction zone, southwestern Japan. *Bull. geol. Soc. Am.* **87,** 1259–68.

O'DAY, M. S. 1974. *The structure and petrology of the Mesozoic and Cenozoic rocks of the Franciscan Complex, Leggett-Piercy area, Northern California Coast Ranges.* Thesis, PhD, Univ. California, Davis, 152 pp.

OGLE, B. A. 1953. Geology of Eel River Valley area, Humboldt County, California. *Bull. Calif. Div. Mines Geol.* **164,** 128 pp.

RICCI-LUCCHI, R. 1975. Depositional cycles in two turbidite formations of northern Apennines (Italy). *J. sediment. Petrol.* **45,** 3–43.

SEELY, D. R., VAIL, P. R. & WALTON, G. G. 1974. Trench slope model. *In:* BURK, C. A. & DRAKE, C. L. (eds). *Geology of Continental Margins,* 249–60. Springer-Verlag, New York.

SNYDER, W. S., DICKINSON, W. R. & SILBERMAN, M. L. 1976. Tectonic implications of space-time patterns of Cenozoic magmatism in the western United States. *Earth planet. Sci. Lett.* **32,** 91–106.

UNDERWOOD, M. B. & BACHMAN, S. B. 1981. Sedimentary facies associations within subduction complexes (this volume).

UNDERWOOD, M. B., BACHMAN, S. B. & SCHWELLER, W. J. 1980. Sedimentary processes and facies associations within trench and trench-slope settings. *In:* FIELD, M. E., BOUMA, A. H. & COLBURN, I. (eds). *Quaternary Depositional Environments on the Pacific Continental Margin,* Pacif. Sec. Soc.econ. Paleontol. Mineral. 211–29.

WALCOTT, R. I. 1978. Geodetic strains and large earthquakes in the axial tectonic belt of North Island, New Zealand. *J. geophys. Res.* **83,** 4419–29.

WALKER, R. G. 1977. Upper Cretaceous resedimented conglomerates at Wheeler Gorge, California: description and field guide—reply. *J. sediment. Petrol.* **47,** 928–30.

—— & MUTTI, E. 1973. Turbidite facies and facies associations. *In: Pacif. Sec. Short Course, Turbidites and Deep-water Sedimentation,* 119–57. Soc. econ. Paleontol. Mineral.

STEVEN B. BACHMAN, Department of Geological Sciences, Cornell University, Ithaca, New York 14853, U.S.A.

The Franciscan Complex of northernmost California: sedimentation and tectonics

K. R. Aalto

SUMMARY: The Franciscan Complex consists, from east to west, of a belt of foliated (textural zone 2 and 3) rocks deformed beneath the Coast Range thrust (CRT), a belt of broken formation (BF) and a belt of sheared olistostromes (SO) consisting of conglomerate, greywacke, limestone, chert, greenstone, plutonic and glaucophane schist blocks in chiefly argillaceous matrix. South of the Klamath River, zone 2 and 3 rocks are of an outer fan or basin-plain facies and have overthrust midfan unfoliated (textural zone 1) BF in which suprafan lobes are preserved as megaboudins. To the north, foliated rocks are in gradational contact with similar BF and two episodes of thrusting were confined to the CRT. Blocks of chert, gabbro and metavolcanic rock were tectonically transported along thrust faults. Greywacke and conglomerate of the SO belt are inner fan deposits. The BF-SO contact is gradational, delineated by the westward appearance of greenstone, chert and plutonic blocks increasing progressively in size and/or abundance. Olistostrome units are intercalated with isoclinally folded, recycled sediments. BF and SO greywackes are commonly feldspathic litharenites with mean parameters (Dickinson 1970): $Q = 54.6$, $C/Q = 0.28$, $F = 21.6$, $P/F = 0.97$, $L = 23.8$ and $V/L = 0.82$. Sediments were deposited in a trench slope basin and intercalated with olistostromes derived from elevating portions of the accretionary basement. Progressing upwards in the local sequence north of the Klamath River, the gradational SO-BF transition, decrease in overtuned beds and a retrograding fan sequence reflect basin filling, diminished faulting and lessened relief concomitant to basin elevation.

Geological setting

The Franciscan Complex is exposed throughout the Coast Ranges geological province (Bailey *et al.* 1964). It is juxtaposed against ultramafic rocks of the western Klamath Mountains belt (Irwin 1977) at the Coast Range thrust, along which Klamath rocks may have overriden the Franciscan Complex by tens of kilometres (Fig. 1, Blake & Jones 1977). Immediately beneath the thrust, Franciscan rocks have suffered cataclasis. For field mapping, the degree of textural reconstitution of greywacke sandstone is assessed following the scheme of Blake *et al.* (1967): textural zone 1 shows no foliation, zone 2 has platey cleavage with bedding features preserved, and zone 3 has well-developed foliation largely obliterating bedding features, and quartz segregation laminae exceeding 1 mm in thickness. Units with zone 3 greywackes are formally termed 'South Fork Mountain Schist' (see Lanphere *et al.* 1978, for a review of stratigraphic nomenclature).

In the region portrayed in Fig. 1, textural grade beneath the Coast Range thrust decreases gradationally northward. To the south, zone 3 metavolcanic and metasedimentary rocks immediately below the Coast Range thrust have overthrust zone 1 broken formation. The Redwood Mountain and Patricks Point outliers (Fig. 1, Irwin 1966) represent klippen of zone 2 and 3 rocks that are thrust over a western belt of Franciscan mélange with zone 1 greywacke (Aalto 1976, 1978; Talley 1976). Talley (1976) subdivides zone 3 rocks of the Redwood Mountain outlier into graphite and lawsonite schist members. These members are petrologically different from zone 3 rocks beneath the Coast Range thrust immediately to the east (Monsen & Aalto 1980). North of the Klamath River (Fig. 1, location D dashed line), zone 2 rocks are in gradational contact with zone 1 broken formation with the degree of textural reconstitution increasing progressively upwards towards the Coast Range thrust. The nature of the transition between the northern and southern regions has not been determined, but it is likely that the zone 3-zone 1 thrust is somewhere overlapped by the Coast Range thrust as one moves north.

The zone 1 Franciscan Complex may be subdivided into an eastern belt of solely broken formation and a western belt of mélange (terminology of Hsü 1974) containing sedimentary, metamorphic and igneous tectonic blocks dispersed in an argillaceous matrix. The boundary between these two units is approximately located in Figs 1 (lines labelled B), 2 & 3.

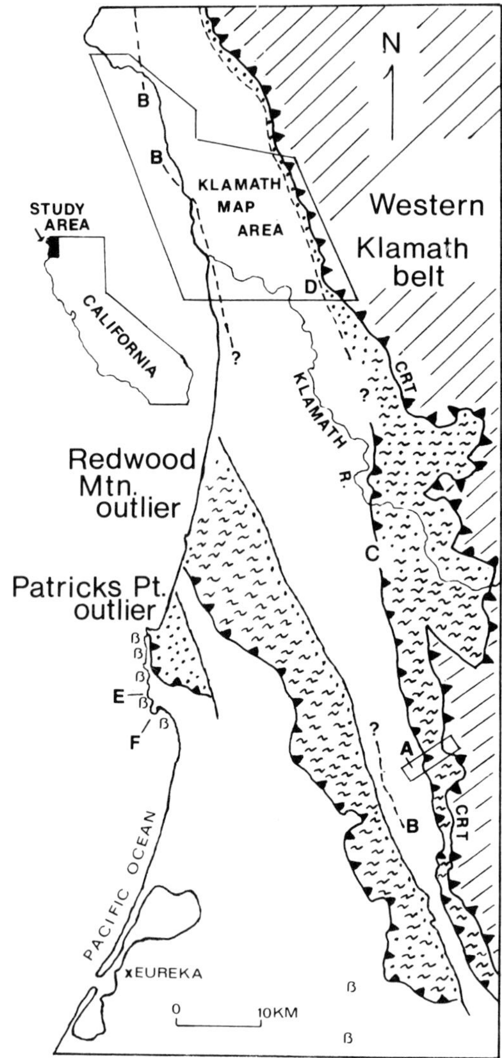

FIG. 1. Location map showing distribution of textural zones and regional geology. CRT is the Coast Range thrust fault. Zone 1 Franciscan basement: no pattern; zone 2: dots; zone 3: wavy dashes; undifferentiated zone 2/3: mixed pattern. Letters A–F refer to features discussed in text. Beta (β) denotes approximate locations of exotic glaucophane schist blocks within zone 1 terrane Sedimentological data of Table 1 are derived from all complete beds exposed in place along Highway 299 to Area A, the coast from the Trinidad map area (Aalto 1976) to Crescent City and in the outlined Klamath map area (Fig. 2). Regional geology from Aalto (1976), Monsen (1979, location A), Shimamoto (1976), Talley (1976), Young (1978), the Weed and Redding 1:250,000 sheets of the Geologic Map of California (1964) and near the Klamath River mouth, BSc thesis work of Ensrud (1978), Jepsen (1979), Lind (1979) and Lozinsky (1979).

Structure

All units of the Franciscan Complex strike NW and dip predominantly to the NE, whether in fault or depositional contact. From east to west (ignoring the zone 2/3 klippe), units become structurally lower.

Zone 2/3 rocks below Coast Range thrust

Franciscan zone 2/3 metasedimentary and metavolcanic rocks beneath the Coast Range thrust exhibit at least two periods of deformation. Primary S_1 foliation of metasedimentary rocks in all regions beneath the thrust is parallel to bedding. Associated with this foliation are F_1 folds whose axial planes are approximately parallel to the foliation. Axial planes of F_2 folds are orientated at an angle to S_1 foliation and are associated with poorly developed S_2 strain-slip cleavage.

The zone 3-zone 1 boundary thrust in the south dips gently to the east (Fig. 1, locations A–C). A marked contrast exists between isoclinically folded zone 3 schists with abundant quartz veins and pods, and the subjacent zone 1 argillite and greywacke, which is mylonitized and overturned but unfoliated (Monsen 1979). Tectonic blocks of greenstone and ribbon chert up to tens of metres in maximum dimension are dispersed within the gouge zone of the zone 3-zone 1 thrust fault. They have no obvious source in the upper or lower thrust plates. Monsen & Aalto (1980) believe that they were tectonically transported along the zone 3-zone 1 thrust fault from a mélange terrane now underlying the western Klamath Mountains belt. Similar tectonic blocks increase in abundance as one nears the Coast Range thrust north of the Klamath River (Figs 2 & 3). They are most concentrated in zone 2 metasediments (Fig. 4A). However, some are found in the underlying zone 1 broken formation as much as 1000 m west of the zone 1-zone 2 boundary (Fig. 2). Blocks of chert and greenstone can exceed 30 m in maximum dimension. The cherts are brecciated and partially recrystallized, but contain radiolaria. Greenstone blocks are brecciated and consist of silicified felsic crystal tuff, porphyritic felsite and basalt. Chloritization and quartz veining are common and blocks have foliated alteration rinds parallel to their long dimensions. Highly deformed,

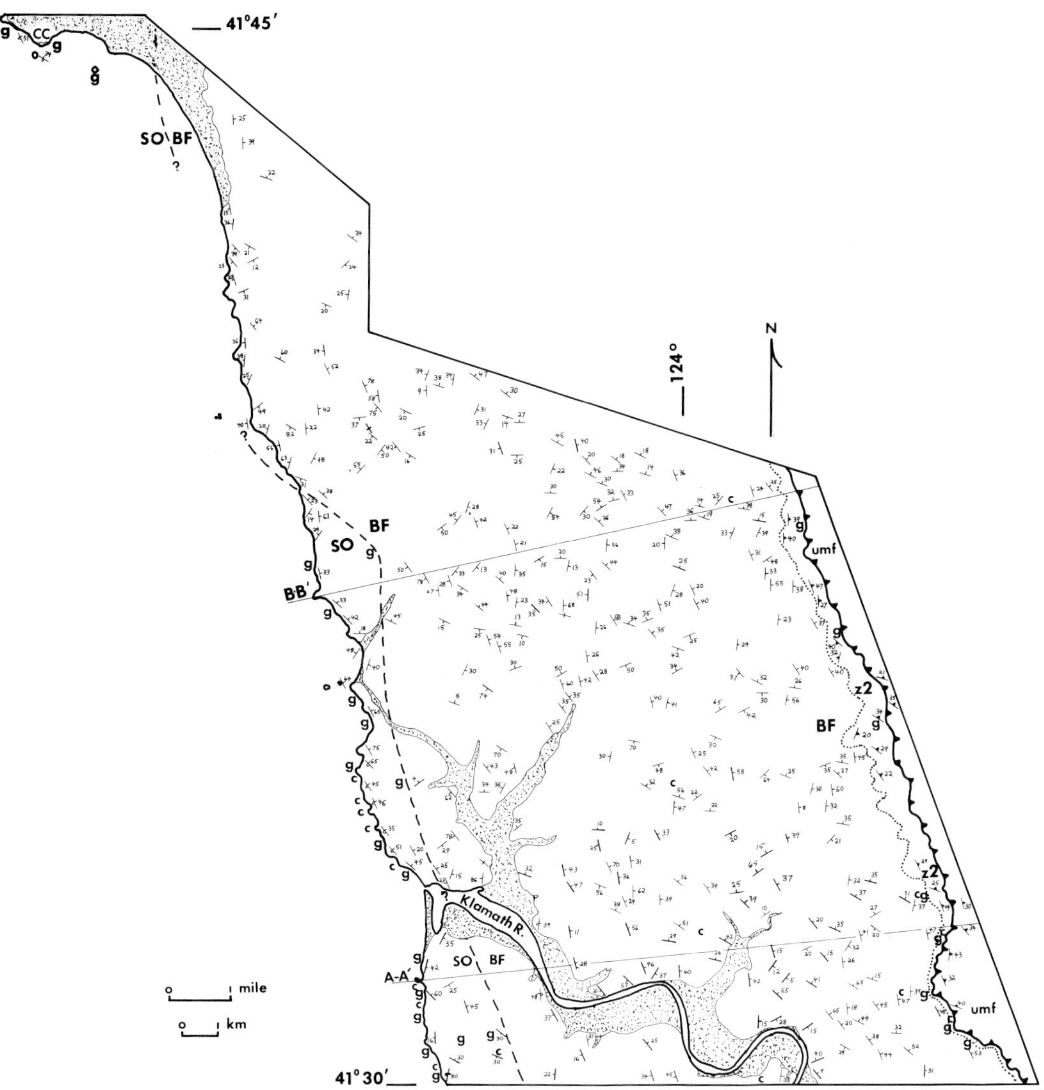

FIG. 2. Klamath map area. From east to west: Umf = ultramafic rock of the western Klamath belt which is bounded by the Coast Range thrust (solid line), z2 = textural zone 2, Franciscan metasedimentary rock (western contact dotted) containing boudins of greenstone (g) and the chert (c), BF = textural zone 1 Franciscan broken formation belt (western contact dashed), and SO = textural zone 1 Franciscan sheared olistostrome belt. Quaternary fluvial sediment and marine terrace cover is depicted by stipple near the Klamath River and Crescent City (CC).

saussuritized, albitized cumulate gabbro blocks exist at location D (Fig. 1) in zone 1 sedimentary rocks.

It is likely that at least two periods of thrust faulting account for many of these geological relationships. In both the north and south, a first period of motion along the Coast Range thrust cataclastically deformed Franciscan rocks and produced the S_1 foliation and F_1 folds. To the south, a second period of thrust faulting transported zone 3 rocks westward over zone 1 broken formation. To the north, this second period of motion was largely confined to the Coast Range thrust; however, subsidiary reverse faults within the lower plate may have provided pathways for transport of the chert, greenstone and gabbro blocks. The immense blocks distant from the thrust were probably let

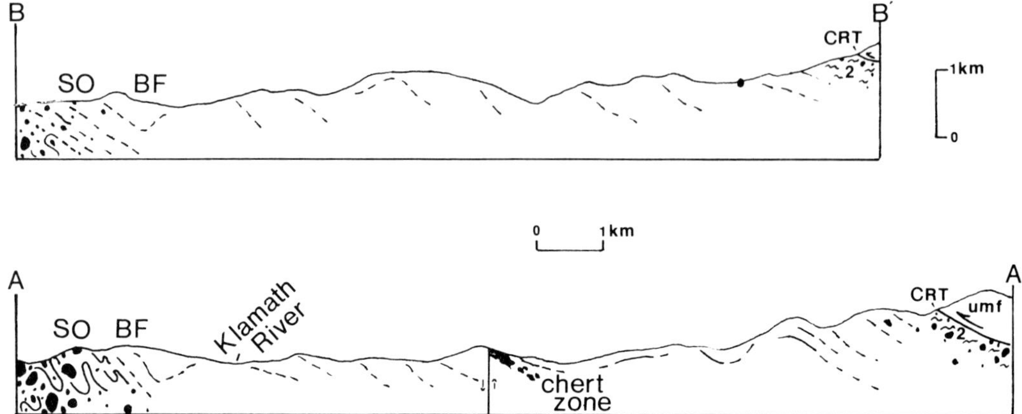

FIG. 3. Topographic and geological cross-sections, Klamath map area. Lines of section shown on Fig. 2. CRT = Coast Range thrust, 2 = textural zone 2 Franciscan metasedimentary rock. Other letter symbols as in Fig. 2. Distribution of greenstone, chert and plutonic boudins shown in black. The position of a linear chert zone within the broken formation belt is indicated on A-A′. Sense of motion and orientation of fault adjacent to this zone is conjectural due to poor exposure.

down during Holocene land degradation. In both regions, the second motion may have produced the S_2 cleavage and F_2 folds.

Zone 2/3 outliers

The thrust fault beneath the Redwood Mountain and Patricks Point outliers is marked by a zone of mylonitized argillite and serpentinite up to 100 m thick, containing dispersed tectonic blocks of brecciated chert, greenstone, conglomerate and greywacke derived from the lower plate of zone 1 mélange (Aalto 1978). Talley (1976) describes complex thrust faulting and emplacement of exotic glaucophane schist, amphibolite and greenstone blocks associated

FIG. 4. (A) Textural zone 2 Franciscan Complex, Klamath map area. Foliated argillite and greywacke containing a large greenstone boudin (under hammer). Exposure within 100 m of the Coast Range thrust in central part of area.
(B) Franciscan Complex of the sheared olistostrome belt, Klamath map area. Block of highly indurated, sheared tuff (labelled T) and dark grey argillite (A) set in the weakly indurated argillaceous matrix (Mx) of a sheared olistostrome unit. Note the veining and near vertical foliation in the foreground block. Brecciated appearance of tuff due to offset of complexly intercalated tuff and argillite. Indurated blocks are transported to the beach in debris flows generated by the failure of olistostrome matrix. Greenstone block (V) at right. Notebook in lower right for scale. Exposure along the coast approximately 1 km north of the southern boundary of the map area.
(C) Franciscan Complex of the Trinidad area. Sheared olistostrome containing boudinaged blocks of greenstone (intercalated tuff and argillite, labelled V), greywacke (S) and limestone (L) dispersed in foliated argillite. The arrow above the elongated tail of the limestone block depicts a block fragment severed by matrix intrusion during deformation. The olistostrome unit is intercalated with turbidites which dip into the cliff. Dashed lines show bedding in the turbidites, offset at lower right. The entire sequence is overturned. Notebook for scale. Exposure located approximately 1.5 km north of Elk Head (Fig. 1, location E).
(D) Franciscan Complex of the broken formation belt, Klamath map area. Boudinaged classic proximal turbidites (facies C). In upper right some boudins are bounded by compressional shears and have assumed phacoidal form. 30 cm rule for scale. Exposure midway along coastal exposure of the BF belt (1 km south of Damnation Creek).
(E) Franciscan Complex of the sheared olistostrome belt, Klamath map area. Overturned interstratified turbidities and diamictite containing angular clasts of sandstone, greenstone (at pen point) and chert (left of pen point). The contact (above pen) between the overturned T_{abe} bed and diamictite is depositional. Unsheared diamictites with unoriented clases exist in adjacent areas. Turbidites are overbank deposits (facies E) and intercalated diamictites (facies A) debris flow deposits. Exposure along coastal cliffs approximately 2 km NNW of the Klamath River mouth.

with serpentinite within the Redwood Mountain outlier. The internal thrusting described by Talley has no counterpart in area A (Fig. 1) to the east, supporting the contention that a direct correlation of zone 3 rocks of the two regions cannot be made (Monsen & Aalto 1980). Zone 2 metasedimentary rocks of the Patricks Point outlier have been overturned by drag folding during thrusting (Aalto 1978).

Along the eastern margin of the Redwood Mountain outlier, zone 1 Franciscan rocks have been emplaced over zone 2 greywacke and argillite by reverse faulting (Lind 1979; Talley 1976). Such faulting presumably postdates emplacement of the klippe and is likely to be related to Cenozoic compression (Silver 1971; Simila *et al.* 1975). Monsen & Aalto (1980) and Talley (1976) suggest that the faults on the

eastern margins of both outliers may have a significant strike-slip component, thus explaining the lack of continuity between zone 3 rocks west of the Coast Range thrust. Such motion is in accord with Cenozoic plate motions (Dott 1979; Herd 1978).

Zone 1 rocks

The zone 1 broken formation terrane (Figs 1 & 2) consists of highly deformed, boudinaged individual greywacke beds and successions of beds in sheared argillite (Fig. 4D). Beds are commonly upright, strike N30–40°W and dip 25–35° either NE or SW (Fig. 5). This pattern is indicative of a chevron style of folding, but fold hinges are rarely exposed. This folding appears to be unrelated to the events induced by motion on the Coast Range thrust.

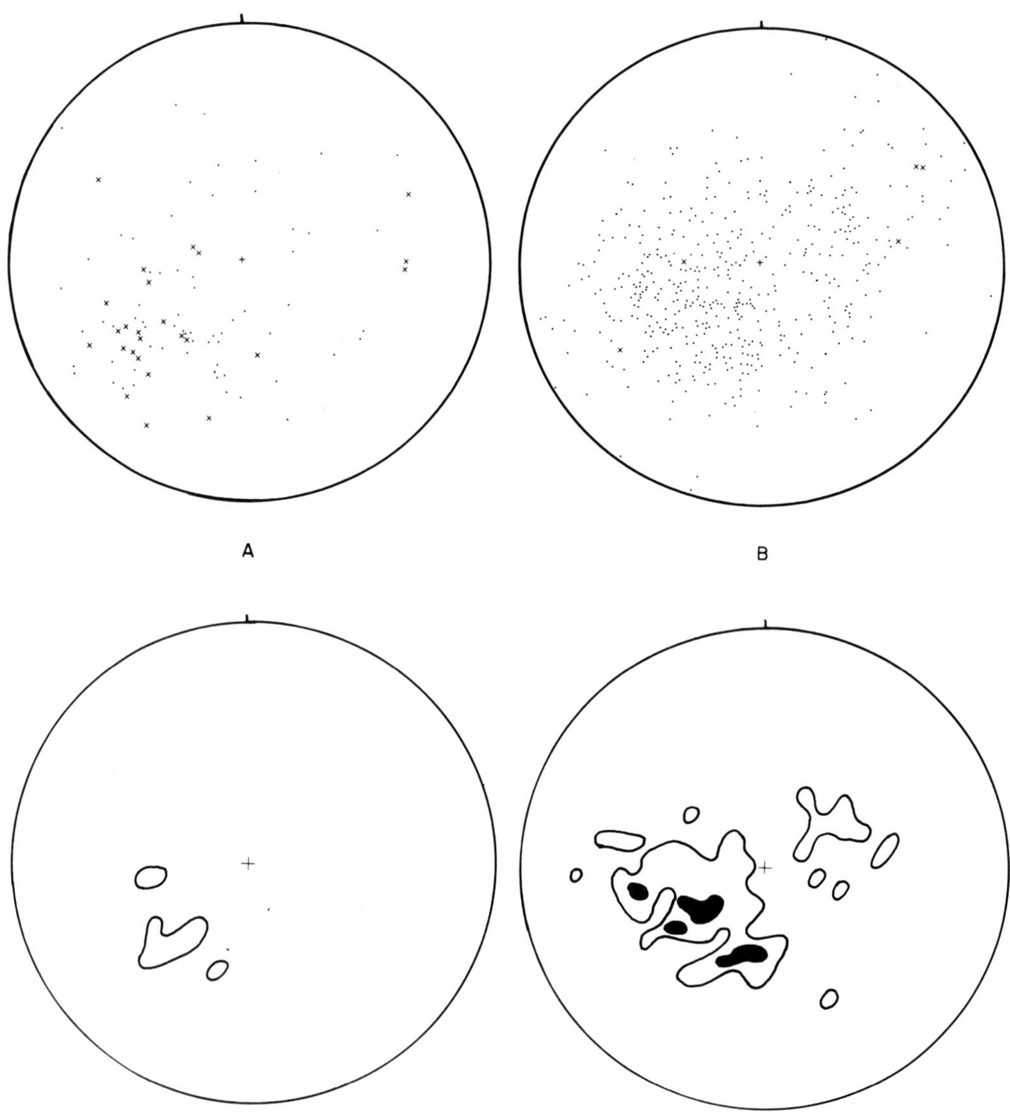

FIG. 5. Poles to bedding within: (A) the sheared olistostrome belt ($n = 102$), and (B) the broken formation belt ($n = 467$) of the Klamath map area (Fig. 2). Upright beds are depicted by dots and overturned beds by x's. Contours show 3 (encircled white) and 6 (encircled black) points per unit area.

In the south-central part of the zone 1 broken formation belt in the Klamath map area (Figs 2 & 3), several elongate bodies of red ribbon chert are arranged in a line trending NNW. Where observable, western (stratigraphically lower) contacts between the chert and surrounding broken formation are either thrusts or high-angle faults. Upper contacts in the two northern localities dip gently NE and could be thrust faulted. However, the presence of intercalated chert and argillite at the upper contact in the northernmost locality suggests conformity.

The contact between the eastern broken formation and western mélange belts within zone 1 Franciscan is apparently gradational. In the Klamath map area (Figs 2 & 3), the contact is marked by appearance of blocks of greenstone, chert and conglomerate which increase in size and abundance to the SW. The blocks are dispersed in a highly fragmented argillaceous matrix with penetrative shear-fracture fabric. The degree of fragmentation has lessened the rock strength and mélange commonly fails by debris flow (Fig. 4B, Aalto 1977). The topography is consequently more subdued than that of the broken formation belt.

Mélange units vary in lateral extent and thickness. Thinner units are clearly subaqueous debris flow (Middleton & Hampton 1973) deposits for they are interstratified with turbidites (Fig. 4C, E). These deposits provide evidence of an olistostrome mode of emplacement of mélange units hundreds of metres in thickness, which are intercalated with sections of broken formation composed of turbidites and associated redeposited sedimentary rocks. In most regions, however, post-emplacement structural activity has in most cases obliterated clear stratigraphic evidence of olistostromes. The degree of deformation within the mélange (herein termed sheared olistostrome) belt is clearly greater than in the eastern belt of broken formation. Isoclinal folding and overturning of beds in the sheared olistostrome belt is common (Figs 4C, E & 5). Axial planes of isoclinal folds commonly strike NW and dip 40–50° NE, and axes plunge SE at shallow angles.

Blocks in olistostrome units may be angular, rounded or tectonically elongated. There is no doubt that blocks have been sheared subsequent to sedimentary mixing for they are commonly bounded by compressional shears and traversed by extensional shear fractures. Argillaceous matrix commonly intrudes into cracks and fractures of disintegrating blocks (Fig. 4C). Irregular blocks show random orientation of slickensided shear surfaces, and elongate blocks have a preferred orientation parallel to the dominant NW strike and NE dip of the Franciscan Complex. Recumbent or isoclinal folds exist in olistostrome matrix material in the vicinity of larger blocks, which commonly have rinds of mylonitized argillite. Such observations imply tectonic deformation. As Hsü (1974) suggests, however, such deformation may be imposed on olistostromes and one must look to other criteria to discover the initial mode of mixing of blocks. The gradational contact between the broken formation and sheared olistostrome belts and the depositional contacts of thin mixed units described above support the interpretation of initial sedimentary mixing.

Immediately to the west of the Patricks Point and Redwood Mountain outliers, glaucophane schist and associated eclogite and amphibolite occur in isolated tectonized blocks within the melange (sheared olistostrome?) units (Fig. 1). The petrology and structural setting of these blocks have been studied in some detail by Shimamoto (1976), who feels that they represent the northern continuation of a zone of such blocks extending as far south as Santa Barbara (Coleman & Lanphere 1971). They have undergone higher grade metamorphism than the rocks with which they are associated, are commonly spheroidal, have chlorite-actinolite-talc rinds and in some localities are associated with serpenitized peridotite bodies.

Petrology

Zone 2/3 rocks

Beneath the Coast Range thrust, zone 3 rocks are predominantly metapelitic, with subordinate amounts of metagreywacke, metachert, metatuff and metabasalt. The metamorphic mineral assemblage is of the glaucophanitic-greenschist facies in most of the region north of area A (Fig. 1, Monsen & Aalto 1980). Young (1978) reports small amounts of lawsonite within zone 3 rocks to the south of area A (Fig. 1), and it is likely that lawsonite content diminishes northward, implying a decrease in metamorphic grade beneath the thrust. Such a contention is supported by the absence of zone 3 rocks in the Klamath map area (Fig. 2).

Medium to dark grey, foliated greywacke and argillite of textural zone 2 crop out north of the Klamath River immediately beneath the Coast Range thrust, in a thrust plate along the eastern margin of the Redwood Mountain outlier (Talley 1976; Young 1978), and in the Patricks

Point outlier. Point-count analysis shows no significant compositional difference between greywacke of the Patricks Point outlier and subjacent zone 1 Franciscan (Aalto 1976). North of the Klamath River (Fig. 2), slightly foliated zone 2 rock is in gradational contact with zone 1 broken formation and of the same composition. Grain components are the same as those of zone 1 greywacke discussed below. However, pumpellyite is more common in plagioclase and chlorite more pervasive in the matrix.

Zone 1 greywacke and argillite

Point counts of 300 grains per thin section show no statistically significant compositional difference between zone 1 greywacke of the broken formation ($n = 22$) and sheared olistostrome ($n = 16$) belts in the Klamath map area, thus they will be discussed together. Aalto (1976) and Young (1978) provide descriptions of zone 1 greywacke to the south. Although most greywackes are lithic arkoses or feldspathic litharenites (*sensu* Folk 1968), the range of composition is wide. No consistent regional trends of composition exist.

Zone 1 sandstones are commonly texturally immature to submature, with poor sorting, angular to subangular grains and an average of 34% matrix and 6% calcite cement. Monocrystalline quartz commonly has undulose extinction, although non-undulose grains of inclusion-free, idiomorphic volcanic quartz exist in nearly every sample. All varieties of polycrystalline quartz described by Young (1976) exist: grains which are polygonized, with elongated original crystals, new crystals and polygonal crystals (although those with new crystals are most common). Chert may contain ghosts of radiolaria and is commonly recrystallized or replaced by calcite. Feldspars include perthite, untwinned albite and trace amounts of K-feldspar which are commonly extensively replaced by sericite or calcite. Twinned plagioclase (An_{17-58}) is less altered but commonly contains pumpellyite. Volcanic rock fragments include felsitic, microlitic, lathwork and vitric grain types (Dickinson 1970) which are commonly partially to wholly replaced by chlorite or calcite. Metasedimentary grain components include quartz-mica schist and phyllite. Sedimentary grains include siltstone and claystone. Microphaneritic grains contain quartz and plagioclase. Minor grain components include detrital zircon, clinopyroxene, muscovite, hornblende, serpentinite and accessory haematite, leucoxene and goethite.

X-ray analysis of 15 samples indicates that the matrix composition of sandstones and the composition of argillite within the sheared olistostrome and broken formation belts is identical, consisting chiefly of varying proportions of quartz and sodic plagioclase silt, illite/sericite and chlorite. The average specific gravity of zone 1 greywackes is 2.66 (range: 2.52–2.72; standard deviation: 0.46), based upon analysis of 50 samples from the Klamath map area.

The detrital framework mode proportions of Dickinson (1970) may be used to assess provenance, although values to some degree reflect reduction of unstable components by diagenetic matrix generation. For the average zone 1 greywacke, Q is moderate (54.6), F is low (21.6), L is low to moderate (23.8), C/Q is low (0.28), P/F is high (0.97) and V/L is moderate (0.82). Additional framework modes (Dickinson & Suzcek 1979) include Qm (39.5), Qp (15.1), P (21.0), K (0.5), Lv (19.5), Ls (4.3) and Lt (38.9). Ternary plots of framework modes proposed by Dickinson & Suzcek (1979) suggest a mixed subduction complex/dissected arc source.

Zone 1 conglomerate

Conglomerate among zone 1 Franciscan rocks is most common in the sheared olistostrome belt. Counts of 100 points made on granules and small pebbles show no consistent regional variation in composition throughout the area portrayed in Fig. 1. Indeed, greater variation exists between two samples taken at exposures less than 100 m apart than throughout the entire region. A study of 10 samples taken south of the Klamath River (Fuller 1977) and six from the Klamath map area indicates that the average conglomerate contains 38% red and green chert (range: 25–51), 21% black chert (range: 8–37), 19% vein quartz (range: 4–34), 9% dacite prophyry and quartz keratophyre (range: 1–41), and less than 4% each of basalt, andesite, tonalite, diorite, quartzarenite, greywacke and limestone. Following Bailey *et al.* (1964) and Cowan & Page (1975), clasts can be considered either intra- or extraformational, depending upon whether or not they could have been derived by recycling Franciscan rock. The most common intraformational lithologies include red and green chert, basalt, andesite and greywacke. Likely extraformational lithologies include quartzarenite, black chert, limestone and diorite. Vein quartz, felsitic volcanic rock and tonalite could fall in either category, since they occur locally within olistostrome blocks. A tectonic landmass, such as an uplifted, eroding subduction zone complex, could have supplied any of the extraformational (as well as the intraformational) clast components.

Composition of mélange blocks

For purposes of field mapping, block types are recognized throughout the sheared olistostrome belt in average amounts as follows: clastic sedimentary (conglomerate and sandstone, $\bar{x} = 50\%$), bedded chert ($\bar{x} = 12\%$), greenstone ($\bar{x} = 38\%$), partially recrystallized limestone (trace), plutonic (trace) and exotic, glaucophane-bearing metamorphic (trace). There are no significant lateral variations of block composition, as determined by field study and by tallying the distribution of blocks visible on 1:12 000 air photos, other than the restriction of exotic blocks to the south (Fig. 1).

Greenstone blocks commonly contain the metamorphic mineral assemblage quartz ± albite + chlorite ± epidote ± pumpellyite ± carbonate. Felsic tuffs range from those composed nearly entirely of crystals, commonly fragmented quartz and plagioclase, to those composed entirely of vitric, microlitic or lathwork lithic fragments in a chloritic matrix. Redeposited (epiclastic?) tuff contains clasts of crystal and crystal-lithic tuff in a chloritized argillaceous matrix. In some highly indurated, foliated blocks, wisps and lenses of tuff have intruded into argillite and later been offset, suggesting ductile deformation followed by development of foliation (Fig. 4B). Such deformation preceded sedimentary mixing in an olistostrome. Epiclastic breccia contains angular clasts in varying proportions of greywacke, argillite, greenstone and chert in a tuffaceous matrix. Basalt with flow banding and pillow structure is commonly composed of microcrystalline plagioclase laths (An_{60-65}) with accessory clinopyroxene. Quartz keratophyre contains quartz and plagioclase (An_{3-5}) phenocrysts in a groundmass of plagioclase, epidote, chlorite and chalcedony.

Plutonic blocks include plagiogranite containing quartz, plagioclase (An_{13-16}) and hornblende; diabase; serpentinized peridotite and highly altered cumulate gabbro containing chiefly albitized plagioclase and clinopyroxene.

Blocks were derived from a variety of source terranes, supporting the contention (Alvarez *et al.* 1980; Blake & Jones 1978) that a diverse assemblage of oceanic crustal plate and arc(?) remnants accreted to the North American plate margin during Franciscan accumulation.

Palaeomagnetic study of nine chert samples from a single block with red radiolarian chert comformably overlying pillow greenstone at the mouth of the Klamath River suggests a low latitude derivation (Jeff Johnston, Woodward-Clyde Consultants, pers. comm. 1980). The chert magnetization directions are tightly grouped with an average inclination of −52°, corresponding to latitude 33°. Since the beds were likely deposited during the predominantly normal polarity period of the Late Mesozoic it is probable they formed south of the equator. Chert was sampled at intervals of approximately 1, 10 and 16 m above the greenstone, representing a depositional interval of 3–18 Ma at sedimentation rates of 1–5 $mm10^{-3}$ yr, and it is unlikely all samples originated in a reversed epoch. Furthermore, most published tectonic models call for a northward oceanic plate motion in the Pacific basin during the Late Mesozoic and a limestone block in the central Franciscan belt some 150 km SSE of this block was shown to be derived from the southern hemisphere (Alvarez *et al.* 1980).

Sedimentology

Depositional setting of the different units within the Franciscan Complex is determined by recording bedding features and measuring short stratigraphic sections in areas in which beds have survived deformation. Table 1 provides a comparison of bed features of the zone 1 sheared olistostrome belt, zone 1 broken formation belt and zone 2/3 metasedimentary rocks throughout the region portrayed in Fig. 1. Features of coarser-grained beds survive deformation better, thus coarser facies data are summarized in separate columns of Table 1. Facies terminology of Walker & Mutti (1973) is employed in classifying classical turbidites. Terminology of Aalto (1976) better describes bedding features among conglomeratic sediment gravity flow deposits and thus is used in this analysis.

Sheared olistostrome belt

Sandstones and conglomerates within the sheared olistostrome belt have coarser beds, characterized by more proximal features, than those of the broken formation belt. Bedded diamictites (Fig. 4E) and abundant conglomeratic sediment gravity flow deposits containing Aalto (1976) divisions II, III and/or IV (facies A of Walker & Mutti 1973) suggest a slope channel or inner fan depositional setting. The

TABLE 1. *Sedimentological comparison of bed types of the zone 1 sheared olistostrome belt, zone 1 broken formation belt and zone 2–3 Franciscan Complex rocks, calculated with and without distal facies data*

Facies (after Aalto 1976*; Walker & Mutti 1973)	Sheared olistostrome belt (N = 1047) (%)		Broken formation belt (N =1354) (%)		Zone 2 and 3 rocks (N = 470) (%)	
Facies A: bedded diamictite	Trace	Trace	—	—	—	—
Facies A: sandy conglomeratic beds with divisions II, III and/or IV	17	34	1	1	Trace	1
Facies B: thick, massive to laminated sandstone beds with divisions V and/or VI	10	19	16	24	5	22
Facies C: sandstone beds with T_{ae} or T_{abe} sequences	24	47	49	75	16	77
Facies D, E and G: thinly bedded laminated sandstones and argillites	49		34		79	

* Division I: massive sandstone or inversely graded pebbly sandstone, II: pebbly sandstone or conglomerate with or without imbrication, III: normally graded pebbly sandstone, IV: diffusely laminated and/or cross-stratified pebbly sandstone, V: massive sandstone, VI: laminated sandstone with dish and pillar structures.

bedded diamictites are intercalated with thinly bedded overbank turbidities of facies E (Walker & Mutti 1973) and possibly represent lateral spills from debris flows which topped a submarine channel margin. The thicker stratigraphic sections which have survived deformation (Fig. 6A, Aalto 1976) have high sandstone/shale ratios and lack well-developed vertical trends in bed thickness or grain-size distribution. Channelling is common beneath thicker, coarser-grained beds.

Broken formation belt

The eastern belt of zone 1 broken formation contains far fewer conglomeratic beds and more classic proximal turbidities and beds with Aalto (1976) divisions V and/or VI (Fig. 4D, facies B and C of Walker & Mutti 1973). Coarsening- and thickening-upward sequences (Fig. 6B) are preserved within megaboudins surrounded by sheared argillite. The megaboudins are lenticular in shape, tens of metres thick and hundreds of metres long. They consist of many separate boudinaged beds and short sequences of beds bounded by compressional shears, indicating both extension and compression. Bed types (Table 1) and vertical trends suggest a middle to outer fan depositional setting. The megaboudins are commonly surrounded by argillite and thinly bedded sandstone with base-cut-out Bouma sequences which, though commonly highly sheared, exhibit characteristics of outer fan turbidities, hemipelagic and pelagic deposits. Each megaboundin may represent part of a suprafan lobe which prograded over outer fan deposits. Migration of middle fan distributary channels due to channel and lobe aggradation led to cessation of deposition of coarser-grained sediment and resumption of hemipelagic and pelagic sedimentation. Thus each lobe was encased in argillaceous sediment and deformed as a separate entity.

Regional synthesis

In the Klamath map area (Fig. 2), the easternmost exposures which gradationally underlie zone 2 rocks are fine-grained and appear to be outer fan and basin-plain deposits. As one progresses from west to east across the Franciscan Complex in this region, sediment facies become progressively more distal, suggesting a retrograding fan sequence if sediments were deposited within a single basin. Because of the gradational nature of the zone 1–zone 2 contact in the north no sedimentological distinction can be made between zone 1 and zone 2 sedimentary rocks.

To the south, Aalto (1976) and Monsen & Aalto (1980) have demonstrated that there is a marked difference in depositional setting of zone 1 versus zone 2/3 rocks. The comparative absence of coarse facies bed types (Table 1), low sandstone/shale ratio and existence of thickening- and coarsening-upward sections suggest an outer fan/basin plain depositional setting for zone 2/3 Franciscan. The zone 1

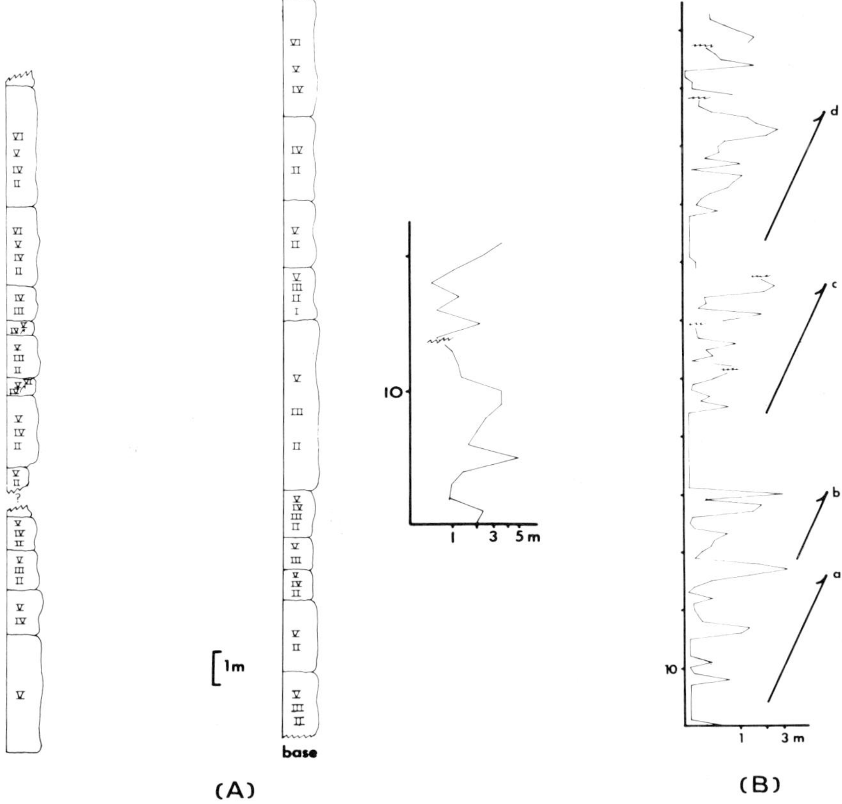

FIG. 6. (A) Stratigraphic section at Crescent City (Battery Point Lighthouse), Klamath map area. Bedding features are depicted by Roman numerals following the Aalto (1976) model (see Table 1). The graph portrays thicknesses of successive beds, from base to top.
(B) Graphic portrayal of stratigraphic section in the broken formation belt, Klamath map area. The graph portrays thicknesses of successive beds, from base to top of the section. The slanted arrows depict four coarsening- and thickening-upward sequences (a–d). Section located near the midpoint of the coastal exposure of the BF belt (3/4 km north of Damnation Creek).

greywackes over which these rocks are thrust faulted are much coarser inner to middle fan deposits (Aalto 1976). This sedimentological contrast supports the contention that the outliers are klippen, and suggests that zone 3 rocks of location A (Fig. 1) have been transported by faulting some distance from a distal Franciscan terrane now underlying the Klamath block.

Sedimentary tectonics and geological history

The Franciscan Complex is considered to have originated by accretion at a convergent plate margin. Differing tectonic models for the Franciscan have been proposed by many workers (e.g. Blake & Jones 1974, 1978; Ingersoll 1978; Maxwell 1974; Suppe 1979, Blake *et al.* 1981). It is not the purpose of this paper to review them in detail. I shall present the one of many possible models (Fig. 7) for this region which I feel best explains the geology described above (Fig. 7), and which is largely based upon models of accretionary prism development proposed by Karig & Sharman (1975) and Moore & Karig (1976).

The sheared olistostrome belt is composed of material derived from two sources. Clastic sedimentary rocks of the belt, including much of the olistostrome matrix, were most likely derived from an elevated subduction zone complex which could have been exposed at either location 2 or 3 in Fig. 7. This would account for the quartzofeldspathic composition of sheared olistostrome belt sediment and the abundance

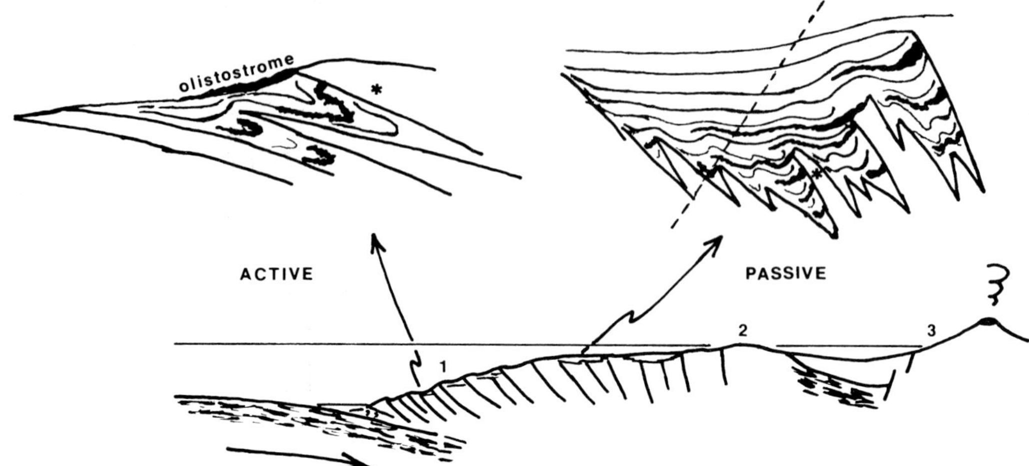

FIG. 7. Sedimentary-tectonic model for the Franciscan complex, Klamath map area. The active mode is characteristic of newly developing trench slope basins in which sediments derived from upslope portions of the accretionary prism (sources 1 or 2) and/or the arc complex (source 3) are intercalated with prism-derived gravity slides (source 1) on small fans affected by continuous compressional faulting. The passive mode is characteristic of older, elevated basins which are filled (sources 2 and/or 3), have diminished relief and are no longer subjected to intense faulting. The asterisk shows the position of an upfaulted sliver of accreted material in the active basin relative to its future position in the passive mode. The dashed line represents the present-day cross-section of the Klamath map area Franciscan.

of non-volcanic lithic components. Intercalation of olistostromes containing previously accreted clastic sedimentary, chert, greenstone and plutonic blocks occurred. Olistostromes were probably derived from structurally elevated portions of the accretionary prism, and were transported downslope to a developing trench slope basin (Fig. 7, active mode). Steep slopes and continuous fault activity account for the isoclinal folding and coarseness of these basin deposits.

As the basin filled and became elevated by continued accretion, fault activity diminished, local relief diminished and olistostromes were less frequently emplaced (Fig. 7, passive mode). The broken formation belt has, by definition, no olistostrome units and contains chiefly middle to outer fan deposits. This more distal facies reflects sediment bypassing of the basin. Diminishing fault activity within the prism accounts for the lack of overturning of beds. Thus, the retrograding fan sequence present across the Franciscan Complex in the Klamath map area resulted from the elevation, filling and eventual sediment bypassing of a trench slope basin (see Underwood & Karig 1979, for a modern example of sediment bypassing on the uppertrench slope).

This model best applies to the Klamath map area, in which there is clear evidence favouring the olistostrome mode of mixing of blocks. Should the model be correct, a restored cross-section of the filled Klamath map area trench slope basin could be drawn along the dotted line shown in Fig. 7 (passive mode). The oldest basin sediments are complexly faulted against slivers of accretionary prism tectonic mélange which constitutes the trench slope basin basement. Thus, to the SW of the Klamath map area one might encounter belts of the actual accretionary basement. Aalto (1978) considered this to be the case in the Trinidad area, in which an isoclinally folded broken formation terrane is situated between mélange terranes each containing exotic glaucophane schist blocks as well as the usual greenstone and chert. While most of these mélange are apparently sheared olistostromes, in places clearly intercalated with turbidites (Fig. 4C), Elk Head (Fig. 1, location E) immediately north of Trinidad is most likely primary tectonically mixed prism material. The headland consists chiefly of overturned pillow greenstone, highly indurated argillite and coarse turbidites (Aalto 1976). All contacts between lithologies are faults. Trinidad Head (Fig. 1, location F; Shireman, pers. comm. 1981) is chiefly an intrusive complex, zoned from east to west, of diabase, gabbro and tonalite with plagiogranite and quartz keratophyre dykes. Both headlands may be part of an accretionary basement slab which separates deformed slope basin rocks. This interpretation is supported by the presence of mélange with serpentinite matrix adjacent to Trinidad Head. Such a mix is more likely tectonic than sedimentary, and thus likely to constitute accretionary basement.

ACKNOWLEDGMENTS: I thank my colleagues and students at Humboldt State University, and especially Clark Blake, Sue Cashman, John Longshore, Susan Monsen and Mike Shimamoto for discussions which contributed to the development of ideas presented in this paper. Also, I thank Sue Cashman, Susan Monsen and John Longshore for reviewing and Jacque Overton for typing the manuscript. Thanks to Lori Dengler for advice on X-ray work. The success of this project was furthered by the cooperation of Simpson Timber Company and Arcata Redwood Company in granting access to their lands. Portions of this study were funded by Humboldt State University Foundation grants #115–75 and 115–77. This study is part of a Franciscan Complex research project supported by the Division of Earth Sciences of the National Science Foundation. NSF Grant #EAR 7809957.

References

AALTO, K. R. 1976. Sedimentology of a mélange: Franciscan of Trinidad, California. *J. sediment. Petrol.* **46,** 913–29.

—— 1977. Franciscan mélange-Quaternary unconformities and terrace stability, Trinidad, California. *Abstr. Prog. geol. Soc. Am.* **9,** 377.

—— 1978. Sedimentology of a mélange: Franciscan of Trinidad, California: a reply. *J. sediment. Petrol.* **48,** 677–9.

ALVAREZ, W., KENT, D. V., SILVA, I. P., SCHWEICKERT, R. A. & LARSON, R. A. 1980. Franciscan Complex limestone deposited at 17° south paleolatitude. *Bull. geol. Soc. Am.* **91,** 476–84.

BAILEY, E. H., IRWIN, W. P. & JONES, D. L. 1964. Franciscan and related rocks, and their significance in the geology of western California. *Bull. Calif. Div. Mines Geol.* **183,** 177 pp.

BISHOP, D. G. 1977. South Fork Mountain Schist at Black Butte and Cottonwood Creek, northern California. *Geology,* **5,** 595–9.

BLAKE, M. C., JR., IRWIN, W. P. & COLEMAN, R. G. 1967. Upside down metamorphic zonation, blueschist facies, along a regional thrust in California. *Prof. Pap. U.S. geol. Surv.* **515,** 1–9.

—— & JONES, D. L. 1974. Origin of Franciscan melanges in northern California. *Spec. Publ. Soc. econ. Paleontol. Mineral. Tulsa*, **19,** 345–57.

—— & —— 1977. Plate tectonic history of the Yolla Bolly junction, northern California: *Guidebook, 73rd Ann. meet, Cordilleran Sec. geol. Soc. Am.* 14 pp.

—— & —— 1978. Allochthonous terranes in northern California?—a reinterpretation. *Pacif. Sec., Soc. econ. Paleontol. Mineral. Pacif. Coast Paleogeogr. Symp.* **2,** 397–400.

COLEMAN, R. G. & LANPHERE, M. A. 1971. Distribution and age of high-grade blueschist, associated eclogites, and amphibolites from Oregon and California. *Bull. geol. Soc. Am.* **82,** 2397–412.

COWAN, D. S. & PAGE, B. M. 1975. Recycled Franciscan material in Franciscan melange west of Paso Robles, California. *Bull. geol. Soc. Am.* **86,** 1089–95.

DICKINSON, W. R. 1970. Interpreting detrital modes of graywacke and arkose. *J. sediment. Petrol.* **40,** 695–707.

—— & SUZCEK, C. A. 1979. Plate tectonics and sandstone compositions. *Bull. Am. Assoc. Petrol. Geol.* **63,** 2164–82.

DOTT, R. H., JR. 1979. Intracontinental plate boundary east of Cape Mendocino, California: comment. *Geology*, **7,** 322–3.

ENSRUD, D. L. 1978. *The geology of area B, sections 8, 9, 16, 17, 20, 21, T 13, N, R 2E, Ship Mountain quadrangle, Del Norte County, California.* Thesis, BSc, Humboldt State Univ. (unpubl.).

FARHOUDI, G. & KARIG, D. E. 1977. Makran of Iran and Pakistan as an active arc system. *Geology,* **5,** 664–8.

FOLK, R. L. 1968. *Petrology of Sedimentary Rocks.* Hemphill's, Austin, Texas. 170 pp.

FULLER, S. 1977. *The origin of the major constituents of the Franciscan melange conglomerates in Humboldt County, California.* Thesis, BSc, Humboldt State Univ. (unpubl.).

HERD, D. G. 1978. Intracontinental plate boundary east of Cape Mendocino, California. *Geology,* **6,** 721–5.

HSÜ, K. J. 1974. Melanges and their distinction from olistostromes. *Spec. Publ. Soc. econ. Paleontol. Mineral. Tulsa,* **19,** 321–33.

INGERSOLL, R. V. 1978. Paleogeography and paleotectonics of the late Mesozoic forearc basin of northern and central California. *Pacif. Sec., Soc. econ. Paleontol. Mineral. Tulsa, Pacif. Coast Paleogeogr. Symp.* **2,** 471–82.

IRWIN, W. P. 1966. Geology of the Klamath Mountains province. *Bull, Calif. Div. Mines Geol.* **190,** 19–38.

—— 1977. Review of the Paleozoic rocks of the Klamath Mountains. *Pacif. Sec., Soc. econ. Paleontol. Mineral. Pacif. Coast Paleogeogr. Symp.* **1,** 441–54.

JEPSEN, K. O. 1979. *Geology of the Klamath Glen area, southern Del Norte County, California.* Thesis, BSc, Humboldt State Univ. (unpubl.).

KARIG, D. E. & SHARMAN, G. F. 1975. Subduction and accretion in trenches. *Bull. geol. Soc. Am.* **86,** 377–89.

LANPHERE, M. A., BLAKE, M. C., JR & IRWIN, W. P. 1978. Early Cretaceous metamorphic age of the South Fork Mountain Schist in the northern Coast Ranges of California. *Am. J. Sci.* **278,** 798–815.

LEATHERS, S. 1978. *Petrology, metamorphic grade and structure of the Redwood Mountain outlier of the South Fork Mountain Schist on the coast of northern California.* Thesis, BSc, Humboldt State Univ. (unpubl.).

LIND, E. K. 1979. *Geology of central belt Franciscan north of Gold Bluffs Beach, California.* Thesis, BSc, Humboldt State Univ. (unpubl.).

LOZINSKY, R. P. 1979. *The geology of the Franciscan Coast Ranges in the vincinity of Klamath, California.* Thesis, BSc, Humboldt State Univ. (unpubl.).

MAXWELL, J. A. 1974. Anatomy of an orogen. *Bull. geol. Soc. Am.* **85,** 1195–204.

MIDDLETON, G. V. & HAMPTON, M. A. 1973. Sediment gravity flows: mechanics of flow and deposition. *In:* MIDDLETON, G. V. & BOUMA, A. H. (co-chairmen). *Turbidities and Deep Sea Sedimentation.* Pacif. Coast Sec., Soc. econ. Paleontol. Mineral. Tulsa, 1–38.

MONSEN, S. A. 1979. *Petrology and structure of the South Fork Mountain Schist and subjacent Franciscan Complex, Pine Ridge, California.* Thesis, BSc, Humboldt State Univ. (unpubl.).

—— & AALTO, K. R. 1980. Petrology, structure and regional tectonics: South Fork Mountain Schist of Pine Ridge Summit, northern California. *Bull. geol. Soc. Am.* **91,** 369–73.

MOORE, G. F. & KARIG, D. E. 1976. Development of sedimentary basins on the lower trench slope. *Geology,* **4,** 693–7.

SHIMAMOTO, M. K. 1976. *Petrology of glaucophane-bearing tectonic blocks from the Weott, Iaqua Buttes, Blue Lake and Trinidad quadrangles, California.* Thesis, BSc, Humboldt State Univ. (unpubl.).

SILVER, E. A. 1971. Transitional tectonics and late Cenozoic structure of the continental margin off northernmost California. *Bull. geol. Soc. Am.* **82,** 1–22.

SIMILA, G. W., PEPPIN, W. A. & MCEVILLY, T. V. 1975. Seismotectonics of the Cape Mendocino, California, area. *Bull. geol. Soc. Am.* **86,** 1399–406.

SUPPE, J. 1979. Structural interpretation of the southern part of the northern Coast Ranges and Sacramento Valley, California: summary. *Bull. geol. Soc. Am.* **90,** 327–30.

TALLEY, K. L. 1976. *Descriptive geology of the Redwood Mountain outlier of the South Fork Mountain Schist, northern Coast Ranges, California.* Thesis, MSc, Southern Methodist Univ. (unpubl.).

UNDERWOOD, M. B. & KARIG, D. E. 1979. Submarine canyons as a key to sedimentation in subduction complexes. *Abstr. Prog. geol. Soc. Am.* **11,** 531.

YOUNG, J. C. 1978. Geology of the Willow Creek quadrangle, Humboldt and Trinity Counties California. *Map Sheet Calif. Div. Mines Geol.* **31.**

YOUNG, S. W. 1976. Petrographic textures of detrital polycrystalline quartz as an aid to interpreting crystalline source rocks. *J. sediment. Petrol.* **46,** 595–603.

WALKER, R. G. & MUTTI, E. 1973. Turbidite facies and facies associations. *In:* MIDDLETON, G. V. & BOUMA, A. H. (co-chairmen). *Turbidities and Deep Sea Sedimentation.* Pacif. Coast Sec., Soc. econ. Paleontol. Mineral. Tulsa, 119–57.

K. R. AALTO, Department of Geology, Humboldt State University, Arcata, California 95521, U.S.A.

Sedimentation, metamorphism and tectonic accretion of the Franciscan assemblage of northern California

M. C. Blake, Jr, A. S. Jayko & D. G. Howell

SUMMARY: The Yolla Bolly terrane of the Franciscan assemblage in northern California is seemingly a typical subduction complex, having undergone penetrative deformation and metamorphism to the high pressure–low temperature blueschist facies. Detailed mapping combined with sedimentological analysis has enabled us to: (1) reconstruct a probable palaeosedimentary environment, (2) analyse the interaction during and after subduction between deformation and metamorphism, and (3) speculate on subsequent deformational history including tectonic accretion of the terrane to North America.

Rocks of the Yolla Bolly terrane consist of three thrust-fault-bounded lithological units: a lower unit of disrupted mudstone and thin-bedded sandstone (broken formation) containing scarce volcanic and radiolarian chert horizons, a middle unit predominantly of thick-bedded to massive sandstone (metagreywacke) that includes several horizons of radiolarian chert, and an upper unit of mudstone and thin-bedded sandstone (broken formation) with numerous intrusive and extrusive volcanic rocks plus rare radiolarian chert. Radiolarians from all three units are of the same age (Tithonian to Valanginian) and together with the sedimentological data, suggest that the rocks represent a continent-derived submarine fan, deposited in a complex transform graben possibly similar to the present-day Gulf of California or basins of the California Continental Borderland, rather than a trench setting.

Metagreywacke containing lawsonite and aragonite yields radiometric ages of approximately 110 Myr and indicate that these rocks were subducted to depths of 20–30 km about 30 or 40 Myr after they were deposited. Shortly after subduction, the rocks were probably involved in a collision that imbricated and tectonically returned the subduction complex to the surface.

The Franciscan assemblage in northern California encompasses three major NW-trending belts. The coastal belt consists almost entirely of sheared sandstone, mudstone, and conglomerate (broken formation) generally interpreted as the youngest (late Cretaceous to Miocene) part of the Franciscan accretionary prism (Bachman 1978; Beutner *et al.* 1980).

The central belt consists largely of tectonic mélange and broken formation and includes most of the high-grade blueschist, eclogite, and exotic limestone knockers in northern California. Palaeomagnetic data suggest that at least some of the terranes within the central belt formed in southern latitudes and were subsequently accreted to North America by large-scale transcurrent plate motions (Alvarez *et al.* 1980).

The eastern belt consists largely of clastic sedimentary rocks locally containing interbedded volcanic rocks and radiolarian chert. All of these rocks have a faint to pronounced metamorphic fabric and contain high-pressure minerals such as lawsonite, aragonite, and glaucophane. The eastern belt contains at least two distinctive tectonostratigraphic units, the South Fork Mountain Schist, and the Yolla Bolly terrane. The South Fork Mountain Schist was originally defined on the basis of metamorphic grade (Blake *et al.* 1967) but subsequent work (Bishop 1977; Worrall 1979; Blake *et al.* unpublished data), indicates that the primary lithology is significantly different from adjacent fossiliferous Franciscan rocks.

This paper briefly describes the lithology, age, metamorphism, and structure of the rocks of the Yolla Bolly terrane and relates these characteristics to plate tectonic models.

Lithology and sedimentology

The Yolla Bolly terrane (Fig. 1) consists of three lithological units separated by thrust faults. The lower unit consists largely of disrupted mudstone and thin-bedded quartzofeldspathic sandstone (broken formation) but also contains coherent lenses of medium- to thick-bedded sandstone and conglomerate plus minor amounts of volcanic rocks and radiolarian chert. The structural thickness is on the order of 1000 m.

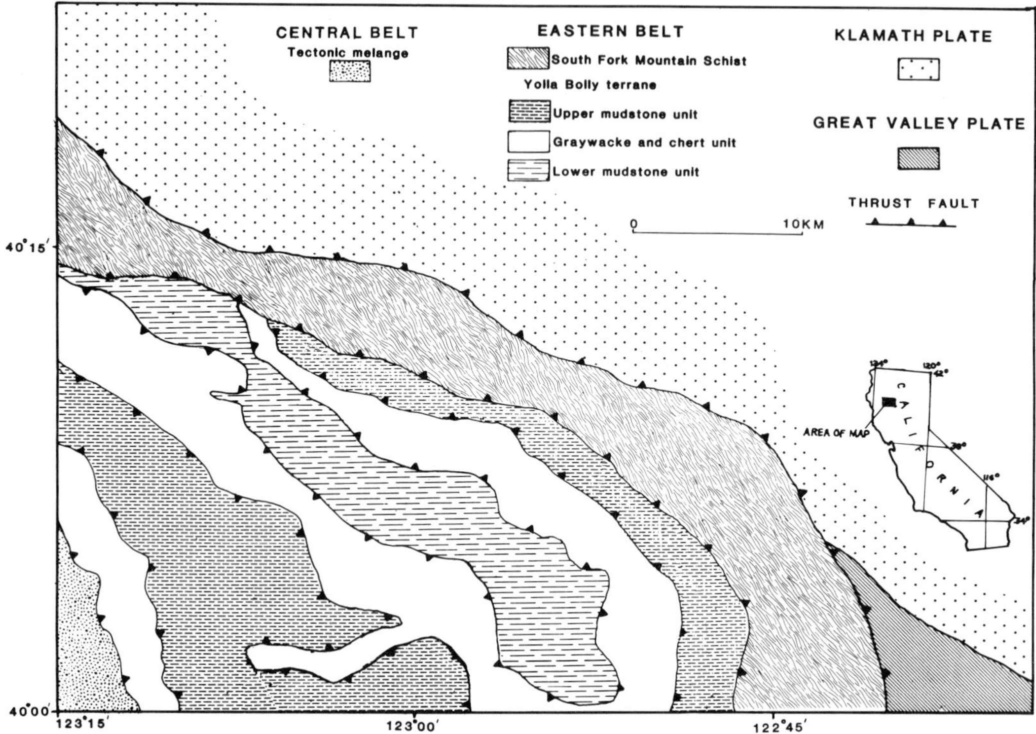

FIG. 1. Generalized geological map of the Yolla Bolly area, northern California. Geology from Worrall (1979) and unpublished mapping by M. C. Blake *et al.*

The middle unit is made up predominantly of thick-bedded to massive quartzofeldspathic sandstone that includes several thick horizons of thin-bedded radiolarian chert. Igneous rocks are rare and consist of small diabase-gabbro sills, largely intrusive into the chert-rich parts of the unit. The thickness of the unit, as deduced from the structure sections is 1000–2000 m.

The upper unit is composed predominantly of mudstone and thin-bedded quartzofeldspathic sandstone, but includes abundant intrusive and extrusive bodies of basalt and quartz keratophyre. Lenses of radiolarian chert are less common than in the other units. The structural thickness is about 1000 m.

Smooth, water-worn exposures in all three units display an abundance of primary sedimentary structures. The observed structures are common characteristics of flysch, and in particular, of turbidites and associated mass-flow deposits. Many beds can be related to the Bouma sequence, and small-scale cross-bedding (ripple-drifted siltstone) is ubiquitous. Sandstones commonly show normal grading and some pebbly beds are inversely graded. The principal type of clast in the coarser grained beds are intrabasinal mud rip-up fragments, though extrabasinal felsic volcanic and granitic clasts are also present. Flame structures, dewatering pillars, and slurry bedding are seen within most sandstone beds, but, surprisingly, dish structures have not been observed. Load structures, flutes and grooves are displayed on the basal surfaces of some beds.

The classification of lithofacies for redeposited mass flow sediments developed by Mutti & Ricci Lucchi (1972), and Ricci-Lucchi (1975) is a convenient means of describing the sedimentary rocks of the Yolla Bolly terrane. Examples of all seven lithofacies (A–G, listed in Underwood & Bachman 1981) can be found in this terrane, though their abundance is highly variable.

The lowest stratigraphic unit is characterized by ripple drifted siltstone (facies E) in association with thin-bedded turbidites (facies D) and minor amounts of massive mudstone (facies G). In the lower part of this unit is a 100–200 m thick interval of sandstone (facies B and C) with interbeds of conglomerate (facies A), thinner

FIG. 2. Examples of coherent strata from the Yolla Bolly terrane, textural zone 1. (A) Lithofacies B, composite (amalgamated) beds of thick-bedded sandstone. (B) Lithofacies D and E thin bedded turbidites. (C) Lithofacies E, ripple-drifted sandstone (Bouma T_{de} and T_{ce} turbidites).

bedded turbidites (facies D and E) and pebbly mudstone and slump folded strata (facies F). Within this sandstone interval there are at least four thinning- and fining-upward megasequences, each 30–50 m thick. Photographs of these facies are shown in Fig. 2.

The above association of lithofacies specifies an environment of principally tranquil, dilute turbidite deposition (overbank?) with episodes of much higher energy turbidite deposition (feeder channels). The conglomerate in basal parts of the inferred channelled sequences is coarser grained than any other conglomerate in the Yolla Bolly terrane. The preponderance of mudstone cut by channels filled with coarse material suggests either an inner fan or slope with feeder channels as a depositional setting. The lower unit is probably the most proximal of the three stratigraphic sequences to be discussed.

The middle unit consists principally of massive bundles of composite (amalgamated) thick-bedded turbidites, fluxoturbidites, or grain-flow deposits (facies B and C), with thin intervals of thin-bedded turbidites (facies D or E). No vertical asymmetric cycles of bedding thicknesses are evident. Because of the abundance of sandstone in this unit and the absence of more ductile interbeds this is the most coherent of the three lithological sequences. Because of this coherency we have been able to map beds of ribbon chert for as much as 20 km along strike (Fig. 3A). Most chert units are 20–50 m thick. The lower contacts of the chert beds are commonly gradational; sandstone passes upward into siliceous shale (1–5 m thick) which grades into pure chert (Fig. 3B). Coarse-grained sandstone usually overlies the chert, commonly separated by a thin (50 cm) basal breccia of chert and mudstone (Fig. 3C, D).

Within the context of a submarine fan model, the above association of lithofacies indicates a middle or supra-fan setting, though the occurrence of interbedded chert is enigmatic. Chert occurs in all three of the stratigraphic units but stands out in the middle unit because of its structural competency. The chert beds must represent periods of diminished to non-existent clastic deposition in areas between places of active deposition in a supra-fan setting, analogous to mud blankets between depositional lobes, or chert deposition may have occurred more uniformly over the entire fan during intervals of high sea-level that trap detrital material in inner shelf, estuarine, and bay settings.

The highest lithological unit is principally ripple-drifted siltstone (facies E) and thin-bedded turbidites (facies D and E). Near the top of this unit are several thinning-upward cycles of greywacke (facies B and D giving way to D and E) which provide evidence of some channellized turbidite deposition. The channel fill is finer grained than the analogous detritus in the lowest stratigraphic unit. However, the principal distinguishing feature of this unit is the abundance of greenstone bodies. Many of the greenstone blocks are essentially knockers in a sheared matrix of facies E siltstone; numerous intrusive rocks, however, of basalt and quartz keratophyre suggest a history of a volcanism possibly related to the depositional environment of this upper unit. The lithofacies associations do not rigidly constrain the depositional environment of this unit; a middle to outer fan fringe setting is possible as is a slope setting. In either case, the intrusive rocks are puzzling.

To summarize, the sedimentological data offer some constraints for the depositional environment of strata in the Yolla Bolly terrane. All the clastic strata derive from mass-flow processes. Deposition probably occurred in a deep-marine setting. Lithofacies associations suggest submarine fan styles of deposition though there are not enough data to specify fan geometries or basin configurations. A possible depositional environment for the strata of the Yolla Bolly terrane would be restricted submarine fans within a region subjected to crustal extension. The lithofacies associations can be explained as basin-trough (middle unit) and basin margin facies (lower and upper units). The bimodal submarine volcanic activity is also consistent with such a regime (Crowell 1976; Lonsdale & Lawver 1980). The occurrence of the thick chert interbeds are problematical, but would seem to fit a restricted basin undergoing extension better than a more dynamic trench setting.

FIG. 3. Depositional relationships between radiolarian chert and turbidites. (A) Through-going chert horizons on south Yolla Bolly Mountain. (B) Interbedded chert and mudstone at the base of a 50 m chert bed along Balm of Gileod Creek. (C) Same chert bed showing overlying coarse-grained sandstone. (D) Same bed showing chert breccia at base of overlying sandstone.

Age

Radiolaria from all three subunits are nearly identical, and indicate a latest Jurassic or earliest Cretaceous age (David L. Jones, pers. comm. 1981). Megafossils were found only in the lowest structural unit and include *Buchia okensis* (upper Tithonian) and *Buchia pacifica* (Valanginian) (Blake 1965). Rb-Sr and $^{40}Ar/^{39}Ar$ measurements on metagreywacke from the Yolla Bolly terrane give metamorphic ages of about 110 Myr (Lanphere *et al.* 1978). It therefore appears that sedimentation occurred about 30–40 Myr prior to blueschist metamorphism.

Structure

At least three periods of folding have been recognized. The earliest is characterized by WNW-trending isoclinal folds (F_1) and associated axial-plane cleavage (S_1). Within the coherent sandstone unit, this cleavage defines a textural zonation that ranges from non-foliated greywacke (textural zone 1) through foliated semischist (textural zone 2) to foliated schist with well-developed quartz segregation (textural zone 3) (Blake *et al.* 1967). Brittle breakage of sandstone beds in the broken formations along the limbs of folds parallel to the tectonic foliation (S_1) is manifested by sheared phacoids or boudins. Small-scale tectonic structures are folded around the F_2 hinges, but seemingly not around F_1 hinges. From these observations we infer that the deformational fabric seen in the broken formation (Fig. 4), and the textural zonation in the meta-greywacke, formed simultaneously during the earliest period of deformation.

A second set of folds (F_2) trend NNE and has a moderately well-developed axial plane schistosity (S_2). These folds are strongly asymmetric showing vergence to the ESE.

The third set (F_3) is characterized by open, NW-trending, and locally overturned folds with vergence from NE to SW. These folds are not accompanied by axial plane cleavage or schistosity.

The thrust faults shown on the geological map (Fig. 1) are located on the basis of the juxtaposition of textural grade and lithological

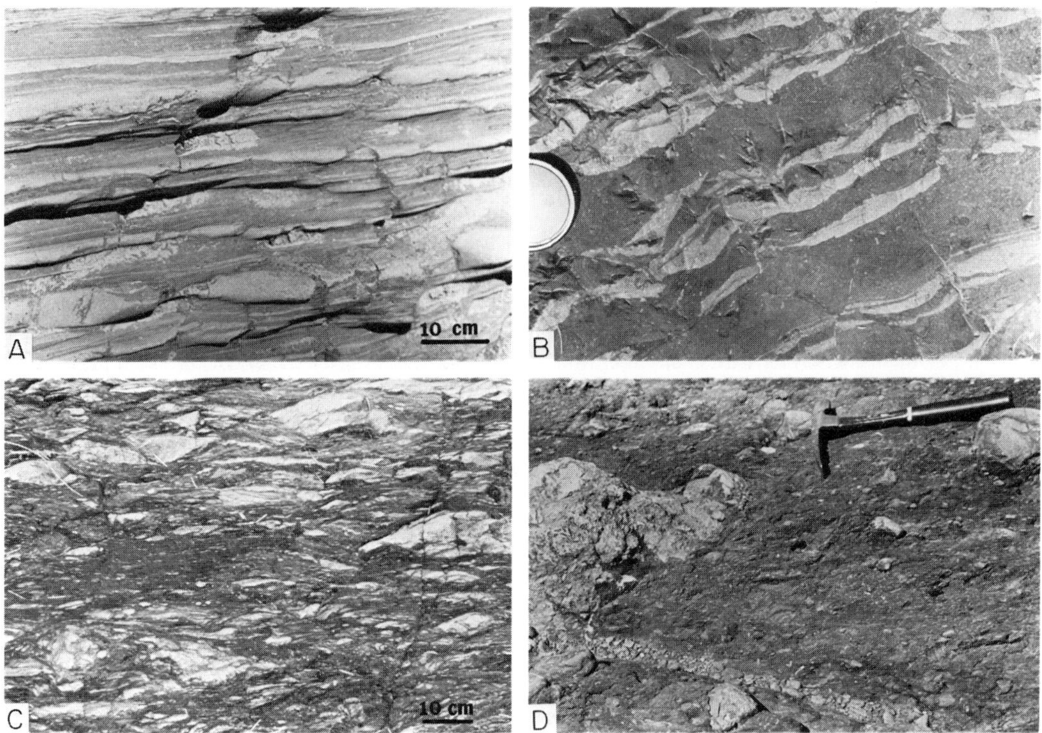

FIG. 4. Examples in the stages of development of broken formation or mélange. (A) Slightly stretched thin-bedded turbidites. (B) Thin bedded turbidites stretched and broken by normal faults. (C) Boudins and phacoids of sandstones in a mud (Te part of a turbidite) matrix, bedding still discernible. (D) Totally disrupted bedding, blocks of sandstone in a homogenized mud matrix.

facies. They appear to have been folded by F_2 and clearly pre-date F_3.

Metamorphism

In addition to the previously described textural zonation, all of the rocks contain high-pressure minerals typical of the blueschist facies. Within the metagreywacke, crystals of lawsonite, phengite, and chlorite have grown parallel to S_1, particularly in the higher textural grades. The greenstones contain, in addition, glaucophane and pumpellyite. Limited oxygen isotope data (Taylor & Coleman 1968) suggest that these rocks formed at temperatures of about 100°–300°C. The widespread presence of aragonite indicates that pressures were on the order of 4–6 kb, corresponding to a depth of 15–20 km.

Conclusions

Keeping in mind the many uncertainties, we suggest the following plate tectonic history. During latest Jurassic and earliest Cretaceous time the clastic sediments were deposited in a restricted basin that was also the site of deposition of thick interbeds of radiolarian chert. Sandstone petrography and the abundance of intrusive and extrusive volcanic rocks suggest that the basin was on a continental margin undergoing rifting, perhaps an environment similar to the present-day Gulf of California or the California Continental Borderland.

Some 30 or 40 Myr later, all of these rocks were involved in a subduction event that produced metagreywacke and broken formation, with the attendant development of high-pressure metamorphic minerals such as lawsonite, aragonite, and glaucophane. Obduction of the subducted complex back to the surface apparently occurred soon after preserving the high-pressure–low-temperature metamorphic mineral assemblage.

References

ALVAREZ, W., KENT, D. V., PREMOLI SILVA, ISABELLA, SCHWEICKERT, R. A. & LARSON, R. A. 1980. Franciscan Complex limestone deposited at 12° south paleolatitude. *Bull. geol. Soc. Am.* **91,** 476–84.

BACHMAN, S. B. 1978. A Cretaceous and early Tertiary subduction complex, Mendocino coast, northern California. *In:* HOWELL, D. G. & MCDOUGALL, K. A. (eds). *Mesozoic Paleogeography of the Western United States.* Pacif. Sec., Soc. econ. Paleontol. Mineral. Tulsa, Symp. Pacif. Coast Paleogeogr. **2,** 419–30.

BEUTNER, E. C., MCLAUGHLIN, R. J., OHLIN, H. N. & SORG, D. H. 1980. *Geologic map of the King Range and Chemise Mountain Instant Study Area, northern California.* U.S. geol. Surv. Map MF-1196A.

BISHOP, D. G. 1977. South Fork Mountain Schist at Black Butte and Cottonwood Creek, northern California. *Geology*, **5,** 595–9.

BLAKE, M. C., JR 1965. *Structure and petrology of low grade metamorphic rocks, blueschist facies, Yolla Bolly area, northern California.* Thesis, Ph.D., Univ. Stanford. 91 pp.

—— , IRWIN, W. P. & COLEMAN, R. G. 1967. Upside-down metamorphic zonation, blueschist facies, along a regional thrust in California and Oregon. Prof. Pap. U.S. geol. Surv. **575–C,** C1–9.

—— , JAYKO, A. S., NEUMANN, R., WILSON, B. & WORRALL, D. M. no date. Geologic map of the Yolla Bolly Wilderness Area, northern California, scale 1:48,000 (unpubl.).

CROWELL, J. C. 1976. Implications of crustal stretching and shortening of Coastal Ventura basin, California. *In: Aspects of the Geologic History of the California Continental Borderland.* Pacif. Sec. Am. Assoc. Petrol. Geol., Misc. Publ. **24,** 365–82.

LANPHERE, M. A., BLAKE, M. C., JR & IRWIN, W. P. 1978. Early Cretaceous metamorphic age of the South Fork Mountain Schist in the northern Coast Ranges of California. *Am. J. Sci.* **278,** 798–815.

LONSDALE, P. & LAWVER, L. A. 1980. Immature plate boundary zones studies with a submersible in the Gulf of California. *Bull. geol. Soc. Am.* **91,** 555–69.

MUTTI, E. & RICCI LUCCHI, F. 1972. Le torbiditi dell'-Appennin settentrionale: introduzione all'analisi di facies. *Mem. Soc. geol. Ital.* **11,** 161–99; English translation by T. H. Nilsen, 1978, *Intern. geol. Rev.* **20,** 125–66; AGI Reprint Ser. 3.

PESSAGNO, E. A., JR 1977. Upper Jurassic radiolaria and radiolarian biostratigraphy of the California Coast Ranges. *Micropaleontology*, **23,** 56–113.

RICCI-LUCCHI, F. 1975. Depositional cycles in two turbidite formations of northern Apennines. *J. sediment. Petrol.* **45,** 3–43.

TAYLOR, H. P. & COLEMAN, R. G. 1968. $^{18}O/^{16}O$ ratios of coexisting minerals in glaucophane-bearing metamorphic rocks. *Bull. geol. Soc. Am.* **79,** 1727–55.

WORRALL, D. M. 1979. *Geology of the south Yolla Bolly area, northern California, and its tectonic implications.* Thesis, Ph.D., Univ. Texas, Austin. 250 pp.

M. C. BLAKE, A. S. JAYKO & D. G. HOWELL, U.S. Geological Survey, Menlo Park, California 94025, U.S.A.

Deformation of partly dewatered and consolidated Franciscan sediments near Piedras Blancas Point, California

Darrel S. Cowan

SUMMARY: Bedded sequences of turbidite sandstone and mudstone and associated chaotic, polymict pebbly mudstone record a post-depositional, pre-metamorphic deformation. Sandstone layers were extended and locally disrupted and fragmented and most clasts in pebbly mudstone were deformed into oblate ellipsoids. On the scale of outcrops and hand-specimens, extension in sandstone was accomplished by: (1) mesoscopically ductile pinch-and-swell behaviour and boudinage; (2) extreme necking, which locally resulted in extension fractures or brecciation; and (3) sets of parallel shear fractures inclined to bedding. On smaller scales, ductile changes in shape were accommodated by both intergranular flow of sand-rich sediment and slip along zones typically 0.1 mm or less in thickness. Microfracturing, granulation, and cataclasis of clastic grains are less commonly observed, even adjacent to microfaults. Mudstone was also mesoscopically ductile but generally it flowed by displacements on penetrative, anastomosing surfaces of slip and was transformed locally into 'scaly clay'. These Franciscan sediments probably contained appreciable pore fluids when they were deformed, since the deformation of unconsolidated sands by homogeneous, intergranular flow rather than widespread cataclasis is fostered by high interstitial fluid pressures. More localized discontinuities, such as extension fractures, shear fractures, and microfaults, formed where sand had more completely dewatered and consolidated. Thus, material response may partly reflect water content and consolidation history. The axially symmetric extension recorded by small-scale structures suggests that sediments were deformed in a shallow-level, laterally unconfined environment on the trench slope, perhaps in response to vertical shortening and horizontal elongation that accompanied gravitational collapse and spreading.

An important question that commonly arises in subduction complexes is whether sediments were deformed while incompletely lithified or after complete consolidation. At active subduction zones, 'soft-sediment' deformation theoretically could occur at the base of the trench slope as sediments are scraped off and accreted, or deep beneath the inner wall of the trench if sediment is hauled down and subducted, or on the inner wall by gravitationally driven sliding or spreading. In the late Mesozoic–early Cenozoic Franciscan subduction complex of the California Coast Ranges, only a few studies, such as those by Kleist (1974) and Bachman (1978), have described soft-sediment structures, probably because much of our attention has focused instead on more highly metamorphosed and multiply deformed terranes where early structures are rarely preserved. The principal objectives of this paper are to describe small-scale structures in Franciscan rocks that are superbly exposed along the shoreline from 2 km north of Piedras Blancas Point to San Simeon Point, California (Fig. 1) and to show that they can be used in conjunction with microfabric to infer the properties and behaviour of sand and mud during deformation. An earlier phase of this study (Cowan 1978) dealt with the origin of polymict, blueschist-bearing olistostromes SE of San Simeon. The results of this research bear on the general problem of whether the deformation accompanied gravity-driven downslope mass transport affecting the shallow levels of a trench slope, or instead was more closely confined within the deeper levels of an accretionary prism and therefore 'tectonic' in character.

The reader should consult Hsü (1968, 1969), Hall (1976), Cowan (1978), and Smith *et al.* (1979) for general background on the geological setting.

Three major rock types are exposed along the shore between Piedras Blancas and San Simeon: well-bedded turbidites comprising 90% sandstone layers displaying a variety of well-preserved Bouma sedimentary structures; lithologically similar but more mudstone-rich rocks in which bedding has been variably disrupted by the early deformation described in detail below; and internally chaotic, polymict pebbly mudstone which locally records the same deformation. The rocks bear a close petrological, structural and metamorphic kinship to the Franciscan exposed SE of San Simeon (Cowan 1978; Hall 1976), even though they are separated by the San Simeon fault zone, along

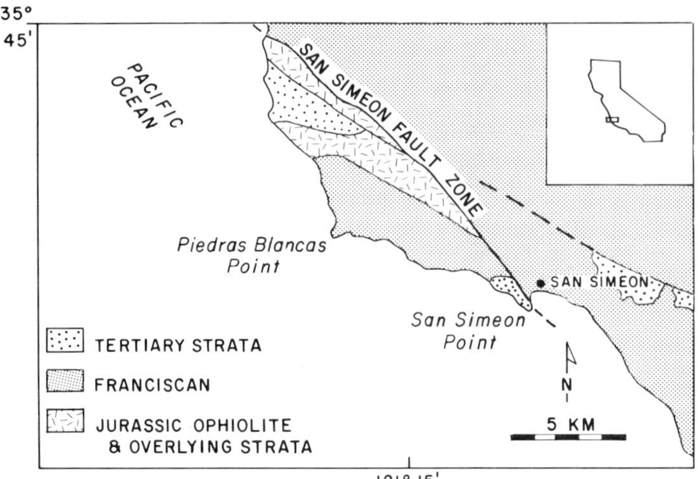

FIG. 1. Generalized geological map of Piedras Blancas–San Simeon area, modified from Hall (1976). 'Jurassic ophiolite and overlying strata' are part of the Great Valley sequence, which formed in a forearc basin associated with the Franciscan subduction complex.

which Hall (1975) has postulated at least 80 km of late Cenozoic right-lateral strike-slip.

Small-scale structures

Primary structures

Some primary sedimentary structures, predating deformation, formed at or immediately below the sediment-water interface in turbidites and are classified here as syn-sedimentary or penecontemporaneous. Fine bedding-parallel laminations, designated S_0, range from <1 mm to several mm in thickness and are defined by changes in grain size and composition. Dark laminae are clay-rich; lighter laminae are sandy. S_0 is prominent in sandstone layers less than 20 cm thick. In mudstone, it is parallel to a weak fissility, which is also syn-sedimentary as defined above. At the base of some sandstone layers, classical examples of disruption accompanying turbidite deposition were observed, including ripped-up mudstone clasts that had already developed bedding-parallel fissility due to compaction at the depositional interface. Some sand irregularly intruded downward or was swirled into less consolidated mud. The dispersal and mixing within associated polymict pebbly mudstones can also be considered as a primary event.

Mesoscopic structures

Most of the Franciscan sedimentary rocks exposed between San Simeon and Piedras Blancas display a pervasive deformation which was superimposed on S_0 and related syn-sedimentary features (Fig. 2). Extensional

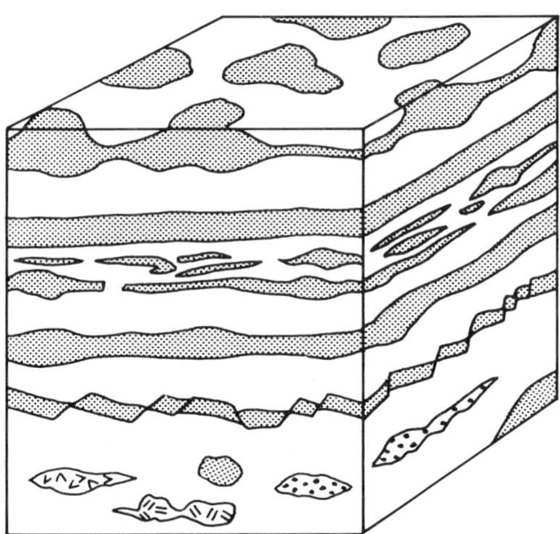

FIG. 2. Diagrammatic summary of mesoscopic structures recording layer-parallel extension, correctly oriented with respect to one another but displayed as they would appear if they were together in one large outcrop. Extended layers are grossly parallel to undeformed beds. Structures are described in the text and actual examples are shown in Figs 3–8. Deformed clasts near bottom are in polymict pebbly mudstone associated with interbedded sandstone (shaded) and mudstone.

FIG. 3. An extended sandstone layer underlain and overlain by more disrupted, thinly bedded, mudstone-rich section. In addition to the necked area near the right edge of the photograph, note the irregularly bulbous protrusions along the upper and lower surfaces of the layer. Layering is parallel to hammer handle, which is 26 cm long.

FIG. 4. Extensional shear fractures developed in a thick sandstone layer. Hammer handle is parallel to bedding laminations within lozenges, which have rotated with respect to unaffected overlying sandstone layers. See text for kinematic interpretations.

Fig. 5. Adjacent sandstone layers exhibiting different mesoscopic deformational behaviour. The hammer rests on part of a layer that flowed ductilely and separated into boudins. The upper surface of the underlying, thicker layer is offset by an extensional shear fracture, recording brittle behaviour, that does not extend into the boudin. Interbedded mudstone also flowed ductilely to accommodate overall flattening.

Fig. 6. Sawn surface of hand specimen, cut normal to layering. White scale bar is 2 cm long. Thicker extended layers are separated by mudstone-rich intervals in which thin sandy laminae are more thoroughly disrupted and disaggregated and locally folded. This sample is representative of the deformation in the thinly layered sections associated with the bed illustrated in Fig. 3.

FIG. 7. Close-up of part of specimen in Fig. 6. Scale bar = 1 cm. Deformation mechanisms vary within a single layer, and from layer to layer. Extension in the prominent bed in centre was accommodated principally by slip along semi-penetrative shear fractures in the strongly necked region, and by intergranular flow at the right edge of the photo. Delicate swirling and incipient disaggregation of thin sandy laminae above involved homogeneous intergranular flow, rather than fracture and cataclasis of individual grains, on a microscopic scale.

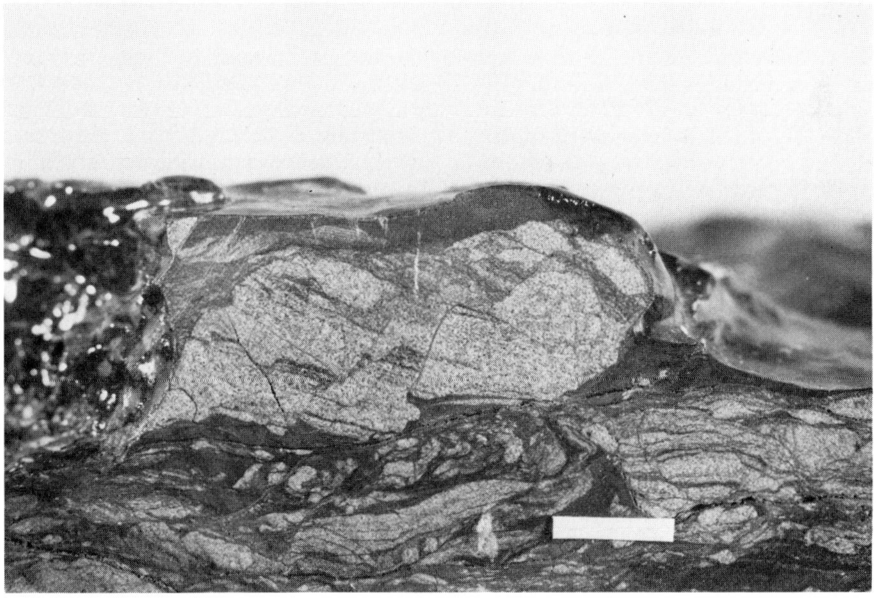

FIG. 8. Another close-up of part of specimen in Fig. 6. Scale bar = 1 cm. Small-scale analogues of extensional shear fracture shown in Fig. 4. Microfractures are antithetic normal faults (see text for discussion). This mesoscopic behaviour is 'transitional' between ductile flow and brittle fracture. See Fig. 11 for a photomicrograph of a typical microfault.

structures including pinch-and-swell, necking, boudinage, and shear fractures are obvious in outcrops (Figs 3, 4 & 5) and hand specimens (Figs 6, 7 & 8). They are most pronounced in pseudo-stratiform sequences of mudstone and sandstone that probably were once bedded turbidites. In some continuous, 500 m long exposures, rocks are variably deformed throughout and structures are not confined to localized, stratigraphically truncated intervals as slump folds typically are (e.g. Grant-Mackie & Lowrie 1964; Gregory 1969; Woodcock 1976).

A general observation on both outcrop and hand-specimen scales is that semi-continuous, thicker sandstone layers are separated by mudstone-rich intervals where thinner sandy layers and laminae are more thoroughly disrupted (Figs 3 & 6). Possible explanations for this disharmonic behaviour are presented below. Although nearly all of the mesoscopic structures described here are evidence of layer instability during extension, they can be divided into two gradational types on the basis of deformational modes observable at this scale. During mesoscopically ductile deformation, sand layers changed shape by mechanisms producing apparently homogeneous flow or by small displacements along internal slip surfaces (Figs 7 & 8). Mesoscopically brittle behaviour involved discontinuities extending completely across layers; extensional shear fractures are a familiar example (Figs 4 & 5). Hsü (1968, 1974) first recognized and illustrated some of these mesostructures in the study area. Cowan (1978) suggested that most of the brittle fractures in olistostromes SE of San Simeon were superimposed on earlier ductile structures (D_1) during a subsequent deformation (D_2). A more extensive study of the bedded sequences at Piedras Blancas indicates that both types of behaviour were largely contemporaneous.

Ductile pinch-and-swell structures, ranging from geometrically ideal to highly irregular, are abundant in semi-continuous layers. Necking is locally extreme, but true boudinage, involving complete separation of a continuous layer into isolated fragments or boudins, is rare. Illustrations of pinch-and-swell structure and boudinage in metamorphic rocks (e.g. Ramsay 1967; Weiss 1972) show that the thickness of most affected layers does not increase during deformation, but some sand layers at Piedras Blancas apparently swelled between thinned necks. Bulbous protrusions resembling sandfilled aneurysms occur along both upper and lower surfaces and thus cannot be explained as load casts or scour-and-fill structures. True swelling attests to mobility of sand within layers and Smith (1977) maintains that 'growing' boudins are evidence for non-Newtonian flow. Some sandy protrusions are hook-shaped or lobate (Fig. 3). Others are partly separated from their parent layers by narrow injected tongues of mudstone. In detail, margins of deformed sand layers range from smooth to irregularly scalloped or hummocky. Pinch-and-swell and associated structures in unmetamorphosed or very low-grade sedimentary rocks are described or illustrated by Horne (1969), Gregory (1969), Corbett (1973), and Kleist (1974), who all interpreted them as evidence for deformation of incompletely consolidated sediments during submarine slumping or sliding. In hand specimens, chaotic zones between extended semi-continuous layers contain variably shaped wispy, elongate, or contorted fragments of sandstone (Fig. 6). Most have highly irregular and locally diffuse boundaries. Many appear to have been injected by mudstone matrix or to have sent tiny dykelets of sand into matrix.

Mesoscopic ductile swelling and necking, and irregular distortion of wispy fragments, were accommodated internally by two principal mechanisms. Delicate S_0 bedding laminations are a useful marker in this regard (Figs 6, 7 & 8). They are either truncated by slip-surfaces (microfaults) less than 1 mm thick or irregularly swirled. Some distinct laminations become progressively vague or distorted and pass into thoroughly churned, homogenous sandstone or siltstone. In most sandy layers, variably oriented microfaults and swirled domains are so intimately associated that they must have formed contemporaneously. Thick (>20 cm) layers of sandstone are generally devoid of S_0 laminations, but similar mechanisms probably operated in them as well. Many outcrops, hand specimens and thin sections of massive, non-laminated fine- to coarse-grained sandstones are crossed by an irregular network of curviplanar dark 'veins', generally <2 mm thick, which I call 'web structure'. They contain more clay and silt than surrounding sandstone and probably served as both dewatering conduits and slip surfaces, since some strands are offset a few millimetres by others.

Mesoscopically brittle structures include extension ('tension') fractures, oriented approximately normal to layering, and extensional shear fractures. The former occur in thin necks between boudins and contain mudstone that flowed from adjacent layers as fractures opened. I arbitrarily classify the internal microfaults described above as extensional shear fractures where they affect both boundaries of a sand layer. They are common structures here as well as in some other terranes that record soft-sediment deformation (e.g. Gregory 1969; Horne 1969; see Weiss 1972, plate 141A for an example in a gneiss). Rast (1956) described analogous fractures, which he termed 'joints', in metamorphic rocks and considered the intervening segments of layers as 'lozenge-shaped boudins'. Nearly all such fractures in the Piedras Blancas–San Simeon area are inclined >45° from the top surface of sandy layers and hence are normal faults if viewed in sections perpendicular to layering. I interpret the faults as Coulomb shear fractures, and displacements along them necessarily resulted in layer-parallel extension. Maximum (σ_1) and minimum (σ_3) principal stresses were oriented respectively normal and parallel to layering at the instant shear fractures formed. Distinctively angular lozenges (Fig. 4) are typically equidimensional rather than elongate indicating that the spacing of fractures is directly proportional to layer thickness. I agree with Whitten (1966) that lozenges formed in this way should be distinguished from true boudins resulting from mesoscopically ductile necking or extension fracturing. Hsü (1968, 1974) suggested that some wedge-shaped lozenges intimately associated with pinch-and-swell and extensional fractures in thickly bedded strata are bounded by 'compressional fractures' (reverse faults), but I found no evidence to support his conclusion when I examined the same exposures.

Rast (1956) pointed out that in his examples, lozenges rotated during slip along fractures. This behaviour is a consequence of both fracture geometry and the fact that faults are confined to sandstone layers and very rarely extend across adjacent mudstone. Fragmented layers are typically overlain and underlain by unfaulted or ductilely extended sandy layers which preserve the local orientation of bedding prior to deformation (Fig. 4). The normal faults are antithetic. S_0 in the fractured layers undergoes a rotation equal to the angle between bedding laminae (S_0) in individual lozenges and S_0 in adjacent beds, while the faults themselves externally rotate in the same sense. (Tyler (1975) analysed an analogous kinematic situation in fractured brittle clasts in ductile matrices.) As a result, the long axes of lozenges become parallel to S_0 in adjacent layers, but bedding laminae within lozenges are noticeably oblique. In spite of the deformation, mesoscopic layering defined by layer boundaries is preserved rather than chaotically obliterated. In any particular bed thicker than about 10 cm, most or all fractures are inclined in the same sense, but faults within thinner layers commonly have opposing dips (see Fig. 7 in Cowan 1978). In some continuous sequences up to 50 m thick, I noted several layers that are each characterized by either sense of rotation. Gregory (1969) described the same situation in lower Miocene strata in New Zealand.

Another less common structure records a deformational mode distinct from the brittle and ductile end-members described above. Necks in some sandstone layers are chaotic breccias consisting of angular to subrounded fragments surrounded by a structureless matrix of sandstone. Extension was accommodated partly by brittle fracture, but fragments rotated randomly within a homogeneously flowing sand which was derived from the same part of the deforming layer.

There are two remarkable aspects of the deformation in the Piedras Blancas–San Simeon area. In my earlier study (Cowan 1978) I showed that clasts in olistostromes SE of San Simeon were variably flattened into oblate ellipsoids and correspondingly extended in all directions in the XY-plane of the total strain ellipsoid. The strong preferred orientation of deformed clasts defines a foliation, S_1. Most of these clasts were mesoscopically ductile, but external changes in shape were accommodated by brecciation, granulation, and cataclastic flow on a smaller scale. Bedded sequences and pebbly mudstones at Piedras Blancas experienced exactly the same type of total strain. Axially symmetric layer-parallel extension, accommodated by pinch-and-swell structure and shear fractures, is apparent in all cross-sections viewed normal to layering (Fig. 2). There is no detectable lineation, defined by either the long axes of boudins or regions of extreme necking, as would be expected if a unique direction of maximum elongation ($X > Y$) had persisted throughout the deformation. Surf-eroded exposures approximately parallel to deformed layering show irregularly rounded, slightly elevated hummocks of sandstone and depressions of mudstone.

Equally striking, in view of the abundance of extensional structures, is the near absence of folds and other evidence for layer-parallel shortening. In intermittent seacliff exposures from Piedras Blancas to a point 20 km SE, I found one 3 m thick interval of disharmonic, mesoscopic folds in the midst of a regularly bedded sequence of sandstone and mudstone. Their geometry and setting recall folds interpreted as submarine slump-related structures elsewhere (e.g. Grant-Mackie & Lowrie 1964;

Gregory 1969; Rupke 1976; Woodcock 1976). A compaction fissility in sandstone and siltstone developed parallel to bedding laminae prior to folding, but a later, weaker fissility formed in the mudstone cores of some folds. A single, 20 cm thick, tightly folded sandstone layer occurs in the midst of deformed strata at Piedras Blancas Point. In both localities, the later fissility is nearly parallel to the axial surfaces of the folds, which are themselves approximately parallel to layering in over- and underlying strata. On a hand-specimen scale, shreds of sandy layers <1 cm thick in highly disrupted mudstone-rich intervals locally define isolated fold closures (Fig. 7).

Microscopic structures

Mesoscopically ductile flow of sand was accommodated on a microscopic scale by either relative intergranular movements or cataclasis. Cataclasis in sandstones produces microscopically distinctive, poorly sorted gouges in which overall grain size has been reduced and angular remnants of sand grains are dispersed in a dark, irresolvable matrix (e.g. Dunn *et al.* 1973; Aydin 1978). In some sandstones, granulation is confined to isolated, planar zones of slip less than 2 mm, but commonly averaging 0.5 mm in thickness. Other sandstones, particularly those displaying mesoscopic web structure, comprise not only a network of planar zones but also irregularly shaped domains of granulated material that invaded pristine sand (Fig. 9). However, in spite of the drastic changes in thickness of extended, semi-continuous layers and the irregular distortion of delicate, wispy laminae only a few millimetres thick, most samples examined petrographically show little evidence for cataclasis or granulation of individual sand grains. The modest alteration of original clay matrix to chlorite and sericite does not mask the well-preserved, easily recognizable detrital texture of the sandstones. Sand grains must have merely rotated and jostled their neighbours, even in some regions of the most intense distortional strain. For example, in the sharply tapering edges of several flattened clasts of sandstone in pebbly mudstone, I could detect no greater incidence of grain breakage or microcracking than in undeformed layers or clasts, where they are insignificant.

Extension and flow are also recorded by microscopic analogues of the pinch-and-swell structure, injection features, and irregular swirling noted in hand-specimens. Detrital sand-size clasts of quartz and feldspar in wisps that are only one or two grains thick are

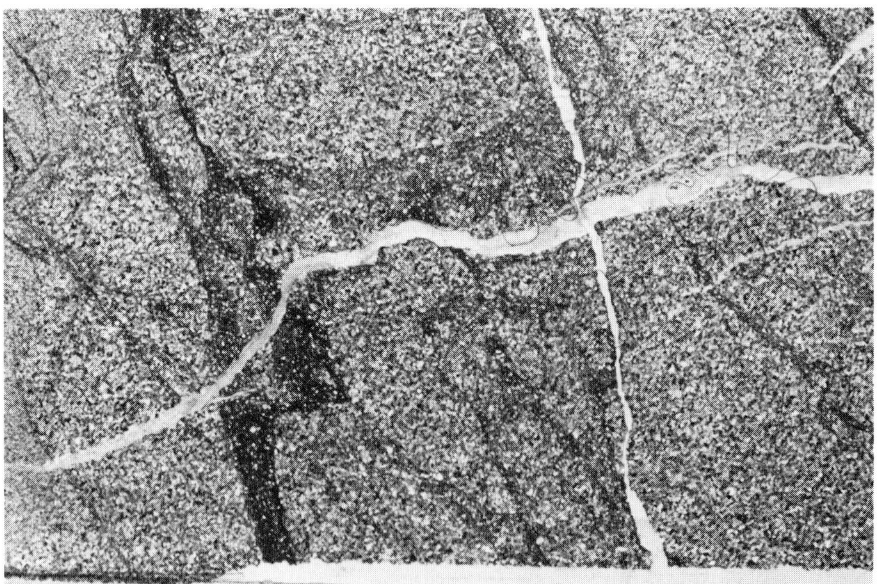

FIG. 9. Photo of a thin section from an elongated sandstone clast in pebbly mudstone; field is 20 × 30 mm. Dark zones consist of cataclastic 'gouge' in which sand grains have been fractured and comminuted. Non-cataclastic flow was probably an important mechanism in the remainder of the specimen. Veins are filled with calcite.

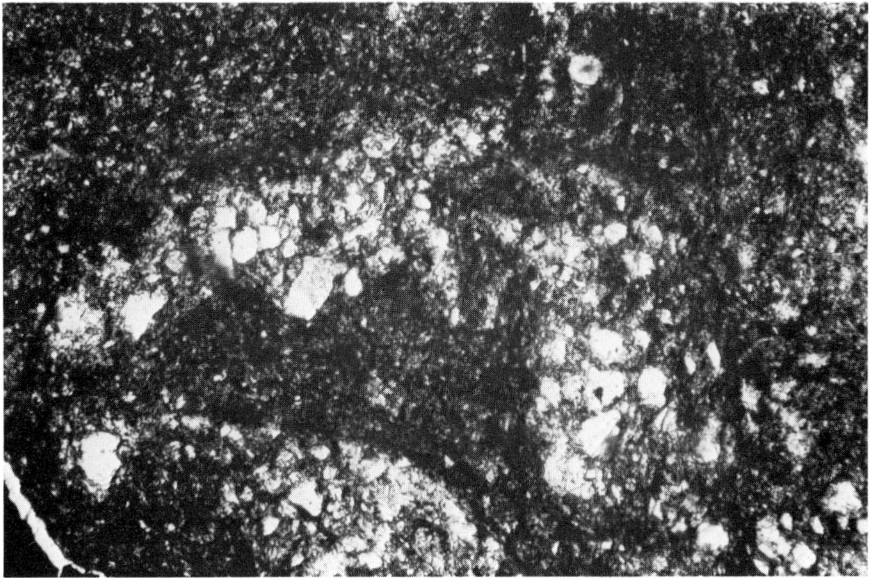

FIG. 10. Photomicrograph of disrupted sandy laminae in siltstone. Swirling and local disaggregation and mixing were accomplished without microfracturing and granulation. Field is 1.4 × 2 mm.

undeformed, as are isolated grains floating in mudstone that has flowed extensively (Fig. 12). D. S. Wood (pers. comms. 1978, 1979) found identical microscopic relations in the Gwna olistostrome on Anglesey, Wales, and emphasized the complete lack of intragranular deformation in clasts of what are now orthoquartzites. I should stress here that none of the microscopic fabrics and textures associated with mesoscopic ductile flow and transposition-related fragmentation in low-grade metamorphic rocks, such as domainal slaty cleavage (Hobbs *et al.* 1976), metamorphic differentiation or *in situ* intragranular plastic strain in quartz, are evident at Piedras Blancas and San Simeon.

Microfaults are narrow slip surfaces that are typically less than 0.1 mm in thickness along which displacements appear to average a few millimetres. In thin section, they are defined by either films of dark brown, irresolvable matrix as thin as 0.01 mm or thicker zones, averaging 0.1 mm, in which platy grains of clays and phyllosilicates are oriented parallel to zone boundaries (Fig. 11). Unlike cataclastic zones, sand grains within and adjacent to these microfaults are not fractured or internally strained. Where faults emerge from sandy layers into silt- or clay-rich layers, they commonly become less distinct as they flatten into parallelism with overall layering; corners of sandy layers are rounded rather than sharp. Some faults merge with bedding-parallel zones of slip.

Mudstones

In undeformed bedded sections, dark brown siltstones and claystones are finely laminated and fissile. The bedding-parallel orientation of detrital clays and micas is probably a primary settling or compaction fabric. Mudstone in deformed sequences (e.g. Figs 3 & 6) is weakly fissile and breaks less readily along irregular surfaces that grossly parallel mesoscopic layering but closely conform in detail to the surfaces of deformed sand layers. Apparently, this fabric is most strongly developed adjacent to sand layers >5 cm thick, where it is possible to break mudstone into progressively thinner lenticular, but non-polished, chips. In thin sections cut normal to layering, these mudstones display vaguely lenticular domains that are defined by slightly different preferred orientations of fine-grained platy materials. Some domain boundaries are gradational but most are narrow planar shear (slip) zones or micro-kinkbands. Other groups of individual micro shear-zones <0.05 mm thick form lacy, anastomosing networks. Similar microstructures in experimentally and naturally deformed clays are described by Morgenstern & Tchalenko (1967), Tchalenko (1968, 1970) and Maltman (1977).

FIG. 11. Photomicrograph of a microfault similar to those illustrated in Fig. 8. Field is 1.4 × 2 mm. Separation of dark, mud-rich laminae is about 0.5 mm in this section. Fault zone is ~0.05 mm wide and individual anastomosing strands contain a film of dark brown clay. Note the absence of cataclastic effects adjacent to fault.

Birefringent phyllosilicates within each zone are strongly oriented parallel or <30° to the zone boundaries—an apparently diagnostic criterion for slip noted by Morgenstern & Tchalenko (1967) and Tchalenko (1968).

Black scaly mudstone or *argille scagliose* is matrix in some deformed pebbly mudstones (Cowan 1978). Hand specimens cleave readily parallel to a pronounced mesoscopic foliation and samples can be broken into lenticular chips that invariably have shiny, polished and locally striated surfaces in contrast to the weakly fissile brown mudstone described above. In this section, a primary settling or compaction fabric defined by silt and sand laminae and oriented phyllosilicates is locally reoriented by planar kinkbands, or, more commonly, irregular to swirled patterns of flow; a superimposed set of anastomosing slip surfaces, intersecting in acute angles <60° is probably responsible for the distinctive foliation.

In summary, mesoscopic flow in deformed mudstone was accommodated by the disruption of a primary sedimentary fabric along a variety of semi-penetrative, microscopic structural discontinuities that are reflected in a crude, layer-parallel mesoscopic foliation. Although very fine-grained, presumably detrital phyllosilicates are strongly reoriented in narrow shear zones, there is no evidence for the post-disruption development of a true slaty cleavage defined by microscopically penetrative planar preferred orientation of inequant detrital or metamorphic minerals. In fact, petrographic observations suggest recrystallization of detrital clay was negligible. Of four mudstone samples analysed by X-ray diffraction (S. Johnson, analyst), one of scaly argillite from San Simeon contains an expandable mixed-layer (montmorillonite-illite) clay and three from Piedras Blancas contain a non-expandable 10 Å clay (presumably illite); all contain chlorite and kaolinite as well.

State of sediments during deformation

Several authors, notably Woodcock (1976), Helwig (1970), Horne (1969), Kleist (1974), and Gregory (1969), have recently addressed the long-standing but vexing problems of how to differentiate between structures formed during gravity-driven slumping of semi-consolidated ('soft') sediments and those resulting from tectonic deformation of consolidated (hard) rocks. Much of the debate concerns the geometry of mesoscopic folds and their relation to adjacent strata. In some literature (e.g. Williams *et al.* 1969; Potter & Pettijohn 1977) there is a tendency to interpret most soft-sediment structures as 'penecontemporaneous' with deposition (hence load or slump phenomena) and to consider those structures postdating lithification as 'tectonic'. Two concepts are involved here: the degree of consolidation or physical state at deformation, and the overall environment of deformation. These should be considered independently, if for no other reason than incompletely lithified sediments could potentially be affected by what all would agree is tectonic defomation as they are carried beneath the toe of the trench slope. My analysis of physical state is based mainly on microfabrics rather than mesoscopic structures, which I feel are more useful as a record of local deformational environment.

Sandstone

At outcrop and hand-specimen scales, both sand layers in originally bedded sequences and sandstone clasts in pebbly mudstones show two contrasting types of behaviour. In the parlance of rock deformation (Griggs & Handin 1960), most have deformed by mesoscopically homogeneous, cohesive *ductile* flow, and some have failed by *brittle* extension and shear fractures. Three deformation mechanisms accommodated cohesive flow at smaller scales. Displacements along closely spaced microfaults are actually observable without a microscope (Fig. 8) and might be considered 'transitional' behaviour. Cataclastic flow, involving microfracturing, granulation, and general reduction in grain size, was also locally important. However, much of the mesoscopic distortion was accomplished by intergranular movements without concomitant cataclasis.

These observations can be interpreted in light of experimental deformation of sandstone and unconsolidated sand. In triaxial compression tests on water-saturated, porous, consolidated sandstones, Handin *et al.* (1963) found that deformational behaviour was strongly dependent on effective confining pressure (P_e), equal to (confining pressure − fluid pressure). At high P_e (low fluid pressures), samples deformed ductilely by wholesale cataclastic flow on the microscale. At progressively lower P_e, grain breakage became localized and finally restricted to narrow, through-going, sharply defined shear zones as the samples experienced transitional, then brittle behaviour. When confining pressure equalled fluid pressure ($P_e = 0$) samples were shortened up to 25% without any evidence for cataclastic flow or granulation, even along shear zones. Handin *et al.* (1963) deduce that frictional resistance to sliding and flow is counteracted by the 'cushioning' effect of high fluid pressures favouring purely intergranular adjustments. Other experiments on dry lithified sandstone (Handin *et al.* 1963) and unconsolidated sands (Borg *et al.* 1960; Friedman *et al.* 1980) show that localized or pervasive cataclastic flow is the predominant deformation mechanism over a wide range of confining pressures (50–200 MPa). From these results, it is clear that much of the sand in the study area was deformed while it was not only water-rich but also under low effective confining pressures, since cataclastic mechanisms were impeded in regions of extreme mesoscopic flow and adjacent to microfaults.

Other evidence suggests that some sands were not only fluid-rich but unconsolidated as well. Individual sand layers or remnants of layers commonly display several types of behaviour, even within a single hand specimen. For example, intergranular cohesive flow may give way to small offsets along closely spaced microfaults, especially in necked regions (Fig. 7). In outcrops, layers that have deformed ductilely are generally interspersed with and adjacent to layers recording brittle failure along through-going fractures (Fig. 5). Most of these structures were probably contemporaneous, since they are intimately associated throughout substantial volumes of sediment. What accounted for these abrupt, local changes in behaviour? Overall confining pressure and strain rate were probably homogeneous and constant, at least on a mesoscopic scale. Local differences in pore pressure could have been responsible. I suggest that the most likely explanation is that sands were already variably dewatered and consolidated at the time of deformation. Unconsolidated, water-rich sands, little modified since deposition, deformed by non-cataclastic, intergranular flow; semi-penetrative microfaults, general cataclasis, and finally brittle shear fractures developed in progressively more consolidated materials. I consider all of these structures to have formed in 'soft' sediment. Behavioural differences reflect local variations in water content (and perhaps P_e), since early consolidation or lithification basically involves compaction, reduction in porosity, and dewatering. If, instead, differences in P_e were solely responsible, all of the layers would have to have been consolidated equally and fluid pressure then selectively raised in small domains to promote intergranular flow. I postulate that early, progressive changes in water content in consolidating sands can have as much influence on behaviour and mechanism as do changes in pore pressure once sands have passed through some critical, but as yet experimentally undetermined, threshold of lithification.

Mudstone

It is much more difficult to decipher the physical state of associated mudstones on the basis of behaviour and mechanisms alone. Boudinage, pinch-and-swell structures, and flattened clasts (Cowan 1978) in the study area all indicate ductility (or material property) contrasts; hence, mud flowed more readily and suffered greater permanent strains on a mesoscopic scale than did sand and other rock types (metabasalt, chert, and metamorphic rock clasts). Other features pointing to mobile mud are the greater degree of stratal disruption in disharmonically deformed mudstone-rich intervals (Figs 3 & 6) and mudstone-filled fissures and re-entrants in sandy layers on all scales. Although mesoscopic ductility of mudstone was complemented by both brittle and ductile behaviour in sand, what I would interpret as a type of cataclastic flow was more widespread in deformed mudstone on a microscopic scale. It seems everywhere to have developed crudely oriented, semi-penetrative discontinuities such as individual or multiple, anastomosing slip surfaces, kink-bands, and swirled zones, while most sand flowed homogeneously and non-cataclastically. These discontinuities are actually expressed mesoscopically as 'scaly' foliation. If sands were indeed water-rich as concluded above, then it is certain muds were also water-rich when they developed the suite of structures described here. For example, Maltman (1977) deformed soft, ductile argillaceous sediments with water contents up to 40% and found that flow is microscopically accommodated on semi-penetrative shear zones and other discontinuities. However, data show that gravitationally compacted muddy sediments dewater rapidly and are essentially consolidated after burial of about 250–300 m (Skempton 1970; Rieke & Chilingarian 1974). I infer that, at the time of deformation, mudstone had undergone a proportionally greater decrease in void ratio and water content than interlayered sands, which served as permeable reservoirs for water escaping from muds. The high fluid pressures so generated prevented consolidation in sands and promoted intergranular flow within. I observed exactly these differences in consolidation in cores from sites 496 and 497 obtained on the inner slope of the Mid-America Trench off Guatemala on DSDP Leg 67 (see von Huene *et al.* 1981); stiff but pliant muds are interbedded with sand layers <2 cm thick that were loose enough literally to be poured out of the core barrel. But it seems paradoxical that stiff mud would behave more ductilely than unconsolidated sand, as the mesoscopic structures at Piedras Blancas and San Simeon imply. When most argillaceous materials are experimentally deformed, they have an interesting response that may resolve the paradox. Conventional graphs of differential stress versus percentage strain (e.g. Skempton 1964) show curves steadily climbing to a 'peak strength' at low total strain and then falling to a lower 'residual strength', or constant stress difference, at which permanent deformation continues indefinitely. Skempton (1964) and Tchalenko (1970) correlate this drop in strength with the development of complex microscopic shear zones and kink bands. In other words, a modest amount of ductile strain in the Franciscan muds, accompanied by the development of some of the discontinuities observable in thin section, may have fundamentally altered their properties and allowed them to achieve greater total strains than interlayered, but more 'competent', sands.

Even if these inferences about the properties of Franciscan mudstones are correct, it is risky to assume that the microstructures described above are diagnostic of a partly consolidated physical state. Morgenstern & Tchalenko (1967) and Tchalenko (1968, 1970) imply that localized microscopic shear-zone structures and induced preferred orientation develop over a wide range in water content. Although mesoscopic behaviour and micromechanisms could depend on not only water content (degree of consolidation) but also fluid pressure, composition, total strain, and strain rate, there are not yet enough experimental data to evaluate the influence of each of these variables.

Summary

The microscopic behaviour of sand suggests that it ranged from unconsolidated to semi-lithified at the time of deformation. Even though it was a water-rich, 'soft' sediment, it developed a variety of small-scale structures that may reflect local differences in water content or fluid pressure. It commonly displays evidence for an 'early ductility' that contrasts with the ductile flow and associated microstructures typical of metamorphic regimes. Wood (pers. comm. 1978), Schuster (1980), Davies & Cave (1976), and Lambert (1965) all have used microfabrics, in concert with other evidence, to infer that sediments were disrupted or deformed while soft, while Gregory (1969),

Horne (1969), and Helwig (1970) invoked differences in water content and consolidation to explain variations in mesoscopic structural style in semi-lithified sediments that were deformed by slumping. My conclusions about physical state were anticipated by Kleist (1974), whose interpretation that Franciscan Coastal Belt sediments in northern California (see Bachman 1981) were deformed while soft was based largely on the geometry of mesoscopic structures.

Environment of deformation

Modern convergent plate margins

We can now try to establish the environment of deformation without having made the *a priori* assumption that soft-sediment, or syn-consolidational, deformation is in some way connected with slumping or sliding. Fig. 12 depicts a variety of hypothetical deformational environments that can be evaluated using the mesoscopic structures at Piedras Blancas and San Simeon. It is an idealized cross-section of a simple accretionary wedge largely based on marine geophysical and DSDP drilling data from active margins (e.g. Seely *et al.* 1974; Karig & Sharman 1975; Hamilton 1977; Moore *et al.* 1979; and others) seasoned with certain deductions about the nature and distribution of deformation in a dynamic subduction zone (e.g. Moore & Karig 1976; Cowan & Silling 1978). Environments fall into two categories depending on scale of observation. Viewing the wedge as a whole, I propose that it can be divided into an accretionary substructure and an overlying superstructure. Each of these large-scale domains in turn comprises smaller, more localized domains of deformation, such as fault zones and conceivably zones of intense flow or folding. The key distinction between substructure and superstructure is that in the former, planar zones of slip or flow are confined; they root within or beneath the wedge and need not intersect the seafloor. In contrast, the superstructure is the realm of delapsion (Hoedemaeker 1973) and is necessarily restricted to shallow levels of the inner trench-slope where laterally unconfined deformation can be partly accommodated on unrooted slip surfaces that must intersect the seafloor. The following discussion elaborates these concepts.

In the substructure, most deformation occurs as a direct consequence of the descent of oceanic lithosphere; panels of sediment are scraped off and accreted at the front of the wedge or subducted beneath the inner wall of the trench. Localized structures (Fig. 12) may include active thrusts at and near the toe of the trench slope accommodating accretion (Karig *et al.* 1980), arcward- or trenchward-dipping faults within the wedge that might represent reactivated basal thrusts (Moore & Karig 1976), and orderly folds (White & Ross 1979) or chaotically churned zones within inter-thrust panels. In the superstructure, deformation results directly from gravity-driven, downslope transport accomplished by a spectrum of mass-movement processes (see Nardin *et al.* 1979 for a recent

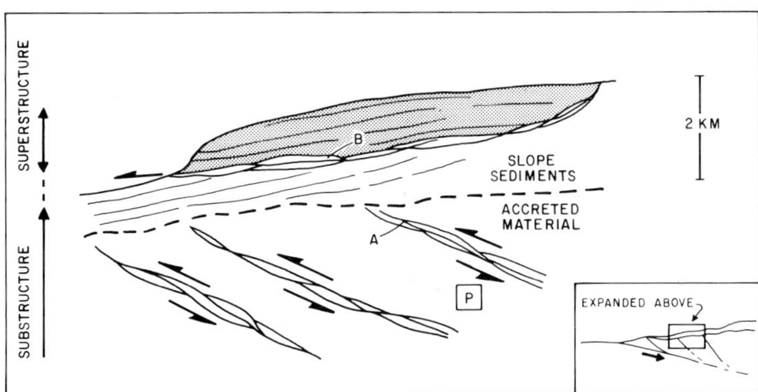

FIG. 12. Diagrammatic cross-section of part of the inner wall of a trench at an active convergent plate boundary. See text for discussion. Horizontal dashed line coincides with the approximate boundary between layered sediments and acoustically complex (presumably accreted) material as shown on a typical reflection profile. Wavy, anastomosing lines represent arcward dipping reflectors, interpreted as fault zones, and a hypothetical detachment zone at the base of a slide (shaded). Deformation at Piedras Blancas may have occurred within the shaded area as semi-consolidated sediments gravitationally collapsed and spread laterally.

review of nomenclature). Slides (including slumps) move en masse along localized, basal slip zones and may suffer internal deformation as well. Slides grade into mass flows (Dott 1963; Morgenstern 1967) as sediment becomes more thoroughly dissaggregated and mixed with water. Although many slide sheets, ranging from 0.1–1 km in thickness, have been documented seismically on passive continental margins (see numerous references in Woodcock 1979), it is not yet known whether they are abundant or extensive on modern trench slopes. Von Huene (1972), Coulbourn & Moberly (1977), and Karig *et al.* (1980) invoke mass-transport mechanisms to explain some irregular topographic features on the inner walls of the Aleutian, Peru-Chile, and Sunda trenches respectively. In view of their prevalence at passive margins, it is likely that shallow-level slide and mass flow deposits will prove to be at least as common on relatively steep, seismically active, and topographically complex trench slopes. They may be confined to the thin (<2 km) apron of sediments draped over some trench slopes (e.g. off Peru-Chile as shown in Kulm *et al.* 1977, and above the Mid-America Trench as illustrated by Ibrahim *et al.* 1979), although slide detachment surfaces could theoretically form at shallow depths on the inner wall regardless of the composition or consolidation history of sediments if shear stresses on the potential detachment exceed the shearing resistance of the material. Sliding would be favoured where sediments are semi-lithified and fluid pressures are high (e.g. Booth 1979).

It would be ideal if one could identify a unique site or trajectory of deformation for any part of the Franciscan, let alone the section exposed in the study area, in Fig. 12. However, this task is complicated by the probability that some key deformation styles are common to both substructure and superstructure, and the smaller our scale of observation, the more equivocal our conclusions regarding large-scale setting are apt to be. For example, the suite of mesoscopic structures developed at A and B in Fig. 12 may be identical since both sites are in fault zones, but the immediate causes of displacements along them are utterly different. Woodcock (1976) also cautions that the geometry and style of mesoscopic folds alone cannot be used to differentiate slide-related from more closely confined 'tectonic' deformation. Some of the following conclusions about both large- and small-scale deformational environments are necessarily speculative in view of these ambiguities.

Significance of axially symmetric extension

The most important constraint on the deformational environment is the type of total strain recorded by small-scale structures in both bedded sequences and pebbly mudstones. The stratal disruption in bedded sequences is due to boudinage, pinch-and-swell structure produced by necking, extension fractures, and shear fractures (Fig. 2). All of these structures involve the extension of sandy layers and concomitant ductile flow in associated less-competent muds or mudstones. Unlike many examples of boudinage, however, there is no detectable preferential direction of extension at Piedras Blancas; extension apparently occurred with equal ease in all directions in the plane of the layering and is therefore axially symmetric with respect to the normal to layering. An imaginary ellipsoid describing the total strain at the scale of hand specimens and outcrops would have the XY-plane (where $X \geqslant Y \geqslant Z$ are the principal axes) parallel to layering, and $\lambda_1 = \lambda_2 > 1$ (where λ_1, λ_2, λ_3 are maximum, intermediate, and minimum quadratic elongations). The bulk mesoscopic strain is qualitatively recorded by the distortion of clasts in pebbly mudstones, both at Piedras Blancas and south of San Simeon (Cowan 1978), most of which were deformed into flattened oblate ellipsoids. At mesoscopic and macroscopic scales, the sediments behaved as a continuum, but the discontinuities accommodating the strain are obvious in most hand specimens and thin sections.

It is emphasized that the total strain is a true triaxial flattening ($\lambda_1 = \lambda_2 > 1$; $\lambda_3 < 1$) involving finite extension in the XY-plane and not 'apparent' flattening produced by a volume loss accompanying compaction. Gravitational compaction is a uniaxial strain involving only vertical shortening in the Z-direction (Fig. 13) but there are no corresponding elongations parallel to X and Y because compacting beds are laterally confined within a sedimentary basin. Thus, the oblate shape of a total strain ellipsoid is not in itself proof that extension parallel to X or Y has occurred, and Ramsay & Wood (1973) and Sanderson (1976) showed that an oblate shape inherited from compaction can persist even where plane tectonic strains, involving no elongation in the Y-direction, are superposed. Dewatering and volume loss were clearly important at Piedras Blancas, but the structures prove that compaction-related changes in the Z-direction were either compensated or succeeded by true extension in the XY-plane. I estimate that layers and clasts locally suffered tectonic extensions as great as 75–100%.

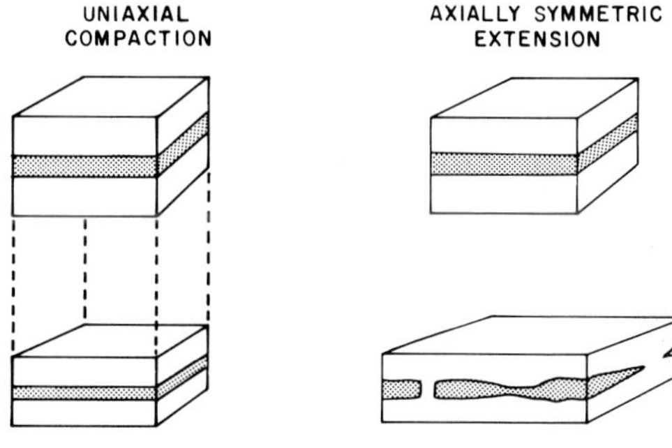

FIG. 13. Volume changes in dewatering sediments are ordinarily accomplished by vertical shortening without corresponding horizontal elongation, since layers are confined in sedimentary basins. Sediments in the study area were laterally unconfined and able to extend in all directions parallel to layering while shortening in a direction normal to layering.

I suggest that the only plausible large-scale setting compatible with the substantial axially symmetric extensions documented here and in Cowan (1978) is the superstructure. First, consider the alternative, using a small volume element *P* in the substructure, shown in cross-section in Fig. 12. Even when the effects of volume change are taken into account, deforming *P* would require an elongation of at least 50% normal to the plane of the figure and parallel to the strike of the accretionary wedge (assuming *Z* is in the plane of the section). Strains of this magnitude are doubtful, because the slice of *P* is effectively confined in this direction by the wedge itself. The substructure is analogous in this regard to classical on-land orogenic belts, such as the Canadian Rockies and Appalachian Valley and Ridge province, where most of the tectonic deformation is considered to be plane strain with negligible elongations parallel to orogenic trends. Hossack (1979) reviews estimates of strike elongations in the Appalachians and parts of the Caledonides and suggests that regional values of 10% are typical. In contrast, the shallow-level, laterally unconfined setting of the superstructure can accommodate regional, axially symmetric extensions. Deformed sediments in the study area can be thought of as having spread laterally, and to do this they must have been in or above a detachment or slip surface that intersected the ocean floor. The low average slope of the ocean floor constrains spreading to be approximately horizontal and compensatory shortening to be vertical. In this environment (Fig. 12), there are no lateral restrictions, other than topography, to impede strike elongations. The situation is somewhat analogous to shallow-level décollements in on-land orogens, above which detached masses can deform independently from confined, subjacent terranes. One could argue that the pronounced *XY*-elongations in the study area are a local anomaly that developed in an overall field of plane 'tectonic' strain in the substructure, but I discount this hypothesis because all of the deformed Franciscan rocks exposed along the coast for 45 km SE of Piedras Blancas appear to record the same type, but different amounts of strain (Cowan 1978).

Several small-scale deformational environments in the superstructure are seemingly compatible with this type of strain. In the simplest case shown in Fig. 12, lateral spreading may: (1) accompany penecontemporaneous sliding within a few metres of the sediment-water interface; (2) occur within the basal detachment zone, which is really just a fault zone in a special, shallow-level setting; (3) occur somewhere above the basal slip surface in a zone of distributed flow accommodating gravitational collapse. However, I believe (1) can be eliminated if the pervasive deformation in the Piedras Blancas–San Simeon region is compared with other examples of penecontemporaneous 'slumping' (e.g. Gregory 1969; Corbett 1973; Woodcock 1976), where slump folds and stratal disruption occur in stratigraphically confined intervals that typically have sharply truncated upper contacts.

A choice between environments (2) and (3) is more difficult. In a gross sense, faults are localized zones of slip characterized by both intense strain relative to adjacent rocks and non-coaxial deformation paths involving a major component of progressive simple shear (e.g. Ramsay & Graham 1970). Ramberg (1977, 1978, 1980) argues, on the basis of theory and experiment, that horizontal lengthening in gravitationally spreading and collapsing sheets or nappes is compensated by vertical shortening and that penetrative 'pure shear' (coaxial progressive strain) is important in this regime. Unfortunately, it is not possible to establish whether a strain path was coaxial or non-coaxial from the shape of the total strain ellipsoid alone (Hobbs *et al.* 1976), but I believe that both the geometry of the small-scale structures in the study area and their overall setting favour the 'collapse', rather than 'fault-zone' hypothesis. First, the same style of deformation is pervasive at mesoscopic and larger scales over a large area. The overall impression is that the heterogeneity in amount of strain reflects local differences in the ratio of mud to sand and perhaps in consolidation history. It does not appear that deformation is sharply localized in 'shear zones' separating thrust sheets or fault blocks that have been juxtaposed or even modestly displaced relative to one another. Secondly, the symmetry of the strain itself makes a fault-zone origin suspect. Faults are generally considered to result in plane strain with negligible extensions normal to slip directions, but triaxial flattening is theoretically possible if rocks in the fault zone are somehow elongaged along strike. Finally, the preponderance of structures recording layer-parallel extension and the near absence of folds in both incipiently and intensely deformed beds and layers (Fig. 2) suggest that layering was oriented at low angles, or possibly parallel, to the XY-planes of both infinitesimal and finite strain ellipsoids throughout the deformation. For example, delicately thinned layers in regions of extreme necking did not rotate with respect to either their parent layers or nearby less-deformed layers. If correct, this relationship between the XY-plane and layering would indicate a coaxial strain path. In summary, the simplest hypothesis permitted by these observations and interpretations is that lateral, horizontal spreading was distributed through a gravitationally collapsing, vertically shortening package of partly consolidated sediments, rather than restricted to a narrow zone of progressive simple shear at its base. If both spreading and layering were approximately horizontal, then it is probable that collapse affected a thin apron of slope sediments mantling the inner wall of a trench, rather than sediments that had been accreted or otherwise deformed and rotated from their originally horizontal orientation.

A model for deformation at Piedras Blancas

I propose that a piedmont glacier or an ice cap is an appropriate analogue for deformation at Piedras Blancas. Ramberg (1977, 1980) has stressed the geometric, kinematic, and dynamic similarities between ice caps and Caledonian nappes in Scandinavia. Both can be visualized as flowing outward in response to the gravitational potential induced by the surface slope of the moving mass. Ice and rock in effect collapse and spread laterally under their own weight. Ramberg emphasizes the internal deformation, or 'collapse strain', distributed throughout these bodies and suggests that lateral displacements (forward movements) are due primarily to vertical shortening balanced by horizontal elongation rather than to sliding along a basal detachment or glide surface. Fig. 14, from Hooke & Huddleston (1978), summarizes the progressive strain inside a glacier. It illustrates a two-dimensional case, but the same deformation paths would apply in ice caps or ice sheets even though substantial elongations occur normal to the plane of the diagram. There is a domain of coaxial strain ('pure shear' in Hooke & Huddleston 1978), accommodating vertical collapse and laterally unconfined strain, in the upper part of the glacier, and the strain path progressively changes to non-coaxial in a basal zone characterized by high shear strains. I visualize the penetrative, mesoscopic triaxial flattening in Piedras Blancas sediments as having occurred in the domain of coaxial deformation, perhaps in the upper part of the larger slide. The inherently low strength of water-rich, incompletely consolidated modern slope sediments deposited on surface slopes as great as 7° (see seismic profiles in Hamilton 1977) would invite collapse. The main reason for drawing an analogy between slides on the inner wall and glaciers is to raise the possibility that signficant and distinctive deformation can occur *within* a moving mass as it spreads and collapses. The model may also apply to thin submarine slides on continental margins in general, although very little is known about the internal structure of deposits in this setting. The overall style of deformation described here may be part of a spectrum developed as slides transform into progressively more fluid-like sediment gravity flows (see Dott 1963 and Morgenstern 1967).

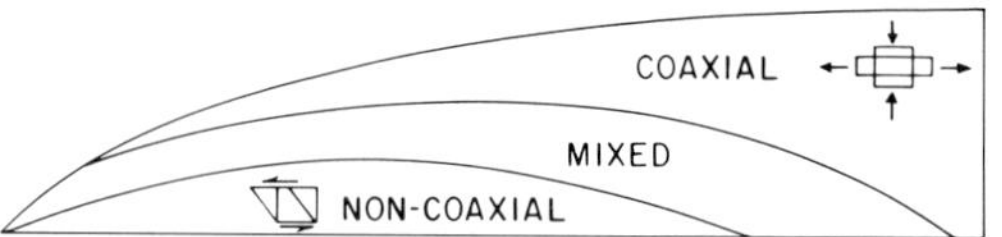

FIG. 14. Profile of the lower part of an active glacier, modified from Hooke & Huddleston (1978). An internal 'collapse strain' (Ramberg 1977) is distributed through the spreading ice, but the nature of the deformation path varies throughout as shown. This model may apply to gravitationally spreading sediment on a trench slope, as discussed in the text. Non-coaxial shear strain near the base probably reflects frictional resistance to sliding.

Conclusions

An important conclusion of this paper is that Franciscan sediments were deformed while in an unconsolidated to semi-lithified state. Similar interpretations elsewhere in the Franciscan, and in many other terranes as well, rely heavily on the geometry of mesoscopic structures thought to be compatible with 'soft-sediment slumping', but I feel that deformational mechanisms are a less equivocal criterion. The axially symmetric extension recorded by a variety of small-scale structures is an unusual type of strain requiring a laterally unconfined deformational environment. Vertical shortening and compensatory horizontal elongation are a possible consequence of gravitational collapse and spreading at a shallow level on the trench slope. Although I am applying these conclusions only to the Franciscan in the study area, my general approach could be useful in other parts of the complex. The suggestion that stratal disruption and fragmentation occurred in response to gravitational spreading may be controversial, since these features have also been ascribed to faulting both in the Franciscan (e.g. Cowan 1974; Bachman 1981) and in other ancient subduction complexes (e.g. Moore & Karig 1980). All available structural data must be precisely described and evaluated in order to test these different hypotheses.

It is appropriate to ask whether the conventional distinctions 'tectonic versus soft-sediment', 'tectonic versus sedimentary', and 'tectonic versus gravity-driven' have any meaning in a trench slope setting. The results of this study argue for abandoning the first and allowing 'tectonic' to denote deformational environment rather than the physical state of sediments. The second may separate processes at the sediment-water interface from all those beneath, but it does not accommodate many mass-transport processes. The third distinction implies that there is some minimum scale beyond which gravity assumes an important role. Although gravitational forces are obviously required to propel thin, surficial slide sheets downslope, Elliott (1976) and Hamilton (1977) suggest that they are also directly responsible for displacements along imbricate, arcward dipping thrusts within accretionary wedges. 'Tectonic deformation' is a useful concept if it can be identified with the large-scale setting of a suite of structures. My proposed distinction between substructure and superstructure is influenced by Dennis's (1972) treatment of slides and thrusts, Hoedemaeker's (1973) definitions of tectonism and delapsion, and Woodcock's (1976) discussion of tectonic and slump structures, all of which embody the connotation that 'tectonic' deformation is confined on a large scale and can be conceptually differentiated from laterally unconfined sliding, slumping, and spreading.

ACKNOWLEDGMENTS: My interest in soft-sediment deformation was stimulated by discussions in the field with D. S. Wood, F. A. Gibbons, and M. T. Brandon. It was catalysed by my participation on Leg 67 of the Deep Sea Drilling Project which afforded the chance to study sediment cores from the Mid-America Trench off Guatemala. I am indebted to M. T. Brandon for his continued suggestions and criticism.

References

AYDIN, A. 1978. Small faults formed as deformation bands in sandstones. *Paleogeophys.* **116,** 913–30.

BACHMAN, S. B. 1978. A Cretaceous and early Tertiary subduction complex, Mendocino Coast, northern California. *In:* HOWELL, D. G. & MCDOUGALL, K. A. (eds). *Mesozoic Paleogeography of the Western United States,* 419–30. Pacif. Sect. Soc. econ. Paleontol. Mineral. Tulsa.

—— 1981. The Coastal Belt of the Franciscan: youngest phase of northern California subduction (this volume).

BOOTH, J. S. 1979. Recent history of mass-wasting on the upper continental slope, northern Gulf of Mexico, as interpreted from the consolidation states of the sediment. *In:* DOYLE, L. J. & PILKEY, O. H. (eds). *Geology of Continental Slopes.* Spec. Publ. Soc. econ. Paleontol. Mineral. Tulsa, **27,** 153–64.

BORG, I., FRIEDMAN, M., HANDIN, J. & HIGGS, D. V. 1960. Experimental deformation of St. Peter sand: a study of cataclastic flow. *In:* GRIGGS, D. & HANDIN J. (eds). *Rock Deformation.* Mem. geol. Soc. Am., **79,** 133–91.

CORBETT, K. D. 1973. Open-cast slump sheets and their relationship to sandstone beds in an upper Cambrian flysch sequence, Tasmania. *J. sediment. Petrol.* **43,** 147–59.

COULBOURN, W. T. & MOBERLY, R. 1977. Structural evolution of fore-arc basins off South America. *Can. J. Earth Sci.* **14,** 102–16.

COWAN, D. S. 1974. Deformation and metamorphism of the Franciscan subduction zone complex northwest of Pacheco Pass, California. *Bull. geol. Soc. Am.* **85,** 1623–34.

—— 1978. Origin of blueschist-bearing chaotic rocks in the Franciscan Complex, San Simeon, California. *Bull. geol. Soc. Am.* **89,** 1415–23.

—— & SILLING, R. M. 1978. A dynamic, scaled model of accretion at trenches and its implications for the tectonic evolution of subduction complexes. *J. geophys. Res.* **83,** 5389–96.

DAVIES, W. & CAVE, R. 1976. Folding and cleavage determined during sedimentation. *Sediment. Geol.* **15,** 89–133.

DENNIS, J. G. 1972. *Structural Geology*. New York.

DOTT, R. H., JR. 1963. Dynamics of subaqueous gravity depositional processes. *Bull. Am. Assoc. Petrol. Geol.* **47,** 104–28.

DUNN, D. E., LAFOUNTAIN, L. J. & JACKSON, R. E. 1973. Porosity dependence and mechanism of brittle fracture in sandstones. *J. geophys. Res.* **78,** 2403–17.

ELLIOTT, D. 1976. The motion of thrust sheets. *J. geophys. Res.* **81,** 949–63.

FRIEDMAN, M., HUGMAN III, R. H. H., & HANDIN, J. 1980. Experimental folding of rocks under confining pressure, part VIII—forced folding of unconsolidated sand and of lubricated layers of limestone and sandstone. *Bull. geol. Soc. Am.* **91,** 307–12.

GRANT-MACKIE, J. A. & LOWRY, D. C. 1964. Upper Triassic rocks of Kiritehere, SW Auckland, New Zealand. Part I: submarine slumping of Norian strata. *Sedimentology,* **3,** 296–317.

GREGORY, M. R. 1969. Sedimentary features and penecontemporaneous slumping in the Waitemata Group, Whangaparaoa Peninsula, North Auckland, New Zealand. *N.Z. J. Geol. Geophys.* **12,** 248–82.

GRIGGS, D. & HANDIN, J. 1960. Observations on fracture and a hypothesis of earthquakes. *In:* GRIGGS, D. & HANDIN, J. (eds). *Rock Deformation.* Mem. geol. Soc. Am. **79,** 347–64.

HALL, C. A., JR 1975. San Simeon–Hosgri fault system, coastal California: economic and environmental implications. *Science,* **190,** 1291–4.

—— 1976. Geologic map of the San Simeon–Piedras Blancas region. San Luis Obispo County, California. *U.S. geol. Surv. Map,* **MF-784.**

HAMILTON, W. 1977. Subduction in the Indonesian region. *In:* TALWANI, M. & PITMAN, W. C. III (eds). *Island Arcs, Deep Sea Trenches, and Back-Arc Basins.* Am. geophys. Union, Ewing Ser. **1,** 15–31.

HANDIN, J., HAGER, R. V., JR, FRIEDMAN, M. & FEATHER, J. N. 1963. Experimental deformation of sedimentary rocks under confining pressures: pore pressure tests. *Bull. Am. Assoc. Petrol. Geol.* **47,** 717–55.

HELWIG, J. 1970. Slump folds and early structures, northeastern Newfoundland Appalachians. *J. Geol.* **78,** 172–87.

HOBBS, B. E., MEANS, W. D. & WILLIAMS, P. F. 1976. *An Outline of Structural Geology.* New York.

HOEDEMAEKER, P. J. 1973. Olisthostromes and other delapsional deposits, and their occurrence in the region of Moratalla (Prov. of Murcia, Spain). *Scr. Geol.* **19,** 1–207.

HOOKE, R. L. & HUDDLESTON, P. J. 1978. Origin of foliation in glaciers. *J. Glaciol.* London, **20,** 285–99.

HORNE, G. S. 1969. Early Ordovician chaotic deposits in the central volcanic belt of northeastern Newfoundland. *Bull. geol. Soc. Am.* **80,** 2451–64.

HOSSACK, J. 1979. The use of balanced cross-sections in the calculation of orogenic contraction: a review. *J. geol. Soc. London,* **136,** 705–11.

HSÜ, K. J. 1968. Principles of mélanges and their bearing on the Franciscan–Knoxville paradox. *Bull. geol. Soc. Am.* **79,** 1063–74.

—— 1969. Preliminary report and geologic guide to Franciscan mélanges of the Morro Bay—San Simeon area, California. *Spec. Publ. Calif. Div. Mines Geol.* **35.**

—— 1974. Mélanges and olistostromes. *In:* DOTT, R. H., JR & SHAVER, R. H. (eds). *Modern and Ancient Geosynclinal Sedimentation.* Spec. Publ. Soc. econ. Paleontol. Mineral. Tulsa, **19,** 321–33.

IBRAHIM, A. K., LATHAM, G. V. & LADD, J. 1979. Seismic refraction and reflection measurements in the Middle America trench offshore Guatemala. *J. geophys. Res.* **84,** 5643–9.

KARIG, D. E., MOORE, G. F., CURRAY, J. R. & LAWRENCE, M. B. 1980. Morphology and shallow structure of the trench slope off Nias Island, Sunda Arc. *In:* HAYES, D. E. (ed.). *The Tectonic and Geologic Evolution of Southeast Asian Seas and Islands.* Geophys. Monogr. Am. geophys. Union **23,** 179–208.

—— & SHARMAN, G. F. III 1975. Subduction and accretion in trenches. *Bull. geol. Soc. Am.* **86,** 377–89.

KLEIST, J. R. 1974. Deformation by soft-sediment extension in the Coastal Belt, Franciscan Complex. *Geology,* **2,** 501–4.

KULM, L. D., SCHWELLER, W. J. & MASIAS, A. 1977. A preliminary analysis of the subduction processes along the Andean continental margin, 6° to 45°S. *In:* TALWANI, M. & PITMAN, W. C. III (eds). *Island Arcs, Deep Sea Trenches, and Back-arc Basins.* Am. geophys. Union, M. Ewing Ser. **1,** 285–301.

LAMBERT, J. L. M. 1965. A reinterpretation of the breccias in the Meneage crush zone of the Lizard boundary, south-west England. *Q. J. geol. Soc. London,* **121,** 339–57.

MALTMAN, A. J. 1977. Some microstructures of experimentally deformed argillaceous sediments. *Tectonophysics,* **39,** 417–36.

MOORE, G. F. & KARIG, D. E. 1976. Development of sedimentary basins on the lower trench slope. *Geology,* **4,** 693–7.

—— & —— 1980. Structural geology of Nias Island, Indonesia: implications for subduction zone tectonics. *Am. J. Sci.* **280,** 193–223.

MOORE, J. C., WATKINS, J. S. *et al.* 1979. Progressive accretion in the Middle America trench, southern Mexico. *Nature,* **281,** 638–42.

MORGENSTERN, N. R. 1967. Submarine slimping and the initiation of turbidity currents. *In:* RICHARDS, A. F. (ed). *Marine Geotechnique.* 189–220. Urbana.

—— & TCHALENKO, J. S. 1967. Microstructural observations on shear zones from slips in natural clays. *Proc. Geotechnical Conf. Oslo,* **1,** 147–52.

NARDIN, T. R., HEIN, F. J., GORSLINE, D. S. & EDWARDS, B. D. 1979. A review of mass movement processes, sediment and acoustic characteristics, and contrasts in slope and base-of-slope systems versus canyon-fan-basin floor systems. *In:* DOYLE, L. J. & PILKEY, O. H. (eds). *Geology of Continental Slopes.* Spec. Publ. Soc. econ. Paleontol. Mineral. Tulsa, **27,** 61–73.

PAGE, B. M. 1978. Franciscan melanges compared with olistostromes of Taiwan and Italy. *Tectonophysics,* **47,** 223–46.

POTTER, P. E. & PETTIJOHN, F. J. 1977. *Paleocurrents and Basin Analysis,* 2nd edn. Berlin.

RAMBERG, H. 1977. Some remarks on the mechanism of nappe movement. *Geol. För. Stockh. Förh.* **99,** 110–17.

—— 1978. Further on stress equations. *Geol. För. Stockh. Förh.* **100,** 254.

—— 1980. Diapirism and gravity collapse in the Scandinavian Caledonides. *J. geol. Soc. London,* **137,** 261–70.

RAMSAY, J. G. 1967. *Folding and Fracturing of Rocks.* New York.

—— & GRAHAM, R. H. 1970. Strain variation in shear belts. *Can. J. Earth Sci.* **7,** 786–813.

—— & WOOD, D. S. 1973. The geometric effects of volume change during deformation processes. *Tectonophysics,* **16,** 263–77.

RAST, N. 1956. The origin and significance of boudinage. *Geol. Mag.* **93,** 401–8.

RIEKE, H. H. & CHILINGARIAN, G. V. 1974. *Compaction of Argillaceous Sediments.* Amsterdam.

RUPKE, N. A. 1976. Large-scale slumping in a flysch basin, southwestern Pyrenees. *J. geol. Soc. London,* **132,** 121–30.

SANDERSON, D. J. 1976. The superposition of compaction and plane strain. *Tectonophysics,* **30,** 35–54.

SCHUSTER, D. C. 1980. *The nature and origin of the Late Precambrian Gwna melange North Wales, United Kingdom.* Thesis, PhD, Univ. Illinois (unpubl.).

SEELY, D. R., VAIL, P. R. & WALTON, G. G. 1974. Trench slope model. *In:* BURK, C. A. & DRAKE, C. L. (eds). *The Geology of Continental Margins,* 249–60. Springer-Verlag, New York.

SKEMPTON, A. W. 1964. Long-term stability of clay slopes. *Geotechnique. London,* **14,** 77–101.

—— 1970. The consolidation of clays by gravitational compaction. *Q. J. geol. Soc. London,* **125,** 373–411.

SMITH, R. B. 1977. Formation of folds, boudinage, and mullions in non-Newtonian materials. *Bull. geol. Soc. Am.* **88,** 312–30.

SMITH, G. W., HOWELL, D. G. & INGERSOLL, R. V. 1979. Late Cretaceous trench-slope basins of central California. *Geology,* **7,** 303–6.

TCHALENKO, J. S. 1968. The evolution of kink-bands and the development of compression textures in sheared clays. *Tectonophysics,* **6,** 159–74.

—— 1970. Similarities between shear zones of different magnitudes. *Bull. geol. Soc. Am.* **81,** 1624–40.

TYLER, J. H. 1975. Fracture and rotation of brittle clasts in a ductile matrix. *J. Geol.* **83,** 501–10.

VON HUENE, R. 1972. Structure of the continental margin and tectonism at the eastern Aleutian Trench. *Bull. geol. Soc. Am.* **83,** 3613–26.

VON HUENE, R. *et al.* 1981. A summary of Deep Sea Drilling Project Leg 67 shipboard results from the Mid-America Trench transect off Guatemala (this volume).

WEISS, L. E. 1972. *The Minor Structures of Deformed Rocks.* New York.

WHITE, R. S. & ROSS, D. A. 1979. Tectonics of the western Gulf of Oman. *J. geophys. Res.* **84,** 3479–89.

WHITTEN, E. H. T. 1966. *Structural Geology of Folded Rocks.* Chicago.

WILLIAMS, P. F., COLLINS, A. R. & WILTSHIRE, R. G. 1969. Cleavage and penecontemporaneous deformation structures in sedimentary rocks. *J. geol.* **77,** 415–25.

WOODCOCK, N. H. 1976. Structural style in slump sheets: Ludlow Series, Powys, Wales. *J. geol. Soc. London,* **132,** 399–415.

—— 1979. Sizes of submarine slides and their significance. *J. Struct. Geol.* **1,** 137–42.

DARREL S. COWAN, Department of Geological Sciences, University of Washington, Seattle, Washington 98195, U.S.A.

Initiation and evolution of the Great Valley forearc basin of northern and central California, U.S.A.

Raymond V. Ingersoll

SUMMARY: The late Mesozoic Great Valley forearc basin of northern and central California evolved from a residual forearc basin formed on top of young oceanic crust to a composite forearc basin resting on both oceanic and continental crust. Depositional environments preserved in outcrop along the west side of the Sacramento Valley evolved from deep oceanic floor and slope in the Late Jurassic to basin plain in the Early Cretaceous to submarine fan complexes in the Late Cretaceous to slope and shelf in the Palaeogene. The forearc basin widened and enlarged, and was supplied with voluminous sediment primarily derived from the coeval magmatic arc to the east. However, significant quantities of ophiolitic detritus (chert, and sedimentary, mafic volcanic and mafic metavolcanic lithic fragments) occur in the lower Great Valley strata, thus indicating erosion of 'tectonic highlands' formed during arc-arc collision immediately preceeding initiation of the late Mesozoic subduction regime.

The Great Valley forearc owes its geometry and history to the peculiarities of subduction initiation and to the shapes of the continental margin and magmatic arcs extant before initiation of the Great Valley subduction phase. The residual forearc developed on oceanic crust that had formed previously by spreading behind an east-facing intraoceanic arc. After this east-facing arc collided with the west-facing continental-margin arc, a new trench formed west of the suture belt. The shape of the new continental margin was irregular, so that a wide residual forearc basin formed in the Great Valley area, whereas no residual forearc basin formed to the north (Klamath area) or possibly, to the south (southern California borderland area that has been dislocated by Neogene strike-slip motions). The northern coastal promontory (Klamath area) provided much of the ophiolitic detritus within the lower part of the Great Valley strata. Sediment derived from the Klamaths was transported southward directly into the forearc basin, as well as westwards and then southwards into the trench west of the forearc basin. The Great Valley forearc basin is preserved and displayed so well today due to its mode of initiation (residual forearc formed in a previously backarc area) and its mode of termination (northward migration of a triple junction that converted a convergent margin into a transform margin).

The Great Valley forearc basin of northern and central California (Fig. 1) is probably the most thoroughly studied and best-understood ancient basin of its kind in the world (Ingersoll 1978a, 1979a). All three major components of the arc-trench system (Franciscan subduction complex, Great Valley forearc basin and Sierra Nevada magmatic arc) (Dickinson 1970; Dickinson & Seely 1979) are exposed on land essentially in their original relationships. The subduction complex contains the most intricate relations (e.g. Aalto 1981; Bachman 1978, 1981; Cowan 1974) due to intense deformation and metamorphism. Nonetheless, coherent patterns of a westward decrease in age of strata, degree of deformation and grade of metamorphism are apparent (Dickinson & Seely 1979; Dickinson *et al.* 1979; Ingersoll *et al.* 1977). Sediments deposited within the forearc basin (Great Valley Group—new nomenclature that replaces Great Valley Sequence of Bailey *et al.* 1964—and related strata) have been studied petrologically, sedimentologically and stratigraphically (e.g. Dickinson & Rich 1972; Ingersoll 1978b, c, 1979a; Mansfield 1979; Ojakangas 1968). The batholithic roots of the magmatic arc have been dated, and their petrology and chemistry have been studied by numerous workers (e.g. Bateman & Dodge 1970; Evernden & Kistler 1970; Presnall & Bateman 1973). West-to-east increase of K-feldspar content and eastward migration of the volcanic front with time are consistent with Dickinson's (1973, 1975) model of arc evolution.

The three components of the late Mesozoic arc-trench system evolved concurrently from their beginnings in the Late Jurassic (Kimmeridgian-Tithonian) through the Cretaceous, for which their relations are most clear. Subduction that created this linked system has been terminated in the Neogene by the northward migration of the Cape Mendocino triple junction (Atwater 1970; Crowell 1979), following a complex Palaeogene history (Dickinson *et al.* 1979).

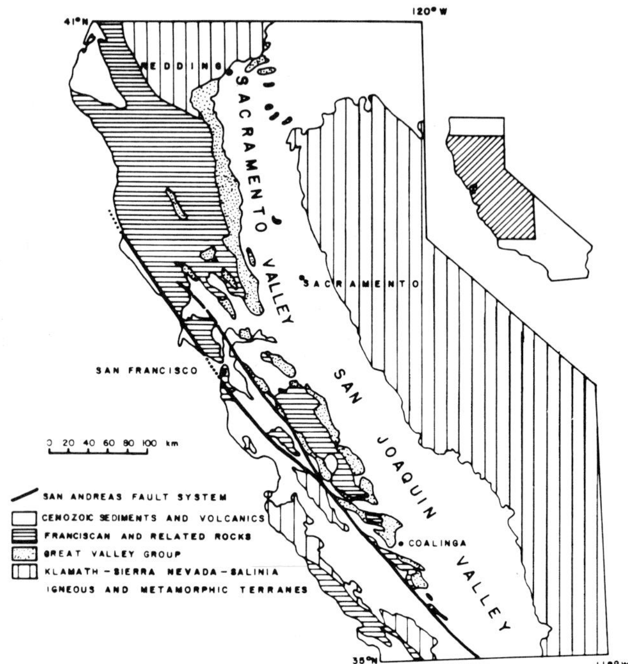

FIG. 1. Location map of northern and central California, showing principal components of the late Mesozoic arc-trench system and geographic locations. The Sierra Nevada igneous and metamorphic terranes represent the roots of the magmatic arc, and the Franciscan complex represents the highly deformed subduction complex formed landward of the trench. The Great Valley Group is primarily Upper Cretaceous in the San Joaquin Valley and is Upper Jurassic to Upper Cretaceous in the Sacramento Valley. Small outcrops of uppermost Cretaceous that nonconformably overlies Sierra Nevada basement occur along the east side of the Sacramento Valley, but are too small to show at this scale. The Great Valley Group rests nonconformably on Klamath basement near Redding. Coast Range ophiolite underlies the Great Valley Group at other surface exposures (after Ingersoll 1978a).

Thus, sedimentation and deformation in the subduction complex, sedimentation in the forearc basin, and intrusion, extrusion, metamorphism and erosion within the magmatic arc all occurred concurrently and resulted in a closely linked triad of major tectonostratigraphic units (Dickinson 1977).

The purpose of this paper is to outline in broad terms the history of this elegant forearc basin. The reader is referred to the numerous references for details that are touched on only briefly here.

Initiation of forearc basin

Schweickert & Cowan (1975) have outlined the most likely scenario for the initiation of the Franciscan–Great Valley–Sierra Nevada arc-trench system. During pre-Kimmeridgian Jurassic time, an east-facing (subduction down to the west) intraoceanic arc approached the western margin of North America where a west-facing (subduction down to the east) continental-margin arc resided (Figs 2 & 3). Batholithic rocks of this continental-margin arc occur today in the eastern Sierra Nevada in California and Nevada. Intervening oceanic lithosphere was consumed beneath both of these arc-trench systems. At the same time that the east-facing intraoceanic arc consumed oceanic lithosphere on its east side, backarc spreading produced new oceanic lithosphere on its west side. Upon collision of the two arc-trench systems in the Kimmeridgian (Nevadan orogeny), a new trench formed west of the suture belt and the Franciscan–Great Valley–Sierra Nevada system initiated (Fig. 3b). In the process of forming a new west-facing continental-margin arc superimposed on a former east-facing intraoceanic arc, oceanic crust that had formed previously in a backarc setting became

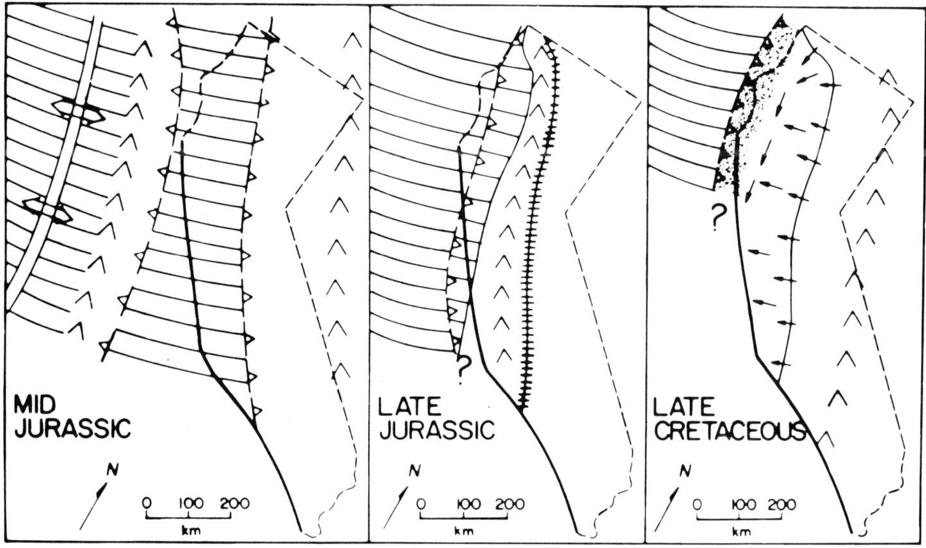

FIG. 2. Schematic palaeogeographic-palaeotectonic maps for the Great Valley forearc. The following symbols indicate various tectonic and geographic features: dashed lines—present California state outline and shoreline; heavy solid lines—Neogene San Andreas fault zone; dashed lines with thrust symbols—trenches; double solid lines with diverging arrows—spreading centre; mountains—magmatic arcs; light solid lines—palaeoshorelines; parallel light solid lines—oceanic crust; stitched line—suture belt; stippled pattern—subduction complex; single-stemmed arrows—sediment dispersal directions. See text for discussion. See Ingersoll (1978a) for additional maps.

the basement (Coast Range ophiolite of Bailey *et al.* 1970) upon which the Great Valley forearc formed (Fig. 3b).

The continental margin created by the combination of backarc spreading, arc-arc collision and initiation of a new subduction zone was irregular in shape. Trenches tend to have smooth, curving shapes in plan view, so that the irregular continental margin formed indentations (oceanic crust trapped in the forearc area) and promontories (continental crust in the forearc area) (Fig. 2). In the Great Valley area, a wide belt of oceanic crust (ophiolite) was trapped behind the trench, so that a residual forearc basin was created (terminology of Dickinson & Seely 1979). To the north (Klamath area), a promontory of accreted arc terranes existed directly behind the new trench. Therefore, no forearc basin formed and sediment derived from this area was transported either directly into the trench and was accreted, or was transported southwards and northwards into adjoining forearc basins and trenches. There may also have been a promontory at the south end of the Great Valley forearc basin, but Neogene transform-related deformation has been so extensive that reconstruction of the late Mesozoic configuration is difficult. However, the best available evidence (Howell *et al.* 1977) is consistent with forearc deposition primarily on continental crust and accreted material (constructional rather than residual forearc basin) along the late Mesozoic continental margin south of the Great Valley area. Thus, the Great Valley area owes its unique characteristics (continuous record of sedimentation from Late Jurassic through the entire Cretaceous) to its specific plate-tectonic history as well as inherited features. The pre-subduction shape of the continental margin controlled the location of the deep residual forearc basin (e.g. Dickinson & Seely 1979).

Depositional environments and sedimentary history of the earliest stages of the Great Valley forearc are obscure. The basal part of the Great Valley Group accumulated on the Coast Range ophiolite (Bailey *et al.* 1970) in a backarc setting. The local occurrence of andesitic volcanics directly above the ophiolite (Evarts 1977) supports this interpretation. It is difficult to know what proportion of the lower part of the Great Valley Group accumulated before initiation of the Great Valley forearc basin (Ingersoll 1978a). Basal deposits include radiolarian cherts likely to have formed in the backarc area before the influx of voluminous suture- and

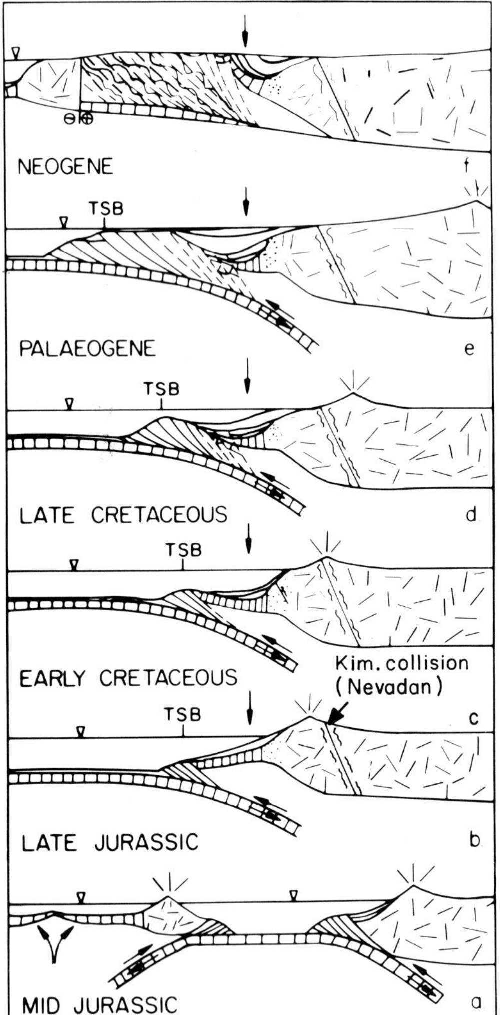

FIG. 3. Schematic cross-sections of northern California from formation of Coast Range ophiolite behind an east-facing intraoceanic arc (a) to termination of Great Valley forearc by conversion to transform margin (f). Vertical arrows indicate location of present outcrops of Great Valley Group along west side of Sacramento Valley. These are used as fixed reference point for each cross-section. Stippled pattern indicates upper-slope discontinuity. TSB is trench-slope break. Foothill suture belt shown by diagonal and wiggly lines. Plus and minus symbols indicate motion on San Andreas fault zone (f). See text for discussion (after Dickinson & Seely 1979, Ingersoll 1979a, and Schweickert & Cowan 1975).

arc-derived detritus (Fig. 3a) (Pessagno 1973). Most of the contacts between ophiolite and basal Great Valley Group are tectonic (Bailey *et al.* 1970; Ingersoll 1978a), but where the contact is depositional, deep oceanic (including basin-plain) deposits seem to predominate. Soon after initiation of the Great Valley phase of subduction, great thicknesses of mud with incised conglomerate-filled channels began accumulating. These deposits represent slope deposition within the newly created residual forearc basin. A wide slope existed from shorelines along the east and north (Klamath area) sides of the basin down to the trench to the west (Ingersoll 1978a). Direct access to the trench thus was provided for suture- and arc-derived detritus.

The Stony Creek Formation along the west side of the Sacramento Valley includes the Tithonian (Upper Jurassic) and Neocomian (Lower Cretaceous) parts of the Great Valley Group (Ingersoll *et al.* 1977). Locally, Kimmeridgian (Upper Jurassic) breccias and ophiolitic detritus occur below the Tithonian. Kimmeridgian conglomerates primarily consist of mafic metavolcanics probably derived directly from the underlying ophiolite (Bertucci 1980). The majority of the Stony Creek Formation includes conglomerate and sandstone clasts consisting of chert, metachert, argillite and other sediments, and varieties of volcanic, plutonic and metamorphic detritus. The indicated provenance is one of combined suture-derived detritus (uplifted and accreted ophiolites, deep-marine sediments and arc terranes), and arc-derived detritus (felsic, intermediate and mafic volcanics) (Bertucci 1980; Ingersoll 1979a, b). Polycrystalline quartz (including chert and metachert) is more prevalent to the north, and volcanic and metavolcanic detritus increases to the south within the Sacramento Valley. The ratio of polycrystalline quartz to total quartz decreases up-section within the lower part of the Great Valley Group. These relations suggest that the Klamath area was the source of most of the suture-derived material, and that following erosion of the sutured terranes ('tectonic highlands' formed by the previous arc-arc collision), detritus eroded from the newly formed continental-margin arc (primarily represented by the Sierra Nevada area today) began accumulating within the forearc basin. Thus, there is a transition upward within the Great Valley Group from locally derived ophiolitic detritus to distantly derived suture-belt detritus to distantly derived magmatic-arc detritus.

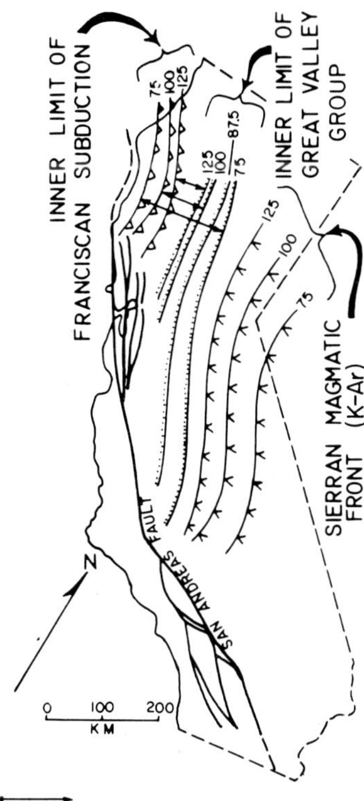

FIG. 4. Schematic map illustrating secular increase in width of Great Valley forearc basin. Approximate positions of migratory boundaries of the forearc basin are shown for times early in Early Cretaceous (125 Ma), in mid-Cretaceous (100 Ma) and late in the Late Cretaceous (75 Ma). Positions of western (outer) boundary at migratory trench-slope break marking inner limit of active subduction are inferred from easternmost extent of successively younger strata within the Franciscan complex. Positions of eastern (inner) boundary at migratory shoreline marking easternmost extent of successively younger strata present in subsurface of modern Great Valley. Positions of the magmatic front marking western limit of igneous belt are inferred from westernmost limits of radiometric dates for Sierra Nevada plutons (Evernden & Kistler 1970) (after Ingersoll *et al.* 1977).

Evolution of forearc basin

The Cretaceous development of the Great Valley forearc consisted of enlargement and filling of the forearc basin concurrently with and as a result of the combined effects of westward and upward growth of the Franciscan subduction complex, and the eastward migration of the Sierran magmatic arc (Figs 3b–e & 4) (Dickinson & Seely 1979; Ingersoll 1978a, 1979a; Ingersoll *et al.* 1977). Thus, the forearc basin (and the arc-trench gap) widened through time as the trench migrated westwards and the magmatic front migrated eastwards. Intense deformation and low-temperature, high-pressure metamorphism characterized the subduction complex (e.g. Bailey *et al.* 1964; Ernst 1970), whereas extrusion, intrusion and high-temperature, low-pressure metamorphism characterized the magmatic arc (e.g. Dickinson 1970; Dickinson & Rich 1972; Evernden & Kistler 1970). Between these two terranes, the relatively undisturbed Great Valley Group accumulated in an enlarging basin.

Some form of mid-Cretaceous deformation likely occurred within the forearc basin as evidenced by subtle warping of Lower Cretaceous strata, local serpentinite protrusions on the east side of the subduction complex, local mid-Cretaceous disconformities, and thrust or reverse faulting within the basin (Brown & Rich 1967; Dickinson & Seely 1979; Ingersoll *et al.* 1977; Jones & Bailey 1973; Jones & Irwin 1971; Jones *et al.* 1969; Peterson 1967). Originally low-angle faults near the base of the Great Valley Group in the northern Sacramento Valley connect with the Coast Range fault to the north-west and may connect in the subsurface with presumed reverse faults underneath the centre of the modern Great Valley (Dickinson & Seely 1979; Ingersoll *et al.* 1977). This intrabasinal faulting may reflect increased crustal telescoping along the upper-slope discontinuity (*sensu* Karig & Sharman 1975) during the mid-Cretaceous.

With the exception of the above-mentioned deformation, the forearc basin evolved systematically as the arc-trench gap widened, and sedimentation continued at a rapid pace (average sedimentation rate for the entire basin during the Late Cretaceous was 280 m/my) (Ingersoll 1979a). By the end of the Cretaceous, the forearc basin was filled to near sea-level in many places (Fig. 3e) and Sierran arc-derived sediment again was able to traverse the arc-trench gap and be incorporated into the subduction complex (Ingersoll 1978a). The basin had evolved into a composite forearc formed on top of accreted material, and both oceanic and continental crust (Dickinson & Seely 1979).

Increased rates of westward motion of North

America relative to the Pacific basin and possible changes in the direction of spreading during the Palaeogene led to the rapid eastward migration of the magmatic arc and possible shearing within the subduction complex (Coney 1976, 1978; Dickinson 1979; Dickinson & Snyder 1978; Dickinson *et al.* 1979; Graham 1979; Howell *et al.* 1977; Nilsen & Clarke 1975). Throughout the Palaeogene, the Great Valley area remained a forearc basin, but with decreased influence of the magmatic arc which was far to the east (Fig. 3e) (Dickinson *et al.* 1979). A wide shelf extended from the eroded roots of the Cretaceous arc in eastern California and Nevada to the trench-slope break along the coast.

Northward migration of the Mendocino triple junction during the Neogene has converted the subducting margin into a transform margin (Atwater 1970; Crowell 1979), and has disrupted palaeotectonic relations, especially along the south-western margin of the former forearc basin (Figs 1 & 3f). The subduction complex has risen above sea-level with the cessation of subduction, probably due, in part, to isostatic adjustment. In the process, the west side of the forearc basin has been eroded and the Great Valley Group has been exposed (Fig. 3f).

Depositional environments along the west side of the Sacramento Valley evolved from deep oceanic before initiation of Great Valley subduction to basin plain and slope in the latest Jurassic to basin plain and outer fan in the Early Cretaceous to complex fan and slope in the Late Cretaceous to slope, shelf and non-marine in the Palaeogene (Fig. 3) (Ingersoll 1978a). These changes in depositional environment were responses to the tectonic events described above as well as the sequential filling of the basin, although eustatic sea-level rise in the Cretaceous and fall in the Palaeogene (e.g. Hancock & Kauffman 1979) had modifying effects (Ingersoll 1980). Submarine sediment-dispersal systems (shelf-slope-channels-fans) enlarged concurrently with basin growth. Thus, individual submarine fan deposits tend to be thinner and less areally extensive in the Lower Cretaceous than in the Upper Cretaceous. By the Late Cretaceous and Palaeogene, submarine canyons had incised the wide shelves so that gravity flows were localized and large fan systems developed (Ingersoll 1978a, 1979a). The remnant deep basin was filled during this time by the westward progradation of slope-shelf-shoreline (deltaic) sequences (Dickinson & Seely 1979; Dickinson *et al.* 1979). During the Neogene, the former forearc basin has been the site of primarily non-marine deposition with local marine conditions owing to sea-level fluctuations. Deep basins formed in the southern part of the San Joaquin Valley due to wrench tectonics associated with Neogene development of the San Andreas fault system (Blake *et al.* 1978; Crowell 1979; Harding 1976).

Petrographic data provide further insight regarding forearc evolution. The Great Valley Group consists of suture-derived and immature arc-derived sedimentary and volcanic material near the base (Stony Creek Formation). The overlying Lodoga Formation (Aptian-Albian) contains increased proportions of monocrystalline quartz, potassium feldspar, mica and metamorphic lithic fragments, suggesting that overlying volcanics, sediments and sutured material had been eroded through to underlying plutonic terranes by the mid-Cretaceous (Dickinson & Rich 1972; Ingersoll 1979b). Intrabasinal tectonism during the mid-Cretaceous (discussed above) may have resulted in reworking of some previously deposited sediments, thus increasing the quartz content of the Lodoga Formation (Dickinson & Rich 1972; Ingersoll *et al.* 1977).

During the Late Cretaceous, alternating episodes of intrusion-extrusion and erosion in the magmatic arc resulted in interbedded volcanic-lithic-rich and arkosic (plutonic-derived) sandstones. Sandstones in the San Joaquin Valley contain higher proportions of mica and metamorphic lithic fragments and more felsic volcanics than do those of the Sacramento Valley (Ingersoll 1978c). These characteristics can be explained by the fact that the magmatic arc formed on top of true continental crust (including older metamorphic terranes) to the south and crust with more oceanic affinities to the north.

By the end of the Cretaceous, arkosic sandstones (monocrystalline quartz and alkali feldspars with significant quantities of mica) accumulated throughout the forearc region (Ingersoll 1978c). These deposits represent erosion of the batholithic roots of the magmatic arc (Dickinson & Rich 1972). Similar arkosic sandstones were deposited throughout the Palaeogene as the active magmatic arc remained far to the east of the Great Valley area (Dickinson *et al.* 1979) (Fig. 3e). The following upward petrological trends within the Great Valley Group are the result of the arc evolution described above (Ingersoll 1978c, 1979b): increasing monocrystalline quartz, decreasing polycrystalline quartz, increasing total feldspar, increasing potassium feldspar, decreasing total lithics, increasing mica and decreasing sedimentary lithics.

Conclusions

The tectonic, sedimentological and petrological history outlined above suggests some general principles regarding the evolution of arc-trench systems. Arc-trench gaps tend to widen with time by the prograde growth of the subduction complex and the retrograde migration of the magmatic front (Dickinson 1973). Forearc basins may initiate as trapped segments of oceanic crust behind newly formed trenches, and grow to be composite wide basins overlapping the former continental margins as well as the growing subduction complex (Dickinson & Seely 1979). Depositional environments tend to evolve from deep oceanic to base-of-slope to slope and shelf to non-marine (Ingersoll 1978a). Low in the stratigraphic fill of the forearc, there may be ophiolitic and suture-derived sediments; these are commonly overlain by volcaniclastic and plutonic-metamorphic detritus. Arkosic potassium-rich detritus predominates at the top of the section (Dickinson & Rich 1972; Ingersoll 1978c).

These models must be applied to other forearc basins with caution. The models owe many of their features to information obtained from the Great Valley forearc; therefore, they may not be applicable to all other cases. Some of the variables that may require changes in these models when applied to other areas include the following: mode of initiation of subduction, geometry and composition of pre-existing continental margin, location along that margin, characteristics of adjoining forearcs, rate and orientation of subduction, intensity of volcanism and plutonism, and climate. The mechanism by which subduction is terminated (triple-junction migration, arc-arc collision, continental collision, etc.) determines the final characteristics of the basin. Each forearc has experienced a unique history and must be understood on its own terms. Nonetheless, knowledge of the detailed history of the Great Valley forearc provides insights to the possible history of others.

ACKNOWLEDGMENTS: I thank W. R. Dickinson for numerous dicussions regarding Great Valley geology throughout my involvement in this study. Dickinson, S. A. Graham and J. K. Leggett provided helpful reviews of the manuscript. The Research Allocations Committee of the University of New Mexico provided financial support.

References

AALTO, K. R. 1981. The Franciscan Complex of northernmost California: sedimentation and tectonics (this volume).

ATWATER, T. 1970. Implications of plate tectonics for the Cenozoic tectonic evolution of western North America. *Bull. geol. Soc. Am.* **81,** 3513–35.

BACHMAN, S. B. 1978. A Cretaceous and early Tertiary subduction complex, Mendocino coast, northern California. *In:* HOWELL, D. G. & McDOUGALL, K. (eds). *Mesozoic Paleogeography of the Western United States.* Pacific Sec., Soc. econ. Paleontol. Mineral. Pacif. Coast Paleogeogr. Symp. **2,** 419–30.

—— 1981. The coastal belt of the Franciscan: youngest phase of northern California subduction (this volume).

BAILEY, E. H., IRWIN, W. P. & JONES, D. L. 1964. Franciscan and related rocks, and their significance in the geology of western California. *Bull. Calif. St. Min. Bur.* **183,** 177 pp.

——, BLAKE, M. C., JR & JONES, D. L. 1970. On-land Mesozoic oceanic crust in California Coast Ranges. *Prof. Pap. U.S. geol. Survey.* **700-C,** 70C–81C.

BATEMAN, P. C. & DODGE, F. C. W. 1970. Variations of major chemical constituents across the central Sierra Nevada batholith. *Bull. geol. Soc. Am.* **81,** 409–20.

BERTUCCI, P. F. 1980. *Petrology and provenance of Upper Jurassic-Lower Cretaceous Great Valley Sequence conglomerate, northwestern Sacramento Valley, California.* Thesis, MS, Univ. Calif., Davis, 143 pp.

BLAKE, M. C., JR, CAMPBELL, R. H., DIBBLEE, T. W., JR, HOWELL, D. G., NILSEN, T. H., NORMARK, W. R., VEDDER, J. G. & SILVER, E. A. 1978. Neogene basin formation in relation to plate-tectonic evolution of the San Andreas fault system, California. *Bull. Am. Assoc. Petrol. Geol.* **62,** 344–72.

BROWN, R. D., JR & RICH, E. I. 1967. Implications of two Cretaceous mass transport deposits, Sacramento Valley, California: comment on a paper by Gary L. Peterson. *J. sediment. Petrol.* **37,** 240–8.

CONEY, P. J. 1976. Plate tectonics and the Laramide orogeny. *Spec. Publ. New Mex. geol. Soc.* **6,** 5–10.

—— 1978. Mesozoic-Cenozoic Cordilleran plate tectonics. *Mem. geol. Soc. Am.* **152,** 33–50.

COWAN, D. S. 1974. Deformation and metamorphism of the Franciscan subduction zone complex northwest of Pacheco Pass, California. *Bull. geol. Soc. Am.* **85,** 1623–34.

CROWELL, J. C. 1979. The San Andreas fault system through time. *J. geol. Soc. London,* **136,** 293–302.

DICKINSON, W. R. 1970. Relations of andesites, granites, and derivative sandstones to arc-trench tectonics. *Rev. Geophys. Space Phys.* **8,** 813–60.

—— 1973. Widths of modern arc-trench gaps proportional to past duration of igneous activity in associated magmatic arcs. *J. geophys. Res.* **78,** 3376–89.
—— 1975. Potash–depth (K–h) relations in continental margin and intra-oceanic magmatic arcs. *Geology,* **3,** 53–6.
—— 1977. Tectono-stratigraphic evolution of subduction-controlled sedimentary assemblages. *In:* Talwani, M. & Pitman, W. C. III (eds). *Island Arcs, Deep Sea Trenches and Back-arc Basins.* Am. geophys. Union, M. Ewing Ser. **1,** 33–40.
—— 1979. Cenozoic plate tectonic setting of the Cordilleran region in the United States. *In:* Armentrout, J. M., Cole, M. R. & Terbest, H., Jr (eds). *Cenozoic Paleogeography of the Western United States.* Pacific Sec., Soc. econ. Paleontol. Mineral. Pacif. Coast Paleogeogr. Symp. **3,** 1–13.
—— , Ingersoll, R. V. & Graham, S. A. 1979. Paleogene sediment dispersal and paleotectonics in northern California. *Bull. geol. Soc. Am.* **90,** pt. I, 897–8, pt. II, 1458–528.
—— & Rich, E. I. 1972. Petrologic intervals and petrofacies in the Great Valley Sequence, Sacramento Valley, California. *Bull. geol. Soc. Am.* **83,** 3007–24.
—— & Seely, D. R. 1979. Structure and stratigraphy of forearc regions. *Bull. Am. Assoc. Petrol. Geol.* **63,** 2–31.
—— & Snyder, W. S. 1978. Plate tectonics of the Laramide orogeny. *Mem. geol. Soc. Am.* **151,** 355–66.
Ernst, W. G. 1970. Tectonic contact between the Franciscan melange and the Great Valley sequence, crustal expression of a late Mesozoic Benioff zone. *J. geophys. Res.* **75,** 886–901.
Evarts, R. C. 1977. The geology and petrology of the Del Puerto ophiolite, Diablo Range, central California Coast Ranges. *Oregon Bull. Dep. Geol. Miner. Ind.* **95,** 121–39.
Evernden, J. F. & Kistler, R. W. 1970. Chronology of emplacement of Mesozoic batholithic complexes in California and western Nevada. *Prof. Pap. U.S. geol. Surv.* **623,** 42 pp.
Graham, S. A. 1979. Tertiary paleotectonics and paleogeography of the Salinian block. *In:* Armentrout, J. M., Cole, M. R. & Terbest, H. Jr (eds). *Cenozoic Paleogeography of the Western United States.* Pacif. Sec., Soc. Econ. Paleontol. Mineral. Pacif. Coast Paleogeogr. Symp. **3,** 45–51.
Hancock, J. M. & Kauffman, E. G. 1979. The great transgressions of the Late Cretaceous. *J. geol. Soc. London,* **136,** 175–86.
Harding, T. P. 1976. Tectonic significance and hydrocarbon trapping consequences of sequential folding synchronous with San Andreas faulting, San Joaquin Valley, California. *Bull. Am. Assoc. Petrol. Geol.* **60,** 356–78.
Howell, D. G., Vedder, J. G. & McDougall, K. (eds) 1977. Cretaceous geology of the California Coast Ranges, west of the San Andreas fault. *Pacific Sec., Soc. econ. Paleontol. Mineral. Pacif. Coast Paleogeogr. Field Guide,* **2,** 109 pp.
Ingersoll, R. V., 1978a. Paleogeography and paleotectonics of the late Mesozoic forearc basin of northern and central California. *In:* Howell, D. G. & McDougall, K. (eds). *Mesozoic Paleogeography of the Western United States.* Pacific. Sec., Soc. econ. Paleontol. Mineral. Pacif. Coast Paleogeogr. Symp. **2,** 471–82.
—— 1978b. Submarine fan facies of the Upper Cretaceous Great Valley Sequence, northern and central California. *Sediment. Geol.* **21,** 203–30.
—— 1978c. Petrofacies and petrologic evolution of the Late Cretaceous fore-arc basin, northern and central California. *J. Geol.* **86,** 335–53.
—— 1979a. Evolution of the Late Cretaceous forearc basin, northern and central California. *Bull. geol. Soc. Am.* **90,** pt. I, 813–26.
—— 1979b. Petrofacies and provenance of the lower part of the Great Valley Sequence, Sacramento Valley, California. *Abstr. Progr. geol. Soc. Am.* **11,** 85–6.
—— 1980. Eustatic, tectonic and sedimentary controls of Cretaceous transgressive-regressive sequences, western, central and eastern USA. *Abstr. Progr. geol. Soc. Am.* **12,** 453.
—— , Rich, E. I. & Dickinson, W. R. 1977. *Great Valley Sequence, Sacramento Valley.* Cord. Sect., geol. Soc. Am. Fieldtrip Guidebook, 72 pp.
Jones, D. L. & Bailey, E. H. 1973. Preliminary biostratigraphic map, Colyear Springs quadrangle, California. *U.S. geol. Surv. Map,* **MF-517**.
—— , Bailey, E. H. & Imlay, R. W. 1969. Structural and stratigraphic significance of the Buchia zones in the Colyear Springs–Paskenta area, California. *Prof. Pap. U.S. geol. Surv.* **647A,** 24 pp.
—— & Irwin, W. P. 1971. Structural implications of an offset Early Cretaceous shoreline in northern California. *Bull. geol. Soc. Am.* **82,** 815–22.
Karig, D. E. & Sharman, G. F. 1975. Subduction and accretion in trenches. *Bull. geol. Soc. Am.* **86,** 377–89.
Mansfield, C. F. 1979. Upper Mesozoic subsea fan deposits in the southern Diablo Range, California: record of the Sierra Nevada magmatic arc. *Bull. geol. Soc. Am.* **90,** pt I, 1025–46.
Nilsen, T. H. & Clarke, S. H., Jr. 1975. Sedimentation and tectonics in the early Tertiary continental borderland of central California. *Prof. Pap U.S. geol. Surv.* **925,** 64 pp.
Ojakangas, R. W. 1968. Cretaceous sedimentation, Sacramento Valley, California. *Bull. geol. Soc. Am.* **79,** 973–1008.
Pessagno, E. A., Jr 1973. Age and geologic significance of radiolarian cherts in the California Coast Ranges. *Geology,* **1,** 153–6.
Peterson, G. L. 1967. Implications of two Cretaceous mass transport deposits, Sacramento Valley, California: reply to comment by Brown and Rich. *J. sediment. Petrol.* **37,** 248–57.
Presnall, D. S. & Bateman, P. C. 1973. Fusion

relations in the system $NaAlSi_3O_8$-$CaAl_2Si_2O_8$-$KAlSi_3O_8$-SiO_2-H_2O and generation of granitic magmas in the Sierra Nevada batholith. *Bull. geol. Soc. Am.* **84,** 3181–202.

SCHWEICKERT, R. A. & COWAN, D. S. 1975. Early Mesozoic tectonic evolution of the western Sierra Nevada, California. *Bull. geol. Soc. Am.* **86,** 1329–36.

RAYMOND V. INGERSOLL, Department of Geology, University of New Mexico, Albuquerque, New Mexico 87131, U.S.A.

FOREARC TERRANES IN OROGENIC BELTS

Cretaceous-Palaeogene Flysch Zone of the East Alps and Carpathians: identification and plate-tectonic significance of 'dormant' and 'active' deep-sea trenches in the Alpine-Carpathian Arc

Reinhard Hesse

SUMMARY: The Flysch Zone of the East Alps is interpreted as a Cretaceous-Palaeogene deep-sea trench fill that accumulated over a period of about 70 Ma when the subduction zone associated with the trench was dormant. In Aptian-Albian time the trench had an elongate, at least 100 km (possibly 200 or 300 km) long basin plain located below the calcite compensation level; sediment input was from two lateral sources (small acute-angled fan or broad feeder channel exposed in the Falknis- and Tasna Nappe; submarine canyon facies in the area north of Salzburg). The existence of a basin plain has also been proved for the Campanian by the recognition of sheet-like turbidites that are continuous for 50 km. The small number of lateral sediment sources is attributed to the existence of slope basins that intercepted the downslope movement of turbidity currents. The lack of syndepositional deformation and of volcanism and volcaniclastic detritus show that sediment accumulation in the trench was not accompanied by active subduction.

This 500 km long and 10 (in the east perhaps 80 or 100) km wide trench of the East Alps was connected with the 1000 km long Carpathian trench whose western segment was also not associated with subduction until Oligocene time. The East Carpathian flysch, however, experienced continuing deformation and (relative) eastward migration of the depositional trough from Albian to Oligocene time. Huge Neogene volcanic centres of the Inner Carpathians are associated with the final diastrophic phases both in the West and East Carpathians, whereas in the East Alps there is no volcanism because subduction does not seem to have reached a great enough depth. The East Alpine-West Carpathian segment of this trench system may represent a major Cretaceous-Palaeogene transform fault zone separating Europe and its bordering seas in the north from the Cretaceous Penninic–Pannonian plate in the south. The latter has been eliminated almost entirely from the surface due to Late Cretaceous-Palaeogene subduction at its southern boundary nearly juxtaposing the European plate (Helvetic and Ultrahelvetic domains) with the Austro-Alpine/South Alpine (plus Adriatic?) plate. Postulated strike-slip motion in the west-east trending segment of the trench was followed by convergent motion, when southward subduction (of the European plate) resumed in late Eocene-Oligocene time.

Combination of an eastward facing trench with active subduction and a dormant trench along a transform fault zone suggests an analogy between the Alpine-Carpathian Arc and the modern West Atlantic loop arcs in the Caribbean and South Scotia Seas. The geodynamic link between these loop-arc structures could be the westward drift of major lithospheric plates between which smaller plates remain relatively stationary because they are 'sheltered' behind westward-dipping subduction zones that are more or less anchored in the asthenosphere.

One of the more challenging tasks awaiting sedimentologists in flysch terranes is to determine the tectonic setting of ancient flysch basins. This is a difficult task and can only be done by comparison with modern basins that we think represent flysch environments.

In this paper the tectonic setting of the Flysch Zone of the East Alps and Carpathians will be analysed (with emphasis on the East Alps) using: (1) palaeocurrent analysis and location and composition of the source areas; (2) anatomy and facies trends of individual flysch formations; (3) inferred depositional environments; (4) reconstructed basin geometry including palaeodepth; and (5) the tectonic behaviour (tilting, syndepositional versus post-depositional deformation) of the basin in general.

On this basis the Flysch Zone will be interpreted as an ancient deep-sea trench, which together with its continuation in the West Carpathians was dormant (i.e. not associated with active subduction) during most of its existence, whereas the East Carpathians reflect active Cretaceous-Tertiary subduction.

Geological setting and previous studies

The Flysch Zone of the East Alps (Rhenodanubian Flysch of Oberhauser 1968) flanks the

Northern Calcareous Alps as an approximately 500 km long, 5–20 km wide, almost straight fold belt forming the forested northern foothills of the East Alps (Fig. 1). It consists of a moderately thick sequence of lithologically distinct turbidite formations whose cumulative thickness reaches a maximum of 1.5–2 km (Table 1). Carbonate flysch formations alternate with terrigenous flysch (Fig. 2). This flysch was laid down during a 70 Ma period of apparently continuous deposition (Barremian-Aptian to Palaeogene).

Sedimentological studies in the Flysch Zone (Butt 1977; Faupl 1975, 1976; Freimoser 1973; Hesse 1973a,b, 1974, 1975; Hesse & Butt 1976; von Rad 1973) have revealed three significant findings: (1) Palaeocurrent directions are predominantly parallel or subparallel to strike of the flysch belt but reverse orientation several times (Fig. 3). (2) At two stratigraphic levels sequences of individual turbidites comprising either an entire formation or significant parts of a formation have been traced along strike for distances up to 115 and 50 km, respectively. (3) Green hemipelagic and red pelagic claystones are free of calcium carbonate although the turbidites with which they alternate may be carbonate-turbidites. Deposition was below the calcite compensation level (CCL).

Palaeocurrent analysis, composition and location of the source areas

Reversal of the palaeocurrent directions at certain formation boundaries is accompanied by a change in lithology from one formation to the next (Fig. 4) indicating that the palaeocurrent reversals are related to switching to petrographically different source areas. Carbonate turbidite formations (Barremian-Aptian Tristel Beds, the thin-bedded Upper Turonian to Lower Campanian Piesenkopf Beds, the Campanian Zementmergel Formation and the Upper Campanian-Lower Maestrichtian Hällritzer Formation) have a westerly source although in the Ofterschwang Beds (Albian-Lower Cenomanian) some easterly derived beds occur that are lithologically quite similar to the westerly derived beds. This is the only formation where turbidites of similar petrographic composition but with opposing palaeocurrent directions alternate with one another (von Rad 1964). The carbonate turbidites consist of mostly bioclastic calcarenites and calcilutites (nannofossil limestones), the latter prevailing in the Piesenkopf Beds. Carbonate turbidite facies of this kind are very widespread in the Upper Cretaceous of the Mediterranean Tethys (Alps, Apennines, Carpathians, etc.) and collectively known under the name 'helminthoid flysch' (Caron *et al.* 1979, 1981) after the meandering trace fossil *Helminthoidea*.

The middle to upper Cenomanian-lower Turonian Reiselsberg Sandstone represents the only major influx of massive terrigenous sands into the flysch basin. Lithologically very similar poorly sorted, often mica-rich, poorly cemented sandstones occur in the upper Maestrichtian-Palaeocene Bleicherhorn Formation. These terrigenous flysch formations have an easterly source. Intercalated carbonate turbidites in these two formations, however, display palaeocurrent directions from the west. The only terrigenous flysch formation with a westerly source is the Aptian-Albian Gault Formation whose sand turbidites are glauconitic quartz greywackes with a significant amount (25–30%) of biogenic and terrigenous carbonate fragments.

The abundance of calcarenitic and calcilutitic turbidites of biogenic origin in the Upper Cretaceous helminthoid flysch facies but also in the Lower Cretaceous Tristel Beds suggests widespread carbonate-producing shelf and slope areas which were tapped by submarine canyons. The fine-grained, quartz-rich glauconitic turbidites of the Gault Formation with their substantial biogenic carbonate content probably represent shelf sands that moved for relatively long distances by (long-shore) currents before they became trapped in canyon heads incising the continental shelf. High matrix content (particularly mica content) and poor sorting of the Reiselsberg Sandstone and part of the Bleicherhorn Formation turbidites, on the other hand, suggest a different origin: these sands were probably funneled through delta-systems during flood stages of rivers without much winnowing, reworking and redeposition. For the Reiselsberg Sandstone this is surprising, because the late Cenomanian-Turonian was probably the time of fastest sea-level rise during the Cretaceous (Watts & Steckler 1979) where one would expect little terrigenous detritus to reach the outer shelf (as evidenced by the widespread occurrence of chalk on the European and North American continental shelves). The conclusion to be drawn from this is that tectonic activity caused uplift and steepening of the relief on land and narrowing of the shelves at sea thus facilitating the supply of coarse terrigenous sands to the deep sea. The source terranes for the Upper Cretaceous terrigenous sands must have been composed predominantly of continental crust (Faupl *et al.* 1970;

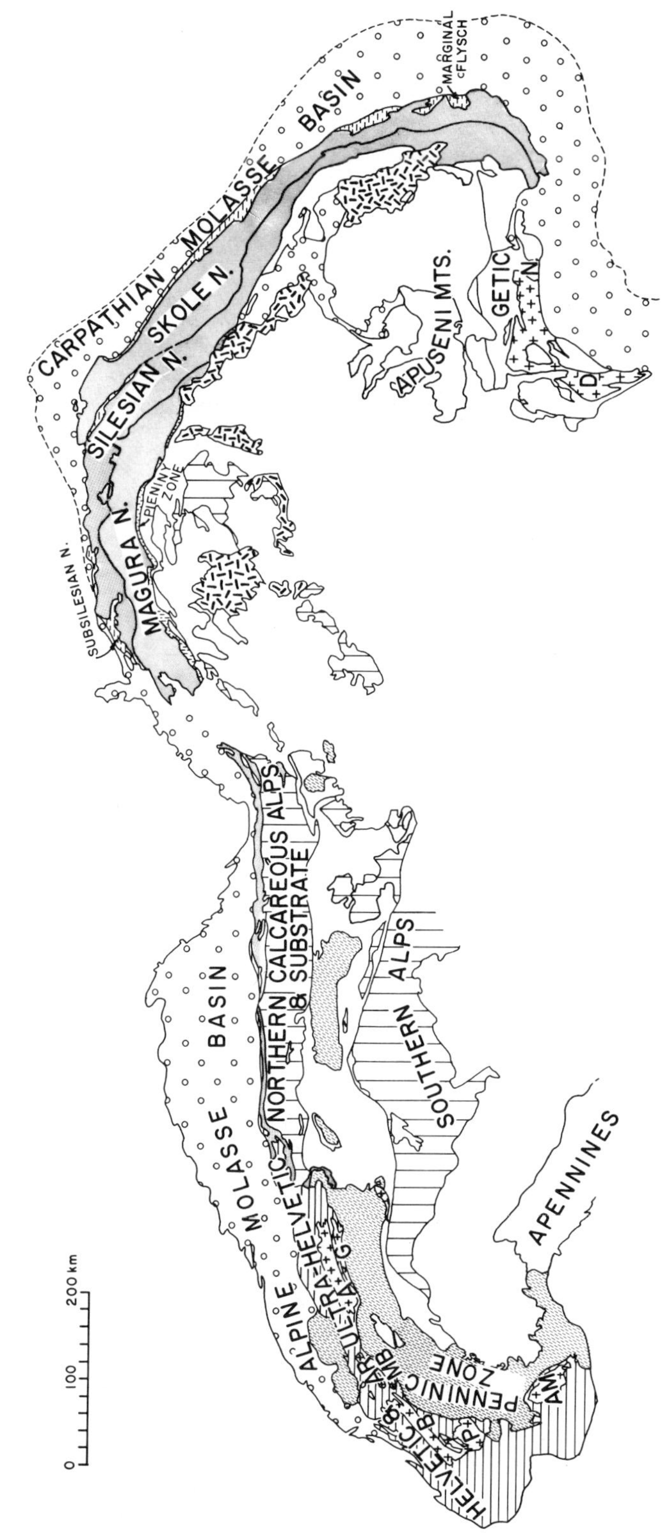

FIG. 1. Flysch Zone of East Alps and Carpathians and adjacent structural units mentioned in text (modified after Tollmann 1969).

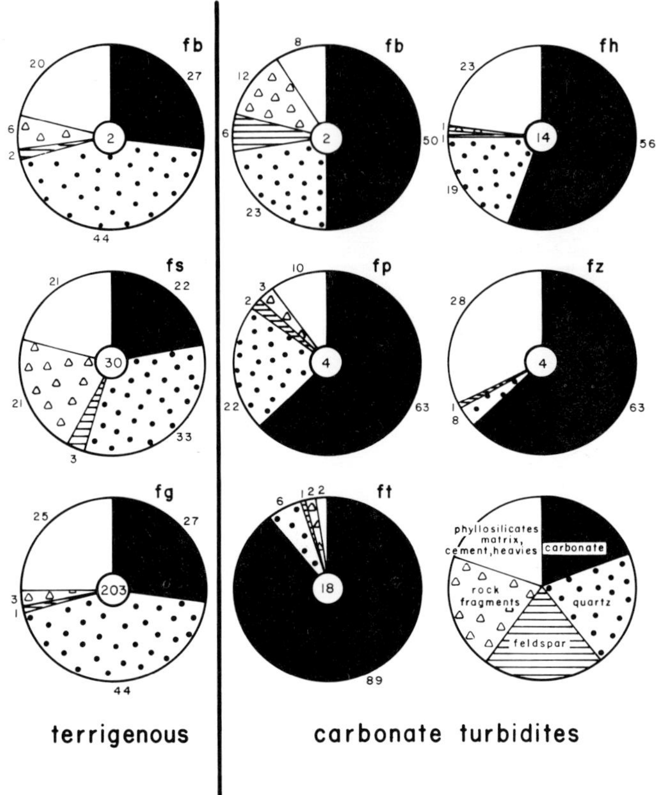

FIG. 2. Average mineralogical composition of turbiditic sandstones and calcarenites of various flysch formations in Bavaria. (Data from Hesse 1973b, p. 175 and von Rad 1973, p. 127; number of samples analysed for each formation in centre of diagram.) Analysis are for basal portions of beds. Formation names: see Fig. 4.

Freimoser 1973) because pebbles and rock fragments of the sands include granite, gneiss, schist and argillite besides Jurassic and earliest Cretaceous sediments (calcarenitic reef-limestone, echinoderm breccia but also pelagic filament limestone, saccocoma limestone, calpionellid limestone, tintinnid and aptychi limestone). Although Faupl *et al.* (1970) and Freimoser (1973) seem to favour a northern source, no conclusive evidence exists as to whether this material was supplied from the north or the south and palaeogeographic considerations make a southern source at least as likely as a northern one (see later section).

One of the puzzles of Alpine orogeny is the scarcity of Mesozoic and Cenozoic volcanism. This is also manifest in the paucity of volcanogenic rock fragments in the terrigenous sediments of the Cretaceous-Palaeogene flysch. More than 1000 petrographic analyses reported by Faupl *et al.* (1970), Freimoser (1973), Grün *et al.* (1972), Hesse (1973a,b), Niedermayr (1966), von Rad (1973) and Wieseneder (1967) have produced nothing but the occasional basaltic or diabasic rock fragment. This has important repercussions on the tectonic interpretation for the Flysch Zone of the East Alps. It is clear that active margin environments with subduction related volcanism are not suitable as tectonic setting for the Flysch Zone of the East Alps and also of the West Carpathians (before Miocene time).

Components derived from basic and ultrabasic rocks indicative of exposed oceanic crust are scarce but not entirely lacking in sediments of the Flysch Zone. Chromspinel has been reported by Faupl (1975) from the Kaumberger Beds in the Vienna Woods (eastern part of the Flysch Zone) and by Prey and Schnabel from sandstones of Reiselsberg type in the southern Vienna Woods and near Ybbsitz (P. Faupl, written comm. 1980). Pebbles of ultramafic

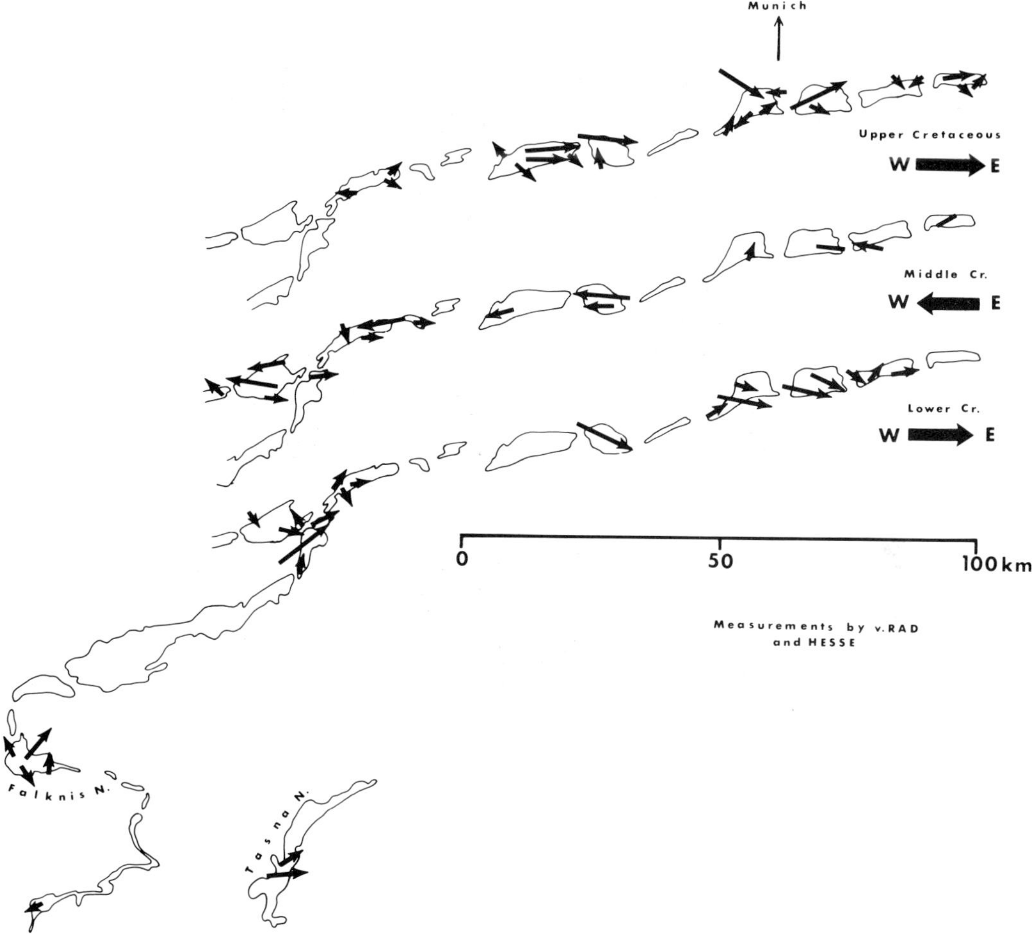

FIG. 3. Reversal of longitudinal palaeocurrent directions for three different times in the Flysch Zone of the East Alps. Current directions for the Falknis and Tasna Nappes are shown only for the Lower Cretaceous because Upper Cretaceous deposits of these nappes are largely pelagic. Note that in middle Cretaceous time (Ofterschwang Beds, Reiselsberg Sandstone) some current directions oppose the general trend from east to west by 180°, a few are at 90°. East-to-west directions in Upper Cretaceous deposits are largely confined to the Bleicherhorn Formation. Length of arrow proportional to number of measurements.

composition, however, are more abundant in the Ultrahelvetic Zone to the north of the Flysch Zone (Dietrich & Franz 1976). In South Penninic units (Verspala Flysch, Arosa Zone) and the Upper Austro-Alpine deposits of the Gosau Basins to the south of the Flysch Zone chromspinel occurs in strata as young as early Campanian according to Woletz (1967). However, Faupl (written comm.) and the author (unpubl. data) found it also in Maestrichtian-Danian strata in the Weyerer Bögen-Gosau and late Campanian strata of the Kössen Gosau basin, respectively.

Composition of the source areas which supplied detritus to the Flysch may thus be established with a reasonable degree of confidence from sandstone and conglomerate mineralogy and petrography if one keeps in mind the possible introduction of bias due to intrastratal solution and selective removal of components during transport and weathering. Locating these source areas, however, is more difficult. Palaeocurrent indicators may be useful to identify the general transport directions, but are not diagnostic to pin-point the exact locations where the detritus was introduced into the

TABLE 1. *Comparison of stratigraphic columns for flysch zones of East Alps and West Carpathians (after Dzulynski* et al. *1959; Hesse 1974; Oberbauser 1980)*

		FLYSCH ZONE						Carpathians
		Vorarlberg	North (sigiswang) Facies	South (Oberstdorf) Facies	Greifenstein Nappe	Vienna Woods Kahlenberg Nappe	Laaber Nappe	Silesian Nappe
Palaeogene	Oligocene							Krosno Beds 600–>2000 m
	Eocene				Greifenstein Beds –500 m		Laaber Beds 2000–3000 m	Menilite Beds 150–300 m; Hieroglyphic Beds 300–80 m
	Palaeocene (Danian)		Bleicherhorn Formation mica-rich greywacke, shale 7200 m	Tratenbach Formation				Ciezkowice Sandstone 150 m; Upper Istebna Beds
Upper Cretaceous	Maestrichtian	Fanola Fm. 500–700 m mica-rich greywacke, calcarenite and shale	Hällritzer Formation					
	Campanian	Planknerbrücke Fm. –550 m Carbonate greywacke and shale	Uppermost Variegated Clayst. carbonate greywacke + shale –400 m		Altlengbach Beds >1000 m; Uppermost Variegated Claystone	Sievering Beds –700 m		
	Santonian	Plankner Fm. 200–>500 m thin-bedded calcisiltite, calcilutite and shale	Plesenkopf Beds thin-bedded siltstone + shale –200 m	Zementmergel Formation calcarenite, carbonatic siltstone thick marl and green shale –200 m	Zementmergel Formation –500 m	Kahlenberg Beds –500 m	Kaumberg Beds 100–300 m	
	Coniacian		Upper Variegated Claystone red and green shale	–20 m				Lower Istebna Beds 1000 m
	Turonian	Schwabbrunnen Fm. >600 m poorly sorted, mica-rich greywacke	600–200 m Reiselsberg Sandstone	poorly sorted, mica-rich greywacke 10–50 m		Reiselsberg Sandstone 0–500 m		
	Cenomanian	Basis Fm.	Offerschwang Bed –100 m siltstone and shale	Lower Variegated Clayst. –200 m		Variegated Claystone		Godula Beds 2000 m
Lower Cretaceous	Albian			Gault Formation 200 m quartz-greywacke, black and green shale		Gault Formation –100 m (in part Bartberg Beds)		Mikuszowice Radiolarite <50 m
	Aptian			Tristel Beds 150 m graded calcarenite, calcilutite green and black shale				Lgota Beds 300 m
	Barremian		substratum unknown					Wernsdorf Shale 200 m
	Hauterivian							Grodischt Sandstone 50 m
	Valanginian							Cleszyn Shale and Limestone 700 m
	Berriasian							

TABLE 2. *Characteristics of turbidites on laterally confined, elongate basin plains*

	Environments with similar characteristics
(1) Palaeocurrent directions: Predominantly longitudinal currents, i.e. current directions parallel to long axis of basin (or parallel to strike of structural belt). Azimuths show bimodal distributions 180° apart, i.e. current reversals are common, between successive formations as well as within individual formations Deposits with opposing current directions are often petrographically distinctly different (different sediment source)	Large deep-sea channels (mid-ocean channels) Perhaps elongate abyssal plains (e.g. Horse-Shoe abyssal plain), otherwise non-existent
(2) Continuity of individual turbidites: up to 100 km or more (East Alps: 115 km, Apennines: 190 km, Ricci-Lucchi 1975a; Caucasus: 200 km, Grossheim 1961)	Deep-sea channels (i.e. Cascadia Channel)
(3) Bed thickness of turbidites: longitudinally continuous layers: cm to more than 10 m (Apennines, Ricci-Lucchi 1975a, Ricci-Lucchi & Valmori 1980)	
(4) Sedimentary structures: base-cut out sequences often starting with b-division (lower parallel lamination). a-divisions (graded or massive) occur occasionally (e.g. felspar-rich marker bed of Gault Fm.). Repeated alternation of b and c-divisions within individual turbidite (possibly due to current reflection from the basin walls). Thickness of e-division dependant on availability of pelitic material, e.g. Gault Fm.: subordinate or absent, Zementmergel Fm: dominant sediment type	
(5) Maximum grain size: typically medium to fine sand. Rarely coarse sand (felspar-rich marker bed of Gault Fm.) or gravel (pebbly mudstone, Gault Fm.; Reiselsberg Sandstone)	
(6) Down-current lithological variations (bed thickness, grain-size, mineral composition, sedimentary structures): minimal. Not detectable on outcrop-scale. Recognizable only within continuous turbidites if traced for long distances (see Fig. 7)	

basin. Ever since palaeocurrent measurements have been used there has been a need to verify by detailed facies analysis the locations of the point sources of sediment supply. If the measured directions are parallel to strike of the structural belt, does this mean that the sources were far away at the basin ends; if they are perpendicular to strike, does this mean we see a nearby lateral source? For the Flysch Zone of the East Alps the absence of sediment sources on parts of the basin margins has been proved by the striking continuity of individual turbidite beds of the Gault Formation, in connection with longitudinal palaeocurrents. Along that 115 km long segment of the Flysch Zone for which the continuity of beds has been established, lateral sediment sources cannot have existed during the time of deposition, because all beds are continuous or pinch out (i.e. the thinner ones) in an easterly direction and no new beds appear (Fig. 5). The same applies to the Zementmergel Formation in which a 190 m thick sequence of carbonate turbidites has been traced bed-by-bed for a 50 km distance (Hesse 1979). Since the two formations have palaeocurrent directions from the west, the point sources where the material was introduced into the flysch basin, therefore, was to the west of the correlated sections.

This, however, should not be misinterpreted to mean that lateral sediment sources did generally not exist along the margins of the flysch

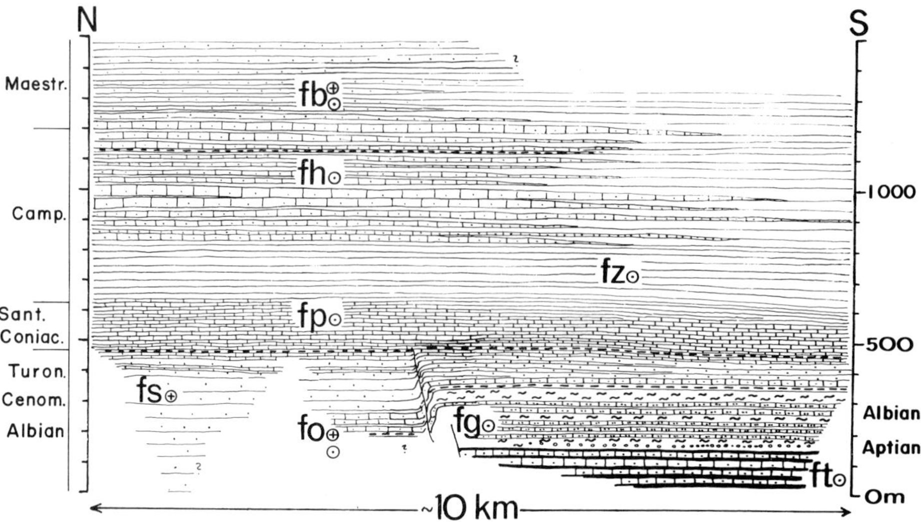

FIG. 4. Palinspastic cross-section for central portion of the Flysch Zone of East Alps in Bavaria (between rivers Lech and Inn). ft: Tristel Beds; fg: Gault Formation; fo: Ofterschwang Beds; fs: Reiselsberg Sandstone: fp: Piesenkopf Beds; fz: Zementmergel Formation; fh: Hällritzer Formation; fb: Bleicherhorn Formation. (In south-eastern Bavaria the latter occurs also in the southern half of the Flysch Zone according to Freimoser 1973.) Circled dots: formations with palaeocurrent directions from the west. Circled crosses: formations with palaeocurrent directions predominantly from east.

basin. They did exist, e.g. for the Gault Formation in the area north of Salzburg, where coarse to very coarse detritus suddenly appears that has no upcurrent equivalent farther west (Hesse 1973a). Another lateral source was established in the west in Graubüuden and Liechtenstein in the Gault Formation of the Falknis and Tasna Nappes (see section on depositional environments). Such lateral sources can have existed in any area at any time, but their existence has to be documented by appropriate facies evidence (occurrence of submarine canyons or deep-sea fans) in addition to palaeocurrents perpendicular to strike.

Anatomy of individual turbidite formations

If the turbidites of individual formations can be traced as continuous layers over major parts of the basin (by 'fingerprinting' them using bed thickness (Fig. 5) and mineralogical composition (Fig. 6)), the anatomy of parts of the basin fill may be unravelled: lithological variations (bed thickness, grain-size, mineralogical composition) may be evaluated quantitatively and gradients be established. Since outcrops of the Gault Formation occur along a narrow belt less than 2 km wide, only longitudinal, downcurrent gradients have been established, whereas for the Zementmergel Formation gradients perpendicular to palaeocurrent direction are also available (unpubl. data). The large number of available thickness data required statistical treatment by a method described by Hesse (1973b). Thickness gradients for the Gault Formation are extremely low and yield an average of -6 mm km^{-1} for 11 beds decreasing downcurrent in thickness and of 4 mm km^{-1} for 12 beds increasing downcurrent (Fig. 7). Gradients for thin-bedded basin-plain turbidites of the Apennines (Ricci-Lucchi & Valmori 1980, table 3) are of the same order of magnitude, whereas thick-bedded basin-plain turbidites show somewhat higher gradients. This com-

FIG. 5. Example of correlation of turbidite sections (Gault Formation, Aptian-Albian, simplified from Hesse 1974). Only the lower half of the formation (i.e. Lower Claystone Member, Lower Greywacke Member, Middle Claystone Member) is shown. Note that bed-by-bed correlation of the sections shown is nearly perfect (except with the most proximal section 225, which contains a larger number of additional thin beds). Bed F2 is a relatively felspar-rich petrographic marker bed ('felspar-bed') which, in a number of outcrops, is amalgamated with underlying bed F1.

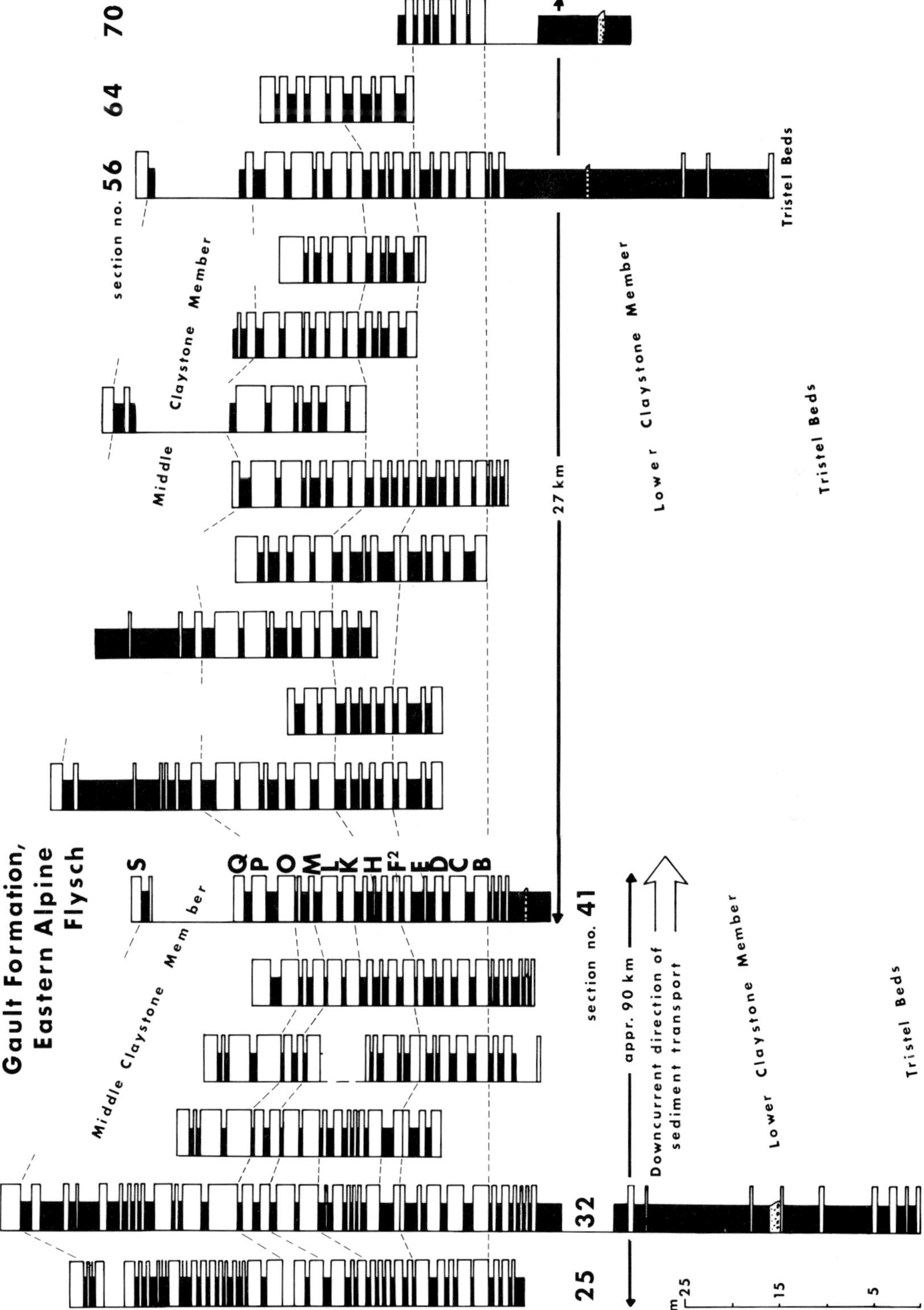

Gault Formation, Eastern Alpine Flysch
section no. 56
64
70
section no. 41
32
25
Middle Claystone Member
Lower Claystone Member
Tristel Beds
27 km
appr. 90 km
Downcurrent direction of sediment transport
S Q P O M L K H F2 E D C B
m 25 15 5

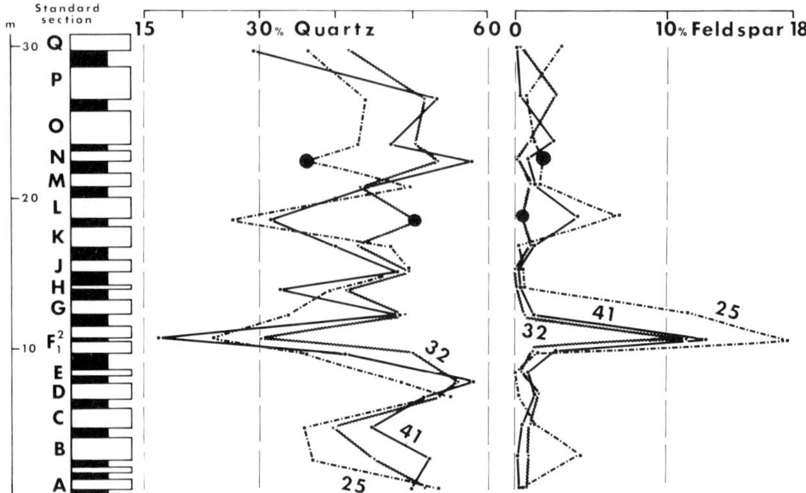

FIG. 6. Bed-by-bed correlation of selected sections (225, 232, 241—shortened to last two digits on figure) of the Gault Formation by mineral composition of basal laminae of every bed. (Simplified from Hesse 1973a, fig. 3: only turbidites A–Q of the Lower Greywacke Member are shown.) Quartz and feldspar contents are plotted versus standard section. Beds that deviate significantly are denoted by full circles (e.g. bed L in section 232, bed N in section 225).

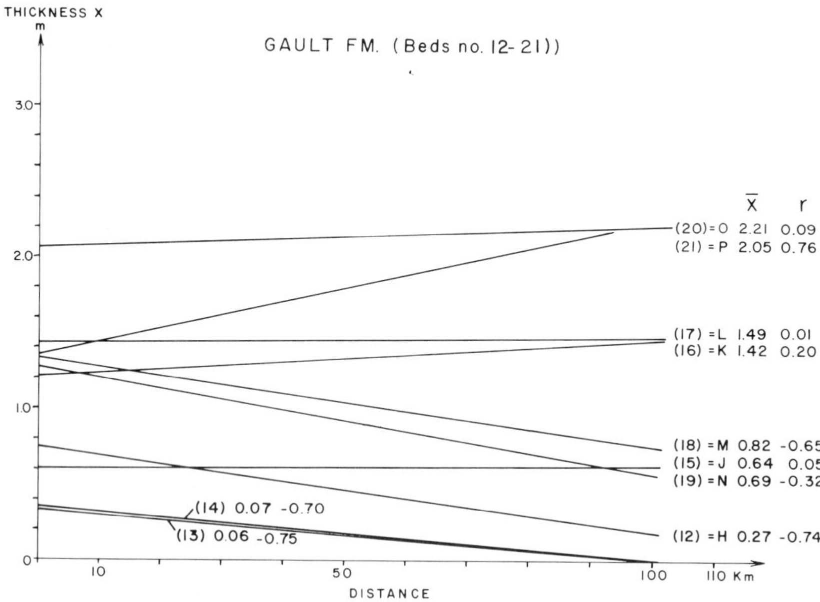

FIG. 7. Vertical and horizontal bed thickness variation of beds no. 12 (−H on Fig. 5) to 21 (−P on Fig. 5) over a downcurrent distance of 106 km. Beds 15–17 and 20, 21 slightly increase downward in thickness with an average gradient of 2.7 m km^{-1}. Average gradient for beds thinning downcurrent is −4.2 mm km^{-1}. Gradients obtained by linear regression analysis. Column to the right: average bed thickness. Note general upward thickening trend of sequence (beds with higher numbers are generally thicker). On the basin plain such a trend reflects a systematic increase in availability of the fine sand-sized material, which may or may not be inherited from a prograding depositional lobe on an adjacent deep-sea fan.

pares well with an average figure of -7 mm km^{-1} for three continuous layers in recent turbidites of the Puerto Rico Trench (gradient determined for a correlation distance of 81 km, Conolly & Ewing 1967) and -3 and 2 mm km^{-1} for three downcurrent thinning and three downcurrent thickening layers, respectively, of the Cascadia Deep-Sea Channel (correlation distance 83 km, Griggs *et al.* 1969). Modern deep-sea trenches and deep-sea channels are clearly environments with very low horizontal gradients for lithological variations in turbidites. Deep-sea channel deposits, however, will show unidirectional palaeocurrents and an association of channel-fill with spill-over facies; they are therefore not possible analogues for the Flysch Zone of the East Alps.

Downcurrent variations in grain-size and mineralogical composition of individual beds display equally low gradients (Hesse 1973a, 1974). For example, the median grain-size at the base of marker bed F2 (characterized by its high detrital felspar content) decreases from more than 500 μm to less than 200 μm over a distance of 90 km in the downcurrent direction. Simultaneously its felspar content decreases from 18 to 11% as a result of hydrodynamic sorting accompanying the grain-size decrease. (The felspar, which is exclusively plagioclase, was enriched in the coarser grain-size fractions of the starting material.)

These low but detectable gradients for downcurrent lithological variations are reflected in the remarkable facies constancy along strike. Across strike, i.e. perpendicular to the prevalent palaeocurrent directions, on the other hand facies varies relatively rapidly. Consequently, two separate stratigraphic columns are in use, one for the southern 'Oberstdorf Facies' and one for the northern 'Sigiswang Facies' (Table 1). Tristel Beds and Gault Formation are restricted to the southern facies (except in the area near Salzburg and at one western locality in the Allgäu, von Rad 1964) either because they have been sheared off the base of the Flysch Nappe during tectonic transport or because the basin was much narrower in Early Cretaceous time, not comprising that part which was later to become the northern facies realm. In the latter case deposition of the Reiselsberg Sandstone (and the Ofterschwang Beds) in the northern facies realm would indicate a mid-Cretaceous widening of the basin and northward overstepping of the turbidite facies by 2–5 km. In the Upper Cretaceous the facies change from north to south is best observed between Hällritzer Formation (in the north) and coeval parts of the Zementmergel Formation in the south. Going from north to south individual turbidites lose their basal structural divisions (mostly b-divisions, some a-divisions) but at the same time gain thickness in the e-division of homogeneous grey marlstone. This suggests that the competence of individual flows to carry sand-sized sediment decreased very slowly in the downcurrent direction but relatively rapidly in the direction perpendicular to flow. The two stratigraphic columns that have been established for the north and south facies in the central segment of the Flysch Zone (in Bavaria) can therefore be applied to areas far to the east in Austria and the west in Allgäu and Vorarlberg (for names of age-equivalent formations see Table 1). Some of the formations can even be recognized in the West Carpathians in Czechoslovakia (Prey 1960, 1965) this being the basis for treating the East Alpine-Carpathian Flysch as one continuous belt.

Inferred depositional environments

Elongate basin plain with limited number of point sources

The long-distance continuity of layers together with the longitudinal palaeocurrent directions and the reversals in current azimuth is conclusive evidence that the depositional environment representing a major part of the flysch basin was a flat, nearly horizontal basin plain. In addition, this was an elongate basin plain. The suggestion of a nearly horizontal basin plain is inescapable—how else could the currents have reversed flow direction? Tilting of the basin floor would be a possibility. However, if current reversals occur repeatedly and even between successive turbidites as in the Ofterschwang Beds, tilting is not a realistic alternative. The long-distance continuity of layers is also best explained by a basin plain environment—by analogy: elongate basin plains and deep-sea channels are the only modern environments for which continuous turbidite layers have been reported (e.g. Conolly & Ewing 1967; Griggs *et al.* 1969; Chough & Hesse 1980; Pilkey *et al.* 1980).

Basin plains may develop in various kinds of turbidite basins including deep-sea trenches. For modern trenches Piper *et al.* (1973) and von Huene (1974) suggested that their trench plains are characterized by an axial leveed channel based on findings in the Aleutian and Peru-Chile trenches. Obviously, in the Gault Formation and in the Zementmergel Formation of the Flysch Zone such a channel is not present. It is questionable, however, whether an axial channel is a necessary element of a deep-sea trench

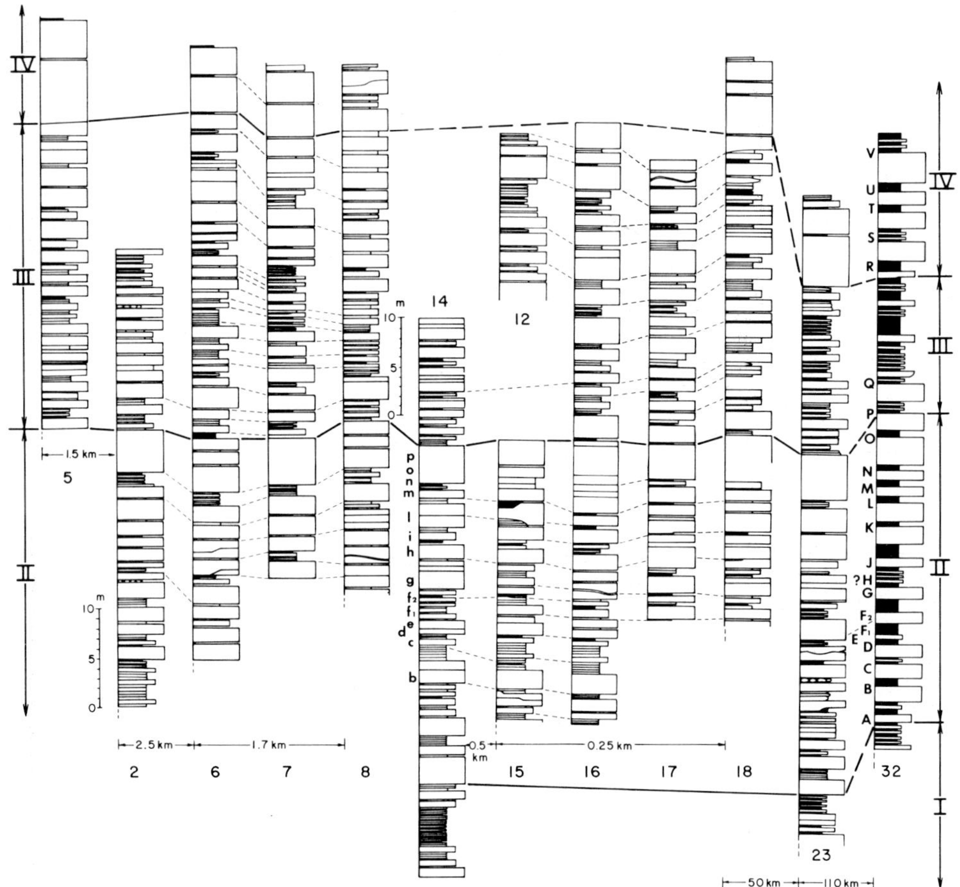

FIG. 8. Correlation of sections, Gault Formation, Falknis Nappe (sections 202, 205 to 208, 212, 214 to 218), Tasna Nappe (223), and, for comparison, Flysch Zone of East Alps (232) (modified from Hesse 1973a, plate 3—first digit of section no. omitted on figure). The Lower thin-bedded Greywacke Member (I, Upper part), the Lower Greywacke Member (II), the Middle thin-bedded Greywacke Member (III) and lower part of Upper thick-bedded Greywacke Member (IV) are shown. The depositional environment of the Falknis and Tasna Nappes is interpreted as an acute-angled deep-sea fan or broad feeder channel. Two groups of turbidites (f2 and n-o-p) are tentatively traced as continuous layers through most sections of the Falknis and Tasna Nappes and are probably also present in section 232 of the Flysch Zone.

Note the general thickening-upward trend from beds 'c' to 'p' and a second similar trend at the top of the sections shown. Similar trends are inherited by the basin plain facies (section 232) of the Flysch Zone (see Figs 5 & 8).

plain, because few trenches have been studied in this regard. In the Mid-America trench no well-defined axial channel was detected during recent deep-sea drilling off Guatemala (von Huene *et al.* 1980, fig. 2). One also can imagine that in narrow trenches the trench walls themselves may act as channel walls.

The characteristics of basin plain turbidites are listed in Table 2. The same kind of evidence (i.e. long distance continuity of layers, palaeocurrent reversals) was used by Ricci-Lucchi (1975 a, b) and Ricci-Lucchi & Valmori (1980) to infer a basin plain environment for much of the Miocene Marnoso-Arenacea Formation of the Northern Apennines. It should be noted that in vertical bed sequences of the Gault Formation certain trends such as a thickening-upward succession can be deciphered (Fig. 5) which might be misinterpreted as a 'prograding fan-lobe trend' if taken out of context. This is to be kept in mind in trend analysis of turbidite sequences: there may be factors other than the geomorphology of a specific depositional environment (such as a

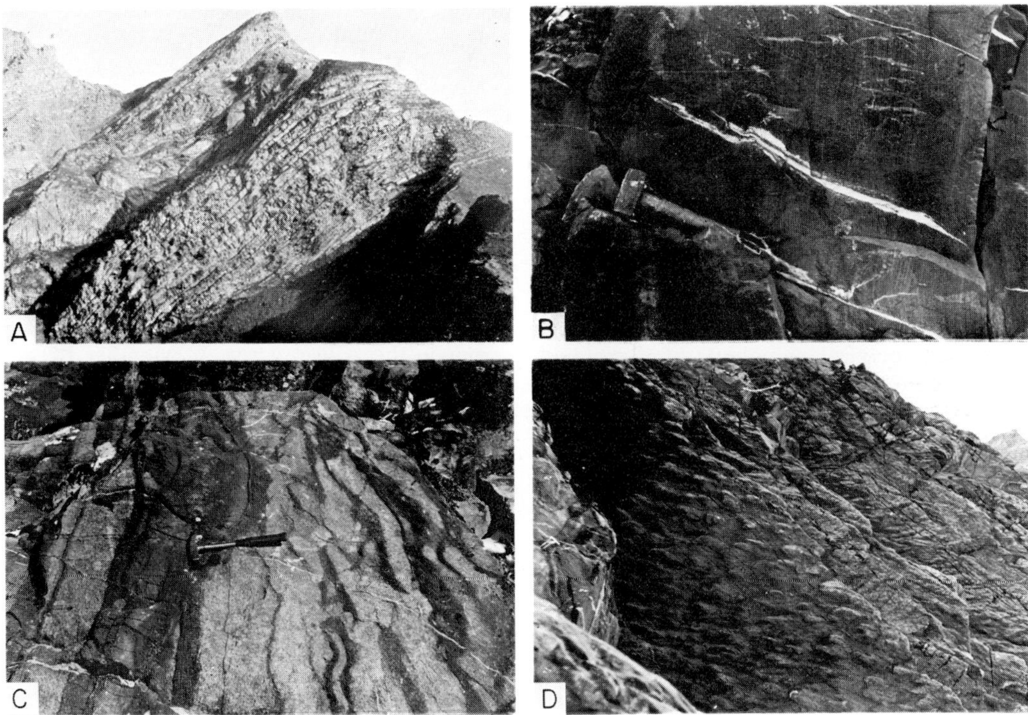

FIG. 9. (A) North wall of Falkinshorn, Liechtenstein, displaying thickening upward sequence of proximal turbidites (lithological members III and IV, see profile 5 on Fig. 8) interpreted as prograding fan lobe. Exposed thickness (between X-X^1) approximately 100 m. Lower 20 m of sequence (below X) separated from overlying sequence by nearly bedding-parallel fault. Detailed bed-by-bed section (profile 5, Fig. 8) measured approximately at centre of picture with the aid of mountain guides J. Gassner and O. Lampert, Triesen, Liechtenstein.
(B) Parallel-laminated and partly wavy-laminated, 80 cm thick turbidite with upper layer of oblique lamination (left side of picture) characteristic of bypassing without deposition of major portion of turbidity current (turbidite facies B_2 of Mutti & Ricci-Lucchi 1975). Gault Formation, Tasna Nappe, Breite Krone, Grisons.
(C) Internal reworking (reactivation) surfaces in thick, coarse-grained turbidite, deep-sea-fan facies. Base of bed to the right. Falknis Nappe, Naafkopf, 2360 m, Liechtenstein.
(D) Protrusions of convolute bedding on upper bedding surface current-remolded into 'false flute casts'. Tasna Nappe, Breite Krone, 2740 m, Silvretta Ranges, Grisons.

deep-sea fan) which control vertical thickness trends ('megarhythms' of Lajoie 1979), etc. These may include climate, sea-level changes and the tectonic behaviour of the source area.

Deep-sea fan in the Gault Formation of the Falknis and Tasna Nappes

In the westernmost outcrops of the Gault Formation in the West-East Alpine border region in Liechtenstein and Graubünden (Switzerland) bed-sequences occur that bear some distinct petrographic and lithological relations with those farther NE in the Flysch Zone but also display some characteristic differences and do not fit a basin-plain depositional model. Although these sections occur in isolated tectonic units of the Falknis and Tasna Nappes separated 50 km from the closest Gault sections in the Flysch Zone, it has been known for some time that the rocks of the three nappes were palaeogeographically linked (e.g. Richter 1933). Palaeocurrent directions (Fig. 3) as well as detailed facies analysis prove that the Falknis and Tasna Gault Formations are the upcurrent equivalents of the Gault Formation in the Flysch Zone.

In the Falknis and Tasna Gault bed-by-bed correlation of sections is generally not possible, although at certain stratigraphic intervals sections have been tied together bed-by-bed over a limited distance (Fig. 8). The formation has

TABLE 3. *Greywacke/claystone thickness ratio, Gault Formation*

Lithological member	Falknis and Tasna Nappes			Flysch Zone		
	Min.	Max.	Av.	Min.	Max.	Av.
IV	11.8	42.7	25.4	0.9	12.5	3.1
II and III	4.7	13.6	7.4	0.6	4.2	1.9
I	1.7	3.8	2.8			0.1

been subdivided into five lithostratigraphic members that are homogeneous lithologically, have limited thickness, can be identified at almost any locality if exposed and correspond to five lithological members of the Gault Formation in the Flysch Zone. For the Falknis and Tasna Nappes these are (I) a Lower thin-bedded Turbidite Member containing a pebbly mudstone and a conglomerate layer, (II) a Lower Turbidite Member of intermediate thickness, (III) a Middle thin-bedded Turbidite Member, (IV) an Upper thick-bedded Turbidite Member (Fig. 9a) and (V) an Upper Claystone Member. The overall thickness of the formation is about the same in the three nappes, except that the Tasna outcrops are stratigraphically incomplete due to faulting. In each member the greywacke-claystone thickness ratio for sections of the Falknis-Tasna Nappe is an order of magnitude higher than for corresponding members in the Flysch Zone (Table 3). This increase in sand-turbidite thickness at the expense of turbiditic and hemipelagic shale or claystone is due to the fact that much of the fine-grained turbiditic sediment bypassed the Falknis-Tasna area and most of the hemipelagic clay deposited was eroded by subsequent turbidity currents. Many of the thinner turbidites of the Falknis-Tasna Nappes pinch out before reaching the Flysch Zone. Therefore, the thin-bedded turbidite members of the Falknis-Tasna Nappes correspond to claystone members in the Flysch Zone, the thick-bedded turbidite members to turbidite members of intermediate thickness. This change in character is gradual. In the westernmost section of the Gault Formation in the Flysch Zone (in the Allgäu area) the middle member is still a thin-bedded turbidite member, as in the Falknis-Tasna Nappes. Further east it loses its turbidites and becomes a claystone member.

For the Falknis-Tasna Gault, bed thickness, grain-size and felspar content of sand-turbidites show a significant increase compared to corresponding members of the Flysch Gault. This is a continuation of the slow gradual upcurrent changes observed within the Flysch Zone. The increase in felspar content and grain-size is reflected in a large number of coarse-grained felspar-bearing sandstone beds, e.g. five in Member II of the Falknis-Tasna Nappe compared to one in Member II of the Flysch Zone (two in the westernmost outcrops in the Allgäu area). In Member IV this effect is even more pronounced, where a large number of felspar-bearing coarse beds occurs in the Falknis-Tasna area but none further east in the Flysch Zone: i.e. the turbidity currents had deposited their coarser load before reaching the basin plain of the Flysch Zone. This is an excellent example of horizontal gradation occurring over a distance of about 50 km. The finer-grained sand fractions which make up the turbidites in the Flysch Zone are present in the Falknis-Tasna area as upper portions of the coarser beds or as individual thin layers. The same effect of horizontal gradation is displayed by the greater abundance and larger diameter of resedimented claystone clasts (intraclasts) in the west.

The number of amalgamated beds is considerably higher in the Falknis-Tasna area than in the Flysch Zone—this being one reason why bed-by-bed correlations were not possible in the former. Lower structure divisions (a-divisions) are much more abundant in the former than in the latter, although this has not been quantified. Similarly, there are sedimentary structures such as low-angle cross-bedding (dune and antidune cross-bedding, Hesse 1973a, plate 9, fig. 2), internal reactivation (reworking) surfaces (Fig. 9c) and small-scale channels (less than 1 m deep, Hesse 1973a, plate 8, fig. 3) in the Falknis and Tasna areas that have been described elsewhere from proximal environments such as deep-sea fan valleys; these are lacking in the Flysch Zone. Rippled surfaces, thin cross-bedded layers at bed-tops (Fig. 9b) and remoulded, stream-lined convolute bedding surfaces ('false flute casts', Fig. 9d), indicative of bypassing have also been observed in the Falknis and Tasna Nappes.

In summary the Gault Formation of the Falknis and Tasna Nappes is distinctly more proximal in character than the Gault Formation of the Flysch Nappe and contains the upcurrent portions of the turbidites whose distal ends on the basin plain of the Flysch Zone extend at least 150 km farther east, perhaps as much as 270 km to the area north of Salzburg. The observed facies associations are similar to Mutti & Ricci-Lucchi's (1975) middle and outer fan facies, characterized by (1) their facies associations B_1 (thick, coarse to medium grained sandstone with occasional granules and pebbles, showing wedge-shaped sets of thick, coarse-grained inclined laminae, plane-parallel and wavy laminations, dish structures and fluid escape pipes, common claystone chips, small-scale cross-lamination and/or convolute lamination), B_2 and E (both similar to B_1 but thinner bedded), C_1 and C_2 ('classical' turbidites with more or less complete turbidite structure-sequences), (2) the absence of larger channels, and (3) the presence of thickening upward cycles (Fig. 8). The main difference in comparison with the Mutti & Ricchi-Lucchi fan model is the presence of only two thickening upward sequences that may be interpreted as prograding fan lobes (members I + II, and members III + IV). This might suggest that the fan was relatively small and consisted of a single channel and depositional lobe probably confined by a funnel-shaped incision at the base-of-slope. 'Feeder-cone' is perhaps a better term for this feature if one considers its function as the main supplier of sediment for the adjacent basin plain. Although it may have existed for 5–8 Ma the rate at which turbidity currents passed through this system was probably not very high

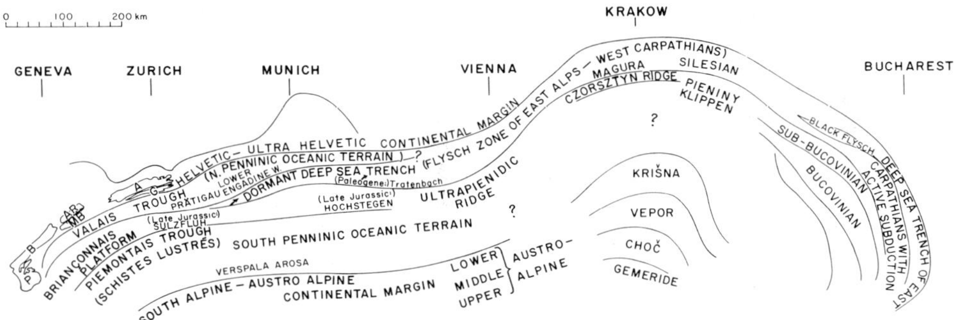

Fig. 10. Inferred schematic arrangement of palaeogeographical zones of Alpine Tethys in Albian time and position of point sources (arrows) for the flysch trench (after Faupl 1978; Gwinner 1971; Hesse 1973a; Hesse & Butt 1976; Lemcke 1970; Sandulescu 1973; Tollmann 1969).

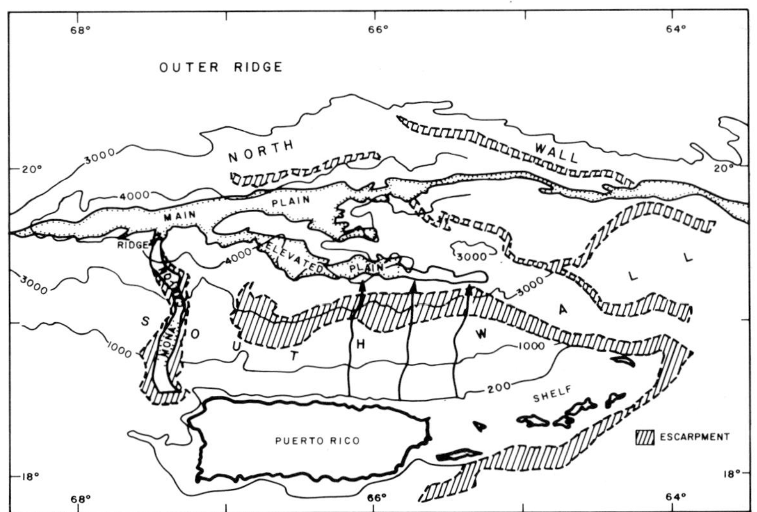

Fig. 11. Bathymetry of the Puerto Rico Trench showing main tributary channel (Mona Canyon) on the south wall (simplified after Conolly & Ewing 1967 and Ewing & Heezen 1955). The Puerto Rico Trench is considered a suitable modern analogue for the Cretaceous deep-sea trench of the Flysch Zone of the East Alps. Isobaths in fathoms.

(up to 50 Ma^{-1}, 10 of which reached the basin plain). The deep-sea fan or feeder cone of the Falknis and Tasna Nappes in Albian time was in all likelihood located on the south side of the SW-NE to W-E oriented fan-basin plain array (Fig. 10), because palaeocurrents are generally toward NE and grain-size decreases in the same direction. A northward decrease in grain-size is especially obvious for the Upper Jurassic Falknis Breccia (Allemann 1956) which may represent an inner fan valley deposit of the same fan. If we look for a modern equivalent of this fan-basin plain assemblage, the modern Puerto Rico Trench seems to provide a close example. It has a main tributary on the south wall (Mona Canyon) which joins with the trench plain (Fig. 11).

The proposed correlation between the Gault Formations of the Falknis and Tasna Nappes and the Flysch Zone was the first example of ancient deep-sea fan deposits that have actually been traced laterally into their basin-plain equivalents (Hesse 1973a). Similar results have been obtained recently by Ricci-Lucchi & Valmori (1980) in Miocene flysch of the Apennines. Contrary to some current models (e.g. Walker 1978), basin-plain turbidites are not restricted to thin-bedded, c-d-e-division distal varieties but include many beds that start with b-divisions and the occasional a-division bed. Some of these may be very thick. This may be a special characteristic of an elongate basin plain fed by a small narrow fan or wide canyon or feeder channel which did not trap major portions of the sediment during downslope transport.

Discussion of the depositional environments in this section has been restricted to two selected formations, for which detailed data are available. The deep-sea fan/basin-plain association of environments may be characteristic for certain other formations of the Flysch Zone, particularly carbonate flysch formations. It may not be applicable, however, to formations such as the Reiselsberg Sandstone which displays invariably proximal characteristics over a distance of at least 200 km along strike. Petrographically similar beds of the Bleicherhorn Formation also display distinctly proximal features. These are the only two formations which also contain significant amounts of fine-grained conglomerates: elsewhere conglomeratic deposits occur only as isolated layers (i.e. pebbly mudstones in the Gault Formation (Hesse 1973b); conglomerates in the Ofterschwang Beds (Stephan & Hesse 1966) and Zementmergel and Hällritzer Formations (Freimoser 1973)). An exception is the coarse boulder conglomerate in the Gault Formation north of Salzburg (Aberer & Braumüller 1958; Hesse 1973a), whose significance as a possible submarine canyon deposit has already been mentioned.

Basin geometry and palaeodepth

The deep-sea fan/basin plain association of depositional environments that has been established repeatedly in time appears to be representative for the Flysch Zone, although other environments may have existed at times (e.g. during Reiselsberg time, see below). This basin plain was relatively narrow compared to its length judging from: (1) the present shape of the flysch belt; (2) the relatively low degree of scatter in the predominantly longitudinal palaeocurrent directions; (3) the successful bed-by-bed correlation in two turbidite formations. If the flysch belt as we see it today was merely a stripe of ocean floor cut randomly out of a broad abyssal plain it would be a miraculous coincidence that its structural orientation is parallel to the sedimentological trends. If the basin plain was part of an originally broad abyssal plain, it would be much more likely to see more scatter and a different pattern in the palaeocurrent directions. The currents not being confined by the basin walls would shift laterally in a similar fashion to fan-lobes on a deep-sea fan and it would be most unlikely to find 50 consecutive beds that are stacked up in an orderly fashion so that each one can be traced for more than 100 km. Sediment volumes contained in individual beds would become unreasonably large, if the shape of the basin plain was square rather than narrow rectangular—to compare two geometric situations that are obviously simplifications but not unrealistic ones, from what we know of modern basin plains. Estimates for individual beds of the Gault Formation range from 1 to 25 km^3 (of solid particles), the lower value applying to an approximately 1 m thick bed (of 2% porosity) extending for 100 km (proved length) times 10 km (assumed basin width), the maximum applying to a bed of 2.5 m average thickness (bed 'AJ', Hesse 1974, fig. 3) extending for 300 km (possible length) times 30 km (possible basin width). These figures compare well with estimates of Ricci-Lucchi & Valmori (1980) for basin-wide, continuous turbidites in the Apennines. If basin width was assumed to be an order of magnitude larger, the thickest beds would contain sediment volumes in the order of 100 km^3—volumes comparable to estimated slump volumes that result from open-slope failure (e.g. 70 km^3 for the slump following the Kwanto earthquake of 1923, Menard 1964) but unlikely for flows that were funnelled through submarine canyon-fan valley systems. Such systems must be capable of containing the entire flow more or less at a time—i.e. the sediment volume plus perhaps 40 times the amount of water in order to make the slurry move.

Palaeodepth of the basin plain was below the calcite compensation level as shown by the absence of calcium carbonate in the green hemipelagic claystone which occurs throughout the stratigraphic column of the Flysch Zone often alternating with carbonate turbidites (Hesse & Butt 1976). Carbonate-free red claystone which occurs at three stratigraphic levels (Albian—lower Cenomanian lower Variegated Claystone, upper Turonian—lower Coniacian upper Variegated Claystone and upper Campanian Uppermost Variegated Claystone) has been interpreted as equivalent to modern brown abyssal clay on the basis of chemical and clay mineralogical analysis (Faupl 1976).

Tectonic behaviour of the basin

There is little if any evidence for syndepositional deformation. If the absence of the Lower Cretaceous Tristel Beds and Gault Formation in the northern facies realm is not due to tectonic erosion at the base of the Flysch Nappe, it may reflect widening of the basin and northward overstepping of the turbidite facies in Ofterschwang and Reiselsberg time. If so, the widening was accompanied by at least 150–550 m deepening (compared to the southern part), this amount being the difference in thickness of the Reiselsberg Sandstone (in the compacted state) between the southern facies (usually less than 50 m) and the northern facies (200–600 m, von Rad 1973). This deepening due to down-faulting (Fig. 4) is the only evidence for possible syndepositional tectonic activity in the Flysch Zone.

Tectonic setting of the Flysch Zone of the East Alps: 'dormant' Cretaceous-Palaeogene deep-sea trench associated with transform faulting

Discussion in previous sections has shown that none of the following types of turbidite basins are adequate modern analogues for the Flysch Zone of the East Alps: (1) active margin basins (i.e. upper slope, mid-slope or lower slope basins, deep-sea trench associated with active subduction); (2) continental rise apron; (3) abyssal cone; (4) abyssal plain; (5) mid-ocean channel; (6) ponded basin on mid-ocean ridge; (7) back-arc basin plain. Among the remaining possible basin types the deep-sea trench with dormant subduction is the tectonic environment best suited to fit the observations (Table 4). Basin geometry (great length compared to nar-

TABLE 4. *Flysch Zone of East Alps: Cretaceous-Palaeocene deep sea trench*

Evidence for trench:
- Dimensions:
  - Length: 500 km (without Carpathians)
  - Width: 10–50 km (widening toward E: possibly > 100 km)
  - Depth: below palaeo-calcite compensation level. Red claystone as equivalent of brown abyssal clay

Depositional environments:
- (a) Elongate, nearly horizontal basin plain:
  - Palaeocurrent directions predominantly parallel to strike (basin axis)
  - Repeated reversal of current directions (change in azimuth by 180° in successive turbidites)
  - Long-distance continuity of individual turbidites
    - Gault Formation: 115 km
    - Zementmergel Formation: 50 km
  - Low gradients of downcurrent change in bed thickness, grain-size, etc. along individual beds (high gradients perpendicular to flow)
- (b) Small to intermediate size deep-sea fan (Falknis-Tasna Nappes)
- Limited number of lateral sediment sources suggests existence of slope basins that intercept access routes of turbidity currents

Lack of contemporaneous subduction:
- Lack of volcanism and volcaniclastic detritus
- Continuous sedimentation without major diastrophism for 70 Ma
- Occurrence of slivers of basic and ultrabasic rocks (ophiolite suite) N and S of Flysch Zone

row width, water depth below the calcite compensation level), association of depositional environments (basin plain plus relatively small deep-sea fans) and limited number of lateral sediment sources, tectonic behaviour (lack of any significant syndepositional deformation), lack of contemporaneous volcanism, and occurrence of possible relics of oceanic crust in adjacent tectonic units are in accordance with a deep-sea trench not associated with subduction. This analogy first suggested by Hsü (1972) and analysed here in detail implies Late Jurassic-earliest Cretaceous formation by subduction of a trench which then became dormant for a period of 50–70 Ma from Barremian-Aptian to Palaeocene or mid-Eocene time.

Although this type of deep-sea trench is not very common, the Puerto Rico Trench provides a modern example. It is 8000 m deep, has a more than 400 km long and up to 30 km wide basin plain with turbidite layers that are continuous for 200 km (Conolly & Ewing 1967). The trench fill of horizontally stratified sediments is up to 1.7 km thick (Ewing & Ewing 1962). In the east half of the trench the basin plain branches into several arms with a combined width of 60 km including the intervening ridge (Fig. 11). At the point of branching, turbidity currents coming from the principal source (Mona Canyon) at the western end of the main plain might theoretically turn southward toward the Puerto Rican shelf, i.e. toward the source area. A similar situation could be envisaged to interpret Faupl's (1975) north-to-south palaeocurrent directions in the Kaumberger Beds of the Vienna Woods without having to resort to an internal source area within the Flysch Zone. The limited number of sources supplying sediment to the main plain of the Puerto Rico Trench (Fig. 11) is typical for modern trenches, which usually have very few access routes for turbidity currents to reach the trench bottom. This is the result of the development of tectonic ridges and ramps on modern trench slopes, which intercept the downslope movement of gravity driven sediment flows trapping them in slope basins. The Puerto Rico Trench was in all likelihood created by pre-Oligocene subduction, but present-day seismic activity indicates sinistral strike-slip motion (Molnar & Sykes 1969) along a major transform fault zone running through the trench slope and forming the northern boundary of the Caribbean plate.

It is of interest that tectonic activity in the source area postulated to derive the terrigenous detritus of the Reiselsberg Sandstone (at a time of rapid global sea-level rise) may have coincided with tectonic activity in the flysch trench proposed by the possible widening and partial downfaulting in Reiselsberg time. This might reflect a pulse of subduction—the only one detected for the 50–70 Ma life-span of the flysch trench. At the same time the postulated Falknis fan ceased to issue sediment to the basin plain. It was apparently cut off from its detrital sediment source either due to uplift or due to formation of a barrier blocking sediment supply or (most likely) due to submergence of the shelf during the Cenomanian transgression. In Late Cretaceous time the Falknis area received pelagic sediment. The coarse clastic sediments of the Reiselsberg Sandstone did not reach the Falknis area because of its elevated position (as a fan). Mass flows transporting these sediments westward bypassed the Falknis area and followed a more northern route supplying Reiselsberg Sandstone as far west as the Wildhaus Syncline in eastern Switzerland (Forrer 1946).

The uniform proximal character of the Reiselsberg Sandstone over a distance of at least 200 km parallel strike could be due to the presence of several lateral sediment sources (von Rad 1973). Alternatively, it might be due to a single stationary source on the trench wall which issued sediment to the trench floor while this was moving past the source like a conveyor belt—a situation which has been implied for fans located on the San Andreas Fault in California (Nelson & Nilsen 1974). This would be an attractive possibility, since it would imply that the flysch trench was in fact the site of (large-scale) strike-slip motion. In this case the Reiselsberg Sandstone would consist of imbricate diachronous sediment lobes rather than synchronous lobes that existed separately side-by-side but coalesced with time. With the available stratigraphic tools at present it will hardly be possible to differentiate between the two possibilities.

Cretaceous palaeogeography of the northern margin of the Tethyan Sea in the Alpine-Carpathian sector and subduction models

Superficially the East Alps and the Carpathians are separated by the depression of the Vienna Basin, but this is a young (Miocene) feature. Geologically the Flysch Zone of the East Alps and the Carpathian Flysch form a continuous belt as shown by stratigraphic and facies relationships (Prey 1960, 1965; Tollmann 1969).

The Carpathian flysch consists of two segments—the West Carpathian segment comprising three separate troughs filled with more or less synchronous turbidite sequences of Upper Jurassic-Lower Cretaceous to Oligocene age, and the East Carpathian segment representing a diachronous succession of troughs which decrease in age from west to east (Contescu 1974). In particular, the middle of the three West Carpathian flysch troughs, the Silesian Trough (Table 1), is linked to the Flysch Zone of the East Alps, whereas the two others, the Skole Trough in the north and the Magura Trough in the south do not seem to have direct equivalents in the East Alps. However, the Pieniny Klippen Belt which borders the Magura Trough (and thus the Carpathian flysch as a whole) in the south, might be traced laterally into the Tiergarten Klippen-Zone of the Vienna Woods (Faupl, pers. comm.). More southerly tectonic units of the Carpathians such as the Križna-Nappe, the Veporids, the Choč-Nappe and the Gemerids, which comprise older Mesozoic sediments and metamorphic rocks, have been correlated with the Lower, Middle and Upper Austroalpine Nappes, respectively, of the Alps (e.g. Tollmann 1969; Stegena *et al.* 1975). Of the structural and palaeogeographic elements north of and under the Carpathian Flysch the Subsilesian Nappe is considered an equivalent of the Ultrahelvetic Buntmergel Formation in Austria (e.g. Ksiazkiewicz 1956).

Like the Cretaceous-Palaeogene flysch of the East Alps the West Carpathian flysch sequences are not interrupted by major diastrophic events or hiatuses except one at the Cretaceous-Tertiary boundary in the Magura Trough whose southernmost portion was stripped off its basement and was displaced southward into the Pieniny Klippen Belt (Birkenmajer 1970). Quite unlike the East Alps and West Carpathians the East Carpathian flysch experienced tectonic deformation from Albian to Miocene time. Both the development of flysch troughs and subsequent diastrophism migrate eastward toward the convex side of the arc (Contescu 1974); a well-expressed diastrophic polarity which is clear evidence for subduction. (For a similar diachronous sequence of flysch deposits in North Africa and Sicily see Wezel 1973.) Thus the Baraolt Nappe is affected in Aptian time; the Ceahlau Nappe in Albian-Cenomanian (-Turonian) time; the Bobu, Teleajen, Macla and Audia Nappes after the end of the Turonian; and the Tarcau and Marginal Folds Nappes in Late Oligocene and Early Miocene times, respectively (Stefanescu 1980).

The Carpathian flysch belt is accompanied on its inner, concave side at a distance of up to 150 km by a broad zone of Neogene calcalkaline volcanic rocks which represent one of the largest volcanic accumulations in Cenozoic Europe. Extrusion of the volcanics coincides with the closure of the flysch troughs in the West Carpathians and with diastrophism in the youngest, outermost depositional trough of the East Carpathians.

It appears justified, therefore, to assume that the deep-sea trench, which received the flysch sediments of the East Alps, continued as a deep furrow around the Carpathian arc dating at least from the Early Cretaceous, changing in character from a conservative to a convergent plate boundary at the West-East Carpathian transition. It was only in Miocene time that the West Carpathian segment became convergent as well, whereas in the East Alps this change occurred after the Palaeocene (probably in late Eocene—early Oligocene time). In the West Carpathians the Miocene subduction zone apparently reached depths sufficient for magma generation by partial melting as evidenced by the vast amounts of Neogene calcalkaline volcanics (Bleahu *et al.* 1973). The deep-sea trench interpretation for the East Alpine—West Carpathian flysch given on the basis of sedimentological evidence from the Alps and the island-arc interpretation for the Carpathians by Bleahu *et al.* (1973) are mutually supporting if one keeps in mind that the trench became eliminated as a topographic feature in the East Alps—West Carpathians at the time when island arc volcanism started in the West Carpathians. In the East Alps evidence for a volcanic arc accompanying the post-mid-Eocene subduction zone is lacking, except for mid-Tertiary volcanism in the South Alps (Dietrich 1976a,b). Subduction may not have been deep enough or, alternatively, the volcanic products may have been lost in tectonic sinks without leaving a trace. In the East Carpathians the subduction zone seems to have been active since the later part of the Early Cretaceous as suggested by the ages of diastrophism, but active volcanism is not documented until Neogene time either (except for some Upper Jurassic volcanism). Again, this must not necessarily mean that subduction did not generate volcanism as the example of the Upper Eocene-Lower Oligocene Tavayannaz Sandstone in the Ultrahelvetic of the West Alps shows—a volcanogenic turbidite sequence composed of andesitic detritus for which not the slightest trace of a volcanic source area exists. Similarly, Palaeogene and older volcanic arcs of the East Carpathians may have been lost in tectonic sinks.

Existence of this Cretaceous-Palaeogene East Alpine-Carpathian deep-sea trench with an actively subducting eastern segment and western segment lacking subduction poses the question of how close to the European continent this feature was in Cretaceous time, in other words—how wide were the intervening oceanic areas? Although one can only speculate on this question, intuitively I would expect they were very broad. This is based on the fact that we do not see continental margin deposits of the southern European margin of any reasonable thickness exposed in the East Alps (except in the Allgaü) or the Carpathians. We have to go to the West Alps to see these sequences in the Helvetic Nappes and the Ultrahelveticum in Switzerland or in the Zone Dauphinoise and the Zone Ultradauphinoise in France. Sequences of similar thickness therefore should be present in the subsurface in the east, under the cover of thrust sheets and nappes, unless the margin was starved.

It has been variously suggested that northern source areas supplied detritus to the Alpine-Carpathian Flysch trench (e.g. Dzulynski *et al.* 1959; Freimoser 1973, a.o.). In the West Carpathians submarine canyons have been detected in drill holes beneath the Carpathian Flysch nappes (Picha 1974); these undoubtedly supplied clastic sediment southward. But has it ever been proved that this material reached the trench? There is evidence it may not have. Why, for example did none of the coarse ultramafic and mafic detritus observed in the Ultrahelveticum in Bavaria (Dietrich & Franz 1976) find its way into the Flysch trench? Its occurrence suggests exposure of (at least) slivers of oceanic crust north of the flysch trench, but this distinctive material does not appear in the trench fill sediments. Where some chromspinel appears in the Vienna Woods Flysch (Faupl 1975), it can probably be derived from a submarine ridge within the Flysch belt itself or may even have a southern source, despite the palaeocurrent directions which point southward (as discussed elsewhere).

Correlation of the Gault Formation between the Falknis and Tasna Nappes and the Flysch Zone established for the Gault Formation places the Flysch Zone in a palaeogeographic position north of the Middle Penninic Falknis and Sulzfluh Nappes and their western equivalent, the Briançonnais Platform (Trümpy & Haccard 1969), but south of the North Penninic Valais Trough, which at its western end (near Bourg St Marucie in the French Savois Alps) comprises an incomplete Late Cretaceous ophiolite sequence (Antoine 1971). Also at the eastern exposed end of this trough in the Lower Engadine Window ophiolites occur between the Bündner Schiefer complex (North Penninic Valais Trough) and the Tasna Nappe (Middle Penninic). Thus the Valais Trough to the north of the Flysch trench appears to have been floored, at least in part, by oceanic crust. Realizing that the entire Prätigau Flysch is part of this trough, 50 km is an absolute minimum for the width of the area between the Flysch trench and the southern continental margin of Europe in the Rhine Valley region. Further east, i.e. east of the Engadine Window, no equivalents of the North Penninic Valais Trough are known. It is possible that the postulated oceanic area between the southern European continental margin and the East Alpine-Carpathian deep-sea trench, of which almost all traces have been lost east of the Engadine Window (except the above-mentioned pebbles and boulders in the Ultrahelveticum), may have been wider in the Alpine sector than in the Carpathian sector, because the Carpathian flysch belt is several times as wide as the East Alpine flysch belt, perhaps at the expense of this postulated oceanic area north of the trench—but this is mere speculation.

South of the Cretaceous-Palaeogene flysch trench a zone of islands and shallow platforms existed which includes rocks of the Sulzfluh-Nappe, possibly the Jurassic Hochstegen Facies in the Tauern Window (Tollmann 1964) and the Briançonnais Platform. Those parts of the platform that are preserved today seemed to be submerged in Early Cretaceous time. However, the abundance of detrital quartz (and felspar) in the Gault Formation indicates that the granitic-gneissic basement of the platform was exposed somewhere. The exposure of silicic basement of the Middle Penninic platform is already visible in the coarse granitic detritus of the Upper Jurassic breccia formations of the Falknis Nappe.

South of this Middle Penninic belt of platforms and islands that were floored by continental crust the South Penninic Piemontais trough (or Schistes lustrés trough) followed: this is the main oceanic realm of the Alpine Tethys. This oceanic area was successively being eliminated from the surface by southward subduction at the South Alpine-Austro-Alpine/South Penninic convergent plate boundary which acted as the main subduction zone in Cretaceous time. It has long been argued that closure of the South Penninic oceanic area in the East Alps was complete by late Campanian time when chromspinel disappears as a heavy mineral from the turbidite sands in the Gosau

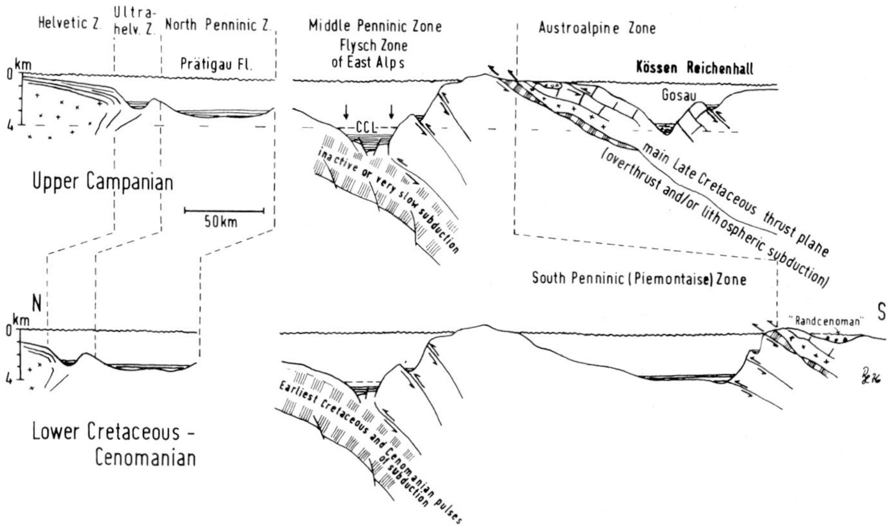

FIG. 12. Hypothetical palinspastic cross-sections (north-south) through the northern Tethys in the western East Alps for Early Cretaceous-Cenomanian and late Campanian times (from Hesse & Butt 1976). Depth ranking of the Kössen and Reichenhall Gosau Basins, the Flysch Zone of the East Alps, and the Ultrahelvetic flysch relative to the CCL based on chemical and mircopalaeontological data. No data were available for the North Penninic Prätigau Flysch. Horizontal distances are absolute minimum figures. Note vertical exaggeration. Continental crust: crosses. Oceanic crust: vertical banding. Mid-slope basins on the southern wall of the flysch trench may be represented by the Triesen Flysch of Leichtenstein, the Tratenbach Beds of Bavaria or perhaps the Laaber Nappe of the Vienna Woods.

Basins (e.g. Tollmann 1978), but—as mentioned—chromspinel is still present in Upper Campanian and Maestrichtian–Danian strata showing that ultramafic rocks, although of unknown origin, were still exposed at that time.

At the same time that subduction proceeded in the South Penninic realm a second plate boundary marked by the flysch trench existed between the North Penninic and the Middle Penninic realm. It must have been generated as a topographic feature by an initial phase of subduction in latest Jurassic-earliest Cretaceous time. During most of the Cretaceous and Palaeogene, however, it acted as a transform fault zone, to become the site of subduction again only after mid-Eocene time.

In Cretaceous time the Alpine Tethys is thus encompassed between two more or less east-west trending plate boundaries (not one, as Dietrich 1976a, b suggested) which enclosed a small plate with both continental (Middle Penninic) and oceanic (South Penninic) crustal domains between them. During this time the northern boundary acted as a transform fault, the southern boundary as a subduction zone. This plate may have extended eastward as far as the western boundary of the Apuseni Mountains or Translyvanian Basin, to include major parts of the predecessor of the present Pannonian Basin before emplacement of the Križna, Veporid, Choč and Gemerid Nappes. By the end of the Cretaceous this former Penninic-plate *sensu lato* had been eliminated from the surface and the Austro-Alpine/South Alpine realm had been juxtaposed with the southern upper slope (or forearc) of the trench (Fig. 12).

The Alpine-Carpathian Arc—a Tethyan analogue of the modern West Atlantic loop arcs in the Caribbean and South Scotia Seas

If deep-sea trenches are anchored in the asthenosphere to some extent by their subduction zones (Kaula 1975; Tullis 1972)—then the Alpine-Carpathian Arc might represent—in a broad sense—an analogue of the present-day Caribbean or South Scotia Arcs. In these modern loop arcs active volcanism is (more or less) restricted to the eastward convex, west dipping subduction zone complexes (Lesser Antilles, South Sandwich Islands) whereas the flanks of the arcs appear to be large transform fault zones marked by islands and, notably in the case of the Caribbean, by the Puerto Rico deep-sea trench which has been inactive as far as subduction is concerned probably at least since the beginning of the Miocene. At these flanks motion is largely strike-slip between the westward moving American plates (Molnar & Sykes 1969) and the relatively stationary Caribbean (Jordan 1975) and (?) South Sandwich plates (Forsyth 1975). On the southern flanks of both arcs slow subduction seems to occur, however, as evidenced by the Tertiary to Recent calc-alkaline volcanism in the South Shetland Islands and structural deformation in the South Caribbean Basin (Case 1974).

The analogy implies strike-slip motion between Europe and the East Alpine-West Carpathian deep-sea trench, which would be sinistral because in Cretaceous time the Eurasian plate can hardly have moved eastward at a time when opening in the North Atlantic had not yet started. Eastward (absolute) plate motion of the Eurasian plate in the Tertiary is also unlikely because of the constraints placed on it by the opening of the marginal seas in the West Pacific and the present-day motions of the plates adjacent to West Pacific subduction zones (Chase 1978). The Penninic-plate *sensu lato* on the other hand would be 'sheltered' behind the East Carpathian subduction zone and would therefore be relatively stationary.

The analogy is of course a very general one. In particular the former southern plate boundary is problematical because it is now juxtaposed with the northern boundary due to the almost complete elimination of the intervening plate, which except in the West Alps, is only exposed in tectonic windows. The evolution of this southern boundary in the Alps was discussed by Trümpy (1975), and in the Eastern European Alpine system by Burchfiel (1980). Laubscher (1971) envisaged 300 km of right lateral movement which would have occurred along a transform fault zone in the area of the southern plate boundary. Envisaging the Cretaceous palaeogeography of the Alpine-Carpathian region in terms of a reasonable modern analogue may not solve all the problems but it may help to spur our imagination and provide some constraints for a more realistic picture. The geodynamic link between the three loop-arc structures could be the westward drift of major lithospheric plates between which smaller plates remain relatively stationary because they are trapped behind westward dipping subduction zones that are anchored in the asthenosphere.

References

Aberer, F. & Braumüler, E. 1958. Über Helvetikum und Flysch in Raum nördlich Salzburg. *Mitt. Geol. Ges. Wien*, **49,** 1–59.

Allemann, F. 1956. Geologie des Fürstentums Liechtenstein. III. Teil-*Selbstverl. Hist. Ver. Fürstentum Liechtenstein, Vaduz*, 224 pp.

Antoine, P. 1971. *La zone de Brèches de Tarentaise entre Bourg-Saint Maurice (Vallé de l'Isère) et la frontière italo-suisse*. These, Univ. Grenoble. 367 pp.

Bleahu, M., Boccaletti, M., Manetti & Peltz, S. 1973. Neogene Carpathian arc: a continental arc displaying the features of an 'island arc'. *J. geophys. Res.* **78,** 5025–32.

Birkenmajer, K. 1970. Pre-Eocene fold structures in the Pieniny Klippen Belt (Carpathians) of Poland. *Stud. geol. Poln.* **31,** 1–77.

—— 1976. The Carpathian orogen and plate tectonics. *Publ. Inst. Geophys. Pol. Acad. Sci. A-2*, **101,** 43–53.

Burchfiel, B. C. 1980. Eastern European Alpine system and the Carpathian orocline as an example of collision tectonics. *Tectonophysics*, **63,** 31–61.

Butt, A. A. 1977. *A comparative study of the depositional environments of Upper Cretaceous rocks in the north part of the Eastern Alps*. Doctorate Thesis. Univ. Tübingen, Germany. 129 pp (unpubl.).

Caron, C., Hesse, R. & Kerckhove, C. 1979. Alpine helminthoid flysch sequences. *Intn. Geol. Correl. Proj. 105: Continental margins in the Alps*. 4th workshop. Field guide. Fribourg, Switzerland. 86 pp.

Caron, C., Hesse, R., Kerckhove, C. van Stuijvenberg, J. & Tassé, N. 1981. Comparison preliminaire des flysches à helminthoides sur trois transversales des Alpes. *Eclog. geol. Helv.* **74** (in press).

Case, J. E. 1974. Major basins along the continental margin of northern South America. *In:* Burk, C. A. & Drake, C. L. (eds). *The Geology of Continental Margins*, 733–41. Springer-Verlag, New York.

Chase, C. G. 1978. Extension behind island arcs and motions relative to hot spots. *J. geophys. Res.* **83,** 5385–7.

Chough, S. K. & Hesse, R. 1980. The Northwest Atlantic mid-ocean channel of the Labrador Sea: III. Head spill vs. body spill deposits from turbidity currents on natural levees. *J. sediment. Petrol.* **50,** 227–34.

CONNOLLY, J. R. & EWING, M. 1967. Sedimentation in the Puerto Rico Trench. *J. sediment. Petrol.* **37,** 44–59.

CONTESCU, L. R. 1974. Geologic history and palaeogeography of Eastern Carpathians: Example of Alpine geosynclinal evolution. *Bull. Am. Assoc. Petrol. Geol.* **58,** 2436–76.

DEWEY, J. F., PITMAN, W. C., III, RYAN, W. B. F. & BONNIN, J. 1973. Plate tectonics and evolution of the Alpine system. *Bull. geol. Soc. Am.* **84,** 3137–80.

DIETRICH, V. J. 1976a. Evolution of the Eastern Alps: a plate tectonics working hypothesis. *Geology,* **4,** 147–52.

—— 1976b. Plattentektonik in den Ostalpen—eine Arbeitshypothese. *Geotekton. Forsch. Stuttgart,* **50,** 1–84.

DIETRICH, V. & FRANZ, U. 1976. Ophiolith-Detritus in den santonen Gosau-Schichten (Nördliche Kalkalpen). *Geotekton. Forsch. Stuttgart,* **50,** 85–109.

DZULYNSKI, S., KSIAZKIEWICZ, M. & KUENEN, PH.H. 1959. Turbidites in flysch of the Polish Carpathians. *Bull. geol. Soc. Am.* **70,** 1089–118.

EWING, J. I. & EWING, M. 1962. Reflection profiling in and around the Puerto Rico Trench. *J. geophys. Res.* **67,** 4729–39.

EWING, M. & HEEZEN B. C. 1955. Puerto Rico Trench geophysical data. *In:* POLDERVAART, A. (ed.). *Crust of the Earth.* Spec. Pap. geol. Soc. Am. **62,** 255–68.

FAUPL, P. 1975. Schwermineralien und Strömungsrichtungen aus den Kaumberger Schichten (Oberkreide) des Wienerwald-Flysches, Niederösterreich. *Neves Jahrb. Geol. Paleontol. Montshofte.* **9,** 528–40.

—— 1976. Vorkommen und Bedeutung roter Pelite in den Kaumberger Schichten (Oberkreide) des Wienerwald-Flysches. Niederösterreich. *Neves Jahrb, Geol. Paleontol. Montshofte.* **8,** 449–70.

—— 1978. Zur räumlichen und zeitlichen Entwicklung von Breccien-und Turbiditserien in den Ostalpen. *Mitt. Ges. Geol. Bergbaustud. Österr.* **25,** 81–110.

—— , GRÜN, W., LAUER, G., PAPP, A., SCHNABEL, W. & STURM, M. 1970. Zur Typisierung der Sieveringer Schichten im Flysch des Wiener Waldes. *Jahrb. geol. B.-A. Wien,* **113,** 73–158.

FORRER, M. 1946. Über den Flysch der östlichen Wildhausermulde. (Vorläufige Mitteilung). *Eclog. geol. Helv.* **39,** 132–40.

FORSYTH, D. W. 1975. Fault plane solutions and tectonics of the South Atlantic and Scotia Sea. *J. geophys. Res.* **80,** 1429–43.

FREIMOSER, M. 1973. Zur Stratigraphie, Sedimentpetrographie und Faziesentwicklung der Südostbayerischen Flyschzone und des Ultrahelvetikums zwischen Bergen/Obb. und Salzburg. *Geologica Bavar.* **66,** 7–91.

GRIGGS, G. B., CAREY, A. G., JR. & KULM, L. D. 1969. Deep-sea sedimentation and sediment-fauna interaction in Cascadia Channel and on Cascadia Abyssal Plain. *Deep Sea Res.* **16,** 157–70.

GROSSHEIM, V. A. 1961. The feasiblity of long distance, horizon-by-horizon correlation of flysch sections ('teleconnection'). *Izvestiya Akad. Nauk. SSSR,* (1962), 41–8.

GRÜN, W., KITTLER, G., LAUER, G., PAPP, A. & SCHNABEL, W. 1972. Studien in der Unterkreide des Wienerwaldes. *Jahrb. Geol. B.-A. Wien.* **115,** 103–85.

GWINNER, M. P. 1972. *Geologie der Alpen.* Schweizerbart, Stuttgart. 477 pp.

HESSE, R. 1973a. Flysch-Gault und Falknis-Tasna-Gault (Unterkreide): Schrittweiser Ubergang von der distalen zur proximalen Flyschfazies auf einer penninischen Trogebene der Alpen. *Geologica et Palaeontologica, Sonderband,* **2,** 55 pp., 12 plates. Elwert-Verlag, Marburg.

—— 1973b. Lithostratigraphie, Petrographie und Entstehungsbedingungen des bayerischen Flysches: Unterkreide. *Geologica Bav.* **66,** 148–222.

—— 1974. Long-distance continuity of turbidites: Possible evidence for an Early Cretaceous trench abyssal plain in the East Alps. *Bull. geol. Soc. Am.* **85,** 859–70.

—— 1975. Turbiditic and non-turbiditic mudstone of Cretaceous flysch sections of the East Alps and other turbidite basins. *Sedimentology,* **22,** 387–416

—— 1979. Bavarian Alps (Trauchgau-Ammergau). *In:* CARON, C., HESSE, R. & KERCKHOVE, C. (eds). *Alpine helminthoid flysch sequences. Intn. Geol. Correl. Proj. 105.* Continental margins in the Alps. 4th workshop, field guide, 65–86, Fribourg, Switzerland.

—— , & BUTT, A. A. 1976. Paleobathymetry of Cretaceous turbidite basins of the East Alps relative to the calcite compensation level. *J. Geol.* **84,** 505–33.

HSÜ, J. K. 1972. Alpine flysch in a Mediterranean setting. *Repts 24th Int. Geol. Congr.* **6,** 67–74, Montreal.

HUENE, R. VON 1974. Modern trench sediments. *In:* BURK, C. A. & DRAKE, C. L. (eds). *The Geology of Continental Margins,* 207–11, Springer-Verlag, New York.

—— , AUBOUIN, J., AZEMA, J., BLACKINTON, G., CARTER, J. A., COULBOURN, W. T., COWAN, D. S., CURIALE, J. A., DENGO, C. A., FAAS, R. W., HARRISON, W., HESSE, R., HUSSONG, D. M., LADD, J. W., MUZYLOV, N., SHIKI, T., THOMPSON, P. R. & WESTBERG, J. 1980. Leg 67: The Deep Sea Drilling Project Mid-America Trench transect off Guatemala. *Bull. geol. Soc. Am.* **91,** 421–32.

JORDAN, T. H. 1975. The present-day motions of the Caribbean Plate. *J. geophys. Res.* **80,** 4433–9.

KAULA, W. M. 1975. Absolute plate motions by boundary velocity minimizations. *J. geophys. Res.* **80,** 244–8.

KSIAZKIEWICZ, M. 1956. Geology of the Northern Carpathians. *Geol. Rdsch.* **45,** 369–411.

LAJOIE, J. 1979. Origin of megarhythms in flysch sequences of the Québec Appalachians. *Can. J. Earth Sci.* **16,** 1518–23.

LAUBSCHER, H. P. 1971. Das Alpen-Dinariden-

Problem und die Palinspastik der südlichen Tethys. *Geol. Rdsch.* **60,** 813–33.

LEMCKE, K. 1970. Epirogenetische Tendenzen im Untergrund und in der Füllung des Molassebeckens nördlich der Alpen. *Bull. Ver. Schweiz. Petrol.-Geol. u. Ing.* **37,** 25–34.

MENARD, H. W. 1964. *Marine Geology of the Pacific.* McGraw-Hill, New York. 271 pp.

MOLNAR, P. & SYKES, L. R. 1969. Tectonics of the Caribbean and Middle America regions from focal mechanisms and seismicity. *Bull. geol. Soc. Am.* **80,** 1639–84.

MUTTI, E. & RICCI-LUCCHI, F. 1975. Turbidite facies and facies associations. *In:* MUTTI, E. *et al.* (eds). *Examples of Turbidite Facies and Facies Associations from Selected Formations of the Northern Appenines.* 9th Int. Sedimentol. Congress Nice, Field trip Guide All, 21–36.

NELSON, C. H. & NILSEN, T. H. 1974. Depositional trends of modern and ancient deep-sea fans. *In:* DOTT, R. H. JR. & SHAVER, R. H. (eds). *Modern and Ancient Geosynclinal Sedimentation.* Spec. Publ. Soc. econ. Palaeontol. Mineral. Tulsa, **19,** 69–91.

NIEDERMAYR, G. v. 1966. Beiträge zur Sedimentpetrographie des Wienerwald-Flysches. *Verh. geol. Bundesanst.* 1/2, 106–41.

OBERHAUSER, R. 1968. Beiträge zur Kenntnis der Tektonik und Paläogeographie während der Oberkreide und des Paläogens im Ostalpenraum. *Jahrb. geol. Bundesanst.* **111,** 115–45.

OBERHAUSER, R. (ed.) 1980. *Der Geologische Aufbau Österreichs.* Springer-Verlag, Berlin. 700 pp.

PICHA, F. 1974. Ancient submarine canyons of the Carpathian miogeosyncline. *In:* DOTT, R. H. JR & SHAVER, R. H. (eds). *Modern and Ancient Geosynclinal Sedimentation.* Spec. Publ. Soc. econ. Paleontol. Mineral. Tulsa, **19,** 126–7.

PILKEY, O. H., LOCKER, S. D. & CLEARY, W. J. 1980. Comparison of sand-layer geometry on flat floors of 10 modern depositional basins. *Bull. Am. Assoc. Petrol. Geol.* **64,** 841–56.

PIPER, D. J. W., VON HUENE, R. & DUNCAN, J. R. 1973. Late Quaternary sedimentation in the active eastern Aleutian trench. *Geology*, **1,** 19–22.

PREY, S. 1960. Gedanken über Flysch und Klippenzonen in Österreich anläßlich einer Exkursion in die polnischen Karpathen. *Verh. geol. Bundesanst.* 197–214.

PREY, S. 1965. Vergleichende Betrachtungen über Westkarpaten und Ostalpen im Anschluß an Exkursionen in die Westkarpaten. *Verh. geol. Bundesanst.* 1/2, 69–107.

RAD, U. VON 1964. *Mineralbestand und Ablagerungsbedingungen der Flyschsedimente im Allgäu.* Doctoral Thesis, Technical Univ. München. 131 pp. (plus appendix) (unpubl.).

—— , 1973. Zur Sedimentologie und Fazies des Allgäuer Flysches. *Geologica Bav.* **66,** 92–147.

RICCI-LUCCHI, F. 1975a. Depositional cycles in two turbidite formations of northern Apennines (Italy). *J. sediment. Petrol.* **45,** 3–43.

—— , 1975b. Miocene paleogeography and basin analysis in the Periadriatic Apennines. *In:* SQUYRES, C. (ed.). *Geology of Italy,* 5–11. Petrol. Explor. Soc. Libya.

—— & VALMORI, E. 1980. Basin-wide turbidites in a Miocene, over-supplied deep-sea plain: a geometrical analysis. *Sedimentology,* **27,** 241–70.

RICHTER, M. 1933. Alter and Stellung der südbayerischen Flyschzone. *Centralbl. Mineral. Geol. Palaeontol,* 496–508.

SANDULESCU, M. 1973. Essai de reconstitution des éléments preparoxismaux alpins des Dacides (internides) orientales. *Rev. Roumain Géol. Géophys., Géograph, Ser. Géolog.* **17,** 145–56.

STEFANESCU, M. 1980. Time of flysch deposition in the Eastern Carpathians. *Ann. Inst. Géol. Géophys.* **56,** 133–42.

STEGENA, L., GECZY, B. & HORVATH, F. 1975. Late Cenozoic evolution of the Pannonian basin. *Tectonophysics,* **26,** 71–90.

STEPHAN, W. & HESSE, R. 1966. *Erläuterungen zur Geologischen Karte von Bayern 1:25,000, Blatt Nr. 8236 Tegernsee.* Bayerisches Geologisches Landesamt München, 304 pp.

TOLLMANN, A. 1964. Zur Fortsetzung des Brianconnais in den Ostalpen. *Mitt. geol. Ges. Wien,* **57,** 469–78.

—— , 1969. Die tektonische Gliederung des Alpen-Karpaten-Bogens. *Geologie*, **18,** 1131–55.

—— , 1978. Plattentektonische Fragen in den Ostalpen und der plattentektonische Mechanismus des mediterranen Orogens. *Mitt. österr. geol. Ges.* **69,** 291–351.

TRÜMPY, R. 1975. Penninic-Austroalpine boundary in the Swiss Alps: a presumed former continental margin and its problems. *Am. J. Sci.* **275A,** 209–38.

TRÜMPY & HACCARD, D. 1969. Réunion extraordinaire de la Société Géologique de France: Les Grisons. 14 au 21 Septembre 1969. *C. r. Somm. Séances Soc. geol. France*, **9,** 329–96.

TULLIS, T. E. 1972. Evidence that lithospheric plates act as anchors. *Eos,* **53,** 522.

WALKER, R. G. 1978. Deep-water sandstone facies and ancient submarine fans: Models for exploration for stratigraphic traps. *Bull. Am. Assoc. Petrol. Geol.* **62,** 933–66.

WATTS, A. B. & STECKLER, M. S. 1979. Subsidence and eustasy at the continental margin of eastern North America. *In:* TALWANI, M., HAY, W. & RYAN, W. B. F. (eds). *Deep Drilling Results in the Atlantic Ocean: continental margins and paleoenvironment.* Am. geophys. Union, M. Ewing, Ser. **3,** 218–34.

WEZEL, F. C. 1973. Diacronismo degli eventi geologici Oligo-Miocenici nelle Maghrebidi. *Riv. Mineral. Sicil.* **24,** 219–32.

WIESENEDER, H. 1967. Zur Petrologie der ostalpinen Flyschzone. *Geol. Rdsch.* **56,** 227–40.

WOLETZ, G. 1967. Schwermineralvergesellschaftungen aus ostalpinen Sedimentationsbecken der Kreidezeit. *Geol. Rdsch.* **56,** 308–20.

REINHARD HESSE, McGill University, Department of Geological sciences, 3450 University Street, Montreal, Quebec H3A 2A7, Canada and Technische Universität, München, Federal Republic of Germany.

The anatomy of a Lower Palaeozoic accretionary forearc: the Southern Uplands of Scotland

J. K. Leggett, W. S. McKerrow & D. M. Casey

SUMMARY: The Southern Uplands of Scotland consists of an accretionary prism which had a prolonged (over 45 Ma) development off the SE coast of ancient North America (Laurentia) during the late Ordovician and Silurian. To the north, a forearc basin occupied the Midland Valley. An arc massif of older metamorphic rocks in the Grampian Highlands, and capping calc-alkaline volcanics, supplied much of the sediment to the trench. In its early development (late Ordovician) the accretionary complex incorporated slivers of ocean-floor material as well as thick turbidites into a steep lower trench slope. Later, in the Silurian, a trench slope break emerged shedding sediment north into the Midland Valley forearc basin. By that time only turbidites and black shales were being accreted. The Southern Uplands is dominated by coherent strata, and, despite intense imbrication, is devoid of mélange. We conclude that the accretionary prism was the result of high sediment input to the trench and very slow, oblique, subduction of the Iapetus Ocean eastwards below Scotland. Décollement surfaces developed during accretion preferentially utilised a highly incompetent black shale layer near the base of the subducting sequence. Prior to the inception of accretion (in the Caradoc), we postulate sediment subduction, and possibly subduction erosion, during the Cambrian and early Ordovician.

The Southern Uplands is a *c*. 12,000 km^2 area of southern Scotland, comprising imbricated and folded Ordovician and Silurian rocks of deep-water origin. Recent reconstructions suggest that the area, and its continuation in Ireland, is a fossil subduction complex preserved on the southern margin of Laurentia (Early Palaeozoic North America) which resulted from accretion above an oceanic plate subducting to the NW (Mitchell & McKerrow 1975; McKerrow *et al*. 1977; Leggett *et al*. 1979a, b; Weir 1979). In this paper we reconstruct the sedimentary and tectonic history of this fossil forearc from Lower Palaeozoic regional geology; we then review the evidence for an accretionary origin for the Southern Uplands, extending our comparisons with modern accretionary forearcs more than we have hitherto.

The reconstructed Northern Iapetus active margin in the British Isles (mid-Ordovician to Early Devonian)

Deformation and metamorphism within the Appalachian-Caledonian orogen are attributed by most geologists to tectonic events associated with the closure of an Early Palaeozoic ocean, the Iapetus (e.g. numerous references in Harris *et al*. 1979). Evidence for the existence of this ocean comes from faunal provinciality, and the distribution of arc-type igneous rocks and ocean floor-type rocks (such as ophiolites) along the orogen. The suture along which the ocean closed is recognised chiefly by contrasts on either side in the faunas, volcanic polarities, structural styles, sedimentary histories and basement characteristics. In the British Isles the suture runs SW to NE from the Shannon estuary in western Ireland through the Solway Firth south of the Southern Uplands, probably curving north under the present North Sea to run between Norway and Shetland (Phillips *et al*. 1976; Thirlwall 1981a). Early Palaeozoic Britain NW of this line formed part of Laurentia (Fig. 1). Differences between individual transects across the Appalachian-Caledonian orogen suggest that the continental margins of the Iapetus were irregular and that individual portion of the northwestern Iapetus margin. Dewey & Kidd 1974). In Fig. 1 we show an interpretation of the late Caradoc palaeogeography of the New England to Greenland portion of the north-western Iapetus margin. Evidence of a magmatic arc in Scotland indicates that subduction was occurring during (also before and after) the Ordovician under the British sector of the Laurentian margin (Phillips *et al*. 1976)). There is no evidence for subduction under the Appalachian or Greenland sectors (e.g. Henriksen & Higgins 1976; Williams 1978). The British sector may have been separated from these other sectors by transform faults (Fig. 1; McKerrow, in press).

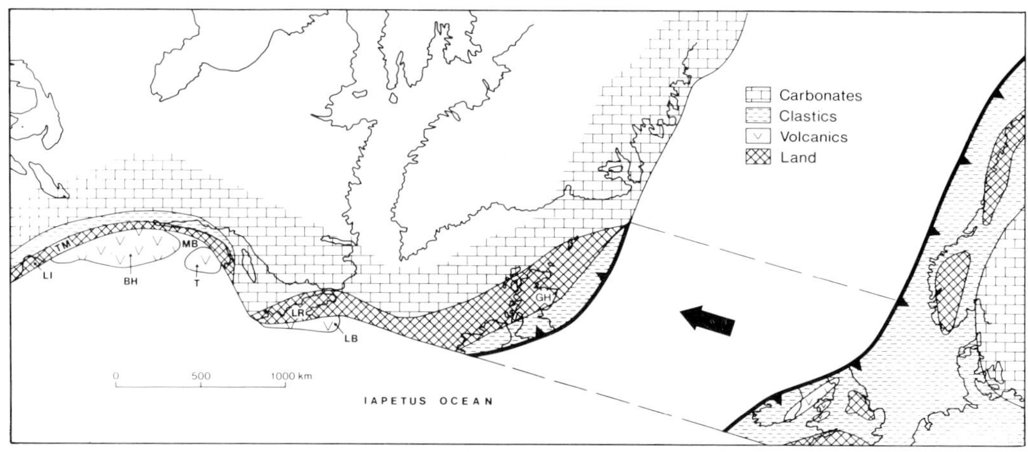

FIG. 1. Possible configurations of the southeastern Laurentian margin from Long Island (LI) to Greenland in late Ordovician times (late Caradoc) after McKerrow (in press). The trench along the NW subducting British sector is shown with a thick line, and is probably separated from the inactive Greenland and North American sectors by transform faults. Carbonates (including shallow marine siliciclastic strata) demark the stable Laurentian shelf. Land areas along the margin represent deformed terranes produced during the Taconic and Grampian orogenies. It has been suggested that Taconic deformation was produced in the mid-Ordovocian by impingement of east-facing intra-oceanic arcs (BH: Bronson Hill, T: Tetagouche, LB: Lushs Bight) on the Laurentian margin (McKerrow in press, and references therein). Grampian deformation may also have been produced by impingement of a volcanic arc on the Laurentian margin, in latest Cambrian times (McKerrow in press, and see section on pre-middle Ordovician history of British sector). Configuration of southeastern Iapetus margin also shown; scale does not refer to width of the ocean, which is unknown. TM: Tactonic Mountains, MB: Matepedia Basin, LR: Long Range, Newfoundland, GH: Grampian Highlands.

FIG. 2. Caledonian (pre-Early Devonian) geological provinces of northern British Isles, and interpretation in terms of northwards subduction from line of Iapetus suture. Outcrops too small to show ornament clearly are generalized. Major faults are dashed where continuation is inferred.
Accretionary complex: NB: Northern Belt, CB: Central Belt, SB: Southern Belt. Faults: KF: Kingledores Fault, OBF: Orlock Bridge Fault, SGSZ: Slieve Glah shear zone. Places: SA: Slieve Aughty; SBh: Slieve Bernagh; SE: Shannon estuary.
Upper slope/forearc basin: basement only exposed in the SW Midland Valley, consisting of the Arenig Ballantrae ophiolite complex (BO), which is probably allochthonous on Precambrian crystalline crystalline basement (see text). A slope/shelf succession (Llanvirn to late Silurian) overlies the ophiolite uncomfortably; this succession and inliers to the NE (containing Llandovery to Early Devonian strata) record the history of a forearc basin formed behind the growing accretionary complex south of the Southern Upland Fault.
Volcanic arc: dashed ornament is mid-Ordovician basalt, andesite and rhyolite of the Tyrone igneous complex (T) and Charlestown inlier (C); more diffuse dashed ornament is the South Mayo Trough, a thick Ordovician-Silurian sequence including early Ordovician basalts, interpreted as an intra-arc rift zone (Ryan & Archer 1977). Solid ornament is Early Devonian (?in part late Silurian–Thirlwall 1981a) andesite. OH: Ochil Hills; OF: Ochil Fault. Possible subduction-related granitoid intrusions in metamorphic frontal arc/arc massif not shown.
Frontal arc/arc massif: most rocks herein were deformed prior to the accretionary phase in latest Cambrian; structural grain shown (highly illustrative). Relation of Great Glen Fault, Leannan Fault (LnF) and Leck Fault (LkF) and of Moine Thrust and Loch Skerrols Thrust (LST), from Johnson *et al.* (1979). Note position of Connemara Dalradian (CD) with respect to putative continuation of Southern Upland Fault, and possible equivalents of basalts, cherts, slates and greywackes of the South Connemara Group (SCG) to Ordovician of Northern Belt of Southern Uplands. For discussion of ophiolitic slivers of the Highland Boundary Complex (HBC) see text.

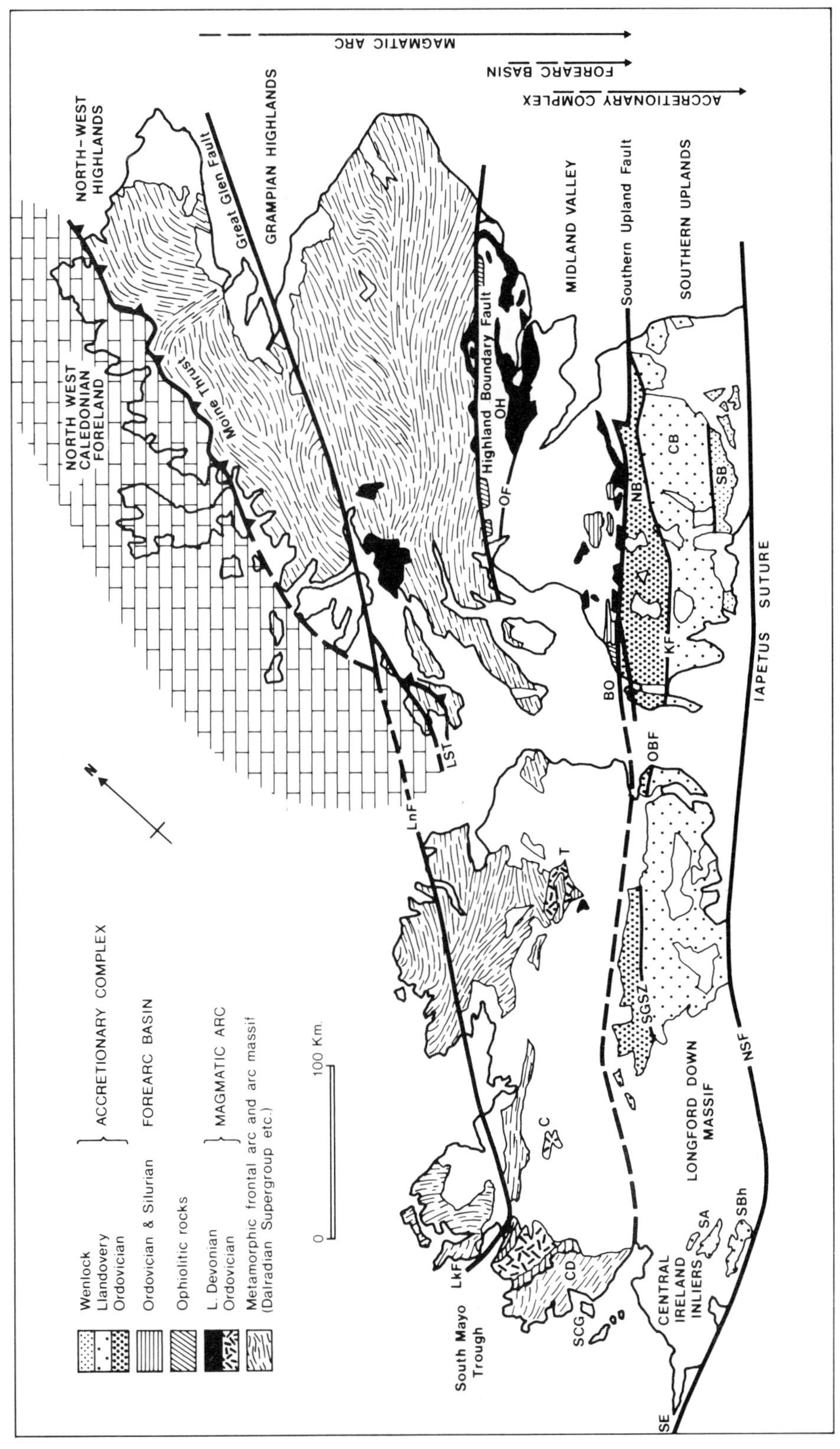
MAGMATIC ARC
FOREARC BASIN
ACCRETIONARY COMPLEX
NORTH-WEST HIGHLANDS
GRAMPIAN HIGHLANDS
Great Glen Fault
MIDLAND VALLEY
Southern Upland Fault
SOUTHERN UPLANDS
NORTH WEST CALEDONIAN FORELAND
Moine Thrust
Highland Boundary Fault
OH
OF
NB
CB
SB
KF
BO
IAPETUS SUTURE
LST
LnF
OBF
N
T
SGSZ
C
NSF
LONGFORD DOWN MASSIF
SBh
SA
CENTRAL IRELAND INLIERS
CD
SCG
LkF
South Mayo Trough
SE
Wenlock
Llandovery
Ordovician
ACCRETIONARY COMPLEX
Ordovician & Silurian
FOREARC BASIN
Ophiolitic rocks
L. Devonian
Ordovician
MAGMATIC ARC
Metamorphic frontal arc and arc massif (Dalradian Supergroup etc.)
0
100 Km.

FIRTH OF FORTH
FIRTH OF CLYDE
MIDLAND VALLEY
SOLWAY FIRTH
Granites and Granodiorites
Upper Palaeozoic and Triassic
Upper Slope / Forearc Basin
Accretionary Complex
Ballantrae Ophiolite Sequence
Lower Palaeozoic
0 5 10 15 km
0 5 10 miles

Since Dewey's (1969, 1971) formative plate tectonic reconstruction numerous authors have supported northwards subduction under Scotland (Mitchell & McKerrow 1975 and references therein), and we can divide northern Britain into structural regions (Fig. 2) using the terminology of Dickinson & Seely (1979). An accretionary complex (mid-Ordovician-late Silurian) occupies the Southern Uplands—Longford Down Massif—Central Ireland Inliers. A forearc basin (late Ordovician—Early Devonian) occupies the southern part of the Midland Valley. A metamorphic frontal arc/arc massif (pre-early Ordovician) occurs in the Grampian Highlands and analogous areas in Ireland. A magmatic arc is represented by volcanic rocks in South Mayo (early Ordovician), County Tyrone (Caradoc), the Midland Valley and parts of the Grampians (?late Silurian and Early Devonian), and by certain contemporaneous granitoids in the Grampian Highlands (early Ordovician to Early Devonian). The British portion of the Appalachian-Caledonian orogen should be viewed as just one transect across the Laurentian margin; even within the small area of the British Isles some geological provinces are discontinuous (e.g. South Mayo Trough and Midland Valley forearc basin, Fig. 2).

Accretionary complex

Interpretation of the Southern Uplands as an accretionary complex (Leggett *et al.* 1979a)

FIG. 3. Geological map of the Southern Uplands, updated from Leggett *et al.* (1979a, b).
Major faults: shown as thick lines where mapped (sources below), dashed where continuations are inferred from Geological Survey Maps, dashed with a question mark where continuation of strike fault is uncertain or where a wrench fault has been deduced from displacements in outcrops of basalt-chert-graptolitic shale (basal lithologies). Black areas are imbricate zones of basal lithologies; almost all of these come from mapped areas in source references below, but some inferred from belts of outcrop of basal lithologies on Geological Survey maps.
Stratigraphic sequences shown in Fig. 4 are located by numbers 1 to 10 and I to X. Two traverses are shown, across the central Southern Uplands and the west coast. Correlative sequences shown using roman numerals on west coast; these can be used to trace some of the major reverse faults between the two traverses. For over 80 years the Southern Uplands has been divided into three belts (Peach & Horne 1899). The Northern Belt corresponds to our tracts 1, 2, and 3; the Central Belt to tracts 4–8; and the Southern Belt to tracts 9–10.
Structural profiles shown in Figs 8–11 are located by double lines and bold letters. Composite profiles are labelled at both ends.
Place names (towns and villages shown with an asterisk): Af Glen Afton dam, B Biggar, Ba Ballantrae, BH Bennane Head, BHd Burrow Head, C. Coulter, CB Clanyard Bay, CI Craighead Inlier, Cm Crossmichael, CmI Carmichael Inlier, Co B Coldingham Bay, CP Corsewall Point, Cr Craigmichan, D. Dumfries, DB Drumbreddan Bay, DL Dobbs Linn, E. Ettrickbridgend, Ed Eddleston, EI Eastfield Inlier, Ey Eyemouth, FB Float Bay, G Girvan, Gk Glenkiln, Gl Glenluce, GoF Gatehouse of Fleet, GPt Gipsy Point, H Hawick, Ha Hartfell, HHI Hagshaw Hills Inlier, I Innerleithen, K Kirkcudbright, L Leadburn, La Lauder, LH Lammermuir Hills, Lm Langholm, Lw Lesmahagow, LwI Lesmahagow Inlier, M. Moffat, MB Morroch Bay, Me Melrose, MG Mull of Galloway, MH Moorfoot Hills, Mo Moniaive, N Noblehouse, ND Nithsdale, NS Newton Stewart, P Portayew, Pe Peebles, PhI Pentland Hills Inlier, Pp Portpatrick, Ps Portslogan, R. Riccarton Junction, S Stobo, SP Siccar Point, Sq Sanquhar, Sn Straiton, St Stranraer, SV Stinchar Valley, T Foot of Talla Water (Tweedsmuir, Talla region), Th Thornhill, W Whithorn, WH Wrae Hill.
Faults: BF Bush Fault, BlF Berryfell Fault, BL Blairbuies Line, CF Carcow Fault, CPF Carmichael-Pentland Fault, CT Culcronchie Thrust, DCF Dove Cove Fault, DF Drumblair Fault, EVF Ettrick Valley Fault, FL Fardingmullach Line, GAF Glen App Fault, GBF Gillespie Burn Fault, GF Grassfield Fault, GRSB Glan Reif Shatter Belt, HBF Harwood Burn Fault, HF Hyndlee Fault, HL Hawick Line, HtF Hermitage Fault, HtL Hartfell Line, KF Kingledores Faults, KT Killantringan Thrust, LF Langlee Fault, LKF Loch Ken Fault, LL Leadhills Line, LRF Loch Ryan Fault, MSB Mosspaul Shatter Belt, MVF Moffat Valley Fault, PNT Pool Ness Thrust, PT Pibble Thrust, RL Riccarton Line, SF Stinchar Fault, SUF Southern Upland Fault, TdF Talmond Fault, TF Tudhope Fault, TL Talnotry Line, WF Wolfhopelee Fault, YLF Yad Linn Fault.
Devonian granites: CCG Cairnsmore of Carsphairn, CFG Cairnsmore of Fleet, CG Criffel, LDG Loch Doon, SG Spango.
Sources: Clarkson *et al.* (1975); Cook & Weir (1979); Craig & Walton (1959); Floyd (1976); Fyfe & Weir (1976); Gordon (1962); Kelling (1961); Leggett *et al.* (1979a, b); Lumsden *et al.* (1967); Peach & Horne (1899); Rust (1965); Toghill (1970b); Walton (1955, 1956, 1961); Warren (1964); Weir (1968, 1977); Welsh (1964); Williams (1962).

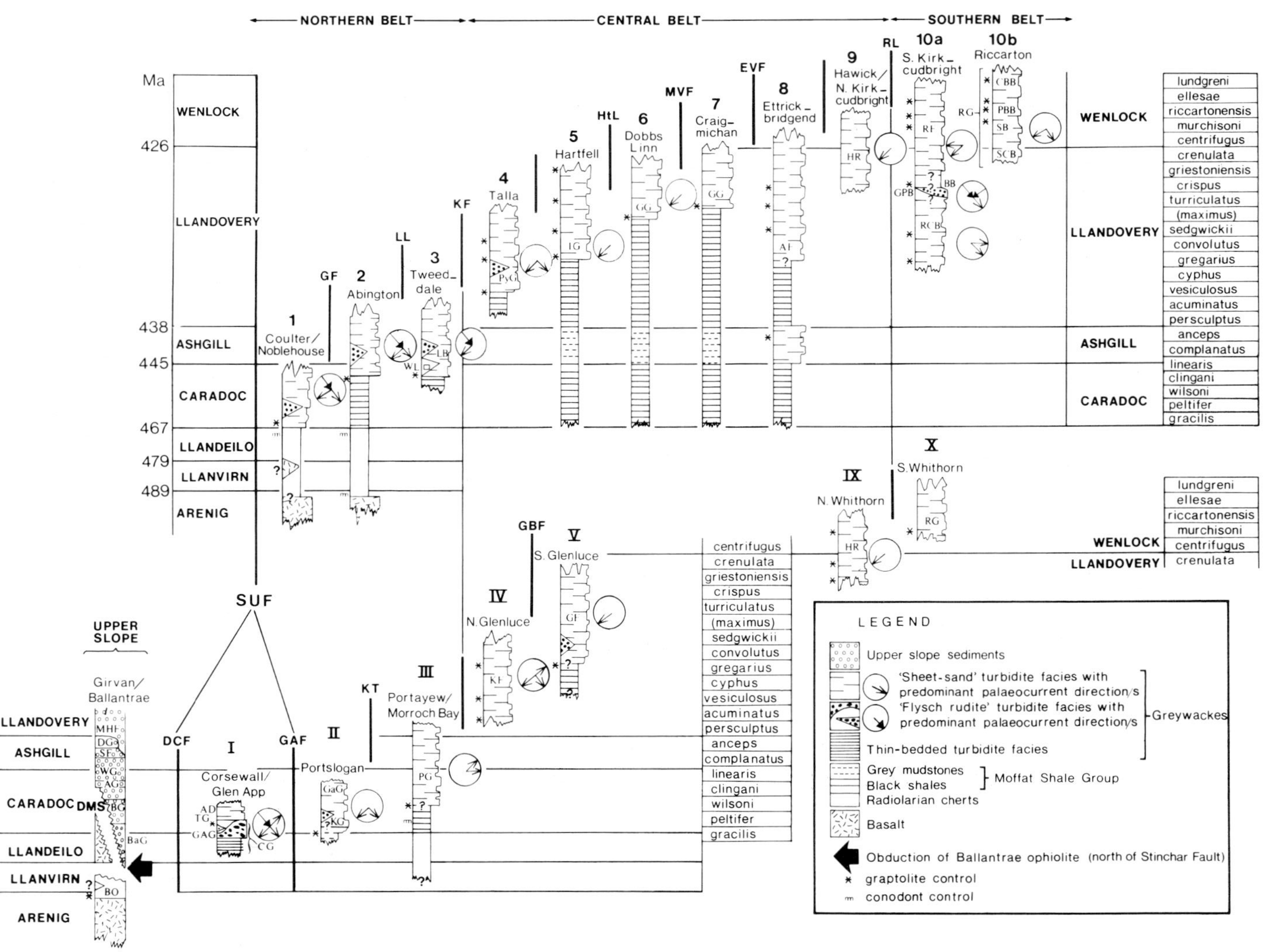
NORTHERN BELT
CENTRAL BELT
SOUTHERN BELT
Ma
426
438
445
467
479
489
WENLOCK
LLANDOVERY
ASHGILL
CARADOC
LLANDEILO
LLANVIRN
ARENIG
1 Coulter/Noblehouse
2 Abington
3 Tweeddale
4 Talla
5 Hartfell
6 Dobbs Linn
7 Craigmichan
8 Ettrickbridgend
9 Hawick/N. Kirkcudbright
10a S. Kirkcudbright
10b Riccarton
GF
LL
KF
HtL
MVF
EVF
RL
lundgreni
ellesae
riccartonensis
murchisoni
centrifugus
crenulata
griestoniensis
crispus
turriculatus
(maximus)
sedgwickii
convolutus
gregarius
cyphus
vesiculosus
acuminatus
persculptus
anceps
complanatus
linearis
clingani
wilsoni
peltifer
gracilis
SUF
DCF
GAF
KT
GBF
I Corsewall/Glen App
II Portslogan
III Portayew/Morroch Bay
IV N. Glenluce
V S. Glenluce
IX N. Whithorn
X S. Whithorn
UPPER SLOPE
Girvan/Ballantrae
DMS
LEGEND
Upper slope sediments
'Sheet-sand' turbidite facies with predominant palaeocurrent direction/s
'Flysch rudite' turbidite facies with predominant palaeocurrent direction/s
Greywackes
Thin-bedded turbidite facies
Grey mudstones
Black shales
Moffat Shale Group
Radiolarian cherts
Basalt
Obduction of Ballantrae ophiolite (north of Stinchar Fault)
graptolite control
conodont control

hinges on four lines of evidence:

(i) The Iapetus suture immediately to the south of the area and the contemporaneous magmatic arc to the north indicate a forearc setting in mid-Ordovician through late Silurian times.

(ii) Rocks preserved in the area are interpreted as ocean floor and trench deposits (basalts, metalliferous sediments, cherts, black shales, and greywackes).

(iii) The structural arrangement of the rocks preserved resembles the accretionary prism model of Seely *et al.* (1974).

(iv) Deformation appears to have occurred progressively during the evolution of the area, rather than in a climactic event.

Forearc basin

The Silurian succession in the Girvan area (Fig. 3) is marine and all inliers NE of Girvan have Silurian successions of Llandovery marine sediments and Wenlock terrestrial sediments. Rocks older than Silurian are only seen at Girvan, where an Ordovician marine succession, interpreted as an upper slope and shelf sequence, rests unconformably on the early Ordovician Ballantrae ophiolite complex (Fig. 4).

Sediments deposited during the latest Caradoc indicate a NW-inclined palaeoslope (Ingham 1978, p. 170). This may be related to the first emergence of a trench slope break in the Southern Uplands accretionary complex. Ensuing sediments in the Girvan area are varied Llandovery marine strata up to *c.* 3 km thick (Fig. 4). These are largely deeper water shelf deposits having both southerly and northerly provenance (Cocks & Toghill 1973).

Llandovery strata in inliers to the east of Girvan are turbiditic, and derived almost entirely from the south (summary in Leggett 1980a). Wenlock sediments include clasts of

---

FIG. 4. Time-stratigraphic diagram showing representative stratigraphic sequences for fault slices in two traverses across the Southern Uplands from the Southern Upland Fault in the north to the unconformable Upper Palaeozoic cover in the south. Adapted from Leggett (1980a), and Leggett *et al.* (1979b). Time-scale after McKerrow *et al.* (1980). Numbers in circles and boxes refer to locations in Fig. 3. Iapetus suture lies c. 20 km to SE of youngest fault slice. Greywackes divided into three broad facies types (see text), with representative predominant palaeocurrent directions beside columns. Details of grey mudstones in pelagic facies simplified: thin black seams occur in the Ashgill grey mudstones, and grey mudstones are intercalated in the upper division of the Moffat Shale (see Leggett 1978, fig. 3). Graptolite control shown only in turbidite facies: correlations in pelagic facies accurate to within one graptolite zone unless a question mark is present, in which case horizon is known to within ± 2 or 3 graptolite zones. Otherwise no exact temporal correlations intended. Solid vertical lines represent faults labelled as in Fig. 3. We follow Clarkson *et al.* (1975) in grouping the *symmetricus* (now termed *rigidus*), *linnarssoni* and *ellesae* (formerly termed *rigidus*) zones together as the '*C. ellesae*' zone. Lumsden *et al.* (1967) grouped the same three middle Wenlock zones together under '*C. linnarssoni*' (see Fig. 9). Conodonts in cherts closely associated with some of the basalts in the Coulter-Noblehouse sequence suggest a Llanvirn-Llandeilo age for the volcanicity (Lamont & Lindström 1957; Bergström 1971). We have shown the possible presence of Arenig basalts at the base of the sequence because basalts of this age occur both to the north of the Coulter-Noblehouse tract (in the Ballantrae ophiolite sequence) and to the south (in the Afton-Abington sequence). Biostratigraphic data within Ballantrae ophiolite sequence and in unconformable cover indicate emplacement within Arenig or Llanvirn (summary in Bluck 1978).

Lithostratigraphic units: AD Albany division, AF Abbotsford Flags, AG Ardwell Group, BB Balmae Beds, BaG Barr Group, BG Balclatchie Group, BO Ballantrae ophiolite, CBB Caddroun Burn Beds, CG Corsewall Group, DG Drummock Group, DMS Downan Mafic Sequence, GaG Galdenoch Group, GAG Glen App Group, GF Garheugh Formation, GG Gala Group, GPB Gipsy Point Beds, HR Hawick Rocks, IG Intermediate Group, KF Kilfillan Formation, KG Kirkcolm Group, LB Lowther Beds, MHF Mulloch Hill Formation, PBB Penchrise Burn Beds, PG Portpatrick Group, PyG Pyroxenous Group, RCB Raeberry Castle Beds, RF Ross Formation, RG Riccarton Group, SB Shankend Beds, SCB Stobs Castle Beds, SF Shalloch Formation, TG Tappins Group, WG Whitehouse Group, WL Wrae Limestone.

Sources for stratigraphic and sedimentological data: upper columns: 1, 2, 3, Peach & Horne (1899); Eckford & Ritchie (1931); Lamont & Lindström (1957); McKerrow *et al.* (1977); Leggett *et al.* (1979a, b), Leggett (1980b) 4, 5, 6, D. Casey (unpublished data); Peach & Horne (1899); Walton (1955), 7, 8, Peach & Horne (1899); Toghill (1970a). 9, Peach & Horne (1899); Warren (1963, 1964). 10a, Craig & Walton (1959, 1962); Walton (1968); Clarkson *et al.* (1975). 10b, Warren (1963, 1964). Lower columns: 1, Walton (1956, 1961); Kelling (1961, 1962); Williams (1962). 2, 3, Kelling (1961, 1962); Welsh (1964); Kelling & Welsh (1970). 4, 5, Gordon (1962). 9b, 10c, Rust (1965); Girvan-Ballantrae sequence from Williams (1962); Lamont & Lindström (1957); Bergström (1971); Ingham (1978).

cleaved greywacke (McGiven 1968), clearly deformed long before any supposed end-Silurian Caledonian orogeny. The emerging trench slope break at this time must have been fringed by shallow-water deposits, because comminuted shelly debris occurs in the Llandovery turbidites (Rolfe 1961).

Uplift caused the forearc basin to become emergent in late Llandovery or early Wenlock times. Terrestrial red beds, including thick alluvial fan conglomerates, overlie the turbidites disconformably. East of the Girvan area the only sedimentary transition between marine and terrestrial strata occurs in the Lesmahagow inlier (Fig. 3). During the Wenlock, movements on the ancestral Southern Upland Fault controlled sedimentation in the forearc basin. Alluvial fan complexes prograded northwards into the basin three times, reflecting movements on the Southern Upland Fault or other faults in the emergent northern part of the accretionary complex (McGiven 1968; Fig. 6). These may be related to rapid uplift associated with discrete phases of offscraping in the south.

The Ballantrae ophiolite is a complex faulted terrane exposed in the SW corner of the Midland Valley (Fig. 3). It largely comprises mafic volcanic and volcaniclastic rocks; serpentinite belts and local bodies of ultramafic rock, gabbro, dolerite, trondjemite and amphibolite are also present. Radiolarian cherts, black shales and debris-flow conglomerates occur above and within the mafic sequences.

Bamford (1979 and references therein) records a 6–14 km thick layer of rocks with seismic velocities of 5.8–6.0 km s^{-1}, interpreted as Lower Palaeozoic, across the Midland Valley. These are underlain by >6.4 km s^{-1} rocks, interpreted as pre-Caledonian high-grade Precambrian metamorphic rocks (Bamford 1979). This interpretation is strongly supported by granulite xenoliths discovered in Upper Palaeozoic volcanic vents across the Midland Valley (Graham & Upton 1978). Laboratory measurements of velocities do not appear to be adequate at present to discriminate ophiolitic lithologies from Lower Palaeozoic sediments in the 5.8–6.0 km s^{-1} layer; high proportions of serpentinite in the ophiolites are likely to provide a further complication (e.g. Powell 1978). Therefore, we do not know what proportion of the basement of the Midland Valley forearc basin comprises ophiolitic rocks of the kind exposed at Ballantrae.

Frontal arc/arc massif

North-west of the Highland Boundary Fault in Scotland and along strike in Ireland (Fig. 2) thick late Proterozoic and Cambrian sediments (Dalradian Supergroup) were deformed and metamorphosed in the early Ordovician Grampian orogeny about 500 Ma ago (summary in Johnson *et al.* 1979). Plate tectonic explanations for the Grampian orogeny are many and varied but almost all relate the orogeny to subduction along the northwestern Iapetus margin. The earliest stratigraphic sequence in the Southern Uplands, immediately south of the Southern Upland Fault, contains an *N. gracilis* Zone fauna (Fig. 4). Accretion must have started after this time. Therefore the Grampian events were clearly over before accretion commenced in the Southern Uplands. In relation to the mid Ordovician-late Silurian construction of the Southern Uplands accretionary complex the Grampian belt forms a joint frontal arc and arc massif.

Magmatic arc

North of the Southern Upland Fault and its hidden continuation in Ireland (Fig. 2) arc-type magmas were erupted at intervals from late Cambrian through to Early Devonian (summary in Stillman & Francis 1979). Contemporaneous granitoids intrude the late Precambrian—Cambrian (Grampian) metamorphic rocks and Proterozoic high-grade gneissic basement of the arc massif (summary in Brown 1979).

No arc volcanics of definite Silurian age are exposed, but indirect evidence for arc eruptions throughout the Ordovician and Silurian comes from the Southern Uplands: abundant andesitic detritus occurs in Ordovician and Silurian greywackes (Kelling 1962; Sanders & Morris 1978; Walton 1956), and numerous metabentonitic seams occur in pelagic and hemipelagic sediments (Cameron & Anderson 1979; Leggett 1979).

By far the most voluminous volcanic rocks exposed in the magmatic arc are calc-alkaline extrusives and pyroclastics, mostly andesites but also including basalts and rhyolites, interbedded in Lower Old Red Sandstone (ORS) terrestrial deposits of possible Early Devonian age (Thirlwall 1981a). In the Ochil Hills (Fig. 2) these lavas are up to 3 km thick. Systematic changes in Sr, K and other elements in the lavas across the region are similar to trends observed across modern magmatic arcs (Thirlwall 1981a). These ORS volcanics pose a problem in dating the closure of the Iapetus Ocean (McKerrow & Cocks 1976; Thirwall 1981a). Either the sector of Iapetus between Scotland and England was still open in the Early Devonian, subducting to the NW under Scotland, or the Lower ORS is in part of late Silurian age. Lower ORS dating on palaeontological grounds is uncertain and K-Ar ages of Lower ORS lavas are either Silurian or Early Devonian depended on the time-scale chosen (Thirlwall 1981a).

The Southern Uplands accretionary complex

Reconstruction of the evolution of the complex

Stratigraphy

In this section we review the geological evidence used to reconstruct the Southern Uplands accretionary complex. Fig. 3 shows the main reverse faults recognised to date in the Southern Uplands, and Fig. 4 shows representative stratigraphic sequences from the tracts defined by these strike faults. We have described the stratigraphic sequences at length, and discussed their correlation across the Southern Uplands, in an appendix to this paper (Supplementary Publication No. 18038 of the Geological Society of London, copies obtainable from the Geological Society Library and the British Lending Library at Boston Spa). Our intention here is to draw attention only to the main points. The diagrams show that each tract is characterized by a different sequence, and that progressively younger strata are found in sequences towards the SE (site of the Iapetus). The accretionary prism model of Seely *et al.* (1974) predicts such a configuration for subduction complexes. The viability of this interpretation in the case of the Southern Uplands is strengthened by the predominant younging of strata NW (continentwards) within each tract, by the oceanic/trench affinities of strata in the sequences (see later) and by evidence for a contemporaneous magmatic arc NW of the region.

Some of the reverse faults can be traced along the entire outcrop (e.g. the Riccarton Line); one, the Kingledores Fault, can be traced into the Irish continuation of the belt (Fig. 2). Most, however, are of uncertain continuity, and several cannot be traced more than a few tens of kilometres, suggesting that some tracts may be discontinuous (Fig. 3). Similarly, stratigraphic sequences can be traced within tracts for varying distances. For example, sequence 5 at Hartfell (Fig. 4) can be traced west across the ground south of Moniaive through the aureole of the Cairnsmore of Fleet granite, to the Wigtownshire coast south of Glenluce, a distance of 100 km (Fig. 3). In contrast sequence 7 at Craigmichan (Fig. 4) is only known to extend for a distance of *c.* 30 km in the region east of Moffat (Fig. 3).

Following the accretionary prism model we propose that each fault-bounded tract represents a discrete packet of accreted ocean floor and/or trench sediment offscraped above the northwards subducting Iapetus ocean crust. The differences between stratigraphic sequences contained in these tracts (e.g. sequences 3 & 4, Fig. 4) suggest a large geographic separation between some of the sequences prior to accretion (McKerrow *et al.* 1977). The apparent lack of continuity of some of the tracts implies original offscraping of lenticular packets during accretion.

Study of Fig. 3 reveals that there are rather more major reverse faults than there are defined stratigraphic sequences (e.g. faults west of the type areas for sequences 2 and 3 in the central part of the Northern Belt and faults traversing the area of sequence III on the Galloway Coast). Clearly, therefore, the different stratigraphic sequences should be used as only a general guide to the configuration of original accretionary tracts; the areas of outcrop of sequences in some cases contain more than one original sliver of off-scraped ocean floor and/or trench sediment. This conclusion has been confirmed wherever areas have been mapped in detail. The best example is in the Langholm region, where Lumsden *et al.* (1967) have mapped several major strike faults which separate greywacke sequences of different age (progressively younger towards the SE) within tract 10 (see Figs. 3, 4 & 10Q).

Ordovician reconstruction

Fig. 5 shows a reconstruction of the Southern Uplands accretionary complex in the late Ordovician. The evolutionary history can be followed using stratigraphic evidence summarized in Figs 3 and 4. In the following account we refer to lettered labels on Figs 3–6.

The Ballantrae ophiolite (BO) north of the Stinchar Fault (SF) was emplaced on the Laurentian continental margin in Arenig or early Llanvirn times (e.g. Dewey 1974) and is overlain unconformably by a Llanvirn to Ashgill upper slope succession (Ingham 1978). In the southern part of the ophiolite the Downan mafic sequence (DMS Fig. 4) between the Stinchar Fault and Dove Cove Fault (DCF) was probably emplaced in the late Llandeilo or early Caradoc (*gracilis* Zone, Barrett *et al.* in press and unpublished fieldwork).

Basement below the ophiolite is not exposed, but boulders in the upper slope succession (Benan Conglomerate) suggest a Proterozoic/

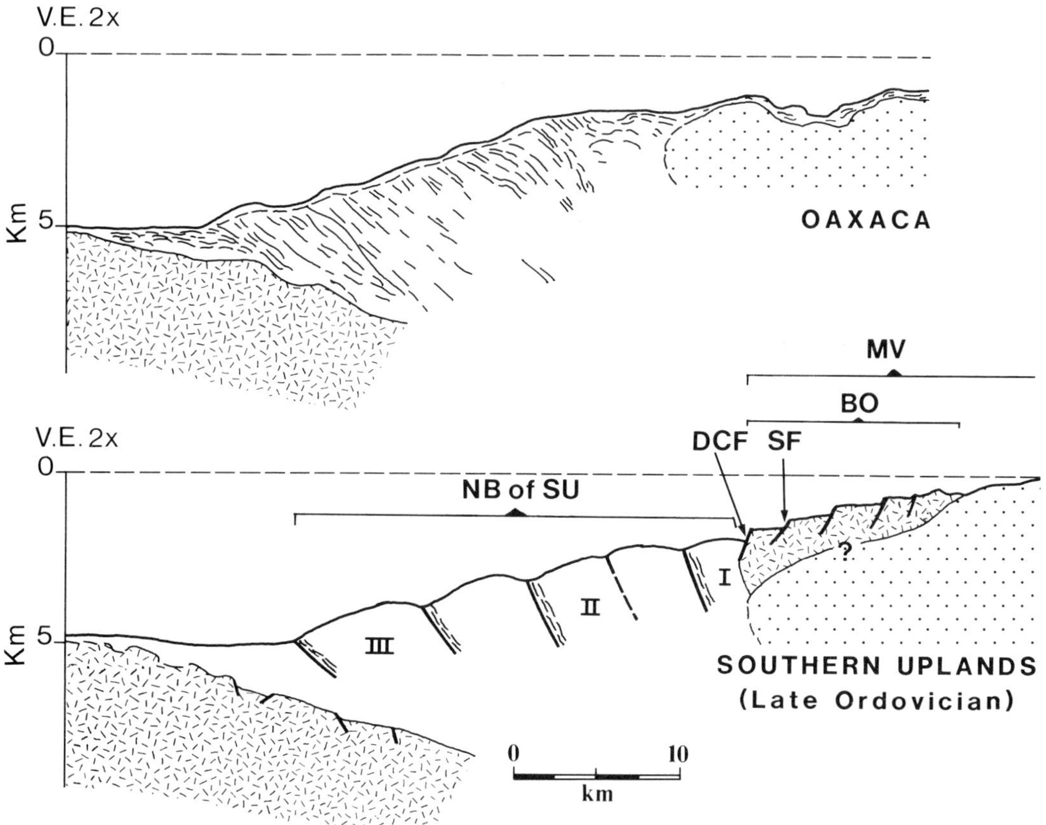

FIG. 5. Reconstruction of Southern Uplands accretionary prism in late Ordovician, compared with modern Middle American accretionary complex off Mexico at same scale. Southern Uplands reconstructed on a line along west coast, Oaxaca (Mexico) profile from Moore *et al.* (1979). Continental basement dotted; subducting oceanic crust and ophiolitic rocks in oblique dashed ornament. For further explanation see text. BO Ballantrae ophiolite, DCF Dove Cove Fault, MV Midland Valley, NB of SU Northern Belt of Southern Uplands, SF Stinchar Fault.

Cambrian metamorphic/granitic basement (dotted ornament in Fig. 5) under the SW Midland Valley (MV) (Longman *et al.* 1979). Seismic refraction evidence indicates that this type of basement may terminate at the Southern Upland Fault (Bamford 1979).

South of the Southern Upland Fault, of which the Dove Cove and Stinchar faults are braiding strands (Fig. 3), greywacke tracts of the Southern Uplands began to be accreted in the late Llandeilo or Caradoc. This we know because all graptolites recorded in the Coulter/Noblehouse and Corsewall/Glen App sequences (sequences 1 and I, Figs 3 & 4) are from the *gracilis* Zone. The remaining tracts of the Ordovician Northern Belt of the Southern Uplands (NB of SU) were accreted by late Ordovician time. Accretionary slices are well-defined by décollement surfaces which are the major strike faults of Fig. 3; along these faults basal lithologies (basalt, chert, black shale) have been imbricated (dashed ornament in Fig. 5). The positions of sequences I–III (Figs 3 & 4) are shown in Fig. 5; note that there are more fault-bounded tracts (probable discrete accreted slices) than there are defined sequences (see faults mapped on Fig. 3 both on the west coast and in the area east of the Loch Ryan Fault).

The Oaxaca profile in Fig. 5 makes an interesting comparison. Continental basement is overlain by an upper slope succession lower Miocene at the base. Below the thin blanket of sediment on the lower slope, landward-dipping reflections are accreted trench deposits. Deep-sea drilling shows that the earliest accretion was late Miocene (Moore *et al.* 1979, 1981). The main points to note in comparing the two profiles in Fig. 5 are:

(i) Clear reverse strike faults in the Southern Uplands (Fig. 3) suggest that individual packets of accreted strata were morphologically well defined (cf. Nankai Trough profiles, Moore & Karig 1976). Seismic records suggest less definition of discrete accreted slices under the Oaxaca lower slope (Shipley *et al.* 1981).

(ii) On both margins the earliest upper slope sediments are more than 10 Ma older than the first accretion of trench clastics.

(iii) The Ordovician part of the Southern Uplands accretionary complex was constructed in *c.* 20–30 Ma (from *gracilis* times to sometime between latest Caradoc and end Ordovician: see constraints of graptolite biostratigraphy in Fig. 4, sequences 1–3). The Oaxaca accretionary complex was constructed in *c.* 10 Ma (since late Miocene: Moore *et al.* 1979).

(iv) Ordovician accretion in the Southern Uplands included ocean floor pelagic sediments and possibly slivers of the underlying oceanic basement: pelagic-hemipelagic sediments and basalts are preserved in all the Ordovician sequences (see Fig. 4). Compared to this 'efficient' accretion in the Ordovician Southern Uplands, the Oaxaca accretion seems to be 'inefficient': pelagic-hemipelagic ocean floor deposits and much of the trench sediment seems to have been subducted below the accretionary complex (Watkins *et al.* 1981).

Silurian reconstruction

By late Silurian times all tracts exposed today (Figs 3 & 4) had been accreted. Later Silurian sequences (?Ludlow) may lie between the Upper Palaeozoic overstep and the Iapetus suture (Figs 2 & 3).

Uplift caused by the prolonged accretion had produced during latest Ordovician and early Silurian a trench slope break, which shed sediment north to the Midland Valley (MV) forearc basin (Leggett 1980a). The trench slope break, known as Cockburnland (C), was bounded by the Southern Upland Fault (SUF), which was active during Silurian sedimentation in the forearc basin (McGiven 1968). The horizontal arrow north of the fault in Fig. 6 shows the area in which inliers preserving the succession in this basin are exposed. The base of the forearc basin succession is not exposed east of the Girvan-Ballantrae area (Fig. 3). The thickest exposed forearc basin succession (at Girvan) is *c.* 3 km. Basement may comprise Proterozoic granulites at depth, as previously discussed. The forearc succession in the north of the Midland Valley is buried under Upper Palaeozoic cover.

Ophiolitic rocks of the Highland Boundary Complex (HBC) crop out at intervals along the Highland Boundary Fault (HBF) on the north border of the Midland Valley (Fig. 2). These relate to the earlier history of the margin prior to the onset of accretion in the Southern Uplands. They date from the Cambrian (Henderson & Robertson 1981) or early Ordovican (Bluck *et al.* 1980). The frontal arc/arc massif was exposed in the Grampian Highlands (GH).

In the Sumatra forearc the accretionary complex is also emergent. The trench slope break forms the Mentawai Islands, one of which is Nias (Fig. 5), on which both accretionary complex and lower slope basin sediments have been mapped (Moore & Karig 1980). In the forearc basin up to *c.* 6 km of sediment have accumulated behind the growing accretionary complex (G. F. Moore *et al.* 1981, and references therein). Basement under the basin may be ophiolitic (Curray *et al.* 1979) or older accreted material (Karig *et al.* 1980).

The main points to note in comparing the two profiles in Fig. 6 are:

(i) Accretion of thick clastics from the

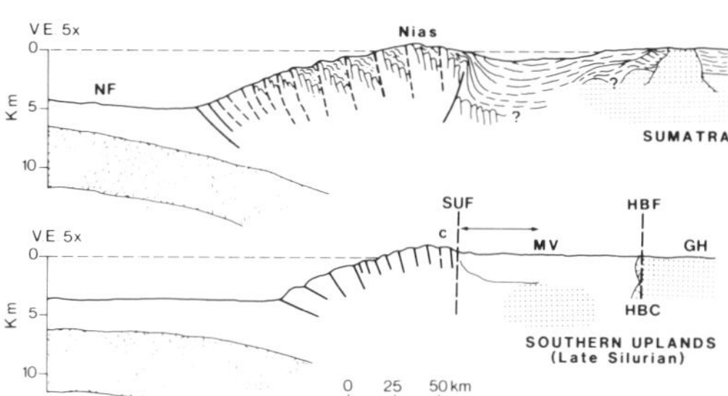

FIG. 6. Reconstruction of Southern Uplands accretionary prism in late Silurian, compared with modern Sundra arc accretionary complex off Sumatra at same scale. Southern Uplands reconstructed on a central section, Sumatra profile from Karig *et al.* (1980). Continental basement dotted; subducting oceanic crust in oblique dashed ornament. For further explanation see text.

C Cockburnland, GH Grampian Highlands, HBC Highland Boundary Complex, HBF Highland Boundary Fault, MV Midland Valley, NF Nicobar Fan, SUF Southern Upland Fault.

Nicobar Fan (NF, Fig. 5) caused the rapid post-Miocene growth of the Sumatran accretionary complex. Greywacke stratigraphy in the Central and Southern Belts of the Southern Uplands suggests that a similar thick ocean floor clastic sediment pile was being accreted in the Silurian along the northwestern Iapetus margin in Britain (Leggett 1980a and see later).

(ii) The similar tectonic/palaeogeographic role of the Southern Upland Fault (SUF) and the reverse fault/flexure behind the Sumatran trench slope break: both controlled forearc basin sedimentation (Karig *et al.* 1980; McGiven 1968).

(iii) The Sumatran accretionary complex developed in 20 Ma since early Miocene (Karig *et al.* 1980); the Southern Uplands accretionary complex developed in 45–50 Ma from mid-Ordovician to late Silurian (Leggett *et al.* 1979a). Plate convergence along the Sunda arc off Sumatra is relatively fast (Karig *et al.* 1980), so the implication of the prolonged development of the Southern Uplands accretionary complex is that convergence between the northern Iapetus oceanic plate and the Laurentian continent was slow.

Lithological and palaeogeographic analysis

Analysis of the Southern Uplands stratigraphic sequences (Fig. 4) allows reconstruction of environments on the Iapetus ocean floor and along its northwestern margin. The basalts, cherts and black shales have been interpreted as ocean floor lithologies (Leggett 1978, 1979) and the greywackes as both ocean floor and trench deposits (Leggett 1980a, b). In this section we briefly describe the lithologies in upward succession, stressing the reasons for these interpretations. Further details can be found in the source references.

(1) Oceanic volcanics

Basalts exposed at the base of sequences 1 and 2 (Fig. 4) occupy strip-like outcrops up to 200 m thick and 4 km long, fault-bounded at the base. There is only minor interdigitation of basalts in overlying pelagic sediments. Further small basalt outcrops are found in the southern part of the Northern Belt along strike from the type area of sequence 3. There are no basalts in the Central or Southern Belts.

Massive basalt dominates outcrops, though pillow lavas, thin lava flows and basalt breccias also occur. Most of the basalts are non-vesicular or sparsely vesicular, suggesting eruption under at least several kilometres of water (e.g. Jones 1969).

Sequence 1 contains alkali basalts and sequence 2 predominantly tholeiites. The least-mobile trace elements, Ti, Zr, Y and Nb, indicate that the basalts are of oceanic derivation (Lambert *et al.* 1981). If so, they must have been slivered off the top of subducting oceanic basement during accretion. This process has been described in modern trenches (see, e.g. Kulm *et al.* 1981). However, in some cases where contacts are not exposed in the field we cannot be sure that the Southern Uplands basalt bodies are not flow units intercalated within a pelagic sediment section.

Minor Caradoc volcanic rocks in the Northern Belt crop out in the Tweeddale area and east of Sanquhar in the Bail Hill area (Fig. 3). The Tweeddale lavas are blocks of peralkaline rhyolite (Thirlwall 1981b) in a debris flow deposit. Intercalated ash beds and blocks of limestone with shallow water fossils led Leggett (1980b) to interpret this debris flow unit (the Wrae Limestone; sequence 3, Fig. 4) as an olistostrome derived by submarine sliding from a carbonate-capped oceanic volcano. Part of such a volcano may be preserved in the Bail Hill region, where several square kilometres of varied igneous rocks with broadly dioritic affinities have been interpreted as an accreted oceanic seamount (Hepworth *et al.* 1981).

(2) Ocean-floor pelagic-hemipelagic sediments

The earliest pelagic sediments are Fe-rich mudstones. These have intimate field relations with the basalts, occurring in cross-cutting veins, interstitial pods and lenses, and beds between and above flows. Fe contents of up to 40%, Mn contents up to 3%, low Al and Ti contents, and Ce-depletion in a REE pattern from a supra-basaltic sample indicate a syn-magmatic origin from hydrothermal-exhalative solutions (Leggett & Smith 1980). These deposits, apparent analogues of hydrothermal metalliferous sediments from modern spreading centres and basal ocean-floor sediment sections, support the conclusion that slivers of oceanic basement have been preserved in the Northern Belt.

Radiolarian cherts and siliceous mudstones overlie the basalts. These are red, grey and green microcrystalline deposits, commonly bedded on a centimetre-to decimetre-scale, but locally massive. Radiolarite layers are common, and sedimentary structures include crude grading (suggesting redeposition), parallel lamination, syn-sedimentary faulting and chaotic

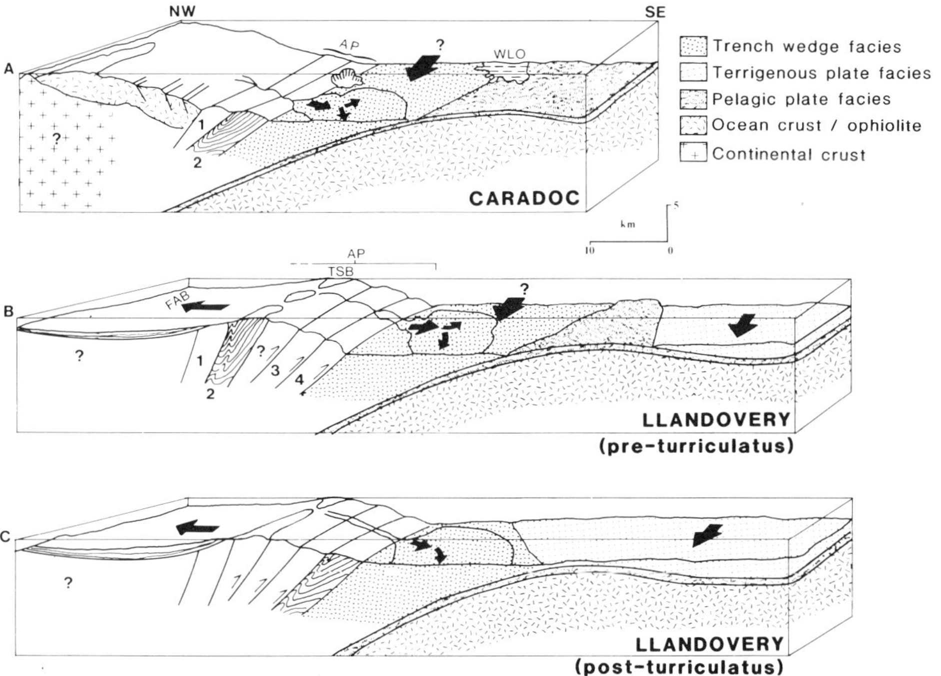

FIG. 7. Schematic cross-sections across the forearc region of the north Iapetus margin in southern Scotland at various times in the Ordovician and Silurian (after Leggett 1980a). Representative sediment dispersal directions shown (see references, Fig. 4). Major accreted slices numbered as in upper columns of Fig. 4.
(A) *Stage 1: Caradoc.* Trench wedge facies overlie pelagic plate facies; nascent accretionary prism (AP) oceanwards of obducted ophiolite on Laurentian continental crust. Source of Wrae Limestone olistostrome (WLO), thought to be an oceanic island (Leggett 1980b), not shown.
(B) *Stage 2: Early and middle Llandovery.* Emergent trench slope break (TSB) delimits two basinal provinces: the lower trench slope and trench to the south, and forearc basin (FAB) to the north. Elevated area on subducting oceanic plate (?outer trench high: see text) delimits two turbidite facies provinces, one in the trench and another on the ocean floor. Question mark in one of the accreted slices represents the probable existence of a fourth Ordovician slice between sequences 2 and 3 (Leggett *et al.* 1979a).
(C) *Stage 3: Late Llandovery.* Oceanic basement high overwhelmed, uniform trench-ocean floor flysch blanket ensues.

slumped bedding. Radiolarians are primitive spumellarians, and are preserved within diagenetic balls of silica 300–400 μm in diameter (Leggett 1978, 1979).

The radiolaria are of no biostratigraphic value, and dates are provided by graptolites (Peach & Horne 1899) and conodonts (Lamont & Lindström 1957). These extend from mid-Arenig to late Llandeilo, about 25 Ma. The thickest continuous chert section known is only *c.* 60 m. Notwithstanding the intense imbrication in the outcrops of basal lithologies (see later) and the limited number of faunas on which the dating is based, it appears that the cherts accumulated very slowly, at rates perhaps as low as several m Ma^{-1} and certainly equally as slow as siliceous pelagic deposits in modern ocean basins.

The cherts are overlain by the pelagic-hemipelagic Moffat Shale Group, comprising black shales and grey mudstones, which accumulated from latest Llandeilo/early Caradoc to late Llandovery (Fig. 4). Biostratigraphic control is excellent in these deposits, which contain abundant graptolites. The maximum exposed thickness is *c.* 100 m (Fig. 4, sequence 7), and the duration of sedimentation *c.* 40 Ma. Most of the Moffat Shale is finely laminated, carbon-rich shale devoid of benthos, which clearly accumulated under euxinic conditions. Interbedded grey mudstones, especially common in the Ashgill (Fig. 4), reflect more oxic conditions. Euxinification has been ascribed to productivity fluctuations arising from eustatic transgressions (Leggett 1978, 1980c; Leggett *et al.* 1981). The most significant aspect of the Southern Uplands pelagic deposits is that they represent some 60 Ma of geological time in which less than 200 m (after compaction) of sediment was deposited. Such prolonged slow accumulation can only have occurred in a wide ocean basin.

(3) Trench and ocean floor turbidite facies

Greywacke turbidites in the Southern Uplands are dominated by monotonous medium- to thick-bedded, usually acyclic, sequences and massive sandstones. These are distinguished in Fig. 4 as the 'sheet sand' facies class. Other, less common, facies include rudites (disorganized and organized) and thin-bedded turbidites, usually with low sandstone-shale ratios. The distribution of these facies classes is depicted in Fig. 4. This can only be done in an illustrative way because biostratigraphic control is generally poor in the greywackes, and detailed lithostratigraphy is generally lacking.

Fig. 7 is a reconstruction of the history of clastic sedimentation along the northwestern Iapetus margin using stratigraphic information summarized in Fig. 4 and previously published facies analysis (Leggett 1980a). Ordovician greywackes are interpreted as trench deposits on the basis of:

(i) common rudites (Fig. 4, sequences 1–3 and I–III) which map out as lenticular bodies along the length of the Northern Belt and have predominantly lateral (NW to SE) palaeocurrents, suggesting localized point sources (Kelling & Holroyd 1978; Fig. 7A);

(ii) immature petrography (e.g. Kelling 1962; Walton 1965), suggesting direct input of arc- and ophiolite-derived detritus without trapping in a forearc basin;

(iii) evidence for flow confinement and low depositional gradients, provided by axial palaeocurrents most commonly flowing NE to SW but with some also flowing SW to NE (Fig. 4, sequences 1–3 and I–III). Rare palaeocurrent indicators suggesting currents from the SE (Kelling 1962) perhaps represent flows which rebounded off the outer trench slope.

The Ordovician trench interpretation is supported by the stratigraphy of the Ordovician strata preserved in the Central Belt. Caradoc deposits SE (i.e. outboard or oceanwards) of Caradoc greywackes of the Northern Belt are entirely pelagic (sequences 4–8, Fig. 4). The same applies to Ashgill strata, except that one as yet unstudied greywacke unit occurs oceanwards of the Northern Belt greywackes in sequence 8 (see discussion in Leggett 1980a).

Biostratigraphic control in the Ordovician greywackes is poor. Graptolite faunas have been recorded from sequences 1 and 3 (see Fig. 4 and caption): in both cases only one graptolite zone is represented. If the Ordovician greywackes are trench deposits the greywackes in each sequence are not likely to extend any higher than these zones. The residence time of turbidite sections in modern trenches on accretionary margins (e.g. Moore 1979) is not likely to be longer than the 3–5 Ma average duration of an upper Ordovician graptolite zone (McKerrow *et al.* 1980).

In the mid-Llandovery, greywackes appear in sequences both 'outboard' (sequences 8, 10a) and 'inboard' (sequences 4, 5, 6) of pelagic facies. This suggests the presence of an elevated area slowly accumulating pelagic sediment (black shale), separating two turbidite provinces (Fig. 7B). Leggett (1980a) interpreted the elevated area as an outer trench high. By this interpretation greywackes in the 'inboard' sequences are trench deposits and greywackes in the 'outboard' sequences are abyssal plain (ocean floor) turbidites. Perhaps more likely, in view of the duration (several graptolite zones) of the strata in sequences 4 and 5 (Fig. 4), is that the high represents an intra-oceanic basement feature, such as the Ninetyeast Ridge in the Indian Ocean, separating an 'inner' trench-ocean floor turbidite accumulation and an 'outer' abyssal plain accumulation.

After *turriculatus* times (sequence 7, Fig. 4) the high was submerged (sequences 5, 8, 9, 10a and b in Fig. 4) and was covered by a uniform sediment blanket (Fig. 7C). The thick greywackes of these later sequences contain interbedded graptolitic horizons and accumulated for long periods. Clarkson *et al.* (1975) have proposed a true thickness of *c.* 5 km for the strata in sequence 9 (Fig. 4) in the Kirkcudbright area (Fig. 3). This supports the interpretation of later Silurian greywackes as a Nicobar Fan-type accumulation (Fig. 6). Facies considerations also support this view: most beds are derived axially, are relatively quartzose, and devoid of appreciable rudite bodies (summary in Walton 1965).

A final category of trench and ocean floor clastics comprises the debris flow/olistostrome deposits. The Wrae Limestone olistostrome has

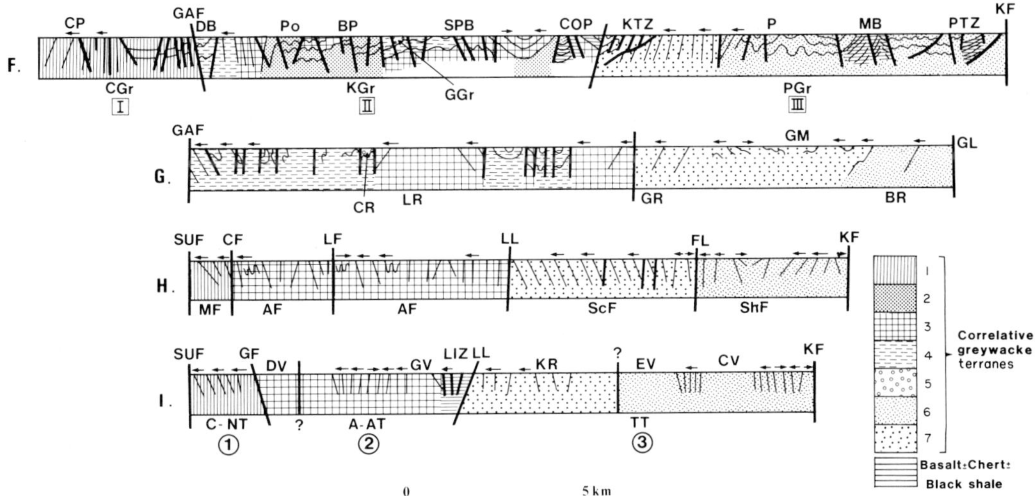

FIG. 8. Structural profiles in the Northern Belt of the Southern Uplands. Lines of section are on Fig. 3. Major reverse faults, ie. those bounding tracts containing considerably different stratigraphic sequences and considered to be décollement surfaces dating from initial offscraping during accretion, are shown extending above and below line of section. Illustrative bedding attitudes shown with thin lines in top two-thirds of section; arrows above section are representative sedimentary younging data. North-west on the sections is on the left. F. the Rhinns of Galloway (from Kelling 1961). G. the region east of Loch Ryan (from Welsh 1964). H. the region west of Nithsdale (from Floyd 1976). I. the Clyde valley region (from Leggett *et al.* 1979a).
Faults: labelled as in Fig. 3, plus KTZ—Killantringan Thrust Zone; LIZ—Leadhills Imbricate Zone; PTZ—Portayew Thrust Zone.
Greywacke units (lettering below sections: units named by authors in papers from which sections are taken). For details of correlations see Figs 3, 4 and text).
1. Corsewall Group (CGr); Marchburn Formation (MF); Coulter-Noblehouse Tract (C-NT).
2, 3, 4. Kirkcolm Group (KGr)—divided into Lower Barren (ii), Upper Barren (iv), and intervening Meta-clast (iii) divisions (profile F); Lochryan Rocks (LR)—divided into Lower (iii) and Upper (iv) divisions (profile G); Afton Formation (AF); Afton-Abington tract (A-AT).
5. Galdenoch Group (GCr); Cairnerzean Rocks (CR).
6, 7 Portpatrick Group (PGr)—divided into a lower Acid-clast division (6) and an upper Basic-clast division (7); Glenwhan Rocks (GR); Boreland Rocks (BR); Scar Formation (ScF): Shinnel Formation (ShF); Tweeddale tract (TT).
Locations: BP Broad Port; COP Cave Ochtree Point; CP Corsewall Point; CV Clyde Valley; DB Dally Bay; DV Duneaton Valley; EV Elvanwater valley; GM Glenwhan Moor; GV Glengonnar Valley; KR Kirkton Rig; MB Morroch Bay; P Portpatrick, Po Portobello; SPB Salt Pans Bay.

already been mentioned. Other such Caradoc carbonate debris flows occur in the Northern Belt, though volumetrically of minor importance and poorly exposed (summary in Leggett 1980b). These may represent olistostromes derived from the upper slope, where Caradoc limestones are known (e.g. Craighead Limestone in the Ardwell Group, Williams 1962; AG in Fig. 4). They may equally well derive from carbonate-capped oceanic islands, an interpretation favoured by the volcanic blocks and tuffaceous interbeds in the Wrae Limestone olistostrome.

(4) Lower trench slope sediments

We believe that the present level of erosion in the Southern Uplands is probably too deep to have preserved extensive lower slope sediments. Mapping in more recent accretionary complexes suggests that large differences in style of deformation and age between adjacent terranes are characteristic of lower slope sediment sections resting on accretionary basement (e.g. Bachman 1981; Moore & Karig 1976; see also J. C. Moore *et al.* 1981). There is no evidence in the Southern Uplands for significant age differences between contiguous greywacke sections, as would perhaps be expected between accretionary basement and a younger mantling slope sediment pile. Possible minor exceptions are an area of Wenlock rocks juxtaposed with Llandovery greywackes in the

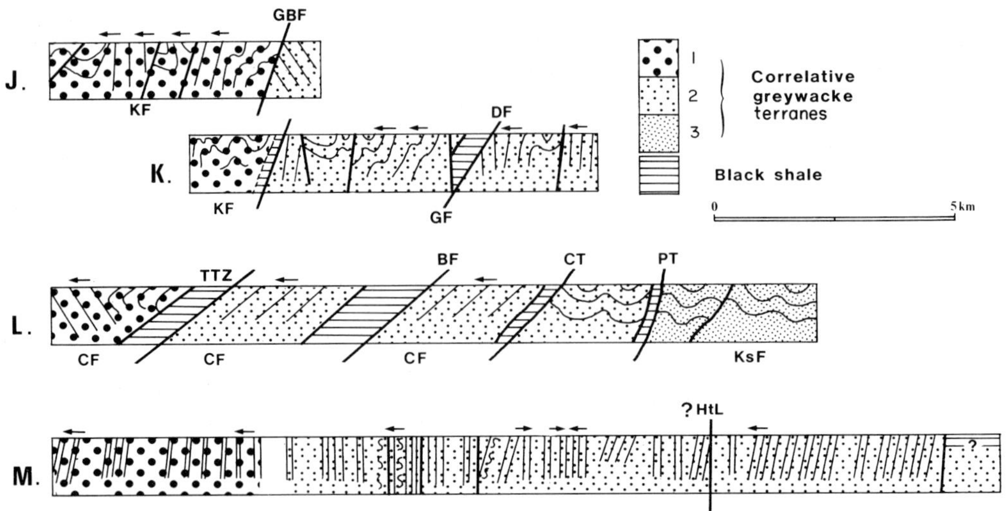

FIG. 9. Structural profiles of the Central Belt of the Southern Uplands. Details as in Fig. 8. North-west on the sections is to the left. J. west Wigtownshire coast, north (from Gordon 1962). K. west Wigtownshire coast, south (from Gordon 1962). L. region west of Cairnsmore of Fleet granite (from Cook & Weir 1979). M. region north of Moffat (from unpublished Institute of Geological Sciences field logs of the 1976 British Gas pipeline section).
Faults: labelled as in Fig. 3, plus BF Blairbuies Fault, TTZ Talnotry Thrust Zone.
Greywacke units (lettering below sections: units named by authors in papers from which sections are taken). For details of correlation see Figs 3, 4 and text.
1. Kilfillan Formation (KF); Craignell Formation (CF) in part. 2. Garheugh Formation (GF); Craignell Formation (CF) in part. 3. Knockeans Formation (KsF). Correlation between profiles K and L is tentative (see Cook & Weir 1979); correlation between profile L and unnamed greywacke terranes in profile M is extremely tentative.

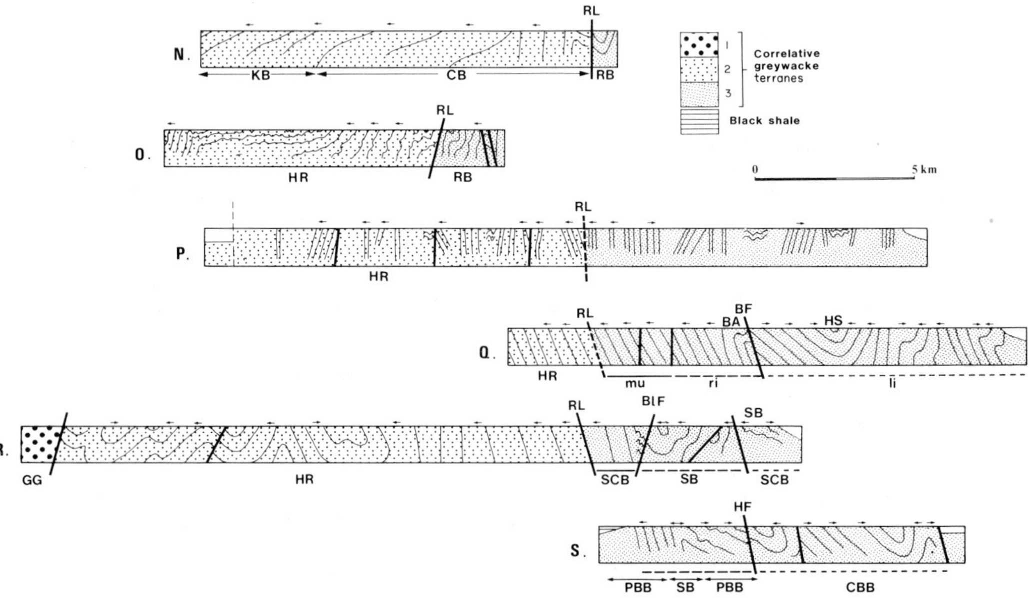

continuation of the Central Belt in the Longford-Down Inlier of Ireland (Leggett *et al.* 1979b), and the anomalous inlier at Coldingham on the NE coast in Scotland. At Coldingham abundant flat-lying structures of possible submarine slide origin contrast markedly with the steeply dipping greywacke terranes to the north and south. Other such candiates for lower slope sediment remnants may come to light with future mapping, but we do not anticipate that any will be large.

(5) Irish sequences

Stratigraphy of sequences in the Irish continuation of the Southern Uplands, summarized by Leggett *et al.* (1979b), is broadly similar to that of sequences described here. A major difference is greater interbedding of basaltic volcanics, both in pelagic and clastic successions, in the equivalent of the Northern Belt in Cavan (NW corner of the Longford-Down massif, Fig. 2). Greywacke facies and petrology are broadly analogous, with more rudites and more immature, volcaniclastic detritus in Ordovician than in Silurian rocks (Leggett *et al.* 1979b). In detail, however, sequences in individual tracts do not correlate between mapped areas in the west and east of the Longford-Down massif. As in the Southern Uplands, it appears that most tracts do not extend along the belt for more than *c.* 100 km.

Structure

Structural profiles of the Northern, Central and Southern Belts (Figs 8–10 respectively) illustrate the megascopic structure of the Southern Uplands. Reverse strike faults and belts of homoclinal, steeply-dipping NW-younging greywackes dominate. Major strike faults are those which bound tracts of differing stratigraphy (Figs 3 & 4) and have indeterminable though presumably very large displacements. Major faults so defined we interpret for reasons already outlined as original décollement surfaces during accretion of the tracts, which infers that they were initiated as low-angle thrusts. This view is supported by parallelism or near-parallelism of bedding and most of the major strike faults (Figs 8–10). Rotation within a growing accretionary complex, coupled with probable oversteepening resulting from eventual impingement against the European margin, may explain the steep dips of these faults today (McKerrow *et al.* 1977).

Basal lithologies are imbricated along the major faults in irregular zones (Fig. 3) from several tens of metres to more than 1 km wide (Leggett *et al.* 1979a; Weir 1979). 'Line', a term used by Floyd (1976) and Weir (1979) for some of the major faults, derives from the prominence of linear outcrops of basal lithologies (basalt, chert and graptolitic shale) marking the traces of imbricate zones on Geological Survey maps (Fig. 3).

Within tracts, steeply dipping homoclinal zones in which beds young NW are predominant. F_1 folds, both megascopic and mesoscopic, are chiefly upright, asymmetric, open to isoclinal, and SE-verging, with variably plunging NE-SW axes (summary in Stringer & Treagus 1980). Axial planes are predominantly inclined steeply NW or SE (e.g. Anderson & Cameron 1979). Widespread slump folds occur in the western part of the Longford-Down inlier (Morris 1979).

Cleavage transects F_1 folds for a distance of up to 50 km north of the Iapetus suture, and is rotated clockwise with respect to F_1 axial traces by up to 20° (Phillips *et al.* 1979). Phillips *et al.* (1979) suggest that the non-axial planar cleavage results from dextral shear parallel to the regional strike during brittle deformation. Stringer & Treagus (1980) following Borradaile (1978) suggest that it is a result of folding of planes non-orthogonal with respect to bulk strain axes. In non-coaxial strain histories cleavage formation can be delayed relative to fold formation as a result of intergranular move-

FIG. 10. Structural profiles of the Southern Belt of the Southern Uplands. Details as in Fig. 8. North-west on the sections is to the left. N. coast section east of Whithorn (from Rust 1965). O. Kirkcudbrightshire coast (from Craig & Walton 1959). P. Region south of Moffat (a composite section from unpublished Institute of Geological Sciences field logs of the 1975 and 1976 British Gas pipeline sections. Q. Langholm region (from Lumsden *et al.* 1967). R. Hawick region (from Warren 1964). S. Riccarton Junction region (from Warren 1964).
Faults: labelled as in Fig. 3.
Greywacke units (lettering below sections: named by authors in papers from which sections are taken; for details of correlation see Figs 3, 4 and text).
1. Gala Greywackes (Llandovery) (GG). 2. Kirmaiden Beds (KB); Carghidown Beds (CB); Hawick Rocks (HR). 3. Riccarton Beds (RB); Stobs Castle Beds (SCB); Shankend Beds (SB); Penchrise Burn Beds (PBB); Caddroun Burn Beds (CBB). In the Riccarton Beds, and subdivisions theoreof, units containing diagnostic graptolite zones are shown: mu, *murchisoni* (solid line); ri, *riccartonensis* (long dashes); li, *linnarssoni* (short dashes).

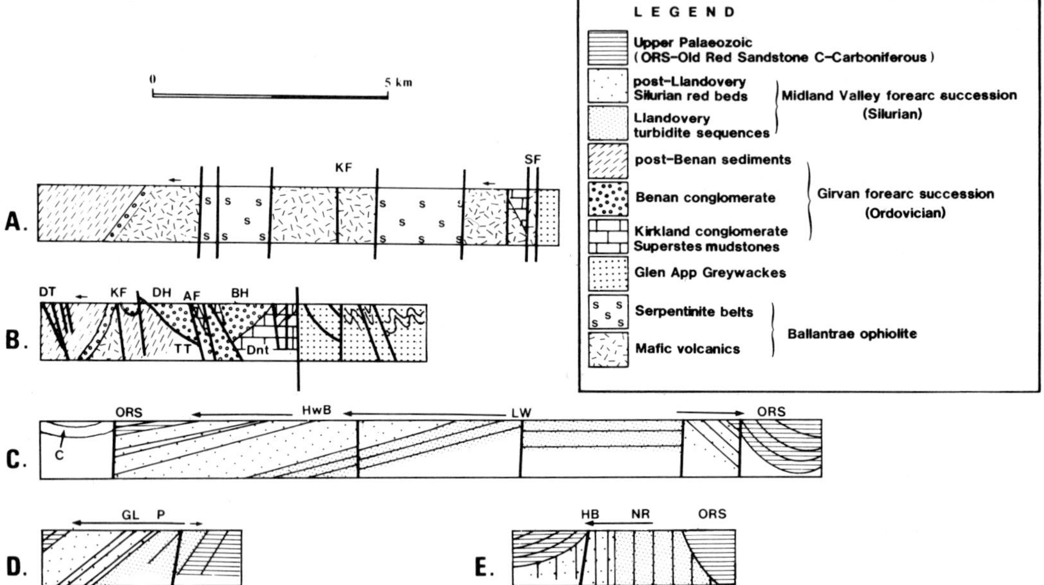

FIG. 11. Structural profiles of the Midland Valley. Lines of section are on Fig. 3. Illustrative bedding attitudes shown with thin lines, faults with thick lines. Arrows are representative sedimentary younging data. North-west on the sections is to the left. (A) The Ballantrae ophiolite (from Dewey 1974). (B) The Stinchar valley region (from Williams 1959—vertical scale exaggerated several times). (C) The Lesmahagow inlier (from Geological Survey one inch sheet 23). (D) The Hagshaw Hills inlier (from Rolfe 1961). (E) The Pentland Hills inliers (from Tipper 1974).
Faults: AF Assel Fault, DT Doune Thrust, DnT Dupin Thrust, KF Knocklauch Fault, SF Stinchar Fault, TT Tormitchell Thrust.
Localities: BH Benan Hill, DH Dinvin Hill, GL Glenbuck Loch, HB Henshaw Burn, HwB Hareshaw Burn, LW Logan Water, NR North Esk Reservoir, P Parisholm, PH Parisholm Hill.

ment within beds (Borradaile 1978). Such movements are highly likely in the early phases of deformation in subduction complexes, where interstitial fluid pressures are likely to be high (e.g. Cowan 1981).

The transecting S_1 cleavage, and probable primary variation in F_1 fold axis plunges (Stringer & Treagus 1980) has caused confusion leading to publication of overly complex polyphase deformation histories (Anderson 1980, commenting on Weir 1979 and references therein). Most recent structural interpretations have related all main folds and strike faults to a compound, diachronous D_1 event produced during accretion (Anderson & Cameron 1979; Eales 1979; Leggett *et al.* 1979a, b).

Post-D_1 folding becomes progressively more common towards the suture (Phillips *et al.* 1979). In the eastern part of the Longford-Down massif profile (Fig. 2) F_2 folds are gentle, south-verging flexures and F_3 folds are north-verging monoclines; both are of Caledonide trend (NW-SE axes), have associated crenulation cleavages, short limbs of < 20 m, and do not affect the regional outcrop (Anderson & Cameron 1979). In the southern part of the Southern Uplands similar F_2 folds are concentrated in a linear belt 2–3 km wide (Stringer & Treagus 1980). Phillips *et al.* (1979) ascribe these relatively minor post-D_1 deformation effects to final closure of the Iapetus.

Wrench faults which displace D_2 structures are common. These are orientated NNW, N and NE (Fig. 3) and are predominantly sinistral with displacements of up to 1 km (summary in Weir 1979). Anderson & Cameron (1979) relate these to closure of Iapetus. Intrusive rocks of probable Early Devonian (post-closure) age have exploited faults in the Southern Uplands. Porphyrites and lamprophyres commonly intrude wrench faults; some of these are sheared, indicating continuing fault movements (Anderson & Cameron 1979; Weir 1979). Small linear granite stocks have the same trends as reverse faults and wrench faults (Geological Survey one-inch sheets 15 and 24).

Structural geology appears more complex in the western than in the eastern part of the Longford Down massif and the Southern Uplands (Morris, unpublished data). In particular there is common pre-cleavage overturning of strata (Morris 1979).

Prehnite-pumpellyite facies metamorphism occurs in basalts and basic-clast greywackes in Longford-Down (Oliver 1978) and the Southern Uplands (Oliver & Leggett 1980). Assemblages of the greenschist and zeolite facies have not been found. Detrital prehnite indicates sedimentary recycling of metamorphic minerals and suggests that the burial metamorphism derives from 'tectonic' burial associated with accretion (Oliver & Leggett 1980).

Detrital glaucophane in late Llandeilo or Caradoc greywackes from Cavan (Sanders & Morris 1978) and the west coast of the Southern Uplands (Kelling 1962; Walton 1956) may indicate pre-existing (?Cambrian or early Ordovician) subduction terranes, eroded to deep levels, exposed along the Laurentian margin. Glaucophane-bearing clasts are exposed in local mélange outcrops in the Ballantrae ophiolite, but their geological significance is as yet unclear (summary in Bluck 1978).

Structure of the Midland Valley

The structure of the post-ophiolite Ordovician and Silurian strata of the Midland Valley is varied, though much less complicated than structure in the Southern Uplands. Northwards-directed, SE-inclined thrusts in the Girvan succession (Williams 1962, Fig. 11b) can only be dated as 'post-Ordovician'. Weir (1979) relates this thrusting to northwards obduction of the whole ophiolite plus sedimentary cover on to Laurentian continental basement sometime towards the end of the Silurian.

Folding varies in intensity along the Midland Valley. Strata in the Lesmahagow inlier are folded into a gentle open asymmetric anticline (Fig. 11C). The Hagshaw Hills rocks form a tight asymmetric anticline with a reverse-faulted axis (Fig. 11D). In the Pentland Hills and in much of the Girvan area, strata are steeply inclined without obvious fold axes. Cleavage is absent. Ingham (1978) ascribes local intense folding in thin-bedded sandstones of the Ardwell Group (Fig. 4) near Girvan (Fig. 3) to slumping.

Folding of forearc strata occurred at some stage during deposition of the Lower Old Red Sandstone (ORS), because ORS beds are folded with Silurian in the Lesmahagow and Hagshaw Hills inliers, but are unconformable on Silurian in the Pentland Hills inlier (Fig. 11). Southwards vergence of the folds and northwards tectonic transport on the thrusts (Fig. 11) suggest compression from the south. Final closure of the Iapetus ocean during deposition of the ORS (?latest Silurian or Early Devonian—see earlier discussion) is the simplest explanation for this deformation.

Discussion

Mechanism of offscraping

The level of initial décollement during accretion of individual tracts in the Southern Uplands was fundamentally influenced by the black shales of the Moffat Shale Group, which provided an incompetent level in the subducting ocean floor/trench sequences. Black shales are thin or are absent in the entirely Ordovician sequences of the Northern Belt (Fig. 4). Décollement occurred at the top of, or within, chert sections, or locally within basalts in sequence 1 where no black shales are developed. Cherts and basalts are also preserved locally below thin black shales in the other Ordovician tracts. In the Central Belt the thickness of the Moffat Shale Group increases due to the diachronism of the greywackes, and all sequences were detached at the top, within, or at the base of the black shales (Fig. 4). Underlying ocean floor lithologies were presumably subducted or accreted at deeper levels: an example of selective subduction (Moore 1975). Décollement occurred within greywacke sequences in the southern part of the Central Belt and the Southern Belt. The ocean-floor clastic pile subducting at that time was probably too thick to enable offscraping of complete sections.

During detachment of each packet of strata (tract) folding and low-angle thrusting were essentially synchronous. Folding was probably accomplished mostly during the early stages of accretion and uplift. The intensity of imbrication, both in zones along tract-bounding major faults and within greywacke terranes, indicates that deformation within tracts was accomplished by continued strike faulting (reverse or thrust motion) after strain-hardening and during rotation of tracts within the prism. We analyse the relationship between folding and faulting in more detail elsewhere (Leggett & Casey, in press).

Controls on structural style

Accretionary complexes dominated by coherent strata tend to develop above modern subduction zones along which sediment input is high and convergence is fast (Cowan & Silling 1978; Moore 1979). The Southern Uplands is devoid of mélange, despite intense imbrication.

Sediment input to the subduction zone during the accretionary phase was undoubtedly high. Stratigraphy of greywacke sequences in the Central and Southern Belts shows that for much of Silurian time thick ocean-floor clastics were being fed into the Scottish subduction zone. Rates calculated from published estimates of thickness in greywacke units (excluding compaction effects) are *c.* 165–185 m Ma^{-1} (Leggett 1980a). Slow convergence is indicated by the relatively slow development of the Southern Uplands accretionary complex compared with the Sumatran margin (Fig. 5 and previous section). So the essentially coherent nature of accreted strata in the Southern Uplands can readily be explained in terms of what Moore (1979) considers the two most important controls in structural style in modern accretionary complexes.

Other parameters were undoubtedly important. The effect of stratigraphy in the subducting sequences has already been described. Oblique convergence (Thirlwall 1981a) may further have facilitated offscraping of coherent strata.

Controls on sedimentation

Most sediment in the early Silurian forearc basin was derived from the south, off the rising trench slope break of the Southern Uplands, accretionary complex. Yet sand finding its way to the trench and ocean floor at the same time compositionally reflects the arc/orogen hinterland to the north (e.g. Walton 1965). This introduces two problems: why does the frontal arc/magmatic arc not contribute more sediment to the Midland Valley forearc basin, and how does the sediment bypass the forearc basin to reach the ocean floor? Unfortunately, Lower Palaeozoic strata under the northern part of the Midland Valley are buried beneath Upper Palaeozoic cover. ORS sediment in the Strathmore syncline along the northern border of the Midland Valley are some 10 km thick (Armstrong & Paterson 1970). This immense thickness indicates a substantial depression, possibly inherited from Silurian time and in some way linked to the lack of northerly-derived material in the southern part of the Silurian forearc basin.

The volcaniclastic nature of Silurian flysch in sequences accreted in the Llandovery is not such a puzzle. Much of the sediment may have been derived from previously accreted Ordovician flysch (Fig. 4, currents from the NW). Additionally, forearc basins along modern active margins are rarely continuous for long distances (e.g. Dickinson & Seely 1979), and the Midland Valley forearc basin may have been bypassed further to the NE, in the portion of the margin now buried under the North Sea. Most Silurian greywackes show a dominance of currents from the NE (Fig. 4; Walton 1965) i.e. parallel to the trench axis (Fig. 7).

Pre-middle Ordovician history of the northwestern Iapetus margin in Britain

Volcanic and plutonic evidence already outlined indicates that Iapetus oceanic crust was subducting under the British sector of the Laurentian margin in late Cambrian and early Ordovician times. By implication, the requirement for a subduction-related mechanism for the Grampian orogeny (Dewey 1971; Phillips *et al.* 1976) supports this. Accretion did not begin in the Southern Uplands until mid-Ordovician, and there is no evidence for an appreciable older accretionary terrane under the Midland Valley. What happened in the forearc before onset of accretion in the Southern Uplands?

By analogy with processes observed on modern active margins (summary in Scholl *et al.* 1981) there are three possible explanations. First, the margin may have been non-accretionary; any sediment accumulating on the trench or coming into the trench on the down-going ocean plate may have been subducted. Secondly, the subducting ocean plate may have mechanically eroded the over-riding Laurentian margin prior to the Southern Uplands accretionary phase. Thirdly, strike-slip movements on major faults sub-parallel to the margin may have removed older (Cambrian and early Ordovician) forearc elements.

There is evidence for missing early Ordovician stratigraphy. As already outlined, the upper slope succession on the Ballantrae ophiolite includes thick northerly derived conglomerates of Llanvirn and Llandeilo age. According to Williams (1962) some of these are of shelf- and some of slope-origin. Age-equivalent sediments to the south, in the early accreted tracts in the Southern Uplands, are pelagic cherts, siliceous mudstones and thin metabentonites. Facies intermediate between these conglomerates and cherts are not preserved. Furthermore, the earliest accreted slice immediately south of the ophiolite was not emplaced until *gracilis* times, more than 10 Ma after deposition of upper slope sediment began on top of the ophiolite (Figs 4 & 5). The missing stratigraphy may be explained by strike-slip removal of Llanvirn-Llandeilo accreted tracts and slope sediment, but sediment subduction is equally likely.

Subduction erosion may also have occurred. Longman *et al.* (1979) cite dated granitic and metamorphic boulders in a Llanvirn conglomerate from the supra-ophiolitic succession as evidence for a Cambrian-early Ordovician magmatic arc under the SW Midland Valley. No scale is shown on their diagrammatic reconstruction, but clearly the arc-trench gap would be far too narrow (a few tens of kilometres) to allow subduction from the line of the Southern Upland Fault, no matter how steep the subduction zone. Although strike-slip removal of a pre-middle Ordovician forearc terrane is again a possibility, mechanical erosion of the overriding Laurentian plate during Llanvirn-Llandeilo times may also be responsible. Lower ORS andesites are also anomalously close to the suture in the southern Midland Valley, giving an arc-trench gap of only *c.* 70 km (Fig. 2). Thirlwall (1981a) invokes a steep subduction zone associated with the late stages of closure to explain this. Subduction erosion during latest Silurian and/or Early Devonian may equally well be responsible, and would perhaps explain the lack of Early Devonian marine strata in the southern part of the accretionary complex if the Lower ORS is indeed of Early Devonian age.

There is no evidence for the missing Cambrian–early Ordovician forearc terrane in Ireland, and total strike-slip displacement of several 100 km would be necessary to remove it to the submerged extension of the Caledonides on the British continental shelf (assuming oblique ESE-WNW convergence at this time—Fig. 1 and Thirwall 1981a). We suspect that stratigraphic links across the Southern Upland Fault do not allow this amount of movement along the south border of the Midland Valley after the mid-Ordovician. The unique late Llandeilo-early Caradoc flysch rudites at Corsewall Point and in Glen App (sequence I, Fig. 4) occur immediately south of the Ballantrae ophiolite and contain clasts 90% of which can be matched with lithologies exposed in the ophiolite (Walton 1956; Kelling & Holroyd 1978). Petrography can similarly be linked between boulders in Wenlock conglomerates of the Midland Valley forearc succession and lithologies exposed in the Northern Belt, south of the Southern Upland Fault. So any strike-slip movements must have occurred in Cambrian and early Ordovician times.

Clearly the geological significance of the Ballantrae ophiolite bears heavily on this problem. The ophiolite has several atypical features. Differentiated lavas, as well as tholeiites, occur in the pillow lava and massive basalt sequences, which are anomalously thick compared with normal ocean crust (Lewis 1976). Voluminous monomict basaltic breccias are intercalated in the mafic sequences (Lewis & Bloxam 1977); rounded pebbles therein and intense vesicularity in pillows indicate eruption in relatively shallow water. There is no appreciable sheeted dyke complex.

Bluck *et al.* (1980), following Lewis & Bloxam (1977), use these features to interpret the mafic sequences in the ophiolite as part of an island arc assemblage. Dating construction of the ophiolite (from gabbro and trondjemite) and emplacement (from amphibolite and an associated middle Arenig olistrostrome), Bluck *et al.* (1980) show only a limited time gap between the two. This, they argue, indicates that elements of a short-lived marginal basin, closure of which emplaced the ophiolite on the Laurentian margin, are also included. Barrett *et al.* (in press) stress the lack of evidence for what should have been an appreciable magmatic arc terrane south of the ophiolite, such as those associated with Ordovician ophiolites in Newfoundland (Fig. 1). They prefer to interpret the Ballantrae ophiolite as accreted slivers of ocean floor volcanic edifices. This explanation is consistent with the atypical features of the ophiolite and explains its forearc setting (Laurentian ophiolites in Newfoundland are demonstrably of back-arc origin: see Dewey 1974 and Williams 1978). It also explains the intense disruption of the ophiolite by high-angle reverse faults. Intermittent accretion of high-standing ocean crust slivers, possibly in a regime of overall sediment subduction, would also explain the division of the ophiolite into Arenig mafic sequences north of the Stinchar Fault and a late Llandeilo or early Caradoc sequence (Downan mafic sequence, Fig. 4) to the south. Volcanism on the ocean plate just prior to accretion in the Arenig would explain the narrow time gap between construction of the ophiolite and emplacement.

Geochemistry of mafic rocks does not help with the debate as to whether the ophiolite was generated in an arc, composite arc/marginal basin, or ocean floor environment. On balance, arguments from regional geology best support an open-ocean origin for the ophiolite.

Finally, how is the latest Cambrian/early Ordovician Grampian orogeny explained? Evidence for a Cambrian/early Ordovician arc under the SW Midland Valley makes an arc/continent collision one of the more attractive theories for the Grampian Orogeny in Scotland (Longman *et al.* 1979). However, this does not readily explain the Grampian deformation in the west of Ireland (Connemara Dalradian, Fig.

2) where the Midland Valley-type basement rocks appear to be missing. It is possible that they have been removed by strike-slip faulting.

In summary, pre-middle Ordovician rocks in southern Scotland and northern Ireland give a tantalizingly incomplete picture of subduction-related events along the Laurentian margin. Sediment subduction, erosion, and strike-slip translation by their very nature inevitably result in 'lost' stratigraphy and other complexities. For this reason we do not expect to be able to reconstruct pre-middle Ordovician events as confidently as we do mid-Ordovician to late Silurian events.

Effects of closure of the Iapetus

Compared with the intense deformation which produced nappes in Norway, Greenland, and Shetland, the Southern Uplands seems to have escaped closure-related tectonism when the Laurentian and European continents collided. All major (D_1) structures (reverse faults and associated folds) feasibly relate to accretion (Anderson & Cameron 1979; Eales 1979; Leggett *et al.* 1979a; Weir 1979). Anderson & Cameron (1979) ascribe the D_2 deformation, which is mesocopic only, to closure. Other possible effects, summarized by Johnson *et al.* (1979), are thrust motion during initial movements on the Highland Boundary Fault and along the Moine Thrust, transcurrent movement on faults between these two, and folding in the Midland Valley.

Relative lack of closure-related deformation in the Southern Uplands could be due to diachronous collision, which according to Phillips *et al.* (1976) occurred earlier in the NE and produced considerable dextral strike-slip along the suture. Thirlwall (1981a), on the other hand, uses geochemical data in Lower ORS volcanics to argue for oblique ESE-WSW convergence and for a swing in the suture line to north-south under the North Sea. This is an attractive hypothesis, allowing head-on collision, and associated nappe formation, in Norway, Greenland and Shetland, and producing plate motion almost parallel to the northern British portion of the margin. Consequent slow subduction under the Southern Uplands would allow preservation of primary structures in the accretionary complex. Vein arrays near the suture indicate both dextral and sinistral movements (Anderson, pers. comm. 1981) and this topic requires further investigation.

Despite this evidence for oblique relative plate motion we suspect that, except along the suture, there are too many similarities in geology either side of the major faults exposed (Fig. 2) to allow strike-slip displacements on the scale of those in Cenozoic western North America. Across the Great Glen Fault, notwithstanding limited post-Early Devonian pre-Carboniferous strike-slip displacement (summary in Johnson *et al.* 1979) stratigraphy, structure and metamorphism in Precambrian rocks are very similar. The Highland Boundary Fault is not traceable west of the Midland Valley, and seems to be a local feature. Early Palaeozoic strike-slip movements may well have occurred along the Laurentian margin in Britain, but the fault or faults responsible are not obvious. Further assessment of this possibility must await palaeomagnetic work.

Basement below southern Scotland

Seismic velocities indicate that the basement below the Midland Valley comprises 6–14 km of Lower Palaeozoic rocks, already discussed, and a layer of pre-Caledonian high-grade metamorphic rocks extending to 20 km depth (Bamford *et al.* 1977; Bamford 1979). This metamorphic layer, with seismic velocities of 6.4 km s^{-1} abuts against a 6.3 km s^{-1} layer at the Southern Upland Fault. This 6.3 km s^{-1} layer under the Southern Uplands is overlain by *c.* 14–16 km of 5.8–6.0 km s^{-1} rocks interpreted as greywackes, and may be pre-Caledonian basement but this is not certain (Bamford 1979, p. 94). Gneissic xenoliths in Upper Palaeozoic vents confirm crystalline basement below the Midland Valley, but have not as yet been found in Carboniferous vents in the Southern Uplands. Strogen (1974) has found such xenoliths in Carboniferous vents immediately along strike from the Southern Uplands in central Ireland. This might indicate a local prong of continental basement at depth, perhaps explaining the greater overturning of strata in the western part of Longford-Down massif (Leggett *et al.* 1979b). It could equally well indicate continental basement below the whole accretionary terrane.

Evidence already outlined indicates that the Southern Uplands rocks derive originally from an ocean-floor/trench environment and cannot have been deposited *in situ* on a continental substrate. If the 6.3 km s^{-1} layer below the Southern Uplands is older continental material, we face the intriguing possibility that the entire fossil accretionary complex is allochthonous. If that did prove to be the case, the most likely explanation would be obduction of the accretionary terrane, possibly including the Ballantrae ophiolite and its Ordovician cover (cf. Weir 1979), over thinned continental basement, as a result of closure of Iapetus. Such a process need involve northwards translation of only several tens of kilometres. Recent seismic reflection work in the Southern Appalachians shows that already-deformed terranes can be translated several hundreds of kilometres over crystalline basement during collision (Cook *et al.* 1979).

Further indications of anomalous crustal structures below the Southern Uplands are the lack of a clearly defined Moho (Bamford 1979) and the occurrence of a block of rock of high electical conductivity extending well into the mantle (Hutton *et al.* 1980). However, there is no evidence in seismic work (Bamford 1979) or gravity work (Powell 1971) that a thick relic of Iapetus ocean crust has been preserved in the present crustal profile (Dewey 1971), as seems to have happened below the Newfoundland sector of the orogen (Sheridan 1974, figs 3 & 5). This does not negate the subduction-accretion explanation for southern Scotland. We suggest that it is entirely possible for an accretionary complex to be generated by subduction of an ocean of which all other traces were obliterated on closure.

Further resolution of crustal structure below southern Scotland awaits COCORP-type deep seismic reflection profiling.

ACKNOWLEDGMENTS: We are grateful to Mike Audley-Charles for reviewing the manuscript, and to Bernard Anderson, Alistair Robertson, and the participants on the 1980 Geological Society field trip to the Southern Uplands for many helpful discussions. JKL & DMC thank NERC for financial assistance during fieldwork.

References

ANDERSON, T. B. 1980. Deformation sequences in the Southern Uplands. *Scott. J. Geol.* **17**, 78–80.

ANDERSON, T. B. & CAMERON, T. D. J. 1979. A structural profile of Caledonian deformation in Down. *In:* HARRIS, A. L., HOLLAND, C. H. & LEAKE, B. E. (eds). *The Caledonides of the British Isles—reviewed.* Spec. Publ. geol. Soc. London, 263–7.

ARMSTRONG, M. & PATERSON, I. B. 1970. The Lower Old Red Sandstone of the Strathmore region. *Rep. No.* 70/12, *Inst. Geol. Sci.*

BACHMAN, S. B. 1981. The Coastal Belt of the Franciscan: youngest phase of northern California subduction (this volume).

BAMFORD, D. 1979. Seismic constraints on the deep geology of the Caledonides of northern Britain. *In:* HARRIS, A. L., HOLLAND, C. H. & LEAKE, B. E. (eds). *The Caledonides of the British Isles—reviewed.* Spec. Publ. geol. Soc. London, 93–6.

—— , NUNN, K., PRODEHL, C. & JACOB, B. 1977. LISPB–III. Upper crustal structure of Northern Britain. *J. geol. Soc. London,* **133**, 481–8.

BARRETT, T. J., JENKYNS, H. C., LEGGETT, J. K. & ROBERTSON, A. H. F. 1981. Age and origin of Ballantrae ophiolite and its significance to the Caledonian orogeny and Ordovician time scale: Comment. *Geology*, in press.

BERGSTRÖM, S. M. 1971. Conodont biostratigraphy of the Middle and Upper Ordovician of Europe and eastern North America. *In:* SWEET, W. C. & BERGSTRÖM, S. M. (eds). *Symposium on Conodont Biostratigraphy.* Mem. geol. Soc. Am. **127,** 83–162.

BLUCK, B. J. 1978. Geology of a continental margin 1: The Ballantrae Complex. *In:* BOWES, D. R. & LEAKE, B. E. (eds). *Crustal Evolution in Northwestern Britain and Adjacent regions.* Geol. J. Spec. Issue No. 10. 151–62.

—— , HALLIDAY, A. N. AFTALION, M., & MACINTYRE, R. M. 1980. Age and origin of Ballantrae ophiolite and its significance to the Caledonian Orogeny and Ordovician time scale. *Geology*, **8**, 492–5.

BORRADAILE, G. J. 1978. Transected folds: a study illustrated with examples from Canada and Scotland. *Bull. geol. Soc. Am.* **89,** 481–93.

BROWN, G. C. 1979. Geochemical and geophysical constraints on the origin and evolution of Caledonian granites. *In:* HARRIS, A. L., HOLLAND, C. H. & LEAKE, B. E. (eds). *The Caledonides of the British Isles—reviewed.* Spec. Publ. geol. Soc. London, 645–51.

CAMERON, T. D. J. & ANDERSON, T. B. 1979. Silurian metabentonites in County Down, Northern Ireland. *Geol. J.* **15,** 59–75.

CLARKSON, C. M., CRAIG, G. Y. & WALTON, E. K. 1975. The Silurian rocks bordering Kirkcudbright Bay, South Scotland. *Trans. R. Soc. Edinburgh,* **69**, 313–25.

COCKS, L. R. M. & TOGHILL, P. 1973. The biostratigraphy of the Silurian rocks of the Girvan district, Scotland. *J. geol. Soc. London*, **129,** 209–43.

COOK, D. R. & WEIR, J. A. 1979. Structure of the Lower Palaeozoic rocks around Cairnsmore of Fleet, Galloway. *Scott. J. Geol.* **15,** 187–202.

COOK, F. A., ALBAUGH, D. S., BROWN, L. D., KAUFMAN, S., OLIVER, J. E. & HATCHER, R. D., JR. 1979. Thin-skinned tectonics of the crystalline Appalachians: COCORP seismic reflection profiling of the Blue Ridge and Piedmont. *Geology*, **7**, 563–7.

COWAN, D. S. 1981. Deformation of partly dewatered and consolidated Franciscan sediments near Piedras Blancas Point, California (this volume).

—— & SILLING, R. M. 1978. A dynamic, scaled model of accretion at trenches and its implications for the tectonic evolution of subduction complexes. *J. geophys. Res.* **83,** 5389–96.

CRAIG, G. Y. & WALTON, E. K. 1959. Sequence and structure in the Silurian rocks of Kirkcudbrightshire. *Geol. Mag.* **96,** 209–20.

—— & —— 1962. Sedimentary structures and palaeocurrent directions from the Silurian rocks of Kircudbrightshire. *Trans. Edinb. geol. Soc.* **19,** 100–19.

CURRAY, J. R., MOORE, D. G., LAWVER, L. A., EMMEL, F. J., RAITT, R. W., HENRY, M. & KIECKHEFER, R. 1979. Tectonics of the Andaman Sea and Burma. *Mem. Am. Assoc. Petrol. Geol.* **29,** 189–98.

DEWEY, J. F. 1969. Evolution of the Appalachian/Caledonian orogen. *Nature, London*, **222,** 124–9.

—— 1971. A model for the Lower Palaeozoic evolution of the early Caledonides of Scotland and Ireland. *Scott. J. Geol.* **7,** 219–40.

—— 1974. Continental margins and ophiolite obduction: Appalachian Caledonian System. *In:* BURK, C. A. & DRAKE, C. L. (eds). *The Geology of Continental Margins*, 933–50. Springer-Verlag, New York.

—— & KIDD, W. S. F. 1974. Continental collisions in the Appalachian Caledonian orogenic belt. Variations related to complete and incomplete suturing. *Geology*, **2,** 543–6.

DICKINSON, W. R. & SEELY, D. R. 1979. Structure and stratigraphy of forearc regions. *Bull. Am. Assoc. Petrol. Geol.* **63,** 2–31.

EALES, M. H. 1979. Structure of the Southern Uplands. *In:* HARRIS, A. L., HOLLAND, C. H. & LEAKE, B. E. (eds). *The Caledonides of the British Isles—reviewed.* Spec. Publ. geol. Soc. London, 269–73.

ECKFORD, R. J. A. & RITCHIE, M. 1931. The lavas of Tweeddale and their position in the Caradocian sequence. *Summ. Prog. geol. Surv. G.B.* Part 3, 46–57.

FLOYD, J. D. 1976. *The Ordovician rocks of West Nithsdale.* Thesis, PhD. Univ. St Andrews (unpubl.).

FYFE, T. B. & WEIR, J. A. 1976. The Ettrick Valley Thrust and the upper limit of the Moffat Shales in Craigmichan Scaurs (Dumfries and Galloway Region: Annandale and Eskdale District). *Scott. J. Geol.* **12,** 93–102.

GORDON, A. J. 1962. *The Lower Palaeozoic rocks around Glenluce, Wigtonshire.* Thesis, PhD. Univ. Edinburgh (unpubl.).

GRAHAM, A. M. & UPTON, B. G. J. 1978. Gneisses in diatremes, Scottish Midland Valley: petrology and tectonic implications. *J. Geol. Soc. London,* **135,** 219–28.

HARRIS, A. L., HOLLAND, C. H. & LEAKE, B. E. (eds). 1979. *The Caledonides of the British Isles—reviewed.* Spec. Publ. geol. Soc. London. Scottish Academic Press, Edinburgh. 768 pp.

HENDERSON, W. G. & ROBERTSON, A. H. F. 1981. The Highland Border rocks and their relation to marginal basin development in the Scottish Caledonides. *Geol. Soc. Newsl. London*, **10,** (2), 23–4.

HENRIKSEN, N. & HIGGINS, A. K. 1976. East Greenland Caledonian fold belt. *In:* ESCHER, A. & WATT, W. S. (eds). *Geology of Greenland*, 182–246. Geological Survey of Greenland.

HEPWORTH, B. C., OLIVER, G. J. H. & MCMURTRY, M. J. 1981. Sedimentology, volcanism, structure and metamorphism of the Lower Palaeozoic accretionary complex; Bail Hill–Abington area of the Southern Uplands of Scotland (this volume).

HUTTON, V. R. S., INGHAM, M. R. & MBIPOM, E. W. 1980. An electrical model of the crust and upper mantle in Scotland. *Nature, London*, **287,** 30–2.

INGHAM, J. K. 1978. Geology of a continental margin 2: middle and late Ordovician transgression, Girvan. *In:* BOWES, D. R. & LEAKE, B. E. (eds). *Crustal Evolution in Northwestern Britain and Adjacent Regions.* Geol. J. Spec. Issue No. 10, 163–76.

JOHNSON, M. R. W., SANDERSON, D. J. & SOPER, N. J. 1979. Deformation in the Caledonides of England, Ireland and Scotland. *In:* HARRIS, A. L., HOLLAND, C. H. & LEAKE, B. E. (eds). *The Caledonides of the British Isles—reviewed.* Spec. Publ. geol. Soc. London, 165–86.

JONES, J. G. 1969. Pillow lavas as depth indicators. *Am. J. Sci.* **267,** 181–95.

KARIG, D. E., LAWRENCE, M. B., MOORE, G. F. & CURRAY, J. R. 1980. Structural framework of the fore-arc basin, NW Sumatra. *J. geol. Soc. London*, **137,** 77–91.

KELLING, G. 1961. The stratigraphy and structure of the Ordovician rocks of the Rhinns of Galloway. *Q. J. geol. Soc. London,* **117,** 37–75.

—— 1962. The petrology and sedimentation of the Upper Ordovician rocks of the Rhinns of Galloway, south-west Scotland. *Trans. R. Soc. Edinburgh*, **65,** 107–37.

—— & HOLROYD, J. 1978. Clast size, shape, and composition in some ancient and modern fan gravels. *In:* STANLEY, D. J. & KELLING, G. (eds). *Sedimentation in Submarine Canyons, Fans and Trenches,* 138–59. Dowden, Hutchinson & Ross, Stroudsburg, Pennsylvania.

—— & WELSH, W. 1970. The Loch Ryan Fault. *Scott. J. Geol.* **6,** 266–71.

KULM, L. D., RESIG, J. M., THORNBURG, T. M. & SCHRADER, H.-J. 1981. Cenozoic structure, stratigraphy and tectonics of the central Peru forearc (this volume).

LAMBERT, R. ST. J., HOLLAND, J. G. & LEGGETT, J. K. 1981. Petrology and tectonic setting of some Ordovician volcanic rocks from the Southern Uplands of Scotland. *J. geol. Soc. London*, **138,** 421–36.

LAMONT, A. & LINDSTRÖM, M. 1957. Arenigian and Llandeilian cherts identified in the Southern Uplands of Scotland by means of conodonts etc. *Trans. geol. Soc. Edinburgh,* **17,** 60–70.

LEGGETT, J. K. 1978. Eustacy and pelagic regimes in the Iapetus Ocean during the Ordovician and Silurian. *Earth planet. Sci. Lett.* **41,** 163–9.

—— 1979. Oceanic sediments from the Ordovician of the Southern Uplands. *In:* HARRIS, A. L., HOLLAND, C. H. & LEAKE, B. E. (eds). *The Caledonides of the British Isles—reviewed.* Spec. Publ. geol. Soc. London, 495–8.

—— 1980a. The sedimentological evolution of a

Lower Palaeozoic accretionary fore-arc in the Southern Uplands of Scotland. *Sedimentology,* **27,** 401–17.

—— 1980b. Palaeogeographic setting of the Wrae Limestone: an Ordovician submarine slide deposit in Tweeddale. *Scott. J. Geol.* **16,** 91–104.

—— 1980c. British Lower Palaeozoic black shales and their palaeo-oceanographic significance. *J. geol. Soc. London,* **137,** 139–56.

—— & CASEY, D. M. 1981. The Southern Uplands accretionary prism: implications for controls on structural development of subduction complexes. *In:* DRAKE, C. L. & WATKINS, J. S. (eds). *Continental Margin Processes*. Spec. Publ. Am. Assoc. Petrol. Geol. (in press).

—— , MCKERROW, W. S., COCKS, L. R. M. & RICKARDS, R. B. 1981. Periodicity in the early Palaeozoic marine realm. *J. geol. Soc. London,* **138,** 167–76.

—— , MCKERROW, W. S. & EALES, M. H. 1979a. The Southern Uplands of Scotland; a Lower Palaeozoic accretionary prism. *J. geol. Soc. London*, **136,** 755–70.

—— , MCKERROW, W. S., MORRIS, J. H., OLIVER, G. J. H. & PHILLIPS, W. E. A. 1979b. The north-western margin of the Iapetus Ocean. *In:* HARRIS, A. L., HOLLAND, C. H. & LEAKE, B. E. (eds). *The Caledonides of the British Isles—reviewed.* Spec. Publ. geol. Soc. London, 499–511.

—— & SMITH, T. K. 1980. Fe-rich deposits associated with Ordovician basalts in the Southern Uplands of Scotland: possible Lower Palaeozoic equivalents of modern active ridge sediments. *Earth planet. Sci. Lett.* **47,** 431–40.

LEWIS, A. D. 1976. *The geochemistry and geology of Girvan-Ballantrae ophiolite and related Ordovician volcanics in the Southern Uplands of Scotland.* Thesis, PhD. Univ. Wales (unpubl.).

LEWIS, A. D. & BLOXAM, T. W. 1977. Petrotectonic environments of the Girvan-Ballantrae lavas from rare earth element distributions. *Scott. J. Geol.* **13,** 211–22.

LONGMAN, C. D., BLUCK, B. J. & VAN BREEMEN, O. 1979. Ordovician conglomerates and evolution of Midland Valley. *Nature, London*, **280,** 578–81.

LUMSDEN, G. I., TULLOCH, W., HOWELLS, M. F. & DAVIES, A. 1967. The geology of the neighbourhood of Langholm. *Mem. geol. Surv. Scotland.* Sheet 11. 225 pp.

MCGIVEN, A. 1968. *Sedimentation and provenance of post-Valentian conglomerates up to and including the basal conglomerate of the Lower Old Red Sandstone in the southern part of the Midland Valley of Scotland.* Thesis, PhD. Univ. Glasgow (unpubl.).

MCKERROW, W. S. 1979. Ordovician and Silurian changes in sea level. *J. geol. Soc. London,* **136,** 137–45.

—— 1981. The Northwest margin of the Iapetus Ocean during the early Palaeozoic. *In:* DRAKE, C. L. & WATKINS, J. S. (eds). *Continental Margin Processes.* Spec. Publ. Am. Assoc. Petrol. Geol. (in press).

—— & COCKS, L. R. M. 1976. Progressive faunal migration across the Iapetus Ocean. *Nature, London,* **263,** 304–6.

—— , LAMBERT, R. ST.J. & CHAMBERLAIN, V. E. 1980. The Ordovician, Silurian and Devonian time-scale. *Earth planet. Sci. Lett.* **51,** 1–8.

—— , LEGGETT, J. K. & EALES, M. H. 1977. Imbricate thrust model of the Southern Uplands of Scotland. *Nature, London,* **267,** 237–9.

MITCHELL, A. H. G. & MCKERROW, W. S. 1975. Analogous evolution of the Burma orogen and the Scottish Caledonides. *Bull. geol. Soc. Am.* **86,** 305–15.

MOORE, G. F., CURRAY, J. R. & EMMEL, F. J. 1981. Sedimentation in the Sunda Trench and Forearc region (this volume).

—— & KARIG, D. E. 1980. Structural geology of Nias Island, Indonesia: implications for subduction zone tectonics. *Am. J. Sci.* **280,** 193–223.

MOORE, J. C. 1975. Selective subduction. *Geology*, **3,** 530–2.

—— 1979. Variation in strain and strain rate during underthrusting of trench deposits. *Geology*, **7,** 185–8.

—— & KARIG, D. E. 1976. Sedimentology, structural geology, and tectonics of the Shikoku subduction zone, southwestern Japan. *Bull. geol. Soc. Am.* **87,** 1259–68.

—— & WATKINS, J. S. *et al.* 1979. Progressive accretion in the Middle America Trench, Southern Mexico. *Nature, London,* **281,** 638–42.

—— & —— 1981. Tectonic processes along the Middle America Trench inner slope (this volume).

MORRIS, J. H. 1979. Lower Palaeozoic soft-sediment deformation structures in the western end of the Longford-Down inlier, Ireland. *In:* HARRIS, A. L., HOLLAND, C. H. & LEAKE, B. E. (eds). *The Caledonides of the British Isles—reviewed.* Spec. Publ. Geol. Soc. London, 513–6.

OLIVER, G. J. H. 1978. Prehnite—pumpellyite facies metamorphism in County Cavan, Ireland. *Nature, London*, **274,** 242–3.

—— & LEGGETT, J. K. 1980. Metamorphism in an accretionary prism: prehnite-pumpellyite facies metamorphism of the Southern Uplands of Scotland. *Trans. R. Soc. Edinburgh,* **71,** 235–46.

PEACH, B. N. & HORNE, J. 1899. The Silurian rocks of Britain, 1, Scotland. *Mem. geol. Surv. Scotland,* 749 pp.

PHILLIPS, W. E. A., STILLMAN, C. J. & MURPHY, T. 1976. A Caledonian plate tectonic model. *J. geol. Soc. London,* **132,** 579–609.

—— , FLEGG, A. M. & ANDERSON, T. B. 1979. Strain adjacent to the Iapetus suture in Ireland. *In:* HARRIS, A. L., HOLLAND, C. H. & LEAKE, B. E. (eds). *The Caledonides of the British Isles—reviewed.* Spec. Publ. geol. Soc. London, 257–62.

POWELL, D. W. 1971. A model for the Lower Palaeozoic evolution of the southern margin of the early Caledonides of Scotland and Ireland. *Scott. J. Geol.* **7,** 369–72.

—— 1978. Geology of a continental margin 3: gravity and magnetic anomaly interpretation of the Girvan-Ballantrae district. *In:* Bowes, D. R. & Leake, B. E. (eds). *Crustal Evolution in Northwestern Britain and Adjacent Regions.* Geol. J. Spec. Issue No. 10, 177–82.

Rolfe, W. D. I. 1961. The geology of the Hagshaw Hills Silurian Inlier. *Trans. Edinb. geol. Soc.* **18,** 240–69.

Rust, B. R. 1965. The stratigraphy and structure of the Whithorn area of Wigtonshire, Scotland. *Scott. J. Geol.* **1,** 101–33.

Ryan, P. D. & Archer, J. B. 1977. The South Mayo Trough: a possible Ordovician Gulf of California-type marginal basin in the west of Ireland, *Can. J. Earth Sci.* **14,** 2453–61.

Sanders, I. S. & Morris, J. H. 1978. Evidence for Caledonian subduction from greywacke detritus in the Longford-Down inlier. *J. Earth Sci. R. Dubl. Soc.* **1,** 53–62.

Scholl, D. W., Von Huene, R., Vallier, T. L. & Howell, D. G. 1981. Sedimentary masses and concepts about tectonic processes at underthrust ocean margins. *Geology*, **8,** 564–8.

Seely, D. R., Vail, P. R. & Walton, G. G. 1974. Trench Slope Model. *In:* Burk, C. A. & Drake, C. L. (eds). *The Geology of Continental Margins*, 249–60. Springer-Verlag, New York.

Sheridan, R. E. 1974. Atlantic continental margin of North America. *In:* Burk, C. A. & Drake, C. L. (eds). *The Geology of Continental Margins.* 391–407. Springer-Verlag, New York.

Shipley, T. H., Ladd, J. W., Buffler, R. T. & Watkins, J. S. 1981. Facies belts of the Middle America Trench and forearc region, southern Mexico: results from Leg 66 DSDP (this volume).

Stillman, C. J. & Francis, E. H. 1979. Caledonide volcanism in Britain and Ireland. *In:* Harris, A. L., Holland, C. H. & Leake, B. E. (eds). *The Caledonides of the British Isles—reviewed.* Spec. Publ. geol. Soc. London, 557–77.

Stringer, P. & Treagus, J. E. 1980. Non-axial planar cleavage in the Hawick Rocks of the Galloway area, Southern Uplands, Scotland. *J. struct. Geol.* **2,** 129–48.

Strogen, P. 1974. The sub-Palaeozoic basement in central Ireland. *Nature, Phys. Sci.* **250,** 262–3.

Thirlwall, M. F. 1981a. Implications for Caledonian plate tectonic models of chemical data from volcanic rocks of the British Old Red Sandstone. *J. geol. Soc. London,* **138,** 123–38.

—— 1981b. Peralkaline rhyolites from the Ordovician Tweeddale lavas, Peebleshire. *Geol. J.* **16,** 41–4.

Tipper, J. C. 1974. *The Marine Silurian of the North Esk Inlier.* Thesis, PhD. Univ. Edinburgh (unpubl.).

Toghill, P. 1970a. Highest Ordovician (Hartfell Shales) graptolite faunas from the Moffat area, south Scotland. *Bull. Br. Mus. nat. Hist., Geol.* **19,** 1–26.

—— 1970b. The south-east limit of the Moffat Shales in the upper Ettrick valley region, Selkirkshire. *Scott. J. Geol.* **6,** 233–42.

Walton, E. K. 1955. Silurian greywackes in Peebleshire. *Proc. R. Soc. Edinb.* **B65,** 327–57.

—— 1956. Two Ordovician conglomerates in South Ayrshire. *Trans. geol. Soc. Glasgow*, **22,** 133–56.

—— 1961. Some aspects of the succession and structure in the Lower Palaeozoic rocks of the Southern Uplands of Scotland. *Geol. Rdsch.* **50,** 63–77.

—— 1965. Lower Palaeozoic rocks: stratigraphy, palaeogeography and structure. *In:* Craig, G. Y. (ed.) *The Geology of Scotland*, 161–227. Oliver & Boyd, Edinburgh.

—— 1968. Some rare sedimentary structures in the Silurian rocks of Kirkcudbrightshire. *Scott. J. Geol.* **4,** 335–69.

Warren, P. T. 1963. The petrography, sedimentation and provenance of the Wenlock rocks near Hawick, Roxburghshire. *Trans. geol. Soc. Edinburgh,* **19,** 225–55.

—— 1964. The stratigraphy and structure of the Silurian rocks south-east of Hawick, Roxburgshire. *Q. J. geol. Soc. London*, **120,** 192–222.

Watkins, J. S., Moore, J. C. *et al.* 1981. Accretion underplating, subduction and tectonic evolution—Middle America Trench, southern Mexico: results from Leg 66 DSDP. *In: Oceanologica Acta*, in press.

Weir. J. A. 1968. Structural history of the Silurian rocks of the coast west of Gatehouse, Kirkcudbrightshire. *Scott. J. Geol.* **4,** 31–52.

—— 1977. The Ettrick Valley Thrust and the upper limit of the Moffat Shales in Craigmichan Scaurs: a reply. *Scott. J. Geol.* **13,** 75–7.

—— 1979. Tectonic contrasts in the Southern Uplands. *Scott. J. Geol.* **15,** 169–86.

Welsh, W. 1964. *The Ordovician rocks of North-West Wigtonshire.* Thesis, PhD. Univ. Edinburgh (unpubl.).

Williams, A. 1959. A structural history of the Girvan district, S.W. Ayrshire. *Trans. R. Soc. Edinburgh*, **63,** 629–97.

—— 1962. The Barr and Lower Ardmillan Series (Caradoc) of the Girvan district, southwest Ayrshire. *Mem. geol. Soc. London*, **3,** 267 pp.

Williams, H. 1978. Geological development of the northern Appalachians: its bearing on the evolution of the British Isles. *In:* Bowes, D. R. & Leake, B. E. (eds). *Crustal Evolution in Northwestern Britain and Adjacent Regions.* Geol. J. Spec. Issue No. 10, 1–22.

Jeremy K. Leggett, Department of Geology, Royal School of Mines, Imperial College, London SW7 2BP.

W. Stuart McKerrow, David M. Casey, Department of Geology and Mineralogy, Parks Road, Oxford OX1 3PR.

Sedimentology, volcanism, structure and metamorphism of the northern margin of a Lower Palaeozoic accretionary complex; Bail Hill–Abington area of the Southern Uplands of Scotland

B. C. Hepworth, G. J. H. Oliver & M. J. McMurtry

SUMMARY: NE-SW faults in the Bail Hill–Abington area of the Northern Belt of the Southern Uplands define blocks up to 3.2 km wide. The strata, folded and locally overturned, young predominantly to the NW but successive blocks to the SW contain progressively younger sequences. Analogous configurations occur in modern accretionary margins. The oldest rocks are Arenig basalts, dolerites, cherts and brown mudstones underlying red shales, possibly Llanvirn, and black fossiliferous shales and cherts of Llandeilo and Caradoc age.

Trench sediments overlying pelagic sequences represent a range of depositional mechanisms. Rudites and associated fine-grained lithologies of lateral origin relate to a lower trench slope canyon system, whilst axially transported sands, originating on the lower trench slope, were deposited by turbidity currents and related flows.

The Bail Hill Volcanic Group (Upper Llandeilo) represents a mildly alkaline seamount in the Iapetus Ocean, with volcanic activity spanning the transition from pelagic plate to trench sedimentation before accretion.

Faults, initially low-angle thrusts, and bedding were rotated through the vertical within the accretionary complex, pre-dating or accompanying slaty cleavage development. Soft sediment deformation, two fold phases and a kink-band set are recognized. Imbricate fault zones located in incompetent pelagic sequences are tentatively equated with tectonic mélanges of other accretionary complexes.

Index minerals, illite crystallinity and 'vitrinite' reflectance establish metamorphic grade as a zeolite to prehnite-pumpellyite facies.

The Southern Uplands of Scotland are familiar as a Lower Palaeozoic accretionary complex, based on evidence summarized in Leggett *et al.* (1981). The Northern Belt (Peach & Horne 1899) consists entirely of Ordovician sediments and volcanics and forms the northernmost zone of the complex. Caledonian deformation produced a series of fault-bounded blocks (Fig. 1) of steeply dipping strata (Mitchell 1974; McKerrow *et al.* 1977).

Stratigraphy

Detailed stratigraphical sequences for the area are published elsewhere (McMurtry 1980a; Hepworth, in preparation). Formation distribution and age relationships are shown schematically in Fig. 2. New geological names are introduced in italics. The pelagic succession is described in this section in a conventional manner, starting with the oldest lithologies. Trench sediments are classified into formations based on lithofacies and are described in the next section, followed by a description of the *Bail Hill Volcanic Group* (McMurtry 1980a). The stratigraphical correlations of Williams *et al.* (1972) are used throughout.

Raven Gill Formation (70 m)

This lowest unit in the Northern Belt consists of massive basalts, dolerites, blue-grey radiolarian cherts and brown mudstones (type section—Raven Gill NS 920199). The mudstones contain conodonts (Lamont & Lindström 1957) and graptolites (Peach & Horne 1899) dated as Lower Arenig, and inarticulate brachiopods less specific in age. The igneous rocks include spilitic lavas displaying vesicular and trachytic textures. A 0.4 m wide shear zone separates this formation from tectonically lower but stratigraphically higher turbiditic sandstones to the south.

Post-Lower Arenig, Pre-Moffat Shale strata (minimum thickness 20 m)

Conodonts in red shales from the Leadhills Imbricate Zone 4 km ENE of Abington, the Fardingmullach Line in west Nithsdale (Floyd 1975) and at Morroch Bay in the Rhinns of Galloway, have been dated as Llandeilo (Lamont & Lindström 1957). Faunas of upper Arenig or Llanvirn age (*D. hirundo* to *D. murchisoni* zones) have not been found in the Southern Uplands. Black shales exposed along the northern sides of the *Eller* and *Howcon Faults* are underlain by green and grey radiola-

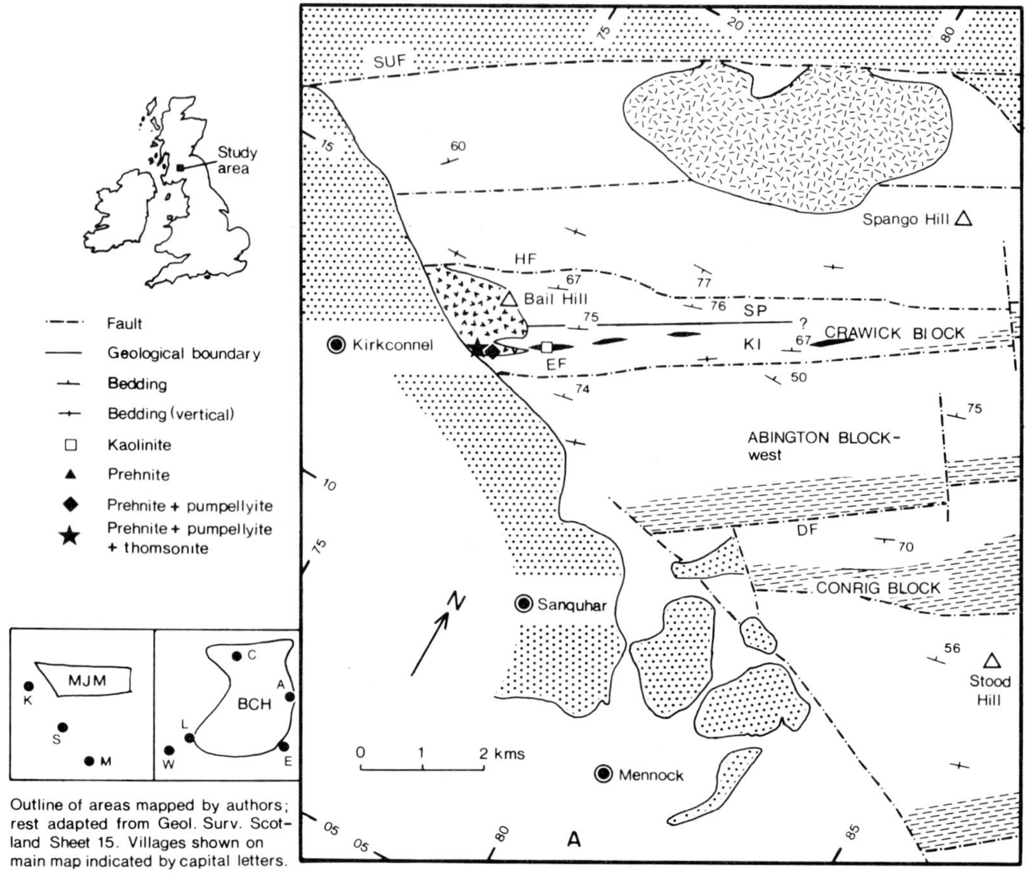

Fig. 1.(a) Map of the Ordovician rocks of the Bail Hill–Abington area. The Leadhills Imbricate Zone extends from the Abington area south-westwards into the Bail Hill district where it is inferred structurally to underlie the Conrig Block. The Duntercleuch Fault is interpreted as a splay from the Leadhills Imbricate Zone. The Spango Granite is Devonian in age.

Formations

AB Abington Formation
CF Crawfordjohn Formation
CR Conrig Formation
EV Elvan Formation
GC Glencaple Formation
GU Guffock Formation
KI Kiln Formation
MB Mill Burn Member
RG Raven Gill Formation
SP Spothfore Formation

Faults

CF Crawfordjohn Fault
DF Duntercleuch Fault
EF Eller Fault
FF Fardingmullach Fault
HF Howcon Fault
SBF Shield Burn Fault
SUF Southern Uplands Fault
WDF Wellgrain Dod Fault

Structural abbreviations

CJB Crawfordjohn Block
FIZ Fardingmullach Imbricate Zone
GF Glencaple fold belt
LIZ Leadhills Imbricate Zone
MBB Mill Burn Block
SBIZ Shield Burn Imbricate Zone

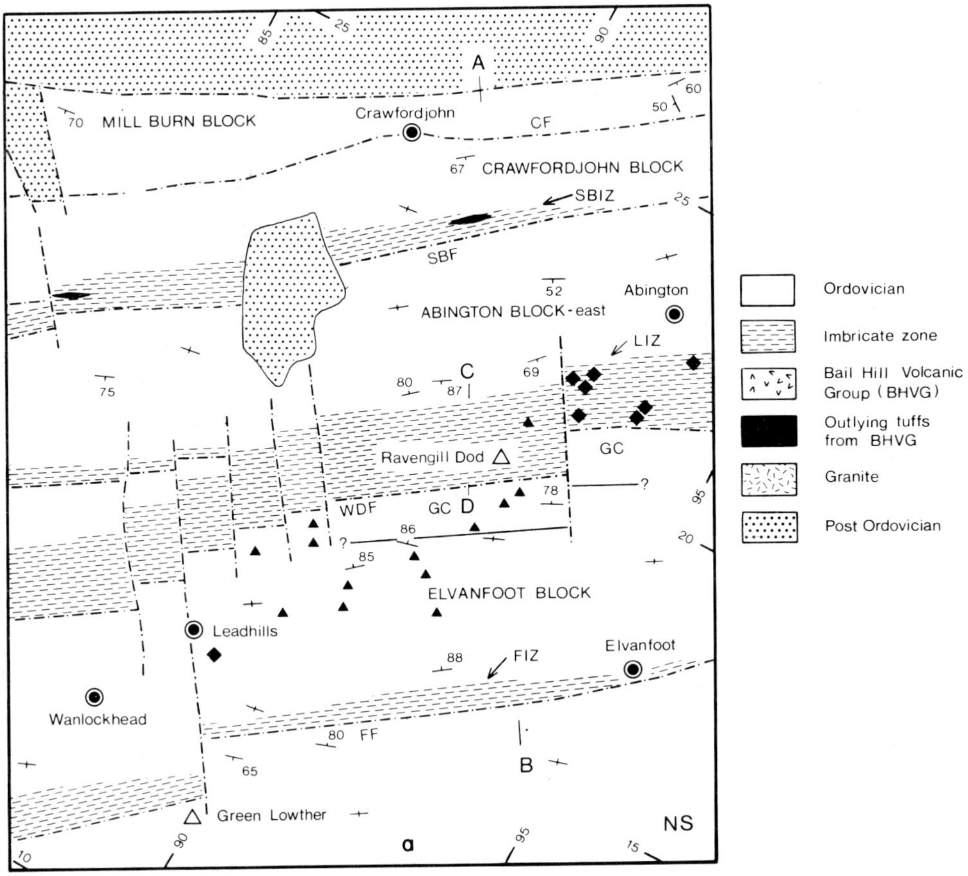

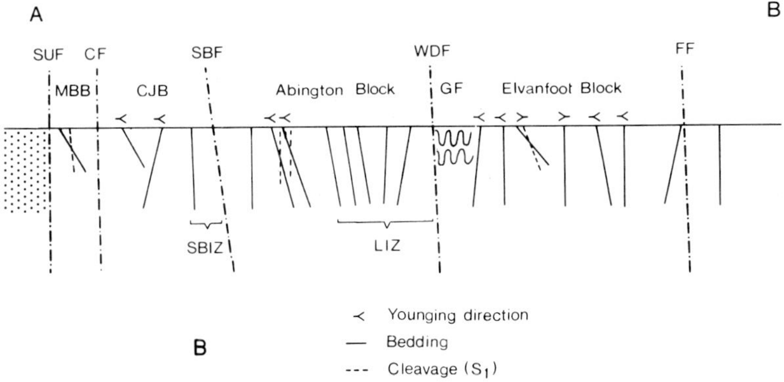

FIG. 1.(b) Schematic structural profile across the Abington district. Stratigraphical sequences are cut up into a series of NE-SW orientated blocks by major reverse faults. Line of section and scale given on main diagram.

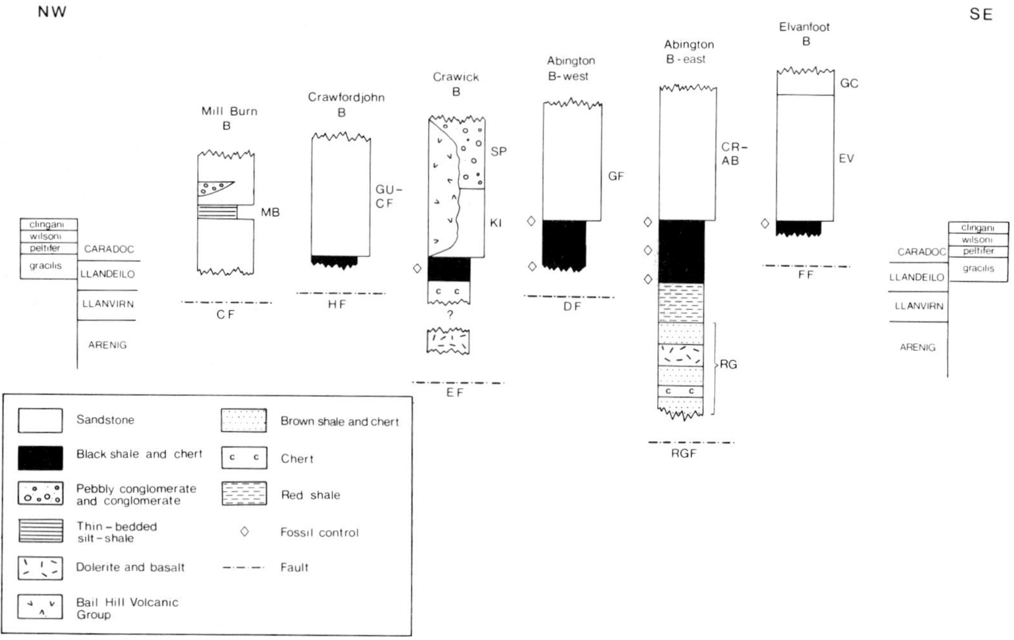

FIG. 2. Stratigraphical sequences of the structural blocks in the Bail Hill–Abington area. For abbreviations see Fig. 1. Based on McMurtry (1980a) and Hepworth (unpublished data).

rian cherts which have an estimated minimum thickness of a few tens of metres; Leggett *et al.* (1979) described a 100 m thick continuous section in the Coulter-Noblehouse area (equivalent to the *Mill Burn Block*, Fig. 2), consisting of red cherts, siliceous mudstones and turbiditic sandstones. These unfossiliferous chert sequences may represent the Upper Arenig-Llanvirn time span (e.g. Peach & Horne 1899).

Moffat Shales (*c.* 20 m)

Across the Northern and Central Belts (Peach & Horne 1899) black shale sedimentation commenced during *N. gracilis* times (Fig. 2). The black shales, with interbedded radiolarian cherts, are typically exposed to the north of faults, with younger graptolite zones and/or faunas cropping out in successive fault blocks to the SE (Lapworth 1878; Walton 1965; McKerrow *et al.* 1977). The shales pass conformably up into turbidites at the base of the *Elvan Formation*.

Clastic sediments

Turbiditic sandstones conformably overlie the Moffat Shales and have been subdivided into formations on the basis of sedimentology and petrography. Modal analysis (Fig. 3) shows petrographic compositions ranging from quartz-rich (e.g. *Glenflosh Formation,* McMurtry 1980a) to ferromagnesian-rich (e.g. Elvan Formation). Graptolite evidence from the Moffat Shales (Fig. 2) shows a progressive south-eastward delay in the onset of clastic sedimentation, as estabished for the Northern and Central Belts (e.g. Lapworth 1878).

Sandstone sequences attain a maximum thickness of 3200 m in the *Elvanfoot Block*; this value probably reflects tectonic thickening by faults rather than the original depositional thickness. Assuming a moderate sedimentation rate for trench sediments of 1000 m Ma^{-1} (Piper *et al.* 1973) and a post-depositional compaction of 50% (cf. Moore *et al.* 1980), the clastic sediments in the Elvanfoot Block might have been deposited in less than 6.5 Ma. This figure is compatible with the time range of 6.5 Ma calculated by Carter *et al.* (1980) for the *D. clingani* zone. It is possible that the other, thinner turbidite formations were likewise deposited within the space of a single graptolite zone.

Sedimentology

Sedimentary facies within the clastic formations were produced by sedimentary processes operating within a trench environment (Piper 1972; Leggett 1980). The sedimentary facies have

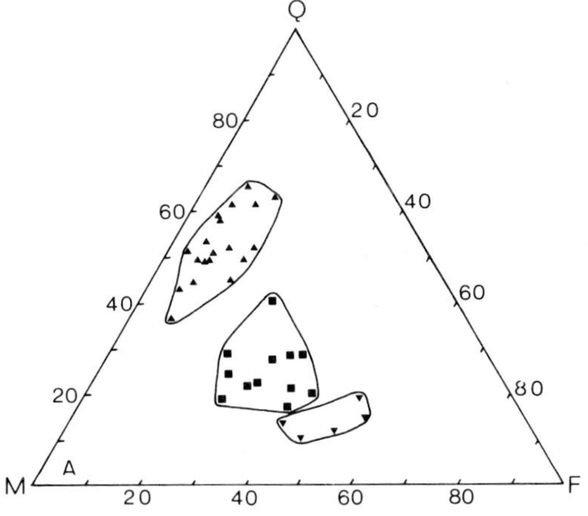

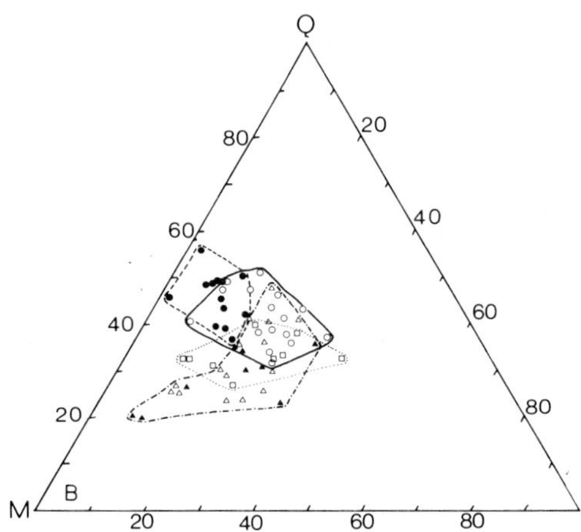

FIG. 3.(A) Modal analyses of sandstones; triangles = Abington Formation; squares = Glencaple Formation; inverted triangles = Elvan Formation. Q = quartz + acid + metamorphic fragments. F = feldspar + amphibole + pyroxene + basic fragments. M = matrix + mica + sedimentary fragments. Analysed by B.C.H.
(B) Modal analyses of sandstones; filled circles = Glenflosh Formation; open circles = Kiln Formation; squares = Guffock Formation; open triangles = Spothfore Formation (matrix); filled triangles = Spothfore Formation (clasts). Analysed by M.J.M.

been separated into those of lateral and those of axial origin (Table 1 and Fig. 4). Stratigraphy and field relations are not implied in the discussion, unless stated, and the lithologies are described in a progressively downcurrent direction (i.e. becoming more distal).

Laterally transported sediments

The Crawick Block (Fig. 2) includes four facies associations which define a large-scale thickening and coarsening-upward cycle (Ricci-Lucchi 1975). Pelagic black Moffat Shales at the base of the block pass upwards into the thin-bedded *Kiln Formation* (McMurtry 1980a), in which two facies associations define upper and lower (*Sheil Rig* and *Butt Hill*) members of the formation. The Kiln Formation is succeeded conformably by the rudaceous *Spothfore Formation* (McMurtry 1980a). Both formations are laterally impersistent (Fig. 1).

Spothfore Formation (300–800 m)

The Spothfore Formation (type section—upper Spothfore Burn 767138, Fig. 1a) consists of rudites, coarse- to fine-grained sandstones and siltstones. Clasts range in size up to several metres and are all of intrabasinal origin, with sediments constituting 95% and volcanic rocks 5% of the total. The rudites are generally disorganised, suggesting emplacement by debris flows (Middleton & Hampton 1973), although local lamination and imbrication indicate currents flowing towards the SE, transverse to the postulated trench (McMurtry 1980a). Thin-bedded sandstones and siltstones interbedded with the rudites were also deposited by currents flowing towards the SE.

Laterally discontinuous sequences of red and green chert with dimensions of several tens of metres both across and along strike crop out within the Spothfore rudites. The cherts show randomly orientated folds (782153) in contrast to nearby rudites which are unfolded; bedding is oblique to that in rudites (780153) and the contacts are sharp and irregular. The cherts are interpreted as large gravity-emplaced (exotic) blocks within the rudites.

Graptolite faunas in shale clasts in the rudites are from low in the *N. gracilis* zone and are older than *in situ* faunas from the Crawick Block. This suggests the clasts (including the large chert blocks) were derived from the north, originating from earlier accreted sediments and/or lower slope deposits.

The Spothfore Formation was formed in an inner fan environment, related to a canyon system incised across the lower slope of the accretionary complex.

Kiln Formation (700–900 m)

The thin-bedded Kiln Formation (type section in the Kiln Burn between 768142 and 773133) comprises six members. The Butt Hill and Sheil Rig Members represent distinct

TABLE 1. Summary of sedimentological characteristics of selected facies from the Bail Hill–Abington area. 1 = Spothfore Formation; 2 = Sheil Rig Member; 3 = Butt Hill Member; 4 = Elvan Formation; 5 = Abington Formation thick-bedded facies; 6 = Abington Formation medium-bedded facies; 7 = Guffock Formation; 8 = Glencaple Formation. *Superscripted number* (1) Describes sand layer only. (2) D = distribution grading, I = inverse grading. (3) After Walker & Mutti (1973). (4) After Mutti & Ricci-Lucchi (1975). (5) Direction given from source to depositional site

Characteristic facies	1	2	3	4	5	6	7	8
Bed thickness (i) range	0.6–5m	mm–cm	mm–cm	0.4–4m	0.2–4m	15–120cm	0.3–0.6m	10–80cm
(ii) average	–	–	–	105cm	110m	32cm	30cm	28cm
Lateral wedging	–	yes	no	–	rare	no	no	no
Amalgamation	yes	–	–	yes	yes	rare	–	rare
Sharp tops [1]	yes	yes	yes	yes	no	no	no	no
Sharp bases [1]	yes	yes	yes	yes	yes	yes	yes	yes
Sole marks	no	no	no	no	no	no	yes	no
Grain size (i) typical	CS-boulder	MS	FS–MS	VCS	VCS	CS	CS	CS
(ii) maximum	5m	–	–	granule	pebble	VCS	–	pebble
Graded [2]	no	no	rarely	no	D, I	D	D	D
Bouma sequences	no	no	no	no	ae	ae	ae	a–e
Mudst./sandst. clasts	<2m	no	no	<2cm	<0.3m	yes	yes	yes
Soft sed. deformation	no	no	yes	rare	no	no	no	yes
Facies [3]	A1	E	–	B	C	C	C	C[4]
Currents [5]	SE	SE	–	–	–	SW	SW	SW
Associated facies [3]	A2, E, F	A1, C2[4]	C2[4]	C2[4], E	B2, D	D, E	D, E	C1[4], B2, D, G
Mode of deposition	Debris flow	Reworked turbidite	"Overbank" turbidite	High conc. grav. flow	High conc. turbidite	Turbidite	Turbidite	Low conc. turbidite

sedimentary facies within which four tuffaceous members crop out (see below).

The Kiln Formation occurs between the underlying Moffat Shales and the overlying Spothfore Formation; deposition of these sediments thus spans the transition from an area of pelagic to a more proximal environment of channel-mouth sedimentation.

Sheil Rig Member. This upper member underlies and is interbedded with rudites of the Spothfore Formation. It is dominated by lenticular, cross-laminated, fine-grained sandstones and siltstones and is considered to be of a proximal nature. The sandy laminae are less than 3 cm thick, are moderately sorted and contain abundant matrix (15%). Palaeocurrents indicate transport to the SE (i.e. transverse to the trench). Similar lithologies have been described from the Hecho Group in the Pyrenees and ascribed to a proximal environment (Mutti 1977).

Butt Hill Member. This lower member overlies pelagic black Moffat Shales and grey and green radiolarian cherts at the base of the Kiln Formation. Units of fine-grained, parallel-laminated sandstone or siltstone alternate every few centimetres with units of fine-grained siltstone or shale. Occasional well-sorted sands and silts with sharp tops and bases are interpreted as turbidites reworked by bottom currents. Both the Sheil Rig and Butt Hill Members have intercalations of thicker-bedded graded sandstones which show transport towards the SW (i.e. parallel to the trench), indicating that deposition of the thin-bedded lithologies took place within an open trench environment, rather than in a canyon or channel incised into the lower slope.

Axially transported sediments

Although palaeocurrent data are generally sparse the dominant transport direction was towards the SW, the thick sedimentary sequences retaining their distinctive sedimentological dispersal patterns parallel to the present-day strike.

Elvan Formation (2500 m)

The Elvan Formation (type section in Longcleuch from 913175 to 921172) is a thick-bedded facies rich in basic igneous clasts and ferromagnesian minerals (Fig. 3a). Massive and pebbly sandstones constitute over 60% of the formation. Beds are characteristically thick and do not show Bouma sequences. Deposition took place from high-concentration sediment gravity flows.

There is a general consensus that similar deposits formed in proximal fan environments (e.g. Walker 1966; Stanley *et al.* 1978). The lack of associated canyon-wall deposits (slump facies, etc.), channelling and material coarser than granule size, together with the lateral and

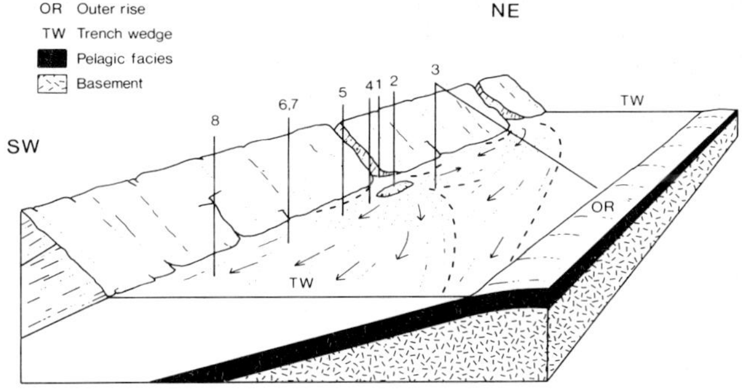

FIG. 4. Schematic diagram of the distribution of sedimentary facies shown by the sandstone sequences. No stratigraphical correlations are implicit unless stated in the text. Details of individual facies are shown in Table 1 using the same numbering system.

vertical continuity of the lithology, precludes deposition of the Elvan Formation in an inner fan environment. Deposition in the trench near a canyon mouth is considered likely, but the lack of knowledge of sand body geometry precludes discrimination between a broadly channellized distributary system and a non-channellized sheet-flow accumulation.

Abington and Glenflosh Formations (2800 m)

The quartz-rich (Fig. 3) *Abington* and Glenflosh Formations are lateral equivalents (Fig. 2), and in the following descriptions will be referred to as the Abington Formation. The Glenflosh Formation has a type section in road and stream cuttings flanking Glenflosh Hill (795140); the Abington Formation has type sections at Craighead Quarry (919238) and road cuttings north of Abington (930243) and has been subdivided into a thick-bedded and medium-bedded sandstone facies.

Thick-bedded sandstones. These are of local development within the Abington Formation and display thinning and fining-upward sequences on a scale of a few metres, a feature typical of distributary channels in a mid-fan environment (Walker & Mutti 1973). Massive, thick-bedded and very coarse sandstones occur, similar to lithologies of the Elvan Formation. However, the dominant lithology of the Abington Formation contrasts with the Elvan Formation in showing internal sedimentary structures, including distribution and inverse grading. It is suggested that this facies represents deposition from more 'mature' flows (turbulence being the major sediment support mechanism) than the Elvan Formation lithologies. Hence these deposits are inferred to occur downcurrent from massive sandstones (Fig. 4) within a broadly channellized distributary system.

Medium-bedded sandstones and shales. The dominant lithology of the Abington Formation consists of interbedded sandstone/shale couplets, 10–100 cm thick (typically 30 cm), which are plane-parallel at outcrop scale. Distribution grading (T_{ae} Bouma intervals) is common and T_b, T_c and T_d divisions are occasionally observed in the upper parts of units.

T_{ae} sequences are generally considered typical of a 'proximal' environment (Walker 1976). Similar sequences have been described as passing downcurrent and laterally into more complete Bouma sequences (Elmore *et al.* 1979). The lack of conspicuous channels in this facies suggests deposition as sheet flows, analogous to a mid-fan, non-channellized area.

Glencaple Formation (*c.* 1100 m)

This formation conformably overlies the more thickly bedded Elvan Formation (Fig. 2) and together the two formations define a first order thinning-upward cycle (Ricci-Lucchi 1975). The type section is in Glencaple Burn (924197–924193). Frequent incursions of the massive Elvan facies and similar petrographies indicate a conformable transition between the two formations.

Characteristically $T_{a\text{-}e}$ Bouma divisions are present. Most beds begin with a graded T_a division of the distribution variety, although delayed, symmetrical and inverse coarse-tail grading are observed in the thicker beds. Similar complete vertical sequences of Bouma divisions have been described from outer fan non-channellized depositional lobes (Mutti & Ricci-Lucchi 1975) and mid-fan sparsely channelled depositional lobes (Walker & Mutti 1973).

Medium-bedded turbidites showing relatively complete Bouma sequences have been documented to pass upcurrent into both massive, coarse-grained, thick-bedded, channellized deposits (e.g. Walker 1966; Stanley *et al.* 1978) and T_{ae} sequences (Elmore *et al.* 1979).

This facies is suggested to be the most distal of all the sandstone lithologies described, formed from low-concentration flows in the trench. The presence of thick-bedded Elvan lithologies indicates sporadic incursions of high-concentration flows, resulting from local progradations of the channellized mid-fan in a down-fan direction. The transition from the Elvan Formation to the Glencaple Formation is attributed either to a lateral shift in the axis of deposition of the coarse, thick-bedded facies, or to a general decrease in the volume of turbidite flows into the trench.

Discussion

The clastic sediments have been interpreted as trench deposits principally because of the dominance of linear sediment dispersal patterns (NE-SW) and the lack of a source to the SE (Piper 1972). In the Northern Belt of the Southern Uplands axial sediment transport and deposition was principally by turbidity currents. No evidence has yet been found to suggest that the deposition of the bulk of the sandstones was related directly to an axial channel, such as in the Chile Trench (Schweller & Kulm 1978). In contrast, apart from the channel-related thick-bedded facies of the Abington Formation, the majority of the sediments were deposited by sheet flows. Retention of distinctive sedimentological characteristics over 20 km down-basin and parallel-bounding surfaces to most sandstone units suggest individual flows of large volume occupying the whole width of the trench, and extending for considerable distances over the flat trench-floor. However, the sequences of massive coarse sandstones in the Elvan Formation indicate that other mechanisms such as high-concentration sediment gravity flow can produce thick sedimentary bodies.

The transition from lateral to axial flow in the trench is thought to have resulted from deflection of the turbidity currents by the outer slope (Piper 1972). This transition was apparently accomplished at the depositional site of the Elvan Formation, which is the coarsest-grained and thickest-bedded of the laterally persistent lithologies. The lateral consistency of sedimentological characteristics of most formations suggests a single stable source for each of them, operational over a considerable time span.

Markedly different petrographies for sandstone sequences of identical age (i.e. Abington and Elvan Formations) indicate the simultaneous operation of two different point sources transporting sediments into two distinct and geographically separated basins.

Nowhere in the area can a lateral transition be observed from proximal to distal facies, as described from deep-sea fans and basins (e.g. Stanley *et al.* 1978) and established for turbidite models (Mutti & Ricci-Lucchi 1975). Contrasting lithofacies can be recognized at separated localities, which individually are comparable with components of the fan model. These form thick sequences, suggesting that sediment supply originated from widely separated points, as in the Chile Trench (Schweller & Kulm 1978).

Volcanic rocks

The Bail Hill Volcanic Group crops out at the western edge of the area and constitutes a range of mildly alkaline lavas and pyroclastics (McMurtry 1980a,b). Early basalts were erupted on to pelagic black shales of *N. gracilis* age. The basalts contain phenocyrsts of clinopyroxene and bytownite, are locally vesicular and amygdaloidal, although more commonly show autobrecciated textures indicative of eruption under water. These basalts are succeeded by middle-stage hawaiite and mugearite lavas and pyroclastics containing phenocrysts of oligoclase/andesine, pargasitic amphibole, biotite/phlogopite and apatite.

The presence of gabbroic and dioritic xenoliths in some lithologies and the relationships of mineral compositions between the coarse-grained xenoliths and volcanic phenocrysts would suggest that a large intrusive body underlay the volcano in Ordovician times and periodically extruded the material that now constitutes the Bail Hill Volcanic Group. This body was subducted along with layers 2 and 3 of the Iapetus Ocean crust. The presence of extrusive and shallow intrusive material (not found at outcrop) as clasts in an infilled volcanic neck would suggest that extrusive activity was lithologically more diverse than present outcrops indicate.

Four outlying pyroclastic members are interbedded with turbiditic sediments of the Kiln Formation (Fig. 1). The thickest, the *Stoodfold Member* (McMurtry 1980a) is petrographically similar to the middle-stage activity of the Bail

Hill volcano and comprises lithic and crystal tuffs, volcaniclastics and agglomerates interbedded with siltstones containing flattened pumiceous material.

The volcano is believed to have originated on the outer rise as a mildly alkaline seamount, its activity spanning the transition from pelagic plate to trench environment, and to have become extinct prior to burial by rudites of the Spothfore Formation. The volcano was detached from the downgoing Iapetus Ocean plate and accreted (McMurtry 1980a).

Structure

The Bail Hill–Abington area consists largely of NW younging, typically vertical beds with little evidence of the 'flat belts' of Craig & Walton (1959), although megascopic folds are found infrequently in all blocks. Overturning of bedding occurs close to the Southern Uplands Fault.

Faulting

Outcrop distribution in the Northern Belt is primarily controlled by large strike faults, recognized on the basis of structural and palaeontological data, trending ENE-WSW and separating sedimentary sequences (predominantly sandstones) up to 3.2 km thick. Lithologies above the faults are strongly imbricated into linear outcrop belts, surrounded to the NW and SE by younger sandstones, thereby defining inliers.

One such inlier, the Leadhills Imbricate Zone (Leggett *et al.* 1979), displays internal organization of the fault slices at the base of the *Abington Block*. Progressively older rocks crop out towards the centre of the inlier, in the sequence *C. peltifer* black shales, Llandeilo red shales and Arenig mudstones and cherts (Fig. 5). To the NW of the Arenig strata imbricate thrusts of approximately equal displacement bring to the surface fault slices consisting of black shales, cherts and thin wedges of Abington Formation turbidites. Imbrication is intense, with a minimum of 11 fault slices recognized in a 700 m poorly exposed transect across strike. Towards the NW margin of the inlier the fault slices die out, and black shales are inferred to pass up into turbiditic sandstones of the Abington Formation.

Both the Leadhills and the Fardingmullach Imbricate Zones can be traced SW for 15 km into Nithsdale (Floyd 1975) and for 80 km thence to the coast in the Rhinns of Galloway (Kelling 1961, 1962) where they are exposed at Killintringan and Morroch Bay respectively. Other faults such as the *Howcon* and *Eller*

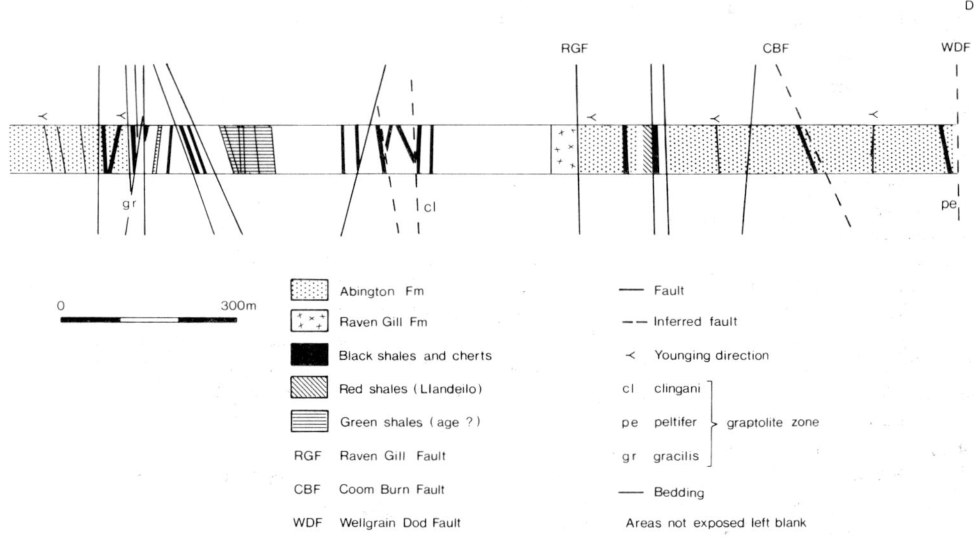

FIG. 5. Structural profile through the Leadhills Imbricate Zone; location of line of section is shown on Fig. 1(A). The sandstones of the Abington Formation indicated on the extreme left of the diagram are continuous for 2.8 km north-westwards up to the Shield Burn Fault. Strata range in age from Arenig to upper Caradoc. Faults are recognized on the basis of structural and palaeontological data.

Faults (McMurtry 1980a) and the *Duntercleuch Fault* and Leadhills Imbricate Zone merge towards the NE, thereby reducing the number of structural segments recognized in sections across strike (Fig. 1). Weir (1979) considered the faults to show a listric geometry, with translation towards the SE.

It is envisaged that movement initially occurred along low-angle thrusts, thereby juxtaposing lithologies originally deposited tens of kilometres apart (McKerrow *et al.* 1977). Subsequent rotation of bedding and thrusts through the vertical took place within the accretionary complex (Eales 1979). The imbricate zones are envisaged as laterally extensive fault zones underlying largely coherent blocks several kilometres in thickness. Innumerable brittle fractures and faults sub-parallel to strike served as planes along which displacement occurred; thus a zone of intense tectonic dislocation facilitated underthrusting of structural blocks towards the NW.

The amount of thickening of the strata by faulting can be estimated from the 1700 m wide Leadhills Imbricate Zone which consists of repetitions of the 180 m thick pre-upper Llandeilo lithologies plus slivers of turbiditic sandstones from the overlying Abington Formation. The presence of Llandeilo and Arenig lithologies in the Leadhills Imbricate Zone suggests that *décollement* occurred within Iapetus Ocean Layer I in the Northern Belt at a lower horizon than in the Central Belt where no units older than Caradoc (*N. gracilis* zone) are found.

Folding

Main deformation (D_1)

The recognition of a deformational chronology is hindered by the variable and limited development of slaty cleavage (S_1). In the *Mill Burn* and Abington Blocks, S_1 maintains a statistically vertical attitude, despite variations in bedding orientation ranging from dipping steeply to the north, to SE (overturned) dipping strata. Rotation of bedding and the major faults from the horizontal to the vertical is considered to be either a pre- or syn-S_1 event.

Folds, generally without the development of a slaty cleavage, are found in all blocks and are preferentially developed in thin- to medium-bedded lithologies (including cherts). Axial surfaces are parallel to the general strike; fold axes plunge towards the SW at angles between 16°–90°. Wavelength of folds varies from a few centimetres up to metres. Folds are generally isolated in development. Hinges of major F_1 folds are rarely seen in the field, possibly due to the formation of axial-planar strike faults associated with inlier development. Where unfaulted, fold pairs verge towards the SE.

Post-D_1 *deformation*

Later folding (F_2) has only been recognized in the Mill Burn Block, immediately south of the Southern Uplands Fault. Inclined plunging folds of sinistral vergence show a transecting S_1 cleavage, and possibly represent D_0 soft sediment deformation. This cleavage is locally deformed by F_2 folds of dextral vergence which plunge southwards at angles of about 40°.

A conjugate kink-band set has been identified from the Leadhills Imbricate Zone. Single sets of kink-bands have been found in the Abington Block.

Discussion

An implication of the accretionary complex model is that deformation is likely to be diachronous across the width of the Southern Uplands (McKerrow *et al.* 1977). Hence D_1, D_2 etc. defined in any one block are not necessarily contemporaneous. South-easterly vergence of early fold phases across the Southern Uplands is consistent with underthrusting towards the NW (Anderson & Cameron 1979). This uniformity of fold vergence direction is seen in modern accretionary complexes, and is dominantly oceanward (J. C. Moore & Karig 1976).

The major fault-bounded blocks are interpreted as individual accreted packets (McKerrow *et al.* 1977). The thickness of each fault block falls within the values of 2–8 km inferred for the accretionary packages of the Shikoku subduction zone (J. C. Moore & Karig 1976, fig. 11). Eales (1979) considered the main fold phase to be partly coeval with thrusting.

In contrast to many other accretionary complexes, tectonic mélanges are absent in the Southern Uplands. We consider that their place is taken in the Northern Belt by imbricate zones located in the incompetent pelagic sequences. Both tectonic mélanges and imbricate zones are associated with strong fault movements (Cowan 1974; Connelly 1978).

Metamorphism

Index Minerals

Spilites from the Leadhills Imbricate Zone

Spilitized lavas, dolerites and breccias from the Leadhills Imbricate Zone (Fig. 1) show recognizable igneous textures but the rocks are altered although unfoliated. Calcic feldspar has altered to albite which itself is partially replaced by mixtures of chlorite (Fig. 6a), carbonate, prehnite and green pumpellyite (Fig. 6b). Phenocrysts of clinopyroxene are unaltered or are partly replaced by chlorite along cracks and edges. Sphene has largely replaced opaque oxides. Amygdales contain variable mixtures of radiating sheaves of chlorite, quartz, carbonate, albite, sericite, pumpellyite and prehnite. None of the fine-grained basalts have recognizable prehnite or pumpellyite, though they are commonly riddled with veins of carbonate ± chlorite ± albite ± quartz ± prehnite ± pumpellyite. The secondary minerals indicate prehnite-pumpellyite facies metamorphism. Comparison of these pumpellyites with those from elsewhere in the world (Fig. 6b) suggests that the metamorphism is of a grade similar to Vancouver Island (field b of Kuniyoshi & Liou 1976), with temperatures between 300 and 360°C and pressures exceeding 1–3 kbar (Nitsch 1971).

Veins in the Elvan Formation

Feldspar detritus in these sandstones is albite. An abundance of prehnite + calcite + quartz veins indicates appreciable mobility of Si, Ca and A1, supporting the metamorphic origin of the albite by the breakdown of detrital calcic plagioclase. These veins may have formed in hydraulic fractures of metamorphic origin (cf. Norris & Henley 1976). Pumpellyite, however, has not been identified in prehnite veins—this assemblage does not define a particular metamorphic grade but does indicate a grade lower than the greenschist facies and, because zeolites have not been found, presumably prehnite-pumpellyite facies.

Lavas of the Bail Hill Volcanic Group

The *in situ* rocks of the Bail Hill Volcanic Group contain original calcic plagioclase, occasionally altered to K-feldspar (possibly adularia) or, more frequently, partially replaced by carbonate, chlorite and white mica.

Amygdales in the lavas contain mixtures of carbonate, thomsonite, prehnite, pumpellyite and albite. The lava matrix sometimes contains pumpellyite and always chlorite. Veins of carbonate and analcite are common. The presence of thomsonite and analcite indicates that the Bail Hill Bolcanic Group has been metamorphosed under conditions of the zeolite facies. The non-albitization of the calcic feldspar corroborates this.

The presence of prehnite and pumpellyite in the same rocks suggests that pressures of over 1 kbar may have been reached (Ernst 1974).

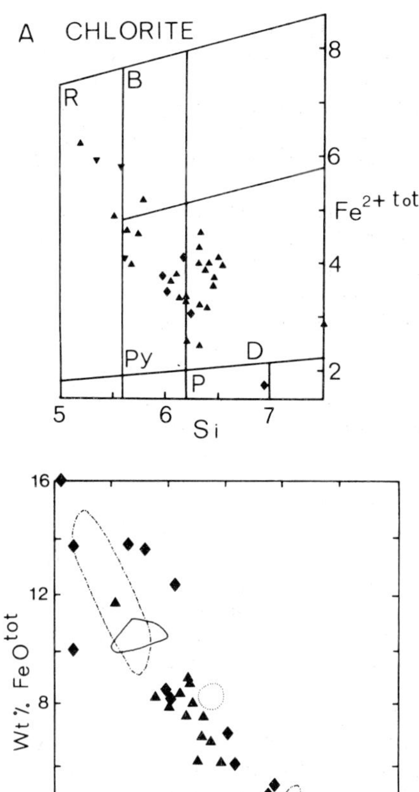

FIG. 6.(A) Chlorite probe analyses. Filled triangles = Leadhills Imbricate Zone spilite; diamonds = Bail Hill Volcanic Group; inverted triangles = Elvan Formation. R = ripidolite; B = brunsvigite; Py = pycnochlorite; D = diabantite; P = penninite (after Hey 1954). Analysed by B.C.H. and G.J.H.O. at Edinburgh and Glasgow Universities.
(B) Pumpellyite composition in terms of wt % FeO^{tot} and wt % Al_2O_3. Diamonds = Bail Hill Volcanic Group; triangles = Leadhills Imbricate Zone spilite. Dot-dash = composition field Vancouver Island (a), after Kuniyoshi & Liou (1976); dotted line = composition field Vancouver Island (b), after Surdam (1969); solid line = composition field, Wakatipu I and II (Kawachi 1976); dash-dot-dot line = composition field, Wakatipu III (Kawachi 1976). Analysed by B.C.H. and G.J.H.O. at Edinburgh and Glasgow Universities.

The presence of analcite implies that metamorphic temperatures did not exceed 200°C, so that the analcite + quartz = albite + water reaction did not occur (Liou 1971). The thomsonite + quartz = wairakite reaction at 280–316°C (Coombs *et al.* 1959) has not taken place, although it is to be remarked that the Bail Hill Volcanic Group is quartz-free. The zeolite metamorphic minerals may have formed soon after extrusion of the volcanics as a result of deuteric alteration, or at some time later under conditions of burial metamorphism.

Siltstones from the Stoodfold Member

These siltstones contain small white rounded kaolinite patches up to 2 mm in diameter, which are distributed through the bedding laminae. The patches formed after the sediment had been partially or wholly compacted, hence their non-flattened nature and random distribution.

Volcaniclastics associated with the siltstones show fresh calcic plagioclase, although some have been partially altered to carbonate and white mica; amphiboles have altered to mixtures of chlorite, carbonate and iron oxides.

The lack of albitization and presence of kaolinite in sediments in the same block as the zeolite-bearing Bail Hill Volcanic Group confirms the lower grade of metamorphism in the Crawick Block compared with the Elvanfoot Block. A stability maximum between 115 and 200°C has been suggested for kaolinite (Dunoyer de Segozac 1970) and similar conditions are inferred for the Crawick Block.

Illite crystallinity

Turbiditic sandstones throughout the area show considerably more than 10% matrix (Fig. 3) suggesting that they have undergone recrystallization with unstable (volcanigenic) detritus breaking down to form more matrix (Cummins 1962). X-ray diffraction analyses of the $<2\ \mu m$ fraction of shales, siltstones and sandstones show quartz + feldspar + chlorite + illite mineralogies which are present in all specimens examined from the Crawfordjohn, Abington and Elvanfoot Blocks. Probe analyses of detrital and metamorphic chlorites show a range of compositions, including ripidolite, brunsvigite, pycnochlorite and diabantite (Fig. 6a). All illite crystallinity (IC) values of $<2\ \mu m$ grain-size fractions fall within the anchizone (Kübler 1968) which has been equated with the prehnite-pumpellyite facies (Kisch 1974).

Under burial metamorphic conditions IC might be expected to decrease within a fault block in the direction of younging (i.e. towards the NW). No significant patterns of IC distribution emerge across the Abington and Elvanfoot blocks, so it is inferred that metamorphism continued after the succession had been cut up into fault-bounded blocks and rotated to verticality. Illite crystallinity has not yet been measured in the blocks north of the Abington Block.

Reflectance

The mean maximum reflectance ($\bar{R}_0$ max) of 'vitrinite' (graptolite) fragments in black shales has been measured for the Abington and Elvanfoot Blocks. Values range from 3.91 to 8.22%, equivalent to meta-anthracite coal rank and the prehnite-pumpellyite facies (Kisch 1974). No significant variation occurs in the area. On the basis of one set of determinations from one locality giving $\bar{R}_0$ max = 5.75%, shales at the base of the Elvan Formation are considered to be of the same metamorphic grade as those at the base of the stratigraphically lower Abington Formation to the NW. This finding agrees with conclusions drawn on the basis of IC.

Discussion

The Bail Hill Volcanic Group has been metamorphosed to a lower grade than the spilites of the Leadhills Imbricate Zone. It is not known whether the change in grade from zeolite to prehnite-pumpellyite facies is coincident with the fault separating the Crawick Block from the Crawfordjohn Block (Figs 1 & 2), or whether it occurs within one of the blocks. On this point, it is significant that illite crystallinity values do not vary in sections across the Leadhills Imbricate Zone to the top of the underlying Glencaple Formation. Given the present-day thickness of the turbidite formations, in excess of 3 km, burial metamorphism must have occurred at the base of the sedimentary pile in the trench whilst the top of the sequence was still unlithified. Metamorphism throughout a sandstone succession would only have been possible after thrusting under the previously accreted wedge, within the accretionary complex. It is considered that metamorphism was a continuous process, initiated in the trench environment and progressively increased in grade as thrusting occurred and bedding rotated towards the vertical.

The low grade of metamorphism in the Bail Hill–Abington area is similar to that found in the matrix of mélanges and sandstones of other subduction complexes such as those on Kodiak

Island (Connelly 1978), the Franciscan Complex of California (Bailey *et al.* 1964; Cowan 1978) and the Oyo Complex of Nias Island (Moore & Karig 1980). Further interpetation of the area awaits more thorough investigations of younger and tectonically active accretionary complexes. The inferred tectonic history presented here may be useful in developing a more detailed model to explain the complicated relationships between deformation and metamorphism at other accretionary margins.

ACKNOWLEDGMENTS: We thank Professor E. K. Walton and Dr. J. A. Weir for much helpful discussion and for critically reading the manuscript. We are also grateful to the Grant Institute of Geology, University of Edinburgh, the Organic Geochemistry Unit of the University of Newcastle upon Tyne and the Department of Geology, University of Glasgow for use of analytical facilities; and to Mrs S. Johnson, Mrs J. Galloway and J. Jukes for typing the manuscripts. Photographic work was produced by Mr J. Allen.

We gratefully acknowledge financial support from the N.E.R.C.; G.J.H.O. was financed in addition by the University of St. Andrews. Special thanks are given to the reviewers for improvements to the manuscript.

References

ANDERSON, T. B. & CAMERON, T. D. J. 1979. A structural profile of Caledonian deformation in Down. *In:* HARRIS, A. L., HOLLAND, C. H. & LEAKE, B. E. (eds). *The Caledonides of the British Isles—reviewed.* Spec. Publ. geol. Soc. London, **8,** 263–7.

BAILEY, E. H., IRWIN, W. P. & JONES, D. L. 1964. Franciscan and related rocks, and their significance in the geology of western California. *Bull. Calif. Div. Mines Geol.* **183,** 89–112.

CARTER, C., TREXLER, J. H., JR. & CHURKIN, M., JR. 1980. Dating of graptolite zones by sedimentation rates: implications for rates of evolution. *Lethaia,* **13,** 279–87.

CONNELLY, W. 1978. Uyak Complex, Kodiak Islands, Alaska: A Cretaceous subduction complex. *Bull. geol. Soc. Am.* **89,** 755–69.

COOMBS, D. S., ELLIS, A. J., FYFE, W. S. & TAYLOR, A. M. 1959. The zeolite facies, with comments on the interpretation of hydrothermal syntheses. *Geochim. cosmochim. Acta,* **17,** 53–107.

COWAN, D. S. 1974. Deformation and metamorphism of the Franciscan subduction zone complex northwest of Pacheco Pass, California. *Bull. geol. Soc. Am.* **85,** 1623–34.

—— 1978. Origin of blueschist-bearing chaotic rocks in the Franciscan Complex, San Simeon, California. *Bull. geol. Soc. Am.* **89,** 1415–23.

CRAIG, G. Y. & WALTON, E. K. 1959. Sequence and structure in the Silurian rocks of Kirkcudbrightshire. *Geol. Mag.* **96,** 209–20.

CUMMINS, W. A. 1962. The greywacke problem. *Liverpool Manchester geol. J.* **3,** 51–72.

DUNOYER DE SEGONZAC, G. 1970. The transformation of clay minerals during diagenesis and low-grade metamorphism: A review. *Sedimentology,* **15,** 281–346.

EALES, M. H. 1979. Structure of the Southern Uplands. *In:* HARRIS, A. L., HOLLAND, C. H. & LEAKE, B. E. (eds). *The Caledonides of the British Isles—reviewed,* Spec. Publ. geol. Soc. London, **8,** 269–73.

ELMORE, R. D., PILKEY, O. H., CLEARY, W. J. & CURRAN, H. A. 1979.Black Shell turbidite, Hatteras Abyssal Plain, western Atlantic Ocean. *Bull. geol. Soc. Am.* **90,** 1165–76.

ERNST, W. G. 1974. Metamorphism and ancient continental margins. *In:* BURK, C. A. & DRAKE, C. L. (eds). *The Geology of Continental Margins,* 933–50. Springer-Verlag, New York.

FLOYD, J. D. 1975. *The Ordovician rocks of West Nithsdale.* Thesis, Ph.D. Univ. St. Andrews (unpubl.).

HEY, M. H. 1954. A new review of the chlorites. *Mineralog. Mag.* **30,** 277–92.

KAWACHI, Y. 1975. Pumpellyite-actinolite and contiguous facies metamorphism in part of Upper Wakatipu district, South Island, New Zealand. *N.Z. J. Geol. Geophys.* **18,** 401–41.

KELLING, G. 1961. The stratigraphy and structure of the Ordovician rocks of the Rhinns of Galloway. *Q.J. geol. Soc. London,* **117,** 37–75.

—— 1962. The petrology and sedimentation of Upper Ordovician rocks of the Rhinns of Galloway. *Trans. R. Soc. Edinburgh,* **65,** 107–37.

KISCH, H. J. 1974. Anthracite and meta-anthracite coal ranks associated with 'anchimetamorphism' and 'very-low-stage' metamorphism, I, II, and III. *Proc. Kon. Ned. Akad. Wet. Amsterdam, Ser. B,* **77,** 81–118.

KÜBLER, B. 1968. La cristallinité de l'illite et les zones tout à fait supérieures du métamorphisme. *Etages Tectoniques,* 105–22. A la Baconnière, Neuchâtel, Switzerland.

KUNIYOSHI, S. & LIOU, J. G. 1976. Burial metamorphism of the Karmutsen volcanic rocks, northeastern Vancouver Island, British Columbia. *Am. J. Sci.* **276,** 1096–119.

LAMONT, A. & LINDSTRÖM, M. 1957. Arenigian and Llandeilian cherts identified in the Southern Uplands of Scotland by means of conodonts, etc. *Trans. geol. Soc. Edinburgh,* **17,** 60–70.

LAPWORTH, C. 1878. The Moffat Series. *Q.J. geol. Soc. London,* **34,** 240–346.

LEGGETT, J. K. 1980. The sedimentological evolution of a Lower Palaeozoic accretionary fore-arc in the Southern Uplands of Scotland. *Sedimentology,* **27,** 401–17.

——, MCKERROW, W. S. & EALES, M. H. 1979. The Southern Uplands of Scotland: A Lower Palaeozoic accretionary prism. *J. geol. Soc. London,* **136,** 755–70.

—— , McKerrow, W. S. & Casey, D. M. 1981. The anatomy of a Lower Palaeozoic accretionary prism: The Southern Uplands of Scotland (this volume).

Liou, J. G. 1971. Synthesis and stability relations of prehnite, $Ca_2Al_2Si_3O_{10}(OH)_2$. *Am. Mineral.* **56,** 507–31.

McKerrow, W. S., Leggett, J. K. & Eales, M. H. 1977. Imbricate thrust model of the Southern Uplands of Scotland. *Nature, London,* **267,** 237–9.

McMurtry, M. J. 1980a. *The Ordovician rocks of the Bail Hill area, Sanquhar, South Scotland: Volcanism and sedimentation in the Iapetus Ocean.* Thesis, Ph.D. Univ. St. Andrews (unpubl.).

—— 1980b. Discussion of: Evidence for Caledonian subduction from greywacke detritus in the Longford-Down inlier. *J. Earth Sci. R. Dubl. Soc.* **2,** 209–12.

Middleton, G. V. & Hampton, M. A. 1973. Sediment gravity flows: mechanics of flow and deposition. *In*: Middleton, G. V. & Bouma, A. H. (eds). *Turbidites and Deep Water Sedimentation.* Pacif. Sec. Short Course, Soc. Econ. Paleontol. Mineral. 1–38.

Mitchell, A. H. G. 1974. Flysch-ophiolite sequences: polarity indicators in arc and collision-type orogens. *Nature, London,* **248,** 747–9.

Moore, G. F., Billman, H. G., Hehanussa, P. E. & Karig, D. E. 1980. Sedimentology and paleobathymetry of Neogene trench-slope deposits, Nias Island, Indonesia. *J. Geol. Chicago,* **88,** 161–80.

Moore, G. F. & Karig, D. E. 1980. Structural geology of Nias Island, Indonesia: Implications for subduction zone tectonics. *Am. J. Sci.* **280,** 193–223.

Moore, J. C. & Karig, D. E. 1976. Sedimentology, structural geology and tectonics of the Shikoku subduction zone, southwestern Japan. *Bull. geol. Soc. Am.* **87,** 1259–68.

Mutti, E. 1977. Distinctive thin-bedded turbidite facies and related depositional environments in the Eocene Hecho Group (South-central Pyrenees, Spain). *Sedimentology,* **24,** 107–31.

—— & Ricci-Lucchi, F. 1975. Turbidite facies and facies associations. *In:* Mutti, E., Parea, G. C., Ricci-Lucchi, F., Sagri, M., Zanzucchi, G., Ghibaudo, G. & Jaccarino, S. (eds). *Examples of Turbidite Facies and Facies Associations from Selected Formations of the Northern Apennines.* IX Int. Congr. Sedim. (Nice), Field Trip A-II, 21–36.

Nitsch, K. H. 1971. Stabilitätsbeziehungen von Prehnit- und Pumpellyithaltigen Paragenesen. *Contrib. Mineral. Petrol.* **30,** 240–60.

Norris, R. J. & Henley, R. W. 1976. Dewatering of a metamorphic pile. *Geology,* **4,** 333–6.

Peach, B. N. & Horne, J. 1899. The Silurian rocks of Britain, 1, Scotland. *Mem. geol. Surv. Scotland,* 749 pp.

Piper, D. J. W. 1972. Trench sedimentation in the Lower Palaeozoic of the Southern Uplands. *Scott. J. Geol.* **8,** 289–91.

—— , von Huene, R. & Duncan, J. R. 1973. Late Quaternary sedimentation in the active eastern Aleutian Trench. *Geology,* **1,** 19–22.

Ricci-Lucchi, F. 1975. Depositional cycles in two turbidite formations of northern Apennines (Italy). *J. sediment. Petrol.* **45,** 3–43.

Schweller, W. J. & Kulm, L. D. 1978. Depositional patterns and channelized sedimentation in active eastern Pacific trenches. *In:* Stanley, D. J. & Kelling, G. (eds). *Sedimentation in Submarine Canyons, Fans, and Trenches,* 311–24. Dowden, Hutchinson & Ross, Stroudsburg, Pennsylvania.

Stanley, D. J., Palmer, H. D., & Dill, R. F. 1978. Coarse sediment transport by mass flow and turbidity current processes and downslope transformations in Annot Sandstone canyon-fan valley systems. *In:* Stanley, D. J. & Kelling, G. (eds). *Sedimentation in Submarine Canyons, Fans, and Trenches,* 85–115. Dowden, Hutchinson & Ross, Stroudsburg, Pennsylvania.

Surdam, R. C. 1969. Electron microprobe study of prehnite and pumpellyite from the Karmutsen Group, Vancouver Island, British Columbia. *Am. Mineral.* **54,** 256–66.

Walker, R. G. 1966. Shale Grit and Grindslow Shales: transition from turbidite to shallow water sediments in the Upper Carboniferous of northern England. *J. sediment. Petrol.* **36,** 90–114.

—— 1976. Facies Models 2. Turbidites and associated coarse clastic deposits. *Geoscience Can.* **3,** 25–36.

—— & Mutti, E. 1973. Turbidite facies and facies associations. *In:* Middleton, G. V. & Bouma, A. H. (eds). *Turbidites and Deep Water Sedimentation.* Pacif. Sec., Short Course, Soc. econ. Paleontol. Mineral. 119–57.

Walton, E. K. 1965. Lower Palaeozoic rocks: stratigraphy, palaeogeography and structure. *In:* Craig, G. Y. (ed). *The Geology of Scotland,* 161–227. Oliver & Boyd, Edinburgh.

Weir, J. A. 1979. Tectonic contrasts in the Southern Uplands. *Scott, J. Geol.* **15,** 169–86.

Williams, A., Strachan, I., Bassett, D. A., Dean, W. T., Ingham, J. K., Wright, A. D. & Whittington, H. B. 1972. A correlation of Ordovician rocks in the British Isles. *Spec. Rep. geol. Soc. London,* **3,** 74 pp.

Barry C. Hepworth, Grahame J. H. Oliver & Michael J. McMurtry, Department of Geology, University of St. Andrews, Fife, Scotland.

FACIES, PETROLOGY AND MODELS

Sedimentary facies associations within subduction complexes

Michael B. Underwood & Steven B. Bachman

SUMMARY: Sedimentation patterns within modern subduction zones are complex and variable, and do not necessarily follow models of submarine fan sedimentation. Environmental reconstructions within ancient subduction complexes should follow modern analogues as closely as possible and consider several criteria, including turbidite facies associations, vertical depositional cycles, regional palaeocurrent patterns, and in many cases, structural style and sandstone petrology.

Important variables in trench sedimentation include the volume and texture of sediment entering the trench and the distribution of major sediment sources, especially large submarine canyons. Sediment transport in the trench is commonly longitudinal, although locally, such as at the mouth of a submarine canyon, flow may be at a high angle to the continental margin.

Patterns of trench-slope sedimentation depend largely upon the topography of the slope. In general, coarse sediment is either trapped behind tectonic ridges (within slope basins) or bypasses the slope via submarine canyons. Background sedimentation is dominated by hemipelagic settling. Current directions are commonly at a high angle to the margin, but longitudinal flow may occur within large elongate slope basins. Major facies associations include submarine-canyon, slope, mature-slope-basin, and immature-slope-basin.

Classification schemes for turbidites and related resedimented deposits (e.g. Walker & Mutti 1973; Mutti & Ricci-Lucchi 1975; Ricci-Lucchi 1975) are applicable to deep-sea deposits regardless of the geometry of the depositional system. Turbidite facies are commonly used to describe specific facies associations, such as slope, submarine fan, and basin plain (e.g. Walker & Mutti 1973; Bouma & Nilsen 1978; Ingersoll 1978a; Walker 1978; Pescadore 1978). In these models, analogies are drawn between stratigraphic sequences of turbidites and modern depositional geometries. However, submarine fan models (Normark 1970, 1978; Nelson & Kulm 1973; Nelson & Nilsen 1974; Nilsen 1980) are not applicable to all modern sedimentary basins, particularly elongate troughs (e.g. Hsü *et al.* 1980; Pilkey *et al.* 1980). Moreover, turbidite facies associations, depositional cycles, and submarine-fan facies associations are not necessarily equivalent or interchangeable.

Meaningful environmental reconstructions within ancient subduction complexes depend upon well-established modern analogues, and should, as closely as possible, be patterned after the geometry of those analogues. For this reason, we first discuss the diversity of sedimentary processes and depositional systems within modern subduction zones, where several types of sediment bodies, including submarine fans, are developed. The sedimentological characteristics of several examples of ancient trench and trench-slope deposits are also considered. We then model turbidite facies associations for the major depositional settings. Because many trench and trench-slope environments have not been adequately sampled, our facies associations in some cases are largely inferential, and designations are based upon the types of deposits logically expected for a given topography and sediment input. In this paper, turbidite facies terminology is used in only the most general sense (Table 1), following the classifications of Walker & Mutti (1973) and Ricci-Lucchi (1975).

The deformation associated with subduction and/or accretion commonly hampers the environmental reconstruction of ancient trench and trench-slope deposits. However, the scale of structural dismemberment, in many cases, is such that identifiable depositional cycles and facies changes are preserved. Nevertheless, several criteria must generally be considered before contrasting sedimentary environments (e.g. trench floor versus slope basin) can be recognized within structurally complex terranes (e.g. Bachman 1978, 1981; Moore 1979; Moore & Karig 1980; Moore *et al.* 1980).

Trench-floor deposits

Surface samples and Deep Sea Drilling Project (DSDP) cores show that modern trench sediments generally consist of varying proportions

TABLE 1. *Classification of turbidites and associated deposits, after Walker & Mutti (1973) and Ricci-Lucchi (1975)*

Facies
A — thick to massive, lenticular beds of coarse sandstone, pebbly sandstone, and conglomerate
B — massive to thick-bedded, medium- to coarse-grained sandstone
C — thick-bedded, classical sandy turbidites, with graded beds at base
D — thinner-bedded and finer-grained turbidites, including lutite turbidites; basal graded beds absent
E — thin- to medium-bedded, coarse- to fine-grained sandstone, with irregular wedge-shaped beds and cross-stratification
F — chaotic deposits formed by slumps and slides
G — hemipelagic mudstone and shale

of sandy turbidites and interbeds of mud (Ross 1971; Kulm, von Huene *et al.* 1973; Piper *et al.* 1973; Kulm & Fowler 1974; von Huene 1974; Karig, Ingle *et al.* 1975; J. C. Moore & Karig 1976; Moore, Watkins *et al.* 1979; von Huene, Aubouin *et al.* 1980; McMillen & Haines 1981; McMillen *et al.* 1981). Sedimentary sequences underlying the trench turbidites commonly include abyssal pelagic clay or biogenic ooze and hemipelagic sediments (Piper *et al.* 1973; Schweller & Kulm 1978; Moore, Watkins *et al.* 1979). This sequence would not be expected, however, if large volumes of terrigenous turbidites are deposited on the abyssal floor seaward of the trench, as with the Bengal Fan (Curray & Moore 1974), or if the trench floor is starved of coarse terrigenous material, as is the Japan Trench (Langseth, Okada *et al.* 1978; Arthur, von Huene *et al.* 1981).

The geometry of trench fill is dependent

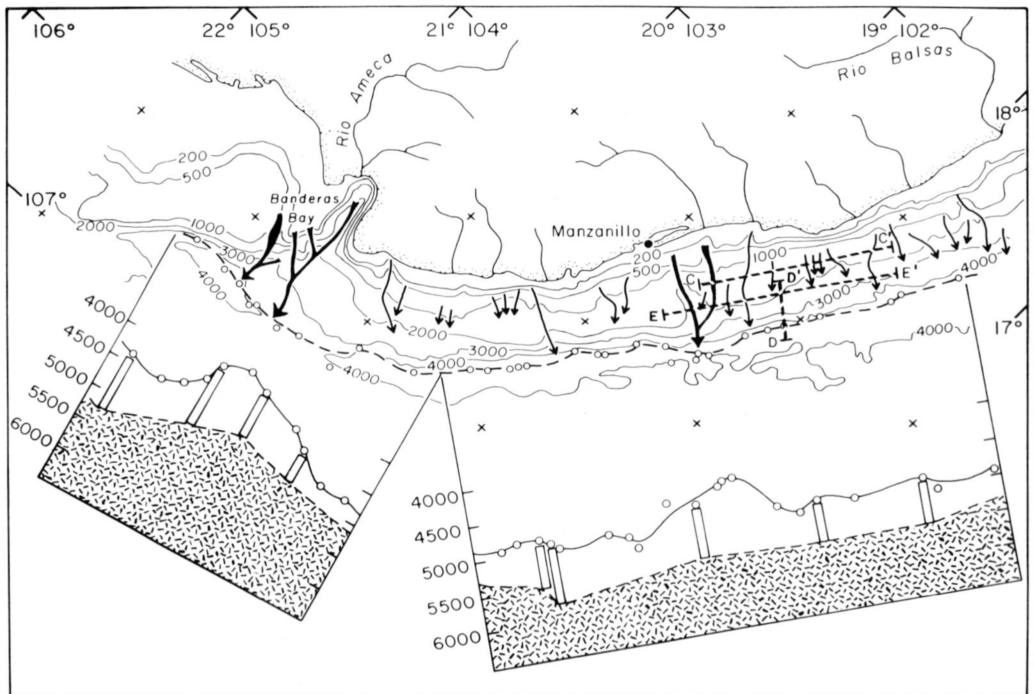

FIG. 1. Bathymetric map and trench-axis profile of the Middle America Trench from Banderas Bay to Rio Balsas, Mexico. General bathymetry (in metres) is modified from Fisher (1961). Axial profile incorporates data from Fisher (1961), Ross & Shor (1965), and unpublished reflection records from the following sources: Scripps Institution of Oceanography cruises *Scan-11*, *Cocotow-4*, *Iguana-1*, *Iguana-5*, Lamont-Doherty Geological Observatory cruise *Vema-28*, and *Glomar Challenger* transit to Deep Sea Drilling Project Leg 66 drilling sites. Axial-depth values are in corrected metres. Sediment thickness was calculated using $V = 2.0$ km s^{-1}, such that 0.5 s penetration = 500 m of sediment. See Fig. 2 for explanation of symbols. Modified from Underwood & Karig (1980).

upon the interaction between sediment supply and convergence rate, the distribution and relative size of submarine canyons along the trench slope, and the bathymetric relief on the downgoing oceanic plate (Schweller & Kulm 1978; Underwood *et al.* 1980). The great bulk of terrigenous material reaches the trench floor via large submarine canyons that are deeply incised into the trench slope (Underwood & Karig 1980). The sand/mud ratio of trench deposits, in many cases, is thus dependent upon the proximity of submarine canyons to the site of deposition. After sediment reaches the floor of the trench, axial transport involving distances of hundreds of kilometres is indicated (Scholl 1974; Kulm *et al.* 1977; Schweller & Kulm 1978).

Submarine canyons and sediment bypassing

Large submarine canyons are prominent features along the landward slope of the northern Middle America Trench (Figs 1 & 2; Fisher 1961; Underwood & Karig 1980; Shipley *et al.* 1980; McMillen & Haines 1981; McMillen *et al.* 1981) as well as the Oregon-Washington continental margin (Carlson & Nelson 1969; Barnard 1978; Kulm & Scheidegger 1979) and the southern Chile Trench (Scholl *et al.* 1970; Hayes 1974; Prince *et al.* 1980). These canyons have been responsible for funnelling terrigenous sediment from the continental shelf directly to the trench floor, thereby bypassing the trench slope. Sediment bypassing was particularly effective during low stands of sea-level associated with Pleistocene glacial epochs (e.g. Scholl *et al.* 1977; Barnard 1978). Data from current-meter studies indicate that at least one of the canyons off Mexico (Rio Balsas Canyon) is still active (Reimnitz & Gutierrez-Estrada 1970; Shepard *et al.* 1979); by analogy, other prominent canyon systems that head on the narrow continental shelf (Figs 1 & 2) are probably active as well.

Sediment bypassing of the slope off Mexico is also supported by an axial profile of the Middle America Trench, which shows a strong correlation between the thickness of sediment and the location of large submarine canyons (Figs 1 & 2; Underwood & Karig 1980). These data suggest that the submarine canyons act as the

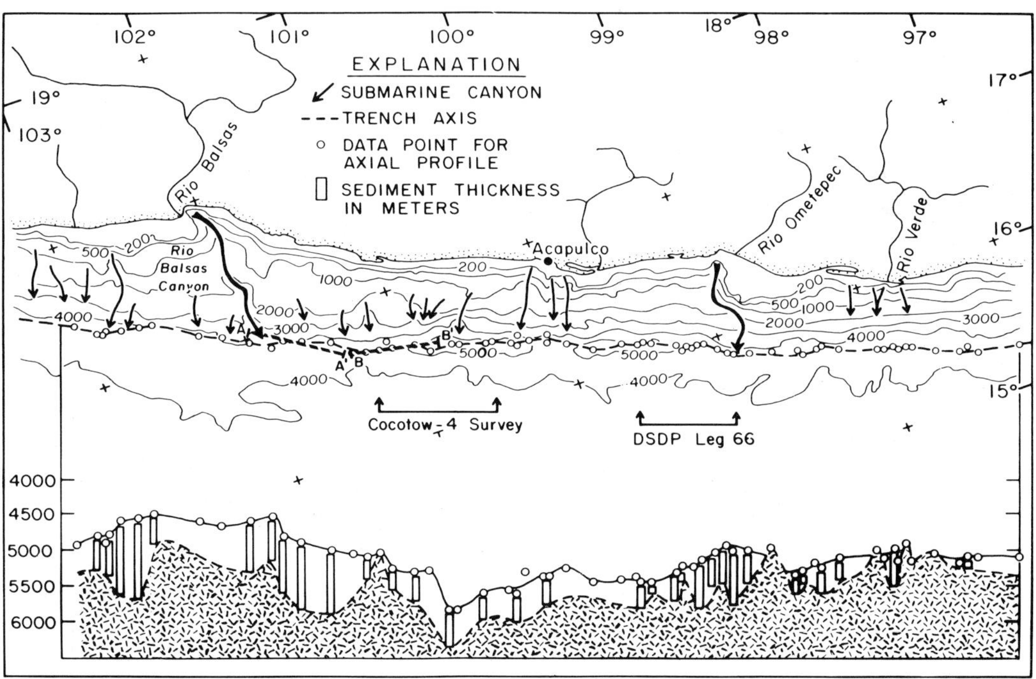

FIG. 2. Bathymetric map and trench-axis profile of the Middle America Trench from Rio Balsas to Tehuantepec Bay, Mexico. Depth calculations and bathymetry are the same as for Fig. 1. Data sources, in addition to those cited in Fig. 1, include Karig *et al.* (1978), Shipley *et al.* (1980), and unpublished reflection profiles from the University of Texas Marine Lab (DSDP Leg 66 site survey). Modified from Underwood & Karig (1980).

primary conduits for sediment entering the trench. Sediment thickness decreases in both directions away from point sources at the canyon mouths, indicating that bi-directional transport probably occurs on the trench floor. Seismic reflection data at the mouth of Rio Balsas Canyon show a thick wedge of sediment that is channellized along its upper surface; these appear to be distributary channels of a submarine fan (Underwood & Karig 1980).

Rates of sedimentation in the Middle America Trench (Ross 1971; Moore, Watkins *et al.* 1979) are not high enough to allow seaward progradation of submarine fans over the outer trench slope (Underwood *et al.* 1980). As a result, individual channels initially orientated perpendicular to the trench axis either shift their courses in response to the axial depth gradient or are deflected by the outer trench slope. Channellized transport on the trench floor appears to be very localized, as channels are no longer evident within a distance of approximately 30 km from the mouth of Rio Balsas Canyon (Underwood & Karig 1980; Underwood *et al.* 1980).

In trench systems with higher rates of sedimentation and lower convergence rates (e.g. the Oregon-Washington margin), submarine fans have built out over the seaward trench slope, masking the trench as a bathymetric feature (Kulm & Fowler 1974; Schweller & Kulm 1978). The Astoria Fan off Oregon, for example, has been used as a model of submarine-fan sedimentation (Nelson & Kulm 1973; Nelson & Nilsen 1974); in this case, the effects of tectonics on trench sedimentation appear to be minimal.

Large submarine canyons do not provide the only source of sediment to reach the trench floor. Secondary sources include background settling of hemipelagic debris and locally-derived mass flows, including slumps and slides (e.g. Piper *et al.* 1973; von Huene 1974; Moore *et al.* 1976; Karig *et al.* 1981). High pore water pressures resulting from tectonically induced dewatering at the base of the trench slope, combined in some cases with oversteepened slopes, provides a mechanism for recurrent slope failure (e.g. Carson 1980). Mass flows thus initiated may transport reworked slope sediments or accreted trench deposits to the trench floor, commonly via small submarine canyons that head on the lower slope (e.g. Karig *et al.* 1981). In general, hemipelagic deposition and locally-derived mass flows provide a significant source of trench sediment only if the terrigenous supply is low. Moreover, once slumps and debris flows reach the trench floor, redistribution of the sediment by longitudinal flow mechanisms may be expected (Piper *et al.* 1973).

Axial transport

Axial channels are prominent features in the eastern Aleutian Trench (Piper *et al.* 1973; von Huene 1974) and the southern Chile Trench (Scholl *et al.* 1970; von Huene 1974; Schweller & Kulm 1978; Underwood *et al.* 1980), and are locally developed in the Japan Trench (Arthur, von Huene *et al.* 1981), the northern Sunda Trench (G. F. Moore, pers. comm. 1980), and the Middle America Trench off Oaxaca (Shipley *et al.* 1980; McMillen *et al.* 1981). The spectacular channel off southern Chile is continuous for over 1000 km (Schweller & Kulm 1978); however, the extent of individual flow units has not been documented. The longitudinal distance and continuity of axial transport is thus poorly known.

Non-channellized longitudinal flow almost certainly occurs along trench floors in many cases. Axial channels are absent in the Middle America Trench north of Acapulco, for example (Ross & Shor 1965; Karig *et al.* 1978), but textural data and sedimentary structures suggest that turbidity currents transport sediment away from the mouths of submarine canyons (Ross 1971). Moreover, the continuity of the trench wedge, which maintains a thickness of over 300 m north of Rio Balsas Canyon (Figs 1 & 2), suggests that axial transport occurs for distances of 200–300 km between major point sources. Other trenches that do not display axial channels include the central Aleutian Trench (Scholl 1974), the Kuril-Kamchatka Trench (Scholl 1974), northern California (Silver 1971), Oregon-Washington margin (Kulm & Fowler 1974; Barnard 1978), the Gulf of Oman (White & Klitgord 1976), the Peru Trench (Schweller & Kulm 1978), the Ecuador Trench (Lonsdale 1978), and the Middle America Trench off Guatemala (Seely *et al.* 1974; Ladd *et al.* 1978; Ibrahim *et al.* 1979).

The presence or absence of an axial channel is probably dependent upon several variables, such as the axial depth gradient, the texture of sediment reaching the trench floor, the rate of sedimentation, and the type of mass flow mechanism involved in axial transport. Where channels are lacking, coarse sediment is probably transported as sheet-like turbidites, in a manner analogous to sand-layer deposition within flat-floored basins along the Atlantic margin (e.g. Bennetts & Pilkey 1976; Pilkey *et al.* 1980). Individual sand layers in these abyssal-plain basins are continuous for distances of up to 500 km (Elmore *et al.* 1979). Sand-layer thickness and the percentage of sand gradually decreases away from the source (Pilkey *et al.* 1980). Similar sheet-like flows in trenches are generally confined by the seaward and landward trench slopes, but in some circumstances, sand bodies may migrate up or overtop the outer trench slope (Damuth 1979; J. C. Moore *et al.* 1981).

Most trenches appear to contain a continuous wedge of turbidites. In some cases, however, sediment supplies are not large enough for sediment bodies to prograde over basement highs on the downgoing oceanic plate, such as seamounts, aseismic ridges, and fault blocks (e.g. Scholl 1974; Kulm *et al.* 1977). In the northern Middle America Trench, basement topography has created silled basins south of Rio Balsas Canyon (Fig. 3). Depth to the trench floor increases by over 200 m from the north side of an exposed basement ridge to the south side (Figs 2 & 3). South-directed turbidity currents emanating from Rio Balsas Canyon apparently filled the trench virtually to the crest of the basement ridge; subsequent flows could then spill over the ridge and continue down the trench axis. Farther south, near the DSDP Leg 66 drilling sites, closely spaced reflection profiles show that axial transport is locally blocked by basement highs (Shipley *et al.* 1980); low rates of sedimentation result in discontinuous and apparently isolated ponds of sediment (Fig. 2).

Ancient analogues

In spite of post-depositional deformation, sedimentary sequences up to hundreds of metres in thickness are preserved within ancient subduction complexes, such as the Coastal Belt Franciscan Complex of northern California (Bachman 1978, 1981). Inferred trench sediments within the Coastal Belt include all turbidite facies (Table 1). However, massive channellized sandstone (facies B) is locally dominant and closely associated with thin-bedded deposits of probable channel overspill origin. Chaotic fine-grained deposits (facies F) may represent hemipelagic material slumped off the base of the trench slope. Turbidite facies associations are consistent with upper- and mid-fan depositional settings, but outer-fan facies associations are generally lacking. Palaeocurrent data indicate that flow directions were parallel to the inferred continental margin; deposition may therefore have occurred within an axial channel on the trench floor. Elsewhere within the Franciscan Complex, mid- to outer-fan facies associations, consisting mostly of facies C and D turbidites, occur together with inner-fan deposits (Aalto 1976). These data suggest that trench fans were at least locally developed within the Franciscan trench.

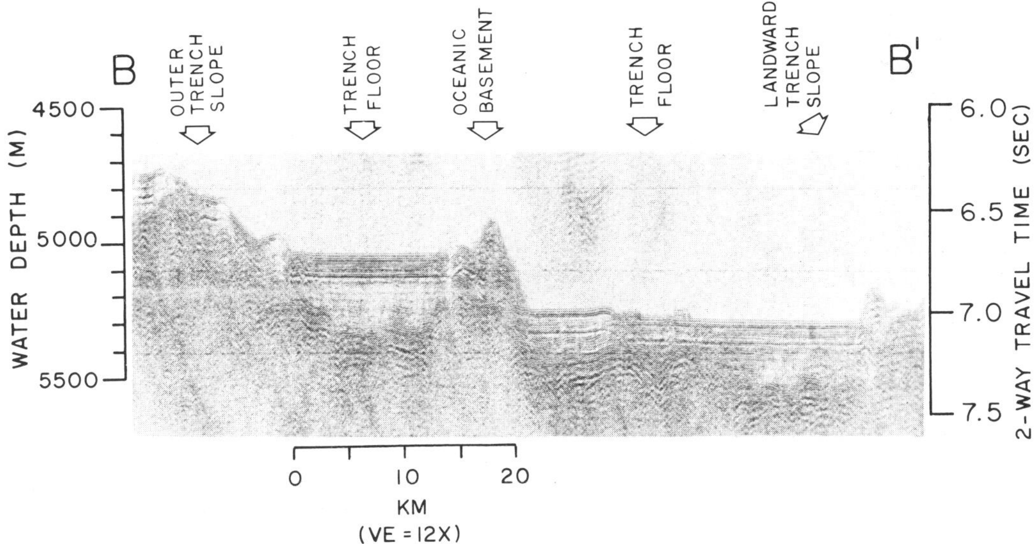

FIG. 3. Single-channel seismic-reflection (air-gun) profile across the floor of the Middle America Trench south of Rio Balsas Canyon. Location of trackline B-B′ is shown in Fig. 2. Seismic profile is from DSDP Leg 66 transit.

Sedimentological data from other ancient subduction complexes demonstrate the diversity of possible transport processes and depositional geometries within trenches. Ordovician greywackes in the Southern Uplands of Scotland, for example, are interpreted as accreted trench deposits and consist primarily of facies B and C sandstones (Leggett *et al.* 1979; Leggett 1980). Localized conglomeratic units may represent point sources within a largely non-channellized trench-floor transport system (Leggett 1980). Permian to Jurassic strata in New Zealand include both mélanges and coherent packets of strongly channellized thick-bedded sandstone and conglomerate (facies A & B) with associated levee deposits and overbank turbidites (Carter *et al.* 1978). Some of the channel-fill sequences probably accumulated in channels orientated parallel to the trench axis; localized point sources are also likely (Carter *et al.* 1978). Tectonically juxtaposed packets of accreted trench turbidites on Barbados are laterally continuous and up to 500 m thick; facies associations are consistent with deposition within middle and outer portions of one or more submarine fans (Speed 1981; Pudsey & Reading 1981). Inferred trench sediments on Kodiak Island, Alaska, were probably deposited within a basin-plain environment; dominant facies include fine-grained facies D turbidites and interbeds of facies G hemipelagic shale (Nilsen & Moore 1979). Regional facies relations and palaeocurrent data suggest that the basin-plain deposits represent long-distance axial transport and distal sedimentation within a major trench-floor system in SE Alaska (Nilsen & Bouma 1977; Nilsen & Moore 1979; Nilsen & Zuffa 1981).

Sedimentary facies model

Based upon the available data from both modern and ancient trench deposits, we propose a conceptual model for trench sedimentation which includes four major types of sediment bodies and corresponding facies associations. The four trench-floor facies associations are trench-fan, axial-channel, non-channellized, and starved-trench (Fig. 4).

Trench fans, in most cases, are not large enough to prograde over the outer trench slope, and the transition from more proximal to more distal facies associations takes place in a direction that is parallel to the trench axis. Submarine fans distorted by restricted basin geometries have been recognized within the geological record (e.g. Pescadore 1978). Trench-fan deposits include all turbidite facies and submarine-fan facies associations, and palaeocurrent patterns range from radial to longitudinal (Fig. 4). Vertical cycles should include typical thinning- and fining-upward sequences associated with channel migration and abandonment, as well as thickening- and coarsening-upward cycles associated with progradation of outer-fan depositional lobes (e.g. Ricci-Lucchi 1975; Mutti *et al.* 1978; Walker 1978).

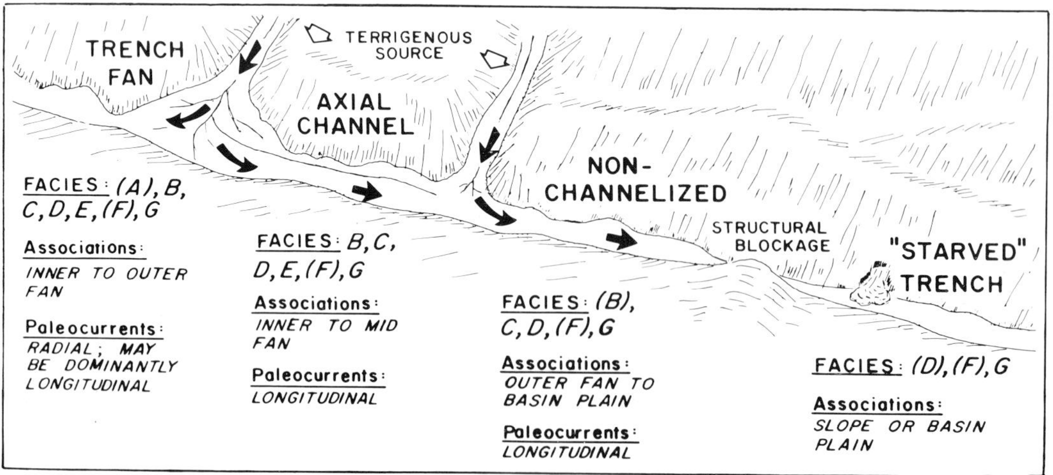

FIG. 4. Conceptual diagram showing the turbidite facies and palaeocurrent patterns predicted for sediment bodies on the trench floor. See Table 1 for definition of facies terminology. Minor turbidite facies are shown in parentheses, and equivalent and/or similar associations, including submarine-fan facies associations, are also indicated. Adapted from Underwood *et al.* (1980).

Trenches containing large axial channels do not fit well into a model of submarine-fan sedimentation. Thick sequences of channellized sandstone (facies B & C) are likely in such cases; these deposits would pass sharply into associated fine-grained levee and overbank deposits (facies E, with D & G). Palaeocurrents reflect longitudinal flow (Fig. 4). Facies associations are similar to those of inner- to mid-fan facies, but more distal fan facies associations are notably lacking. Basin-floor channels of this type have been described in the Alps and Apennines by Sagri (1979).

Evidence from seismic reflection data suggests that non-channellized sheet flow is common in modern trenches. Facies associations related to this style of sedimentation are probably similar to non-channellized outer-fan and basin-plain facies associations, although palaeocurrent patterns would be dominantly longitudinal (Fig. 4). Facies C and D turbidites and interlayered hemipelagic mudstones (facies G) are expected, but well-developed vertical cycles are unlikely. Analogies with modern abyssal-floor basins suggest a decrease in both sandstone/shale ratios and sand-layer thickness with increasing distance from the sediment source (Pilkey *et al.* 1980).

Trenches may be starved of coarse clastic material either because no terrigenous sands reach the trench or axial transport on the trench floor is blocked. The sedimentary facies associations expected with a 'starved' trench are similar to those of the slope or basin plain (Fig. 4); deposits are dominated by facies G hemipelagic muds and fine-grained facies D turbidites, along with possible facies F slump bodies.

Trench-slope deposits

The structure and morphology of forearc regions are diverse and variable (Seely 1979; Dickinson & Seely 1979). Large sedimentary basins such as the Aleutian Terrace (Marlow *et al.* 1973; Scholl 1974) and the Mentawai Trough, off Sumatra (Karig *et al.* 1980) are commonly located between the active magmatic arc and the trench-slope break. Several smaller basins or benches may develop on the upper trench slope, such as off Oregon-Washington (Barnard 1978; Kulm & Scheidegger 1979) and Peru-Chile (Coulbourn & Moberly 1977; Moberly *et al.* 1981; Kulm *et al.* 1977; Kulm *et al.* 1981). Models of forearc-basin sedimentation are well-established. Sedimentary sequences range from non-marine, deltaic, and shelf deposits to large complexes of coalescing submarine fans (Dickinson 1971, 1974; Ingersoll 1978b; Dickinson & Seely 1979). In this paper, forearc basins are considered only in so far as they affect sedimentation along the lower trench slope.

Sedimentary basins along the lower slope are generally elongate and bounded by tectonically active ridges (e.g. von Huene 1972, 1979; Kulm & Fowler 1974; Carson *et al.* 1974; G. F. Moore & Karig 1976; White & Klitgord 1976; Karig *et al.* 1979; Arthur, von Huene *et al.* 1981). Locally, slope basins may contain: (1) hemipelagic deposits or remobilized material slumped off adjacent bathymetric highs; (2) turbidites of shallow-water origin transported beyond the trench-slope break via submarine canyons; or (3) debris-flow deposits containing blocks of the underlying accretionary complex (G. F. Moore & Karig 1976). Any small basin or terrace is a potential site for the accumulation of sand-sized detritus, and sand layers have been cored in lower-slope basins off Kodiak Island (Kulm, von Huene *et al.* 1973; Howell & von Huene 1978; M. Hampton, pers. comm. 1979), New Zealand (Lewis 1981), and Oregon-Washington (Kulm, von Huene *et al.* 1973; Kulm & Fowler 1974; Barnard 1978; Kulm & Scheidegger 1979). Because of upslope blockage of sediment transport by tectonic ridges and deposition within basins located higher on the slope, the chances of coarse sediment reaching a lower-slope basin decreases with increasing distance down the trench slope (Underwood *et al.* 1980).

Where basins are not present, the apron of sediments covering the lower trench slope generally consists of hemipelagic muds and occasional thin beds of fine sand and silt (Ross 1971; Kulm, von Huene *et al.* 1973; Barnard 1978; Langseth, Okada *et al.* 1978; Moore, Watkins *et al.* 1979; Kulm & Scheidegger 1979; Krissek *et al.* 1980; von Huene, Aubouin *et al.* 1980; McMillen & Haines 1981; McMillen *et al.* 1981; Arthur, von Huene *et al.* 1981). In many cases, the slope muds overlie much coarser trench sediments that were accreted at the base of the lower slope (e.g. J. C. Moore & Karig 1976; Moore, Watkins *et al.* 1979). Tectonic loading and oversteepened slopes may cause sediment failure (slumps and slides) within the apron of slope muds (e.g. Hampton *et al.* 1978).

Upslope trapping

Small submarine canyons play an important role in the distribution of sediment along the trench slope by providing an effective transport route for coarse detritus. Most of the canyons on the landward slope of the northern Middle America Trench head downslope of the continental shelf (Figs 1 & 2; Fisher 1961), and, as a result, they probably do not receive a direct supply of terrigenous detritus with the present position of sea-level. It is likely, however, that sand-sized material entered the canyons during global low stands associated with Pleistocene glacial epochs. Moreover, unconfined mass movements (sediment creep, slumps, slides, debris flows) may carry coarse debris beyond the shelf break (Field & Clarke 1979; Nardin *et al.* 1979). Small submarine canyons can then intercept and funnel downslope the turbidity currents generated by mass movements occurring upslope or on the outer shelf.

Tectonic ridges and other bathymetric highs on the trench slope commonly cut off the paths of submarine canyons and block the transport of turbidites and other mass flows. The mechanics of upslope trapping are well-documented in the northern Middle America Trench, where detailed local bathymetry shows that the small canyons typically coalesce downslope into fewer and larger canyons (Karig *et al.* 1978; Underwood *et al.* 1980). As an individual canyon approaches a ridge, such as the trench-slope break, the canyon channel is blocked, the canyon gradient decreases, and sediment becomes ponded behind the ridge in a slope or forearc basin. Reflection profiles display far fewer submarine canyons below the trench-slope break (Underwood *et al.* 1980). In general, only the largest canyons continue across the active ridges of the trench slope (Underwood & Karig 1980). Approximately 80% of the submarine canyons off Mexico end without reaching the trench floor; most terminate above the trench-slope break (Figs 1 & 2). Where a greater number of active ridges are present, such as the Sunda Trench (G. F. Moore & Karig 1976; Karig *et al.* 1979), the effects of structural blockage of sediment transport are more pronounced, and the chances of submarine canyons maintaining their channels greatly diminishes downslope.

Ancient analogues

In general, differentiation of trench-floor and trench-slope deposits within ancient subduction complexes is difficult on the basis of sedimentary facies analysis alone, except in cases such as Barbados, where a clear lithological distinction exists between thin-bedded biogenic marls (slope deposits) and accreted trench turbidites (e.g. Speed 1981). Structural style is perhaps the most commonly used criterion for distinguishing between these two types of units, as trench deposits are generally more highly deformed than associated slope sediments (e.g. Bachman 1978; Moore & Karig 1980).

Perhaps the most complete record of lower-slope sedimentation is exposed on Nias Island, off Sumatra (Moore 1979; Moore & Karig 1976, 1980; Moore *et al.* 1980). Stratigraphic sequences exposed on Nias define an overall coarsening- and thickening-upward trend associated with basin uplift and the progradation of a submarine fan complex (Moore *et al.* 1980). Basal slope strata overlie mélange (accreted trench deposits) and consist of hemipelagic marls and lutite turbidites (facies D & G). There is a pronounced increase upsection in coarser facies A, B, and C deposits, and small-scale vertical cycles suggest that deposition of coarser facies occurred within outer- to inner-fan environments (Moore *et al.* 1980). The stratigraphic record on Nias demonstrates the influence of tectonic processes on lower-slope sedimentation. Apparently, individual basins, initially isolated from sources of coarse detritus, were uplifted and eventually linked with downcutting submarine canyons which headed off the coast of Sumatra (Moore 1979; Moore *et al.* 1980). Continued uplift and deposition allowed progradation of submarine fans and the infilling of the slope basins.

Several examples of possible slope-basin deposits occur within the Franciscan Complex of California (Howell *et al.* 1977; Underwood 1977; Bachman 1978, 1981; Smith *et al.* 1979). These strata are relatively undeformed and generally dominated by thick sequences of channellized facies B sandstone, with associated deposits of levee/overbank origin (facies E, with minor D & G). Some sections in the Coastal Belt contain abundant facies C & D turbidites, but distal fan-facies associations are generally lacking (Bachman 1978, 1981). Facies associations suggest deposition within mid- to outer-fan channels and depositional lobes, or their facies equivalents (Howell *et al.* 1977; Bachman 1978; Smith *et al.* 1979). The predominance of thick-bedded sandstone can be attributed to selective trapping of the coarse-grained parts of gravity flows within restricted slope basins, while more turbulent fine-grained fractions were able to bypass the basins and settle farther down the trench slope (Bachman 1978; Smith *et al.* 1979).

Ancient trench-slope deposits on Kodiak Island consist primarily of thick sequences of facies G mudstone (Nilsen & Moore 1979). Chaotic deposits (facies F) and thick beds of channellized sandstone and conglomerate (facies A & B) crop out locally and probably represent slump deposits and canyon/channel fill, respectively (Nilsen & Moore 1979). Trench-slope deposits on New Zealand contain deposits of low energy bottom currents (contourites) in addition to the dominant hemipelagic mudstones and lutite turbidites (Carter *et al.* 1978).

Sedimentary facies model

In our facies model for trench-slope deposits (Fig. 5), submarine-canyon and slope facies associations are generally the same as for tectonically inactive environments (e.g. Whitaker 1974; Kelling & Stanley 1976). Slope deposits are dominated by facies G hemipelagic mudstones and thin-bedded, fine-grained facies D turbidites (Fig. 5). The occurrence of chaotic deposits associated with slope failure (facies F) may be more common within subduction complexes, however, because of the related tectonic activity. Channel lag deposits from submarine canyons consist largely of coarse facies A, B, & C deposits; in large canyons, levee and overbank deposits (facies D, E, & G) may represent local deposition outside the channel thalweg. Slumps off canyon walls (facies F) may also become interstratified with canyon fill. Inactive canyons are filled largely with hemipelagic slope deposits and fine-grained turbidites (facies D & G). Palaeocurrents from canyon fill should be dominantly at a high angle to the margin (Fig. 5).

Because of upslope trapping, trench-slope basins near the base of the landward slope generally receive only fine-grained sediments derived from hemipelagic settling and dilute density flows (facies D & G). Slumps and slides derived from tectonically active bounding ridges are also likely, but such chaotic deposits would consist primarily of remobilized fine-grained slope sediments. In our facies model, basins that receive only fine-grained material are classified as 'immature' slope basins (Fig. 5).

Progressive growth of the accretionary prism generally causes uplift of trench-slope basins

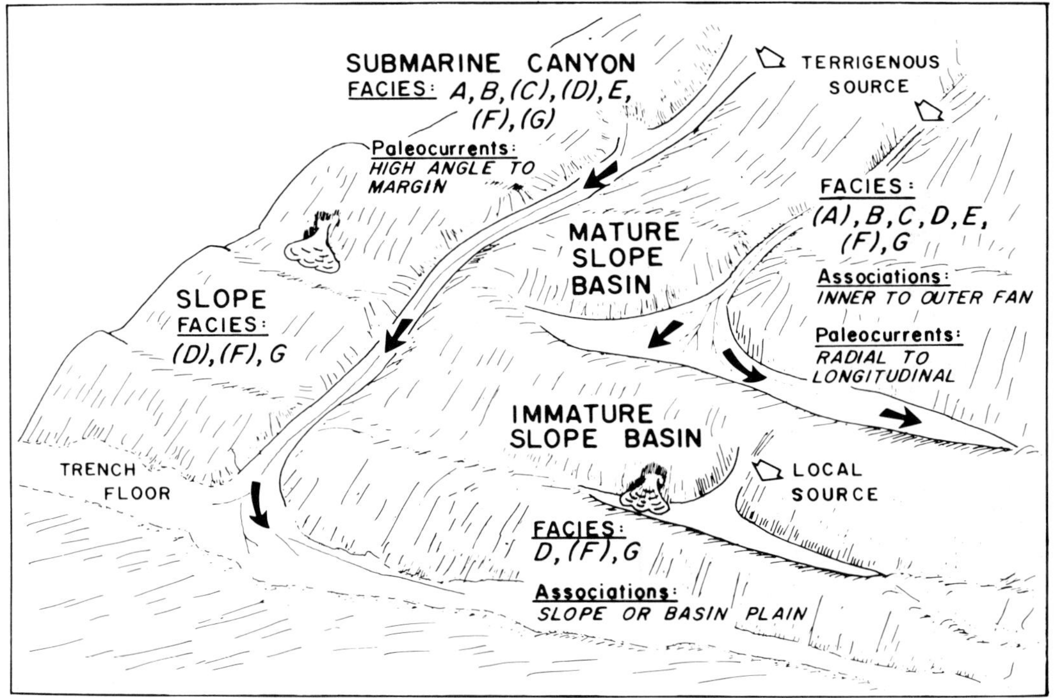

FIG. 5. Conceptual diagram showing the turbidite facies and palaeocurrent patterns predicted for trench-slope settings. The terminology follows that of Fig. 4 (see Table 1). The terms 'mature' and 'immature' slope basin refer to the presence or absence, respectively, of a source of coarse terrigenous sediment, and are not necessarily indicative of any particular bathymetric position on the trench slope. Adapted from Underwood *et al.* (1980).

through time; this is associated with a decrease in tectonic activity with increasing distance from the active basal slope (Seely *et al.* 1974; Karig & Sharman 1975). As a result, bounding thrust ridges may become inactive during the uplift history of individual slope basins, thereby increasing the width of the basins through time (G. F. Moore & Karig 1976). The combined effect of these processes may be the sudden influx of coarse terrigenous detritus into previously isolated slope basins, as basin uplift is accompanied by progressive downcutting of submarine canyons (e.g. Moore *et al.* 1980). In our facies model, we define any such basin that receives a direct supply of coarse clastic material as 'mature', regardless of its actual position on the trench slope (Fig. 5).

The facies associations for mature slope basins are varied and depend largely upon the geometry of the basin. All of the submarine-fan facies associations may be present in large slope basins. In contrast, small narrow basins may selectively trap coarse material and allow fine-grained sediment to bypass the basin. Palaeocurrent patterns may also vary from radial to longitudinal (Fig. 5).

Discussion

The sedimentary facies models presented in this paper are somewhat speculative, due largely to the general lack of detailed sampling within modern subduction zones. The available data clearly indicate, however, that over-simplified models cannot account for the observed complexity and diversity of sediment bodies and patterns of sedimentation. We have attempted to stress the variety of possible facies associations and identify some of the variables that affect sedimentation within trench-floor and trench-slope environments.

Turbidite facies terminology is an important tool in the analysis of sedimentary strata, including strata exposed within ancient subduction complexes. However, turbidite facies associations within subduction complexes may be other than those of a submarine fan. The facies associations that do not fit well into the slope/fan/basin-plain model of deep-basin sedimentation include axial-channel, non-channellized (sheet-flow) trench-floor, starved-trench, immature-slope-basin, and certain associations within mature slope basins, such as small basins that selectively trap coarse detritus. Submarine fans also occur within subduction zones, but we believe that the use of terminology linked to submarine-fan morphologies is clearly not appropriate in all cases. The identification of submarine-fan deposits within subduction complexes should depend upon the documentation of appropriate palaeocurrent patterns, vertical cycles, and fan facies associations that are consistent with submarine-fan progradation or regression. In the absence of such documentation, facies association terminology should be linked with the geometries of the other types of sediment bodies observed within modern trenches, such as axial channels.

Our facies models have other implications for environmental reconstructions within accretionary margins, including the identification and differentiation of separate tectonostratigraphic units. Because of the complexity and variability of sedimentation that is possible within a single trench, general lithologies and turbidite facies associations can change dramatically within a relatively short distance. For example, structural blockage of axial transport paths by a seamount on the downgoing oceanic plate could allow thick sequences of sandy turbidites to accumulate on one side of the blockage, with hemipelagic muds dominating the other side. Subsequent accretion and uplift of the seamount and trench fill could expose sections of strata that might appear to represent completely unrelated geological histories. The resulting juxtaposition of sandy turbidites, basalts, and mudstones could be erroneously attributed to large-scale tectonic events associated with accretion, rather than the original depositional geometry that existed on the trench floor. In other cases, it may be nearly impossible to differentiate between trench fill and trench-slope deposits. For example, abyssal sediments, trench sediments, and slope sediments could all be very similar if the trench is starved of coarse clastic material. If tectonostratigraphic designations are based primarily upon sedimentary facies relations, then all of the resulting strata would be included in the same unit, and the designations would have little genetic meaning.

In addition to the predicted lithological variations and complexities, significant differences in sandstone petrology could result for trench-floor and slope-basin deposits that accumulate in different sites along the length of an accretionary margin. For example, both regional and local changes in source rocks and onland drainage patterns could introduce sands of differing composition into the submarine canyons that funnel detritus into different sedimentary basins. Moreover, long-distance axial transport in the trench could juxtapose detritus of one composition against sediments of a different provenance that were transported via local submarine canyons. It is therefore important to integrate palaeocurrent data with data on sandstone petrology before reliable provenance determinations can be made.

We suggest that normal sedimentary processes, rather than large-scale tectonic events, may be responsible in some cases for lithological and compositional changes recorded within many subduction complexes and accretionary margins. Clearly, the possible effects of penecontemporaneous or subsequent tectonism must be considered, but the possibility of relating observed changes or differences to a logical system of sediment transport and deposition cannot be ignored. The separation of strata into meaningful tectonostratigraphic units should depend upon several criteria, such as contrasts in sedimentary facies associations, sandstone petrology, structural style, palaeontology, and general physical properties.

ACKNOWLEDGMENTS: We would like to thank the following people for valuable contributions to this paper: M. Hampton, D. Howell, D. Karig, G. F. Moore, J. C. Moore, W. Normark, and W. Schweller. The manuscript was greatly improved by critical reviews from C. Ando, J. Leggett, G. F. Moore, W. Schweller, and the Italian and English sedimentologists who served as anonymous referees.

References

AALTO, K. R. 1976. Sedimentology of a melange: Franciscan of Trinidad, California. *J. sediment. Petrol.* **46,** 913–29.

ARTHUR, M., VON HUENE, R. *et al.* 1981. Sedimentation in the Japan Trench. *Initial Rep. Deep Sea drill. Proj.* **57,** U.S. Govt Printing Office, Washington, in press.

BACHMAN, S. B. 1978. A Cretaceous and early Tertiary subduction complex, Mendocino coast, northern California. *In:* HOWELL, D. G. & MCDOUGALL, K. (eds). *Mesozoic Paleogeography of the Western United States.* Pacif. Sec. Soc. econ. Paleontol. Mineral. Pacif. Coast Paleogeogr. Symp. **2,** 419–30.

BACHMAN, S. B. 1981. The Coastal Belt of the Franciscan: youngest phase of northern Californian subduction (this volume).

BARNARD, W. D. 1978. The Washington continental slope: Quaternary tectonics and sedimentation. *Mar. Geol.* **27,** 79–114.

BENNETTS, K. R. & PILKEY, O. H. 1976. Characteristics of three turbidites, Hispaniola-Caicos Basin. *Bull. geol. Soc. Am.* **87,** 1291–300.

BOUMA, A. H. & NILSEN, T. H. 1978. Turbidite facies and deep-sea fans—with examples from Kodiak Island, Alaska. *Proc. Offshore Tech. Conf.* **1,** 559–70.

CARLSON, P. R. & NELSON, C. H. 1969. Sediments and sedimentary structures of the Astoria submarine canyon-fan system, northeast Pacific. *J. sediment. Petrol.* **39,** 1269–82.

CARSON, B. 1980. Sediment dewatering associated with subduction: experimental results. *Abstr. Progr. geol. Soc. Am.* **12,** 399.

——, YUAN, J., MYERS, P. B. & BARNARD, W. D. 1974. Initial deep-sea sediment deformation at the base of the Washington continental slope—a response to subduction. *Geology,* **2,** 561–4.

CARTER, R. M., HICKS, M. D., NORRIS, R. J. & TURNBULL, I. M. 1978. Sedimentation patterns in an ancient arc-trench-ocean basin complex: Carboniferous to Jurassic Rangitata orogen, New Zealand. *In*: STANLEY, D. J. & KELLING, G. (eds). *Sedimentation in Submarine Canyons, Fans, and Trenches,* 340–61. Dowden, Hutchinson & Ross, Stroudsburg, Pennsylvania.

Coulbourn, W. T. & Moberly, R. 1977. Structural evidence of the evolution of forearc basins off South America. *Can. J. Earth Sci.* **14,** 102–16.

CURRAY, J. R. & MOORE, D. G. 1974. Sedimentary and tectonic processes in the Bengal deep-sea fan and geosyncline. *In*: BURK, C. A. & DRAKE, C. L. (eds). *The Geology of Continental Margins,* 617–28. Springer-Verlag, New York.

DAMUTH, J. E. 1979. Migrating sediment waves created by turbidity currents in the northern South China Basin. *Geology,* **7,** 520–3.

DICKINSON, W. R. 1971. Clastic sedimentary sequences deposited in shelf, slope, and trough settings between magmatic arcs and associated trenches. *Pacif. Geol.* **3,** 15–30.

—— 1974. Sedimentation within and beside ancient and modern magmatic arcs. *In*: DOTT, R. H. & SHAVER, R. H. (eds). *Modern and Ancient Geosynclinal Sedimentation.* Spec. Publ. Soc. econ. Paleontol. Mineral. Tulsa, **19,** 230–9.

DICKINSON, W. R. & SEELY, D. R. 1979. Structure and stratigraphy of forearc regions. *Bull. Am. Assoc. Petrol. Geol.* **63,** 2–31.

ELMORE, R. D., PILKEY, O. H., CLEARY, W. J. & CURRAN, H. A. 1979. Black Shell turbidite, Hatteras Abyssal Plain, western Atlantic Ocean. *Bull. geol. Soc. Am.* **90,** 1165–76.

FIELD, M. E. & CLARKE, S. H. JR 1979. Small-scale slumps and slides and their significance for basin slope processes, southern California Borderland. *In*: DOYLE, L. J. & PILKEY, O. H. (eds). *Geology of Continental Slopes.* Spec. Publ. Soc. econ. Paleontol. Mineral. Tulsa, **27,** 223–30.

FISHER, R. L. 1961. Middle America Trench: topography and structure. *Bull. geol. Soc. Am.* **72,** 703–20.

HAMPTON, M. A., BOUMA, A. H., CARLSON, P. R., MOLNIA, B. F. & CLUKEY, E. C. 1978. Quantitative study of slope instability in the Gulf of Alaska. *Proc. Offshore Tech. Conf.* **4,** 2307–18.

HAYES, D. E. 1974. Continental margin of western South America. *In*: BURK, C. A. & DRAKE, C. L. (eds). *The Geology of Continental Margins*, 581–90. Springer-Verlag, New York.

HOWELL, D. G., VEDDER, J. G., MCLEAN, H., JOYCE, J. M., CLARKE, S. H. JR & SMITH, G. 1977. Review of Cretaceous geology, Salinian and Nacimiento blocks, coast ranges of central California. *In*: HOWELL, D. G., VEDDER, J. G. & MCDOUGALL, D. (eds). *Cretaceous Geology of the California Coast Ranges, West of the San Andreas Fault.* Pacif. Sec., Soc. econ. Paleontol. Mineral. Tulsa, Pacif. Coast Paleogeogr. Field Guide, **2,** 1–46.

—— & VON HUENE, R. 1978. Trench-slope basins off Kodiak Island, a modern analog for some Late Cretaceous rocks of central California. *Abstr. Progr. geol. Soc. Am.* **10,** 109–10.

HSÜ, K. J., KELTS, K. & VALENTINE, J. W. 1980. Resedimented facies in Ventura Basin, California, and model of longitudinal transport of turbidity currents. *Bull. Am. Assoc. Petrol. Geol.* **64,** 1034–51.

IBRAHIM, A. K., LATHAM, G. V. & LADD, J. 1979. Seismic refraction and reflection measurements in the Middle America Trench offshore Guatemala. *J. geophys. Res.* **84,** 5643–9.

INGERSOLL, R. V. 1978a. Submarine fan facies of the Upper Cretaceous Great Valley sequence, northern and central California. *Sediment. Geol.* **21,** 205–30.

—— 1978b. Paleogeography and paleotectonics of the late Mesozoic forearc basin of northern and central California. *In*: HOWELL, D. G. & MCDOUGALL, K. (eds). *Mesozoic Paleogeography of the Western United States.* Pacif. Sec., Soc. econ. Paleontol. Mineral. Pacif. Coast Paleogeogr. Symp. **2,** 471–82.

KARIG, D. E., CARDWELL, R. K., MOORE, G. F. & MOORE, D. G. 1978. Late Cretaceous subduction and continental margin truncation along the northern Middle America Trench. *Bull. geol. Soc. Am.* **89,** 265–76.

—— , INGLE, J. C. *et al.* 1975. *Initial rep. Deep Sea drill. Proj.* **32,** 927 pp. U.S. Govt Printing Office, Washington.

——, LAWRENCE, M. B., MOORE, G. F. & CURRAY, J. R. 1980. Structural framework of the forearc basin, NW Sumatra. *J. geol. Soc. London,* **137,** 77–91.

——, MOORE, G. F., CURRAY, J. R. & LAWRENCE, M. B. 1981. Morphology and shallow structure of the trench slope off Nias Island, Sunda arc, *in special SEATAR volume.* Am. geophys. Union, in press.

—— & SHARMAN, G. F., III 1975. Subduction and accretion in trenches. *Bull. geol. Soc. Am.* **86,** 377–89.

——, SUPARKA, S., MOORE, G. F. & HEHANUSSA, P. E. 1979. Structure and Cenozoic evolution of the Sunda Arc in the central Sumatra region. *In*: WATKINS, J. S., MONTADERT, L. & DICKERSON, P. W. (eds). *Geological and Geophysical Investigations of Continental Margins.* Mem. Am. Assoc. Petrol. Geol. **29,** 223–37.

KELLING, G. & STANLEY, D. J. 1976. Sedimentation in canyon, slope, and base-of-slope environments. *In:* STANLEY, D. J. & SWIFT, D. J. P. (eds). *Marine Sediment Transport and Environment Management,* 379–435. Wiley Interscience, New York.

KRISSEK, L. A., SCHEIDEGGER, K. F. & KULM, L. D. 1980. Surface sediments of the Peru-Chile continental margin and Nazca plate. *Bull. geol. Soc. Am.* **91,** 321–31.

KULM, L. D. & FOWLER, G. A. 1974. Oregon continental margin structure and stratigraphy: a test of the imbricate thrust model, *In*: BURK, C. A. & DRAKE, C. L. (eds). *The Geology of Continental Margins*, 261–83. Springer-Verlag, New York.

—— & SCHEIDEGGER, K. F. 1979. Quaternary sedimentation on the tectonicaly active Oregon continental slope. *In*: DOYLE, L. J. & PILKEY, O. H. (eds). *Geology of Continental Slopes.* Spec. Publ. Soc. econ. Paleontol. Mineral. Tulsa, **29,** 247–64.

—— , SCHWELLER, W. J. & MASIAS, A. 1977. A preliminary analysis of the subduction processes along the Andean continental margin, 6° to 45°S, *In:* TALWANI, M. & PITMAN, W. C., III (eds). *Island Arcs, Deep-sea Trenches, and Back-arc Basins.* Am. Geophys. Union, M. Ewing Ser. **1,** 285–301.

—— , VON HUENE, R. *et al.* 1973. *Initial Rep. Deep Sea drill. Proj.* **18,** U.S. Govt Printing Office, Washington, 1077 pp.

KULM, L. D., RESIG, J. M., THORNBURG, T. M. & SCHRADER, H.-J. 1981. Cenozoic structure, stratigraphy and tectonics of the central Peru forearc (this volume).

LADD, J. W., IBRAHIM, A. K., MCMILLEN, K. J., LATHAM, G. V., VON HUENE, R., WATKINS, J. S., MOORE, J. C. & WORZELL, J. L. 1978. Tectonics of the Middle America Trench, offshore Guatamala. Int. Symp. *Guatamala 4 February Earthquake and Reconstruction Process*, **1,** Guatamala City, May 1978.

LANGSETH, M., OKADA, H. *et al.* 1978. Near the Japan Trench—transects begins. *Geotimes*, **23,** 22–6.

LEGGETT, J. K. 1980. The sedimentological evolution of a Lower Palaeozoic accretionary fore-arc in the Southern Uplands of Scotland. *Sedimentology*, **27,** 401–17.

——, MCKERROW, W. S. & EALES, M. H. 1979. The Southern Uplands of Scotland: a Lower Palaeozoic accretionary prism. *J. geol. Soc. London*, **136,** 755–70.

LEWIS, K. B. 1981. Quaternary sedimentation of the Kikurangi oblique subduction and transform margin. *In:* BALLANCE, P. F. & READING, H. G. (eds). *Sedimentation in Oblique-slip Moblile Zones.* Spec. Publ. int. Assoc. Sediment. **4,** 171–89. Blackwell Scientific Publications, Oxford.

LONSDALE, P. 1978. Ecuadorian subduction system. *Bull. Am. Assoc. Petrol. Geol.* **62,** 2454–77.

MARLOW, M. S., SCHOLL, D. W., BUFFINGTON, E. C. & ALPHA, T. R. 1973. Tectonic history of the central Aleutian arc. *Bull. geol. Soc. Am.* **84,** 1555–74.

McMillen, K. J., Enkeboll, R. H., Moore, J. C., Shipley, T. A. & Ladd, J. W. 1981. Sedimentation in different tectonic environments of the Middle America Trench, Southern Mexico and Guatemala (this volume).

—— & Haines, T. R. 1981. Late Quaternary sediments of the southern Mexico margin. *In*: Moore, J. C., Watkins, J. S. *et al.* (eds). *Initial Rep. Deep Sea drill. Proj.*, **66,** U.S. Govt Printing Office, Washington.

Moberley, R., Shepherd, G. L. & Coulbourn, W. T. 1981. Forearc and other basins, continental margin of northern and southern Peru and adjacent Ecuador and Chile (this volume).

Moore, D. G., Curray, J. R. & Emmel, F. J. 1976. Large submarine slide (olistostrome) associated with Sunda arc subduction zone, northeast Indian Ocean. *Mar. Geol.* **21,** 211–26.

Moore, G. F. 1979. Petrology of subduction zone sandstones from Nias Island, Indonesia. *J. sediment. Petrol.* **49,** 71–84.

——, Billman, H. G., Hehanussa, P. E. & Karig, D. E. 1980. Sedimentology and paleobathymetry of Neogene trench-slope deposits, Nias Island, Indonesia. *J. Geol.* **88,** 161–80.

—— & Karig, D. E. 1976. Development of sedimentary basins on the lower trench slope. *Geology*, **4,** 501–4.

—— & —— 1980. Structural geology of Nias Island, Indonesia: implications for subduction zone tectonics. *Am. J. Sci.* **280,** 193–223.

Moore, J. C. & Karig, D. E. 1976. Sedimentology, structural geology, and tectonics of the Shikoku subduction zone. *Bull. geol. Soc. Am.* **87,** 1259–68.

——, Watkins, J. S. *et al.* 1979. Progressive accretion in the Middle America Trench, southern Mexico. *Nature. London*, **281**, 638–42.

Moore, J. C. *et al.* 1981. Facies belts of the Middle America Trench and forearc region, southern Mexico: results from Leg 66 DSDP (this volume).

Mutti, E., Nilsen, T. H. & Ricci-Lucchi, F. 1978. Outer fan depositional lobes of the Laga Formation (upper Miocene and lower Pliocene) east-central Italy. *In*: Stanley, D. J. & Kelling, G. (eds). *Sedimentation in Submarine Canyons, Fans and Trenches*, 210–23. Dowden, Hutchinson & Ross, Stroudsburg, Pennsylvania.

—— & Ricci-Lucchi, F. 1975. Turbidite facies and facies associations. *In*: *Examples of turbidite and facies associations from selected formations of the northern Apennines*, 21–36. 9th Int. Congr. Sedimentology, Nice, Guidebook.

Nardin, T. R., Edwards, B. D. & Gorsline, D. S. 1979. Santa Cruz Basin, California Borderland: dominance of slope processes in basin sedimentation. *In:* Doyle, L. J. & Pilkey, O. H. (eds). *Geology of Continental Slopes*. Spec. Publ. Soc. econ. Paleontol. Mineral. Tulsa, **27,** 209–22.

Nelson, C. H. & Kulm, L. D. 1973. Submarine fans and deep-sea channels. *In*: *Turbidites and Deep Water Sedimentation*. Pacif. Sec., Short Course, Soc. econ. Paleontol. Mineral. 39–78.

—— & Nilsen, T. H. 1974. Depositional trends of modern and ancient deep-sea fans. *In*: Dott, R. H. & Shaver, R. H. (eds). *Modern and Ancient Geosynclinal Sedimentation*. Spec. Publ. Soc. econ. Paleontol. Mineral., Tulsa, **19,** 69–91.

Nilsen, T. H. 1980. Modern and ancient submarine fans: discussion of papers by R. G. Walker and W. R. Normark. *Bull. Am. Assoc. Petrol. Geol.* **64,** 1094–101.

—— & Bouma, A. H. 1977. Turbidite sedimentology and depositional framework of the Upper Cretaceous Kodiak formation and related stratigraphic units, southern Alaska. *Abstr. Progr. geol. Soc. Am.* **9,** 1115.

—— & Moore, G. W. 1979. Reconnaissance study of Upper Cretaceous to Miocene stratigraphic units and sedimentary facies, Kodiak and adjacent islands, Alaska. *Prof. Pap. U.S. geol. Surv.* **1093,** 1–34.

Normark, W. R. 1970. Growth patterns of deep-sea fans. *Bull. Am. Assoc. Petrol. Geol.* **54,** 2170–95.

—— 1978. Fan valleys, channels, and depositional lobes on modern submarine fans: Characters for recognition of sandy turbidite environments. *Bull. Am. Assoc. Petrol. Geol.* **62,** 912–31.

Pescadore, T. 1978. The Irpinids: A model of tectonically controlled fan and base-of-slope sedimentation in southern Italy. *In*: Stanley, D. J. & Kelling, G. (eds). *Sedimentation in Submarine Canyons, Fans, and Trenches*, 325–39. Dowden, Hutchinson & Ross, Stroudsburg, Pennsylvania.

Pilkey, O. H., Locker, S. D. & Cleary, W. J. 1980. Comparison of sand layer geometry on the flat floors of ten modern depositional basins. *Bull. Am. Assoc. Petrol. Geol.* **64,** 841–56.

Piper, D. J. W., von Huene, R. & Duncan, J. R. 1973. Late Quaternary sedimentation in the active eastern Aleutian Trench. *Geology*, **1,** 19–22.

Prince, R. A., Schweller, W. J., Coulbourn, W. T., Shepard, G. L., Ness, G. E. & Masias, A. 1980. *Bathymetry of the Peru-Chile Continental Margin and Trench*. Geol. Soc. Am. Map Ser.

Pudsey, C. J. & Reading, H. E. 1981. Sedimentology and structure of the Scotland Group, Barbados (this volume).

Reimnitz, E. & Guiterrez-Estrada, M. 1970. Rapid changes in the head of the Rio Balsas submarine canyon system, Mexico. *Mar. Geol.* **8,** 245–58.

Ricci-Lucchi, F. 1975. Depositional cycles in two turbidite formations of northern Apennines (Italy). *J. sediment. Petrol.* **45,** 3–43.

Ross, D. A. 1971. Sediments of the northern Middle America Trench. *Bull. geol. Soc. Am.* **82,** 303–22.

—— & Shor, G. G. Jr 1965. Reflection profiles across the Middle America Trench. *J. geophys. Res.* **70,** 5551–71.

Sagri, M. 1979. Upper Cretaceous carbonate turbidites of the Alps and Apennines deposited below the calcite compensation level. *J. sediment. Petrol.* **49,** 23–8.

Scholl, D. W. 1974. Sedimentary sequences in the north Pacific trenches. *In:* Burk, C. A. & Drake, C. L. (eds). *The Geology of Continental*

Margins, 493–504. Springer-Verlag, New York.

—— , Christensen, M. N., von Huene, R. & Marlow, M. S. 1970. Peru-Chile trench sediments and sea-floor spreading. *Bull. geol. Soc. Am.* **81,** 1339–60.

—— , Marlow, M. S. & Cooper, A. K. 1977. Sediment subduction and offscrapping at Pacific margins. *In*: Talwani, M. & Pitman, W. C., III (eds). *Island Arcs, Deep Sea Trenches, and Back-Arc Basins.* Am. geophys. Union, M. Ewing Ser. **1,** 199–210.

Schweller, W. J. & Kulm, L. D. 1978. Depositional patterns and channelized sedimentation in active eastern Pacific trenches. *In*: Stanley, D. J. & Kelling, G. (eds). *Sedimentation in Submarine Canyons, Fans, and Trenches*, 311–24. Dowden, Hutchinson & Ross, Stroudsburg, Pennsylvania.

Seely, D. R. 1979. The evolution of structural highs bordering major forearc basins, *In:* Watkins, J. S., Montadert, L. & Dickerson, P. W. (eds). *Geological and Geophysical Investigations of Continental Margins.* Mem. Am. Assoc. Petrol. Geol. **29,** 245–60.

—— , Vail, P. R. & Walton, G. G. 1974. Trench slope model. *In*: Burk, C. A. & Drake, C. L. (eds). *The Geology of Continental Margins*, 249–60. Springer-Verlag, New York.

Shepard, F. P., Marshall, N. F., McLoughlin, P. A. & Sullivan, G. G. 1979. Currents in submarine canyons and other sea valleys. *Stud. Geol. Am. Assoc. Petrol. Geol.* **8,** 1–173.

Shipley, T. H., McMillen, K. J., Watkins, J. S., Moore, J. C., Sandoval-Ochoa, J. H. & Worzel, J. L. 1980. Continental margin of Guerrero and Oaxaca, Mexico. *Mar. Geol.* **35,** 65–82.

Silver, E. A. 1971. Transitional tectonics and late Cenozoic structure of the continental margin off northernmost California. *Bull. geol. Soc. Am.* **82,** 1–22.

Smith, G. W., Howell, D. G. & Ingersoll, R. V. 1979. Late Cretaceous trench-slope basins of central California. *Geology*, **7,** 303–6.

Speed, R. C. 1981. New views on the geology of Barbados. *Second Latin Am. Geol. Congr.*

Underwood, M. B. 1977. The Pfeiffer Beach slab deposits, Monterey County, California: possible trench-slope basin. *In*: Howell, D. G., Vedder, J. G. & McDougall, K. (eds). *Cretaceous Geology of the California Coast Ranges, West of the San Andreas Fault.* Pacif. Sec. Soc. econ. Paleontol. Mineral., Pacif. Coast Paleogeogr. Field Guide, **2,** 57–69.

—— , Bachman, S. B. & Schweller, W. J. 1980. Sedimentary processes and facies associations within trench and trench-slope settings. *In*: Field, M. E., Bouma, A. H. & Colburn, I. (eds). *Quaternary Depositional Environments on the Pacific Continental Margin*, 211–29. Pacif. Sec. Soc. econ. Paleontol. Mineral., Tulsa.

von Huene, R. 1972. Structure of the continental margin and tectonism at the eastern Aleutian Trench. *Bull. geol. Soc. Am.* **83,** 3613–26.

—— 1974. Modern trench sediments. *In*: Burk, C. A. & Drake, C. L. (eds). *The Geology of Continental Margins*, 207–11. Springer-Verlag, New York.

—— 1979. Structure of the outer convergent margin off Kodiak Island, Alaska, from multichannel seismic records. *In:* Watkins, J. S., Montadert, L. & Dickerson, P. W. (eds). *Geological and Geophysical Investigations of Continental Margins.* Mem. Am. Assoc. Petrol. Geol. **29,** 261–71.

—— , Aubouin, *et al.* 1980. Leg 67: The Deep Sea Drilling Project Middle America Trench transect off Guatemala. *Bull. geol. Soc. Am.* **91,** 421–32.

Walker, R. G. 1978. Deep-water sandstone facies and ancient submarine fans: Models for exploration for stratigraphic traps. *Bull. Am. Assoc. Petrol. Geol.* **62,** 932–66.

—— & Mutti, E. 1973. Turbidite facies and facies associations. *In*: *Turbidites and Deep Water Sedimentation.* Pacif. Sec., Short Course, Soc. econ. Paleontol. Mineral. 119–58.

Whitaker, J. H. McD. 1974. Ancient submarine canyons and fan valleys. *In*: Dott, R. H. & Shaver, R. H. (eds). *Modern and Ancient Geosynclinal Sedimentation.* Spec. Publ. Soc. econ. Paleontol. Mineral., Tulsa, **19,** 106–25.

White, R. S. & Klitgord, K. 1976. Sediment deformation and plate tectonics in the Gulf of Oman. *Earth planet. Sci. Lett.* **32,** 199–209.

Michael B. Underwood & Steven B. Bachman, Department of Geological Sciences, Kimball Hall, Cornell University, Ithaca, New York 14853, U.S.A.

Composition of modern deep-sea sands from arc-related basins

J. Barry Maynard, Renzo Valloni & Ho-Shing Yu

SUMMARY: Petrographic and chemical examination of modern deep-sea sands shows a clear distinction between active and passive plate-tectonic settings. Among active settings, sands from forearc basins of island arcs and basins from strike-slip continental margins can be distinguished, but those from other arc-related settings overlap considerably in composition. In particular, sands from continental margin subduction zones and those from the backarc of island arcs appear to be indistinguishable.

Chemically, arc-related sands are very close in composition to ancient greywacke sandstones. For instance, sodium almost always exceeds potassium. Thus it is not necessary for sodium to be added to sands diagenetically to make greywackes, as has been suggested.

The influence of tectonics on the composition of sandstones has long been a central feature of sandstone petrology. The geosynclinal model of global tectonics stimulated much thought on this subject. Probably the best-known ideas are those of Krynine (e.g. 1942). Subsequently such research declined, but the plate model for global tectonics has produced a new wave of interest (e.g. Crook 1974; Schwab 1975; Potter 1978; Dickinson & Suczek 1979). Much of this work has relied on data from ancient sediments, where diagenetic changes and difficulty in deciphering the tectonic setting limit the usefulness of the conclusions. Much more work is needed on recent sediments to establish a baseline against which ancient sediments can be compared.

We have previously reported (Valloni & Maynard 1981) the results of our work on the petrography of a set of recent deep-sea sands from piston cores. In this paper we shall compare this petrographic information with bulk chemistry, and attempt to relate these to tectonics in arc-related settings.

Plate tectonic models of depositional basins

First, it is necessary to define the terms used. The classification of sedimentary environments using plate tectonics has been extensively discussed by Dickinson (1974), Dickinson & Seely (1979) and Reading (1978, chapter 14). These ideas are summarized in Table 1. The most important subdivision is into passive and active settings, as originally pointed out by Inman & Nordstrom (1971). In the classification used here, passive settings are synonymous with their 'trailing-edge coast'. We have subdivided their 'collision coasts' and 'marginal sea categories'. The petrography of sands from the passive settings and the collision-related settings has been described in some detail by Dickinson & Valloni (1980). In this paper, therefore, we concentrate more on the arc-related settings, and introduce the chemical properties of the sediments.

Deep-sea sands are found in most of the environments listed in Table 1, but are most common in strike-slip and subduction settings and the later stages of intercontinental rifting. Of the subduction settings, the continent side of a subducting continental margin (the backarc of continental margin arcs) does not contain deep-sea sands. Instead, many of the world's big rivers derive their sediments from these areas (e.g. the Amazon), so that the sands described by Potter (1978) may largely fill this category. Thus the important environments from the standpoint of this paper are basins of island arcs and basins lying seaward of continental margin volcanic arcs or strike-slip plate boundaries.

Petrography

About 80 samples of deep-sea sands, mostly from Lamont-Doherty piston cores supplemented by some DSDP material, were studied petrographically (Fig. 1). The details of the methods used and the depository for all of the original data are given in Valloni & Maynard (1981). For petrography, the sands were sieved to isolate the fine sand fraction, because of the strong effect of grain size on feldspar content (e.g. Field & Pilkey 1969, fig. 3). The results are summarized in Table 2 and Fig. 2.

The samples separate well into passive (trailing edge) and active settings on the basis of the

TABLE 1. *Plate tectonic classification of sedimentary environments (after Reading 1979, chapter 14; Dickinson & Valloni 1980, fig. 2)*

- I Spreading-related or passive settings
  - (A) *Intracratonic rifts* (East Africa). Mostly filled by alluvial fans and lakes
  - (B) *Failed rifts* or aulacogens (Benue Trough). Thick sequence of deposits ranging from deep-sea fan to fluvial
  - (C) *Intercontinental rifts*
    - 1 Early (Red Sea). Evaporites and some clastics
    - 2 Late (Atlantic). Early stage sediments along margins overlain by continental shelf or deltaic deposits, passing seaward into oceanic crust overlain by turbidites and finally pelagics. Often termed *trailing-edge*
- II Active settings
  - (A) Continental collision-related
    - 1 *Remnant ocean basins* (Bay of Bengal). Thick turbidite fan deposits eventually piled into imbricate thrust sheets
    - 2 *Late orogenic basins* (sub-Himalayas). Variety of terrestrial and shallow marine deposits, but dominantly fluvial. Termed *peripheral foreland basins* if on the subducting plate
  - (B) *Strike-slip fault-related settings* (California). Thick sequences of deep-marine to fluvial sediments in small basins, mostly derived from adjacent uplifts
  - (C) Subduction-related settings
    - 1 *Continental margin magmatic arcs* (Andes)
      - (a) Forearc. Thin to thick deposits ranging from turbidites to fluvial. Termed *leading-edge* in some classifications
      - (b) Backarc. Very thick, mostly terrestrial, accumulations. Forms much of the molasse facies of older models. Sometimes termed *ensialic backarc* or *retroarc*
    - 2 *Intra-oceanic arcs* (Japan, Aleutians)
      - (a) Forearc. Thin turbidite deposits plus pelagics
      - (b) Backarc. Thin to thick accumulations of clastics near the arc, passing into pelagics and possibly into terrigeneous clastics again at the continental margin

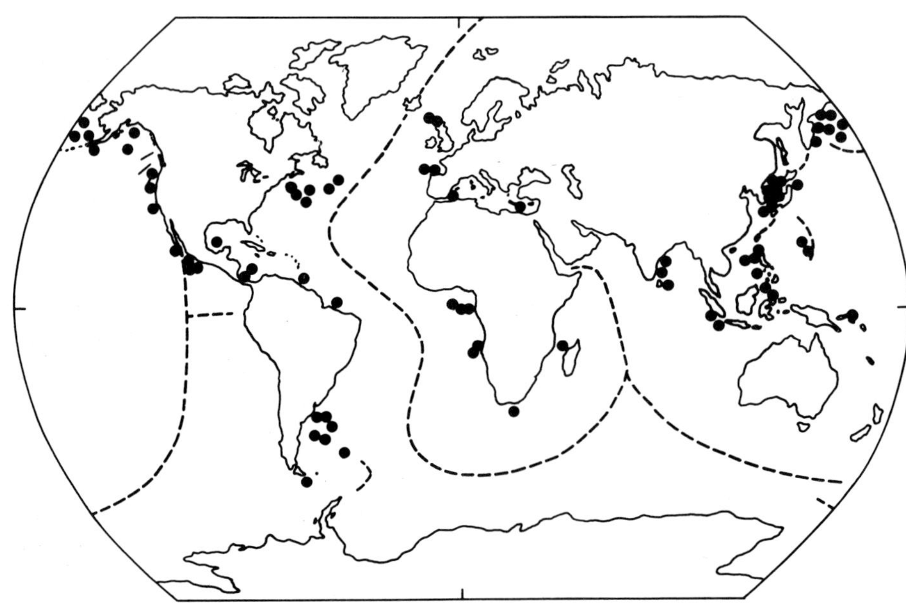

FIG. 1. Sample location map.

primary petrographic variables quartz (Q), feldspar (F), and fine-grained rock fragments (L). Samples with more than 40% quartz are almost exclusively from basins adjacent to passive continental margins (Fig. 2). In addition,

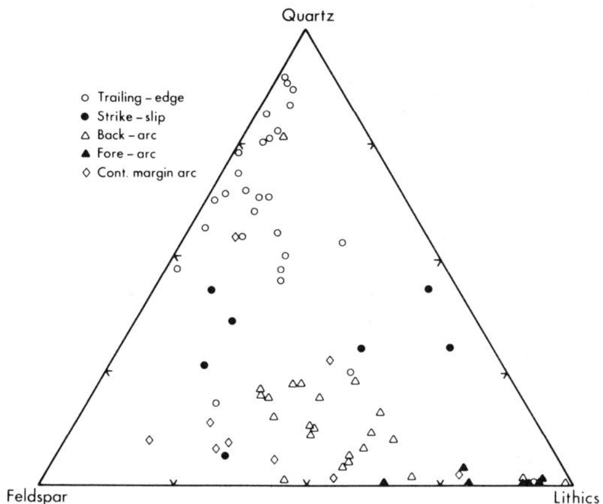

FIG. 2. Framework constituents of modern deep-sea sands from various tectonic settings.

samples from intraoceanic forearc basins are easily distinguished: they are composed almost entirely of volcanic rock fragments. Samples from the other three settings have extensive areas of overlap on triangular plots (Fig. 3).

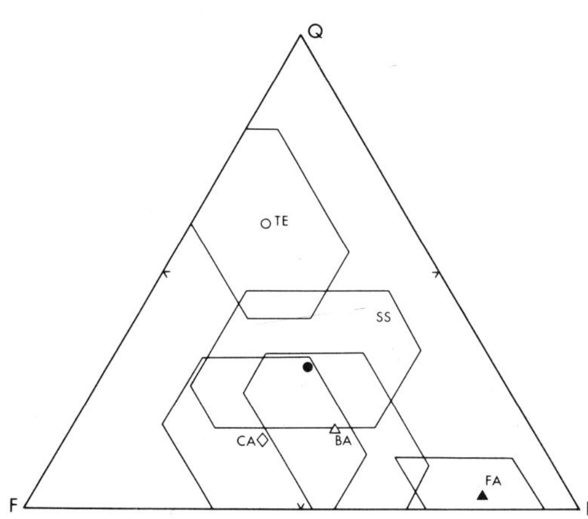

FIG. 3. Range of variation in framework petrography of deep-sea sands. The polygons show one standard deviation about the mean. (See Ingersoll 1978 for a discussion of this representation.) TE = trailing-edge; SS = strike-slip; CA = continental margin arc; BA = backarc of island arc; FA = forearc of island arc.

Supplementary variables such as the type of feldspar or rock fragments have often been used as further discriminants. Fig. 4 shows that

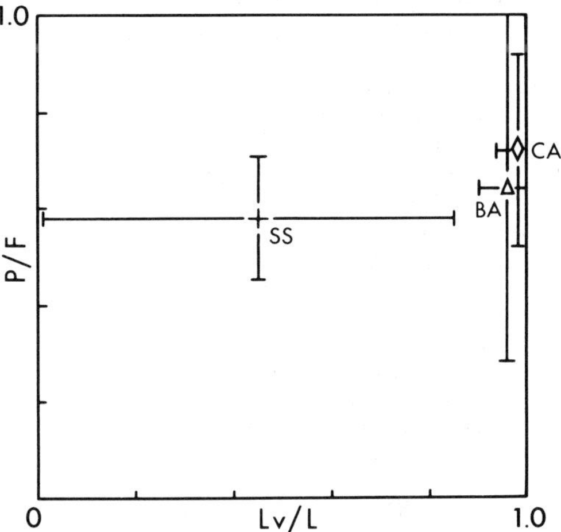

FIG. 4. Petrography of deep-sea sands from tectonic settings that overlap on Fig. 3. Note the very close similarity of samples from the backarc of island arcs (BA) to those from the forearc of continental margin arcs (CA). P/F is the ratio of plagioclase to total feldspar; Lv/L the ratio of volcanics to total lithics.

all three of these settings have about the same proportion of plagioclase in the feldspar, but that the strike-slip related settings have a significantly lower proportion of volcanic grains in the rock fragments. In ancient rocks, this difference may translate to less matrix. Thus, of the five settings considered, three have a reasonably distinct petrography, but basins lying on the ocean side of magmatic arcs on continental margins seem to contain sands almost identical in composition to those found in the backarc basins of island-arcs.

All of the arc-related sands share the characteristics of low quartz content and a high proportion of volcanic lithic fragments. Valloni & Maynard (1981) suggested several subdivisions of this petrographic type. They identified:

Volcanic-vitric sands (Vv). These sands have virtually no quartz, and are made up almost entirely of volcanic grains having a vitric texture or glass shards. They are almost entirely confined to forearc basins of island arcs (Fig. 5). *Volcanic-lithic sands* (Vp and Vk). These sands contain about equal amounts of feldspars and lithic fragments, with volcanic grains again making up almost all of the lithic fragments.

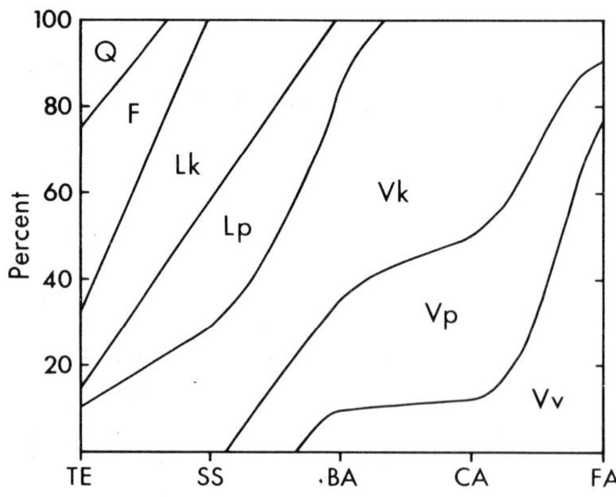

Fig. 5. Distribution of petrographic types by tectonic setting. See text for abbreviations.

ity: the forearc and backarc basins contain very different sands for a given arc-type. To what extent this difference between backarc and forearc sands also applies to continental margin arcs we cannot say, having no backarc sands of this type. Studies of shelf sands of Indonesia would provide the appropriate contrast. This result suggests that in general more elaborate tectonic subdivisions will not be reflected in petrography, although within a given basin such variations may be detectable (e.g. Moore 1979).

Another limitation is the surprising similarity in composition of sands from the backarc of island arcs and the forearc of continental margin arcs. These two settings are likely to be preponderant in ancient arc-related terranes, and their close similarity is discouraging for the use of petrography to distinguish them. Why this pattern occurs is not clear to us.

Table 2. *Petrology of modern deep-sea sands*

Tectonic setting	No. of samples	Q	F	L	C/Q	P/F	Lv/L
Trailing-edge (TE)	29	61	26	13	0.10	0.13	0.13
Strike-slip (SS)	7	31	36	33	0.31	0.58	0.45
Continent margin-arc (CA)	8	16	53	31	–	0.72	0.98
Oceanic arc							
(1) Backarc (BA)	27	16	34	50	0.19	0.64	0.89
(2) Forearc (FA)	9	3	16	81	–	0.90	0.99

Heavy minerals are particularly abundant (3–6%). Vp refers to a sub-variety with almost all plagioclase feldspar, Vk to one with appreciable (>20%) K-feldspar. These sands are characteristic of both the backarc basins of island arcs and the forearc basins of continental margin arcs.

Lithofeldspathic sands (Lp and Lk). These sands are significantly richer in quartz than the preceding varieties. Feldspar and lithic fragments are again present in subequal amounts, but volcanic grains no longer predominate. Two sub-types are distinguished based on the predominance of plagioclase (Lp) or K-feldspar (Lk). Among the quartz grains, polycrystalline varieties are abundant (20–30%). These sands are most common in strike-slip settings, although they are also found off passive continental margins. Two quartz-rich types (Q and F) can also be distinguished, but they are confined to passive settings (Fig. 5).

The distribution of sand types is somewhat curious, because there is apparently no relationship to the type of island arc. That is 'mature' island arcs like Japan and 'primitive' arcs like the Marianas, seem to be producing the same sands. The big difference is in polar-

How do these results compare with ancient sediments? Dickinson & Suzcek (1979) found similar compositions in ancient sandstones. Their arc-related samples fall into the same part of the QFL triangle as ours (Fig. 6), except for our samples from intra-oceanic forearc basins. Probably such sands are rarely preserved in sandstones on the continents. Their 'continental block provenance' corresponds to our passive settings. We have few samples from the area they designate 'recycled orogen provenances' but sands of this composition make up the bulk of the samples reported from modern big rivers by Potter (1978). Crook (1974) has presented data from analysis of over 300 ancient turbidites that differ in some respects from our modern sands (Fig. 7). He divided these rocks into quartz-rich, quartz-intermediate, and quartz-poor varieties, as shown. He suggested that these divisions correspond to trailing-edge continental margins, leading-edge continental margins (both strike-slip and subduction), and island arcs. Our modern sands show both the passive margin sands and the arc-related sands crossing these divisions. Instead, a division at about 40% quartz separates the passive and active settings, with

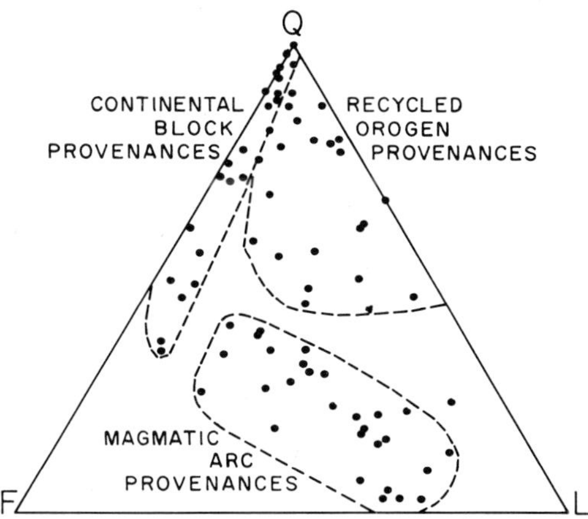

FIG. 6. Petrography of sands as related to tectonic setting of the source area, largely from ancient samples, from Dickinson & Suczek (1979, fig. 9). The distribution matches closely that of Fig. 2, except compositions assigned to recycled orogen provenances are rare in our samples.

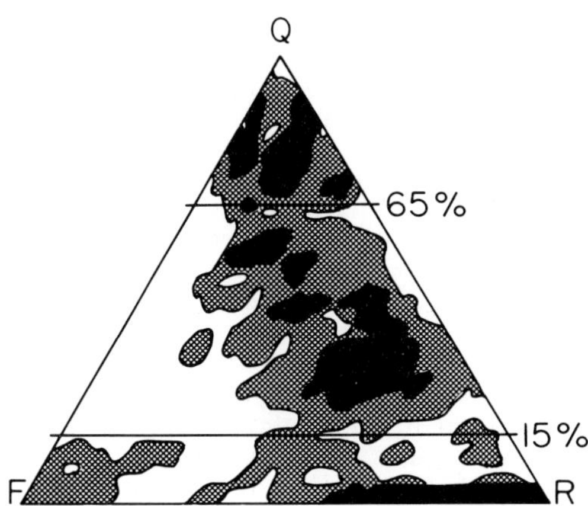

FIG. 7. Petrography of ancient turbidite sandstones, after Crook (1974). The QFR triangle differs from the QFL triangle used in the other figures in shifting the values somewhat towards the R corner (all polymineralic grains counted as rock fragments), as discussed in Valloni & Maynard (1981).

no possible further subdivision of the active settings based on quartz. Some of the difference between our results and his are probably attributable to our use of only fine sands, which might give higher feldspar. It seems unlikely that all of the difference can be explained by such analytical variables, however. This contrast between recent and ancient sands needs further study, especially considering the similarity of Potter's (1978) river sands to Crook's pattern.

Bulk chemistry

Most of these samples were also analysed for bulk chemistry. The object was to make a comparison between the composition of beds of turbidite sand and the immediately overlying mud. These can then be compared with ancient turbidite sand-shale pairs to see whether there is exchange of components between the sand and shale beds during diagenesis. Also we wished to see if the tectonic subdivisions previously described are associated with distinctive chemical compositions. During diagenesis, and low-grade metamorphism, much of the type of petrographic information described earlier becomes obscured. This is particularly true of the types of rock fragments, many of which are converted to matrix. Therefore it would be particularly helpful if chemical analyses could be used, because these processes seem to produce little change in bulk chemistry, except for water and CO_2. In addition to the modern sands and associated muds, 129 samples of ancient turbidite sandstones and shales were analysed. Age, location, and chemical analyses for these samples are presented in the appendix. Determinations were made using atomic absorption following fusion of the sample with $LiBO_2$ and dissolution in HNO_3 (Yu 1979).

The set of ancient turbidite sands and shales that we used is unfortunately rather limited. The sampling is biased towards Cambro-Ordovician rocks of the Appalachian Basin. Also, virtually all of the samples we collected turned out to be from active tectonic settings, based on our interpretation of the literature, probably either forearc to a continental margin (e.g. Southern Uplands of Scotland) or from backarc basins of island arcs (Cambrian of Appalachian Basin?). Usually, this kind of assignment to subvarieties of active settings is very tentative, so we have not attempted it. On the other hand, the average sandstone composition from these samples is close to that reported by Pettijohn (1963, table 12) for a similar sampling of greywackes, so we feel that our samples are representative. There is no comparable data set for the shales: ours appears to be the first compilation of chemical data on shales of this type.

As with petrography, the results show that passive settings are quite distinct (Table 3, Figs

TABLE 3. *Bulk chemistry of modern deep-sea sands and associated muds*

Tectonic setting	SiO_2	Al_2O_3	Fe_2O_3	MgO	CaO	K_2O	Na_2O
1 Sands							
Trailing-edge	77.9	9.8	2.9	1.3	4.1	2.0	1.9
Strike-slip	67.8	15.6	3.7	2.3	3.6	2.9	3.9
Continental margin arc	69.5	14.1	3.9	1.9	4.4	2.6	3.6
Backarc	68.8	14.4	4.5	2.4	4.4	2.0	3.6
Forearc	61.5	15.2	7.7	3.8	6.7	1.4	3.8
2 Muds							
Trailing-edge	65.9	13.7	5.3	2.8	8.2	2.6	1.5
Strike-slip	65.8	14.4	6.8	3.4	4.9	2.0	2.7
Continental margin arc	66.1	16.9	6.4	3.2	3.0	2.5	2.4
Backarc	68.0	14.9	6.5	3.1	2.8	2.3	2.5
Forearc	68.9	12.1	7.2	3.0	4.9	1.5	2.6

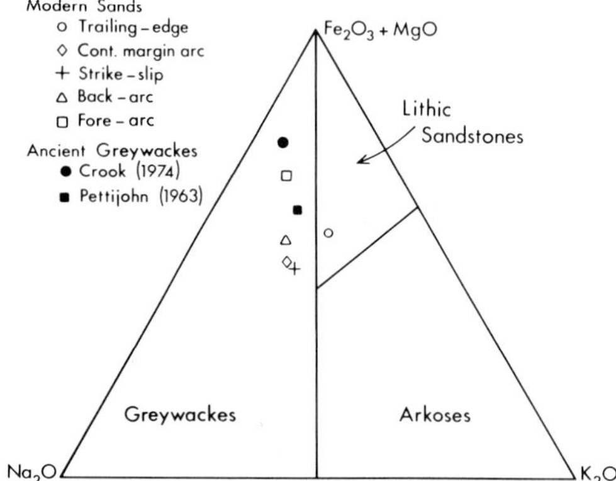

FIG. 8. Comparison of the chemical composition of modern deep-sea sands with ancient greywackes. Sands from trailing-edge continental margins fall outside the compositional range for greywackes, but the arc-related samples all have compositions typical of greywackes. Fields from Blatt *et al.* (1972, fig. 9–2).

8 & 9). The sands are particularly rich in SiO_2 and the K_2O/Na_2O ratio is almost always greater than one. The muds are even more enriched in K_2O (Fig. 10), but have SiO_2 values similar to the other muds. Sands and muds from forearc basins of island-arcs are also distinctive, having extremely low K_2O/Na_2O ratios and very high Fe_2O_3 + MgO (Table 3). The other three settings, however, have sands and muds of very similar composition. Samples from strike-slip basins and continental margin-arc basins are chemically identical, and intra-oceanic backarc basins are distinguished only by slightly lower K_2O/Na_2O and slightly higher Fe_2O_3 + MgO (Table 3). Figs 9 and 10 show, however, that there is a large amount of overlap. Thus it seems unlikely that bulk chemistry can be used to distinguish among types of arc-related sandstones in ancient rocks.

Greywackes, the typical sandstone type of turbidites, are exceptionally rich in Na_2O compared with other sandstone types (e.g. Pettijohn *et al.* 1972, p. 211). It is not known, however, whether this high Na_2O reflects Na-rich source rocks or is a diagenetic feature. For instance, both primary and secondary sources of Na_2O have been proposed for the albite-rich Charny greywackes of Quebec (Middleton 1972; Lajoie 1973; Lajoie *et al.* 1974). Diagenetic addition of Na_2O could be either a local process, caused by exchange of K^+ in sands for Na^+ in interbedded shales, or could involve introduction of Na^+ by migrating pore fluids. A comparison of the chemistry of modern and ancient turbidite sands should show whether the high Na_2O can reasonably be interpreted as a primary feature or whether diagenetic additions seem to be called for.

Fig. 9 shows that sands from the four active settings are similar in chemical composition to ancient greywacke sandstones, particularly in the K_2O/Na_2O ratio. The shales, however, show a pronounced enrichment in K_2O compared with their modern counterparts (Table 4). Does this reflect diagenetic changes? Fig. 11 shows a comparison between the compositions of the sands and their associated muds or shales. This diagram shows that the increase in K_2O/Na_2O in the ancient shales is not matched by a corresponding decrease in the sands. Thus the sands could not have acted as a source of K^+ during diagenesis. In fact, the close similarity in composition of the recent and ancient turbidites

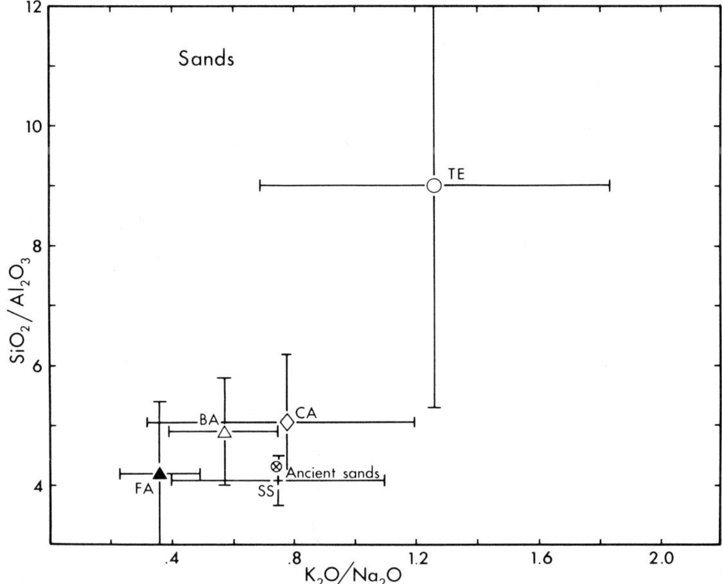

FIG. 9. Chemistry of deep-sea sands. Note the extensive overlap of the active tectonic settings. The coincidence of arc-related sands and ancient greywackes suggests that diagenetic additions are not necessary to explain the composition of these sandstones. Error bars are one standard deviation each side of the mean.

TABLE 4. *Comparison between modern deep-sea sands and ancient turbidites*

	SiO_2	Al_2O_3	Fe_2O_3	MgO	CaO	K_2O	Na_2O	K_2O/Na_2O
1 Sands								
Modern*	68.8	14.5	4.4	2.2	4.3	2.2	3.7	0.60
Ancient†	70.4	14.2	5.4	2.2	2.6	2.1	3.1	0.68
Ancient‡	66.9	14.4	4.6	2.1	7.7	2.0	2.5	0.78
2 Shales								
Modern*	67.3	15.3	6.5	3.2	3.1	2.3	2.5	0.92
Ancient‡	61.3	18.3	6.5	3.1	5.1	4.0	1.4	2.9

* Average of SS, BA and CA samples.
† Data of Pettijohn (1963).
‡ One hundred and twenty-nine samples of sand and shale from ancient turbidites (Appendix).

suggests that no diagenetic additions are necessary to explain the chemistry of these sandstones.

How then can the chemistry of the shales be explained? Note the much higher K_2O content of the older shales. This pattern is not confined to this type of shale, but seems to be a worldwide secular trend in shale composition (Weaver 1967; Ronov & Migdisov 1971; Van Moort 1972). Some preliminary work in our laboratory suggests that this high K_2O in Lower Palaeozoic shales is due to detrital K-feldspar, and so is a primary rather than a diagenetic effect. A similar conclusion probably also applies to the older sandstones. Note in Fig. 11 that the Lower Palaeozoic sandstones have a higher K_2O/Na_2O ratio than younger ones. This trend again argues against diagenetic addition of Na_2O, but it also suggests that Lower Palaeozoic clastics have an anomalously large K_2O content. There are many reports (Bowie *et al.* 1966; Odom 1975; Stablein & Dapples 1977; Swett & Smit 1972) of highly feldspathic sands of this age from environments not normally associated with arkoses (e.g. beaches and tidal flats). Interpretations of their origin vary, but the abundance of this type of occurrence suggests some non-uniform event in the Earth's history.

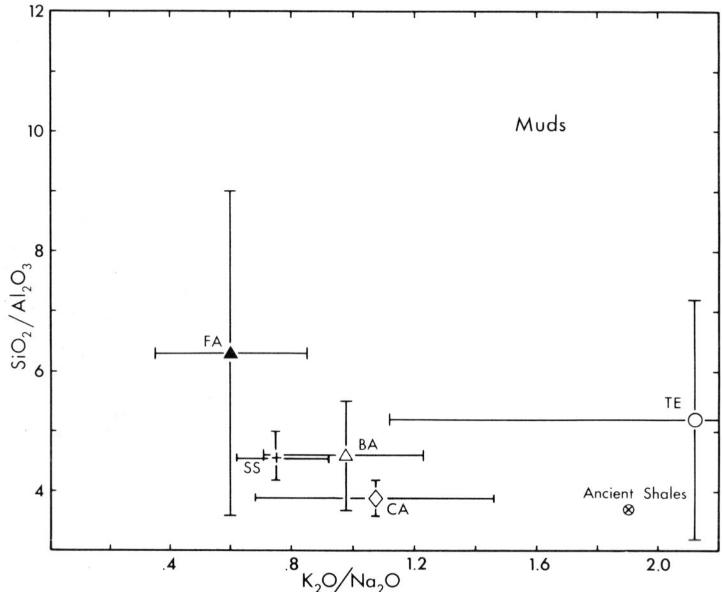

FIG. 10. Chemistry of muds interbedded with deep-sea sands. As with the sands, the arc-related settings show extensive overlap. Note the strong enrichment of ancient shales in K_2O.

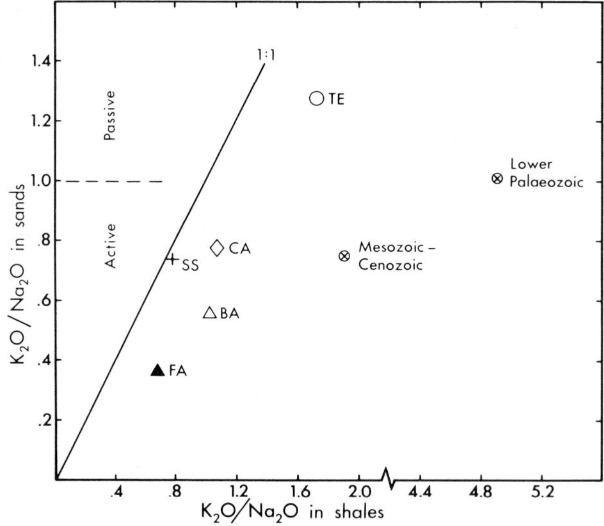

FIG. 11. Comparison of the compositions of turbidite sands and associated muds or shales shows that older sands do not become depleted in K_2O relative to the shales. The very high K_2O/Na_2O in Palaeozoic shales is due to detrital K-feldspar, and so is a depositional rather than a diagenetic effect.

Conclusions

Tectonics was found to exert a strong influence on the composition of modern deep-sea sands, but only broad categories are distinguishable. For example, active and passive settings are quite distinct in both petrography and bulk chemistry. Quartz contents higher than 40%

and K_2O/Na_2O greater than 1.0 indicate sands from passive (trailing-edge) continental margins. Subdivisions of active settings are much less distinct, with the forearc of continental margin arcs (leading-edge margins) having almost identical sands to those from the backarc of island arcs.

The bulk chemistry of sands from active settings is quite similar to the reported composition of average greywacke. From this we infer that it is not necessary to have post-depositional changes in composition to make greywackes. Comparison of some ancient turbidite shales with modern turbidite muds supports this conclusion. High K_2O in Lower Palaeozoic shales of this type appears to be a primary rather than a diagenetic effect.

ACKNOWLEDGMENTS: We thank our many colleagues who have discussed these ideas with us, particularly Paul Potter, Bill Dickinson, and Chris Suczek. Maynard wishes to express grateful appreciation to the staff of Grant Institute of Geology, University of Edinburgh, for help with preparation of the manuscript. Research supported in part by NSF Grant EAR79–03919. The Lamont-Doherty Core Collection is maintained through NSF Grants OCE76–18049 and ONR contract CN14-75-C-0210.

References

BLATT, H., MIDDLETON, G. V. & MURRAY, R. 1972. *Origin of Sedimentary Rocks*, Prentice-Hall, New Jersey, 634 pp.

BOWIE, S. H. U., DAWSON, J., GALLAGHER, M. J. & OSTLE, D. 1966. Potassium-rich sediments in the Cambrian of northwest Scotland. *Trans. Instn Ming Metall.* **78,** B125–45; **79,** B60–9.

CROOK, K. A. W. 1974. Lithogenesis and tectonics: the significance of compositional variation in flysch arenites (greywackes). *Spec. Publ. Soc. econ. Paleontol. Mineral. Tulsa*, **19,** 304–10.

DICKINSON, W. R. 1974. Plate tectonics and sedimentation. *Spec. Publ. Soc. econ. Paleontol. Mineral. Tulsa*, **22,** 1–27.

DICKINSON, W. R. & SEELY, D. R. 1979. Structure and stratigraphy of forearc regions. *Bull. Am. Assoc. Petrol. Geol.* **63,** 2–31.

—— & SUCZEK, C. A. 1979. Plate tectonics and sandstone composition. *Bull. Am. Assoc. Petrol. Geol.* **63,** 2164–72.

—— & VALLONI, R. 1980. Plate settings and provenance of sands in modern ocean basins. *Geology*, **8,** 82–6.

FIELD, M. E. & PILKEY, O. H. 1969. Feldspars in Atlantic continental margin sand off the southeastern U.S. *Bull. geol. Soc. Am.* **80,** 2097–102.

INGERSOLL, R. V. 1978. Petro-facies and petrologic evolution of the late Cretaceous forearc basin, northern and central California. *J. Geol.* **86,** 335–52.

INMAN, D. L. & NORDSTROM, C. E. 1971. On the tectonic and morphologic classification of coasts. *J. Geol.* **79,** 1–21.

KRYNINE, P. D. 1942. Differential sedimentation and its products during one complete geosynclinal cycle. *Ann. Primero congresso Panamericano de Ingeniera de Minas y Geologia (Santiago, Chile) Geologia 1st Pt.* **2,** 537–61.

LAJOIE, J. 1973. Albite of secondary origin in Charny Sandstones, Quebec: discussion. *J. sediment. Petrol.* **43,** 575–6.

—— HEROUX, Y. & MATHEY, B. 1974. The Precambrian shield and Lower Paleozoic shelf: the unstable provenance of the Lower Paleozoic flysch sandstones and conglomerates between Beaumont and Bic, Quebec. *Can. J. Earth Sci.* **11,** 951–63.

MIDDLETON, G. V. 1972. Albite of secondary origin in Charny Sandstones, Quebec. *J. sediment. Petrol.* **42,** 341–9.

MOORE, G. F. 1979. Petrography of subduction zone sandstones from Nias Island, Indonesia. *J. sediment. Petrol.* **49,** 71–84.

ODOM, I. E., 1975. Feldspar grain-size relations in Cambrian arenites, upper Mississippi Valley. *J. sediment. Petrol.* **45,** 636–51.

PETTIJOHN, F. J. 1963. Chemical composition of sandstones. *Prof. Pap. U.S. geol. Surv.* **440–S,** 21 pp.

—— POTTER, P. E. & SIEVER, R. 1972. *Sand and Sandstone.* Springer-Verlag, New York, 628 pp.

POTTER, P. E. 1978. Petrology and composition of modern big river sands. *J. Geol.* **86,** 423–49.

READING, H. G. (ed). 1978. *Sedimentary Environments and Facies.* Blackwell Scientific Publications, Oxford, 576 pp.

RONOV, A. B. & MIGDISOV, A. A. 1971. Geochemical history of the crystalline basement and the sedimentary cover of the Russian and North American platforms. *Sedimentology*, **16,** 137–87.

SCHWAB, F. L. 1975. Framework mineralogy and chemical composition of continental margin-type sandstone. *Geology*, **3,** 487–90.

STABLEIN, N. K. & DAPPLES, E. C. 1977. Feldspars of the Tunnel City Group (Cambrian), western Wisconsin. *J. sediment. Petrol.* **47,** 1512–38.

SWETT, K. & SMIT, D. E. 1972. Paleogeography and depositional environments of the Cambro-Ordovician shallow marine facies of the North Atlantic. *Bull. geol. Soc. Am.* **83,** 3223–48.

VALLONI, R. & MAYNARD, J. B. 1981. Detrital modes of modern deep-sea sands as related to plate tectonics: a first approximation. *Sedimentology*, **28,** 75–83.

VAN MOORT, J. C. 1972. The K_2O, CaO, MgO and CO_2 contents of shales and related rocks and their implications for sedimentary evolution since the Proterozoic. *24th Int. geol. Congr., Sect.* **10,** 427–39.

Weaver, C. E. 1967. Potassium, illite and the ocean. *Geochim. cosmochim. Acta*, **31,** 2182–96.

Yu, H. 1979. *Three aspects of sandstone diagenesis: compaction and cementation of quartz arenites and chemical changes in graywackes.* Dissertation, PhD, Univ. Cincinnati (unpubl.).

J. Barry Maynard, Department of Geology, University of Cincinnati, Cincinnati, Ohio 45221, U.S.A.

Renzo Valloni, Istituto Petrografia, Universita di Parma, 43100 Parma, Italy.

Ho-Shing Yu*, Department of Geology, University of Cincinnati, Cincinnati, Ohio 45221, U.S.A.

* *Present address*: Pennzoil International, P.O. Box 2967, Houston, Texas 77001, U.S.A.

Appendix

Bulk chemistry of some ancient turbidite sands (sd) and Shales (sh)

		# of samples	SiO_2	Al_2O_3	Fe_2O_3	MgO	CaO	K_2O	Na_2O	SiO_2/Al_2O_3	K_2O/Na_2O	Inferred tectonic setting
Pico, Repetto Fmns.	Sd	3	66.6	15.0	2.62	0.88	8.84	3.14	2.94	4.44	1.07	Active
Pliocene												
Ventura Basin, Cal.	Sh	4	67.5	15.9	5.47	2.51	3.11	2.92	2.54	4.25	1.15	
Marnoso Arenacea	Sd	1	60.5	13.9	2.61	2.60	16.5	1.89	2.07	4.35	0.91	Active
Lower Miocene												
Muraglione Pass, Toscana, Italy	Sh	3	49.2	16.7	5.27	5.85	18.9	2.88	1.14	2.95	2.53	
Macigno	Sd	2	66.8	14.5	3.48	2.32	8.20	2.21	2.56	4.61	0.86	Active
Oligocene												
Pieve pelago, Italy	Sh	4	66.4	15.0	5.33	3.28	4.37	3.11	1.97	4.43	1.58	
Loreto Fmn	Sd	1	58.4	16.6	4.76	2.43	15.4	0.88	2.19	3.52	0.40	Active
Oligocene												
Magallenes Prov., Chile												
Prealpine Flysch	Sd	4	62.4	13.7	3.21	2.92	14.8	1.32	1.67	4.55	0.79	Active
Oligocene												
Val d'Illiez Switzerland	Sh	7	58.7	16.4	5.61	3.34	12.5	2.46	1.09	3.59	2.26	
Cretaceous sands of Tierra del Fuego	Sd	13	66.8	16.4	6.60	2.72	2.26	1.80	3.49	4.07	0.52	Active
Magallenes Prov., Chile												
Lewes River Group	Sd	4	63.9	15.5	5.75	2.67	6.28	2.00	3.79	4.14	0.53	Active
Laberge Grp.												
Triassic-Jurassic												
Whitehorse Yukon												
Escuminac Fmn.	Sd	1	68.3	10.5	4.13	2.00	12.3	1.98	0.77	6.50	2.57	Passive
Devonian												
Miguasha, Quebec	Sh	2	66.7	16.0	6.72	3.00	3.64	3.00	0.94	4.18	3.19	
Gala Greywacke	Sd	3	67.3	15.0	5.35	2.18	1.67	3.61	4.91	4.49	0.73	Active
Silurian												
Innerliethen, Scot.	Sh	2	67.7	15.7	5.43	3.23	1.20	3.97	2.65	4.31	1.50	
Aberystwyth Grit	Sd	3	71.8	15.0	6.70	1.80	2.03	1.35	1.41	4.78	0.96	Active
Silurian												
Aberystwyth, Wales	Sh	7	66.7	17.4	7.22	2.02	1.25	4.19	1.29	3.83	3.25	
Frenchville Fmn	Sd	1	57.1	11.9	2.70	1.15	19.3	3.57	4.39	4.82	0.81	Active
Silurian												
Stockholm, Maine	Sh	3	65.7	12.5	5.80	3.12	8.01	3.07	1.77	5.26	1.73	
Martinsburg Fmn	Sd	16	83.0	11.4	3.93	0.63	0.29	0.48	0.30	7.25	1.60	Active
Ordovician												
Lehigh Gap, Penn.	Sh	17	58.7	25.9	6.26	1.66	0.36	6.51	0.53	2.27	12.3	
Martinsburg Fmn	Sd	4	73.5	11.3	5.80	1.71	4.11	2.06	1.51	6.50	1.36	Active
Ordovician												

Williamsport, Md.	Sh	4	61.9	21.8	7.53	2.07	0.66	5.22	0.82	2.84	6.36	
Utica Fmn. Ordovician	Sd	4	69.1	15.0	4.51	2.15	5.72	1.84	1.70	4.62	1.08	Active
St Antoine de Tilley, Quebec	Sh	3	63.3	16.8	6.41	4.37	3.31	4.72	1.06	3.77	4.45	
Normanskill Fmn Ordovician	Sd											
Hudson Valley, N.Y.	Sh	4	56.9	19.6	8.00	3.66	7.14	4.09	0.71	2.90	5.76	Active
Sillery Grp. Cambrian	Sd	5	69.1	15.2	6.38	2.48	2.84	1.43	2.63	4.55	0.54	Active
St Lawrence Valley, Quebec	Sh	4	54.6	25.8	9.74	2.44	0.64	5.42	1.37	2.12	3.95	
Average sand		15	66.9	14.4	4.6	2.1	7.7	2.0	2.5	4.65	0.78	
Average shale		13	61.3	18.3	6.5	3.1	5.1	4.0	1.4	3.35	2.86	
Pettijohn's average greywacke (volatile-free basis)		61	70.4	14.2	5.4	2.2	2.6	2.1	3.1	4.96	0.68	

Initiation of subduction zones: implications for arc evolution and ophiolite development

D. E. Karig

SUMMARY: The initial location of trenches with respect to continental margins is a critical factor in the nature of the forearc region and its subsequent evolution, as well as for the interpretation of ophiolites. Although recent studies of ancient margins and highly evolved arcs suggest that wide strips of oceanic crust are commonly trapped between the trench and volcanic arc, review of young active arcs does not support this conclusion. Neogene trenches have formed predominantly very close to the crustal interface between oceanic crust and continental or older island arc crust. A few arcs have developed along transform zones, but none can be identified as having formed by breakage within an oceanic plate. Apparently the high strength of normal oceanic lithosphere causes the initial fracture to utilize pre-existing zones of weakness or pre-stressing, such as the downbowed and fractured crustal interfaces along passive margins. There seems to be no physical basis for the initial rupture to form smooth, large radius arcs, thereby trapping oceanic crust in re-entrants. Moreover, oceanic crust or ophiolites originating adjacent to a continental or arc margin ought to betray that heritage by a high terrigenous or volcaniclastic component in their overlying sediment. Large ophiolite sheets that can be interpreted as forearc slabs may develop in areas with relatively unusual plate geometries, as where oceanic spreading zones cut diagonally across the forearc. Disrupted ophiolites could, in many cases, represent slices of oceanic crust and upper mantle accreted to and responsible for the growth of oceanic arcs. Evidence for accreted slices in the Mariana arc and the range of forearc geometries among oceanic arcs allow construction of an evolutionary sequence in which an initial steep and narrow trench shape is differentiated into upper and lower sections by intermittent accretion of oceanic crustal slices. As a result, the slope apron on the upper section rotates arcward and can rise to form a basin as is displayed in the southern Middle America arc.

Evolution of the Earth's crust beneath trench-arc systems remains one of the more contentious topics in the geological sciences. The problem bears strongly not only on the original geometry of arc systems and their subsequent evolution, but also on the development of ophiolites. Models of convergent plate margins commonly assume that much of the arc, in particular the forearc, is underlain by oceanic crust (Dewey & Bird 1970; Mitchell & Reading 1971). A major reason for this assumption is the widespread belief that the initial crustal fracture leading to trench formation occurs within oceanic crust. Support for such a model is also derived from the occurrence of ophiolites that cap structurally imbricated stacks of accreted material (Dewey 1976) and sometimes form the seaward flanks of ancient forearc basins (Dickinson & Seely 1979).

A recent corollary to this model identified the trench slope break as the seaward edge of the trapped oceanic crustal slab, which thus serves as a mechanical buttress against which an accretionary prism is built (e.g. Hamilton 1977). As a result of this process, the trench slope break would not migrate seaward as accretion proceeds but could even migrate landward if there were compressional deformation within the forearc basin (Dickinson & Seely 1979; Seely 1979). These authors imply that broad initial forearc regions, often related to embayments in the continental margin, are typical or at least very common.

Studies of the morphotectonic geometries of contemporary arc systems have led to somewhat different conclusions concerning the initiation and evolution of arc systems (Karig 1974, 1977; Karig *et al.* 1980). The systematic increase in the width of the accretionary prism or inner trench slope with increasing maturity of the arc system suggests an arc evolution involving a significant upgrowth and outgrowth of the trench slope and slope break from an initial position generally along interfaces between oceanic crust and either continental or older arc crust. This had led me (Karig 1977) to consider trapping of oceanic crust beneath forearcs as a relatively unusual situation.

In this paper I approach the problem of initiation and early evolution of arcs by examining examples of young arc systems (less than 10 Myr old), where the initial characteristics can still be observed. I also use data from older active arc systems in which the early evolution is relatively well constrained by geological and geophysical data.

Examples of arc initiation

Most identifiable young arc systems occur along pre-existing interfaces between oceanic and continental crust, or between oceanic and island arc crust, at least in a regional sense. Commonly these interfaces represent rifted, passive margins (e.g. New Hebrides). Although only back arc rift margins have been utilized during the past 10 Myr, passive continental margins bordering main ocean basins must also be assumed as likely sites. Often crustal interfaces represent rejuvenated convergent margins (e.g. northern Middle America). Less commonly, arc initiation appears to have utilized pre-existing transform boundaries (e.g. McQuarie). With the possible exception of the Mussau Trough (which might also have been a transform) I cannot identify any young arc that has originated within an oceanic lithospheric plate.

Plate convergence within marginal basins has often been assumed to differ from that in oceanic basins, primarily to explain the emplacement of some ophiolites (e.g. Dewey 1976). Such assumptions are not supported by observations in active arc systems. Crustal convergence occurs by subduction, initiated along a basin flank (e.g. Shikoku, Solomons, New Hebrides), even when the basin crust is quite young.

These regional observations, however, are not adequate to show what might occur on a more local scale; one which is appropriate to the understanding of ophiolites and subduction complexes. Description of several better known examples of initial or early arc geometries (Fig. 1) serves that purpose and as a basis for discussion of the possible constraints on arc initiation.

New Hebrides arc

The present New Hebrides trench has developed along the rear (western and southern) flank of an older arc, apparently as a result of an arc polarity reversal that occurred about 6 Myr BP (Falvey 1978; Carney & MacFarlane 1978). Regional tectonic reconstruction (Falvey 1978) indicates that the trench developed along the rifted flank of an early Tertiary back arc basin. The section of the arc south of 21°S, and swinging eastward into the Hunter-Matthew fracture zone appears more primitive and may have developed from that transform (Karig & Mammerickx 1972), possibly by the migration of the pole of relative plate motion (e.g. Dewey 1975). As a modification to this scheme, Falvey (1978) suggests that the change in boundary type occurs at a triple junction, from which a spreading ridge heads northward into the Fiji Basin.

The inner trench slope of the entire New Hebrides arc is steep and narrow in comparison to those of older arc systems (Karig & Sharman 1975). Although there are local variations in slope morphology (Ravenne *et al.* 1977; Daniel 1978), there is no clear tectonic division into lower and upper slope sections as recognized in the larger Tonga of Mariana arcs. Local sedimentary aprons on the upper slope surround the larger active volcanoes but even these are lacking in the Hunter-Matthew section of the arc, where the inner slope is narrower still, and displays irregular, unsedimented benches (Karig & Mammerickx 1972).

Recent refraction and detailed seismological studies in the central section of the arc (Ibrahim *et al.* 1980; Isacks *et al.* 1981) indicate that the arc is built of crust some 30 km thick, having a velocity structure intermediate between oceanic and continental in character, and similar to that of other oceanic arcs (e.g. Murauchi *et al.* 1968; Shor *et al.* 1971). This type of crust appears to underlie the upper part of the trench slope (Fig. 2), whereas low-velocity material, suggestive of accreted material, underlies the lowermost slope. In between lies an area of crust, 20–30 km wide, where the velocity structure is ambiguous or unconstrained. Beneath this area could lie marginal basin crust trapped behind the trench, thin arc crust representing the normal-faulted, transitional zone between arc and marginal basin, or accreted slices of oceanic crustal rocks.

The sum of acquired geophysical data defines a narrow, steep trench slope constructed primarily of igneous material in an unknown structural geometry. This lack of a wide forearc basin or terrace, is consistent with the very narrow flattened section in the shallow section of the seismic zone (e.g. Isacks & Barazangi 1977). Ophiolitic slices, or even narrow ophiolite sheets, could lie beneath part of the trench slope, but any such ophiolite sheet ought to carry a distinctive cover of volcaniclastic strata, as it would have developed immediately west of the pre-reversal, Miocene volcanic arc.

Philippine archipelago

The Philippine archipelago has been an area with a very complicated pattern of rapidly

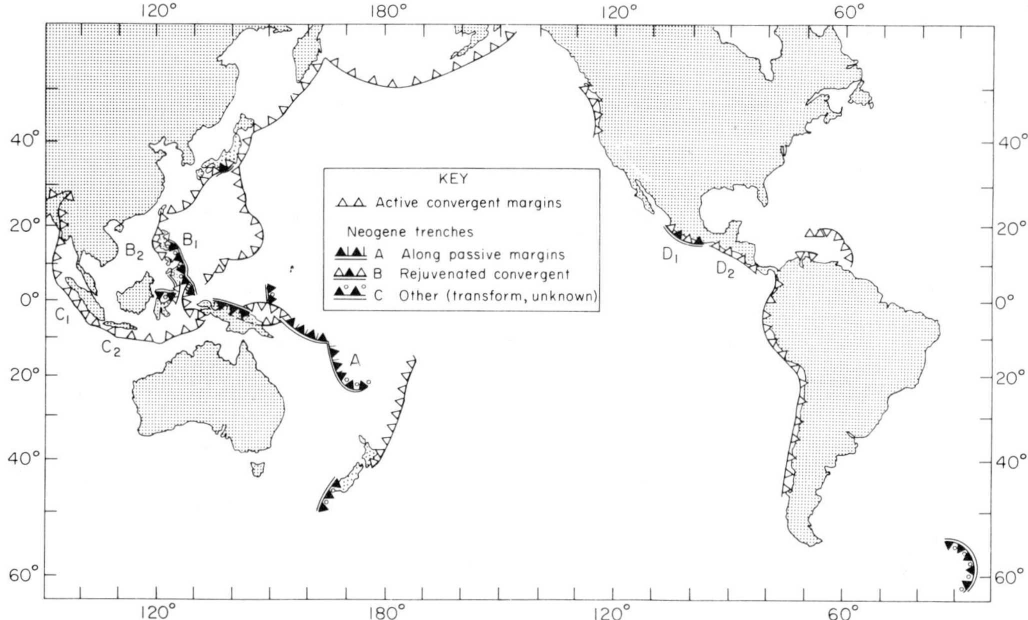

FIG. 1. Active convergent margins of the Indo–Pacific and Atlantic regions, noting arc systems either initiated or re-activated within the past 10 Myr. Systems used as examples in the text are indicated and identified by letter: (A) Vanuatu (New Hebrides); (B) Luzon; (C) Sunda; (D) Middle America.

changing plate consumption during the Neogene, including formation of several new convergent margins (Fig. 1). Most is known about the margins bounding Luzon, where repeated arc polarity reversals are suspected. Subduction has been occurring along the western margin (Manila Trench; Fig. 3) since some time in the Miocene, but a new trench is now developing along the eastern coast (Karig 1977; Bowin *et al.* 1978). This eastern zone appears to be the northward propagating extension of the Philippine Trench (Fig. 3), all of which is quite young and which is extending southward as well (Karig 1975; Hamilton 1979). Profiles across the Philippine Trench at widely scattered locations show a steep inner trench slope, without a significant accretionary prism (Karig 1975; Hamilton 1979). This trench seems to have developed along or near the pre-existing margin of archipelagic crust, although in most areas the data are not capable of determining whether or not a narrow strip of oceanic crust might be trapped west of the site of trench initiation.

The nascent trench east of Luzon has clearly developed at the base of the slope between 16 and 17°N, where the most extensive surveying has been undertaken (Karig, unpublished results of SI0 *Antipode IV* cruise). In that region an extremely steep slope (averaging $> 10°$) extends from the crest of the Sierra Madre Range to a very small prism of accreted sediments at the slope base (Karig 1977). Because this slope marks the eastern boundary of a Cretaceous–Palaeogene subduction complex (Murphy 1973; Karig 1973), the nascent trench would appear to have rejuvenated an older convergent margin. If the descending oceanic lithosphere continues beneath this slope with the geometry implied by its inclination at the trench, there is no room for a slice of trapped oceanic crust. Nor does the non-magnetic character of the slope suggest that any such oceanic crust exists.

Subduction has persisted along the western margin of Luzon since early or mid-Miocene, producing a well-developed accretionary prism of moderate size (e.g. Ludwig *et al.* 1967; Fig. 3). A discussion of this margin bears on the problem of arc initiation and the origin of ophiolites because that section of the arc between 15°N and 16°N exposes the Zambales ophiolite, in a position east of the present forearc basin (Fig. 3). This large ophiolite is one of the youngest and, at least in its upper structural levels, one of the least deformed examples known. As important, it retains a recognizable tectonic position within an active arc.

Tectonic studies of this body are only begin-

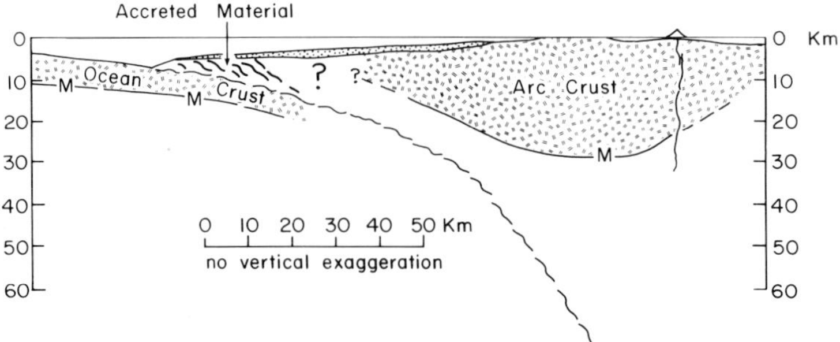

FIG. 2. Crustal and tectonic section of the central New Hebrides arc: an example of a relatively young (6 Myr) arc system. This section is constrained by data from Isacks & Barazangi (1977); Ibrahim *et al.* (1980); Luyendyk *et al.* (1974); Ravenne *et al.* (1977); and unpublished seismological data of B. L. Isacks. This section illustrates the relatively narrow and steep forearc area, and the limited area beneath which oceanic crust might be trapped.

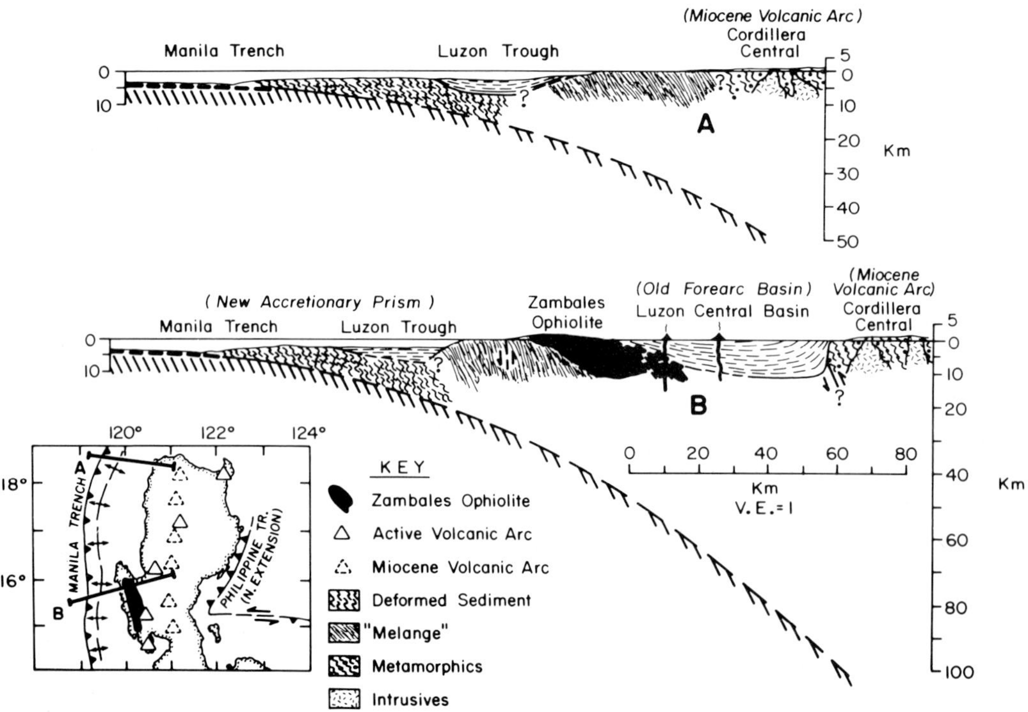

FIG. 3. Tectonic sketch map and two schematic tectonic sections across western Luzon, where the Zambales ophiolite slab occurs. The southern section crosses the arc through the Zambales ophiolite and the Luzon Central Basin (an old forearc basin) and shows its relationship to the present forearc basin (Luzon Trough). The northern section lacks the older forearc and ophiolite but exposes a deformed terrane including disrupted ophiolitic fragments. Also shown on the map is the northward propagating end of the Philippine Trench, along the east coast of Luzon. Data are from Hayes & Ludwig (1967); Ludwig (1970); Schweller & Karig (1979); Cardwell *et al.* (1980) and various unpublished industrial and Philippine governmental sources.

ning, but preliminary mapping and analysis of available data define the ophiolite as a gently east-dipping sheet that forms the western flank of the Luzon Central Basin (Irving 1950; Schweller & Karig 1979; de Boer *et al.* 1980). The strata filling that basin and covering the ophiolite begin with tuffs and tuffaceous pelagic limestone of late Eocene to Oligocene age (Hashimoto *et al.* 1977; Garrison *et al.* 1979), changing upward to lithic and volcaniclastic turbidites and related sediments. A shallowing of the depth of deposition through time, from several kilometres for the basal strata to sea-level during the Pliocene, is reasonably certain (Ingle 1975; Garrison *et al.* 1979). One surprising observation is the relative lack of deformation in the upper levels of the ophiolite and in its sediment cover that might be associated with trench initiation. At most we note a few widely spaced folds and a submarine unconformity across which there may be differential tilting.

The Zambales ophiolite is most easily interpreted as a crustal sheet, formed in an oceanic (or back-arc) basin west of Palaeogene Luzon, and trapped between the initial position of the Manila Trench and a volcanic arc (Cordillera Central) sometimes early in the Miocene. This tectonic setting, together with subsurface stratigraphic and magnetic data from unpublished oil company reports would indicate that the ophiolite and the Luzon Central Valley formed an initially broad forearc basin, and evolved much in the manner suggested by Dickinson & Seely (1979).

The Zambales region would appear to contradict the major premise of this study, but several observations suggest that an unusual plate geometry was responsible for its creation. First, the ophiolite neither occupied an embayment in the mid-Tertiary margin of Luzon nor continues along strike to the north or south. To the north, the magmatic (frontal) arc is flanked instead by an assemblage of highly disrupted ophiolitic and sedimentary fragments (e.g. Fanning 1912). Secondly, magnetic anomalies obtained during oil company surveys over the eastern flank of the ophiolite and the Central Luzon Valley, as well as preliminary studies of dykes near the top of the ophiolite (e.g. J. F. Violette, written comm. 1980) indicate that the spreading ridge which generated the ophiolite trended north-easterly. If, as suspected, the ophiolite is autochthonous with respect to central Luzon, the rift zone along which the ophiolite was generated somehow had to be resolved against another type of plate boundary in that vicinity.

Sunda arc

Although the present subduction geometry along the western margin of Sumatra appears to date from the late Oligocene (Karig *et al.* 1980), the large amount of subsurface and onland data from the Nias transect (Kiechefer *et al.* 1980: Moore & Karig 1980) lead to a number of conclusions pertinent to the initiation and evolution of that subduction pulse. The Oligocene trench seems to have developed along an irregular continental margin, disrupted by strike-slip faults (Karig *et al.* 1980). Basins, trapped within the re-entrants, can be identified on multichannel reflection profiles, but most of these appear to overlie the continental slope. Refraction data from the area east of Nias (Kieckhefer *et al.* 1980) indicate that that slope is underlain by continental crust or more likely, an older subduction complex (Karig *et al.* 1980).

Extensive palaeodepth and structural control from offshore wells and from the island of Nias demonstrate that the trench slope evolved from a small, deep terrace stage in the early Miocene to one with a broad upper slope, near sea-level, by the end of the mid-Miocene (Karig *et al.* 1980). One observation pertinent to the evolution of arcs is that Miocene slope strata, deposited on the upper trench slope with a significant initial dip, now lie on the trenchward flank of the forearc basin, and have been rotated arcward into a nearly horizontal orientation.

In the section of the Sunda arc including Java and Bali, refraction studies have been interpreted as favouring trapped oceanic crust beneath the forearc basin (Curray *et al.* 1977), but this interpretation is vulnerable on several grounds. The thickness and velocity of seismic units beneath the forearc basin are quite variable and are significantly different from those of typical ocean crust. They bear at least as much resemblance to arc crust (Shor *et al.* 1971) or to subduction complexes (Eaton *et al.* 1970). In particular, the 4.5 to 5 km s^{-1} material can be as easily or better interpreted as dewatered on slightly metamorphosed sediments as basaltic second layer (e.g. Kieckhefer *et al.* 1980).

A Cretaceous–Palaeogene subduction complex, comprising mélange and slices of oceanic crustal lithologies, is exposed in several areas near the south coast of Java (Ketner *et al.* 1976; Hamilton 1979). Subsurface data from the area offshore reveal a regional unconformity, cut at sea-level during the late Oligocene to early Miocene (Bolliger & de Ruiter 1976). Together, these data suggest that at least the inner (northern) side of the forearc basin is underlaid by a subduction complex that was emergent during the initiation of the Neogene subduction event, rather than by oceanic crust trapped behind the new trench. In short, the data from the eastern Sunda arc do not support the presence of an oceanic crustal slab beneath that forearc basin.

Middle America arc

Finite plate reconstructions imply that consumption of oceanic crust has been a general feature of the west coast of Mexico and Central America since well back into the Mesozoic. Nevertheless, the response of this margin has varied greatly, both spatially and temporally, affording several glimpses of the initiation and evolution of trench slopes.

Initiation of renewal of subduction during the late (?) Miocene along the Middle America Arc north of the Tehuantepec Ridge followed removal of a pre-existing continental margin by an as yet poorly defined process (Karig *et al.* 1978a). The narrow steep trench slope is clearly divided into upper and lower slope sections. Single channel seismic reflection and magnetic profiles (Karig *et al.* 1979a) indicated approximately 25–30 km of outgrowth by accretion of sediments directly against the old continental basement. Deep sea drilling (Moore, Watkins *et al.* 1979, 1981) and associated multichannel reflection profiles (Shipley *et al.* 1980, 1981) confirmed this geometry and more clearly outlined the existence of an upper slope apron or incipient forearc basin over the contact between accreted sediments and continental crust. Slope parallel strata in this slope apron provide another example of initially inclined basal units in the forearc setting. A well-defined trench slope break lies 10–15 km seaward of the edge of the continental crust, well within the accretionary prism. In this rejuvenated arc there is again no trapped oceanic slab. Although accretion began against a buttress of continental basement, the trench slope break has already begun to migrate outboard and is rising.

The character of the arc south of the Tehuantepec Ridge contrasts strongly with the juvenile aspect to the north, but merits discussion because it has served as the model of forearc development by Seely *et al.* (1974) and Seely (1979). This arc has been cited as a well-constrained example where a wide slab of oceanic crust was initially trapped behind a trench as the basement for a large forearc basin (Seely 1979). Although many of the well and seismic data used in that interpretation have not been made public, the available information from other studies suggest that evolution of this arc is open to alternative interpretations.

The basement beneath the outer half of the forearc basin emerges on the Nicoya, Osa and Burica peninsulas of Costa Rica and Panama where it has been mapped as consisting of slices of oceanic basalt with zones of disruption which contain mélanges of oceanic crustal rocks and sediments (e.g. Dengo 1962; Henninsen & Weyl 1968; Galli-Olivier 1979). The term ophiolite has been applied to this complex but the apparent degree of disruption seriously detracts from interpretation of this basement as a trapped slab of oceanic crust.

Uplifted basement masses are delineated beneath and seaward of the shelf edge by many seismic reflection and magnetic profiles, and although interpreted by Seely (1979) as diapirs rising from the oceanic slab, might as well be viewed as thrust slices (Ladd *et al.* 1978). This interpretation would be more consistent with structural nature of the exposed terrane.

In the Nicoya complex, highly deformed cherts and limestones of Jurassic to Late Cretaceous age are structurally mixed with the oceanic igneous rocks (Galli-Olivier 1979; Schmidt-Effing 1979) and probably represent sediments of the ocean basin. These highly deformed rocks are overlain by deep water siliceous limestones and shales of late Cretaceous and Palaeocene age (Galli-Olivier 1979: Harrison 1953), which appear from their position, degree of deformation, and sedimentological character to be trench slope sediments. This unit is dominated by hemipelagic sediments near the base of the section in seaward areas but form a transgressive clastic sequence futher east and upsection (Galli-Olivier 1979). This latter facies is several kilometres thick and has reasonably been interpreted as upper slope or forearc basin strata (Galli-Olivier 1979). In places a conglomerate basal sequence is observed to unconformably overlie basalts of the subduction complex (Galli-Olivier 1979). The geological relationships in the Nicoya and related complexes indicate that subduction was operative by Coniacian time.

Combination of palaeodepth information from the borehole on the shelf edge off Guatamala (Seely 1979) with the sedimentary data from the emergent subduction complex along tectonic strike to the SE suggests that, rather than oceanic sediments, basal strata of the forearc section are Coniacian slope apron sediments, rotated arcward from an original seaward dip. They overlie disrupted oceanic crustal slices and mélange that were very likely accreted to the trench slope. Deposition at abyssal depths and subsequent shallowing would characterize lower slope sediments equally as well or better than they would oceanic sediments. Overlying and to the east of these strata are turbidites which could reflect the increase in depositional area on a widening and flattening upper slope or forearc basin. The distribution of Late Cretaceous sediments suggests that the trench was initially in the vicinity of the shelf-edge high and that the forearc region was steep and narrow, similar to that of young oceanic arcs. With accretion of oceanic crustal slices, together with oceanic sediment, the trench migrated westward, the trench slope flattened, and as the trench slope break rose, a basin developed.

Sites of trench initiation

Physical contraints and definitive characteristics

As important as the documentation of the initial geometry of arc systems is the identification of the characteristics of potential sites of subduction that constrain the position of the initial rupture zone. It is likewise necessary to know what characteristics of the pre-existing crust and sediment cover might be used to identify and differentiate loci of trench initiation. These characteristics are relevant not only to the structure of arc systems but also to the origin of ophiolite sheets.

The difficulty in spontaneously initiating subduction within oceanic plates, because of the high stress levels that would be required to fail the lithosphere, have been discussed by McKenzie (1977) and by England & Wortel (1980). Thus any pre-existing zone of weakness, or of suitable pre-stressing in the lithosphere would provide a preferential setting for subduction. Review of young arcs indicates that these settings include both interfaces in crustal type and pre-existing lithospheric plate boundaries. Of the former, I have identified the passive margins, created by spreading in either main ocean or back-arc basins and previously active convergent margins. Of pre-existing plate boundaries, transforms appear to be utilized most commonly.

Passive margins

Extensive geophysical studies of the Atlantic and Red Sea margins (e.g. Grow *et al.* 1979; Lehner & de Ruiter 1977; Lowell & Genik 1972; Fairhead 1973) indicate that the landward edge of oceanic crust abuts densified and structurally thinned continental (transitional) crust, which thickens and acquires typical continental characteristics over a width of 50–100 km (Grow *et al.* 1979).

During the subsequent evolution of passive margins, the sediment load of the continental rise prism commonly results in the downbowing of the landward flank of the oceanic crust and lithosphere (Grow *et al.* 1979; Watts & Steckler 1979; Lehner & de Ruiter 1977). This broad flexure involves dips of only 1° or less but it may well be responsible for major faulting and even decoupling of the crustal interface during the early stages of the margin evolution (Watts & Steckler 1979; Turcotte *et al.* 1977).

The basal strata overlying the initially formed oceanic crust along passive continental margins show a significant terrigenous component for at least several tens of kilometre seaward of the crustal interface. This terrigenous input may be expressed as very rapidly deposited, landward thickening clastic wedges (Tucholke & Mountain 1979; Klitgord & Grow 1980), at times mixed with basalt (Moore 1973), or by more subtle dispersion of the terrigenous component in hemipelagic deposits (Whitmarsh, Weser, Ross *et al.* 1974; Thiede 1974). There is adequate evidence that the basal strata are not typically oceanic in character.

A second kind of passive margin forms as a result of back-arc spreading. Regardless as to whether or not this process of crustal formation is identical with spreading within main ocean basins, and produces identical crust, there are significant differences in the sediment cover of the two settings. Because back-arc spreading begins near the volcanic chain, the first sediments deposited on initial basin crust are rapidly deposited volcaniclastics (Karig & Moore 1975; DSDP Staff, Leg 60, 1978). As the basin widens, the arcward flank receives a large apron of volcaniclastic debris, which appears to depress and possibly fault the area of initial basin crust (Karig 1972; Mrozowski & Hayes 1979) in a manner similar to that of the loaded passive margins. The rear or continental flank of these basins is rapidly separated from the volcanic arc and either collects a small apron of debris from the adjacent remnant arc or becomes an apparently typical passive continental margin, depending on the configuration of regional tectonic elements.

The point to be made is that the area of initial crust in marginal basins, like that of passive margins, tends to become downflexed and possibly faulted during the evolution of those basins. Marginal basin flanks are also the sites of quite distinctive sedimentation, which should easily differentiate that setting not only from normal ocean basins but from other parts of marginal basins (Karig & Moore 1975). These differences ought to be clearly expressed in the basal strata of ophiolite and in forearc basins where marginal basins are involved.

The structural geometry of both types of passive margins suggests that the interface between oceanic and transitional crust is the most suitable zone for the initiation of subduction. Not only is there a good possibility of lithospheric weakness along that interface, but the basin crust is already downflexed into a suitable geometric configuration. This argument is entirely qualitative, but accords with the commonly observed initiation of young arcs near that interface.

This model is easily visualized in a two-dimensional setting, but it has become commonly accepted that most passive margins have developed along highly irregular lines. This in turn has led to an intuitive model which assumes that the initial rupture surface of a subduction zone cannot follow the intricacies of such a margin, but instead cuts across oceanic re-entrants in order to develop the smooth curvature that characterizes arc systems (Karig 1977; Dickinson & Seely 1979; Seely 1979). These re-entrants would then be loci of forearc basins built on oceanic crust, and following a collision event, could become large ophiolites. A more careful analysis suggests that this model is incorrect and without a firm physical basis.

First, there is no reason why an initial break could not closely follow an irregular crustal interface. Motivation for the idea of a smooth break was that, because mechanical flexures of oceanic lithosphere have half-wavelengths on the order of 100 km, the initial geometry of the descending plate would not be able to reflect small or sharp irregularities along the pre-existing passive margin. However, the oceanic lithosphere is more correctly visualized as being initially slipped beneath the upper plate at a very low angle rather than being bent down along that boundary. Slippage could readily occur along a tooth-like pattern of subduction and transform segments, analogous to the pattern of floating ice-fingers (see Weeks & Anderson 1958) used by Wilson (1966) to illustrate plate convergence. The flexure of oceanic lithosphere at the outer swell would simply reflect the average shape of the initial boundary. Small irregularities in the subduction zone pattern would quickly be smoothed and masked by preferential sedimentation and accretion in indentations along the inner trench slope (Karig *et al.* 1981). Larger offsets would probably be preserved as trench-trench transforms, an example of which occurs near 16°N along the Philippine Trench.

Secondly, it is not at all clear that most rifted margins are irregular on the scale inferred by the 'trapping' model. For example, if, as is generally assumed, the crustal interface of the North American passive margin is marked by the East Coast Magnetic Anomaly (e.g. Klitgord & Behrendt 1979), then that interface would have curvatures similar to those of some arcs.

In short, it is not evident why large oceanic slabs would be trapped by initiation of subduction along passive margins. Small sheets are more logical products but in any case, trapped oceanic crust in this setting ought to betray its origin in the terrigenous or volcaniclastic character of the capping sediments and possibly in anomalous crustal structure.

Inactive subduction margins

Subduction can also be reinitiated along previously convergent, but deactivated continental or arc margins. The location of renewed subduction appears to be close to the seaward margin of the older subduction complex, probably controlled by the geometry and mechanical conditions during and following cessation of the previous subduction episode.

After subduction ceases, the geometry of the convergent margin does not appear to change significantly, because the shallow morphotectonic framework of that margin is primarily controlled by the elastic loading of the oceanic lithosphere (Karig *et al.* 1976). If both plates were completely elastic, cessation of convergence should affect the geometry only by the removal of deeper level stresses through thermal resorption of the descending slab. In margins with well-developed accretionary prisms, this effect would be uplift, primarily in the upper sections of the trench slope. Continued deposition in the trench would decrease the mass imbalance, but would cause additional crustal depression of the region. Departure from perfectly elastic conditions results in elastic-plastic bending at a hinge area that generally forms near the trench (McAdoo *et al.* 1978), and should lead to local uplift seaward of that hinge.

To the extent that these conclusions concerning preservation of subduction zone geometries are correct, renewal of subduction would be expected to utilize the previous slip surface. This has apparently occurred along eastern Luzon, and possibly along the Sunda and eastern Aleutian arcs (Karig *et al.* 1976). Where this mechanism is operative, no oceanic slab should be stranded beneath the new forearc region.

Tectonic disruption of a quiescent convergent boundary might also occur, leading to a number of possible configurations in which large expanses of oceanic crust could become stranded beneath the area later to be converted to a forearc region. One such pattern is a spreading ridge that rifts the upper plate and older or pre-existing subduction complex. An excellent example is offered by the Gulf of California. A slight migration of the Pacific–North American pole could lead to renewal of subduction along Baja California, leaving the mouth of the Gulf of California trapped as a very large oceanic sheet in the forearc area.

Strike-slip disruption of a quiescent subduction margin could also disrupt the simple convergent plate geometry. Whether such strike-slip faults are confined to the upper plate or not and how they affect the subduction interface is a problem that has not yet been addressed.

Transform boundaries

Although it is almost certain that some oceanic arc systems were initiated along transforms or fracture zones, it is less obvious that this process traps a 60–100 km wide belt of typical oceanic crust between the trench and volcanic arc (e.g. Dickinson & Seely 1979). Long, active transform faults are more promising loci of trench initiation than are fracture zones because they are through-going lithospheric ruptures, along which pole migration could produce suitable relative motions.

However, neither the longer transforms nor major fracture zones are simple fault boundaries, but rather consist of a structurally complex zone that can exceed 50 km in width (e.g. Bonatti *et al.* 1979). Within such zones, slices of oceanic crust are apparently interfaulted against comminuted mixtures of crustal and mantle and oceanic sedimentary rocks, some of which may qualify as mélanges (Choukroune *et al.* 1978; Saleeby 1977). If subduction were initiated along a long transform, the resultant forearc area might well be floored with oceanic material but most likely with very atypical structural characteristics.

Conclusions and implications

Implications for the structure and evolution of arc systems

Profiles across very young arc systems demonstrate that the modal configuration of arc initiation includes a steep inner trench slope, resulting in a narrow forearc region and the lack of a forearc basin. Entrapment of strips of oceanic crust beneath the inner trench slope, especially those wider than several tens of kilometres, appears from that perspective to be a relatively unusual event. A positive correlation in width and height of various accretionary prisms with the width of the upper flattened section of the seismic zone continues to provide strong evidence for the upward and outward growth of the forearc region. However, it is clear that the forearc does not merely expand with a fixed geometry.

The well-controlled history of the Sunda arc illustrates a strong upward growth of the accretionary prism and outward migration of the trench but with changes in tectonic style since the early Miocene (Karig *et al.* 1980). The lesser migration of the trench slope break requires that the lower trench slope broadens and possibly flattens as the arc evolves. This broadening of the lower slope can be observed in the evolutionary spectrum of arc systems, from a 20 km width in the northern Middle America arc (Karig *et al.* 1978b; Shipley *et al.* 1981) to a maximum observed width of 250 km in the Makran (Farhoudi & Karig 1977).

In arcs dominated by accretion of sediments, lateral growth of the accretionary prism with age certainly involves onlap and transgression of the old continental or arc slope, as the tectonic dam provided by the trench slope break rises (e.g. Dickinson & Seely 1979). However, use of the old slope as a reference (Karig 1977), as well as more detailed studies of individual arcs, shows that there is also an outward migration of the trench slope break. The trench slope break initially forms near the base or seaward edge of the old slope (and probably marks that boundary) but it is much oversimplified to extrapolate that assumption and to identify the trench slope break as a buttress in wider, more evolved arcs involving sediment accretion (e.g. Hamilton 1977).

Evolution of oceanic arc systems, where the sediment influx is low and oceanic crust is principally involved is more difficult to determine, largely because internal structures cannot be observed on reflection profiles. The increase in size of these forearc regions with increased duration of subduction points toward some form of cumulative accretion. Contrary to the characteristics expected of such a model, recent deep sea drilling results in this type of arc have generally been interpreted as supporting subsidence and even tectonic erosion of the forearc area (DSDP Staff, Leg 60 1978). These conclusions, particularly in the Mariana arc, are open to serious question. Counter-evidence for relative uplift of the trench slope break in that arc can be cited from the nature of dredged and drilled sediments (Karig 1971; Fischer, Heezen *et al.* 1971; DSDP Staff, Leg 60 1978), from deformation along a surveyed submarine canyon (Karig 1971), and from dredging of serpentinized ultramafics from structural highs along the trench slope break (Evans & Hawkins 1979). South of the drilling transect, the trench slope appears to consist of long basement ridges, convex toward the trench and uplifted relative to sediments trapped behind (Karig & Ranken, in preparation). These ridges are most easily interpreted as intermittently accreted slices of oceanic crust and upper mantle, similar to those outlined in the Peru–Chile (Kulm *et al.* 1973) and southern Middle America (Ladd *et al.* 1978) arcs.

Combining morphology and apparent structure of very young and immature arcs such as the New Hebrides with the concept that slices of oceanic crust can be accreted to the inner slope leads to a model that is compatible with most data from the range of active oceanic arcs (Fig. 4). In this model, accretion of oceanic slices to the base of a steep initial trench slope would result in broadening of that slope and uplift of at least its trenchward section. This in turn would enhance the development of a slope apron by increasing the width over which deposition could occur and by decreasing the gradient of the upper slope. The assumption that the Tonga and Mariana arcs are more evolved examples in this evolution implies that a trench slope break develops, dividing the inner slope into upper and lower sections. The upper slope section could continue to rotate arcward until it develops into a forearc basin, as displayed along the southern Middle America arc. Basal strata in these forearc basins are interpreted as arcward-rotated, originally trenchward-dipping slope deposits, overlying either pre-existing crust, accreted material, or possibly even early arc volcanics.

Emplacement of large, coherent ophiolite sheets has been attributed to many processes, only a few of which are consistent with recent tectonism observed in the oceans. The nature of the thrust sheets capped by most such ophiolites

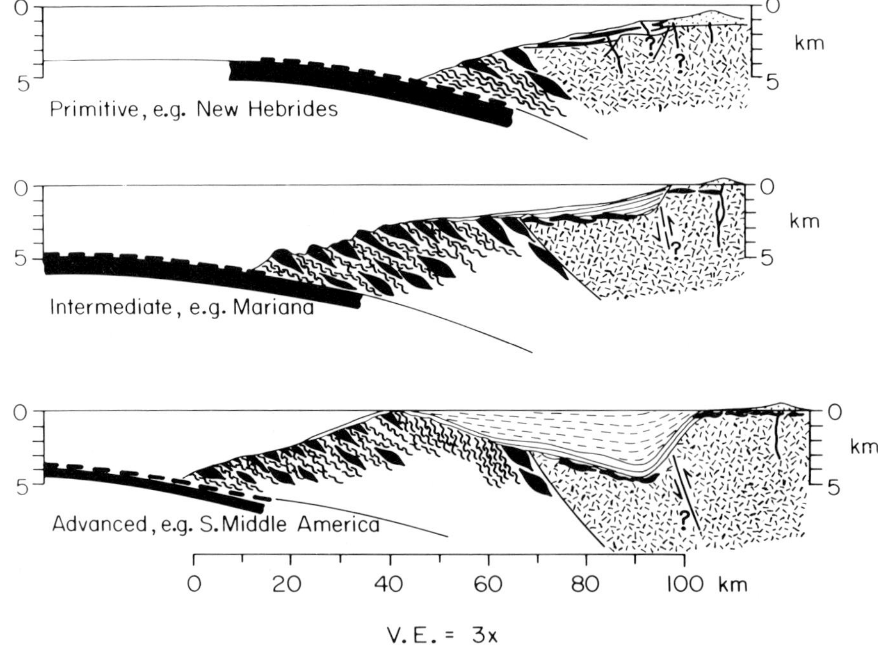

FIG. 4. Schematic evolution of oceanic arc systems based on the assumption that structural variations in such arcs of different ages are equivalent to temporal variations. Growth is by intermittent accretion of relatively large slices of oceanic crust and upper mantle, manifested in emergent terrane as disrupted ophiolite. Uplift, rather than lateral growth, dominates, converting an originally steep undifferentiated slope into an upper section (forearc basin) and a steep lower slope.

(e.g. Dewey 1976), coupled with a realistic model of arc-continent collisions as is now occurring in the Banda Arc (e.g. von der Borch 1979), strongly imply that these ophiolites occupied a forearc setting during part of that evolution. The Zambales ophiolite is one of these forearc sheets, uplifted before collision with and emplacement on to a continental margin. This sheet is autochthonous with respect to interior Luzon, and closely fits the models of Seely (1979), Dickinson & Seely (1979) and Hamilton (1977).

Quite a different picture emerges from the foregoing description of young arc systems. The lack of wide forearc regions in young arcs and their common initiation along crustal interfaces argues strongly against trapping of wide bands of oceanic crust behind the site of trench initiation, and beneath forearc basins. The apparent contradiction might be resolved if large ophiolite sheets did evolve in a forearc setting but in relatively unusual circumstances.

Anomalous plate boundary geometries, in particular those in which oceanic spreading has transected the forearc region (Fig. 5) do occur along convergent margins at present, and could account for several perplexing characteristics commonly observed in ophiolites. Upper plate

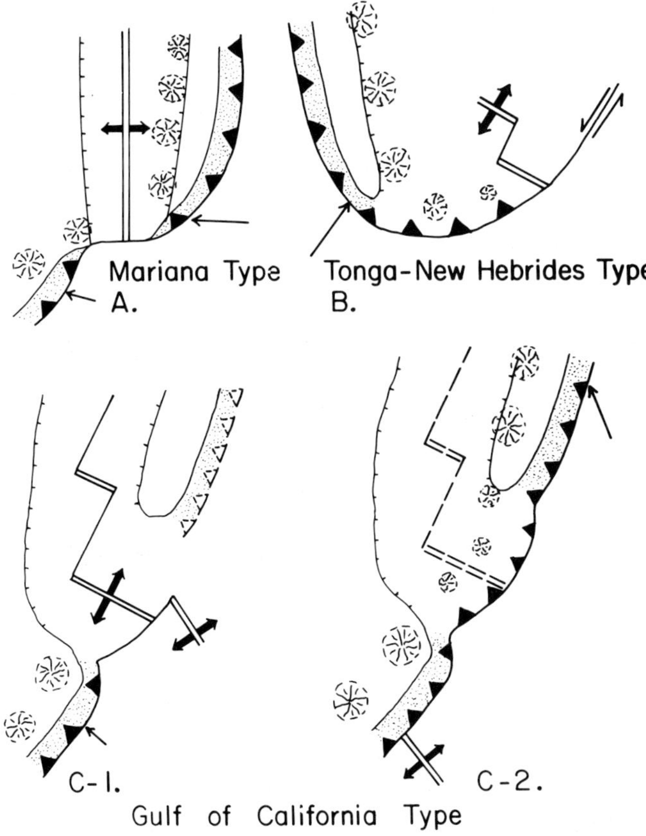

FIG. 5. Several examples of plate geometries along convergent margins in which transverse upper plate spreading zones could lead to creation of ophiolite sheets in a forearc setting. (A) In this case, modelled after the southern end of the Mariana arc, forearc rifting results from the extension of a back-arc spreading zone to the trench. Cessation of back-arc spreading could lead to the creation of a volcanic chain across very young 'oceanic' crust. (B) In a related case, drawn from the southern New Hebrides arc, upper plate rifting separates trench from transform segments of plate boundary. Initiation of subduction and of a volcanic arc would follow the north-easterly migration of the triple junction. (C) In a different set of circumstances, as could evolve in the Gulf of California, a main ocean basin spreading zone could rupture a pre-existing arc system, creating a large area of oceanic crust within the arc or continental crust (1). Migration of poles of relative motion might subsequently reinitiate subduction along the margin, trapping oceanic crust in and behind the forearc area (2).

rifting could be contemporaneous with subduction (southern Mariana) or could be followed by subduction, as would be the situation if a slight pole shift re-initiated subduction along Baja California and across the Gulf of California. In either case, oceanic crust would underlie the entire forearc region as well as the superimposed volcanic arc. Not only could the age difference between the oceanic (ophiolitic) crust and the arc volcanics be very small, but the interval between creation of the crust and emplacement on a continental margin could be very short. Forearc rifting could also account for spreading oblique to the margin and for the common lack of terrigenous basal strata.

Many so-called ophiolites have been structurally disrupted, either before or during emplacement on continental margins. Some of these, especially those disrupted before emplacement, could represent slices of far-travelled oceanic crust accreted to inner trench slopes. The great difference in tectonic significance between forearc ophiolite sheets, initially autochthonous with respect to the upper plate and disrupted oceanic slices, allocthonous with respect to both plates supports the recommendation of the ophiolite Penrose Conference (Conference Participants 1972) to restrict the term ophiolite to structurally integral sheets. It is also a blatant plea for increased study of the sediments associated with ophiolites and of post-magmatic structures of these bodies.

References

BOLLIGER, W. & DE RUITER, P. A. C. 1976. Geology of the south-central Java offshore area. *Proc. Indonesian Petrol. Assoc., 4th Ann. Conv., 1975,* **1,** 67–81.

BONATTI, E., CHERMAK, A. & HONNOREZ, J. 1979. Tectonic and igneous emplacement of crust in oceanic transform zones. *In: Deep Drilling Results in the Atlantic Ocean: ocean crust,* Am. geophys. Union, M. Ewing Ser. **2,** 239–48.

BOWIN, C., LEE, R. S., LEE, C. S. & SCHOUTEN, H. 1978. Plate convergence and accretion in the Taiwan-Luzon region. *Bull. Am. Assoc. Petrol. Geol.* **62,** 1645–72.

CARDWELL, R. K., ISACKS, B. L. & KARIG, D. E. 1980. The spatial distribution of earthquakes, focal mechanism solutions, and subducted lithosphere in the Philippine and northeastern Indonesian islands. *Geophys. Monogr. Am. geophys. Union,* **23,** 1–35.

CARNEY, J. N. & MACFARLANE, A. 1978. Lower to middle Miocene sediments on Maewo, New Hebrides, and their relevance to the development of the outer Melanesian arc system. *Aust. Soc. Explor. Geophys.* **9,** 123–30.

CHOUKROUNE, P., FRANCHETEAU, J. & LEPICHON, X. 1978. In situ structural observations along transform fault A in the FAMOUS area, Mid-Atlantic Ridge. *Bull. geol. Soc. Am.* **89,** 1013–29.

Conference Participants 1972. Ophiolites. *Geotimes,* **17,** 24–5.

CURRAY, J. R., SHOR, G. G. JR., PAITT, R. W. & HENRY, M. 1977. Seismic refraction and reflection studies of crustal structure of the eastern Sunda and western Banda arcs. *J. geophys. Res.* **82,** 2479–89.

DANIEL, J. 1978. Morphology and structure of the southern part of the New Hebrides island arc system. *J. Phys. Earth,* **26,** 5181–90.

DSDP Staff, Leg 60 1978. Leg 60 ends in Guam. *Geotimes,* **23,** 19–22.

DE BOER, J., ODOM, L. A., RAGLAND, P. C., SNIDER, F. G. & TILFORD, H. R. 1980. The Bataan orogene: eastward subduction, tectonic rotations, and volcanism in the western Pacific (Philippines). *Tectonophysics,* **67,** 251–82.

DENGO, G. 1962. Tectonic-igneous sequence in Costa Rica. *Mem. geol. Soc. Am.* 133–61.

DEWEY, J. F. 1975. Finite plate evolution: implications for the evolution of transforms, triple junctions, and orogenic belts. *Am. J. Sci.* **275-A,** 260–84.

—— 1976. Ophiolite obduction. *Tectonophysics,* **31,** 93–120.

—— & BIRD, J. M. 1970. Mountain belts and the new global tectonics. *J. geophys. Res.* **75,** 2625–47.

DICKINSON, W. R. & SEELY, D. R. 1979. Structure and stratigraphy of forearc regions. *Bull. Am. Assoc. Petrol. Geol.* **63,** 2–31.

EATON, J. P., O'NEILL, M. E. & MURDOCK, J. N. 1970. Aftershocks of the 1966 Parkfield-Cholame, California, earthquake: a detailed study. *Bull. seismol. Soc. Am.* **60,** 1151–98.

ENGLAND, P. C. & WORTEL, R. 1980. Some consequences of the subduction of young slabs. *Tectonophysics,* **47,** 403–15.

EVANS, C. & HAWKINS, J. W. 1979. Mariana arc-trench system: petrology of 'seamounts' on the trench-slope break. *Trans. Am. geophys. Union,* **60,** 968.

FAIRHEAD, J. D. 1973. Crustal structure of the Gulf of Aden and the Red Sea. *Tectonophysics,* **20,** 261–7.

FALVEY, D. 1978. Analysis of palaeomagnetic data from Viti Levu, Fiji. *Aust. Soc. Explor. Geophys.* **9,** 117–23.

FANNING, P. R. 1912. Geologic reconnaissance of northwestern Pangasinan. *Philippine J. Sci.* **7A,** 255–82.

FARHOUDI, G. & KARIG, D. E. 1977. Makran of Iran and Pakistan as an active arc system. *Geology,* **5,** 664–8.

FISCHER, A. G., HEEZEN, B. C. *et al.* 1971. *Init. Rep. Deep Sea drill. Project,* **6,** U.S. Govt Printing Office, Washington.

Galli-Olivier. C. 1979. Ophiolite and island-arc volcanism in Costa Rica. *Bull. geol. Soc. Am.* **90,** 444–52.

Garrison, R. E., Espiritu, E., Horan, L. J. & Mack, L. E. 1979. Petrology, sedimentology, and diagenesis of hemiplegic limestone and tuffaceous turbidites in the Aksitero Formation, Central Luzon, Philippines. *Prof. Pap. U.S. geol. Surv.* **1112,** 16.

Grow, J. A., Bowin, C. O. & Hutchinson, D. R. 1979. The gravity field of the U.S. Atlantic continental margin. *Tectonophysics*, **59,** 27–52.

Hamilton, W. 1977. Subduction in the Indonesian region. *In:* Talwani M. & Pitman, W. C. (eds). *Island Arcs, Deep Sea Trenches and Back-arc Basins* Am. Geophys. Union, M. Ewing Ser. **1,** 15–32.

—— 1979. Tectonics of the Indonesian region. *Prof. Pap. U.S. geol. Surv.* **1978,** 345 pp.

Harrison, J. V. 1953. The geology of the Santa Elena Peninsula in Costa Rica, Central America. *7th Pacif. Sci. Congr.* 102–14.

Hashimoto, W., Matsumaru, K., Kurihara, K., David, P. P & Balce, G. R. 1977. Larger foraminiferal assemblages useful for the correlation of the Cenozoic marine sediments in the mobile belt of the Philippines. *Geol. Paleo. South-east Asia,* **18,** 103–24.

Hayes, D. E. & Ludwig, W. J. 1967. The Manila Trench and West Luzon Trough—II. Gravity and magnetics measurements. *Deep Sea Res.* **14,** 545–60.

Henninsen, D. & Weyl, R. 1968. Ozeanische kruste in Nicoya-komplex von Costa Rica (Mittelamerika). *Geol. Rdsch.* **57,** 33–47.

Ibrahim, A. K., Pontoise, B., Latham, G., Larue, M., Chen, T., Isacks, B., Recy J. & Louat, R. 1980. Structure of the New Hebrides arc-trench system. *J. geophys. Res.* **85,** 253–66.

Ingle, J. C., Jr 1975. Summary of late Paleogene-Neogene insular stratigraphy, paleobathymetry, and correlations, Philippine Sea and Sea of Japan region. *In:* Karig, D. E., Ingle, J. C., Jr *et al.* (eds). *Init. Rep. Deep Sea drill Proj.* **31,** 317–856. U.S. Govt Printing Office, Washington.

Irving, E. M. 1950. Review of Philippine basement geology and its problems. *Philippine J. Sci.* **79,** 267–307.

Isacks, B. L. & Barazangi, M. 1977. Geometry of Benioff zones: lateral segmentation and downwards bending of the subducted lithosphere. *In:* Talwani, M. & Pitman, W. C. III (eds). *Island Arcs, Deep Sea Trenches and Back-arc Basins,* Am. geophys. Union, M. Ewing Ser. **1,** 99–114.

—— , Cardwell, R. K., Chatelain, J.-L., Barazangi, M., Marthelot, J. M., Chinn, D. & Louat, R. 1981. Seismicity and tectonics of the central New Hebrides island arc. *Am. geophys. Union, Ewing Series,* **4.**

Karig, D. E. 1971. Structural history of the Mariana arc system. *Bull. geol. Soc. Am.* **82,** 323–44.

—— 1972. Remnant arcs. *Bull. geol. Soc. Am.* **83,** 1057–68.

—— 1973. Plate convergence between the Philippines and the Ryukyu islands. *Mar. Geol.* **14,** 153–68.

—— 1974. Evolution of arc systems in the Western Pacific. *Ann. Rev. Earth planet. Sci.* **2,** 51–75.

—— 1975. Basin genesis in the Philippine Sea. *In:* Karig, D. E., Ingle, J. C. Jr *et al.* (eds). *Init. Rep. Deep Sea drill. Proj.* **31,** 857–79. U.S. Govt Printing Office, Washington.

—— 1977. Growth patterns on the upper trench slope. *In:* Talwani, M. & Pitman, W. C. III (eds). *Island Arcs, Deep Sea Trenches, and Back-arc Basins.* Am. geophys. Union, M. Ewing Ser. **1,** 175–86.

—— , Caldwell, J. G. & Parmentier, E. M. 1976. Effects of accretion on the geometry of the descending lithosphere. *J. geophys. Res.* **81,** 6281–91.

——, Cardwell, R. K., Moore, G. F. & Moore, D. G. 1978a. Late Cenozoic subduction and continental truncation along the northern Middle America Trench. *Bull. geol. Soc. Am.* **89,** 265–76.

—— , Lawrence, M. B., Moore, G. F. & Curray, J. R. 1980. Structural framework of the forearc basin, northwest Sumatra. *J. geol. Soc. London,* **137,** 77–91.

—— & Mammerickx, J. 1972. Tectonic framework of the New Hebrides island arc system. *Mar. Geol.* **12,** 187–205.

—— & Moore, G. F. 1975. Tectonically controlled sedimentation in marginal basins. *Earth planet. Sci. Lett.* **26,** 233–8.

—— & Ranken, B. in preparation. Tectonic elements of the southern Mariana arc.

—— & Sharman, G. F. 1975. Subduction and accretion in trenches. *Bull. geol. Soc. Am.* **86,** 377–89.

Ketner, K. B., Kastowo, Modjo, S., Naeser, C. W., Obradovich, J. D., Robinson, K. Suptandar, T. & Wikarnd, 1976. Pre-Eocene rocks of Java, Indonesia. *J. Res. U.S. geol. Surv.* **4,** 605–14.

Kieckhefer, R. M., Shor, G. G. Jr & Curray, J. R. 1980. Seismic refraction studies of the Sunda Trench and forearc basin. *J. geophys. Res.* **85,** 863–89.

Klitgord, K. D. & Behrendt, J. C. 1979. Basin structure of the U.S. Atlantic margin. *Mem. Am. Assoc. Petrol. Geol.* **29,** 85–112.

—— & Grow, J. A. 1980. Jurassic seismic stratigraphy and basement structure of western Atlantic margin quiet zone. *Bull. Am. Assoc. Petrol. Geol.* **64,** 1658–80.

Kulm, L. D., Scheidegger, K. F., Prince, R. A., Dymond, J., Moore, T. C. & Hussong, D. M. 1973. Tholeiite basalt ridge in the Peru Trench. *Geology,* **1,** 11–14.

Ladd, J. W., Ibrahim, A. K., McMillen, K. J., Latham, G. V., von Huene, R. E., Watkins, J. S., Moore, J. C. & Worzel, J. L. 1978. Tectonics of the Middle America Trench. *In: Int Symp. Guatamalan Earthquake and Reconstruction Process*, Feb. 4, 1976, 2.

LEHNER, P. & DE RUITER, P. A. C. 1977. Structural history of Atlantic margin of Africa. *Bull. Am. Assoc. Petrol. Geol.* **61,** 961–81.

LOWELL, J. D. & GENIK, G. J. 1972. Sea-floor spreading and structural evolution of Southern Red Sea. *Bull. Am. Assoc. Petrol. Geol.* **56,** 247–59.

LUDWIG, W. J. 1970. The Manila Trench and West Luzon Trough—III: seismic refraction measurement. *Deep Sea Res.* **17,** 553–71.

—— , HAYES, D. E. & EWING, J. I. 1967. The Manila Trench and West Luzon Trough—I: bathymetry and sediment distribution. *Deep Sea Res.* **14,** 533–44.

LUYENDYK, B. P., BRYAN, W. B. & JEZEK, P. A. 1974. Shallow structure of the New Hebrides island arc. *Bull. geol. Soc. Am.*, **85,** 1287–300.

MCADOO, D. C., CALDWELL, J. G. & TURCOTTE, D. L. 1978. On the elastic-perfectly plastic bending of the lithosphere under generalized loading with application to the Kuril trench. *Geophys. J. R. astron. Soc.* **54,** 11–26.

MCKENZIE, D. P. 1977. The initiation of trenches: a finite amplitude instability. *In:* TALWANI, M. & PITMAN, W. C. III (eds). *Island Arcs, Deep Sea Trenches and Back-arc basins.* Am. geophys, Union, M. Ewing Ser. **1,** 57–61.

MITCHELL, A. H. G. & READING, H. G. 1971. Evolution of island arcs. *J. Geol.* **79,** 253–84.

MOORE, D. G. 1973. Plate edge deformation and crustal growth, Gulf of California structural province. *Bull. geol. Soc Am.* **84,** 1883–906.

MOORE, G. F. & KARIG, D. E. 1980. Structural geology of Nias Island, Indonesia: implications for subduction zone tectonics. *Am. J. Sci.* **280,** 193–223.

MOORE, J. C., WATKINS, J. S. *et al.* 1979. Progressive accretion in the Middle America Trench, southern Mexico. *Nature. London,* **281,** 638–42.

—— , —— *et al.* 1981. Facies belts of the Middle America Trench and forearc region, southern Mexico: results from Leg 66 DSDP (this volume).

MROZOWSKI, C. L. & HAYES, D. E. 1979. The evolution of the Parece Vela basin, Eastern Philippine Sea. *Earth planet. Sci. Lett.* **46,** 49–67.

MURAUCHI, S., DEN, N., ASANO, S., HOTTA, H., YOSHII, T., ASANUMA, T., HAGIWARA, K., ICHIKAWA, K., SATO, T., LUDWIG, W. J., EWING, J. I., EDGAR, N. T. & HOUTZ, R. E. 1968. Crustal structure of the Philippine Sea. *J. geophys. Res.* **73,** 3143–71.

MURPHY, R. W. 1973. The Manila Trench-West Taiwan fold belt: a flipped subduction zone. *Bull. geol. Soc. Malaysia*, **6,** 27–42.

RAVENNE, C., PASCAL, G., DUBOIS, J., DUGAS, F. & MONTADERT, L. 1977. Model of a young intraoceanic arc: the New Hebrides island arc. *Int. Symp. Geodynamics in South-West Pacific,* 63–78.

SALEEBY J. B. 1977. Fracture zone tectonics, continental margin fragmentation and emplacement of the Kings-Kaweah ophiolite belt, southwest Sierra Nevada, California. *In:* COLEMAN, R. G. & IRWIN, W. P. (eds). *North American Ophiolite Volume. Bull. Oregon, Dep. Geol Miner. Ind.* **95,** 141–60.

SCHMIDT-EFFING, VON R. 1979. Alter und Genese des Nicoya-Komplexes, einer ozeanischen Palaokruste (Oberjura bis Eozan) im sudlichen Zentralamerika. *Geol. Rdsch.* **68,** 457–94.

SCHWELLER, W. J. & KARIG, D. E. 1979. Constraints on the origin and emplacement of the Zambales ophiolite, Luzon, Philippines. *Abstr. Progr. geol. Soc. Am.* **11,** 512–3.

SEELY, D. R. 1979. The evolution of structural highs bordering major forearc basins. *Mem. Am. Assoc. Petrol. Geol.* **29,** 245–60.

—— , VAIL, P. R. & WALTON, G. G. 1974. Trench slope model. *In:* BURK, C. A. & DRAKE, C. L. (eds). *The Geology of Continental Margins*, 249–60. Springer-Verlag, New York.

SHIPLEY, T. H., MCMILLEN, K. J., WATKINS, J. S., MOORE, J. C., SANDOVAL-OCHOA, J. H. & WORZEL, J. L. 1980. Continental margin of Guerrero and Oaxaca, Mexico. *Mar. Geol.* **35,** 65–82.

—— , LADD, J. W., BUFFLER, R. T. & WATKINS, J. S. 1981. Tectonic processes along the Middle America Trench inner slope (this volume).

SHOR, G. G. JR., KIRK, H. K. & MENARD, H. W. 1971. Crustal structure of the Melanesian area. *J. geophys. Res.* **76,** 2562–86.

THIEDE, J. 1974. Sediment coarse fractions from the western Indian Ocean and the Gulf of Aden, Deep Sea Drilling Project, Leg 24. *In:* FISCHER, R. L., BUNCE, E. T. *et al.* (eds). *Init. Rep. Deep Sea drill. Proj.* **24,** 561–766. U.S. Govt Printing Office, Washington.

TUCHOLKE, B. E. & MOUNTAIN, G. S. 1979. Seismic stratigraphy, lithostratigraphy and paleosedimentation patterns in the North American basin. *In: Deep Drilling Results in the Atlantic Ocean: continental margins and paleoenvironment,* Am. geophys. Union, M. Ewing Ser. **3,** 58–86.

TURCOTTE, D. L., AHERN, J. L. & BIRD, J. M. 1977. The state of stress at continental margins. *Tectonophysics,* **42,** 1–28.

VON DER BORCH, C. C. 1979. Continental island arc collision in the Banda Arc. *Tectonophysics,* **54,** 169–93.

WATTS, A. B. & STECKLER, M. S. 1979. Subsidence and eustacy at the continental margin of eastern North America. *In: Deep Drilling Results in the Atlantic Ocean: continental margins and paleoenvironment.* Am. geophys. Union, M. Ewing Ser. **3,** 218–34.

WEEKS, W. F. & ANDERSON, D. L. 1958. Sea ice thrust structures. *J. Glaciol. London*, **3,** 173.

WHITMARSH, R. B., WESER, O. E., ROSS, D. A. *et al.* 1974. *Initial Rep. Deep sea drill. Proj.* **23,** 11–31. U.S. Govt Printing Office, Washington.

WILSON, J. T. 1966. Are the structures of the Caribbean and Scotia arc regions analogous to ice rafting? *Earth planet. Sci. Lett.* **1,** 335–8.

D. E. KARIG, Department of Geological Sciences, Cornell University, Ithaca, New York 14853, U.S.A.